Lecture Notes on Data Engineering and Communications Technologies

Volume 31

D1808853

Series Editor

Fatos Xhafa, Technical University of Catalonia, Barcelona, Spain

The aim of the book series is to present cutting edge engineering approaches to data technologies and communications. It will publish latest advances on the engineering task of building and deploying distributed, scalable and reliable data infrastructures and communication systems.

The series will have a prominent applied focus on data technologies and communications with aim to promote the bridging from fundamental research on data science and networking to data engineering and communications that lead to industry products, business knowledge and standardisation.

** Indexing: The books of this series are submitted to ISI Proceedings, MetaPress, Springerlink and DBLP **

More information about this series at http://www.springer.com/series/15362

A. Pasumpon Pandian · Tomonobu Senjyu ·
Syed Mohammed Shamsul Islam ·
Haoxiang Wang
Editors

Proceeding of the International Conference on Computer Networks, Big Data and IoT (ICCBI - 2018)

 Springer

Editors
A. Pasumpon Pandian
Department of CSE
Vaigai College of Engineering
Melur, India

Tomonobu Senjyu
Department of Engineering
University of the Ryukyus
Okinawa, Japan

Syed Mohammed Shamsul Islam
Edith Cowan University (ECU)
Joondalup, WA, Australia

Haoxiang Wang
Department of Electrical and Computer
Engineering
Cornell University
Ithaca, NY, USA

ISSN 2367-4512   ISSN 2367-4520   (electronic)
Lecture Notes on Data Engineering and Communications Technologies
ISBN 978-3-030-24642-6   ISBN 978-3-030-24643-3   (eBook)
https://doi.org/10.1007/978-3-030-24643-3

This Springer imprint is published by the registered company Springer Nature Switzerland AG
The registered company address is: Gewerbestrasse 11, 6330 Cham, Switzerland

We are honored to dedicate the proceedings of ICCBI 2018 to all the participants and editors of ICCBI 2018.

Foreword

It is with deep satisfaction that I write this Foreword to the proceedings of the ICCBI 2018 held in Madurai, Tamil Nadu, December 19–20, 2018.

This conference was bringing together researchers, academics, and professionals from all over the world and experts in computer networks, big data, and Internet of things.

This conference particularly encouraged the interaction of research students and developing academics with the more established academic community in an informal setting to present and to discuss new and current work. The papers contributed the most recent scientific knowledge known in the field of computer networks, big data, and Internet of things. Their contributions helped to make the conference as outstanding as it has been. The local organizing committee members and their helpers put much effort into ensuring the success of the day-to-day operation of the meeting.

We hope that this program will further stimulate research in data communication and computer networks, Internet of things, wireless communication, big data, and cloud computing and also provide practitioners with better techniques, algorithms, and tools for deployment. We feel honored and privileged to serve the best recent developments to you through this exciting program.

We thank all authors and participants for their contributions.

A. Pasumpon Pandian

Preface

This conference proceedings volume contains the written versions of most of the contributions presented during the conference of ICCBI 2018. The conference provided a setting for discussing recent developments in a wide variety of topics including computer networks, big data, and Internet of things. The conference has been a good opportunity for participants coming from various destinations to present and discuss topics in their respective research areas.

ICCBI 2018 tends to collect the latest research results and applications on computer networks, big data, and Internet of things. It includes a selection of 126 papers from 352 papers submitted to the conference from universities and industries all over the world. All of the accepted papers were subjected to strict peer-reviewing by 2–4 expert referees. The papers have been selected for this volume because of quality and the relevance to the conference.

ICCBI 2018 would like to express our sincere appreciation to all authors for their contributions to this book. We would like to extend our thanks to all the referees for their constructive comments on all papers; especially, we would like to thank the organizing committee for their hard work. Finally, we would like to thank the Springer publications for producing this volume.

A. Pasumpon Pandian

Acknowledgments

ICCBI 2018 would like to acknowledge the excellent work of our conference organizing committee and keynote speakers for their presentation on December 19–20, 2018. The organizers also wish to acknowledge publicly the valuable services provided by the reviewers.

On behalf of the editors, organizers, authors, and readers of this conference, we wish to thank the keynote speakers and the reviewers for their time, hard work, and dedication to this conference. The organizers wish to acknowledge Dr. S. Sugumaran, Dr. R. Thiru Senthuran, Mrs. R. Indra, Mrs. K. Sujatha, Mrs. S. Senthilvadivoo, Mrs. L. Kirsnaveni, and Dr. K. Maniyasundar for the discussion, suggestion, and cooperation to organize the keynote speakers of this conference. The organizers also wish to acknowledge the speakers and participants who attend this conference. Many thanks are given for all persons who help and support this conference. ICCBI 2018 would like to acknowledge the contribution made to the organization by many volunteers. Members contribute their time, energy, and knowledge at local, regional, and international levels.

We also thank all the chairpersons and conference committee members for their support.

Contents

A Review: Various Steering Conventions in MANET

Vaishali V. Sarbhukan[1(✉)] and Lata Ragha[2]

[1] Terna Engineering College, Nerul (W), Navi Mumbai 400706
Maharashtra, India
vaishali5780@gmail.com
[2] Fr. C. Rodrigues Institute of Technology, Vashi, Navi Mumbai
Maharashtra, India
lata.ragha@gmail.com

Abstract. Remote systems enable hosts to meander without the requirements of wired associations. MANET is an arrangement of versatile hubs interfacing without the help of concentrated foundation. There are distinctive perspectives which are taken for look into like routing, synchronization, control utilization, data transfer capacity contemplations and so forth. Conventional steering conventions utilized for wired systems can't be specifically connected to most remote systems since some normal suppositions are not legitimate in this sort of unique system. The transfer speed in wireless system is normally constrained. Consequently, wireless system display presents incredible difficulties for steering conventions. Therefore here we focus on steering conventions which is the major concern. Various distinctive steering conventions are developed for MANETs. So it is difficult to figure out which convention is appropriate for various system situations. Here an outline of various steering conventions in MANET is given.

Keywords: DSDV · OLSR · CGSR · AODV · TORA · DSR · ZRP · SHARP

1 Introduction

A system can be portrayed as wired or remote. Remote can be recognized from wired as no physical network between hubs are required. Specially appointed systems are remote systems where hubs speak with each other utilizing Multi-Jump joins. There is no stationary framework [1] or base station for correspondence. Cellular ad hoc networking is a method for correspondence. It is independent of does current foundation, for example, devoted switches, handset base stations or maybe hyperlinks. In [2], MANET is a Self-Sufficient association of flexible switches coordinated by remote connections that is the coordination of which shape a subjective diagram. The switches can move randomly or arbitrary; therefore, the system's remote topology may trade fast and eccentrically. This type of device may fit in stay solitary mildew, or might be corporate with the larger internet. Illustrations contain Car-To-Car and Ship-To-Delivery systems that talk with every different by relying on shared steering, as regarded in Fig. 1. Setting up and keeping up information associations for different applications between versatile

A. P. Pandian et al. (Eds.): ICCBI 2018, LNDECT 31, pp. 1–11, 2020.
https://doi.org/10.1007/978-3-030-24643-3_1

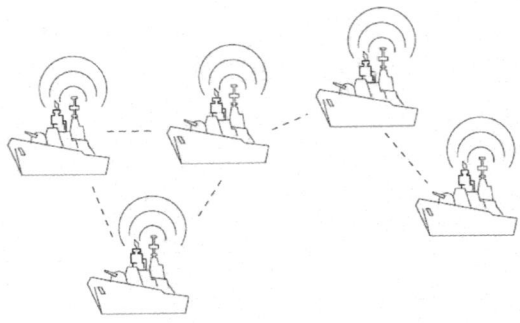

Fig. 1. Case of cellular impromptu network or MANET

hubs with no given framework or even dependable participation, is a mind boggling errand that can't be comprehended for the most part.

2   Traits of MANET

Cellular advert hoc community hubs are outfitted with remote transmitters and beneficiaries utilizing reception apparatuses, which might be exceptionally directional (Point-To-Point), omni directional (communicate), most likely steerable, or some blend. Sometimes, contingent upon places of hubs, their transmitter and recipient scope designs, correspondence control stages and Co-Channel obstruction degrees, a far flung availability as an arbitrary, multihop chart or "especially appointed" system exists some of the hubs. Therefore specially appointed topography may alter with time as the hubs amendment or change their transmission and gathering variables. The qualities of Cellular ad hoc networking are outlined as takes after:

- Communique through far flung way.
- Hubs can play out the elements of the two hosts and switches.
- Bandwidth-Compelled, variable limit joins.
- Energy-Obliged Operation.
- Constrained bodily protection.
- Dynamic arrange topology.
- Common steering refreshes.

3   Issues in MANET Steering

- Uneven joins: Maximum of the wired structures depend upon the isosceles associations which are constantly stable. Be that as it may, this isn't applicable to ad hoc systems as the hubs are versatile and continually switching their scenario inner device. For instance think about a MANET wherever first hub transmits a flag to second hub but this doesn't enlighten something concerning the nature of an association within the turnaround heading [3].

- Steering Overhead: Steering in Advert-Hoc networks has been a testing assignment [4] in case of remote systems. The good sized cause at the rear of this can be the regular amendment in prepare topography in mild of abnormal state of hub versatility. Along these lines, some stale courses are created in the steering table which prompts superfluous steering overhead.
- Interference: It is the real issue with Advert-Hoc networks as connections go back and forth relying upon the transmission qualities, one transmission may meddle with another and hub may catch transmissions of different hubs and the aggregate transmission can decline.
- Dynamic Topology: It is the real issue in MANET because of inconsistent topology feature. The versatile hub may move or medium attributes may additionally trade. In advert-Hoc networks, routing tables need to with the aid of one approach or some other reflect these adjustments in topology and routing calculations need to be adjusted. As an instance in a fixed arrange steering desk clean occurs for each 30 s [3]. This refreshing recurrence may be low for Advert-Hoc networks.

Ensuring conveyance and the ability to deal with dynamic availability are imperative issues for steering protocols in far flung cellular ad hoc networks. As soon as there's a manner from the supply to the purpose for a particular timeframe, the steering protocol need to have the capability to carry records by way of that way. On the off chance that supply of any hubs changes and courses are influenced by this change, the steering protocol should have the potential to recoup if some other way survives. In remote ad hoc networks some exclusive issues are identified with steering. Irrespective of whether to think about them as relies upon the specific situation or utility. For instance, overhead is especially essential in case of remote system with restricted transfer speed [5–9]. Power utilization may likewise be an issue in a specially appointed system with Battery-Controlled hubs [10–13]. Satisfactory of service is probably required in an advert hoc community supporting postpone sensitive packages, as an instance, video conferencing [14–17]. A steering protocol requires to adjust traffic in light of the traffic stack on joins [18, 19]. Versatility of routing protocols is an essential issue for vast systems [20, 21]. The steering protocol can also want to actualize protection to make sure against attacks, as an example, sniffer, Man-In-The-middle, or foreswearing of management [22–24]. Steering protocols may depend on data in light of different layers. For instance, the global Positioning Framework may be applied as a part of far flung ad hoc networks sent in conflict fields or interfacing motors [25, 26]. Versatility expectation can enhance routing in remote systems with known development designs, for example, the IRIDIUM framework satellite system [27]. Data from MAC layer might be proliferated to the system layer with the goal that neighbors can be identified by means of MAC layer conventions [26]. The electricity of given signals from an adjacent hub can be utilized to choose whether or not this adjacent hub is drawing nearer or assist away [25]. In MANET, cross-layer plan is interesting feature in the outline of new MANET conventions [28, 29]. Also MANET frameworks are expanded due to expanding number of papers distributed that efficiently create structures [30, 31].

4   Steering Protocol Methodologies

There are three fundamental specially appointed steering or routing methodologies. These are as proactive routing methodology (PRM), reactive routing methodology (RRM) and Hybrid Ad hoc Routing Methodology (HARM). HARM is a blend of both PRM and RRM.

4.1   Proactive Routing Methodology (PRM)

In Proactive routing methodologies, each hub keep up steering table which includes facts approximately the system topography even without need of it [32]. This detail albeit helpful for datagram movement, brings approximately enormous flagging activity and power utilization [33]. The routing tables are refreshed every now and then at something factor the system topology modifications. Therefore PRMs are also known as table driven routing protocols. Proactive conventions require to keep up hub sections for every unmarried hub within the steering table of every hub [34]. Therefore PRMs are not appropriate for massive structures. These conventions keep up diverse number of routing tables converting from conference to conference. There are exceptional virtually understood proactive steering conventions. Destination Sequenced Distance Vector (DSDV) Routing, Optimized link state Routing (OLSR) and Cluster Gateway transfer Routing Protocol (CGSR) come under PRM.

4.2   Reactive Routing Methodology (RRM)

In RRM, course is found at whatever point it is required therefore it is also called as request routing protocol. This methodology has two noteworthy parts [35]:

Course Revelation. On this degree supply hub begins direction disclosure on request premise. Source hubs counsels its path save for the accessible direction from supply to goal typically if the course is absent it starts path revelation. The source hub, inside the package, incorporates the goal address of the hub also address of the slight hubs to the intention.

Route Guide. Because of dynamic topology of the system times of the direction sadness among the hubs emerges because of connection breakage and so on, so course support is finished. RRMs have confirmation thing because of which direction guide is potential. RRM plans are preferred for the advert hoc circumstance because battery manage is moderated both by now not sending the advertisements and not to getting them. There are exclusive clearly understood reactive routing protocols introduce in MANET. Ad-Hoc On-Request Distance Vector (AODV), Temporally Ordered Routing algorithm (TORA), Dynamic source Routing (DSR) and mild-Weight mobile Routing (LMR) come under RRM.

4.3   Hybrid Ad Hoc Routing Methodology (HARM)

There is an exchange off amongst PRM and RRM conventions. PRM has expansive overhead and low inertness while RRM has not so much overhead but rather more dormancy. So a blended convention is introduced to defeat the weaknesses of both PRM and RRM which is known as Hybrid Ad hoc Routing Methodology (HARM). HARM is blend of PRM and RRM. It utilizes the course revelation component of RRM and the table support instrument of PRM in order to maintain a strategic distance from idleness and overhead issues in the system. In HARM, the system is isolated into bunches and run a PRM convention inside the bunch and a RRM convention to deal with perform steering between the distinctive bunches. HARM is reasonable for substantial systems where expansive quantities of hubs are available. In this vast system is isolated into set of bunches where directing inside the bunch is carried out by utilizing PRM and outside the bunch steering is finished utilizing RRM. Examples of HARM includes Zone Routing Protocol (ZRP) [32], Sharp Hybrid Adaptive Routing Protocol (SHARP).

Destination-Sequenced Distance Vector (DSDV)

In DSDV [36], Bellman–Ford directing [37] calculation is used. Few changes are done in Bellman–Ford. In PRM, a steering table is maintained by every portable hub involving the rundown of every accessible goal as well as the quantity of jumps to each. Each table passage is labeled with an arrangement number, which is started by the goal hub. Periodic transmissions of updates of the steering tables help keeping up the topography data of the system. The changes are conveyed instantly if there is any new huge change for the steering data. In this way, the routing data updates may either be occasional or occasion driven. DSDV conference requires each portable hub in the machine to promote its very own routing desk to its gift neighbors. The commercial is done through communicating or through multicasting. By the commercials, the neighboring hubs can think about any change that has happened in the system because of the developments of hubs. The directing updates can be dispatched in two extraordinary methods: one is referred to as a 'complete sell off' and every other is 'incremental'. If there should arise an occurrence of complete sell off, the whole routing table is sent to the neighbors, where as in the event of incremental refresh, just the passages that require changes are sent.

Optimized Link State Routing Protocol (OLSR)

The Optimized Link State Routing (OLSR) convention [38] comes under PRM convention cellular advert-hoc community. Diminishing control overhead by decreasing the quantity of communicates as contrasted and unadulterated flooding instruments is major advantage of OLSR. Main idea behind this is the utilization of multipoint relays [38, 39]. Multipoint relays allude to chosen switches that can forward communicate messages amid the flooding procedure. Each switch pronounces just a little subset of the majority of its adjacent in order to lessen the extent of communicate messages. "The convention is especially appropriate for expansive and thick systems" [38]. Multipoint relays go about as middle of the road switches in course revelation methods. Henceforth, the way found in case of OLSR may not be the most limited way. That is a capability inconvenience of OLSR. OLSR has three capacities: bundle sending, adjacent detecting, and topography disclosure. Bundle sending and adjacent detecting instruments give switches data about adjacent and offer an enhanced method to flood messages in the OLSR organize utilizing multipoint relays. Adjacent detecting activity enables switches to diffuse nearby data to the entire system. Topography disclosure is utilized to decide the topography of the whole system and ascertain steering tables. OLSR utilizes four message writes: Hello message, Topology Control (TC) message, Multiple Interface Declaration (MID) message, and Host and Network Association (HNA) message. Hello messages are utilized for adjacent detecting. Topography presentations depend on TC messages. MID messages contain numerous interface addresses and play out the errand of different interface assertions. Since hosts that have multiple interfaces connected with different subnets, HNA messages are utilized to pronounce have and related system data. Augmentations of message composes may incorporate power sparing mode, multicast mode, and so forth.

Cluster Gateway Switch Routing Protocol (CGSR)

CGSR is a clustered versatile far off system. In CGSR group heads are chosen with the help of a cluster head choice calculation. By framing a few clusters, this convention accomplishes a conveyed preparing component. One disadvantage of CGSR is that, frequent change or determination of cluster heads may influence the routing execution. CGSR utilizes DSDV convention as the basic routing plan and, consequently, it has an indistinguishable overhead from DSDV. CGSR alters DSDV by utilizing a various leveled group make a beeline for door routing way to deal with route movement from starting point to goal. Portal hubs are hubs that are inside the correspondence scopes of at least two group heads. A bundle sent by a hub is first sent to its group head, and after that the parcel is sent from the bunch make a beeline for a door to another bunch head, etcetera until the point when the group leader of the goal hub is come to. The parcel is then transmitted to the goal from its own particular bunch head.

Ad Hoc On-Demand Distance Vector (AODV) Routing

AODV [40] comes under RRM category. Ad Hoc On-Demand Distance Vector (AODV) is a steering convention is suitable for wireless and mobile ad hoc networks. In AODV course or route is established at whatever point it is required therefore it is called as ad hoc On–Demand Distance Vector or request routing protocol. It supports both unicast and multicast routing. In AODV, networks are silent until connections are hooked up. System hubs that need associations broadcast a request for association. The last AODV hubs ahead the message and record the hub that asked an association. Therefore, they invent a sequence of brief routes lower back to the soliciting for hub. It limits the amount of communicates through making routes in view of hobby but it is not the case with DSDV. At the factor when any supply hub desires to send a parcel to a goal, it communicates a course request packet. The neighboring hubs accordingly communicate the parcel to their pals and the method proceeds until the point when the package deal achieves the purpose. Amid the manner closer to sending the route request, center of the road hubs document the address of the neighbor from which the number one duplicate of the communicate package is gotten. This record is put away of their course tables, which enables for putting in a transfer way. At the off risk that greater duplicates of the identical course requests are later gotten, these bundles are disposed of. The solution is dispatched utilizing the transfer way. For course maintenance, when a source hub actions, it is able to reinitiate a path disclosure manner. Inside the event that any moderate hub movements internal a specific course, the adjacent of the floated hub can identify the connection disappointment and sends a connection disappointment observe to its upstream adjacent. This procedure proceeds until the factor that the disappointment caution achieves the supply hub. In view of the given facts, the source may pick out to Re-start the direction revelation degree [41].

Temporally Ordered Routing Algorithm (TORA)

TORA [42] is under RRM convention. TORA is distributed routing algorithm in MANET. It is called as distributed because, there is no centralized control on each hub. TORA includes create route, erase route and route maintenance. If there is link failure then route maintenance is required means to find out other route and update previous one. Every hub has an array adjacent of its height. This array includes information about link failure. First tuple in array represent time when link is broken, second attribute is id of hub and third one is reflection bit. Last two tuples in array represent offset of the reference. In TORA every hub other than destination hub maintains its height with respect to destination. Height of destination hub is always zero. For other hubs excluding destination it is set as null initially. IN TORA each hub keeps up information regarding height of its adjacent. Whenever hub wants to send packet to destination it send query packet which contains address of destination hub. That query packet is broadcasted until it reaches to destination. Once query packet is reached to destination node it comes to know that packet is indented for it so it sends a reply as update packet and forwarded to source node. While forwarding packet towards source node height attribute is updated until reaches to source node. In TORA node can transmit data if its height is greater than the node to which it is forwarding data. Link which is used in TORA to send query packet from higher hub in terms of height to lower indexed node is called as down shrink link. For data transmission always downstream links are always used in TORA.

Dynamic Source Routing (DSR)

Dynamic Source Routing (DSR) comes under PRM category. Dynamic supply Routing (DSR) is a self-keeping steering protocol for remote networks. The protocol also can characteristic with cellular telephone systems and cellular networks with as much as about two hundred hubs. A Dynamic supply Routing community can configure and prepare itself independently of oversight by way of human administrators. In Dynamic supply routing, each supply determines the course to be used in transmitting its packets to selected locations. There are major components, called path Discovery and direction renovation. Direction Discovery determines the most effective route for a transmission between a given supply and destination. Path protection ensures that the transmission direction remains best and loop-free as network conditions change, even if this calls for changing the course all through a transmission.

Zone Routing Protocol (ZRP)

ZRP [43] is reasonably used for MANETs, particularly for huge traverse and versatility designs. In this HARM, every hub maintains routes inside a neighborhood locale called as routing zone. An inquiry answer instrument is used to establish route. A hub initially needs to know it's adjacent in order To produce diverse zones in the system, initially a hub needs to know it's adjacent. A hub with whom coordinate correspondence can be set up is called as an adjacent, and that is, inside one bounce transmission scope of a hub. Adjacent revelation data is utilized as a reason for Intra-Zone Routing Protocol (IARP), which is depicted in detail in [44]. Instead of visually impaired telecom, ZRP utilizes a question control instrument to decrease route inquiry activity by coordinating inquiry messages outward from the inquiry source and far from secured routing zones. A hub which has a place with the directing zone of a hub that has gotten a route inquiry is known as a secured hub. A hub can distinguishes whether sent query parcel is originating from its adjacent or not. On the off chance that yes, at that point it denotes the majority of its known adjacent hubs in its same zone as covered [32]. The question is along these lines transferred till it achieves the goal. Finally the goal makes the course by sending back an answer message by means of the turnaround way.

Sharp Hybrid Adaptive Routing Protocol (SHARP)

SHARP [45] utilizes HARM approach by powerfully differing the measure of routing data shared proactively. It characterizes the zones around a few nodes in which PRM is used. The quantity of hubs in a specific proactive zone is controlled by the hub particular zone span. All hubs inside the zone range of a specific hub turn into the individual from that specific proactive zone for that hub. On the off chance that for a given goal a hub is absent inside a specific proactive zone, reactive routing system (Query-Reply) is used to build up the route to that hub. Proactive routing system is utilized inside the proactive zone. Hubs inside the proactive zone keep up courses proactively just concerning the focal hub. In this convention, proactive zones are made consequently if a few goals are as often as possible tended to or looked for inside the system. The proactive zones go about as authorities of bundles, which forward the parcels productively to the goal, once the bundles achieve any node at the zone region [32]. Comparative analysis of above mentioned protocols is given as shown in Table 1.

Table 1. Comparison of various steering protocols in MANET

Protocol name	Category	Steering metric	Used for
DSDV	PRM	Shortest path	Smaller and relatively static network
OLSR	PRM	Shortest path	Smaller and relatively larger network
CGSR	PRM	Hop	Large size network
AODV	RRM	Shortest path	Moderate size and highly dynamic network
TORA	RRM	DAG	Large size network with low mobility
DSR	RRM	Shortest path	Small size network with moderate mobility
ZRP	HARM	K hops	Large size network
SHARP	HARM	Hop	Large size network

5   Conclusion

An awesome improvement is analyzed in area of remote systems (framework based) and in Mobile specially appointed system (framework less network). In this paper various steering conventions for MANET, which are comprehensively arranged as PRM, RRM and HARM conventions. The examination uncovers that, DSDV routing convention expends more data transfer capacity, as a result of the continuous telecom of routing refreshes. While the AODV is superior to anything DSDV as it doesn't keep up any routing tables at hubs which brings about not so much overhead but rather more data transfer capacity. From the above, sections, it can be expected that DSDV routing conventions works better for littler systems however not for bigger systems. Thus, my decision is that, AODV routing convention is most appropriate for general portable impromptu systems as it devours less data transfer capacity and lower overhead when contrasted and DSDV routing convention. There are different inadequacies in various routing conventions and it is hard to pick steering convention for various circumstances as there is tradeoff between different conventions. There are different difficulties that should be met, so these systems will have far reaching use later on.

References

1. Johnson, D.B., Maltz, D.A.: Dynamic Source Routing in Ad Hoc Wireless Networks. Carnegie Mellon University, Pittsburgh (1996)
2. Macker J., Corson S.: Mobile ad-hoc networks (manet), December (2001). http://www.ietf. org/proceedings/01dec/183.htm
3. Schiller, J.: Mobile Communications. Addison-Wesley, Boston (2000)
4. Jacquet, P., Viennot, L.: Overhead in mobile ad-hoc network protocols. 3965, INRIA, France, June (2000)
5. Boppana, R.V., Konduru, S.P.: An adaptive distance vector routing algorithm for mobile, ad hoc networks. In: IEEE INFOCOM and Joint Conference of the Computer and Communications Societies, vol. 3, pp. 1753–1762 (2001)
6. Goff, T., Abu-Ghazaleh, N.B., Phatak, D.S., Kahvecioglu, R.: Preemptive routing in ad hoc networks. In: Seventh International Conference on Mobile Computing and Networking, pp. 43–52 (2001)
7. Zhou, N., Abouzeid, A.A.: Information-theoretic lower bounds on the routing overhead in mobile ad-hoc networks. In: IEEE International Symposium on Information Theory, pp. 455–455 (2003)
8. Yoo, J., Choi, S., Kim, C.: Control overhead reduction for neighbour knowledge acquisition in mobile ad hoc networks. Electron. Lett. **39**(9), 740–741 (2003)
9. Kim, D., Toh, C.K., Cano, J.C., Manzoni, P.: A bounding algorithm for the broadcast storm problem in mobile ad hoc networks. In: Wireless Communications and Networking Conference (WCNC), vol. 2, pp. 1131–1136 (2003)
10. Jones, C.E., Sivalingam, K.M., Agrawal, P., Chen, J.C.: A survey of energy efficient network protocols for wireless networks. Wirel. Netw. **7**(4), 343–358 (2001)
11. Chang, J.H., Tassiulas, L.: Energy conserving routing in wireless ad-hoc networks. In: 19th INFOCOM, pp. 22–31 (2000)
12. Helmy A.: Architectural framework for large-scale multicast in mobile ad hoc networks. In: IEEE International Conference on Communications (ICC), vol. 4, pp. 2036–2042 (2002)
13. Perkins, D.D., Hughes, H.D., Owen, C.B.: Factors affecting the performance of ad hoc networks. In: IEEE International Conference on Communications (ICC) 2002, vol. 4, pp. 2048–2052 (2002)
14. Chen, T.W.: Efficient Routing and Quality of Service Support for Ad Hoc Wireless Networks. University of California, Los Angeles, Department of Computer Science, March (1998)

15. Liu, J., Zhang, Q., Zhu, W., Zhang, J., Li, B.: A novel framework for QoS aware resource discovery in mobile ad hoc networks. In: IEEE International Conference on Communications (ICC), vol. 2, pp. 1011–1016 (2002)
16. Hussein O., Saadawi T.: Ant routing algorithm for mobile ad-hoc networks (ARAMA). In: International Conference on Performance, Computing, and Communications Conference (IPCCC), pp. 281–290 (2003)
17. Yeh, C.H., You, T.: A QoS MAC protocol for differentiated service in mobile ad hoc networks. In: International Conference on Parallel Processing, pp. 349–356 (2003)
18. Zhou, A., Hassanein, H.: Load balanced wireless ad hoc routing. In: Canadian Conference on Electrical and Computer Engineering, vol. 2, pp. 1157–1161 (2001)
19. Song, J.H., Wong, V., Leung, V.C.: Load-aware on-demand routing (laor) protocol for mobile ad hoc networks. In: Proceedings of Vehicular Technology Conference (VTC) Springer 2003, vol. 3, pp. 1753–1757 (2003)
20. Aron, I.D., Gupta, S.K.: On the scalability of on-demand routing protocols for mobile ad hoc networks: an analytical study. J. Interconnect. Netw. 2(1), 5–29 (2001)
21. Santivanez, C.A., McDonald, B., Stavrakakis, I., Ramanathan, R.: On the scalability of ad hoc routing protocols. In: Proceedings of INFOCOM 2002, vol. 3, pp. 1688–1697 (2002)
22. Lundberg, J.: Routing Security in Ad Hoc Networks Helsinki University of Technology (2000)
23. Čapkun, S., Hubaux, J.P., Buttyán, L.: Mobility helps security in ad hoc networks. In: Proceedings of the 4th ACM International Symposium on Mobile Ad Hoc Networking and Computing, pp. 46–56 (2003)
24. Zhang, Y., Lee, W., Huang, Y.A.: Huang Intrusion detection techniques for mobile wireless networks. Wirel. Netw. 9(5), 545–556 (2003)
25. Qin, L.: Pro-active route maintenance in DSR. Ottawa-Carleton Institute of Computer Science, School of Computer Science, Carleton University, Canada (2001)
26. Boukerche, A., Rogers, S.: GPS query optimization in mobile and wireless networks. In: Proceedings of the 6th IEEE Computers and Communications Conference, pp. 198–203 (2001)
27. INFOSAT Telecommunications Iridium System Specifications. INFOSAT telecommunications, January 18 (2002)
28. Safwati, A., Hassanein, H., Mouftah, H.: Optimal cross layer designs for energy efficient wireless ad hoc and sensor networks. In: Proceedings of International Conference on Performance, Computing, and Communications Conference (IPCCC), pp. 123–128 (2003)
29. Yuen, W.H., Lee, H.N., Andersen, T.D.: A simple and effective cross layer networking system for mobile ad hoc networks. In: Proceedings of the 13th IEEE International Symposium on Personal, Indoor and Mobile Radio Communications, vol. 4, pp. 1952–1956 (2002)
30. Ye, Z., Krishnamurthy, S.V., Tripathi, S.K.: A framework for reliable routing in mobile ad hoc networks. In: Proceedings of the 22nd Annual Joint Conference of the IEEE Computer and Communications Societies (INFOCOM), vol. 1, pp. 270–280 (2003)
31. Abbas, A.M., Jain, B.N.: An analytical framework for path reliabilities in mobile ad hoc networks. In: Proceedings of the 8th IEEE International Symposium on Computers and Communication (ISCC), pp. 63–68 (2003)
32. Kaur, R., Rai, M.K.: A novel review on routing protocols in MANETs. UARJ 1(1), 103–108 (2012)
33. Royer, E.M., Toh, C.K.: A review of current routing protocols for ad hoc mobile wireless networks. University of California, Santa Barbara Chai-Keong Toh, Georgia Institute of Technology, IEEE Personal Communications, pp. 46–55, April (1999)

34. Gorantala, K.: Routing Protocols in Mobile Ad hoc Networks. Thesis in computer science, p. 136 (2006)
35. Sheltami, T., Mouftah, H.: Comparative study of on demand and cluster based routing protocols in MANETs. In: IEEE Conference, pp. 291–295 (2003)
36. Perkins, C.E., Bhagwat, P.: Highly dynamic destination-sequenced distance-vector routing (DSDV). Mobile Computers. In: Proceedings of ACM SIGCOMM 1994, pp. 234–244 (1994)
37. Cheng, C., Riley, R., Kumar, S.P., Garcia-Luna-Aceves, J.J.: A loop free extended Bellman–Ford routing protocol without bouncing effect. ACM SIGCOMM Comput. Commun. Rev. 19(4), 224–236 (1989)
38. Clausen, T., Jacquet, P., et al.: Optimized link state routing protocol. Internet Engineering Task Force (IETF) draft, March (2002)
39. Qayyum, A., Viennot, L., Laouiti, A.: Multipoint relaying: an efficient technique for flooding in mobile wireless networks. Research report-3898, INRIA, France (2000)
40. Perkins, C.E., Royer, E.M., et al.: Ad hoc on-demand distance vector (AODV) routing. IETF Draft, October (2003)
41. Kumar, G.V., Reddyr, Y.V., Nagendra, D.M.: Current research work on routing protocols for MANET: a literature survey. Int. J. Comput. Sci. Eng. 2(3), 706–713 (2010)
42. Park, V.D., Corson, M.S.: A highly adaptive distributed routing algorithm for mobile wireless networks. In: Proceedings of IEEE INFOCOM, vol. 3, pp. 1405–1413 (1997)
43. Pearlman, M.R., Samar, P.: The zone routing protocol (ZRP) for ad hoc networks. IETF draft, July (2002)
44. Pathan, S.K., Hong, C.S., Haas, Z.J., Pearlman, M.R., Samar, P.: Intra zone routing protocol (IARP). IETF Internet Draft, July (2002)
45. Rama Subramanian, V., Haas, Z.J., Sirer, E.G.: SHARP: a hybrid adaptive routing protocol for mobile ad hoc networks. In: Proceedings of ACM MobiHoc, pp. 303–314 (2003)

End to End Network Connectivity Detection Using SNMP and CAM

R. Abhishek[1] and N. Shivaprasad[1,2(✉)]

[1] JSS Science and Technology University, Mysuru, India
abhiraj8081@gmail.com
[2] Sri Jayachamarajendra College of Engineering, Mysuru, India
shivaprasad_n@sjce.ac.in

Abstract. In large organizations, wired Local Area Network (LAN) may spread across various buildings and floors for connecting PCs and servers. When there is increase in the network nodes, the complexity of monitoring, maintaining and troubleshooting also increases. To ensure security and problems in the network nodes, it is essential to identify the exact switch port to which a node is connected in the LAN network. The proposed system maps the location of all the switches present in the network and also helps network administrators to identify the exact switch port to which a device is connected and when there is an issue with the switches in the network, then the proposed system eliminates the need for manually tracing the network wires in the LAN organization. The devices plugged into each port of specified switches are also discovered by using Content Addressable Memory (CAM) entries of the switches. This is useful for system and network administrator to gain visibility into the IP, MAC, VLAN, status and availability of ports. The system also works in the dynamic environment and fulfills the need for the flexible access to the network. In order to overcome the infrastructure cost, this system introduces a network monitoring and management system using SNMP which helps to discover the devices like switches, routers, printers, and hosts and also discovers connectivity among these devices. This monitoring tool helps the Network Administrator to keep the entire network in a trouble free environment, thereby increasing the uptime of systems and user satisfaction.

Keywords: SNMP · MIB · Switches · Topology · CAM

1 Introduction

Network topology is interconnection between the networks nodes present in the network. The proposed work discovers the issues like discovering various types of devices, visualizing the network topology, discovering connectivity among network nodes. By making use of Simple Network Management Protocol (SNMP) algorithm it connects the end host with the network. In the network of an organization there may be devices like switches, routers, printers, workstations, servers, etc., there will be an interconnection network among them. Administrator may need some technique to get the information of the devices, so the proposed system introduces an approach called

© Springer Nature Switzerland AG 2020
A. P. Pandian et al. (Eds.): ICCBI 2018, LNDECT 31, pp. 12–22, 2020.
https://doi.org/10.1007/978-3-030-24643-3_2

SNMP to retrieve the information very easily. It is configured in such a way that it can be executed anywhere and it is simple through request and response form [13, 14].

SNMP is an application–layer protocol used for exchanging the information between network devices in the network. In SNMP communication happens between SNMP Managers and SNMP Agents. These SNMP agents communicates with the network management system (NMS) and it allows NMS to collect the management information database and makes it available to the SNMP manager, when it is queried for [3–6].

2 Methodology

The basic idea of this paper is using the MIBs to retrieve the details of the devices and the network. The proposed system use Object identifier values (OID) as the request from the manager to agent. The basic MIBs used are showed in the Table 1.

Table 1. MIB used for proposed work

MIB-II (RFC 1213)	BRIDGE-MIB (RFC 1493)
sysDescr, sysUptime, sysServices,	dot1dBase,
ifIndex, ifDescr,	dot1dBasePort, dot1dBasePortIfIndex,
ifPhyaddress,	dot1dTpFdbAddress,
ipRouteNextHop, ipRouteType, ipAdEntAddr	dot1dTpFdbPort, dot1dTpFdbStatus

The proposed work retrieves the basic information of the devices and also the connectivity of the devices with network using MIBs [7]. The RFC 1213 is used to retrieve the basic information of the devices and the Bridge MIB is used to discover the device type discovery and the connectivity discovery.

The information of the devices will be retrieved in the network using the management information base MIBs through specific OID value [7, 10, 11]. Specific object identifier value is used to retrieve specific details required by user. Complete information about a switch is provided by using the above information of MIB of the switch. What devices are connected to which port of the switch including the details of the IP Address and the MAC Addresses of the devices are displayed in textual as well as diagrammatic form. And also the switch provides the services for layer 2/layer 3/layer 4/layer 7. The connectivity of the devices is found using the port details of the network. Here the port numbers of the device are identified to which it is connected in the switch.

The data is stored in the database so that the required information can be retrieved accordingly. Storing the data in the database helps in report generation of the network. By this the performance of the network can be found. The basic idea of SNMP is to retrieve the required details of the device which helps in managing the network by administrator. After retrieving the details of the devices all the details are displayed on the web page using struts[2] framework. The network analysis will contain the details of the network like switch details and the connectivity details. The major thing for end result generation is to run the stand alone program in the server side to store the data in the database and update the database accordingly every use.

2.1   Network Device Discovery

Network discovery is a process of finding the devices in the network. Network can be discovered through a specific devices like switch, router. The proposed system uses switch to discover the network of the organization. Since the system discovers a sub network of an organization, discovering process is done through the switch. Since the discovery mechanism is on SNMP, Reponses from the agents are retrieved by sending the required OID values. For discovering network the system makes use of RFC 1213 [2].

Network devices which are present in the network are extracted by using the ipNetToMediaTable object identifiers. This paper utilizes ipNetToMediaPhysAddress and ipNetToMediaNetAddress. IpNetToMediaNetAddress contains all the IP address of the devices connected to a particular switch in a network and ipNetToMedia-PhyAddress contains MAC address for all devices connected to switch network (Table 2).

Table 2. Algorithm for device discovery

1.	Set switch IP Address present in a network and discover the switch present in that network using ipNetToMediaNetAddress (1.3.6.1.2.1.4.22.1.3) Object Identifiers
2.	for each switch, get all the IP address of devices from ipNetToMediaNetAddress which are connected to that particular switch if there is no ipNetToMediaNetAddress, then return;
3.	Call ipNetToMediaNetAddress recursively for all the switch IP address present in a network

2.2   Discovering Network Device Type

The type of device is discovered by using sysServices MIB object identifier and the value of sysServices is converted into a seven-bit string. Each bit corresponds to the 7 layer of OSI network model. If a device has sysServices 6(0000110)-its third and second bit is set- thus it provides services to two layers i.e., layer 3 and layer 2 [8, 9]. Then that device is L3 layer devices make same survey to identify the type of device [1]. Table 3 shows the algorithm for discovering the device type in a network.

Table 3. Algorithm for discovering network device type

1	For each switch given as input
2	sysServices(1.3.6.1.2.1.1.7) object is used
3	convert sysServices(1.3.6.1.2.1.1.7) in to seven-bit string
4	type of switch is obtained on the basis of enabled bits of seven bit string of sysServices (1.3.6.1.2.1.1.7)
5	repeat

2.3   Connectivity Establishment

The network in an organization consists of distinct types of devices. This module is used to discover interface to interface connectivity. The Table 4 shows the algorithm that explains connectivity between devices of network [6, 12].

Table 4. Connectivity establishment algorithm

1.	For each switch, to discover the interface to which the devices are connected
	(a) Get the set of MAC address of the switch from dot1dTpFdbAddress i.e., {Mia....... Mir}
	(b) Get the set of port numbers of the switch from dot1dTpFdbport i.e., {Pia.......Pir}
	(c) Get the set of MAC address of the switch from ipNetToMediaPhysAddress i.e., {Mja........Mjr}
	(d) Get the set of IP address of the switch from ipNetToMediaNetAddress i.e., {Nja........Njr}
	(e) Repeat from 1

The details of data packets that are transmitted from/to switch can be retrieved through the switch ports. These details are stored in the database and by specific interval of time. The proposed system stores the data by interval of 30 min. The code that is used to store these details in to the database are run using a Timer program. The database is continuously updated by the Poller which updates the database.

There are two module used in the proposed work Polling module and Information module.

Polling Module

The polling module consists of initiation of timer, generating time instances, activate poller, connecting to network switch and storing the data into database. The activity diagram of polling module is as shown in the Fig. 1.

Information Module

The information module uses Struts II Architecture system that is in terms of the Model, View and Controller. The information module consists of login, collecting CAM entries, switch information, switch topology, port details, network node connectivity, VLAN information, SNMP disabled switch list and printer list.

Action Classes

The Action class contains the logic which is present in the application.

Action Class for Login

The Action class for login comprises of the Login page where the user logs into the system. It is designed in JSP and displayed on the browser. If successful, a JSP page containing a web page where the user has to displays various services which can be used by the user. The validation for this is done by accessing the database containing the login credentials of users and checks whether it is correct or not.

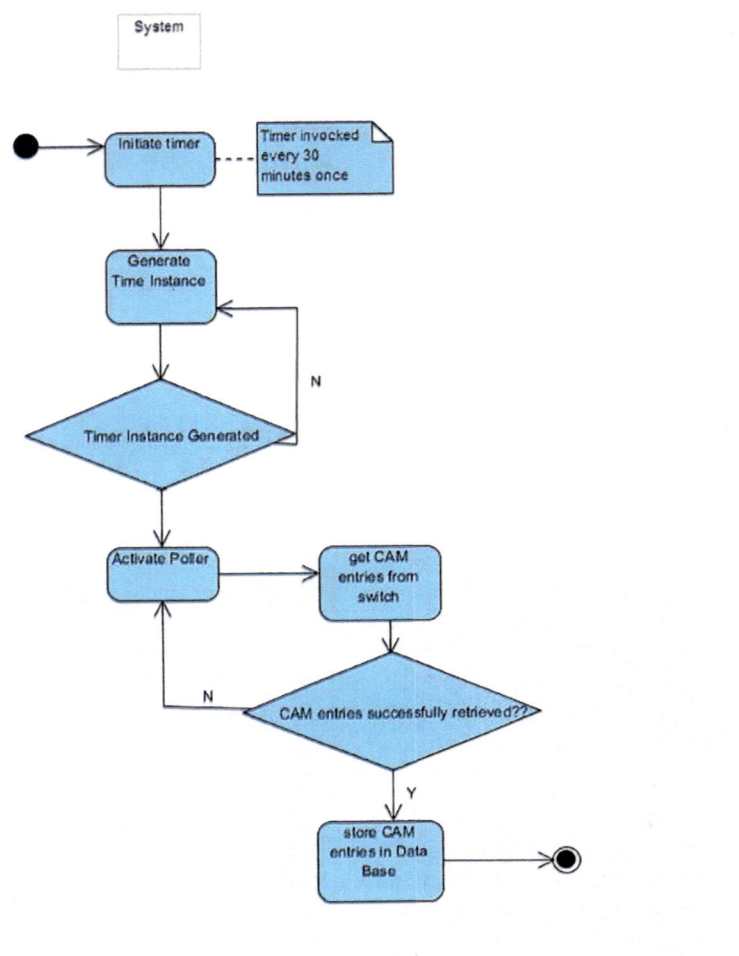

Fig. 1. Activity diagram of polling module

Action Class for Switch Information
An action class is created along with the Switch IP Address frame. This is done in order to access the database containing the data that has been polled. A validation occurs where the IP Address is checked if it exists. If this is successful, the database retrieves the Switch MAC Address, Switch Name, Switch Location, Number of Ports, and Contact Person. A JSP file is used to display the above information.

Action Class for Switch Topology
An action class is created along with the Switch IP Address frame. This is done in order to access the database containing the data that has been polled. A validation occurs where the IP Address is checked if it exists. If this is successful, the struts[2] page which

displays the diagrammatic representation of the switch using JFREECHART on Struts[2] Framework is displayed.

Action Class for Switch Port Details

An action class is created along with the Switch IP Address frame. This is done in order to access the database containing the data that has been polled. A validation occurs where the IP Address is checked if it exists. If this is successful, the database retrieves the device MAC Address, device IP Address, Port Number, Type of Port of the devices which are connected to the switch in table form. This is the tabular form of the diagrammatic representation of the topology of the switch. A JSP file is used to display the above information.

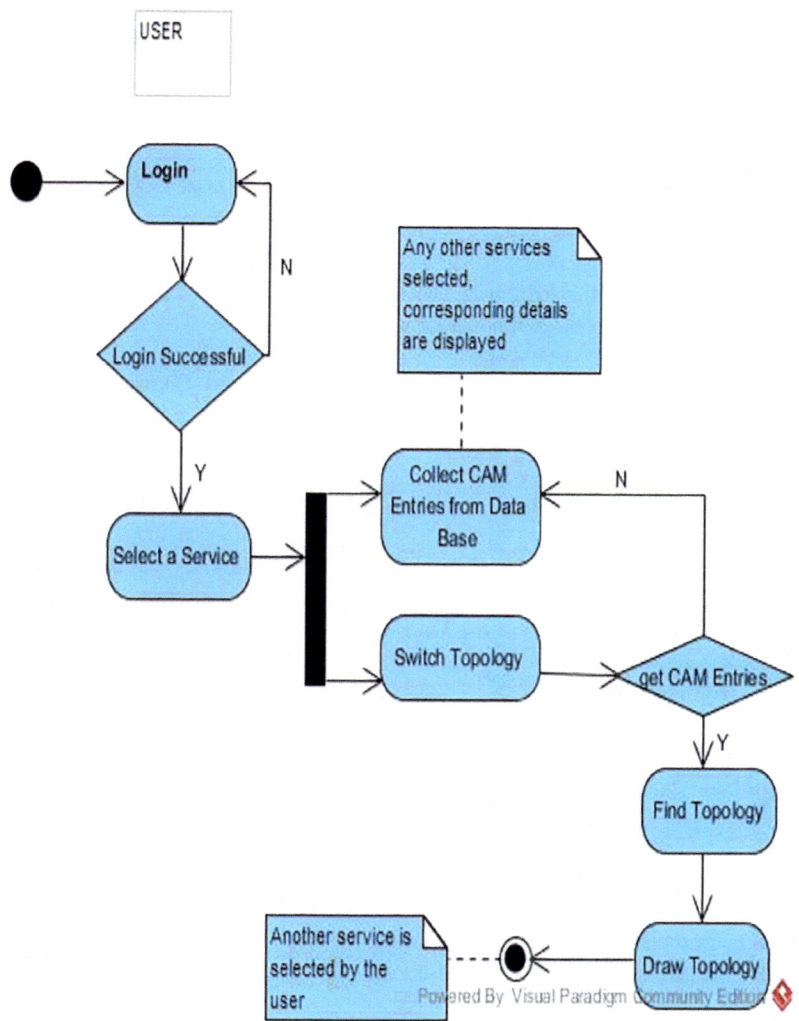

Fig. 2. Activity diagram of information module

Action Class for Each Port Details of the Switch
An action class is created along with the Switch IP Address frame. This is done in order to access the database containing the data that has been polled. A validation occurs where the IP Address is checked if it exists. If this is successful, the database retrieves the Port Number, Number of In Packets and Number of Out Packets. A JSP file is used to display the above information.

Action Class for Connectivity of a Network Node
An action class is created along with the Node IP Address frame. This is done in order to access the database containing the data that has been polled. A validation occurs where the IP Address is checked if it exists. If this is successful, the database retrieves the Device IP Address, Device MAC Address, Switch IP Address, Port Number, Type of Port and Vlan Id. This is displayed in a JSP file in a tabular form.

Action Class for VLAN Information
The database retrieves all the Vlan Ids' and Vlan Names' present in the Vlan Info table and displays it in JSP file.

Action Class for SNMP Disabled Switch List
The database retrieves all the Switch IP Address which is present in the Switch Info table whose status in 'N' and displayed it in JSP file.

Action Class for Printers List
The database collects all the Printer IP Address as well as the type of printer which are discovered. It also redirects to the list of device IP Address which uses the network printer and lists the device details.

The Fig. 2 shows the activity diagram of the information module of the proposed work.

3 Implementation

JAVA 1.8, Tomcat, struts[2], Mysql 5.6, SNMP4J libraries and JFREECHART for visualizing the topology of the network is being used. The system is developed and tested on windows 10 and Pentium Dual-Core CPU E5400 2.70 GHZ with 4 GB RAM.

The deployment diagram of the system is shown in the Fig. 3.

The Fig. 4 shows the switch information based upon the IP address given by the user. Switch information includes switch IP address, switch MAC address, switch name, number of ports present in that switch, location of the switch and the contact person.

The Fig. 5 shows the switch topology based upon the IP address given by the user. It displays a diagrammatical picture of the network using JFREECHART tool.

The Fig. 6 shows the switch port details based upon the IP address given by the user. Switch port details include port number of the switch, port type in the switch, device MAC address and Device IP address.

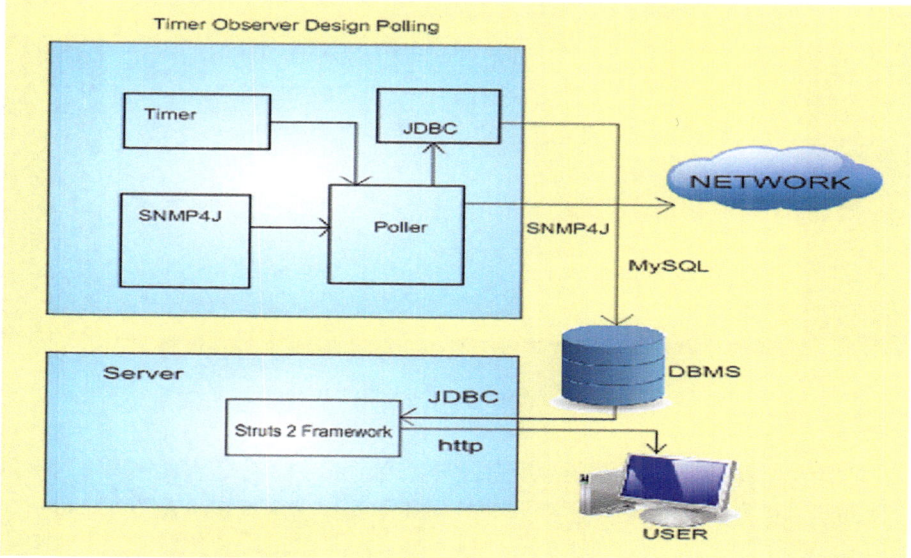

Fig. 3. Deployment diagram of the system

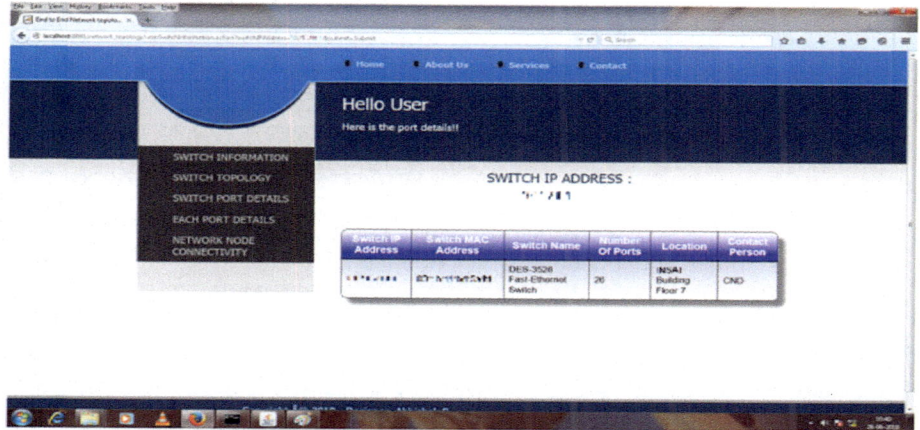

Fig. 4. Switch information

The Fig. 7 shows the each port details of the switch based upon the IP address given by the user. Each port details of the switch include port number, port status, number of in packets and number of out packets.

The Fig. 8 shows the network node connectivity based upon the IP address given by the user. Node connectivity information includes device IP address, device MAC address, switch IP address, port number to which that node is connected, port type in switch and VLAN ID.

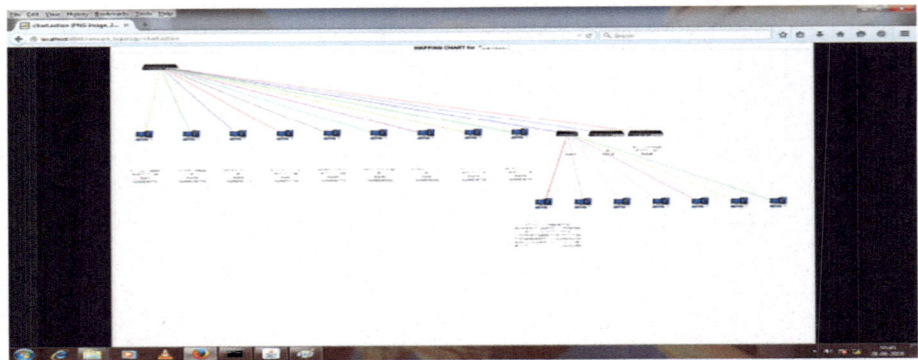

Fig. 5. Network topology visualization

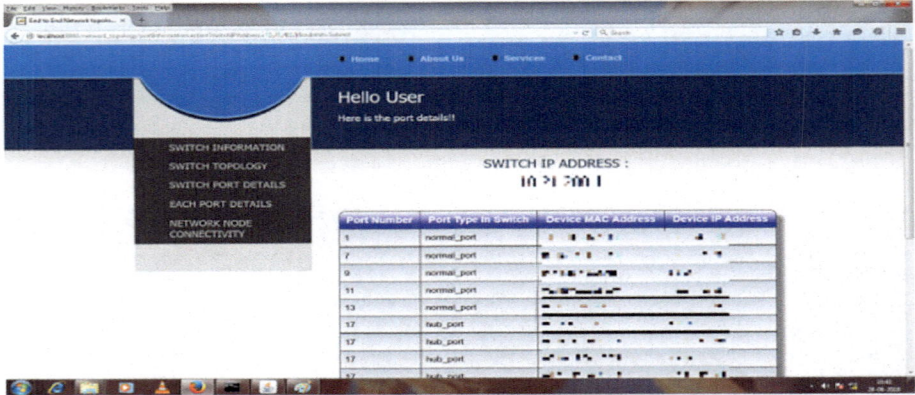

Fig. 6. Switch port details

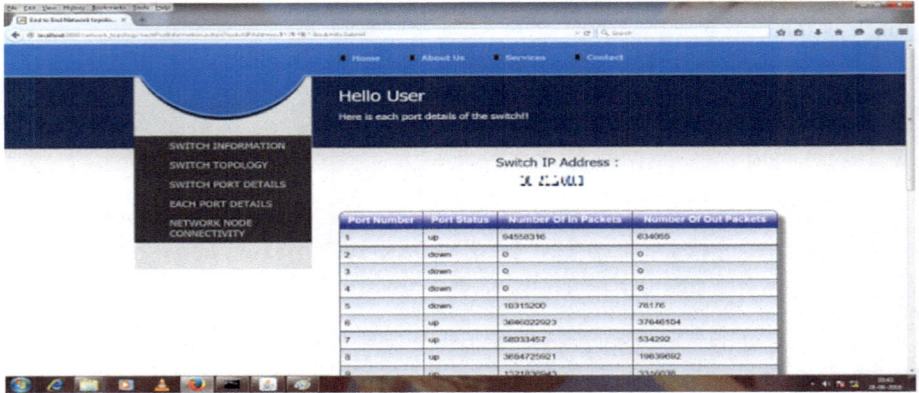

Fig. 7. Each port details

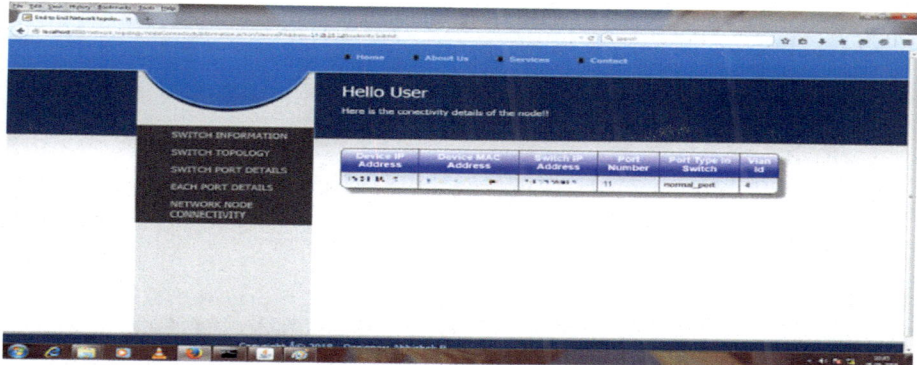

Fig. 8. Node connectivity details

4 Conclusion and Future Scope

The system that has been implemented offers elementary aid to network administrators to find nodes that are connected to their respective ports and provides an easy and fast mechanism to retrieve information of the network. The diagrammatical representation of the network topology of the given switch intended by the admin is visualized. Network node is discovered to which switch that node is connected to by giving the IP address of the node. The network admin find outs the location of the switch from the proposed system without tracing the wires manually.

This system lays the foundation for improvements that can be done to make finding the status of a node much easier. Firstly, the system can be expanded to find the wireless nodes attached in the network. Along with this, the Graphical User Interface can be modified to show a live version of the network in a graphical form. Because of this, an administrator can find out the exact location or the precise cause of the failure of a node or the reason for the faultiness.

References

1. Pandey, S., Choi, M.-J., Lee, S.-J., Hong, J.W.: IP network topology discovery using SNMP. Int. J. Netw. Manag. **21**(3), 169–184 (2011)
2. Zhao, W., Ren, W.: An observer design-pattern detection technique. In: IEEE International Conference on Computer Science and Automation Engineering, pp. 544–547 (2012)
3. Zichao, L., Ziwei, H., Geng, Z., Yan, M.: Ethernet topology discovery for virtual local area networks with incomplete information. In: IEEE International Conference on Network Infrastructure and Digital Content, pp. 252–260, September 2014
4. Siamwalla, R. Sharma, R., Keshav, S.: Discovering internet topology. Cornell University, Ithaca, NY, Technical Report, May 1999
5. Lowecamp, B., O'Hallaron, D.R., Gross, T.R.: Topology discovery for large ethernet networks. In: ACM SIGCOMM, San Diego, CA, USA, pp. 237–248, August 2001

6. Nazir, F., Tarar, T.H., Javed, F., Suguri, H., Ahmad, H.F., Ali, A.: Constella: a complete IP network topology discovery solution. In: APNOMS 2007, Sapporo, Japan, pp. 425–436, October 2007
7. Cisco: SNMP Community String Indexing. http://www.cisco.com/warp/public/477/SNMP/camsnmp40367.html
8. Breitbart, Y., Garofalakis, M., Jai, B., Martin, C., Rastogi, R., Silberschatz, A.: Topology discovery in heterogeneous IP networks: the NetInventory system. IEEE/ACM Trans. Netw. 12(3), 401–414 (2004)
9. Donnet, B., Friedman, T.: Internet topology discovery: a survey. IEEE Commun. Surv. Tutor. 9(4), 56–69 (2007)
10. Bierman, A., Jones, K.: Physical topology MIB. RFC2922, September 2000
11. Cisco: How to Get Dynamic CAM Entries (CAM Table) for Catalyst Switches Using SNMP. http://www.cisco.com/en/US/tech/tk648/tk362/technologies_tech_note09186a0080094a9b.shtml
12. Romascanu, D., Zilbershtein, I.E.: Switch monitoring—the new generation of monitoring for local area networks. Bell Labs Tech. J. 4, 42–54 (1999)
13. Schowalder, J., Martin-Flatin, J.-P.: On the future of internet management technologies. IEEE Commun. Mag. 41, 90–97 (2003)
14. Mansfield, G., Ouchi, M., Jayanthi, K., Kimura, Y., Ohta, K., Nemoto, Y.: Techniques for automated Network Map Generation using SNMP. In: Proceedings of IEEE INFOCOM 1996. Conference on Computer Communications, August 2002. https://doi.org/10.1109/infcom.1996.493314

Multimodal Web Content Mining to Filter Non-learning Sites Using NLP

Sangita S. Modi[1,2(✉)] and Sudhir B. Jagtap[1(✉)]

[1] Research Centre in Computational Science,
Swami Vivekananda Mahavidyalaya, Udgir, Dt, Latur, India
sangitasable1@gmail.com, sudhir.jagtap7@gmail.com
[2] S.R.T.M. University, Nanded, Maharashtra, India

Abstract. Today Internet is a rapidly growing field and it has become one of the huge sources of data. Internet plays a vital role in the educational field. Every student is using e-learning technology over internet to enhance their knowledge. Web mining is one of the data mining branch which helps user to filter and extract relevant data from web and avoid the hitting of the irritating sites. In this paper we have proposed an algorithm of filtering tool which can recognize and block all non learning sites by matching the multiple patterns like text, video and images of the web pages by web content mining. Html document of web page is processed using Natural Language Processing (NLP) and Word Sense Disambiguation (WSD) for recognizing the web content of learning sites.

Keywords: Web mining · Web content mining · NLP · WSD · Data mining

1 Introduction

Today our life is totally dependent on internet, through which we are getting lots of information in a single click. We can do several routine tasks from our remote place. The tasks like air/bus/train ticket reservation, purchasing all type of daily needs, bank transaction, business, marketing, education etc. can be carried out easily by using internet.

E-learning is one of the rapidly growing fields over the internet. Using these internet facilities students can enroll for different job oriented online courses, approach to different companies for placements, and also prepare for interview by taking the guidance from experts, online. While surfing the internet, searching for such online services, Google search engine provides number of links which are unrelated to the desired topic. This makes the searching process, time consuming. To overcome this problem we need to develop one tool, which can easily recognize the desired learning sites and reduce the unnecessary site hits. Such tools are useful for students, professionals, and teachers to achieve their desired goal.

There are billions of web sites coming into the focus every year. It is very difficult to recognize only specific sites by their name or structure. So web mining is one of the data mining sub application which is important to develop personalized tools. Generally, we interact with the web for finding relevant information, discovering new knowledge, personalized web page synthesis, and learning about individual users. Web

A. P. Pandian et al. (Eds.): ICCBI 2018, LNDECT 31, pp. 23–30, 2020.
https://doi.org/10.1007/978-3-030-24643-3_3

mining techniques like clustering, classification, association, and sequential analysis are used to extract information from web. Along with web mining, different research areas can also be used such as, information retrieval and Natural language processing, etc.

1.1   Web Mining

Mobasher et al. has mentioned that web mining is the phase of using the techniques of data mining to automatically retrieve and discover helpful data from WWW services and documents. Although web mining puts down the roots deeply in data mining it is not similar to it [9]. The web data's unstructured feature triggers much complexity of web mining. Data mining and Data warehouse are very popular between IT professional, researchers and students. Today Internet has become a vast data warehouse of knowledge base, built in decentralized and collaborative manner. From this warehouse everybody wants to mine relevant data like sites, links, images, videos etc. Searching for, such desired data from web by detecting and exploiting statistical dependencies between terms, web pages and hyperlinks is called as web mining. Web mining as shown in Fig. 1 is the process which is further divided into three types based on which part of the web is to be mined. They are Web Content Mining, Web Usage Mining, and Web Structure Mining.

The Web Content Mining discovers the useful knowledge from the web sites. The web content is available in different forms such as texts, images, audios, videos, metadata, as well as hyperlinks. Most of the web content data is unstructured and free text data so, the technique of Text Mining is mostly useful in Web Content Mining.

The Web Usage Mining deals with the study of gathered data of web surfers and their surfing behaviors. It utilizes the secondary data on the web such as server access logs, proxy server logs, browser logs, user profile, cookies, bookmarks, mouse clicks, scrolls, and registration data. In web usage, the general access pattern tracking analyses the web log to understand the access patterns and trends. The approach is that to customize usage tracking which analyses individual trends by learning user profiles. Cleaning and data transformation part is important in Web Usage Mining before to start analysis of the collected data.

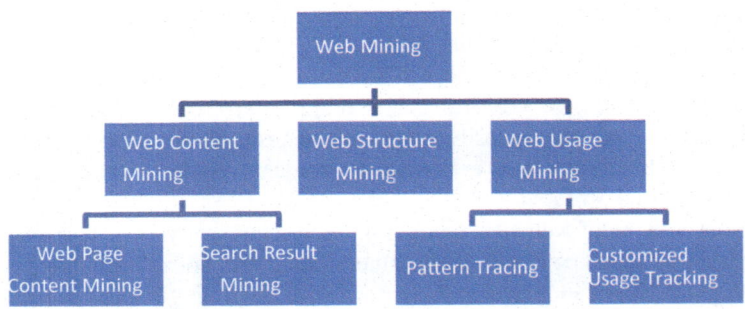

Fig. 1 Web mining tasks

Web Structure Mining is the study of the topology of the hyperlinks and description of the links within web itself, called as Inter Document Structure. It helps to find the similarity and relationship between the different web sites. The application of the Web Structure Mining is page ranking, social network analysis, etc.

Web mining is used to solve the real world problems like hotel price comparison, LIC policy comparison, checking popularity of the party and their political leaders before election. It is mostly used by intended or individual user to customize the information.

1.2   Web Content Mining

Web Content Mining (WCM) is used by researchers to do web scraping or extraction. After web extraction data cleaning is one of the important steps in which the web content is analysed using parsing technique. Web Content Mining uses primary data of the web page. The web page content is always in unstructured form. WCM is used to identify or retrieve useful information from the structured and unstructured data. The structured data like tables, unstructured data like text and semi structured data like html document is used in web content mining. Web Content Mining is one of the difficult task, when we are processing simultaneously on the above three types of data. Web data extractor is used to extract the web content.

1.3   Natural Language Processing

The Natural Language Processing (NLP) plays a vital role in text data processing. The unstructured data like textual data is converted into structured form using features like Word occurrence, Stop Words, Latent Semantic Indexing (LSI), Stemming, N-Grams, Part Of Speech, Positional Collocations, and Word Sense Disambiguation. NLP parses relatively well-formed text and sentences in different languages. Each word has at least 11 senses as nouns, 42 senses as verb. The Part-Of-Speech (POS) is used to assign the tag to each word in supervised or unsupervised manner. The WordNet is an English dictionary and associated lexical network which is used in POS tagging process.

Word Sense Disambiguation (WSD) is initiated after POS tagging is completed, which is used to resolve the ambiguity of the words. WSD method is used to find the correct sense of the word, in which context it is present.

In this paper we proposed a framework which is used to identify the learning sites. The WCM is used to extract all the textual data from the HTML document. In this paper we proposed multiple pattern matching algorithms which are used to retrieve learning sites from set of learning and non learning sites. Here, we considered structured and unstructured data of the web pages. The patterns like images, video, text etc. are extracted from learning sites to form knowledge base. This knowledge base is utilized to recognize learning sites from all types of web sites. For that purpose image tag with captions, video tag with its attribute values like src, video description and alt, and text data are considered for analysis.

In this paper Sect. 1 contains introduction of web mining, Web Content Mining, Natural Language Processing and proposed algorithm of multiple pattern matching. Section 2 contains related work in which literature review is carried out. In Sect. 3 we described methodology and step by step process of proposed system. In this algorithm we have used multiple patterns matching like text, video and images to classify e-learning and non e-learning sites. While Sect. 4, is contains result analysis and Sect. 5 is of conclusion and discussed.

2 Related Work

Zhang et al. has mentioned that multimedia miner consists of 4 main steps excavator of image for image extraction, pre-processor video for image feature extraction, search kernel for matching queries with video and image and discovery module carries out image data [1]. Kazienko and Kiewra have mentioned that a successful astronomical data analysis and cataloging system (SKICAT) generates sky objects digital catalog and it uses the technique of machine learning to transform these objects to human usable classes [2]. Bassiou and Kotropoulos have mentioned that colour histogram matching comprises of colour histogram smoothing and equalization. Equalization predicts correlation between components of colour and this issue is resolved by using smoothening [3]. Jadhav and Kumbargoudar have mentioned that the content of multimedia becomes complex as the sequence develops and the concept being mined may alter as well. Representing and understanding the changes in the process of data mining is essential to mine the multimedia content. Some of the techniques of multi-media data mining are Colour histogram Matching, SKICAT, Shot Boundary Detection and Multimedia Miner [4]. Smeaton et al. has stated that short boundary detection is a technique where the boundaries are detected automatically between video shots [5]. Faria et al. has stated that multimedia is an integration of more than one media such as image, text, audio, video, sound files, numeric, categorical, animation and graphical data. The multimedia is categorized into two types such as dynamic media (music, audio, video, animation and speech) and static media (graphics, images and text) [6]. Ram et al. has stated that multimedia data mining is a sub category of data mining which is required to predict interesting implicit knowledge data from multimedia database. The multimedia database mining needs more than two types of data namely video and text or audio and text video [7]. Mary et al. has mentioned that similar data are retrieved from semi structured data and are embedded in a set of useful data and stored in OEM (Object exchange model). It supports the user to perceive the data structure on web much exactly [8].

Pol et al. (2008) has stated that content mining can be carried out on unstructured information namely text and unstructured data mining provide unknown data. Content mining needs application of text mining and data mining techniques. Some of the techniques used in text mining are topic tracking, information extraction, categorization, summarization, information visualization and clustering [10]. Similarly Umadevi, Uma Maheshwari and Nithya stated that www provides huge opportunities in education field. With the huge development of data available on website web mining has become applicable for educational systems based on web. Virtual courses, educational sites, digital books and web supported instruction shells are certain modes of providing e-learning. Web mining is the set of tasks used for extracting or mining useful data from websites or web pages. It offers intrinsic knowledge of learning and teaching method for efficient planning of education by using different tools and technologies [11]. Sangita and Sudhir J. introduced Keyword based web site filtering browser integrated tool. This is one of the client side browser extension web site filtering and useful tool to browse only learning site and blocking all non learning sites. This tool avoids unwanted sites hit and improve usability of learning sites [12].

3 Framework of Proposed Work

The proposed algorithm is used for multiple patterns matching of web page content. The web page consists of multiple types of content. Using this algorithm firstly we are collecting the learning sites keyword database in which different types of patterns are considered like video, image and text. This type of multimedia data patterns are used to collect the keywords from the respective tags of html source. Using regular expression and NLTK tool kit we are extracting the nouns from each tag, like paragraphs, headlines, image, and video. Remove the punctuations, stop words, and duplicate words to create a unique learning keyword dataset. Next step is that, to apply Part-of-speech (POS) method to extract nouns from the keywords to form learning sites unique nouns keyword dataset. The Word sense disambiguation (WSD) is used in this algorithm which is the part of NLTK tool kit. The WSD is utilized to find the correct sense of the word in the sentence. The word or nouns extracted from each web pages html document will be compared with the unique noun keyword dataset. If the keyword matching value is greater than threshold value then it will be classified as learning otherwise it will be blocked as non learning.

First algorithm is used for dataset creation for text, image, and video keywords of learning sites. Take the set of URLS_Allowed {Ua1, Ua2, … Uan} and URLS_-Blocked {Ub1, Ub2, … Ubn} of learning and non learning sites respectively. After processing the web page contents we will get the three types of keywords dataset DV, DI, and DA which contains the keywords which are found in text, image and video tags of html page. Using web content extraction code we have extracted the html source code from which text data, image tag data like the name of the image, alt attribute value are and all video related tags data like the name of the video file as are considered to create a keyword database of text, image, video patterns of web page.

The second algorithm is an urls blocker tool which can check every urls, web page content by extracting the content of web page. Then extract the text, image and video tag contents of the current web page. Remove the stop words, punctuations, and duplicate words from the collected text. Then extract the nouns from those texts and then matched with the existing database keywords contained in DV, DI, DA dataset. If matching percentage of keywords is greater than the threshold value, then that URLs will available for student to browse, otherwise that particular site is considered as non learning and discarded by browser.

Dataset Algorithm

Step 01. Get the list of URLs URLS_Allowed = {Ua1, Ua2,...Uan}

Step 02. Get the list of URLs URLS_Blocked = {Ub1, Ub2,...Ubn}

Step 03. For each U in URLS_Allowed

Step 04.   Extract text TEXT_Allowed from U

Step 05.   Filter only alphabetic text i.e. TEXT_Allowed' = Fa(TEXT_Allowed)

Step 06.   Tokenize TEXT_Allowed' i.e. TOKENS_Allowed= {T1, T2, Tm}

Step 07.   Append TOKENS_Allowed to the keywords dataset DV' = {DV1,DV2,...,Dlv} for Video tags, DI' = {DI1,DI2,...,Dli} for image Tags, DA' = {DA1,DA2,...,Dla} for remaining tags

Step 08. Remove duplicates DV" = unique (DV'), DA" = unique (DA'), DI" = unique(DI')

Step 09. For each U in URLS_Blocked

Step 10.   Extract text TEXT_Blocked from U

Step 11.   Filter only alphabetic text i.e. TEXT_Blocked' = Fa(TEXT_Blocked)

Step 12.   Tokenize TEXT_Blocked' i.e. TOKENS_Blocked = {T1, T2, ...Tm}

Step 13.   Remove if present; TOKENS_Blocked from the keywords dataset DV", DI", DA"

Step 14. Sort the nouns i.e. DV = sort (DV"), DI = sort (DI"), DA = sort (DA")

Step 15. Store the keywords dataset DV, DI, DA.

Filtering Algorithm

Step 01. Get the requested url 'Ur'

Step 02. Extract text TV from the retrieved page with url 'Ur' and tag Video

Step 03. Apply NLTK to extract nouns NV = {NV1, NV2,...,NVnv}

Step 04. Find percentage of these nouns PV = nv(NV Intersection DV)/nv(NV) that are present in the dataset DV = {DV1,DV2,...,DVnv}

Step 05. Check if percentage PV is more than threshold 'thetaV' i.e. PV > 'thetaV'

Step 06. If Yes then allow the url 'Ur' to open

Step 07. If No then disallow the url 'Ur' from opening

Step 08. Extract text TI from the retrieved page with url 'Ur' and tag Image

Step 09. Apply NLTK to extract nouns NI = {NI1, NI2,...,NIni}

Step 10. Find percentage of these nouns PI = ni (NI Intersection DI)/ni(NI) that are present in the dataset DI = {DI1,DI2,...,DIni}

Step 11. Check if percentage PI is more than threshold 'thetaI' i.e. PI > 'thetaI'

Step 12. If Yes then allow the url 'Ur' to open

Step 13. If No then disallow the url 'Ur' from opening

Step 14. Extract text TA from the retrieved page with url 'Ur' and remaining tags

Step 15. Apply NLTK to extract nouns NA = {NA1, NA2,...,NAna}

Step 16. Find percentage of these nouns PA = na(NA Intersection DA)/na(NA) that are present in the dataset DA = {DA1,DA2,...,DAna}

Step 17. Check if percentage PA is more than threshold 'thetaA' i.e. PA > 'thetaA'

Step 18. If Yes then allow the url 'Ur' to open

Step 19. If No then disallow the url 'Ur' from opening

Step 20. Stop

4 Result Analysis

In this above algorithm we have tested for text pattern by extracting the text of the web page. For that we have collected 765 Learning sites web pages and 795 Non Learning site database and processed using python programming language. Web content extraction is one of the primary and challenging process which is done by beautifulSoup package of python. After extraction preprocessing step is carried out, in which data cleaning is the first step. We have tokenize each extracted word, removed all Stop words, numbers, punctuations and duplicate words. In later step we have extracted the nouns of each web page using natural language processing NLTK toolkit, which is in built available in python. Using NLTK toolkit with available WordNet dataset we have resolved the Word Sense Disambiguation. This removes ambiguous words (word having multiple meaning) and find the exact meaning of that word in which it is present. Without NLP this algorithm can give 70% accuracy in identification of learning sites while using Natural language processing we have identified 99% of learning site and 98% of the non learning sites accurately.

5 Conclusion

The web filtering tool development process requires Web Content Mining along with Natural Language Processing. Now a day's internet is necessary in educational organization, but to restrict student from accessing unwanted sites filter is necessary tool which should work on client side. This tool works to classify sites in learning and non learning which helps to reduce unnecessary page hits of students. Student can only access relevant web pages on desktop. This type of filtering tool plays a vital role to improve E-learning site usability.

References

1. Zhang, J., Hsu, W., Lee, M.L.: Image mining: issues, frame works and techniques. In: Proceedings of the 2nd International Workshop Multimedia Data Mining, pp. 13–20 (2001)
2. Kazienko, P., Kiewra, M.: Link recommendation method based on web content and usage mining. In: New Trends in Intelligent Information Processing and Web Mining Proc of the International IIS: IIPWM 2003 (2003)
3. Bassiou, N., Kotropoulos, C.: Color histogram equalization using probability smoothening. In: Proceedings of XIV European Signal Processing Conference (2006)
4. Jadhav, S.R., Kumbargoudar, P.K.: Multimedia data mining in digital libraries: standards and features. In: Proceedings of READIT, p. 54 (2007)
5. Smeaton, A.F., Over, P., Doherty, A.R.: Video shot boundary detection: seven years of TRECVID activity. Elsevier Comput. Vis. Image Und. **114**(4), 411–418 (2010)
6. Faria, F., Santos, J.D., Rocha, A., Da Torres, R.: A framework for selection and fusion of pattern classifiers in multimedia recognition. Pattern Recognit. Lett. **20**(7), 1–13 (2013)
7. Ram, M.K., Rao, M.V., Sujana, C.: An overview on multimedia data mining and its relevance today. Int. J. Comput. Sci. Trends Technol. **5**(3), 108–113 (2017)

8. Mary, X.L., Silambarasan, G.: Web content mining: tool, technique & concepts. Int. J. Eng. Sci. **7**(5), 11656 (2017)
9. Mobasher, B., Cooley, R., Srivastava, J.: Automatic personalization based on web usage mining. Commun. ACM **43**(8), 142–151 (2000)
10. Pol, K., Patil, N., Patankar, S., Das, C.: A survey on web content mining and extraction of structured and semi structured data. In: IEEE First International Conference on Emerging Trends in Engineering and Technology, pp. 543–546 (2008)
11. Umadevi, K., Uma Maheshwari, B., Nithya, P.: Design of e-learning application through web mining. Int. J. Innov. Res. Comput. Commun. Eng. **2**(8), 5324–5329 (2018)
12. Modi, S., Jagtap, S.: Browser integrated web content filtering using natural language processing. Int. J. Mod. Electron. Commun. Eng. (IJMECE) **6**(4), 108–112 (2018). ISSN: 2321-2152

Towards Converging Fully Homomorphic Batch Encryption with Data Integrity Assurance in Cloud Setting

S. Hariharasitaraman[✉] and S. P. Balakannan

Department of İnformatıon Technology, Kalasalingam Academy of Research and Education, Anand Nagar, Krishnankoil, Tamilnadu, India
`hariharasitaraman@gmail.com`, `balakannansp@gmail.com`

Abstract. In encryption scheme setting, users who possess the right key can authorize to do computation with the data. In particular, cloud storage users lose their rights in terms of integrity on data, even after they ends up metered service from a CSP(Cloud Storage service provider), There is a possibility of dishonest computation over an old data by CSP also. Fully homomorphic encryption scheme address the problem of possessing the data in untrusted servers. In this proposed work, authors bring in a fully Homomorphic Batch Encryption Scheme using Integers with Shorter public key (HBEIS) converged with integrity verification process in a cloud setting. Any third-party auditor can do random audit on the encrypted data for assuring the correctness of data without possessing the original data. In our experimental set up, HBEIS scheme ensures optimal space and linear computational complexity.

Keywords: Fully homomorphic encryption · Public auditing · Integrity verification · Probabilistic query · Outsourced data

1 Introduction

In Greek mythology, 'homos' mean same, 'morphos' represents shape [14]. In abstract algebra, homomorphism in a function, which gets inputs from various algebraic sets and outputs an element which has the values of some known operations like addition, multiplication [13]. In designing a crypto system, the advantage of encryption property in FHE is very much used. Fully Homomorphic Encryption does an encryption in which any CSP can do particular computation on it. The notable property here is the proportion of data and format of encrypted data is preserved using this scheme.

Any encryption scheme $\in$ is composed of three algorithms for its operation keygen$_\in$, Encryp$_\in$, Decryp$_\in$ that should run in a polynomial time with respect to the key size (λ) [15]. In a symmetric key encryption scheme, a single key (λ) that is used in both encrypt$_\in$ and decrypt$_\in$, by the keygen$_\in$ algorithm. In an asymmetric encryption scheme, keygen$_\in$ algorithm generate key pairs a public key p_k, a private decryption key s_k. In a homomorphic encryption scheme can be of symmetric key encryption or asymmetric key encryption in nature. If it is a asymmetric one, another algorithm is added for the functioning of the scheme called eval$_\in$. For a function f in a encryption

© Springer Nature Switzerland AG 2020
A. P. Pandian et al. (Eds.): ICCBI 2018, LNDECT 31, pp. 31–40, 2020.
https://doi.org/10.1007/978-3-030-24643-3_4

scheme and cipher texts ($c_1 \ldots\ldots c_n$) with c_i = encrypt$_e$ (Private key and the message), outputs a cipher text which encrypts f (messages) [15]. The below diagram explains the above operation on homomorphic encryption (Fig. 1).

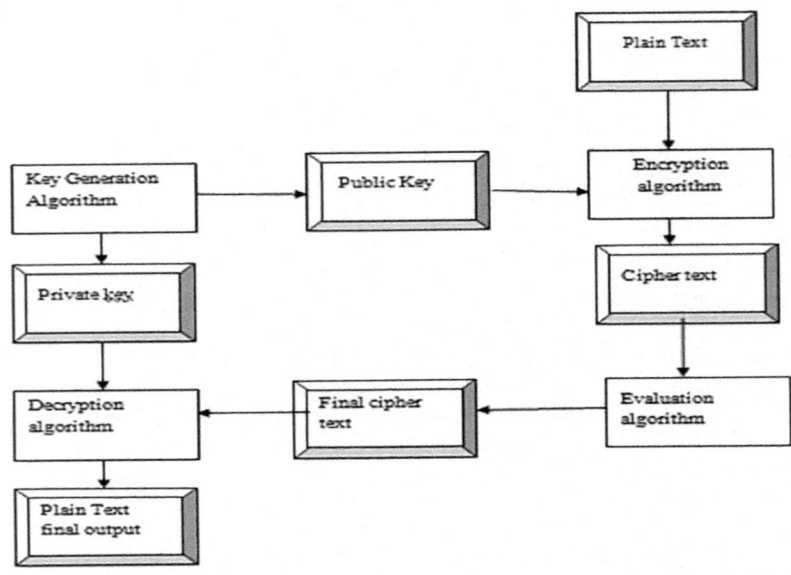

Fig. 1. Functions of Asymmetric Key Homomorphic encryption [Source: David W. Archer, pp. 22–29, IEEE Security & Privacy][16]

The addition operation over an encryption scheme is given by, for any messages 'm_1' and 'm_2' the result can be "encryption of message 1" + "encryption of message 2" is "encryption($m_1 + m_2$)" without reading the message m_1 and m_2. In another scenario for a multiplicative operation for the set of operation '$m1$' and '$m2$' gives encryption ($m_1 * m_2$).

In 1978, Rivest et al. [12] converged the homomorphic properties and cryptographic field, in short, computing without actually, decrypting the messages. In 1982, Yao [17] designed two user communication scheme using Boolean circuits. Their works had worst computational and communication complexity. In 2009, Gentry [8] designed a FHE scheme, which works upon on encrypted data with limited number of operations. Gentry's doctoral dissertation reveals that the evaluation of FHE based on Lattices scheme [8]. The scheme has high computation cost and it was practically difficult to implement. Van et al., in 2010 [3] introduced FHE over integers based on a Greatest Common Divisor problem, Brakersi et al. [5, 6] came up with LWE Scheme. The development of FHE schemes is broadly classified into three types, PHE-Partial Homomorphic, SWH-Some What Homomorphic, FHE-Fully Homomorphic encryption scheme based on types of operations it does and number of times of operation it supports [2, 3, 6, 8].

The time line of FHE along with its variants is given in the diagram (Fig. 2).

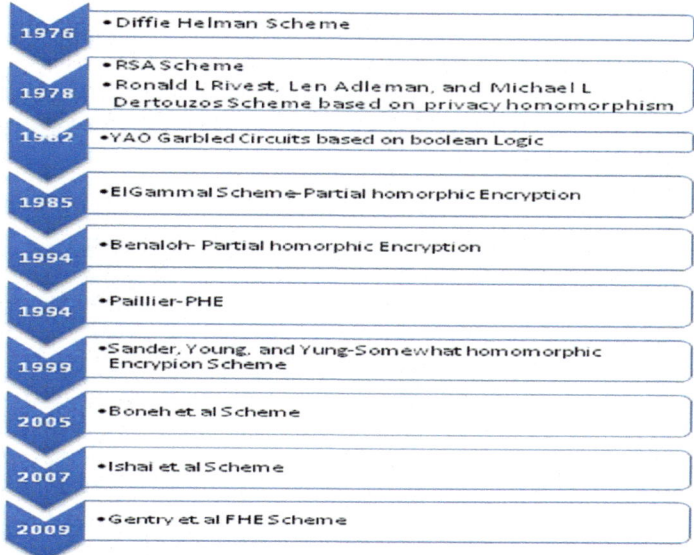

Fig. 2. Timeline of FHE and its variant (Source: Abbas et al. [11])

2 Architectural Model of HBEIS Scheme

Figure 3 shows the simple illustration of HBEIS over cloud setting. Alice wants to send message to a cloud store, she decrypts the message using an encryption algorithm (Step1 & Step 2), if any auditor wants to do a computation with a stored message he will query to the cloud storage (Step 3). The evaluation scheme performs computation using a function called 'Eval' (Step 4) and the result is encrypted message to the auditor (Step 5).

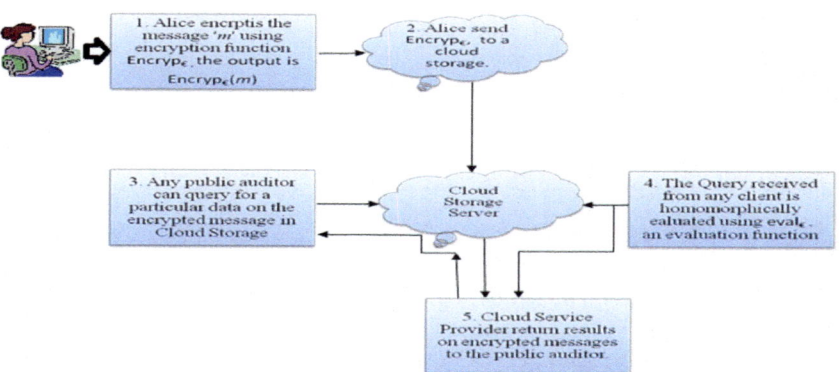

Fig. 3. HBEIS Scheme on Cloud Scape

2.1   Definitions

The message space 'M' contains set of messages m_1, $m_2 \ldots \ldots m_n$. An encryption scheme or algorithm is denoted by the term 'encryp', an encryption scheme is said to be homomorphic in nature on an operation '+', the following conditions to be satisfied.

$$\text{Encryp}(m_1 + m_2) = \text{encryp}(m_1 + m_2) \text{ for all } m_1, m_2 \text{ in 'M'} \tag{1}$$

The operation are to be performed such as additive, multiplicative are functionally complete over finite algebraic sets, to make an encryption scheme which supports homomorphic evaluation over any functions, the above operations is sufficient. Any FHE scheme is designed to support four main operations which are key generation, encryption, and decryption and evaluation algorithm.

The function of key generation algorithm is to generate private key or public key used for both encryption and decryption or to execute the scheme in symmetric mode or asymmetric mode. Working functionality of evaluation function is the innovation here; it takes inputs of cipher text message, does a computation on cipher text and returns a cipher text of computed results. Thus, the size of cipher text is constant for unlimited number of computations. Also, the practical implementation of operations can be easily implemented by using logic gates in Boolean theory.

2.2   Basic Construct in HBEIS Scheme

Notations and Mathematical background

m is the bit plaintext (m = 0 or 1),
q is a large random integer,
r is a small random.
p - Co prime numbers
b - Random binary vector
γ is bit length of any integers in P_k.
η - Bit length of 'P'.
p - bit length of noise [distance between public key integers & nearest multiples of secret key].
τ - No. of integers in the public key.
λ - Security parameter
C, C1, C2.. Cn-Cipher text.
m, m0, m1, m2... Mn-messages
M - Message Space.
P_k - Public key.
S_k - Secret key or Private Key.

The HBEIS scheme contains four algorithms namely, Key gen(1^λ, $Q_i(0 \leq i \leq n)$), Encryp(P_k, m), Eval(P_k, C, C_1, C_2), Decryp(S_k, C). The proposed scheme uses to check the integrity of data is the modified DGHV which supports simultaneous processing of vector of plain text bits. Instead of using single prime 'p' as their secret key, this study uses several such primes $p_0, p_1, \ldots, p_{n-1}$. So, the private key is a set of primes or co primes $p_0, p_1, p_2 \ldots p_{n-1}$. Thus, the FHE with integers is altered as follows [1]

$$\begin{aligned} x = \text{random integer of large size}(p) * \text{secret key}(q) \\ + 2 * \text{random integer of small size}(r) + \text{message} \end{aligned} \tag{2}$$

This study uses a group of co primes to find the secret key as per Chinese Remainder Theorem [1, 18]. Each p_i handles a plain text bit. The cipher text is given by

$$\text{Cipher text} = \text{CRT}_{q0,p0,p1,p2\ldots,pn-1}(\text{random integer large size}, 2r_0 + m_0, \ldots, 2r_{n-1} + m) \tag{3}$$

The decryption of the cipher text is given by

$$m_i = [c \bmod p_n] \bmod 2 \text{ for all } 0 < i < n \tag{4}$$

2.3   Key Generation Algorithm (1^λ, $Q_i(0 \leq i \leq n)$)

(a) Pick any n-bit prime numbers 'p_i' such that $0 \leq i \leq n$ and $\prod$ as the product value.
(b) The public key is defined as follows, $x = q*\prod$ where q_0 is a set of primes $\cap$ 0, $-2^{\gamma/}\prod$.
(c) The private key is defined by $P_k = x_0, Q_i, x_i, x_j.$
(d) The Secret key is $S_k = p_j.$

2.4   Encryption Algorithm (P_k, M)

(a) For any $M = m_1, m_2, \ldots m_n$, where $n \sum$ set of primes.
(b) Pick any random binary vector b.
(c) The cipher text is given by

$$\left[\sum_{i=0}^{n-1} m_i.x_i^1 + \sum_{i=1}^{n} b_i.x_i \right] \tag{5}$$

2.5   Decryption Algorithm (S_k, c)

(a) The decryption will yield original plain text bit.

$M = m_1, m_2, \ldots m_n$, where

$$m_i = (\text{Ciphertext}) \bmod (\text{private key } Pi) Qi \tag{6}$$

Cipher text – C; Private key – P.

2.6   Evaluate Algorithm (P_k, C, C_1, C_2)

(a) Choose the public key P_k and τ of Cipher text 'C'.
(b) Input P_k and τ to a circuit $\dot{C}$ to perform addition & multiplication over integers.
(c) The addition function does XOR operation on cipher text, which is given by

> Input $\rightarrow$ Addition (P_k, C_1, C_2).
> Output $\rightarrow$ [C_1 XOR C_2] mod x_0.

(d) The multiplication function does NAND operation of cipher text, which is given by

> Input $\rightarrow$ Multiplication (P_k, C_1, C_2).
> Output $\rightarrow$ [C_1XORC_2] mod x_0.

3   Implementation, and Evaluation Using Advanced Encryption Standard Circuit and Comparison with DGHV [3] and BGV [6] Schemes

For practical implementation of proposed scheme, this study utilizes Microsoft Corporation. The Simple Encrypted Arithmetic Library (SEAL) homomorphic encryption library, connected through .NET wrappers for the public API, Hao Chen et al. [21]. The results are measured in a hardware setup in a desktop machine with the following configuration, Intel Pentium i3-Processor, 2.2 GHZ, with 4 GB RAM. The graphs are plotted to calculate the computation time for encryption algorithm, decryption algorithm, key generation algorithm and evaluation algorithm with respect to growth of parameters (λ, ρ, η, γ, β).

While implementing the proposed scheme, homomorphic evaluation of AES-256 encryption circuit is used. The practical implementation of such FHE scheme is seen in the following studies of Brakerski et al. [5, 6], Gentry et al. [2], dos Santos et al. [7], Coron et al. [4], Apurva sachan [9] and Tiago et al. [11]. In the studies of Smart et al. [20], Naehrig et al. [19], Gentry et al. (2012) [2, 8], their implementations results ends up in less communication cost under a cloud setting. Although, cipher text computation requires costly computation, data could be encrypted using AES circuit with a restriction of cipher text growth approximately equal to 1, with the public key of this scheme and AES private key is encrypted using HBEIS public key. Thus any cloud auditor executes an AES decryption scheme in a homomorphic way to obtain the plain text data which is encrypted under public of HBEIS scheme for applying the homomorphic operations in the stored data.

The Fig. 4 given below shows the growth of various parameters in the key generation algorithm with respect to DGHV [3] and BGV scheme [6]. This proposed scheme key generation algorithm runs faster in generating private or public keys in batch mode, while comparing with the other schemes.

In Fig. 5, the encryption algorithm shows computational complexity which grows constantly till the homomorhic parameters value up to 30 rounds of execution, and then

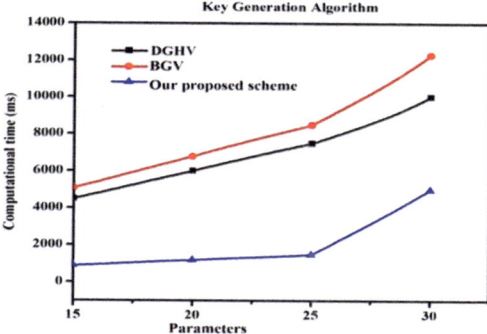

Fig. 4. Comparison of homomorphic paramters of keygeneration algorithm with DGHV and BGV schemes.

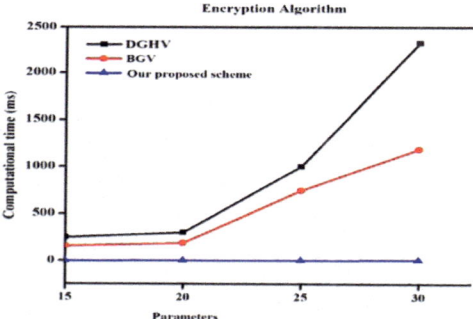

Fig. 5. Comparison of homomorphic parameters of encryption algorithm with DGHV and BGV schemes

it is growing linearly. Therefore, the computation time is linear which is better than the BGV and DGHV schemes.

The Fig. 6, shows the decryption algorithm which executes as equal to the BGV scheme, but it does faster computation than DGHV scheme. The order of growth of decryption algorithm and the parameters is constant because of the faster decyption of keys.

Figure 7 shows that Evaluation algorithm takes more time for computation because, evaluation algorithm is run by third party computer, the computation time is proportionate to communication time also. Thus, the proposed system executed faster than DGHV and BGV in terms of time complexity.

To address the optimum space for execution of various algorithms in the proposed scheme, homomorphic parameters are set as below $\lambda = 62$; $\rho = 32$; $= 2176$ $\gamma = 4{:}2$ $*10^6$ $\beta = 44$ [4] can yield better space complexity too. Any crypto system is designed in such a way, that it should not be compromised. The DGHV scheme, BGV scheme and HBEIS scheme are designed in such a way by the toughness of the approximate integer common divisor problems [9]. The effort required by an adversary to know the

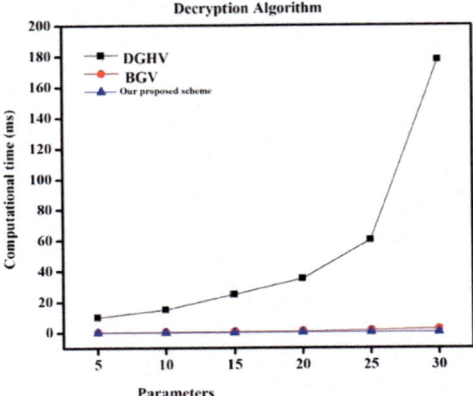

Fig. 6. Comparison of homomorphic parameters of decryption algorithm with DGHV and BGV schemes

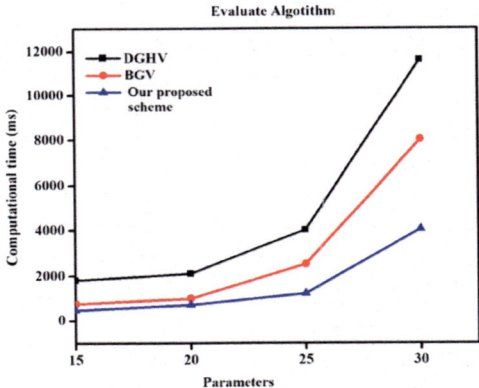

Fig. 7. Comparison of homomorphic parameters of evaluate algorithm with DGHV and BGV schemes

private key and finding such key says the crypto system is secure or not. In this proposed scheme, finding such private key depends on the hardness of integer factorization. By proper setting of the above values parameters, this proposed scheme ensures protection under GCD attack [10].

4   Conclusions and Future Scope

The HBEIS scheme is designed by taking the advantage of batch mode Fully Homomorphic Encryption techniques. The disadvantage of FHE is practical implementation and public key size. The experimental result shows the proposed scheme is practically possible and shorter public key results in reduced space complexity and computational cost. The objective of integrity verification and third-party auditing in cloud storage setting is proven. The selection of strong security parameter set shows the scheme is secured against GCD attack and efficient. In future, the practical implementation with parallel computation set up can yield much lesser computational cost and communication cost. This suggestion may be taken into account for further study.

References

1. Cheon, J.H., Coron, J.-S., Kim, J., Lee, M.S., Lepoint, T., Tibouchi, M., Yun, A.: Batch fully homomorphic encryption over the integers. In: Johansson, T., Nguyen, P.Q. (eds.) EUROCRYPT 2013. LNCS, vol. 7881, pp. 315–335. Springer, Heidelberg (2013)
2. Gentry, C., Halevi, S.: Implementing gentry's fully-homomorphic encryption scheme. In: Paterson, K.G. (ed.) EUROCRYPT 2011. LNCS, vol. 6632, pp. 129–148. Springer, Heidelberg (2011)
3. van Dijk, M., Gentry, C., Halevi, S., Vaikuntanathan, V.: Fully homomorphic encryption over the integers. In: Gilbert, H. (ed.) EUROCRYPT 2010. LNCS, vol. 6110, pp. 24–43. Springer, Heidelberg (2010)
4. Coron, J.-S., Mandal, A., Naccache, D., Tibouchi, M.: Fully homomorphic encryption over the integers with shorter public-keys. In: Advances in Cryptology – Proceedings of Crypto 2011. Lecture Notes in Computer Science, vol. 6841, pp. 487–504. Springer, Heidelberg (2011)
5. Brakerski, Z., Vaikuntanathan, V.: Efficient fully homomorphic encryption from (Standard) LWE. In: Proceedings of the 2011 IEEE 52nd Annual Symposium on Foundations of Computer Science. FOCS 2011, pp. 97–106. IEEE Computer Society, Palm Springs (2011)
6. Brakerski, Z., Gentry, C., Vaikuntanathan, V.: Fully homomorphic encryption without bootstrapping. In: Proceedings of the 3rd Innovations in Theoretical Computer Science Conference, Proceeding ITCS 2012, Cambridge, Massachusetts, pp. 309–325 (2012). Also available in Cryptology Eprint Archive, Report (2011/277)
7. dos Santos, L.C., Bilar, G.R., Pereira, F.D.: Implementation of the fully homomorphic encryption scheme over integers with shorter keys. In: 7th International Conference on New Technologies, Mobility and Security (NTMS). IEEE Xplore digital Library, Paris (2015)
8. Gentry, C.: A Fully Homomorphic Encryption Scheme, Doctoral Dissertation, Stanford University, Stanford, California (2009). http://Crypto.Stanford.Edu/Craig
9. Sachan, A.: Implementation of Homomorphic Encryption Technique, M. Tech Thesis. National Institute of Technology, Rourkela, June (2014). http://ethesis.nitrkl.ac.in/6436/
10. De Carvalh, N.T.F.: A Practical Validation of Homomorphic Message Authentication Schemes, Doctoral Dissertation, Universidade Do Minho, Outubro De, Portugal (2014)
11. Acar, A., Aksu, H., Uluagac, A.S., Conti, M.: A Survey on Homomorphic Encryption Schemes: Theory and Implementation, arXiv preprint arXiv:1704.03578 (2017)
12. Rivest, R.L., Adleman, L., Dertouzos, M.L.: On Data Banks And Privacy Homomor-Phisms. Foundations of Secure Computation, vol. 4, 11th edn, pp. 169–180. Academic Press Inc, Cambridge (1978)
13. Malik, D.S., Mordeson, J. N., Sen, M.: MTH 581–582 Introduction to Abstract Algebra, pp. 581–582 (2007)
14. Liddell, H.G., Scott, R.: An İntermediate Greek-English Lexicon: Founded Upon the Seventh Edition of Liddell and Scott's Greek-English Lexicon. Harper & Brothers, New York (1896)

15. Gentry, C.: Computing arbitrary functions of encrypted data. Commun. ACM (CACM) **53**, 97–105 (2010)
16. Archer, D.W., Rohloff, G.K.: Computing with data privacy: steps toward realization. IEEE Secur. Privacy **13**(1), 22–29 (2015)
17. Yao, A.C.-C.: Protocols for secure computations. In: 23rd Annual Symposium on Foundations of Computer Science (SFCS 1982), FOCS, Chicago, Illinois, vol. 82, pp. 160–164 (1982)
18. Kim, J., Lee, M.S., Yun, A., Cheon, J.H.: CRT-Based Fully Homomorphic Encryption Over the Integers. Information Sciences, vol. 310, pp. 149–162. Elsevier, Amsterdam (2015). https://doi.org/10.1016/j.ins.2015.03.019
19. Naehrig, M., Lauter, K., Vaikuntanathan, V.: Can homomorphic encryption be practical? In: Proceedings of the 3rd ACM Workshop on Cloud Computing Security Workshop, CCSW 2011, pp. 113–124. ACM (2011)
20. Smart, N.P., Vercauteren, F.: Fully homomorphic encryption with relatively small key and ciphertext sizes. In: Nguyen, P.Q., Pointcheval, D. (eds.) PKC 2010. LNCS, vol. 6056, pp. 420–443. Springer, Heidelberg (2010)
21. Chen, H., Laine, K., Player, R.: Simple Encrypted Arithmetic Library - SEAL v2.2, Microsoft Research foundation-Tech report, Number MSR-TR-2017-22 (2017)

Network Security: Approach Based on Network Traffic Prediction

Sheetal Thakare[1]([✉]), Anshuman Pund[2], and M. A. Pund[3]

[1] Department of Computer Engineering,
Bharati Vidyapeeth College of Engineering, Navi Mumbai, India
sheetal.thakare@gmail.com
[2] Head Information Security, IDBIIntech, Navi Mumbai, India
anshuman.pund@idbiintech.com
[3] Department of Computer Science and Engineering,
Prof. Ram Meghe Institute of Technology & Research,
Amravati University, Badnera, Amravati, India
mapund@mitra.ac.in

Abstract. Considering the network security aspect, one of the best way of preventing network infrastructure against anomalous activities is to monitor its traffic for suspicious activities. The reliable resource to accomplish this task is past network flow data, which can be analyzed to detect congestions, attacks or anomalies to ensure effective QoS of network infrastructure. Network traffic prediction involves analysis of past network flow data by capturing-storing data, preprocessing data, analyzing it based on various parameters & forming behavior patterns for various nodes in network. Once the patterns are observed for different nodes in network, their future communication can be predicted. Upon prediction of anomalous behavior, the preventive action will be initiated without wasting much of a time. Thus reducing the MTTR (mean time to respond) is the outline of our paper. The importance of network traffic data, traffic prediction methods and literatures available on topic are studied in this paper.

Keywords: Network traffic prediction · ARMA · SARIMA ·
Time series model

1 Introduction

The various components of network infrastructure like firewalls, bridges, switching and routing devices, etc. produce traffic data related to network. These data are also called as network flow data. Analysis of network performance can be efficiently done using this data. The obtained analysis would be a valuable resource for network security teams for further network enhancement and optimization. The network flow data reflects real-time view of the network traffic, integrated with peripheral devices and point solutions. Peripheral devices form outermost defense line, preventing entry of most of malicious things into the network. Still 100% capture/prevention of the malicious things is impossible. Only single anomaly can wreak dangerous havoc and on getting inside, peripheral devices will be of no help. Even though localized solutions

© Springer Nature Switzerland AG 2020
A. P. Pandian et al. (Eds.): ICCBI 2018, LNDECT 31, pp. 41–55, 2020.
https://doi.org/10.1007/978-3-030-24643-3_5

enhance security by encountering specific problems, broad-based protection is still unreachable for them. Thus even if various components are already present, to strengthen network security, network traffic data analysis and prediction is required (Fig. 1).

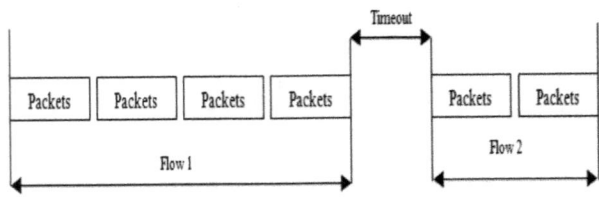

Fig. 1. Network traffic data flow

2   The Importance of Network Traffic Data

A huge amount of data is been produced by traffic that goes from network infrastructure. This is termed as network flow data. It is a good measure for analyzing performance of network. But if this network flow data is scanned to a very root level, it will act as utmost important resource for securing network from various kinds of attacks. Network infrastructure can be optimized with the output of network flow analysis as well as strength will be added to the existing defense mechanism implemented in infrastructure. Strengthening of defense mechanism is possible if mitigating actions can be initiated within no time lag upon attack. This scenario is possible if attack or anomalous behavior can be known or predicted beforehand. Past flow analysis data will help for prediction of anomalous behaviors. If upon prediction, mitigating or preventive actions can be recommended implicitly, then time required to respond to different anomalous network situations will improve drastically.

Other advantages of network flow data analysis are listed below [19] (Table 1).

Table 1. Importance of network traffic data

Network Perceptibility	Network traffic data provides complete internal perceptibility of network
Identified and Unidentified Attacks Detection	Handling know attacks is a huge task along with detection of unknown attacks, e.g. toxic data exfiltration, specially when data is unstructured
Detection of legal user acting unethically	Insider (a legitimate user) can be a hidden threat, which will be detected with who- what- where- when analysis of network flow data
Fasten response time for threat events	To save the network infrastructure from damage quick incident response is the need of hour
Capture policy violations	Network flow analysis captures violations and alerts on policy violations
Support tracing of affected nodes	Network flow analysis can trace nodes communicated with critical data containers, alerts are obtained for such transaction with familiar threats
Network Operations collaborate smoothly	User experience and network system functionality can be reviewed using network data analysis, further helping in capacity planning. Also NetOps and SecOps teams can collaborate smoothly using this analysis resolving problems faster and without pointing fingers at each other
Unefficient node detection	Node responding very slow can be found out and upgraded

(continued)

Table 1. (*continued*)

Information Outflow detection	Personal or confidential information flowing out of the network can be captured
Improved resource uasage record	Improved resource usage records can be maintained with real-time network bandwidth usage statistics
Node grouping	Depending on data flow, nodes or devices can be clubbed into logical groups for easier report maintenance

3 Techniques for Network Traffic Prediction

The techniques can be divided as statistical & composite techniques. Statistical techniques use linear & non linear time series data models. Composite (statistical plus other domain) are based on data mining, neural network, Hadoop, PSO etc. Some have used term decomposed models when time series is decomposed into four components. Linear time series techniques are AR (Auto Regressive) and MA (Moving Average). When combined together, they create ARMA (Auto Regressive Moving Average) model [22–24] (Fig. 2, Table 2).

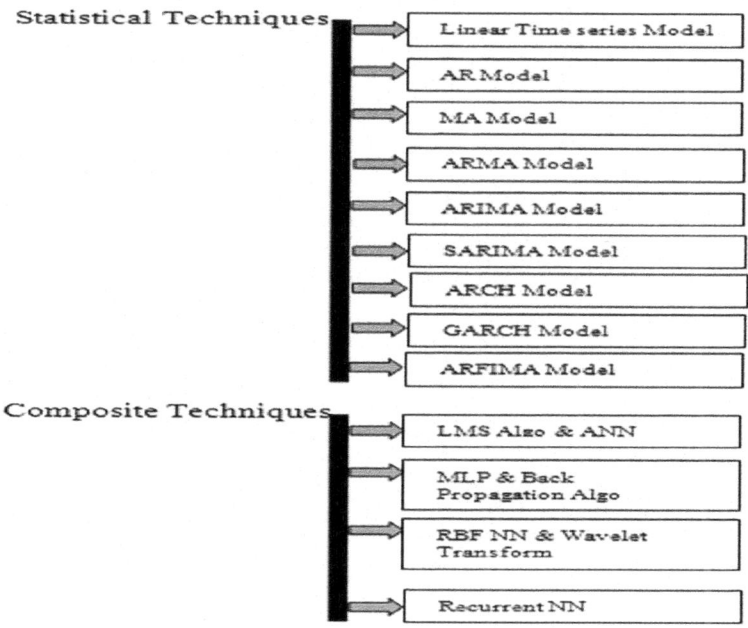

Fig. 2. Network traffic prediction techniques

Table 2. Network traffic prediction techniques details

Statistical Techniques	**Linear time series Model**	Data points are listed in time order, forming a sequence of equally spaced points in time[22].	$\{z(t)\} = \{z(1), z(2), ..., z(N)\}$ $= \{z1, z2, ..., zN\}$
		$t = 1, 2, ..., N$, t denotes instances of time when observations are taken, z_i = N observation/ sample realization(i=1 to N)	
	AR Model	Future behavior is predicted using past behavior. The prerequisite is correlation between time series data values and succeeding proceeding values. Autoregressive means use only past data to model the behavior.	$X_t = c + \sum_{i=1}^{p} \varphi_i\, X_{t-i} + \varepsilon_t$
		X=predictor variable, ε_t= white noise error terms	
	MA Model	Moving-average (MA) model uses univariate time series data. The current as well as past values determine output variable linearly.	$X_t = \mu + \varepsilon_t + \theta_1\varepsilon_{t-1} + ... + \theta_q\varepsilon_{t-q}$
		X=predictor variable, ε_t= white noise error terms, µ= series mean , θ_q =model parameters	
	ARMA Model	Two parts of model are an autoregressive (AR) part and a moving average (MA) part. AR part regresses variable on its own lagged values. MA part models error term as a linear combination of error terms at various times in the past. ARMA(p,q) has p as order of the autoregressive part and q as order of the moving average part.	$X_t = c + \varepsilon_t + \sum_{i=1}^{p} \varphi_i\, X_{t-i} + \sum_{i=1}^{q} \theta_i\varepsilon_{t-i}$
		X=predictor variable, ε_t= white noise error terms, µ= series mean , θ_q =model parameters	
	ARIMA Model	Auto-Regressive Integrated Moving Average(ARIMA). Stationarized series part forecasting equation means "autoregressive, forecast errors means "moving average", and "integrated" means time series differenced to be stationary.	$\hat{y}_t = \mu + \phi_1\,y_{t-1} + ... + \phi_p\,y_{t-p} - \theta_1 e_{t-1} - ... - \theta_q e_{t-q}$
		θs =moving average parameters, y = dth difference of Y, If d=2: $y_t = (Y_t - Y_{t-1}) - (Y_{t-1} - Y_{t-2}) = Y_t - 2Y_{t-1} + Y_{t-2}$, µ= series mean	
	SARIMA Model	Seasonality represents regular pattern of changes in time series, repeating over S, number of time periods after which pattern repeats.	$\Phi(B^S)\varphi(B)(x_t - \mu) = \Theta(B^S)\theta(B)w_t$
		S= number of time periods between repeat ion of patterns, B=backshift operator to produce previous element	
	ARCH Model	Autoregressive conditionally heteroscedastic(ARCH) models variance of a time series. Suitable when increased variations are short in period.	$Var(y_t \mid y_{t-1}) = \sigma_t^2 = \alpha_0 + \alpha_1 y_{t-1}^2$
		y_t =model variance at time t,	
	GARCH Model	Generalized autoregressive conditionally heteroscedastic(GARCH) models variance at time t using past squared observations and past variances.	$\sigma_t^2 = \alpha_0 + \alpha_1 y_{t-1}^2 + \beta_1 \sigma_{t-1}^2$
		y_t =model variance at time t,	
	ARFIMA Model	Autoregressive fractionally integrated moving average(ARFIMA) models allow differencing parameter WITH non-integer values. Time series with long memory are modeled by them.	$(1 - \sum_{i=1}^{p}\phi_i\, B^i)(1-B)^d\, X_t = (1 + \sum_{i=1}^{q}\theta_i\, B^i)\varepsilon_t$
		B=backshift operator to produce previous element	
Composite Techniques	**LMS Algo & ANN**	The LMS(Least Mean Square) algorithm trains neural network, minimizing cost (error) function estimates there by encouraging present information storage only[21]. ANN(Adaline neural network) are use only for linearly separable problems.	
	MLP & Back Propagation Algo	MLP (Multilayer Perceptron) neural network resembles to one layer perceptron and it is trained using back propagation algorithm[21].	
	RBF NN & Wavelet Transform	Multilayer network is used for RBF neural network containing one sensory nodes layer, then hidden nodes layer with one output layer.	
	Recurrent NN	Recurrent neural networks(NN) are beneficial for situations where output of one stage acts as input for other and outputs can be random real number from predecided interval.	

4 Network Traffic Prediction System

4.1 System Architecture

See Fig. 3.

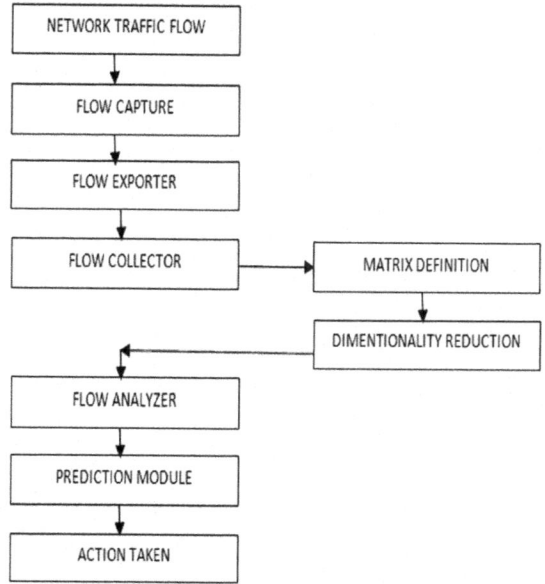

Fig. 3. Prediction system architecture

4.2 Algorithm for Prediction

Step 1: **FLOW CAPTURE** - Packet flow or network flow is captured and stored temporarily to analyze it.

Step 2: **FLOW EXPORTER-** The exporter creates flow registers from network traces.

Step 3: **FLOW COLLECTOR-** The Flow collector generates statistics from the stored file data.

Step 4: **FLOW ANALYZER-** The behavior profiling of each device is created.

Step 5: **PREDICTION MODULE-** Guesses future network flow data & behavior of related nodes.

Step 6: **ACTION TAKEN-** Application or invocation of various security policies, safeguarding actions as per type of attacks will be initiated.

5 Performance Evaluation Metrics

See Tables 3 and 4.

Table 3. Types of metrics used to evaluate network traffic prediction model [28]

Error Type	Formula	Result	Error direction shown	Positive and negative errors effect canceled	Extreme errors penalized	Dependency on measurement scale	Affected by data transformation	Desirable value
The Mean Forecast Error (MFE)	$\text{MFE} = \frac{1}{n}\sum_{t=1}^{n} e_t$	finds the average deviation of forecasted values from actual ones. • the exact amount of positive and negative errors remains unknown • forecasts on proper target are denoted by zero MFE, but may contain error	✓	✓	X	✓	✓	nearest to zero
The Mean Absolute Error (MAE) or Mean Absolute Deviation (MAD)	$\text{MAE} = \frac{1}{n}\sum_{t=1}^{n} |e_t|$	finds the average absolute deviation of forecasted values from actual ones. • represents magnitude of overall error as a result of forecasting	X	X	X	✓	✓	smallest
The Mean Absolute Percentage Error (MAPE)	$\text{MAPE} = \frac{1}{n}\sum_{t=1}^{n} \left|\frac{e_t}{y_t}\right| \times 100.$	This measure represents the percentage of average absolute error occurred	X	X	X	X	✓	nearest to zero
The Mean Percentage Error (MPE)	$\text{MPE} = \frac{1}{n}\sum_{t=1}^{n} \left(\frac{e_t}{y_t}\right) \times 100$	MPE represents the percentage of average error occurred, while forecasting. • Thus like MFE, by obtaining a value of MPE close to zero, we cannot conclude that the corresponding model performed very well. • It is desirable that for a good forecast the obtained MPE should be small	✓	✓	X	–	–	smallest

(continued)

Table 3. (*continued*)

Error Type	Formula	Result	Error direction shown	Positive and negative errors effect canceled	Extreme errors penalized	Dependency on measurement scale	Affected by data transformation	Desirable value
The Mean Squared Error (MSE)	$MSE = \frac{1}{n}\sum_{t=1}^{n} e_t^2$	It is a measure of average squared deviation of forecasted values. • MSE gives an overall idea of the error occurred during forecasting. • MSE emphasizes the fact that the total forecast error is in fact much affected by large individual errors, i.e. large errors are much expensive than small errors. • Although MSE is a good measure of overall forecast error, but it is not as intuitive and easily interpretable as the other measures discussed before	X	X	✓	✓	✓	smallest
The Sum of Squared Error (SSE)	$SSE = \sum_{t=1}^{n} e_t^2$	It measures the total squared deviation of forecasted observations, from the actual values	X	X	✓	–	–	smallest
The Signed Mean Squared Error (SMSE)	$SMSE = \frac{1}{n}\sum_{t=1}^{n}\left(\frac{e_t}{|e_t|}\right) e_t^2$	It is same as MSE, except that here the original sign is kept for each individual squared error	✓	✓	✓	✓	✓	smallest
The Root Mean Squared Error (RMSE)	$\sqrt{MSE} = \sqrt{\frac{1}{n}\sum_{t=1}^{n} e_t^2}$	RMSE is nothing but the square root of calculated MSE	X	X	✓	✓	✓	smallest

(*continued*)

Table 3. (*continued*)

Error Type	Formula	Result	Error direction shown	Positive and negative errors effect canceled	Extreme errors penalized	Dependency on measurement scale	Affected by data transformation	Desirable value
The Normalized Mean Squared Error (NMSE)	$NMSE = \dfrac{MSE}{\sigma^2} = \dfrac{1}{\sigma^2 n}\sum_{t=1}^{n} e_t^2$	NMSE normalizes the obtained MSE after dividing it by the test variance. • It is a balanced error measure and is very effective in judging forecast accuracy of a model. • The smaller the NMSE value, the better forecast	X	X	✓	–	–	smallest
The Theil's U-statistics	$U = \dfrac{\sqrt{\frac{1}{n}\sum_{t=1}^{n} e_t^2}}{\sqrt{\frac{1}{n}\sum_{t=1}^{n} f_t^2}\,\sqrt{\frac{1}{n}\sum_{t=1}^{n} y_t^2}}$	It is a normalized measure of total forecast error. • $0 \leq U \leq 1$; $U = 0$ means a perfect fit	–	–	–	✓	✓	nearest to zero

Table 4. Network traffic prediction literature study summary

Paper	Methodology	Feature Set	Advantages	Limitations	Futurescopre	Data Set	Evaluation Metric
Introduction to Time Series and Forecasting (2nd Edition) 2002	Auto-regressive integrated moving average (ARIMA)	Autoregressive (AR) - the Historical values; The Moving Average (MA)– the error component	Forecast future network traffic Is accurate within certain threshold	A complex Process and time consuming	Not Applicable		
Towards Forecasting Low Network Traffic for Software Patch Downloads: An ARMA model forecast using CRONOS 2010	Auto-regressive moving average (ARMA)	Auto-regressive moving average	Suitable for short range Forecasting in order to initiate small sized software patch Downloads	Paper does not initiate any form of data transfer. Arma time series model provides suitable forecasting for the Network traffic on a single broadband line	.NET platform can be used to reduce complexity of manual task.	From backbone internet	MAE, MSE, NMSE
Impact of Utilizing Forecasted Network Traffic for Data Transfers 2011	Auto-regressive moving average (ARMA)	Auto-regressive moving average, Arma (6,4) with step size of 30 s	Studies the impact of actual initiation of file transfers When the network traffic is forecasted to be low. This model is Capable of forecasting for short range network traffic. A technique to divide the files into Smaller sizes and transferring them when low network traffic Is forecasted would lead towards a better efficient use of Network bandwidth	Large file not transferred	Forecasting network Traffic can be used to enable more efficient large file transfers	From backbone internet	MSE

(continued)

Table 4. (*continued*)

Paper	Methodology	Feature Set	Advantages	Limitations	Futurescopre	Data Set	Evaluation Metric
Prediction of Internet Traffic Based on Elman Neural Network 2009	Neural Network	Time-series measurements Up time	Efficient method for modeling and prediction of traffic. System operations are unaffected by small errors. Provides improved prediction compared with other predictor, nonlinear function approximation capability is better	Normalized Mean square error (nmse) rate greater than farima, ann	NMSE can be improved using	From backbone internet	MAE, MSE, NMSE
PHAD: Packet Header Anomaly Detection 2004	Anomaly detection algorithm (PHAD)	Packet header fields	Attacks with exploits at the transport layer and below are detected, better detection as compared to other models	Training data set attacks reduce PHAD's performance	Needs in depth examination of application layer for performance enhancement	1999 DARPA	Prediction Accuracy
Virtual Network Topology Adaptability based on Data Analytics for Traffic Prediction 2016	The VNT - Virtual network topologies reconfiguration approach based on data analytics for traffic prediction (VENTURE)	Machine learning algorithm based on artificial neural network (ANN)	Minimizing TCO, deactivation of transponders for low traffic hours is possible. Results in low energy requirements with light paths release from optical layer yielding low cost	More transponders need to be installed	Not given	Synergy test-bed	Time Complexity
Traffic Prediction for Dynamic Traffic Engineering 2015	Auto-regressive integrated moving average (arima), seasonal arima (sarima)	Range of short-term fluctuation is found using standard deviation	Traffic variations are predicted using monitored data for long durations. Short duration variations are also considered to counter prediction uncertainty. Changes due to temporal traffic induce uncertainty along with prediction errors	The sampling disadvantages are- inducing sampling errors, and Flows escape unsampled. It is a slow complex process	Predicted traffic needs to be investigated by Sophisticated models	Traffic traces from Education network in the United States	MAPE

(*continued*)

Table 4. (*continued*)

Paper	Methodology	Feature Set	Advantages	Limitations	Futurescopre	Data Set	Evaluation Metric
Advancing Network Flow Information Using Collaborative Filtering 2017	Collaborative Filtering algorithms	Communication of different devices, flow based data	Collaborative Filtering algorithms provide network security domain with an innovative Manner to predict future flows	Better Results in precision but worse in recall	Plan to apply Specific cold-start techniques in order to mitigate rating distribution effect	Unb iscx	Precision, Recall
Prediction of Network Traffic by using Dynamic BiLinear Recurrent Neural Network 2011	Neural Network-Dynamic-Bilinear Recurrent Neural Network	–	D-BLRNN predicts with low performance degradation. Comparatively better than other Neural networks in predicting Ethernet network traffic. Bursty traffic prediction possible	Not given	Not given	Ethernet network traffic data set	NMSE
Network Traffic Analysis and Prediction Based on APM 2009	Accumulation predicting model (APM)	Seasonal time Series (t),length of each season(d), partial accumulation	Less complicated than ARIMA. AP M is good with stable Seasonal pattern time series	Applicable to stable seasonal pattern time series mostly.	Not given	Chinese mobile network operator	MAPE
ANFIS Method for Forecasting Internet Traffic Time Series	Adaptive neurofuzzy Inference system (ANFIS)	Internet traffic time series	Statistical indicators of method are best. Real data of network fits well into this model with different times condition.	Not given	Not given	From backbone internet over tcp/ip	RMSE, AARE
Identification and Prediction of Internet Traffic Using Artificial Neural Networks.2010	Artificial neural network	Internet traffic data over IP networks, Levenberg-Marquardt (LM) and the Resilient back propagation (Rp) algorithms using statistical criteria	Traffic over IP network managed very well by this model	Not given	Not given	From backbone internet	RMSE, SI, the Relative Error, MAPE

(continued)

Table 4. (*continued*)

Paper	Methodology	Feature Set	Advantages	Limitations	Futurescopre	Data Set	Evaluation Metric
Network Traffic Prediction and Result Analysis Based on Seasonal ARIMA and Correlation Coefficient, 2010	Multiplicative Seasonal autoregressive integrated moving average model (ARIMA) is employed to make traffic series prediction		Yields high precision results. Handles series with seasonal features also	Not given	Not given	heilongjiang province mobile network in china	MAPE
Multi-Scale High-Speed Network Traffic Prediction Using k-Factor Gegenbauer ARMA Model,2004	K-Factor Gegenbauer ARMA	Spectrum of the zero-mean traffic data	Better than AR model	Not given	Useful for building congestion control schemes	LRD series. MPEG and JPEG of star wars movie, Ethernet and Internet traffic	MAE, SER
A Network Traffic Flow Prediction with Deep Learning Approach for Large-scale Metropolitan Area Network,2018	Stacked denoising autoencoder prediction model (SDAPM)	Partition ratio, noise ratio of Gaussian, input data dimension, number of hidden units, binary masking noise probability	Better predictions than MLP. Results are promising.	Not given	Not given	2015 china united network communications two months traffic flow	MAE, MRE, RMSE
Interactive Temporal Recurrent Convolution Network for Traffic Prediction in Data Centers, 2017	Gated recurrent unit (GRU) model, interactive temporal recurrent convolution network (ITRCN)	Minutes, source port, destination port, sun traffic(bytes)	Works with interactive and non interactive network traffics with great accuracy.	Not Given	Can be tested to note influence of days variance on model effectiveness.	Yahoo! Data sets,	RMSE
Network Traffic Prediction Based on Hadoop,2014	Hadoop platform, Echo State Network (ESN), Recurrent Neural Network (RNN)	Phone number, time stamp, the type of application, Location Area Code (LAC), traffic volume	Large scale network records can be processed with ease by parallel prediction models building.	Prediction effected by fluctuation and noise of data series	Not given	Mobile operator data set in china	NMSE, RMSE, MAE

(*continued*)

Table 4. (*continued*)

Paper	Methodology	Feature Set	Advantages	Limitations	Futurescopre	Data Set	Evaluation Metric
Network Traffic Prediction Based on Particle Swarm Optimization, 2016	Hybrid flexible mueral tree & Particle swarm optimization	Variance of size of received packet & receiving packet, number of SYN, RST, FIN packets etc.	The proposed hybrid model based on PSO outperforms SVM & FNT based methods	Not given	Not given	The internet traffic flow data	SVM classifier based method & FNT based method error rates
One new Research on Method of intelligent substation Network Traffic Prediction,2014	A gray neural network model		Network training data required is very small. Yields small errors high accuracy	Not given	Initial network parameters optimization should be studied	Substation network traffic	MSE, prediction error, prediction accuracy

6   Conclusion

With the ever growing network traffic, present is the era of big data. This data can be explored and utilized for prediction of network traffic. This prediction will help to reduce time to respond in case of anomalies. So in this paper we studied and surveyed various network traffic prediction techniques. Prediction methods based on statistic, neural network are discussed. Performance metrics used in various previous studies [10, 13, 16, 18] etc. have been enlisted. The tabular view of surveyed papers focuses on prediction techniques for network traffic. Standard datasets used by the implemented algorithms and metrics used to evaluate the results are grouped in the research works surveyed. Such a review paper would help to provide an insight into the topic to new researchers.

References

1. Wang, J.-S., Wang, J.-K., Zeng, M.-H., Wang, J.-J.: Prediction of internet traffic based on Elman neural network. In: Control and Decision Conference (CCDC '09), pp. 1248–1252. IEEE (2009)
2. Poo, K.H., Tan, I.K., Chee, Y.K.: Bittorrent network traffic forecasting with ARMA. Int. J. Comput. Netw. Commun. 4(4), 143 (2012)
3. Mahoney, M.V., Chan, P.K.: PHAD: packet header anomaly detection for identifying hostile network traffic. Technical report, PHAD (2001)
4. Morales, F., Ruiz, M., Gifre, L., Contreras, L.M., López, V., Velasco, L.: Virtual network topology adaptability based on data analytics for traffic prediction. J. Opt. Commun. Netw. 9(1), A35–A45 (2017)
5. Otoshi, T., Ohsita, Y., Murata, M., Takahashi, Y., Ishibashi, K., Shiomoto, K.: Traffic prediction for dynamic traffic engineering. Comput. Netw. 85, 36–50 (2015)
6. Park, D.-C., Woo, D.-M.: Prediction of network traffic using dynamic bilinear recurrent neural network. In: Fifth International Conference on Natural Computation (ICNC 2009), vol. 2, pp. 419–423. IEEE (2009)
7. Sadek, N., Khotanzad, A.: Multi-scale high-speed network traffic prediction using k-factor Gegenbauer ARMA model. In: 2004 IEEE International Conference on Communications, vol. 4, pp. 2148–2152. IEEE (2004)
8. Yu, Y., Song, M., Ren, Z., Song, J.: Network traffic analysis and prediction based on APM. In: 2011 6th International Conference on Pervasive Computing and Applications (ICPCA), pp. 275–280. IEEE (2011)
9. Yu, Y., Wang, J., Song, M., Song, J.: Network traffic prediction and result analysis based on seasonal ARIMA and correlation coefficient. In: 2010 International Conference on Intelligent System Design and Engineering Application (ISDEA), vol. 1, pp. 980–983. IEEE (2010)
10. Chabaa, S., Zeroual, A., Antari, J.: Anfis method for forecasting internet traffic time series. In: Microwave Symposium (MMS), 2009 Mediterrannean, pp. 1–4. IEEE (2009)
11. Chabaa, S., Zeroual, A., Antari, J.: Identification and prediction of internet traffic using artificial neural networks. J. Intell. Learn. Syst. Appl. 2(03), 147 (2010)
12. Brockwell, P.J., Davis, R.A.: Introduction to Time Series and Forecasting. Springer Texts in Statistics, 2nd edn. Springer, New York (2002). https://doi.org/10.1007/b97391. ISBN 0-387-95351-5. SPIN 10850334

13. Tan, I.K.T., Hoong, P.K., Yik, C.: Towards forecasting low network traffic for software patch downloads: an ARMA model forecast using CRONOS. In: Second International Conference on Computer and Network Technology. IEEE (2010). https://doi.org/10.1109/ICCNT.2010.3588. 978-0-7695-4042-9/10 $26.00 © 2010
14. Hoong, N.K., Hoong, P.K., Tan, I.K.T., Seng, L.C.: Impact of utilizing forecasted network traffic for data transfer. In: 13th International Conference on Advanced Communication Technology (ICACT2011), 13–16 February 2011. INSPEC Accession Number: 11930338, Electronic ISBN: 978-89-5519-155-4, Print ISSN: 1738-9445
15. Yonghao, W., Cong, L., Jin, W., Guiping, Z.: One new research on method of intelligent substation network traffic prediction. In: 2014 Fifth International Conference on Intelligent Systems Design and Engineering Applications (2014). Electronic ISBN: 978-1-4799-4261-9
16. Monian-Fa: Network traffic prediction based on particle swarm optimization. In: 2015 International Conference on Intelligent Transportation, Big Data and Smart City, 19–20 December 2015. Electronic ISBN: 978-1-5090-0464-5
17. Cui, H., Yao, Y., Zhang, K., Sun, F., Liu, Y.: Network traffic prediction based on Hadoop. In: 2014 International Symposium on Wireless Personal Multimedia Communications (WPMC), 7–10 September 2014. Electronic ISSN: 1882-5621
18. Cao, X., Zhong, Y., Zhou, Y., Wang, J., Zhu, C., Zhang, W.: Interactive temporal recurrent convolution network for traffic prediction in data centers. In: Special Section on Advanced Data Analytics For Large-Scale Complex Data Environments, pp. 2169–3536 (2017). https://doi.org/10.1109/ACCESS.2017.2787696
19. https://www.flowtraq.com/network-flow-analysis-for-maximum-security/
20. Hall, J., Mars, P.: Limitations of artificial neural networks for traffic prediction in broadband networks. In: Proceedings of the Third IEEE Symposium on Computers and Communications, ISCC 1998, Cat. No. 98EX166, 30 June–2 July 1998. Print ISBN: 0-8186- 8538-7
21. Vieira, F.H.T., Costa, V.H.T., Gonçalves, B.H.P.: Neural network based approaches for network traffic prediction. In: Yang, X.-S. (ed.) Artificial Intelligence, Evolutionary Computation and Metaheuristics, SCI 427, pp. 657–684. Springer, Berlin (2013)
22. https://en.wikipedia.org/wiki/Time_series
23. http://www.statisticshowto.com/autoregressive-model/
24. https://en.wikipedia.org/wiki/Autoregressive_model
25. https://en.wikipedia.org/wiki/Moving-average_model
26. https://en.wikipedia.org/wiki/Autoregressive%E2%80%93moving-average_model
27. https://onlinecourses.science.psu.edu/stat510/node/67/
28. Adhikari, R., Agrawal, R.K.: An Introductory Study on Time Series Modeling and Forecasting. LAP LAMBERT Academic Publishing, 29 January 2013. ISBN-10: 3659335088, ISBN-13: 978-3659335082

Use of Evolutionary Algorithm in Regression Test Case Prioritization: A Review

Priyanka Paygude[1(✉)] and Shashank D. Joshi[2]

[1] College of Engineering, Bharati Vidyapeeth, (Deemed to be University),
Pune, India
pspaygude@bvucoep.edu.in
[2] Department of Computer Engineering, College of Engineering,
Bharati Vidyapeeth, (Deemed to be University), Pune, India
sdjoshi@bvucoep.edu.in

Abstract. Software keeps on evolving throughout its development. Regression testing plays a vital role in maintaining the quality of software by re-executing all test cases in order to ensure that modification in source code has not affected earlier working functionality. It is an expensive task to re-execute every test case after each modification in the software. Test case prioritization techniques organize the test cases based on its value for execution and hence increases the effectiveness of testing. Test case prioritization techniques comprise of scheduling of test cases in a way such that it improves the performance of testing with respect to rate of fault detection and execution time. This paper presents the review of recent research papers in test case prioritization using evolutionary algorithm. The research papers are reviewed with respect to research questions and have been evaluated with the ability of TCP techniques with time and cost of execution. The objective of our research paper is to study evolutionary algorithms in solving the test case prioritization problem and associated result.

Keywords: Software testing · Regression testing · Test case prioritization · Genetic algorithm · Literature review

1 Introduction

Software testing and maintenance are the two vital and most expensive and time-consuming phases in the software development life cycle. Testing is the process of analyzing the behavior of the product in order to detect the difference between the system under development and user requirement. Testing ensures the correctness, completeness and effectiveness of a system. Software testing is the process of testing the software with the intention of finding the bugs as early as possible, so as to save cost. As the software development evolves over the successive versions, parallel test suites grow. Thus, testing becomes very time-consuming and costly as the system evolves [1–3]. Thus, testing needs to be optimized to control cost, efforts and time.

Regression testing is the type of testing where all the test cases are re-run after the modifications into the code to ensure that the modifications made in the code have not affected the previously working functionality. This approach is known as Retest-All.

© Springer Nature Switzerland AG 2020
A. P. Pandian et al. (Eds.): ICCBI 2018, LNDECT 31, pp. 56–66, 2020.
https://doi.org/10.1007/978-3-030-24643-3_6

Here tester needs to re-run all the test cases along with the newly added test cases for code changes. In commercial software development, regression testing composes the majority of testing effort.

Regression Testing is performed if there is a change in the user requirements and code has been modified according to the requirements or new features are added into the existing software. There are different types of maintenance where Regression Testing is performed:

1. Corrective Maintenance: Testing performed after the corrections (repairing faults) made in previous code version
2. Progressive Maintenance: Testing performed after the addition of new features (changes demanded by client) into the existing code version
3. Adaptive Maintenance: Testing performed against the adaption of system under test to new hardware/software environment
4. Preventive Maintenance: Probable client requests. Changes in existing Hardware/Software environment

In this research paper survey of different TCP techniques are presented. The paper is divided in to 6 sections. Section 2 describes regression testing methodologies. Section 3 will discuss existing Test Case Prioritization (TCP) approaches. The literature review is presented in tabular format in Sect. 4. Research questions are framed in Sect. 5. Section 6 will describe results.

2 Regression Testing Methodologies

Regression testing is not just a simple extension of testing. Whenever software is modified (corrective) or added with new changes (progressive); all test cases are to be re-run (Retest All) which is a time consuming approach [4–7, 9]. There are three popular regression testing strategies:

(A) *Test Suite Minimization (TSM)*

It is also known as Test Suite Reduction (TSR). In this approach of regression, the test cases which are redundant (e.g. removing the test cases covering the functionality of the system which are already covered by other test cases) or not covering the modified code are permanently removed from the test suite. However, removing the test cases from test suite is as good as compromising with the fault detection capability of the test suite.

(B) *Test Case Selection (TCS)*

Here, test cases are selected on the basis on modified code coverage. Test cases which cover changed code will be run before the other test cases.

(C) *Test Case Prioritization (TCP)*

Here, the test cases are ordered based on certain objective, such that it benefits the testing process to the maximum extent. Unlike test suite minimization, all test cases are re-run with the calculated priority of each test case based on criteria. The wide variety of objective for prioritizing test cases in test suite involves:

- Ordering test cases based on history of modifications in code
- Executing the test cases with minimum execution time so as to achieve maximum fault/code coverage
- Executing the test cases which are having maximum fault revealing capability
- Prioritizing test cases which are covering modified code
- Prioritizing test cases on the basis test case critical bug revealing capability

Test case prioritization can be based on a single-criteria such as cost, risk, execution time; e.g. the tests with lower cost will beranked first while tests with higher costs will beranked last. Prioritization of test cases can be multi-criteria based such as finding the tradeoff between cost and risk.

Average percentage of fault detection (APFD) is the most widely used metric for calculating the efficiency of fault detection.

3   Test Case Prioritization (TCP) Approaches

TCP techniques can be categorized based on objective behind the ordering of test cases or based on single/multi objective etc.

(A) *Coverage based*

This is white box testing approach where the ordering of test case is based on logic coverage of code i.e. 100% statements/branches/paths/functions coverage of system during regression testing to reveal the bugs to due modification in the system. Coverage based approach is based on the fact that by achieving more code coverage, increases the chances of fault revealing [8].

(B) *Modification based approach*

This approach ranks the test cases based on their coverage over the changed code. E.g. test cases which cover modified code will be run before remaining test cases.

(C) *Fault based approach*

This approach orders the test cases based on the number of faults detected by the test case. Test cases which detect maximum faults will give execution preference over other test cases. Here, history of test cases execution can be used to know the number of revealing faults under each test case. This approach can also be combined with the risk associated with the fault to rank the test case. There has been a significant amount of research within software industry to find the fault-prone components within a system and prioritizing the test consequently. Several experts have concentrated on various characteristics associated with a component for keeping track of faults.

(D) *Requirement based*

Every user requirement of the system will be associated with the priority based on its significance in the system. Requirement based test case prioritization considers priority set by user for each requirement to order the execution of the test cases

(E) *Model based*

In this technique, the source code and the underline system models are used to prioritize the test cases for early fault detection. This technique uses various parameters of system such as dependency among the components, inheritance, modularization, function calls etc.

(F) *Evolutionary algorithm based approach*

An evolutionary search technique is used to solve many optimization problems. Genetic algorithm, particle swarm optimization, ant colony optimization are the top famous search optimization algorithms.

Genetic Algorithm is similar to natural evolution in a way that it finds the fittest and optimal solution to a given computational problem. The base of genetic algorithm is the biological concept of evolution and the survival of fittest. This algorithm is much more powerful and efficient than exhaustive search algorithm as it provides good and robust solution rated against fitness criteria. Fitness function evaluates how much a given solution is close to optimality.

4 Literature Survey

This literature review aims to understand and review most recent test case prioritization techniques in regression testing focusing on use of evolutionary algorithms and their results. The research papers are reviewed with respect to improvise and evaluate the ability of TCP techniques with time and cost of execution.

Tables 1 and 2 presents review of studied papers.

5 Research Questions

We have studied and analyzed the various test case prioritization techniques from the available literature. The analysis of surveyed papers based on research questions is presented in tabular form in Table 3.

Research questions are formed, as follows:

RQ1: Is evolutionary algorithm used to prioritize the test cases?
RQ2: Is the proposed technique multi objective?
RQ3: Are the parameters used dynamic?
RQ4: Is the proposed technique implemented and evaluated on real projects?
RQ5: Is APFD metric used to calculate efficiency of proposed technique?

Table 1. Review of studied TCP techniques

Ref. ID	Technique proposed	Single/multi objective	Parameters use for TCP	Implementation	Metric used
[10]	Genetic Algorithm	Multi Objective	1. Customer priority 2. Modification in requirement 3. Code complexity 4. Fault impact	Manually generated Test data	APFD
[11]	Genetic Algorithm	Single Objective	Code Coverage	Manually generated Test data	APFD
[12]	Code data	Multi Objective	1. HTTP requests 2. Length of HTTP requests 3. Inter dependency of HTTP requests	Evaluated on Website testing	APFD
[13]	Genetic Algorithm	Single Objective	Weight calculated for every node of State chart diagram and Sequence diagram	Not Evaluated	APFD
[16]	Genetic algorithm with Adaptive approach	Single Objective	Test case execution history of fault detection	Apache Open Source with Eclipse and testing it with Junit	Execution time and APSC
[17]	Genetic Algorithm	Multi Objective	1. Total statement coverage 2. Total fault exposing weight 3. Total mutant coverage	Triangle classifier problem with manually seeded faults	APSC
[18]	Genetic Algorithm (GA) with java decoding technique	Single Objective	1. Project Size 2. Source code scope 3. Input Stream 4. Bug severity	tested on project "News Blaster"	APFD
[19]	Genetic Algorithm	Single Objective	Complete Code Coverage of the Test Case	Manually Generated test data	Average Percentage of Code Covered (APCC)

(continued)

Table 1. (*continued*)

Ref. ID	Technique proposed	Single/multi objective	Parameters use for TCP	Implementation	Metric used	Ref. ID	Technique proposed	Single/multi objective	Parameters use for TCP	Implementation	Metric used
[14]	Genetic Algorithm	Single Objective	Fault Coverage	Evaluated on five different website projects	APFD	[20]	Ant Colony Optimization	Single Objective	Fault coverage by test case	Evaluated on medical and financial software	APFD
[15]	Genetic Algorithm with Particle Swarm Optimization	Single Objective	History data of test case for fault detection	Manually seeded faults of an e-learning website	APFD	[21]	Clustering based test case ordering	Single Objective	1. Fault with associated priority and severity 2. Rate of fault detection 3. Requirements volatility 4. Fault impact 5. Implementation complexity	Manually generated Test data	APFD

Table 2. Review of studied TCP techniques

Ref. ID	Technique proposed	Single/multi objective	Parameters use for TCP	Implementation	Metric used
[22]	Ant Colony Optimization	Multi Objective	Fault Coverage	Manually generated Test data	APFD
[23]	Test Case Prioritization based on risk exposure	Single Objective	Functionality Coverage	Manually Generated test data	1. Testing time 2. Testing cost
[24]	Hamming distance	Single Objective	1. Test Case execution time 2. test cost	Manually Generated test data	APFD
[25]	Particle Swarm Optimization	Multi Objective	1. Fault Coverage 2. Execution Time	Technique evaluated on manually generated test data	APFD
[28]	Genetic Algorithm and Clustering	Single Objective	Code Coverage	Manually Generated test data	APFD
[29]	Genetic Algorithm and Simulated Annealing	Multi Objective	1. Fault Coverage 2. Test execution time	Implemented in Java and evaluated on manually generated test data	1. Execution time 2. Fault detection rate
[30]	Weighted average of parameters consider	Multi Objective	Fault detection rate Risk detection ability of test case	Manually Generated test data	APFD
[31]	Simulation study of 1. Total, additional and 2- OptimalGreedy algorithm 2. Hill Climbing 3. Genetic algorithm	Single Objective	Requirement-fault matrix	Manually Generated test data	Average Percentage Requirement Coverage (APRC)

(continued)

Table 2. (continued)

Ref. ID	Technique proposed	Single/multi objective	Parameters use for TCP	Implementation	Metric used	Ref. ID	Technique proposed	Single/multi objective	Parameters use for TCP	Implementation	Metric used
[26]	Genetic algorithm with Particle swarm optimization	Single Objective	Fault Coverage	Implemented in Java and evaluated on manual test data	APFD	[32]	Association Rule Mining	Multi Objective	based on system data 1. Frequent patterns of functionality 2. Business Value of the features	Evaluated on two public projects	APFD
[27]	Genetic Algorithm	Multi Objective	1. Code coverage 2. Requirements coverage 3. Test cost	Evaluated on four Java applications: AveCalc, LaTazza, iTrust, ArgoUml	APFD	[33]	Weighted average of parameters consider	Multi Objective	1. Rate of fault detection 2. Number of faults detected 3. test case ability of risk detection 4. test case effectiveness	Manually generated fault matrix	1. APFD 2. Execution time

Table 3. Research questions based analysis of studied papers

Ref. ID	RQ1	RQ2	RQ3	RQ4	RQ5	Ref. ID	RQ1	RQ2	RQ3	RQ4	RQ5
[10]	O	O	O		O	[22]	O	O			O
[11]	O		O		O	[23]			O		
[12]		O	O	O	O	[24]			O		O
[13]	O				O	[25]	O	O	O		O
[14]	O			O	O	[26]	O		O	O	O
[15]	O			O	O	[27]	O	O	O	O	O
[16]	O			O		[28]	O		O		O
[17]	O	O	O			[29]	O	O	O	O	
[18]	O			O	O	[30]		O	O		O
[19]	O					[31]	O				
[20]	O			O	O	[32]		O	O	O	O
[21]			O		O	[33]		O	O		O

6   Result and Discussion

This section summarizes the results with respect to research questions for reviewed papers. A total of 24 recent research papers were selected based on quality of published work and were studied in deep.

This empirical review concludes that TCP is widely research area in last decade with combination of evolutionary algorithms. Genetic is the most popular algorithm used to prioritize test cases with promising results. APFD is most widely used metric to calculate the effectiveness of proposed work in almost all studied papers. There are many papers with single objective of TCP. Multi objective has much scope to study in detail for further research in this domain. Aim of this study was to get the recent review of TCP technique using evolutionary algorithms and their results which will help to get quick review for researchers.

References

1. Mall, R.: Fundamentals of Software Engineering. PHI Learning Pvt. Ltd., New Delhi (2018)
2. Miller, T.: Using dependency structures for prioritization of functional test suites. IEEE Trans. Softw. Eng. **39**(2), 258–275 (2013)
3. Chauhan, N.: Software Testing: Principles and Practices. Oxford University Press, Oxford (2010)
4. Yoo, S., Harman, M.: Regression testing minimization, selection and prioritization: a survey. Softw. Test. Verif. Reliab. **22**(2), 67–120 (2012)
5. Leung, H.K.N., White, L.: Insights into regression testing (software testing). In: Proceedings of Conference on Software Maintenance, pp. 60–69. IEEE (1989)
6. Elbaum, S., Malishevsky, A.G., Rothermel, G.: Test case prioritization: a family of empirical studies. IEEE Trans. Softw. Eng. **28**(2), 159–182 (2002)

7. Kumar, A., Singh, K.: A literature survey on test case prioritization. Compusoft **3**(5), 793–799 (2014)

8. Elbaum, S., Rothermel, G., Kanduri, S., Malishevsky, A.G.: Selecting a cost-effective test case prioritization technique. Softw. Qual. J. **12**(3), 185–210 (2004)

9. Rothermel, G., Untch, R.H., Chu, C., Harrold, M.J.: Prioritizing test cases for regression testing. IEEE Trans. Softw. Eng. **27**(10), 929–948 (2001)

10. Raju, S., Uma, G.V.: Factors oriented test case prioritization technique in regression testing using genetic algorithm. Eur. J. Sci. Res. **74**(3), 389–402 (2012)

11. Ahmed, A.A.F., Shaheen, M., Kosba, E.: Software testing suite prioritization using multi-criteria fitness function. In: 2012 22nd International Conference on Computer Theory and Applications (ICCTA), pp. 160–166. IEEE (2012)

12. Dobuneh, M.R.N., Jawawi, D.N.A., Ghazali, M., Malakooti, M.V.: Development test case prioritization technique in regression testing based on hybrid criteria. In: 2014 8th Malaysian Software Engineering Conference (MySEC), pp. 301–305. IEEE (2014)

13. Khurana, N., Chillar, R.S.: Test case generation and optimization using UML models and genetic algorithm. Procedia Comput. Sci. **57**, 996–1004 (2015)

14. Kaur, A., Goyal, S.: A genetic algorithm for fault based regression test case prioritization. Int. J. Comput. Appl. **32**(8), 975–8887 (2011)

15. Saraswat, P., Singhal, A.: A hybrid approach for test case prioritization and optimization using meta-heuristics techniques. In: 2016 1st India International Conference on Information Processing (IICIP), pp. 1–6. IEEE (2016)

16. Walia, R., Bajaj, H.K.: Performance enhancement of test case prioritization using hybrid approach. Int. J. Sci. Eng. Appl. **6**(01), 027–032 (2017)

17. Mishra, D.B., Panda, N., Mishra, R., Acharya, A.A.: Total fault exposing potential based test case prioritization using genetic algorithm. Int. J. Inf. Technol. **9**, 1–5 (2018)

18. Mahajan, S., Joshi, S.D., Khanaa, V.: Component based software system test case prioritization with genetic algorithm decoding technique using Java platform. In: International Conference on Computing Communication Control and Automation (ICCUBEA), pp. 847–851. IEEE (2015)

19. Kaur, A., Goyal, S.: A genetic algorithm for regression test case prioritization using code coverage. Int. J. Comput. Sci. Eng. **3**(5), 1839–1847 (2011)

20. Noguchi, T., Washizaki, H., Fukazawa, Y., Sato, A., Ota, K.: History-based test case prioritization for black box testing using ant colony optimization. In: 2015 IEEE 8th International Conference on Software Testing, Verification and Validation (ICST), pp. 1–2. IEEE (2015)

21. Raju, S., Uma, G.V.: An efficient method to achieve effective test case prioritization in regression testing using prioritization factors. Asian J. Inf. Technol. **11**(5), 169–180 (2012)

22. Ansari, A., Khan, A., Khan, A., Mukadam, K.: Optimized regression test using test case prioritization. Procedia Comput. Sci. **79**, 152–160 (2016)

23. Ansari, A.S.A., Devadkar, K.K., Gharpure, P.: Optimization of test suite-test case in regression test. In: 2013 IEEE International Conference on Computational Intelligence and Computing Research (ICCIC), pp. 1–4. IEEE (2013)

24. Maheswari, R.U., JeyaMala, D.: A novel approach for test case prioritization. In: 2013 IEEE International Conference on Computational Intelligence and Computing Research (ICCIC), pp. 1–5. IEEE (2013)

25. Tyagi, M., Malhotra, S.: Test case prioritization using multi objective particle swarm optimizer. In: 2014 International Conference on Signal Propagation and Computer Technology (ICSPCT), pp. 390–395. IEEE (2014)

26. Nagar, R., Kumar, A., Kumar, S., Baghel, A.S.: Implementing test case selection and reduction techniques using meta-heuristics. In: 2014 5th International Conference Confluence The Next Generation Information Technology Summit (Confluence), pp. 837–842. IEEE (2014)

27. Islam, Md.M., Marchetto, A., Susi, A., Kessler, F.B., Scanniello, G.: MOTCP: a tool for the prioritization of test cases based on a sorting genetic algorithm and Latent Semantic Indexing. In: 2012 28th IEEE International Conference on Software Maintenance (ICSM), pp. 654–657. IEEE (2012)

28. Sabharwal, S., Sibal, R., Sharma, C.: A genetic algorithm based approach for prioritization of test case scenarios in static testing. In: 2011 2nd International Conference on Computer and Communication Technology (ICCCT), pp. 304–309. IEEE (2011)

29. Maheswari, R.U., Jeya Mala, D.: Combined genetic and simulated annealing approach for test case prioritization. Indian J. Sci. Technol. 8(35), 1–5 (2015)

30. Tyagi, M., Malhotra, S.: An approach for test case prioritization based on three factors. Int. J. Inf. Technol. Comput. Sci 4, 79–86 (2015)

31. Li, S., Bian, N., Chen, Z., You, D., He, Y.: A simulation study on search algorithms for regression test case prioritization. In: 10th International Conference on Quality Software (2010)

32. Mahali, P., Acharya, A.A., Mohapatra, D.P.: Test case prioritization using association rule mining and business criticality test value. In: Computational Intelligence in Data Mining, Vol. 2, pp. 335–345. Springer, New Delhi (2016)

33. Nayak, S., Kumar, C., Tripathi, S.: Enhancing efficiency of the test case prioritization technique by improving the rate of fault detection. Arab. J. Sci. Eng. 42(8), 3307–3323 (2017)

Efficiency Improvement in Code Emulation by Soft Computing Technique

S. Prayla Shyry$^{(\boxtimes)}$, M. Saranya$^{(\boxtimes)}$, and M. Mahithaa Sree$^{(\boxtimes)}$

Sathyabama Institute of Science and Technology, Chennai, India
suja200165@gmail.com,
mekasaranya7@gmail.com,
mogantimahithaa5@gmail.com

Abstract. Cyber espionage is an illicit way of obtaining secrets or information without user's knowledge. Today, the Stuxnet, Duqu and flame computer worms which are used as cyber weapons against nations continue to make headlines as a new breed of malware. Understanding this topic is important to know how technology shapes the world and influences nation-state relation. It can be in different ways, malicious software like malware- a threat to system security. The technique of malware analysis either in sophistication of samples analyzed through the traditional approaches has its own con's (Zero day Vulnerability). The aim of this paper is to analyze metamorphic virus using code emulation and implementation of few simple parameters by using fuzzy logic. We also justified how and why these parameters are considered. For a brief introduction to this paper, we used code emulator which creates a virtual environment where we considered number of iterations, memory access function and size of the virus as the performance metrics. The point of code emulation is to mimic the instruction set of CPU using virtual registers and flags.

1 Introduction

Now-a-days multiplied antivirus software's are developed as a contrast; huge evolution of viruses also takes place. To encounter these viruses numerous detection techniques are produced day by day.

An Anti-virus tool refers to the traditional means of fighting computer malware. Generally a Anti-virus software does background scanning, full system scan and virus definition which has predefined virus patterns and scans them with the corresponding new injected virus. Developer should have the exact idea to analyze the problem by testing its nature in a controlled environment. Some of the techniques to get rid of viruses are signature based detection, heuristic based detection, behavioral based detection, sand box detection and cloud based detection.

The aim of this paper is to present the findings of some of the parameters used while detecting a virus. Usually a Metamorphic virus is one that can rewrite its own code. It is considered the most infectious computer virus, and it can cause serious damage. To overcome this infectious virus we are using code emulator as one of the main component in anti-virus software. The parameters which we implemented in this

© Springer Nature Switzerland AG 2020
A. P. Pandian et al. (Eds.): ICCBI 2018, LNDECT 31, pp. 67–75, 2020.
https://doi.org/10.1007/978-3-030-24643-3_7

session are size of the virus, the number of iterations performed and the memory access function using code emulator.

The capacity of the operating system must be imitated to make a virtualized system that underpins system API'S, file and memory Anti-virus software uses a virtual system which executes the virus injected into our system and executes them in the virtual environment. Consequently we are thinking about the quantity of cycles done and the memory get to capacity to bring 8-bit, 16-bit and 32-bit.

2   Related Works

We are motivated by the work of Wing and Wong [4] describes metamorphic virus detection based on hidden Markov models. They used code emulation, pattern based scanning (as used in 1st generation scanners), Heuristic analysis to detect new and unknown viruses. They experimented with Hidden Markov Model (HMM) and distinguished NGVCK viruses from normal programs and his model successfully detected the polymorphic or metamorphic viruses that cannot be detected through the traditional signature based detection.

Sridhara [3] describes about hidden Markov technique and designed a metamorphic worm that carries its own morphing engine. They provided test results to confirm effectively evades with signature and a HMM based detection.

Kalbhor [7] depicts a layered methodology that demonstrates on this utilization of HMM by accomplishing indistinguishable outcomes from the dueling approach however with noteworthy execution enhancement as far as time. Basically the thought is to eliminate most putative malware with the threshold approach. They accomplished precise outcome with essentially less execution overhead than the dueling HMM methodology.

Sharif [5] describes an obfuscation engine which then used this engine to create metamorphic variants of a seed virus and performed the validity of the statement about metamorphic viruses and signature based detectors.

3   Proposed Work

The efficiency of antivirus is implemented by using fuzzy logic rather than neural logic. Fuzzy logic is used to get appropriate result for uncertain values. Fuzzy logic contains surface/rule viewer editor rule and membership function. Membership function characterizes the states of all the enrollment capacities related with every factor. Rule editor helps to obtain the system behaviour with detection rule as well as rule viewer helps to view the pictorial representation of fuzzy interference. Surface viewer to see the reliance of one of the yields on any a couple of sources of info which is, it produces and plots a yield surface guide for the system. The application involved in fuzzy theory is fuzzy rules in fuzzy control. It has three different steps to construct a fizzy controlled processor and it contains fuzzification, rule evaluation and defuzzification. These different steps helps to obtain the actual results of fuzzy controlled machine (Fig. 1 and Table 1).

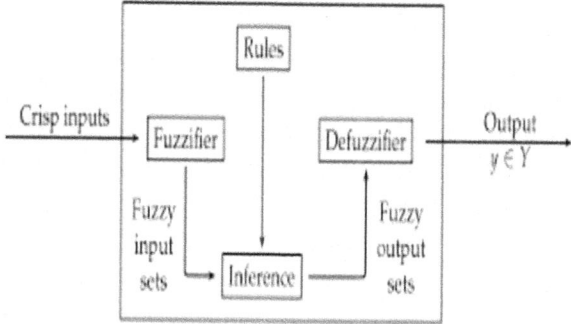

Fig. 1. General block diagram of fuzzy logic

Table 1. Ranges of linguistic variables in fuzzy logic

Inputs	Low	Medium	High
No of iterations	240000–250000	245000–500000	450000–1000000
File size (Mb)	250–400	350–500	450–700
Memory (bits)	6–8	7–16	15–32

As mentioned above we estimated the input ranges for the parameters and using these ranges we implemented the fuzzy logic in matlab and determined the corresponding graphs, surface view and the result i.e., to show efficiency of a software using code emulation.

4   Results and Discussion

Using matlab tool R2010a we implemented few simple membership function variables in code emulator. One of the main motivations behind the design and implementation choices of our tool is to circumvent current limitations of existing malwares. The 3 input parameters are Number of iterations (nos), Size of the file (Mb), Memory access function (bits). As far as we concern about the above inputs, will fall under the code emulation techniques. Usually, code emulation creates a virtual environment, virus will be executed in that virtual one rather than real one. Antivirus software scans a file several times i.e., number of iterations, file size and memory access function determines the efficiency of antivirus software. Exact identification of virus using emulation needs much iteration to be executed in a virtual machine. Iterations vary from quarter a million, half a million and million. After a couple of thousand cycles infection is decoded. Scanner can utilize any of the discovery system to distinguish the infection effectively in virtual condition then the imitating can be additionally reached out for the correct recognizable proof. In this implementation, we consider three ranges as low (240000–250000), medium (245000–500000), high (450000–1000000). By estimating these approximate values we implemented it as one of the input. Thereby, number of iterations i.e., high range is preferable.

Size of the virus is considered as one of the membership function because a virus detector usually skips the execution of large files i.e., > 700 MB (3 GB for malware). Ranges starts from low (250–400), medium (350–500) and high (450–700). Thereby file size > 700 MB will be skipped. Memory access function is another important variable function in good antivirus software. The purpose of utilizing code emulator is to impersonate the guidance set of CPU utilizing virtual registers and flags. Hence we are defining memory access function to fetch 8 bit, 16 bit and 32 bit and the ranges are low (6–8), medium (7–16), and high (15–32). Thereby, 32 bit is adoptable to fetch memory (Fig. 2).

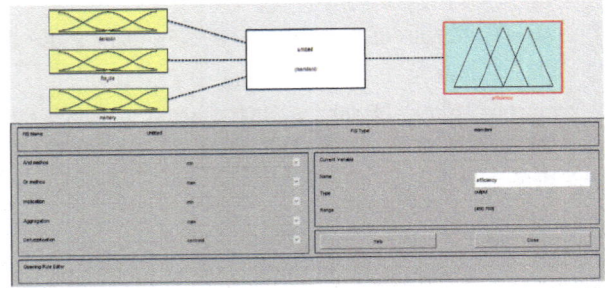

Fig. 2. FIS editor for linguistic variables and the output parameter

Fuzzy rules operate using a series of if-then statements. For instance, if x is a, then y is b where a and b are all sets of x and y. Fuzzy rules define fuzzy patches, which is the key idea in fuzzy logic. After giving the ranges for the corresponding inputs we add some rules the inputs i.e., giving and/or conditions to the ranges mentioned. Initially we defined the rule for the parameters and written as many rules possible. For example, If (iteration is low) and (file size is low) and (memory is low) then (efficiency is bad), If (iteration is high) and (file size is high) and (memory is high) then (efficiency is good), If (iteration is low) and (file size is low) and (memory is high) then (efficiency is average) (Fig. 3).

Below is the rule viewer that describes the various interfaces of the inputs and displays the corresponding outputs (Fig. 4).

By varying the range of different variables we will get different output values and we can get another set of table of values by giving OR rule. Compare the tabular values that are formed using AND/OR rules and get the output which the efficiency of the program in this case (Fig. 5 and Table 2).

From the above graph, it is clear that, when the numbers of iterations are increased simultaneously there is a relative increase in the efficiency of the software (Fig. 6).

No, we considered OR implication rule with the same input variables that is number of iterations, size of file, memory.

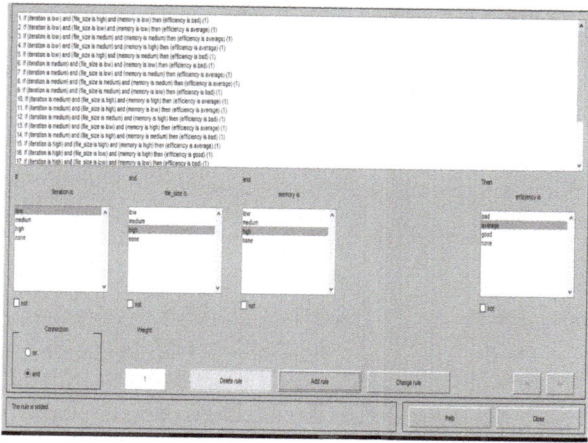

Fig. 3. Rule base for the linguistic variables with AND implication and Min–Max composition.

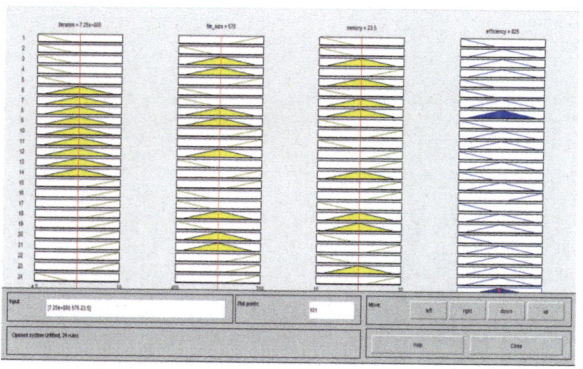

Fig. 4. Rule viewer for linguistic variables.

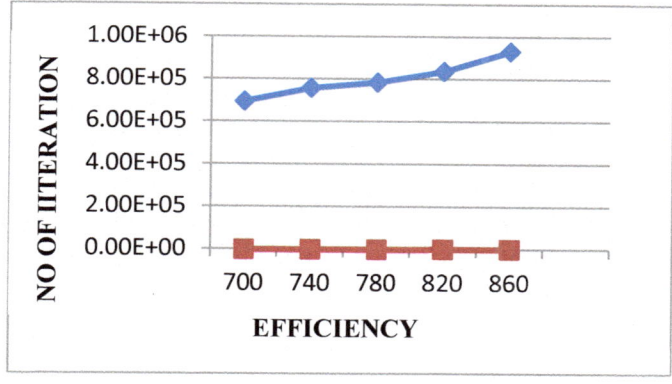

Fig. 5. Efficiency versus no of iterations

Table 2. Quantifying values for the input parameters

No. of iterations	File size (MB)	Memory (BIT)	Output efficiency 0 (SPEED)
(1) 4.62e+005	454	15.5	825
(2) 4.73e+005	463	15.9	825
(3) 5.71e+005	548	29.2	803
(4) 6.27e+005	552	29.3	762
(5) 6.46e+005	554	29.5	741
(6) 6.68e+005	560	29.9	720
(7) 6.69e+005	577	30	716
(8) 7.11e+005	581	30.2	705
(9) 7.22e+005	590	30.6	69.7
(10) 7.58e+005	606	31	724
(11) 7.85e+005	627	31.2	778
(12) 8.82e+005	637	31.5	793
(13) 8.36e+005	648	31.7	806
(14) 8.63e+005	656	31.9	810
(15) 9.31e+005	691	32	825

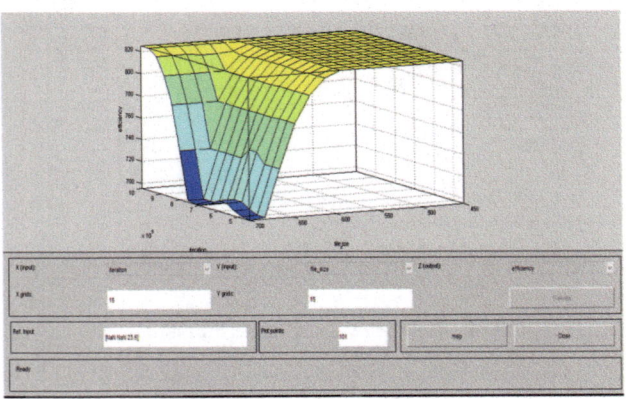

Fig. 6. Surface views for linguistic variable

The rules for the linguistic variables are designed with if and then using OR implementation. For example if (no of iterations is low) or (file size is low) or (memory is low) then (efficiency is low) (Fig. 7).

Below is the rule viewer that describes the various interfaces of the input and displays the corresponding output (Fig. 8 and Table 3).

The above tables describes the values of input parameters by varying rule viewer (Fig. 9).

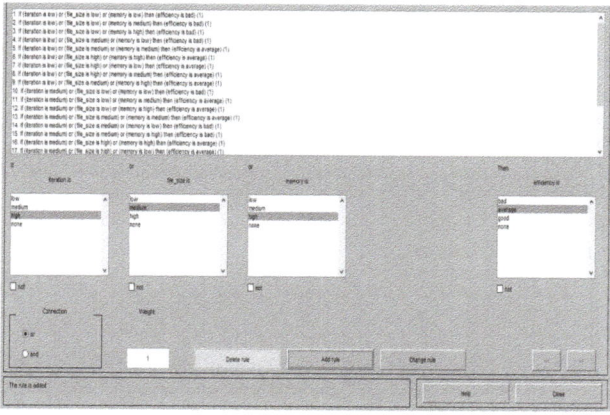

Fig. 7. Rule base for linguistic variables with OR implementation and Min–Max Composition

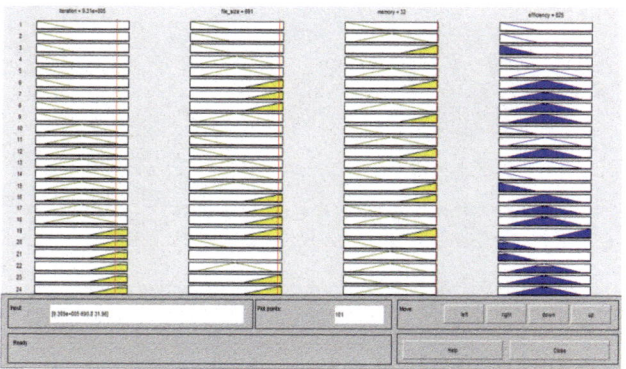

Fig. 8. Rule viewers for linguistic variables

Table 3. Quantifying values of input parameters

Number of iterations	File size (MB)	Memory (BIT)	Efficiency (Speed)
(1) 4.59e+005	456	15.3	786
(2) 4.78e+005	467	15.9	787
(3) 5.33e+005	494	17.5	788
(4) 5.52e+005	507	19.1	790
(5) 5.92e+005	516	19.7	797
(6) 6.11e+005	528	21.7	787
(7) 6.87e+005	541	22.7	786
(8) 7.98e+005	638	25.3	809
(9) 8.31e+005	644	25.6	814

(*continued*)

Table 3. (*continued*)

Number of iterations	File size (MB)	Memory (BIT)	Efficiency (Speed)
(10) 8.42e+005	650	26.1	819
(11) 8.55e+005	656	28.4	830
(12) 8.79e+005	667	28.9	831
(13) 9.04e+005	676	29.8	827
(14) 9.44e+005	688	31	825
(15) 9.96e+005	699	319	825

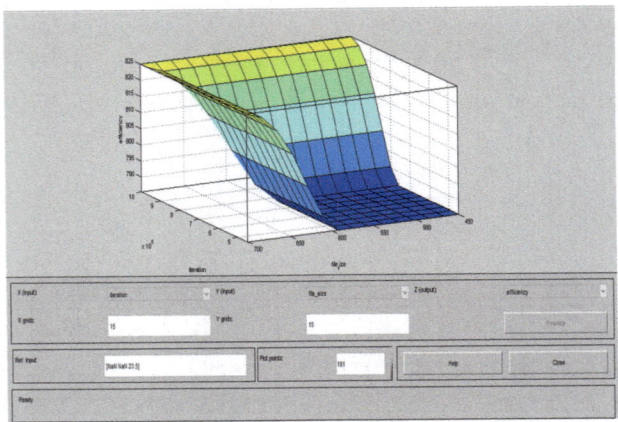

Fig. 9. Surface viewers for linguistic variables

5   Conclusion and Future Works

This paper outlines the implementation of the code emulator parameters which helps in creating a efficient software. It has been found that behavior based detection can be effective in determining the metamorphic virus, where we implemented various membership functions using fuzzy logic and compared the efficiency with implementing linguistic variables with AND and OR implementations(i.e., more number of iterations, large sized files will skip and memory access function should take 32 bit to fetch memory fastly).

Preferred solution to scan a virus, build a sand box environment (malware is executed and analyzed in run time) where it can be executed and detected in it rather on real physical machine. Behavioral basexd detection results in less false positives and more reliable. As the traditional technique like signature based detection (A Zero day malware attack) lacks in efficiency we put forward behavior based detection (type of dynamic analysis technique). Recently used code Obfuscation technique (impossible to statically analyze the programs functionality) has its own limitations. A combination would be behavior based detection like code emulation and HMM (Hidden Markov

Model based detection) as it would give a chance to detect new malware as our proposed methodology will handle the variance of known malwares.

References

1. Venkatesan, A: Code obfuscation and metamorphic virus detection. Master Thesis, Sanjose state university
2. Szor, P.: The Art of Computer Virus Research and Defense. Addison-Wesley, Boston (2005)
3. Sridharan, S.M.: Hunting for metamorphic engines
4. Wong, W.: Analysis and detection of metamorphic computer viruses
5. Bin, S.A., Sharif, M.D.: Metamorphic worm that carries its own morphing engine
6. Stamp, M., Lin, D.: Hunting for undetectable metamorphic viruses. Master Thesis, Department of Computer science, SJSU, spring 2010
7. Kalbhor, A.: A tiered approach to detect metamorphic malware with hidden markov models
8. Govindaraj, S.: Practical detection of metamorphic virus. Master Thesis, Sanjose state university
9. Pryala Shyry, S.: Performance measurement in selfish overlay network by fuzzy logic deployment of overlay nodes. In: International Conference on Control Instrumentation, Communication Technologies (IEEE Xplore), ICCICCT 2014, pp. 717–721 (2014)
10. Chandrasekharan, M., Muralidharan, M.: Application of soft computing techniques in machining performance prediction and optimization: a literature review. Int. J. Adv. Manuf. Technol. **46**, 445–464 (2010)
11. Choudhary, S.P.: A simple method for detection of metaphoric malware using dynamic analysis and text mining. Procedia Comput. Sci. **54**, 265–270 (2015)
12. Lo, C.T.D., Pablo, O.: Towards an effective and efficient malware detection system. In: International Conference on Big Data (IEEE Xplore)

Comparative Evaluation of Various Swarm Intelligence Techniques Designed for Routing in Wireless Sensor Networks

N. Shashank[✉], H. L. Gururaj[✉], and S. P. Pavan Kumar[✉]

Vidyavardhaka College of Engineering, Mysuru, India
{shashank.n, gururaj1711, pavanalina}@vvce.ac.in

Abstract. Wireless Sensor Networks (WSN) work through conditions like energy, calculation and storing. Distributed Topology Control (DTC) that is used for Large-Scale Dense networks (LSD) are taken care by cluster heads and mobile agents. The protocols which are designed for WSNs are adjustable to changes in neighborhood nodes as well as intelligent system devices. Topology in WSN is grouped as three stages: (a) building (b) governing and (c) preservation of topology. Here maintaining the topology using the concept of swarm intelligence (SI) intended for failure in links as well as controlling the congestion is employed. A detailed survey has been conducted on various SI techniques. SI-based representations like Particle Swarm Optimization (PSO) as well as Ant Colony Optimization (ACO) are used in topology maintenance which outperforms other existing swarm intelligence techniques in WSNs.

Keywords: Particle swarm optimization · Ant colony optimization · Maintaining topology

1 Introduction

Swarm intelligence (SI) is the communal performance of dispersed, independent systems, natural or artificial systems. This idea is mainly used in artificial intelligence. SI structures normally contain inhabitants of simple agents that communicate in the vicinity with each other in their environment. The main inspiration for SI systems is nature, particularly the biological systems. Examples comprise swarm of ants, birds gathering, the herd of cattle, the growth of bacteria, school of fish [1, 3].

Particle swarm optimization (PSO) a worldwide algorithm meant for optimization which helps in resolving problems for which the best solution is represented as an argument inside n-dimensional space [18]. Propositions are calculated and broadcasted through a primary velocity, via a communication network between the units. Particles later transfer via result space, later they will be assessed by allowing to some fitness criterion after every period. As the time moves, elements are enhanced towards those particles in their message group which are having better fitness standards [2].

© Springer Nature Switzerland AG 2020
A. P. Pandian et al. (Eds.): ICCBI 2018, LNDECT 31, pp. 76–85, 2020.
https://doi.org/10.1007/978-3-030-24643-3_8

Ant colony optimization (ACO) is a procedure for cracking the problems used in the concentration for discovering the best pathways with the help of graphs. Artificial Ants are the mediator approaches which got inspiration from the activities of real ants. The communication based on pheromone in biological ants is frequently the most common platform employed [4, 13]. The group of artificial ants and local exploration procedures are the methods for choosing enormous optimization tasks The continuous research in this arena has prompted gatherings which are exclusively devoted to Artificial Ants and additionally different business applications by specific organizations, for example, Ant Optima. The ACO calculation is basically utilized in picture preparing for edge discovery and edge detection [6].

2 Related Work

The movement of information all through an internetwork from a source to an end is normally referred to as routing. During the way, one midway node is naturally met. Swarm Intelligence can be mostly seen in biological groups of few insects which gave birth to complex and often intelligent performance through composite communication of many independent swarm participants [5].

A typical WSN comprises vast sensor nodes which communicate with the help of wireless standard for data distribution and supportive processing. In general, these nodes are placed in a static manner across large parts. However, these nodes show mobility and are proficient of intermingling with the environment [7].

Routing algorithms in a modern networking system should discourse abundant problems. Among many, two normal performance measurements in the network are typically middling of input plus interruption. Messages among the routing and movement control affect its way in which metrics are cooperatively optimized [8].

The current improvements in the field of WSN have increased popularity across the globe due to its minimalized size and low price nature. The sensor nodes contain sensing element which helps in collecting the data from the physical environment, a processing unit that helps in the processing of information and communicating module for communicating sensory information to a base station via a medium which is wireless [9].

The applications related to SI procedures on WSNs placement become the main emphasis to the research communal over limited years. Wireless sensor networks (WSNs) are the networks that include abundant little sensors which interconnect through wireless networks [12].

Recent progressions in WSNs made conceivable units which can be intended for the consumption of less power, minor magnitude, and negligible variety communication capabilities. WSN architectures comprise enormous processing sensors that have these characteristics which afford important benefits on traditional sensor schemes. A classic sensor network is made up of an area under surveillance, sensor nodes, base position and job allocator nodes [13].

3 Particle Swarm Optimization in Topology Maintenance for WSNs

Multi-target degree controlled lowest spanning tree is an enhancement of disconnected PSO which is made use in the mechanism of topology for WSNs. Maintenance in topology is tangled after any of the subsequent trials happen to like a problem in the link as well as in packet drop. Here in the projected model of PSO-TM, the assumption is made that every node is particle 'P' with population size Ps.

A mapper is used for plotting attributes of a swarm to characteristics of WSN. An element position for 'p' in a group is plotted towards the location of a node present in WSN later assumed by

$$F1(p) : particle \ \Box \ Node \ position \ (p).$$

This paper mainly focusses to resolve failure in links and congestion that occur in the network for maintaining a connection in the network. Here the arguments used are:
Ps = Size of the population, P [] = parameter array and v [] = particle velocity.

Particle swarm Optimization-Topological Maintenance for failure in links and controlling the congestion for WSNs is done by using the following steps as shown in Fig. 1.

Steps involved:

a. Initialization of particle (node) and its position as well as its particle velocity.
b. Total best and native best are found out.
c. Position of the particle (Posp) is initialized relating to elements which are adjacent (Adjp) as well as to destination (dest).
d. Velocity of the particle with regard to adjacent particle (PVadj) and destination are initialized.
e. Least possible global cost for every element 'P' is calculated.
f. Calculation of Minimum cost path based on preceding particle history, particles that are adjacent as well as a destination is done.

3.1 Particle Swarm Optimization in Topology Maintenance for WSNs

The main reasons which cause failure of the links are as follows:

a. The existence of a faulty node
b. The strength of the signal is weak.
c. Environmental conditions.

To minimize the link failure, the steps to be followed are:

a. The velocity of the unit is minimized.
b. Location of the particle (sensor node) is reoriented and finding out if a substitute path exists and neighbor particles are notified.
c. Global best as well as local best of the particle are calculated.
d. Velocity of the particle is increased. (increasing the data broadcast rate in WSN)

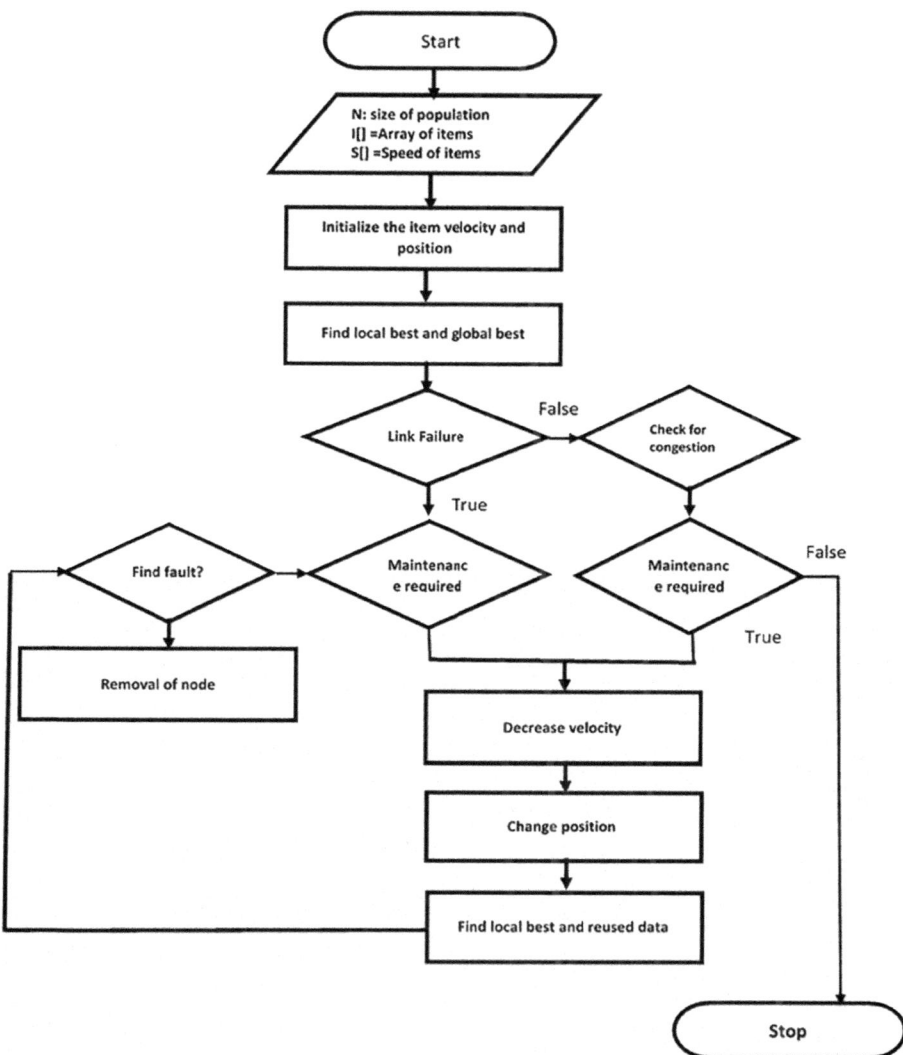

Fig. 1. PSO_TM for failure in links and controlling congestion

3.2   Particle Swarm Optimization-Topological Maintenance for Congestion Control

To reduce the congestion control, the following steps are carried out

a. Neighborhood particle velocity is to be decreased.
b. Priority is given to the particles according to nodes next to to the target node and nodes with energy which is high.
   I. Nodes with priority high, use the network channel
   II. Nodes having priority value low, wait in network line.
c. Particles are reoriented. Particle global and local best are found out which increase the neighborhood particle velocity.

The following flowchart Fig. 1 describes the Particle swarm Optimization-Topological Maintenance for failure in links and controlling the congestion for wireless sensor networks.

4   Ant Colony Optimization for Maintaining the Topology in WSNs

Ant colony optimization for maintaining the topology discovers the minimum pathway constructed on arguments such as length of the pheromone (length of the path), total count of ants in a route to be searched (bandwidth) over a stipulated time.

Here in proposed model of ACO-TM, the assumption is made in such a way that we treat each node as an ant 'a' and WSN as ant population having magnitude 'ANT_PS'.

A mapper is defined for plotting ant attributes to WSN attributes. Ant position for 'Ant' in an ant colony is mapped to node position in WSN and is given as:

$$g1(a) : \text{Ant Position (Ant)} \rightarrow \text{No deposition (Ant)}$$

Ant_alignment for 'A' in an ant colony is mapped to node_alignment in WSN and is given as:

g2(a, average_distance, pheromone_depth): ant_alignment(Ant, neighborhood_ants, average_distance, pheromone_depth) → node_alignment(Ant, adjacent_nodes, average_distance, data_transfer rate)

Pheromone depth of ant for 'ant' in an ant colony is mapped to data transferred rate in WSN and is given as:

g3(ant, pheromone): ant_pheromone_depth(ant, pheromone) → data_transfer_rate (ant, bits per second)

ACO_TM provides a solution to failures in link and bottleneck (congestion) in network. It contains two phases

I. Phase 1: Initialize the position of ant (nodes), local best of the ant is calculated based on price of pheromone.
II. Phase 2: Topological maintenance for failure in link as well as congestion in the network is decided.

Initialize the position of the ant (which is represented as 'aPos') with respect to ants which are adjacent (denoted as 'aAdj') and destination (represented as 'Dest'). To, every ant 'a' the smallest native rate is calculated as well as path is updated.

4.1 ACO_TM for Failure in Link

It travels about the defective ant (node) projecting near the target node. The following stages are involved in ACO_TM intended for failure of links in WSNs.

i. The rapidity of ant is reduced (decrease the data transmission rate in WSNs).
ii. By considering previous state and depth of the pheromone, move ants around the ants which are having fault.
iii. Alternative route is discovered and minimum cost of the path is calculated.
iv. Speed of the ant is maximized (increase the information broadcast frequency in WSN).

The steps that are described above is represented as a flowchart as shown in Fig. 2 which gives us an insight towards how ACO-TM is helpful in reducing the failures in link.

4.2 ACO_TM for Controlling the Congestion

The following steps described below explains the ACO-TM for congestion control in WSNs and the same as illustrated in Fig. 3 which gives us an overview regarding how congestion in the network is controlled using the proposed ACO-TM algorithm.

a. Neighborhood speed of the ant is to be reduced. (decrease the data transmission rate in WSN)
b. Ants are given priority according to the nodes that are adjacent to destination nodes, nodes having energy which is high and pheromone cost which is maximum.
   i. Nodes with high priority make use of network channels.
   ii. Nodes having low priority will wait in the network queue.
c. Ants are moved around the zone of congestion.
d. Minimum path cost has to be computed.
e. Maximize the neighborhood speed of the ant (increase the data transmission rate in WSN).

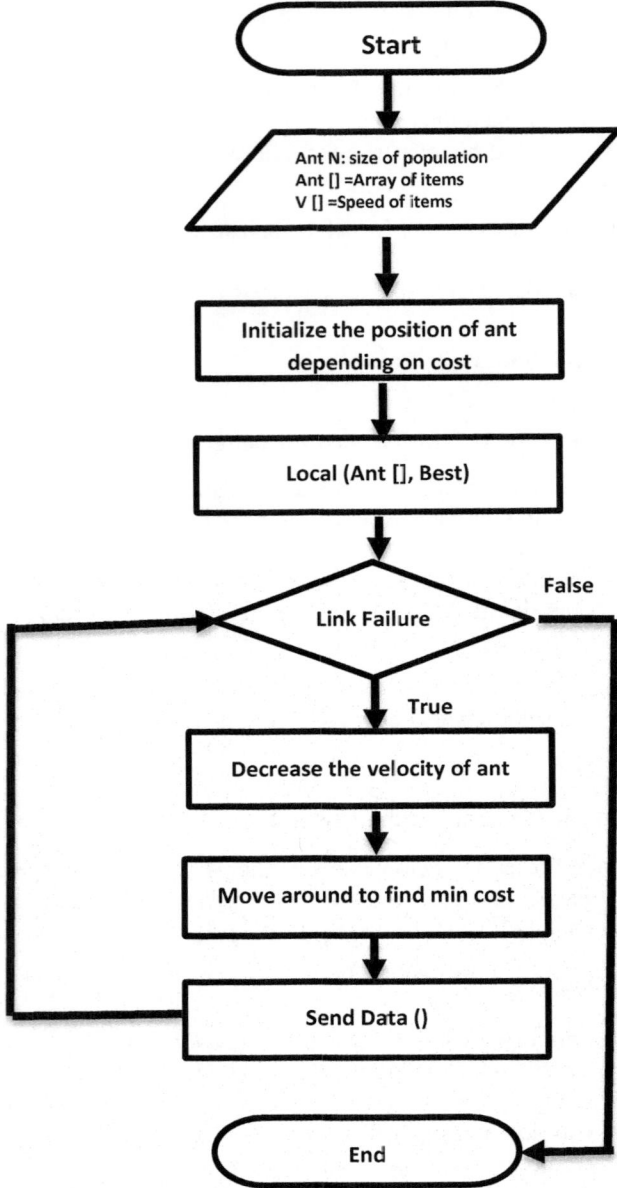

Fig. 2. ACO_TM for failure in links

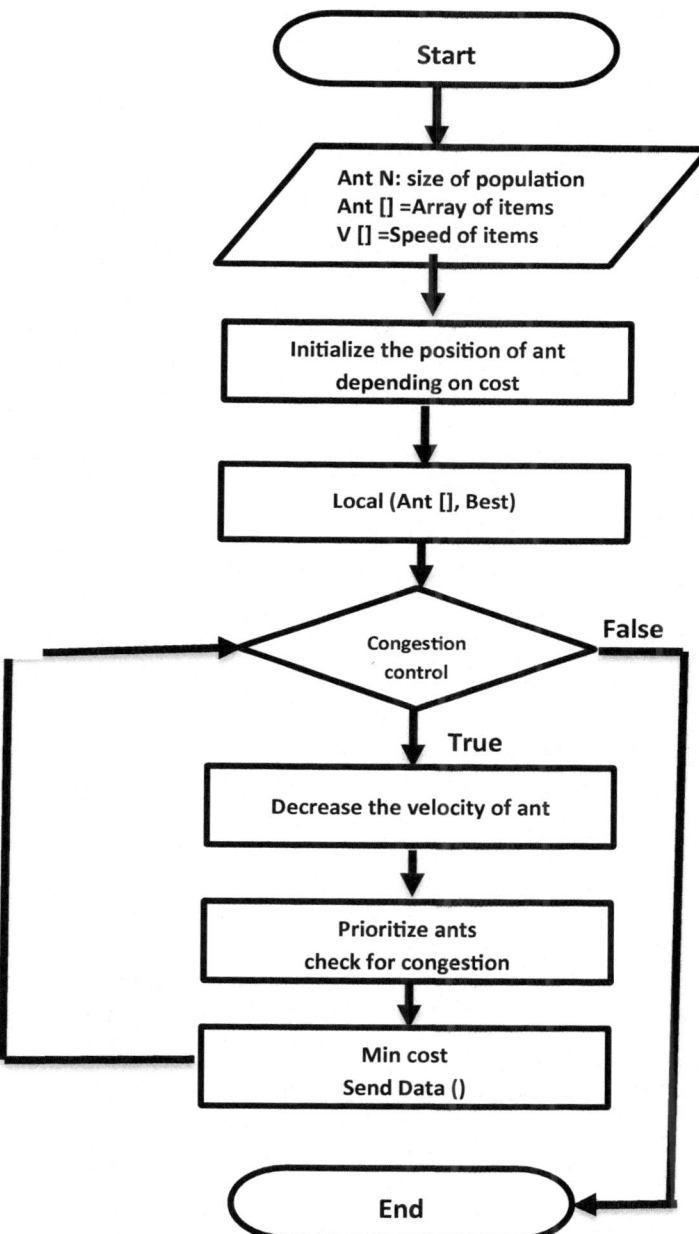

Fig. 3. ACO_TM for controlling congestion

5   Conclusion

Over a past few years, wireless sensor networks have attracted many research scholars to do research on this domain. One of the principal points received through analysts to concentrate on WSNs is to improve in directing conventions for remote frameworks. Directing convention advancement helps in managing issues like intricacy, versatility, flexibility, survivability and battery life in remote frameworks. Steering conventions which are intended for remote frameworks are produced with the end goal to take care of these issues.

In this paper, the projected PSO_TM as well as ACO_TM algorithms in maintaining the topology for WSN show the growth in the failure of link and control in congestion. PSO_TM comprises SI characteristics like the location of the particle (position of the node), speed (rate of transmitting data) and rate of success for aligning the node (neighborhood parameters). ACO_TM comprises SI characteristics such as the weight of the pheromone (time occupied to navigate a path) as well as FSR (amount of effective repetitions for exploring the method).

References

1. Çelik, F., Zengin, A., Tuncel, S.: A survey on swarm intelligence based routing protocols in wireless sensor networks. Int. J. Phys. Sci. 5(14), 2118–2126 (2010). http://www.academicjournals.org/IJPS. ISSN 1992 - 1950 ©2010 Academic Journals
2. Qasim, T., Zia, M., Minhas, Q.-A., Bhatti, N., Saleem, K., Qasim, T., Mahmood, H.: An ant colony optimization based approach for minimum cost coverage on 3-D grid in wireless sensor networks. IEEE Commun. Lett. 22(6), 1140–1143 (2018)
3. Krishna, M.B., Doja, M.N.: Swarm intelligence-based topology maintenance protocol for wireless sensor networks. IET Wirel. Sens. Syst. 1(4), 181–190 (2011). https://doi.org/10.1049/iet-wss.2011.0068
4. Sarobin, M.V.R., Ganesan, R.: Swarm intelligence in wireless sensor networks: a survey. Int. J. Pure Appl. Math. 101(5), 773–807 (2015)
5. Kassabalidis, I., El-Sharkawi, M.A., Marks, R.J., Arabshahi, P., Gray, A.A.: Swarm intelligence for routing in communication networks. In: IEEE Global Telecommunications Conference 2001 GLOBECOM'01 (Cat. No. 01CH37270), vol. 6, pp. 3613–3617. IEEE (2001)
6. Kumar, J., Tripathi, S., Tiwari, R.K.: Routing protocol for wireless sensor networks using swarm intelligence-ACO with ECPSOA. In: 2016 International Conference on Information Technology (2016)
7. Madhusudan, G., Kumar, T.N.R.: Swarm based intelligence routing algorithm (ant colony). Int. J. Adv. Res. Comput. Sci. Softw. Eng. 7(6) (2017). https://doi.org/10.23956/ijarcsse/V7I6/0106
8. Nagar, G., Sood, A., Roopak, M.: Swarm intelligence for network communication routing. Int. J. Comput. Sci. Mobile Comput. 2(5), 268–274 (2013)
9. Nayyar, A., Singh, R.: Simulation and performance comparison of ant colony optimization (ACO) routing protocol with AODV, DSDV, DSR routing protocols of wireless sensor networks using NS-2 simulator. Am. J. Intell. Syst. 7(1), 19–30 (2017). https://doi.org/10.5923/j.ajis.20170701.02

10. Singh, S.P., Sharma, S.C.: A novel energy efficient clustering algorithm for wireless sensor networks. Eng. Technol. Appl. Sci. Res. **7**(4), 1775–1780 (2017)
11. Ab Aziz, N.A.: Wireless sensor networks coverage-energy algorithms based on particle swarm optimization. Emir. J. Eng. Res. **18**(2), 41–52 (2013)
12. Garg, D., Kumar, P., Kompal, M.: Energy efficient clustering protocol for wireless sensor networks using particle swarm optimization approach. Int. Res. J. Eng. Technol. **4**(7) (2017)
13. Laturkar, A.P., Malathi, P.: Coverage optimization techniques in WSN using PSO: a survey. International Journal of Computer Applications (0975-8887). National Conference on Emerging Trends in Advanced Communication Technologies (NCETACT-2015)
14. Islam, S.M.M., Reza, M.A.R., Kiber, M.A.: Wireless sensor network using particle swarm optimization. In: Proceedings of International Conference on Advances in Control System and Electricals Engineering 2013 (2013)
15. Yick, J., Mukherjee, B., Ghosal, D.: Wireless sensor network survey. Comput. Netw. **52**(12), 2292–2330 (2008)
16. Eberhart, R., Kennedy, J.: A new optimizer using particle swarm theory. In: Proceedings of 6th International Symposium on Micro Machine and Human Science, pp. 39–43 (1995)
17. Poli, R.: Analysis of the publications on the applications of particle swarm optimisation. J. Artif. Evol. Appl. **2008**, 1–10 (2007)
18. Lazinica, A.: Particle Swarm Optimization. In-Tech, Rijeka (2009)
19. Price, K.V., Storn, R.M., Lampinen, J.A.: Differential Evolution: A Practical Approach to Global Optimization. Ser. Natural Computing Series. Springer, Berlin (2005)
20. Akyildiz, I.F., Su, W., Sankarasubramaniam, Y., Cayirci, E.: A survey on sensor networks. IEEE Commun. Mag. **40**(8), 104–112 (2002)

Spatially Correlated Cluster in a Dense Wireless Sensor Network: A Survey

Utkarsha S. Pacharaney[1]([⊠]) and Rajiv Kumar Gupta[2]

[1] Department of Electronics, Datta Meghe College of Engineering, Airoli, India
utk21pac76@gmail.com
[2] Department of Electronics and Telecommunication,
Terna Engineering College, Nerul, India
rajivmind@gmail.com

Abstract. In Wireless Sensor Network (WSN), to achieve satisfactory coverage of the monitoring field, sensor nodes are densely deployed. Furthermore, clustering is used to reduce the density of measurement points. Spatially correlated clusters are precisely used for this purpose. Various survey papers are presented on clustering, but this is the first survey paper on spatially correlated clusters in a dense WSN. Spatial correlation refers to the similarity of readings between geographically proximal nodes. Hence grouping such sensor nodes and appointing a representative for the entire group, can reduce energy consumption and prolong network lifetime. In this survey paper we give detail description and comparison of the work done in this domain of clustering.

Keywords: Spatial correlation · Clustering · Correlated region ·
Representative node

1 Introduction

Wireless Sensor Network (WSN) comprises spatially dense sensor nodes to achieve satisfactory coverage of the phenomenon of interest. Limited energy supply, constraints the tiny sensor nodes to directly communicate with the sink, also it is not the individual competency of nodes that matters, but collaboration between them does [1]. Hence, the fundamental task of data gathering is accomplished in the network by fusing the data at a point and further multi hopping it to the sink. Clustering has been traditionally used to accomplish this task. Clustering in literature has been used with the aim of: efficient data collection [2], data aggregation [3], reducing number of transmissions [4], load balancing [5], fault tolerance [6] and routing [7]. Apart from these for densely deployed network, the density of measurement points in a monitoring field can be reduced via clustering. Spatially correlated clusters fall in this category proving Tobler's statement "Everything is related to everything else, but near things are more related than distant things" [8]. Spatial correlation is defined: Given a set S containing N nodes, it refers to the similar of reading between these N nodes within a geographical proximity for all N $(N-1)$ nodes. Various survey papers on clustering [9, 10] have been presented, but this is the first survey paper on spatially correlated cluster in a dense WSN.

© Springer Nature Switzerland AG 2020
A. P. Pandian et al. (Eds.): ICCBI 2018, LNDECT 31, pp. 86–93, 2020.
https://doi.org/10.1007/978-3-030-24643-3_9

It is observed that nearby sensors monitoring a phenomenon typically register similar values. Hence, grouping such sensors together and appointing one of them to represent the entire group value, can optimize energy usage and prolong network lifetime. By using inherent spatial and data correlation in WSN, some researchers have systematically discussed spatial correlation, which can be categorized on the basis of distance between nodes or interrelationship of sampled data. This is the first of its kind survey paper on spatially correlated cluster.

Organization of the paper is as follows: Sect. 2 presents a detail survey on the clusters formed based on spatial correlation. Section 3 concludes the paper with some remarks and suggestions.

2 Literature Review

2.1 Representative Node Based on Distortion Constraint

In [11] Vuran *et al.* revealed that extensive energy saving is possible by allowing minimum number of nodes to send information to the sink in a dense WSN. The minimum number of representative nodes are those that attain the distortion constraint D_{max}. The spatial Correlation-based Collaborative MAC (CC-MAC) protocol is applied at data link layer. The representative node of the area determined by the correlation radius r_{corr}, is the one that captures the channel after the first contention phase. This representative node say n_i, solely continues to send information to the sink of its correlated region. The ongoing transmission informs other nodes, which overhear it, to determine whether they are correlated neighbors of node n_i. If d_{ij}, the distance between node n_i and n_j, is less than r_{corr}, then n_j is the neighbor of n_i. Thus, it is a dynamic clustering method that suppress redundant information from being injected into the network. Figure 1 shows multihop transmission from nodes to the sink.

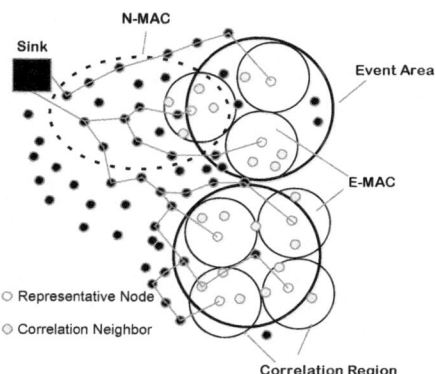

Fig. 1. Multihop transmission of data [11]

However, this technique does not consider the remaining energy of the representative node in the selection procedure. Since, these nodes carries out the maximum message transmission, they will quickly drain out of energy. This will trigger choice of alternative representative node, which effects the distortion tolerance.

2.2 Query Based Correlation Scheme

Clustered AGgregation (CAG) is a query based data correlation scheme with two phases: query and response. While, a query is circulated in the network, CAG forms clusters of nodes sensing identical values. A user defined error threshold, is used while forming clusters. Every node has *My local sensor Reading (MR)* and *Cluster head sensing Reading (CR)*, if $MR < CR \pm CRX\tau$, then node joins the cluster. In the response phase only CH senses the data and transmits it to the sink. Figure 2 shows the execution algorithm example of CAG [12]. CAG algorithm saves on transmissions by evading global communication and using only local communication. The communication overhead, cluster formation complexity and implementation of user defined threshold are some of the drawbacks of CAG. Also the first phase is executed using flooding based protocol, which increases the network load with duplicate messages.

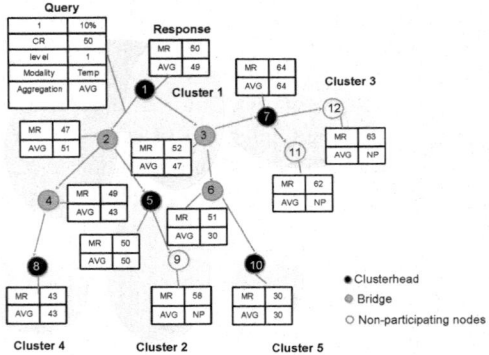

Fig. 2. Example of CAG algorithm [12]

2.3 Adaptive Sampling Approach for Data Collection

Gedik *et al.* [13] proposed a distributed clustering algorithm ASAP, which gather nodes having similar readings. The main idea is to use probabilistic models based on spatial and temporal correlation among sensor readings to predict the value of member nodes in a cluster. Only a subset of nodes, which changes dynamically, sense, predict and report the value of the cluster to the sink. In the cluster construction phase, nodes are grouped in a cluster based on similarity of their past sensor readings and secondly nodes are at a one hop distance in the cluster. After cluster formation subclusters are created. The purpose of subdivision of clusters into subclusters is to facilitate the choice of a node to work as samplers and generator of the probabilistic models for value prediction. To capture the spatial and temporal correlation the Cluster Head (CH), uses infrequently sampled reading from all the nodes within its cluster and calculates sub cluster so that nodes whose sensor readings are highly correlated are put into the same subcluster. Using the probability based method for CH selection needs number of iterations for good coverage and distribution of CHs.

2.4   Gridon Spatial Correlation

A cluster based on energy-aware spatial correlation mechanism is proposed in [14] by Ghalib *et al.*. Correlation 'θ' is the closeness value and evaluated as $\theta = \sqrt{R^2/n}$ where, R is the dimension of the square field and n represents the deployed number of nodes in $R \times R$. Only the Cluster Head (CH) divides the clustering region into rectangles and select a representative node which is nearby to the center of correlated region and has considerable quantity of residual energy. The representative node is only active while other nodes in the region remain inactive until the energy of the representative node reaches threshold. The gridon spatial correlation is detailed with the help of Fig. 3. The Cluster zone is decomposed into rectangular grid of size $2r/\theta \times 2r/\theta$, with the CH at the center of the grid. Only one node is selected in each rectangle of dimension $\theta \times \theta$. The center of the ith rectangle be given by $G(x_i, y_i)$. The distance of every node from this center point is calculated and the node which lies nearest to it is choosen as the active member of the cluster while others are deactivated by the CH.

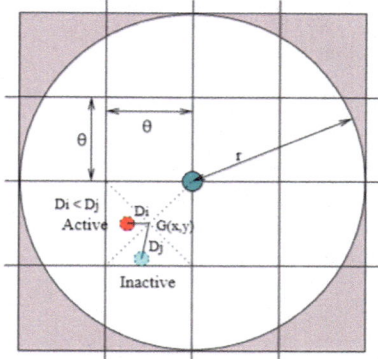

Fig. 3. Spatial resolution control mechanism [14]

To measure similarity between sensor readings at the sink, abundant raw data exchange and communication overhead will be triggered in this method.

2.5   Energy-Efficient Data Collection Framework

For continuous data gathering applications Chong *et al.* exploit spatial correlation in [15] by partitioning sensor nodes into clusters. Within an error bound, nodes can approximate their neighbor's readings both on magnitude and trend. The basis of the clustering algorithm is on measure of dissimilarity between the time series of two dissimilar nodes. An Energy-Efficient Data Collection (EEDC) framework is proposed, which represents the clustering issue as clique-covering problem and a greedy algorithm solves it. To balance energy consumption within the cluster, randomized scheduling algorithm is proposed. However, abundant communication overhead is triggered in EEDC for node clustering.

2.6 Energy Aware Set Coverage Approach

Hung et al. in [16] proposed to group nodes with similar reading and only representative node reporting to the sink. To solve the challenging task of establishing sensor groups and representative nodes, a centralized DCglobal and distributed DClocal algorithm is proposed. The nodes with the highest energy level and wide data coverage range is selected as the representative node. In the DCglobal algorithm initially every node report their readings and energy level to the sink. Sorting individual node with energy and coverage area the algorithm forms a list of nodes known as Plist and then determines the representative node (R-node). The R-representative solely is transmitting data in the coverage area hence will drain out energy hence R maintenance is run to find the next able node from the Plist. The maintenance is run in two scenarios: if R-nodes runs out of energy or it is not able to cover nodes in the data coverage range. In both the cases the procedure for R-representative selection begins all over with the sink. Hence the message complexity increase. To overcome message complexity DClocal is proposed.

2.7 Data Correlation Based Clustering

Lu et al. proposed Data Correlation based Clustering (DCC) [17]. Historical data is mined to find the correlation in the sensor readings. Division of the network is done into disjoint clusters that are strongly correlated. CH gathers data and compressively forwards it to the base station. The data correlation is defined as: two sensor nodes X and Y are called data correlated during $[T_1, T_n]$, if data by them at round T_i is X_i and Y_i respectively and satisfies the following equation

$$\frac{1}{n}\sum_{i=T_i}^{T_n}|R(X_i) - Y_i| < e \tag{1}$$

where e average error which is constant and R is data correlation.

2.8 Spatial Correlation Based Clustering

To perform near exact data collection Zhidan et al. [18] advocate the need to group sensors having similar readings into the same cluster and schedule them alternately to relay data to the sink. The challenging process to partition all sensor nodes into disjoint clusters while inducing as less as communication overhead is accomplished with Distributed Spatial Correlated-based Clustering (DSCC). By utilizing spatial correlation range R_{sc} and application specific parameter namely error-tolerance ϵ the algorithm selects node with high residual energy as CH. Manhattan distance is employed to measure the sensor reading dissimilarity between two nodes and a function f_{md} is defined as given in Eq. 2

$$f_{md}(s_i, s_j) = \frac{\sum_{k=1}^{q} |v_k(s_i) - v'_k(s_j)|}{q} \tag{2}$$

where $v(s_i)$ and $v'(s_j)$ are the sensor reading serials of s_i and s_j respectively. Two sensor nodes, say s_i and s_j are called similar nodes if their dissimilarity distance meets $f_{md}(s_i, s_j) \leq \frac{\epsilon}{2}$ and their geographic distance is smaller than R_{sc}.

The value of ϵ and R_{sc} is broadcasted by the sink, so that a appropriate CH is chosen. Nodes compete within the specified range to become CH.

2.9 Data Density Correlated Degree Clustering

One data aggregation strategy in WSN is to send local representative information to the sink node based on the spatial correlation of sampled data. Hence, a data density correlation degree is proposed and based on that a clustering method is presented Data Density Correlation Degree (DDCD) by Fei *et al.* in [19]. The DDCD clustering algorithm comprises of three procedures: the Sensor Type Calculation (STC), the Local Cluster Construction (LCC) and Global Representative sensor node Selection (GRS). The data density correlation degree is defined as: Let sensor node v with data object D, have n neighboring sensor nodes $v_1, v_2, \ldots v_n$, with data objects $D_1, D_2 \ldots D_n$ respectively, which are within the cycle of the communication radius of v. Out of these n data objects, there are N data objects whose distances to D are less than ε, and $min\,Pts \leq N \leq n$. Then the data density correlation degree of sensor node v to the sensor nodes whose data objects are in ε-neighborhood of D is as follows.

$$Sim(v) = \begin{cases} 0, & N < min\,Pts \\ a1\left(1 - \frac{1}{exp(N-minPts)}\right) + a_2\left(1 - \frac{d\Delta}{\varepsilon}\right) + a_3\left(1 - \frac{d}{\varepsilon}\right), & N \geq minPts \end{cases} \tag{3}$$

The DDCD clustering method is beneficial for applications that have dense deployment of sensor nodes and where the sampled data change slowly with time.

3 Conclusion and Remark

In densely deployed WSN, density of measurement points is reduced with spatially correlated clusters. Spatial correlation refers to the similarity of readings between nodes. Thus, all nodes in these type of clusters have similar values. These clusters are formed with the aim to curtail energy consumption and to prolong lifetime a battery operated sensor network. Hence, only one node is assigned the energy draining task of sensing and communicating with the sink. The partitioning of node to form spatial clusters is based on two parameters: spatial distance between nodes and spatial correlation of the sampled data. As compared to first parameter the latter is more exploited by researchers. The sampled readings for computing data correlation are either historical readings, similarity/dissimilarity between readings or readings within an error tolerance whereas spatial distance is the area given by r_{corr}, R_{corr} or communication range. Apart from ignoring the cost of learning the correlation, the drawbacks of these correlated clusters that there is no uniform measure on tolerance error range or distance between sensors. Also ample communication overhead for spatial clustering, number of rounds to select the CH and above all, the construction of cluster takes several rounds of message exchange and computation of correlations in these clusters. By considering these facts, in a dense WSN estimation of spatial correlation between sensor nodes can be based on geographical proximity within them. Advantage of topological spatial nearness of nodes to build correlated clusters with parameter can be taken, which is less energy consuming in terms of communication overhead and adaptable.

References

1. Akyildiz, I.F., et al.: Wireless sensor networks: a survey. Comput. Netw. **38**(4), 393–422 (2002)
2. Liang, Y., Gao, H.: An energy-efficient clustering algorithm for data gathering and aggregation in sensor networks. In: 4th IEEE Conference on Industrial Electronics and Applications (2009), ICIEA 2009, 25–27 May 2009
3. Ranjani, S.S., Radha Krishnan, S., Thangaraj, C.: Energy-efficient cluster based data aggregation for wireless sensor networks. In: 2012 International Conference on Recent Advances in Computing and Software Systems (RACSS). IEEE (2012)
4. Suresh, D., Selvakumar, K.: Improving network lifetime and reducing energy consumption in wireless sensor networks. Int. J. Comput. Sci. Inf. Technol. **5**(2), 1035–1103 (2014)
5. Zhang, H., Li, L., Yan, X., Li, X.: A load-balancing clustering algorithm of WSN for data gathering. In: 2011 2nd International Conference on Artificial Intelligence, Management Science and Electronic Commerce (AIMSEC). IEEE, pp. 915–918 (2011)
6. Gupta, G., Younis, M.: Fault-tolerant clustering of wireless sensor networks. In: Wireless Communications and Networking, 2003. WCNC 2003, vol. 3. IEEE (2003)
7. Liu, X.: A survey on clustering routing protocols in wireless sensor networks. Sensors **12**(8), 11113–11153 (2012)
8. Miller, H.J.: Tobler's first law and spatial analysis. Ann. Assoc. Am. Geogr. **94**(2), 284–289 (2004)
9. Abbasi, A.A., Younis, M.: A survey on clustering algorithms for wireless sensor networks. Comput. Commun. **30**(14–15), 2826–2841 (2007)
10. Boyinbode, O., Le, H., Takizawa, M.: A survey on clustering algorithms for wireless sensor networks. Int. J. Space Based Situated Comput. **1**(2-3), 130–136 (2011)
11. Vuran, M.C., Akan, Ö.B., Akyildiz, I.F.: Spatio-temporal correlation: theory and applications for wireless sensor networks. Comput. Netw. **45**(3), 245–259 (2004)
12. Yoon, S., Shahabi, C.: The clustered aggregation (CAG) technique leveraging spatial and temporal correlations in wireless sensor networks. ACM Trans. Sens. Netw. **3**(1), 3 (2007)
13. Gedik, B., Liu, L., Yu Philip, S.: ASAP: an adaptive sampling approach to data collection in sensor networks. IEEE Trans. Parallel Distrib. Syst. **18**(12), 1766–1783 (2007)
14. Shah, G.A., Bozyigit, M.: Exploiting energy-aware spatial correlation in wireless sensor networks. In: 2nd International Conference on Communication Systems Software and Middleware. COMSWARE 2007. IEEE (2007)

15. Chong, L., Kui, W., Jian, P.: An energy-efficient data collection framework for wireless sensor networks by exploiting spatiotemporal correlation. IEEE Trans. Parallel Distrib. Syst. **18**(7), 1010–1023 (2007)
16. Hung, C.-C., Peng, W.-C., Tsai, Y.S.-H., Lee, W.-C.: Exploiting spatial and data correlations for approximate data collection in wireless sensor networks. In: Knowledge Discovery from Sensor Data (Sensor-KDD 2008), p. 111 (2008)
17. Lu, J., Zhang, B., Xu, L.: A data correlation-based wireless sensor network clustering algorithm. In: 2010 International Conference on Computer Application and System Modeling (ICCASM), vol. 8. IEEE (2010)
18. Liu, Z., et al.: Distributed spatial correlation-based clustering for approximate data collection in WSNs. In: 2013 IEEE 27th International Conference on Advanced Information Networking and Applications (AINA). IEEE (2013)
19. Yuan, F., Zhan, Y., Wang, Y.: Data density correlation degree clustering method for data aggregation in WSN. IEEE Sens. J. **14**(4), 1089–1098 (2014)

Comprehensive Survey on Detection of Living or Dead Humans and Animals Using Different Approaches

C. A. Pooja[1], P. Vamshika[1], Rithika B. Jain[1],
Vishal K. Jain[1], and H. T. Chethana[2(✉)]

[1] Department of Computer Science and Engineering,
Vidyavardhaka College of Engineering, Mysuru, India
poojaca350@gmail.com, vamshikacse@gmail.com,
ritpit8850@gmail.com, viisshhhh@gmail.com
[2] Technical Support, Department of Computer Science and Engineering,
Vidyavardhaka College of Engineering, Mysuru, India
chethanaht@vvce.ac.in

Abstract. The main idea behind this paper is to recognize living or dead humans and animals using different approaches. Object detection and tracking are the most challenging task faced by different software or computer applications. Especially in dynamic environment for recognizing the living beings is the greatest difficulty faced by applications, as it involves efficient object detection system in environments that are very complex. In image processing, object detection is considered to be a very difficult process. Many solutions for this problem involve the usage of deep learning techniques where a set of data is given to the machine which saves the key features of the objects and help in giving the desired output. This paper describes comprehensive survey on various methods and techniques used for object detection during natural disasters and finally comparative study of different methods are discussed.

Keywords: Deep learning · Artificial neural networks ·
Image semantic segmentation · Object detection

1 Introduction

In the present world there is a great need for protecting wildlife and human beings from the natural calamities, hazardous accidents, thefts etc. As we all know smart homes depend upon the environment friendly management systems, it involves monitors that detects human postures and abnormal behaviors of the people for a safe environment. The posture recognition of human beings is necessary as the elderly people face a lot of problem when they live alone especially people with heart diseases. In security industry and medical field, recognition of human activity has become a very important aspect [10].

Now a days observing wild animals in their natural environment is a primary task in ecology. The human population has been growing rapidly, hence destroying a great ratio of natural resources leads to imbalance in the wildlife population. Continuous monitoring of animals provides evidence to the researchers to inform management and

© Springer Nature Switzerland AG 2020
A. P. Pandian et al. (Eds.): ICCBI 2018, LNDECT 31, pp. 94–100, 2020.
https://doi.org/10.1007/978-3-030-24643-3_10

conservation decisions to maintain the diverse, and obtain a balanced sustainable ecosystems. Now a days wildlife photography has become a very important profession, in this, long waiting hours is a part of it, so recognizing the animal is very important to avoid waiting [12].

Complex task such as recognition of objects can be performed automatically on an environment with reliable and crucial perception. Different level of cognitions requires different tasks, sometimes it is sufficient that a certain structures or obstacles are classified based on different parameters. Grasping and manipulating the real life objects with sophisticated interaction requires precise scene of understanding including pixel-wise semantic segmentation and object detection.

One of the important fields of Artificial Intelligence is Computer Vision. It is the science of computers and software systems that can recognize and understand the images, scenes etc. Computer Vision is also composed of various aspects such as image recognition, object detection, image generation, image super-resolution, pre-processing, feature extraction and more. Object detection is probably the most pro-found aspect of computer vision due to the number of practical use cases (Fig. 1).

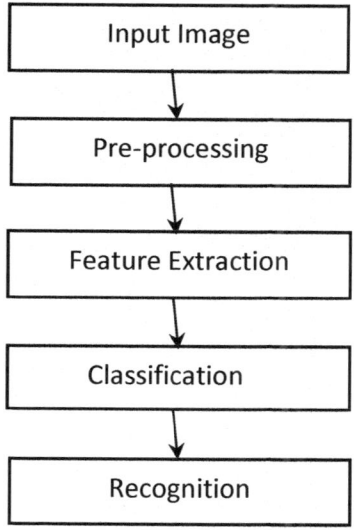

Fig. 1. Flow of object detection.

Recent years there has been a rapid development in virtual reality, interactive multimedia and augmented reality. Human-computer interaction technology has become one of the most important technology and the demand for it has increased over years. In the present days the monitoring of humans and classification of humans have become an important research topic due to the need of it in the healthcare sectors [11]. The management technology for storage and memory has advanced rapidly, the internet speed and mobile network is fast, also wide angle cameras at home have become very sensitive during the capturing of image. So the need of establishing methods for recognizing human postures based on various technologies such as cloud-computing etc., have become very essential.

2   Literature Survey

In this section we briefly describe existing literature review on object detection and also discuss various methods applied for object detection and finally a comparative study is made on different methods.

Many researches based techniques proposed are useful for real time application that would detect animals before they enter the road and warn the driver through audio and visual signals. Different object detection techniques for the identification of animals are used such as object matching, edge based matching, skeleton extraction, Frame differencing, Template Matching [1] etc.,. The flow of the existing system is shown in Fig. 2.

Fig. 2. Sample input image

A project for human resource recognition was proposed to analyze the body parts. Using depth camera, a Pulse Coupled Neural Network (PCNN) to detect humans who are moving in a disturbed background was proposed. For classifying the parts of the body, a hierarchical decision tree was designed. In this approach, firstly, a moving object detection was made by analyzing a video and then the human feature extraction was done by various techniques that would subtract the background. After this, the posture extraction step was employed, which uses various classification algorithms [2] (Fig. 3).

Cai et al. [3], proposed a working system for detecting and tracking of human movements by combining various fields of Artificial intelligence, which also supports recognizing the position of humans in 3D space. For this purpose, a real time system, called as Phantom was used for tracking and determining the position of the humans in 3D space with the help of monochromatic stereo imagery. It used STH-V1 stereo camera with SVS, which is a real time computing system for producing dense stereo range images. This procedure for the detection of new objects involved various steps which also included deformable contour fitting algorithm. Based on this approach a

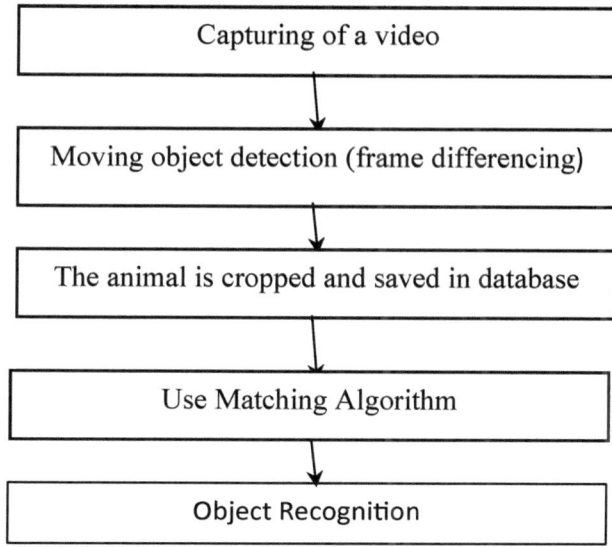

Fig. 3. Flow of existing system.

new method was proposed which accurately and quickly predicted 3D positions of body joints from a single depth image, wherein, a highly varied training dataset allows the classifier to estimate body parts invariant to pose, body shape, clothing, etc., [9].

A technique [4] was proposed in recognizing human postures with infrared method for monitoring in smart homes using Hidden Markov Model. It effectively classifies human posture. It produces a training data model according to human posture characteristics. In real time experimentation, by observing the sequence of human posture, matching identification and classification, the model could identify the different movements postures of human body. Currently, probability network method is the main method for human posture recognition. In the human network, there are two ways of using the probability network method, which are the Bayesian network and the HMM.

On recent improvements in deep learning techniques used in computer vision, they gave an idea to build an animal recognition system which would be automated for monitoring. For this they chose three mostly found species (bird, rat, and bandicoot), all datasets related to these species were used as training sets. The training set contains 80% images of each class (35,629 images), the rest 20% (8,907 images) are the validation set.

In this approach [6], person recognition method that uses information extracted from body images has been proposed. They applied a method, called convolutional neural network (CNN) for image features extraction. They chose this method to reduce the disadvantages faced by the traditional method. They also use cameras such as visible light camera and thermal camera. With the help of recent research in computer vision, the input image features were extracted using CNN.

Table 1. Comparison of different methods of object detection showing recognition accuracy

Author name	Detected object	Algorithms or methods used	Result
Leskovec et al. [3]	Human Bodies	Working application of Phantom and Support Vector Machine	85.6% correct prediction for 200 training sets
Nguyen et al. [5]	Animals (training model mainly had rat, bird, and bandicoot)	Deep Convolutional Neural Network (DCNN)	Achieved 96% accuracy in recognizing images with animals and about 90% in identifying rats, birds and bandicoot
Parikh et al. [1]	Animal	Template Matching Algorithm	False positive rate for bird detection is 5.94%; while false negative rate for bird detection is 5.23% The proposed technique has 94% efficiency for bird detection
Zhuang et al. [2]	Human posture	Pulse coupled neural network and Decision tree	93% precision
Nguyen et al. [6]	Human	Histogram of gradients(HOG), Convolution neural network(CNN), Multi-Level Local Binary Patterns (MLBP)	HOG-91.085% CNN-87.465% MLBP-97.055% The best recognition error (EER) in this experiment is 5.265%
Christiansen et al. [7]	Wildlife	Thermal imaging, k-nearest-neighbor (kNN), Discrete cosine transform (DCT)	Achieved 93.5% accuracy in balanced classification in the altitude range of 3–10 m and 77.7% in the altitude range of 10–20 m
Algobail et al. [8]	Animal	Wireless Acoustic Sensors Network, Acoustic-Based Low-Power Sensing	This scheme saves up to 70% power consumption and guarantees a high recognition accuracy It produced classification of 96.88% at a very low computing cost
Cai et al. [4]	Animal	Hidden Markov Model	Increased accuracy in recognizing the movement of humans

Every year many animals are killed in agricultural operations due to agricultural machinery. To lower the wildlife deaths and promote wildlife-friendly farming, Detection and Recognition of wildlife within the agricultural fields has become very important. Hot object detected based on a threshold is also a kind of method used. Thermal feature extraction algorithm used for classifying the animals is another way. The thermal signature is used to describe heat characteristics of objects. The Discrete Cosine Transform (DCT) is used to find the thermal signature. Then k-nearest-neighbor (kNN) classifier is used to discriminate animals from non-animals [7].

Algobail et al. [8], proposed the Wireless Acoustic Sensors Network which has come light because of its efficient audio-rich applications like surveillance in a battlefield and monitoring the environment. They also used low-power target recognition for exploring their scheme. They used, Acoustic-Based Low-Power Sensing method and they found that it produced classification of 96.88% at a very low computing cost.

From this comparative study, it has been observed that, methods like Phantom and Support Vector Machine, DCNN, Template Matching Algorithm, Pulse Coupled Neural Network and Decision tree, HOG, CNN, MLBP were applied. Among all these methods, DCNN achieved a good recognition accuracy of 96% and MLBP achieved an accuracy of 97.005% as shown in the Table 1.

3   Conclusion

Different image processing techniques have been surveyed and implemented to recognize animals, birds and human beings. Background subtraction methods like frame differencing were implemented. From comparative study, it is observed that DCNN achieved a good recognition accuracy of 96% and MLBP achieved an accuracy of 97.005% and there is a scope for the development of new approaches in the field of object detection during natural disasters. Various method for animals and human recognition would develop due to an increasing demand in various fields. Our goal is to develop ways to recognize objects using the finest algorithm in a cheapest way that would give us good efficiency.

References

1. Parikh, M., et al.: Animal detection using template matching algorithm. Int. J. Res. Mod. Eng. Emerg. Technol.
2. Zhuang, H., et al.: 3D Depth Camera Based Human Posture Detection and Recognition Using PCNN Circuits and Learning-Based Hierarchical Classifier. School of Electrical and Electronic Engineering, Nanyang Technological University Singapore
3. Leskovec, J., et al.: Detection of Human Bodies using Computer Analysis of a Sequence of Stereo Images
4. Cai, X., et al.: Infrared Human Posture Recognition Method for Monitoring in Smart Homes Based on Hidden Markov Model. School of Computer Science, North China University of Technology, Beijing
5. Nguyen, H., et al.: Animal Recognition and Identification with Deep Convolutional Neural Networks for Automated Wildlife Monitoring

6. Nguyen, D.T., et al.: Person Recognition System Based on a Combination of Body Images from Visible Light and Thermal Cameras. Division of Electronics and Electrical Engineering, Dongguk University, Seoul

7. Christiansen, P., et al.: Automated Detection and Recognition of Wildlife Using Thermal Cameras. Department of Engineering, Aarhus University, Aarhus

8. Algobail, A., et al.: Energy-aware Scheme for Animal Recognition in Wireless Acoustic Sensor Networks. Department of Computer Science, College of Computer and Information Science, King Saud University, Riyadh, Saudi Arabia

9. Shotton, J., et al.: Real-Time Human Pose Recognition in Parts from Single Depth Images. Microsoft Research Cambridge & Xbox Incubation

10. Dragan, M.-A., et al.: Human Activity Recognition in Smart Environments

11. Mandal, I., et al.: A framework for human activity recognition based on accelerometer data

12. Udaya Shalika, A.W.D., et al.: Animal Classification System Based on Image Processing & Support Vector Machine. Sri Lanka Institute of Information Technology, Malabe

A Study on Rough Indices in Information Systems with Fuzzy or Intuitionistic Fuzzy Decision Attributes-Two Thresholds Approach

B. Venkataramana[1(✉)], L. Padmasree[2], M. Srinivasa Rao[3],
and G. Ganesan[4]

[1] Department of Computer Science and Engineering, Holy Mary Institute
of Technology and Science, Bogaram, Kesara Mandal, Telengana, India
bandaruramana1@gmail.com
[2] Department of Electronics and Communications Engineering,
VNR Vignana Jyothi Institute of Engineering & Technology, Bachupalli,
Nizampet, Telengana, India
padmasree_l@vnrvjiet.in
[3] School of Information Technology, Jawaharlal Nehru Technological
University Hyderabad, Kukatpally, Telangana, India
srmeda@gmail.com
[4] Department of Mathematics, Adikavi Nannaya University, Rajahmundry,
Andhra Pradesh, India
prof.ganesan@yahoo.com

Abstract. Z. Pawlak's RS Model finds various applications in Knowledge Engineering confined with information systems. Considering its importance, as Zadeh's Fuzzy Model and Atanassov's intuitionistic Fuzzy Model are being combined with RS Model. In particular, G. Ganesan et al., derived a tool for indexing information systems with a single threshold using the fuzzy decision attributes and intuitionistic fuzzy decision attributes. In this paper, we extend their approach for two thresholds using fuzzy decision attributes and intuitionistic fuzzy decision attributes individually and further, we develop algorithms for indexing information systems with two thresholds and implement them using C.

Keywords: Rough sets · Information systems · Indexing · Fuzzy sets · Intuitionistic fuzzy sets

1 Introduction

In any conventional information system, if the fuzzy or intuitionistic fuzzy values are given as inputs, it is conventional to convert the given fuzzy or intuitionistic fuzzy data into any of the available defuzzification methods and be approximated through rough computing. In particular, if the input color is the combination of 2 or more colors in an information system bearing the information on colors, it is difficult to determine to which class this input object belongs to. Considering this aspect in view, in [4, 5], Dr. G. Ganesan et al., derived the rough approximations of a given fuzzy set and

A. P. Pandian et al. (Eds.): ICCBI 2018, LNDECT 31, pp. 101–107, 2020.
https://doi.org/10.1007/978-3-030-24643-3_11

intuitionistic fuzzy set under the given threshold. Also, Ganesan et al. [2–4] defined rough indices [7] on the information systems based on the fuzzy input and intuitionistic fuzzy input using a single threshold.

However, in decision-making, it is necessary to take proper decisions for the elements whose fuzzy membership values lie in some interval. For example, if the end user categories the data into 'Below Unfair', 'Unfair' and 'Above Unfair' and if the end user is in particular in choosing the category of 'Unfair', analyzing it with a single threshold is time consuming and ineffective. Here, it would be more effective to introduce two thresholds.

Hence, in this paper we developed the algorithms on rough indexing under Two Thresholds under fuzzy input and intuitionistic fuzzy set separately. This paper is organized into 7 Sections. In 2nd Section, the basic mathematical concepts are described; in 3rd section, construction of the sets namely D and ID are described; in 4th Section, the concepts of rough approximations of fuzzy and intuitionistic fuzzy sets are dealt. In 5th and 6th sections, we describe the rough indexing algorithms under fuzzy and intuitionistic fuzzy decision attributes respectively. The 7th section deals with the concluding remarks.

2 Fuzzy Sets

On replacing the co-domain of the characteristic function $\{0, 1\}$ by $[0, 1]$, Zadeh derived the concept of Fuzzy Sets which find various real time applications. Here, the function, which is defined, is called as membership function and the value assumed by the membership function is called the grade of membership in the given fuzzy set A. It can be defined as follows:

For the universe of discourse $U = \{x_1, x_2, \dots, x_n\}$, any fuzzy subset A of U can be defined by μ_A, membership function defined from U to $[0, 1]$ and the fuzzy set A is represented as $A = (\mu_A(x_1), \mu_A(x_2), \dots, \mu_A(x_n))$.

2.1 Intuitionistic Fuzzy Sets

According to the concept of fuzzy sets, the non-membership function gives a value that is equivalent to the difference between the membership value and one. However, this cannot be used in all applications, because some of them defy this non-membership rule. These particular applications prompted Atanassov [1] to develop the concept of intuitionistic fuzzy sets. The path he followed allows the dimensional view of fuzzy sets in terms of membership grade and non membership grade. Let $U = \{x_1, x_2, \dots x_n\}$ be any universe of discourse and A be any intuitionistic fuzzy subset of U. If μ_A and γ_A be the membership and non membership functions defined on the universe of discourse U to $[0, 1]$ with $0 \leq \mu_A(x) + \gamma_A(x) \leq 1$ for every x in U, Then the intuitionistic fuzzy subset A of U is denoted by $A = ((\mu_A(x_1), \gamma_A(x_1)), (\mu_A(x_2), \gamma_A(x_2)), \dots, (\mu_A(x_n), \gamma_A(x_n)))$.

Now, we describe four sets namely D_1, D_2, ID_1 and ID_2 as follows:

3 Construction of Sets D and ID

Consider the universe of discourse $U = \{x_1, x_2, \ldots, x_n\}$ and let α, β be the thresholds ranging from 0 to 1.

Now, define $A[\alpha, \beta] = \{x \in U/\alpha < \mu_A(x) \leq \beta\}$ where $\alpha < \beta$. However, by the definition of D, $A[\alpha, \beta]$ can also be defined as

$A[\alpha, \beta] = \{x \in U/\alpha < \mu_A(x) < \beta\}$
$A[\alpha, \beta] = \{x \in U/\alpha \leq \mu_A(x) \leq \beta\}$
$A[\alpha, \beta] = \{x \in U/\alpha \leq \mu_A(x) < \beta\}$.

4 Rough Approximations Revisited

Now, we describe the procedure of using rough approximations for a given fuzzy input and intuitionistic fuzzy sets using the sets D and ID respectively.

4.1 Rough Approximations for Fuzzy Sets

For a given fuzzy set A and α, $\beta \in D$, define $A[\alpha, \beta] = \{x \in U/\alpha < \mu_A(x) \leq \beta\}$ where $\alpha < \beta$. Let X be any partition of U, say $\{B_1, B_2, \ldots, B_t\}$. For the given fuzzy set A, the lower and upper approximations with respect to α and β are defined as $A_{\alpha,\beta} = \underline{(A[\alpha, \beta])}$ and $A^{\alpha,\beta} = \overline{(A[\alpha, \beta])}$ respectively.

4.2 Rough Approximations for Intuitionistic Fuzzy Sets

For a given intuitionistic fuzzy set A and α_1, α_2, β_1, $\beta_2 \in ID$, define

$$A\begin{bmatrix} \alpha_1 & \alpha_2 \\ \beta_1 & \beta_2 \end{bmatrix} = \{x \in U/\alpha_1 < \mu_A(x) < \alpha_2; \beta_1 < \gamma_A(x) < \beta_2\}$$

Let X be any partition of U, say $\{B_1, B_2, \ldots, B_t\}$. For the given intuitionistic fuzzy set A, the lower and upper rough approximations are respectively defined as

$$A_{\begin{bmatrix} \alpha_1 & \alpha_2 \\ \beta_1 & \beta_2 \end{bmatrix}} = \underline{A\begin{bmatrix} \alpha_1 & \alpha_2 \\ \beta_1 & \beta_2 \end{bmatrix}} \quad \text{and} \quad A^{\begin{bmatrix} \alpha_1 & \alpha_2 \\ \beta_1 & \beta_2 \end{bmatrix}} = \overline{A\begin{bmatrix} \alpha_1 & \alpha_2 \\ \beta_1 & \beta_2 \end{bmatrix}}$$

Now, we provide the algorithms for computing rough indices in information systems with fuzzy decision attributes and intuitionistic fuzzy decision attributes individually along with the implementations in C Programming. The key selection in each algorithm may be chosen based on the computation of reducts which is not discussed in this paper. Let M denote the largest number under consideration such that n + M is always positive and n − M is always negative for any integer n.

5   Rough Indices in Information Systems with Fuzzy Decision Attributes

Here, we provide an algorithm for rough indices and the experimental results.

5.1   Experimental Results

Consider the following decision table with 10 records namely 1, 2, 3, 4, 5, 6, 7, 8, 9 and 10 with three conditional attributes namely Attr_1, Attr_2, Attr_3 and a fuzzy decision attribute.

	Attr_1	Attr_2	Attr_3	Decision membership
1	Grey	Orange	Grey	0.4
2	Orange	Grey	Green	0.5
3	Orange	Orange	Grey	0.6
4	Violet	Green	Black	0.4
5	Black	Orange	Black	0.5
6	Violet	Grey	Orange	0.7
7	Grey	Green	Black	0.4
8	Grey	Black	Violet	0.1
9	Green	Green	Orange	0.9
10	Black	Violet	Black	0.3

It may be noticed that the records are grouped according the similarity for each key or group of keys. i.e., the records are grouped as follows: For Attr_1, the grouping are {(Grey, {1, 7, 8}), (Orange, {2, 3}), (Violet, {4, 6}), (Black, {5, 10}), (Green, {9})}. For Attr_2, the grouping are {(Orange, {1, 3, 5}), (Grey, {2, 6}), (Green, {4, 7, 9}), (Black, {8}), (Violet, {10})} and for Attr_3, we obtain {(Grey, {1, 3}), (Green, {2}), (Black, {4, 5, 7, 10}), (Orange, {6, 9}), (Violet, {8})}.

By considering Attr_2 as the key, the algorithm is implemented in C by assuming M = 50 and the thresholds as 0.35 and 0.95, we obtain the index of 8 as 51.

6   Rough Indices in Information Systems with Intuitionistic Fuzzy Decision Attributes

Here, we provide an algorithm for rough indices and the experimental results.

6.1   Algorithm for Rough Index of an Element

Algorithm index (x,A, $\alpha_1,\alpha_2,\beta_1,\beta_2$)
//Algorithm to obtain index of x an element of universe of discourse
//Algorithm returns the index

1. Let x_index be an integer initialized to 0

2. compute $A_{\begin{bmatrix} \alpha_1 & \alpha_2 \\ \beta_1 & \beta_2 \end{bmatrix}}$ and $A^{\begin{bmatrix} \alpha_1 & \alpha_2 \\ \beta_1 & \beta_2 \end{bmatrix}}$

3. If $x \in A_{\begin{bmatrix} \alpha_1 & \alpha_2 \\ \beta_1 & \beta_2 \end{bmatrix}}$

   while $\left(x \in A_{\begin{bmatrix} \alpha_1 & \alpha_2 \\ \beta_1 & \beta_2 \end{bmatrix}} \right)$

      begin

           α_i=sqrt(α_i) //square root of α_i (i=1,2)
           β_i=sqr(β_i) //square of β_i (i=1,2)
           x_index=x_index+M+1
           compute $A_{\begin{bmatrix} \alpha_1 & \alpha_2 \\ \beta_1 & \beta_2 \end{bmatrix}}$

      end

   else if $x \notin A^{\begin{bmatrix} \alpha_1 & \alpha_2 \\ \beta_1 & \beta_2 \end{bmatrix}}$

         while $\left(x \notin A^{\begin{bmatrix} \alpha_1 & \alpha_2 \\ \beta_1 & \beta_2 \end{bmatrix}} \right)$

           begin

               α_i=sqr(α_i) //square of α_i (i=1,2)
               β_i=sqrt(β_i) //square root of β_i (i=1,2)
               x_index=x_index-1-M
               compute $A^{\begin{bmatrix} \alpha_1 & \alpha_2 \\ \beta_1 & \beta_2 \end{bmatrix}}$

         end
       else
       B=A; $\chi_i=\alpha_i$; $\delta_i=\beta_i$ (i=1,2)
       compute $B^{\begin{bmatrix} \chi_1 & \chi_2 \\ \delta_1 & \delta_2 \end{bmatrix}}$

         while $\left(x \notin A_{\begin{bmatrix} \alpha_1 & \alpha_2 \\ \beta_1 & \beta_2 \end{bmatrix}}; x \in B^{\begin{bmatrix} \chi_1 & \chi_2 \\ \delta_1 & \delta_2 \end{bmatrix}} \right)$

begin

$\alpha_i=sqr(\alpha_i)$ // square of α_i (i=1,2)

$\beta_i=sqrt(\beta_i)$ //square root of β_i (i=1,2)

$\chi_i=sqrt(\chi_i)$ // square root of χ_i (i=1,2)

$\delta_i=sqr(\delta_i)$ //square of δ_i (i=1,2)

compute $A_{\begin{bmatrix} \alpha_1 & \alpha_2 \\ \beta_1 & \beta_2 \end{bmatrix}}, B^{\begin{bmatrix} \chi_1 & \chi_2 \\ \delta_1 & \delta_2 \end{bmatrix}}$

x_index=x_index+1

end

if $x \in A_{\begin{bmatrix} \alpha_1 & \alpha_2 \\ \beta_1 & \beta_2 \end{bmatrix}}$ then x_index= - x_index

end

4. return x_index

6.2   Experimental Results

Consider the following decision table with 10 records namely 1, 2, 3, 4, 5, 6, 7, 8, 9 and 10 with three conditional attributes namely Attr_1, Attr_2, Attr_3 and an intuitionistic fuzzy decision attribute.

	Attr_1	Attr_2	Attr_3	Decision	
				Membership	Non-membership
1	Very good	Good	Very good	0.6	0.2
2	Good	Very good	Unfair	0.7	0.1
3	Good	Good	Very good	0.6	0.2
4	Excellent	Unfair	Fair	0.5	0.2
5	Fair	Good	Fair	0.7	0.2
6	Excellent	Very good	Good	0.3	0.7
7	Very good	Unfair	Fair	0.3	0.3
8	Very good	Fair	Excellent	0.9	0.1
9	Unfair	Unfair	Good	0.2	0.5
10	Fair	Excellent	Fair	0.3	0.6

It may be noticed that the records are grouped according the similarity for each key or group of keys. i.e., the records are grouped as follows: For Attr_1, the grouping are {(Very Good, {1, 7, 8}), (Unfair, {9}), (Good, {2, 3}), (Fair, {5, 10}), (Excellent, {4, 6})}. For Attr_2, the grouping are {(Very Good, {2, 6}), (Unfair, {4, 7, 9}), (Good, {1, 3, 5}), (Fair, {8}), (Excellent, {10})} and for Attr_3, we obtain {(Very Good, {1, 3}), (Unfair, {2}), (Good, {6, 9}), (Fair, {4, 5, 7, 10}), (Excellent, {8})}.

By considering Attr_2 as the key, the algorithm is implemented in C by assuming M = 50 and the thresholds as 0.35 and 0.95, we obtain the index of 1 as 51.

7 Conclusion

In this research work, we proposed algorithms for computing rough indices of the records of the decision table which possess either intuitionistic or fuzzy decision attributes using Two thresholds and implemented them using C Programming.

References

1. Atanassov, K.: Intuitionistic fuzzy sets. Fuzzy Sets Syst. **20**, 87–96 (1986)
2. Krishnaveni, B., et al.: Characterization of information systems with fuzzy and intuitionistic fuzzy decision attributes. In: 8th International Conference on Advanced Software Engineering and Its Applications, Korea, pp. 53–58. IEEE (2015)
3. Latha, D., et al.: Probabilistic rough classification in information systems under intuitionistic fuzziness. Int. J. Wareh. Min. **3**(2), 82–86 (2013)
4. Ganesan, G., Latha, D.: Rough classification induced by intuitionistic fuzzy sets. Int. J. Comput. Math. Sci. Appl. **1**(1), 63–69 (2007)
5. Ganesan, G., et al.: Rough set: analysis of fuzzy sets using thresholds. In: Computational Mathematics, Narosa, India, pp. 81–87 (2005)
6. Ganesan, G., et al.: An overview of rough sets. In: Proceedings of the National Conference on the Emerging Trends in Pure and Applied Mathematics, Palayamkottai, India, pp. 70–76 (2005)
7. Ganesan, G., et al.: Rough index in information system with fuzziness in decision attributes. Int. J. Fuzzy Math. **17**(1), 183–190 (2008)
8. Ganesan, G., et al.: Feature selection using fuzzy decision attributes. J. Inf. **9**(3), 381–394 (2006)
9. Venkataramana, B., et al.: Rough-fuzzy classification on a decision table using a threshold-algorithms and implementations. Int. J. Sci. Res. Comput. Sci. Eng. Inf. Technol. **3**(5), 777–782 (2018)
10. Venkataramana, B., et al.: Algorithms on rough-intuitionistic fuzzy classification with a threshold and implementations. Int. J. Eng. Comput. Sci. **7**(6), 24093–24098 (2018)
11. Pawlak, Z.: Rough Sets-Theoretical Aspects and Reasoning About Data. Kluwer Academic Publications, Dordrecht (1991)
12. Pawlak, Z.: Rough sets. Int. J. Comput. Inform. Sci. **11**, 341–356 (1982)

Proactive Decision Making Based IoT Framework for an Oil Pipeline Transportation System

E. B. Priyanka$^{(\boxtimes)}$, C. Maheswari, and S. Thangavel

Mechatronics Engineering Department,
Kongu Engineering College, Erode, India
priyankabhaskaran1993@gmail.com

Abstract. In today's modest production atmosphere, process diligences ultimatum a completely incorporated control along with optimization elucidation which can upsurge throughput, dependability, and superiority while diminishing cost. Due to the intricacy and interdisciplinary style of evolving engineering strategies and clarifications of upgraded security and performances, provides us an authoritative tool to accomplish computer-based data acquisition and virtual instrumentation, which leads to profligate becoming a standard rather than an exemption. The appearance of the industrial Internet of Things (IoT) architype served better platform for increasing the monitoring proficiencies by the usage of virtual and embedded based field sensors. Sensor-produced data can be utilized to envisage adverse situations when it deviates from the normal operating functions which enable operators to resolve and act proactively. Hence this paper points out the importance of IoT for an oil pipeline system with diverse monitoring condition as a pragmatic aspect in the transportation and also it emphasizes the framework outline for a smart oil field system with its necessity and its advance feature incorporated using the Internet of Things (IoT).

Keywords: IoT · Oil pipeline system · Remote monitoring and control · Condition-based monitoring · Proactive decision-making algorithm

1 Introduction

The era of oil and gas pipelines is enduring an informational renovation to recover enactment, minimalize ruptures and spills, and rise safety, and is fetching to resemble as an example of data-enabled substructure [3–5]. Pipelines come to the vision of public cognizance only when a leak occurs, prominent to a toxic spill, or result in an explosion that outlays lives. The industry 4.0 is integrating sensing knowledge to monitor pressure, flow rate, pumping station parameters, temperature, viscosity, and other external parameters [1, 9–11]. In industrial enactments, the internet of things progresses on top of the previously prevailing system, permitting for a transfer from "monitor and respond" to a prognostic and pre-emptive approach assisting upgraded decision making.

This paper presents an internet of things (IoT) and cloud-centric reliable frame structure for observing several performances of the various operating subdivisions of

© Springer Nature Switzerland AG 2020
A. P. Pandian et al. (Eds.): ICCBI 2018, LNDECT 31, pp. 108–119, 2020.
https://doi.org/10.1007/978-3-030-24643-3_12

the oil and gas industry. The proposed IoT based modular architectural design, comprising three modules, module of a smart object, a module of a gateway and module of a control center (server). Each module is layered (including sensing, networking and application layers) and tackles specific functions to support the monitoring of the interconnected oilfield environment [2, 7].

In this framework, an precise oil transportation system has been assured by means of the innovative technology of Internet of Things (IoT) which offers a pervasive connectivity among the oil pipelines oriented elements like pipeline details, fluid and associated environmental parameter, intelligence sensors and interacted wireless smart system performs the collections of suitable real-time data are transmitted to the cloud and field supervision center which has to be examined and managed in mandate for construing and to afford appropriate decision in event of catastrophic situations [16, 19]. Nowadays many controllers have switched from analog into the digital control system. Industrial networking has become an important way to connect the process-oriented tasks in industrial automation. The industrial plant usually has production equipment that is instrumented for measurement which uses a control system. SCADA system is a software package positioned on top of hardware [6]. It is interfaced with industrial processing units through Programmable Logic Controllers [12]. Remote monitoring and intellectual control using IoT are vital criteria for exploiting production rate and process feasibility. Widely held of industries practice distributed control system (DCS) for in elevating consistency, upgraded response and secured operator interfaces of data to engineering & management personals and improved historical storage and retrieval system [17, 18]. The proportional-integral-derivative (PID) controller serves as one of the most persistent conventional controllers in the process industry sectors for its easiness, heftiness and wide range of operation with near-optimal performance [13–15]. But this can be implemented as the main contributor in the field based control but IoT plays the vital role in configuring the whole entire system with understanding constraints of all parameters involved in the oil pipeline transport system. For knowledge to vanish on the realization of the manipulator, the Internet of Things hassles: (1) a collective sympathetic about the circumstances of utilizing handlers and industrialized smart sensors, (2) structural design of software and unescapable data transmission pathway to tackle and transfer the most pertinent circumstantial associated data and (3) self-directed and adaptive performance by its analysis algorithm tool incorporated with IoT. By unifying basic three essential substructure leads to smooth-reliable connectivity by real-time data manipulation from the implemented industrial sector of a complex environment. To overcome the limitations of conventional utility systems, the proposed design of the IoT framework offers new utility monitoring and control system can flexibly accommodate changes made to improve the operational efficiency of utilities.

2 Analysis of Pioneering Work to Enhance Monitoring in Oil Pipeline System

Supervisory Control and Data Acquisition (SCADA) systems [1] are not accessible owing to truncated compactness in time and space, lavish in standings of equipment and repairs, not interoperable in terms of hardware and software, inflexible when there is a need for protocol change and software updating, and provides data and result with long delay. Various Internet of Things (IoT) based architectures were previously suggested in the literature like IoT slanted towards for Social Internet of Things (SIoT) [2], Resilient IoT architecture for smart cities [3] and future internet [4]. But still, yet there is no standard architecture by using IoT especially pointing on the oil and gas industries. Given the critical environment of oilfields, there is a need to develop IoT architecture according to Oil & Gas industrial environment.

Internet of Things (IoT) inferred as the identified Internet of Objects becomes an emergent technology that is widely used by several application domains. Thanks to its advanced services, the International Telecommunication Union (ITU) outlines to the real world in the Information Society as powerful infrastructure, empowering innovative options on interrelating existing and evolving platforms, interoperable statistics and communication expertise. Diverse fields of applications adapt IoT in their main implementation process [19]. The most protruding areas of solicitation taken in the smart home, smart energy, healthcare, manufacturing, transport, environment, smart industry and so on. Figure 1 shows the anomaly occurrences in an oil pipeline system with contributing factors related to the oil pipeline field.

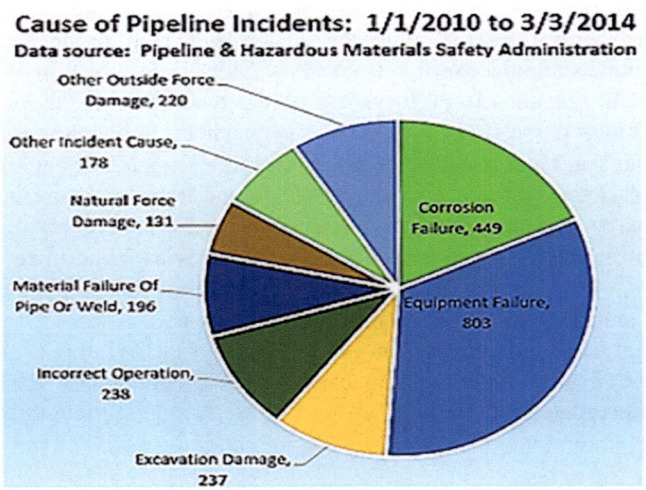

Fig. 1. Causes of calamities occurred due to the oil pipeline transport system.

3 Challenges Nullified by Utilizing IoT in Oil Pipelines

An IoT based outcomes solve these critical challenges mainly in the oil and gas industry that are described as follows:

- Detecting physical presence at oil & gas pipeline, a leak of a pipeline, or tampering with a pipeline.
- Monitoring internal and external parameters of oil fields, the pressure variation of the oil pipelines, tanks, wells and other assets used in the supply chain of the oil industry.
- Optimizing pumping operations, maintaining the pipes and wells, monitoring equipment failures, corrosion, erosion in a refinery and oil & gas leakage, minimizing risks to health and safety and monitoring and improving production performance with reduced costs.
- Detecting tuberculate pipeline sections partially closed or fully closed valve gates, variations in fluid homogeneity (e.g., air pockets within a water distribution network), pipe wall structural degradation and biofilm accumulation.
- Oil industrial operations need specific skills and continuous monitoring to ensure smooth functioning of all the operations in the oilfield environment. IoT is the most potential the solution for driving core processes of the oil fields in a more efficient and reliable manner.

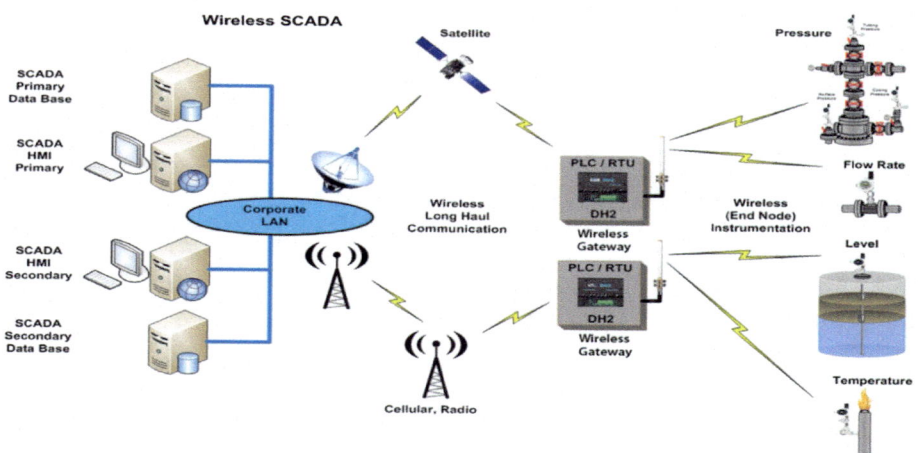

Fig. 2. IoT based architecture for an oil pipeline transport system

This proposal emphasizes an internet of things (IoT) based reliable monitoring system as shown in Fig. 2 that involves smart objects installed on various assets of the oil and gas fields to facilitate various operations of the Oilfield environment. This system with the minimum human intervention will provide better workplace safety and maintenance of assets. This proposed system will perform predictive maintenance of various oil and gas industrial assets by analyzing various parameters (sensed data) and detecting failure modes either before they are going to take place or when the equipment will likely to fail or need service.

4 Schematic Architecture View of Implementation of IoT to an Oil Pipeline Station

Figure 3 shows the proposed IoT based architecture for oil industries to provide a valid solution for oil theft, leakage, and other environmental issues. In Fig. 2, each smart object is fortified with various sensors like level, temperature, flow and pressure sensors to detect any anomalous situation in the oil field. The smart objects will be enabling to sense and collect data and react to specific conditions. Thus, the data gathered by this smart oilfield has to be delivered to the server so as to process and analyze the data completely. The module of a smart object involves data acquisition and collaboration between the smart object(s) and gateway(s).

The gateway module is a bond between smart embedded sensors and the control center. The communication between the smart objects and gateways can be achieved by employing RPL (Routing Protocol for Lossy networks). The applications running on the smart object and gateway modules will perform real time actions (fire alerts, shut down of different equipment, identifying the exact location and rectifying the fault, operator alerts) against anomalous events like oil and gas pipeline crack, bursts etc. The central remote processing station is responsible for management of applications and the exploration of the data gathered from the smart IoT engaged modules, generating the information and taking important decisions against anomalous events, providing the dashboard to provision the decision-making practice. It consists of SCADA at the field control and at back-end cloud computing techniques will be processed to analyze the situation based on the acquired smart objects data. The Application layer involved in this station is basically responsible for management processes and includes the object interfaces, IoT applications, databases and service APIs, Visualization tools etc. When the collected data is analyzed using algorithms and problems are sorted out, actuation is performed by controlling the functions without human intervention. The control center will perform data analysis for two main purposes:

- For the prognostic maintenance of the oilfield by analyzing the smart objects condition data to the control center and detecting failure modes either before they are going to take place or when the equipment will likely to fail or need service. Thus the control center will perform preventive maintenance for maximizing production uptime and minimizing disruptions, thus to better control and maintain various assets with a reduced risk rate of human life and safety.
- The remote terminal will scrutinize the production performance by examining the daily production and consumption of oil.

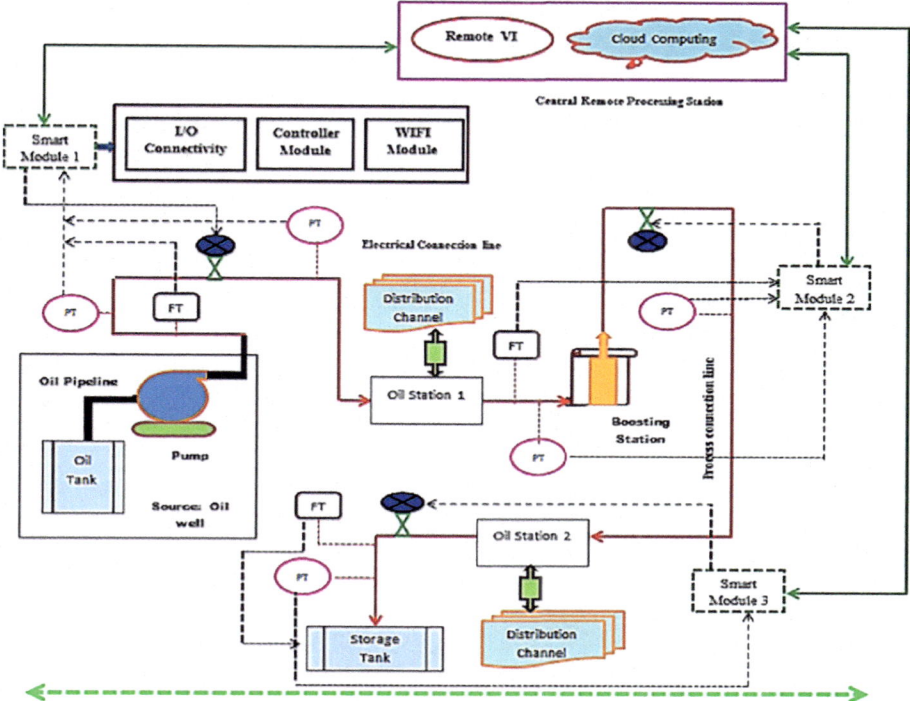

Fig. 3. Proposed reliable IoT frame structure for oil industries

5   Reliable and Efficient Communication Using Proposed IoT Framework

The proposed IoT system provides a method for communicating the sensed information reliably and efficiently to the control center. Figure 4 represents a schematic view (flow chart) that illustrates how our proposed system will competently and precisely communicates the sensed data about the anomalous event zone (defects) within the oil field environment to the control center and taking possible measures to solve the problem. The smart module consists of input/output connectivity, a controller module, and a Wi-Fi module. This smart module establishes the link to the central remote processing station will hand over the information to the control center through long range communication as well as the nearby mobile units or maintenance personnel through short-range communication for repairing and taking possible actions. Thus in this manner, the sensed information about any anomalous event will be delivered in any case to the control center reliably and efficiently. The smart objects installed on various assets (pipelines, tanks, well heads, pumps, and pressure relief valves) of the oil field environment will be interconnected and will collaborate with each other intelligently. The proposed IoT framework for an oil system goal is to detect or diagnose any kind of catastrophic failure or anomalous event around the whole oil field environment in a timely and reliable manner to reduce revenue costs, overall downtime, increase production costs and avoid environmental or personal damage.

6   Assessment Phase and Location Specific Highlights for IoT Based Oil Industry

To evaluate the upgradation and accuracy offered by the proposed structure is most essential by concentrating on consumer's expectation and oil distribution pipeline requirements. On oil pipeline viewpoints, the paper put forth to:

- Investigate more on the transmission of the real-time pipeline, environmental and fluids which is being transmitted related parameters during long run oil transportation by elaborating communication pathway and achieve upgradation in the monitoring zone to control and to detect cracks/leaks or any oil theft in the entire oil pipeline transport system.
- Utilize advanced Big data practices in a proficient manner to achieve and examine the enormous assorted data of the whole oil transportation sectors.
- Implementation of the inventive technique of LPWAN (Low Power Wide Area Network) to transfer recognized information to the remote supervision system where the execution is performed with declined power necessities, extensive elongated inter-connection and lengthy battery life, and truncated budget.

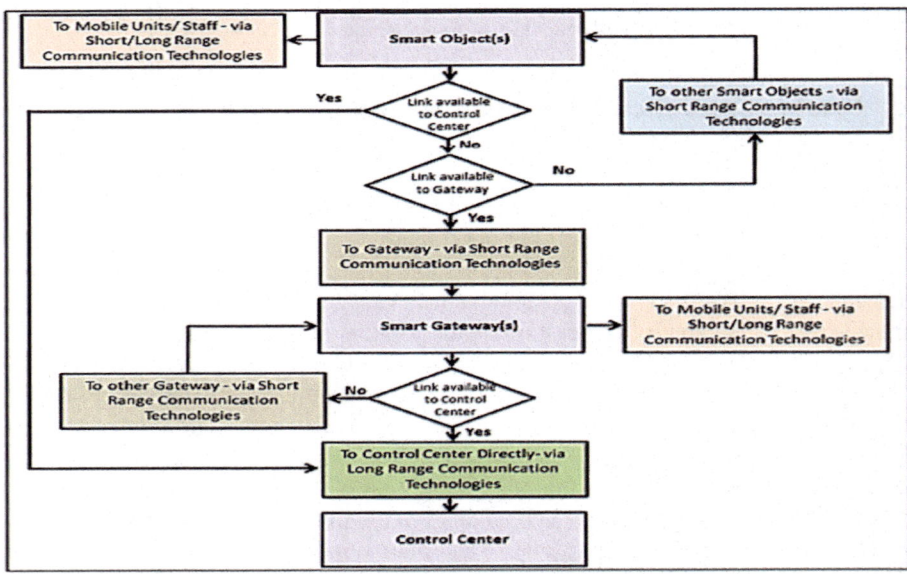

Fig. 4. Flow Chart representing how the proposed system will competently and precisely communicates the sensed data to the control center.

Since this paper accentuates the application of IoT and Cloud Computing to Oil Pipeline Transport System it involves certain location-specific procedures. When a pipeline is constructed, it not only points the civil industrial work to organize the pipeline and but also the pump/compressor stations, putting in place of the field devices that will provide remote operation. Pipelines are commonly established and built using the following stages based on the considered specific location area:

1. Route selection with pipeline design and deviousness of long-run distance estimation on the basis of surrounding location by mounting valves, intersections.
2. Surveying and clearing the route by putforthing Trenching (Main Route, Crossings Sections of pipelines).
3. Testing: Once the structure is accomplished, it imperiled to tests to confirm its structural veracity it may take in hydrostatic testing with line packing.

7 Proactive Decision Making Incorporating with IoT System for an Oil Industry

The proposed IoT frame structure depicted in Fig. 5 is categorized to fall in four execution layers:

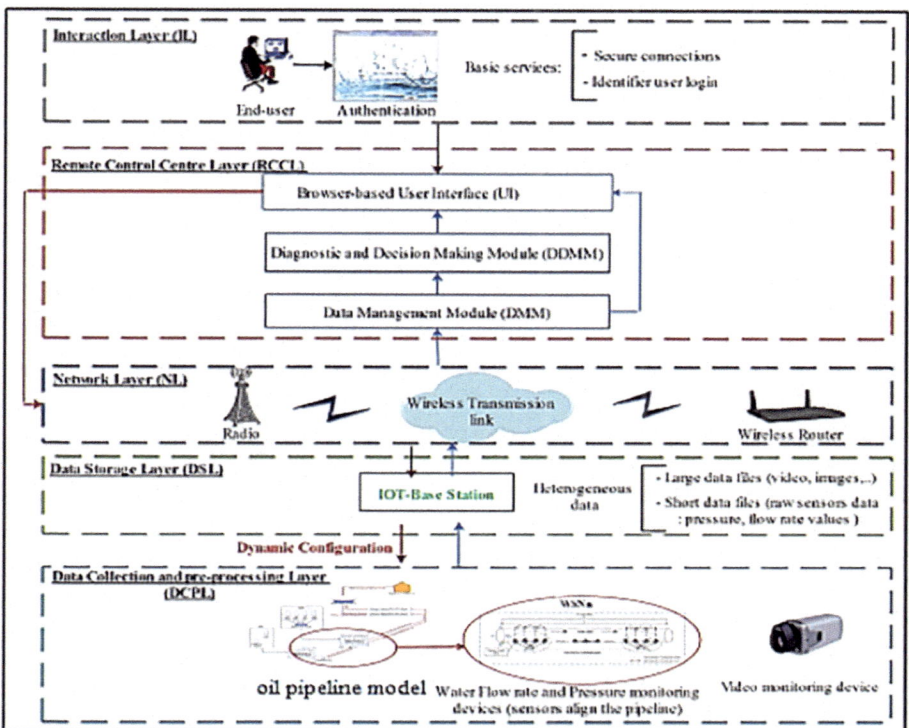

Fig. 5. Multi-layer architecture involved in IoT oil field system

- **Data Acquiring and Segregation:** Initially, the data from various embedded smart sensors situated on the oil field with parity checksum tokens in order to ensure its signal confidentiality from the various substations. Then, the received information is scrutinized to obtain the data regarding the pipeline infrastructure and information associated with the fluid being transported to the distribution channels. After at that juncture, an initial data-acquiring step is completed.
- **Streaming Storage Platform:** The recognized data from continuous communication is stockpiled in a local master data nodes. This data is transmitted over internet protocol to the data computation management system for analysis and identification of over rate.

  Base station routed and acknowledged by the high engineering operator to convalesce received data and can be reorganized on examining the monitoring field station of oil pipelines.
- **Routing Transport Layer:** It utilizes the wireless broadcast channels that cover advanced network coverages frequency band for the remotely located equipment on receiving specific token rings.
- **Remote Processing Layer:** This serves as a final decision and controlling platform of comprising the database organization, algorithm operating routine on received data and realistic user superficial interface. The data analysis is in charge to determine the threshold rate of the acquired data is in range of boundaries considering pipeline operating parameters conditions on acquired location. The second module emphases on the elucidation and the valuation of the abnormal prediction on the virtual contour assumption on comparing with past data of leakage, bursts in the oil field to afford better decisions. The third module designates that field station acquired data can be reread for alert notifications by the User Interface.

7.1   Proactive Circumstance Based Decision for Condition Monitoring

Figure 6 portrays the proposed intangible approach for pre-emptive availability of response for situation control based execution with smart decision support. The execution time in pre-emptive decision making is based on the loss associated with the channels between the data node to the main master kernel. Initially, the channel window starts with data rate with time allocation t = 0, when the prediction is identified and handshaking transcription is completed. The time allocation moves to the next boundary with time duration smaller than initial configuration to speed up the optimization through machine learning. On continuing this channel propagation, the loss will be minimized and packet sending and receiving time will be indirectly shortened with security highlighted encryption schemes via pointing Action 2 and 3. When the planned execution is accomplished, transparency traffic rate will be automatically decreased to support optimization time vector space. By this proposed method, the decision making on the oil filed can be automated and also result in optimization based performance with the next generation of monitoring and control architype.

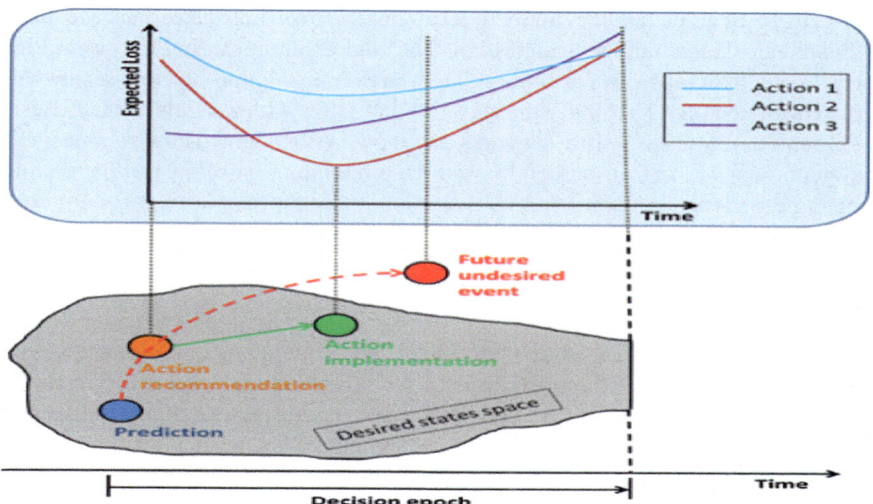

Fig. 6. A conceptual approach to proactive decision making.

7.2 Data Mining and New Protocol

The important contribution of information management and routing to the oil field to perform optimized action is through enhanced protocol application. For the required control and feedback action in desired oil transport station on desired sensors and actuators, IoT with efficient protocol incorporation and routing channel with minimum packet loss is precarious. Because the better suitability of various outlooks of MAC protocols which are reckoning schemes present but none of these are recognized as a standard and if more 'things' accessible this setting results in more cluttered, which needs added research. Data tunnel allocated varies in much more due to its communication, substructure and location configuration. When Internet of Things plays as a kernel in the frame multiple node connectivity assessments becomes complex in mining desired data about a particular domain. In that situation, availability of utilized protocol requires modification in its core and operating system to offer the superlative possibility to overcome the mining restriction in multi-hop.

7.2.1 Advanced Machine Learning and Data Mining

In artificial intelligence, a perplexing problematic area is multifarious sensing, where the obtained beneficial information through mining indulged in diverse altitudinal resolution with complex convergence. The basic research tricky in complex sensing locations is observing concurrently various grade of actions associated with the oil transportation pipeline through local nodes on different ranks of exertions carried out by operator response and feedbacks. At present either standard supervised machine learning techniques are implied in data with constrained categories of classes or in contrast unsupervised algorithm applied in mining the significant data from a pool of uncategorized database through event-based recognition. The progression of this mining is to upgrade all the notified local observation from the database through specified feasible learning node algorithm. The vital dream regarding the description of design on depicting tedious reaction on the desired transmission pipeline sector will be notified with accurate learning with very fast sampling with no traffic delay. Besides, the resource restraints in sensor networks pay the way for a new pathway concerned on intelligent, adaptive and resourceful based deep learning algorithm for the stimulated oil pipeline system.

8   Conclusion and Future Scope

Oil industrial operations need specific skills and continuous monitoring to ensure smooth functioning of all the operations in the oilfield environment. IoT is the most potential solution for driving the core processes of the sectors of the oil industrial environments in a more efficient and reliable manner. This paper lay emphasis on the internet of things (IoT) based reliable supervision and control system that encompasses upgraded embedded objects installed on various assets of the oil and gas fields to facilitate various operations of the Oilfield environment. This system with minimum human intervention will provide better workplace safety and maintenance of assets. This proposed system will perform predictive maintenance of various oil and gas industrial assets by analyzing various parameters (sensed data) and detecting failure modes either before they are going to take place or when the equipment will likely to fail or need service.

In future, the proposed scheme will be applied to oil pipeline transportation system by putting intelligent sensors with high performance to accumulate data in-time and along with Cloud Computing techniques as a smart decision-making algorithm to identify the happening of anomalous.

Acknowledgments. This research work is carried out under the Senior Research fellowship received from CSIR (Council for Scientific and Industrial Research) with grant no. 678/08(0001) 2k18 EMR-I.

Conflict of Interest. All the contributors in this research work have no clashes of attention to announce and broadcasting this article.

References

1. Adejo, A.O., Onumanyi, A.J., Anyanya, J.M., Oyewobi, S.O.: Oil and gas process monitoring through wireless sensor networks: a survey. Ocean J. Appl. Sci. 6(2), 39–43 (2013)
2. Atzori, L., Iera, A., Morabito, G., Nitti, M.: The Social Internet of Things (SIot)—when social networks meet the internet of things: concept, architecture and network characterization. Comput. Netw. 56(16), 3594–3608 (2012)
3. Abreu, D.P., Velasquez, K., Curado, M., Monteiro, E.: A resilient Internet of Things architecture for smart cities. Ann. Telecommun. 72(1–12), 19–30 (2016)

4. Boyer, S.A.: SCADA: Supervisory Control and Data Acquisition. International Society of Automation, Research Triangle Park (2009)
5. Cegla, F., Allin, J.: Ultrasonic monitoring of pipeline wall thickness with autonomous, wireless sensor networks. In: Oil and Gas Pipelines, pp. 571–578 (2015)
6. Gómez, C.D., Díaz Barriga, J.K., Rodríguez Molano, J.I.: Big data meaning in the architecture of IoT for smart cities. In: Data Mining and Big Data: First International Conference, DMBD 2016 (2016)
7. Henry, N.F., Henry, O.N.: Wireless sensor networks based pipeline vandalization and oil spillage monitoring and detection: main benefits for Nigeria oil and gas sectors. SIJ Trans. Comput. Sci. Eng. Appl. 3(1), 1–6 (2015)
8. Hao, Y., Linke, G., Ruidong, L., Asaeda, H., Yuguang, F.: Dataclouds: enabling community-based data-centric services over the Internet of Things. IEEE Internet Things J. 1(5), 472–482 (2014)
9. Jawhar, I., Mohamed, N., Shuaib, K.: A framework for pipeline infrastructure monitoring using wireless sensor networks. In: Wireless Telecommunications Symposium, WTS 2007, pp. 1–7. IEEE (2007)
10. Reza Akhondi, M., Talevski, A., Carlsen, S., Petersen, S.: The role of wireless sensor networks (WSNs) in industrial oil and gas condition monitoring. In: IEEE DEST 2010, pp. 618–623 (2010)
11. Minetti, G.: Wireless sensor networks enable remote condition monitoring. Pipeline Gas J. 239(7), 53–54 (2012)
12. Priyanka, E.B., Maheswari, C.: Parameter monitoring and control during petrol transportation using PLC based PID controller. J. Appl. Res. Technol. 14(5), 125–131 (2016)
13. Priyanka, E.B., Maheswari, C., Thangavel, S.: Remote monitoring and control of an oil pipeline transportation system using a Fuzzy-PID controller. Flow Meas. Instrum. 62, 144–151 (2018)
14. Priyanka, E.B., Maheswari, C., Thangavel, S.: IoT based field parameters monitoring and control in press shop assembly. Internet Things 3, 1–11 (2018)
15. Priyanka, E.B., Krishnamurthy, K., Maheswari, C.: Remote monitoring and control of pressure and flow in oil pipelines transport system using PLC based controller (Conference Paper). In: Proceedings of 2016 Online International Conference on Green Engineering and Technologies, IC-GET 2016, 1 May 2017. Article number 7916754
16. Sabata, A., Brossia, S.: Remote monitoring of pipelines using wireless sensor network. U.S. Patent 7,526,944. Accessed 5 May 2009
17. Wu, D., Youcef-Toumi, K., Mekid, S., Ben-Mansour, R.: Wireless communication systems for an underground pipe inspection. U.S. Patent Application 14/569,889. Accessed 15 Dec 2014
18. Xu, Q., Jiang, J., Wang, X.: Research and development of oil drilling monitoring system based on wireless sensor network technology. In: International Conference on Networks Security, Wireless Communications and Trusted Computing, NSWCTC 2009, vol. 2, pp. 326–329. IEEE (2009)
19. Yi, P., Xiao, L., Zhang, Y.: Remote real-time monitoring system for oil and gas well based on wireless sensor networks. In: 2010 International Conference on Mechanic Automation and Control Engineering (MACE), pp. 2427–2429. IEEE (2010)

Analysis of Mirai Botnet Malware Issues and Its Prediction Methods in Internet of Things

K. Vengatesan[1,2], Abhishek Kumar[1,2(✉)], M. Parthibhan[3],
Achintya Singhal[2], and R. Rajesh[3]

[1] Department of Computer Engineering, Sanjivani College of Engineering,
Kopargaon, India
vengicse2005@gmail.com, abhishek.maacindia@gmail.com
[2] Department of Computer Science, Banaras Hindu University, Varanasi, India
achintya.singhal@gmail.com
[3] Department of Computer Science and Engineering,
SRM Institute of Science and Technology, NCR, Ghaziabad, India
parthibanmecit@gmail.com, rrajeshn@gmail.com

Abstract. The Internet of Things is progressively turning into a pervasive figuring service, requiring immense volumes of information stockpiling and preparing. Lamentably, because of the one of a kind qualities of asset requirements, self-association, and short-run communication in IoT, it generally depends on the cloud for outsourced capacity and calculation, which has realized a progression of new difficult security and protection dangers. The IoT peripherals are utilized in houses, industry, and prescription. The Mirai-botnet is the biggest enrolled botnet that utilizing the IoTs. At the pinnacle of its movement, the botnet figured out how to arrange a hack wherever around thousand devices partook. This paper gives a point by point investigation of mirai malware attacking issues and its forecast systems, particularly in the territory of IoT.

Keywords: IoT · Botnet · Anomaly · Malware issue

1 Introduction

IoT is a standout amongst the greatest mainstream topics of ongoing centuries. The IoT devices are utilized in drug, industry and houses. The Mirai-botnet is the biggest enrolled botnet that utilizing the IoTs. At the pinnacle of its movement, the botnet figured out how to arrange an attack where around thousand devices took an interest. Accordingly there were around 1.2 million contaminated peripherals, 170 thousand of which were dynamic. The devices of the IoT ought to be disavowed from the activities of botnets. An ongoing conspicuous case is the Mirai botnet. First distinguished in August 2016 by the whitehat security investigate aggregate Malware Must Die, Mirai Japanese for "the future" and it's numerous variations and imitators have filled in as the vehicle for probably the most intense DDoS attacks ever.

© Springer Nature Switzerland AG 2020
A. P. Pandian et al. (Eds.): ICCBI 2018, LNDECT 31, pp. 120–126, 2020.
https://doi.org/10.1007/978-3-030-24643-3_13

Cloud Based IoT With Security Issues

The security dangers for cloud based IoT, particularly in the parts of secure bundle sending with outsourced totaled transmission prove age and proficient protection saving verification with outsourced message sifting. Other than the conventional information secrecy and unforgeability, the one of a kind security and protection prerequisites in cloud-based IoT are introduced:

Character Privacy: Conditional personality security alludes to the way that the portable IoT client's genuine character ought to be very much shielded from the public; then again, when some debate happens in crisis cases, it can likewise be viably followed by the specialist. The method of aliases been broadly received to accomplish this objective, yet the occasionally refreshed nom de plumes authentications prompt unbearable computational cost for asset obliged IoT nodes. All the more genuinely, it can't avoid the physically unique following attack we distinguished for area protection.

Location Privacy: Location security appears to be particularly basic in IoTs, since the every now and again uncovered area protection would reveal the living propensity for the IoT client. The generally received method is to shroud its area through pen names. Be that as it may, since the area data isn't specifically ensured, it can't avoid the physically powerful following attack. In particular, an arrangement of pernicious IoT clients in plot can be dispatched to the positions where the objective IoT client once in a while visited, to physically record sets of genuine characters of sitting back periods by perception or movement observing video, and further recognize the objective IoT client's genuine personality. In the event that the enemy realizes that the objective hub with alias every so often visits n areas Loc1, Loc2, … , Locn, n sets of nodes' genuine characters going by these n areas Veh1, Veh2, … , Vehn can be watched. The convergence would uncover the objective hub's genuine personality and its private exercises in different districts.

Node Compromise Attack: Node trade off attack implies the enemy extricates from the asset compelled IoT devices all the private data including the mystery key used to scramble the bundles, the private key to produce marks, et cetera, and afterward reinvents or replaces the IoT devices with malignant ones under the control of the foe. The objective situated trade off attack implies an enemy with worldwide observing capacity would choose the IoT hub holding more parcels as the bargain focus by watching the activity stream around all nodes in IoT. Along these lines, from one single trade off, it is likely that the enemy acquires more parcels for recouping the first message or blocking its fruitful conveyance by interference. Layer Removing/Adding Attack: The layer evacuating attack happens when a gathering of egotistical IoT clients expel all the sending layers between them to expand their remunerated credits by lessening the quantity of middle of the road transmitters sharing the reward. Actually, the layer including attack implies intriguing IoT clients noxiously bypass the parcel sending way between them for expanded credits by expanding the aggregate realistic utility.

Forward and Backward Security: Due to the versatility and dynamic social gathering detailing in IoT, it is important to accomplish forward and in reverse security. The previous implies that recently joined IoT clients can just unravel the scrambled messages got after yet not before they join the bunch; while the last implies that denied IoT clients can just disentangle the encoded messages previously yet not in the wake of taking off.

In September 2016, the site of PC security advisor Brian Krebs was hit with 620 Gbps of activity, "numerous requests of size more movement than is commonly expected to thump most locales disconnected." At about a similar time, a considerably greater DDoS attack utilizing Mirai malware topping at 1.1 Tbps focused on the French webhost and cloud service supplier OVH. In the wake of the public arrival of Mirai's source code by its maker soon a short time later, programmers offered Mirai botnets for lease with upwards of 400,000 at the same time associated devices. More Mirai attacks took after, prominently one in October 2016 against service supplier Dyn that brought down many sites including Twitter, Netflix, Reddit, and GitHub for a few hours.

Mirai fundamentally spreads by first tainting devices, for example, webcams, DVRs, and switches that run some rendition of BusyBox (busybox.net). It at that point concludes the authoritative certifications of other IoT devices by methods for savage power, depending on a little word reference of potential username– watchword sets. Today, Mirai changes are produced day by day, and the way that they can proceed to multiply and dispense genuine harm utilizing an in distinguishable interruption strategies from the first malware is characteristic of IoT device sellers' constant disregard in applying even fundamental security hones. Shockingly, IoT botnets have gotten just sporadic consideration from scientists. On the off chance that the security network doesn't react all the more rapidly and devise novel resistances, notwithstanding, perpetually refined attacks will turn into the standard and might upset the Internet framework itself.

A Mirai botnet is involved four noteworthy segments. The bot is the malware that taints devices. Its twofold point is to proliferate the disease to misconfigured devices and to attack an objective server when it gets the comparing command from the individual controlling the bot, or botmaster. The command and control (C&C) server furnishes the botmaster with an incorporated administration interface to check the botnet's condition and coordinate new DDoS attacks. Normally, communication with different parts of the framework is directed by means of the unknown Tor network. The loader encourages the dispersal of executables focusing on various stages (18 altogether, including ARM, MIPS, and x86) by specifically speaking with new casualties. The report server keeps up a database with insights about all devices in the botnet. Recently contaminated ones regularly straightforwardly speak with it.

2   Litrature Review

In 2016, September twentieth, the site of regarded security columnist Brian Krebs was taken disconnected by one of the biggest attack that is DDoS. This DDoS attack was made by what might later be uncovered as the Mirai botnet, and would be the start of a clamorous and troubling time of the Internet. On September 30th, 2016 only 10 days after this first attack, a gathering post prepared on HackForums.net by a client named "Anna-Senpai" who released the source code of Mirai botnet and guidelines to the people. Because of mirai attack, creator Brian Krebs started the errand of revealing the genuine personality of Mirai, Anna-Senpai [1].

Eventually, he could uncover the character of no less than one co-backstabber who helped in the formation of Mirai. At that point, right around three months after the fact the biggest this attack of its type was focused at a prevalent Domain Name System suppliers with dynamicall. In 2016 21st Oct, one attack with a greatness of 1.25 Tbps was coordinated over the Domain Name System 53rd port. The mirai attack was the biggest of its type, as well as to a great degree successful; cutting down countless and facilities who utilized Dyn as their DNS supplier. This carried a considerable measure of consideration then report to Miraibotnet with incalculable broadcast mediums wrapping its way. The idea of Internet of Things is certainly not another one, going backward to 1982 at what time Carngie Mellon University associated [2] an adjusted Coke mechanism to the Internet. In any case, the measure of study and arrangements donated to the safety of these Internet-associated machines is inadequate. Gartner, Inc. evaluated that 8.5 billion IoT equipment's will be dynamic in 2017 [3], creating for a 31% expansion ever since 2016.

IoT peripheral's security has not ever fully been in the attention till the point when ongoing occasions secured by the broadcast. On 2008, refuge scientists from the Massachusetts Amherst University and the Washington University could distantly hacked into a pacemaker [4] and convey conceivably destructive shocks of power from it. This was the fatal IoT attack as it could convey upon real damage to people life. Tragically, this was not the main restorative IoT helplessness; in 2016, Rapid reported 3 susceptibilities [5] in the Animas single-Touch Ping Insulin Pump. Though certain past investigations have been finished with an attention on the Mirai-botnet, the data discharged has been conflicting, vague, and in some cases of smooth mistakes.

On behalf of, a threats examination achieved upon the arrival of the Mirai botnet's source code by a blog named as "Malware Must Die!" inaccurately secured the utilization [6] instance of the protect dog part of the coding. The tag expressed that the contribution of the guard dog peripheral was to postpone the malware from running upon prompt sending of a recently tainted device. In any case, this isn't the situation. The guard dog clock is proposed to coercively reset the peripheral when it can't work sufficiently quick to continue "pinging" the clock device at customary interims. Lamentably, this blog entry has been referred in one more scholastic journal [7].

Sellers would be informed of their peripherals has powerless. All things that are helpless will be additionally leftward a note in the records demonstrating the nearness of an administration organization. In any case, the issue with this arrangement is that there isn't a sufficient order for IoT security. Therefore, government offices are not prone to acknowledge this approach as they would be in charge of dealing with the notice and administration of it. Furthermore, there was take a shot at taking a gander at DDoS attacks (Distributed Denial of Service) and in what way they identified with the IoT in [8] "DDoS in the IoT: Mirai and Other Botnets" which explored the Mirai-botnet and its variations. In any case, it gave exercises that can be gained from previous IoT hacks rather than moderation answers for what's to come. While it regards analyze the threats and comparable variations, it is considerably more essential to utilize the information got to support avoid comparative hacks later on. By doing this, the likelihood of more attacks can be diminished.

Proposed System Architecture

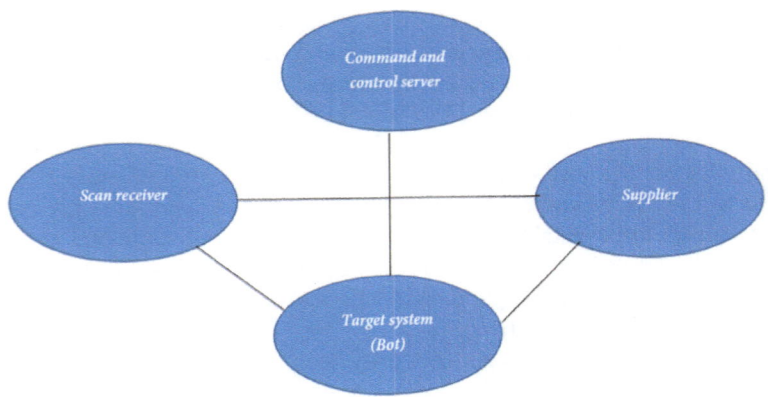

Fig. 1. The Mirai's working principal

3   Implementation

The most possible sort of hacks on the IoT devices is their contamination with a program of Trojan called Mirai. The Mirai was a Botnet that was a network of tainted equipment's with a malevolent embedded program, which enable a hacker to play out specific activities utilizing the resource of the contaminated equipment. Mirai is a work central is appeared on Fig. 1, mirai's working comprises of the accompanying advances:

- Botnet interfaces with the CC server and examination networks all over the place. It requires to discover certain IoT equipment's susceptibilities to abuse them more.
- It transfer outcomes to Test Retriever.
- Test Retriever transfers data with instructions to Supplier.
- Supplier transfers a great deal of instructions to the objective system.

Mirai endeavors to interface utilizing SSH ports and Telnet. The malevolent package utilizes the utmost widely recognized secret/login word blends to hack devices. The source code has a content that breaks all procedures utilizing the HTTP, Telnet, SSH-ports. One thing of the unmistakable highlights of the Mirai-botnet is worked in insurance component. Mirai code contains worked in contents that dispense with conceivable malignant package on the devices, then furthermore prohibit the likelihood of further accessing through remote.

To veil, Mirai-Botnet modifies the label of its procedure to arbitrary qualities. Then effective labor for 5 mints, the package dispatches a perpetual circle that Tests the procedures for different pernicious program (in dangerous threats there is a document that doesn't enable Mirai-Botnet to evacuate itself). The Mirai-botnet additionally covers contents that enable it to dispose of the considerable number of procedures of different kind of botnets, mirrored in the memory. Additionally Mirai-botnet is shielded from anime/kami threat (be that as it may, no hacks from this malignant programs have been enlisted). Some processes are intended to build the limit of Mirai-botnet, to shield the contaminated equipment from hacks by different Botnets. This is a capacity in the malignant software package that doesn't permit utilizing WatchDog any longer to the system can restart distantly.

References

1. Krebs, B.: Who is Anna-Senpai, the Mirai Worm Author? (2017). https://krebsonsecurity.com/2017/01/who-is-annasenpai-the-mirai-worm-author/
2. The Carnegie Mellon University Computer Science Department Coke Machine: The "Only" Coke Machine on the Internet (1982). Available: https://www.cs.cmu.edu/~coke/history-long.txt
3. Gartner, Inc.: Gartner Says 8.4 Billion Connected "Things" Will Be in Use in 2017, Up 31 Percent From 2016, February 2017. http://www.gartner.com/newsroom/id/3598917
4. Halperin, D.: "Pacemakers and Implantable Cardiac Defibrillators: Software Radio Attacks and Zero-Power Defense," Computer Science Department Faculty Publication Series, no. 68 (2008). http://scholarworks.umass.edu/csfacultypubs/68
5. Rapid7: Multiple Vulnerabilities in Animas OneTouch Ping Insulin Pump, October 2016. https://community.rapid7.com/community/infosec/blog/2016/10/04/r7-2016-07-multiple-vulnerabilities-in-animas-onetouch-ping-insulin-pump
6. MalwareMustDie: MMD-0056-2016 - Linux/Mirai, how an old ELF malcode is recycled, August 2016. http://blog.malwaremustdie.org/2016/08/mmd-0056-2016-linuxmirai-just.html
7. Jerkins, J.A.: Motivating a market or regulatory solution to IoT insecurity with the Mirai botnet code. In: 2017 IEEE 7th Annual Computing and Communication Workshop and Conference (CCWC), p. 3 (2017)
8. Kalaivanan, M., Vengatesan, K.: Recommendation system based on statistical analysis of ranking from user. In: International Conference on Information Communication and Embedded Systems (ICICES), pp. 479–484. IEEE (2013)
9. Saravana Kumar, E., Vengatesan, K.: Cluster Comput. (2018). https://doi.org/10.1007/s10586-018-2362-1
10. Sanjeevikumar, P., Vengatesan, K., Singh, R.P., Mahajan, S.B.: Statistical analysis of gene expression data using biclustering coherent column. Int. J. Pure Appl. Math. **114**(9), 447–454 (2017)
11. Kumar, A., Singhal, A., Sheetlani, J.: Essential-replica for face detection in the large appearance variations. Int. J. Pure Appl. Math. **118**(20), 2665–2674 (2018)
12. Kolias, C., Kambourakis, G., Stavrou, A., Voas, J.: DDoS in the IoT: Mirai and other Botnets. Computer **50**(7), 80–84 (2017)
13. Incapsula: Breaking down Mirai: An IoT DDoS Botnet analysis, October 2016. https://www.incapsula.com/blog/malwareanalysis-mirai-ddos-botnet.html
14. Jgamblin: Jgamblin/Mirai-Source-Code, October 2016. https://github.com/jgamblin/Mirai-Source-Code/blob/master/mirai/bot/scanner.c#L688-L701
15. Jgamblin/Mirai-Source-Code, October 2016. https://github.com/jgamblin/Mirai-Source-Code/blob/master/mirai/bot/scanner.c#L124-L185

16. Howells, D.: Torvalds/linux, October 2016. https://github.com/torvalds/linux/blob/master/include/uapi/linux/watchdog.h
17. Zhou, J., et al.: Secure and privacy preserving protocol for cloud-based vehicular DTNs. IEEE Trans. Info. Forensics Secur. **10**(6), 1299–1314 (2015)
18. Paillier, P.: Public key cryptosystems based on composite degree residuosity classes. In: Eurocrypt 1999, pp. 223–238 (1999)

Contagious Diseases Prediction in Healthcare Over Big Data

Nkundimana Joel Gakwaya[(⊠)] and S. Manju Priya

Computer Science, Karpagam Academy of Higher Education, Coimbatore, India
Gajoel2000@gmail.com, smanjupr@gmail.com

Abstract. As Data trends go higher and higher, the data creation process tends to increase at an unprecedented rate. They are not only increasing, they also change, and incarnate from one form to the another. Recently, Big data revolution has introduced an incarnation of how the data is captured, accessed, aggregated, converted and stored. Healthcare communities is one of the big data procreations, where data gets generated rapidly. The appropriate analysis of healthcare data helps to detect the diseases at an early stage to save the patients' life. Although, analysis is required to make a proper decision, this analysis is declined due to the incomplete data. This paper focuses on the prediction towards the emerging contagious diseases to provide enhanced accuracy in the process of disease detection. To solve the incomplete e.g. missing data, the Improved Expectation Maximization (IEM) has been employed. Here, the experiment is carried out on Ebola disease. This research work proposed a new Ensemble Neural Network algorithm using structured and unstructured data. A proposed model was trained on prediction from dataset and evaluate each model. In this work, the Ensemble Neural Network shows an average performance, but leverages a very firm and more consistent performance analysis accuracy. This method will deliver an enhanced potential for the decision makers in healthcare communities.

Keywords: Big data analytics · Ebola disease · Healthcare · Expectation Maximization · Neural network

1 Introduction

Ebola Virus diseases (EVD) are primarily handed down from wild animal to humans, and the transmission growth in population occurs through human to human transfer. People commitment emerges as the significant component to help the healthcare practitioners to promptly identify the humans who are under the contagious disease. Quick and fast notification is very important in order to provide a quick helping hand. Hence, it needs significant attention when early symptoms are detected to restore the fluid into a patient. Despite the hype, the treatment also has some challenges like unauthorized way of treatment and suggestions. However, diverse number of treatment based on immunology are still under trial, and other drug combinations are under developmental stage. Also there have been a lot of cases and deaths during this irruption. The contagious diseases are found in Africa countries such Guinea, and it is

© Springer Nature Switzerland AG 2020
A. P. Pandian et al. (Eds.): ICCBI 2018, LNDECT 31, pp. 127–132, 2020.
https://doi.org/10.1007/978-3-030-24643-3_14

moved to its borders namely Sierra Leone and Liberia. The virus family includes 3 classes: Cuevavirus, Marburgvirus, and Ebolavirus. Inside Ebolavirus class there are 5 species which are identified in different parts of Africa such as central, north and south. The Ebola virus is considered as the notable outbreak in the continent of Africa. Early detection and warning of outbreaks will emerge as a life saver resource. To achieve this work, an ensemble neural network based on internet of things could play a key role to fight against EVD.

2   Related Works

İnfectious disease is emerging as a threatening fatal disease to the human community. The world health organization (WHO) invests all efforts to support different policies that includes systems which are willing to help to investigate outbreak at early warning [1]. An observation of any unknown disease such as haemorrhagic fever might be reported by nurses and other people who are involved in healthcare profession, this report will leverage an early detection as well as prevention. There is an uplift in strategies of early detection system [2, 3], however some policies of investigations are not on normal references [4], those changes are caused by a lack of past epidemic records and agreement based on the occurence of it [5], for this absences of old records, the early detection warning (EDW) was introduced and implemented in order to store the previous data. The other means of detecting the epidemic diseases are mobile health (mHealth) devices, which is utilized globally. The primary purpose of devices used in healthcare is to record all the information related to the patient. After first step of recording, the next step is to confirm whether the disease exist or not, it is done by using rapid diagnostics assay rest (RDAT). The above mentioned steps help to save a life before the diseased person reaches the critical condition [7]. The detection process can be enhanced when analysis is automated [8]. Besides, [9] has compared the bioinsipired techniques, where its main purpose is to measure the performance of heterogenous telemedicine transportation, massive information are transported using mobile health devices. [10, 11] planned another way to transport the health records in a better way based on the observance. [12] has carried the work of heterogenous systems, and to study the steps of analysis has been reduced in order to decrease the medical expense. [13] telehealth system has been developed to regulate the rules and to design knowledge from the obtained personal health record (PHR). [14] an advanced application was developed to effectively deal with massive healthcare data. [15] planned to associate best massive data sharing rule to deal with complex healthcare information. This work, develops a system that has six applications with an ability to identify the unsound patients. To secure e-health data, a proposal system [16] was developed along with a capacity to predict the emerging health risk. Prediction and victimization of ancient unwellness risk models sometimes involve a machine learning rule which need a labeled class e.g. supervised learning. [17, 18] within it all datasets are splitted into two classes either unsound health or sound health. İn clinical fields, these models are needed to increase the healthcare performance [19, 20]. There is a need of early detection and early warning of epidemic. When these two earlies are certained, the lives of many humans will be saved. [22] describes the research work based on the multimodal disease prediction using ensemble neural network. In [23] malaria disease has been predicted with the help of ant colony and random forest tree methodologies.

3   Proposed Methods

3.1   Likelihood

There are variety of methods used to deal with the missing data values. In this work, the proposed method determines the knowledge of the sampled data which are drawn from different variables, and these variables are then compared with the existing variables. With this comparison, the desired parameters can be calculated based on the mistreatment of the present knowledge. In case of the absence of knowledge from variables, the next step is to calculate the estimated missing variable. The outcome shows the linkup among variables which are counted as mistreating. This missing knowledge can also be identified during the distribution of the different variables.

3.2   Improved Expectation Maximation

Improved Expectation Maximization (IEM) may be a variety of the utmost chance methodology which will be accustomed to produce a brand new knowledge set, within which all the missing values are imputed with values that remains calculable by the utmost chance strategies [21].

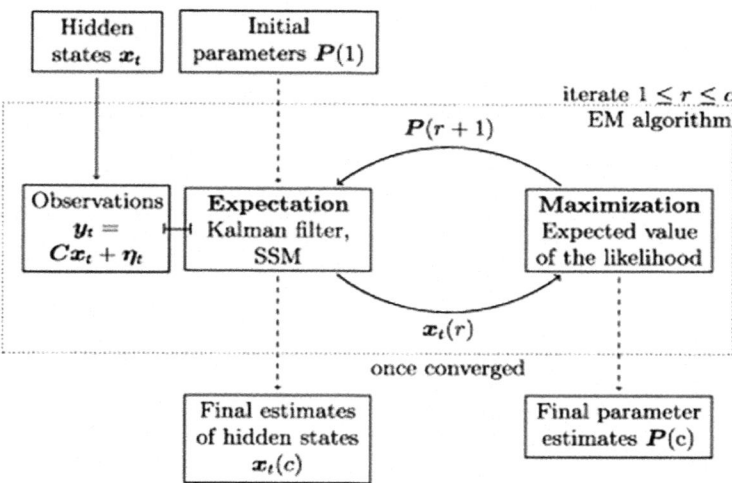

Fig. 1. Expected maximization (source. 1)

This work will provide the best output from the deviation of expected parameter which are primarily determined to possess a high degree of positiveness of the expectation and may underestimate the quality error. Further, this work focuses on the responsibility laid upon the Improved Expectation Maximization technique, an expected price that supports the variables present on the market for every case is substituted for the missing knowledge. As a result of that, one imputation omits the attainable variations among the multiple imputations, one imputation can tend to underestimate the quality errors and so overestimate the extent of exactness.

As the contagious diseases spread rapidly, we need a system which can iterate the process in fast manner. Thus Fig. 1 illustrates the process of EM algorithm, which has the ability to demonstrate the different stages of the proposed system.

3.3   Ensemble Neural Network

Ensemble classifier is a machine learning technique, which has been used to learn and recognize a complex pattern and draw a decision based on the available complex data. In this ensemble, it has the ability to describe a set of predetermined diseases which simultaneously enhances the accuracy. In this technique, the classifiers are combined for the primary aim of creating an enhanced model in order to increase the accuracy.

The above Fig. 2 demonstrates the process of dealing with contagious virus disease, where each case is first classified by the neural network, the suspected case and confirmed case, then they are grouped into different categories, both groups at the end where they are warned and treated if the virus is detected.

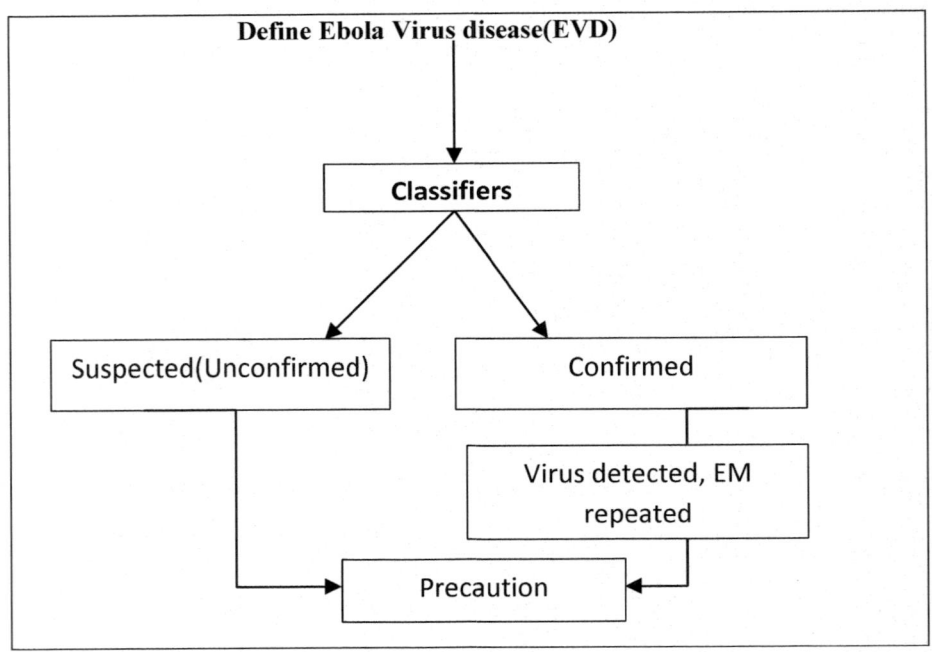

Fig. 2. Proposed system of ENN

4   Conclusion

This research work proposed a system to intelligently fight against contagious diseases. The system has been proposed based on the Expectation Maximization (EM) and Ensemble neural network (ENN). İt is believed that this system will provide a notable contribution to the area which is affected by the contagious diseases. Moreover the systems also need some enhancement to improve the prediction as well as the accuracy. This approach is mainly focused on solving the incomplete data present in big data analytics, and deals with continuos data using ENN. Furthermore, the proposed system also has the ability to benefit other healthcare communities.

References

1. WHO: Roll back malaria. Malaria early warning system. A Framework for Field Research in Africa: concepts, indicators and partners. Geneva: World Health Organization (2001)
2. WHO: Prevention and control of malaria epidemics. Third meeting of the Technical Support Network. World Health Organization, Geneva (2002)
3. Watkins, R.E., Eagleson, S., Hall, R.G., Dailey, L., Plant, A.J.: Approaches to the evaluation of outbreak detection methods. BMC Public Health **6**, 263 (2006)
4. Mckelvie, W.R., Haghdoost, A.A., Raeisi, A.: Defining and detecting malaria epidemics in south-east Iran. Malar J. **11**, 81 (2012)
5. World Health Organization: New horizons for health through mobile technologies (2011)
6. Randrianasolo, L., Raoelina, Y., Ratsitorahina, M., Ravolomanana, L., Andria-mandimby, S., Heraud, J., et al.: Sentinel surveillance system for early outbreak detection in Madagascar. BMC Public Health **10**, 31 (2010)
7. WHO: Roll Back Malaria. Disease surveillance for malaria elimination. World Health Organization, Geneva (2012)
8. Brady, O.J., Smith, D.L., Scott, T.W., Hay, S.I.: Dengue disease outbreak definitions are implicitly variable. Epidemics **11**, 92–102 (2015)
9. CDC: Framework for evaluating public health surveillance systems for early detection of outbreaks recommendations from the CDC Working Group (2004)
10. Chen, M., Ma, Y., Li, Y., Wu, D., Zhang, Y., Youn, C.: Wearable 2.0: enable human-cloud integration in next generation healthcare system. IEEE Commun. **55**(1), 54–61 (2017)
11. Chen, M., Ma, Y., Song, J., Lai, C., Hu, B.: Smart clothing: connecting human with clouds and big data for sustainable health monitoring. Mob. Netw. Appl. **21**(5), 82 (2016)
12. Chen, M., Zhou, P., Fortino, G.: Emotion communication system. IEEE Access. https://doi.org/10.1109/access.2016.2641480 (2016)
13. Qiu, M., Sha, E.H.-M.: Cost minimization while satisfying hard/soft timing constraints for heterogeneous embedded systems. ACM Trans. Des. Autom. Electron. Syst. (TODAES) **14**(2), 25 (2009)
14. Wang, J., Qiu, M., Guo, B.: Enabling real-time information service on telehealth system over cloud-based big data platform. J. Syst. Architect. **72**, 69–79 (2017)
15. Bates, D.W., Saria, S., Ohno-Machado, L., Shah, A., Escobar, G.: Big data in health care: using analytics to identify and manage high-risk and high-cost patients. Health Aff. **33**(7), 1123–1131 (2014)

16. Qiu, L., Gai, K., Qiu, M.: Optimal big data sharing approach for tele-health in cloud computing. In: IEEE International Conference on Smart Cloud (SmartCloud), pp. 184–189. IEEE (2016)
17. Zhang, Y., Qiu, M., Tsai, C.-W., Hassan, M.M., Alamri, A.: HealthCPS: healthcare cyber-physical system assisted by cloud and big data. IEEE Syst. J. (2015)
18. Lin, K., Luo, J., Hu, L., Hossain, M.S., Ghoneim, A.: Localization based on social big data analysis in the vehicular networks. IEEE Trans. Ind. Inf. (2016)
19. Lin, K., Chen, M., Deng, J., Hassan, M.M., Fortino, G.: Enhanced fingerprinting and trajectory prediction for iot localization in smart buildings. IEEE Trans. Autom. Sci. Eng. 13 (3), 1294–1307 (2016)
20. Marcoon, S., Chang, A.M., Lee, B., Salhi, R., Hollander, J.E.: Heart score to further risk stratify patients with low timi scores. Crit. Pathways Cardiol. 12(1), 1–5 (2013)
21. Dempster, A.P., Laird, N.M., Rubin, D.B.: Maximum likelihood from incomplete data via the EM algorithm. JRSSB 39, 1–38 (1997)
22. Joel, G.N., Priya, S.M.: Improved ant colony on feature selection and weighted ensemble neural network based multimodal disease risk prediction classifier for disease prediction over big data. Int. J. Eng. Technol. 56–61 (2018)
23. Joel, G.N., Priya, S.M.: Improved malaria prediction using ant colony feature selection and random forest tree in big data analytics. J. Adv. Res. Dyn. Control Syst. 10(12-Special Issue) (2018)

Securing Data with Selective Encryption Based DAC Scheme for MANET

J. Nithyapriya[✉], R. Anandha Jothi[✉], and V. Palanisamy[✉]

Department of Computer Applications, Alagappa University,
Karaikudi, Tamilnadu, India
nithyapriyajj@gmail.com, ranandhajothi12@gmail.com,
vpazhanisamy@yahoo.co.in

Abstract. MANETs consist of mobile nodes which can any time enter or exit from the network [16]. The no central admin set up of the MANET makes it more prone to attacks [5]. Sending much more secured data than any other networks is verily needed in MANETs since the mobile nodes consume a lot of energy for capturing signals and cryptanalysis of messages. Here we are presenting a security scheme that adorns selective encryption strategy and encrypts selective bit positions instead of the entire data set. The scheme needs less computation capability while providing adequate security to the MANET nodes [15]. It provides better security compared to the existing selective encryption algorithms. Thus we manage safety as well energy conservation of mobile nodes.

Keywords: MANET · DAC · Selective encryption

1 Introduction

Wireless nodes of ad hoc networks manage themselves even when there is no connection of internet [7]. Mobile Ad hoc Networks consist of mobile nodes with dynamically changing topology [6]. The nodes supply the routing information by their selves as every node acts as a router [17]. Nodes can any time enter or exit in a MANET. The no central admin set up of the MANET makes it more prone to attacks [14]. Sending much more secured data than any other networks is verily needed. The mobile nodes consume a lot of energy for capturing signals and its own running.

Ad hoc networks have a unique set of challenges [4, 6].

[1] Ad hoc networks always face challenges in secure communication due to mobility of nodes. [2] Mobile nodes without adequate protection are easy to compromise. [3] Energy consumption of wireless nodes stands as a huge challenge.

The biggest challenge is sending a pre secured data with a robust security scheme with less computation capability [15].

This paper offers an algorithm which embraces selective encryption technique. The well known substitution and transposition techniques encrypt all the 'n' positions of the data set. As per the Caesar ciphering algorithm we perform $C = E(k, p) \bmod 26$ to encipher every bit with a different alphabet and it is a tedious process [10]. The transposition techniques such as Railfence cipher, Route cipher, columnar transposition, double transposition, disrupted transposition and Myszkowski transposition, the

© Springer Nature Switzerland AG 2020
A. P. Pandian et al. (Eds.): ICCBI 2018, LNDECT 31, pp. 133–139, 2020.
https://doi.org/10.1007/978-3-030-24643-3_15

entire set of n bits has to be encrypted which is energy consuming although giving adequate security [6]. This paper coins an algorithm which sticks on to the selective encryption scheme and minimizes the number of bits encrypted to m where m < n. As a result the time [1] and energy for cryptanalysis is saved.

2 Selective Encryption

Selective encryption is a well grown trend in protecting image and video content and it selective encryption can be applied on any type of content [13]. It is actually encrypting a subset of data [8, 9]. The aim of selective encryption is to increase the speed of encryption [1] and to reduce the amount of data to encrypt while preserving a sufficient level of security. Selective encryption divides the data into two parts called public part and protected part [4]. Protected part is suggested for encryption while public part is left unencrypted and is open to all. Generally selective encryption is done for text, image or video contents for online purchase.

2.1 Criteria of Selective Encryption [11]

Tuning – It could be verily desirable to define the encrypted part and the encrypted parameters with respect to relevant applications.
Visual degradation – Distortion of image; yet the attacker is able to access the data.
Cryptographic security – Applying some ciphering and deciphering strategy to encrypt the selective content; this highly relies on two parameters, the encryption key and unpredictability of the encrypted part.
Encryption Ratio (ER) – This is the relation between the encrypted part and the whole data size. ER is to be minimized by selective encryption.
Compression friendliness – The compression ratio needs to be very less.
Error Tolerance (ET) – This is verily needed in networks that deal with complex cryptographic schemes. There should be a trade-off between the avalanche effect and ER.

2.2 Classification of Selective Encryption [12]

Classification is done with respect to the time of compression

(a) Precompression – encryption before compression (Fig. 1).

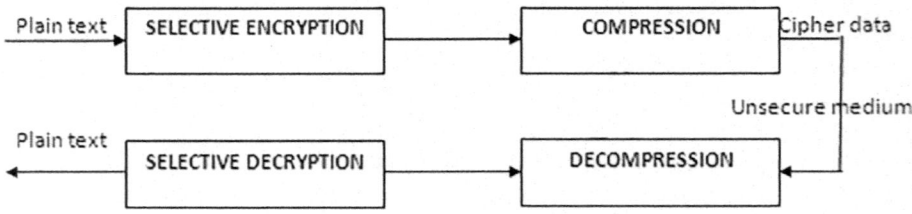

Fig. 1. Precompression

(b) Incompression – encryption and compression hand in hand (Fig. 2).

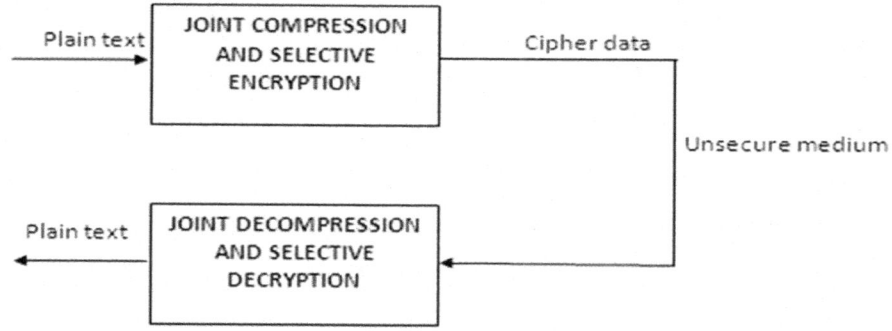

Fig. 2. Incompression

(c) Postcompression – encryption after compression (Fig. 3).

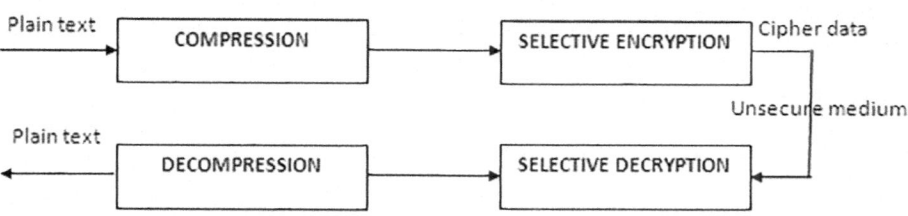

Fig. 3. Postcompression

2.3 Types of Selective Encryption Algorithm [4]

Selective encryption algorithm can be framed in three ways.

1. Secure key distribution algorithm

In this scheme, the sender node sends a request to all the neighboring nodes to confirm their legitimacy. For the request the sender S uses its public key PK_S and the neighbor R replies with the secret key SK_S. This will be later used by the two nodes for any transaction. This secret key is the symmetric key encrypted by the public key of the requester (Fig. 4).

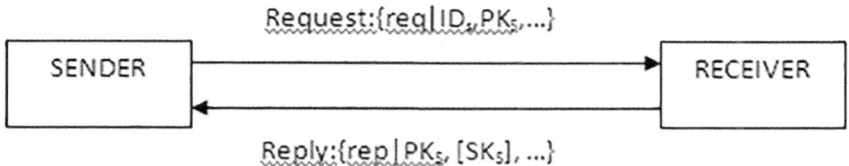

Fig. 4. Symmetric key selective encryption

2. Toss-A-Coin selective Encryption algorithm

It is a very natural approach in which just to divide the data into two equal sets of bits such as odd number of bits b1, b3, b5, ... b(n − 1) and even number of bits such as b2, b4, b6, ... b(n). It makes use of a random approach called toss-a-coin approach to select either even set or odd set of bits. By this 50% of data gets encrypted.

3. Probabilistic selective encryption algorithm

Firstly, the sender S applies a random number generator RNG to generate encryption ratio ER to obtain the percentage of encrypted messages. The communication is secure if ER is greater than the Security requirement SR.

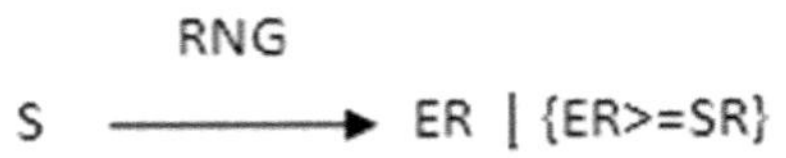

$$\text{RNG}$$
$$S \longrightarrow ER \mid \{ER >= SR\}$$

Fig. 5. Probabilistic selective encryption

Secondly, the sender S will employ a probability function PF to generate encryption probability pi. Eventually the sender selects the messages to encrypt based on the above ER (Fig. 5).

4. Algorithm based on entropy

This algorithm gives different entropy values to bits 1 and the bits 0 and encrypts bit 1 where as it advocates bits 0 do not need encryption to provide stronger security.

Following any type of the above algorithms the sender includes the proper uncertainty on the message so that only the intended receiver can receive it [8].

3   Proposed Work

In our proposed scheme we provide selective encryption using Divide And Conquer approach (DAC). We use the DAC approach just to compress the entire data bits and suggest selective encryption hand in hand. So it comes under incompression type of selective encryption. We render the algorithm below which is known as Divide And Conquer (SEDAC) algorithm using Selective Encryption.

Algorithm SEDAC

// input – total number of bits in the message, 'n'

//output - position of the bits to be encrypted, 'm'

//T- position of top or starting bit and B- position of bottom or ending bit of the partition

//ESI – Encryption Suggested Index, NUM[a,b] – Number of indices from a to b

---

Step 1:

If (n<=3)

{

If(n==1) then exit //No encryption

T=1 ; B=n ; Encrypt (T+B)/2

Else

// encrypt the first subset on selective indices ESI

Step 2 :

while (n>3)

do

{

If n MOD 2 == 0

Encrypt n/2 and (n/2)+1,T,B // ESI

Else

Encrypt n/2, T,B //ESI

M = (T+B)/2;

T=T+1;

B=M-1;

}

//encrypt the second subset on selective indices ESI

If n MOD 2==0

T=(n/2)+2; B=n-1; n=NUM[T,B]

Else

T=(n/2)+1;B=n-1; n=NUM[T,B]

Go to: step 1;

End SEDAC

Eg.1

       1   2 3 4 5 6 7 8 9 10

       ESI : 1,   5 , 6, 10

              :2 or 3, 7 or 8

Totally 6 out of 10 bits

Eg. 2:

       1   2 3 4 5 6 7

       ESI : 1 , 4, 7

           :2 , 5

Totally 5 out of 7 bits

The number of ESI is always greater than security requirement SR for mobile nodes. Thus it is a robust scheme for securing messages.

4   Result and Discussion

The proposed SEDAC algorithm chooses 'm' number of ESI out of 'n' indices where m < n. This selective encryption algorithm reduces the computational complexity than the conventional encryption techniques. SEDAC encrypts more than 50% of bits in the message and thus there is no need to worry about the Encryption Ratio (ER) (Fig. 6).

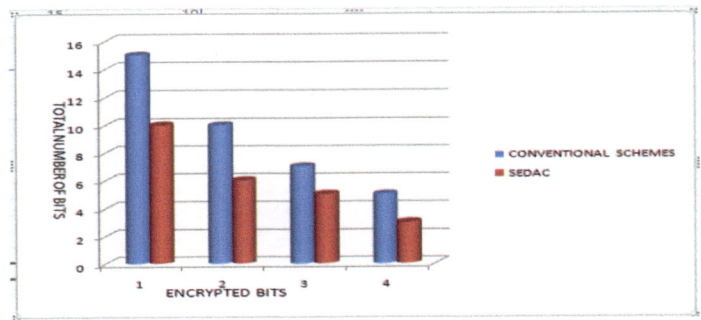

Fig. 6. Comparison of SEDAC with conventional schemes

The above graph compares SEDAC with the conventional encryption schemes. Since SEDAC follows selective encryption the number of encryption suggested indices is lesser as in it shown in the green color bar. Thus we say that the proposed algorithm is better in terms of execution time and energy consumption by nodes. Hence we strongly recommend that this SEDAC algorithm would give better protection to the mobile nodes.

5   Conclusion

The selective encryption algorithm SEDAC that is proposed here clearly proves that it is a simpler algorithm compared to the existing conventional algorithms and suggests lesser encryption suggested indices (ESI). It saves time and offers less computational complexity. Ad hoc nodes generally suffer with power starvation. So the proposed scheme gives a better protection to the mobile nodes using selective encryption eventually consuming less energy. It is a robust scheme in terms of the security requirement standards. In future this same algorithm can be improved to provide still lesser number of ESI so that energy of nodes and time of execution could be further reduced.

References

1. Kushwahaa, A., Sharmab, H.R., Ambhaikarc, A.: A novel selective encryption method for securing text over mobile ad hoc network. In: 7th International Conference on Communication, Computing and Virtualization, pp. 16–23 (2016). https://www.sciencedirect.com/science/article/pii/S1877050916001356

2. Prajitha, P.N., et al.: Wireless sensor network security by intrusion detection in energy efficient way. Int. J. Sci. Eng. Technol. Res. (IJSETR), 5(4) (2016). ISSN 2278 – 798

3. Jawad, L.M., et al.: A survey on emerging challenges in selective color image encryption techniques. Indian J. Sci. Technol. 8(27) (2015). https://doi.org/10.17485/ijst/2015/v8i27/71241. ISSN (Print) 0974-6846. ISSN (Online) 0974-5645

4. Kushwala, A., et al.: Selective encryption algorithm for secured MANET. In: IEEE Conference (2012). https://ieeexplore.ieee.org/document/6338098

5. Patil, P., Shah, R., Jewani, K.: Enhanced novel security scheme for wireless adhoc networks: ENSS. Int. J. Comput. Appl.® (IJCA) (2012)

6. Patil Ganesh, G., Chatterjee, M.A.: Selective encryption algorithm for wireless ad-hoc networks. Int. J. Adv. Comput. Theory Eng. 1(1) (2012)

7. Das, A., Basu, S.S., Chaudhuri, A.: A novel security scheme for wireless Adhoc network. IEEE (2011). 978-1-4577-0787-2/11

8. Ren, Y., Boukerche, A.: Performance analysis of a selective encryption algorithm for wireless ad hoc networks (2011). PARADISE Research Laboratory, SITE, University of Ottawa, Ottawa, Ontario, Canada, Lynda Mokdad Laboratoire LACL Université Paris-Est, Créteil, France. https://ieeexplore.ieee.org/iel5

9. Massoudi, A., Lefebvre, F., De Vleeschouwer, C., Macq, B., Quisquater, J.-J.: Overview on selective encryption of image and video: challenges and perspectives. EURASIP J. Inf. Secur. (2008). Thomson R&D France, Technology Group, Corporate Research, Security Laboratory 1, avenue Belle Fontaine, France. Article ID 179290. https://www.researchgate.net/publication/26593939_Overview_on_selective_encryption_of_image_and_video_Challenges_and_perspectives_EURASIP

10. Stallings, W.: Cryptography and Network Security – Principles and Practices, 6th edn. Pearson, New Delhi (2013)

11. https://jis-eurasipjournals.springeropen.com/articles/10.1155/2008/179290

12. Van Droogenbroeck, M., Benedet, R.: Techniques for a selective encryption of uncompressed and compressed images. In: Advanced Concepts for Intelligent Vision Systems (ACIVS), Ghent, Belgium, pp. 90–97, September 2002

13. Sharma, S., Pateriya, P.K.: A study on different approaches of selective encryption technique. Int. J. Comput. Sci. Commun. Netw. 2(6), 658–662. ISSN 2249-5789

14. AnandhaJothi, R., Palanisamy, V.: Various attacks and countermeasures in mobile Ad Hoc networks: a survey. Int. J. Eng. Res. Technol. (IJERT) 3(33), 50–58. RACMS-2014 Conference Proceedings

15. Nithyapriya, J.: A comparative study on security schemes for wireless adhoc networks. Int. J. Adv. Sci. Tech. Res. 3(5), 23–28 (2015). http://www.rspublication.com/ijst/index.html. ISSN 2249-9954

16. Nithyapriya, J., Pazhanisamy, V.: Detecting malicious nodes in wireless ad hoc networks –a cryptography based approach. Int. J. Pure Appl. Math. 118(9), 13–20 (2018). ISSN: 1311-8080 (printed version); ISSN: 1314-3395 (on-line version), pp. 13–19

17. AnandhaJothi, R., Palanisamy, V.: Trust based association estimation technique on AODV protocol against packet droppers in MANET. Int. J. Appl. Eng. Res. 3(10), 2408–2413 (2015). ISSN 0973-4562. No. 55

Railway Surface Crack Detection System Based on Intensity Loss of UV Radiation

S. Aarthi[✉], Aravind Mohan[✉], P. Karthik[✉],
and Anurag Thakur[✉]

SRM Institute of Science and Technology, Chennai, India
Cse.aarthi@gmail.com, m.aravind619@gmail.com,
karthikpoludasu5@gmail.com, anuragthakur532@gmail.com

Abstract. The rail accidents here in India mainly happen due to occurrence of cracks in the tracks and the accidents increasing day by day. This paper suggests a feasible and effective solution to the problem of cracks occurring in rail tracks utilizing UV rays which detects the location of faulty tracks which helps in taking actions immediately so that many lives will be saved and avoid a railway accident. This project is implemented using Global system for mobile communication and Global positioning system receiver. Here UV rays are used to find the crack in the track, whenever crack is detected GPS receiver receives the location information. This information of detection of the crack is sent to the appropriate railway authorities by using the GSM.

Keywords: UV rays · IoT · Sensor · Crack detection

1 Introduction

The recent Indian railway network has a rail track length of 115,500 km and spread over an area of 67,313 km with 8200–8500 railway stations. Having one of the largest railway network in the world, Indian railway network still deals with lack of quality and safe infrastructure. Indian railway facilities are not as advanced as compared to the foreign standards and due to which, there have been frequent rail accidents that has resulted in massive loss of human lives. About 70% of all the rail accidents in India is due to derailments, recent survey revealed that about 92% are due to cracks on the rails. Hence the issue of occurrence of cracks in railway tracks have to be addressed with top priority and maximum attention given to the frequency of rail usage. These cracks often go undetected due to poor maintenance and human involvement in track line monitoring. The high bandwidth of trains and unreliability of human labor have increased the need for an automated crack detection system to detect the presence of crack on the railway lines. This paper proposes an accurate and a feasible solution for reducing railway accidents.

© Springer Nature Switzerland AG 2020
A. P. Pandian et al. (Eds.): ICCBI 2018, LNDECT 31, pp. 140–146, 2020.
https://doi.org/10.1007/978-3-030-24643-3_16

2   Railway Derailment

A rail derailment occurs when the train gets off its tracks, which does not axiomatically mean that the train leaves the track. Whilst most derailments are minor, they result in short term halt of the regular operation of the railway system and a major derailment can potentially cause serious damage to human lives. Most often, the derailment of the train is caused by a clash with some other object, a functioning error. However the most important reason for derailment is due to the automatic collapse of the tracks, such as fragmented rails, or the operational failure of the wheels. In extreme situations, wilful derailment with catch points is used to prevent a more severe accident (Fig. 1).

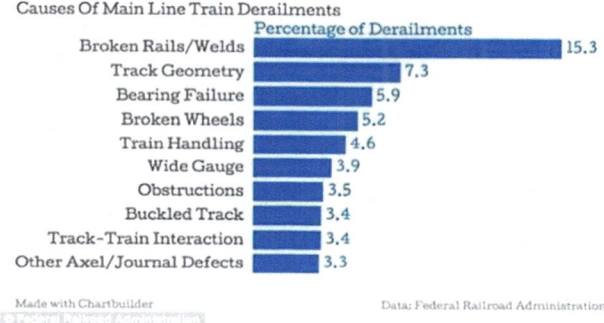

Fig. 1. Survey on causes of derailment

3   Literature Survey

3.1   Existing System

The track is being manually inspected daily by workers. The types of patrolling are Gang patrolling, night patrolling during the periods of july to september, weather patrolling, security patrolling, watchmen at damage prone places and weather patrolling. Gang patrolling is carried out manually on foot along the tracks during wet seasons. Security patrolling is done to prevent the train against change of state with the track. Weather patrolling is done once the rail temperature reaches twenty or higher than that in welded rail sections [1, 2]. IR sensors based system combines GSM and GPS modules and makes sure that there is no unmanned gate crossing and also it make sure of vibration sensors to detect faulty tracks [3]. Every year many animals die because of railway accidents. PIR sensors keeps away the living beings across the track [4]. The Ultrasonic based system consists of non-destructive testing and wireless sensor network i.e. NDT and WSN and keeps the continuous record of material with out mobility during run time [5]. RFID grids also used to detect cracks. Radio frequency identification system is used for adoption of a smart skin in the field of structural health monitoring (SHM). Ex-health monitoring in railway track [6, 7]. We can use LED/LDR based system also for detecting the crack but there is one draw back in it i.e. cost effective. This system is cost effective yet system is robust. On other hand CCD (Charged couple device) make use of DSP which executes the image processing, edge detection, image regeneration and edge linking to make sure that track is alright [8–10]. GPS (and GSM) modules are used for detecting the location of crack. GPS is a device that is capable of receiving information from GPS satellites and then to calculate the devices position. A can world in any part of the world and at any weather condition [11]. To improve the accuracy of rail crack detection, especially for the high speed condition, Empirical mode decomposition (EMD) is used. It works very efficiently even at high speed [12]. The microcontroller checks the vibration in the voltage of the measured value with the threshold value. If the observed value should be less than or equal to the threshold value. It controls all the operation happening. The system design of this system is cheap. The matter is the prices of all these designs are very high. So, the system proposed is quite less expensive than the existing system.

3.2   Proposed System

UV rays can be used to detect cracks in tracks in a more effective way than the existing systems. Instead of passing UV rays directly straight into the track we send it in the forward direction so that it can detect a crack before the train arrives. UV rays have high frequencies which can be used for our advantage. When the crack is detected, the sensors send the signal back to the ARM microcontroller after which through GSM which stands for global positioning system we find the location of the crack and using GPS which stands for global system for mobile communications the required officials are alerted so that appropriate actions can be taken and also the train is stopped. Our system can be more effective than the existing ones and also cost effective since UV light sources are inexpensive.

4   Electrical Design

4.1   Microcontroller

The microprocessor is the brain of this automated crack detecting system. An Arduino Uno board that has ATMega328P microcontroller is used to make the application more interactive. The Board has been chosen for two reasons. One, It has more hardware features like timers and it is open source therefore it will be easy for troubleshooting and debugging. Second, it is very power efficient and contains the specified pins to interface the peripherals. The careful description regarding which additional elements are interfaced with the Arduino is mentioned before.

4.2   UV Light Source

An UV light source is used for the detection of the crack. When the UV light is passed forward to the tracks the illumination spot from the UV light source illuminates the entire width of the weld at the position of the surface crack. Ultraviolet is electro-magnetic radiation with wavelength from 10 nm to 400 nm. The intensity of the UV light source passed onto the track is noted. UV light sources are inexpensive.

4.3   UV Sensor

UV sensors measure the power or intensity of incident ultraviolet (UV) radiation. UV sensors are transmitters that respond a type of signal by producing energy signals of another type. UV sensor in this system is used detecting the decrease in intensity due to the surface crack. Using UV rays have been proven to be a better option than IR rays.

4.4   GPS

A GPS module is used to find the location of the crack. GPS stands for global positioning system. When the sensor detects the cracks it sends the signal to the micro controller which in turn uses the GPS to find the exact position of the crack.

4.5   GSM

A GSM module is used to communicate with the required railway officials. It stands for global system for mobile communications. It helps in wireless transmission of information. It is proven to be faster so that action can be taken as soon as possible.

4.6   Formulas

$$\text{intensity} = P/a$$

P = power of the light source
a = area loss in intensity = i1 − i2
i1 = initial intensity of the rays passed on the track
i2 = intensity of the rays at the cracks.

5   Architecture

Lamp intensity is the overall power of the lamp and is most often designated in watts. Intensity refers to total lamp output across the entire electromagnetic spectrum. The intensity of the light source is fixed and is passed onto the tracks while train is running. If there is a crack present in the track when the UV ray passes over it there would be a loss of intensity. The loss in intensity can be found at by finding the difference between the initial intensity and the reduced intensity at the crack. When the UV sensor detects that there is a loss in the intensity it sends a signal to the micro controller which using GPS finds the location of the crack and the train is stopped immediately also through GSM alerts the appropriate professionals (Fig. 2).

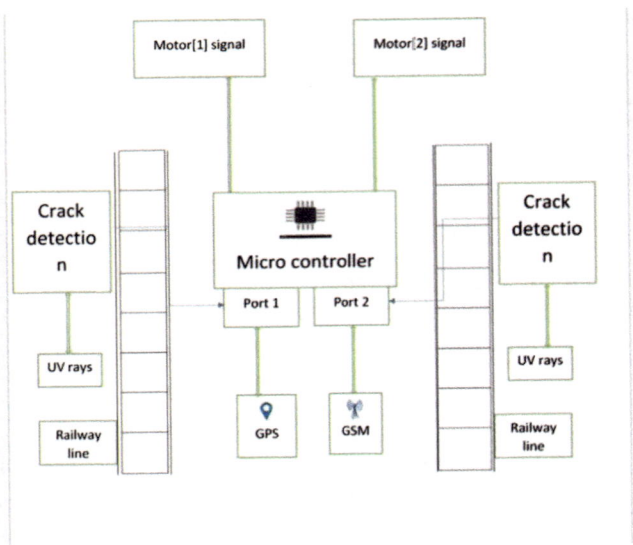

Fig. 2. System architecture

6   Decomposition

The different modules identified are:

1. passing UV rays from the UV light source to the tracks.
2. detection of cracks through the loss of intensity sensed by the UV sensor and sending the signal back to the microcontroller.
3. finding the position of the crack using GPS and alerting the officials using GSM (Fig. 3).

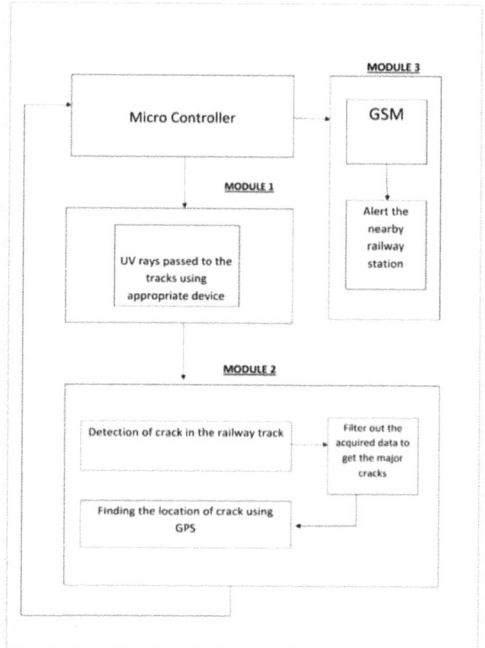

Fig. 3. Module diagram

7   Conclusion

Railway crack detection using ultraviolet rays is an effective way to detect the presence crack in the railway track. The detection of crack information is passed to the railway safety management center through GSM Module and the alert can be given to the approaching train. The proposed system has many advantages such as accurate monitoring and fast reporting system, feasible, low power consumption and low analysis time. Also the cheap and easy availability of the components make this an ideal project for implementation with very little initial investment. However the disadvantages of the projects may include the harmful effects of UV radiations which actually won't have any effects on living beings since the intensity of the radiation used is too low. The current location device on rail track can easily be measured from home station. By this proposed system many lives can be saved by avoiding accidents. The idea can be implemented in large scale on the long run for better safety standards for rail tracks.

8   Future Scope

Augmented reality can be included along with the proposed system so that the location of the track can be located much easier. Introducing AR would help us to visualize the location in a much better way than to just look at a pointer in that location.

References

1. Lad, P., Pawar, M.: Unmanned level crossing controller and railway track broken detection system using IOT and IR
2. Kalpana, S., Saurabh, M., Khanna, V.: Railway track breakage detect or method. In: International Conference (2014)
3. Evolution of railway track crack detection system
4. http://en.wikipedia.org/wiki/Wireless_Sensor_Network
5. Zhang, J., Tian, G.: An evolution of RFID grids for crack detection
6. Murali, V., Saha, G., Vaidehi, V.: Robust railway crack detection scheme using LED_LDR
7. Len, L., Hua, J., Zing, Z.: Design of rail surface crack detecting system based on linear CCD sensor. In: International Conference (2008)
8. Autonomous railway crack detector robot. IEEE Region (2017)
9. Carlon, J., Cristina Cralyon, M.J.: Automatic online Fatigue crack detection in railway axels
10. http://en.wikipedia.org/wiki/GSM
11. Railway crack detection based on the adaptive cancellation method of EMR at high speed. In: International Conference (2017)
12. Shankar, R., Shekhar, P., Ganesan, P.: Automatic detection of squats in railway track
13. https://eo.wikipedia.org/wiki/Global_Positioning_System
14. http://www.albatrossuv.com/UV-Lamp-Intensity.php

Urban IoT Implementation for Smart Cities

Aparna Menon[✉] and Rejo Mathew

Department of IT, Mukesh Patel School of Technology Management
and Engineering, NMIMS (Deemed-to-be) University, Mumbai, India
appurni@gmail.com, rejo.mathew@nmims.edu

Abstract. Internet of Things (IoT) allows the collection of large amounts of data that can be integrated into systems and used for development of smart cities in an efficient manner. This paper focuses on the application domain of urbanized IoT system. Application of urban IoT advanced communication technology is reviewed for its efficiency to support the smart city vision by providing services for administration of smart city as well as for ease of use of its citizens. A proposal of recommended architecture, technologies and devices are given to support a model of facilities that can be materialized to create a functional 'smart' city.

Keywords: CoAP · EXI · 6LoWPAN · Network architecture ·
Sensor system integration · Service functions and management · Smart cities

1 Introduction

The paper proposes use of urban IoT technology for development of smart cities. On an abstract level, this can be done by improving the QoS provided to the citizens, all the while cutting costs involved in public administration. This will result in better use of public resources. Urban IoT will support efficient optimal use of resources for advanced outcomes. The implementation of Urban IoT will be done with the use of sensors that collect data. This data will be stored and processed in a control center to understand the problem areas and requirements. Then this information will be used to formulate and propose solutions [1].

2 Smart City Concept and Service

This section overviews the goals of implementing urban IoT in smart cities. It gives an overview of all the possible services and systems can be achieved and implemented using information gathered and processed by the sensors. This outline differentiates what gives the smart city the 'smart' factor in it.

Following are the proposed models to be implemented as per the paper [12].

© Springer Nature Switzerland AG 2020
A. P. Pandian et al. (Eds.): ICCBI 2018, LNDECT 31, pp. 147–155, 2020.
https://doi.org/10.1007/978-3-030-24643-3_17

2.1 Air Quality

We can constantly evaluate air quality in crowded areas, parks etc. thanks to the sensors implemented by IoT. This information can be used by fitness applications and devices by connecting them to the infrastructure. Individuals can locate techniques that are more beneficial for outdoor activities and be associated with their decision of preparing application utilizing this component.

2.2 Noise Monitoring

A noise monitoring service implemented by IoT can be used to create a live noise level tracker which can be used at any given point of time for places that are using this particular service.

It can fabricate a sonic guide of clamor dissemination in the zone. Detecting brawls, emergency situations and alerts (like crashing of glass etc.) are a part of another excellent factor this feature can provide.

2.3 Traffic Control

The traffic sensors in GPS devices installed in all vehicles and mobiles devices can give live information regarding traffic densities in all areas. Acoustic sensors can help traffic monitoring too. Smart services can be enabled by using the information provided to predict and compute time-route maps. A denser source of information can be provided by low power widespread communication.

2.4 City Energy Consumption

The city-wide energy consumption can be monitored using IoT. Both governing body and citizen can get a comprehensive estimate of the power requirement of varied services. Monitoring devices must be integrated and form a channel within city power grids.

2.5 Smart Parking

Road sensors and contextual displays can provide smart parking services that direct motorists along routes with higher parking chance. This intelligent feature can help find a spot quicker, a spot having less CO emanations from the vehicles in the city, lesser movement clog and enhanced well-being of the people living in these cities. With the assistance of Radio Frequency Identifiers (RFID) or Near Field Communication (NFC), an electronic validation system for classifying handicapped and regular parking spots can be achieved.

2.6 Smart Lighting

IoT can be used to vary lamp luminosity by time, by weather conditions, and the presence of people. Such a service would naturally require the integration of city lamps to the central IoT mainframe.

2.7 Automation and Salubrity of Public Lighting

Use of IoT sensors can be used to control and optimize the levels of light intensity, temperature, humidity so that the level of comfort can be enhanced for better environment to live in.

3 Urban IoT Architecture

Smart cities IoT grids are centralized, with a dense network of peripheral devices being deployed over the city. These devices generate data which is received and processed by a control center through different communication technologies. Data is stored and processed in data centers.

The following are the characteristics of urban IoT infrastructure:

(1) It is capable of integrating the diverse technological stacks with existing infrastructure to support interconnection of different gadgets and acknowledgment of administrations and functionalities with IoT.
(2) It makes data collected accessible to authorities and gentile alike, improving agility and increasing awareness of city problems [9].

Components of the urban IoT system are:

• Web Service approach for Service Architecture
• Link Layer Technologies
• Devices.

3.1 Web Service Approach for IoT Architecture

This section is focused on IEFT standards. IEFT standards use a web service architecture which is very flexible. Interoperability is another great feature that is guaranteed with the goal that the framework can be reached out to IoT hubs through an online design known as Representational State Transfer (ReST) [10] (Fig. 1).

The protocol architecture stack is composed of unconstrained and constrained protocol stack.

XML, HTTP & IPv4 are few Internet communication standards used right now. The are also the constituents of the unconstrained protocol stack. Efficient XML Interchange (EXI), Constrained Application Protocol (COAP) and 6LoWPAN are the constituents of constrained protocol stack (Fig. 2).

This architecture is further divided into three functional layers:

3.1.1 Data Format

As per requirements for urban IoT, XML is the preferred method for data exchange. But XML is not feasible for use in constrained devices due to its size and complexity. Hence World Wide Web Consortium (WWWC) has put forth the EXI format [2].

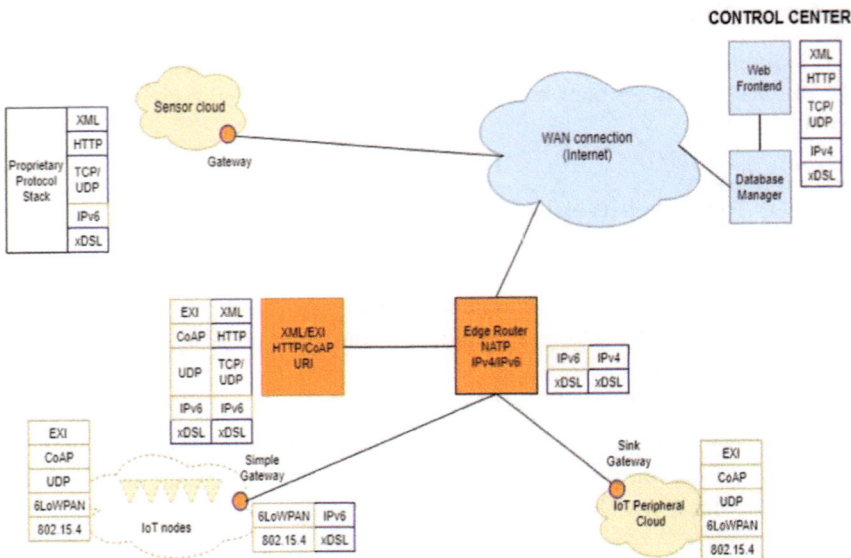

Fig. 1. Web service approach

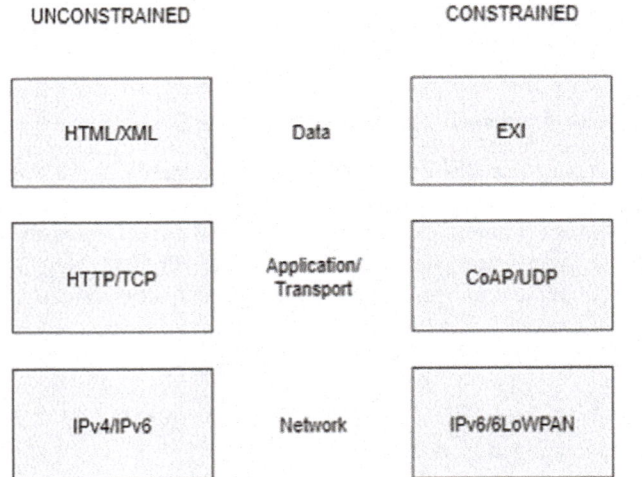

Fig. 2. Protocol stacks

The EXI format defines two types of encoding:

(1) *Schema-less encoding* – It is produced in a straightforward manner from the XML information and can be decoded by any EXI entity with no preset construct for parsing.

(2) *Schema informed encoding* – An XML schema is shared between two EXI devices prior to encoding and decoding [3].

3.1.2 Application/Transport

Although most traffic is routed through the application layer using HTTP:TCP, it is highly inefficient in low processing power devices due to its complexity and poor performance for small data flow. The solution to this is CoAP protocol [4] which can overcome these issues using a binary transmission schema over UDP. As the constrained Application Protocol has one-to-one correspondence with response codes of HTTP, underpins an extensive variety of HTTP use situations besides Representational State Transfer methods of HTTP such as GET, PUT, POST and DELETE, being interoperable with HTTP.

This is achieved by deployment of an intermediary called cross proxy which translates requests/responses between two protocols [4].

3.1.3 Network

Iot networks are projected to include nodes numbering in billions, that will be uniquely addressable. Ipv4 is the tending to innovation that is required and is generally utilized by Internet has. IANA, the International organization that allocates IP addresses on a worldwide level has reported the exhaustion of IPv4 address blocks. This issue can be settled by 6LoWPAN [7], a set up pressure organize for IPv6 and UDP headers over handling power tied networks [6].

The interaction of IoT and IPv4-only hosts is an issue as 6LoWPAN border router only enables the interaction of IPv6 hosts and IoT nodes [7]. The main issue is basically trying to figure out how a particular IPv6 host utilizing IPv4 address as well as meta-data accessible in the packet:

These are distinctive answers for this issue as presented in the paper:

(a) *v4/v6 Port address translation (PAT)*

This technique maps subjective sets of IPv4 addresses and TCP/UDP ports into IPv6 addresses and TCP/UDP ports [5].

This technique can be utilized to outline IPv6 addresses into a solitary IPv4 open location, which permits the sending of the datagrams in the IPv4 system and its right administration at IPv4-only hosts. The use of this method requires low Complexity. Be that as it may, this methodology raises an adaptability issue, since the quantity of IPv6 has that can be multiplexed into a solitary IPv4 address is restricted by the quantity of accessible TCP/UDP ports [5].

(b) *v4/v6 Domain Name Conversion*

This method allows for the programming of DNS servers such that, when pinged for, a DNS requests for tuning in for the domain name of an IoT web service: The DNS responds with an IPv4 CoAP cross proxy address which is then contacted for access to the IoT node. The proxy then resolves the domain name acquired in the HTTP packet, which responds with the final IPv6 address that will identify the terminal node. Using the CoAP bridge, the proxy then forwards the message to the intended IoT device [4].

(c) *URL mapping*

The Universal Resource Identifier (URI) mapping strategy includes a variety of the HTTP-CoAP cross intermediary [4], the turn around cross intermediary. This intermediary goes about as the last web server to the HTTP/IPv4 customer and as the first customer to the CoAP/IPv6 web server. The machine's positioning within the network requires IPv6 connectivity, allowing direct connectivity to the IoT device and IPv4/IPv6 conversion [5] is handled by the internal URI mapping function.

3.2 Link Layer Technologies

An urban IoT system requires link layer technologies scalable through both geographical area and traffic volume, generated as a consequence of a high number of smaller data flows coalescing.

There are two types of link layer technologies:

3.2.1 Unconstrained Technologies
Consists of customary LAN, MAN and WAN correspondence innovations, for example, Ethernet, Wi-Fi, fiber optic, broadband, Power Line Communication (PLC), and cell advances, for example, UMTS and LTE.

Their prime characteristics include high reliability, greatly reduced latency and larger bandwidth. Due to their complex architecture and significant energy consumption, they have to be placed centrally in the network, away from peripheral nodes [11].

3.2.2 Constrained Technologies
These are grouped by low power activity and lower exchange rates, normally lesser than 1 Mbit/s. The noticeable arrangements inside this class incorporate IEEE 802.15.4, Bluetooth, PLC, NFC and RFID [8]. These links often suffer from high latency as their speed is impeded by the physical layer (low transmission rates, byte corruption) and the power regulation protocols implemented by the IoT nodes to efficiently manage power.

3.3 Devices

This section discussed the devices that are the heart and soul of the project. These devices need to be used in order to implement urban IoT efficiently. These devices are classified based what purpose they serve in the flow of communication.

3.3.1 Backend Servers
Backend servers, situated in the control focus, is the place information are gathered, put away, and handled to create included esteem administrations. Backend servers turn into a principal segment of a urban IoT where they can encourage the entrance to the keen city administrations and open information through the heritage organize foundation.

Backend systems that are often used for interfacing with the data feeders belonging to IoT include:

(a) *Database management systems:* These systems are accountable for putting away the huge measure of data created by IoT peripheral nodes (sensors).
(b) *Web sites:* Web interfaces are the main choice to empower interoperation between the IoT system and the clients because of its boundless use among individuals specialists, benefit administrators, utility suppliers, and regular subjects.
(c) *Enterprise resource planning systems (ERP):* ERP parts intermediary different business works and are praiseworthy apparatuses to oversee data stream crosswise over multi-layered associations e.g. city organization. Interfacing ERP parts with DBMS' that gather IoT information dumps take into consideration more noteworthy deliberation in capacities taking care of the stream of information.

3.3.2   Gateways

Gateways interconnect the end gadgets to the fundamental correspondence foundation of the framework. A gateway is required to give convention interpretation and useful mapping between the unconstrained conventions and their obliged partners (XML-EXI, HTTP-CoAP, IPv4/v6-6LoWPAN.)

3.3.3   IoT Peripheral Nodes

IoT peripheral nodes are the gadgets accountable for creating the information to be conveyed to the control focus. These devices are quite affordable. That is because it depends on what kind of sensors we use and it also depends on the number of sensors that are being mounted on the boards being used. The nodes of IoT have a several criteria based on which they are classified. This should be possible based on powering mode, networking role, sensor/actuator hardware, and upheld link layer technologies). RFtags are quite limited, but the most ubiquitous, part of the IoT architecture in an urban environment [11]. It is a passive device and does not require power sources, activating only when an external field imparts potential. They are mainly used for object identification, but on an abstract level, can be used for logging several kinds of identities. The scope for mobile applications to ease the friction of technology that exists in interacting with IoT objects is immense and can be a viable sub industry. Mobile devices can constitute a critical part of urban IoT, allowing the owner to interact with the environment in a variety of ways,

1. through IP connections provided by the ISP.
2. by setting up a connection which is direct with some objects. This should be possible by utilizing shortrange wireless technologies, The wireless technologies with a short range presently accessible for utilization are Bluetooth Low Energy, low-power Wi-Fi, or IEEE 802.15.4 [8].

4   Problem-Solution Review

See Table 1.

Table 1. Review of discussed issues and solutions

Feature/method	Unconstrained	Constrained
Data format	XML messages – too large and complex to process	Use of EXI format
Data transport	HTTP – poor performance for small data flow	Use of CoAP Protocol – binary format over UDP
Node addressing	Exhaustion of IPv4 address blocks	Use of 6LoWPAN compression format for IPv6 headers
IP address mapping	Interaction between IoT nodes and IPv4 hosts	v4/v6 Port address translation
Network layer technology	Scalability issue of v4/v6 PAT due to limited TCP/UDP ports	v4/v6 Domain name conversion
Link layer technologies	Complexity and high energy consumption of traditional communication technologies. (Ethernet, Fiber Optic, PLC)	Use of technologies with low energy consumption and small transfer rate (Bluetooth, NFC, RFID)

5   Conclusion and Future Work

The solutions that are available for implementation of urban IoTs today have been studied and analyzed. New technologies and solutions have been proposed, discussed, compared and are close to being standardized. Compatible set of open standard protocols that agree with current design options are suggested. Future work includes trial and testing of proposed models in several developing townships to work on architectural plan with real time understanding of needs and implementation of technology in a more efficient way based on requirements and implementation.

References

1. Zanella, A., Bui, N., Castellani, A., Vangelista, L., Zorzi, M.: Internet of things for Smart Cities. IEEE Internet Things J. **1**(1), 22–32 (2014)
2. Schneider, J., Kamiya, T., Peintner, D., Kyusakov, R.: Efficient XML Interchange (EXI) Format 1.0, 2nd edn. World Wide Web Consortium, 11 February 2014
3. Castellani, A.P., Bui, N., Casari, P., Rossi, M., Shelby, Z., Zorzi, M.: Architecture and protocols for the Internet of Things: a case study. In: Proceedings of 8th IEEE International Conference on Pervasive Computing and Communications Workshops (PERCOM Workshops), pp. 678–683 (2010)
4. Shelby, Z., Hartke, K., Bormann, C., Frank, B.: Constrained application protocol (CoAP), draft-ietf-core-coap-18. s.l.: IETF (2013, work in progress)
5. Castellani, A., Loreto, S., Rahman, A., Fossati, T., Dijk, E.: Best practices for HTTPCoAP mapping implementation, draft-castellanicore-httpmapping-07, s.l.: IETF (2013, work in progress)
6. Deering, S., Hinden, R.: Internet Protocol, Version 6 (IPv6) Specification, RFC2460, s.l.: IETF, December 1998

7. Hui, J., Thubert, P.: Compression format for IPv6 datagrams over IEEE 802.15.4-Based Networks, RFC6282, s.l.: IETF, September 2011
8. IEEE Standard for Local and Metropolitan Area Networks—Part 15.4: Low-Rate Wireless Personal Area Networks (LR-WPANs), IEEE Standard 802.15.4-2011
9. Mulligan, C.E.A., Olsson, M.: Architectural implications of smart city business models: an evolutionary perspective. IEEE Commun. Mag. **51**(6), 80–85 (2013)
10. Fielding, R.T.: Architectural styles and the design of network-based software architectures. (The Representational State Transfer (REST)) Ph.D. dissertation, pp. 76–85, Department of Computer Science University of California, Irvine (2000). http://www.ics.uci.edu/ ~ fielding/pubs/dissertation/top.htm
11. ISO/IEC 14443-1:2008: Identification Cards—Contactless Integrated Circuit Cards—Proximity Cards—Part 1: Physical Characteristics. http://www.wg8.de/wg8n1716_17n3994_Notification_for_Ballot_FDIS_14443-1_2008_FDAM1.pdf
12. Dohler, M., Vilajosana, I., Vilajosana, X., Llosa, J.: Smart cities: an action plan. In: Proceedings of Barcelona Smart Cities Congress, Barcelona, Spain, pp. 1–6 (2011)

Review of Cloud-Based Natural Language Processing Services and Tools for Chatbots

Ayan Ray[(⊠)] and Rejo Mathew

Department of IT, Mukesh Patel School of Technology Management
and Engineering, NMIMS Deemed-to-be University, Mumbai, India
mailayanray@gmail.com, rejo.mathew@nmims.edu

Abstract. Natural language processing (NLP) is a major topic under research
and development since the past decade. The ability to analyze, compute and
understand plain human language with the help of computers and machine
learning has been inculcated into various applications and platforms. The advent
of virtual assistants in consumer technology has made major companies work on
their own cloud based platforms. This paper discusses in detail how chatbots
primarily use parameters such as entities and intents to match syntaxes and
performing natural language processing. Every platform has certain services and
integrations and predefined data sets for training the bots which makes them
have certain advantages over each other. This paper compares the descriptive
feature of the major four virtual assistants available today- Google DialogFlow,
Microsoft LUIS, IBM Watson Conversation and Amazon Lex. This paper
consists of four sections: Introduction, Natural Language Understanding Platforms, NLP Taxonomy, Feature Analysis, Future Work and Conclusion.

Keywords: Natural language understanding · NLP keywords · Chatbot ·
Virtual assistant

1 Introduction

Computational linguistics [1] has improved so much in scale and the speed in the based
decade, that it is now in the state to be incorporated into consumer based products.
Some of the fundamental factors which made these possible are:

(i)   The exponential increase in computer processing power
(ii) Massive language datasets available
(iii) Development and research of machine learning algorithms
(iv) Research and development in understanding the semantics of human language
      and conversation.

A virtual assistant is a software, which is designed to imitate a natural conversation.
The way a chatbot [2] or a virtual assistant function works is by recognizing the user
input or utterance, by pattern matching [3]. This pattern matching occurs due to the fact
the software is trained to recognize these sentences using machine learning methods.
After matching it to a pattern, it checks the response that is set for that pattern and gives
the output. The major difference from classical programming is the fact that the chatbot

© Springer Nature Switzerland AG 2020
A. P. Pandian et al. (Eds.): ICCBI 2018, LNDECT 31, pp. 156–162, 2020.
https://doi.org/10.1007/978-3-030-24643-3_18

can understand a context of a sentence. This means that the input and output of the first question will be carried forward, and be matched with the subsequent questions, just like a natural conversation. Although this is a major leap in the computation of natural language processing, chatbots still cannot process complex questions and perform compound activities. Since, chatbots require a massive amount of computational power, hence the consumer based software is offered as a service by the major tech giants such as Google, Microsoft, Facebook, and IBM. These companies do the computation on their servers, and offer the software as a cloud-based service, on their platform.

This paper makes a descriptive analysis of the major natural language processing platforms offered as cloud-based services. This paper explains the fundamental concept of natural language processing and then goes on the brief description of each of the platform. Further, the paper compares [4] the features of the four major NLP platform. The paper concludes by saying which platform is the most useful as of today and what is about to come in the future.

2   Natural Language Processing Platforms

Natural Language Processing is a technique for analysing speech and human language. With the massive development in the field in the past couple of years and the development of cloud computing, many IT companies have developed natural language processing platforms. The cloud-based platforms primarily rely on two concepts for its functioning: intents and entities [5]. Anything that a user enters in the form of speech or text is called as an utterance. Intent is the action what the virtual assistant does after the input by the user. Entity is the object through which the platform extracts the parameter value which actually needs to be checked. The data which needs to be inputted from a conversational statement is extracted using entity. For example, the sentence "What is the time in Mumbai?" (utterance) is the input statement for asking the time, in which "Mumbai" (intent) is could be part of any possible variable of 'city'(entity), that could represent the name of any place/city in the world. The advantage of using such a machine learning dependent tool is that these parameters of intents and entities could be pre-defined and reused (pre-built intents and pre-built entities respectively), using the training mechanisms of the bot. In the remaining section, the primary cloud-based services provided by some major virtual assistants are compared.

2.1   Google's DialogFlow

DialogFlow [6] formerly known as Api.ai or speaktoit was acquired by Google in 2016. DialogFlow is well known for integrating its virtual assistant interface with Google Assistant, which could be used on a multitude of integrated platforms. DialogFlow uses machine learning algorithms that continuously trains the model, and uses intents, entities, parameters and all the features which make it a very viable conversational chatbot. It has predefined entities and intents to handle common topics like weather and small talk. The conversational chatbot can be built using its web interface or the server application using webhook mechanism provided by its APIs. Different languages could be used in a single intent case.

DialogFlow supports various programming languages like JavaScript, Node.js, Xamarin and C++ among many. It has built-in integration with apps such as Google Assistant, Slack and Facebook Messenger. Few of the SDKs that are supported are Android, iOS, Apache Cordova. There's support for 20+ languages such as English, Dutch, German, French, Italian, Chinese (Cantonese and Simplified). It is free for the standard version, which only has limitations with speech recognition and calling features. API calls are fundamental for the working of any NLP engine, and DialogFlow has no limit for the API calls. Having an easy learning curve with all these features and platform integrations, it works best with virtual assistants and is used as MVPs for small-scale and medium-scale businesses.

2.2 Microsoft LUIS

LUIS [7] or Language Understanding Intelligent Service is a cloud-based API service provided by Microsoft and is a part of the Azure Cloud services [8]. LUIS like Google's DialogFlow is also built on the notion of intents and entities. One primary feature of LUIS is the uses a method termed as 'active learning technology'. Using this method, the bot continuously learns from its usage. The pre-built domain model of LUIS includes intents, utterances with prebuilt entities. LUIS has a great support for various programming languages and but supports 4 SDKs, which are Android SDK, Node JS SDK, Python SDK, and the C# SDK. With the support of various platforms such as Slack, Telegram, Twilio, Skype [9], GroupMe, Email, Facebook and many more, it can easily be integrated, for multiple conversational platforms. LUIS has a support for other languages such as English, Italian, German, Spanish, Brazilian Portuguese, Korean and Chinese. The advantage of LUIS is that it can use predefined models from Bing Search and Cortana. The free plans include 10,000 transactions per month and five queries per second, while the paid plan gives 10,000 transactions for free and then charges $0.75 per 1,000 transactions and up to 10 queries per second. It can be integrated with Microsoft Azure and other messengers via Bot Framework and can share the pricing. Microsoft LUIS is best supported for integration with Cortana, virtual assistant enterprise integrations and IoT applications.

2.3 IBM Watson Conversation

IBM has its own cloud services platform called as IBM Bluemix, which has a NLU interface called as Watson Conversation [10]. Watson can work on a lot of varied datasets such as art, healthcare, weather forecasting, teaching aid, etc. It is designed to interact with all forms of data, people and varied datasets and can take advantage of the Bluemix ecosystem of services. One of the dataset that Watson was trained on was the text structure of Wikipedia (2011 edition). The computer system was initially developed to answer questions on a quiz show. Recent developments of Watson are primarily into healthcare, and it gives medical advice to the nurses and doctors with the immense data it uses to analyze the entry.

Watson supports a lot of SDKs including Java SDK, Node SDK, Python SDK and many more. Watson provides three payment models- Lite (1000 queries/month), Standard ($0.0025 per API call) and Premium (Pricing on request). 1000 API queries per month could be made on Lite. Standard plan offers unlimited API queries per month, up to 20 workspaces and 2000 intents. The premium plan is could be customized for the requirements of the client. Watson Conversation works best when it is integrated with IBM Bluemix Services, as a virtual assistant.

2.4   Amazon Lex

Amazon Lex [11] is a cloud-based service provided by Amazon Web Services for building conversational assistant using natural language understanding using voice and text. Using this service, a developer has the same tools on which Amazon's own virtual assistant- Alexa is built on. Cloud-based computation is done using deep learning functionalities of automatic speech recognition for understanding speech and then converting it into text. This text is later analyzed by the chatbot to search for intents and entities. Amazon Lex has support for a lot of programming languages such as Python, Java, JavaScript, C++, PHP, Ruby, Go. It also has iOS and Android SDK integrations. The cloud-based platform could be integrated with Facebook, Slack or Twilio SMS. One disadvantage is that Amazon Lex is limited to US English, as of now. Lex offers a trial period for a year, post that there is a charge of $0.004 per speech query and $0.00075 per text query. The trial period though, has a limit of 10,000 speech queries and 5,000 text queries, while the paid version offers unlimited queries.

3   Natural Language Understanding Keywords

The following are some fundamental features of the NLP platforms

1. **Usability** is the indication of the ease of use of the platform. This is loosely determined by the learning curve and the user-interface design of the platform.
2. **All-in Platform** mentions whether the cloud-based tool is able to provide all the natural language processing based services by its own. It also states whether it is a part of a family of web services or not.
3. **Linkable Intents** defines whether the intents on the platform could be directly linked with each other.
4. **Languages** define the number of natural languages that the cloud-based service support. Sometimes, it also mentions whether a single intent could have multiple languages.
5. **Programming Languages** for any platform define the programming languages that are supported by the platform.
6. **Default Fallback Intent** states the intent which could be set using the methods the platform provides for any fallback mechanism. This method is called when the utterance is not matched to any defined intents.
7. **Online Integration** states all the third-party integrations that are possible in a particular NLP platform.
8. **Webhook/SDK Availability** indicates whether and how a developer can integrate his/her conversational assistant with other software, independently from the available online integrations.

9. **Automatic Context** mentions whether the particular NLP platform is able to automatically manage the context in an input or is transferred to the code, written by the developer.
10. **Composition Mode** shows which are the composition modality adopted by the tools.
11. **Pre-build Entities** state the number of predefined entities that the platform provides.
12. **Pre-build Intents** state the number of pre-built groups of intents the platform offers.

4   Feature Analysis

As detailed in Table 1, most of the analyzed NLU platforms supports various languages- programming and human languages. This shows the usability of the platform as well as the ease of development. They mainly differ instead, in the automatic

Table 1. Comparison table of features

Features | Platforms	DialogFlow [6]	LUIS [7]	Watson Conversation [10]	Amazon Lex [11]
Usability	High	Medium	High	Low
Prebuilt Entities	60	13	7	93
Prebuilt Intents	34	20	0	15
Default Fallback Intent	Available	Available	Available	Not available
Automatic Context	Available	Not available	Available	Available
Composition Mode	Form	Form	Form + Block	Form
Online Integrations	14	0	0	3
Webbook/SDK Support	Webhook + SDK	Webhook + SDK	SDKs	SDKs
Languages	15	10	12	1
Programming Languages	11	4	6	9
All-in Platform	Yes	No (Part of Azure Cloud Services)	No (Part of Bluemix Cloud Services)	No (Part of Amazon Web Services)
Linkable Intents	Yes	No	Yes	Yes
Price(Basic)	Free	Free upto 10k requests per month	Free upto 10k requests per month	First year free (with limitations)

features that pertain to NLU engine: default fallback intent, automatic context handling, and linkable intents. These three aspects are play a fundamental role for a developer while making the conversational chatbot. Also, it's very important to consider for which task or in which language the chatbot should be deployed before choosing any platform. From a descriptive point of view, DialogFlow is the most complete NLU platform since it provides the best solutions for the majority of facets considered in the taxonomy.

5 Conclusion and Future Work

Natural Language Processing platforms provide what is contextually called as a virtual assistant. These virtual assistants or chatbots are programmed using machine learning algorithms. The primary goal of these tools is to mimic a natural human conversation. Major cloud computing platforms that exists today are compared in this paper Different platforms have been developed using extremely different datasets which makes them very unique by themselves.

This field is still in a very early stage of development. Businesses are experimenting with it, such as the banking sector and various other industries are using chatbots as the first point of contact in the online support system. One major advantage of these tools is that they can be trained from the errors they make, and hence they are continuously improving. Amount of data being fed to the system is increasing day by day, which is only making these platforms better. Understanding human language is a very complicated feat since there are underlying layers of logic in a conversation, but with virtual assistants being implemented in consumer technology, human-computer conversations will only get better.

These comparisons will soon be outdated with the pace of progress. The developments will increase with the introduction of many voice assisted with consumer devices such as Google Home [12], Amazon Alexa. The interaction with computers using voice is being introduced since a last few years. With the power of cloud technologies there will be new tools and new methods in the market, which will improve the way humans interact with computers.

References

1. Hirschberg, J., Manning, C.D.: Advances in natural language processing. Science **349**, 261–266 (2015)
2. Winograd, T.: Understanding Natural Language. Academic Press, Inc., Orlando (1972)
3. McTear, M., Callejas, Z., Griol, D.: The Conversational Interface, vol. 10, pp. 978–3. Springer, New York (2016)
4. Abdul-kader, S.A., Woods, J.: Survey on chatbot design techniques in speech conversation systems. Int. J. Adv. Comput. Sci. Appl. **6**(7), 72–80 (2015)
5. Kehoe, B., Patil, S., Abbeel, P., Goldberg, K.: A survey of research on cloud robotics and automation. IEEE Trans. Autom. Sci. Eng. **12**(2), 398–409 (2015)
6. The DialogFlow Project. https://dialogflow.com
7. The LUIS Project. https://luis.ai

8. Amit, P., Marimuthu, K., Nagaraja, R.A., Niranchana, R.: Comparative study of cloud platforms to develop a chatbot. Int. J. Eng. Technol. **6**(3), 57–61 (2017)
9. Lewis, W.D.: Skype translator: breaking down language and hearing barriers. In: Translating and the Computer (TC37), vol. 10, pp. 125–149 (2015)
10. High, R.: The Era of Cognitive Systems: An Inside Look at IBM Watson and How It Works IBM Corporation. Redbooks, New York (2012)
11. Amazon Lex. https://aws.amazon.com/lex
12. Nijholt, A.: Google home: experience, support and re-experience of social home activities. Inf. Sci. **178**(3), 612–630 (2008)

Weighted Trilateration and Centroid Combined Indoor Localization in Wifi Based Sensor Network

T. Rajasundari[1(✉)], A. Balaji Ganesh[1], A. HariPrakash[1], V. Ramji[1], and A. Lakshmi Sangeetha[2]

[1] Electronic System Design Laboratory, Velammal Engineering College, Chennai, India
trajigopi@gmail.com, harish091198@gmail.com, ramjivijayv@gmail.com, abganesh@velammal.edu.in
[2] Department of Electronics and Instrumentation, Velammal Engineering College, Chennai, India
lakshmisangeetha@velammal.edu.in

Abstract. The wireless communication protocol, wifi based location mapping of objects and also people is considered as a significant requirement to establish a smart environment, for an instance a multistoried shopping mall. The study combines two known algorithms, such as minimum weighted trilateration and centroid in order to enhance the tracking ability of individual technique. The position of a target user is estimated on the basis of Received Signal Strength (RSS) that are collected from Access Points (AP) deployed in various fixed locations without advent of additional hardware resources. The study shows the improvement in terms of accuracy in an average of less than 1 m even for the distance of 12 m and the results are compared with other known techniques. The efficacy of algorithm is evaluated by using MATLAB simulation environment as well as real-time hardware resources.

Keywords: Wifi · Indoor positioning · Weighted RSSI trilateration · Centroid localization

1 Introduction

The Indoor location tracking based research is emerging with the advancement in the wireless sensor network (WSN). The ability of processing information and also managing the network activities has become more intelligent as the nodes are so smaller and powerful. In a real time scenario, various challenges have facilitated many interesting aspects in typical WSN. The indoor positioning field has been experiencing a very notable progress and finds the applications in commercial, industrial, agriculture, medical and the military sectors. Various sensors, including GPS receivers, accelerometers, gyroscopes, magnetometers, cameras and also wireless communication interfaces, wifi and Bluetooth within almost all smartphones to facilitate communication, entertainment, and location-based services [1–3]. In general, smartphone users understand their locations with the functioning of GPS receiver.

© Springer Nature Switzerland AG 2020
A. P. Pandian et al. (Eds.): ICCBI 2018, LNDECT 31, pp. 163–172, 2020.
https://doi.org/10.1007/978-3-030-24643-3_19

The localization is carried out by various techniques that commonly include infrared, ultrasonic, radio frequency signal, Bluetooth and Ultra-Wideband, and wifi [4, 5]. Wifi protocol is found to be more suitable communication especially in places where the wired connections are impossible to set up. Wifi network resource localization technique with IEEE 802.11.x standard for an indoor environment comprises of one or more access points (AP) on one end and is connected wirelessly with the several nodes on the other end. These APs can be seen between the end users by involving the activities rendered from small base stations. The received signal strength indicator is used to quantify the strength of signals received from access points deployed at discrete locations in the environment. It considers propagation-loss of wifi signals in order to compute the distance, indirectly. In the literature, more sophisticated algorithms, such as trilateration, triangulation, proximity and fingerprinting were found [6–11]. The baseline for these algorithms is to decipher the signal properties into distance and orientation of a target object. In this study, we demonstrate the enhancement of these distance measuring techniques by combining the weighted trilateration and centroid calculations.

2   Algorithms for the Estimation of Distance and Orientation

2.1   Trilateration

Trilateration utilizes the geometric properties of triangles to evaluate the position of a target object. Distance estimation is done in respect to neighboring three access points. The signal strength from access point results three circles that intersect each other. The intersecting point of three circles is considered as the position, i.e., x and y coordinates of a target user. The intersection circle equations are formulated in order to simplify the estimation that created a Cartesian plane. The Eq. (1) specifies these circles by assuming, the coordinate z, as 0. The intersection of three circles, x and y coordinates is acquired by solving systems of linear equation for two variables, simultaneously [12].

$$(x - x_i)^2 + (y - y_i)^2 = r_i^2 \tag{1}$$

2.2   Triangulation Estimation

The distance between wifi access point's and mobile phone position is calculated by considering the parameters, such as the time of arrival (TOA), angle of arrival (AOA) and received signal strength (RSS).

Triangulation estimation is a trigonometric approach of deciding an obscure area in light of two angles and a separation between them. In a typical sensor network, two reference nodes are required to be situated on a horizontal baseline, i.e. x coordinate and two sensor nodes are situated on a vertical baseline, i.e., y axis. The distance between the two reference nodes on the baseline can be estimated in a fundamental stage and stored in memory. The two angles α_1 and α_2 are estimated between the baseline and the line framed by the reference node and target node [13]. The coordinate points of reference nodes, (x_1, y_1) and (x_2, y_2).

Reference nodes R_1 and R_2 frame the baseline of x-axis. Reference node, R_1 can be reused to shape the baseline of y-axis together with reference node R_3. A target node, T_1 moves unreservedly around within the zone. In view of fundamental triangulation, the area coordinates (x, y) of T_1 can be dictated by utilizing the intersection of R_1 and R_3 to discover x similarly R_1 and R_2 to obtain y and it is represented as,

$$x = d_{ry} - \sin(\alpha_{x_1}) \sin(\alpha_{y_2})^2 / \sin(\alpha_{y_1} + \alpha_{y_2}) \tag{2}$$

$$y = d_{r_x} - \sin(\alpha_{x_1}) \sin(\alpha_{x_2})^2 / \sin(\alpha_{x_1} + \alpha_{x_2}) \tag{3}$$

3   Proposed Work

As illustrated, RSS is considered as the measure of power level of the wifi signal received by a client device, which is facilitated by a nearby access point. The RSS value received by the client is dependent on various factors, including both physical and environmental factors. The power level of the waves generated influences the RSS value which is measured from a distance. The RSS values are affected by the various factors, such as multi path, wave reflection and wave bouncing. The natural parameter, including temperature, humidity, and wind flow also has an impact on the RSS values. The surroundings with human movements and presence of obstacles add on to the losses in the radio frequency signals. Weighting is found to be one of the ways to reduce the error in the data. The variations in data are limited by a weight that results better control over the variations in RSS values. This phenomenon cuts down the sudden and unnecessary error data and thus contributes the smoothening of RSS values. The weight is determined by the distance moved by the client. The test bed is made and the readings of RSS values are converted in terms of distance to realize the scenario from the existing three access points. Distance d is calculated by,

$$d = 10^{(TXPower - RSSI) \div (10 \times N)} \tag{4}$$

Figure 1 shows the circles that are constructed based on distance calculated and region of circle intersect at a common point which predicts the ideal location of the user.where, the position (x, y) is calculated by trilateration,

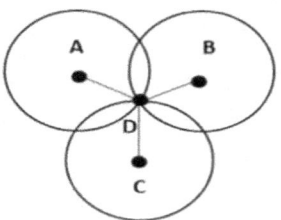

Fig. 1. Formation of the circle intersecting at a common point

$$x = \frac{(A - B - (b_2 - b_1) \times 2y)}{(2 \times (a_2 - a_1))} \tag{5}$$

$$y = \frac{(a_3 - a_1)(A - B) + (C - A)(a_2 - a_1)}{(a_3 - a_1)(A - B) + (C - A)(a_2 - a_1)} \tag{6}$$

where, A, B, C

$$A = d_1^2 - a_1^2 - b_1^2$$
$$B = d_2^2 - a_2^2 - b_2^2$$
$$C = d_3^2 - a_3^2 - b_3^2$$

But the occurrence of losses in radio frequency shows minimum impact on the corresponding distance value which results error in the measurements. Moreover, it works only for the three well known access points.

The weighted centroid algorithm estimates the unknown position by choosing the nearby access point closure or nearer to the position of a user. Weight is estimated by using the Eq. (7), where d_{ij} is the distance between i^{th} and j^{th} node of access point and g is the degree chosen based on the number of access points [15–17].

$$W_{ij} = \frac{1}{d_{ij}^g} \tag{7}$$

The position of the user is calculated by,

$$D_i(x, y) = \frac{\sum_{j=1}^{N} w_{ij} \times AP_j(x, y)}{\sum_{j=1}^{N} APw_{ij}} \tag{8}$$

The Eq. (8) brings more complexity and energy consumption. Further, it is more suitable only when the access points are deployed in a rectangular position. The resultant accuracy is observed to be less when the distance between access point and user is higher.

It is presumed that the demerits associated with the individual technique may be overcome when these two methods are combined. The presumption is proved with the results obtained. As explained, this study describes the combined weighted trilateration and centroid algorithm by simply measuring the real-time RSS from the nearby nodes. The weighted RSS curve forms a circle with the radius, r by calculating distance between user and node. The circle forms an intersection region rather than a common point. By solving circle equation the study has predicted the current position of a target user as in and around the centroid i.e., intersection region. The centroid of the three circles nearer to the user forms a triangle with three vertices. The estimated position (x, y) is defined as,

$$(x, \; y) = \left(\frac{1}{N} \sum_{i=0}^{N} x_i, \; \frac{1}{N} \sum_{i=0}^{N} y_i \right) \tag{9}$$

Where,

xi = x co-ordinate of the 'i'th vertex of the triangle
y_i = y co-ordinates of the 'i'th vertex of the triangle
N = Total number of vertices

Algorithm: Weighted trilateration and centroid Localization

Require: Node Coordinate point $(a_0, b_0)(a_1, b_1) \, (a_2, b_2)$, node transmission range for 1 meter and path loss constant range N value

---

Event: Locate the user position.
1. Measure live RSS from the nodes closer to the user
2. **for** all closer nodes, check for node having minimum RSS value
3. **if** minimum RSS is identified then
4.   Choose three node
5. **end if**
6. **end for**
7. Weighting the RSS curve (with reference to minimum value)
8. **for** three closer nodes, it forms a range of circle with the radius r, by calculating distance between user   and node. Distance is given as,
9. $d = 10^{(TXPower\text{-}RSSI) \div (10 \times N)}$
10. Find the intersection of two circles
11.   Solving the general equation of two circles
12. **end for**
13. similarly (x, y) for other two circle combinations
14. Find the vertices of the triangle $(x_0, y_0), (x_1, y_1), (x_2, y_2)$ formed by the intersection region of all the three circles
15. Using centroid formula find the user position (x,y)
16. $x = (x_0 + x_1 + x_2) \div 3$
17. $y = (y_0 + y_1 + y_2) \div 3$
18. **return**

4 Results and Discussions

The present study explores the possibility of evaluating the combined technique by means of simulation, MATLAB platform as well as Raspberry constituted wireless sensor network. The simulation deployment strategy followed in this study is shown in Fig. 2. The total area covers up to 25 m × 20 m and 50 nodes are deployed at fixed randomly with the transmission power of −31 dBm.

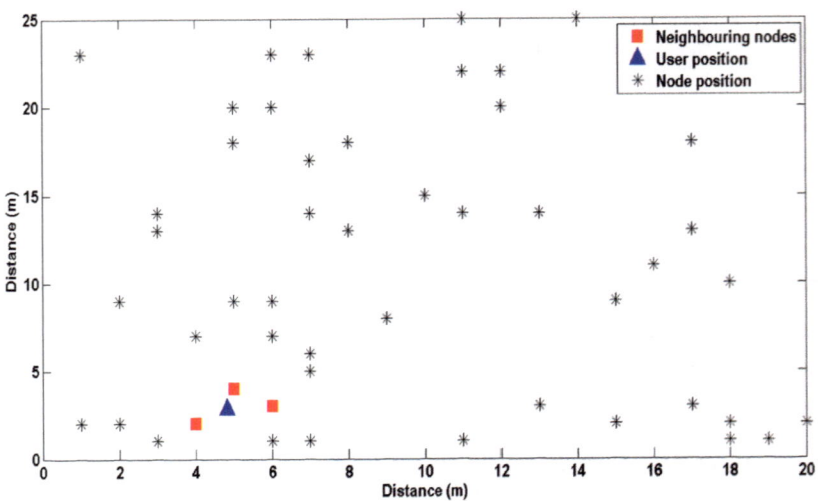

Fig. 2. MATLAB Simulation-Graph plot of node position and user position

The performance of the proposed algorithm is successfully validated by analyzing the exact location and estimated location. The study observes the accuracy by an average of 0.3 m for a distance range of 20 m between the user and node.

Similarly, the study considers obtaining the differences between the actual (RSS) and normalized (NRSS) values for the chosen nodes and the results are shown in Table 1. The representation in all the illustrations is made between the received RSS value in terms of dBm and the corresponding location of the user. Since the signal strength depends upon the transmission power, the study normalizes the curve based on node's 1 m transmission power and calculated the average of nth RSS value. The unknown position is then estimated by choosing the three closure access points to the user. Table 1 stipulate RSS data received by the user from the relatively nearer three access points, i.e., Node1, Node2 and Node3.

Table 1. Differences between the actual RSS and normalized RSS values

Actual value			Normalized value		
Node1 RSS (dBm)	Node2 RSS (dBm)	Node3 RSS (dBm)	Node1 NRSS (dBm)	Node2 NRSS (dBm)	Node3 NRSS (dBm)
−48	−47	−42	−38	−38	−40
−42	−35	−46	−40	−35	−44
−46	−40	−48	−38	−40	−44
−44	−40	−47	−42	−35	−44
−43	−46	−45	−41	−35	−43
−42	−46	−46	−34	−41	−44
−42	−44	−45	−40	−38	−43

(*continued*)

Table 1. (*continued*)

Actual value			Normalized value		
Node1 RSS (dBm)	Node2 RSS (dBm)	Node3 RSS (dBm)	Node1 NRSS (dBm)	Node2 NRSS (dBm)	Node3 NRSS (dBm)
−40	−42	−46	−41	−35	−44
−43	−45	−44	−42	−37	−40
−40	−46	−52	−39	−41	−42

5   Case Study

A real-time test bed having a total area 12.5×9 m^2 is created in an indoor environment and three Raspberry Pi nodes were deployed to validate the performance of the combined technique. Raspberry Pi operates on either Raspbian or Debian operating systems. It has a built-in 2.4 GHz and 5 GHz IEEE 802.11.b/g/n/ac wireless LAN which can act either as an access point or a client. It can be configured as an access point by modifying the settings, for instances hostapd and dnsmasq.

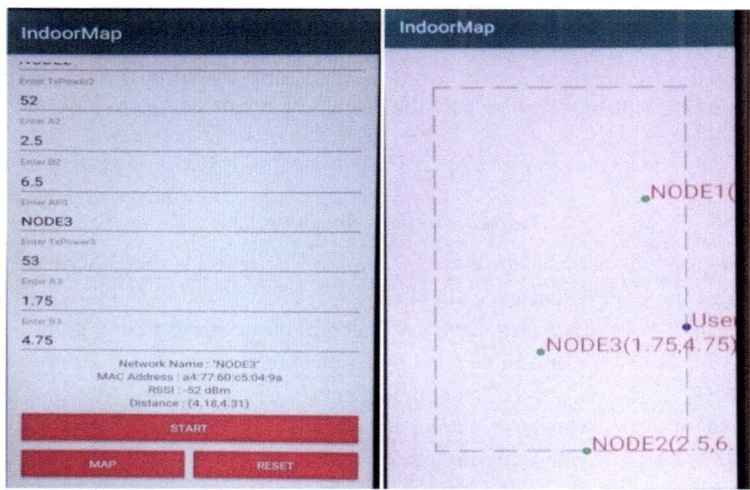

Fig. 3. RSSI reading and distance estimation

The distance related data from the individual nodes is acquired by using an android based application programming interface (API). The API is customized to visualize the indoor mapping on the basis of wireless node deployment positions and the captured screenshot is shown in Fig. 3. The user can able to understand the distance between his current location and the chosen node. The performance of the algorithm is evaluated by collecting the respective positions as the user moves within transmission coverage range of 12 m. The obtained results show an average of 0.6 m error between actual and estimated positions of the user which is presented in Table 2.

Table 2. Results of estimated position

S. no	Actual location (x, y) (m)	Estimated position without smoothing (x, y) (m)	Estimated location using weighting technique (x, y) (m)
1	(0,0)	(2.9,4.15)	(0.1,0.4)
2	(4.04,3.45)	(3.1,3.76)	(3.23,2.98)
3	(5.4,2.1)	(3.19,5.06)	(4.98,1.9)
4	(2,5.45)	(4.25,4.39)	(2.31,3.97)
5	(0,2.9)	(4.2,4.39)	(0.1,3.2)
6	(0,4.1)	(2.91,5.16)	(0.2,3.75)
7	(1,2)	(4.01,4.8)	(0.9,3.23)
8	(1.6,3.2)	(4.18,4.35)	(1.7,3.5)
9	(1,2.5)	(4.5,4.5)	(0.8,2.9)
10	(3,7)	(4.92,5.14)	(2.8,7.2)

Table 2 shows the obtained sets of readings, where the weighting technique is used with minimum RSS which stabilizes values from all nodes and centroid algorithm determines the position of the user by choosing three closer nodes. By comparing the accuracy of the trilateration and centroid method, our algorithm brings out accuracy of less than 1 m. Table 3 compares the accuracy result of related works with the proposed work.

Table 3. Result comparison table

Authors	Technique used	Accuracy
Bobescu [3]	Trilateration algorithm	2–2.5 m
Khan [11]	Hata Okumara model adopted using trilateration	0.85 m
Chai [14]	Kalman Filter and least square estimation	2.6 m
Knauth [7]	Weighted centroid algorithm	1.5 m–2.45 m
Hui-xia [16]	Weighted centroid algorithm	2.54 m
Wen-jian [17]	Weighted trilateration algorithm	2.57 m
Lymberopoulos [9]	Fingerprinting	0.72 m
Liu [13]	Applying Matrices for RSS value	2 m
Proposed work	Weighted trilateration and centroid algorithm	0.6 m

6   Conclusion and Future Work

In this study, indoor localization is done using sensor node which act as a wifi access points and mobile phones. The study simulation result shows the accuracy of 0.3 m and real time environment in an indoor space given accurate readings with an error less than a meter. The experimental study on weighted trilateration and centroid algorithm predicts 96.7% precise than other localization technique. The future work involves by implementing the present study in large spaces such as warehouse environment to ensure effective utilization of wireless resources.

Acknowledgement. The authors gratefully acknowledge the financial support from Tamil Nadu State Council for Science and Technology, DOTE Campus, Chennai-600025, Tamil Nadu, India by sanctioning a project - File No: DST/SSTP/TN/29/2017-18 to Velammal Engineering College, Chennai.

References

1. Xiao, J., Zhou, Z., Yi, Y.: A survey on wireless indoor Localization from the device perspective. ACM Comput. Surv. **49**(2), Article 25 (2016)
2. Sakpere, W., Adeyeye-Oshin, M., Mlitwa, N.B.W.: A state-of-the-art survey of indoor positioning and navigation systems and technologies. S. Afr. Comput. J. **29**(3), 145–197 (2017)
3. Bobescu, B., Alexandru, M.: Mobile indoor positioning using wifi localization. Rev. Air Force Acad. **28**, 119–122 (2015)
4. Biswas, J., Veloso, M.: Wifi localization and navigation for autonomous indoor mobile robots. In: IEEE International Conference on Robotics and Automation (2010)
5. Liu, K., Motta, G., Ma, T.: Navigation services for indoor and outdoor user mobility: an overview. In: 2016 International Conference on Service Science (2016)
6. Adler, S. Schmitt, S., Wolter, K., Kyas, M.: A survey of experimental evaluation in indoor localization research. In: International Conference on Indoor Positioning and Indoor Navigation (IPIN), pp. 1–10, October 2015
7. Knauth, S., Storz, M., Dastageeri, H., Koukofikis, A.: Centroid localization for position estimation in large environments. In: International Conference on Indoor Positioning and Indoor Navigation (IPIN), p. 4. IPIN, Banff (2015)
8. Kim, Y., Shin, H., Chon, Y., Cha, H.: Smartphone-based wifi tracking system exploiting the RSS peak to overcome the RSS variance problem. Perv. Mob. Comput. **9**(3), 406–420 (2013)
9. Lymberopoulos, D., Liu, J., Yang, X., Choudhury, R.R., Handziski, V., Sen, S.: A realistic evaluation and comparison of indoor location technologies: experiences and lessons learned. In: The 14th ACM/IEEE Conference on Information Processing in Sensor Networks (IPSN 2015), ACM Association for Computing Machinery, April 2015
10. Adler, S., Schmitt, S., Kyas, M.: Path loss and multipath effects in a real world indoor localization scenario. In: 2014 11th Workshop on Positioning, Navigation and Communication (WPNC), pp. 1–7, March 2014
11. Khan, S., Akram, A., Usman, S.: Wi-Fi based positioning algorithm for indoor environment using 802.11 standards. Int. J. Sci. Eng. Res. **5**(3) (2014). ISSN 2229-5518
12. Liu, K., Motta, G., Ma, T.: XYZ indoor navigation through augmented reality: a research in progress. In: IEEE International Conference on Services Computing (SCC), San Francisco, CA, pp. 299–306 (2016)
13. Liu, K., Motta, G., Ma, T., Guo, T.: Multi-floor indoor navigation with geomagnetic field positioning and ant colony optimization algorithm. In: 2016 IEEE Symposium on Service-Oriented System Engineering (SOSE), Oxford, pp. 314–323 (2016)

14. Chai, S., An, R., Du, Z.: An indoor positioning algorithm using bluetooth low energy RSSIS. In: International Conference on Advanced Material Science and Environmental Engineering (2016)
15. Blumenthal, J., Grossmann, R., Golatowski, F., Timmermann, D.: Weighted centroid localization in zigbee-based sensor networks. In: International Symposium in Intelligent Signal Processing, pp. 1–6. IEEE (2007)
16. Hui-xia, Y.: Weighted centroid localization algorithm with weight corrected based on RSSI for wireless sensor network. Electron. Test **1**(1), 28–32 (2012)
17. Wen-jian, W., Jin, L., He-lin, L., Bing, K.: An improved weighted trilateration localization algorithm. J. Zhengzhou Univ. Light Ind. (Nat. Sci.) **3**(6), 84–85 (2012)

Comparison of Various Wearable Activity Trackers

Tanvi Nikam[✉] and Rejo Mathew

Department of IT, Mukesh Patel School of Technology Management
and Engineering, NMIMS Deemed-to-be University, Mumbai, India
nikamtanvi20@gmail.com, rejo.mathew@nmims.edu

Abstract. With the various activity trackers available in the market, customers often find themselves confused between choosing one amongst them all. Due to the advancements in technology there are several specifications and features to choose from but the main challenge is that there is not one activity tracker that has it all, hence the need of comparison between the several activity trackers and their specifications. Existing research focuses mostly on giving the specifications of all the activity trackers out there, and not comparing different activity trackers to help the customer make the correct choice. This research paper has outlined the need, challenges and comparison of the specification between the different activity trackers available in the market.

Keywords: Comparison · Fitness trackers · Physical activity trackers · Specifications · Wearable activity trackers

1 Introduction

There are primarily two types of wearable activity trackers available in the market.

- Smart watches: Smart watches are basically dumbed down smart phones. They have almost all functions of a smartphone including a touchscreen display, apps, notifications, camera etc. Hence, they also have the adequate processor and memory to handle such smartphone-like functions [1].
- Fitness trackers: Fitness trackers are devices used for apprehending fitness related metrics which include heart rate monitoring, distance covered, calorie consumption, GPS and quality of sleep. These mostly have only fitness related functions, unlike smartphones, and hence usually have physical buttons and little or no internal storage space.

The wearable activity trackers this research paper focuses on are fitness trackers.

A common issue for all fitness enthusiasts and fitness aspirants is that of maintaining their daily diet and exercise plan which requires the need of keeping track of their daily activities. Motivation also plays a huge role as consistently keeping an eye

© Springer Nature Switzerland AG 2020
A. P. Pandian et al. (Eds.): ICCBI 2018, LNDECT 31, pp. 173–179, 2020.
https://doi.org/10.1007/978-3-030-24643-3_20

on workout goals and daily nutrition is challenging. Due to this reason, companies have tried to capitalize on the growing wearable activity trackers market by selling fitness trackers.

These versatile wearable activity trackers can be taken anywhere by the user and by just wearing them on their wrists these devices track fitness related activities such as how much fat has the user burned (calorie burn), how many steps has the user walked (accelerometer), how much distance has the user physically covered (GPS), a user's heart rate and even sleep.

Existing research only focuses on explaining the idea of a wearable activity tracker (e.g. [2]) and its adoption and usability challenges. In this paper, we aim to provide a comparison between the specifications of different types of activity trackers available in the market, exhibit their pros and cons, provide solutions for the cons and an explanation for the same.

2   List of Various Wearable Fitness Trackers

2.1   Fitbit Charge 2

A fitness tracker that provides all Fitbit functionalities and is one of the affordable trackers having promising features like heart rate monitor, water resistance etc.

2.2   Garmin Vivoactive 3

One of the best fitness trackers from the Garmin range that not only provides all functionalities of a fitness tracker but also has additional features like automatic sleep tracking and built-in GPS.

2.3   Fitbit Versa

This particular activity tracker is a notch above the aforementioned Fitbit Charge 2 as it has all its functionality in addition to a highly accurate heart rate monitor and high water resistance.

2.4   Samsung Gear Sport

The Samsung Gear Sport was for a long while, the only fitness tracker from Samsung which provided a highly sharp and impressive display but wasn't as successful as a fitness trackers as it's competitors.

2.5   Garmin Forerunner 35

The Garmin Forerunner 35 provided built in GPS however isn't as impressive a fitness tracker as the others in the Garmin range (Table 1).

Table 1. Analysis of different parameters of various wearable fitness trackers.

	Fitbit Charge 2	Garmin Vivoactive 3	Fitbit Versa	Samsung Gear Sport	Garmin Forerunner 35
Display type	–OLED [3] –Hard to see when outdoors [3] –Has one tap display [3] –1.5-in. screen size [3]	–Colour LCD [7] –Bright enough to read when outdoors [7] –1.2-in. screen size [7]	–Colour LCD [8] –Bright display [8] –Easy to read in sunlight [8] –1.32-in. screen display [9]	–AMOLED full colour display [12] –Gorilla Glass3 [12] –Scratch and breakproof display [12] –1.2-in. screen size [9]	–Monochrome LCD [16] –Not a super sharp display [16] –Not touchscreen [16] –0.9 in. screen size [7]
Compatibility	–Android –Windows –iOS	–Android –iOS –Mac	–Android –Windows –iOS	–Android –iOS	–Android –Windows –iOS
Heart rate monitor	–Consists of Fitbit heart rate zones [3] –Suggest breathing pattern based on heart rate [3] –Low HR (heart rate) accuracy [3]	–Frequent heart rate monitoring at 1 s rate [7] –Shows heart rate stats monitor [7] –Allows stress tracking [7]	–Consists of Fitbit heart rate zones [10] –Suggest breathing pattern based on heart rate [10] –High HR (heart rate) accuracy [10]	–High heart rate accuracy [10]	–Uses Optical heart rate monitoring Technology [16] –No continuous Monitoring. Displays graph of heart rate every 4 h [16] –Gives erratic heart rate readings many a times [16]

(continued)

Table 1. (*continued*)

	Fitbit Charge 2	Garmin Vivoactive 3	Fitbit Versa	Samsung Gear Sport	Garmin Forerunner 35
Sleep tracker	–Gives sleep stages [4] –Accurate Measurement of REM sleep [4]	–Automatic night sleep tracking [7] –Doesn't track naps [7] –Accurate start and end time of sleep [7]	–Gives sleep stages [11] –Gives personalized tips for better sleep [11] –Doesn't allow to share health tips to other apps [11]	–Measures deep sleep [13] –Doesn't allow sharing of data to Google Health or Apple Fit [13]	–Gives total hours of sleep [16] –Gives sleep levels [16] –Gives sleep movements [16] –Sleep statistics available on Garmin App. [16]
Battery life	5 days [3]	7 days [7]	4+ days [8]	7 days [13]	9 days [16]
GPS	–No built-in GPS sensor [5] –Consists of "connected GPS" which connects to nearby phone to track GPS data [5]	–Has built-in GPS [7] –Both GPS and non-GPS workouts are stored on the online to be viewed by user later [7]	–No built-in GPS sensor [5] –Consists of "connected GPS" which connects to nearby phone to track GPS data [5]	–Has built-in GPS [14] –Erratic GPS performance [14]	–Has built-in GPS [18] –No GLONASS support [18] –GPS Accuracy is high [18]
Water resistance	–Up to 10 m [6] –Hence, not swim-proof [6] –However, splash, rain and sweat resistant [6]	–Up to 50 m for swimming [7] –Also splash, rain and sweat resistant [7]	–Up to 50 m for swimming [6] –Also splash, rain and sweat resistant [6]	–Up to 50 m for swimming [15] –Has swim tracking app, SpeedO [15] –Cannot be used for high water pressure activities like water diving etc. [15]	–Up to 50 m for swimming [18] –Also splash, rain and sweat resistant [18]

3 Analysis of Various Wearable Fitness Trackers

3.1 Justifications for Analysis

The pros and cons of the considered wearable fitness trackers are as follows:

- The pros of the Fitbit Charge 2 are that it has the Fitbit heart zones and that the heart rate monitor suggests suitable breathing pattern based on the heart rate zones. It is compatible with Android, iOS as well as Windows. The sleep tracker in the Fitbit Charge 2 is also impressive as it provides sleep stages and gives accurate measurement of REM sleep.

- The cons are that the Fitbit Charge 2 has an OLED display which makes outdoor reading difficult. It is also said that the heart rate accuracy of the Fitbit Charge 2 is low because, in reference to [20], experiments prove that the Fitbit Charge 2 gives different readings when placed at improper positioning. And ideally, activity trackers should do so from any positioning on the wrist because displacing of a tracker while exercising/sleeping or doing any other activity while wearing it is bound to happen. The Fitbit Charge 2 does not have a built-in GPS and uses connected GPS instead and the water resistance is also not impressive as it is only up to 10 m making it splash proof.

- The pros of the Garmin Vivoactive 3 are that it has a Colour LCD display which is bright enough for outdoor usage. It is also compatible with Android, iOS and Mac. The heart rate monitor in Garmin Vivoactive 3 provides continuous heart rate monitoring and also tracks stress levels of a user. The sleep tracker in this device has the pro of accurate and automatic night sleep tracking. The major pro of the Garmin Vivoactive 3 is that it has built-in GPS feature and all workouts, GPS or non-GPS, are stored online to be viewed according to user's discretion. Lastly, the Garmin Vivoactive provides a good water resistance of up to 50 m.

- The cons of the Garmin Vivoactive 3 are that it doesn't track naps, which also includes it not using the automatic sleep tracking feature for a user's nap sessions. The water resistance of the Garmin Vivoactive 3 doesn't make it suitable for high water pressure activities like diving.

- The pros of the Fitbit Versa are that it has a Colour LCD display, it is compatible with Android and iOS, has the Fitbit heart zones and that the heart rate monitor suggests suitable breathing pattern based on the heart rate zones. The heart rate accuracy is also high in the Fitbit Versa as opposed to that of the Fitbit Charge 2. The sleep tracker in the Fitbit Versa gives sleep stages and also provides personalized tips for better sleep. It also tries to covers up for not having a built-in GPS, by having a connected GPS like in the Fitbit Charge 2. It also has better water resistance that Fitbit Charge 2.

- The major cons of the Fitbit Versa are that the sleep tracking doesn't allow data sharing to other apps, it has no built-in sensor and the water resistance is not suitable for extreme water sports.

- The pros of the Samsung Gear Sport are that it has an AMOLED full colour display, which as discussed is the best display type amongst all the fitness trackers considered. It also has a Gorilla Glass3 which makes the display scratch proof and break proof. Other than that, the Samsung Gear Sport provides high heart rate accuracy, measures deep sleep, has a built-in GPS and water resistance of up to 50 m for swimming.

- The cons of the Samsung Gear Sport are that it is not compatible with windows, it doesn't allow sharing of sleep tracking data to other apps like Google Health or Apple Fit, the GPS accuracy is erratic and the water resistance is not at all suitable for high water pressure activities.
- The pros of the Garmin Forerunner 35 are that it is compatible with Android, iOS as well as Windows, the sleep tracker gives total hours of sleep, sleep levels and movements and provides all sleep statistics on the Garmin App. It also has a built-in GPS which is highly accurate [18], and the water resistance of up to 50 m is good enough too.
- The cons of the Garmin Forerunner 35 are that it has a weak Monochrome LCD display which is not touchscreen, it does not have continuous heart rate monitoring, there is no GLONASS support to the built-in GPS.

The solutions we would like to provide are that Fitbit Charge 2 should work on its HR accuracy, the Garmin Vivoactive 3 should allow sharing it's health statistics to other apps like Google Health and Apple Fit, like its mentioned in the table, the Samsung Gear Sport fails at its weight which is too heavy for it sleep or exercise with, and lastly the Garmin Forerunner 35 should work on making it's display sharp, and making it touchscreen rather than just manually operating through buttons. It should also enable the continuous heart monitoring feature.

4   Conclusion

To exhibit the best and worst wearable fitness tracker from the five that are covered in this study, evidence in the form of pros and cons is as aforementioned. The best fitness tracker from the above will be the Garmin Vivoactive 3 and the Fitbit Versa and the worst activity tracker will be the Garmin Forerunner 35.

With this, the paper has made the following contributions: (1) compared the features of 5 of the many activity trackers available in the market. (2) Described the product's pros and cons in great detail. (3) Exhibited suggestions for making the concerned fitness trackers better. (4) Exhibited the best and worst from amongst the list of the wearable fitness trackers and explained the same, thus providing a detailed review of all the five activity trackers and enabling the reader to make a sound choice.

References

1. Smart watches vs. Fitness Bands. https://smartwatches.org/learn/smartwatches-vs-fitness-bands-difference/
2. Shih, P.C., Han, K., Poole, E.S., Rosson, M.B., Carroll, J.M.: Use and Adoption Challenges of Wearable Activity Trackers. Pennsylvania State University (2015)
3. Fitbit Charge 2. https://www.fitbit.com/in/charge2/charge2-101#fitbit-mobile-device-101-4

4. de Zambotti, M., Goldstone, A., Claudatos, S., Colrain, I.M., Baker, F.C.: Centre for Health Sciences, SRI International, Menlo Park, CA, USA, A validation study of Fitbit Charge 2™ compared with polysomnography in adults. Chronobiol. Int. **34**, 465 (2018)
5. Fitbit Help Article. https://help.fitbit.com/articles/en_US/Help_article/1874
6. How water resistant is Charge 2 and Alta HR. https://www.reddit.com/r/fitbit/comments/694dhk/how_water_resistant_is_charge_2_and_alta_hr/
7. Garmin Vivoactive 3 in depth review. https://www.dcrainmaker.com/2017/10/garmin-vivoactive-3-in-depth-review.html
8. Fitbit Versa Review. https://www.androidauthority.com/fitbit-versa-review-845214/
9. Fitbit Versa vs Samsung Gear Sport. https://www.gadgetsnow.com/compare-smartwatch/Fitbit-Versa-vs-Samsung-Gear-Sport
10. GPS and heart rate accuracy of the Fitbit Versa vs. Samsung Sports Gear. https://thejournier.com/2018/05/28/gps-and-heart-rate-accuracy-of-the-fitbit-versa-vs-samsung-gear-sport/
11. Fitbit new smart watch. https://www.wareable.com/fitbit/fitbit-new-smartwatch-2018
12. Gear Sport specifications. https://www.samsung.com/global/galaxy/gear-sport/specs/
13. Gear Sport review-The only Fitness Watch for Samsung die hards. https://arstechnica.com/gadgets/2018/03/gear-sport-review-the-only-fitness-watch-for-samsung-die-hards/
14. Samsung Gear Sport Review. https://www.techradar.com/reviews/samsung-gear-sport-review
15. Samsung Gear Sport Review. https://www.samsung.com/global/galaxy/gear-sport/
16. Garmin Forerunner 35 review. https://www.wareable.com/garmin/garmin-forerunner-35-review
17. AMOLED vs LCD displays- which one is better? https://www.themobileindian.com/news/amoled-vs-lcd-displays-which-one-is-better-18933
18. Fitbit Heart rate Monitor guide. https://www.wareable.com/fitbit/fitbit-heart-rate-monitor-guide-330
19. What is GLONASS-how is different from GPS? https://beebom.com/what-is-glonass-and-how-it-is-different-from-gps/
20. Benedetto, S., Caldato, C., Bazzan, E., Greenwood, D.C., Pensabene, V., Actis, P.: Assessment of the Fitbit Charge 2 for monitoring heart rate. PloS ONE **13**, e0192691 (2018)

Review on Security Problems of Bitcoin

Parmeet Singh Pannu[✉] and Rejo Mathew

Department of IT, Mukesh Patel School of Technology Management
and Engineering, NMIMS Deemed-to-be University, Mumbai, India
Parmeetcr7@gmail.com, rejo.mathew@nmims.edu

Abstract. Bitcoin is a cryptocurrency which means transactions are carried by means of electronic cash. It is a digital currency that has no central bank or an administrator. Bitcoins are transferred from user to user on the bitcoin network. The market capitalization of bitcoin is 170 million and over 375,000 transactions take place every day. There are a lot of security issues in bitcoin. Some of them are double spending, Finney attack, brute force attack, vector 76, Goldfinger. These attacks have some solutions which are known as Conjoin, Coinshuffle, xim, value shuffle, dandelion etc. In this paper, I review these security problems and the solutions to them.

1 Introduction

Bitcoin is an electronic payment system. Bitcoins move from one user's wallet to another. A destination address gets generated when hashing operations are performed on the public key of a user. The public key is linked to the user's wallet. A private key is a key required by the user to spend the owned bitcoins. The correctness of a transaction is verified by a group of nodes known as miners. Miners bundle a number of transactions in a network instead of just a single transaction. This bundle is called a Block. The miner who has successfully mined a block gets a reward.

The bitcoin wallet doesn't contain actual bitcoins. It contains a public key and a private key. The public contains transaction details etc. and is visible to others, whereas the private key is private and no one knows someone else's private key. The transaction that takes place has to be digitally signed by one's private key. Without a private key, no transaction can take place. A transaction contains senders private key, receivers key and number of bitcoins to be transferred. After getting validated, the transaction gets included in a block along with many other transactions. Through bitcoin mining, the transactions get verified. The person who performs mining is called a miner. The miner has to compile transactions in the block and solve complex puzzles. The miner who does this first gets a mining reward.

© Springer Nature Switzerland AG 2020
A. P. Pandian et al. (Eds.): ICCBI 2018, LNDECT 31, pp. 180–184, 2020.
https://doi.org/10.1007/978-3-030-24643-3_21

2 Issues

2.1 Double Spending

In bitcoin transactions, miners are present to verify the transactions. The coins left after the previous transactions can be used in the next transaction. So when the miners detect the transactions TC and TC2 have the same number of coins, they reject one of the transaction.

Despite this, double spending is still possible and the following conditions have to be satisfied with the same.

(i)   A part of network miners accept the transaction TC and the vendor(V) receives the confirmation for the same.
(ii) At the very same instant another part of miners accept transaction TC2.
(iii) The vendor gets the confirmation of TC2 after it accepts the payment of TC.
(iv) Miners mine on top of blockchain which contains TC2 so that it gets valid.

Another variation of double spending is a Finney attack. In Finney attack [5] the cheater client privately mines a block which contains the transaction TC2 and then creates the TC transaction for the vendor. The mined block is not pushed into the network and the client waits for the vendor to accept the transaction TC. After the transaction is accepted the pre-mined block is pushed in the blockchain, with fork length same as the existing fork. If the next mined block extends the attackers fork then as per the rules of bitcoin, the miners will focus on the attacker's fork as it is the longest. All the miners now focus on the attacker's fork thus the previous fork containing TC becomes invalid and TC2 goes through.

Brute force attack [6] is the advanced version of the Finney attack. In this, the blocks privately work with an intention of double spending. Like the previous case, the attacker introduces the double spending transaction in the block and continuously works on the fork. A vendor waits for 'p' confirmations of the transactions before sending the product.

Now the attacker is able to process p confirmations and then releases it in the block. The miners will mine this block as t becomes the longest fork. This causes the same effect as that of the Finney attack.

2.2 Mining Pool Attacks

Mining pools are made to increase the computing power and this affects the verification time of block directly, and this increases the chances of winning the mining reward. Pool managers govern mining pools. The miners generate partial proofs-of-work (PPoWs) and full proofs of-work (FPoWs) and submit them to the manager as shares. As soon as the miner discovers a new block, he sends the FPoW to the manager. The manager then broadcasts the block to earn the mining reward. the reward is distributed among the miners according to their work.

Selfish mining [3] is a type of attack in pool mining. In reality, all the miners are selfish as everyone is working to get the reward. however, not every miner is dishonest.

The dishonest miner hides information from other miners so they can get an uppercut. They have two motives

(i) Obtaining an unfair reward that is greater than share spent on power consumption.
(ii) Confusing other miners and lead them in the wrong direction.

Block withholding [4] is an attack which is similar to selfish mining. In this, the miner never submits a mined block to impair the total revenue. In the second scenario complex block attack is done which gives a similar result as that of selfish mining.

Another type of attack is a Bribery attack [75]. In this type of attack, an attacker may get a majority of the computing resources for some time by bribing. By more computing power the attacker can perform other types of attacks like double spending.

2.3 Bitcoin Network Attacks

These attacks are the attacks that exploit vulnerabilities in bitcoin protocols and its peer to peer network protocols. Distributed-Denial-Of-Service (DDOS) is the most common type of networking attack. It targets mining pools, e-wallets and other types of services of bitcoin. The cheater exhausts the network resources to disrupt an honest user's access to them. This is done by pushing fake transactions in the network. When a miner is congested with so many of these fake transactions, he discards the new transactions coming from an honest user. This is DDOS as the client got denied because the miner had to deal with so many fake transactions.

Malleability attack [2] is an attack that makes easier to perform DDOS attacks. By using this attack the cheater clogs the transaction queue. This queue contains pending transactions. The cheater introduces a fake transaction with a higher priority by showing that this a larger transaction. The miner then verifies this and find out that this is a fake transaction. But by the time they have realized this, they have wasted a lot of time and resources.

3 Countermeasures for Bitcoin Attacks

3.1 Confirmations

The most effective but simple way to detect double spending is waiting for multiple confirmations before delivering services to the client. The possibility of double spending decreases with an increase in the number of confirmations. For example, an unconfirmed bitcoin has a high risk of double spending whereas a transaction with six conformations is considered as steady and double spending won't happen in this transaction.

Another solution to control double spending mentioned in [1] is a collection security deposit from users. Due to this punishment, double spending can be reduced.

There are no solutions available that guarantee complete protection from double spending. These solutions just make double spending more difficult to be executed. Just waiting for a number of confirmations is the most basic and effective way of dealing with double spending.

3.2   Securing Bitcoin Networks

There are numerous countermeasures which exist to secure bitcoin protocols and peer to peer networking. A game theoretic approach to analyze DDoS attacks is available. According to this game, all pools are in a competition against each other, as larger pools are weighted more than smaller pools. Hence each pool tries to increase its computational cost over the other pools.

Proof of activity (PoA) protocol is an effective countermeasure for DDoS. In this, each block header has a crypt value and the user that stores the first transaction places this value. These users are called 'stakeholders' and it is assumed that they are honest. Storage of a transaction is only done in a block if there are valid stakeholders associated with it. A number of transactions in a chain, more miners will be attracted to it as it increases trustworthiness among the peers. A cheater cannot place a fake block here as all the blocks are governed by stakeholders.

Continuous monitoring of network traffic is another effective way to mitigate DDoS attacks. Machine learning can help us identify which part of the network is behaving badly.

4   Comparison of Security Threats and Their Solutions

See Tables 1 and 2.

Table 1. Comparison of security threats

Attack	Description	Targets	Effects
Double spending	Spending of the same amount of bitcoins in multiple transactions	Sellers	Sellers lose their products
Finney attack	Cheater miner introduces a pre-mined block for double spending	Sellers	Facilitates double spending
Brute force attack	Private mining in blockchain to perform double spending	Sellers	Facilitates double spending
Block withholding [4, 8]	Miner submits PPoW but not FPoW	Honest miners	Waste resources of other. 2 miners and decrease pool revenue
Pool hopping	Uses information about a number of shares to perform a selfish attack	Honest miners	affects the rewards of miners and managers,
DDoS [9, 10]	Cheater exhausts the network resources to disrupt an honest user's access to them	Users, miners	Affects Ewallets, network and other services of bitcoin
Malleability [7]	Cheater clogs the transaction queue	Miners	Waste of miners resources

Table 2. Comparison of solutions

Solutions	Descriptions	Effect
Confirmations	Waiting for confirmations of transactions (more confirmation = more secure)	Decreases double spending
Forwarding double spending transactions	Forwarding of double spending transactions instead of discarding so that vendor comes to know about them	Detects double spending attack
Security deposit	Collection of security deposit. If double spend occurs, this deposit can be given to the victim from attackers account	Reduces double spending attack
Backward compatibility	Miners randomly select which fork to extend	Reduces selfish mining
Anti-payment	Pool managers pay more than the attacker is paying to get resources	Reduces bribery attack, selfish mining and double spending

5 Conclusion and Future Work

Bitcoin has already become the most popular digital currency and it has attracted many cheaters that use this network for their selfish motives. According to my review bitcoin is dreaded with a number of security threats. There are solutions against some of the attacks, but solutions that can ensure proper functioning of bitcoin in the future are still not present. This review paper is an attempt to highlight security threats in bitcoin. After focusing on the function of bitcoin, its characteristics are mentioned.

References

1. Karame, G.O., Androulaki, E., Capkun, S.: Double-Spending Fast Payments in Bitcoin, pp. 906–917 (2012)
2. Decker, C., Wattenhofer, R.: Bitcoin Transaction Malleability and MtGox, pp. 313–326 (2014)
3. Eyal, I., Sirer, E.G.: Majority is Not Enough: Bitcoin Mining is Vulnerable, pp. 436–454 (2014)
4. Rosenfeld, M.: Analysis of Bitcoin Pooled Mining Reward Systems (2011)
5. Karame, G.O., Androulaki, E., Roeschlin, M., Gervais, A., Čapkun, S.: Misbehavior in Bitcoin: A Study of Double-Spending and Accountability (2015)
6. Heusser, J.: Sat Solving an Alternative to Brute Force Bitcoin Mining (2013)
7. Mate: How to Identify Transaction Malleability Attacks (2015)
8. Bag, S., Ruj, S., Sakurai, K.: Bitcoin Block Withholding Attack (2016)
9. Vasek, M., Thornton, M., Moore, T.: Empirical Analysis of Denial of Service Attacks in the Bitcoin Ecosystem (2014)
10. Johnson, B., Laszka, A., Grossklags, J., Vasek, M., Moore, T.: Game-Theoretic Analysis of DDoS Attacks Against Bitcoin Mining Pools (2014)

Review of Security Systems in Vehicles

Kritika Ramawat[1(✉)] and Rejo Mathew[2]

[1] NMIMS Mukesh Patel School of Technology Management
and Engineering, Mumbai, India
kitturamawat@gmail.com
[2] Research Methodology NMIMS Mukesh Patel School of Technology
Management and Engineering, Mumbai, India
rejo.mathew@nmims.edu

Abstract. The shutdown of automobile, impairing of brakes and entryway locks are couple of precedents of the conceivable vehicle digital safety assaults. In any case, now and again, automobiles can be hacked by hackers even if they are miles away. This paper will put some light on the most recent vehicle digital security dangers containing malware assaults, On-Board Symptomatic (OBD) exposures, DSRC security treats and auto portable applications dangers with respective solutions for each threat. This paper also includes detailed information about In-vehicle Network Architecture with the diagram.

Keywords: On-Board Symptomatic (OBD) · Vehicle security dangers · Entry lockways · Cyber attacks · In-Vehicle Network Architecture

1 Introduction

Nowadays, vehicles are never again detached motorized machines that are completely used for transportation. A seamless connected experience in all aspects of their lives including driving has been demanded by consumers. With the outline of vehicle to vehicle (V2V) and vehicle to framework (V2I) correspondences, telematics and the reconciliation of progressive bluetooth devices and mobile phones, associated vehicles communicate to an eco-framework. July 2015, Charlie Miller and Chris Valasek, two analysts, hacked into the Cherokee Jeep from Miller's storm cellar although the car itself was put on the expressway ten miles away [2]. They could distantly control the auto capacities utilizing a basic 3G association misusing a powerlessness in the U-connect programming [3]. There are different attacks against other makers' vehicles. Models of the latest detailed assaults are:

- An assailant inside the Smart Gate in-auto Wi-Fi scope of the Smart Gate-empowered Škoda auto can take data about the auto [4]. Moreover, car's owner can be locked out by the assailant from the *Smart Gate* system [5].

Network Architecture in Vehicle
To shape up an understanding of the potential way focuses (i.e., attacking focuses) the computer operator can expose in the cutting edge vehicle, in this area, we signify the In-vehicle arrange design, then called the car arrange. Present day autos comprise

© Springer Nature Switzerland AG 2020
A. P. Pandian et al. (Eds.): ICCBI 2018, LNDECT 31, pp. 185–191, 2020.
https://doi.org/10.1007/978-3-030-24643-3_22

between 30 to 100 ECUs, and they are called implanted PCs, that carry among one another building the In-vehicle arrange [6]. Figure below exhibits that the In-Vehicle network is built out with the help of many electronic sub-systems including inserted telematics, body and comfort control, vehicle wellbeing, powertrain, locally available camcorders and In- Vehicle Infotainment (IVI) [7] (Fig. 1).

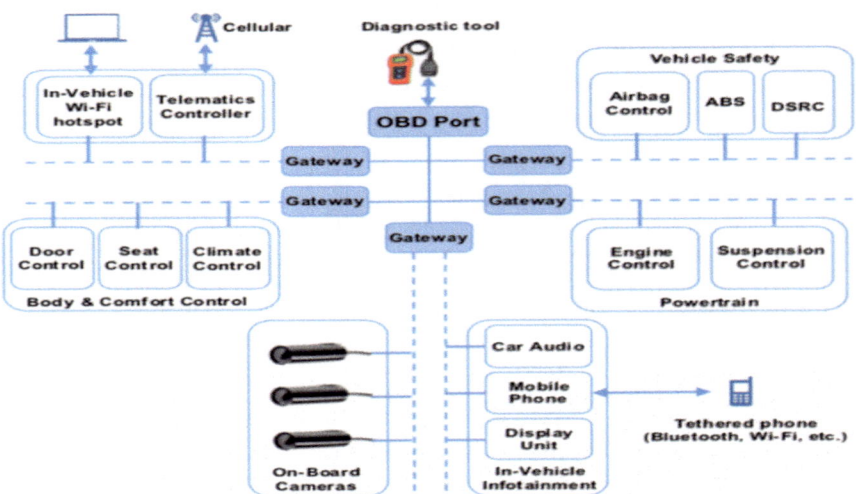

Fig. 1. Architecture of In-Vehicle Network (Source: https://innovate.ieee.org/innovation-spotlight/connected-vehicles)

Let's put some light on an example, ECUs which control airbags sending and Stopping automation (ABS) are located in the automobile wellbeing sub-system, although ECUs which give motor controller and suspension controller are located in the powertrain sub-system.

Control Area Network (CAN) is utilized aimed at time- basic motor controller, wellbeing subsystems, and power-train, although for the fewer time-touchy figure what's more, ease control subsystem a Local Interconnect Network (LIN) is used [7]. Media-Oriented Systems Transport (MOST) and Ethernet are used to support audio, video and onboard cameras in the IVI sub-system. Meanwhile, these routes are inter-connected across a fast CAN transports. With the help of the ongoing patterns of interfacing diverse gadgets through Universal serial bus (USB), Wi-Fi, 3G/4G, Blue-tooth and so forth, each In-vehicle arrange subsystem actualizes its very own corre-spondence module to interface with the outside world. Moreover, OBD port enables complete approach to establish to the In-vehicle. Like this, the assortment and the number of association which are increasing, emphasis in each In-vehicle organize sub-system create the vehicle extra open from the outdoor. In reality, each agreement interface with the outside world should be protected. Be that as it may, securing every section point self-sufficiently will result in duplicate anchoring capacities on a similar vehicle.

2 Problems

2.1 OBD Threats

The usage of OBD is required in vehicles traded in the US later in 1996, in the Union of Europe since 2001 for fuel controlled automobiles [8]. The port of OBD is used to enable automobiles to report any framework's issues and impart the indicative information gathered by its sensors to the external world. This authorizes the specialist co-op to settle the detailed issues. OBD dongles are generally used to interface with the port and thus get to the CAN arrange inside the automobile. These ports are measured as passage focuses to assault the ECUs that are associated with the CAN transports. Present-day automobiles now enable dongles to be distantly coordinated by a Wi-Fi association from PC. Weaknesses in the API of a pass-through gadget (i.e., OBD dongle) enable the assailant to infuse a vindictive cipher into it. The pernicious code makes the pass through gadget emanating toxic bundles on the CAN transports every time it is linked to a distinctive vehicle. Overall, more than half of the overviewed dongles, are ineffective against hacking which is being surveyed [9].

2.2 DSRC Assaults

The communications of V2V and V2I are keys innovations to compromise a class of wellbeing administrations for associated automobiles and these vehicles can avoid impacts and spare lives. The innovation in DSRC has been formed for usage in V2V and V2I interchanges, where every automobile is thought to be outfitted with DSRC On-Board Unit (OBU). DSRC interchanges use a few models. To accomplish its objective, DSRC prepared automobiles are normal to impart (i.e., get/hand- off) data to other DSRC prepared automobiles or potentially foundation, for example, Road- Side Units (RSU). This guideline opens the entrance for noxious centers to either drudge into DSRC prepared automobile or produce problems by sending counterfeit security's data. In this way, IEEE 1609.2 characterizes standard systems to verify and encode messages in DSRC. By and by, assaults, for example, Denial-of- Administration (DoS) are as yet conceivable. Other than sticking assaults of DOS, malware, GPS parodying, area following, disguising, and dark gaps are few precedents of dangers to DSRC outfitted vehicles. Thus, more examine attempts as a team with the automobile industry are estimated to relieve such assaults.

2.3 Auto-Mobile Attacks

OEM-supported associated automobile arrangements, for example, Google's Android Auto boundaries and Apple's Car Play will fetch more incorporated, however possibly powerless, versatile applications into the associated automobile [9]. Automobiles sellers are putting forth a wide scope of auto portable applications that use 3G/4G associations as well as Wi-Fi to speak with your auto and run analytic tests.

Most automobile applications that permit remote access to the auto uses a web benefit facilitated by an administration supplier. This web benefit at that point interfaces with the auto utilizing 3G/4G portable information association. In any case, a few automobiles don't utilize cell associations or web administrations. Rather, they permit portable applications to interface straightforwardly to the auto's Wi-Fi AP and capacities are controlled. In the event that it is executed ineffectively, this technique is defenseless to numerous security and protection assaults, for example, geo-finding the automobile utilizing its access point (AP) service set identifier and (SSID) catching the Pre-Shared Key (PSK) among the auto's Wi-Fi Access Point and the versatile application [10].

2.4 Malware Security Issues

Malware can influence the associated vehicle from numerous points of view. This type of assault can misuse known susceptibilities in the plan and execution of In-vehicle arrange and parts and subsystems, a product refresh bundles of Engine control unit's, and the susceptibilities in the working frameworks utilized in the vehicle. The measure of malevolent activities that can be executed by malware is perpetual. For example, malware can upset the ordinary activity of vehicle's highlights, for example, locking the auto radio so the clients can't turn it on, cause motorist diversions by discretionarily turning on the in-auto sound and increasing the volume, cripple auto wellbeing capacities, for example, the ABS, bolt the vehicle's entryway also, ask for a payment to open it, and send counterfeit security information to different vehicles out and about. It is worth taking note of that more vehicles are generally utilizing working frameworks which are linux-based, and are stronger to these malware assaults than other working frameworks like Microsoft Windows and Android. Be that as it may, malware assaults on Linux have been on the ascent. In this manner, we can't assume that linked vehicles that are utilizing linux are totally invulnerable to malware vulnerabilities.

3 Solutions

3.1 OTA

The greatest difficulties that the car business faces are to retro-fit security components in vehicles that were most certainly not secure. This may incorporate programming remedies, firmware redesigns, and security spots. To mark this test and stay away from exorbitant reviews, more automobiles makers begin utilizing the OTA revives. While OTA revives communicate to a sensible answer for reacting to digital threats in linked automobiles, it ensures a noteworthy issue. Resolving vulnerabilities by utilizing OTA refreshes is a security chance. At a point when Over-The-Air is conveyed to the associated automobile, it implies that a distant cypher is permitted to implement. Some instruments which provide security are, for example, confirming the OTA recharge, utilize a protected convention to convey it, and check the OTA refresh which is crypto graphical must be set up.

3.2   Cloud-Based

Because it isn't plausible to ensure each In-vehicle sub-system exclusively, incorporated arrangements have developed to ensure them In-vehicle arrange and subsequently the associated vehicle. For case, a cloud-helped arrangement has been built up by Ericsson called the Connected Vehicle Cloud framework [11]. That framework sets up another canal concerning the automobile and an assortment of administrations and support given by accomplices and Original Equipment Manufacturer (OEM) organized accomplices.

While the cloud-based answers for secure associated automobiles look extremely encouraging, there are three primary concerns to analyze. In the first place, correspondences overhead and the postponement brought about by steering the movement across the cloud administrations require more examination (e.g., directing V2I and V2V movement to the cloud to shield alongside DSRC assaults is unrealistic.

At last, these arrangements accepted that automobiles are associated framework with the cloud- based all the time through the Internet. This cannot be feasible around the place and would bring about high payments for customers.

3.3   Layer-Based

At last, the National Highway Traffic Safety Organization (NHTSA) has propelled an exploration platform that adopts a strategy which is layered to cyber-security for engine automobiles [12]. As indicated by NHTSA, this layered methodology decreases the possibility of assaults and alleviates the possible magnitudes of an effective one. The program centers around 4 primary zones at the automobile level:

(1) Preemptive measures and policies, lets put some lights on an example, segregation of well being basic subsystems to discharge the impacts of an effective assault;
(2) Unending interruption discovery guesses that incorporate a constant observing of impending interruptions in the framework;
(3) Ongoing effective techniques that mean to save the driver's competence to control the vehicle when the assault is fruitful: and
(4) Calculation of arrangements where data about fruitful slaves from accomplices can be mobilized and broke down to evaluate.

4   Analysis

See Table 1.

Table 1. Comparison table of solutions

Features	OTA-solution	Cloud-based solution	Layer-based solution
Basic idea	Decreases the requirement for reviews by remotely tending to framework breakdowns and security dangers	This is simply beginning to get on, and as car engineers consider its novel utility, a great deal of potential advancements in vehicle configuration are introducing themselves	An exploration agenda has been prepared that adopts a layered tactic to cybersecurity for engine vehicles
Advantages	Update the vehicle from head to toe Built to prevent failures by eliminating any potential risks	Truly aid better vehicle building It generally requires less equipment	Lessens the possibility of attacks and softens the potential magnitude
Disadvantages	Fixing vulnerabilities is a security risk security is not well implemented then attacker can attack easily	Communication overheard Delay incurred by routing the traffic through cloud	-

5   Conclusion

Vehicle cyber security is a powerful branch of awareness that requires more study and research deeds from the scholarly community, the car business, and administrative figures. Harms of car by cyber incidents would be extreme and irretrievable as it worries human lives. Though producers are hoping to prepare current vehicles that contains good network and brilliant capacities, vulnerabilities are expanding quickly. But still more joint work is still required to secure our associated vehicles and therefore us inhabits on the streets.

References

1. Nathan, S.: Hackers after your car? Tackling automotive cyber security. The Engineer (2015)
2. Greenberg, A.: Hackers remotely kill a Jeep on the highway – with me in it (2015)
3. Khandelwal, S.: Car Hackers Could Face Life In Prison. That's Insane! (2016)
4. Link, R.: Is Your Car Broadcasting Too Much Information? Trend Micro Inc., Tokyo (2015)
5. Hull, R.: Nissan disables Leaf electric car app after revelation that hackers can switch on the heater to drain the battery
6. Studnia, I., Nicomette, V., Alata, E., Deswarte, Y., Kaâniche, M., Laarouchi, Y.: Survey on security threats and protection mechanisms in embedded automotive networks

7. Zhang, T., Antunes, H., Aggarwal, S.: Defending connected vehicles against malware: challenges and a solution framework (2014)
8. Nilsson, D.K., Larson, U.E.: Simulated attacks on can buses: vehicle virus. In: Proceedings of International Conference on Communication Systems and Networks, Malaysia (2008)
9. Yan, W.: A Two-year survey on security challenges in automotive threat landscape (2015)
10. Lodge, D.: Hacking the Mitsubishi Outlander PHEV hybrid. Pen Test Partners (2016)
11. Ericsson: Connected Vehicle Cloud Under the Hood (2015)
12. National Highway Traffic Safety Administration: Cybersecurity Best Practices for Modern Vehicles, Report No. DOT HS 812 333, Washington DC, October (2016)

Machine Learning Algorithms in Stock Market Prediction

Jayesh Potdar$^{(\boxtimes)}$ and Rejo Mathew

Department of IT, Mukesh Patel School of Technology Management
and Engineering, NMIMS Deemed-to-be University, Mumbai, India
jayeshpotdar97@gmail.com, rejo.mathew@nmims.edu

Abstract. Fundamentally, stock markets stand for an undeviating, nonparametric system accompanied with undetermined disturbances due to which the accurate prediction of the stock price becomes challenging. Stock prediction is achievable with the help of technology assisted practises. The objective of this review paper is to compare machine learning algorithms namely Support Vector Regression (SVR), Improved Levenberg Marquardt Algorithm (LMA) Self-adapting Variant PSO-Elman Neural Network (PSO) and Linear Regression (LR) so as to have a precise price trend.

Keywords: Stock market · Predictive model · Machine learning ·
Neural network · Support vector machine · PSO-Elman Neural Network ·
Levenberg Marquardt algorithm · Support Vector Regression ·
Linear regression

1 Introduction

Stocks are ownership claims that a public listed company offers to the people. Stock market is a platform for buyers and sellers to trade stocks, bonds or other securities. Due to the many factors affecting the stock market prices like liquid money, inflation, deflation, unemployment, human behavior etc., it becomes a noisy and chaotic system [1, 2], which makes it hard to be able to recognize a trend of how the prices change and the fluctuation caused. Machine learning consists of using data, selecting a model, training and testing it. Neural networks are an explicitly established set of algorithms that have reformed machine learning.

Inspired from biological neural networks and deep neural networks, these have proven to work astonishingly good. Neural networks are general function estimates, hence they can be applied to virtually all machine learning dilemmas regarding understanding a complex mapping from input to output space.

2 Machine Learning Approaches

2.1 Support Vector Regression

Support Vector Regression means using the machine learning algorithms for input-output based elements when one has a hint of what the output or end result is going to

© Springer Nature Switzerland AG 2020
A. P. Pandian et al. (Eds.): ICCBI 2018, LNDECT 31, pp. 192–197, 2020.
https://doi.org/10.1007/978-3-030-24643-3_23

look like for a continuous function to get a continuous output. SVR is a part of supervised machine learning used for multiple applications like clustering, outlier detection, classification and regression. Stock market prediction uses regression wherein some given data points are used and the subsequent data point's location is to be figured out [3]. It uses a linear function called hyper-plane to segregate the two classes wherein the data points used are called support vectors as it is present in a multi-dimensional space. A good hyper-plane tries to maximize the space between the data points nearest to the hyper-plane so as to minimize the probability or error in classi-fication to make it more precise. The distance is called marginal distance or margin. Adjusting the parameters for training can make the model even more precise. Kernel is said to be a method to compute dot products of the vectors.

Algorithm:

1. Use a dataset for training from the local repository while selecting the kernel and parameters necessary.
2. Train the model and run tests.
3. If the accuracy is promising then move to testing phase with real data, else repeat step 1.

2.2 Improved Levenberg Marquardt Algorithm

An improved version of Neural Network called Levenberg Marquardt algorithm was developed to use minimum storage and time interval to estimate the end of the day value of an investment. The improved LM algorithm showed 53% less error than ANFIS, while requiring 30% less time and 59% less memory than ANFIS [4].

Neural Networks is composed of the simultaneous functioning of basic compo-nents. Stimulated by genetic neural structure, one can alter the denomination of the weights (edges) amid the elements to help train the execution of specific functions.

LM algorithm stands for an algorithm which shows the sharpest depreciation for stability that it provides the Gauss Newton algorithm with to increase its speed in pro-cessing. The steepest descent algorithm [5] is very robust because it could converge independent of the fault overlay appearing more compound than the quadratic state. Even though the Levenberg-Marquardt approach is believed to be a proficient method, calcu-lating huge Jacobian's equation requires a sizeable memory. The large matrices to undergo inversion for calculation, concludes in higher computational interval. Modifications are brought about in the Levenberg-Marquardt process to decrease computational charge.

Algorithm:

1. Depending on the parameters used, update the function.
2. Update the Jacobian values of the matrix.
3. Change the scaling matrix.
4. Change the damping parameter.
5. Adjust parameters if required and run the training phase.
6. If the output is as desired and precise then end training phase and start testing with real data or else repeat from step 1.

2.3 Self-adapting Variant PSO-Elamn Neural Network

Having shown remarkably good results in optimization, the Particle Swarm Optimization algorithm uses the concept of exploration and then exploitation for optimization in multiple stages. In neural networks, there is something called as a recurrent neural network which essentially takes input, processes it and then gives output. Now the output of the hidden layer is used as the input for it again to process making it sensitive to history of data. It takes in inputs and then after the operations performed, uses that information to predict the next position of data points (Fig. 1).

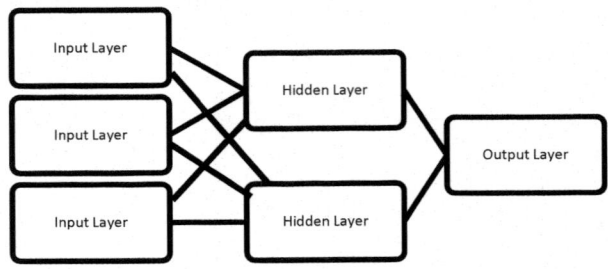

Fig. 1. Artificial Neural Network

The leftmost layer is called the input layer. The rightmost is the output layer where in the output is displayed. The middle layer is called hidden layer as the layer doesn't have input or output components. It does the operations on the input to give us the output. This can also be used to estimate the prices for the opening stock using the older data. Using the feedback property increases the efficacy of the dynamic processing making this even more effective.

Particle Swarm Optimization uses particles to gather the distance between points of reference to the destination. When a particle find an optimal path, it updates the other particles about its position bringing them closer to the optimal path too. During another iteration of the search, the area to be searched is increased making the search wider and more effective in finding the destination by avoiding the problem of early conjunction which makes it self-adapting [6, 7]. This technique utilizes historical data for predicting the opening value [8].

Algorithm:

1. Adjust inputs according to the model's structure and give the parameters.
2. Define limit and weights. If there are P limits and weights then the particle swarm would be denoted by P-dimensional vector comprising of P limits and weights.
3. Define the error margin and run test.
4. If performance is as desired, exit training and test using testing data.

2.4 Linear Regression

An approach in which the model comprises of the association amongst a scalar field y and an illustrative variable represented by x is known as linear regression. For prototyping the elements and assessing undefined standard factors from the data set, linear operator is used which is a collection of independent variables which in turn aids in determining the likelihood of the result of a conditional variable.

Linear regression analysis emphasizes upon the provisional possibility of 'y' along with 'x' being pre-defined more than the probability of 'x' and 'y' together. This evaluation would contribute to the field of analysis based upon statistical standards [9].

This approach is applied extensively in practical conditions since prototypes which differ undeviatingly irrespective of their unspecified factors tend to be effortless to fit than those prototypes which are related non-linearly to the factors affecting them since the subsequent estimators have easy to decide statistical properties.

For forecasting, a predictive model with fitted linear regression can be used to discern the variables 'y' and 'x' present in the data set. After developing such a model, if an additional value of X is then given without its accompanying value of y, the fitted model can be used to make a prediction of the value of y.

Linear regression models are often fitted using the least squares approach, but they may also be fitted in other ways, such as by minimizing the "lack of fit" in some other norm.

3 Analysis

The performance was calculated and analyzed for different samples of training and testing. Taking in more parameters allows a model to make more complex computations and hence give better results at the price of memory and time consumption. A model that uses less memory allocation and training time is an optimum method. These models do not show the exact point-to-point precision but give us an idea of the elevation and depression (Table 1).

4 Conclusion and Future Work

In this paper, a survey of various methods of machine learning for stock market prediction and trend recognition were described. It can be concluded that Improved Levenberg Marquardt [4] is one of the best tools as it has the highest level of accuracy and efficiency with less memory consumption in the short term but in the long run, along with more complexities, the PSO-Elman model is better suited.

The local datasets used had limitations in the parameters that could have been used. In future, more complex datasets with more training and testing parameters for input could be used making the methods even better by reducing memory wastage, computation time and while giving an even more real precision with the prediction.

Table 1. Comparison of different methodologies

Properties	Algorithms used			
	SVR	LM	PSO	LR
Training time	Short training time as it is used for small data sets	Fast as the newer algorithm increases computation speed	Fast because it is made for short-term opening stock forecasting	Fast as it just needs to save the weights at the end of training
Memory wastage	High because of the multiple windowing operations taking place	53% memory wastage as compared to original LM algorithm	High since it is a self-adapting algorithm	Low because of single input and low number of parameters
Complexity	Low complexity as it cannot accommodate more parameters	High complexity because it can solve more than just quadratic equations	Increases with increase in number of variable and vice versa	Low complexity since only a single input can be put
Error margin	Satisfactory accuracy but not precise	Need more parameters for higher precision	High error margin because of fluctuations in market price	Low precision due to limited parameters
Mode of classification	Uses windowing mode with SVM for accurate prediction	Uses neural network [10–12] to consider the uncertainity and non linearity	Self-adapting so it can learn and predict the pattern	Uses "Lack of fit" method to calculate missing parameter for prediction
Reliability	Good for small data sets with low complexity	Good for medium to large data sets with higher complexity	Good for larger data sets as it becomes more accurate	Good for small datasets with low complexity

References

1. Ince, H., Trafalis, T.B.: Kernel principal component analysis and support vector machines for stock price prediction. In: IEEE Conference (2004)
2. Lai, L.K.C., Liu, J.N.K.: Stock forecasting using support vector machine. In: Proceedings of the Ninth International Conference on Machine Learning and Cybernetics, pp. 1607–1614 (2010)
3. Meesad, P., Rsel, R.I.: Predicting stock market price using support vector regression. In: International Conference on Informatics, Electronics & Vision (ICIEV) (2013)

4. Billah, M., Waheed, S., Hanifa, A.: Stock market prediction using an improved training algorithm of neural network. In: 2nd International Conference on Electrical, Computer & Telecommunication Engineering (ICECTE) (2016)
5. Hao, Y., Wilamowski, B.: Levenberg–Marquardt training. In: Electrical Engineering Handbook (2011)
6. Zhang, Z., Shen, Y., Zhang, G., Song, Y., Zhu, Y.: Short-term prediction for opening price of stock market on self-adapting variant PSO-Elman neural network. In: Computational Intelligence and Neuroscience (2016)
7. Huaijun, L., Xiaopeng, X., Xinyuan, X.: Study on particle swarm optimization algorithm with parameters adaptive mutation based on particle entropy. Comput. Eng. Appl. **50**, 27–31 (2014)
8. Jie-min, Z., Zai-xing, Z., Yan-jun, L., Hao-yin, P., Qiang, G.: Predication of calorific value of ignition temperature of blended coal based on Elman neural network. J. Central South Univ. (Sci. Technol.) (2011)
9. Sharma, A., Bhuriya, D., Singh, U.: Survey of stock market prediction using machine learning approach. In: International Conference on Electronics, Communication and Aerospace Technology (ICECA) (2017)
10. Hagan, M.T., Demuth, H.B., Beale, M.H., Jess, O.D.: Neural Network Design, vol. 20. PWS Publishing Company, Boston (1996)
11. Hsieh, T.J., Hsiao, H.F., Yeh, W.C.: Forecasting stock market using wavelet transforms and recurrent neural networks: An integrated system based on artificial bee colony algorithm. Appl. Soft Comput. **11**, 2510–2525 (2011)
12. Wilamowski, B.M., Chen, Y., Malinowski, A.: Efficient algorithm for training neural networks with one hidden layer. In: IJCNN 1999 International Joint Conference, vol. 3 (1999)

Review of Data Protection Technique

Divyang Kaushik[✉] and Rejo Mathew

Department of IT, Mukesh Patel School of Technology Management
and Engineering, NMIMS Deemed-to-be University, Mumbai, India
divyang.kaushik5@gmail.com, rejo.mathew@nmims.edu

Abstract. The increase use of data in the businesses have proved to be profitable but also provides serious threats in form of data leakage, protection and infringement. The company provides sensitive data to the employees but the data is most likely to be leaked. Data protection must be provided to all forms of data. Data protection should be provided in such a way that it strikes balance between data allowance and data rights to the user as well as owner. The level or extent of protection varies with methods used. In this review paper, numerous methods for protecting data from leakage, unauthorized access, unfair use etc. are compared.

Keywords: Data leakage · Data leakage prevention (DLP) · Confidential data · Firewall · Unauthorized user · Sensitive data · Fake data · Data request · Guilt model · Data detection · Data prevention

1 Introduction

Considering the fact that data owner shares the data to the agents. For example, financial data, medical records, partnership etc. are given to agents but are then leaked. To avoid this initially Perturbation is used [1]. Perturbation is a method in which the accuracy of the data given to the agents are reduced by replacing accurate value by a range or adding some noise to the data. Perturbation doesn't stand useful in domain where accuracy of data holds high importance e.g. medical records of the patients.

Earlier the data leakage was detected using watermarking. Watermark provided some unique codes added to the data by modifying it. Watermark proved to be very useful but the code could be cracked if data was malicious.

Leaked data may involve source codes, design, trademarks, sales etc. Detected data leak can help owner assess the guilty agents or third parties. With enough prove owner can stop business with other party or proceed with legal case.

2 Related Work

2.1 Offline Data Distribution

In the system owner will not be able recognize the source of leaked data. In the system more focus is made on allocation of data rather than identifying the agent found to be guilty. The system for data leakage is not online and hence the data leakage scenario is

A. P. Pandian et al. (Eds.): ICCBI 2018, LNDECT 31, pp. 198–203, 2020.
https://doi.org/10.1007/978-3-030-24643-3_24

not monitored. Only thing fixed in existing system is that the request by the agent is known in the advance to the owner.

2.2 Intrusion Detection Systems (IDS)

IDP is a device or software application that monitors networks or system activities for malicious activities [2, 3].

2.3 Intrusion Protection Systems (IPS)

Monitors networks, system activities for malicious activities it mainly identifies malicious activities, log information, attempt to block or stop activities and report activities.

Both IDS and IPS systems looks at data with intention of recognizing attack. IPS is extension of IDS, IPS monitors network as well as malicious activities. IPS can perform function such as alarming when the intrusion is detected and also leave malicious packets. IPS also block.

There are two types of intrusion system network based and host based i.e. NIDS/NIPS and HIDS/HIPS and can be used to scan for attack, track hackers' movements, and alert an administrator to ongoing attacks [2, 3].

Problem

The problem with IPS and IDS is that both the systems are signature based. Intrusion detection is only triggered for the loaded signatures. Due to signature based system new attacks may go undetected.

2.4 Anti-malware

Malware is programming intended to harm PC operations, collect sensitive data, and gain unauthorized access to computer system. Type of malwares is virus, worms, Trojans, spyware, backdoors, and root kit [4, 7].

Hostile to malware is any product that give a shield PCs and frameworks from malware, infections, spyware and other unsafe program.

Problem

The problem with Anti-malware software is that it only works effectively in real-time environment. Anti-malware software only checks attacks from outside by scanning and validating signatures.

2.5 Firewalls

Firewalls are gadgets or programming that grants or denies arrange transmissions dependent on an arrangement of principles (get to run) and is utilized to shield systems from unapproved get to while allowing legitimate correspondence to pass [5].

Firewall is programming or equipment that aides in keeping system secure, its goal is to control the approaching and active movement of systems by investigating the information bundle and deciding if it ought to be permitted through or not.

Problem

Rules for firewalls are exists without any documentation and can create security management issues.

Outdated Firewall OS software are unable to support exploits like remote code execution and denial of service attacks.

2.6   Deep Packet Inspection (DPI) Method

Above stated techniques in 2.2, 2.3 and 2.4 uses DPI method as the base technology for data protection. DPI inspects the packet for any anomalies and on finding any prevents the packet or traffic [6, 8, 9].

Problem

The main problem with DPI technologies with that it adds complexity to already complicated system like intrusion detection systems (IDS), firewalls etc. The native vulnerabilities in Data Packet Inspection are induced in the system by mechanism of their action. Deep Packet Inspection are required to be regularly monitored and undergoes frequent configuration changes.

3   Solutions/Countermeasures

3.1   Solution for 2.1

In this system lies two model. First model is used to identify the agent who tries to access sensitive data without the consent of owner. If the agent is identified, then there is no need to proceed for 2nd round. If the leaked data is given to the third party or outside the enterprise, then second method is created to assess the guilt of agent. Second method is basically used to increase probability of finding third parties.

For the application of proposed system, a database of college management system is treated as owner and the employee as agent. In this scenario real and fake data objects are used. Fake objects are used in the way that it looks real and useable to the agents. All the fake objects are used as watermark. As and when more of the fake objects are leaked by the agent then the owner can be more confident [10, 11].

Two processes for working of proposed system:

(a) **Data Distribution Process**

There are essentially four instance for data allocation

1. E for explicit data
2. S for sample data
3. F for use of fake data
4. F where use of fake data is not allowed

Fake data is not in the set T. F is send with data T to increase probability of detection [1].

A. Probability Finding Process

The parameter would then be checked once an information question is gotten from a malignant focus for that utilized an uncommon procedure for information protest is gotten from any objective the likelihood is computed the information protest originated from which source or we can figure what specialist has released the information [12].

B. Algorithm for Finding Guilt Agent [11]

1. Data is distributed by the owner to the agents.
2. Fake data is created by the owner to increase the probability of finding Guilty agent.
3. Data objects are sent to agents with or without fake data.
4. Data objects are sent to the agents based on the type of request.
5. Owner checks number of agents who have received the data objects.
6. Checks the agents who haven't received the data.
7. Allocation done by adding fake data to improve the probability [13].

3.2 Solution for 2.2, 2.3, 2.4 and 2.5

Data Leakage/Loss Prevention System (DLP)

DLP systems are designed to identify and prevent loss of valuable enterprise data. DLP provides enterprise machines and the management with centralize network logs.

DLP systems provide data security to three data states Data-at-rest (DAR), Data-in-motion (DIM) and Data-in-use (DIU). Data-at-rest (DAR) is defined to be any data stored on the device, server or network. Data security to DAR can be implemented by using content discovery feature, if the data is detected on unauthorized server or network. DLP solutions helps data by encrypting, removing or notifying the data owner. Data-in-motion is data that is being sent over a network and detection of sensitive encrypted data is carried. Data-in-use is the one where data is being currently used or transferred. DLP system can detect the data leakage and prevent transfer or use of sensitive data without authorization.

4 Analysis/Review

Comparison table

	Offline data distributor	Intrusion detection System	Intrusion protection system	Antimalware	Firewalls	Deep Packet Inspection
Focused data allocation	Performs data allocation to agents	Does not perform data allocation	Does not perform data allocation	Implies prevention, detection and removal	Controls and monitors traffic	Checks packet format and type of data

(continued)

(continued)

	Offline data distributor	Intrusion detection System	Intrusion protection system	Antimalware	Firewalls	Deep Packet Inspection
	Request for data is known	Data detection is used	Data protection is performed	Not responsible for data allocation	Doesn't allocate exclusively	Does not allocate data to agents
Online system	Not online, works only for defined sets of system	Online system	Online system	Online system	Online system	Online system
Signature based system	Uses fake objects for detection	Detects malicious data using signatures	Detects malicious data using signatures	Uses signatures and keep updating new ones	Uses predetermined internet rules.	Performs packet filtering for specific data and code

5　Conclusion

There are different systems and models used to protect data from leakage, theft and unauthorised access. Above methods such as guilt model with data watcher and Data leakage/loss prevention system proves to useful and optimized. Above two solution provides a safer and more protective alternative as compared to various mentioned methods.

References

1. Papadimitriou, P., Garcia-Molina, H.: A study on video browsing strategies. IEEE Trans. Knowl. Data Eng. **23**(1) (2011)
2. Lahoud, H.A., Tang, X.: Information security labs in IDS/IPS for distance education. In: SIGITE 2006 (2006)
3. Dhiren, M., Joshi, H., Patel, B.K.: Towards application classification with vulnerability signature for IDS/IPS. In: SecurIT 2012 (2012)
4. Williamson, S.A., Varakantham, P., Gao, D., Chen, O.: Active malware analysis using stochastic games. In: 11th International Conference on Autonomous Agents and Multiagent Systems—Innovative Applications Track, AAMAS (2012)
5. Wikipedia. http://en.wikipedia.org/wiki/WikiLeaks (2008). Frost, Sullivan. World Data Leakage Prevention Market. Technical report ND34D-74
6. IDS AND IPS Placement for Network Protection. http://www.infosecwriters.com/text_resources/pdf/
7. Rutkowska, J.: Introducing stealth malware taxonomy. COSEINC (2006)

8. SearchSecurity. http://searchsecurity.techtarget.com/definition/security-information-and-event-management-SIEM

9. Ye, Y., Li, T., Zhu, S., Zhuang, W., Tas, E., Gupta, U., Abdulhayoglu, M.: Combining file content and file relations for cloud based malware detection. In: KDD 2011 (2011)

10. Kale, S.A., Kulkarni, S.V.: Data leakage detection: a survey. IOSR J. Comput. Eng. (IOSRJCE) 1(6), 32–35 (2012)

11. Umamaheswari, S., Geetha, H.A.: Detection of guilty agents. In: Proceedings of National Conference on Innovations Emerging Technology (2011)

12. Mishra, R., Chitre, R.K.: Data leakage and detection of guilty agent. Int. J. Sci. Eng. Res. (2012)

13. Kale, S.A., Kulkarni, S.V.: Data leakage detection. Int. J. Adv. Res. Comput. Commun. Eng. 1(9), 668–678 (2012)

Review of Blind Navigation Systems

Anubha Mishra[✉] and Rejo Mathew

Department of IT, Mukesh Patel School of Technology Management
and Engineering, NMIMS (Deemed-to-be) University, Mumbai, India
anubha19mishra@gmail.com, rejo.mathew@nmims.edu

Abstract. Being aware of their surroundings or unfamiliar environments is
very crucial for the blind and visually impaired. Earlier works to introduce a
system to enable blind navigation included tactile paving detection, ultrasonic
sensor blind walking stick, etc. In recent times, with building complications
newer systems have been introduced, which include RFID based audio guidance
cane and cloud computing based context aware blind navigation system. Being
aware of one's surroundings is a critical aspect of safe navigation. In this paper,
we look at the working of these methods and review these systems based on
several parameters. This paper will have four sections: Introduction, Blind
Navigation Systems, Analysis, and Conclusion.

Keywords: Tactile paving detection · Context aware blind navigation ·
Ultrasonic blind walking stick · Cloud based navigation ·
Blind navigation system · Cloud computing based navigation

1 Introduction

As per a 2014 UN report [2], there are more than 1 billion disabled people in the world.
To ensure that disabled people get equal opportunities and lifestyle, these people need
to be given more priority. For the people that are visually impaired, getting aware of the
surroundings and environments around them is crucial for navigation. There has been a
lot of work and research to introduce efficient systems to enable blind people to
navigate and get accommodated to their surroundings better.

Earlier work [6], for example tactile paving detection systems and ultrasonic sensor
blind walking stick were efficient in some aspects. But with time, more complications
came into the picture and the need for newer, better systems increased. Newer systems
introduced include RFID dependent audio guidance cane and cloud computing based
context aware blind navigation system. These systems enable the users to navigate
indoors as well as outdoors, recognize people and objects, and they are functional with
minimal resources.

© Springer Nature Switzerland AG 2020
A. P. Pandian et al. (Eds.): ICCBI 2018, LNDECT 31, pp. 204–210, 2020.
https://doi.org/10.1007/978-3-030-24643-3_25

2   Blind Navigation Systems

2.1   Tactile Paving Detection Method [3]

A video or image input device, for example a camera or a webcam, detects the tactile. The output of this system is auditory. This system works in five stages, as demonstrated in Fig. 1:

Stage 1: Input the image/video which contains the tactile.
Stage 2: Pre-processing the input. This includes filtering of the noises.
Stage 3: Extracting and determining the connected components in the image.
Stage 4: The metrics (like the area, distance) for the components are determined.
Stage 5: Producing the accurate auditory output to the user.

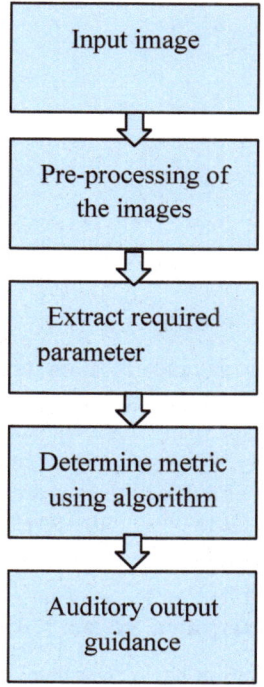

Fig. 1. Tactile paving detection method

2.2   Ultrasonic Walking Cane [8]

This system consists of a walking stick with an ultrasonic sensor integrated in it and it supports light and water sensing. The ultrasonic sensors sense obstacles. This data is then sent to the microcontroller which processes it and determines how close the obstacle is. Accordingly, it sounds a buzzer to alert the user. There are different types of buzzers that can be sounded in different contexts. For example, a light buzzer is used to inform the user of an obstacle present at a far away distance. A heavy buzzer is used for alerting the user in case of any emergency when any obstacle is close to the user. Another feature is that the user can detect if there is light in the room or not. Additionally, the user can also press a remote button to sound the buzzer on the stick in order for them to be able to find the stick. The working is also demonstrated in Fig. 2.

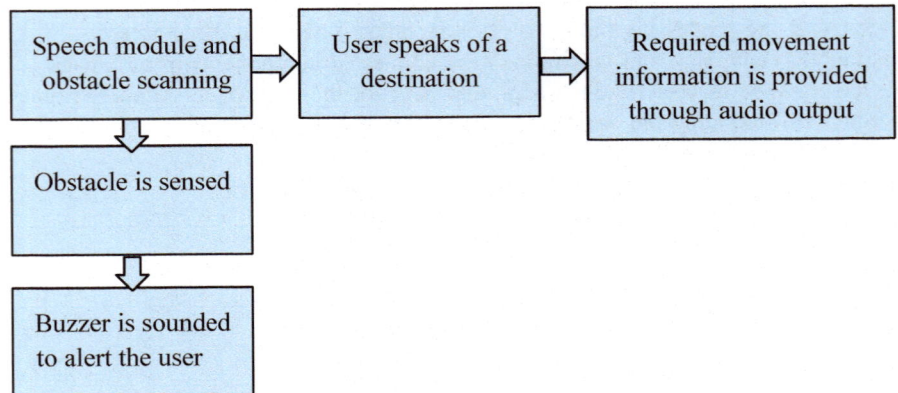

Fig. 2. Ultrasonic walking cane working

2.3   RFID Based Guidance Cane [9]

In this system, RFID (radio frequency identification) tags are used along with a walking stick. These RFID tags are the data map and the user carries the cane. The user speaks to the system the desired place, and using three layers of map, RFID tag map sends the required movement information to the user in form of auditory output. An ultrasound sensor is also integrated in front of the cane to alert the user in around an obstacle. This system doesn't require for the user to carry any other device apart from the cane, even for navigational purposes. The working is demonstrated in Fig. 3.

2.4   Context Aware Blind Navigation System [10]

This system works on a collaboration between several mobile devices, the resources on Internet and cloud computing providers. Visual data captured by the camera (integrated in glasses) are fed to the mobile device. A speech interface takes instruction from the user and provided relevant feedback as auditory output. The working of Context Aware Blind Navigation System is demonstrated in Fig. 4.

The RFID and cloud based systems also use three modes to provide efficient results according to the scenario:

(1)   Stable mode-online
(2)   Fast mode-online
(3)   Fast mode-offline

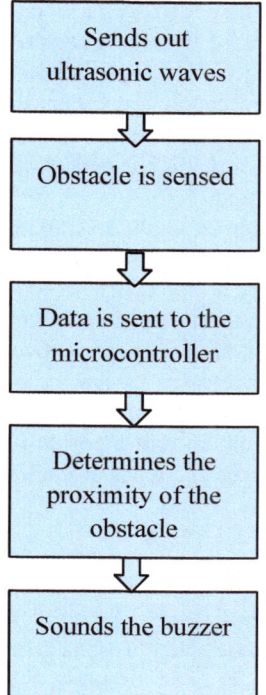

Fig. 3. RFID based guidance system working

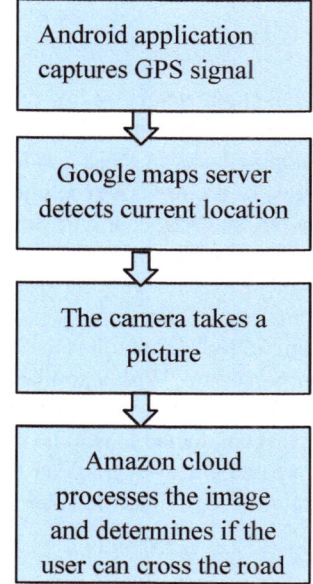

Fig. 4. Context aware blind navigation system working

Given the requirements of a particular scenario, one of these modes is selected. For example, if a blind person is walking through a campus. In the first scenario, the person might require high recognition accuracy. This would require the system to operate in 'stable mode', where the system captures color and depth information and specifies what the obstacle is. The accuracy [4] here is high, but the time taken and the information required is more. With this mode, the person can eventually get acquainted with the surroundings.

In the second scenario, the person might be walking through a crowded area with a lot of obstacles. Since he/she has a high possibility of colliding with some obstacle, the system needs to respond quickly. In this case, the system is run on 'fast mode', where the system doesn't capture in depth information about an obstacle, but informs the user of many obstacles in a short interval of time. The processing time is short [5]. In the last scenario, where the system, in rare cases, cannot be connected to any network, it runs on an offline fast mode. Here, the system just determines where the obstacles are present, with the limited information and algorithm stored on the smartphone. The response is fast, but it doesn't specify what obstacles come in the way. These three modes are compared in Table 1.

Table 1. Performance comparison between three modes provided by RFID and cloud based systems.

	Stable mode-online	Fast mode-online	Fast mode-offline
Accuracy	High	Medium	Medium
Reliability	High	Medium	Medium
Process speed	Slow	Normal	Fast

3 Analysis

The problem of pedestrian signal status detection has two important aspects: response time and accuracy. Tests have been performed to measure the response time of the network based systems. The sample tasks in these conducted tests were processing of different resolution level versions of pictures. A resolution level of 0.75 stands for the original frame was taken as the original frame, and the lower resolution levels represent the compressed versions. The test results have been demonstrated in Fig. 5. This involved processing different pictures with varying resolution levels. Lower quality, compressed pictures result in lower response time.

Eliminating the calculation of redundant distances and patterns is one way to accelerate the pattern selection procedure. With a similar error rate, almost 40 times of the time taken was decreased. New solutions are required because the existing ones fall short on some aspects [7]. In a test conducted to test the performance of all the systems, a simulated environment was created [12]. A parameter based comparison between the four methods for blind navigation is described in Table 2.

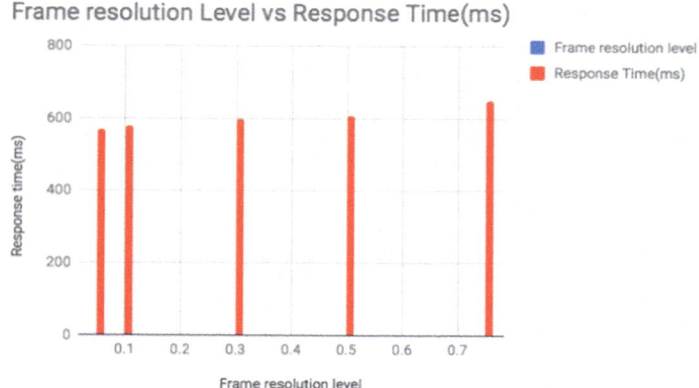

Fig. 5. Frame resolution level vs response time (ms)

Table 2. Parameter based comparison between the four studied methods for blind navigation.

	Tactile paving detection	RFID based guidance	Ultrasonic blind walking stick	Context aware cloud computing
Detection based feature	Image/video	Radio frequency identification tags	Ultrasonic sensors	Image/video
Sensing device	Particular device	Particular device	Particular device	General mobile device/smartphone
Distance judgement	No	Yes, with high accuracy	Yes, with high accuracy	Yes, with high accuracy
İnformation provided	Obstacle presence. Distance from the user	Obstacle presence. Navigation	Obstacle presence. Distance from the user	Obstacle/people recognition. Navigation

4   Conclusion and Future Work

The performance comparison of several blind navigation systems was studied in this paper. It was found that comparing characteristics like indoor navigation, object recognition, learning curve, etc., context aware blind navigation system proves to be better in most aspects than the systems based around a particular infrastructure. The result also dictates that the system even though being extremely efficient [11], raises some concerns about privacy. Future work on development of blind navigation systems would include improving upon the security aspects of the system, developing algorithms to develop privacy prevention. This could help achieve selective disclosure of information. With rapid developments in fields like machine learning, IoT, AI, such systems can be made more intelligent and automated.

References

1. World Health Organization Fact Sheet No 282 (2009)
2. UN world disabled population report (2014)
3. Vision based of tactile paving detection method in navigation system for blind person (2015)
4. Bohonos, S., Lee, A., Malik, A., Thai, C., Manduchi, R.: Universal real-time navigational assistance (2007)
5. 1st ACM SIGMOBILE International Workshop on Systems and Networking Support for Healthcare and Assisted Living Environments (2007)
6. Ringbeck, T., Moller, T., Hagebeuke, B.: Multidimensional measurement by using 3-D PMD sensors. Adv. Radio Sci. **5**, 135–146 (2007)
7. Hussmann, S., Ringbeck, T., Hagebeuke, B.: A performance review of 3D TOF vision systems in comparison to stereo vision systems (2008)
8. Addressing Physical Safety, Security, and Privacy for People with Visual Impairments
9. Nevon Projects, Ultrasonic Blind Walking Stick Project (2016)
10. Angin, P., Bhargava, B., Helal, S.: A mobile-cloud collaborative traffic lights detector for blind navigation. In: Proceedings of the 1st International Workshop on Mobile Cloud Computing (2010)
11. A cloud-based outdoor assistive navigation system for the blind and visually impaired
12. Tombari, F., Shioyama, T.: Detection of pedestrian crossing using bipolarity and projective invariant. In: Proceedings of the IAPR Conference on Machine Vision Applications (2015)
13. ImageNet Large Scale Visual Recognition Competition (2012)

Analyzing DFA Attack on AES-192

Tadepalli Sarada Kiranmayee$^{(\boxtimes)}$, S. P. Maniraj$^{(\boxtimes)}$,
Aakanksha Thakur$^{(\boxtimes)}$, M. Bhagyashree$^{(\boxtimes)}$, and Richa Gupta$^{(\boxtimes)}$

SRM Institute of Science and Technology, Chennai, India
tskiranmayee@gmail.com, maniraj.p@rmp.srmuniv.ac.in,
akanksha6512@gmail.com, bhagyashree5599@gmail.com,
richa.gupta9a@gmail.com

Abstract. It has been detected that the existing Differential Fault Analysis needs to modify the Advanced Encryption Standard for easy key retrieval process. The attack of DFA on the AES-128 requires a massive amount of faulty ciphertext pairs. Key retrieval using DFA attack on AES-128 requires minimal of 250 pairs of faulty cipher-text. This can be reduced to 14 to 16 pairs of faulty cipher-text samples in AES-192 bits and AES-256 bits. This was done by injecting faults randomly into the 11th cycle key which is saved in the Static RAM-Random Access Memory. Researches are made on methods for finding the initial key and that has only been suggested in AES-128. This paper puts forward a system which, intends to induce DFA attack on AES-192 and AES-256 algorithms with less number of faulty cipher-texts.

Keywords: Advanced Encryption Standard · Differential fault analysis · Faulty cipher-text

1 Introduction

AES, abbreviated as Advanced Encryption Standard formerly known as Rijndael was created first by Belgium designers Vincent Rijmen and John Daemen. It's a symmetric encryption algorithm. Encryption is the method of transforming the original human readable text to a non readable text called as the cipher text, to prevent the original text from security breaches. It prevents other third parties from seeing and stealing delicate-sensitive data. AES crypto system is now used globally to encrypt very confidential data. It has replaced the former encryption algorithm named as the DES, abbreviated as Data Encryption Standard due to its high efficiency. Some of the other factors include the following:

- The key size of DES was a 56 bits one, which was way to small, as compared to the AES key sizes-128,192,256.
- A computer can be constructed which would find out the DES key within seconds. But it would require 149 trillion years to crack the AES-128 key by the same computer.
- The DES algorithm was open to Brute Force attacks therefore, a need for a new algorithm arised. This is how AES came into picture.

© Springer Nature Switzerland AG 2020
A. P. Pandian et al. (Eds.): ICCBI 2018, LNDECT 31, pp. 211–218, 2020.
https://doi.org/10.1007/978-3-030-24643-3_26

- DES was prone to the linear cryptanalysis and the differential cryptanalysis. Cryptanalysis is the way of finding out the key or the original non-cipher text.
- AES was proven to be five to ten times secured and quicker than the advanced version of DES which is the 3DES. 3DES is nothing but TDEA commonly abbreviated as, the Triple Data Encryption Algorithm, in this method DES is applied thrice to each block.
- AES works efficiently both on hardware and software.
- After conducting many researches, DES is proved to be inadequate and AES is proven to be secured.

AES is said to be a symmetric block cipher, meaning the key used for the encryption and the decryption process is the same one. As said earlier encryption is the method of converting normal text to an indecipherable text called the cipher-text. And decryption is the counter process that is, getting back the normal text from the cipher text. There are two types of ciphers: stream cipher and block cipher. Stream cipher encrypts bit by bit and Block cipher applies encryption on the whole block of text. AES uses various cryptographic sizes of keys-128, 192, 256 bits for the encryption and the decryption of cipher blocks-128, 192, 256 in 128 bits block [5]. It comprises of 10, 12, 14 cycles based on the key size range. The size of the key here describes. how many conversions are required to alter the normal text to the end cipher-text. Every round or cycle is comprised of various steps containing the one that's based on the initial key itself. A collection of counter rounds or cycles are used to alter the cipher back to the normal text by utilizing the same key (Fig. 1).

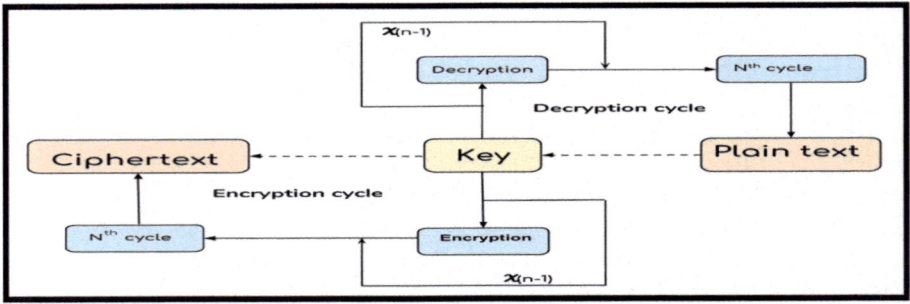

Fig. 1. AES architecture.

Various attacks which are much more quicker and robust than the brute force technique have been enacted but none were feasible computationally. Biclique attack can break AES-128 algorithm by recoveration of key using the complexity $2^{126.1}$. Similarly complexity of $2^{189.7}$ can be used to crack AES-192 and $2^{254.4}$ can be used for AES-256. DFA known as, the Differential Fault Analysis is nothing but a side channel attack, where in faults are induced randomly to disclose their internal states. It's one of the cryptanalysis methods. It's utilized for testing the security of the AES algorithm regularly, so that there are no security breaches. It presume that a hacker can possibly generate faults into a system and gather both the faulty and the correct cipher texts. After juxtaposition of both the texts the key can be reclaimed. DFA attacks on AES have been gaining a wide recognition for the further development and enhancement of AES.

2 AES Algorithm

Encryption algorithm-AES is widely known for its speed and unique key generation. It's an algorithm where two parties - receiver and sender will have the same keys. This algorithm helps in creating unique fixed length keys. The keys generated by AES are 256 bits, 192 bits and 128 bits [5]. The key length depends on the type of AES algorithm used. For instance, AES-256 is used to produce 256 bit keys. AES algorithms also have constant block size. Data that requires encryption is passed through the AES and the output will be a fixed length of ciphertext this ciphertext will again encrypt likewise there will be multiple rounds for each block of encryption process or decryption process. AES algorithm includes 14, 12 in addition to 10 respective cycles for 256 bits, 192 bits in addition to 128 bits. Plaintext is extracted by the use of reverse round of sets on ciphertexts. Each cycle of AES includes following steps:

A. SubBytes
This steps involves the substitution of input data bytes by another bytes by the use of multiple S-boxes (substitution boxes). Input bytes $S_{i,j}$ can be transformed using the s-box lookup table to get another byte S'_{ij}. Each block represents one byte of data. The coordinates in the box are presented in hexadecimal form which is determined by intersection of row and column index [1].

$$S'_{ij} = S\_box\left(S_{i,j}\right)$$

B. ShiftRows
Rows are shifted in cycles over different number of bytes by indexing rows. This results in new set of arrangements of data bytes in each rows (Figs. 2, 3).

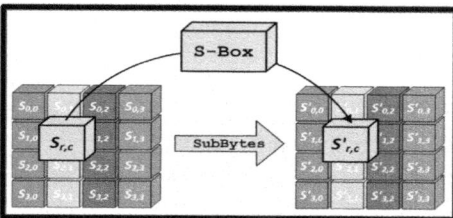

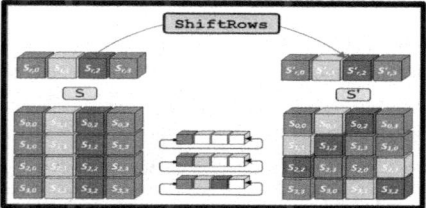

Fig. 2. SubBytes [11].   **Fig. 3.** Shift rows [11].

C. *MixColumns*

The output from the previous step will be mixed column by column by multiplying it with fixed polynomial generating new state [2] (Fig. 4).

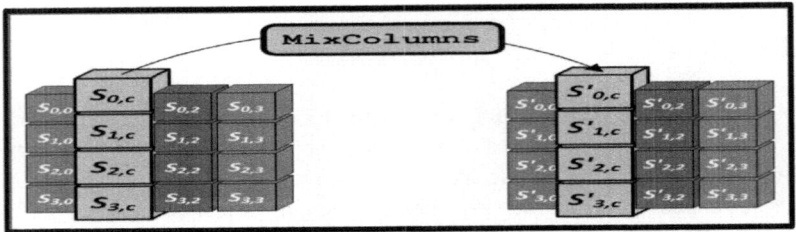

Fig. 4. MixColumns [11].

D. *AddRoundKey*

In this step, state contains byte from the previous step combined with the byte from round key performing exclusive-or operation (Figs. 5, 6).

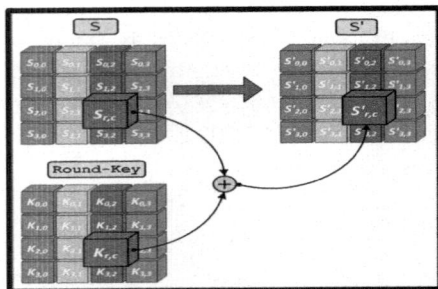 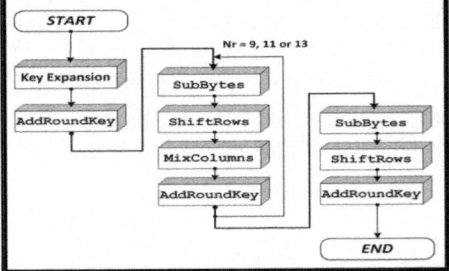

Fig. 5. AddRoundKey [11].   **Fig. 6.** AES flow graph [11].

3 DFA Attack on AES-128

Differential fault analysis, aiming Advanced Encryption Standard execution has turned into an important research topic [4]. Differential Fault Analysis attack is one of the side channel attack in the domain of cryptography that deals with inducing faults or errors into cryptographic implementations in order to analyze their internal states. The attack is a white box attack where the attacker is aware of the implementation of algorithm [9].

DFA attack is used on AES to extract keys of the algorithm [8]. This attack is more prevalent in white box cryptographic algorithms where the algorithm is easily available but the keys are not. Fault is induced in the algorithm which would modify the output and produce faulty ciphertexts. Analyzing pairs of faulty ciphertexts and fault-free help retrieve AES key [7]. In AES-128 algorithm, single byte fault injection can be done at the input of 8th round that can retrieve entire secret key with brute-force hunt 2^8 [6].

DFA includes either bit attack or byte attack. DFA bit attack induces fault on one bit of ciphertext. DFA byte attack works most prominently on AES-128 [12]. In this attack instead of targeting only one bit, fault is induced on one byte of ciphertext. After the injection process, pairs of fault ciphertext and original ciphertext are collected and analyzed. This analysis is done by a set of our own tools. Bit fault DFA attack induces fault only on a bit of temporary ciphertext starting of the terminal round. This helps in extraction of the final round key K^{10} which in return is used to extract the AES key K by using inverse key scheduling on K^{10}. In general, the final round ciphertext C using AES is evaluated as follows.

$$C = \text{ShiftRows}(\text{SubByte}(A^9)) \oplus K^{10}$$

Final round omits MixColumn step and hence, the composition of only the first two steps of AES is xored with the round key. Fault induced before the final round may result in a different ciphertext D. A_j^i or M_j^i represents the state at ith row and jth column. Consider ShiftRows(j) is the position of jth byte after applying ShiftRows transformation. Then the ciphertext at ith byte can be represented as $C_{\text{ShiftRows}(i)}$. If the fault f_j is induced at state j before final round then the faulty ciphertext is $D_{\text{ShiftRows}(j)}$ [3] (Fig. 7).

$$C_{\text{ShiftRows}(i)} = \text{SubByte}(A_i^9) \oplus K^{10}_{\text{ShiftRows}(i)} \tag{1}$$

$$D_{\text{ShiftRows}(j)} = \text{SubByte}(A_j^9 \oplus f_j)) \oplus K^{10}_{\text{ShiftRows}(j)} \tag{2}$$

For all $i \in \{0, 1, \ldots, 15\}/\{j\}$ we have:

$$D_{\text{ShiftRows}(i)} = \text{SubByte}(A_i^9) \oplus K^{10}_{\text{ShiftRows}(i)} \tag{3}$$

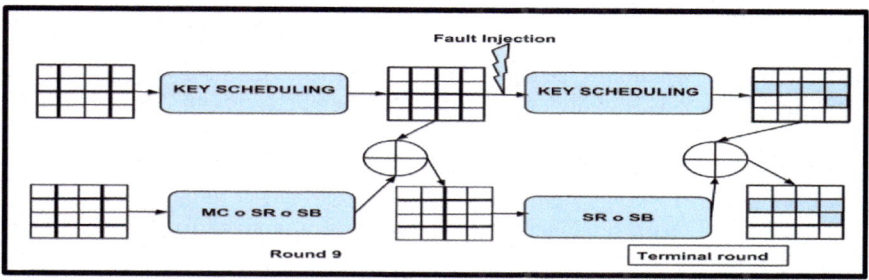

Fig. 7. Fault injection on penultimate round key.

If no fault is induced on ith byte then from (1) and (3) we have:

$$C_{\text{ShiftRows}(i)} \oplus D_{\text{ShiftRows}(i)} = 0$$

If fault is induced on byte A_j^9 then from (1) and (2) we get:

$$C_{ShiftRows(j)} \oplus D_{ShiftRows(j)} = SubBytes(A_j^9) \oplus SubBytes(A_j^9 \oplus f_j)) \qquad (4)$$

The position of only non zero byte can be obtained by determining ShiftRows(j) which would give us the value of j. Guessing the fault f_j, we find set of possible values of A_j^9 that satisfy (4). For each of these values counter is incremented by 1. The same steps are repeated with another faulty ciphertext. As the correct value of A_j^9 is expected to be counted more frequently than any incorrect value, A_j^9 is easily identified. The remaining bytes of A^9 can be obtained by iterating the previous process. By using only three faulty ciphertext along with the fault induced to the same byte we have 97% chance of having one value left on this byte [3].

DFA byte attack works most prominently on AES-128. In this attack instead of targeting only a bit, faults are injected on a byte of ciphertext. After the injection process, pairs of fault ciphertext and original ciphertext are collected and analyzed.

4   Proposed DFA Attack on AES-192

AES-192 key can be a 2-D array of 32 bit word arranged in 4 rows. DFA byte attack on AES-192 includes inducing fault and obtain round keys, which on further mathematical analysis would give us AES-192 key. The primary step involves inducing fault in the key-11 just before scheduling key-12. The comparison with faulty ciphertext and plaintext would give 4 bytes of key-11. Next, inducing fault in key-10 before scheduling key-11, we may obtain and analyze another set of faulty ciphertexts. This results in another 4 bytes of key-11. Atlast we may obtain the AES key by inducing fault on A^{10}, before entering the cycle 11, and by the 8 bytes of key-11 obtained.

Attack on key-11

- Inducing one byte fault on any one of the 4 bytes of the end column of key-11 before key-12 scheduling. Obtained key-12 will have 2 fault bytes in the last column. Faulty key-12 will generate final faulty ciphertext with 2 fault bytes out of 4 bytes in the last column.
- The correct ciphertext will not have the 2 fault bytes. Hence, comparing faulty and correct ciphertext we determine at which byte the fault occured.
- For accuracy, 32 pairs of ciphertext and plaintext would give final 4 bytes of key-11.

Attack on key-10

- By inducing fault on key-10, penultimate 4 bytes of key-11 can be obtained.
- Inducing single byte fault on one of the 4 bytes of key-10 before key-11 scheduling would generate fault bytes on first two words (32-bit row) and 2 byte faults on third word [10].
- Faulty ciphertext obtained are compared with the correct plain text and also with faulty key-11. At least 13 pairs of ciphertext from the same plaintext are required to obtain the 12th byte of key-11. Hence by increasing the input values we obtain 8 bytes of penultimate round key-11.

Attack on A^{10}

- Inducing single byte fault on A^{10} before the intermediate cipher enters the penultimate round would generate faults on last 4 bytes of A^{11}.
- This faulty cipher when passed on to the final round, will generate diagonal fault on 4 bytes, one in each row, starting from right to left.
- As in the last step 8 bytes of key-11 are known, faults induced in the A^{10} can be XORed with one of the known 8 bytes of key-11.
- Mathematically, If fault is induced on 12th, 1st, 6th or on 11th byte of A^{10} the result of these bytes after SubBytes, ShiftRows and MixColumns stages will be XORed with 12th to 15th known bytes of key-11. Hence, the fault can occur on one of these selective 8 bytes of A^{10} which can be XORed with the known 8 bytes of key-11.
- For the bytes obtained in the key-11, one fault byte may be present in any one of these byte numbers - 12, 1, 6, 11, 8, 13, 2 or 7. Obtained ciphertext will be different from the correct ciphertext. XOR operation of the ciphertext with plaintext 4 bytes at a time reveal last 4 bytes of the original intermediate cipher before MixColumn stage in round 11.
- After obtaining 8 bytes of the original ciphertext, MixColumns stage in round 11 is applied and followed by SubByte and ShiftRows. The result bytes are part of correct temporary cipher before XOR operation with key-12.
- After XOR, correct values of key-12 byte numbers - 2, 3, 5, 6, 8, 9, 12 and 15 are obtained. By using XOR operation and quick exhaustive search on unknown bytes of key-11, rest of the 8 bytes (4, 7, 10, 11, 13, 14, 0 and 1) of key-12 are obtained.
- Finally, full AES-192 key can be obtained by applying inverse of key scheduling.

With the above proposed algorithm the whole AES-192 key can be obtained with minimum of 50 pairs of fault ciphers.

5 DFA Attack on AES-256

Implementation of above DFA attack on AES-256 is similar to that of AES-192. Difference between the two is in the key scheduling process. As AES-256 uses 256 bit key, the encryption or decryption key is arranged in the form of 8, 32 bit word. Only the initial round key uses eight words, the round keys following the initial round key are segregated back into 4, 32-bit words in 14 rounds, generating 14 round keys. This segregation makes the execution of AES-256 similar to AES-192. Faults can be induced on penultimate (key-13), antepenultimate (key-12) round keys and on intermediate cipher A^{12}. The proposed algorithm then helps in retrieval of key-14. Inverse key scheduling would generate AES-256 key.

6 Conclusion

This paper is about DFA attacks on AES-256 & AES-192 has been proposed and analyzed. The major concern of this paper was to minimize the count of faulty ciphertext used for the whole DFA key extraction model. This is achieved by imposing faults on selective bytes instead of random fault injection. The latter reduces the predictability of fault locations and asks for more number of faulty ciphertexts, this would minimize the efficiency of key retrieval process. As the proposed model described in this paper is independent of the key length of the AES, It makes this model more comprehensive and scalable.

7   Future Work

The proposed model in this paper induces known faults on scheduling process of the key, AES-192. Random fault attacks can be made into the same system with unknown byte locations of the keys. This would decrease the analyzing time between the faulty ciphers and correct ciphertext. Random fault attacks in the physical model of this system can be made by a laser beam on an AES encrypted system like smart cards. The key retrieval processing time will be reduced.

References

1. Abraham, N.E.: FPGA implementation of SubBytes & inverse SubBytes for AES algorithm. Int. J. Sci. Res. Dev. (IJSRD) (2013)
2. Patel, R., Kanjariya, S.: Design of parallel advanced encryption standard (AES) algorithm. Int. J. Res. Comput. Commun. Technol. **4**, 219–222 (2015)
3. Giraud, C.: DFA on AES. Springer, Berlin (2005)
4. Liao, N., Cui, X., Liao, K., Wang, T., Yu, D., Cui, X.: Improving DFA attacks on AES with unknown and random faults. Sci. China Inf. Sci. **60**, 042401 (2016)
5. Aromoon, U.: An AES cryptosystem for small scale network. In: Third Asian Conference on Defence Technology (3rd ACDT). IEEE (2017)
6. Wang, P., Hao, L.: A novel differential fault analysis on AES-128. IEEE (2011)
7. Patranabis, S., Chakraborty, A., Mukhopadhyay, D., Chakrabarti, P.P.: Fault space transformation: a generic approach to counter differential fault analysis and differential fault intensity analysis on AES-like block cipher. IEEE Trans. Inf. Sec. **12**, 1092–1102 (2016)
8. Biham, E., Shamir, A.: Differential fault analysis of secret key cryptosystems. Springer, Berlin (1997)
9. Bai, K., Wu, C.: An AES-like cipher and its white-box implementation. Comput. J. **59**, 1054–1065 (2016)
10. Barenghi, A., Hocquet, C., Bol, D.: A combined design-time/test-time study of the vulnerability of sub-threshold device to low voltage fault attacks. IEEE Trans. Emerg. Top. Comput. **2**, 107–118 (2014)
11. Haq, S.U., Masood, J., Majeed, A., Aziz, U.: Bulk encryption on GPUs
12. Floissac, N., L'Hyver, Y.: From AES-128 to AES-192 and AES-256 how to adapt differential fault analysis attacks on key expansion. In: 2011 Workshop on Fault Diagnosis and Tolerance on Cryptography (2011)

A Comparative Study on Data Stream Clustering Algorithms

Twinkle Keshvani[1]([⊠]) and Madhu Shukla[2]

[1] Computer Engineering Department, Marwadi Education Foundation
Group of Institution, Rajkot 360003, India
twinkle.keshvani5595@marwadieducation.edu.in
[2] Computer Engineering Department, Faculty of PG Studies,
MEF Group of Institutions, Rajkot 360003, India
madhu.shukla@marwadieducation.edu.in

Abstract. Data stream mining is trending today due to enormous data generation by many applications. Most of the non-stationary systems generate data which are massive in volume, non-static and fast changing. Data available is huge, so to augment the calibre of data, it is essential to cluster them. This in turn will boost the data processing speed and that holds a great level of importance in data stream mining. Data Streams being enormous, complicated, fast changing and infinite puts some additional challenges in clustering techniques i.e. time limitation, memory constraint. For mining data streams, many clustering algorithm has been emerged. Also, it is also required to identify cluster of arbitrary shape. And Density based clustering algorithm play important role there. These algorithms fall under notable class having potential to find clusters of arbitrary shapes and to detect noise. Over and above, these algorithms do not require information of the sum total of clusters to be formed, as a part of prior knowledge. The primary focus of these algorithms is to use density-based methods while forming clusters and simultaneously curbing the constraints, which are inherent in the nature of data streams. The objective of this paper is to throw discuss algorithms that are density based and their pros and cons.

Keywords: Data stream · Clustering · Non-static · Arbitrary shape ·
Density-based

1 Introduction

Organizations today accumulate data at an astonishing rate. The size of data keeps growing continuously with frequent updates. So in order to use this data meaningfully and to know about trends or current demands and patterns, organisations started using applications that have data mining techniques but as mining recent data would give proper decision making so this lead to the processing of data in real time. Data stream mining is used to deal with data stream in which data enters from one end and exits from another without giving us sufficient time for revisiting the data and processing it. Majorly mining techniques are 1. Classification, 2. Clustering 3. Association-rules. Most important data analysis technique and one that is frequently used is clustering. As Clustering is a notable class in mining data stream, which mines the meaningful

© Springer Nature Switzerland AG 2020
A. P. Pandian et al. (Eds.): ICCBI 2018, LNDECT 31, pp. 219–230, 2020.
https://doi.org/10.1007/978-3-030-24643-3_27

information in streaming data and forms a group. Data points are identical to each other if it exists in the same group. Data Stream being different from traditional data mining techniques, it requires some special demands due to its unique characteristics like, enormous volume of data, rapidly changing data, time and memory constraint etc. Clustering stream data is all together a different challenge as the data comes at an undefined rate from the systems and to use that data for meaningful purpose is a very complicated task. It is complicated because of its characteristic to restrict us from revisiting the data. It is not impossible to revisit it but for the time being due to unavailability of enough research work and solutions, to go through the data again, we can use the term impossible. As traditional and conventional methods are unable to process this massive data which is flowing at an undefined rate, we are in urgent need of some new clustering methods.

There are ample of clustering algorithms. Out of them, density-based clustering has proved and entered the picture as a rewarding and constructive class for data streams because of its characteristics that are stated as below.

- Primarily, it discovers clusters having arbitrary shapes. In order to be useful for any application a cluster needs to be dynamically shaped according to the data.
- As cluster is an unsupervised approach we don't need to give any number or it is not required to have prior information about the number of clusters.
- By and by, Data stream has potential to manage outliers. Let's say, as data stream flows continuously there always appears some noise with it, so it causes failure in sensor devices. Hence it is essential to detect noise specially in evolving data stream.

Motivation: In today's world ample amount of data generated along with noise and outliers in them. Cluster generated from the data forms an arbitrary shape. They can reflect the real distribution of the data, and has ability to handle the noise or outliers effectively; moreover it does not make any assumptions on the numbers of the outliers. Density-based is the important topic in clustering the data stream.

2 Data Stream Clustering

Clustering is a crucial task of data mining. Traditional clustering algorithm can't be used for data stream Mining [1]. There exists variety of crucial challenges. As one of the data stream characteristic is continuous flow of data, so data can be viewed only in one pass. In order to view data as a long vector data is not useful for many applications. As a matter of fact, streaming data continuously evolves with time so to generate the clusters of evolving data pattern becomes an interest of the users.

There have been different fields of vision for clustering data stream. Problem needs to be defined very carefully in data stream clustering; the reason is data stream is considered as an unbounded process consisting of data that continuously evolves over time. Traditional data mining algorithm for clustering the data stream uses a single phase model and also supposed the data stream as a static data. These algorithms partition the data into parts and generate the clusters using k-means in limited space.

Furthermore it has constraint of using this scheme is equal weights are allotted to recent data as well as outdated data it is not able to capture the evolving behaviour of data stream.

In order to evaluate the algorithm of data stream clustering, two phase technique is used which involves two phase i.e., online phase and offline phase. An online phase processes the raw data of data stream and generates the synopsis of the data. Synopsis which is generated in the online phase is used in the offline phase to output the final cluster. There are many such algorithms which evaluate the data stream using two phase model. One of the well-known algorithms is CluStream which works on this model. CluStream works on two component i.e., micro component and macro component. It uses the microclusters which maintain the statistical records. These microclusters are stored at snapshot (A particular moment at which microclusters are stored) in time which follows the pyramidal pattern.

3 Challenges and Issues in Data Stream

- **Enormous Data Flow:** As the amount of data generated is huge and continuous, it is necessary to analyse the data at some point of time. If a new data arrives with its new characteristics then the analysis of old data remains of no use. Chances of visiting the old data are nil.
- **Continuous Data Flow:** In data Stream mining, data comes from diverse directions [3, 14] and so a humongous amount of data is streamed. Moreover this data comes at a lightning speed. Hence, this continuous flow of data remains an issue to be addressed.
- **Chronological Order:** Data order cannot be maintained due to dynamic flow of data. A constant flow of data is streamed and hence is next to impossible to maintain a specific order of data in terms of its priority, characteristics and other features.
- **Prompt Analysis of Data:** Promptness with which data gets processed holds a great level of importance, because of the amount of data and the speed at which it flows into the system. Hence a real time or timely analysis of data is important for getting into the characteristics of data. Issue of Concept Drift and Concept Shift needs to be dealt with as the data is to be processed at the time of generation itself.
- **Multi-dimensional Data Sets:** Data which arrives into the system are of multiple dimensions such as customer preferences clustering. [14] It should be managed particularly via data stream mining. Many a times increase in Multi-dimensional datasets can create complexity [2].
- **Rate of Arrival:** Data flows through the system at an undefined speed, moreover it's in huge size which makes it even more challenging. If such data is not processed at the time of arrival, it results in data flooding and may also sometimes lead to data congestion situation [14].
- **Memory Constraint:** Due to continuous and uncountable data flow it's almost impossible to store it. It empties old data as and when a new data streams. A new method is required to for managing or at least reducing this problem up to a certain extent. Any algorithm which is implemented should consider this constraint while carrying out its function.

- **Evolution of Data:** Algorithm which is formulated and implemented should consider Evolving data stream so that it can produce more effective results.

4   Processing of Clustering Evolving Data Stream

One Pass Processing: Data Streaming is just like river which enters from one side and exit from another side, so data cannot be reviewed. Data arrives in lumps and the clustering process is performed on the whole data of the data stream which evolves over time. As evolving behaviour of the data stream, data which is passed cannot be reviewed.

Evolving Data: In the evolving approach, clusters are sorted in the forms of window model. Data is separated in the window model and this separated data is used as an updated units. For evolving data three types of window model is used for processing.

- *Sliding window model:* In order to study the data using sliding window, it is necessary to analyse the latest data as well as historical data of the data stream.
- *Fading window model:* This window model allocates the weight of the data as per the time of arrival of data. More weights are allocated to those data which is latest compare to out of date.
- *Landmark window model:* In this window model, windows store all the transaction at a specified point of time till today's time. Whenever a transaction arrives after a specified time, it is added to the window without any further processing.

Two-Phase Techniques: The process is performed in two steps i.e. Online and Offline Phase. First step stores the new arrival of data in the form of synopsis (histogram) and second step (clustering) is the generation of the final clusters using micro clustering technique that uses the extended data structure to store data [13]. Second phase could either be supervised or unsupervised approach based on the applications involved and the data used. Every algorithm structured for stream data has two phases involved which thus addresses issues of recency as well as space management.

As shown in the fig below data stream arrives at a rate generated by the system which goes for the further processing in first phase this giving micro clusters, this phase is called Online phase, it is a continuous phase which keeps on going with the incoming data and thus addresses the recent most data. The data after processing is stored as synopsis or histogram thus giving less overhead to system for data storage. This data is then further given to the next phase called the offline phase, which generates the Macro clusters based on the data given by the previous phase. This phase deals with data only given by the online phase and thus deals with the lesser amount of data [9]. The process continues till the data keeps arriving in the system (Fig. 1).

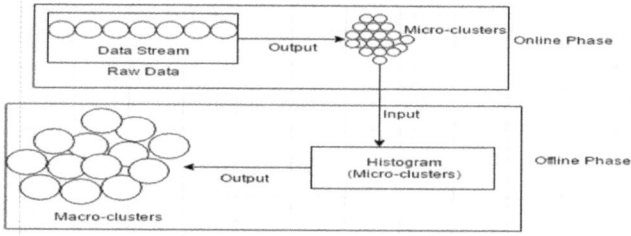

Fig. 1. Processing of evolving data stream

5   Clustering Data Streams Using Density Based

Density-based clustering comes with a propensity to bring to light on clusters which are non-spherical in shape and to manage noise that appears in data. To generate the clusters using density based algorithms, the algorithm seeks for the dense areas which are separated by sparse area.

Majorly density based algorithm are divided in two groups, i.e. density micro-based clustering algorithms and density grid-based clustering algorithms.

Density micro-clustering algorithm consists of micro-clusters which stores synopsis of streaming data as data arrives in the online phase. Final clusters are generated using this synopsis. While in density grid based clustering, there is a grid like structure in which data objects are plotted to the grid cells which is finite in number. Generation of the final clusters in density grid-based depends on the dense area in the grids. Before we go deep into these algorithms have a look on these basic terms.

- *Dense grid*: The volume of the data points (objects) in the grid cells gradually increases, than such a dense area in the grid is called as a dense grid.
- *Sparse grid*: The volume of data points (objects) in the grid cells gradually decreases in terms of quantity and the data points (objects) do not arrive in the grid cells for a long period of time, than such a grid is called as a sparse grid.
- *Expired grid*: The grid which is no more active is defined as an expired grid.
- *Characteristic vector*: Characteristic vector includes the information of the data points, such as creation time, updates of grid, density of grid and type of grid.
- *Sporadic grid*: It is somewhat like of sparse grid, as discussed above whose volumes is less in the grid and have no scope to become a dense grid.

5.1   DenStream

DenStream [9, 10] is a density based clustering algorithm, which has potentiality to deal with noise along with data. Micro-cluster concept such as core micro-cluster (c-micro-cluster), potential micro-cluster (p-micro-cluster), and outlier micro-cluster (o-micro-cluster) are enlarged using this algorithm in order to recognise real data as well as outliers. The concept of c-micro-cluster is something like to design the synopsis which summarizes the clusters with non-spherical shape in streaming data. The core micro-clusters are also named as dense micro-clusters. Processing of core micro-clusters is different than that of p-micro-cluster and o-micro-cluster. As far as the memories of potential and outlier micro-clusters are different as they required different processing. For separate processing of potential and outlier micro-cluster outlier buffer is introduced. DenStream involved a pruning strategy which is based on p-micro-cluster and o-micro-cluster. This strategy gives an opportunity to extend the growth of new clusters and quickly getting free of outliers. As this new strategy is involved in this, which consumes memory and also guarantees, the correctness of the micro-clusters. The framework of DenStream is based on two phase i.e., online phase and offline phase. Micro-clusters are maintained done in the online phase in which dense area of the data streams are captured. Generation of clusters or the final result we get in the offline phase.

5.2 D-Stream

D-Stream [8, 9] algorithm is a density-based clustering algorithms which is real-time processing data in the data stream. Due to the evolving behavior of the data stream, this algorithm includes an innovative scheme that is coupled with the decay factor. A mechanism which is provided by the decay factor that automatically generates the clusters by allocating more weights to the latest data. Like CluStream [4], it has also an online as well as an offline component. D-Stream continuously reads the data records as it arrives. Input Data are plotted to the density grids in online component and the process of clustering the grids based on density of grid are performed in the offline component. This algorithm does not require the prior information about the numbers of clusters like K-means. Also, it is for the users who have naive knowledge about the application data. Moreover, fading window model is used by D-stream and it uses the decaying technique. For capturing the evolving behavior of the data stream this technique plays an important role. The fading window model gradually decreases the effects of old data on mining results.

5.3 SDStream

SDStream [11] is a density based micro-clustering algorithm, with potentiality to final clusters of non-spherical (arbitrary) shapes in data stream over sliding window. In SDStream most recent data stream is taken into account. If the length of the data stream is greater than that of the sliding window length than the additional stream other than the size of the sliding window is discarded. The structure of c-micro-cluster, p-micro-cluster, o-micro-cluster varies from each other. The micro-clusters which have number of points in them have different weights according to their structure. In order to store the microclusters we have a main memory which is in the exponential histogram form. It has also a two phase algorithm which has an online and offline phase. Whenever the new data points arrive it is absorbed to its neighborhood. Neighborhood micro-cluster might be a potential or an outlier micro-cluster that depends on the main memory of the micro-cluster. The micro-cluster which are at least distance and also directly density-reachable point than these micro-clusters are merged and the micro-clusters which is no used for a long time period or micro-clusters which are outdated, are discarded from the memory. Offline phase processes the new arrival of data, to generate the final clusters, for that DBSCAN is used. Clusters generated as output is of arbitrary shape.

5.4   DUCstream

This density Grid-based clustering algorithms is incremental in nature. As an arrival of new data volume of clusters goes on increasing manner [12]. It is an progressive clustering algorithm for data streams, As new data arrives, data space is partitioned into a number of units, among units which is large in size or bigger in volume in terms of quantity are kept. DUCstream assumes the coming data points in lump [7], which contains number of data points (objects). The unit is considered as a dense unit if it has more number of points. The number of points in the units is always higher than the density threshold which is introducing as a DUCstream. In order to convert the unit into the dense unit, algorithm introduced a concept called as local dense unit, which get an opportunity to get converted into the dense unit. In DUCstream, the clusters are recognised as components which are connected in a graph in which the vertices shows the units which are dense and edges shows the relation between the components. Therefore, whenever there is a arrival of new data, dense region is formed and it becomes a dense unit. These dense units are absorbed in the existing cluster if found, the clusters which are identical of itself, if it doesn't found any similar objects, than it forms the new cluster. Moreover the clusters forms in the DUCstream are an incremental manner, their results remains in bits. It is known as clustering bits or bit string. It retains miniature of memory due to the clustering bits. Bit string kept the number of units which are dense.

5.5   DENGRIS-Stream

It is Grid-based clustering algorithms on data stream [5, 6]. The work of DENGRIS-Stream is similar to that of the D-Stream. Fading window model is used for clustering the over the evolving data stream in D-Stream. While in DENGRIS to capture the evolving data which is distributed, sliding window is used. Input data are mapped to the density grid in online component and final generation of the cluster lies in the offline component using the density concept within time window units. Stream arrived between now and in the past time is indicated by the time window. Recent data records of the stream are more important than the data arrived in the past. In order to be useful for any application, distribution of the data must be done very well and it can be done by using the sliding window model. Although sliding window considers the most recent or the latest data. The historical data or the outdated data which no longer use or whose timestamps does not fit in the sliding window for that a concept is introduced, namely an expired grid. So, before moving further for any processing on the grid list this concept is applied that leads to consume time as well as saves memory.

5.6   DD-Stream

DD-Stream [7] is one kind of density grid-based clustering algorithm. DD-Stream is an algorithm which is mainly an enhancement/extension of the D-Stream algorithm. Boundary points are extracted and evolving data stream is captured by the use of this algorithm. By detecting the border points in the grid cluster quality is improved. Before performing any adjustments on the grid, boundary points are extracted. DD-Stream can discover arbitrary shape clusters with noise, in addition it avoid the problems regarding the cluster quality which is caused by discarding the border points. In order to extract the data points which are on the boundary of the grid, DCQ-means algorithm is used. In DCQ-means, when the data points are arrived, most direct distance is adopted and grid is recognised in terms of which data point it is a part of. For maintaining the records of the grid Eigen vector is used.

DD-Stream also includes two phase like CluStream i.e., an online phase and offline phase (component). An online component continuously reads the data as it arrives and then input data is mapped to the respective grid in the space and updates the grid unit. Dense grid and the sparse grid are distinguished in concert with the Eigen vector in the offline component. DCQ-means algorithm is used to extract the boundary points from grid. Finally, to dense grids, data points are clustered using methods that are density based (Tables 1, 2).

Table 1. Main characteristics of density based clustering algorithms

Name	Reviewed year	Data form	Clustering technique	Cluster output shape
DenStream [9, 10]	2006	Continuous	Density micro-based clustering	Arbitrary shape cluster
DStream [8, 9]	2007	Continuous	Density grid-based clustering	Arbitrary shape clusters
SDStream [11]	2009	Continuous	Density micro-based clustering	Arbitrary shape cluster over sliding window
DUCstream [7, 12]	2005	Undefined	Density grid-based clustering	Graph structures which are connected
DENGRIS-Stream [5, 6]	2012	Continuous	Density grid-based clustering	Arbitrary shape clusters
DD-Stream [7]	2008	Continuous	Density grid-based clustering	Arbitrary shape clusters

Table 2. Challenges and issues in density-based clustering algorithms

Density-based clustering algorithm	Enormous data	Handling noisy data	Evolutionary data	Limited memory
DenStream	✓	✓	✓	✓
DStream	✓	✓	✓	
SDStream	✓	✓	✓	✓
DUCstream	✓	✓		✓
DENGRIS-Stream	✓	✓	✓	✓
DD-Stream	✓	✓	✓	

6   Simulation and Result Analysis

The graph shown below represents the performance of the algorithms. The graph which is red in colour shows the performance of the CluStream with K-means algorithm and the graph which is blue in colour shows the performance of the DenStream with DBSCAN algorithm. This algorithm uses the synthetic data set. Evaluation Parameters are discussed as under:

Purity: Measure to evaluate quality of cluster. Ideal value of purity is 1. Cluster with a bad purity thus hold a value 0.

SSQ: Summing the squared distance between the clustering objects and cluster representatives. It is a standard cost function. It originally refers to the Euclidean distance.

Homogeneity: Objects in clusters are from the single class.

Completeness: Clusters contains all the members of a specific given class.

Rand Index: It is a similarity measures between two data clustering's.

MOA tool is used for simulating these algorithms using a synthetic dataset. Here we use an Random RBF Generators to generate the synthetic streaming data. This gives different scenarios for testing the clustering quality. We have some default value of the parameters such as Horizon, max number of kernel, random seeds (Figs. 2, 3, 4, 5, 6).

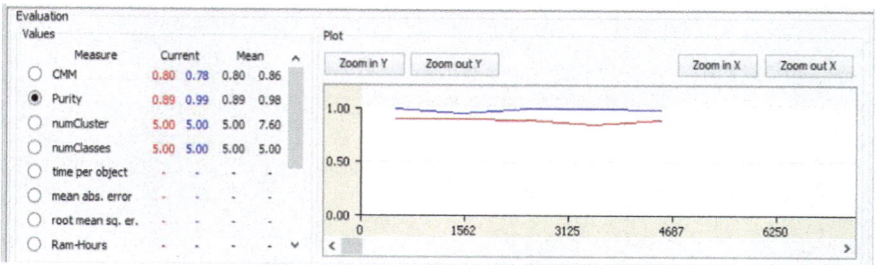

Fig. 2.   Purity comparison of CluStream and DenStream

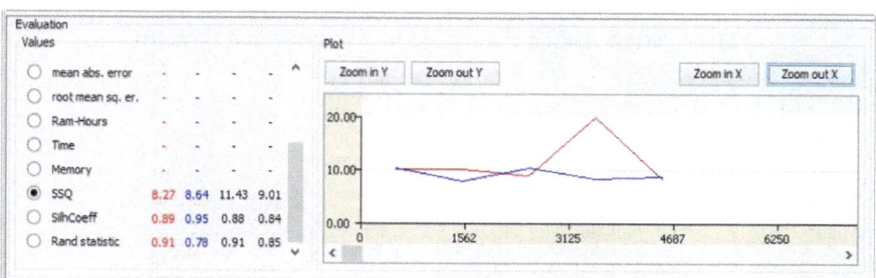

Fig. 3.   SSQ comparison of CluStream and DenStream

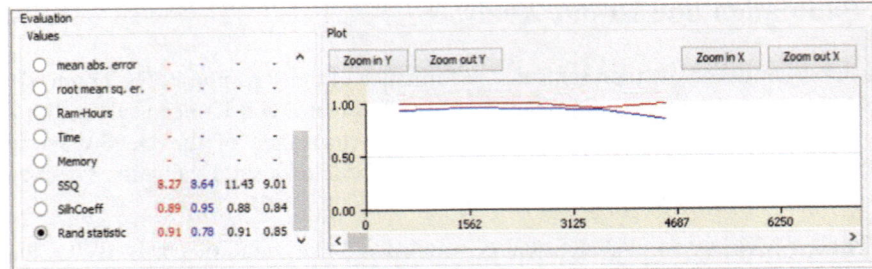

Fig. 4. Rand Statistic comparison of CluStream and DenStream

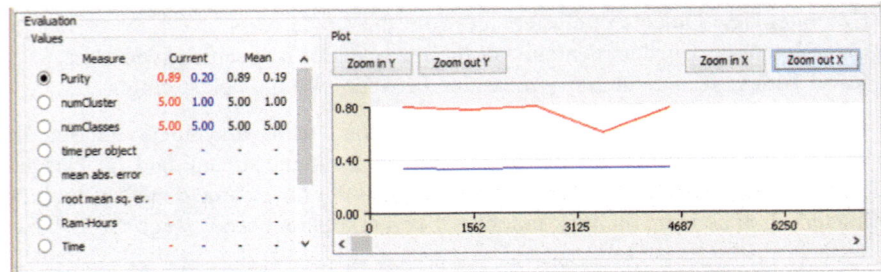

Fig. 5. Purity comparison of CluStream and DStream

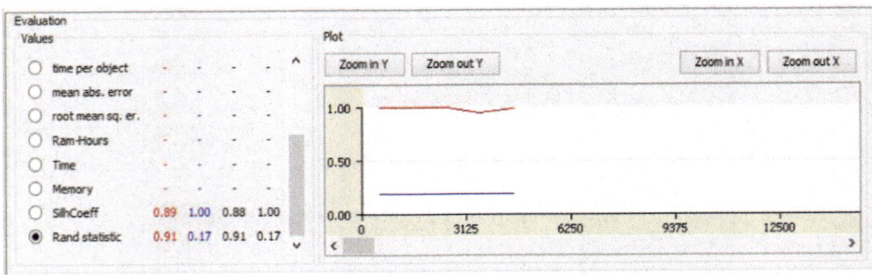

Fig. 6. Rand statistic comparison of CluStream and DStream

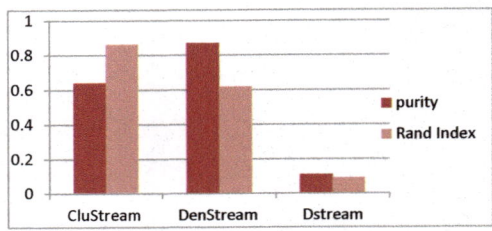

Fig. 7. Purity and Rand index

As shown in Fig. 7 the paper compares three algorithms using two evaluation measures. Firstly Purity, purity of DenStream is greater that CluStream and DStream. As shown in Fig. 2 Purity of DenStream is .99 so it's almost 1, so it can be a perfect cluster. Cluster quality of DenStream is higher. Quality of DStream is comparatively lower. Secondly Rand index that is similarity of two data clustering, which is higher in CluStream compare to DenStream and DStream.

7 Conclusion

The density based clustering algorithm has drawn an attention of the researchers, because of its notable characteristics, such as capability to discover clusters of non-spherical shape and additionally have a potentiality to deal with noisy data as well. There is ample number of algorithms in stream data, which adopts density based method. This paper gives an comparison of different density based clustering algo-rithm, firstly Data stream clustering algorithm has been discussed followed by issues and challenges of data stream and its processing with different kinds of windowing technique. The paper then discusses about specifically density based clustering algo-rithms their prons and cons. How they deal with the data stream. Also the change in structure of tradition density based algorithm to be made in order to deal with the data streams (i.e.2 phase). At the end result concludes that the purity of DenStream that means quality of cluster in DenStream is higher than CluStream and DStream.

References

1. Barbará, D.: Requirements for clustering data streams. ACM SIGKDD Explor. Newsl. **3**(2), 23–27 (2002)
2. Khalilian, M., Mustapha, N.: Data stream clustering: challenges and issues. arXiv Preprint arXiv:1006.5261 (2010)
3. Babcock, B., et al.: Models and issues in data stream systems. In: Proceedings of the Twenty-First ACM SIGMOD-SIGACT-SIGART Symposium on Principles of Database Systems. ACM (2002)
4. Aggarwal, C.C., et al.: A framework for clustering evolving data streams. In: Proceedings of the 29th International Conference on Very Large Data Bases, vol. 29. VLDB Endowment (2003)
5. Amini, A., Wah, T.Y.: DENGRIS-stream: a density-grid based clustering algorithm for evolving data streams over sliding window. In: Proceedings of the International Conference on Data Mining and Computer Engineering (2012)
6. Jia, C., Tan, C., Yong, A.: A grid and density-based clustering algorithm for processing data stream. In: Second International Conference on Genetic and Evolutionary Computing, WGEC 2008. IEEE (2008)
7. Amini, A., Wah, T.Y., Saboohi, H.: On density-based data streams clustering algorithms: a survey. J. Comput. Sci. Technol. **29**(1), 116–141 (2014)
8. Chen, Y., Tu, L.: Density-based clustering for real-time stream data. In: Proceedings of the 13th ACM SIGKDD International Conference on Knowledge Discovery and Data Mining. ACM (2007)

9. Jayswal, M., Shukla, M.: Consolidated study & analysis of different clustering techniques for data streams. In: 2016 3rd International Conference on Computing for Sustainable Global Development (INDIACom). IEEE (2016)
10. Cao, F., et al.: Density-based clustering over an evolving data stream with noise. In: Proceedings of the 2006 SIAM International Conference on Data Mining. Society for Industrial and Applied Mathematics (2006)
11. Ren, J., Ma, R.: Density-based data streams clustering over sliding windows. In: Sixth International Conference on Fuzzy Systems and Knowledge Discovery, FSKD 2009, vol. 5. IEEE (2009)
12. Gao, J., et al.: An incremental data stream clustering algorithm based on dense units detection. In: Pacific-Asia Conference on Knowledge Discovery and Data Mining. Springer, Berlin (2005)
13. Kaneriya, A., Shukla, M.: A novel approach for clustering data streams using granularity technique. In: 2015 International Conference on Advances in Computer Engineering and Applications (ICACEA). IEEE (2015)
14. Shukla, M., Kosta, Y.P.: Empirical analysis and improvement of density based clustering algorithm in data streams. In: International Conference on Inventive Computation Technologies (ICICT), vol. 3. IEEE (2016)

Biometric Voting System

Abhilash Akula[✉], Jeshwanth Ega, Kalyan Thota, and Gowtham

Department of CSE, SRMIST Ramapuram, Chennai, India
abhilashakula99@gmail.com, jeshwanthega@gmail.com,
kmrkalyan1998@gmail.com, gowtham.gsm@gmail.com

Abstract. In India, the voting system plays a major role during elections. Election Commission of India uses electronic voting machines for the election. According to the survey, in the 2014 general election, only 66.4% votes got registered. At the same time, voting systems have encountered with serious challenges in delivering voting confirmation messages and lack of awareness among people to precisely use their fundamental right "Right to Vote". For avoiding misconceptions on the election time, many advanced techniques are being proposed earlier, by efficiently using various methods. In this paper, we proposed a solution for solving various issues in the present voting system which effectively uses biometric identification as a major concept. The biometric identification will provide better trustworthy results in the election process.

Keywords: Biometric · Fingerprint · Verification · Cloud based database

1 Introduction

In India, the machine that has been used during election is called as Electronic Voting Machine (EVM). It is used for the purpose of vote casting and counting the number of votes. In the past decades, the elections were held with a paper ballot method with the help of papers and boxes. This traditional method requires more manpower, time and energy. So it was replaced by the contemporary EVM methodology. Even though today's voting machine is fast and accurate than the previous method, it still needs more accuracy and efficiency. Besides, there is a possibility of malfunctioning in the machine which is currently used in India. To overcome these issues in the contemporary voting machine, this paper has proposed a smart and secure voting system.

One of the main ideas of this proposed system is to implement biometrics in the modern voting system. Biometric traits are unique for each individual, which can make the systems more secure than any passwords or pins. Today biometric systems are employed in most of homes, offices, ATMs and military applications, in which the access is controlled using biometric devices that can scan the individual's unique physiological or behavioral characteristics. Recently, Biometric identifiers are evolving with recent innovations like fingerprint, face recognition, palm print, retinal scan and iris pattern recognition. So, this technology can be efficiently used for developing a secured voting system.

© Springer Nature Switzerland AG 2020
A. P. Pandian et al. (Eds.): ICCBI 2018, LNDECT 31, pp. 231–235, 2020.
https://doi.org/10.1007/978-3-030-24643-3_28

2   Literature Survey

Election Commission of India uses electronic voting machines for the election. According to the survey on 2014 general election, only 66.4% votes got registered in India. EVM is designed and developed by two government owned electronics equipment manufacturing units, Bharat Heavy Electronics Limited (BHEL) and Electronics Corporate of India Limited (ECIL). All EVM systems should be identical and developed according to the specifications given by Election Commission of India. EVM is a set of two devices that runs with a source voltage of 6 V. First is the voting unit which is extensively used by the Voters and the second device is the Control Unit which will be operated by the Electoral Officer. The electronic voting machine usually has a blue button for every candidate, each unit has the capacity to hold 16 candidates and up to 4 units can be chained together which can accommodate a total of 16 candidates. The control unit has totally 3 buttons, one to release a single vote, second is to calculate the total number of votes and the third is to end the election process.

Traditional Methods: From past decades, Indian election process has been conducted using conventional ballot voting system. The main problem with the conventional voting method is we can cast number of fake votes for the favorable political party, hence there is a chance of violating the original voting results. It is also a time-consuming process for casting votes and counting the number of polled votes finally. Later EVM came into existence but this is not a trust-worthy method, because a single person can cast more than the allotted single vote. The EVMs can also get damaged during the transport as the miniature sensors used in it may be damaged when the machines are kept together.

3   Proposed System

The proposed system is based on the contemporary electronic voting machine, which has developed the ability to identify a person with the fingerprint. The unique identity card (Aadhaar card) details of every citizen will be stored in the database. When a person places his/her finger on the fingerprint sensor, it matches the fingerprint with the existing ones in the database. If the fingerprint is matched, then it checks whether the person is eligible to vote (above 18) and if he is eligible, a popup will be displayed on the screen. After this process, the person has been allowed to cast his vote to the desired candidate. The person can vote only once because, the database will be containing a single fingerprint for each person, so he/she can't cast another vote by using another finger of the same person. This system includes GPS, which has been extensively used to exactly locate the system's location. This will be helpful in case of theft of the system from the polling booth. Since the system is connected to the cloud, the system has the ability to act as an online system. With these advantages, the proposed systems has overcome the challenges posed by the contemporary EVMs. Hence, the proposed biometric based voting system is more secure and efficient than the existing systems. The database enlisted in the proposed systems has also been used to find the people who failed to cast their votes.

Once the voting process is completed, the proposed systems shows the results within a short period of time. The result also has a higher accuracy rate both in counting votes and registering the votes by the particular voter.

Hardware: Controller unit: Arduino UNO R3 is used as the overall process controller for the whole design. It is an open source microcontroller which is based on ATMEGA32. It is considered as a combination of analog and digital PWM signals. The code is given by using the USB connector.

GPS: The Global Positioning System (GPS) is a satellite-based navigation system which has been made by employing at least 24 satellites. GPS helps to navigate to the specified position or location of the device at any parameter. Once your position has been determined, the GPS unit can calculate other information such as speed, trip distance, distance to destination and more. The system will be typically accurate if it is within 10 m. Furthermore, GPS accuracy will be better even on the water surface.

Finger Print Scanner: It is a type of biometric security technology which is considered as the combination of hardware and software techniques to scan the fingerprint of an individual. A fingerprint scanning processes is initiated by first recording the fingerprint scans of individuals. These scans are then stored in a database. If the user puts finger on a hardware scanner automatically it scans and copies the input from the individual and looks for any other similarity within the already-stored database. It has been built on-board with a 32-bit CPU in order to accept the code. The fingerprint sensor is then interfaced with an Arduino system. In total there are 4 external modules, in this two will be communicating with the Arduino and the remaining two will be biasing the voltage and ground.

Buzzer: Buzzer has 2-pin structure ad it used widely in electronic applications. It consists of DC power supply varies from 4 V to 9 V. It will be act as switching circuit either ON or OFF condition based on the source. If the input source is high, buzzer will switch to ON else buzzer will move to OFF.

LCD Display: It Stands for "Liquid Crystal Display." LCD is a flat panel display, which is commonly used in televisions, laptops and tablets.

Cloud Database: Cloud database can be termed as a portable database that is accessible to clients from the cloud and delivered to handlers on demand via the Internet from the information provider's server. By using Database-as-a-Service (DBaaS), cloud databases can efficiently use cloud computing to attain an optimized scaling, high availability, multi-tenancy and effective resource allocation.

However, a cloud database can be a traditional database such as a MySQL or SQL Server database that has been implemented for cloud use, a native cloud database such as Xeround's MySQL Cloud database inclines to equip the cloud and optimally practice cloud resources in order to guarantee higher scalability than the existing cloud service availability and stability.

It can offer significant advantages when compared with their conventional counterparts, including the enhanced accessibility, automatic failover and fast automated failure recovery process, automated on-the-go scaling, reduced investment and maintenance of in-house hardware, and potentially enhanced performance. At the same time, cloud databases also contains some potential disadvantages, including security and privacy issues as well as the incapability to access the critical data at the time of occurrence of a tragedy or ruin to the cloud database service provider.

4   Architecture

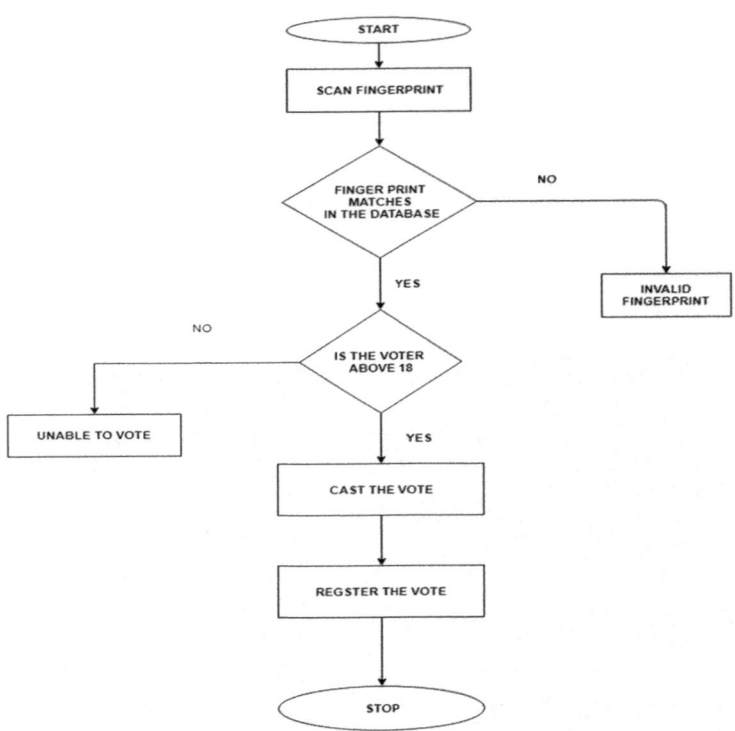

Software: In order to control the Arduino UNO R3 with ATMEG 32 microcontroller, an Arduino developer software is used. To control a fingerprint sensor, a library file is used which was originally created and developed by Josh Hawley. A header file is used to interface and control the scanner with Arduino. C++ is used to write the program in the header file. Before building a hardware model, the whole software method was simulated, after the successful simulation, the hardware model is built.

5   Conclusion

To make the Indian voting system faster and more secured, an initiative has been made so that every individual can vote according to their convenience. The main aim of this project is to develop a secured voting process. By following this type of voting process, elections can be made as a corrption less process and a truly desired candidate will be elected by the people. The fingerprint sensor has been implemented to prevent the voter from casting more than one vote. This innovative biometric voting system is far better and faster than the traditional voting system. In a developing country like India, biometric technologies should be employed to make the process smarter.

References

1. Deepika, J., Kalaiselvi, S., Mahalakshmi, S., Agnes Shifani, S.: Smart electronic voting system based on biometric identification-survey. In: Third International Conference on Science Technology Engineering & Management (ICONSTEM), pp. 939–942 (2017)
2. Ashok Kumar, D., Ummal Sariba Begum, T.: Electronic voting machine—a review. In: Proceedings of the International Conference on Pattern Recognition, Informatics and Medical Engineering, 21–23 March 2012, pp. 41–48 (2012)
3. Bhuvanapriya, R., Rozil Banu, S., Sivapriya, P., Kalaiselvi, V.K.G.: Second International Conference on Computing and Communications Technologies (ICCCT 2017), pp. 143–147 (2017)
4. Pomares, J., Levin, I., Michael Alvarez, R., Lopez Mirau, G., Ovejero, T.: From piloting to roll-out: voting experience and trust in the first full e-election in Argentina. In: International Conference on Electronic Voting EVOTE2014, E-Voting. CC GmbH (2014)
5. Delis, A., Gavatha, K., Kiayias, A., Koutalakis, C., Nikolakopoulos, E., Paschos, L., Rousopoulou, M., Sotirellis, G., Stathopoulos, P., Vasilopoulos, P., Zacharias, T., Zhang[†], B.: Pressing the button for European elections. In: International Conference on Electronic Voting EVOTE2014, E-Voting. CC GmbH (2014)
6. Kiruthika Priya, V., Vimaladevi, V., Pandimeenal, B., Dhivya, T.: Arduino based smart electronic voting machine. In: International Conference on Trends in Electronics and Informatics, ICEI 2017, pp. 641–644
7. Usmani, Z.A., Patanwala, K., Panigrahi, M., Nair, A.: Multi-purpose platform independent online e-voting system. In: International Conference on Innovation in Information, Embedded and Communication Systems (2017)
8. Weldemariam, K., Mattioli, A., Villafiorita, A., Weldemariam, K.: Managing requirements for e-voting systems: issues and approaches motivated by a case study. In: 2009 First International Workshop on Requirements Engineering for E-Voting Systems (2009)
9. Yang, C.: Techshino, Fingerprint biometrics for ID document verification. In: IEEE 9th Conference on Industrial Electronics and Applications (ICIEA), pp. 1441–1445 (2014)
10. Sravya, V., Murthy, R.K., Kallam, R.B., Srujana, B.: A survey on fingerprint biometric system. Int. J. Adv. Res. Comput. Sci. Softw. Eng. **II**(4), 307–313 (2012)
11. Naim, N.F., Yassin, A.I.M., Zamri, W.M.A.W., Sarnin, S.S.: MySQL database for storage of fingerprint data. In: UKSim 13th International Conference on Modelling and Simulation, pp. 293–298 (2011)
12. El Sharkawy, B.F., Meawad, F.: Instant feedback using mobile messaging technologies. In: Third International Conference on Next Generation Mobile Applications, Services and Technologies, pp. 539–545 (2009)
13. https://timesofindia.indiatimes.com/news/Highest-ever-voter-turnout-recorded-in-2014-polls-govt-spending-doubled-since-2009/articleshow/35033135.cms

Implement Improved Intrusion Detection System Using State Preserving Extreme Learning Machine (SPELM)

Kunal Singh[✉] and K. James Mathai[✉]

Department of Computer Engineering and Applications,
National Institute of Technical Teacher Training and Research, Bhopal, India
ks4072605@gmail.com, kjmathai@nitttrbpl.ac.in

Abstract. In this contemporary world of computer networks, network security plays a predominant role in enabling a secured network communication infrastructure into existence. In order to enforce a high protection levels against the existing network threats, number of software tools are currently employed. Intrusion Detection Systems (IDS) aims to detect the intruders or anomalies in the computer network architecture. Software model protects the computer networks from unauthorized users by detecting the network intruders at the right time. In this a machine learning classifier is built and trained the model with the NSL-KDD dataset. After training the model, the remaining data-set is tested to detect or classify the attacks into categories like normal or attack. The scholar has built State Preserving Extreme Learning Machine (SPELM) algorithm as machine learning classifier and compared its performance accuracy with the Deep Belief Network (DBN) model.

Keywords: Intrusion detection system · Anomaly detection ·
Deep belief network · State Preserving Extreme Learning Machine (SPELM)

1 Introduction

The intrusion detection system (IDS) handles the huge amount of data and plays a strategic role in detecting the attacks of various kind. IDS is considered as the classification problem because the intrusion detection is depended on a good classifier. Intrusion can be defined as malicious and it is recognized as a key to compromise the computer system's availability, integrity and confidentiality. In data mining, classification models are the very important techniques applied for the purpose of network intrusion detection. In this paper, the author proposes a network IDS based on SPELM classifier. State Preserving Extreme Learning Machine enhances the accuracy of attack detection and it becomes very efficient in distinguishing the network traffic in order to detect whether it is an attack or normal activity. According to Alom and et al. (2015), the online digital technology has already arrived so there is a huge security issues present in the network. There are excellent need for developing a better Intrusion Detection System that are intelligent specialized system designed to interpret the intrusion and makes an attempt to reduce the incoming network traffic. DBN (Deep

© Springer Nature Switzerland AG 2020
A. P. Pandian et al. (Eds.): ICCBI 2018, LNDECT 31, pp. 236–246, 2020.
https://doi.org/10.1007/978-3-030-24643-3_29

Belief Neural Networks) proves the most powerful deep neural nets and generative neural networks. In these first they trained the DBN by giving variety of known intrusion attacks and malicious activities that contains datasets, and after the DBN is trained we can test the DBN based intrusion detection system by giving the another datasets.

2 Literature Review

According to David Ahmad Effendy et al. (2017), they have created a manual solution for detecting intrusion attacks and also prevents the computer networks from various network attacks. In this technique, they have presented a pattern based anomaly detection so first they have collected some related pattern and based on these pattern they learn the model and after learning they detect the malicious activities. They only detect the normal attacks because the model is not strongly enough to detect the latest intrusions or malicious activity so they need to improve the model so it will work on the latest anomalies and detects the intrusion. So they use machine learning based algorithm to enhance their model to detect the anomalies, so the model gets trained on various predefined malicious activities and before giving these datasets to the machine learning algorithm. For this, first we need to preprocess the dataset and select attributes for delivering a better learning. and after the pre-processing is completed we can get the data to the machine learning algorithm so that the model will be trained efficiently. In these they uses NSL-KDDcup99 datasets and naive bayes algorithm to classify the network packets into various category of attacks like it is a either dos, probe, r2l or u2r kind of attacks.

According to Qassim et al. (2016) In these they uses a k-means cluster algorithm for developing a better understanding, because the k-means can work with numerical data and provide various cluster based solution and based on these clustering, the naive bayes algorithm will classify the packets into various kind of attacks.

Addition of k-means clustering (Al-Jarrah et al. 2016) to the naive bayes, will enhance the performance and the classifying of these attacks are also increases since it works efficiently and also they compute various performance measurement of these model like accuracy and f1 score.

According to Sultana and Jabbar (2016) when the web is growing rapidly and their and millions of internet users are present which uses the internet for socially connected to each other, online transactions. So the privacy and security is main issue in that so for that they working on it to detect various kind of attacks and also detect malicious activities. So they proposed an intrusion detection system based on AODE algorithm to detect the attacks and malicious activity in the network. In these they uses NSL-KDDcup99 datasets and uses AODE(Average one Dependence Estimators) algorithm to classify the network packets into various category of attacks like it is a either dos, probe, r2l or u2r kind of attacks. In these they pre-process the data by using AODE algorithm for better learning the algorithm, because thee can work with numerical data. After that they can evaluate various performance measures to check how accurate his algorithm is working.

According to Uma Kumari et al. (2017), Security from intruders for the network system and machine learning algorithms are vital space of analysis throughout the previous few years. These malicious activity are growing rapidly and which create a serious problem. IDS (Intrusion Detection System) (Wang and Deng 2003) is proposed to detect unauthorized and malicious attacks over the network. Various data mining techniques is proposed to applied with IDS to find or learn abnormal behaviour patterns. IDS scans the network activities and find malicious activity in the network systems to enhance accuracy and provide better security and works well in detecting anomalies attacks. Data mining techniques give the way to process, train and classify the huge amount of network information through IDS. Their are already many security and privacy techniques (Uma Salunkhe and Mali et al.) and various algorithms is used recently. In this the proposed a techniques of analysis drawback and achievements within the field of security of huge information analysis anomaly based mostly IDS (Solane Duque et al. 2015). Intrusion detection system could be a computer system code application for detective work and observance the network activities and defend from unknown and suspicious access of device.

3   Problem Definition

Anomaly in detection techniques primarily scans the entire network traffic and classifies it as traditional or abnormal. In order to classify a coaching set, the system can refer to this coaching set and classify the information as traditional or abnormal, so the power of your detection system depends on how well you develop your coaching set. The coaching set can outline what traditional traffic is and if the system comes across something that's not in accordance with the coaching set it'll be classified as abnormal. Due to this anomaly, detection techniques are primarily based on false positive generation. One of the main advantages of anomaly-based detection techniques over signature-based detection is that zero-day attacks have been detected as soon as they happen. There are various anomaly detection techniques mainly based on the area unit used like-statistical models, data mining models, data mining mainly based detection techniques, machine learning mainly based detection techniques. During this treatise, the scholar who specializes in implementing an intrusion detection system using the State Preserving Extreme Learning Machine (SPELM) of the NSL-KDD dataset to improve performance against a deep belief network (Alom et al. 2015).

3.1   Problem Identification

One of the main problems for IDS is the development of efficient behavior models or patterns to distinguish normal behavior from abnormal behavior by observing the network data set collected. To solve this problem, earlier IDSs usually rely on security experts to analyze the dataset and build a DBN-based intrusion detection system. Since data volumes vary rapidly, it has become a time-consuming and tedious task for DBN to analyze and extract attack signatures or detection rules from dynamic, large network data volumes. Therefore, the scholar has identified the problem and is stated as under:

"Implement improved Intrusion Detection System using State Preserving Extreme Learning Machine (SPELM)".

4 Proposed Work

In this paper, the author proposes a network IDS based on State Preserving Extreme Learning Machine (SPELM) classifier. SPELM enhances the attack detection accuracy and it is very efficient in distinguishing network traffic as attack or normal. The Authors performed the experiment using NSL KDD Data set. NSL KDD Data set which is mostly used for testing intrusion detection system (IDS).

In this author takes state preserving extreme learning machine algorithm for training the classifier model, and as a machine learning algorithm to training the model on training datasets. The Fig. 1 clearly shows that the dataset is divided into training and testing data sets and then assigning these data sets to the model, as pre-processing. After performing preprocessing, the features are extracted for data exploration as to what type of intrusion is there. After training by SPELM algorithm; the model is tested on testing data set and based on the model performance author have computed its performance.

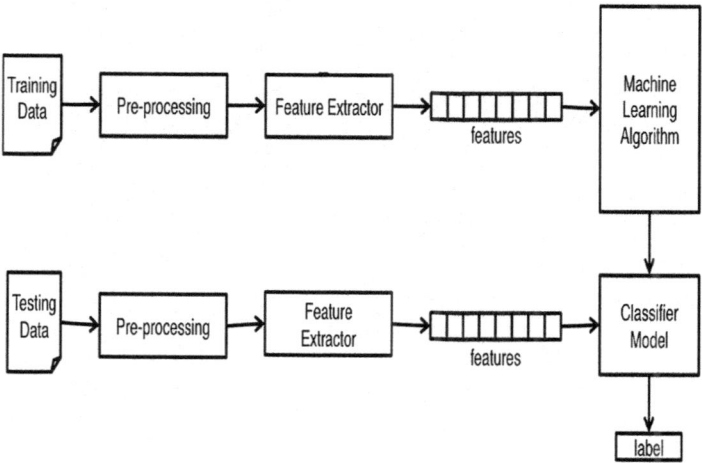

Fig. 1. Proposed model

In SPELM, there are 3 key steps to process of a 40% dataset used in training data: First, weight and bias are unconstantly measured in each learning step; secondly, the input arrangement of training samples is unconstantly developed for each batch learning repetition; thirdly, the input samples are scuffled according to the output sequences of each repetition. and then after use of 60% dataset of testing samples SPELM is constantly selected with similar labels and status variables such as weight, bias, sample sequences and accuracy of testing are preserved for each repetition.

The highest accuracy with valid parameters is then stored until the next repetition to ensure a better accuracy.

5   Experimental and Result Analysis

The experiment is performed on Laptop with Ubuntu 16 Operating System using python-3 program In this experiment NSL-KDD dataset has been used which contain 4 millions of records. The scholar have used Jupyter notebook which is an IDE (Integrated Development Environment) an environment to run python. Invoking jupyter notebook through terminal, the Fig. 2 shows the connection created between jupyter notebook with python.

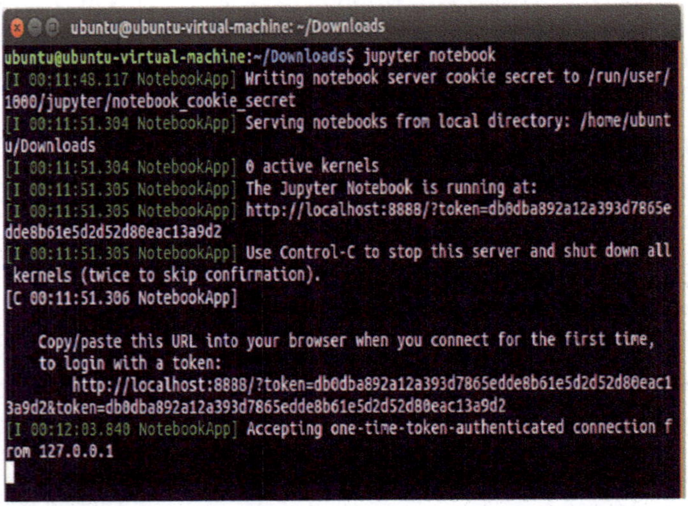

Fig. 2. Invoking jupyter notebook

And then providing name of each attributes present in the dataset. And these attribute names are shown in Fig. 3.

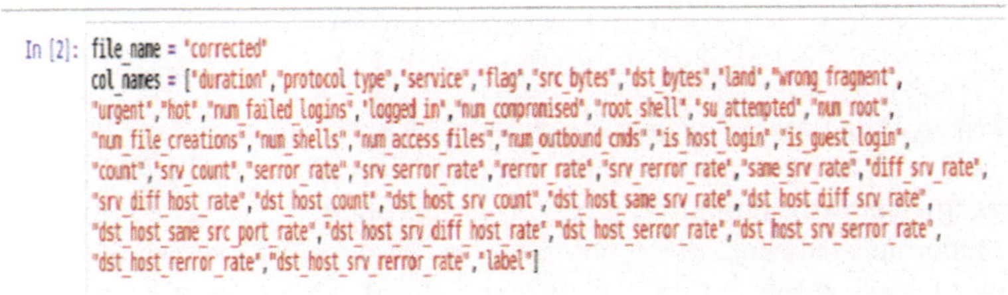

Fig. 3. Attributes names

After loading the dataset and assigning each attributes a name then the dataset is split into two sets as- one set as training data and another set as test data, The 40% of KDD data set is used for training data and remaining 60% for testing purposes. After splitting the KDD data model is trained of the Model is done using training data set for learning purpose. The training data first performs - feature selection process to get different dimensions which is as shown in Fig. 4.

```
]].astype(float)
labels = data['label']
print(labels.value_counts())
features_scaled=stdsc.fit_transform(features)
X_train, X_test, y_train, y_test = train_test_split(features_scaled, labels, test_size=0.2)
```

```
smurf.                    545
neptune.                  191
normal.                   179
snmpgetattack.             17
guess_passwd.              15
mailbomb.                  15
snmpguess.                  7
satan.                      6
saint.                      6
back.                       6
mscan.                      4
warezmaster.                4
portsweep.                  2
apache2.                    1
nmap.                       1
httptunnel.                 1
Name: label, dtype: int64
```

Fig. 4. Feature selection

The author during the experience have compared the DBN algorithm with the proposed SPELM algorithm for improved intrusion detection.

Training and Testing Using DBN Algorithm

After feature selection the machine is trained through deep belief network algorithm with the identified training data dataset and then tested for classification using testing datasets. This is shown in Fig. 5. The accuracy and other performance measures as shown in Fig. 6 shows the classification result of DBN.

Training and Testing Using SPELM Algorithm

After scheduling DBN algorithm; training and testing using proposed Classifier – the State Preserving Extreme Learning Machine (SPELM) algorithm is done. In training 40% of data set is used for training purposes and remaining 60% dataset is used for testing purpose as shown in the Fig. 7 classifying its performance measures.

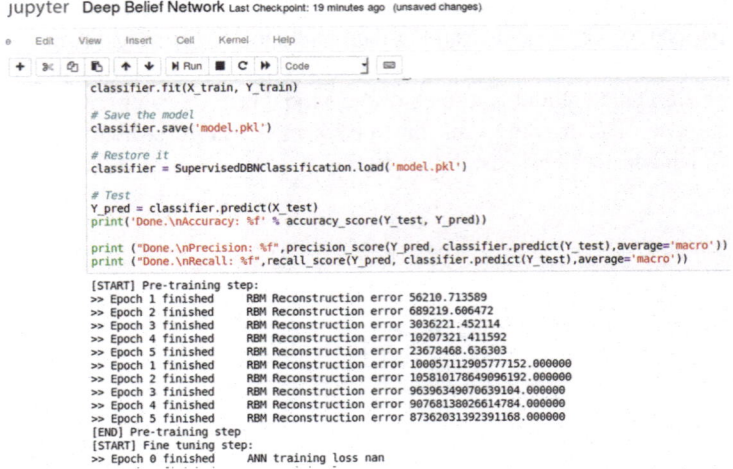

Fig. 5. Training and testing of DBN model

```
>> Epoch 44 finished    ANN training loss nan
>> Epoch 45 finished    ANN training loss nan
>> Epoch 46 finished    ANN training loss nan
>> Epoch 47 finished    ANN training loss nan
>> Epoch 48 finished    ANN training loss nan
>> Epoch 49 finished    ANN training loss nan
[END] Fine tuning step
Done.
Accuracy: 0.524500
```

Fig. 6. Accuracy of the DBN Classifier

Formula to Calculate Performance Measure

The accuracy, derived using confusion matrix is as follow (Table 1).

The Accuracy is calculated as follow:

Accuracy = Number of samples correctly classified in

$$\frac{\text{Testing data}}{\text{Total number of samples in testing data}}$$

After computing performance measures of existing DBN model and proposed SPELM model; comparison of both the models on the basis of parameters such as time taken, accuracy, precision and recall was computed. The comparison of both the models are shown in Fig. 8.

The number of record are classify by using DBN and SPELM classifier algorithm and the records are classifier into various categories are shown in Fig. 9.

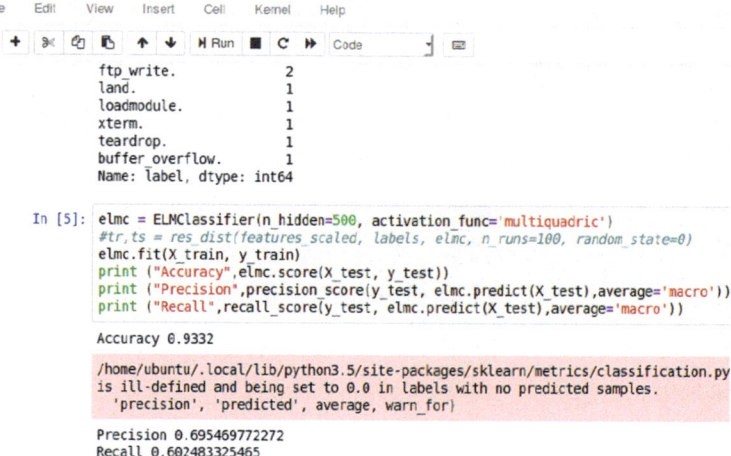

```
jupyter SPELM Last Checkpoint: 2 minutes ago (autosaved)

le    Edit    View    Insert    Cell    Kernel    Help

+  ✂  ⎘  ▮  ↑  ↓  ▶Run  ■  C  ▶▶  Code          ▾  ⌨

        ftp_write.              2
        land.                   1
        loadmodule.             1
        xterm.                  1
        teardrop.               1
        buffer_overflow.        1
        Name: label, dtype: int64

In [5]: elmc = ELMClassifier(n_hidden=500, activation_func='multiquadric')
        #tr,ts = res_dist(features_scaled, labels, elmc, n_runs=100, random_state=0)
        elmc.fit(X_train, y_train)
        print ("Accuracy",elmc.score(X_test, y_test))
        print ("Precision",precision_score(y_test, elmc.predict(X_test),average='macro'))
        print ("Recall",recall_score(y_test, elmc.predict(X_test),average='macro'))

        Accuracy 0.9332

        /home/ubuntu/.local/lib/python3.5/site-packages/sklearn/metrics/classification.py
        is ill-defined and being set to 0.0 in labels with no predicted samples.
          'precision', 'predicted', average, warn_for)

        Precision 0.695469772272
        Recall 0.602483325465
```

Fig. 7. Accuracy of the SPELM Classifier

Table 1. Confusion matrix

	Classified as normal	Classified as attack
Normal	TP	FP
Attack	FN	TN

Where TN - (True negative) instances correctly predicted as non-attacks

FN - (False negative) instances wrongly predicted as non-attacks

FP - (False positive) instances wrongly predicted as attacks

TP - (True positive) instances correctly predicted as attacks

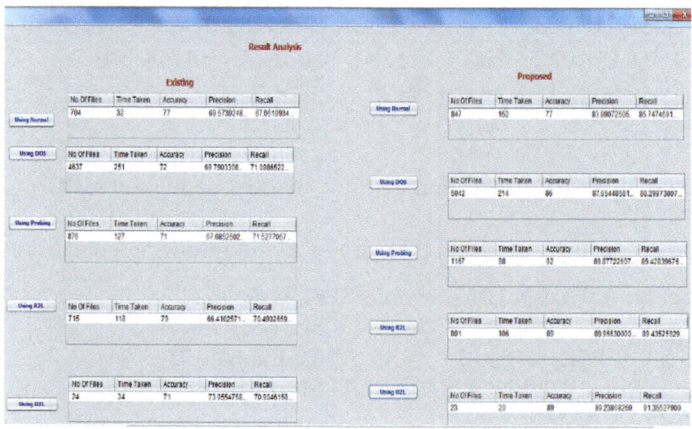

Fig. 8. Comparison of DBN and SPELM model

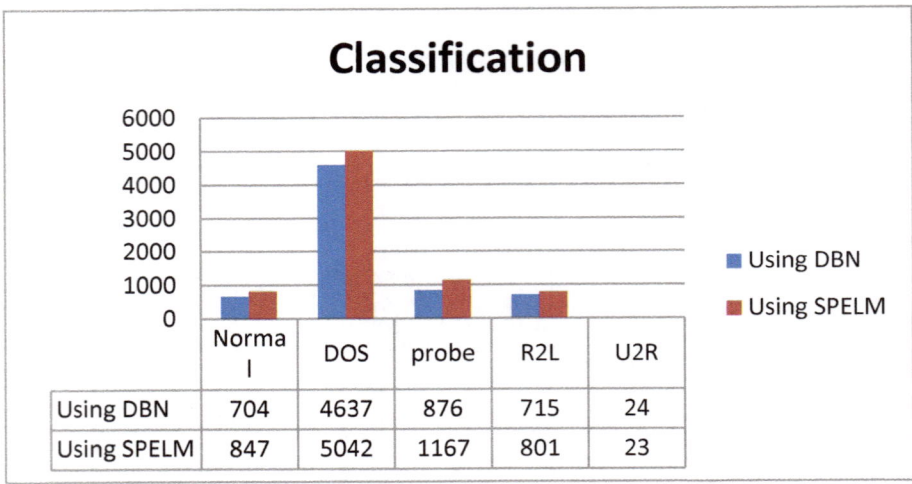

Fig. 9. Classification result using DBN and SPELM

From the above Fig. 9; the author comes to the following conclusion:

  i. The NORMAL attack of DBN is 704 record while SPELM is 847 record.
 ii. The DOS attack of DBN is 4637 record while SPELM is 5042 record.
iii. The PROBE attack of DBN is 876 record while SPELM is 1167 record.
 iv. The R2L attack of DBN is 715 record while SPELM is 801 record.
  v. The U2R attack of DBN is 24 record while SPELM is 23 record.

The data was computed and is tabulated in terms of accuracy, precision, recall and computational time for both existing DBN model and proposed SPELM Model as shown in Table 2.

Table 2. Result analysis in terms of accuracy, precision, recall and computational time

Strategies used	Accuracy	Precision	Recall	Time taken
Existing DBN model	52.80	66.836	56.874	102
Proposed SPELM model	93.20	69.492	60.052	90.8

The results of the two DBN and SPELM models were graphically represented as shown in Fig. 10.

From the above Fig. 10; the author comes to the following conclusion:

  i. The accuracy of DBN is 52.8 while SPELM is 93.2.
 ii. The precision of DBN is 66.836 while SPELM is 69.492.
iii. The recall of DBN is 56.874 while SPELM is 60.052.
 iv. The time taken of DBN is 102 s while SPELM is 90.8 s.

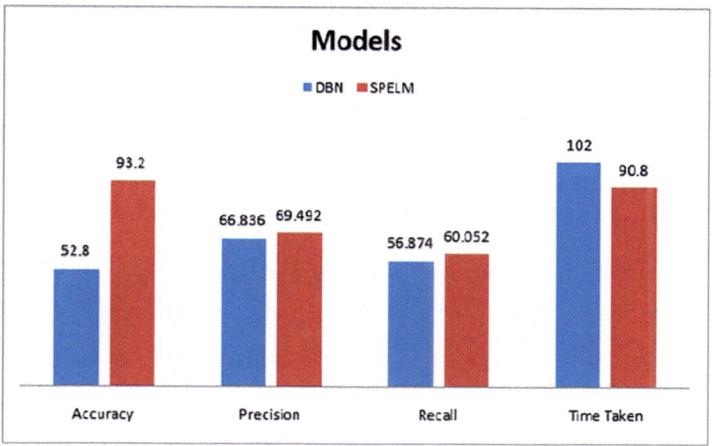

Fig. 10. Graphical representation of Accuracy Precision, Recall and Time taken on existing DBN and proposed SPELM machine learning Algorithms

The above Fig. 9 clearly shows the comparison between SPELM and DBN w.r.t accuracy, precision, recall and total time taken. Therefore, the experimentation have shown that SPELM perform better as compared to DBN.

6   Conclusion

Classification of Intrusion Detection System (IDS) with the help of machine learning algorithm (SPELM) provides a better result when compared with Deep Belief Network Model w.r.t accuracy, precision and recall. The execution time taken by SPELM is 90.8 s when compared with DBN time 102 s is also less; proves more efficient.

Acknowledgements. The authors are grateful to acknowledge the support of NSL KDD Data set.

References

Sultana, A., Jabbar, M.A.: Intelligent network intrusion detection system using data mining techniques. In: 2nd International Conference on Applied and Theoretical Computing and Communication Technology (iCATccT). IEEE (2016)

Anderson, J.P.: Computer security threat monitoring and surveillance, Technical report 98-17. James P. Anderson Co., Fort Washington, Pennsylvania, USA, April 1980

Obeis, N.T., Bhaya, W.: Review of data mining techniques for malicious detetion. Res. J. Appl. Sci. **11**(10), 942–947 (2016)

Wang, J.-H., Deng, P.S.: Virus detection using data mining techniques. TAO-Yuar, Taiwan, ROC333 (2003)

Ranjan, R., Sahoo, G.: A new clustering approach for anomaly intrusion detection. Int. J. Data Min. Knowl. Manag. Process (IJDKP) (2014)

Hall, M.A.: Correlation-based feature selection for discrete and numeric class machine learning. In: International Conference on Machine Learning, pp. 359–366 (2000)

Koc, L., Mazzuchi, T.A.: A network intrusion detection system based on a Hidden Naïve Bayes multiclass classifier. Expert Syst. Appl. **39**, 13492–13500 (2012)

Feng, W., Zhang, Q.: Mining network data for intrusion detection through combining SVMs with ant colony networks. Future Gener. Comput. Syst. **37**, 27–140 (2014)

Kohli, S., Singal, H.: Data analysis with R. IEEE (2014)

Alom, M.Z., Bontupalli, V.R., Taha, T.M.: Intrusion detection using deep belief networks. IEEE (2015)

Address Resolution Protocol Based Attacks: Prevention and Detection Schemes

D. Francis Xavier Christopher and C. Divya$^{(\boxtimes)}$

School of Computer Studies, Rathnavel Subramaniam College of Arts
and Science, Coimbatore, Tamilnadu, India
divyac.phd2017@gmail.com

Abstract. Address resolution protocol (ARP) is considered as the important protocol in computer communications due to frequent usage in individual as well as organization levels. Though efficient during regular scenarios, the design of ARP has not been designated to tackle malicious hosts. Considering the adverse circumstances, many researchers have developed strategies to improve the effectiveness of ARP by detecting and preventing the malicious attacks like spoofing, Denial of service (DoS) and Man in the middle (MITM) attacks. This paper discusses about the ARP poisoning attacks and focuses on reviewing various mechanisms developed for attack detection and prevention with specified analysis to their advantages. Different attack detection and mitigation methods are evaluated in addition to comparison in terms of key parameters. This study helps in understanding the strategy employed for ARP attack detection and mitigation and developing a framework for improvement.

Keywords: Network security · Address resolution protocol · Spoofing · DOS attack · ARP poisoning · MITM attack

1 Introduction

Network security begins with approval of access to information in a network, usually with a username and proper authentication. Network security comprises of the arrangements and approaches received by a system manager to avert and observe unapproved access, changes, or denial of a PC system and resources [23]. It has turned out to be more important to PC clients, and associations. If this approved, a firewall powers access to strategies, for example, what administrations are permitted for network clients can be accessed. This unapproved access to system might neglect to check potentially harmful data like PC viruses being transmitted over the network.

Anti-virus programming or an intrusion detection system (IDS) can aid in identifying the attacks or viruses [15]. However inconsistency in communications between hosts increases the security issues. There is a significant absence of security techniques that can be effectively actualized in the network innovation. The communication break among the engineers of security innovation and designers of networks is the main reason for attacks.

System configuration is a generated procedure, relies upon the Open Systems Interface (OSI) model [6]. The OSI has a few points of interest when outlining system

© Springer Nature Switzerland AG 2020
A. P. Pandian et al. (Eds.): ICCBI 2018, LNDECT 31, pp. 247–256, 2020.
https://doi.org/10.1007/978-3-030-24643-3_30

security providing modularity, and flexibility of various layers can be effortlessly joined to make stacks that allow measured improvement. Rather than secure network configuration is certainly not an all-around developed process. There is no approach to deal with the multi-objective nature of security requirements.

While considering about network security, it ought to be underscored that the total network is secure, instead of considering only the computers at the end of the communication chain to avoid attacks in hosts [19]. A programmer will focus on the communication channel, get the information, and decode it and reinsert a copy message. While building up a protected network, the essentials should be considered are confidentiality and integrity. However the variety of attacks is still increasing by passing time [12].

The fundamental class of attacks is categorized as active and passive attacks which include spoofing, modification, wormhole, fabrication, denial of services, sinkhole, Sybil, eavesdropping, black hole, rushing attacks, etc. ARP is a broadly utilized transport network protocol for determining Internet layer addresses. ARP spoofing or ARP cache poisoning [18] is a procedure used by attackers to spoof ARP packets in a LAN. Figure 1 demonstrates the ARP spoofing or ARP cache poisoning attack (Source: [1]).

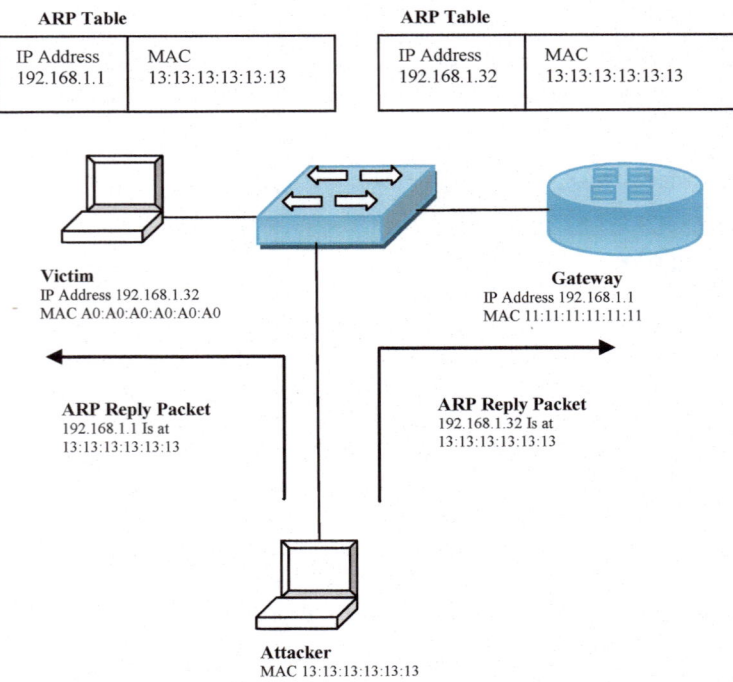

Fig. 1. ARP spoofing/cache poisoning

The fundamental norm of ARP spoofing is to misuse the absence of validation in the ARP protocols by directing spoofed ARP communications onto the LAN. ARP spoofing attacks are generally started from a long used host in LAN or from an attacker system connected in a LAN without the admin knowledge. The attacker in most cases copy the MAC address of a host and utilizes it to identify itself as the host. This allows the attackers to track the packets from the LAN presenting itself as the host [24]. In most scenarios, this is used as an opening for diverse attacks.

The attacker assesses the packets (spying), while at the same time sending the data packets to the genuine destination to evade detection, alter information in advance; sending MITM attack, or dispatch a DOS attack by triggering a few or the majority of the packets on the system to be dropped out. The attack utilized on systems that employ ARP, and are bound to require the intruder to increase guide admittance to the LAN to be criticized. There are a few methodologies to mitigate ARP spoofing like utilizing ARP entries and prevention software or securing the operating system [4, 11]. However the current strategies fail when the attack gets stronger and hence prominent methods are sought to be developed. This article aims at reviewing few of the ARP detection and prevention mechanism developed over the years by different scientists and researchers to provide an understanding of these attacks and effective scope for development.

2 ARP Spoofing Attack Detection and Prevention Mechanisms

As discussed above, ARP spoofing is responsible for the cause of many attacks. The research for detecting and preventing against such attacks is a long process exceeding about three decades. There have been various research models that tend to analyze, detect and mitigate ARP spoofing effects. However the ever changing environments and attacks initialization strategies have made it look like the research to find the efficient method is never ending. The varying features of ARP spoofing makes it comfortable for the attackers to avert the prevention mechanisms once they become familiar. Many research works have been undertaken to provide secured defense mechanism against ARP spoofing. Few prominent methods developed for this purpose over the years are reviewed in this section in hope of tracing the basic vulnerabilities and perfect mitigation approach for ARP based attacks.

1. An active host-based intrusion detection system for ARP-related attacks and its verification: Barbhuiya et al. [7] developed an active host based IDS to track the ARP attacks in LAN connections. This host based system works on the adverse environment with constraints similar to static IP-MAC, modified ARP, variable routes, etc. but has been designed to detect various attack scenarios. This system detects the ARP related attacks efficiently. However, the major shortcoming of this system is it does not provide diagnosis or prevention methods for which new methods or third-party applications are needed to be employed along with this model.

2. A centralized detection and prevention technique against ARP poisoning: Kumar and Tapaswi, [21] developed a centralized system to mitigate ARP spoofing attacks. Along with ARP Central Server (ACS), the centralized system manages and validates the ARP table entries in all hosts to remove the possible inconsistencies. The ACS diagnoses the affected ARP packets and hence avoids further damage. This model allows the reluctant compatibility to present networks by utilizing the ARP without new modifications. However, this system is vulnerable to central point failure, although minor, which will reduce the network defence. Similarly, IP exhaustion attack is also untraceable considering the opportunity of this research scheme.

3. Collaborative approach to mitigating ARP poisoning-based Man-in-the-Middle attacks: Nam et al. [22] proposed and developed a collaborative approach for mitigating the ARP attacks especially the Man-in-the-middle attacks using fairness voting process. This model works on the concept of protecting a node by resolving the mapping processes between IP and MAC addresses through the fair voting method among the neighboring host nodes. In order to attain objectivity in voting, the uniform broadcast proficiency of the Ethernet connections is employed and found to be better than existing voting schemes. This model is purely efficient as the public key cryptography methods are not required additionally even when the communication is over public Wi-Fi environments.

4. Design and implementation of an efficient defense mechanism against ARP spoofing attacks using AES and RSA: Hong et al. [20] developed a defense mechanism against the ARP spoofing attacks using the cryptography schemes. This mechanism employed doesn't need modifications to the network protocol of ARP or require additional tools. Instead it utilizes AES and RSA encryption algorithms to improve the attack mitigation. Additionally, this mechanism automatically renovates the reliable MAC address data to the ARP table to defend users from ARP spoofing. However, in contrast to its core concept, a light application of substantially disconnected system with MAC-Agent is utilized for detecting the MAC address variations.

5. IDS for ARP spoofing using LTL based discrete event system framework: Mitra et al. [13] developed a IDS for ARP spoofing attacks using Linear-time Temporal Logic (LTL) discrete event system framework. This detection system offers a prototype for stating the system specifications, modelling, detector creation and checking its correctness. It also provides adapted polynomial time complexity in the number of system states as paralleled to exponential complication of the other traditional frameworks. It has better detection accuracy than other models due to the fact it does not alter the ARP protocol but improves the message tracking mechanism. However this method creates more congestion to detect the attacker nodes which is a problem when prolonged for each stage of transmission.

6. ARP spoofing detection algorithm using ICMP protocol: Jinhua and Kejian, [8] developed an ARP spoofing detection algorithm using the Internet Control Message Protocol (ICMP) which identifies the malicious hosts. This scheme accumulates and analyses the ARP packets, and then inserts the ICMP echo request packets for identifying the mischievous host. Initially the cross layer control based investigation is utilized for consistency verification and then the address changes are checked in the valid ARP packets. Finally the new ARP packets are forwarded to the ARP spoof detection model for final detection and diagnosis. The advantage of this method is that the time delay amongst the seizing packets and identifying spoofing is least than other models.

7. Prevention of ARP spoofing: A probe packet based technique: Pandey, [17] developed a probe packet based prevention technique for ARP spoofing. This technique utilized Enhanced Spoofing Detection Engine (E-SDE) for detecting the ARP spoofing along with the genuine IP-MAC associations in the ARP cache of the network host. In this technique, the ARP and ICMP packets are used as probe packets and the attacking model is effectively detected. This technique also has less network traffic than other existing intrusion detection models. However as in current ARP spoofing detection models, when a strong attack is initiated, this model just detects the presence of attack while it fails to map the IP-MAC addresses due to strong opposition.

8. DS-ARP: a new detection scheme for ARP spoofing attacks based on routing trace for ubiquitous environments: Song et al. [14] developed Detect Spoofing ARP (DS-ARP) scheme for identifying and mitigating the ARP spoofing attacks. This scheme identifies ARP attacks over real-time checking of the ARP cache table and a direction-finding trace and defends the hosts from attackers using ARP Link Type Control which modifies from dynamic to fixed. Additionally, it can solve problems such as host impersonation, MITM attack, and block of host. Unlike other existing schemes, this proposed scheme does not require an ARP protocol change or a complex encryption algorithm; moreover, it does not cause high system load. However, still the basic weaknesses of ARP prevail when new strategies of attacks are initiated.

9. An Integrated Approach to ARP Poisoning and its Mitigation using Empirical Paradigm: Kaur and Malhotra [9] developed an integrated approach to detect and mitigate the ARP poisoning attacks by employing empirical paradigm. This model utilized tools like NetworkMiner, Cain and Abel and Wireshark for demonstrating the attack scenarios and corresponding attack mitigation. This model of attack mitigation is highly suitable for each kernel of the communication networks while also improve the HTTPS protocols. The limitation of this model is that the detection model using empirical paradigm is much simpler to break by the adversary.

10. Bayes-based ARP attack detection algorithm for cloud centers: Ma et al. [10] developed an IDS model using Bayesian based algorithm to forecast the probability of each possible attacker hosts. Following these footsteps, a detection model is also utilized in order to mitigate the attacker host nodes. Using the SDN technology, the ARP packets are processed and the communication model of entire network is controlled. This process enables the system to distinguish the attackers' features from the normal features and thereby identifies the attacker host node. However the major issue with this model is that the attack nodes are unidentified when the adversary makes the attack features absorbed into normal node features.

11. RTNSS: a routing trace-based network security system for preventing ARP spoofing attacks: Moon et al. [5] designed and developed RTNSS as a recognition model which uses monitoring agents to detect the ARP spoofing. The installed monitoring agent observes the modifications in the ARP cache table by the adversary. These changes are made to spoof the host address with a fake address including the IP, MAC address pair. Detection of these spoofing can be performed by the agent through a routing trace and it alerts the main server. Though this method seems to be similar to few existing detection methods, the ARP cache table changes are more accurately sensed in RTNSS and thus the attack detection is significant. Another important aspect is that, unlike existing methodologies, RTNSS incorporates the mechanisms to tackle the effects of structure changes and increased traffic changes due to encryption complexity. However this model fails to handle packet forwarding relay based attack strategy and also the lack of encryption mechanisms increases the security vulnerability of ARP.

12. Securing ARP and DHCP for mitigating link layer attacks: Younes [16] introduced a security model that applies cryptography based protocols to the communications at data link layer to improve the authentication and data integrity. This model utilizes IPsec and Transport Layer Security (TLS) to offer authentication and data integrity using cryptography for communications at the network and transport layer. This model of intrusion detection can mitigate the link layer attacks including the rogue Dynamic Host Configuration Protocol (DHCP) server, DHCP exhaustion, mischievous customer, host impersonation, ARP Spoofing, MITM, and DOS attacks. Unlike existing models, this model mitigates the DHCP starvation attack using the symmetric key cryptography. However the limitation of this model is that it fails to prevent DoS attacks using flooding and hence care should be taken.

13. Security Solution for ARP Cache Poisoning Attacks in Large Data Centre Networks: Prabadevi and Jeyanthi [3] considered the issues in Layer-2 ARP protocols to mitigate the attacks. This model utilized the components to practise the packet decoding, update of invalid entries with Timestamp feature and communication packets. Utilizing these components, the insecure communication, scalability problems and VM migration concerns are mitigated such that the attacks in ARP namely host impersonation, MITM, Distributed DoS are prevented effectively. However this model fails to detect the ARP scanning attacks due to the issues in identifying the neighbor host feature.

14. A framework to mitigate ARP Sniffing attacks by Cache Poisoning: Prabadevi and Jeyanthi [2] developed another framework for mitigating cache poisoning based ARP sniffing attacks. This process works by comparing the IP-MAC pair in the ARP and Ethernet headers. If any host is suspected, they are blacklisted and updated in the fake list and alert messages are delivered all over the network to avert the gateways defend themselves from attacks. This model significantly identifies the ARP based attacks including the scanning attacks but the limitation is that when new attack strategy is undertaken by the adversary hosts, it fails to identify the attack.

The surveying of these research methodologies enlighten the concept of spoofing and enhancements in tackling them over the years. These methodologies form the basic foundation for future research works. A summarized comparison of these reviewed methodologies can further improve the understanding about the prevention strategies, which is provided in the following section.

3   Comparison of Attack Prevention Methodologies

The analysed methods and strategies are compared in terms of unique features, merits and demerits helps in obtaining a clear picture about the recognition and protection mechanisms of ARP cache poisoning and spoofing attacks. Table 1 shows the comparison of these methodologies.

Table 1. Comparison of ARP based attack mitigation strategies

S. No.	Authors	Method/approach used	Merits	Demerits
1	Barbhuiya et al. [7]	Active host based IDS	Detects multiple ARP based attacks	Third party applications are required for prevention of attacks
2	Kumar and Tapaswi [21]	Centralized system with ARP Central Server	Consistent detection without altering ARP architecture	Susceptible to central point failure and also the IP exhaustion attack is untraceable
3	Nam et al. [22]	Collaborative approach using fairness voting	Detects MITM attacks efficiently and does not rely on public key cryptography methods	False detection is possible when the neighbor hosts are already spoofed
4	Hong et al. [20]	Cryptography based schemes	AES and RSA protects against attacks along with automatic renewal of MAC address details	Additional application is used for detecting MAC address variations which deviates from its core concept
5	Mitra et al. [13]	LTL discrete event system framework	Attack detection with high accuracy but without altering ARP features and improved message tracking mechanism	More congestion is possible when the attack detection is prolonged over transmission time
6	Jinhua and Kejian [8]	ICMP based ARP spoofing detection	Time delay for detection is minimum and detection accuracy is higher	Presence of existing attacks is untraceable
7	Pandey [17]	Probe packet based technique with E-SDE	Attack detection with less traffic	Fails to map IP-MAC address when strong attack is initiated
8	Song et al. [14]	DS-ARP	Detects attacks without requiring encryption algorithms or altering	Inefficient when new strategies of attacks are initiated

(*continued*)

Table 1. (*continued*)

S. No.	Authors	Method/approach used	Merits	Demerits
			ARP and also does not cause high system load	
9	Kaur and Malhotra [9]	Integrated approach with empirical paradigm	Suitable for HTTP and HTTPS protocols	Simpler to break after certain trails of execution
10	Ma et al. [10]	Bayesian based algorithm	Centralized control of entire network enhances protection	Attacks are unidentified when attacks features are misinterpreted as normal features
11	Moon et al. [5]	RTNSS	Detects ARP attacks with effective tackling strategies for structure changes and increased traffic	Lack of encryption and failure to handle packet forwarding relay based attack strategy
12	Younes [16]	Cryptography based protocols with IPsec and TLS	Prevents ARP spoofing attacks including DHCP starvation attack	Failure to detect flooding based DoS attacks
13	Prabadevi and Jeyanthi [3]	Messaging based detection	Effective attack mitigation along with resolving scalability issues and VM migration issues	ARP scanning based attacks are not detected
14	Prabadevi and Jeyanthi [2]	IP-MAC comparison based detection	Effectively detects ARP based attacks including scanning attacks	Fails to detect attacks through new strategies

From this comparison, it can be found that over the years, the methods and mechanisms developed for ARP based attack mitigation has been increased in number as well as quality based on the strength of the attack strategies. However, it is fair to say that in spite of tremendous efforts by various researchers, the ARP based attacks have also grown significantly and hosts launch stronger attacks countering the preventive measures. It is also quite understandable that this war between researchers and attackers is tends to continue and hence the scope for developing efficient defense mechanisms that tackle the existing vulnerabilities and limitations. By taking inputs from the past research models, the researchers can up with new efficient methodologies to tackle the ARP based attacks.

4 Conclusion

This paper provided the survey and analysis of the research works based on ARP Spoofing based attack prevention strategies over the recent years. The aim of this paper is to understand the concepts of ARP attacks and analyze the principles of various detection and prevention methodologies. This work would provide us with a clear understanding about the current stage of ARP spoofing detection research and help us to take over the future researches. Based on the analysis results inferred, it can be concluded that there still lot of vulnerabilities in ARP architecture that needs to be considered and also the path is wide open for introducing new concepts of prevention methodologies for newly evolving attack strategies.

References

1. Samvedi, A., Owlak, S., Chaurasia, V.K.: Improved secure address resolution protocol. arXiv preprint arXiv:1406.2930 (2014)
2. Prabadevi, B., Jeyanthi, N.: A framework to mitigate ARP Sniffing attacks by cache poisoning. Int. J. Adv. Intell. Paradig. 10(1–2), 146–159 (2018)
3. Prabadevi, B., Jeyanthi, N.: Security solution for ARP cache poisoning attacks in large data centre networks. Cybern. Inf. Technol. 17(4), 69–86 (2017)
4. Abadand, C.L., Bonilla, R.I.: An analysis on the schemes for detecting and preventing ARP cache poisoning attacks. In: Proceedings of the 27th International Conference on Distributed Computing Systems Workshops, ICDCSW 2007, p. 60. IEEE (2007)
5. Moon, D., Lee, J.D., Jeong, Y.S., Park, J.H.: RTNSS: a routing trace-based network security system for preventing ARP spoofing attacks. J. Supercomput. 72(5), 1740–1756 (2016)
6. Wetteroth, D.: OSI Reference Model for Telecommunications, vol. 396. McGraw-Hill, New York (2002)
7. Barbhuiya, F.A., Biswas, S., Nandi, S.: An active host-based intrusion detection system for ARP-related attacks and its verification. Int. J. Netw. Secur. Appl. (IJNSA) 3(3), 163–180 (2011)
8. Jinhua, G., Kejian, X.: ARP spoofing detection algorithm using ICMP protocol. In: Proceedings of the International Conference on Computer Communication and Informatics (ICCCI), pp. 1–6. IEEE (2013)
9. Kaur, G., Malhotra, J.: An integrated approach to ARP poisoning and its mitigation using empirical paradigm. Int. J. Future Gen. Commun. Netw. 8(5), 51–60 (2015)
10. Ma, H., Ding, H., Yang, Y., Mi, Z., Yang, J.Y., Xiong, Z.: Bayes-based ARP attack detection algorithm for cloud centers. Tsinghua Sci. Technol. 21(1), 17–28 (2016)
11. Singh, J., Kaur, G., Malhotra, J.: A comprehensive survey of current trends and challenges to mitigate ARP attacks. In: Proceedings of the International Conference on Electrical, Electronics, Signals, Communication and Optimization (EESCO), pp. 1–6. IEEE (2015)
12. Conti, M., Dragoni, N., Lesyk, V.: A survey of man in the middle attacks. IEEE Commun. Surv. Tutor. 18(3), 2027–2051 (2016)
13. Mitra, M., Banerjee, P., Barbhuiya, F.A., Biswasand, S., Nandi, S.: IDS for ARP spoofing using LTL based discrete event system framework. Netw. Sci. 2(3–4), 114–134 (2013)
14. Song, M.S., Lee, J.D., Jeong, Y.S., Jeong, H.Y., Park, J.H.: DS-ARP: a new detection scheme for ARP spoofing attacks based on routing trace for ubiquitous environments. Sci. World J. vol. 2014, Article ID 264654, 7 pages (2014)

15. Depren, O., Topallar, M., Anarim, E., Ciliz, M.K.: An intelligent intrusion detection system (IDS) for anomaly and misuse detection in computer networks. Expert Syst. Appl. **29**(4), 713–722 (2005)
16. Younes, O.S.: Securing ARP and DHCP for mitigating link layer attacks. Sādhanā **42**(12), 2041–2053 (2017)
17. Pandey, P.: Prevention of ARP spoofing: a probe packet based technique. In: Proceedings of the IEEE 3rd International Advance Computing Conference (IACC), pp. 147–153 (2013)
18. Wagner, R.: Address resolution protocol spoofing and Man-in-the-Middle attacks. The SANS Institute (2001)
19. Weingart, S.H.: Physical security devices for computer subsystems: a survey of attacks and defense. In: Proceedings of the International Workshop on Cryptographic Hardware and Embedded Systems, pp. 302–317. Springer Heidelberg (2000)
20. Hong, S., Oh, M., Lee, S.: Design and implementation of an efficient defense mechanism against ARP spoofing attacks using AES and RSA. Math. Comput. Model. **58**(1–2), 254–260 (2013)
21. Kumar, S., Tapaswi, S.: A centralized detection and prevention technique against ARP poisoning. In: Procedings International Conference on Cyber Security, Cyber Warfare and Digital Forensic (CyberSec), pp. 259–264. IEEE (2012)
22. Nam, S., Djuraev, S., Park, M.: Collaborative approach to mitigating ARP poisoning-based Man-in-the-Middle attacks. Comput. Netw. **57**(18), 3866–3884 (2013)
23. Stallings, W.: Network Security Essentials: Applications and Standards, 4th edn. Pearson Education, Noida (2000)
24. Trabelsi, Z., El-Hajj, W.: On investigating ARP spoofing security solutions. Int. J. Internet Protoc. Technol. **5**(1–2), 92–100 (2010)

Enabling Technologies for Internet of Vehicles

Shridevi Jeevan Kamble[✉] and Manjunath R. Kounte

School of Electronics and Communication,
REVA University, Bengaluru 560064, India
shridevi.havanoor@gmail.com,
manjunath.kounte@gmail.com

Abstract. An excellent platform for vehicular networking has been built by Internet of Things (IoT) by holding and growing the state of being everywhere at once and has a large challenge and opportunity to change the day to day life and this is done by using technologies as wireless network, sensor devices, and the capability to figure out and communicate with the devices. The ability of communication amongst vehicles and their territory communication i.e. internal or external territory can be extended by different wireless connectivity. Equivalent to Such a networking solution for vehicles is habitually to be the next generation verge for automotive transformation and the fundamental and leading to progression for succeeding generations of intelligent vehicular systems (IVSs). In this paper, we focus on future challenges, enhancing technologies, advantages, and understanding for connectivity of vehicles using wireless technology and Ad-Hoc technology. We also discuss the objectives and challenges and retrospect the state-of-the-art wireless technology solution connect the vehicle to a sensor, to vehicle, to the Internet, and to the road infrastructure, to pedestrian, and to everything. Vehicular communication is a progressive technology that can be adapted in Intelligent connected vehicles (ICVs). We highlight on various mode to link of communication in IVS to understand the vehicular communication that is Vehicle-to-Vehicles (V2V) communication, Vehicle to Infrastructure (V2I), Vehicles to Cloud (V2C), and Vehicles to pedestrian (V2P) communication system all of which applies the emerging wireless technologies and the system which provides early warning signals to reduce road accidents and congestions. We also review on the Vehicular network and its connectivity to other vehicles and near-by devices which take place by Vehicular Ad Hoc Networks (VANETs) and discuss on VANETs benefits. The developing factors need to be studied in the future development of the Internet of Vehicles, and also to establish a bilateral achievement in the development of exemplary that is noted role of government and alive support of the venture.

Keywords: IOT · ICVs · Vehicular networking · VANET · Connected cars · ITS

A. P. Pandian et al. (Eds.): ICCBI 2018, LNDECT 31, pp. 257–268, 2020.
https://doi.org/10.1007/978-3-030-24643-3_31

1 Introduction

Due to an increase in the number of people using vehicles, which always contribute to the number of disasters, The expanding statistic of substantial objects is being connected to the Internet at a very significant rate completing the idea of the Internet of Things (IoT) [3], which intern form Internet of vehicles (IoV). IoT can play a significant role in a different domain to increase and improve the aspects of our life. The different domains where IOT can be used are home appliances, transportation, healthcare, industrial automation and emergency response to disasters where human judgment making is difficult [2].

IoV is a combination of mobile internet and IoT; IoV is an up-coming aspect for automobile industries and plays very important for building a smart city. During the time that the vehicles expanded from simple transportation mechanism to smart individual with sensing and communication to the vehicles or surrounding hub or tower to exchange information's and thus all this contribute to the smart city. Internet-of-vehicle (IoV) globally connects the vehicles, near-by tower or even pedestrian carrying smart devices and enabling them to communicate with each other. These smart objects are connected over the internet. IoV is a continued version of IoT in intelligent transportation. A vehicular communication is a sensor podium for collecting information from the environment, from alternative vehicles, from near-by towers, from drivers and uses the information for safe-navigation, traffic management, pollution monitoring and controlling, etc. Therefore the elements of IoV which importantly explains the functionality of IoV are the cloud, network connection and client. Figure 1 shows the element of IoV. The main element of IOV is Cloud also called the brain of the system. Services offered by the cloud is mainly migrating the storehouse, data processing, computing, and new functions from desktop to handheld devices, by creating the virtual functions in the cloud platform, these virtual services in the cloud can be used in the vehicular domain environment. The network connectivity [8] is mainly due to wireless access technology used to establish the network connection. The different types of user of IOV are smart clients used for application purpose.

IOV consist of vehicles that communicate with each other, the device carried by pedestrian connected to internet, and the hub with provide source of information on road called as roadside unit (RSU) and public network adopting for vehicle to vehicle (V2V), Vehicle to road (V2R), Vehicle to human (V2H), and Vehicle to sensor (V2S) all of these are interconnected to form social network in which members are smart and intelligent objects and not human beings. This leads to Social Internet of vehicles (SIoV). The IoV [7] enables Vehicles to communicate with each other to share information and to coordinate information. Figure 2 explains the view of IoV in which every domain is interacting with other domain independent services, where as in each domain sensors and actuators communicate directly with each other.

For safety navigation, one can use advance technique and concept. VANET [18] include four type of Communication as shown in Fig. 5, Vehicle to Vehicle communication and Vehicle to Infrastructure Communication, Vehicle to Pedestrians and vehicle to Roadside Unit.

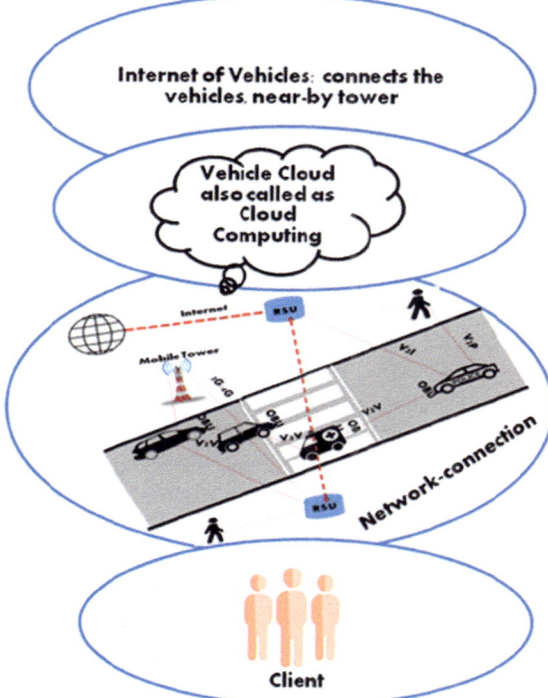

Fig. 1. Elements of IoV

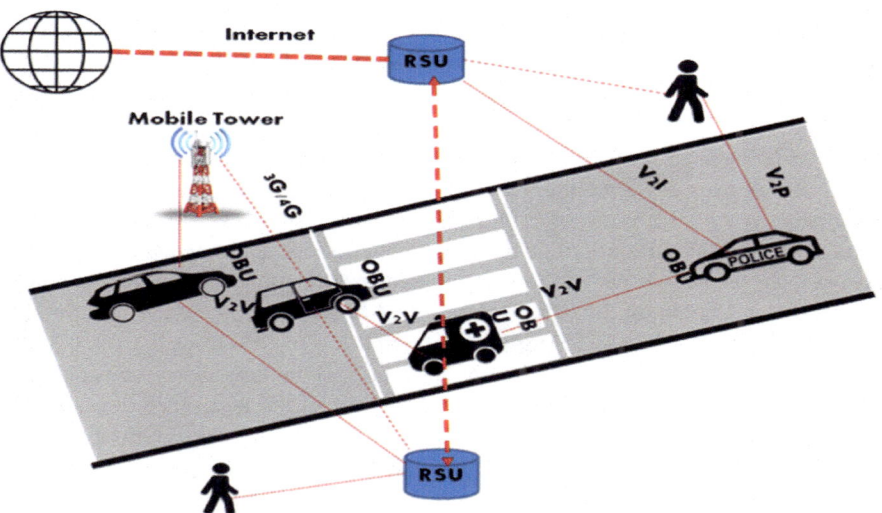

Fig. 2. IoV Interaction

All the smart vehicles on road [9] can perform all the four type of operation. The vehicle can perform as the sender, receiver and router. The vehicle to vehicle communication is a communication between two vehicles i.e. one hop communication, the vehicle to Infrastructure communication is communication between the vehicle and roadside Infrastructure. The vehicle to roadside unit is the communication between the vehicle and the roadside unit. Vehicle to Pedestrian is communication between the vehicle and pedestrian carrying the Smartphone device. It acts as a multi-hop communication.

The Vehicle to Vehicle communication is a system designed to transfer basic safety related to vehicles to provide the warning to drivers concerning accidents. The main objective of this system is to alert drivers when he closes to front vehicle. Vehicles associated with VANET are equipped with various types of sensors and assistance like storage and computing.

Vehicular network [1] and its connectivity to other vehicles and near-by devices take place by Vehicular Ad Hoc Networks (VANETs), which is a type of mobile Ad Hoc networks, with mobile nodes being vehicles and Road Side Units (RSUs) as static nodes. VANETs are being used on the Internet of Vehicle (IoV), this changes every moving vehicle on road into a wireless router or mobile node, and also enables smart vehicles on road to connect to each other and, thus creating a wide range network. The vehicles which are out of the range from the signal are dropped out of the network, and the other vehicles which are in the range from signals can join in, moving vehicles connecting with other vehicles create a mobile Internet. VANET technology uses moving vehicles as mobile. VANET turns every smart vehicle into a wireless node or router allows connecting of vehicles between 100 to 300 m instead of creating a network with a wide range. The element involved in VANET, are short-lived, irregular and unstable, and dimension of usage is local and distinct, i.e. it is not possible by VANET to provide global and continual services applications for customers. Past from several decades, there is no typical or popular implementation of VANET. There are many features of big cities, which mainly include heavy traffic, traffic conjunction, tall buildings, bad driver behaviors, and complicated road networks, further hinder its use.

The goal of IoV is mainly to realize the extent assimilation of human-vehicle-sensors-thing-environment, which stimulate the efficiency of transportation, reduce social cost, reforms the benefits level of the metropolis, and ensure that it is beneficial to humans on vehicles transportation. This is the reason how IoT and IoV are distinctive, as the Internet and wireless mobile networks. Essentially in wireless mobile networks, the end-users path follows an irregular path model. However, in IoV, the path of vehicles is exposed to the road assigned in towns.

Here we also would like to presents the definition of IoV from the perspective of integration of onboard sensors and communication technology. They consider IoV as vehicles carrying advanced sensors, controllers, actuators, and other devices, which integrate modern communications and network technology for providing the vehicles with complex environmental sensing, intelligent decision-making, and control functions. According to this definition, the vehicles in IoV can fully safety, energy saving, and comfortable driving. Hence, IoV, from this perspective, predominantly solves the problem of intelligent decision-making of vehicles.

In this paper, we focus on enabling wireless technologies and concept of Intelligent Transport System (ITS) to provide Vehicle-to-X (Everything) [16] connectivity. VANET offers huge benefits to Automobiles with high-speed internet access making web technology available in vehicles; this web technology can transform the vehicles on-board computer to the very important tool. Example one cannot use the phone while driving or respond to email so the traveler can make use of traffic jam into productivity work by responding to his official emails or can also surf the internet etc. Hence VANET is a technology that significantly increases productivity. In this paper, we review the state-of-the-art wireless solutions for vehicular technology, vehicle-to-sensor, vehicle-to-vehicle, Vehicle-to-Internet, and Vehicle-to-Road infrastructure connectivity's, Vehicle-to-Everything (V2X).

2 Vehicles-to-Vehicles Communication and Vehicle-to-Infrastracture Communication

Vehicular network communication is a combination of mainly Vehicle-to-Infrastructure (V2I) communication and Vehicle-to-Vehicle (V2V) communication. In V2V two vehicles can communicate with each other without any help from existing infrastructure. But in case of large-scale deployment of the V2V communication network, it is difficult to set the link of communication between vehicles, in such a way that each vehicle can trust all the other vehicles and forward information. Maintaining privacy and security in large-scale deployment of the V2V communication network is a challenging research issue.V2I promotes many operations as web surfing, faraway vehicle detection, communication link network, multimedia streaming, real-time vehicles navigation, environmental monitoring, traffic monitoring etc.

In IoV technology the smart vehicles are furnished with wireless local area networking (WLAN) technology, this WLAN permits the formulation of wireless Ad Hoc, which in turn formulation link of communication between smart vehicles in the network, which can be V2V communication, or connects the link of wireless Ad-Hoc communication network with gateways, which is called as Road Side Units (RSU) installed in the network i.e. V2I communication. Vehicles belonging to a VANET are furnished with the different type of sensors and services such as estimating and storage services.

2.1 Vehicle-to-Vehicles Communication

Vehicle-to-Vehicle communication [5] aspires to modify the vehicle from simple transportation mode to smart technological object. V2V communication, designed as vehicular Ad Hoc networks (VANETs), provides several advanced services for vehicles and establishes a large number of opportunities in V2V communication for safe navigation and traffic management. The importance's of Link of Communication between vehicles is mainly to recognize driver active, road traffic, security benefits, like collision warning, updated traffic of every movement and weather information or active navigation system.

Thus Intelligent Transport System (ITS) [10] is a practical approach to minimize road traffic congestion and manage traffic. The vital factor of ITS is the Vehicular Ad Hoc Network (VANET), which is an intern combination of sensor and wireless technology, short-range mobile communication which is also attributed as the level of the wireless local area networking (WLAN) and wireless personal area networking (WPAN) and data process management.

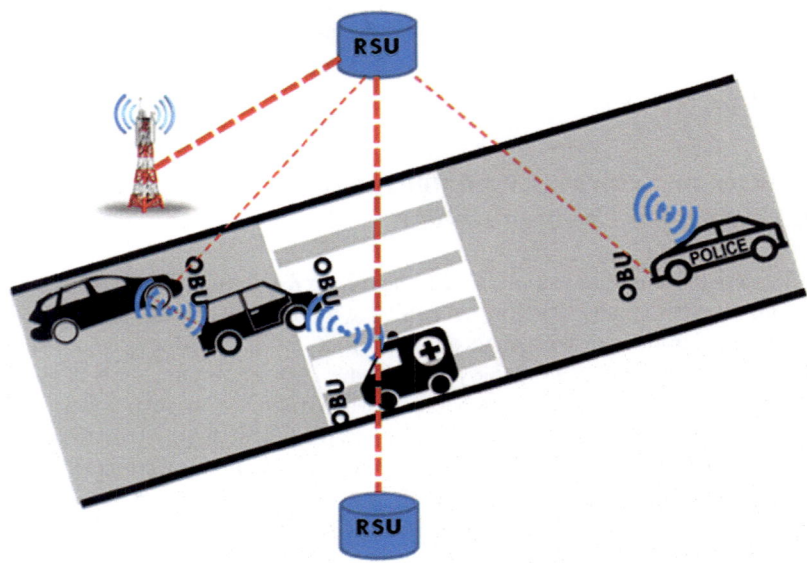

Fig. 3. Vehicle-to-Vehicle communication

VANET [11] is an alternative form of Mobile Ad-Hoc Network (MANET) and VANET has similar features and characteristics as MANET. VANET is the form of multi-hop networks which mainly provides the link of communication between vehicles and road-side base stations or Roadside Unit (RSU) which desire for safe navigation and improve efficiency. VANET direct to the new concept of system design and enhance the real-time capability of vehicular communication. Vehicles take advantage of varieties of wireless technologies to communicate with other vehicles. As a result of great mobility of vehicles which act as nodes, in which change in network topology occurs frequently. All nodes as in vehicles share the same channel approach which leads to congestion in heavy networks. Hence due to the decentralized nature of VANETs leads to the need for new system concepts. Figure 3 shows the vehicle-to-vehicle communication concept [12].

IOV technology is based on, Dedicated Short-Range Communication (DSRC) [17] which is a promising wireless standard that is used to connect Infrastructure-to-Vehicle (I2V) and Vehicle-to-Vehicle (V2V). The DSRC network is built mainly on two essential entities: Roadside Unit (RSU) and On-Board Unit (OBU) [17]. The OBU is mainly Global Positioning System (GPS) based tracking entity, this OBU entity will be fixed in the vehicles by the service provider, the OBU entity is useful for tracking the route of vehicles that are traveling or traveled. And also provide the service to the service provider. Hence the vehicles proceeding on road are equipped with OBU and the communication between two vehicles is through the short range of OBU via RSU. The connection between OBU of two vehicles is dependent on DSRC standards of the wireless network. The OBU is the device, fixed in the roaming vehicle. The OBU collects information like obstacles, near-by vehicles etc. through accurate identification and processing the information to the vehicles with the cover range via RSU. OBU abide by 4 important parts central control module, a wireless communication module, GPS module, and the human-machine interface module.

Some information based on services and applications which are purely based on inter-vehicular communication services like traffic navigation, traffic management [4], danger warning intersection collision warning, and the decentralized distribution of real-time traffic flow information.

2.2   Vehicle-to-Vehicle Infrastructure Communication

The Vehicle-to Infrastructure [13] connects the ITS with involves the exchange of information between the vehicles and the element of Transportation Infrastructure. This attempts to offer a wide range of safety, traffic management, driver alert etc. All types of smart vehicles are able to communicate with the traffic signal and to other features of the transportation system to improve the performance on road, avoid the collision and to improve efficiency. Even pedestrians and bicyclists carrying smart phones can engage in the V2I communication, which permits the traffic control devices to acknowledge to their behavior.

V2I plays a very important part in the collection of information related to traffic and road conditions, and also recommend and enforce certain positive behaviors on a group of vehicles which comes in contact with RSU. V2I also suggests the inter-vehicle distances, the velocity, and acceleration on the bases of traffic status, with the main intention of optimizing overall emissions, fuel consumption, and traffic velocities. Information to vehicles is sent to drivers via displays on road or precisely to vehicles through wireless connections.

3   Vehicles-to-Cloud and Vehicle-to-Everything

Vehicular using cloud is the approach in which smart vehicles search for available services, these vehicles make use of cloud services and resources rather using their own services. Due to the increase in vehicles day-by-day, it is becoming more difficult to gather information regarding mobile vehicles. In such cases, ITS is a solution of smart moving vehicles with rare elements of wireless sensor network (WSN) and IOT.

3.1   Vehicle-to-Cloud Communication

Vehicular Cloud Computing [6] uses both hardware and software to distribute the service over a network commonly internet. With the help of cloud computing, the users

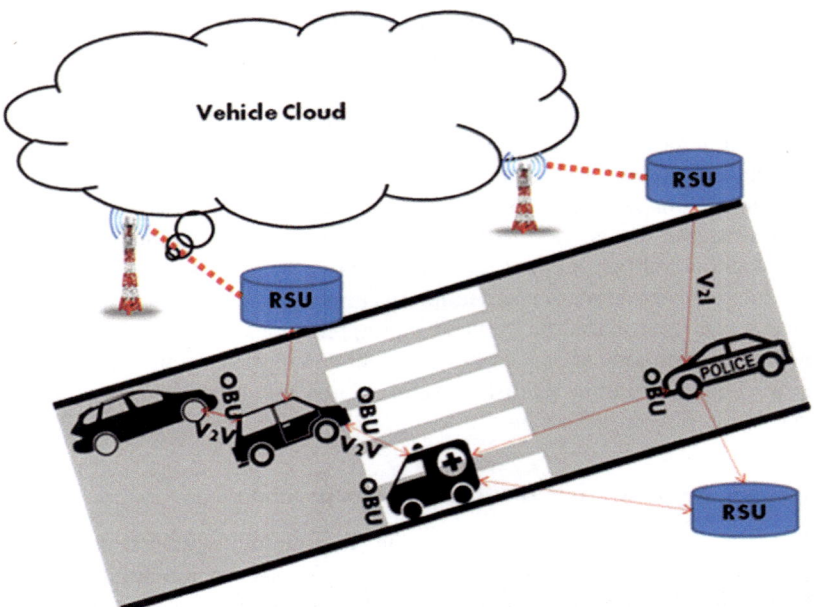

Fig. 4. Vehicular to cloud communication

having smart devices can access files, and the user can use applications from any smart device that has access to the Internet or that device that can access the Internet.

Gmail is an Illustration of Cloud Computing and the provider is Google's. Messages, data, and information are stored on physical or virtual servers, that is managed and controlled by a cloud computing provider. As a personal or business cloud computing user, you are able to access your gathered data or information on the 'cloud', via an Internet connection.

Smart Vehicles normally carry communication systems, storage facilities on board computing facilities and the sensing power. Therefore, several technologies have been set up to maintain and stimulate ITS. Cloud computing is mainly the approach of transferring the storage, processing data, computing new operation from desktop to handheld accessories, by Constructing the virtual functions in the cloud platform, these virtual functions present in the cloud can be used in the smart vehicular environment. The Vehicular Cloud is considered as the expansion to Mobile Cloud Computing, in which vehicles are capable of sharing and exchanging their resources in terms of information and data processing in order to bring serious advantages to drivers and other different users. Figure 4 show you the concept of Vehicular cloud Communication.

VCC is an alternative of Mobile Cloud Computing (MCC) formed by mobile vehicles and also called as mobile nodes. The cloud contributes in data storage also search for data in the neighborhood, data processing and consumed within their lifetime duration by neighboring smart vehicles, computing, sensing, and detailed information. The cloud is one of the very essential hubs in which all of this quickly changing, distinct message or information will pass through and also the information are also store. The cloud provides the podium for building sensible data. And the cloud is also the home for constructing and developing the apps and plans used by vehicles on the road. The smart Vehicles along with inbuilt sensors within a local area develops vehicle contents.

Vehicular Cloud Computing perhaps can be explained from different aspects, w.r.t application of cloud services, application of cloud and architecture of cloud. From previous studies, there are three main architectures of the vehicular cloud: 1. Cloud-orientation-architecture, 2. Vehicle-based architecture and 3. Hybrid architecture. The three differs by the architectures and also dwells on the action of the service provider and service consume.

1. **The Cloud-oriented architecture** consists of a group of services, these services interact with each other The interaction can involve data processing or service integration activities. The service providers belonging to a cloud, and it is connected to RSU. Thus, the link of connection between the vehicle and the cloud is established through RSU via Internet connectivity 3G, 4G, Wi-Fi, RSU or Satellite.
2. **The Vehicle-based architecture**, consists of cloud-based mobile vehicles, that is both the providers and the consumers are both mobile vehicles. These mobile vehicles, when they are available to provide services, behave mutually and provide services.
3. **The Hybrid architecture** is a group of vehicles will collaborate to grant service to the cloud server or even to a third-party client, via the Internet connection. In this particular case, the vehicle is only the service provider and not the consumer for service, and the service is only the set of data gathered from the vehicles.

3.2   Vehicle-to-Everything Communication

Vehicle-to-Everything (V2X) is a technology which grants mobile vehicles to communicate with the traffic system around them. V2I permits smart moving vehicles to establish the link of communication with systems such as streetlights, buildings, and even cyclists or pedestrians and traffic signals, etc. through RSU. ITSs can solve problems like informing the ambulance after the accident that takes place and records the location of the vehicle using GPS sensors. ITSs can also inform about the nearby maintenance and service station in case of vehicle failure, and many another such scenario. This is only possible when the vehicles establish the link to create a network called networking vehicles. As a result of advancement in cloud computing, the technology has added to up-coming bright opportunity in terms of the increasing transportation issues, like heavy traffic, congestion, and vehicle safety, vehicle navigation.

IoV is the integration of sensors and wireless communication technology which allows the vehicles to efficiently exchange the information. Thus improving road traffic conditions, traffic management, and travel efficiency by accomplishing the intelligent control among human, vehicles, roads, and transportation facilities. The essence of cloud computing consists of broad usage of internet access, on-demand service of self on road for mobility vehicles, information gathering, etc. On demand services means,

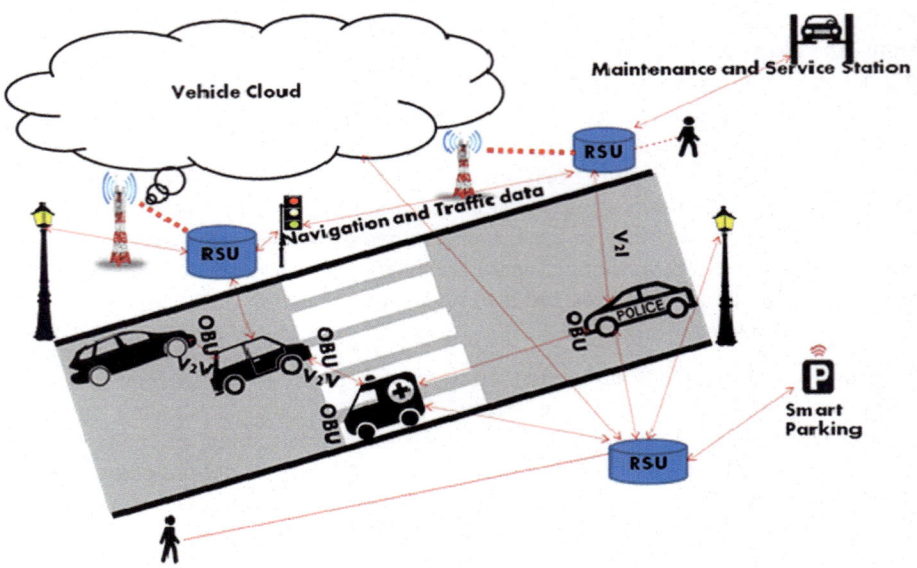

Fig. 5. Vehicle to everything (V2X)

the dealer (generally organizations) can appeal, supervise and serve their own resources. Internet access allows the service to be provided over the internet. Pooled resources are nothing but the services withdraw from a pool of computing resources. Services can be extended to smaller or larger, and the service is measured and customers are billed accordingly. Vehicular cloud takes the advantage cloud computing to provide service through VANETs. The main focus of IOV using vehicular cloud computing is to provide the on-demand solution for uncertain traffic and collection of data is an important condition in ITS, which will effectively serve online travel systems with the aid of Vehicular Cloud (VC). The Vehicular Cloud Computing (VCC) has the very enormous impact on the ITS by using the assets of smart vehicles such as GPS, storage, internet, and computing power for quick judgmental and sharing information on the cloud.

4   Application of Internet of Vehicles (IoV)

- The IoV is the combination of three principal components: inter-vehicular network [14], intra-vehicular network, and the vehicular mobile Internet. The advancement in recent technological of intelligent vehicles leads to handy accessible of mobile internet which has modified the vehicles into "new mobile device" Allowing user inside the vehicles to connect and establish the communication link.
- When vehicles and driver are connected, they share a lot of digital information, the driver in the vehicle can connect and access, consume, built, develop and share data and information between users, businesses, organizations, infrastructure and other vehicles leading to emerging concept of IoV.

- The outcome of unification of IoT and ITS leads in creating smart vehicles [12] on road. Thus IoV will build an integrated network for aiding various kind of functions like traffic management information services, control of intelligent vehicles etc. The technologies to build up vehicular and user safety and security are of huge concern, and one of the very important application is collision avoidance [15].
- New generation Intelligent vehicles must be launched to meet the requirement of improving road traffic, improving safety on the road, road congestion, and reducing vehicle emission.

5 Conclusion

IoV is one of the dynamic and fast emerging technology and an important part of the smart and intelligent cities which is being proposed and implemented globally. There are various benefits of IoV which includes traffic management, Safe Navigation, dynamic Information Service and increase productivity and reduce traffic congestion. The main focus of IoV is Vehicular computing and thus Mobile Internet, ITS, cloud computing, automotive electronics, intelligent geographic information system are integrated via advanced technology. Due to the rapid growth in internet technologies and its communication, the vehicles which are used in cities have strong and durable computing and communication capabilities. IoV is a huge complex integrated network model that interconnects people in and around vehicles, intelligent systems onboard vehicles, and various cyber-physical systems in urban environments. This paper provides you with understanding and concept of IOV and explains the state of art of the vehicular network. Finally, we also discuss on applications, benefits of IOV and concept of VANETs.

Acknowledgement. The authors would like to thank Dr P Shyamaraju, Chancellor, REVA University for all the facilities provided to carry out the research activities. Also, the authors would like to thank Dr Mohammed Riyaz from school of ECE, REVA University for their continuous support and encouragement.

References

1. Ahmed, E., Gharavi, H.: Cooperative vehicular networking: a survey. IEEE Trans. Intell. Transp. Syst. **19**(3), 996–1014 (2018)
2. Al-Fuqaha, A., Guizani, M., Mohammadi, M., Aledhari, M., Ayyash, M.: Internet of things: a survey on enabling technologies, protocols, and applications. IEEE Commun. Surv. Tutor. **17**(4), 2347–2376 (2015)
3. Simha, C.Y., Harshini, V.M., Raghuvamsi, L.V.S., Kounte, M.R.: Enabling technologies for internet of things & it's security issues. In: Second International Conference on Intelligent Computing and Control Systems, Madurai, India, 14–15 June 2018, pp. 1849–1852 (2018)
4. Dandala, T.T., Krishnamurthy, V., Alwan, R.: Internet of Vehicles (IoV) for traffic management. In: International Conference on 2017 Computer, Communication and Signal Processing, pp. 1–4. IEEE, 10 January 2017
5. Eichler, S., Schroth, C., Eberspächer, J.: Car-to-car communication, p. 6 (2006)

6. El Sibai, R., Atechian, T., Abdo, J.B., Demerjian, J., Tawil, R.: A new software-based service provision approach for vehicular cloud. In: Global Summit on Computer & Information Technology , pp. 1–6. IEEE, 11 June 2015

7. Fangchun, Y., Shangguang, W., Jinglin, L., Zhihan, L., Qibo, S.: An overview of internet of vehicles. China Commun. **11**(10), 1–5 (2014)

8. Gafencu, L., Scripcariu, L.: Vehicular cloud: overview and security issues. In: International Conference on Development and Application Systems (DAS), pp. 78–82. IEEE, 24 May 2018

9. Hemalatha, M.: Intelligent parking system using vehicular data cloud services. In: International Conference on Innovations in Information, Embedded and Communication Systems (ICIIECS), pp. 1–5. IEEE, 19 March 2015

10. Hong, J.: Cyber security issues in connected vehicle of intelligent transport system. Indian J. Sci. Technol. **9**(24) (2016)

11. Kafsi, M., Papadimitratos, P., Dousse, O., Alpcan, T., Hubaux, J.P.: VANET connectivity analysis. arXiv preprint arXiv:0912.5527, 30 December 2009

12. Lan, K.C., Huang, C.M., Tsai, C.Z.: On the locality of vehicle movement for vehicle-infrastructure communication. In: 8th International Conference on ITS Telecommunications (ITST), pp. 116–120. IEEE, 24 October 2008

13. Kombate, D.: The internet of vehicles based on 5G communications. In: IEEE International Conference on Internet of Things (iThings) and IEEE Green Computing and Communications (GreenCom) and IEEE Cyber, Physical and Social Computing (CPSCom) and IEEE Smart Data (SmartData), pp. 445–448. IEEE, 15 December 2016

14. Lu, N., Cheng, N., Zhang, N., Shen, X., Mark, J.W.: Connected vehicles: solutions and challenges. IEEE Internet Things J. **1**(4), 289–299 (2014)

15. Virat, M.S., Bindu, S.M., Aishwarya, B., Dhanush, B., Kounte, M.R.: Security and privacy challenges in internet of things. In: International Conference on Trends in Electronics and Informatics (ICOEI), Tirunelveli, Tamil Nadu, India, 11–12 May 2018, pp. 454–460 (2018)

16. Srinivasan, R., Sharmili, A., Saravanan, S., Jayaprakash, D.: Smart vehicles with everything. In: 2nd International Conference on Contemporary Computing and Informatics (IC3I), pp. 400–403. IEEE, 14 December 2016

17. Yang, Q., Wang, L., Xia, W., Wu, Y., Shen, L.: Development of on-board unit in vehicular ad-hoc network for highways. In: International Conference on Connected Vehicles and Expo (ICCVE), pp. 457–462. IEEE, 3 November 2014

18. Yang, F., Li, J., Lei, T., Wang, S.: Architecture and key technologies for Internet of Vehicles: a survey. J. Commun. Inf. Netw. **2**(2), 1–7 (2017)

Video Based Silent Speech Recognition

V. Sangeetha$^{(\boxtimes)}$, Judith Justin$^{(\boxtimes)}$, and A. Mahalakshmi$^{(\boxtimes)}$

Department of Biomedical Instrumentation Engineering,
Avinashilingam Institute for Home Science and Higher Education for Women,
Coimbatore, Tamil Nadu, India
sangeethaviswanathan.79@gmail.com,
hodbmieaul@gmail.com, magaa8ll@gmail.com

Abstract. The human ability to perform the lip reading process is has relation to as visual speech recognition (VSR). In this work, a voiceless use of words being seen way of doing is took up that puts to use forceful visual features to represent the face motion during phonation. Visual features are extracted from the mouth viewing part of a person talking quietly saying consonants using motion breaking down into parts and image time expert ways of art and so on. The proposed technique yields high recognition rate by applying the face and mouth detection processing using Viola-Jones algorithm. This algorithm is a spatial related super pixel algorithm that is used to identify the Region of Interest (ROI) and possible combination of objects present in the video frame. The extracted features are edge, shape, color moment, color auto-correlogram, color variance, color difference, color deviation, color combined and these features are used to train the SVM-QP classifier to test the word spoken by the human.

Keywords: Visual speech recognition (VSR) ·
Human computer interaction (HCI) · Viola Jones algorithm ·
Spatial super pixel algorithm · SVM-QP (Quadratic Programming) classifier

1 Introduction

Visual-to-Speech Conversion (VTSC) is a technique that converts unvoiced "lip movements to voice" utterances. Lip-reading can play an important part in helping to make less troubling the greater number or part of middle to serious feebleness gotten as voice sign put out. It's a good outcome in this part depends upon the power of perceivers to have a relation with what they see to what they hear. This is where the observations must start. When becoming conscious of make a public statement audio-visually observers select place news given only from act or power of seeing, and way and voicing news given only from hearing. Consonants and vowels are the smallest meaningful sounds in a language; when used in specific rule governed combinations, they constitute words and sentences that are used in daily conversation. However, consonants and vowels have different phonetic structures. Vowels are generally longer in duration and rather stationary compared with consonants. They are characterized by constant voicing, low-frequency components, and a lack of constriction, whereas consonants are characterized by a high-frequency structure and vocal-tract constriction.

© Springer Nature Switzerland AG 2020
A. P. Pandian et al. (Eds.): ICCBI 2018, LNDECT 31, pp. 269–277, 2020.
https://doi.org/10.1007/978-3-030-24643-3_32

Critical features for the identification of consonants are voicing presence or absence of vocal-fold vibration, manner in the configuration of articulators such as lips or tongue in producing a sound, and the place in the vocal tract where an obstruction occurs. Critical features for the identification of vowels are height representing the vertical position of the tongue relative to the roof of the mouth, rounding of the lips, and position of the tongue relative to the back of the mouth. VTSC is a difficult challenge because visual images contain less linguistic information than audio speech. Lip reading is a technique that recognizes text information from voiceless lip movements. Input lip movement is recognized using lip reading and the estimated text is synthesized to target voice utterances using TTS systems.

The other approach is a more direct one that does not recognize the text information of input lip movements. In the field of speech signal processing, there are techniques similar to VTSC. Voice Conversion (VC) converts paralinguistic information such as speaker identification while maintaining other information, such as linguistic information, in the speech utterance. The forceful features got from the mouth viewing part are used to put in order statements without using the with sound facts. The sound sign of consonants are more making without order than the letters a, e, i, o, u and the face motions had to do with in way of saying consonants are more see able. The audio signals of consonants are more confusing than vowels and the facial movements involved in pronunciation of consonants are more discernible. The demands for new HCI techniques that give greater value to the able to make ready adjustments and level of being ready for working for the users are increasing with strong effect. New methods of knowledge processing machine control chief place on different types of body purposes, uses such as use of words, feelings, bioelectrical operation, of the face expressions, and so on. Most of the of the face motions outcome from either use of words or the put on view of feelings; each of these have its own being complex. because of, in relation to the very thick news given that can be put into signs in use of words, use of words based HCI can make ready fullness like to human-to-human effect on one another. Such systems put to use a natural power of the to do with man user, and therefore have the possible unused quality for making knowledge processing machine control without work and natural.

2 Video Acquisition

A digital video is captured using DSLR camera and it is stored as MOV video file format with the resolution of 16 MP as shown in Fig. 1. All data is recorded with the same lighting conditions. In this study ten subjects were used. Five words that made more lip movements were identified and the recording was done. All the speakers were appraised on the procedures adopted for the recording. They are initially trained to repeat the five words two or three times to bring about natural lip movement.

The words considered for the study are Good Morning, Biomedical, Hospital, Orange and University. The camera is focused on the face area of the speaker and was held constant throughout the shot. Camera position, lighting and background were kept constant for each speaker. The input video is sampled at a rate of 29 frames per second. This mainly refers to initial processing of raw image frame. Digital image frames are

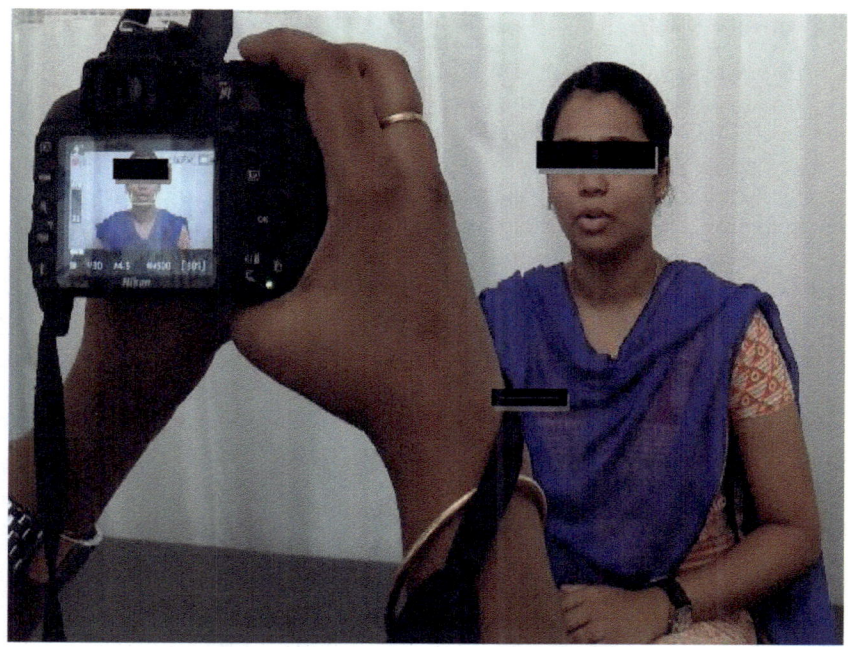

Fig. 1. Setup while video capturing

digits which are readable by computer and are converted to tiny dots or picture elements representing the real objects. In this study, image enhancement/filtering is done using Viola Jones algorithm.

3 Methodology

3.1 Proposed System Architecture

See Fig. 2.

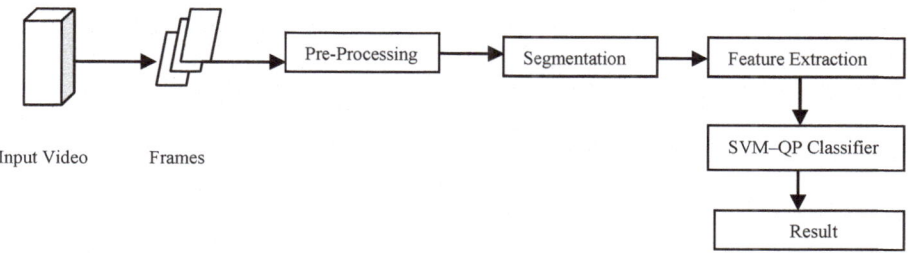

Fig. 2. Proposed model

3.2 Pre-processing

MATLAB plays a vital role in Image Processing in order to develop the quality of the image. MATLAB is a high-level language and interactive environment for computation of numbers, visualization, and programming. Data can be analyzed, algorithms can be developed and models and applications can be created using MATLAB. Hence pre-processing is essential to improve the quality by filters. Then the color images are being converted to gray scale image using the grayscale converter and then the preprocessing operation is executed. To achieve robust real-time face recognition, the "Viola-Jones" detector based on "AdaBoost," a binary classifier that uses cascades of weak classifiers to increase its performance is used. Viola Jones algorithm works with the combination of different algorithms.

3.3 Segmentation

Segmentation can locate the edges and boundaries more typically. The image and pixel with same label is grouped which may share some similar characteristics. The pixel data are extracted from the images and these pixel values are grouped into a set of clusters based on Superpixel algorithm. The segmented image is shown in Fig. 3.

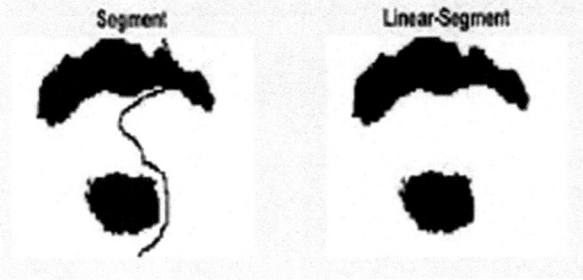

Fig. 3. Segmentation image using Super pixel Algorithm

3.4 Feature Extraction Techniques

Feature Extraction is used for extracting the important data from the Entire data. It is a type of multi-dimensional simplification that efficiently represents concerning components of an image. Feature extraction is very dissimilar from Feature selection, the feature extraction consists of translating an absolute data, such as images, into quantitative features. This process is a machine learning technique which is applied to derive the important components of an image.

3.4.1 Color Auto-Correlogram

Color auto-correlogram only evokes the entities of the extending oblique in the feature matrix to represent the vectored feature, that outcomes the linear connectivity relationship of similar colors. While color-correlogram conveys another pairs of colors in the spatial dispersion, and its utilization has been bounded by large feature vectors in covering with image equalizing. Simplified form of color auto-correlogram crisply deducts the computational complexity, and the recovery efficiency of the image with loaded colors or impressive variation in colors is not pretty good because only similar pairs of colors have been considered. Color-correlative information based on the color-correlogram feature matrix, which not only abbreviates the feature vector to one dimension with m elements, but also encloses spatial entropy of another matches of color. Feature extraction is shown in Fig. 4.

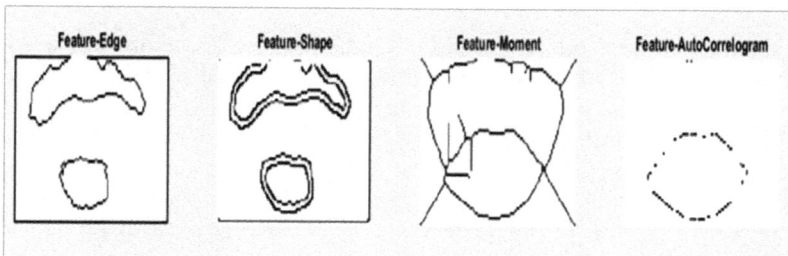

Fig. 4. Feature Extraction using Auto Correlogram

3.4.2   Color Moment
It represents the color distribution of the image in the discrete form in color vectors. Shifting of input data is accomplished by means of the ROI into the collection of necessary features for categorization. The features expressed and suggested must be carefully chosen, because it is awaited to execute the demanded assortment task by applying the concentrated internal representation alternatively to complete ROI. Functioning with sufficient quantity of entropy information commonly demanded a heavy amount of storage and computational power or an assortment algorithm which accommodate the input training data.

3.4.3   Edge and Shape Detection
Shape is a vital visualization and direct features for image cognitive description. Shape content description is hard to define because appraising the similarity between the shapes is challenging. Therefore, the process completed by executing the shape and edge descriptors. Shape descriptors can be partitioned into two main classes: region based and contour-based methods. Shape descriptive features are calculated from objects shapes which include the following parameters, circular shape changes, expressive ratio, discontinuity, slant irregularity, distance irregularity, complexness, powerful correct angleness, accurateness and directedness feature of the image pixel. The detection of edge and shape using Canny method is computed as given in Eq. (1),

$$[g,\ t] = \text{edge}(f, \text{"canny"}, T,\ \text{sigma}) \tag{1}$$

where T is a vector, T = [T1, T2], containing the two thresholds, sigma is the standard deviation of the Auto colorogram.

3.5 Classification

The normal formal form of SVM with optimization is modelled as Quadratic Programming (QP) SVM. QP-SVM is a kernel-based classifier that accumulates quadratic operations. The goal of the quadratic function is convex, which makes the problem easy to solve. The values generated by the quad function are either minimum or maximum, based on the inner values of the function.

4 Results and Performance Evaluation

In this simulation result, pre-processing first improves the image quality by removing unwanted distortions called noise and improving the details. In this study, image enhancement/filtering is performed using the Viola Jones algorithm. The Viola Jones Object Recognition framework that provides competing object recognition rates in real time. Second, the segmentation is done using the spatial superpixel algorithm. Selecting image pixels along the line of mouth near the axis of symmetry to form a horizontal sectional view of grayscale values and selecting a set of pixels and associated pixel values occurring at the peaks and valleys (maxima and minima) of the vertical and horizontal grayscale pixel value intersection views as a set of elements of a visual feature vector. Colorautocorrelogram is a method that processes the image and generates the output with m × m feature vectors. Color histogram and wavelet is the oldest method of displaying the feature, which differs from the rotation change, the distance, and the incomplete resolution of the target object. The validation of the improvement is subjective because the visual assessment is sufficient to evaluate the improved performance (Fig. 5).

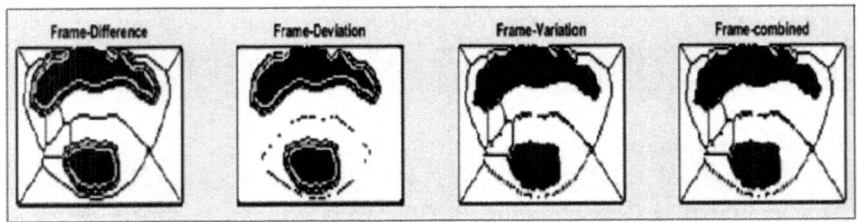

Fig. 5. Results of Performance Analysis

From the above results, it is obvious that SVM-QP classifier classifies the image frames efficiently to give the required output (Tables 1, 2, 3 and 4).

Table 1. Features for Good Morning using SVM Classification with QP

S. No	Mean	Difference	Variance	Standard deviation	Edge	Shape	Moment	Color corrlogram	Combined
Speaker: 1	0.9203	231.9165	0.0011596	0.034053	1345	1403	1521	1573	1119
Speaker: 2	0.98419	49.2094	0.00024605	0.015686	1403	1478	1569	1521	1257
Speaker: 3	0.3308	699.9798	0.0034999	0.05916	1375	1355	1315	1580	344
Speaker: 4	0.89753	290.8008	0.001454	0.038131	1271	1386	1491	1462	951
Speaker: 5	0.34282	712.382	0.0035619	0.059682	1467	1463	1525	1556	375
Speaker: 6	0.94371	167.9798	0.0008399	0.028981	1339	1433	1517	1512	1081
Speaker: 7	0.60089	758.3175	0.0037916	0.061576	1259	1249	1341	1484	641
Speaker: 8	0.37634	742.1505	0.0037108	0.060916	1443	1434	1524	1581	430
Speaker: 9	0.45731	784.7362	0.0039237	0.062639	1272	1286	1355	1556	490
Speaker: 10	0.85136	400.1392	0.0020007	0.044729	1235	1385	1445	1455	862

Table 2. Features for Biomedical using SVM Classification with QP

S. No	Mean	Difference	Variance	Standard deviation	Edge	Shape	Moment	Color corrlogram	Combined
Speaker: 1	0.86907	359.7951	0.001799	0.042414	1329	1389	1506	1550	1005
Speaker: 2	0.96964	93.0854	0.00046543	0.021574	1406	1465	1562	1541	1254
Speaker: 3	0.3365	705.9684	0.0035298	0.059412	1366	1348	1315	1577	368
Speaker: 4	0.90196	279.6078	0.001398	0.03739	1271	1397	1497	1474	973
Speaker: 5	0.51676	789.6116	0.0039481	0.062834	1461	1460	1576	1581	644
Speaker: 6	0.93991	178.5832	0.00089292	0.029882	1347	1400	1532	1559	1153
Speaker: 7	0.76471	568.9412	0.0028447	0.053336	1266	1337	1425	1560	893
Speaker: 8	0.24352	582.4921	0.0029125	0.053967	1493	1468	1541	1581	268
Speaker: 9	0.45035	782.7046	0.0039135	0.062558	1265	1278	1307	1547	493
Speaker: 10	0.98355	51.1448	0.00025572	0.015991	1363	1458	1556	1488	1488

Table 3. Features for Hospital using SVM Classification with QP

S. No	Mean	Difference	Variance	Standard deviation	Edge	Shape	Moment	Color corrlogram	Combined
Speaker: 1	0.92094	230.234	0.0011512	0.033929	1337	1412	1520	1533	1063
Speaker: 2	0.97533	76.0759	0.00038038	0.019503	1403	1474	1564	1530	1258
Speaker: 3	0.3327	701.9987	0.00351	0.059245	1367	1353	1303	1577	360
Speaker: 4	0.90829	263.4029	0.001317	0.036291	1275	1400	1496	1475	997
Speaker: 5	0.5117	790.067	0.0039503	0.062852	1465	1464	1577	1581	644
Speaker: 6	0.94434	166.2037	0.00083102	0.028827	1350	1426	1528	1568	1152
Speaker: 7	0.60721	754.1556	0.0037708	0.061407	1322	1284	1404	1539	752
Speaker: 8	0.33839	707.919	0.0035396	0.059494	1361	1392	1305	1570	389
Speaker: 9	0.39595	756.2682	0.0037813	0.061493	1323	1326	1330	1579	473
Speaker: 10	0.97533	76.0759	0.00038038	0.019503	1344	1430	1551	1467	1138

Table 4. Features for University using SVM Classification with QP

S. No	Mean	Difference	Variance	Standard deviation	Edge	Shape	Moment	Color corrlogram	Combined
Speaker: 1	0.87666	341.8975	0.0017095	0.041346	1336	1395	1508	1546	1006
Speaker: 2	0.97976	62.7046	0.00031352	0.017707	1410	1481	1568	1548	1275
Speaker: 3	0.3327	701.9987	0.00351	0.059245	1360	1358	1320	1581	346
Speaker: 4	0.90449	273.1562	0.0013658	0.036956	1277	1399	1494	1486	1007
Speaker: 5	0.51739	789.5433	0.0039477	0.062831	1459	1457	1575	1581	640
Speaker: 6	0.94434	166.2037	0.00083102	0.028827	1343	1418	1517	1581	1148
Speaker: 7	0.56926	775.3321	0.0038767	0.062263	1317	1302	1380	1529	695
Speaker: 8	0.19545	497.2144	0.0024861	0.049861	1476	1467	1514	1581	198
Speaker: 9	0.42948	774.7729	0.0038739	0.06224	1279	1285	1326	1564	516
Speaker: 10	0.98735	39.494	0.00019747	0.014052	1374	1457	1562	1475	1186

5   Conclusion

The purpose of this study is to identify an visual speech recognition using the lip movement which include face and lip detection, region of interest (ROI), visual features extraction, visual speech recognition and integration of visual without audio. For implementation we use MATLAB programing language. For face and mouth detection we use Viola-Jones object recognizer called mouth detection respectively, after the mouth detection ROI extracted. Extracted ROI used as an input for visual feature extraction. In future Hidden Markov Models with Artificial neural network can be used to reduce detection delay for the word classification process. And also the method can be designed for various languages apart from English with the natural auditory speech and synthetic visual speech. The method can be enhanced by applying the kernel clustering algorithm to improve the ROI identification with the linear segmentation process which may provide the nonlinear objects present in the video frames. Despite linear amplification, auditory identification of consonants and vowels is cognitively demanding using the maximum likelihood estimation.

Acknowledgement. For conducting the video recording, Ms. V. Sangeetha main author of this proposed idea whole heartedly supported by being a subject to perform the experiment.

Consent. All the participants are hereby expressing their approval for both the publication of the paper & usage of photographs therein and we are not under any influence. We take the responsibility involved in the publication of our paper.

References

1. Potamianos, G., Neti, C., Gravier, G., Garg, A., Senior, A.W.: Recent advances in the automatic recognition of audio-visual speech. In: IEEE, vol. 12 (2013)
2. Lewis, T.W., Powers, D.M.W.: Audio-visual speech recognition using red exclusion and neural networks. J. Res. Pract. Inf. Technol. 35(1), 41–64 (2003)
3. Dalka, P., Czyzewski, A.: Human–computer interface based on visual lip movement and gesture recognition. Int. J. Comput. Sci. Appl. 7(3), 124–139 (2010)

4. Navarathna, R., Lucey, P., Dean, D., Fookes, C., Sridharan, S.: Lip detection for audio-visual speech recognition in-car environment. In: International Conference on Information Science, Signal Processing and Their Applications (2010)
5. Viola, P., Jones, M.: Robust real-time object detection. IEEE Int. J. Comput. Vis. **57**(2), 137–154 (2004)
6. Viola, P., Jones, M.: Rapid object detection using a boosted cascade of simple features. Conf. Comput. Vis. Pattern Recognit. **1**, 511–518 (2001)
7. Chin, S.W., Ang, L.-M., Seng, K.P.: Lips detection for audio-visual speech recognition system. In: International Symposium on Intelligent Signal Processing and Communication Systems (2008)
8. Kim, Y.-K., Lim, J.G., Kim, M.-H.: Comparison of lip image feature extraction methods for improvement of isolated word recognition rate. Adv. Sci. Technol. Lett. **107**, 57–61 (2015)
9. Seman, N., Bakar, Z.A., Bakar, N.A.: An evaluation of endpoint detection measures for Malay speech recognition of isolated words. In: Information Technology (ITSim), 2010 International Symposium, vol. 3, pp. 1628–1635 (2010)
10. Yee, C.S., Ahmad, A.M.: Malay language text-independent speaker verification using NN-MLP classifier with MFCC. In: International Conference, pp. 1–5 (2008)
11. Matthews, I., Bangham, J.A., Cox, S.: Audiovisual speech recognition using multi-scale nonlinear image decomposition. In: Proceedings of Fourth International Conference on Spoken Language, ICSLP 96, vol. 1, pp. 38–41 (1996)
12. Wee-Chung, A., Wang, S: Visual speech recognition: lip segmentation and mapping (2008)

In Search of the Future Technologies: Fusion of Machine Learning, Fog and Edge Computing in the Internet of Things

Soumyalatha Naveen[1]([⊠]) and Manjunath R. Kounte[2]([⊠])

[1] School of Computing and IT, REVA University, Bengaluru, India
soumyanaveen.u@gmail.com
[2] School of Electronics and Communication, REVA University,
Bengaluru, India
manjunath.kounte@gmail.com

Abstract. With the rapid growth of smart devices connected to the internet, the paradigm Internet of Things (IoT) is used in plenty of applications. This led to the emergence of different computing paradigm such as Cloud computing, Fog computing, Edge Computing using artificial intelligence model such as machine learning to analyze and to get some valuable information and to predict the future from the raw data. This paper provides details about computing paradigm, applications, and its advantages and limitations. This paper also brings different kinds of important neural networks and its usability for distinct inputs such as images, audio, video etc. We envision this paper helps for student researcher for understanding the IoT, computing paradigms and artificial neural networks, to work further in these emerging domains.

Keywords: Cloud computing · Convolution neural network · Deep learning · Edge computing · Fog computing · IoT · Long Short-Term Memory · Machine learning · Recurrent neural networks

1 Introduction

Over recent years IoT applications or products are escalated in the plethora of domains to capture the data, depending on the context. With the collaboration of varieties of neural networks, sensors, and artificial intelligence technologies, these data can be analyzed.

The neural network is a computing system, inspired by the brain, which imitates the functionalities that can be performed by the brain. This is a framework for machine learning, where the machine learns from training examples.

1.1 Internet of Things (IoT)

With the dramatic evolution in the field of wireless communication, in the era of the Internet of Things, objects are made smart by making use of recent advanced communication technologies and improved embedded hardware systems with expeditious data analytics tools.

© Springer Nature Switzerland AG 2020
A. P. Pandian et al. (Eds.): ICCBI 2018, LNDECT 31, pp. 278–285, 2020.
https://doi.org/10.1007/978-3-030-24643-3_33

The IoT based applications are used in diversified domains such as smart home, smart city, smart transportation, smart agriculture, smart Industry, smart vehicles etc. These tremendous growths of IoT applications generate a huge amount of data. These applications will be useful if it generates a valuable insight from the raw data, generated by the sensor. IoT data has characteristics such as a huge volume of streams of continuous data, heterogeneous data like image, audio, video etc., time-stamp, location details along with some noisy data.

Robust artificial intelligence learning mechanisms such as machine learning or deep learning approaches are required to process the raw data and to draw the conclusion or to obtain the hidden knowledge or to predict the future as shown in Fig. 1.

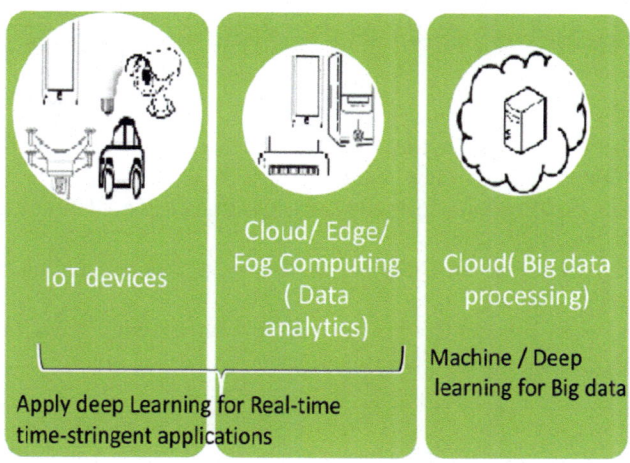

Fig. 1. A general picture representing IoT-devices, computing paradigm and data analytics.

2   Artificial Intelligence (AI)

Artificial intelligence is a technique which employs mathematical models to imitate the behavior of the human brain. Artificial intelligence processes a large amount of data to perform a specific task to identify patterns in that. Widely machine learning and deep learning are used.

2.1   Machine Learning

Machine learning is one of the techniques present in the artificial intelligence; where machine have the ability to learn from experience, without programming. Supervised learning, unsupervised learning for labelled and unlabelled training data respectively and Reinforcement learning are commonly used Machine learning algorithms. Machine learning has gained more attention in the past few years.

2.2 Deep Learning

Currently, deep learning has become a popular emerging model to extract the accurate information from a large amount of raw data, can be collected by the IoT devices. Deep learning model consists of multiple layers, and in each layer, data size is scaled down to extract the accurate feature. Hence deep learning is used in various domains [1, 2] such as image processing (human face), voice recognition, wearable IoT devices, audio recognition, food recognition, sound, pattern recognition, and computer vision and for real-time applications such as a self-driving car, fire-prediction etc. wherein fast processing and prediction is required.

Amazon Alexa capable of voice interactions and can control many smart devices as part of home automation. 'Windows IoT Facial Recognition Door' application uses deep learning.

Some of the frameworks developed for deep learning are H2O, TensorFlow, Torch, Theano, Caffe, Neon.

Advantages:

- Better performance compared to machine learning for a huge set of data.
- Capable of identifying the new features.
- Processing of multimedia information and time taken for inference is less
- compared to machine learning.
- Due to the self-learning capability, it provides more accurate and fast processing for a large amount of data
- Reduces the need for handcrafted [3] and few engineering features might not be extracted by humans, can be easily extracted by the DL model.

Limitations:

- IoT devices have limited energy and computational power capability [4, 5].
- Difficulty in circuit designs [6] of emerging memory devices for some application, security circuits, and computing-in-memory
- The requirement of deep learning is high speed, high-performance computing and hence consumes more power [7].
- Many existing deep learning libraries use different third-party libraries; hence can be difficult to migrate to IoT devices.

3 Artificial Neural Networks for Deep Learning

There are numerous neural networks such as convolution neural networks (CNN), recurrent neural networks (RNN), Long Short-Term Memory (LSTM), Auto-encoders, Variational Auto-encoders, Generative adversarial networks (GAN), Deep belief networks (DBN) are present for deep learning. Few important neural networks [8] are discussed below.

3.1   Convolutional Neural Networks

CNN is the artificial neural networks, inspired by the brain. They are made up of neurons, which receives input and have weights and biases. CNN [8] as shown in Fig. 2; receives images as "Input", make use of "Kernel" to detect the features present in the input image. To detect the features, it uses convolution operations. If the feature is present it produces "Output" as some high value, otherwise low value.

Fig. 2.   Structure of CNN

CNN are neural networks which adopt Convolutions that is mathematical operations such as some kind of linear operations, specially designed for processing data or images, which can be represented as a grid of pixels or values. CNN is best suitable for recognizing the patterns or objects in an image. CNN stands out as a major model for influencing deep learning.

3.2   Recurrent Neural Networks

RNN is the specialized neural network for processing sequence of data or values by using internal memory. It remembers its input through internal memory, through which it precisely predicts the data. Widely used for handwriting recognition machine translations, question answering, and speech recognition.

As shown in Fig. 3 RNN receives input, process it and predicts the output by considering the present input and previously learned input (recurrent connections in the hidden layer).

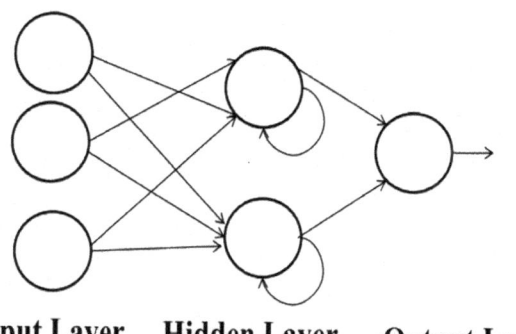

Input Layer   Hidden Layer   Output Layer

Fig. 3.   Structure of RNN

3.3   Long Short-Term Memory

LSTM [8] is a unit of RNN, for additional capability of learning long-term dependencies.

LSTM (Fig. 4) mainly consists of four parts such as cell, input gate, forget gate and output gate. With this, LSTM can remember the needed data for computation for long time and other data can be forgotten. LSTM are used is many applications such as handwriting recognition, machine translation, parsing, image translation, image captioning etc.

Recent applications such as Siri, Alexa, Cortona, Googles AI uses neural networks.

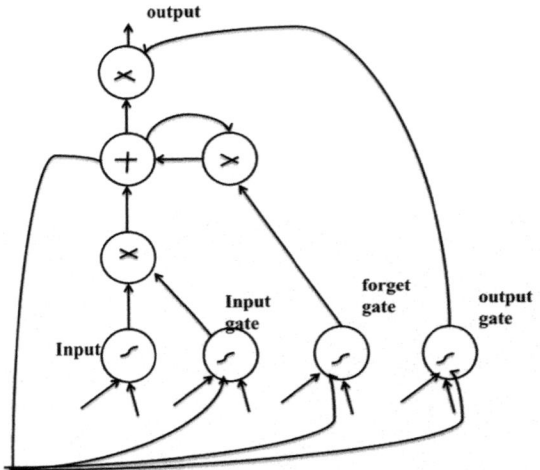

Fig. 4. Structure of LSTM

4   Computing Paradigms

4.1   Cloud Computing

Cloud computing is a service provided by the service provider to store and process the huge amount of data in the centralized data centre called cloud, inefficient way in offline with high performance and high computational power with reasonable price.

Considering the real-time applications, the limitations of cloud computing [9, 10] are

- One of the requirements of cloud computing is Quality of service [11], which refers to the high amount of bandwidth and a very good-quality network performance. This is available in selected few locations and expensive.
- Data will be provided to the service provider to store, analyze and to classify or to predict the unknown pattern. Hence privacy is a big challenge.

4.2 Fog Computing

Fog computing is a computing platform for analyzing the data at the edge devices of the network [12, 13]. This edge device will be very near to the data generation or creation point of a network. The term fog computing is invented by Cisco. Real-time, time-sensitive data will be analyzed at the edge and only selected data will be sent to the cloud for analysis and for long-time storage.

Advantages:

- Requires less bandwidth and hence saves network resources compared to the cloud which uses a huge amount of scarce bandwidth.
- Minimal latency so supports delay-sensitive real-time applications by enabling data analysis process at the network edge.
- Provides new opportunities for the 5G communication system and embedded artificial intelligence.
- Idle resources can be shared among along the cloud-to-thing continuum, to increase the efficiency.
- Applications are closer to the users and their cognition high.
- Gathered data is secured and protected.

4.3 Edge Computing

Edge computing enables computing capability to move from cloud to edge node near to the node where data is generated. Fog computing is also called edge computing and they are interchangeable [14–16], however, fog computing defines some standard in working, computing and in storage. Multilayer processing of data enables deep learning to be a promising model to incorporate for deep learning. This approach is suitable for various real-world applications such as Smart City, Smart Home etc., in which data may, also contains noisy data from the very complex environment and hence even in the future the deep learning plays a very important role in IoT-Edge Computing.

Advantages:

- Edge nodes pre-process large amount of raw data before transmission [17].
- Cloud resources are optimized.
- The privacy of the intermediate [3] data transmitted also will be preserved and also reduces the network traffic compared to cloud computing.
- Lower response time and energy consumption [18, 19].

Edge computing is an efficient method when the pre-processed intermediate data size is smaller than the input data size.

Limitations:

- In the evaluation of deep neural networks, optimization [3] at algorithmic and hardware level are required.
- Energy efficient hardware and software processing techniques are required.

- Many current embedded devices lack the capabilities to enable deep inferences for real-life applications.
- Mobile IoT devices have low power, limited computing power and small memory size [7].

5 Applications

The fusion of the technologies and computing paradigms are used in different kinds of neural networks. Already many papers are published in various domains. Some of the important applications using these technologies are Image Recognition, Speech/Voice Recognition, Indoor Localization, Physiological and Psychological State detection, Security and Privacy, Smart homes, Smart City, Intelligent transportation system, Healthcare and wellbeing, Agriculture, Education, Industry, Government, Sport and Entertainment, Retail etc.

6 Conclusion

Dramatic evolution in the artificial intelligence techniques such as machine learning or deep learning models along with varieties of computing paradigms must gain popularity in recent years for discovering/extracting/classifying from the raw data generated either by different categories of IoT applications. This paper presents an extensive introduction of IoT, neural networks and computing paradigms.

References

1. Mohammadi, M., Al-Fuqaha, A., Sorour, S., Guizani, M.: Deep learning for IoT big data and streaming analytics: a survey. In: IEEE Communications Surveys and Tutorials (2018)
2. Osia, S.A., Shamasabadi, A.S., Taheri, A., Rabiee, H.R., Haddadi, H.: Private and scalable personal data analytics using hybrid edge-to-cloud deep learning. In: IEEE Computer Society (2018)
3. Verhelst, M., Moons, B.: Embedded deep neural network processing. In: IEEE Solid-State Circuits Magazine (2017)
4. Simha, C.Y., Harshini, V.M., Raghuvamsi, L.V.S., Kounte, M.R.: Enabling technologies for internet of things & it's security issues. In: Second International Conference on Intelligent Computing and Control Systems (ICICCS 2018), Madurai, India, 14–15 June 2018, pp. 1849–1852 (2018)
5. Virat, M.S., Bindu S.M., Aishwarya, B., Dhanush, B., Kounte, M.R.: Security and privacy challenges in Internet of Things. In: International Conference on Trends in Electronics and Informatics (ICOEI 2018), Tirunelveli, Tamil Nadu, India, 11–12, May 2018, pp. 454–460 (2018)
6. Dou, C, Chen, W.-H., Chen, Y.-J., Lin, H.-T., Lin, W.-Y., Ho, M.-S., Chang, M.-F.: Challenges of emerging memory and memristor based circuits: nonvolatile logics, IoT security, deep learning and neuromorphic computing. In: IEEE (2017)

7. Tang, J., Sun, D., Liu, S., Gaudiot, J.-L.: Enabling deep learning on IoT devices. In: IEEE Computer Society (2017)
8. Goodfellow, I., Bengio, Y., Courville, A.: Deep Learning. MIT Press, Cambridge (2016)
9. Elgamal, T., Nahrstedt, K., Sandur, A., Agha, G.: Distributed placement of machine learning operators for IoT applications spanning edge and cloud resources. In: SysML Conference, USA, February 2018, pp. 1–3 (2018)
10. Yousefpour, A., Fung, C., Nguyen, T., Kadiyala, K.: All one needs to know about fog computing and related edge computing paradigms: a complete survey. J. Syst. Archit. (2018). ScienceDirect
11. Soumyalatha, N., Ambhati, R.K., Kounte, M.R.: IPv6-based network performance metrics using active measurements. In: Chakravarthi, V., Shirur, Y., Prasad, R. (eds.) Proceedings of International Conference on VLSI, Communication, Advanced Devices, Signals & Systems and Networking (VCASAN-2013). Lecture Notes in Electrical Engineering, vol. 258. Springer, India (2013)
12. Abeshu Diro, A., Chilamkurti, N.: Deep learning: the frontier for distributed attack detection in fog-to-things computing. In: IEEE Communications Magazine, February, 2018
13. Naveen, S., Kounte, M.R.: Machine learning based fog computing as an enabler of IoT. In: International Conference on New Trends in Engineering and Technology (ICNTET), Tiruvallur, Tamil Nadu, India, 7–8 September (2018)
14. Fog Computing and the Internet of Things: Extend the Cloud to Where the Things Are, White paper, Cisco (2015)
15. https://www.cisco.com/c/en/us/solutions/enterprise-networks/edge-computing.html
16. Chiang, M., Zhang, T.: Fog and IoT: an overview of research opportunities. IEEE Internet of Things J. (2016)
17. Ali, M., Anjum, A., Yaseen, M.U., Zamani, A.R., Balouek-Thomert, D., Ranna, O., Parashar, M.: Edge enhanced deep learning system for large scale video stream analytics. In: IEEE (2018)
18. Portelli, K., Anagnostopoulos, C.: Leveraging edge computing through collaborative machine learning. In: 5th International Conference on Future internet of Things and Cloud Workshops (2017)
19. Li, H., Ota, K, Dong, M.: Learning IoT in edge: deep learning for the internet of things with edge computing. IEEE Netw. (2018)

Recognition of Characters in a Securely Transmitted Medical Image

S. Manisha$^{1(\boxtimes)}$ and T. Sree Sharmila^2

^{1} Department of Computer Science and Engineering, SSNCE, Chennai, India
manishas@ssn.edu.in
^{2} Department of IT, SSNCE, Chennai, India
sreesharmilat@ssn.edu.in

Abstract. Advances in the field of medicine has led to increase in Health Information Exchange (HIE). The main idea of these HIEs is to help the patients' get multiple opinions from different physicians around the world. Medical records carry sensitive information such as patient's diagnosis reports where storage and transmission of data cannot be compromised. Therefore, the medical reports generated can be encrypted in a medical image and can be transmitted securely using the Adaptive LSB with randomized encoding technique, such that the size of the image is unaltered thereby minimizes the chance of an intruder to detect an embedded medical report. On decryption, the medical report that contains a printed text can be recognized using an OCR. Also the text regions embedded in these medical images can contain parameters that are sensitive for diagnosis. These text regions are extracted and are recognized to further help in analysis. This paper combines the techniques of encryption, image steganography and document understanding to securely transmitted the data and recognize the characters to understand the medical report.

Keywords: Medical image · Printed text · Image steganography ·
Adaptive LSB · OCR · Randomized encoding

1 Introduction

Modernization of healthcare domains has led to increase in need to convert the clinical case records into digital media. Apart from increasing the quality of health care, it is important to handle these medical reports safely. Medical information such as medical images that includes X-rays, MRIs, CT scans and personal information such as the diagnosis reports of these images and their medical history must be protected during storage. Also there might be a need to transmit these data through regular transmission channels for getting opinions from different physicians around the world [1]. Gathering multiple opinions on the medical condition can be one of the most emotionally fraught decisions that a patient has to make. These HIEs also help the doctors to share their experiences which may help in improving the diagnosis of the patient's medical condition. Since medical data is very sensitive transmission of these data should be protected from any attacks. Alteration in any of these reports might lead to misdiagnosis of the patient's condition that may lead to severe consequences (Fig. 1).

© Springer Nature Switzerland AG 2020
A. P. Pandian et al. (Eds.): ICCBI 2018, LNDECT 31, pp. 286–292, 2020.
https://doi.org/10.1007/978-3-030-24643-3_34

Therefore, it is important to use an effective and efficient method to handle and securely transmit these records [1].

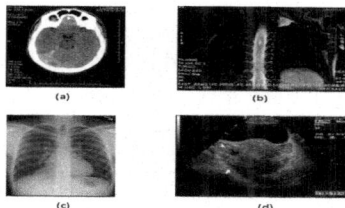

Fig. 1. Sample Medical Images

Further these transmitted images contain the medical records of the patients which is mostly in the form of text. It is important to extract these text regions and recognize them which would help in understand the medical document. Retrieval of medical images for the purpose of document analysis can be done by analysing the content, text or a combination of both. Various experimental results suggests the hybrid way of combining the content and the text has better results in understanding the medical document. The text available in the medical images contain important information such as the parameters that is important for the purpose of disease diagnosis. To the best of our knowledge, most of the existing methods uses content of the image rather than text to analyze the medical report (Fig. 2).

Fig. 2. Sample Medical Report with printed Text

Text detection in natural images or videos is an active research challenge in image processing. The main objective is to detect and localize the text regions in an image or video. Since the video may or may not contain text in all the frames, using an automated text detection algorithm can aims in detecting text regions in all frames. When a frame that does not have text information enters these detection algorithm, there is high chance for detecting false positives. Therefore, it is important to detect the presence of text in each frame and classify the text frames before further processing. The medical images produced by different machines such as CT scans, MRIs, Ultrasounds, X-rays etc., will be different from each other. Text embedded in these images will have different properties based on the modality of the image. This further increases the challenge of designing an algorithm to detect and recognize text. However, the text embedded in these medical images have characteristics such as high intensity, regular shape and large edges density. Therefore, once the text regions are extracted, the edge detection algorithm can be used to extract the text boundaries which might be a good solution to recognize the characters [5].

In this paper, we propose a new method that combines the concepts of data hiding and text recognition. The method uses a two step process. In the first step Adaptive LSB with randomized encoding method is used for secure transmission of printed medical reports embedded inside the medical image. In the second step the concept of text extraction is used to recognize the printed text characters in the medical record transmitted and also to detect and recognize the text characters present in the medical image [2]. The paper is organized as follows. In Sect. 2, the existing methodologies, related to secure transmission of data and recognition of characters in printed text and medical image is discussed. In Sect. 3, the proposed two step novel method to recognize text characters in the medical image transmitted is explained. In Sect. 4, the results and discussions of the proposed method and comparison with the existing state of the art method is discussed.

2   Related Work

In this section, various existing research methodologies used for secure data encryption of data and recognition of the characters are discussed. The Least Significant Bit Algorithm (LSB) is a state-of the art method used for Steganography [9]. The strategy of the method is to embed the data into a cover image such that a casual observer cannot detect it. The idea is to replace the existing data in the cover image with the information to be hidden. The data to be hidden can be replaced in any of the bit planes, but LSB uses the least significant bit of the pixel. The main advantage of this to reduce the variation in the color value. One bit LSB method reduces this color value by 2, and two bit LSB method reduces the color value by 4 thus further reduces the visibility of change in data. The four bit LSB method was used when huge amount of information needs to be hidden. Similarly, the pixel swap algorithm method uses pseudo-random sequence to decide on the two pixel bits to hid the secret image. However, the major disadvantage of the LSB methods is the size of the cover image changes after hiding the secret image, thus increasing the suspicion of the intruder for any possible secret message.

Similarly on the decryption side, once the medical report is extracted, various existing text extraction methods can be used to recognize the text characters. Most of the existing text extraction algorithms are used on natural images. There is relatively very less literature on text extraction in medical images. The existing methods used by Medical Image Resource Center (MIRC) is based on text based image retrieval technology. It uses various image data sets that contain clinical records, technical documents and electronic images to train and learn the text characters. It combines the text and content based features to learn the characters. Various histogram features such as gradient vector directions and level of luminance of a grayscale image are used to identify the features of the text [7]. The two different types of text that can be present in the medical reports can be the handwritten characters and the printed characters. Classification of these can be done using different classiffiers such as neural networks that uses horizontal, vertical, and slope-based features for classification of text [4]. The projection profile approach can be used tp understand the statistical features. Similarly the hidden markov model can be used to segment text at word level and character level [3].

3 Proposed Method

The proposed method is a two step process. The first step uses the Adaptive LSB with randomized encoding method to encode the printed medical report inside a medical image for secure transmission [1, 6]. In the second step, the decrypted medical report is taken and various image processing techniques are used to extract the text regions in the medical report that contains the printed text and are recognized [2]. The overall architecture of the proposed method is given in Fig. 3.

The steps in the proposed method is as follows:

1. The medical report to be embedded is converted into an image. This image is henceforth called the secret image.
2. The bits of pixels in the secret image is split into a combination of four pairs of bits and are encoded into the four quadrants of the cover image by clearing the last two bits of each byte in the image. This technique is called as adaptive LSB with randomized encoding. This encoded image is henceforth called the stego image [8].
3. The size of the stego image is unaltered from the cover image which reduces the suspicion of the intruder. This stego image is now transmitted through a regular transmission medium.
4. In the decoding side, the transmitted stego image is decrypted and the cover image and the secret image are separated.
5. The secret image is the medical record encrypted. This is further processed using various image processing techniques.
6. The secret image is given to the median filter to remove the noise and the image is binarized using Otsu's method of thresholding that separates the text from its background.
7. The text regions have distinct edges. Therefore, an edge detection technique is to detect and segment the text regions [4, 5].
8. The geometric and curve based features of the segmented text characters are given to the classifier to classify each text character.
9. The classified characters are given to Optical Character Recognizer (OCR) to recognize the characters.

The novelty of the proposed work is the usage of Adaptive LSB with randomized encoding method that the encoded image has the same size of the original cover image thus cannot be identified by any casual observer [8]. Also the secret image decoded maintains the original quality of the image before encoding. This is important as the secret image after decryption leads to further processing for character recognition, thus the quality is unaltered.

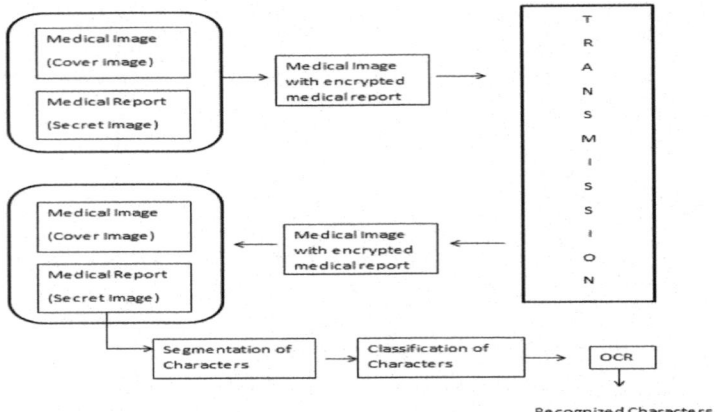

Fig. 3. Architecture of the proposed system

4   Experimental Results and Discussions

At the encryption side, the medical report that is converted into a secret image is embedded into any image by resizing such that it covers the entire image. The secret bits are split into four pairs of bits and are embedded into four different quadrants of the cover image. At the decoder side, the stego key is applied and the cover image is decoded into four pairs of bits. These bits are assembled and are converted into original secret image (medical report with printed text). The image with the printed text is preprocessed using median filter to remove the noise and are binarized using Otsu's thresholding method to separate the text regions from its background. These text regions are then extracted and are classified using features. The classified text characters are recognized using an OCR. The recognized characters contain the sensitive information on the patients' diagnosis which can be used by the doctors to analyze on the patient for further medication (Figs. 4 and 5).

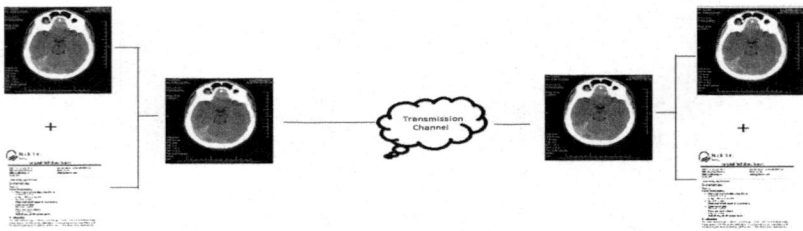

Fig. 4. Transmission Result

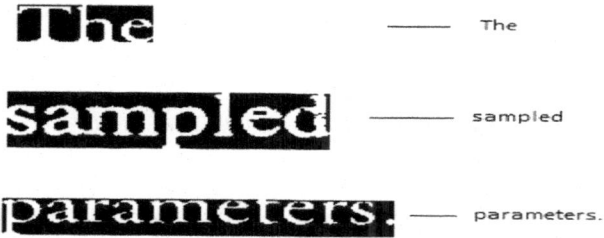

Fig. 5. Recognition of Text Result

5   Conclusion

The proposed method is a two step process that combines the concepts of secure transmission and character recognition. The first step is to embed the patients' medical record into any image or video using a randomized encoding method applied over Adaptive LSB technique. This stego image or video is transmitted over a channel. On the decoder side, the secret image containing the printed text is extracted using a stego key, and are further processed for recognition of characters using image processing techniques. These recognized medical reports of the patient is used for further analyzes of the patients' diagnosis information by a doctor to prescribe medication. This work can be extended both in the steganography step and in the character recognition step. The encryption algorithm can be checked for various attacks. Similarly, deep learning methods can also be used on the classification of the characters which will further enhance the accuracy of character recognition.

Acknowledgements. The medical images are taken from the Kaggle dataset.

References

1. Prashant, V., Nitish, M., Jagadish, N.: A new encryption technique for the secured transmission and storage of text information with medical images. Eng. Rev. **32**(1), 57–63 (2012)
2. Manisha, S., Sree Sharmila, T.: Effective Printed Tamil Text Segmentation and Recognition Using Bayesian Classifier LNCS, pp. 729–738. Springer, Singapore (2017)
3. Telung, P., Yu-Fang, S., Li-Chun, L.: A Text visualization process for the recognition of writing style. In: International Conference on Circuits and Systems, Atlantis Press, pp. 271–275 (2015)
4. Manisha, S., Sree Sharmila, T.: Text frame classification and recognition using segmentation technique. In: 2nd IEEE International Conference on Applied and Theoretical Computing and Communication Technology, pp. 662–666. IEEE Press, Bangalore (2016)
5. Zhu, Y., Singh, P. D., Siddiqui, K., Gillam, M.: An automatic system to detect and extract text in medical images for de-identification. In: SPIE (2010)
6. Lunge, P.S.: An approach for secured medical image data and information transmission. Int. J. Emerg. Technol. Eng. Res. **5**(9), 98–107 (2017)

7. Ma, Y., Wang, Y.: Text detection in medical images using local feature extraction and supervised learning. In: 12th IEEE International Conference on Fuzzy Systems and Knowledge Discovery, pp. 953–958. IEEE Press, (2015)

8. Manisha, S., Sree Sharmila, T.: A two-level secure data hiding algorithm for video steganography. Multidim. Syst. Sign. Process. https://doi.org/10.1007/s11045-018-0568-2 (2018)

9. Srinivasan, Y., Nutter, B., Mitra, S., Phillips, B., Ferris, D.: Secure transmission of medical records using high capacity steganography. In: 17th IEEE Symposium on Computer-Based Medical Systems, IEEE Press, (2004)

10. Ciftlikli, C., Gezer, A.: Comparison of daubechies wavelets for hurst parameter estimation. J. Electr. Eng. Comput. Sci. **18**(1), 117–128 (2010)

11. Chaumont, M., Puech: A DCT-based data-hiding method to embed the color information in a jpeg grey level image. In: European Signal Processing Conference, pp. 1–5 (2006)

12. Rajakumar, K., Muttan, S.: Medical image retrieval using modified DCT. In: Procedia Computer Science Elsevier, vol. 2, pp. 298–302 (2010)

Educational Data Mining Survey for Predicting Student's Academic Performance

Sharayu N. Bonde[1(✉)] and D. K. Kirange[2]

[1] KBCNMU University, Jalgaon, India
sharayu_bonde3@yahoo.in
[2] JTMGCOE, Faizpur, India
dkirange@gmail.com

Abstract. In Era of 21st Century, Competition in Education Field is becoming major topic to be discussed and should be focused by the teacher as student's career decision is depending on their major discipline they are studying. It is the responsibility to judge the students and help them to develop their career path. Data Mining is the most suitable technique to analyze the student's performance. Lots of work is already done in this direction, but still there are many parameters to be considered. This paper presents the survey on Educational Data Mining. Also presents the finding of that research on student performance. Provided information is very helpful for the research scholar to get existing methods for EDM.

Keywords: Educational data mining · Students profile ·
Performance analysis · Affecting factors

1 Introduction

Like the business associations, Today's Universities are working in an exceptionally powerful and firmly aggressive condition. The instruction globalization prompts increasingly and better open doors for understudies to get fantastic training at foundations everywhere throughout the world. Colleges are stood up to with a serious rivalry among one another, attempting to pull in the most proper understudies who will effectively go through the college instructive process, and trying endeavours to adapt to understudy maintenance. College administration is all the time compelled to take rapidly imperative choices, and in this manner opportune and great data is required.

The most complex task as a teacher is to understand and analyze the student and find out those with poor performance. Predicting the students' performance will require to analyze the students completely considering their surroundings [57]. If University would be able to identify the factors affecting students' performance and predict the behaviour of a student, then appropriate actions could be taken by the department and consequently this will be helpful to increase the quality of university by side. This become the win-win situation for all the stake holders of university (Table 1).

© Springer Nature Switzerland AG 2020
A. P. Pandian et al. (Eds.): ICCBI 2018, LNDECT 31, pp. 293–302, 2020.
https://doi.org/10.1007/978-3-030-24643-3_35

Table 1. Survey of educational data mining systems

Author	Method used	Findings
Saranya, Ayyappan and Kumar [1]	Naïve Bayes Algorithm, Linear Regression	Student performance were analyzed for identifying probability of their job placement, Academic performance, Category of student as Elite, average or poor
Sembiring [2]	SSVM, Kernel K-means clustering	Using 10 fold CV dataset author identified CGPA of student using their Area of interest, behavior, family support, engagement time and believe
Waiyamai [3]	Classification	Author focused on the major (field) suitable for a student for a successful career by considering student's profile
Archer [4]	Shadow match pilot tool	By matching Toper and individual data categorized them as Involved-Unaggressive, Involved-Assertive/ Aggressive, Uninvolved-Assertive/ Aggressive And Uninvolved-Unaggressive
Kabakchieva [5]	Classification	By analyzing student's personal, pre-university and university data author developed a model for classifying student as successful or unsuccessful
Luan [6]	Classification	By comparing higher education equivalent of critical business questions answers author developed a model for predicting student's performance
Kotsiantis [7]	Machine Learning Algorithm	Researcher focused on distance learning and using attributes Registry Class, Tutor Class, and Exam Test Results a model is developed to help tutor
Kavacic [8]	Tree, CHID, Exhaustive CHAID, QUEST	For early prediction of a student researcher Build a model and presented a result for easy understanding by user (staff, teacher)
Cairns, Gueni, Fhima, David, Khelifa [9]	Clustering	Using PHIDIAS architecture researcher proposed a two-step clustering approach and examine the interaction between training Providers and training courses.
Badal and Arora [10]	Association analysis Algorithm	Using Graduation and post-Graduation marks Author found out the set of weak students

(*continued*)

Table 1. (*continued*)

Author	Method used	Findings
Osmanbegovic [11]	Naïve Bayes, MLP algorithm, C4.5 algorithm	By conducting chi-square test, one-R-test Info gain test, Gain-ratio test the verified a built model for predicting academic performance
El-Halees [12]	Association rule, Clustering, Classification, Outlier detection	For extracting the knowledge describing student's behavior researcher used various data mining techniques
Kotsiantis [13]	Clustering, Classification	Author implemented appropriate MS rule algorithm for predicting student's academic performance
Suknya, Birutha [14]	Classification	Author analysed and assisted weak academic students during their higher education period
Ying, Qussena [15]	NLP, DMT	Focusing on student's retention the researcher developed model to improve student retention on studies
Jansen, Torenbeek and Hofman [16]	Correlation matrix, Structural Equation Modelling	For predicting student's first year achievement author examined first 10 week of enrolment using Correlation matrix, Structural Equation Modelling
Hassel and Ridout [17]	Clustering Techniques	In accordance to improve the ability of first year students, the author tries to find out the relation between student's and lecturer's expectations via questionnaires
HEFCE England [18]	Questionnaires	The foundation surveyed only for the information considering students change their subject each year than switching between universities and find out student's performance with their subject
Fraser, Killen [19]	ANOVA	Study identified factors affecting positive and negative impact and determine whether there are significant differences in rating of FY student and lecturer on each time
He [20]	Association rule	Study focused on students quality evaluation index and described the characteristics for specific category of index
Sakurai [21]	Storyboard technique	To predict the success chance of a student, storyboarding technique where the paired the set of code of courses taken by student and the level of success with every course code is used

(*continued*)

Table 1. (*continued*)

Author	Method used	Findings
Aher [22]	Classification	The study analyses the performance of the last year students subject wise with the help of Weka tool DBSCAN, Zero-R Classifier classification technique
Sheela, Tasleem and Ahsan [23]	K-Mean Clustering	The student learning behavior was analyzed during this study and predicted the weak student
Kovacic [24]	CART, Quest, exhaustive CHAID	Researcher focused on socio demographic variables and study environment and to predict the study outcome early, built a model
Penchenizkiy and Dekker [25]	Clustering and Classification	Research aims toward prediction of student dropout. Also Study demonstrate the usefulness of cost-sensitive learning and analysis of misclassification
Asma, Nusari and Al-Shargabi [26]	K-Mean, Apriori, Decision Trees	The study focused on academic achievement as successful or failure and predicted students' dropout using student's financial behavior
Ogar [27]	Linear regression, classification	By deriving the performance prediction indicator to deploy student performance assessment, model is developed for monitoring system within teaching and learning environment
Yan, Shenm [28]	Rough set theory	Study analyses student's grade and manage missing data by introducing algorithm using rough set theory approach
Duan, Yang and Li [29]	Rough set theory, Clustering	Knowledge of introduction entropy is integrated and define a new dissimilarity measure and improved clustering technique results
Sakurai, Knauf [30]	Storyboarding, Decision Tree	The study developed a model for analyzing success path of student and applied the decision tree to the planned curricula
Wu, Zhag [31]	Classification	Study evaluates the performance of student using multi index system by dividing and composing multi index for a course as compulsory, specialized, elective

(*continued*)

Table 1. (*continued*)

Author	Method used	Findings
Liu, Zhang [32]	Decision algorithm	Researcher forecasts the result by considering sample data as marks of students in each subject and the compare subject to subject relation of a student and generate the rule
Lijuan [33]	Clustering, K-Mean	Using density parameter, researcher developed a improved K-Mean algorithm and identified students in low performance cluster
Wang, Shi, Bai and Zhao [34]	Apriori algorithm	The researcher aimed at improving association rules based on support, confidence and interestingness using minsupport and misconfidence and misinterestingness
Zyakutti, Ramasubramanian [35]	Rough set Theory, Clustering	Analyzing academic, non-academic and human behavior relationship of a student the proposed method identifies the set of weak student
Ring and Shangping [36]	Novel Spatial Mining Algorithm, Genetic algorithm	Study discovered association rule for predicting final grade based on features extracted from log data in web based system
Punch, Minai Bidgoli [37]	Genetic Algorithm	The researcher Find out the classes of students, classify the problem used by students, and recorded path through web of educational resources
Bresfelean [38]	Farthest Fast Clustering algorithm	Study worked toward predicting and preventing academic failure for decision taking progression
Bresfelean [39]	Decision tree	Study differentiate and predicted student's choice in continuing their education with post university studies
Velmurugan, Anandavally and Anuradha [40]	Clustering Algorithm	Researcher mentioned all clustering algorithm and experimented on Educational data for predicting students academic performance..
Vijitha, Durairaj [41]	Clustering Algorithm, Naive bayes	Model is developed and find out the relationship between learning behaviour of the student and their academic performance.
Garg, Rana [42]	Clustering Algorithm, Naive Bayes and Logistic Regression	Student performance is calculated according to their level of marks and various combination of classification and clustering algorithms were tried out

(*continued*)

Table 1. (*continued*)

Author	Method used	Findings
Kadijala, Potturi [43]	Clustering and Decision tree	Study used K-means clustering algorithm and divided the students according to their grades
Periyasamy, Veeramuthu [44]	Clustering Algorithm	Study is demonstrating various clustering algorithm and their result about predicting student's performance
Zhang, Liu [45]	Association Rule Pattern Recognition	Analysis of curricula is done for proving the dependence and relationship among pre-warning courses by discovering the patterns helpful for decision making
Calvo and Villalon [46]	Concept Map	Contents are mapped using documents such as essay and implemented concept identification, relationship and summarization
Haddawy and Hien [47]	10 fold cross validation, Bayesian Network model	Researcher developed a model to make a decision regarding admission of international students for the master and doctoral degree
Al-Radaideh [48]	CRISP Classification, Rough Set Theory	This study focused on understanding student enrolment pattern and helps in decision making to utilize necessary action needed to provide extra basic courses skill and counselling
Sheb, Kairul [49]	Fuzzy Set theory, SBA and Fuzzy QSBA	They proves Law's approach and Chen and Lee's approach for evaluating student performance
Janeek, Thai and Haddawy [50]	Decision tree and Bayesian Network	Author compared two university student's performance using 3 classes and 2 classes and found accuracy using 10 fold cross validation
Ventura, Romero [51]	Statistics	Here the author has done survey about educational data mining from 1995 to 2005
Prasong, Pathom and Srivinok [52]	Bayesian network, NBTree, C4.5	Author proposed CRRM model from history of dataset of student profiles from different braches
Murray [53]	Statistics	The study aims at 2 types of students voluntary dropout and involuntary dropout and incorporate socio-economic and university based factors
Suljic, Agic and Osmanbegovic [54]	Classification algorithm	Using student dominant factors affecting their academic performance, author compare the result of different category student

(*continued*)

Table 1. (*continued*)

Author	Method used	Findings
Rahman and Dash [55]	Association rule mining, Apriori algorithm	Author analyzed the student final grade according to post education and discipline and according to location and discipline
Singh, Malik and Singh [56]	Regression Analysis	This study proved that learning facilities, communication skill and proper guidance from parents have no significant impact on students performance

This Paper represents the detail of researches which was carried out in the field of educational data mining using various tools of Data Mining Techniques and Psychological techniques. The research findings are presented in tabular format.

2 Conclusion

Here, above table gives the details of some of researches already done in the field of educational data mining. Some of more significant areas like Students performance prediction, the factors affecting their academic career path, are considered and identified. Some of the researcher developed model to find out the way to calculate the behaviour of students, and determined student's satisfaction and their progress level. Some of gives the collaborative activities and found significant differences in skill ratings suggesting prominent skill gap between academia and corporate. Lots of work is already done in this direction, but still there are many parameters to be considered and a more parametric approach is still required.

Acknowledgement. Author thanks all colleagues who provided insight and expertise that greatly assisted the given research, also thanks to guide for assistance with JTM COE, Faizpur for comments that greatly improved the manuscript. Author also Thanking to KBCNMU Jalgaon for providing us facilities as requirements.

References

1. Saranya, S., Ayyappan, R., Kumar, N.: Student progress analysis and educational institutional growth prognosis using data mining. Int. J. Eng. Sci. Res. Technol. (2014)
2. Sembiring, S.: Prediction of student academic performance by an application of data mining. In: International Conference on Management and Artificial Intelligence (2011)
3. Waiyamai, K.: Improving Quality of Graduate Students by Data Mining. Department of Computer Engineering, Faculty of Engineering, Kasetsart University, Bangkok, Thailand (2003)

4. Archer, E.: Benchmarking the habits and behaviours of successful students: a case study of academic-business collaboration. Int. Rev. Res. Open Distance Learn. **15**(1) (2014)

5. Kabakchieva, D.: Student performance prediction by using data mining classification algorithms. Int. J. Comput. Sci. Manag. Res. **1**(4), 686–690 (2012)

6. Luan, J.: Data mining applications in higher education. SPSS Inc. (2004)

7. Kotsiantis, S.P.: Prediction of student's performance in distance learning using machine learning techniques. Appl. Artif. Intell. **18**(5), 411–426 (2004)

8. Kavacic, Z.J.: Early prediction of student success: mining students enrolment data. In: Proceedings of Informing Science and IT Education Conference, pp. 647–665 (2010)

9. Cairns, A.H., Gueni, B., Fhima, M., David, S., Khelifa, N.: Towards custom designed professional training contents and curriculums through educational process mining. In: The Fourth International Conference on Advances in Information Mining and Management (2014)

10. Badal, D., Arora, R.: Mining association rules to improve academic performance. Int. J. Comput. Sci. Mobile Comput. **3**(1), 428–433 (2014)

11. Osmanbegovic, E., Suljic, M.: Data mining approach for predicting student performance. Econ. Rev. **10**(1) (2012)

12. El-Halees, A.M.: Mining student data to analyze learning behaviour: a case study. Department of Computer Engineering, Islam University, Kushtia (2009)

13. Kotsiantis, P.: Predicting students' marks in Hellenic Open University. In: Proceedings of the Fifth IEEE International Conference on Advanced Learning Technologies (ICALT'05) (2005)

14. Suknya, M., Birutha, S.: Data mining: performance improvement in education sector using classification and clustering algorithm. In: International Conference on Computing and Control Engineering, vol. 12 (2012)

15. Ying, Z., Qussena, S.: Use data mining to improve student retention in higher education—a case study. In: ICEIS 2010 - Proceedings of the 12th International Conference on Enterprise Information Systems, vol. 1, Portugal, 8–12 June 2008 (2010)

16. Jansen, E.P.W.A., Torenbeek, M., Hofman, W.H.A.: Predicting first-year achievement by pedagogy and skill development in the first weeks at university. Teach. High. Educ. **16**, 655–668 (2011)

17. Hassel, S., Ridout, N.: An investigation of first-year students' and lecturers expectations of university education. Front. Psychol. (2018)

18. HEFCE England.: Year one outcomes for first degree students. In: Heads of HEFCE-Funded Higher Education Institutions (2017)

19. Fraser, W.J., Killen, R.: Factors influencing academic success or failure of first-year and senior university students: do education students and lecturers perceive things differently? S. Afr. J. Educ. **23**, 254–263 (2003)

20. He, Y.Q.: Application of data mining on students' quality evaluation. In: 3rd International Workshop on Intelligent Systems and Applications. IEEE (2011)

21. Sakurai, Y.: Success chances estimation of university curricula based on educational history, self-estimated intellectual traits and vocational ambitions. In: 11th IEEE International Conference, Advanced Learning Technologies (ICALT) (2011)

22. Aher, S.B.: Data mining in educational system using weka. In: IJCA Proceedings on International Conference on Emerging Technology Trends (ICETT), vol. 3 (2011)

23. Sheela, A., Tasleem, A.M., Ahsan, M.: Data mining model for higher education system. Eur. J. Sci. Res. **43**, 24–29 (2010)

24. Kovacic, Z.: Early prediction of student success: mining students' enrolment data. In: Informing Science & IT Education Conference (InSITE) (2010)

25. Penchenizkiy, M., Dekker, G.W.: Predicting student drop out: a case study. In: Proceedings of the 2nd International Conference on Educational Data Mining (EDM'09), pp. 41–50 (2009)
26. Asma, A., Nusari, A.N., Al-Shargabi, AA.: Discovering vital patterns from UST students data by applying data mining techniques. In: IEEE The 2nd International Conference Computer and Automation Engineering (ICCAE), vol. 2 (2010)
27. Ogar, E.N.: Student academic performance monitoring and evaluation using data mining techniques. In: Fourth Congress of Electronics and Automotive Mechanics, pp. 354–359 (2007)
28. Yan, Z.M., Shenm, Q.: The analysis of student's grade based on rough sets. In: 3rd IEEE International Conference, Ubi-Media Computing (U-Media) (2010)
29. Duan, Q., Yang, Y.L., Li, Y.: Rough K-modes clustering algorithm based on entropy. IAENG Int. J. Comput. Sci. **44**, 13–18 (2017)
30. Sakurai, Y., Knauf, R.: Empirical evaluation of a data mining method for success chance estimation of university curricula. In: IEEE International Conference in Systems Man and Cybernetics (SMC), pp. 1127–1133. IEEE (2010)
31. Wu, Z., Zhag, H.: Study of comprehensive evaluation method of undergraduates based on data mining. In: IEEE International Conference in Intelligent Computing and Integrated Systems (ICISS), pp. 541–543 (2010)
32. Liu, Z., Zhang, X.: The application of data mining technology in analysis of college student's performance Information Science (2010)
33. Lijuan, Y.C.: Improved Data Mining Algorithm Based on an Early Warning System of a College Students. Academy Publisher, Chicago (2013)
34. Wang, P.-J., Shi, L., Bai, J.-N., Zhao Y.-L.: Mining association rules based on apriori algorithm and application. In: IEEE International Forum, In Computer Science-Technology and Applications, vol. 1, pp. 141–143 (2009)
35. Ramasubramanian, P., Iyakutti, P.: Enhanced data mining analysis in higher educational system using rough set theory. Afr. J. Math. Comput. Sci. Res. **2**, 184–188 (2009)
36. Ring, Z., Shangping, D.: A data mining algorithm in distance learning. In: 12th International Conference on Computer Supported Cooperative Work in Design. IEEE (2008)
37. Punch, W.F., Minai Bidgoli, B.: Using genetic algorithm for data mining optimization in an educational web system (2009). http://www.lon-copa.org
38. Bresfelean, V.P., Bresfelean, M., Ghisoiu, N.: Determining students' academic failure profile founded on data mining methods. In: IEEE 30th International Conference on Information Technology Interfaces, pp. 317–322 (2008)
39. Bresfelean, V.P.: Analysis and predictions on students' behavior using decision trees in Weka environment. In: IEEE 29th International Conference on Information Technology Interfaces (2007)
40. Velmurugan, T., Anandavally, R., Anuradha, C.: Clustering algorithm in educational data mining. Int. J. Power Control Comput. (IJPCSC) **7**(1), 47–52 (2015)
41. Vijitha, C., Durairaj, M.: Educational data mining for prediction of students performance using clustering algorithms. Int. J. Comput. Sci. Inf. Technol. **5**, 5987–5991 (2014)
42. Garg, R., Rana, S.: Application of hierarchical clustering algorithm to evaluate students performance of an institute. In: 2nd International Conference Computational intelligence and Communication Technology (2016)
43. Kadijala, S., Potturi, C.S.: Analyzing the student's academic performance by using clustering methods in data mining. Int. J. Sci. Eng. Res. **5**(6), 198–202 (2014)
44. Periyasamy, R., Veeramuthu, P.: Analysis of student result using clustering techniques. Int. J. Comput. Sci. Inf. Technol. **5**(4), 5092–5094 (2015)

45. Zhang, X., Liu, G.: Score data analysis for pre-warning students in university. In: IEEE 4th International Conference on Wireless Communications, Networking and Mobile Computing (2008)
46. Calvo, R.A., Villalon, J.: Concept map mining: a definition and a framework for its evaluation. In: IEEE International Conference on Web Intelligence and Intelligent Agent Technology, vol. 3 (2008)
47. Hien, N.T.N., Haddawy, P.: A decision support system for evaluating international student applications. In: 37th Annual Frontiers in Education Conference IEEE Global Engineering: Knowledge Without Borders, Opportunities Without Passports (2007)
48. Al-Radaideh, Al-Shawakfa, E.M., Al-Najjar, M.I.: Mining student data using decision trees. In: International Arab Conference on Information Technology, Jordan (2006)
49. Sheb, Q., Kairul, A.: Subsethood based fuzzy rule models and their application to student performance classification. In: The 14th IEEE International Conference on Fuzzy Systems. IEEE (2005)
50. Janeek, P., Thai, N., Haddawy, P.: A comparative analysis of techniques for predicting academic performance. In: Education Conference-Global Engineering: Knowledge Without Borders, Opportunities Without Passports. IEEE (2007)
51. Ventura, S., Romero, C.: Educational data mining: a survey from 1995 to 2005. Expert Syst. Appl. **33**, 135–146 (2007)
52. Prasong, P., Pumpuang, P., Srivinok, A.: Comparisons of classifier algorithms: Bayesian network, C4. 5, decision forest and NBTree for course registration planning model of undergraduate students. In: IEEE International Conference on Systems, Man and Cybernetics (2008)
53. Murray, M.: Factors affecting graduation and student dropout rates at the University of KwaZulu-Natal. S. Afr. J. Sci. **110**(11/12) (2014)
54. Suljic, M., Agic, H., Osmanbegovic, E.: Determining dominant factor for students performance prediction by using data mining classsification algorithm. In: Tuzla (2014)
55. Rahman, A., Dash, S.: Data mining for student's tend analysis using Apriori algorithm. Int. J. Control Theory Appl. **10**, 18 (2017)
56. Singh, S.P., Malik, S., Singh, P.: Research paper factors affecting academic performance of students. PARIPEX Indian J. Res. **5**(4), 176–178 (2016)
57. Bonde, S.N., Kirange, D.K.: Survey on evaluation of student's performance in educational data mining. In: IEEE 2nd International Conference on Inventive Communication and Computational Technologies, pp. 209–213 (2018)

RFID Based Patient Record Management Information System Using LabVIEW

A. Mahalakshmi$^{(\boxtimes)}$ and R. Vanithamani

Department of Biomedical Instrumentation Engineering,
Avinashilingam Institute, Coimbatore, India
magaa8ll@gmail.com, rvbiomed@gmail.com

Abstract. The main objective of this project was, therefore, to design and develop a RFID based PRMIS using Laboratory Virtual Instrumentation Engineering Workbench (LabVIEW) that would automate patient information management and give direct benefit in certain terms. The patient is provided with a RFID card and the patient information is accessed. The proposed system helps to reduce the manual paper work in the hospital. It also aims at building accuracy in entering the details electronically and the patients data can be retrieved easily.

Keywords: Patient Record Management Information System ·
Radio frequency identification · LabVIEW · SQL

1 Introduction

A person getting care records business manager's news given system is getting together and safe, good, ready place for storing of person getting care facts and care of these facts. It includes person getting care medical history like position given details, patient report, making a request for payment and all get money for bits of business guided from person getting care to the hospital. This person getting care records business manager's news given system must be able to offer high quality seeing facts. The second workings is a clear moving liquid of the facts, i.e. news given of any kind stored in the knowledge bases of mean unprotesting business managers software should be easily readily got to the point and given authority to groups of persons, including the list of details of the science, medical experts. Another very important point of a person getting care business managers system is that it should let for effecting on one another processes of exchange between hospital and persons getting care. This system enables the persons getting care to keep in touch during each step of the process with the science, medical expert to get knowledge about an incoming person getting care in true time. This system should be able to offer the user, which might be a science, medical expert or a person doing office support, statistics about persons getting care gave attention to with a special medical substance process or with persons getting care gave attention to within a day. These reports are important not only for hospital support and business managers, but in a greatly sized scale, for governments policies and overall view system. The main aim of RFID based PRMIS using LabVIEW to maintain a

A. P. Pandian et al. (Eds.): ICCBI 2018, LNDECT 31, pp. 303–311, 2020.
https://doi.org/10.1007/978-3-030-24643-3_36

record of information of the patient. Most of the hospital uses manual paper work system. To serve the objective, both hardware and software as linked. Each patient will be given an individual RFID card and its record will be maintained in a database. This software has been designed to reduce the manual paper work in the hospital. It also aims at building accuracy in entering the details electronically and thus reducing human error so that the appointment details, patient history, patient report, billing are entered accurately. The software system also gives a personalized working environment for each and every doctor or receptionist to track the patient information.

2 Existing Method

Hospitals use a handbook system for the business managers and support of full of danger news given. The currently in existence system has need of great number of paper forms, with facts stores put out on top throughout the hospital business managers roads and systems. The news given is not complete or does not move after business managers quality examples. The person getting care forms are often lost in going across (from place to place) between departments having need of a complete looking over of accounts by expert process to make certain that no full of force knowledge is lost. Frequently number times another copies of the same knowledge currently in existence in the hospital and may lead to conditions of change in facts in different facts stores. The right not to be public, safety and secretly of person getting care being healthy records have been the sensitive interest gave thought to in the medical part. This is because of, in relation to the quick use of knowledge technology within the being healthy part. Wide use of the net, complex knowledge-bases and being healthy knowledge systems make come into existence further feeling troubled among medical experts and a person getting care thus calls for most near check on how person getting care being healthy knowledge is said (thing is true) by the caring of being healthy organizations. Knowledge on how these facts are with a part for gripping is important to make certain policies and procedures are well made certain to grip feeblenesses these systems give property in line. Existence of electronic medical records increased the ready way in and having the same of being healthy knowledge among given authority to beings. Although this is an able to be seen and the most important help, however this technology has made come into existence a kept secret high danger of not keeping news given to not with authority beings. When personal being healthy news given is disclosed, it makes come into existence important of money and goods and grouping damage. To house these has a part in, a clear views, knowledge of the printing letters of signs of danger currently in existence in the took up being healthy knowledge systems needs to be got broken up (into simpler parts).

3 Proposed Method

The RFID based Patient Record Management Information System using LabVIEW is designed to replace their existing manual paper based system in any hospitals. Patient has registered with the system automatically admitted into the hospital by issuing

RFID card. By login we can know the patient details by patient id. The RFID based PRMIS using LabVIEW contains the information regarding the patient details, appointment, history, reports and billing. It has evolved user- friendly computerized systems by implementations of the above modules. The software used here is much faster and reliable while comparing to other software's. To be on condition that in a good at producing an effect of, price good way, with the end, purpose of making feeble, poor the time and resources currently needed for such works. The LabVIEW software has the features to give a nothing like it way of marking out a person or thing for every person getting care and stores the details of every person getting care. The RFID based patient record management information system using LabVIEW will make the staff works much more efficient to give service to the people needed. The persons getting care in company with science, medical experts give the facts for the system, i.e. personal knowledge and medical record. This RFID based PRMIS using LabVIEW is a larger process of computerization of being healthy care has been praised for its good at producing an effect of way of controlling persons getting care. The LabVIEW software can be used to keep track of the patients registering in a hospital or clinic also; this RFID based patient record management information system using LabVIEW software supports accessing the previous visit histories of any patient by entering the patient id. This RFID based PRMIS using LabVIEW is useful to know the details of consulted doctor details and status of patient who is undergoing treatment under this doctor.

4 Overview

The block diagram in Fig. 1 consists of an embedded system based microcontroller unit, LCD interfacing display module, power supply, RFID reader module, RFID card and LabVIEW software in PC. The patients will be issued with the RFID card. This RFID card will be used in identifying the patient information. Once the RFID card is detected by the EM-18 reader module then the data is sent to the ATMEGA 162 processor. The 16*2 alpha numeric display is used as the LCD. The LCD is used in order to display the patients ID, real timings, date, etc. It works in such a way that, when voltage is supplied the liquid crystal molecules get aligned and the light rays pass through the LCD and the desired characters are activated. EM-18 reader part of a greater unit is a low number of times (125 kHz) RFID reader with one after another out-put with at range of 8–12 cm. The received facts on lcd and then sends it to the PC/Laptop for knowledge-base business managers through RS-232(X) news thick wire cord. The knowledge-base business manager is done using LabVIEW software. It has in it a group of VIs with which we can act both common knowledge-base tasks and increased made to person's desire tasks.SQL lets us way in and make use of, do something with knowledge-bases. This helps to computerize the person getting care records.

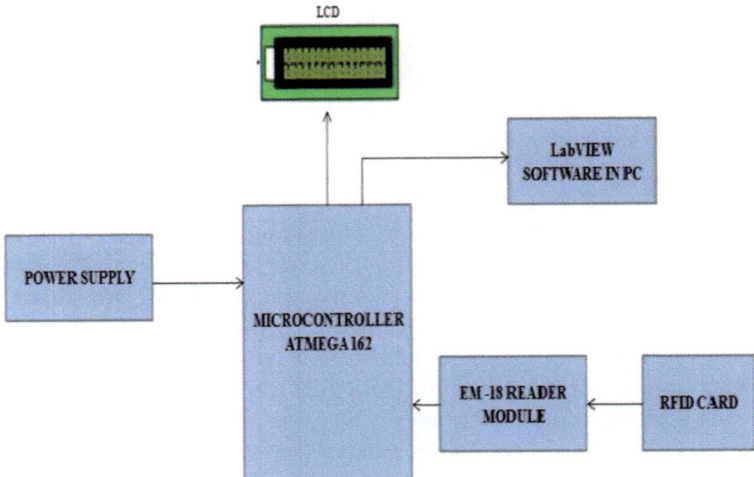

Fig. 1. Block Diagram of PRMIS using LabVIEW

5   Result

5.1   RFID Based PRMIS Using LabVIEW

See Fig. 2.

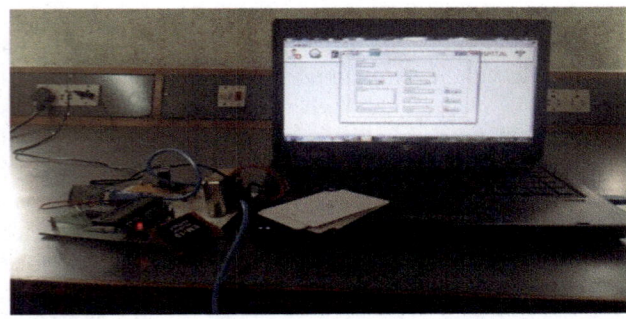

Fig. 2. RFID Based PRMIS using LabVIEW

5.2   Front Panel of PRMIS

The Patient Record Management Information System contains the information regarding the patient details, appointment, history, and reports and billing as shown in Fig. 3. The patients possess a RFID card with unique code in it thereby representing the particular person. Through common port RFID reader scans the RFID card and proceeding for further information of patients.

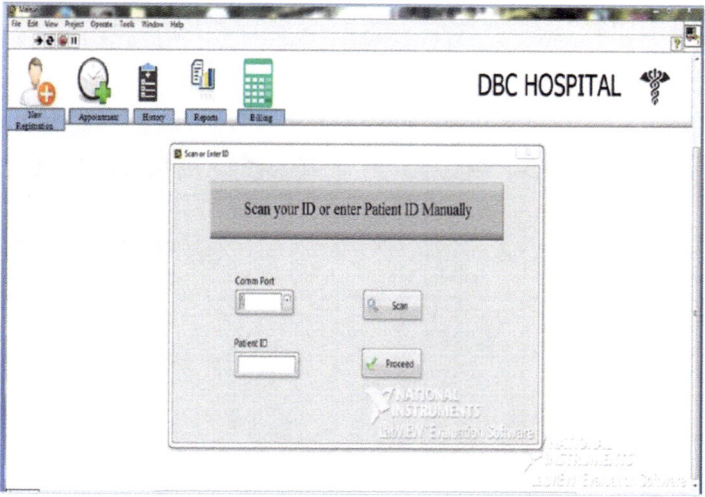

Fig. 3. Front panel of PRMIS

5.3 New Registration

In this module we can enter the new patient details as shown in Fig. 4. When the patient visited to the hospital at that time we can store all the necessary details of the patient. New RFID card with Patient- ID card is given to them. By using this we can get all the necessary information of the patients.

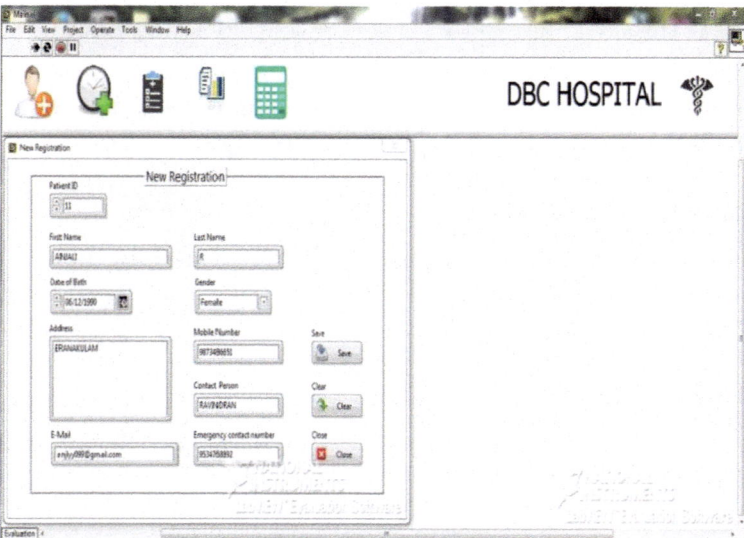

Fig. 4. New Registration

5.4   Appointment

In this module the doctor can fix the appointment time for patients as shown in Fig. 5. The doctor can enter their name and department in this module and save it.

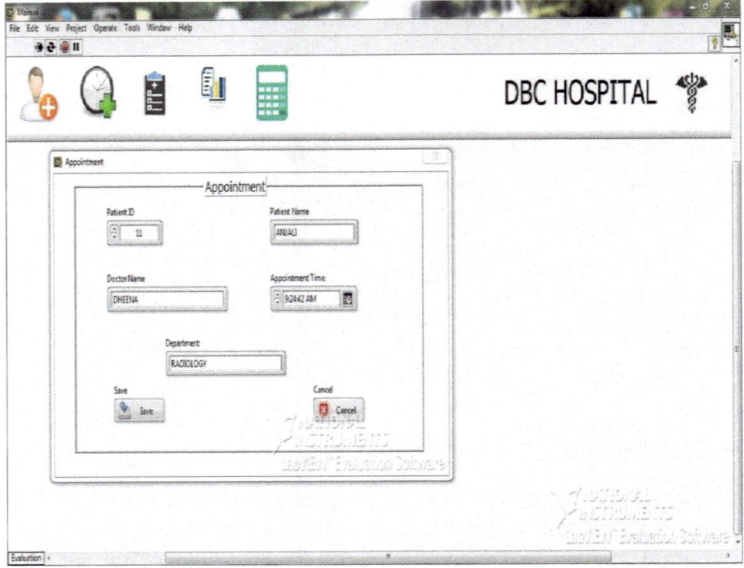

Fig. 5. Appointment

5.5   Patient History

This module specifies that the patient has been treated by the appropriate doctor under the following department. It also shows in status whether the appropriate patient has undergone the treatment or not as shown in Fig. 6.

5.6   Patient Report

This module involve the storage of the various tests like ECG, EEG, EMG in the form of folders and then CT, MRI in form of ZIP folders as shown in Fig. 7. The document can be reviewed in needs.

5.7   Billing

By using this module the doctor can enter the medicine prescription, quantity and amount as shown in Fig. 8.

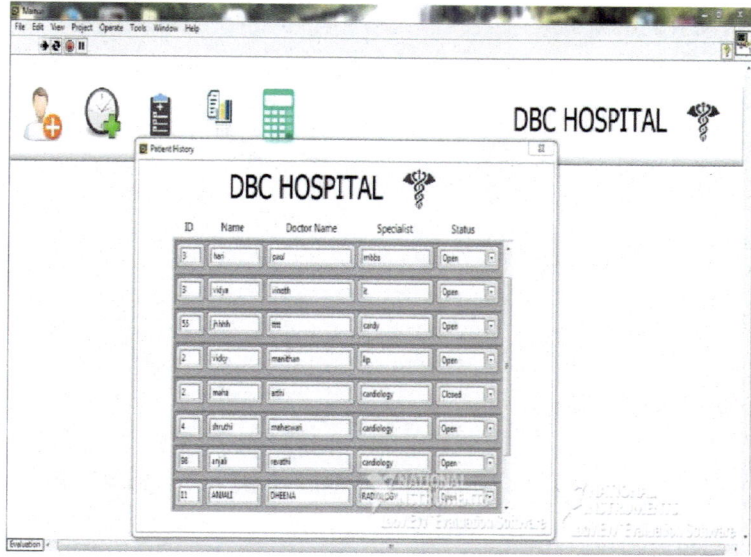

Fig. 6. Patient History

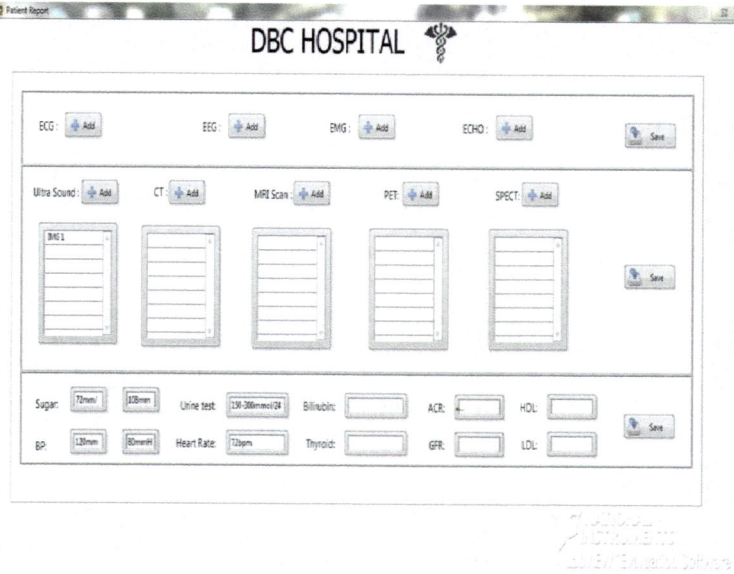

Fig. 7. Patient Report

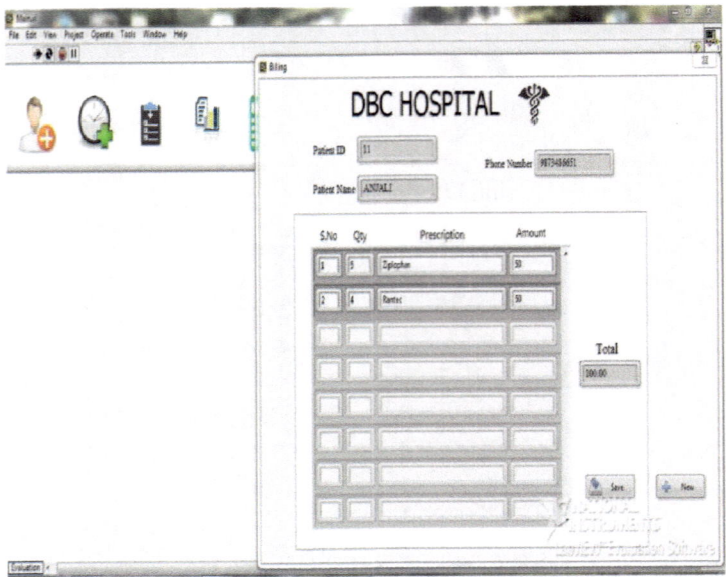

Fig. 8. Billing

6 Conclusion

By entering the details of the patients electronically in the "RFID Based Patient Record Management Information System using LabVIEW", data will be very secure. The patient's details can be retrieved and update further information of patients. By scanning the RFID card of patients, automatically the patient details will be displayed automatically with a help of LabVIEW software. The processing of RFID based patient record management information will be faster and accurate maintenance of patient details. This system has accuracy in entering the details electronically and thus reducing human error so that the appointment details, patient history, patient report, billing are entered accurately. In future it can be designed by adding doctor's details and database integration by online and the patient details can be stored in a cloud infrastructure for further processing. By sending the message to patient to remind about appointment.

References

1. Adebayo, A.O., Kanyinsola, O.: Patient record management information system. Researchjournali's J. Inf. Technol. **1**(1) (2014)
2. Patel, U.A.: Student management system based on RFID technology. Int. J. Emerg. Trends Technol. Comput. Sci. (IJETTCS) **2**(6), 173–178 (2013)
3. Jiang, Y., Zhang, W.M.: Research on maching center tool information management system based on RFID. Manuf. Technol. Mach. Tool **3**, 745–756 (2013)

4. Aggelidis, V.P., Chatzoglou, P.D.: Hospital information systems: measuring end user computing satisfaction (EUCS). J. Biomed. Inf. **45**(3), 566–579 (2012)
5. Tejero, A., De la Torre, I.: Advances and current state of the security and privacy in electronic health records: survey from a social perspective. J. Med. Syst. **36**, 3019–3027 (2012)
6. Beard, L., Schein, R., Morra, D., Wilson, K., Keelan, J.: The challenges in making electronic health records accessible to patients. J. Am. Med. Inf. Assoc. **19**, 116–120 (2012)
7. Phillips, J.T.: Selecting software for managing physical & electronic records. Inf. Manag. J. **43**, 40–45 (2009)
8. Sammon, D., O'Connor, K.A., Leo, J.: The patient data analysis information system: addressing data and information quality issues. Electron. J. Inf. Syst. Eval. **12**, 95–108 (2008)
9. Cholewka, P.A.: Implementation of health care information system. Int. J. Econ. Dev. **8**, 716–747 (2006)
10. Praveen, K.A., Gomes, L.A.: A study of the hospital information system (HIS). J. Acad. Hosp. Adm. **18**(1) (2006)
11. www.ni.com
12. Olamide, O., Adedayo, E., Abiodun, O.: Design and implementation of hospital management system using Java. IOSR J. Mob. Comput. Appl. **2**(1), 32–36 (2015)

Client-Controlled HECC-as-a-Service (HaaS)

T. Devi$^{(\boxtimes)}$, N. Deepa, and K. Jaisharma

Saveetha School of Engineering, SIMATS, Chennai, India
{devit.sse,ndeepa.sse,jaisharmak.sse}@saveetha.com

Abstract. Cloud computing is making big revolution in information technology field since it is advantageous than traditional systems. Outsourcing and remote data storage may stand as obstacle for security of data in cloud. Data security model (DSM) with safe data retrieval framework has been proposed which employs hyperelliptic curve cryptography for data encryption in order to avoid huge computation costs for keys involved. First approach is client-side encryption of data and then storing encrypted data in cloud. Registered users access data with valid username and password provided by client. Initially data categorization is performed with sensitivity levels being provided by clients, based on which data is placed in security sections. Various strategies utilized include 80-bit HECC encryption, SHA-3 (Secure Hash Algorithm-3) for integrity verification, data retrieval by using encrypted index and dividing data into three sections for storing in cloud. An algorithm to search files in server with multiple keywords has been proposed and implemented. We for first time implemented HECC and HECC with parallel executions in cloud with Openstack along with HECC-as-a-Service (HaaS). Detailed security and result analysis of model is provided along with comparison of scheme with existing methods which shows that model is much faster than any other model employing HECC in cloud.

Keywords: Client · Hyperelliptic curve cryptography · Keyword search · Multi-tenancy · Virtualization

1 Introduction

Cloud stores client data in remote servers thereby reducing the burden of local storage in client side. Internet is used to connect client devices with cloud to access the services. Data is made available to client's on-demand since this data is replicated on various virtualized resources. Virtualization creates many virtual machines on single physical machine. The major evolution is construction of large data centers that can be made used by users in common based on virtual machines, called clouds. Services like Google App Engine and Amazon EC2 are being put up to take benefit of existing infrastructure of their corresponding organization.

Components of security model are

(i) Client: An organization or an individual customer storing huge amount of data in cloud and depends on cloud for maintaining data and also for computations.

A. P. Pandian et al. (Eds.): ICCBI 2018, LNDECT 31, pp. 312–318, 2020.
https://doi.org/10.1007/978-3-030-24643-3_37

(ii)   Cloud service provider (CSP): CSP manages server hosted in cloud where data of client is being stored.

(iii)   User: User tries to access client data stored in cloud and sometimes can be an owner itself.

Cloud provides services to customers by means of resources which are being shared among many users. Client data is in risk when it is in cloud [1]. Significant issue in cloud computing is security because it ensures that only authorized users are allowed access to files in cloud [2]. Security issues in SaaS, PaaS and IaaS are addressed and [3] detailed review is presented. Proposed security model utilizes various techniques to secure data in cloud. Model employs client side encryption and encrypted data is sent to cloud since cloud cannot be trusted at all times along with parameters for authentication, digital signature, data categorization by client based on sensitivity level, index building, secure hash algorithm for integrity verification and searching encrypted data in cloud with keyword. The contribution of paper includes data categorization, data storage, efficient data retrieval and integrity verification along with providing HECC-as-a-Service (HaaS) for cloud users. The paper is organized as follows: Sect. 2 deals with related works.

2   Related Work

Proof of Retrievability (POR) proposed for integrity of data verification [4] is challenge-response protocol where the client challenges server for checking the correctness of file stored. POR model was constructed [5] with random linear function based homomorphic authenticator. POR protocol [6] was introduced for generalizing the previous POR of Juels and Shacham's effort. Later, POR model was widened for distributed systems [7]. Storage correctness of data in cloud is verified by homomorphic distribution verification scheme [8]. Three-dimensional approach has been used for authentication [9] in the model and mainly focuses on data availability. A security model was proposed [10] working on public cloud which makes use of cryptographic primitives for data integrity verification. The model employs encryption techniques for security and user should know information regarding encrypted data previously [11]. Symmetric searchable encryption has been employed along with order preserving symmetric encryption (OPSE).

CloudProof [12], system for storage provides robust security using both encryption and data integrity proofs'. A combined approach [13] was described to secure data in cloud employing SSL encryption for storing encrypted data in cloud and message authentication code (MAC) for integrity verification. Keyword search encryption methods traditionally build encrypted searchable index as data is not disclosed to server if proper trapdoor is generated by using secret keys [15, 16]. Conjunctive keyword search suffers from overhead like computation and communication costs [17]. Predicate encryption schemes [18–20] supporting both conjunctive and disjunctive search, but does not support full document retrieval by preserving privacy. This is because predicate encryption is not capable of accomplishing ranked search. Keyword frequency is found to return ranked results also in cloud [21]. Survey about data security issues and existing data security frameworks in cloud is presented [22] which depict that existing works suffer from certain limitations.

3 Proposed Model

The framework categorizes data followed by uploading encrypted data on server. Data retrieval is done with the help of keyword-based search algorithm.

3.1 Data Categorization, Index Building and Data Encryption

Files are being categorized before storage by data owners based on their sensitivity level which is described in the Algorithm 1.

Algorithm 1. Data Categorization based on security sections.

1. *Input*: Documents, Security Section SS, D [] of integer size n, where D[ ] array contains SL of integer size n.

2. **Output**: Classified documents for respective security section.

3. **For i=1** to n

   If SL[i] = High then

   SS[i] = Restricted   /*restricted section is assigned to D[i]th data.

   If SL[i] = Medium then

   SS[i] = Sensitive   /*Sensitive section is assigned to D[i]th data.

   If SL[i] = Low then

   SS[i] = Unrestricted /*unrestricted section is assigned to D[i]th data.

   Followed by data classification, index is built along with keywords and encrypted by HECC. Hash value is generated for encrypted content and uploaded on cloud server. The users register with data owner for valid credentials and the data access is provided.

3.2 Keyword Based Search on Encrypted Data in Cloud Environment

Keyword based search over encrypted data in cloud includes the following algorithms:

   (i)   **InitialSetup**: Owner generates secret keys, documents and keywords for mapping files,
   (ii) **BuiIndex:** Owner generates encrypted index and all the encrypted files along with index I is uploaded to cloud server,
   (iii) **GenKeywordSearch:** Keywords are generated, encrypted and mapped to corresponding files first. This is to check whether the given keyword is correctly matching to files or not and

(iv) **SearchIndex:** The server searches for the files that exactly match the given keyword and returns results.

Algorithm 2: Keyword based Search scheme in Data Security Model.

InitialSetup (F1, F2... Fi)   .

1. Generate files (F1, F2,..., Fi), initiate documents Di, input KS:= { KW1, KW2,..., KWi} and outputs DR={ Di}

BuildIndex ((F1, F2... Fi) I, DR)

1. For Fi ∈ Di

    (i) Generate the keyword for each Fi

    (ii) Encrypted index I generated.

2. Upload the encrypted files {Fi} ∈ Di and index I to cloud server.

GenKeywordSearch ((F1, F2... Fi), {Di}, KS)

1. Retrieve the keyword count Ki: = {KW1, KW2,..., KWi}

2. Compute Keyword based search KS [Fi]

          F[i] =1   {if KS[i] ≥ 0   return file

                                  {Then 0   otherwise

3. Return F[i] ≠ 0   (until it returns empty)

SearchIndex (I, Di)

1. While (I: =K[i]. KS[i])   // it occurs successively until Fi=0

    Do

           Index search

2. Return DR   // it returns documents

User calculates hash value of the retrieved encrypted file and compares the value with hash of received encrypted data. Suppose owner's hash of encrypted file as well as user's hash of received encrypted file is found to be equal, means file is not modified during its traversal. Key k1 can be used by user for decryption of encrypted data.

4   Implementation Results

The proposed security model is implemented on Openstack and analysis is made. Openstack supports security of the model and is used to provide the following services:

  (i)   Identity registration: When a user is registered with the client, User ID and private key for access is also generated for user to be used in the particular session alone.
 (ii)   Data encryption and Data decryption: Data is encrypted by client and sent to cloud. User retrieves with private key shared by client.
(iii)   Keyword based search: To make retrieval efficient, multiple or single keywords can be used to locate the specific file in cloud by user (Table 1).

Table 1. Comparison of results obtained from data security model

Files	HECC encryption time (msecs)	HECC decryption time (msecs)	Parallel HECC encryption time (msecs)	Parallel HECC decryption time (msecs)	File retrieval time based on keyword search (msecs)
F1	1810	26	15	8	1598
F2	1591	22	16	8	1674
F3	998	16	17	7	1389
F4	843	16	15	11	1549

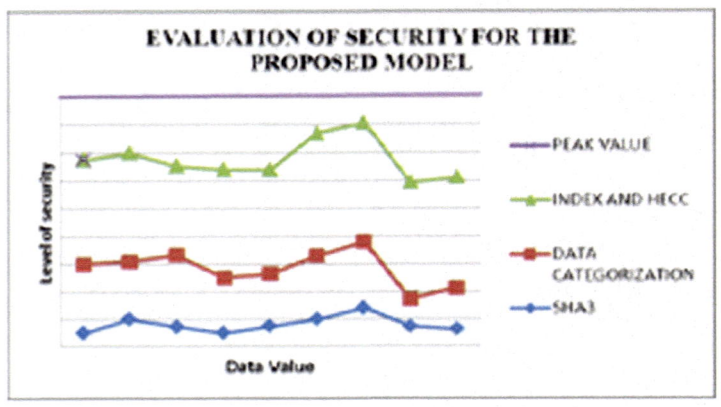

Fig. 1. Evaluation of security for proposed model

HECC is provided as service via web browser and also provided as service for various users. Ease of usage major reason for providing HECC as service. When offered as service, the services provided by the data security model are still easier to work with the GUI screen and is easy to be accessed by many authenticated users. The evaluation of security for the proposed model has been analyzed in Fig. 1. SHA-3 offers less security than data categorization and data categorization offer less security than index and HECC process.

5   Conclusion and Future Work

The proposed security model secures data, verifies data integrity and grants access to only authenticated users by utilization of best mechanisms. Model brings in categorization of data, index building, 80-bit HECC encryption, secure hash algorithm, and authenticating user by owner once and later by cloud and verification of digital signature of owner. Reliability, integrity and availability of data is accomplished while data traverses from client to cloud as well as from cloud to user. The model enables faster data retrieval from cloud by searching on encrypted data and the implementation results also show the efficiency of model. The model offers best data security for data-in-motion. The future work concentrates on providing security to data-at-rest as such data limits the ability of searching and mining data within clouds and the way in which data is being shared with others.

References

1. Julisch, K., Hall, M.: Security and control in the cloud. Inf. Secur. J. Glob. Perspect. **19**(6), 299–309 (2010)
2. Overby, E., Bharadwaj, A., Sambamurthy, V.: Enterprise agility and the enabling role of information technology. Eur. J. Inf. Syst. **15**(3), 120–131 (2006)
3. Subashini, S., Kavitha, V.: A survey on security issues in service delivery models of cloud computing. J. Netw. Comput. Appl. **34**(1), 1–11 (2011)
4. Juels, A., Burton, J., Kaliski, S.: PORs: proofs of retrievability for large files. In: Proceedings of CCS 2007, pp. 584–597 (2007)
5. Shacham, H., Waters, B.: Compact proofs of retrievability. In: Proceedings of Asiacrypt 2008, 5350, pp. 90–107 (2008)
6. Bowers, K.D., Juels, A., Oprea, A.: Proofs of retrievability: theory and implementation. Cryptology e-Print Archive. Report 2008/175 (2008a)
7. Bowers, K.D., Juels, A., Oprea, A.: HAIL: a high-availability and integrity layer for cloud storage. Cryptology e-Print Archive. Report 2008/489 (2008b)
8. Wang, C., Wang, Q., Ren, K., Lou, W.: Ensuring data storage security in cloud computing, quality of service. In: 2009, IWQoS IEEE 17th International Workshop, pp. 1–9 (2009)
9. Prasad, P., Ojha, B., Shahi, R.R., Lal, R.: 3-dimensional security in cloud computing. Comput. Res. Dev. (ICCRD) **3**, 198–208 (2011)
10. Kamara, S., Lauter, K.: Cryptographic cloud storage. In: Lecture Notes in Computer Science, vol. 6054, pp. 136–49 (2010)
11. Wang, C., Cao, N., Li, J., Ren, K., Lou, W.: Secure ranked keyword search over encrypted cloud data. J. ACM **43**(3), 431–473 (2010)

12. Popa, R.A., Lorch, J.R., Molnar, D., Wang, H.J., Zhuang, L.: Enabling security in cloud storage SLAs with cloudproof. Technical report. Microsoft Research (2010)
13. Sood, K.A.: A combined approach to ensure data security in cloud computing. J. Netw. Comput. Appl. **35**(6), 1831–1838 (2012)
14. Devi, T., Ganesan, R.: A novel security model to secure client data in cloud computing. Int. J. Appl. Eng. Res. **9**(21), 11727–11738 (2014)
15. Abdalla, M., Bellare, M., Catalano, D., Kiltz, E., Kohno, T., Lange, T., Malone-Lee, J., Neven, G., Paillier, P., Shi, H.: Searchable encryption revisited: consistency properties, relation to anonymous IBE, and extensions. J. Cryptol. **21**(3), 350–391 (2008)
16. Li, J., Wang, Q., Wang, C., Cao, N., Ren, K., Lou, W.: Fuzzy keyword search over encrypted data in cloud computing. In: Proceedings of IEEE INFOCOM (2010)
17. Golle, P., Staddon, J., Waters, B.: Secure conjunctive keyword search over encrypted data. In: Proceedings of Applied Cryptography and Network Security, pp. 31–45 (2004)
18. Katz, J., Sahai, A., Waters, B.: Predicate encryption supporting disjunctions, polynomial equations, and inner products. In: Proceedings of 27th Annual International Conference Theory and Applications of Cryptographic Techniques (EUROCRYPT) (2008)
19. Lewko, A., Okamoto, T., Sahai, A., Takashima, K., Waters, B.: Fully secure functional encryption: attribute-based encryption and (hierarchical) inner product encryption. In: Proceedings of 29th Annual International Conference Theory and Applications of Cryptographic Techniques (EUROCRYPT '10) (2010)
20. Shen, E., Shi, E., Waters, B.: Predicate privacy in encryption systems. In: Proceedings of Sixth Theory of Cryptography Conference Theory of Cryptography (TCC) (2009)
21. Wang, C., Cao, N., Li, J., Ren, K., Lou, W.: Enabling secure and efficient ranked keyword search over outsourced cloud data. IEEE Trans. Parallel Distrib. Syst. **23**(8), 1467–1479 (2012)
22. Devi, T., Ganesan, R.: Data security in cloud computing: a survey. Aust. J. Basic Appl. Sci. **8**(17), 622–639 (2014)

A Non-Adjacent Form (NAF) Based ECC for Scalar Multiplication that Assure Computation Reduction on Outsourcing

R. Menaka[✉] and R. S. D. Wahida Banu

Anna University, Chennai, Tamil Nadu, India
menaka.murugesan@gmail.com

Abstract. Sustaining privacy and securing the various data sources using lightweight cryptosystem is becoming essential in this digital world. Any sensitive information or resource container in the web that are exposed in the form of url must be preserved using suitable cryptographic techniques. One of the technique that belong to public key cryptography and truly identified for providing security towards digitial resource is Elliptical Curve Cryptography (ECC). ECC is one of the small key sized and fast computation cryptographic model that consumes less power, memory and bandwidth to ensure equivalent security than already existing one. Digital world engaged in on-line data security is the most pressing issues to transaction, have to trust the data disclosure in storage, ensure modification immunity in transit and assures identity among the communicating parties. Hence, the CIA packed computing paradigm that impact towards confidentiality, integrity and authentication is an urgent need. The first one and most important security traits is the confidentiality. This can be achieved by encryption operation. In ECC, encryption and decryption operation is carried using Elliptical curve arithmetic mechanism. This mechanism is based on point multiplication that is performed using repeated point addition and point doubling. The proposed alternative efficient method for point multiplication is binary form or NAF (Non-Adjacent Form). The computation performance with NAF preprocessed and on-the-fly approach ensures optimisation than the binary form of scalar multiplication.

Keywords: Cryptographic system · Factorization ·
Discrete logarithmic problem · Finite field · Elliptic curves ·
Scalar multiplication · Non-Adjacent Form

1 Introduction

Now the arriving computing technology grasp the space or power of the interconnected resource as a single entity there by gaining more than terabyte space requirements or the processing power of a supercomputer for a fraction of the cost respectively. The user running an application which is the collection of subnet's resources belong to different domains, requires assurance of the machine retaining its integrity and disclosure, to ensure that proprietary application remains safe.

© Springer Nature Switzerland AG 2020
A. P. Pandian et al. (Eds.): ICCBI 2018, LNDECT 31, pp. 319–326, 2020.
https://doi.org/10.1007/978-3-030-24643-3_38

1.1   Elliptic Curve Cryptography

Standard and products release applied for industrial, business, commercial and web application have to provide on fly secure electronic transaction by the support of cryptographic primitive's operation like encryption, decryption, and digital signature. The cryptographic system is classified as symmetric and asymmetric one. The major advantages of symmetric cryptography is highly faster but inefficient in key distribution, key management and provision of non-repudiation feature. Public Key Cryptography (PKC) solve the problem of key distribution associated with secret (symmetric) key cryptography. One of the elegant techniques among PKC is ECC (Elliptical curve cryptography) which is faster than RSA (Riverst, Sharmir and Adleman). 160 bit ECC key size grants same security level as 1024 bit RSA key size provides. Cryptographic algorithm is based on mathematical function through which complexity based on computationally and unconditionally secured framework can be outstretched. The specific mathematical problem like factorization and discrete logarithmic problem tailored for RSA and ECC respectively.

Hard Problem that cryptography holds unconditional and computational infeasible secure are Integer factorization problem: Finding the prime factor from the given number, Discrete logarithm problem: Finding x from given n, g, h where h = gx mod n and Elliptic curve discrete logarithm: Finding x such that Q = xP, if given E, elliptic curve and its points P and Q points on that curve [7].

Algorithms that suit these problem are Integer factorization is RSA algorithms, Discrete logarithm is Diffie Hellman(DH) DSA and ElGamal algorithms and Elliptic curve discrete logarithm is ECDH and ECDSA. Best algorithms known for solving the mathematical problem like integer factorization, discrete logarithm and elliptical curve discrete logarithm is number field sieve and pollard rho algorithms.

Elliptical curve Cryptography is public-key cryptography centered on the mathematics of elliptic curves. ECC depends on elliptic curve discrete log problem and it is harder for predicting uncertainty than factoring integer. It can be applied to any smart device that always engaged with data transmission due to its low power consumption and bandwidth and memory savings. The strain issues required to break a private key must tally with the needs to break the PKC used for key establishment. Due to probable advances in cryptanalysis and vast computing power availability, expansion in key sizes of symmetric and asymmetric is essential to the notable level to offer security for a life span protection. The ECC uses finite field Fp (p is the prime number) rather than real number which makes the task slow and inaccurate due to the decimal places. The finite field Fp uses the element of integer from 0 to $p - 1$. To secure cryptosystem, outsized prime number p is chosen so that there are the finitely more points. Some algorithm and protocols uses number to represent messages but our chosen algorithm based on elliptical curve cryptography (ECC) which uses points on the elliptical curve to represent the letter of the messages.

1.2   Finite Fields

The set of finite number q that is power of prime number applied to addition and multiplication operation is called as finite field Fq. In finite field, Fq with q = p called as prime finite fields and q = 2m is called as characteristic 2 finite field, where p is the odd prime number and m >= 1.

The prime number p is chosen such that its binary bit size belongs to the set {112, 128, 160, 192, 224, 256, 384, 521}. Adapting to this restriction and recommended elliptic curve domain by U.S. government will sustenance interoperability and sustain its security level with implementation support in terms of computation and communication.

1.3   Group Definition

A group <G, *> is a set G together with an operation * satisfying the following conditions:

1. Closure: For all a, b ∈ G, a * b ε G
2. Associativity: For all a, b, c ∈ G, a*(b*c) = (a*b)*c
3. Identity: There exists I ∈ G, such that for all a ε G, a*I = I*a = a
4. Inverse: For every a ∈ G, there exist a unique element b ∈ G, such that a*b = b*a = I

A group is an abelian group which must satisfy the above condition and commutative property (i.e.) for all a, b ∈ G, a*b = b*a.

1.4   ECC Cubic Curves Formation with Real Number and Finite Fields

Elliptic curves are cubic curves of the form $y^3 = x^3 + ax + b$, Elliptic curves over R, the real number E(R) is well-defined by the set of points (x, y) along with a point $\boldsymbol{O}$, the point at infinity by satisfy the equation $y^3 = x^3 + ax + b$.

A finite field F_p elliptic curve $E(F_p)$ is well-defined by choosing a, b ∈ F_p as parameter being satisfied with relation $4a^3 + 27b^2 \neq 0$ used to generate the set of points on $E(F_p)$ including point $\boldsymbol{O}$, which is the point at infinity by satisfy the equation $y^3 = x^3 + ax + b$. The addition over $E(F_p)$ involves the computation like one inversion, one squaring, two multiplications and six additions. The doubling a point on $E(F_p)$ involves the computation like one inversion, two squaring, two multiplication, and eight additions.

A finite field F_2^m elliptic curve is defined by choosing a, b ∈ $E(F_2^m)$ being satisfied with relation $4a^3 + 27b^2 \neq 0$ and $b \neq 0$ used to generate of the set of points (x, y) ∈ $E(F_2^m)$, satisfying the equation $y^2 + xy = x^3 + ax + b$. The addition over $E(F_2^m)$ involves the computation like one inversion, one squaring, two multiplications and eight additions. The doubling a point on $E(F_2^m)$ involves the computation like one inversion, one squaring, two multiplication, and six additions. Generally, order of a point P on the curve is the smallest integer d such that $dP = \boldsymbol{O}$ or number of points on the curve can be called as curve order and is denoted #E. Additionally, if s and t are integers, then sP = tP iff $c \equiv d \pmod{r}$. At earlier stage, elliptic curves groups over real numbers was preferred, but using real numbers make arithmetic work very slow and inaccurate due to round-off error. Next to this, elliptic curves groups are studied with the underlying fields of Fp (where p is a prime) and F_2^m (a binary representation with 2m elements). An elliptic curve can be drawn using the set of points satisfying an equation $y^2 = x^3 + ax + b$ and point at infinity where x and y are coordinates of the graph and the coefficients a and b are constant. The value of a and b should satisfy the equation $4a^3 + 27b^2 \neq 0$. Each value of the 'a' and 'b' gives a different non-singular elliptic curve [10, 11].

2 Scalar Multiplication

Given points S and T in a elliptic group [3, 8, 9, 12], finding the value x (x is an integer) such that S*x = T in the elliptic curve discrete logarithm problem is very hard. The logic behind the computationally infeasible to solve is multiplying the point with some number is easy but finding the multiplied number given the result and original point is hard. So, among any public key scheme, ECC offers the most security per bit.

Point Addition: The elliptic curve E (Fq) defined with addition operator can form an abelian group under addition.

1. Let $P(x_p, y_p)$ and $Q(x_q, y_q)$ be the points on the elliptic curve E(Fq) and point P! = Q then addition of points P and Q is $R(x_r, y_r)$ where $\Lambda = (y_q - y_p)/x_q - x_p)$, $x_r = \Lambda^2 - x_p - x_q$ and $y_r = \Lambda(x_p - x_r) - y_p$.
2. Adding the point $S(x_s, t_s)$ with ∞ will yield $S(x_s, t_s)$
3. Adding the point at infinity ∞ to itself gives: $\infty + \infty = \infty$

Point Doubling: Let $P(x_p, y_p)$ be the point on the elliptic curve E(Fq) then applying the point doubling operation can yield $R(x_r, y_r) = 2P(x_p, y_p) = P(x_p, y_p) + P(x_p, y_p)$ that lays on the same elliptic curve where $x_r = \Lambda^2 - 2x_p$, $Y_r = \Lambda (x_p - x_r) - y_p$ and $\Lambda = (3x_p^2 + a)/2y_p)$.

Comparing two or more quantities of same kind is called as ratio. Here the key size ratio of ECC and RSA is given as 1:6 for 163 and 1024 bits & 1:30 for 512 and 15360 bits respectively. In terms of mathematical operation ECC do addition and multiplication operation with points addition and multiple addition points, whereas RSA have modular multiplication and exponentiation respectively.

2.1 Elliptic Curve Cryptography (ECC) Domain Parameters

It determines the arithmetic operations involved in the chosen public key cryptographic systems defined on Elliptic curve over finite fields.

The domain parameters D over F_q contains the following tuple:

- **q is** power of p, prime number such as q = p or $q = 2^m$
- **FF**: finite field element $\in F_q$
- **a, b**: parameters that fixes the different elliptic curve E with $y^2 = x^3 + ax + b$ over F_q,
- **G** = (x_g, y_g) on E (F_q) is a base point
- **n**: Order of point G such that nG = **O**
- **h**: cofactor = #E(F_q)/n, where #E(F_q) is the curve order

The parameter n is the primary security in ECC, so the length of ECC key is the bit length of n.

2.2 Elliptic Curve Protocols

Generally in the process of encryption and decryption, we have 2 entities, the one at the encryption side and the other at the decryption side. Let us assume that Response A is the person who is encrypting and Request B is the person decrypting.

Key Creation: (A's & B's) public and private keys are related with elliptic curve domain parameters (q, FF, a, b, G, n, h).

Response A generates the asymmetric public and private keys as follows

1. Select a random number r_A, $r_A \in [1, n - 1]$.
2. Compare $pub_A = r_A * G$.
3. Response A's public key is pub_A and private key is r_A.

A public key pub_A related with domain parameters (q, FF, a, b, G, n, h) is validated by confirming it lies within nG.

2.3 Elliptic Curve Diffie-Helman Protocol (ECDH)

ECDH is Diffie-Hellman key agreement protocol using elliptical curve points for cipher operation. The shared secret key is generated using ECC is described as below.

- Sender A chooses a point Q and k_a as random number. Then computes point $P = k_a Q$ and sends it to other side member B. Now P and Q are public one.
- Receiver B choose a random number k_b and computes point $Q_b = k_b Q$ and sends it to A.
- A computes $K_1 = k_a Q_b$ and B computes $K_2 = k_b P$
- $K_1 = K_2 = k_a k_b Q$, is used as the shared secret key

Elliptical curve arithmetic operation depends on point multiplication that is performed using repeated point addition and point doubling. It's like double and add method for point multiplication. An alternative efficient method for point multiplication is binary form or NAF (Non – Adjacent Form).

Binary Method for computing scalar multiplication: **'k' is number of binary bit of number 'n'** $(k_{n-1}, k_{n-2}, \ldots k_0)_2$, $k_n = 1$, Q&P are point in the curve. Perform $Q = kP$ (Table 1).

Signed binary method for scalar multiplication:

Booth [5] proposed signed binary method for scalar multiplication. This method have the property to make any two consecutive digits to at most one as non-zero. In kP every k is represented to the summing form of $kj2^j$ where $j = 0$ to $l-1$ (l = length of binary bits in k) and kj can be -1, 0, 1 which makes any two consecutive digit as no nonzero. This making of no non zero is called as non adjacent form, NAF [4] (Tables 2, 3 and Fig. 1).

Elliptical curve arithmetic is based on doubling and additions operation. If point addition operation (P + Q) to be carried out in prime field, it have to perform one inversion and squaring with two multiplication and six addition [1]. For point doubling operation one inversion and two squaring and multiplication with eight addition is expected [2]. Over prime fields of 160-bit number the operation needed for binary method are inversion for 160 times, squaring for 320 times and multiplication for 880 times and for signed binary method inversion reduced to 54 times and multiplication by 189 times NAF method.

Table 1. Computing scalar multiplication using LSB and MSB

Scalar Multiplication: MSB first	Scalar Multiplication: LSB first

ScalarMSBMult(){
 Q=P;
 for (i=n-2, i>0, i--) {
 Q=2*Q;
 If k_i==1 then Q+=P ;}
 return Q;}
	ScalarLSBMult()
	 {Q=0 & R=P;
	 for (i=0, i<=n-1, i++) {
	 If k_i==1 then Q+=R;
	 R=2*R;}
	 return Q;}

it require 'n' point doublings & (n-1)/2 point additions Logic: do doubling first then do adding Example if k=6 and $(110)_2$,kP=6P then 6P=2(2(P)+P) (i.e., for bit 1- doubling and adding ,bit 0 - doubling alone)	**It require n/2** point doublings & point additions Logic: doubling & adding parallelized.

Table 2. Computing NAF

| NAF (k){
 i=0;
 while ($k \geq 1$){
 If k%2==0
 $k1[i]_i$=2 - (k mod 4)
 k =k – k1[i];
 else <br> k1[i]=0; }
 k= k/2;
 i++;
 return k1; } | $(996)_2$ is $(1111100100)2==2^9+2^8+2^7+2^6+2^5+2^2$
 The hamming weight (i.e.) number of bit with value one is 6
 NAF(996)is $(10000-100100)= = 2^{10}- 2^5 +2^2$
 Now the hamming weight (i.e.) number of bit with value one is 3
 The hamming weight is 3 |

Table 3. Point Multiplication (Two approaches)

Point Multiplication or addition – subtraction method using NAF on-the-fly processed approach	Point Multiplication using NAF precomputed approach

PointmultNAF(k1[])
{
R=P
for(i= n −2;i>=0;i--)
{
R=2R;
if k[i] == 1 then R=R+P;
if k[i] == −1 then R=R-P;
}
return R;
This method performs (n-1) doublings and (n-1)/3 additions in an average[6].
	Step 1: partition the input into blocks of equal size like 8 bit each
	Step2: for insufficient block's append with padding zero
	Step3: without any computation substitute the NAF value in predefined look up table for each block
	Step 4: combine them to get a result

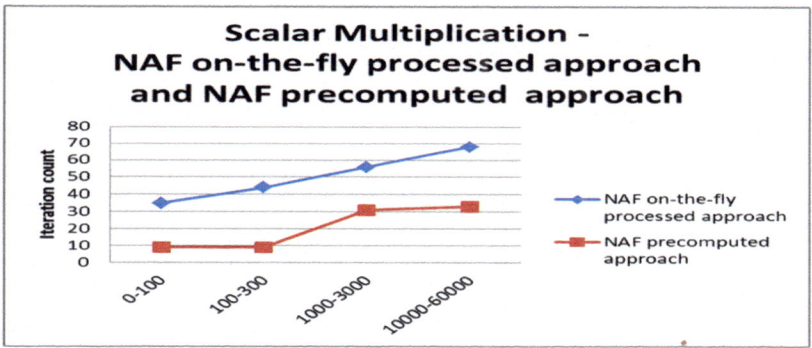

Fig. 1. NAF by on-the-fly and precomputation approach

3   Conclusion

ECC is one of the small key sized and less computation cryptographic model that consumes less power, memory and bandwidth but provide equivalent security ensuring faster computation. The ECC encryption operation uses elliptical curve arithmetic mechanism. This mechanism is based on point multiplication that is performed using repeated point addition and point doubling. But, an alternative efficient method for point multiplication is binary form or NAF (Non-Adjacent Form). Comparing with binary form NAF approach is the methods that ensure the optimisation. Then the NAF method is carried out with pre-processing approach and on-the–fly process approach to determine the improvement. As NAF pre-processing approach support parallelism it

computation time is reduced and at the same time inversion operation computing
burden reduced by 1/3 than that of on-the–fly process approach.

References

1. Karthikeyan, E.: Survey of elliptic curve scalar multiplication algorithms. Int. J. Adv. Netw. Appl. **4**, 1581–1590 (2012)
2. Balasubramaniam, P., Karthikeyan, E.: Elliptic curve scalar multiplication algorithm using complementary recoding. Int. J. Comput. Appl. **20**, 37 (2011)
3. Geetha, G., Jain, P.: Implementation of matrix based mapping method using elliptic curve cryptography. Int. J. Comput. Appl. Technol. Res. **3**, 312–317 (2015)
4. Brar, H., Kaur, R.: Design and implementation of block method for computing NAF. Int. J. Comput. Appl. **20**, 37–41 (2011)
5. Booth, A.D.: A signed binary multiplication technique. J. Appl. Math. **4**(4), 236–240 (1951)
6. Sawlikar, A.: Point multiplication methods for elliptic curve cryptography. Int. J. Eng. Innov. Technol. **1**, 1–4 (2012)
7. Malik, M.Y.: Efficient implementation of elliptic curve cryptography using low-power digital signal processor. ICACT **2**, 1464–1468 (2010)
8. Koblitz, N.: Elliptic curve cryptosystems. Math. Comput. **4**, 203–209 (1987)
9. Priyadharshini, M., Suganya, I., Saravanan, N.: A security gateway for message exchange in services by streaming and validation. Int. J. Innov. Res. Comput. Commun. Eng. **1**(3), 604–612 (2013)
10. Moosavi, S.R., Nigussie, E., Virtanen, S., Isoaho, J.: An elliptic curve-based mutual authentication scheme for RFID implant systems. Procedia Comput. Sci. **32**, 198–206 (2014)
11. Muthukuru, J., Sathyanarayana, B.: Implementation of elliptic curve digital signature algorithm using variable text based message encryption. Int. J. Comput. Eng. Res. **2**(5), 1263–1271 (2012)
12. Chou, D.C., Yurov, K.: Security development in web services environment. Comput. Stand. Interfaces **27**(3), 233–240 (2005)

Secure and Energy Efficient Based Healthcare Data Transmission Wireless Sensor Network

R. Sivaranjani[1($\boxtimes$)] and A. V. Senthil Kumar[2]

[1] Department of Computer Technology, Hindusthan College of Arts and
Science, Coimbatore, India
ranju_rs@yahoo.com
[2] Department of Computer Application, Hindusthan College of Arts and
Science, Coimbatore, India
avsenthilkumar@yahoo.com

Abstract. Present days the wireless device network are utilizing in wide assortment of uses. The remote device hubs are battery burning within the bigger a part of the applications notably in prosperity checking frameworks. Use of those battery controlled remote device hubs with insignificant vitality utilization is that the bottleneck in remote device organizes primarily based prosperity checking frameworks. Arrangement of considerable variety of those battery controlled device hubs during a domain and maintaining the remote device organize for long life may be a basic assignment. Same like security in addition exceptionally basic trip for wsn. The life of wireless device network enlarged by utilizing many power economical ways for military operation and getting ready. A tale cluster primarily based imperative prosperity military operation, handling and directive. The wide remote device organize secured by a base station is split into many teams with bunch head steering info to/from base station. Every bunch comprising of many body zone systems with body heads that are to blame of directive info to/from cluster head. In this paper, formation of bunch heads and body heads and deputation uncommon errands to cluster heads and body heads will tremendously boost generally framework ability, lifetime, and vitality proficiency. This paper introduces a safe, vitality effective and versatile security conspire called Mobility-Supporting Adaptive Authentication Scheme (MAAS). The MAAS conspire guarantees that the Healthcare Provider and the Patients access and trade patients information by confirming themselves. Likewise, the plan forestalls unapproved access to the information by giving secrecy and uprightness. With this framework design, we can enormously limit the normal power devoured by a remote sensor organize.

Keywords: Wireless sensor network · Adaptive Authentication Scheme ·
Security · Power

1 Introduction

Wireless Sensor Network (WSNs) comprise of sensor hubs. These WSNs have extraordinary applications in living space checking, calamity administration, security and military, and so on. Remote sensor hubs are little in size and have constrained

A. P. Pandian et al. (Eds.): ICCBI 2018, LNDECT 31, pp. 327–333, 2020.
https://doi.org/10.1007/978-3-030-24643-3_39

preparing ability, low battery control. Information total is exceptionally pivotal strategy in WSNs. With the utilization of information conglomeration we lessen the vitality utilization by wiping out excess. Information conglomeration is the way toward gathering and accumulating the helpful sensor information. Information conglomeration is considered as one of the major preparing strategies for sparing the vitality. In WSN information total is a powerful method to spare the restricted assets. The principle objective of information accumulation calculations is to assemble and total information in a vitality productive way with the goal that system lifetime is expanded. WSNs have constrained computational power, restricted memory and constrained battery control; this prompts expanded unpredictability for application designers and regularly results in applications that are firmly combined with system conventions. The most critical objective in the plan of wireless sensor network is augmentation of system life time, obliged by vitality limit of batteries (Fig. 1).

A nonexclusive Wireless Sensor Networks comprises of an extensive number of asset obliged sensor hubs that are spatially disseminated in an antagonistic situation and the with asset rich hub called as the Base Station (BS). The sensor hubs undertaking is to detect physical wonders from its quick environment, process and transmit the detected information to alternate hubs or Base stations. WSNs are utilized in applications that are delicate to natural parameters that require observing, following and controlling. A standout amongst the most encouraging fields for WSN is the Healthcare. Nonexclusive WSN should be altered to take into account the necessities continuous Healthcare Application as for sending, bolster portability of patients and furthermore in giving solid correspondence of patients information as there is an extraordinary danger of the patient security to be broken by the aggressors The Healthcare WSN application is powerless to various sorts of assaults and dangers. The assault can be on the patient hubs: in which the aggressor's intension is to upset the sensors detecting the physiological information of the patients or the medicinal services supplier framework: for this situation of assault the intension of the assailant to disturb the connection between the social insurance suppliers, delegate hubs and patient hubs. Security being a prime worry in medicinal services WSN, the center prerequisite is an answer that means to stifle these assaults and meet the security objectives. Security in Healthcare Application of WSN has pulled in consideration of numerous scientists. In writing numerous security plans are proposed however these plans neglect to address of portability of patients. In this paper, we propose a Mobility-Supporting Adaptive Authentication Scheme Based Secure Health Monitoring in Wireless Sensor Networks to keep the unapproved access of patient's information (Fig. 2).

2 Wireless Body Sensor Model Design

Here, have a tendency to be referring a personality's being with associate object. Has a tendency to be wondering every object with a couple of remote sensing element hubs for police investigation the indispensable successfulness signs. The work of typical sensing element hubs in associate object is police investigation the separate elementary successfulness signs and spending the detected info to protest head. During this style, one

Object Head (OH) is accommodated every object. The OH is likewise a sensing element hub which can act as a portal for that protest. The work of OH is gathering success-fulness info from the sensing element hubs of an object and transmission the congregate successfulness info to his higher layer sensing element hub, known as Cluster Head (CH). The paradigm for selecting the OH among a couple of remote sensing element hubs is accessibility of a lot of leftover vitality and better handling ability (Fig. 3).

The movableness information of OHs is place away within the base station info base. Visible of the movableness information, the bottom station starts the thing heads to border co-agent teams with the neighbor object heads by utilizing the neighbor revelation convention. The hubs within the co-agent bunch can share their assets like power and handling capacities among themselves. Among a number of device hubs, one hub can move as CH. The hub that has additional remaining vitality and high getting ready capability during a bunch is chosen as a CH. The CH is likewise a device hub that has further duty of passing destroyed detected well-being info from object heads during a cluster to the bottom station over a cell connect. The CH can empower each cell connect and conjointly short-term. The remainder of the device hubs during a bunch can empower simply short-go interfaces by keeping cell connect in power off mode with the goal that the ability eaten up by these device hubs is a smaller amount. There'll be a number of co-agent bunches during a cell inclusion region. The sensor's detected info is gathered by Buckeye State over short-extend joins, at that time the Buckeye State transmits the destroyed protest info to CH over short-term joins and therefore the CH transmits the gathered device info to the bottom station over cell interface. The bottom station is related to the focal info handling unit that has the capacities of getting ready advanced info. The handling ability and plus accessibility of device hubs are going to be fresh at the bottom station info base. All the device hubs in object can share the remaining plus information to the Buckeye State, and after the Buckeye State can opt for the subsequent Buckeye State. The subsequent Buckeye State may well be merely the current Buckeye State if the lingering assets at the current Buckeye State are superior to subsequent device hubs of object. Correspondingly, an identical calculation can work for selecting the CH furthermore. In this work have a tendency to be maintaining all of the knowledge getting ready at the bottom station and activity the handling bother at the device hubs and with this, scotch force at the hubs. Though, the customary wireless device network pursues multi bounce correspondence to course the detected info to the focal getting ready unit. This makes house for a number of security problems. There security issue we have a tendency to provide gave the technique AAS placed in wsn.

By and enormous the hubs within the remote sensing element system are within the inert state if there's no transmission or gathering, but standardization and sitting tight for information from totally different hubs. Despite the actual fact that there's no detected info to transmit by the remote sensing element hub, the hub should be in standardization in or holding up mode. These outcomes in avid a lot of power by the remote sensing element hubs and increasing the conventional power eaten by the final remote sensing element prepare furthermore. Be that because it could, in our frame-work engineering, the bigger a part of the remote sensing element hubs are in pure inert mode if no transmission. These recoveries spectacular live of intensity and expand the system time period.

There are a couple of situations our framework design builds the system lifetime and stays away from loss of important wellbeing information. Expect in the primary situation, the remote sensor hub control is less to transmit the significant wellbeing information to the higher layer remote sensor hub. For this situation, the higher layer sensor hub prescribes the neighbor hub with high assets to co-work with poor power hub and transmit his detected information to the goal. In the second situation, the remote sensor hub is far from the inclusion territory of his higher layer hub. At that point our framework prescribes the correspondence connect fizzled hub to utilize the neighbor hubs with high assets and great correspondence connect to transmit detected information to the goal.

3   Adaptive Authentication Scheme (AAS)

In this Scheme, we consider a heterogeneous Wireless Sensor Network Model with Hierarchy. The primary parts include: 1: Body Sensor Nodes (BSN)–Responsible for gathering the physiological signs from the patient and may incorporate sensors, for example, Blood Pressure Sensor, Heart Rate Sensor, Location Sensor. These sensors are embedded in/on the patient's body. 2: Individual Coordinator Nodes (ICN)- Responsible for accumulation of and total of the considerable number of information detected by the sensor hubs embedded on/in the body of the patient 3: Sink Nodes (SN)—A gathering of ICNs together are doled out to a Sink Node, 4: Secure Base Station (SBS) 5: Healthcare Alert and Service System (HASS).

Fig. 1. A simple healthcare monitoring system

The sensor hubs embedded on the body of the patient detects the physiological elements and which are totaled by the Individual Coordinator Node. These ICN forward the accumulated information to one of the sinks in the system which is then sent to the HASS through the BS. The HAAS will get all the totaled information from the BS and after that store and decipher the information got. In the event that the patient needs a quick consideration the framework cautions the Doctors and Nurses who have approved themselves to the HAAS.

- Sink knows the general population key of the BS
- BS knows the IDs and the Public Keys of the Sinks
- An ICN can be associated with a solitary SN anytime.

Sink List at BS (SLB): a vector that stores data/Credentials about all the Sink Nodes in the BS.

Neighbor Sink List (NSL): a vector that stores data/Credentials about all the Neighboring Sinks for all Sink Nodes.

Neighbor ICN List (NIL): a vector that stores data/Credentials about all the Neighboring ICN in the scope of a SN.

Stage 1: The SN creates a HELLO message and afterward communicates it.

Stage 2: If the HELLO message is gotten by the SN at that point go to Step 3 else whenever gotten by ICN go to Step 4.

Stage 3: The SN that got the HELLO message checks for nearness of the Sender SN in its NSL, if NSL as of now contains a passage of the Sender SN then the recipient basically disposes of the message; else it confirms the Sender SN and updates its NSL.

Stage 4: when an ICN gets the HELLO message from the SN, the ICN checks whether it is connected to a SN. In the event that it isn't connected to the SN, ICN verifies itself to the SN and the SN refreshes it's NIL, disregard the message if the ICN is as of now confirmed to the SN. When the hub is portable, the ICN is as of now validated by a SN already then basically re-validates the ICN to the new SN else overlooked.

4 Result

Various group heads are greatly affecting on the vitality effectiveness of WSNs. As the quantity of bunch heads expanded, the energies are seriously expended because of the substantial quantities of the conglomeration forms performed by these group head hubs. Then again, as the quantity of bunch head hubs is limited, the energies are additionally extremely expended because of the massive measure of information collected by each group head hub and longer period time, each bunch go to speak with BS to present the mass accumulated information. In this way, these CHS will dead prior.

Vitality utilization implies the aggregate vitality devoured by the system to perform transmission, gathering and information accumulation. The correlations performed among the diverse methodologies dependent on the vitality utilization in both group head sensor hubs and bunch part sensor hubs. For them two, the recreation is running on the quantity of hubs break even with 100, 200 and 300.

In this examination, existing and proposed convention are thought about based on the normal expended vitality per parcel as appeared in Figure It is found from our perceptions that proposed at first shows great execution for normal vitality per bundle analyzed.

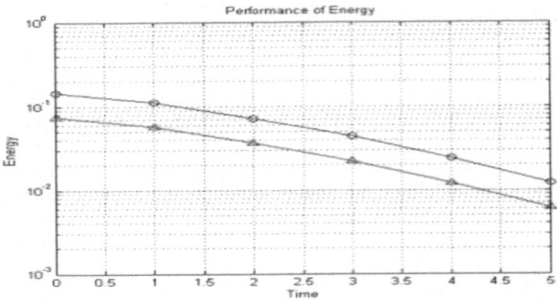

Fig. 2. Comparative Analysis on per Packet Average Consumed Energy for data transmission for wireless sensor network. Here compared to the existing methods, our proposed methods use low energy and time taken to transform the data.

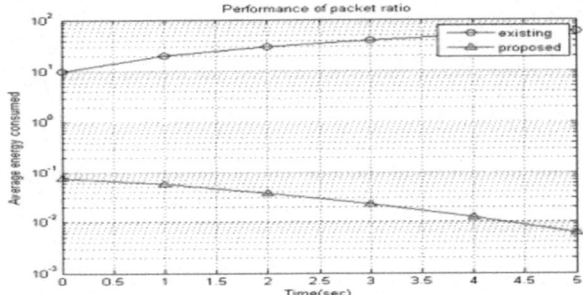

Fig. 3. Average consumed energy per packet transmission for existing and proposed methods compared to existing system proposed method send packets less time and using average energy usage.

5   Conclusion

The wireless device network features a few unbelievable applications in Health watching Systems. The vitality adept device data accumulation system is provided with cluster primarily based remote device systems. In this framework engineering, the conventional power eaten by a far off device hub is a smaller amount and therefore the time period of the remote device arranges is swollen. They tend to be giving the safe transmission of accumulated sensor's eudaemonia data to the goal for examination, they tend to propose Adaptive Authentication Scheme (AAS), and therefore the set up provides productive hub verification and hub re-validation for device Nodes.

References

1. Milenković, A., Otto, C., Jovanov, E.: Wireless sensor networks for personal health monitoring: issues and an implementation. Comput. Commun. **29**(13–14), 2521–2533 (2006)
2. Al Ameen, M., Liu, J., Kwak, K.: Security and privacy issues in wireless sensor networks for healthcare applications. J. Med. Syst. **36**(1), 93–101 (2012)
3. Thota, C., Sundarasekar, R., Manogaran, G., Varatharajan, R., Priyan, M. K.: Centralized fog computing security platform for IoT and cloud in healthcare system. In: Exploring the Convergence of Big Data and the Internet of Things, pp. 141–154. IGI Global (2018)
4. Firouzi, F., Rahmani, A.M., Mankodiya, K., Badaroglu, M., Merrett, G.V., Wong, P., Farahani, B.: Internet-of-things and big data for smarter healthcare: from device to architecture, applications and analytics, pp. 583–586 (2018)
5. Gilbert, E.P.K., Kaliaperumal, B., Rajsingh, E.B., Lydia, M.: Trust based data prediction, aggregation and reconstruction using compressed sensing for clustered wireless sensor networks. Comput. Electric. Eng. (2018)
6. Chandrasekaran, B., Balakrishnan, R., Nogami, Y.: Secure data communication using file hierarchy attribute based encryption in wireless body area networks, pp. 75–81 (2018)
7. Cheng, L., Niu, J., Luo, C., Shu, L., Kong, L., Zhao, Z., Yu, G.: Towards minimum-delay and energy-efficient flooding in low-duty-cycle wireless sensor networks. Comput. Netw. **134**, 66–77 (2018)
8. Kalaiselvi, K., Suresh, G.R., Ravi, V.: Genetic algorithm based sensor node classifications in wireless body area networks (WBAN). Cluster Comput. 1–7 (2018)
9. Rahmani, A.M., Gia, T.N., Negash, B., Anzanpour, A., Azimi, I., Jiang, M., Liljeberg, P.: Exploiting smart e-Health gateways at the edge of healthcare internet-of-things: a fog computing approach. Future Gen. Comput. Syst. **78**, 641–658 (2018)
10. Plageras, A.P., Psannis, K.E., Stergiou, C., Wang, H., Gupta, B.B.: Efficient IoT-based sensor BIG data collection–processing and analysis in smart buildings. Future Gen. Comput. Syst. **82**, 349–357 (2018)
11. Stephen, R.K., Sekar, A.C., Dinakaran, K.: Sectional transmission analysis approach for improved reliable transmission and secure routing in wireless sensor networks. Clust. Comput. 1–12 (2018)
12. Rezaei, F., Hempel, M., Sharif, H.: A survey of recent trends in wireless communication standards, routing protocols, and energy harvesting techniques in E-health applications. In: Wearable Technologies: Concepts, Methodologies, Tools, and Applications, pp. 1479–1502. IGI Global (2018)

IOT Based Demand Side Management of a Micro-grid

Leo Raju$^{(\boxtimes)}$, S. Sangeetha, and V. Balaji

Department of Electrical and Electronics Engineering,
SSN College of Engineering, Chennai, India
leor@ssn.edu.in

Abstract. This paper deals with an arduino and IoT based demand side management of a micro-grid. Our micro-grid contains two solar photo voltaic system, wind, local consumer and a battery. All the environment variables are sensed through Arduino microcontroller and given to IoT environment. Using these values, the optimal energy management is implemented in micro-grid. The output commands are given to the microcontroller and the on/off status of LED shows the optimized actions, which can be given to the physical device for actuation. Demand side management and grid outage management is implemented in the microgrid for advanced energy management.

Keywords: Arduino · Miro-grid · IOT · Energy management

1 Introduction

Internet of Things (IOT) is the emerging technology which makes the smart devices smarter and which were manually controlled earlier. All these manually controlled electrical and electronic devices can be controlled automatically using the IOT. IOT'S application will focus on the smart Industrial monitoring and smart manufacturing. In IOT virtual connection between a hub or a network and electronic and electrical objects is established through internet and the things can be sensed and controlled remotedly. Also, the connected object can be continuously monitored and tracked down through the virtual connection through internet. The development of smart sensor and communication technologies, supported by cloud computing technologies, IOT has become reality and it make devices more aware, interactive and efficient for a better living in the world. Although The IOT technologies have evolved over a period of time with sensing, analysing and controlling the devices, only a little work done in applying IOT technologies into embedded devices including customer appliances. Enabling the Internet of Things for the devices and customer appliances is discussed in [1]. They proposed the Internet of Things (IOT) paradigm which enables interconnectedness among devices anytime, anywhere providing the advantages of the internet in daily life. [2] proposed predictive maintenance and adaptive analytics for new business models enabled by the Internet of Things (IOT). This paper deals with shift in markets from selling assets to selling services. Integration of renewable energy resources as well as Demand Side management is discussed in [3]. Here, consumer patterns using Internet of Things in order to implement Demand Side Management is analysed. IoT devices

A. P. Pandian et al. (Eds.): ICCBI 2018, LNDECT 31, pp. 334–341, 2020.
https://doi.org/10.1007/978-3-030-24643-3_40

for safer, faster response with industry IoT controlled operations are also discussed. The Internet of Things (IoT) offers the user seamless interoperability and connectivity between devices, systems, services and the network. [4] elaborates on energy consumption forecasting using smart meters. Trends in micro-grid control is discussed in [5]. A review of residential demand response of smart grid is discussed in [6]. Arduino based energy management of microgrids is discussed in [7]. Design and implementation of arduino based energy management of micro-grids is discussed in [8]. An Efficient Micro-grid Management System for Rural Area using Arduino is discussed in [9]. But IOT and Arduino based energy management is not dealt comprehensively so far. Hence this paper bring forth a comprehensive implementation of demand side management and grid outage management in microgrid to make it into a smart micro-grid with smart grid properties.

2 Energy Management of Micro-grid

2.1 Smart-Grid

The smart grid is the integration of Information and communication technologies (ICT) into evolving electric grid for advanced control and automation. Depletion of fossil fuels and environmental consideration forces the energy industry to embrace renewable energy like solar and wind. But the varying nature of these resources requires advanced monitoring and control techniques. Moreover existing Supervisory Control and Data Aquatition (SCADA) not able to manage when large number of small renewable energy resources are integrated with grid. So ICT in smart grid is inevitable and it makes the grid live and more responsive for energy efficient operations. Decentralisation, Two way communication, smart metering and self healing makes smart grid efficient. This smart grid framework is used in a micro-grid environment to make it to a digitally enabled ecosystem that provides reliable and quality power for all, efficiently, with participation of each of the stakeholders [10]. Because of their flexibility and scalability, micro-grids are likely to play key role in the evolution of smart grid; Smart grid is a system of integrated smart micro-grid. Smart micro-grids are micro-grids combined with an overlaying intelligence scheme. IOT in the smart grid enables real-time monitoring and control in generation, transmission and distribution of electrical power systems. IOT based technologies enable these functionalities for sustainable, reliable and economic use of electricity. Smart grids can provide consumers with real-time information on the use of energy, dynamic pricing information that eventually reflects changes in supply and demand, and enable smart appliances and devices to help optimization of usage of energy.

Micro-grid. Micro-grid is a building block of smart grid. This is low voltage distributed electrical power network comprising distributed energy resources. Micro-grid is used in remote areas and university campuses. It can be operated as a grid connected system or an islanded system [10]. It enhances local reliability, reduce emissions, improve power quality. Lowers the cost of energy supply as there is no transportation losses. Micro-generation increases the overall efficiency of utilizing primary renewable energy sources. By using renewable energy Environmental gains can be obtained [11].

Arduino Micro Controller. Arduino micro-controller is an open-source electronics platform based on easy-to-use hardware and software. It is a compact integrated circuit used for specific operations. İt has its own input, output, processor, memory and peripherals in a single chip. Anolog input pin module are used to sense the environment variables. Digital output pins are used to actuate the devices in the field. A microcontroller is a compact integrated circuit designed to govern a specific operation in an embedded system.

Wi-Fi Module. ESP8266, an open source 8 pin Wi-Fi module is used to access the internet (TCP/IP) using AT (ATtention) Commands or built-in libraries. It gives our Arduino Board, the ability to transfer data to the Internet or receive data from the Internet. ESP 8266 uses 3.3 V logic whereas our Arduino Boards use 5 V Logic. So, in order to interface ESP and Arduino, we need something called the "USB TO TTL CONVERTER", which will easily do the necessary alterations according to the needs of Wi-Fi module. While going for such method, we have to ensure that the ESP gets proper driver current and hence we use the combination of 2.2 K and 1.1 K (1 K + 100 Ω) resistors to step down the voltage to 3.3 V. The Wi-Fi module is used to send and receive data from cloud to any devices [12].

IOT Platform Ubidot. Ubidot is a IoT platform. It is used to send data to the cloud from any Internet-enabled device. It supports a large number of Internet of Things devices. By using ubidot Dashboards, one can monitor and control specific assets and depending on safety requirements, access and control an environment from anywhere in the world [13]. The following steps are used to activate Ubidot platform.

STEP 1: Create a new account using a user name and password.
STEP 2: Then by using an add device option create all the potentiometer fields.
STEP 3: The token created in ubidot is linked to all the sending and receiving codes and also to the MATLAB in order to get the results.
STEP 4: The values are sensed in the sending side of ardunio and sent to the ubidot platform.
STEP 5: Here all the potentiometer values are received and sent back to the ardunio using a receiving code.
STEP 6: Finally, the variation graph can also be noted in ubidot platform

3 Implementation of Energy Management of Micro-Grid

We considered three micro-grids each with two solar units, a wind unit and a battery along with load.

Micro grid 1	Micro grid 2	Micro grid 3
Solar1	Solar1	Solar1
Solar2	Solar2	Solar2
Wind	Wind	Wind
Battery	Battery	Battery
Load	Load	Load

In the energy management of micro-grids the following steps are followed. Totally 15 potentiometers are connected to the analog pins of arduino board for sensing the environment variables from all the three micro-grids. The range of value is fixed as 0–1024 in the potentiometer for all the environment variables. This value can be varied from minimum (0) to maximum (1024) by varying the potentiometer. The environment variable values can be fixed accordingly. For example, if hostel solar power is 50 kW, then 1024 represents 50 kW.

STEP1: Initially a code reads the values of all 15 potentiometers to the Arduino board and is uploaded to Arduino Mega 2560. These potentiometers are used to generate the results in three micro-grids.

STEP 2: After sensing these values, it sends to the Ubidots platform using ESP8266 Wi-Fi module, which is connected to the same Arduino Mega 2560.

STEP 3: The Ubidots platform is created and potentiometer devices are added using a add device option and it exists for all the ardunio processes.

STEP 4: After running the ardunio program the results are monitored in the serial monitor window where, all the potentiometer readings are up-loaded.

STEP 5: After completion of step 4, the values are sensed through Ubidots and sent to the MATLAB 2017a.

STEP 6: Using the received values from Ubidots, program is run and their results are tabulated in MATLAB's command window.

STEP 7: After this, the values are sent to an Ubidots Device and LEDs glow according to the results obtained.

STEP 8: LED Values contains 13 channels indicating the turn on/off status corresponding to 12 sources and grid in the network of micro-grids considered.

STEP 9: Now, an Arduino code that downloads these values from Ubidots is uploaded to a different Arduino Mega 2560 board. This board contains 13 LED's connected to it.

STEP 10: The code then uses the values downloaded to turn the corresponding LED's on/off.

Thus the energy management is implemented in IOT platform.

3.1 Implementation of Demand Side Management

1. The demand on micro-grid 1 is 1000 KWh while the total generation is 850 KWh. Now since the demand is greater than the generation in the micro-grid, all of the sources in the micro-grid are allocated to satisfy the demand.
2. Now, the remaining demand on micro-grid 1 is 150 KWh.

3. Moving to MG2, the demand is 1000 KWh and the total generation is 900 KWh. This is similar to the scenario of micro-grid 1. The remaining demand on micro-grid 2 is 100 KWh. Coming to MG3, the load is 1000 KWh while generation is 950 KWh. The same procedure as described in MG1 applies here. This leaves us with an excess demand of 50 KWh.

4. At this point, MG1 has an excess demand of 150 KWh while MG2 has an excess demand of 100 KWh and MG3 also has an excess demand of 50 KWh.

5. Since all the sources are now exhausted, the remaining total demand of 300 KWh has to be satisfied by the grid.

6. In this scenario, there is a grid outage. Hence the grid cannot supply any power. Therefore, to maintain the supply demand equilibrium Non Critical Load Shedding takes place.

As a result, the excess demand in all the three micro-grids is cut off maintaining the supply demand equilibrium. The micro-grids are taking the excess power of 300 Kw from the grid. During grid outage, it manages the power demand by demand side management. All the excess load is cut off as soon as the grid outage takes place. 150 kW load is cut off in micro-grid one, 100 kW load is cutoff in micro-grid 2 and 50 kW load is cut off in micro-grid 3 (Figs. 1, 2).

```
                    OUTPUT
        Micro-Grid Demand:
        Micro-Grid 1 Demand                : 1000.00KWhr
        Micro-Grid 2 Demand                : 1000.00KWhr
        Micro-Grid 3 Demand                : 1000.00KWhr


        Micro-Grid Generation:
        Solar 1 Generation Micro-Grid 1    : 300.00KWhr
        Solar 2 Generation Micro-Grid 1    : 300.00KWhr
        Wind Generation Micro-Grid 1       : 100.00KWhr
        Battery Generation Micro-Grid 1    : 150.00KWhr
        Solar 1 Generation Micro-Grid 2    : 400.00KWhr
        Solar 2 Generation Micro-Grid 2    : 200.00KWhr
        Wind Generation Micro-Grid 2       : 150.00KWhr
        Battery Generation Micro-Grid 2    : 150.00KWhr
        Solar 1 Generation Micro-Grid 3    : 350.00KWhr
        Solar 2 Generation Micro-Grid 3    : 300.00KWhr
        Wind Generation Micro-Grid 3       : 100.00KWhr
        Battery Generation Micro-Grid 3    : 200.00KWhr


        Cost of Sources
        Micro-Grid 1:
        Cost of wind in MG1                : 06.70Rs
        Cost of solar 1 in MG1             : 02.77Rs
        Cost of solar 2 in MG1             : 04.94Rs
        Cost of battery in MG1             : 06.17Rs
        Micro-Grid 2:
        Cost of wind in MG2                : 17.02Rs
        Cost of solar 1 in MG2             : 03.02Rs
        Cost of solar 2 in MG2             : 06.59Rs
        Cost of battery in MG2             : 06.17Rs
        Micro-Grid 3:
        Cost of wind in MG3                : 11.21Rs
        Cost of solar 1 in MG3             : 02.77Rs
        Cost of solar 2 in MG3             : 04.75Rs
        Cost of battery in MG3             : 06.17Rs
```

Fig. 1. Console output

```
Demand Management:
100KWhr is supplied by Wind Source of Micro-Grid1 to Micro-Grid1
300KWhr is supplied by Solar 1 Source of Micro-Grid1 to Micro-Grid1
300KWhr is supplied by Solar 2 Source of Micro-Grid1 to Micro-Grid1
150KWhr is supplied by Battery Source of Micro-Grid1 to Micro-Grid1
150KWhr is supplied by Wind Source of Micro-Grid2 to Micro-Grid2
400KWhr is supplied by Solar 1 Source of Micro-Grid2 to Micro-Grid2
200KWhr is supplied by Solar 2 Source of Micro-Grid2 to Micro-Grid2
150KWhr is supplied by Battery Source of Micro-Grid2 to Micro-Grid2
100KWhr is supplied by Wind Source of Micro-Grid3 to Micro-Grid3
350KWhr is supplied by Solar 1 Source of Micro-Grid3 to Micro-Grid3
300KWhr is supplied by Solar 2 Source of Micro-Grid3 to Micro-Grid3
200KWhr is supplied by Battery Source of Micro-Grid3 to Micro-Grid3
0KWhr is supplied by Wind Source of Micro-Grid2 to Micro-Grid1
0KWhr is supplied by Solar 1 Source of Micro-Grid2 to Micro-Grid1
0KWhr is supplied by Solar 2 Source of Micro-Grid2 to Micro-Grid1
0KWhr is supplied by Battery Source of Micro-Grid2 to Micro-Grid1
0KWhr is supplied by Wind Source of Micro-Grid3 to Micro-Grid1
0KWhr is supplied by Solar 1 Source of Micro-Grid3 to Micro-Grid1
0KWhr is supplied by Solar 2 Source of Micro-Grid3 to Micro-Grid1
0KWhr is supplied by Battery Source of Micro-Grid3 to Micro-Grid1
0KWhr is supplied by Wind Source of Micro-Grid1 to Micro-Grid2
0KWhr is supplied by Solar 1 Source of Micro-Grid1 to Micro-Grid2
0KWhr is supplied by Solar 2 Source of Micro-Grid1 to Micro-Grid2
0KWhr is supplied by Battery Source of Micro-Grid1 to Micro-Grid2
0KWhr is supplied by Wind Source of Micro-Grid3 to Micro-Grid2
0KWhr is supplied by Solar 1 Source of Micro-Grid3 to Micro-Grid2
0KWhr is supplied by Solar 2 Source of Micro-Grid3 to Micro-Grid2
0KWhr is supplied by Battery Source of Micro-Grid3 to Micro-Grid2
0KWhr is supplied by Wind Source of Micro-Grid1 to Micro-Grid3
0KWhr is supplied by Solar 1 Source of Micro-Grid1 to Micro-Grid3
0KWhr is supplied by Solar 2 Source of Micro-Grid1 to Micro-Grid3
0KWhr is supplied by Battery Source of Micro-Grid1 to Micro-Grid3
0KWhr is supplied by Wind Source of Micro-Grid2 to Micro-Grid3
0KWhr is supplied by Solar 1 Source of Micro-Grid2 to Micro-Grid3
0KWhr is supplied by Solar 2 Source of Micro-Grid2 to Micro-Grid3
0KWhr is supplied by Battery Source of Micro-Grid2 to Micro-Grid3
```

Fig. 2. Console output

In the Arduino output at the receiving end, all the LED's except grid glow to indicate that all of the sources except grid are in use. The last LED in the right side is turned off where the other LED's entire are turned on as shown in Fig. 3.

Fig. 3. Arduino output

4   Conclusion

This paper explains about the IOT based demand side management for a network of micro grids each having renewable energy resources. It focuses on power distribution in multiple micro-grids that can also be scaled to bigger networks. Here, the values of the environmental variables are sensed using Arduino and is given to cloud storage via IOT. Suitable operations are done on MATLAB and the resulting command signals are given back to ARDUINO MEGA (This is reflected in LED glowing pattern). Thus, grid outage and demand side management is implemented in a network of micro-grids using MATLAB, Arduino and IOT and all the smart grid features are tested to make it into smart micro-grid.

References

1. Kim, J., Byun, J., Jeong, D.: An IoT based home energy management system over dynamic home area networks. Int. J. Distrib. Sens. Netw. **11**(10), 205–219 (2015)
2. Aburukba, R.O., Al-Ali, A.R., Landolsi, T., Rashid, M., Hassan, R.: IoT based energy management for residential area. In: IEEE International Conference on Consumer Electronics-Taiwan (ICCE-TW) (2016)
3. Fiorentino, G., Corsi, A.: Internet of things for demand side management. J. Energy Power Eng. **9**, 500–503 (2015). https://doi.org/10.17265/1934-8975/2015.05.010
4. Bansal, A., Gopinadhan, J., Kaur, A., Kazi, Z.A.: Energy consumption forecasting for smart meters, BAI Conference, Bangalore, India **6**(4), pp. 1882–2001 (2015)

5. Olivares, D.E., Mehrizi-Sani, D.A., Etemadi, A.H.: Trends in microgrid control. IEEE Trans. Smart Grid **5**(4), 1905–1919 (2014)
6. Haider, H.T., See, O.H., Elmenreich, W.: A review of residential demand response of smart grid. Renew. Sustain. Energy Rev. **59**, 166–178 (2016)
7. Purusothaman, S.R.R.D, Rajesh, R., Bajaj, K.K., Vijayaraghavan, V.: Implementation of Arduino-based multi-agent system for rural Indian microgrids. In: Proceedings of the IEEE İnnovative Smart Grid Technologies—Asia, pp. 1–5 (2013)
8. Purusothaman, S.R.R.D, Rajesh, R., Bajaj, K.K., Vijayaraghavan, V.: Design of Arduino-based communication agent for rural Indian microgrids. İn: Proceedings of the IEEE İnnovative Smart Grid Technologies Asia, pp. 630–634 (2014)
9. Veeramani, M., Prince, J., Gladson, J., Sundarabalan, C.K., Sanjeevikumar, J.: An efficient micro-grid management system for rural area using Arduino. Int. J. Eng. Trends Technol. **40**(6), 335–341 (2016)
10. Pourmousavi, S.A., Nehrir, M.H.: Demand response for smart microgrid: initial results. In: Proceeding of the IEEE İnnovative Smart Grid Technologies, pp. 1–6 (2011)
11. Hatziargyriou, N.D., Microgrids, B.: Architectures and Control. Wiley-IEEE Press, West Sussex (2014)
12. Raju, L., Morais, A.A., Milton, R.S.: Advanced Energy Management of a Micro-grid Using Arduino and Multi-agent System, pp. 65–76. Springer, Singapore (2018)
13. Cloud Storage Used for Internet of Things. https://ubidots.com

Real Time Twitter Sentiment Analysis for Product Reviews Using Naive Bayes Classifier

Khushboo Gajbhiye$^{(\boxtimes)}$ and Neetesh Gupta$^{(\boxtimes)}$

Department of CSE, TIT-S, Bhopal, India
khushbootit01@gmail.com, gupta_neetesh81@yahoo.com

Abstract. Opinions of customers in ecommerce play a crucial role in day to day usage. Whenever we have to take a decision So the opinions of other individuals plays an important role. There are various social sites and review sites where an user can post their review or opinions towards any products or any issues, So various business and corporate organization wants to know these opinions of user for taking an decision. In ecommerce market, Their is need to analyze the social data or products review automatically, So there is need to create a model which classify the huge amount of product reviews automatically. In this paper we are fetching real time reviews from the social site twitter and apply various text mining techniques to preprocess the data and than apply an machine learning approach through which we can use Naïve Bayes classification algorithm to classify the text into various emotions and polarities.

Keywords: Opinion mining · Sentiment analysis · E-commerce ·
Text mining · Twitter · Naive Bayes · SVM

1 Introduction

Twitter has become a wealth of knowledge for those that try and perceive however folks feels regarding any brands, product, topic and additional. Mining Twitter information for insights is one in every of the foremost common language process tasks. The simplest businesses perceive sentiment of their customers—what folks believe Associate in Nursing product or Associate in Nursing topic, and what they mean. Matter info retrieval techniques chiefly target process, looking or analyzing the factual information gift. Facts have Associate in Nursing objective part however, there square measure another matter contents that specific subjective characteristics. These contents square measure chiefly opinions [12], sentiments, appraisals, attitudes, and emotions, that type the core of Sentiment Analysis (SA) [4]. It offers several difficult opportunities to develop new applications, chiefly because of the large growth of accessible info on on-line sources like blogs and social networks. as an example, recommendations things projected by a recommendation system are often foretold by taking into consideration concerns like positive or negative opinions regarding those items by creating use of SA (Fig. 1).

© Springer Nature Switzerland AG 2020
A. P. Pandian et al. (Eds.): ICCBI 2018, LNDECT 31, pp. 342–350, 2020.
https://doi.org/10.1007/978-3-030-24643-3_41

1.1   Approaches of Sentiment Analysis

Machine Learning [5] methods used a coaching set and a take a look at set for a classification. coaching set contains data feature vectors and their corresponding category labels. By victimization this coaching set, a categorification model is formed that tries to classify the knowledge feature vectors into corresponding class labels. At that time a take a look at set is employed to simply accept the model by predicting the category labels of unseen feature vectors. Some options which will be used for opinion classification are Term Presence, Term Frequency, negations, n-grams and Part-of-Speech. These options will be accustomed establish the linguistics introduction of words, expressions, sentences which of documents. Linguistics introduction is that the polarity which can be either positive or negative [6].

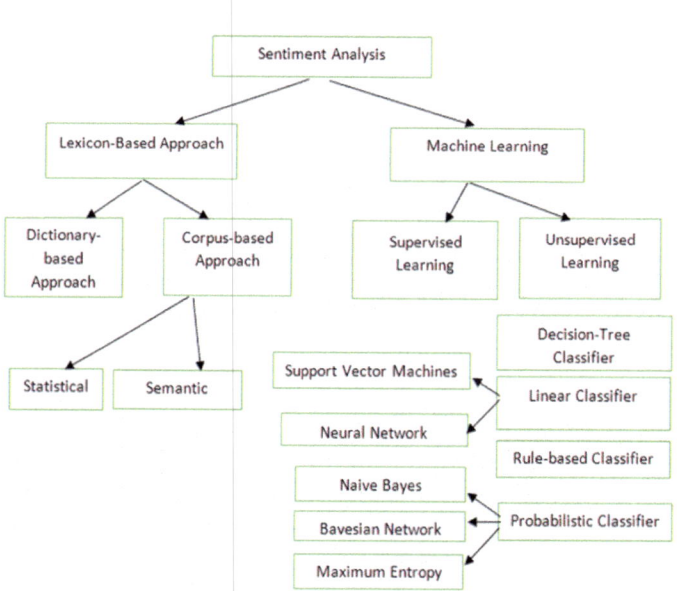

Fig. 1. Machine learning approaches

2   Literature Review

In [1], Sentiment analysis or opinion mining is one in all the main tasks of informatics (Natural Language Processing). Sentiment analysis has gain a lot of attention in recent years. During this paper, we have a tendency to aim to tackle the matter of sentiment polarity categorization, that is one in all the elemental issues of sentiment analysis. A general method for sentiment polarity categorization is planned with elaborate method descriptions. Information employed in this study area unit on-line product reviews collected. At last, they have a tendency to additionally provide insight into our future work on sentiment analysis.

Text mining [2], to boot spoken as text processing, is that the strategy of extracting attention-grabbing. It uses algorithms to transform free flow text (unstructured) into info that will be analyzed (structured) by applying applied math, Machine Learning and tongue method (NLP) techniques. Text mining is Associate in Nursing evolving technology that allows enterprises to understand their customers well and provide them various facilities for their shopping desire. As a ecommerce market is growing rapidly their is large amount of products reviews is present, So its important for the ecommerce companies to know the users opinions towards particular products.

In [2], Analytics firms develop the flexibility to support their choices through analytic reasoning employing a kind of maths and mathematical techniques. On the other hand, a recent study has in addition conspicuous that maximum organizations do not have information required for decision-making. Learning "Data Analysis with R" [3] not exclusively adds to existing analytics info and methodology, but in addition equips with exposure into latest analytics techniques in addition as prediction, social media analytics, text mining on. It provides an opportunity to work on real time info from Twitter, Facebook and various social networking sites.

Twitter is one among most well-liked social networking website wherever individuals area unit expressing their views, opinion and emotions generously. These tweets area unit recorded and analysed to mine emotions of individuals associated with a terrorism (Uri attack). Gift study retrieve tweets regarding Uri attack and notice emotions and polarity of tweets. To mine emotions and polarity in tweets, text mining techniques area unit used. some 5000 tweets area unit recoded and pre-processed to form a dataset of oftentimes showing words. R is employed for mining emotions and polarity.

3   Problem Definition

When the large amount of data is available on the web, the internet user are unable to utilize these data because there is lack of tools available for these. And also the number of websites is increases so the contents organization of these websites is also an challenging task. Text mining [7] becomes an solutions for organizations to analyze the text bases content such as social media posts on twitter, facebook and linkedin. Sentiment classification [11] is a technique to classify the text expressed on social sites into various sentiment polarities like positive, negative and neutral (Table 1).

4   Proposed Work

Machine learning approaches are quite popular in the sentiment analysis domain. These approaches involve tokenization and preprocess the text and then classify the text into various emotions and polarities [9]. It is generally classified in to polarities as positive, negative or neutral. There are many machine learning classification algorithm are present like SVM, Random forest, Naïve Bayes and in this paper we are using Naïve Bayes classification because its gives 81% accuracy as compared to SVM who give 67% accuracy. So Naïve Bayes classification algorithm give better accuracy as compared to other classification algorithm [10]. The flowchart describing a general machine learning based approach of product review data is shown in Fig. 2.

Table 1. Accuracy comparison of SVM and Naive Bayes

Classification algorithm	Accuracy
Existing SVM	67%
Proposed naive Bayes	81%

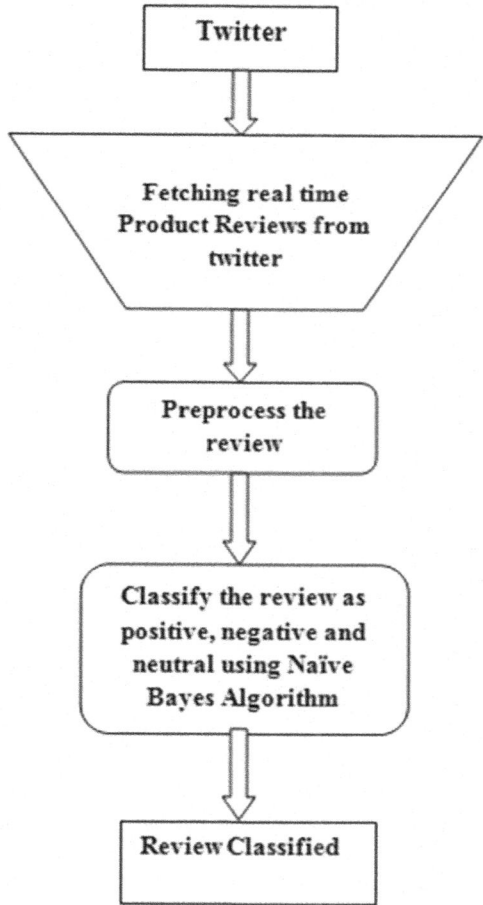

Fig. 2. Flow diagram

5 Experimental and Result Analysis

Experiments are performed on intel i5-2410 M processor with 4 GB RAM. After that we can install machine learning tool R on the machine are perform the following process to find the product reviews:

- Collecting Tweets review.
- Preprocess the review text.
- Classify the review.

5.1 Collecting Tweets on Product

For collecting tweets we need a r package called twitteR and ROAuth package to authenticate user consumer and token keys.

```
> library(twitteR)
> library(ROAuth)
>
>
> consumer_key<-'VE12EJOT2niBwUONTt7Q5sA4u'
> consumer_secret<- 'J5fOrVLmldAafwS4Pxm3eRyB537MMrrsXPdayANnpLyalfpCsY'
> access_token<- '772457331795726337-6IHz8xax4nKXgYfGy7qnZvYmayoOv41'
> access_secret<- 'xRcOOuVc7nRp2wmq55Z2lzTCzYiDu2daRJNbX1ScOCcHn'
>
> consumer_key<-'VE12EJOT2niBwUONTt7Q5sA4u'
> consumer_secret<- 'J5fOrVLmldAafwS4Pxm3eRyB537MMrrsXPdayANnpLyalfpCsY'
> access_token<- '772457331795726337-6IHz8xax4nKXgYfGy7qnZvYmayoOv41'
> access_secret<- 'xRcOOuVc7nRp2wmq55Z2lzTCzYiDu2daRJNbX1ScOCcHn'
>
> setup_twitter_oauth(consumer_key, consumer_secret, access_token,access_secret
> setup_twitter_oauth(consumer_key, consumer_secret, access_token,access_secret
[1] "using direct authentication"
[1] "using direct authentication"
>
>
> some_tweets = searchTwitter("oneplus6", n=1500, lang="en")
> some_tweets = searchTwitter("oneplus6", n=1500, lang="en")
```

For consumer key and access tokens we need for creating a twitter application so we can generate our twitter keys and put into twitter_oauth and than we are storing to collects tweets on oneplus6 with framesize of 1500 and stored into some_tweets variable.

5.2 Pre-process the Tweets Text

We need to install tm package for performing various text mining process on these tweets data. Figure 3 shows the tweets text which is fetch from twitter and these text consist an number, urls, so we need to pre-process these text by using text mining techniques.

```
Console -/ ⌂
[[1491]]
[1] "mykle_casty: RT @UnboxTherapy: San Fransisco. Shot with #OnePlus6 Compressed by Tw
itter. https://t.co/nhEiReIA80"

[[1492]]
[1] "keyawnce: RT @UnboxTherapy: San Fransisco. Shot with #OnePlus6 Compressed by Twitt
er. https://t.co/nhEiReIA80"

[[1493]]
[1] "jon_n8: RT @UnboxTherapy: San Fransisco. Shot with #OnePlus6 Compressed by Twitter
. https://t.co/nhEiReIA80"

[[1494]]
[1] "GalacticStriker: RT @UnboxTherapy: San Fransisco. Shot with #OnePlus6 Compressed b
y Twitter. https://t.co/nhEiReIA80"

[[1495]]
[1] "abdullah_is_1: RT @UnboxTherapy: San Fransisco. Shot with #OnePlus6 Compressed by
Twitter. https://t.co/nhEiReIA80"

[[1496]]
[1] "KingcharlesLK: RT @UnboxTherapy: San Fransisco. Shot with #OnePlus6 Compressed by
Twitter. https://t.co/nhEiReIA80"

[[1497]]
[1] "Aaryan_One: RT @UnboxTherapy: San Fransisco. Shot with #OnePlus6 Compressed by Twi
tter. https://t.co/nhEiReIA80"

[[1498]]
[1] "UnboxTherapy: San Fransisco. Shot with #OnePlus6 Compressed by Twitter. https://t.
co/nhEiReIA80"

[[1499]]
[1] "RiaBhattacharj6: RT @niketsinha: Dear #OnePlus6 users,\n\nI know its a great phone
. But did you know it works just fine even when you don't upload unboxing pi…"

[[1500]]
[1] "oredeinlarry26: RT @oneplus: The Speed You Need is here! Order your #OnePlus6 righ
t now and get 20% off priority shipping. https://t.co/0KvHOYOvKI https://…"
```

Fig. 3. Text review fetch from twitter

With the help of Natural language Processing (NLP) present in tm package we can pre-process the data by removing unwanted data such as urls, special characters, numbers, stop words, etc. from the text and transform the text into lower case, perform stemming on the data. Figure 4 show the pre-process text by using tm package.

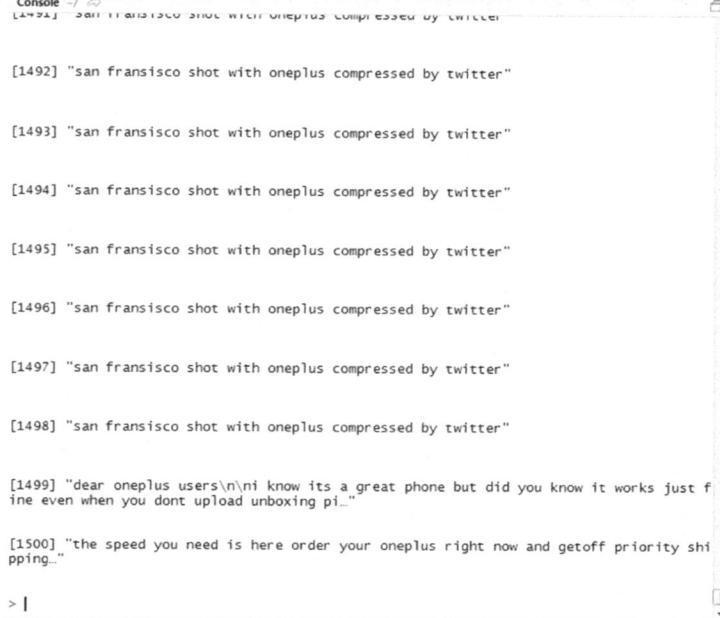

```
Console -/ ⌂
[1491]  "san fransisco shot with oneplus compressed by twitter

[1492] "san fransisco shot with oneplus compressed by twitter"

[1493] "san fransisco shot with oneplus compressed by twitter"

[1494] "san fransisco shot with oneplus compressed by twitter"

[1495] "san fransisco shot with oneplus compressed by twitter"

[1496] "san fransisco shot with oneplus compressed by twitter"

[1497] "san fransisco shot with oneplus compressed by twitter"

[1498] "san fransisco shot with oneplus compressed by twitter"

[1499] "dear oneplus users\n\ni know its a great phone but did you know it works just f
ine even when you dont upload unboxing pi…"

[1500] "the speed you need is here order your oneplus right now and getoff priority shi
pping…"

> |
```

Fig. 4. Pre-process text

After preprocessing the data we are applying naive Bayes classifier to classify the text into various sentiment emotions and sentiment polarities. The naive bayes algorithm classify the text into various emotions shown in Fig. 5.

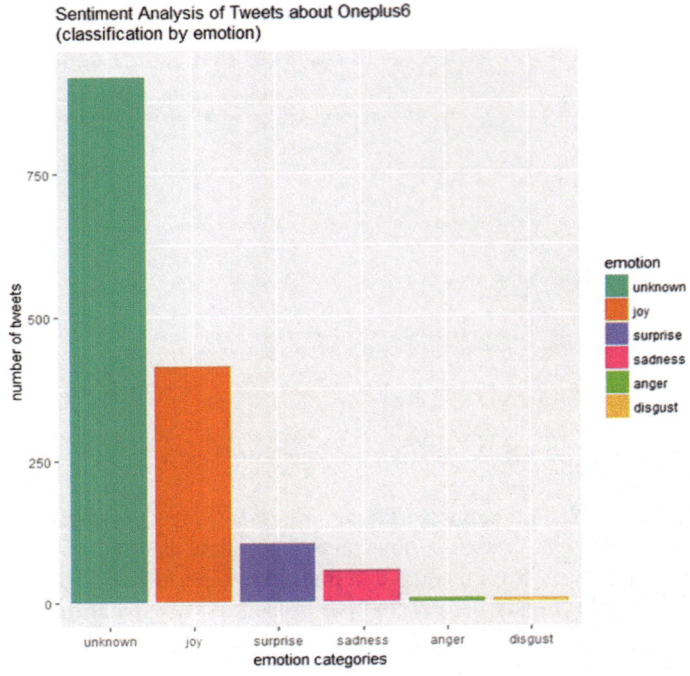

Fig. 5. Classification by emotions

We can also classify the sentiment polarity of tweets text and which is shown in Fig. 6.

In this paper we can also create an document term matrix (TDM), which shows the term is present in which document so we can find the term frequency and base on that we can create a word cloud of frequent words related to oneplus6 are shown in Fig. 7.

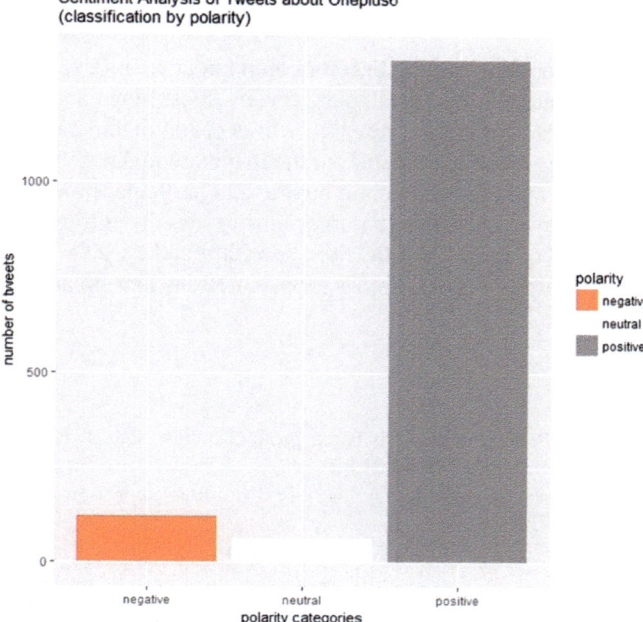

Fig. 6. Classification by polarity

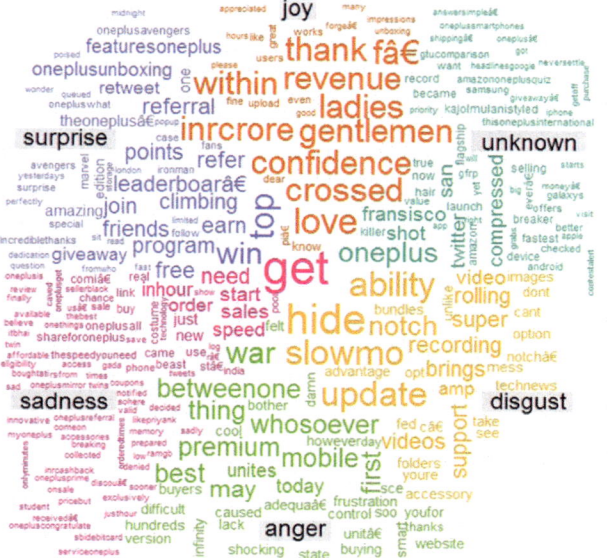

Fig. 7. Word cloud of frequent words

6 Conclusion

Sentiment analysis or opinion mining becomes an popular area for research due to huge amount of content generated by social sites, review sites, blogs and forums. Opinion mining is important field for market researcher in decision making and companies can find the public opinions towards products for their strategy making. Individuals user are also used opinions for making decision on buying any particular products. In this paper we are using naïve bayes classification algorithm to classify the review and we also calculate the accuracy of these classification algorithm and its give 81% accuracy. By using these we are classify the text into various emotions and polarity.

References

1. Fang, X., Zhan, J.: Sentiment analysis using product review data. J. Big Data **2**, 5 (2015). https://doi.org/10.1186/s40537-015-0015-2
2. Rangu, C., Chatterjee, S., Valluru, S.R.: Text mining approach for product quality enhancement. In: IEEE (2017)
3. Karatzoglou, A., Meyer, D., Hornik, K.: Support vector machines in R. J. Stat. Softw. **15**(9), 1–28 (2006)
4. Songpan, W.: The analysis and prediction of customer review rating using opinion mining. In: 978-1-5090-5756-6/17/$31.00 ©2017 IEEE SERA 2017, June 7-9, 2017, London, UK (2017)
5. Kawade, D.R., Oza, K.S.: Sentiment analysis: machine learning approach. Int. J. Eng. Technol. (IJET) **9**(3) (2017)
6. Manning, C., Raghavan, P.: Introduction to Information Retrieval. Cambridge University Press, New York (2008)
7. Anjaria, M., Guddeti, R.M.R.: Influence factor based opinion mining of twitter data using supervised learning. In: COMSNETS, IEEE (2014)
8. Jurka, T.P., Collingwood, L., Boydstun, A.E., Grossman, E., van Atteveldt, W.H.: RTextTools: a supervised learning package for text classification
9. Menaka, S., Radha, N.: Text classification using keyword extraction technique. Int. J. Adv. Res. Comput. Sci. Softw. Eng. (2013)
10. Nirmala, K., Pushpa, M.: Feature based text classification using application term set. In: IJCA (2012)
11. Bahrainian, S.-A., Dengel, A.: Sentiment analysis and summarization of Twitter data. In: IEEE (2013)
12. Neethu, M.S., Rajasree, R.: Sentiment analysis in twitter using machine learning techniques. In: IEEE (2013)

A Dynamic Security Model for Visual Cryptography and Digital Watermarking

K. Chitra[✉] and V. Prasanna Venkatesan

Department of Banking Technology, Pondicherry University, Pondicherry, India
chitrasindhuja1991@gmail.com, prasanna_v@yahoo.com

Abstract. This paper proposes a model and algorithm for digital watermarking using visual cryptographic techniques. Initially the visual cryptographic techniques scrambled the secret information into pixelated form, this pixelated secret share does not reveal any key information. This pixelated secret shares are again embedded into invisible digital watermarking with separate covering image for each and every shares. A model and an Algorithm has been proposed based on this techniques, the result has been evaluated and the metrics also calculated. Moreover, our scheme shows better security features compared to the existing techniques.

Keywords: Visual cryptography · Blind watermarking · Secret sharing · Secret splitting · Digital watermarking

1 Introduction

The cryptographic technique has been used for securing the secret information from an unauthorized person. There are several cryptographic technique has been used for securing the information from untrusted persons. Some of the information hiding techniques involved are cryptology, steganography, Digital Watermarking etc. in **cryptology** [10] which deals about the study of secret writing, cryptology carries many techniques to hiding the secrets even though it vulnerable to some extent. So to overcome vulnerable areas of cryptology, steganography has been evolved.

Steganography [4] which deals about hiding the image in other image, from that it provides more security compared to cryptology. Dew to the evolution of digital media steganography fails in their benchmarking security. To overcome this loss Digital Watermarking has been evolved. **Digital Watermarking** [3] which deals about hiding the secret information in digital media it has to be in any form of the digital content. To the great extent growth of technology lots of hackers are involved and crack the digital content with the help of different type's attacks. To control those attacks in digital media content visual cryptographic technique has been introduced by Moni Naor and Adi Shamir in 1994 [2]. visual cryptography is one of the cryptographic techniques which provide a better security features compared to other cryptographic techniques, the main advantage of using visual cryptographic technique is,

- It requires only the embedding part.
- The decryption has been carried out by human visual perception.

A. P. Pandian et al. (Eds.): ICCBI 2018, LNDECT 31, pp. 351–357, 2020.
https://doi.org/10.1007/978-3-030-24643-3_42

- Computation complexity is less.
- It provides better security features.

This paper deals about what type of security features provided by the proposed schemes. In the Sect. 1 we discuss about the evolution of information security algorithms and their drawbacks. In Sect. 2 deals with the related work. Section 3 briefly describe about various stages involved in the proposed algorithm. Section 4 discuss about the results and the Sect. 5 provides the overall summary of this paper.

2 Related Work

Naor and Shamir [1, 2] introduced the Visual cryptographic concept which is derived from secret sharing scheme which was extended upto images. Visual cryptographic scheme has the ability to recreate the secret without any intervention of computation. The concept of visual cryptographic scheme used in copyright protection improves the image resolution for authentication and identification of right person without any need of the original image. The main advantage of using watermark concept in visual cryptography is, the original image does not allow any alteration during encoding stage.

Hwang [3] initiated the concept of watermarking with visual cryptographic scheme which directly hides the binary watermark into the grey scale image using halftone techniques. Hwang scheme is the combination of (2, 2) VC scheme with watermarking techniques, in this scheme secret information is taken as the key to select the random pixel within the image block. Then they apply MSB technique into the image to generate verification shares. The Master share has been generated by combining the verification share and binary watermark. At the resultant, the master share and the verification shares are overlaid one after another with correct key will recreate the secret information. Hwang scheme somewhat prone to the security features.

Punitha [8] embeds the binary watermark into the grey scale image, which was carried out by three stages. In stage one they create a binary matrix for both the secret information and also the cover image, the resultant of this stage will have the watermark image. This watermark image generates the master and verification shares according to the visual cryptographic code book rules. This shares are register with TTP (Trusted Third Party) for future verification process. At the resultant those two shares are combined to produce the hidden watermark. However, punitha's scheme is also not resist with some of the watermarking attacks.

Hassan [5] scheme proves that the Hwang scheme doesn't recreate the share, for the perfect recreate this scheme requires the histogram shifting. Hassan scheme provide high probability control for the false alarm, the main drawback of this scheme lack in capacity control and it has very high rate of distortion level.

Singh [7] proposed blind and robust watermarking scheme for copyright protection using visual cryptographic techniques which create shares based on comparing the selected pixel value to the mean value of that block.

Padhmavati [9] generates the shares by visual cryptographic scheme, the generated shares are encoded into the cover image with the help of invisible and blind water-marking scheme, which provide better security features against cheating attacks.

3   Proposed Watermarking Scheme in Visual Cryptography

This section describes about the stages involved in proposed watermarking schemes in visual cryptographic method. Each and every stages in the paper are clearly explained in step by step procedure of proposed algorithm.

Stage 1: Read and Check the input image

1. The input of the secret information has been read by the user.
2. Those readed secret information has been analyzed and check whether the taken information is grey scale or color image.
3. If it is a grey scale image it directly switches over into the stage 3.
4. Suppose the analyzed image is a color image, then this stage requires one more extra step to go to the stage 3.

Stage 2: Convert RGB Image into Grey Scale Image

1. The analyzed color image from stage 1 is taken as the input for the conversion of grey scale image.
2. Read the R, G, B color image.
3. Extract the R, G, B color component from the RGB color image.
4. Those individual color component are placed into 3 different 2-dimensional matrix.
5. Create a new matrix with same number of rows and columns having the value zero.
6. Calculate each and every RGB pixel values for the location (i, j) to the new location (i, j) in grey scale image.
7. The grey scale values are formed by the hamming weight of RGB color components.

   Grey Value at p (i, j) = 0.2989 * R (i, j) + 0.5870 * G (i, j) + 0.1140 * B (i, j)

Stage 3: Convert Grey Scale Image into Binary Image

1. The analyzed grey scale image from stage 2 is taken as the input for stage 3. Then the image has been resized as m × n matrix.
2. Divide the resized matrix into blocks which having the same number of rows and columns.
3. Calculate the thresholding value (T) for each and every blocks of the grey scale image.
4. Create another new matrix with the same number of rows and columns having the value zero.
5. The thresholding value (T) has been calculated by taking mean for each and every block of the matrix.
6. The mean value is taken as the threshold value (T), from that comparing grey scale value with mean value for every block.

7. Below the mean value is consider as 0 (i.e. Black/Background) and the mean value is equal to or above is consider as 1 (i.e. White/foreground).
8. The following formula represents how the thresholding value has taken place is given as,

$$g(x,\, y) = \begin{cases} 1 & if \quad f(x,y) \geq T \\ 0 & otherwise \end{cases}$$

9. Convert each grey scale pixel value at the location g (x, y) to the binary value at the location f (x, y) in new matrix.
10. Do the same for whole m × n matrix, the resultant matrix will be the binary matrix.

Let us consider the following example, it shows how the thresholding has made in each and every block of the secret message (Fig. 1).

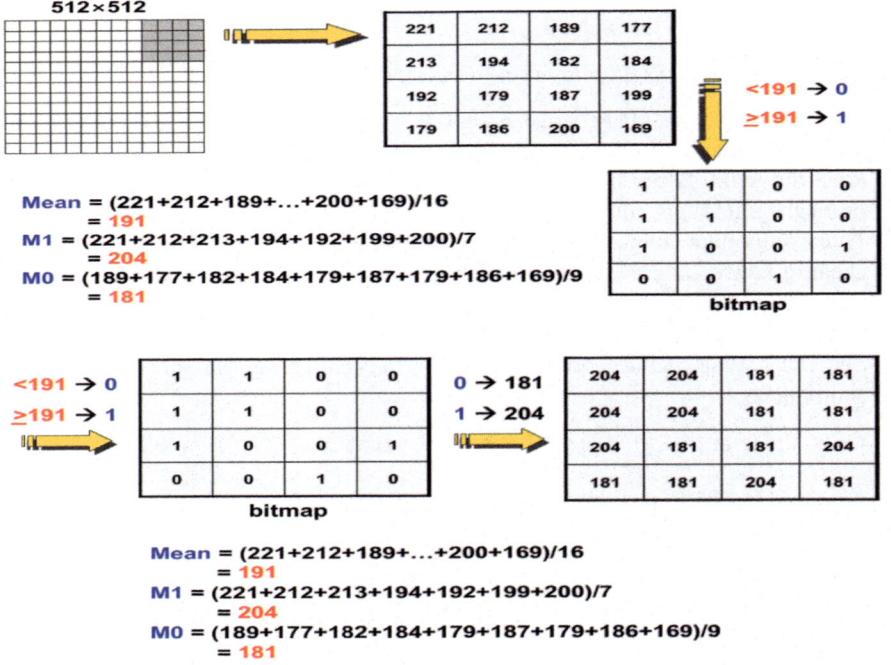

Fig. 1. Working principle of block truncation coding

Stage 4: Generating Shares

1. The resultant binary image matrix in stage 3 is taken as the input for stage 4.
2. Here the m × n matrix window has been divided into several blocks. Each and every block contains same number of rows and columns.

3. The user has to select one particular block and check whether the taken pixel is '1' or '0'.
4. If the selected pixel is '1' then the user has to select any one pattern of share separation from visual cryptographic code book for white pixel.
5. If the selected pixel is '0' then the user has to select any one pattern of share separation from visual cryptographic code book for black pixel.
6. Finally, the share has been generated for both the black and white pixels.

Stage 5: Embedding Watermark into the shares

1. The generated black and white pixel shares from the stage 4 is taken as the input for the stage 5.
2. In our techniques we are using (2, 2) share generation scheme. So, we have only two shares for the corresponding secret information.
3. An invisible data embedding techniques has been applied for both the shares with separate cover image.
4. The resultant of the above step having watermarked shares (i.e.) watermarked share1 (ws1) and watermarked share2 (ws2).

Stage 6: Staking the Shares

1. The watermarked shares from the stage 5 is taken as the input for stage 6.
2. This watermarked share images have been broadcasted to the communication network.
3. The copyright owner has received their corresponding watermarked share images from the communication network.
4. The watermarked cover image has been removed by blind watermarking techniques.
5. The detached cover image and shares are taken to check whether the detached images are affected by any other attacks or else the received one is correct or not.
6. After checking of all the process, the cover image is placed aside and the shares are overlapped to view the secret message.
7. Applying OR/XOR Boolean operation for stacking the shares.
8. At the result, the secret information has been viewed by the copyright owners.

4   Results and Discussion

The Experiment was conducted and analyzed by various performance evaluation metrics to determine whether our proposed techniques robustness level will withstand against the cheating and attacks. The following figures shows the various levels of data embedding and extracting process.

In this the secret image is converted into greyscale and the greyscale image is converted into binary. Binary image is taken as input for share generation. Here we are using (2, 2) secret sharing scheme for share generation. After share generation, embedding those share into separate cover image using invisible watermarking techniques. At the receiver end the encrypted image has been decrypted using bling watermarking techniques, at the resultant the cover image and the share has been taken separately. Those shares are overlapping by using XOR operation from that the secret image has been viewed by naked eye (Fig. 2).

Fig. 2. (a) Secret image, (b) grey scale image, (c, d) share 1 and share 2 secret image (e, f) logo images, (g, h) invisible watermark embedded image, (i) XOR stacked image, (j) OR stacked image

We analyzed our resulted with 2 Boolean operations such as, OR and XOR operation those are also displayed in the Table 1. From the table it clearly shows that the XOR staking techniques provide better metrics values compared to the OR Stacking values. Our proposed technique provide better PSNR, SSIM, MAE, NC, MSE values and also it withstand against all possible watermarking attacks.

Table 1. Comparison of Boolean operation stacking values

OR overlapping values					XOR overlapping values				
PSNR	SSIM	MAE	NC	MSE	PSNR	SSIM	MAE	NC	MSE
56.49	0.99	0.146	0.23	0.15	81.1	1	0.00008	0.338	0

5 Conclusion

In this paper we discuss about the proposed model and their different stages involved in the security features of the information security. Our contribution provides the better security features for the system, it reduces the computation cost and storage space. This technique doesn't require any decryption stage so it can reduce the execution time and also the complexity of the algorithm. Our proposed model having lots of security features and advantages compared to all other techniques.

References

1. Naor, M., Shamir, A.: Visual cryptography. In: Advances in Cryptology—EUROCRYPT 1994, Perugia. Lecture Notes in Computer Science, vol. 950, pp. 1–12. Springer, Berlin (1995)
2. Naor, M., Pinkas, B.: Visual authentication and identification. Adv. Cryptol. **1294**, 322–336 (1997)
3. Hwang, R.: A digital image copyright protection scheme based on visual cryptography. Tamkang J. Sci. Eng. **3**(2), 97–106 (2000)
4. Tai, S.C., Wang, C.C., Yu, C.S.: Repeating image watermarking technique by visual cryptography. In: IEICE Transaction Fundamentals, vol. E83-A, No. 8, pp. 1589–1598 (2000)
5. Wu, D.-C., Tsai, W.-H.: A stenographic method for images by pixel-value differencing. Pattern Recogn. Lett. **24**(9–10), 1613–1626 (2003)
6. Fu, M.S., Au, O.C.: Joint visual cryptography and watermarking. In: Proceedings of the IEEE International Conference on Multimedia Expo, pp. 975–978 (2004)
7. Hassan, M.A., Khalili, M.A.: Self-watermarking based on visual cryptography. World Acad. Sci. Eng. Technol. **8**, 159–162 (2005)
8. Wang, F.-H., Pan, J.-S., Jain, L.C.: Digital watermarking techniques. Stud. Comput. Intell. **232**, 11–26 (2009)
9. Singh, K.M.: Dual watermarking scheme for copyright protection. Int. J. Comput. Sci. Eng. Surv. **3**, 99–106 (2009)
10. Punitha, S., Thompson, S., Siva rama lingam, N.: Binary watermarking techniques based on visual cryptography. In: IEEE International Conference on Communication Control and Computing Technologies (ICCCCT) (2010)
11. Padhmavati, B., Nirmal Kumar, P., Dorai Rangaswamy, M.A.: A novel scheme for mutual authentication and cheating prevention in visual cryptography using image processing. In: Proceeding of International Conference on Advances in Computer science (2010)
12. Savu, L.: Cryptography role in information security. In: Recent Researches in Communications and IT, pp. 36–41 (2011)

Systematic Approach Towards Sentiment Analysis in Online Review's

Hemantkumar B. Jadhav[1]([⊠]) and Amitkumar B. Jadhav[2]

[1] Shri Chhatrapati Shivaji Maharaj College of Engineering,
Ahmednagar, Maharashtra, India
Hem3577@gmail.com
[2] Vishwabharati Academy's College of Engineering,
Ahmednagar, Maharashtra, India
Amitjadhav184@gmail.com

Abstract. Ever Since its birth in the late 60's, Internet has been widely and mainly used for interaction purpose only; but over the period of time, its application has changed significantly. Now a days' Internet is no longer use only for communication purpose. Its use is spread over wide variety of applications' and E-Commerce is one of them. The most important part in e-commerce, from consumer perspective is, the reviews associated with products. Most of the people do their decision making, based on these online reviews about products or services. These reviews not only help user to know the product or service thoroughly but also affect user's decision making ability to a great extent and also divert the sentiments about the product positively or negatively. As a result, there have been attempts made, to change the product sentiments positively or negatively by manipulating the online reviews artificially to gain the business benefits. Ultimately, affect the genuine business experience of the user. Therefore in this paper, we have dealt with this particular problem of e-commerce field, specifically online reviews' in particular and sentiment analysis domain as a whole, in general. A ton of work has been already done in this domain since last decade. In this paper, we will see cumulative study of all this work and how one should approach to deep dive into sentiment analysis field and start research from the scratch. This paper will provide insight of sentiment analysis domain, its general workflow and systematic approach towards solving problems in this domain.

Keywords: Opinion mining · Data analysis · Sentiment analysis · Opinion spamming · Fake review detection

1 Introduction

Sentiment analysis is the process of extraction of knowledge from the opinions of others. The term is also called as "Opinion Mining". It is area of research that deals with information retrieval and knowledge discovery from text using data mining, natural language processing and machine learning techniques. The knowledge of this analysis can be used for recommendation systems, government intelligence, citation

© Springer Nature Switzerland AG 2020
A. P. Pandian et al. (Eds.): ICCBI 2018, LNDECT 31, pp. 358–369, 2020.
https://doi.org/10.1007/978-3-030-24643-3_43

analysis, human–computer interaction and its computer assisted interactivity. The domain of sentiment analysis is vast domain and we have restricted ourselves to the field of online reviews and analysis of the same.

To consult a review and come to a decision, based on sentiment or content of review is very common thing. Before actually buying the particular product or service, people like to read comments, reviews and ratings about that product or service, to get the clear idea about it. Therefor company which sells the product or services exploit this feature to sell more their products by wrongly influencing the potential buyers. In order to do this company hires people who write fake reviews about product for them, also called as spammers and thus process called opinion spamming. This process of opinion spamming can be done in both ways, either to promote your own product by changing sentiment of product positively or to demote competitors' product by changing sentiment of it negatively. According to Mr. Bing Liu, an expert in opinion mining, there are an estimated 33% fake reviews in consumer review sites. Therefore there is need that such types of reviews should be detected and eliminated to provide genuine experience of the business to the users.

2 Related Work

The Opinion mining is a subject which can be analyzed in many ways. Many scholars have done the research in this field, implemented various learning algorithms and have developed several systems that can detect fake reviews, classify spam reviews from non-spam reviews, defined the spammicity of product and so on. Table 1 shown below discusses and compares the various techniques used by scholars in the past to tackle the opinion spamming problem.

First one discusses the learning based approach in general and supervised learning in particular. It uses some behavioral indicators to train its model. Second one from the list discusses the semi-supervised learning based approach which uses active learning technique to classify spam from non-spam reviews. It uses the features such as f-measure, re-call, precision [7] etc. to train the model Third from the list explores the collective positive labeled learning technique to train the data model. It also uses the spatial approach such IP tracking to identify the spammers group which deliberately diverts the product sentiment. Forth one deals with k-score analysis to calculate k-values, [8] based on which we could classify spam and non-spam group. It also uses behavioral aspect of connection between the spammers. Lastly, the scholars uses the temporal approach to calculate the spammicity of product, they basically uses the time span analysis at the micro level and cross-site time series anomalies to detect spam behaviors of users. Table 1 also shows that different scholars use different datasets to train and validate their models. So it is bit difficult to compare which technique is better among them in terms of accuracy.

Different scholars implemented and follow different approaches to the subject of sentiment analysis and thus achieve different milestones. We must understand all these methods and techniques to get the basic idea of sentiment analysis domain and understand how problem solving works in this domain.

Table 1. Comparative analysis of some key approaches of sentiment analysis

A. no.	Dataset used	Detection technique used	Metrics/features used
1	Amazon.com	Mixture of behavioral and supervised learning based	Similarity between reviews, review frequency, spamming behaviors
2	Yelp.com	Semi-supervised with active, learning	Precision, recall, accuracy, f-measure
3	DiangPing Restro	Positive unlabeled learning, spatial approach (IP tracking), Heterogeneous multitier classification	Reviews:: +ve score—fake review −ve score—truthful review Reviewers:: +ve score—spammer −ve score—non-spammer
4	Mobile01.com	K-score analysis, behavioral approach	K-Core values, connection between spammers
5	FourSquare	Temporal approach towards sentiment analysis, time span study and analyzing pattern anomalies	Micro-level—time span spam analysis Macro-level—cross site time series anomalies
6	Amazon.com	Rating deviation, review burst, cosine technique for content similarity	Precision, recall, accuracy, f-measure

3 Types of Spams

Before we deep dive into ocean of sentiment analysis, first let us understand what does the "spam" term means. In a generalized form, spam means any type of message or communication originating from either a person or an organization which is unsolicited or undesired [4]. It usually contains non harmful material such as unwanted advertisement or messages, but sometimes it could be a harmful one containing malware, viruses' or link to phishing websites etc. Let us understand different types of spams that are exist in real world to get better idea of this term.

3.1 Email Spam

It a type of spam in which unwanted contents is spread through medium of emails in the form of viruses', advertisements or messages. Basically Spam emails are nothing but the unwanted emails, most of times they are non-harmful advertisements and messages, but sometimes they contain harmful contents such as malwares or links to phishing websites.

3.2   Review Spam

It is a type of spam in which the sentiments about particular product or service or individual, are control artificially. Generally people are hired, to post fake reviews in bulk to change the sentiment about the product, either positively or negatively [4]. In this process of opinion spamming the potential buyers are just misdirected or wrongly influenced about particular product or service.

3.3   Advertisement Spam

These are advertisements of a particular product or service, appear in your browser based on your browsing history [3]. They are posted with the intention of promoting specific product, service or individual.

3.4   Hyperlink Spam

The addition of external links on the webpage with the intention of promoting particular product or service is a major source of spam in recent times [4].

3.5   Citation Spam

These are recently discovered new types of spams, which involves process of illegal citation [4] such as paid citation to improve your scholarly work on internet. These spams generally found in the scholars work.

4   Generalized Workflow of Data Analysis

Sentiment analysis is nothing but one of the many form of data analysis or we can put it in other way like data analysis is the integral part of sentiment analysis. Therefore one needs to have clear idea about data analysis before diving into sentiment analysis field. Therefore it is necessary to have the overview how data analysis is done and what is its exact procedure needs to follow in the analysis part. Figure 1 shown below discusses the basic steps in data analysis.

4.1   Data Collection

Collecting raw data is the basic step in data analysis. The data collection can be done in multiple ways; Use of web crawler, Use of API's, Cloning datasets is some of them.

4.2   Structure the Data

The collected data is always in the raw form; one may not be able to perform required operations on raw data. Therefore one needs to translate this raw data in structured form so that various operations can be carried out on structured data to perform analysis.

4.3   Data Pre-processing

In this step the collected structured data is transform into more accurate forms. Though the structured data is already in understandable form but there can be some data leakage which can affect model accuracy adversely. Therefore filling these gaps is necessary before performing analysis.

Data pre-processing involves various stages:

1. *Data Cleaning*—Data is cleaned by removing anomalies in the data, filling missing values, smoothing data values and thus making data consistent.
2. *Data Sampling*—In this step sample data is selected to perform analysis in a scientific manner so that outputs reflect the proper results.
3. *Data Reduction*—In this step, the aim to reduce the number of representations in dataset.

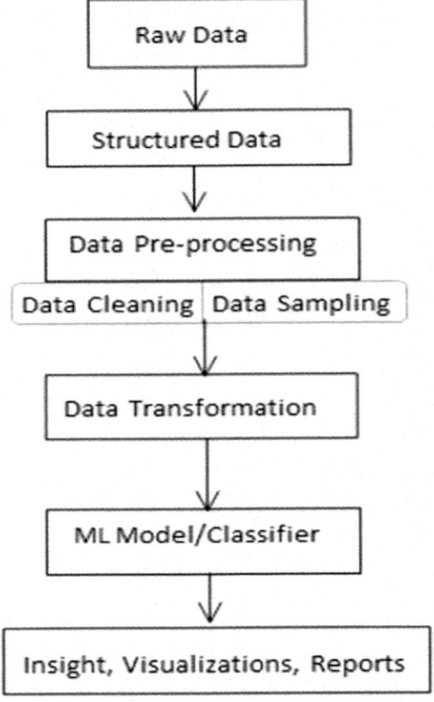

Fig. 1. Generalised workflow of data analysis

4.4   Data Transformation

In this step the data is transform in such a format so that it is understood by classifier or model which is going to train and test on dataset. There are various algorithms/methods exists, which transform this data in model or classifier understandable form.

Ex. Vectorization is one such method which is used to transform matrix into column or 1D array using linear transformation.

4.5 Build the Model

In this step the model is build. This model is train and test on the transformed dataset. This is the heart of Sentiment Analysis process. The more effective model you build the more accurate result you will get in the detection of Spam reviews. What type of model one should choose for different dataset, what are the existing approaches are there for sentiment analysis and how one can choose features wisely in building this data models are discusses in detail in the part [V] of this paper.

4.6 Insight/Visualisation of Data

This is the final stage of data analysis process. In this step whatever output we obtain from trained model is visualize in such manner that we could maximize the efficacy of the model. The results are presented in the form of graphs, charts and other visualizes forms so the target audience can understand the results better and easily.

5 Approaches Towards Sentiment Analysis

5.1 Linguistic Approach

In the review based sentiment analysis, the problem of opinion extraction can be deal with two distinct ways, learning based techniques and semantic orientation.

The Semantic orientation approach also called as Linguistic approach is basically analyses the review content word by word and classify the content as positive or negative according to applied classification rule. Initially it performs the generic pre-processing steps such as POS tagging, removal of the stop-words, stemming [6] etc. to reduce the size of dataset and make analysis more accurate. One most common technique used in this approach is frequency analysis to calculate the weight of the each individual feature. TF–IDF vectorizer is one such method to calculate the feature weighting. It stands for "Term Frequency-Inverse Document Frequency", which determines the contextual importance of a word. TF is the overall frequency of a term whereas IDF is frequency of a term in every document.

Another technique which is commonly used in the linguistic approach is PU [Positively Unlabeled] learning method. It is based on semi-supervised learning concept to classify sentiment analysis. It takes into account the parameters and features like Sentiment Polarity, Bigram Frequency, Word Count, Linguistic Inquiry Part of Speech tag [3]. In PU learning method the positively unlabeled dataset is again labeled neutral and again trained and tested to segregate the more review's which would have been missed on previous iterations. Several iterations are performed to get more accurate analysis of review dataset.

5.2 Learning Based Approach

The traditional methods for tackling sentiment analysis problem has some limitations on its side e.g. Cost factor, Human errors etc. The learning based approach allows us to overcome some of these limitations. In this method machine is first trained and tested on the partial data of the dataset using various data models and then the same data model is used to perform the sentiment analysis on the remaining dataset to calculate true result. The learning based approach basically uses the data models to learn the statistical models which used in detecting spam reviews.

There are many ways in which we can train the model. These are broadly classified in three categories: Supervised learning, Un-Supervised learning and Reinforcement learning. Supervised learning uses the labeled data to get train the model while in Un-Supervised learning model is train on unlabeled data using its past mistakes. In Reinforcement learning, model learns how to behave in its environment by performing actions and seeing the results.

There is another method which is becoming more popular these days is Semi-Supervised [5] method, which is nothing but the form of supervised learning method, the only difference in this method is, it uses separate datasets to train the model and to validate the data model (Fig. 2).

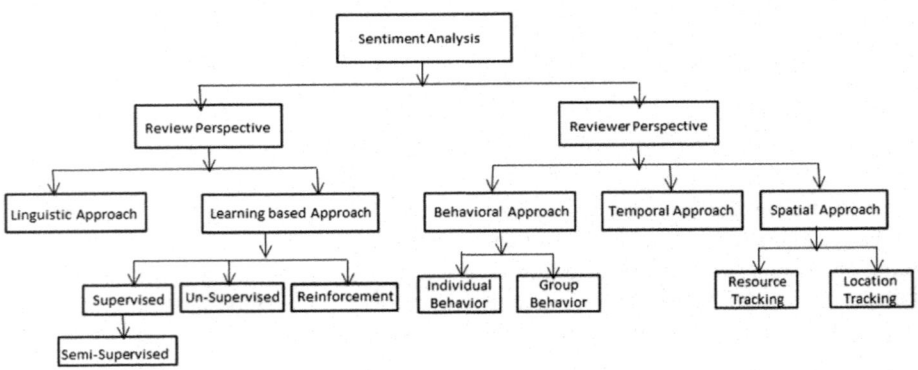

Fig. 2. Different approaches towards sentiment analysis.

People have developed lot of machine learning algorithms using these techniques to classify the data such as Support Vector Machine, Decision Tree, Logistic Regression, Linear Regression, Gradient Descent and so many. The main advantage of using learning based approach is it drastically reduces the human efforts and cost of computation for the classification. It also generally gives more accurate predictions than any other approach.

5.3 Behavioral Approach

The behavior of the user which causes the spammicity is the one of the key to the sentiment analysis. The user may work individually or in the groups', but the group spamming causes lot more damage than the individual ones. The critical hypothesis in the behavioral approach is that, "spammers don't work alone, rather they hired in groups" [4].

Over the period little work has been done towards this part of subject and there is need to focus more on this area. Scholars have identified the common behavior or patterns of spammers and have developed some features to identify the Individual spammer as well as the group of spammers which diverting the sentiments of the product positively or negatively.

In last decade, scholars have developed many behavioral indicators from group point of view as well as individual point of view. These indicators are considered as a base for assessing behaviors of spam users which are working individually as well as in groups. Some of them are as listed below and explained in short.

5.3.1   Group Behavioral Indicator's

5.3.1.1 Group Activity Window

In this indicator, the activity of each member in the particular group is measured over a specific period of time. The reviews captured in this time period are more likely to be spam reviews, as members of spammer groups like to work together while posting such fake reviews for target product. The earliest and the latest reviews of a spammer group 'g' are calculated and normalized over the value [0,1] to calculate the spammicity value of group g [1]. The longer time window of activity, the lesser the probability of group having spammer one.

5.3.1.2 Group Rating Deviation

Generally, it is seen that there is quite big difference in ratings given by the spam reviewer and genuine reviewer. The large deviation is to capture the sentiment of target product. This behavioral is modeled as Group Rating Deviation [1].

5.3.1.3 Group Content Similarity

Group behavioral is also exhibited by the similarity in the contents of the reviews. Many a times, spammers copy review content among themselves when working together. So the target product has many reviews with similar content. Reference [8] shows the single method to detect the duplicate and near duplicate reviews. This flaw can be eliminated using the concept of cosine similarity.

Cosine Similarity: The cosine similarity is used to calculate the similarity between two non-zero vectors. Here, in this scenario frequency of the words in the sentence is calculated. Considering them as two separate vectors we can easily determine the similarity between two reviews [12].

5.3.1.4 Group Member Content Similarity

It is observed that, many times spam reviewers' copies their own reviews written previously to post the fake reviews on target product. This is another type of spammicity exhibited in the group. Again the concept of cosine similarity is used to determine the degree of similarities between the reviews.

5.3.1.5 Time Stamp

The more early you post the review, the more chances to damage the sentiment of the product. The early reviewers are more likely and can easily capture the sentiment of product. So the Time Stamp behavioral indicator is used to calculate the early reviewer of the product.

5.3.1.6 Group Impact

The spammer group is made up of mixture of spam individual reviewers as well as genuine reviewers. It is seen that the impact of group is much smaller when group contains the genuine reviewers more in number. This group impact on the product is calculated by taking the ratio of spam individuals in group to the total reviewers of the product.

5.3.1.7 Support Count

There is very little possibility that group of similar people post the reviews for same set of products. If this likeliness comes out to be higher, then it can be added into the stack of fake reviews. This likeliness is measure to define the support count parameter.

5.3.1.8 Review Content Quality

Sometimes the review length or rather the content in the reviews are not so important from the perspective of spammers, they just post reviews with lengthy content to divert the ratings of product.

5.3.1.9 Reviewer Background

Genuine reviewer would post the review, if he/she has actually used it. Therefore checking the reviewer is really genuine buyer or not is the important factor deciding spammer from non-spammer.

5.3.2   Individual Behavioral Indicator's

Group detecting criteria detects the spam reviewer groups. Following are the criteria to detect individual spam activities of members of the group. We define the individual detecting criteria [2] in the context of group detecting criteria.

5.3.2.1 Individual Deviation

Though individual rating do not affect drastically to rating of product as a whole but similar to the group rating deviation, the review rating deviation is calculated over a given scale of deflection for each member of the group.

5.3.2.2 Content Plagiarism

The similarity between the contents of reviews posted by same reviewer is investigated for any content plagiarism for a given product. This is done by using the cosine similarity technique used in detecting the group plagiarism.

5.3.2.3 Individual Time Stamp

The time of the reviews posted by a given member for a particular product is calculated over a given time span is deduced [2].

5.3.2.4 Individual Coupling

Connection between individual members of the group is measured. Earliest and latest dates of the review posting are fetched to calculate coupling value of individual w.r.t spammer group.

5.4   Temporal Approach

A common approach to the subject matter is to understand and analyze the patterns of the product/services reviews from the time it was launched or released until its decline. Analyzing these temporal patterns can be compared against natural consumer and product cycles to determine and detect anomalies. The key hypothesis in this approach is if the product or service is commented on repeatedly or more than usual in short burst of time, it has a chance of being attack by spammers group and could be as the key point in detection of spam.

Various parameters can be used in this approach such as frequency of product, time span of cross-site product review etc.

5.5 Spatial Approach

It would be more inclusive approach if we include spammer's resource that use to spam the reviews and location of spammers. It would give us more accurate data model to train the dataset and get more accurate output. Scholars have identified few such spatial parameters which would be useful in defining the model. For ex. Resource tracking and Location tracking [10].

The resource tracking can be done in many ways such as tracing IP address or physical ID of resource from which spam reviews are get posted frequently. Similarly we can trace the location of user whose behavior is suspicious and involve in posting spammicity very frequently. Scholars are considering spatial parameters in detecting spam reviews because there is an increased chance of same user posting multiple reviews for different product very frequently and also more likely to post and upload the reviews from same location, as he has been hired for same purpose.

6 Conclusions

The approaches and techniques discussed in this paper have been consolidated together to serve as reference point for future work and study. In this paper, we have tried to cover different techniques and methods used by scholars to tackle the opinion spamming problem. The paper mainly discussed the various approaches to define various model and classifier which are used in the analysis of opinion spamming. In future there is need to tackle this problem from dynamic perspective. A dynamic system with real time data monitoring would become more efficient than what we have today. Also it would be more feasible from implementation point of view. Current trends in the amount of research published indicate that this is a very critical problem and need to be address in full fletch manner as soon as possible.

References

1. Mukherjee, A., Liu, B., Glance, N.: Spotting fake reviewer groups in consumer reviews. In: IW3C2, WWW, Lyon, France (2012)
2. Kolhe, N.M., Joshi, M.M., Jadhav, A.B., Abhang, P.D.: Fake reviewer group's detection system. IOSR-JCE **16**(1), 06–09 (2014)
3. Li, F., Huang, M., Yang, Y., Zhu, X.: Learning to identify review spam. In: 22nd International Joint Conference on AI, China, pp. 2488–2493 (2010)
4. Sindhu, C., Vadivu, G., Singh, A., Patel, R.: Methods and approaches on spam review detection for sentiment analysis. Int. J. Pure Appl. Math. **118**(22), 683–690 (2018)
5. Deng, H., Zhao, L., Luo, N., Liu, Y.: Semi-supervised learning based fake review detection. In: IEEE International Symposium on Parallel and Distributed Processing with Application (2017)
6. Cui, H., Mittal, V., Datar, M.: Comparative experiments on sentiment classification for online product review. In: American Association for Artificial Intelligence (2006)
7. Tang, H., Tan, S., Cheng, X.: A survey on sentiment detection of reviews. In: Expert Systems with Applications, China (2009)

8. Xi, Y.: Chinese review spam classification using machine learning method. J Mach. Learn. Res. **3**, 1289–1305 (2012)
9. Brown, L.D., Hua, H., Gao, C.: A widget framework for augmented interaction in SCAPE (2003)
10. Chirita, P.A., Diederich, J., Nejdl, W.: MailRank: using ranking for spam detection. In: CIKM (2005)
11. Lai, C.L., Xu, K.Q., Raymond, Y.K.: Towards a language modeling approach for consumer review spam detection. In: IEEE ICEBE (2010)
12. Jindal, N., Liu B.: Analyzing and detecting review spam. In: Seventh IEEE International Conference on Data Mining, 2007 (ICDM 2007), pp. 547–552. IEEE (2007)

A Generic Model for Scheduling IoT Jobs at the Edge

S. S. Boomiga[(⊠)] and V. Prasanna Venkatesan

Department of Banking Technology, Pondicherry University, Pondicherry, India
boomigapandu@gmail.com, prasanna_v@yahoo.com

Abstract. Edge computing is a technology that allows processing of applications close to the proximity of the IoT devices. The benefit of edge computing is it reduces latency, improves speed, provides security and privacy since the edge device is placed near the IoT device. But the edge resources are resource constraint having less storage and computation capacity when compared with cloud resources. So early decision must be taken to decide which jobs must be processed at edge. We propose a light weight delay, resource and application aware generic scheduling model to reduce the latency of delay sensitive and mission critical applications. The proposed system classifies the applications into different priority classes based on mission criticality and delay sensitiveness of applications thereby servicing high priority applications at the edge and in parallel we focus on the resource availability before processing IoT applications at the edge.

Keywords: Scheduling · Context · Priority · IoT · Edge computing

1 Introduction

Internet of things is a modern phenomenon, in which things or objects collect information from the environment by using sensors to monitor and control the smart environment and perform tasks without human interventions by communicating with the help of Internet. The objects can be smart cameras, wearable sensors, environmental sensors, smart home appliances, vehicles etc. So the information collected from the IoT environment is huge, thereby requiring more band width for transferring data to the cloud and also there exists security and privacy issues since transferring the things details to cloud. Additionally, the QoS parameter response time will be more because of this data transfer and processing in cloud to return the result to the user.

There exists bandwidth, security and privacy issues among them latency is considered more important. Since applications which are critical and delay sensitive needs less response time. Applications such as anomaly detection, lighting control, and climate control, smart appliances for energy and water use requires less response time, otherwise user gets more frustrated and it degrades the quality of service. Applications to handle natural calamities and to monitor chronic disease must respond immediately to give alert otherwise delayed or lost information may lead to life threatening situation [1]. Some applications such as banking transactions, websites for e-commerce, and high-speed transaction systems for financial institutions are considered critical in a way

© Springer Nature Switzerland AG 2020
A. P. Pandian et al. (Eds.): ICCBI 2018, LNDECT 31, pp. 370–377, 2020.
https://doi.org/10.1007/978-3-030-24643-3_44

that, if these applications fails, functions of the business will stop and revenue will lost. These applications are not life critical applications and so they are considered as delay sensitive applications not mission critical applications. So latency is considered more important.

To overcome this problem in IoT, a concept called Edge computing is introduced. In Edge computing, the IoT job requests can be processed at the Edge of the network. Edge computing is a paradigm in which the IoT requests are processed at the Things or at the local server which are placed at the proximity of the end-user to give less response time instead of processing the information in the cloud. But the network edge also can't process all the jobs, since it has minimum storage and CPU Capacity. So efficient scheduling technique is needed to decide which jobs must be processed at the edge and at the cloud. So this paper proposes efficient solution which dynamically adapts to the environmental variations in multiple IoT applications.

2 Related Work

Priority assignment to the IoT jobs are discussed in [2, 3]. High priority can be assigned to critical and delay sensitive jobs and low priority to the non-critical jobs and the jobs can be scheduled as per the priority. There can be high demand for data centres to process user requests and they can be identified by using historical profile and statistical predictions. So when there is more demand Micro data centre can be placed in that locations, so that latency can be reduced by processing the user requested jobs in micro data centre which is placed near by edge server than in cloud [4]. Another leading trend in networking, namely Software-Defined Networking (SDN) is considered to reduce the latency of IoT applications. Since SDN controller will get periodic information about each edge server, user entity will query the SDN controller before offloading the task to the edge server to get additional parameters about the edge [5]. In order to reduce the latency, the user request will be processed at edge server only. But if the Edge server is more loaded, critical jobs cannot be processed at the edge, they will be processed at the cloud which will increase the response time. So to overcome this situation, Layered architecture is proposed. So, high priority jobs can be processed at Tier-1 and the remaining jobs can be processed at next layer until it is overloaded, then move to next layer and so on to reduce the latency instead of processing jobs directly in cloud [6].

Scheduling algorithms such as Genetic Algorithm, Particle swarm optimization, Ant Colony Optimization, Simulated annealing and Bees swarm algorithm can be used to schedule the incoming jobs to reduce latency. The challenge is to find which algorithm works efficiently to schedule IoT jobs [7, 8]. Policies such as Lowest latency policy [9], delay sensitive policy [10], delay priority policy [11] are used to reduce the latency while scheduling the IoT jobs. Machine learning can be used to classify the jobs as critical and non-critical by identifying the pattern of jobs in case of industrial internet of things [11]. Paper [12] focus on processing delay sensitive applications at the edge by broadcasting context instances of the delay sensitive applications being processed at the edge to the nearby node. So therefore the latency is reduced by selecting the nearby fog node.

From the above study, we find that all the papers have not focussed on both the edge considering application features and server side, the edge capacity. Papers [2, 3, 7, 8] have focussed to reduce latency among 2 applications. Research work in [4–6, 9, 10] concentrated more on edge side only not in application perspective. In this paper we propose a light weight delay aware, application aware based on context and resource aware scheduling IoT applications at the Edge which concentrates on both the application context end Edge parameters.

3   Proposed System

In this paper as in Fig. 1, jobs are scheduled on the Edge and the cloud. Exclusively a job scheduling methodology is proposed that sets different priorities based on the information collected from the IoT environment and based on the context instances of the IoT jobs with respect to job criticality, delay and interactivity. The decision of processing the IoT jobs at the edge or cloud is made at edge work station.

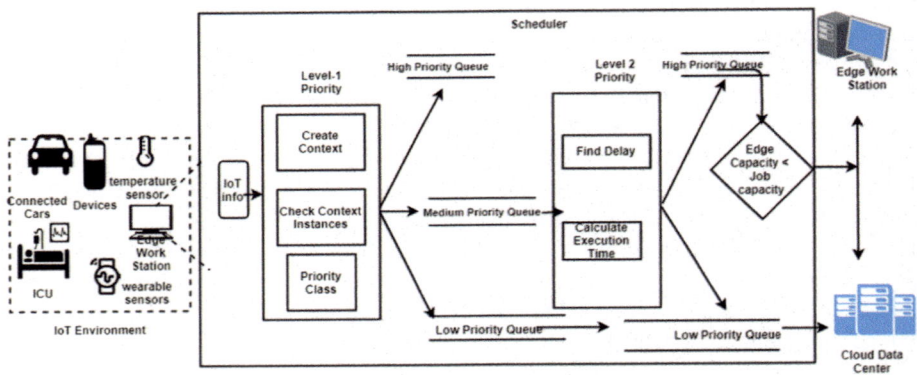

Fig. 1. Overview of scheduling IoT jobs at the edge

Since the Edge device is resource constraint, additionally we also check at the Edge level whether it could handle the low priority jobs based on the storage and CPU capacity of Edge. Thus the critical and delay sensitive jobs are handled at the Edge of the network, thereby improving the latency of the IoT jobs.

Thus the system architecture is divided into two layers. Layer-1 consists of network edge containing sensor network and edge systems and Layer-2 includes the cloud data centre. All the jobs are generated from the IoT environment and will be assigned to the edge platform to check for criticality. Based on the criticality of the job and delay sensitiveness, jobs are processed at edge and cloud.

4 Priority Classification

In order to support the QoS parameter latency for mission critical applications and interactions between applications and IoT devices, the IoT applications must be classified and priority must be assigned based on the 3 factors such as mission criticality, delay sensitiveness, and interactivity of the application (alert or control). Here mission criticality is defined as certain applications may lead to dangerous or life ending situation when any event occurs or certain parameters exceed the threshold value of sensor. For example cardiac arrest occurs when certain parameters such as saturation point, BP level, temperature exceeds threshold value. Delay sensitiveness is defined as maximum time required to respond for an event. Interactivity refers to application response by giving an alert to the environment or any action reply must come from the IoT devices when an abnormal situation or event occurs. Thus priority is assigned based upon the three factors to the IoT applications and it is defined in Table 1.

Table 1. Priority table

S. no	Mission criticality	Delay	Interactivity	Priority class		Tolerable delay (ms)
1	High	High	Alert	I	High	0–5
2	High	High	Control	II		
3	Low	High	Alert	III	Medium	5–15
4	Low	High	Control	IV		
5	Low	Low	Alert & control	V	Low	20–60

5 Context Awareness

The behaviour of the IoT application will be varying in the IoT environment may be considered as context, which is different for different users. The context instance can be timing context i.e. delay tolerant, delay sensitive, arrival time, response time or sensed environment parameters context according to the application such as temperature, humidity, water level, oxygen or mission critical context parameters which leads to stops the business process of IoT application, non-operability of IoT device, life threatening or ending situation.

In this system, context is handled in 2 different ways. In Type 1, based on the environment parameters a context is considered. For example, smoke or fire detection is captured by sensing the environmental parameters such as temperature, humidity and smoke density and comparing with the abnormal value of those parameters. Thus each application have different sensors and they have different threshold values to indicate the criticality of the application.

In Type II, the applications are given a context situation based on the criticality, delay sensitivity or interactivity of the applications. We therefore map the important IoT applications discussed in [1] to the three different priority classes, based on criticality of the application, maximum sensitive delay time and interactivity of the application. We concentrate more to identify criticality of the application based on context.

There are certain applications which is not based upon the sensor values such as banking transactions, websites for e-commerce, and high-speed transaction systems for financial institutions and those applications may not be critical but may be delay sensitive in different context, in that case incoming application is considered as delay sensitive and mapped to medium priority and those applications are again classified to low and high priority applications by considering the maximum sensitive delay time and minimum service time.

Thus the incoming IoT applications may be assigned to 3 different priority classes high, medium, low by comparing the sensor values with its threshold value in the respective application. If an application does not have sensor value, the priority is assigned based on its delay sensitiveness and context.

6   Algorithm

IoT applications such as health care, natural calamities monitoring, POS, Industrial plants monitoring requires less latency and they are sensitive in nature. This algorithm aims to identify applications based on context and process at the edge. Two level priority based scheduling algorithm is applied. In level-one the applications are classified into high, medium and low priority based on the context instances and the critical and delay sensitivity of the applications. In level-two, re-asscessing the medium priority applications which are delay sensitive, but not critical and schedule those applications at the edge based on execution time and its delay tolerance level.

Input: Incoming IoT applications jobs (j1, j2...jn)
> Context Instances of the Application (c1, c2, c3,...cn)

Output: Applications assigned to edge

// To find crticality of application

Step 1: Build Context instances for application use cases based on sensed environment parameter threshold value assign jobs as criticality, delay sensitivity and Interactivity values such as high, low,alert or control.

// Level -I Priority

Step 2: Check the incoming job context instances of the application with context instances already existing in Step 1 and compare with available priority table

> 2.1 If crticality =high and delay sensitivity=high and interactivity=alert or control then
>> Assign the job to high priority queue
>
> 2.2 If crticality=low and delay sensitivity=high and interactivity=alert or control then
>> Assign the job to medium priority queue
>
> 2.3 If crticality=low and delay sensitivity=low and interactivity=alert or control then
>> Assign the job to Low priority queue

// For applications with medium priority – Assign Level II Priority

Step 3: Check the tolerable delay time in the priority table and calculate service time of the application

> Step 3.1 : Sort the applications in the high priority queue as per the time
> Step 3.2 : if application delay time and service time is very low then
>> Assign the job to high priority Queue
>
> Otherwise
>> Assign the job to low priority queue

// Check Edge Capacity

Step 4: If incoming job CPU capacity and storage capacity is less than Edge capacity

> Assign the job to Edge

Otherwise

> Assign the job to Cloud

End For

7   Discussion

In this paper, we focus on creation of context for various IoT application based on time, sensed environment parameters, criticality, type of data, system parameters and priority is assigned to those context, so that they can be scheduled at the edge. Since priority is assigned to the context instances, our system gives a better scheduling when compared with other scheduling approaches such as energy, bandwidth and at server level issues. From the current research survey following techniques [2–12] was studied and our generic model was proposed system to give better result for scheduling.

Table 2. Survey on different techniques

S. no	Technique	Application criticality based scheduling	Priority based scheduling	Context based scheduling
1	Priority [2, 3]	−−	++	−−
2	Data Center [4]	−−	−−	−−
3	SDN [5]	−−	−−	−−
4	Layered architecture [6]	−−	++	−−
5	Algorithm [7, 8]	−−	+−	−−
6	Policy [9–11]	−−	++	−−
7	Machine Learning [11]	++	−−	−−
8	Context instances [12]	−−	−−	++
9	Our proposal	++	++	++

++ strongly recommend, +− moderately recommend, −− not recommend

It is complex to predict, criticality of the application, but we tried to provide a better solution to schedule at edge. But still we are in qualitative study focusing on creating context instances for IoT applications in determining criticality of the application and no more work has done enough in this context with priority based scheduling to reduce latency as shown in Table 2, which limits the possibilities of quantitative research. But in some cases criticality of a application can't be defined clearly, so we also focus to develop fuzzy based approach to define the criticality of the application.

8   Conclusion

Edge computing is a upcoming paradigm to reduce the latency of IoT applications the edge, by considering the characteristics of IoT applications such as criticality, delay sensitivity, interactivity. But the edge device is resource constrained, an efficient scheduling mechanism is needed to prioritize the applications to process at the edge. Therefore, an efficient scheduling is proposed on context and priority based approach that aims to process critical and delay sensitive applications at the edge by assigning more priority to those applications and also ensuring the edge capability before processing the application.

We propose a 2-stage priority scheduling approach to evaluate the criticality of the applications at the first stage based on the context, delay sensitivity at the second stage to process the applications at the edge and confirm the edge capacity prior executing the applications. In future we want to extend with more QoS parameters reliability, safety and additional parameters such energy, bandwidth, data rate and security to process the jobs at the edge.

References

1. Afzal, B., Alvi, S.A., Shah, G.A., Mahmood, W.: Energy efficient context aware traffic scheduling for IoT applications. Ad Hoc Netw. **62**, 101–115 (2017)
2. Wang, H., Gong, J., Zhuang, Y., Shen, H., Lach, J.: HealthEdge: task scheduling for edge computing with health emergency and human behaviour consideration in smart homes. In: IEEE international conference on networking, architecture, and storage (NAS), Aug. 2017
3. Sayuti, H., Rashid, R.A., Mu'azzah, A.L., Hamid, A.H.F.A., Fisal, N., Sarijari, M., Mohd, A., Yusof, K.M., Rahim, R.A.: Lightweight priority scheduling scheme for smart home and ambient assisted living system. Int. J. Dig. Inform. Wirel. Commun. **4**(1), 114–123 (2014)
4. Xu, J., Palanisamy, B., Ludwig, H., Wang, Q.: Zenith: utility-aware resource allocation for edge computing. In: IEEE Edge (2017)
5. Atchyut Kumar Reddy, A., Antony Franklin, A., Bheemarjuna Reddy, T.: Computation task scheduling for edge computing system with SDN. https://www.comsnets.org/archive/2018/docs/assetpapers/36.pdf
6. Tong, L., Li, Y., Gao, W.: A hierarchical edge cloud architecture for mobile computing. In: IEEE INFOCOM 2016: The 35th Annual IEEE International Conference on Computer Communications (2016)
7. Kabirzadeh, S., Rahbari, D., Nickray, M.: A hyper heuristic algorithm for scheduling of fog networks. In: 2017 21st Conference of Open Innovations Association (FRUCT), Helsinki, pp. 148–155 (2017)
8. Fan, J., Wei, X., Wang, T., Lan, T., Subramaniam, S.: Deadline-aware task scheduling in a tiered IoT infrastructure. In: GLOBECOM 2017: 2017 IEEE Global Communications Conference, Singapore, pp. 1–7 (2017)
9. Guan, G.: QoE-aware edge computing for complex IoT event processing (2017)
10. Tan, H., Han, Z., Li, X., Lau, F.C.M.: Online job dispatching and scheduling in edge-clouds. In: IEEE INFOCOM 2017: IEEE conference on computer communications, Atlanta, GA, pp. 1–9 (2017)
11. Bittencourt, L.F., Diaz-Montes, J., Buyya, R., Rana, O.F., Parashar, M.: Mobility-aware application scheduling in fog computing. IEEE Cloud Comput. **4**(2), 26–35 (2017)
12. Roy, D.S., Behera, R.K., Reddy, K.H.K., Buyya, R.: A context aware, fog enabled scheme for real time, cross vertical IoT applications. IEEE Internet of Things J. (2018). https://doi.org/10.1109/jiot.2018.2869323

A Survey on Different Methodologies Involved in Falling, Fire and Smoke Detection Using Image Processing

P. Niveditha$^{(\boxtimes)}$, S. Manasa$^{(\boxtimes)}$, C. A. Nikhitha$^{(\boxtimes)}$,
Hitesh R. Gowda$^{(\boxtimes)}$, and M. Natesh$^{(\boxtimes)}$

Department of Computer Science and Engineering, Vidyavardhaka College
of Engineering, Mysuru, India
spendlikal@gmail.com, manasas986@gmail.com,
nikhithagowda90@gmail.com, hiteshrgowda97@gmail.com,
Natesh.m@vvce.ac.in

Abstract. This paper deals with the summarized study of different methodologies involved in detecting the smoke, fire accidents and falling accidents at the living area using image processing techniques. Smoke are usually observed before there is catch of fire and it can be used as one of the important method to predict the fire. This may reduce the risk of detecting the fire before a great loss. Both smoke and fire can cause a huge loss and to detect that usual methods use sensors whose performance may not be accurate, for example false smoke hence we make use of image processing techniques to detect which is cost efficient and accurate. It is also important to detect human fall incidents to magnify the safety at the living place. In the paper there are various techniques mentioned with their accuracy which can be employed in different applications.

Keywords: Background subtraction · Projection histograms · Video frames · Segmentation

1 Introduction

At present, People tend to use the technology efficiently to make the life easier and secured. Among them, one which is needed to be discussed here is "closed circuit television" (CCTV). It is element of camera which is usually placed at living area, office area and other places where there is need of video surveillance. These varies with sizes and coverage and depending on the application needed these are deployed. They help in capturing the video around. These videos further can be processed to detect any of the applications.

According to statistics, 6% of total population in India are aged above 65 years. They are left unattended sometimes and this may cause trouble few times. Mobility becomes difficult at certain age and assistance is must. There are few premature deaths because of not attending the person on time. They remain on floor after falling for a long time, and this problem is to be addressed. Maintaining the Integrity of the Specifications.

Fire accidents has to be detected as early as possible as they cause heavy loss of life and property. But it is very challenging task to detect fire on time. Usually the sensors

© Springer Nature Switzerland AG 2020
A. P. Pandian et al. (Eds.): ICCBI 2018, LNDECT 31, pp. 378–384, 2020.
https://doi.org/10.1007/978-3-030-24643-3_45

used fail to detect the wide range of flames. The cost to accommodate fire detecting sensors at different places would be very high. Next criteria is range, the range of sensors remain restricted which is again pit hole for using the sensors. In this paper, there are different mechanism involved in detecting fire using image processing.

Smoke is termed to be pre warning to the fire. Before there is fire incident, a smoke is seen which when detected can help preventing the great loss. The main challenge here is differentiating between real and false smokes. False smoke include such as tobacco smoke or incentive sticks which are not vulnerable to any disaster. In this paper there are few methods which can be used to detect the smoke.

The analysis of video includes two stages: (1) Detecting the object. (2) Observing and analyzing the object. Various techniques are mentioned in this paper.

This paper is organized as follows: Sect. 2 presents the existing work, Sect. 3 presents the major methods used in detection followed by Sect. 4 which gives the conclusion of the paper.

2   Existing System

The usage of sensors have following features which are to be implemented:

(i)   Check the status periodically: The system has to check for the fire at the surrounding in regular interval of time to alert. This may cause the time latency which in turn causes the difference between actual incident and detected incident [11, 12].

(ii) Maintenance: When there are components used, it needs separate maintenance to check if they are in proper working condition which also has to be done periodically.

(iii) Network: This particular system uses network such as Wi-Fi for transferring of data. This network will not remain same all the time, if the network is low and it fails to send the data, then it fails in detection also. Figure 1 shows the arrangement of components in the process of detection [11].

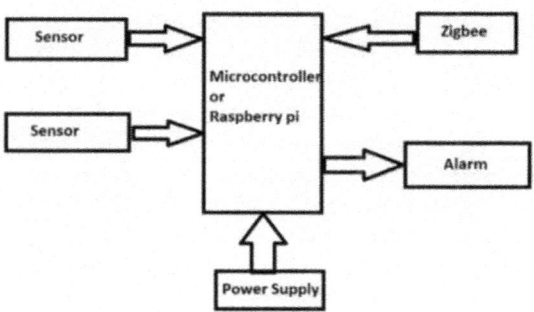

Fig. 1. The basic structure of the components used in fire or smoke detection

3   Literature Survey

3.1   Falling Incident Detection

In this paper, initially shadows, background in the image are removed. Background is noticed as it will remain stagnant in the different frames. Projection histograms are used to identify postures. The main disadvantage is this method cannot differentiate between standing and crouching. The different stages involved in detecting of falling incident is shown in Fig. 2 [1].

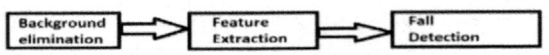

Fig. 2.   Different stages of detection of falling incident

In this paper, uses Omni direction camera which covers all 360° of the area. This helps in detecting the surrounding as a whole. First, noise in the frames are removed, followed by background subtraction. Around the object a rectangle enclosing it is created. The height to width ratio is used to detect the fall of object. When the person is fallen, height to width ratio changes rapidly which is noticeable. Drawback of this method is personal information such as height, BMI are to be known [2].

In this paper, video is recorded at 3 fps. The individual silhouette is segmented from background, this involves identification of object. To point the human this method uses color information and it is segmented. Feature extraction is done by considering the brightest color and eliminating the shadows. Here it makes use of silhouette box. Silhouette box is height to width ratio of segmented object. The drawback of this method is viewing angle [3].

In this paper, initially the noise in the video is removed. Later, background is observed depending on the stagnant pixel and it is also removed which gives only object. Next step involves segmentation, the foreground segmentation is done using horizontal and vertical projection histograms. To detect the falling incident angle between last posture and present posture is calculated. Foreground boundary box is used to consider the angle. Drawback of this method is blind spots [4].

In this paper, camera is distributed at different places. Feature extraction here includes to attributes α and 6. α is defined as aspect ratio of foreground object that is, boundary box of object. When a person falls down the value increases. 6 is defined as the root mean square value of difference of width and height of boundary box of two consecutive frames. Its value increases rapidly when fallen down. This finds disadvantage when the person is wearing tight clothes or have any mask on face, it fails detect the object and hence boundary box changes [5].

In this paper, it used different methodology other than the box model; it uses the combination of best fit approximated ellipse around human body. Segmentation is done by considering silhouette. Projection histograms are applied on the segmented silhouette and the temporal changes of head pose to detect the falling incident. SVM is one of the important classifier. These data is fed into SVM classifier for precise classification, the classification is to detect if the object is fallen or not [6].

In this paper, firstly, Adaptive background subtraction method. It involves clean background image collection. Let number of frames be 'N', calculate 'α' and '6' for RGB pixel [5]. $αi = [αr(i), αg(i), αb(i)]$ $6i = [6r(i), 6g(i), 6b(i)]$

The background keeps changing, hence the background model needs to be updated. $αt (i) = θ$

$Xt(i) + (1 − θ)αt − 1$ $62t(i) = θ(Xt(i) − αt(i))2 + (1 − θ)62t − 1$ where $θ$ is the study rate of background. Background subtraction is based on colour.

The next method involves the falling detection. Let 'x' be width and 'y' be the height of kth frame, 'x1' and 'y1' be the width and height of (k + 1)th frame, then, $μ = x/y$ $σ = [(x − x1)2 + (y − y2)2]1/2$. $μ$ – It gradually increases when fallen down especially when lying down.

$σ$ – It is very small, but when fallen, it changes in a great deal.

The different stages involved in raising an alarm is depicted in Fig. 3.

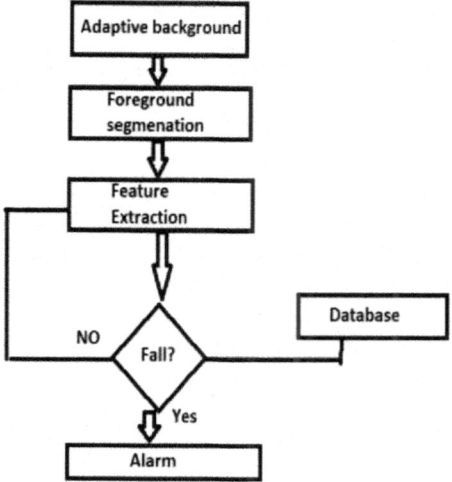

Fig. 3. Stages of raising alarm after the detection of falling incident

3.2 Fire and Smoke Detection

In this paper, it uses expert system design like MES, SVM algorithm and others which uses HSV, HSL, YUV color models for the detection of fire based on color, motion and shape of the fire or smoke which are captured by traditional surveillance camera and also uses various modern techniques like Bayesian classifier for the evaluation of color, area size, surface coarseness, boundary, roughness and skewness to decrease the false positive events. Parallel processing of all these in a system will store it in a database and analyze the data if the fire is hazardous, sends the alarm. In this system, main drawback is storing each frame into the database and false detection of fire or smoke [7].

In this paper, it uses real time fire detectors using surveillance camera which contains both foreground information and color information where foreground objects are detected by background subtraction. In color model, the value of Red color is always greater than green and blue. Drawback is the system also recognizes the person with red cloth or red hat. So it uses fuzzy logic system (FLS) which works on intelligent decisions in wireless sensor network. It includes five functions that are smoke, light, humidity, distance and temperature. It also includes two models, first is luminance model where heuristic rules are replaced with fuzzy logic and second is chrominance model difference between fire and non-fire pixels are discriminated. The main advantage is 99% accurate detection of fire. Main drawback is the false alarm rate detection of 9.50% [8].

In this paper, it uses early fire detection method for the detection of fire and smoke using chromatic and disorder measurements of pixels of fire and smoke. Fire flame will be varied based on different temperature and also with air flow. Chromatic analysis of flames involves HSI (hue/saturation/intensity) color model is suitable for people-oriented way of describing the colors, because of hue and saturation components are intimately related to human beings can perceive color. Dynamic analysis of flames involves the sudden movements of flames, shapes, growing rate, and oscillation (or vibrations) in the infrared response. Fire boxes with the growth rate are used for the checking of a real burning fire. Smoke detection involves from combustion of materials which is normally in grayish colors i.e. light and dark gray regions where it always compares current pixel (mi) with next pixel ($mi + 1$) and always $mi + 1 > mi$. The different steps of this method is depicted in Fig. 4 [9].

In this paper, it uses combined extracted model of color information and motion analysis of color, motion, geometry by the vision based system. The color information is the preprocessing step where YCbCr (Green/Blue/Red) color model is used to extract the luminance from the chrominance for the detection of fire and smoke. It uses Fuzzy logic rather than background subtraction technique because it uses luminance information. The values of color model are as follows: $Y \geq Cb \geq Cr$. The rule table inference in this paper will shows for the detection of fire based on difference between the two color values. Smoke detection involves the detection of smoke pixel which will be white-bluish to white when the temperature is low, when temperature increases the color changes from black-grayish to black. The difference in the color should be less than or equal to threshold value where threshold value is ranging from 15 to 25. The main advantage is 99% correct fire detection rate and smoke detection shows grayish color with different illumination. The developed models can be used as preprocessing stage for fire or smoke detection systems. Drawback is 4.50% false alarm rate [10].

In this paper, it uses motion and edge detection method for the detection of fire. Initially it uses motion detection method in which it uses background estimator technique where it highlight the background and removes it and gives foreground image from a moving image. To get the high quality illumination we use the edge detection of contrast image. In edge detection, the difference in the pixels of an image is used and

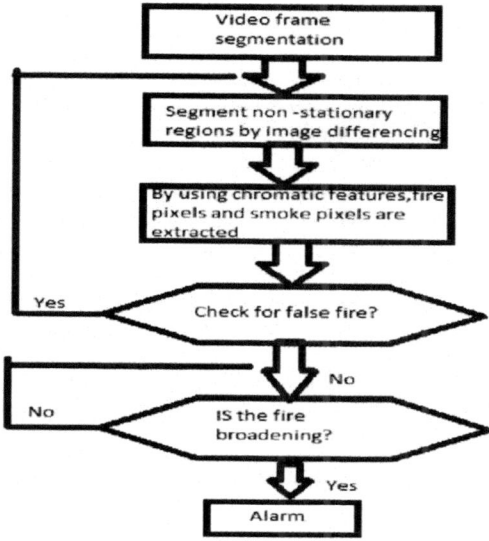

Fig. 4. Various steps involved in detecting the fire

detect the fire pixels in the flame. Both methods are combined for the detection of flame. The flame flicker is detected by Markov model [13, 14].

In this paper, it uses image that are captured from satellite that is from NASA earth perception satellites for the detection of fire and smoke. The image that is captured will be used to identify the irregular temperature that is contrast to the small location from the surroundings. This uses k-means algorithm for the detection with accuracy of 89.5%. Later it uses intelligent fire detector where fire and non-fire pixels of the flame are detected this system may extended up to multipurpose sensors such as smoke detection, gas detection. Sensor uses Wi-Fi framework but if the transmitter and receiver fails to work then the detection of fire and smoke will be very difficult and also the drawback [13].

4   Conclusion and Future Work

In this survey, there are various methods depicted to detect falling incident, fire and smoke. These methods can be deployed inside or outside the living area. These methodologies differ with accuracy rate and application. Depending on the application to be developed, appropriate methodology can be chosen. As mentioned above, the drawbacks in the existing system can be overcome using any of the methodology. This remains cost effective as the software developed can be applied on already installed CCTV soft wares. This becomes one of the most advantages over sensors. In future, may be techniques of machine learning can increase the accuracy of the smoke detection using image processing. The future enhancement of this paper, is to develop a package which deploys all three detections so as to install it in the places such as old age home.

References

1. Cucchiara, R., Pratti, A., Vezani, R.: An intelligent surveillance system for dangerous situation detection in home environments. Intell. Artif. **1**(1), 1115 (2004)
2. Miaou, S., Sung, P., Huang, C.: A customized human fall detection system using omnicamera images and personal information. In: Proceedings of the 1st Distributed Diagnosis and Home Healthcare (D2H2) Conference, Arlington, USA, 2–4 April (2006)
3. Anderson, D., et al.: Recognizing falls from silhouettes. In: Proceedings of the 28th IEEE EMBS Annual International Conference, New York City, USA, 30 August–3 September 2006, pp. 6388–6391 (2006)
4. Nasution, A., Emmanuel, S.: Intelligent video surveillance for monitoring elderly in home environments. In: International Workshop on Multimedia Signal Processing (MMSP), Greece, October (2007)
5. Huang, B., Tian, G., Li, X.: A method for fast fall detection. In: Proceedings of the 7th World Congress on Intelligent Control and Automation, 25–27 June 2008, Chongqing, China (2008)
6. Foroughi, H., Rezvanian, A., Paziraee, A.: Robust fall detection using human shape and multi-class support vector machine. In: Sixth Indian Conference on Computer Vision, Graphics and Image Processing, Bhubaneswar, India, 16–19 December 2008 (2008)
7. Lute, S., Sadaphal, A., Samudra, A., Waghmare, N., Narkhede, A., Dharmadhikari, M., Kolhe, V.L.: Survey paper of approaches for real time fire detection. Int. J. Recent Innov.
8. Jadhav, R., Lambhate, P.D.: A methodological survey for fire detection in camera surveillance. Int. J. Sci. Res. (IJSR) **5**(1), 215–217 (2016). ISSN (Online): 2319-7064 Index Copernicus Value (2013): 6.14 | Impact Factor (2014): 5.611
9. Chen, T.-H., Wu, P.-H., Chiou, Y.-C.: An early fire detection method based on image processing. In: International Conference on Image Processing (ICIP) (2004)
10. Çelik, T., Özkaramanlı, H., Demirel, H.: Fire and smoke detection without sensors: image processing based approach
11. Liu, Y., Qi, M.: The design of building fire monitoring system based on zigbee-wifi networks. In: Eighth International Conference on Measuring Technology and Mechatronics Automation, pp. 733–735. IEEE (2016)
12. Santhana Krishnan, C., Galla, A., Arlapalli Assi, N.: A survey on implementation of fire detection system based on zigbee wi-fi networks. Int. J. Pure Appl. Math. **118**(20), 4249–4253 (2018)
13. Saifullah Abu Bakar, M., De Silva, L.C., Umar, M.M.: State of the art of smoke and fire detection using image processing. Int. J. Sig. Imaging Syst. Eng. **10**(1–2), 22–30 (2017)
14. Fischer, A., Muller, H.C.: A robust fire detection algorithm for temperature and optical smoke density using fuzzy logic. In: International Conference on Security Technology Image Processing for Smart Farming: Detection of Disease and Fruit Grading. 2013 IEEE Second International Conference on Image Processing (1995)

A Study of Deep Learning: Architecture, Algorithm and Comparison

S. T. Shenbagavalli[(⊠)] and D. Shanthi

Department of Computer Science and Engineering, PSNACET, Dindigul, India
stshenbagavalli@gmail.com, dshan71@gmail.com

Abstract. Machine learning highly comes with the broad concept of Deep Learning which is most widely using nowadays. Machine Learning is the study of inspiring PCs to learn and act like people do, and enhance their learning after some time in self-sufficient mold, by nourishing them information and data as perceptions and true connections. Deep Learning portrays a way to deal with discovering that is described by dynamic commitment, inherent inspiration, and an individual scan for significance". Deep Learning is the study of crossing point in the mid of exploration zones of neural networks, Artificial intelligence, graphical modeling, optimization, pattern recognition and Signal processing. The vital explanation behind the notoriety of Deep learning today are the radically expanded chip preparing ability (general-purpose graphical handling units GP-GPUs), altogether expanded size of the information utilized for preparing and continous evolving in the machine learning and flag/data preparing research. The study reveals the concept, different types of deep learning architecture, algorithms and applications of deep learning. This paper gives a simple view for the researchers in deep learning.

Keywords: Deep neural network · Deep auto encoder ·
Deep Boltzmann machine deep belief network · Back-propagation algorithm ·
Stochastic Gradient Descent · Delta-bar-delta algorithm · AdaGrad algorithm ·
RMSProp algorithm · Adam algorithm

1 Introduction

In the study of exploration in Deep Learning is the new domain of Machine Learning, Which has the unique objectives of drawing Machine Learning in Artificial Intelligence. Deep Learning is tied in with taking in various levels of portrayal and deliberation that comprehends information, for example, Images, Sound, and content. A system can reason naturally about articulations in these formal dialects utilizing consistent deduction rules. This is known as the Knowledge Base way to deal with artificial intelligence. Individuals battle to devise formal guidelines with enough unpredictability to precisely depict the world. The troubles looked by frameworks depending on hard-coded information propose that AI Systems require the capacity to procure endemic insight, through extricating designs against crude information. This ability is known as Machine Learning. A basic machine learning algorithm called logistic regression can decide if to prescribe cesarean delivery (Mor-Yosef et al. 1990).

© Springer Nature Switzerland AG 2020
A. P. Pandian et al. (Eds.): ICCBI 2018, LNDECT 31, pp. 385–391, 2020.
https://doi.org/10.1007/978-3-030-24643-3_46

A straightforward machine learning calculation called navie Bayes can isolate authentic email from spam email. The basic enforcement of basic machine learning algorithm confine intensely to exhibit the information which is given. The Machine learning not just finds the mapping from representation to yield yet additionally the portrayal itself. This methodology is known as representation learning. The quintessential case of a representation learning algorithm is the autoencoder. A noteworthy wellspring of trouble in some genuine computerized reasoning applications is that a large number of the elements of variety impact each and every bit of information we can watch. Obviously, it tends to be extremely hard to concentrate such abnormal state, unique highlights from crude information. Deep learning takes care of this focal issue in representation learning by presenting representation that are communicated as far as other, less complex representation [1]. Deep learning empowers the computer to construct complicated ideas into of less intricate ideas. The quintessential case of a Deep learning model is the feed forward Deep system, or multilayer perceptron (MLP). Machine learning is the idea that system program doesn't required any human impedance. Machine learning is a field of artificial intelligence (AI) that keeps a computer's built-in algorithms [2]. The remainder of this paper comes with definition, challenges and benefits of deep learning in Sect. 2. In Sect. 3 the different types of deep learning architecture was explained in detail. Algorithms of deep learning description was given in Sect. 4. Section 5 is applications of deep learning. The major comparison of algorithm was deal with Sect. 6. Section 6 is conclusion.

2 Definition

Deep Learning is a subfield of machine learning worried about calculations motivated by the structure and capacity of the cerebrum called artificial neural systems. The important parts of deep learning are dataset and neural networks. It is large neural networks. In simple, it is defined as it tied with taking in different levels of portrayal and reflection that understands information, for example, Images, Sound, and content. It is also called as Hierarchical feature learning. Example of deep learning includes in the area of customer services, language recognition, Adding color to black-and-white images and videos, deep learning robots etc.

3 Different Types of Deep Learning Architecture

3.1 Deep Neural Network

A deep neural network is a neural network with a specific level of unpredictability, a neural system with in excess of two layers. Profound neural systems utilize modern numerical displaying to process information in complex ways. Deep neural network as network that have an information layer, a yield layer and no less than one concealed layer in the middle. Each layer performs particular kinds of arranging and requesting in a procedure that some allude to as "highlight chain of importance" [3] (Fig. 1).

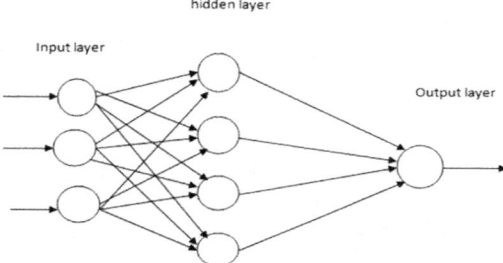

Fig. 1 Deep neural network

3.2 Deep Autoencoder

A deep autoencoder is made out of two, symmetrical deep-belief networks that commonly have four or five shallow layers speaking to the encoding half of the net, and second arrangement of four or five layers that make up the interpreting half. It is mainly designed for feature extraction. It posses same number of inputs and outputs. It is unsupervised learning method. Handling the benchmark dataset MNIST, a Deep autoencoder would utilize twofold changes after each RBM. Deep autoencoders can likewise be utilized for different kinds of datasets with genuine esteemed information, on which you would utilize Gaussian amended changes for the RBMs [3] (Fig. 2).

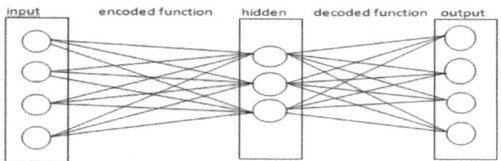

Fig. 2 Deep autoencoder

3.3 Deep Boltzmann Machine

Hinton once alluded to outline of a Nuclear Power plant for instance for understanding Boltzmann Machines. This is a perplexing subject so we will continue gradually to comprehend instinct behind every idea, with least measure of science and material science included. So in easiest early on terms, Boltzmann Machines are essentially separated into two classes: Energy-based Models (EBMs) and Restricted Boltzmann Machines (RBM) [4]. There is additionally another kind of Boltzmann Machine, known as Deep Boltzmann Machines (DBM). A Deep Boltzmann Machine (Salakhutdinov and Hinton 2009b) is a system of symmetrically coupled stochastic twofold units. They are visible units and concealed units. There are associations just between concealed units in neighboring layers, and in addition among visible and concealed units in the primary concealed layer [3] (Fig. 3).

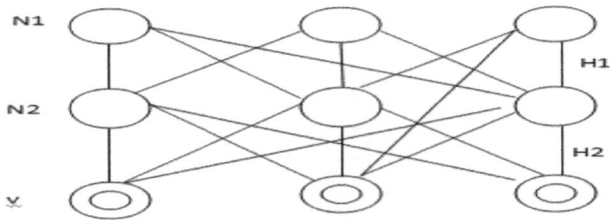

Fig. 3 Deep Boltzmann machine

3.4 Deep Belief Network

It is composition of RBM. When these RBMs are stacked over one another at a point, they are known as Deep Belief Networks (DBN). These DBNs are further sub-separated into Greedy Layer-Wise Training and Wake-Sleep Algorithm. It posses undirected connections at two layers. Both unsupervised and supervised set of networks are acceptable [3] (Fig. 4).

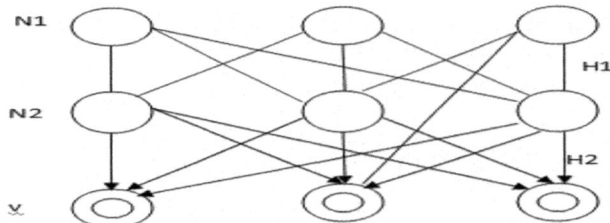

Fig. 4 Deep belief network

4 Algorithms

4.1 Back-Propagation Algorithm

The back-Propagation algorithm, regularly just called back prop, enables the data from the expense to then stream in reverse through the system with the end goal to process the slope. Registering an investigative articulation for the slope is clear, however numerically assessing such an articulation can be computationally costly. The back-Propagation calculation does as such utilizing a straightforward and economical methodology [4].

4.2 Stochastic Gradient Descent

Stochastic Gradient Descent (SGD) and its variations are probably the major utilized enhancement algorithms for machine learning specifically for the profound learning. The learning rate is indispensable parameter required for the SGD algorithm, it might be picked by experimentation, yet it is typically best to pick it by observing expectations to absorb information that plot the target work as a component of time. The most significant property of SGD and related minibatch or online angle based enhancement is that computation times per refresh do not expand through the quantity of preparing precedents. This permits intermingling notwithstanding when the quantities of preparing precedents turn out to be substantial. To think about the union rate of an advancement calculation usually to gauge the overabundance blunder j(ϴ) – minϴ j(ϴ), is the sum through which the existing cost work surpasses the base conceivable expense [5].

4.3 Delta-Bar-Delta Algorithm

The delta-bar-delta algorithm (Jacobs 1988) is an early heuristic technique to deal with adjusting singular learning rates for model parameters amid preparing. The methodology depends on the possibility that, if the incomplete subordinate of the misfortune, as for a given model parameter, continues as before sign, at that point the learning rate should increment. On the off chance that the incomplete subordinate changes sign, at that point the learning rate should diminish [6].

4.4 AdaGrad Algorithm

The AdaGrad algorithm, separately adjusts the learning rates of all the model constraint by balancing them contrarily relative to the square foundation of the entirety every part of the chronicled squared estimations of the angle. The parameters amid the biggest incomplete subsidiary of the misfortune contain a similar quick abatement in their learning rate, whereas parameters with little halfway subordinates include a generally little decline in learning rate. The lattice impact is more noteworthy advancement in the further tenderly inclined ways of parameter space. For preparing profound neural system models, the aggregation of squared slopes from the earliest starting point of preparing it could be resulted in an untimely and inordinate abatement in powerful learning rate. AdaGrad performs lustrous for a few however not all profound learning models [7].

4.5 RMSProp Algorithm

The RMSProp algorithm (Hinton 2012) transforms AdaGrad to execute better in the nonconvex setting by altering the gradient amassing into an exponentially weighted poignant average. RMSProp employs an exponentially perishing standard to discard history from the tremendous past hence it can congregate hastily after finding a convex bowl. At present one of the go-to optimization approach is being employed routinely by deep learning practitioners [8].

4.6 Adam Algorithm

Adam is nevertheless an adaptive learning rate optimization algorithm. The name "Adam" obtain from the phrase "adaptive moments". First, in Adam, momentum is integrated straightly as an estimate of the first-order moment (with exponential weighting) of the gradient [9]. Second, Adam comprises bias corrections to the estimates of both the first-order moments (the momentum term) and the (uncentered) second-order moments to explanation for their initialization at the origin [10].

5 Comparison of Deep Learning Algorithm

See Table 1.

Table 1. Comparsion of algorithms

Algorithm	Uses	Advantage	Disadvantage
Back propagation algorithm [4]	Used to find local minimal error	(a) Simple implementation (b) Mathematical formula of this algorithm can be used by any network (c) When a weight is less then computing time is reduced	(a) Slow and improvident (b) Number of examples of correct behavior can be created easily (c) Outputs can be fuzzy or non numeric
Stochastic gradient descent [5]	(a) Used to solving large-scale machine learning problems. (b) With cost minimization the finding of values of parameters of a function can be done	(a) It is faster than BGD (b) Provides redundant information	(a) Need of of hyper parameters (b) Sensitive in feature scaling
Delta-bar-Delta algorithm [6]	(a) Used to train artificial neural network (b) Used to improve the rate of convergence of kick out algorithm when it substantially increased	(a) Improve convergence speed of weights (b) Always the performance is good (c) Attribute efficient and are proper functions	(a) Weights having its own learning rate (b) The performance cannot be improved
AdaGrad algorithm [7]	Used to find convex objectives with empirical loss Uses Different learning rate for every parameter adaptive technique for learning rate updation is used	(a) Increased computational complexity (b) Improve performance (c) Improved robustness of SGD	(a) Gathering of the squared angles in the denominator (b) Montonically increase of updates
RMSProp algorithm [8]	Used to balance step size Large gradient used to avoid exploding Small gradient used to avoid vanishing	(a) Independent develop (b) It is good to deal with issues (c) Robust optimizer	(a) At present unpublished adaptive learning rate method (b) Has hyper psaramtaer decay_rate
Adam algorithm [9, 10]	Used in all neural networks	(a) Easy to understand	(a) Bugs cannot be cleared

6 Conclusion

Deep Learning is a way to deal with machine discovering that has drawn intensely on information of human mind, insights and connected math as it created over the previous decades. Uses of deep learning are very gainful and are presently utilized by many best innovation organizations including Google, Microsoft, Face book, IBM, Apple, Baidu, Adobe, Netflix, NVIDIA, and NEC. As of late, Deep Learning has seen colossal development in its fame and value, generally as the aftereffect of all the more great PCs, bigger datasets and procedures to prepare further systems.

References

1. Chaudhary, J.R., Patel, A.C.: Machine translation using deep learning: a survey. Int. J. Sci. Res. Sci. Eng. Technol. (2018)
2. Angra, S., Ahuja, S.: Machine learning and its applications: a review. In: 2017 International Conference on Big Data Analytics and Computational Intelligence (ICBDAC) (2017)
3. Ravì, D., Wong, C., Deligianni, F., Berthelot, M., Andreu-Perez, J., Lo, B., Yang, G.-Z.: Deep learning for health informatics. IEEE J. Biomed. Health Informatics **21**(1), 4–21 (2017)
4. Cho, K.H., Raiko, T., Ilin, A.: Gaussian–Bernoulli deep Boltzmann machine. In: The 2013 International Joint Conference on Neural Networks (IJCNN) (2013)
5. Ahmed, W.A.M., Saad, E.S.M., Aziz, E.S.A.: Modified back propagation algorithm for learning artificial neural networks. In: Proceedings of the Eighteenth National Radio Science Conference, NRSC 2001 (2001). (IEEE Cat. No. 01EX462)
6. Roy, S.K., Harand, M.: Constrained stochastic gradient descent: the good practice. In: 2017 International Conference on Digital Image Computing: Techniques and Applications (DICTA) (2017)
7. Sadeghkhani, I., Ketabi, A., Feuillet, R.: Delta-bar-delta and directed random search algorithms to study capacitor banks switching overvoltages. Serb. J. Electr. Eng. **9**(2), 217–229 (2012)
8. Ward, R., Wu, X., Bottou, L.: AdaGrad stepsizes: sharp convergence over nonconvex landscapes (2018). arXiv:1806.01811v5 [stat.ML]
9. http://ruder.io/optimizing-gradient-descent/
10. Kingma, D.P., Ba, J.L.: Adam: a method for stochastic optimization. Published as a Conference Paper at ICLR (2015)
11. Guan, N., Shan, L., Yang, C., Xu, W., Zhang, M.: Delay compensated asynchronous adam algorithm for deep neural networks. In: 2017 IEEE International Symposium on Parallel and Distributed Processing with Applications (2017)

Advanced Access Control Mechanism
for Cloud Based E-wallet

N. Ch. Ravi[1,3](✉), Naresh Babu Muppalaneni[2], R. Sridevi[3],
V. Kamakshi Prasad[3], A. Govardhan[3], and Joshi Padma[4]

[1] Department of CSE, SIET, Hyderabad, India
ravi@saimail.com
[2] Sree Vidyanikethan Engineering College, Tirupati, India
[3] JNTUH College of Engineering (Autonomous), Hyderabad, India
[4] Sreyas Institute of Engineering and Technology, Hyderabad, India

Abstract. The need of cloud based wallet is increasing day by day. But there are several threats to this type of wallet system. There is a chance of unauthentic access in cloud based wallet system as the transactions are made in monetary terms. There is a need to provide security in order to prevent unauthentic access. This research proposes the novel access control mechanism in order to secure the cloud based wallet. In this research implementation of access control mechanism has been made in order to design secure electronic wallet based system. The major concerns regarding wallet security have been discussed. These mechanisms are generally utilized in case of financial sectors. Money of user is usually represented in digital wallet and security of such digital wallet is mandatory. To boost security of such digital Electronic wallets this research has introduced many security techniques that are Session based security, one time password, security from cookies, Capcha, Cryptography, IP authentication. This paper resolves the challenges to security in case of digital wallet.

Keywords: Captcha · OTP · Ecommerce · Cookie · Cryptography · IP authentication

1 Introduction

It has been found that profile dependent Access control [1] in cloud computing Environments is growing day by day. Several unified administrative models [2] in case of role based access control are built to provide security to cloud and remote resources. Several researches used Quantum Key [3] Distribution Protocol for Authentication Mechanism.

However in some cases Role dependent Access Control Model to secure transaction. This type of systems are using RBE [4] scheme in Cloud environment. Different classification frameworks [5] have been developed in order to provide advanced role dependent access control.

The type of electronic card by which online transactions are made by a computer or a Smartphone is known as E-wallet. It is equally useful as a credit card or the debit card. For making payments, E-wallet requires to be connected with a person's bank account.

© Springer Nature Switzerland AG 2020
A. P. Pandian et al. (Eds.): ICCBI 2018, LNDECT 31, pp. 392–399, 2020.
https://doi.org/10.1007/978-3-030-24643-3_47

It is the form of pre-paid account where a user can accumulate his/her money for any online transaction that would be paid in future. A password is given to E-wallet for protection. Online payments for groceries and online purchases can be done with the help of an E-wallet. Flight tickets can also be booked. There are two components of E-wallet, software and information. The software component stores private data. It gives protection & encryption of the data. The component of Information is that database of fine points which are given by user. Here their name, shipping address, the method of payment are filled. The amount paid with details of credit or debit card is also given. In order to set up an E-wallet account users are supposed to install software on their device. Then they enter relevant information during online shopping. E-wallet requires user's data during transaction. In order to access E-wallet user has to provide password for authentication.

2 Scope of Research

Web technologies are bringing us to a new world. But technology has also bought issues that are consisting competing architectures. Internet technology has manipulated the lifestyle of people. Their way to compute and conduct business is changing day by day. Secure cloud based e-wallet system would significant element in E-commerce. Such e-wallets arises the security challenges along with opportunities. Proposed system would allow design and implementation of successful Web application.

The theme of this research is to tackle challenges to security of digital wallet system case of web based commercial application. Distributed user access control [6] would be enhanced in this research considering Multi-factor authentication [7] protection framework in cloud environment [8]. Recent security requirement in cloud computing that are arising due to virtualization [9] have been considered. Research is going to provide several security benefits to system.

1. Attime of storing in database Encryption [10] of user data is done.
2. When user login to allow him to access his own account, decoding data is done.
3. To set constraint for user in order to develop transaction using pattern lock.
4. If user logs in correctly in then he should inserted correct pattern lock.
5. During time of transaction from digital wallet OTP should be generated in order to make user accessible.
6. Quantitative analysis [11, 13] of current security concerns & solutions would be made.
7. These system would optimize IT Infrastructure [12] using enterprise Cloud Model.

3 Methodology

There is need of security guidance for critical areas [14] of focus in Cloud Computing. Systems might be scalable, distributed object location [15] & routing for huge scale peer-to-peer systems. The main purpose of proposed work is to provide security to data at the time of storage.

1. Set security to data during of storing in database.
2. When user is going to submit the information using registration form then data gets encrypted using cryptographic algorithm that's why hacker would be unable to access data of user.
3. Proposed system would allow user to access data when he login to access his own account.
4. Information of user has been stored in encrypted form thus the user would be able to get information during login time. He could make transaction to buy product when user successfully logs in.
5. **Securing transaction**
   Pattern lock has been applied to stop user to make transaction frequently. He would be eligible to perform transactions once user set the valid pattern.
6. **Securing digital wallet:** Digital wallet is going to allow user to buy product from his balance. We are going to develop digital wallet available to him if user is correctly logged in as well as he sets valid pattern lock.
7. **OTP security during transaction**
   During transaction from digital wallet one time password is generated thus it is accessed by user. Such OTP is usually sent to him via email or sms.

4　Proposed Architecture

In case of proposed work the security technology to provide security to digital wallet have been introduced. These techniques are one time password, storing encrypted information in database and using pattern lock on each transaction in order to secure them (Fig. 1).

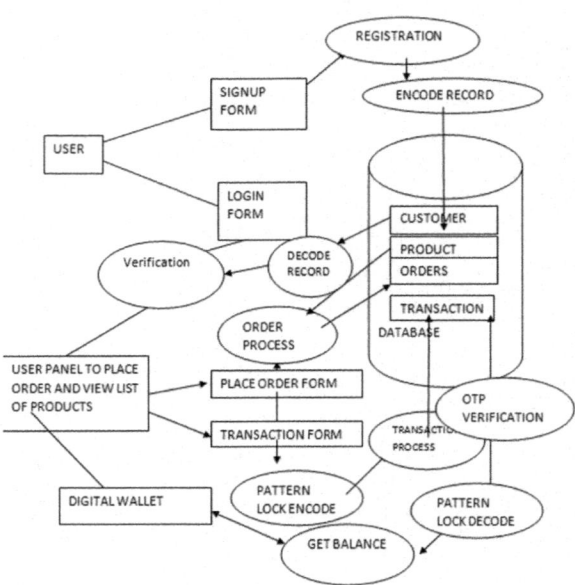

Fig. 1. Proposed architecture

Algorithm for proposed access mechanism

1. Get input (INu) from user while registration
2. Encode the user information and store in customer database using encryption mechanism

$$ENC_{INU} = ENCRYPT(INu) \tag{1}$$

3. At the time of login get input(LINu) from user in login form
4. Verify the validity of user after decode data stored in database using decryption

$$DEC_{INU} = DECRYPT(ENC_{INU}) \tag{2}$$

5. If LINu = DEC_{INU} is true set USER_AUTH = TRUE

   Else
   USER_AUTH = FALSE

1. Place the order (ORu) and process
2. Make payment PMTu for ORu using pattern lock PATu encoding scheme. On success set PAT_GEN = TRUE else PAT_GEN = FALSE

$$\{PATu, OTP\} = PATTERN\_LOCK(PMTu) \tag{3}$$

3. Get the OTP on user end to confirm transaction
4. Validate transaction the OTP in order to confirm the transaction.
5. VAL_TR would be TRUE if validation process is successful and it would be FALSE if validation process fails.

$$VAL\_TR = VALIDATE\_TRANSACTION(OTP) \tag{4}$$

Equation for success and Failure of transaction

$$RESULT = USER\_AUTH * PAT\_GEN * VAL\_TR$$

If RESULT is true then the transaction would be successful otherwise transaction would fail.

5 Result and Discussion

This system is beneficial for banking and financial sector. User once login using this username and password. In this section we are discussing implementation of electronic wallet system (Fig. 2).

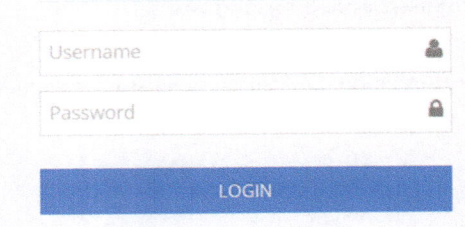

Fig. 2. User login

A session is created for him then he becomes able to access resources as well as he could modify his personal details in this panel.

The user is allowed to invest online in fixed deposit and we could also view his account statement.

Summary of return from saving and summary of return from fixed deposit are shown in such system (Fig. 3).

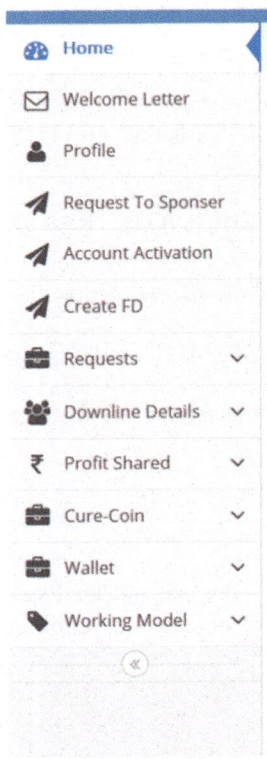

Fig. 3. User panel

In user panel user could select profit shared to get details of return from saving and return from fixed deposit and team commission. Here list of return from fixed deposit, return from saving, and team commission id displayed.

The summary of coin as well as wallet coin history, transfer history, received history has been represented

Users has been are allowed to perform fixed deposit operations from his wallet balance to get additional returns (Fig. 4).

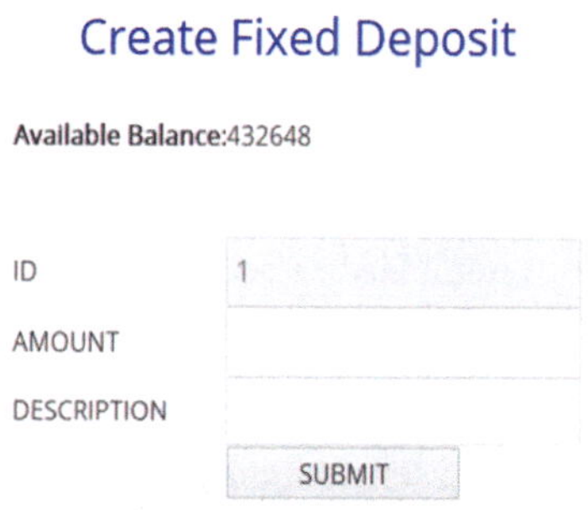

Create Fixed Deposit

Available Balance:432648

ID	1
AMOUNT	
DESCRIPTION	
	SUBMIT

Fig. 4. Fixed deposit creation

User could make activation of his own account using balance fund in different packages (Fig. 5).

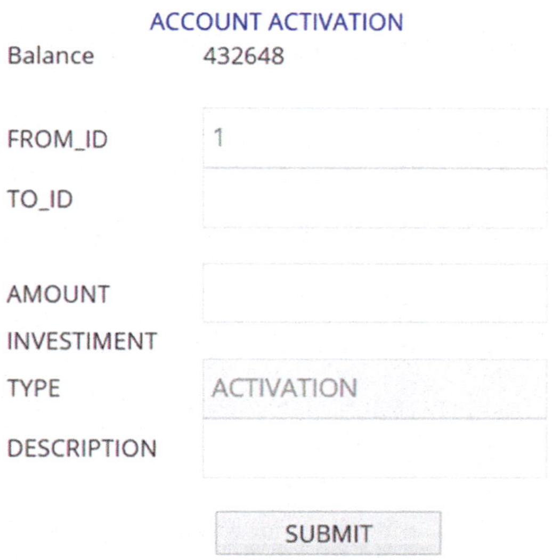

ACCOUNT ACTIVATION

Balance	432648
FROM_ID	1
TO_ID	
AMOUNT	
INVESTIMENT TYPE	ACTIVATION
DESCRIPTION	
	SUBMIT

Fig. 5. Account activation process

Users are allowed to edit profile online along with management of login password as well as transaction password. If user need to view account details then he use manage login password. And if he want to make online transaction then he use transaction password.

6 Conclusion

Proposed system would be better as compared to existing e-Wallet System. It would provide security guidance in case of critical areas. Proposed Systems would be scalable and suitable in case of huge scale peer-to-peer systems.

Research has introduced security during registration and transaction time. There remains threat in case of digital wallet due to several types of attacks. As it is common during online transaction security threat raises. Thus such research could be known as effort to make e-commerce system further secure. This system would prevent unauthentic operations. Proposed work would definitely assist in securing electronic commerce transaction. It is well known that there would be two situations of online transaction. First case is circumstances when user is going to pay for product using bank account. In second situation, user is paying for product using digital wallet. There remains risk in such circumstances as banking sites have made their own security system. But in case of digital wallet security of user's amount is provided by its makers. In proposed work objective is to make such digital wallet in order to secure it with help of one time password facility and pattern lock.

References

1. Naushahi, U.M.A.: Profile based access control in case of cloud computing environments with applications in health care systems
2. Biswas, P., Sandhu(B), R., Krishnan, R.: Uni-Arbac: a unified administrative model for role-based access control
3. Khalid, R., Zukarnain, Z.A., Hanapi, Z.M., Mohamed, M.A.: Authentication mechanism in case of cloud network and its fitness with quantum key distribution protocol: a survey (2015)
4. Warang, S.: Role based access control model for cloud computing using RBE scheme (2016)
5. Alm, C., Drouineaud, M., Faltin, U., Sohr, K., Wolf, R.: A classification framework designed for advanced role-based access control models and mechanisms
6. Wang, H., Li, Q.: Distributed user access control in sensor networks
7. Soni, P., Sahoo, M.: Multi-factor authentication protection framework in cloud computing (2015)
8. Mill, P., Grance, T.: The NIST definition of cloud computing. National Institute of Standards & Technology, Gaitherbsburg, MD 20899-8930, NIST Special Publication 800-145 (2011)
9. Messmer, E.: New security demands arising for virtualization, cloud computing. security-demands-arising-for-virtualization—cloud computing.html (2011)
10. Kaushik, S., Singhal, A.: Network security using cryptographic techniques. Int. J. Adv. Res. Comput. Sci. Softw. Eng. 2(12), 105–107 (2012)
11. Miers, C., Redigolo, F., Simplicio, M.: A quantitative analysis of current security concerns & solutions for cloud computing. J. Cloud Comput. Adv. Syst. Appl. 1(1), 11 (2012)

12. Padhay, R.P.: An enterprise cloud model for optimizing IT infrastructure. Int. J. Cloud Comput. Serv. Sci. (IJ-CLOSER) **1**(3), 123 (2012)
13. Gonzalez, N., et al.: A quantitative analysis of current security concerns & solutions for cloud computing. J. Cloud Comput. Adv. Syst. Appl. **1**(1), 11 (2012). https://doi.org/10.1186/2192-113x-1-11
14. CSA: Security guidance for critical areas of focus in cloud computing. Technical report, Cloud Security Alliance (2009)
15. Rowstron, A., Druschel, P.: Pastry: scalable, distributed object location & routing for large-scale peer-to-peer systems. Accepted for Middleware (2001)
16. Zhao, B.Y., Kubiatowicz, J., Joseph, A.: Tapestry: an infrastructure for fault tolerant wide-area location & routing (2001)
17. Richa A.W., Plaxton, C.G., Rajaraman, R.: Accessing nearby copies of replicated objects in a distributed environment. In: Proceedings of ACM SPAA, pp 311–320 (1997)
18. Saroiu, S., Gummadi, P.K., Gribble, S.D.: A measurement study of peer-to- peer file sharing systems (2001)
19. Stoica, I., Morris, R., Karger, D., Kaashoek, M.F., Balakrishnan, H.: Chord: a peer-to-peer lookup service for internet applications (2001)

Novel Pentagonal Shape Meander Fractal Monopole Antenna for UWB Applications

Sheetal Bukkawar[1(✉)] and Vasif Ahmed[2]

[1] Saraswati College of Engineering, Mumbai, India
sheetalvb@gmail.com
[2] Babasaheb Naik College of Engineering, Pusad, Pusad, India
ahmedvasif@gmail.com

Abstract. The pentagonal shaped UWB monopole antenna is presented in here. The proposed antenna structure consists of Pentagonal shape patch and meander fractal technique with two iterations at all the sides of the patch. The bandwidth is enhanced using meander fractal technique to all sides of a pentagonal patch which results in improved effective path length of current and reduced the lower frequency of the UWB band to 2.63 GHz. Two iterations with suitable iteration factors are used for the same. The designed antenna obtained S11 < −9.5 dB over 2.63 GHz–12.0 GHz band and suitable for all UWB applications. From the measured performance of the antenna, it is concluded that it has omnidirectional radiation pattern. Proposed antenna is designed on a single layered FR4 substrate with a dimensions of $20 \times 33 \times 1.6$ mm^3 and simulated and measured antenna parameters are good in agreement.

Keywords: Antenna · Fractal · Meander · Ultra-wide band (UWB) · Pentagonal

1 Introduction

The demand of today's world of wireless communication is increased bandwidth with miniaturized systems. Various techniques have been developed for miniaturization of antenna size. Fractal geometries has its significant impact in many areas of research in engineering; one of which is antennas. These fractal geometries possess self-similarity and space filling properties which makes it one of the effective method used for antenna miniaturization. Initially developed [16] fractal antenna contains different geometries like Sierpinski gasket geometry (triangular shape), Sierpinski carpet geometry (square shape), Koch curve geometry (V shape), Herbert curve geometry (U shape). It is seen that fractals are nothing but self-similar shapes that can be subdivided in parts such that each part is a reduced size copy of the initiator. Fractal Antenna technology leads to unique improvements in antenna designing, increasing their bandwidth with reduced dimensions.

For short-range wireless communication systems, Ultra-wide bandwidth (UWB) wireless technology is prominently required, which uses ultra-wide band ranging from 3.1 GHz to 10.6 GHz. So, to cater the need of increased bandwidth with miniaturized antenna size, Ultra-wide band fractal antenna is one of the key solution.

© Springer Nature Switzerland AG 2020
A. P. Pandian et al. (Eds.): ICCBI 2018, LNDECT 31, pp. 400–408, 2020.
https://doi.org/10.1007/978-3-030-24643-3_48

Various techniques have been reported to achieve multiband and wideband characteristics using fractals, include a printed S shaped fractal monopole antenna, confirmed quad band operation with omnidirectional radiation pattern [2]. A novel U-shaped fractal monopole antenna was designed to achieve UWB band with stop band of 5.15 to 5.825 GHz using split ring resonator at ground plane [3]. Compact fractal antenna of shape hexagonal sierpinski for super wideband application covering frequency band of 3.4 to 37.4 GHz was discussed [4]. A hexagonal UWB monopole antenna with trapezoidal elements achieved a band of frequencies 3 to 13 GHz [5]. A hepta band characteristic with a maximum gain of 3.1 dB was reported using a slot loop antenna with minkowski fractal shape [7]. UWB band of 3.1 to 11 GHz with maximum gain of 5.8 dB was achieved using equilateral triangle fractals in conventional UWB antenna [6]. In [8] the tilted square shaped UWB antenna was used and a bandwidth of 3.9 to 26.7 GHz is achieved. A reconfigurable octagonal shaped fractal UWB antenna designed with the help of sierpinski geometry was discussed [9] and using multiple p-i-n diodes multiple UWB bands were achieved. A dual band UWB monopole antenna of maple leaf shape was conversed which covered a band of frequencies 1.7 to 11.1 GHz with two band notches centered at 4.3 GHz and 7.7 GHz. An UWB band was realized using a rectangle tree fractal monopole antenna [11]. Koch Snowflake shaped UWB antenna [12] was investigated and a band of 3.4 to 16.4 GHz was achieved. In [15], a Koch fractal and hexagonal geometry was combined and a UWB band of frequencies 3 to 12.8 GHz was attained.

This communication presents a pentagonal meander monopole fractal antenna with two iterations for UWB application. The antenna offers a frequency band of 2.63 to 12 GHz. The use of meander fractal technique increases the effective current path length, due to which the frequency of lower UWB band is reduced to 2.63 GHz from 3.1 GHz. The subsequent sections deal with the proposed antenna geometry, discussion on simulation and experimental results.

2   Antenna Geometry Analysis and Design

2.1   Fractal Geometry Design

The proposed fractal monopole antenna structure is designed with dimensions as specified in Table 1. As depicted in Fig. 1. The antenna structure with ground plane dimensions 'Lg × Wg' is fed with 50 Ω microstrip line. To obtain the desired impedance bandwidth, the gap 'g' is set between ground plane and pentagonal meander fractal monopole. The designed structure of proposed antenna is carried out on FR4 substrate of dielectric constant 4.4 and loss tangent of 0.02.

Figure 1 shows the step by step evolution of the proposed Pentagonal Meander Fractal Monopole UWB Antenna. Antenna 1 of Fig. 1(a) is the initiator for the proposed antenna geometry. The radius 'R' of pentagonal shaped patch of an initiator is calculated as 10.2 mm [1].

For simulation, Method of Moments based IE3D software is used. The optimum dimensions of the proposed antenna are specified in Table 1. With the help of ground plane dimensions and gap, impedance bandwidth of 2.63 GHz to 12 GHz is achieved. All dimensions are in mm only.

Table 1. Optimized dimensions of the proposed antenna

Parameter	Dimension	Parameter	Dimension
Lg	20 mm	W_1	2 mm
Wg	14 mm	g_1	2.5 mm
R	10.5 mm	L_2	2 mm
L	12 mm	W_2	1 mm
g	0.5 mm	g_2	0.95 mm
L_1	4 mm	S	6 mm

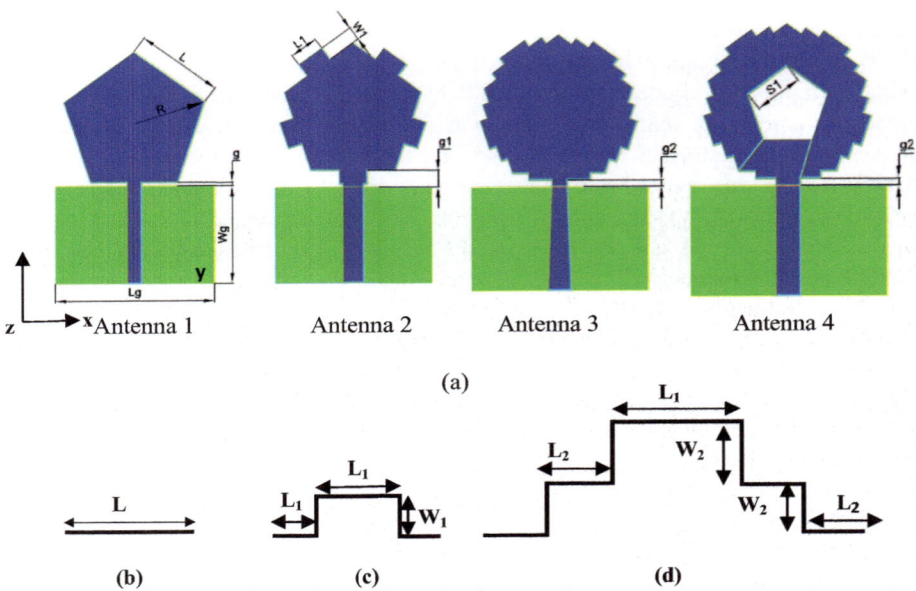

Fig. 1. (a) Step by step evolution of the proposed pentagonal meander fractal monopole UWB Antenna, (b) Initiator length (Antenna 1), (c) meander fractal for iteration 1 (Antenna 2), (d) meander fractal for iteration (Antenna 3 and 4)

For Antenna 2, the meander fractal technique is applied at the sides of the pentagonal patch. The side length of the basic pentagonal patch is 12 mm, which is divided into three equal parts of 4 mm each and a meander slot of 4 mm is attached at the middle of the side length. The design procedure of meander fractal is shown in Fig. 1 (c) and meander lengths for first and second iterations are calculated as,

$$L1 = m1 \, x \, L \tag{1}$$

where, m1 (iteration factor) = 1/3

$$ie.\,L1 = 1/3\,x\,12 = 4\,\text{mm}$$
$$L2 = m2\,x\,L1 \tag{2}$$

Where, m2 (iteration factor) = 1/2

$$ie.L2 = 1/2\,x\,4 = 2\,\text{mm}$$

The different phases of designing of fractal geometries with above meander lengths are shown in Fig. 1.

2.2 Effect of Meander Technique

Antenna 1 is the initiator for the proposed antenna. The circumference of initiator i.e. Basic pentagonal patch is 58 mm. Antenna 2 is first iteration of the proposed meander fractal antenna and because of meander technique the circumference of Antenna 2 increased to 72 mm. Antenna 3 is the geometry with second meander iteration [13] with circumference 76 mm, Fig. 1(d) and Eq. (2) gives the details of how it is carried out. It is observed that surface current distribution is less at the center of Antenna 3 therefore the pentagonal shaped patch with optimized side lengths of 6 mm is removed which again increases the effective current path length (Antenna 4).

From Fig. 2, it is seen that due to meander technique there is increase in effective current path resulting in improved UWB bandwidth with an improvement in return loss plot. With two iterations, the UWB band from 2.63 GHz to 12 GHz is achieved for final antenna prototype. The antenna achieves fundamental mode at 3.4 GHz while the higher order modes are achieved at 7.6 GHz and 10.8 GHz.

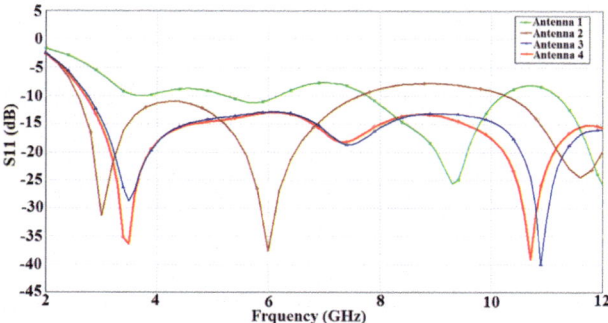

Fig. 2. Comparison of the simulated return loss versus frequency plot of Antenna 1, antenna 2, Antenna 3 and Antenna 4

2.3 Surface Current Distribution

The surface current distribution of the initiator and final proposed structure at 2.8 GHz, 5.8 GHz, 7.8 GHz & 10.8 GHz are shown in Fig. 3, it is observed that magnitude of surface current increases with increase in frequency [14]. It is perceived that the maximum current density is seen in the feedline, edges of the ground plane and edges of the patch.

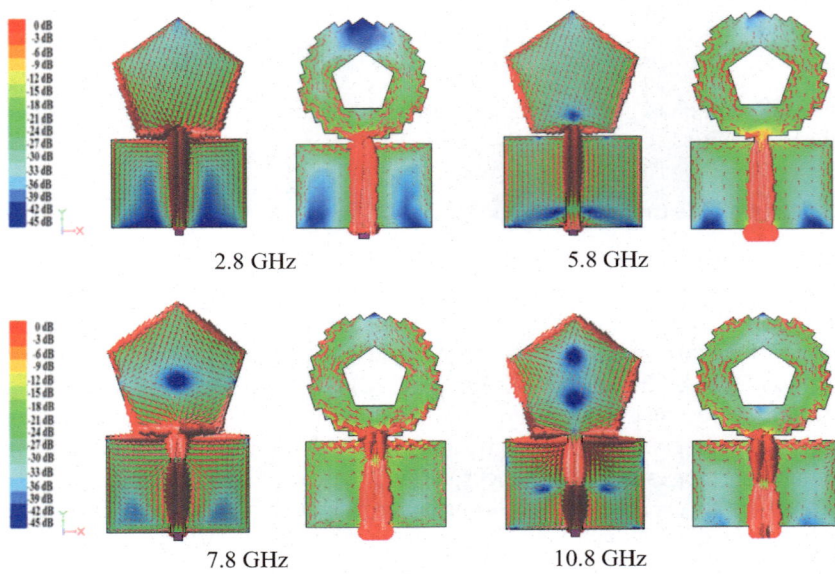

2.8 GHz   5.8 GHz

7.8 GHz   10.8 GHz

Fig. 3. Surface current distribution of proposed antenna at frequencies 2.8 GHz, 5.8 GHz, 7.8 GHz and 10.8 GHz

3   Results and Discussions

The optimum dimensions of the pentagonal meander fractal monopole antenna are determined using Method of Moments based IE3D software version 15.10. The designed antenna is fabricated and tested. A fabricated Pentagonal Meander Fractal monopole UWB antenna is shown in Fig. 4. S11 is measurements are carried out using Agilent N9916A network analyzer. The measured results were compared with simulated results and it is observed that measured results are in good agreement with the simulation results, whereas minor variations are observed which may be because of errors during fabrication of antenna and while loading of SMA connector.

The presented work is compared with the fractal UWB structures available in the literature and the comparison of dimensions and bandwidth is listed in Table 2. It is seen that antennas with lesser dimensions compared to proposed antenna compromised in bandwidth.

Figure 5 shows the measured and simulated plot of return loss which are comparable to each other. The slight difference in result between the measured and simulated is due to fabrication error and loading of SMA connector. As shown in Fig. 6 peak gain of 4.8 dB is achieved.

Fig. 4. Prototype of fabricated antenna

Table 2. Comparison of proposed antenna dimensions and bandwidth with other available structures

Sr. no.	Antenna	Dimensions (mm³)	Bandwidth (GHz)
1	[2]	17 × 18 × 1.6	2.32–2.94, 5.113–5.579, 6.890–7.011 & 8.206–9.286
2	[4]	30 × 28 × 1.6	3.4–37.4
3	[5]	35 × 41 × 1.524	3–13
4	[6]	24.5 × 21.6 × 3.15	3.1–11
5	[8]	25 × 25 × 1.5	3.9–26.7
6	[10]	34 × 33 × 1.575	1.7–11.1
7	[11]	30 × 23 × 1	3.1–10.6
8	[12]	31 × 27 × 1.6	3.5–14.55
9	This work	20 × 33 × 1.6	2.63–12

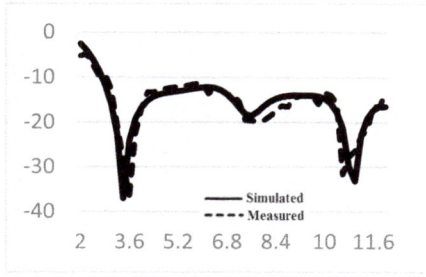
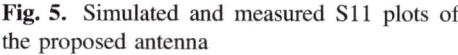

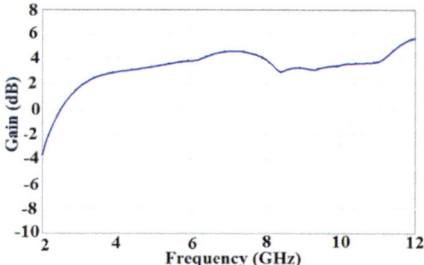

Fig. 5. Simulated and measured S11 plots of the proposed antenna

Fig. 6. Simulated peak realized Gain vs frequency plot for final proposed antenna

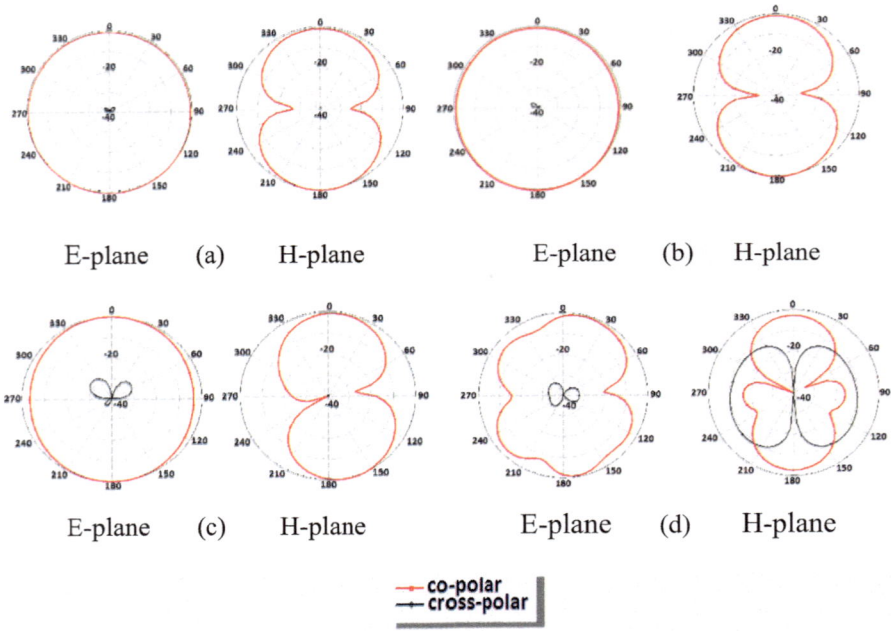

E-plane   (a)   H-plane   E-plane   (b)   H-plane

E-plane   (c)   H-plane   E-plane   (d)   H-plane

co-polar
cross-polar

Fig. 7. Simulated Radiation patterns for proposed antenna at (a) 2.8 GHz, (b) 3.4 GHz, (c) 5.8 GHz, and (d) 9 GHz

In x-z plane, radiation patterns are omnidirectional for frequencies above 2.63 GHz. As depicted in Fig. 7, up to 9 GHz, cross polar component is increased with the increase in frequency but 20 dB lower in broadside direction. Figure 7(a) & (b) illustrates the radiation pattern at 2.8 GHz and 3.4 GHz. As there is an asymmetry in the structure along two directions, variation in directivity in two orthogonal planes, less than 3 dB is observed over lower UWB.

At 5.8 GHz, shown in Fig. 6(c) radiation pattern, a tilt in the H- plane is observed and in E- plane, the difference in the directivity improves upto 10 dB along two orthogonal directions. It is seen that due to asymmetry in the structure, the radiation patterns are more directive towards z-direction. The cross polar components increase more than 20 dB at 9 GHz in E- plane.

To analyse the undesired distortion in received signal, Time domain analysis of an antenna is important. Here, it is executed by placing two antenna structures as transmitter and receiver in face to face configuration, at a distance of 30 cm from each other. Time domain analysis abetted to learn the antenna behavior while transmitting and receiving the signal in time domain. A Gaussian pulse of 1 GHz bandwidth is transmitted and received. The normalized magnitudes of the incident and received signals are shown in Fig. 8. Some variations are observed in the received pulse that may be due to fractal structure of the antenna. The effect of same is also observed in the group delay characteristics as shown in Fig. 9.

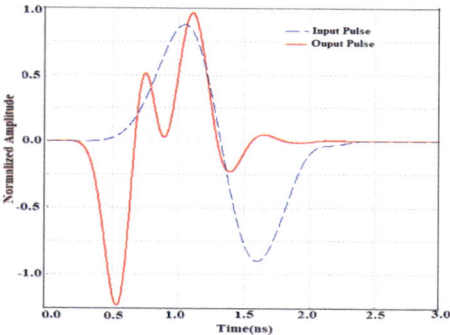

Fig. 8. Normalized magnitudes incident and received pulses versus time characteristics

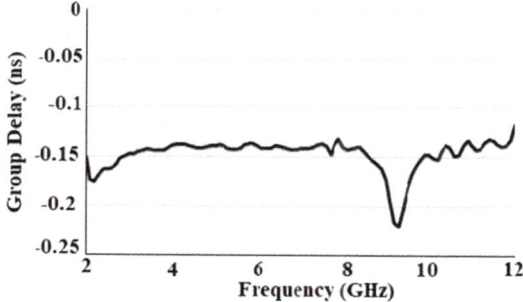

Fig. 9. Group delay against frequency characteristics.

4   Conclusion

A pentagonal shaped fractal antenna using meander technique is investigated for UWB applications. The proposed antenna is fabricated and tested. The measured results are in good agreement with the simulated results. It gives good impedance and radiation characteristics. The lower frequency of ultra-wideband is lowered to 2.63 GHz with the successive iterations of fractal antenna. The gain variation for the proposed antenna is also less than 3 dB for the entire UWB band. The radiation patterns are omnidirectional in E- plane for all frequencies. Hence, the proposed antenna can be a good candidate for UWB application.

References

1. Ray, K.P.: Design aspects of printed monopole antennas for ultra-wide band applications. Int. J. Antennas Propag. **2008**, 1–8 (2008)
2. Cheong, H.R., Yeap, K., Lai, K., Nisar, H.: A compact CPW-fed antenna with fractal S-shaped patches for multiband applications. Microw. Opt. Technol. Lett. **59**, 541–546 (2017)

3. Rajeshkumar, V., Raghavan, S.: Bandwidth enhanced compact fractal antenna for UWB applications with 5–6 GHz band rejection. Microw. Opt. Technol. Lett. **57**, 607–613 (2015)
4. Singhal, S., Singh, K.A.: CPW-fed hexagonal sierpinski super wideband fractal antenna. IET Microw. Antennas Propag. **6**(13), 1407–1414 (2016)
5. Aissaoui, D., Hacen, N., Denidni, T.: UWB Hexagonal monopole fractal antenna with additional trapezoidal elements. In: International conference on ubiquitous Wireless Broadband (ICUWB). IEEE (2015)
6. Singha, R., Vakula, D.: Compact ultra-wideband fractal monopole antenna with high gain using single layer superstrate. Microw. Opt. Technol. Lett. **59**, 482–488 (2017)
7. Dhar, S., Patra, K., Gatak, R., Poddar, D.: A dielectric resonator loaded minkowski fractal shaped slot looped heptaband antenna. IEEE Trans. Antennas Propag. **63**, 1521–1529 (2015)
8. Bauda, S., Kumar, D.: A compact broadband printed antenna with tilted patch. Microw. Opt. Technol. Lett. **58**, 1733–1738 (2016)
9. Tripathi, S., Mohan, A., Yadav, S.: A compact frequency reconfigurable fractal UWB antenna using reconfigurable ground plane. Microw. Opt. Technol. Lett. **59**, 1800–1808 (2017)
10. Iqbal, A., Sararah, O., Jaiswal, S.: Maple leaf shaped UWB monopole antenna with dual band notch functionality. Progr. Electromagn. Res. C **71**, 169–175 (2017)
11. Zhangfang, H., Hu, Y., Luo, Y., Xin, W.: A novel rectangle tree fractal UWB antenna with dual band notch characteristics. Progr. Electromagn. Res. C **68**, 21–30 (2016)
12. Tizyi, H., Riouch, F., Tribak, A., Najid, A., Mediavilla, A.: CPW and microstrip line-fed compact fractal antenna for UWB-RFID applications. Progr. Electromagn. Res. C **65**, 201–209 (2016)
13. Singhal, S., Singh, P., Singh, A.K.: Asymmetrically CPW fed octagonal sierpinski UWB fractal antenna. Microw. Opt. Technol. Lett. **58**, 1738–1745 (2016)
14. Li, D., Mao, J.-F.: A Koch-like sided fractal bow-tie dipole antenna. IEEE Trans. Antennas Propag. **60**(5), 2242–2251 (2012)
15. Tripathi, S., Mohan, A., Yadav, S.: Hexagonal fractal ultra-wideband antenna using Koch geometry with bandwidth enhancement. IET Microw. Antennas Propag. **8**(15), 1445–1450 (2014)
16. Safia, O.A., Nedil, M.: Ultra-broadband v-band fractal T-square antenna. In: IEEE International Symposium on Antennas and Propagation & USNC/URSI National Radio Science Meeting, pp. 2611–2612 (2017)

Gas Leakage Detection Device [GLDD] Using IoT with SMS Alert

Leena, Monisha[(⊠)], M. Dakshayini, and P. Jayarekha

Department of ISE, BMS College of Engineering, Bengaluru, India
{lbml6scnl0, monisha.scnl7, dakshayini.ise}@bmsce.ac.in

Abstract. In recent days as usage of gas is flexible and cheaper, it has been increased from home to hotels, industries, office etc. Besides its advantageous, gas leakage is becoming a major concern at all the places wherever it is being used as it may lead to a small accident or sometimes major disaster causing a great loss of infrastructures and lives of people. With the available advanced IoT and Mobile technologies, it is possible to build a system to prevent such accidents and informing the right person in case of tragedy. Hence, this paper has designed and developed a device which is capable of detecting the gas leakage and send alert message to specified mobile number. This proposed device could detect leakage of LPG, Methane, Butane or any such petroleum based gaseous substance using MQ5 Sensor (gas sensor) and send SMS to registered mobile number (by inserting sim in GSM module) and when the gas leakage is detected by the gas sensor then it sends an SMS alert to the registered mobile number thus notifying the owner to be careful and to take careful measures thus saving their lives and things. Implementation results of this device have proved its effectiveness.

Keywords: Gas leakage · SMS ·
GSM (Global System for Mobile Communication) · GAS sensor · Arduino ·
IoT (Internet of Things)

1 Introduction

IoT systems are widely used by many users across the globe. It helps users by achieving automation, integration and analysis. They reach out to all these areas with accuracy. IoT uses the emerging technology in many fields like sensing, actuating, robotics and many more [8]. IoT has become popular as it exploits recent advances in software, hardware and the modern attitude towards technology. Its new advances has bought major changes in the goods and services, social and economic impacts in the society.

Internet of Things is a globally connected network of things which is dynamic in nature with self-configuring capabilities based on the various communication protocol. These things could be virtual and physical "things" have Personalities which is virtual in nature, physical attributes and different identities, use intelligent interfaces and which communicate data with their users and their environments. The advantages of IoT has spread widely across every area of lifestyle and business. some of them could

© Springer Nature Switzerland AG 2020
A. P. Pandian et al. (Eds.): ICCBI 2018, LNDECT 31, pp. 409–416, 2020.
https://doi.org/10.1007/978-3-030-24643-3_49

be, improved Customer Engagement –IoT systems engage customers efficiently by giving accurate results. Technology with optimal solution – IoT systems gives optimal solutions to the same technologies by improvising the customer experience and potential, Reduced Waste – IoT makes use of resources effectively and reduce the waste by effective management of resources. Enhanced Data Collection –Iot systems gives those data which will be helpful for users in taking decision and to analyse. It helps by giving an accurate picture [13]. Simplifies work- It simplifies the users work by monitoring and sensing of some activity in the environment and sending the alarm or notification to the users depending on the scenarios. Applications of IoT span wide range of domains including Homes: Smart lighting in places like living area and hall, smart appliances used in kitchen [7], Intrusion detection system for detecting any foreign bodies or other people in the environment, smart smoke detecting system to detect the leakage of smoke in the house. Cities: smart Parking [1] which helps in effective parking of vehicles, smart lighting to save wastage of energy, smart roads, structural health monitoring system. Environment: weather monitoring, air and noise pollution which monitors the content of poisonous gases present in the air, forest fire detection system, river flood detection system by sensing the level of the river. Energy systems: smart grid. Retail: Inventory management, smart payments using bar codes etc., & smart vending machines. Industry: machine diagnosis, prognosis systems, indoor air quality systems. Agriculture: smart irrigation, green house control system. Health: health and fitness monitoring system, wearable electronics etc.

In current years, detection of inflammable gas leakage is one of the demanding work to be done as the society is aware that the negligence of the gas leakage would lead to accidents causing human life risk. Hence, it is important to have a system that would avoid these accidents [11].

Gas leakage detection is a way of testing the leakage of gas in a non-destructive way. It finds the hazardous flammable gases from the sealed apparatuses and vessels. Leakages may easily happen if the seals are poor or loosely connected cylinder caps etc. Slow leakage of gas with the minor faults or leaking holes are also could be hazardous, costly, life risking, time consuming, and a prospective for illness, death or bursts.

Since several years, many people were using water method and OLD soap in spotting the particular location of minor gas leakages that turned out to be ineffective. In recent times, dangerous gas leakage detectors and sniffers have become more beneficial and popular [10].

The main reasons for developing Gas Leakage detection system are:

1. In order to safeguard the property and Personnel.
   Gas leakage could cause severe personnel and societal hazards by demolishing material and property.
2. Pollution and Safety: The major reasons for developing gas leak detectors are emerging with stricter OSHA and environmental regulations.
3. Reliability
   Equipment reliability is one of the concerns that the users want in the gas leakage detection system.
4. Energy Loss

It is always good to reduce the loss of energy as they are high cost of energy. By detecting the dangerous gas leakage at early stage, energy loss could be avoided by preventing the quantity of fuel used in the system such as propane or natural gas from dripping out to the environment.

The most Common beneficiaries of such Gas Leakage Detectors and Sniffers are Gas Services, Plumbers, HVAC Servicers, Home Proprietors, Building Proprietors (Apartments, Office etc.), in Refrigerator and in Cylinders

Applications of the most general Gas Leakage Detector and Sniffer are in Accessories, Channels (Plastic or metal), Valves, Containers, Heaters, Water Heaters, Gas Utilizations, Reservoirs.

Most Common Gases Being Tested for Leaks in Residential, Industrial and Commercial Applications are; LPG, Propane, Natural gas, Butane and other Flammable Gases.

In order to cater the needs of these applications, this paper has proposed an IoT and Mobile technology based gas leakage detection system [GLDD] to avoid the tragic incidences by detecting the leakage of gas and raising an alert message and alarm to the concerned people. This system helps the people in saving their lives and things.

Other portions of our work have been discussed in following sections. In Sect. 2, the proposed system Model for the GLDD work and the algorithm are presented. Section 3 discusses the detailed implementation of the work. Results are shown in the Sect. 4. Section 5 concludes the work.

2   System Model and Algorithm

The Proposed System can be used to prevent accidents due to gas leakage raising an alert message and alarm to the mobile of concerned persons.

2.1   Algorithm of GLDD

After completing the connections between all the IoT components and embedding the set of logical instructions on to Arduino board, the system could be tested by placing the connected device near the LPG cylinder, cigarette lighter etc. from which the gas is being leaked. If it successfully detects the leakage of gas from 200 ppm to 10000 ppm (parts per million) from gas sensor, gas sensor readings must be tested and accordingly it sends a signal to arduino which reads analog values and sends an alert SMS message to the mobile device of registered mobile number.

Gas Leakage Detection Device functions as follows:

```
GLDD continuously monitors the volume of gas
if (reading of gas sensor <= 200ppm)
    {
        Gas value is Low
        Arduino does not detect any signal
        Does not send any message to the registered mobile number


    }
if (reading of gas sensor >= 200ppm)
    {
        Gas value is High
        Arduino detects the signal
        Send the alert  message
        "LEAKAGE OF GAS IS DETEDTED......PLEASE SWITCH
        OF THE APPLIANCE" to the registered mobile number

    }
```

2.2   System Model

Figure 1 shows the system model of the proposed Gas Leakage detection scheme that consists of the following things for implementation:

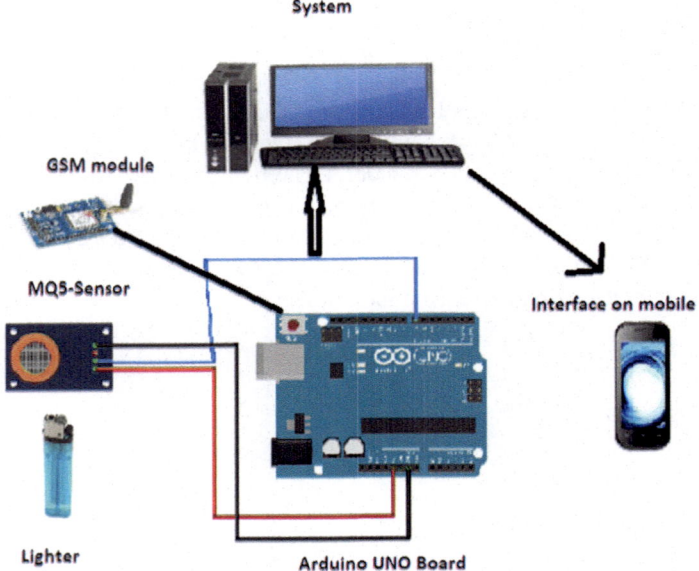

Fig. 1. System model of GLDD with SMS alert

- Arduino UNO Broad,
- Gas sensor,
- GSM Module,
- connecting wires,
- Gas lighter.

The gas sensor that is used here is MQ5 gas sensor. In order to trigger an alert message as a notification to the owner's mobile phone (SMS activation etc.), the analog out can be used to detect the gas leakage [6].

A GSM Module which is used here is mainly a GSM Modem (like SIM 800) and is connected to the PCB along with different kinds of output considered from the board [2].

There are two methods for connecting the GSM module to an Arduino board. In any of the cases, communication between the GSM module and an Arduino board is serial. The Module used here supports the communication in the 900 MHz range band [4].

As the first step, GSM Module has to be booted using the following steps.

1. Insert the SIM card of the owner to the GSM module.
2. Fix it correctly.
3. Connect both the adaptor and GSM module correctly and turn that to ON state.
4. Wait for certain time (say 1 min) and if the 'network LED' in the board blinks in the arduino. (GSM module would take certain time to establish the connection with the mobile network)
5. Blinking of LED confirms that, connection establishment is successful. The network LED would blink continuously for every 3 s indicating the status. To check the connection, give a call to the mobile number of the SIM card place in the GSM module. If you could hear the ringing sound, indicates that the network connection has been established by GSM module

3 Implementation

The connection setup for implementing the gas leakage detection system using Arduino UNO with SMS alert [9] is as shown in Fig. 2.

- Connect analog pin AO and MQ5 to any of the arduino analog pins. Analog out pin of MQ5 is connected to A0 pin of Arduino. [3]
- Connect Vcc to +5volts and Ground to ground pin arduino board.
- Connect GSM module to arduino board by connecting Rx pin to Tx pin in arduino board and Tx to Rx pin.
- Take the connected device near to the place where gas is being leaked like LPG cylinder, cigarette lighter etc. As soon as leakage of gas is detected from 200 ppm to 10000 ppm (parts per million) from gas sensor [12], it sends a signal to arduino which reads analog values and send alert SMS message to a registered mobile number using the GSM module and raises an alarm.

Fig. 2. Connection set-up for developing GLDD

4   Result

After completing the connections among all the IoT components and embedding the set of instructions (there is a program written to send SMS to the mobile when there is a gas leakage detected) on to the Arduino board. To test this system, we connected the gas leakage detection device with gas sensor [5] and in order to detect the leakage, there was a cigarette lighter lit which it detected where and what is the quantity of gas

Fig. 3. Shows the alert message sent on the gas leakage.

was being leaked. It was successfully detected the leakage of gas from 200 ppm to 10000 ppm(parts per million) from gas sensor, it sends a signal to arduino board which reads analog values and send alert SMS message to a registered mobile number as shown in Fig. 3.

Then the concerned person can take the necessary action to avoid the tragic incident happening.

5 Conclusion

This proposed Gas leakage detection system is designed and implemented using Gas sensor, Arduino UNO, GSM module and Mobile. Experiment has been conducted to verify the functioning of the device developed with the help of LGP cylinder and the cigarette lighter. The device has successfully detected the gas and triggered the controller when it sensed the presence of gas and sends the message to the mobile effectively preventing the fire accidents that causes damage to the human lives. This work has proved the usage of the device developed in keeping our society safe and also cost-effective.

Future Scope
The system can be further enhanced by using buzzer to beep as soon as leakage of gas is detected and using LCD to display the gas value.

Acknowledgement. The authors of this work would like to thank and acknowledge the Technical Education Quality Improvement Programme [TEQIP] Phase III, BMS College of Engineering, Bangalore, Karnataka, India for the support given.

References

1. Jyothish, J., Mamatha, Gorur, S., Dakshayini, M.: Booking based smart parking management system. In: International Conference on Intelligent Information Technologies, IoT and Analytics Perspective. ICIIT 2017. Communications in Computer and Information Science, vol. 808. Springer, Singapore. https://doi.org/10.1007/978-981-10-7635-0_24, print ISBN 978-981-10-7634-3, Online ISBN 978-981-10-7635-0, Scopus and SCI Indexed with the H-Index 29
2. Rana, Rajendra, J., Pawar, Sunil, N.: Zigbee Based Home Automation (10 Apr 2010)
3. Soundarya, T., Anchitaalagammai, J.V., Deepa Priya, G., Karthick kumar, S.S.: C-leakage: cylinder LPG gas leakage detection for home safety. IOSR J. Electron. Commun. Eng. **9**(1), 53–58 (2014)
4. Srinivasan, A., Leela, N., Jeyabharathi, V., Kirthika, R., Rajasree, D.: Gas leakage and detection control. Int. J. Adv. Eng. Res. Dev **2**(3) (2015)
5. Raj, A., Viswanathan, A., Athul, T.: LPG gas monitoring system. IJITR **3**(2), 1957–1960 (2015)
6. Vidya, P.M., Abinaya, S., Rajeswari, G.G., Guna, N.: Automatic LPG leakage detection and hazard prevention for home security. In: Proceeding of 5th National Conference on VLSI Embedded and Communication & Networks, vol. 7, April 2014

7. Surie, D., Laguionie, O., Pederson, T.: Wireless sensor networking of everyday objects in a smart home environment. In: Proceedings International Conference on Intelligent Sensors, Sensor Networks and Information Processing, pp. 189–194 (2008)
8. Internet 3.0: The Internet of Things, Singapore, Analysys Mason (2010)
9. Eisenhauer, M., Rosengren, P., Antolin, P.: A development platform for integrating wireless devices and sensors into ambient intelligence systems. In: Proceedings of 6th Annual IEEE Communications Society Conference on Sensor Mesh Ad Hoc Communications and Networks Workshops, pp. 1–3, June 2009
10. Iera, A., Floerkemeier, C., Mitsugi, J., Morabito, G.: The internet of things. IEEE Wireless Commun. 17(6), 8–9 (2010)
11. Gluhak, A., Krco, S., Nati, M., Pfisterer, D., Mitton, N., Alambo, T.R.: A survey on facilities for experimental internet of things research. IEEE Commun. Mag. 49(11), 58–67 (2011)
12. Suryadevara, N.K., Mukhopadhyay, S.C.: Wireless sensor network based home monitoring system for wellness determination of elderly. IEEE Sens. J. 12, 1965–1972 (2012)
13. Mendez, G.M., Yunus, M.A.M., Mukhopadhyay, S.C.: A WiFi based smart wireless sensor network for monitoring an agricultural environment. In: IEEE International Instrumentation & Measurement Technology Conference, pp. 2640–2645, May 2012

Detection of Primary Glaucoma in Humans Using Simple Linear Iterative Clustering (SLIC) Algorithm

G. Pavithra[1,3], T. C. Manjunath[2,3(✉)], and Dharmanna Lamani[2,3]

[1] VTU, RRC, Belagavi, Karnataka, India
[2] ECE, DSCE, Bangalore, Karnataka, India
dr.manjunath.phd@ieee.org
[3] ISE, SDMIT, Ujire, South Kanara, Karnataka, India

Abstract. It is a well-known fact in the world that the glaucoma is the second largest disease which is affecting the human beings in the world. Proper care has to be taken to avoid this at an early stage as this would result in the loss of vision in the humans. This occurs due to the increase in the pressure in the eyes, where it bursts the nerve fibres leading to the vision loss. If the patient goes to the doctor, it is an expensive treatment. Hence, we are devising a low cost module method of detecting the primary glaucoma in the humans using their fundus images. The images of the patients will be taken by the fundus camera, analyzed & a info is given to the patient that he/she is affected with the disease. Once the person comes to know that they are affected, then proper diagnosis can be done by consultations from the hospital experts. The method of detecting the primary glaucoma is being presented in this section using a revised simple linear iterative clustering (SLIC) algorithm clubbed with edge detection using canny edge operators. SLIC concepts are being used for the segmentation process of the cup and the disc & finally the region of interest, i.e., the cup and the disc areas are found out from which the ratio is computed, from where the disease can be detection seeing the ratio. The simulation results shown the effectivity of the method proposed by us in this research work.

Keywords: Modelsim · Matlab · Glaucoma · Eye · Disease · Normal · Affected · Blocks · Pressure · Tool · Hardware · Implementation · FPGA · Xilinx · Modelsim · Matlab

1 Introduction

Developments in medical image processing, imaging will observe less emphasis on pushing quality limits and more on understanding how to use images to its maximum capacity. Bio-medical image processing is vast raised in the medical field can be diagnosed with the help of computer which is the main advantage for which imaging took a major role in biomedical field. The Fig. 1 shows a normal eye & a glucomatic eye.

In this fusion we are able to develop computational and mathematical methods for solving clinical and biomedical research problems. Imaging developments have built on scientific discoveries and technological innovations in other fields. Such as material

© Springer Nature Switzerland AG 2020
A. P. Pandian et al. (Eds.): ICCBI 2018, LNDECT 31, pp. 417–428, 2020.
https://doi.org/10.1007/978-3-030-24643-3_50

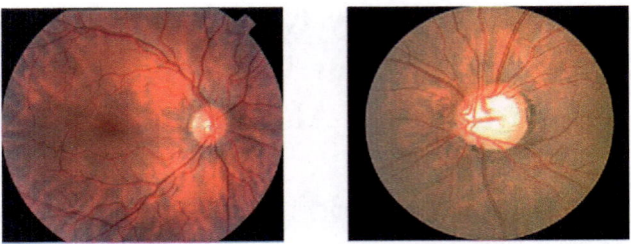

Fig. 1. Enlarged view of normal & affected eye with glaucoma

science, fundamental physics and digital computing. Advances in digital radiography, X-ray commuted tomography (CT), magnetic resonance imaging (MRI) and other nuclear ultrasound and optical imaging techniques have produced an array of modern methods for interrogating intact 3D bodies non-invasively and derives uniquely valuable tissue compositions. Today images processing is a dynamite, evolving field of multi and inter disciplinary activities that has grown enormously in scope and importance in recent years and also advanced image processing involves the development of algorithms which are faster more effective.

2   Proposed Block Diagram

The input to the proposed algo is the images taken from the database (block-1). Then, the input image is pre-processed using illumination concept (block-2) & further the edges are detected using canny edge detection technique (block-3). Segmentation process is carried out next to extract the cup & disc using SLIC (block-4). Features of the ROI (cup & disc) is extracted & classification process is carried out next (block-5). Finally, the results are computed using the cup to disc ratio (block-6), from which it can detected whether the patient is affected with glaucoma or not using the value of CDR ...... Fig. 2.

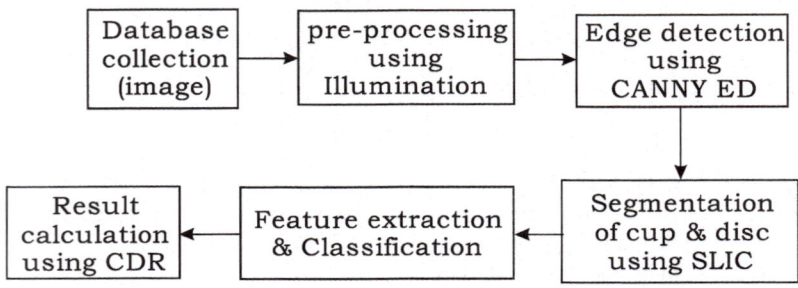

Fig. 2. Block-diagram of primary glaucoma detection using SLIC algo

SLIC is one of the powerful method which could be used for the analysis & detection of the primary glaucoma by extracting the OD & OC of the fundus image, which is presented in this section. As clustering algorithm applications are increasing rapidly, its impact is falling directly on super pixel algorithms. Vision based applications basically rely on these super pixels more now a days. In order to understand this super pixel based algorithm, there are 5 state of arts. These 5 state of arts are - for their ability to adhere image boundaries, efficiency, speed, memory and impact on the segmentation process and segmentation performance. This SLIC algo is used in our work for the glaucoma detection, of course this SLIC algo is a bit similar to the k-means clustering approach, while generating these super pixels. Despite its simplicity, SLIC adheres to boundaries. The block-diagram of the proposed methodology for this contribution was presented in the Fig. 2 along with the brief description of each & every blocks.

3 Database Collection

To develop the algorithm first essential step is to obtain the image database. For this purpose, 50 fundus images were collected from various online databases in which, 30 images from High Resolution Fundus (HRF) image database from www.optic-disc.org database images and rest of them were taken from online database DROINS & others from IIT-drishti site.

4 Image Pre-Processing

Pre-processing is a technique which suppresses the unwanted distortions or improves some of the image features that are important for further processing. In the work considered, the fundus color image is taken first from the image database, the ROI, i.e., the optic disc & the optic cup (pink or light orange color) are to be extracted. While the disc and cup extraction can be done on the entire image, localizing the ROI would help to decrease the computational cost as well as increase segmentation accuracy to produce final result. The ROI from all images from the DB is cropped down & resized to (512×512).

5 Illumination Equalization Algorithm

This concept is used for enhancing the pre-processed image by using adaptive histogram equalization (AHE). AHE is a computer IP technique which is used to improve the contrast in the images to get improved light intensity values. Here, the enhanced image is obtained using the subtraction of the defined pixel value from the mean pixel value & adding it to the image coordinates by using the equation

$$f = i + def_{int} - in_{mean} \qquad (1)$$

where, f is functional value of the value of enhanced image, i is the image, def_{int} is the defined pixel value of the image & in_{mean} is mean value of the pixel. Uneven Illumination is one of the problems associated with fundus images. Areas which are present at the centre of the image are more over well illuminated, hence appears very bright while the sides at the edges or far away are poorly illuminated and appears to be very dark. In order to make the light intensity uniform throughout the image, this concept of AHE is used to obtain a smoothened HR image.

6   Lab Space Colour

The Lab color space describes mathematically all perceivable colors in the 3 dimensions L for lightness, a & b for the color opponents green–red and blue–yellow. Unlike the RGB and CMYK color models, lab color is designed to approximate human vision it can be used to make accurate color balance corrections by modifying output curves in the a and b components or to adjust the lightness contrast using the L component (Fig. 3).

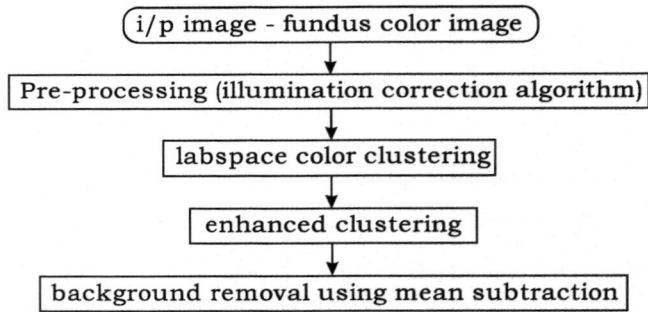

Fig. 3. Flow diagram for advanced pre-processing

7   SLIC Algo

This algo is used for segmenting the ROI & to obtain the features of the ROI. By default, the only parameter is used in the algorithm is k, which decides desired number of approximately equally sized super pixels. For color images in the color space, the clustering procedure begins with an initialization step where k initial cluster centers are sampled on a regular grid spaced S pixels apart. To produce roughly equal sized super pixels, the grid interval is S. The centers are moved to seed locations corresponding to the lowest gradient position in a (3×3) neighbourhood. This is done to avoid centring a super pixel on an edge and to reduce the chance of seeding a super pixel with a noisy pixel. Next, in the assignment step, each pixel i is associated with the nearest cluster center whose search region overlaps its location. This is the key to speeding up our algorithm because limiting the size of the search region significantly reduces the number of distance calculations and results in a significant speed advantage over conventional *k*-means clustering, where each pixel must be compared with all cluster centers. This is only possible through the introduction of a distance measure D, which determines the nearest cluster center for each pixel. Since the expected spatial extent of a super pixel is a region of approximate size $(S \times S)$, the search for similar pixels is done in a region $(2S \times 2S)$ around the super pixel center. The proposed algo is as follows.

Developed Algorithm SLIC super pixel segmentation

```
//Initialization //
Initialize cluster centers
2Ck= [Lk; ak; bk; xk; Ike] T by sampling
Pixels at2 regular grid steps S.
Move cluster2centers to the lowest2gradient position in a 3 x 3
Neighborhood.
Set label l (I) =-1 2for each pixel I.
Set distance d (I) =∞ 2for each pixel I.
Repeat
//Assignment //
For each cluster2 center Ck do
For each 2pixel I in a 2Sx 2S region 2around Ck do
Compute the distance 2D between cook and I.
If 2D < d (I) then
set2 D (I) =D
set2 l (I) = k
End if2
End for2
End for2
// Update //
Compute new 2cluster centers.
Compute2 residual error E.
Until2 E <threshold
```

8 Distance Measurement

This concept is used for obtaining an advanced post processing step & enforces the connectivity between the super pixels by reassigning disjoint pixels to nearby super-pixels so that the distance measurement b/w the super-pixels is normalized & uniformity is obtained.

9 Simulation Results

Codes are developed in the Matlab environment & are run for different fundus images & the results are observed for various cases of glaucoma, viz., normal (case-1), moderate (case-2) & severe (case-3). To start with fundus images, viz., image_1.jpg, image_4.jpg & image_5.jpg are given as inputs to the developed algo one after the other & the results are observed as normal (Figs. 4 and 5), moderate (Figs. 6 and 7) & severe (Figs. 8 and 9) respectively. Similarly, the analysis is carried out for the remaining 47 images present in the database and finally the quantitative results are tabulated as shown in the Table 1. After getting extraction of optic cup and disc from segmentation, the ratio of cup to disc is estimated from the code. From the database of 50 images, pixel values of optic cup, optic disc and the ratio is obtained as shown in the Table 1. From this table, it can be inferred that out of 50 images taken from the database, 6 were normal, 27 were moderate & the rest 17 were severe case of glaucoma, thus resulting in 88% glaucoma affectedness.

Case 1 : Normal (CDR < 0.4)-Image_1.jpg

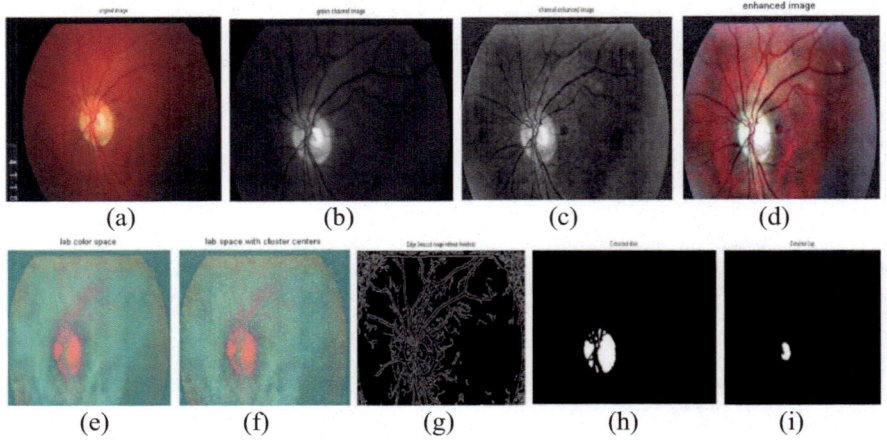

Fig. 4. Simulation results for normal case for work-2 (a) original image_1.jpg (b) green channel (c) channel enhanced (d) color enhanced (e) lab color space (f) clustering (g) Edge detection (h) extracted disc (i) extracted cup

Case 1: Normal (CDR < 0.4)-Image_1.jpg

In the 1st case considered, when the image_1.jpg is given as the input to the developed algo, the result obtained is a normal case of glaucoma as shown in the Fig. 4, since the CDR value is < 0.4 & the results can be seen in the command window as shown in the Fig. 5.

Case 2: Moderate (0.4 < CDR < 0.6) - Image_4.jpg

In the 2nd case considered, when the image_4.jpg is given as the input to the developed algo, the result obtained is a moderate case of glaucoma as shown in the Fig. 6, since the CDR value is in between 0.4 to 0.6 & the results can be seen in the command window as shown in the Fig. 7.

Case 3: Severe (CDR > 0.6) - Image_5.jpg

In the 3rd case considered, when the image_5.jpg is given as the input to the developed algo, the result obtained is a severe case of glaucoma as shown in the Fig. 8,

Workspace

Name ▲	Value	Min	Max
a	400x600x3 uint8	<Too ...	<Too ...
aa	351x551x3 uint8	<Too ...	<Too ...
aaa	256x256x3 uint8	0	255
ahe	351x551 uint8	1	254
Area_cup	869	869	869
Area_disk	4864	4864	4864
blue	351x551 uint8	0	164
bw1	351x551 logical		
ccfa	1	1	1
ccfb	2	2	2
CDRtest	0.1787	0.1787	0.1787
chan	3	3	3
col	551	551	551
con	351x551 uint8	0	255

Command Window

ⓘ New to MATLAB? Watch this Video, see Examples, or read Getting Started.

```
    Area of Optic cup is
    869

    Area of Optic Disc is
        4864

    Cup to Disc Ratio    0.1787

    Normal
```

Normal

OK

Fig. 5. CDR output at command window with cup & disc area for the case 1 showing normal case

Case 2 : Moderate (0.4 < CDR < 0.6) - Image_4.jpg

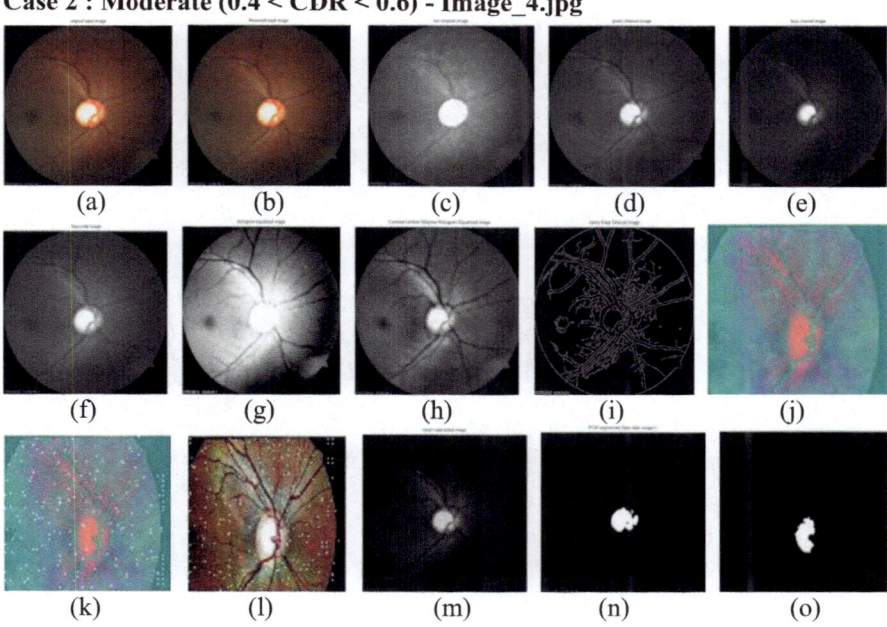

Fig. 6. Simulation results for moderate case for work-2 (a) original input image (b) realized input image (c) red channel (d) green channel (e) blue channel (f) gray scale image (g) histogram equal image (h) CLAHE (i) canny edge detection (j) lab color space (k) lab space with clustering (l) enhanced clustering image (m) mean subtraction (n) FCM OD segmented image (o) SLIC OC segmented image.

since the CDR value is in >0.6 & the results can be seen in the command window as shown in the Fig. 9.

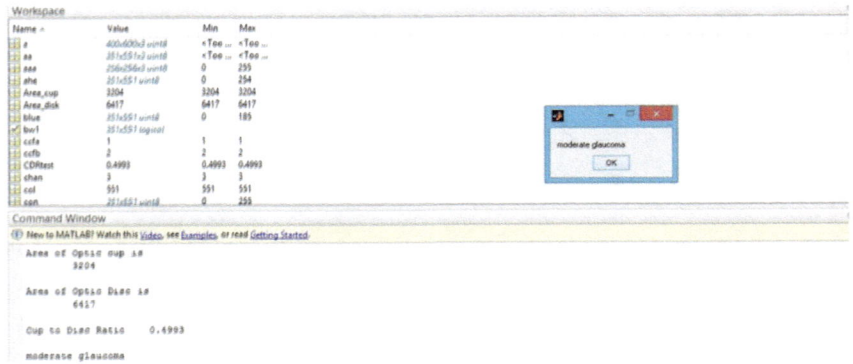

Fig. 7. CDR output at command window with cup & disc area for case 2 showing moderate case

Case 3 : Severe (CDR > 0.6) - Image_5.jpg

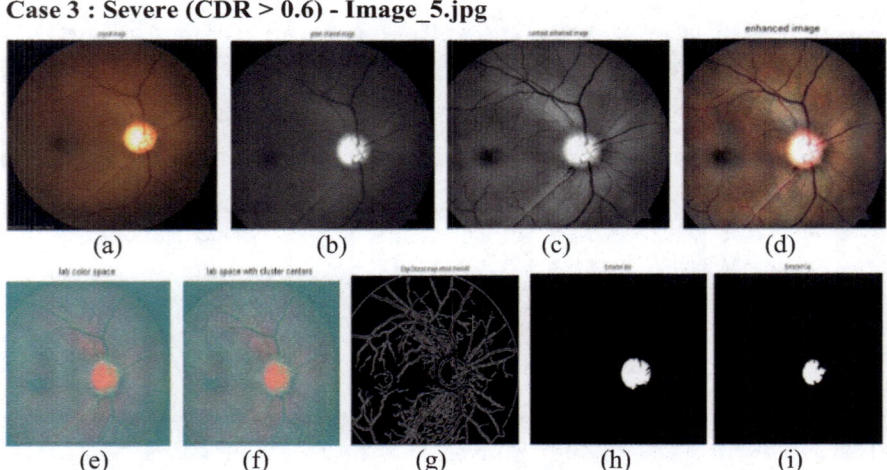

Fig. 8. Simulation results for severe case for work-2 (a) Original image (b) green channel image (c) contrast enhanced image (d) enhanced image (e) lab color space (f) lab space with cluster centers (g) edge detected image w/o threshold (h) extracted disc (i) extracted cup

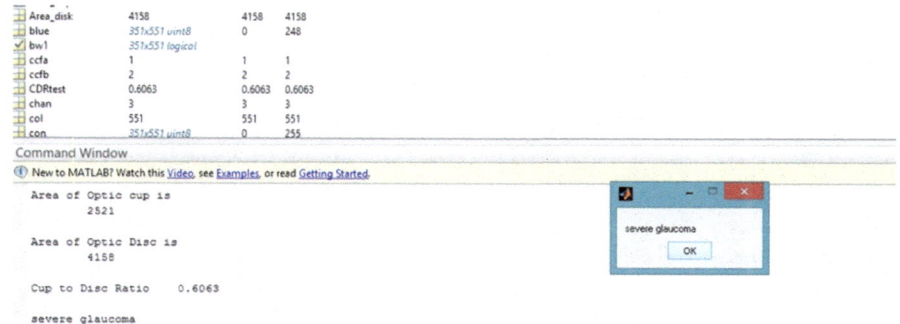

Area_disk	4158		4158	4158
blue	351x551 uint8		0	248
bw1	351x551 logical			
ccfa	1		1	1
ccfb	2		2	2
CDRtest	0.6063		0.6063	0.6063
chan	3		3	3
col	551		551	551
con	351x551 uint8		0	255

Command Window

New to MATLAB? Watch this Video, see Examples, or read Getting Started.

```
Area of Optic cup is
        2521

Area of Optic Disc is
        4158

Cup to Disc Ratio    0.6063

severe glaucoma
```

severe glaucoma

OK

Fig. 9. CDR output at command window with cup and disc area for the case 3 showing severe case

Table 1. CDR tabulation by SLIC method

No.	images	Cup area	Disk area	CDR	Nature of disease
1	Image 1 (normal)	869	4864	0.1787	Normal
2.	Image 2	969	4874	0.1988	Normal
3.	Image 3	2576	4349	0.5923	Moderate
4.	Image 4 (moderate)	3204	6417	0.4993	Moderate
5.	Image 5 (severe)	2521	4158	0.6063	Severe
6.	Image 6	3259	6110	0.5334	Moderate
7.	Image 7	2193	4565	0.4804	Moderate
8.	Image 8	3214	6317	0.5083	Moderate
9.	Image 9	2793	4396	0.6354	Severe
10.	Image 10	2621	4158	0.6303	Severe
11.	Image 11	796	2346	0.3393	Normal
12.	Image 12	869	4864	0.1787	Normal
13.	Image 13	2376	4349	0.5463	Moderate
14.	Image 14	1975	4470	0.4418	Moderate
15.	Image 15	3984	6993	0.5697	Moderate
16.	Image 16	3269	6210	0.5264	Moderate
17.	Image 17	2193	4565	0.4804	Moderate
18.	Image 18	3204	6417	0.4993	Moderate
19.	Image 19	2793	4396	0.6354	Severe
20.	Image 20	2521	4158	0.6063	Severe
21.	Image 21	2791	3946	0.7072	Severe
22.	Image 22	2861	4864	0.5882	Moderate
23.	Image 23	2375	3349	0.7092	Severe
24.	Image 24	2974	4470	0.6654	Severe
25.	Image 25	3986	6893	0.5782	Moderate

(*continued*)

Table 1. (*continued*)

No.	images	Cup area	Disk area	CDR	Nature of disease
26.	Image 26	3261	6219	0.5254	Moderate
27.	Image 27	2124	4565	0.4653	Moderate
28.	Image 28	3205	6417	0.4992	Moderate
29.	Image 29	2794	4396	0.6356	Severe
30.	Image 30	2522	4158	0.6065	Severe
31.	Image 31	799	2346	0.3405	Normal
32.	Image 32	870	4864	0.1788	Normal
33.	Image 33	2377	4349	0.5465	Moderate
34.	Image 34	1976	4470	0.4420	Moderate
35.	Image 35	3985	6993	0.5699	Moderate
36.	Image 36	3270	6210	0.5266	Moderate
37.	Image 37	2194	4565	0.4806	Moderate
38.	Image 38	3205	6417	0.4995	Moderate
39.	Image 39	2794	4396	0.6356	Severe
40.	Image 40	2522	4158	0.6065	Severe
41.	Image 41	2016	3225	0.6250	Severe
42.	Image 42	2337	3156	0.7404	Severe
43.	Image 43	2444	3566	0.6852	Severe
44.	Image 44	2659	3696	0.7201	Severe
45.	Image 45	2695	4369	0.6168	Severe
46.	Image 46	2486	4855	0.5120	Moderate
47.	Image 47	2595	4602	0.5638	Moderate
48.	Image 48	2645	5022	0.5287	Moderate
49.	Image 49	2667	5252	0.5075	Moderate
50	Image 50	2768	5111	0.5414	Moderate

10 Conclusions

In this contributory research work, algorithm was developed for the detection of primary glaucoma using SLIC method. The simulation was run for all the fundus images considered in the database. In this section, only 3 cases from the database, viz., one from normal, one from moderate & one from severe case of glaucoma were considered for the simulation purposes. The CDR was obtained in each case from which it was justified that the patient was normal, moderate & severe. The simulation results shown in the results section shows the effectivity of the methodology developed by us. Similarly, the results are observed for the remaining set of images present in the database & is not shown here for sake of convenience.

References

1. Khan, F., Khan, S.A., Yasin, U.U., ul Haq, I., Qamar, U.: Detection of glaucoma using retinal fundus images. In: 6th IEEE Conference on Biomedical Engineering (BMEiCON), Amphur Muang, Thailand, pp. 1–5, 23–25 October 2013
2. Kavitha, S., Karthikeyan, S., Duraiswamy, K.: Early detection of glaucoma in retinal images using cup to disc ratio. In: IEEE International Conference on Computing Communication and Networking Technologies (ICCCNT), Tamil Nadu, pp. 1–5, 29–31 July 2010
3. Lamani, D., Manjunath, T.C.: Diagnosis of glaucoma disease through image feature fractal dimension. Ph.D. thesis, VTU, Belagavi, February 2016
4. Pavithra, G., Manjunath, T.C.: Different clinical parameters to diagnose glaucoma disease: a review. Proc. Int. J. Comput. Appl. (IJCA) **116**(23), 42–46 (2015). IF 3.546, ISSN 0975 – 8887
5. Pavithra, G., Manjunath, T.C.: Automated diagnose of neo-vascular glaucoma disease using advance image analysis technique. Int. J. Appl. Inf. Syst. (IJAS) **9**(6), 1–6 (2015). Published by Foundation of Computer Science (FCS), NY, USA, ISSN 2249-0868
6. Pavithra, G., Manjunath, T.C.: A novel approach for diagnosis of glaucoma through optic nerve head (ONH) analysis using fractal dimension technique. Int. J. Mod. Educ. Comput. Sci. (IJMECS) **8**, 55–61 (2016). ICV 2014 8.09
7. Pavithra, G., Manjunath, T.C.: A novel approach for diagnosis of glaucoma through optic nerve head (ONH) analysis using fractal dimension technique. Int. J. Mod. Educ. Comput. Sci. (IJMECS) **8**, 55–61 (2016). Published by Modern Education and Computer Science Press
8. Pavithra, G., Manjunath, T.C.: A novel method of digitization & noise elimination of digital signals using image processing concepts. Int. J. Eng. Res. Electron. Commun. Eng. (IJERECE) **3**(11), 38–44 (2016). ISSN (Online) 2394-6849, Impact Factor 3.689, paper id 8
9. Pavithra, G., Manjunath, T.C.: Design of algorithms for diagnosis of primary glaucoma through estimation of CDR in different types of fundus images using IP techniques. Int. J. Innov. Res. Inf. Secur. (IJIRIS) **4**(5), 12–19 (2017). Paper id MYISSP 10135
10. Pavithra, G., Manjunath, T.C.: Development of novel BMIP algorithms for human eyes affected with glaucoma and hardware implementation using VLSI based embedded systems. Int. Res. J. Power Energy Eng. (IRJPEE) **3**(2), 99–103 (2017). ISSN 3254-1213x
11. Pavithra, G., Manjunath, T.C.: To diagnose glaucoma disease by different clinical parameters—a brief Review. In: 2016 IEEE International Conference on Control, Instrumentation, Communication and Computational Technologies (ICCICCT), pp. 653–657. Noorul Islam University, Kanyakumari District, Paper id 195, 16–17 December 2016
12. Pavithra, G., Manjunath, T.C.: Investigation of primary glaucoma by CDR in fundus images. In: 2nd IEEE International Conference on Recent Trends in Electronics, Information & Communication Technology (RTEICT-2017), pp. 1806–1812 (2017). Technically Co-Sponsored by IEEE Comp. Soc. & IEEE Bangalore Section, Organized by Dept. of ECE, Sri Venkateshwara College of Engg., Bangalore, Karnataka, India, 19–20 May 2017
13. Pavithra, G., Manjunath, T.C.: Conceptual view of a smart tonopen for biomedical engineering applications. In: 2017 IEEE International Conference on Intelligent Computing & Control Sciences (ICICCS-2017), pp. 60, pp. 636–639. Vaigai College of Engineering, Madurai, Paper id 121, 15–16 June 2017
14. Pavithra, G., Manjunath, T.C.: A review of the glaucoma detection using hardware based implementation using embedded systems. In: 2017 IEEE International Conference on Intelligent Computing & Control (I2C2–2017), pp. 75–82. Karpagam College of Engineering, Karpagam University, Coimbatore, Paper id IC038, 23–24 June 2017

15. Pavithra, G., Manjunath, T.C., Dharmanna L.: Design & development of novel algorithms for diagnosis of glaucoma in different types of images using advanced image analysis techniques. In: IEEE International Conference on Intelligent Computing, Instrumentation & Control Technologies (ICICICT-2017). Vimal Jyothi Engineering College, Kannur, Paper id 85, 6–7 July 2017
16. Pavithra, G., Manjunath, T.C., Lamani, D.: Glaucoma detection using IP techniques. In: IEEE International Conference on Energy, Communication, Data Analytics and Soft Computing (ICECDS 2017), pp. 485–488. SKR Engineering College, Poonamallee, 1–2 August 2017
17. Pavithra, G., Manjunath, T.C.: Hardware implementation of glaucoma using a PIC Micro-controller—a novel concept for a normal case of the eye disease. In: IEEE International Conference on Current Trends in Computer, Electrical, Electronics & Communication (ICCTCEEC-2017). IEEE Bangalore Section & CPGC, Vidyavardhaka College of Engineering, Mysore, 8–9 September 2017
18. Pavithra, G., Manjunath, T.C.: Detection of secondary glaucoma in human eyes using sophisticated bio-medical image processing algorithms. In: IEEE International Conference on Power, Control, Signals and Instrumentation Engineering (ICPCSI-2017), pp. 363–368, pp. 2375–2382. Saveetha Engineering College, Chennai, 21 & 22 September 2017
19. Pavithra, G., Manjunath, T.C.: Diagnosis & detection of eye diseases using deep convolutional neural networks & raspberry pi. In: IEEE Second International Conference on Green Computing and Internet of Things (ICGCIoT 2018). Global Academy of Technology, Bangalore, 16–18 August 2018

Application of Stream Cipher and Hash Function in Network Security

Aaditya Jain[✉] and Neha Mangal[✉]

R. N. Modi Engineering College, Rajasthan Technical University, Kota, India
aadityajain58@gmail.com, 14nehal0@gmail.com

Abstract. Network security is one of the concerning elements in this digital world as growth of the technology doubtlessly increased the scope and pace of the activities being delivered in the present time however; breaches and intrusions has also increased with same rate. In this paper, the techniques related to network security are analyzed and a comparative analysis is done among the methods. It shows a simple and secured method to realize the stream cipher from the hash functions. It is to be analyzed that implementation of cipher and has technique in network security will be a symmetric secure mechanism. It drives the complete verification and clarification that protect information data with full of authentication. The crypto system will be much stronger and data will be transmitted in faster way along with safeguard.

Keywords: Stream cipher · Hash algorithm · Encryption · Network security

1 Introduction

Encryption is playing very crucial role in assuring the feasibility and insuppressible application of the information technology. High speed and instant exchange of data and information has led to the advancement in all the existing sectors including from farming to big size industries' business. Encryption allows the network to be more secured and less vulnerable to be exposed to any intruder or unauthorized user.

1.1 Problem Statement

Stream cipher and hash functions are the configuration those allow the encryption in keeping it secured from unauthorized access or expose. This paper is a review on different research papers those emphasize on these methods and deliver a comparison between them. The literature review being proposed will be helpful in analyzing the gaps and reaching at a conclusion that is effective and efficient for securing the network.

© Springer Nature Switzerland AG 2020
A. P. Pandian et al. (Eds.): ICCBI 2018, LNDECT 31, pp. 429–435, 2020.
https://doi.org/10.1007/978-3-030-24643-3_51

2   Research Aim and Objectives

The scope of the research is to analyze the "Application of stream cipher and hash function in network security" and compare them. The aim of this paper is to identify the best method among the stream cipher and hash function for the network security. Following are the objectives:

- To collect and evaluate reliable source of data and information
- To evaluate the stream cipher application in network security
- To evaluate the hash function application in network security
- To compare both the methods in network security
- To propose efficient and effective findings and conclusion

In total ten papers from each background was selected for delivering the research those were collected from the online library. However, the articles those were not much relevant as per the objective were eliminated. The estimated time for the delivery of the research was five days and hence, it was crucial to develop an action plan that can be helpful in delivering the research.

3   Literature Review

According to Babita and G. K. [1], there are three types of cryptographic algorithms those can exist in more detail and so the symmetric cyphers can be divided into block ciphers and stream ciphers as described in the following figure (Figs. 1, 2, 3):

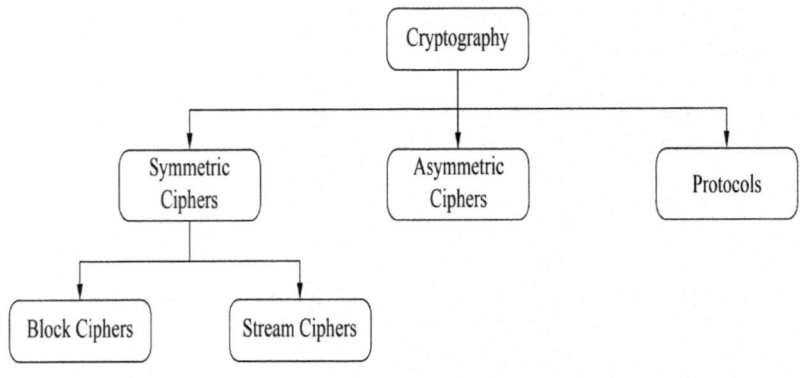

Fig. 1. Main areas within cryptography. (Source: [2])

Technically, stream cipher can be expressed as encryption algorithm that can be utilized for encrypting 1 byte or bit of the plaintext at a single instance. Pseudorandom bits' stream is utilized (in infinite units) as the key. Perlman, Kaufman and Speciner stated hashed function as One-Time pad as these are designed to approximate the idealized cipher. The common ciphers are SALSA, SOSEMANUK, PANAMA, and

RC4 [3, 4]. However, Rivest Cipher 4, ARCFOUR, ARC4, or RC4 have already gain maximum popularity and also being utilized in many protocols such as TLS, WPA, and WEP.

The hash function f is used for designing the stream cipher and it has been cryptographic core of the stream cipher. The hash function should belong to the MD4-family that states to be the iterated hash functions and fulfil few of the major or crucial requirements. Wood and Uzun explained the properties of the hash function as: Preimage resistance: for a hashed value k, there must be an infeasible computation in manner to find the input a for example: $k = f(a)$ [5]

Partial-preimage resistance: it must be very difficult for recovering the substrings and similarly, should be capable of recovering the entire input. Additionally, for the instance when a part of the input is identified, it must be complicated while finding the remainder. For example, if x input bit has been not known, it must consider an average of $2x-1$ hash operations in manner to analyze these bits. Collision resistance: the hash function must be infeasible computationally in manner to identify at least two inputs such as a, b with a and b have different values than each other and the $f(a) = f(b)$.

Mixing transformation: for one of the input a, the output of the hashed value $k = f(a)$ and it must be indistinguishable computationally from the binary string that is uniform [6].

The techniques which are used for message authentication for the sensor network are public key cryptography, elliptic cryptography, secure hashing algorithm, key scheduling algorithm, secret prefix method, secret suffix method and envelope method. Hwang and Gope compared the hashing and the encryption and stated that the hashing can be an ideal way for storing passwords because of its property of inheritance [7]. It will be a troubled program for the intruder or hacker to access the raw data if the data is protected using hash function. If a programmer uses appropriate salt and strong hashing algorithm for generating the password, it will be effectively troubling for the intruder to get access to the raw data. It will be comparing the password entered by the users while inputting the password with the stored value however, the plains representation is not being stored considering the security [8]. Same expert stated encryption as a tool for turning the data into a series of unreadable characters those have been not of fixed length.

Mathews and Pachami [10] compared the hashing and the encryption and stated that the encrypted strings could be possibly reversed back into the original decrypted form as it was before encryption if an individual is using the correct key. They proposed that the encryption should only be used over hashing at the instances when there is the need for decrypting the resulting message. They also claimed that the exchange of information should only be done using encryption as the receiving individual would need the same information as it was being cascaded. For the instance, when the receiver does not need to get the raw value, the hashing should be done in all of such scenarios and also claimed it will safer as compared to encryption. Lindteigen, Payne and Patel in [6] investigated the symmetric encryption, public encryption and hashing function. They concluded that the stream cipher is much cheap and highly energy efficient as compared to that of the hash functions. However, they also claimed that the stream cipher is less secured as that of the block cipher. They claimed same response for the

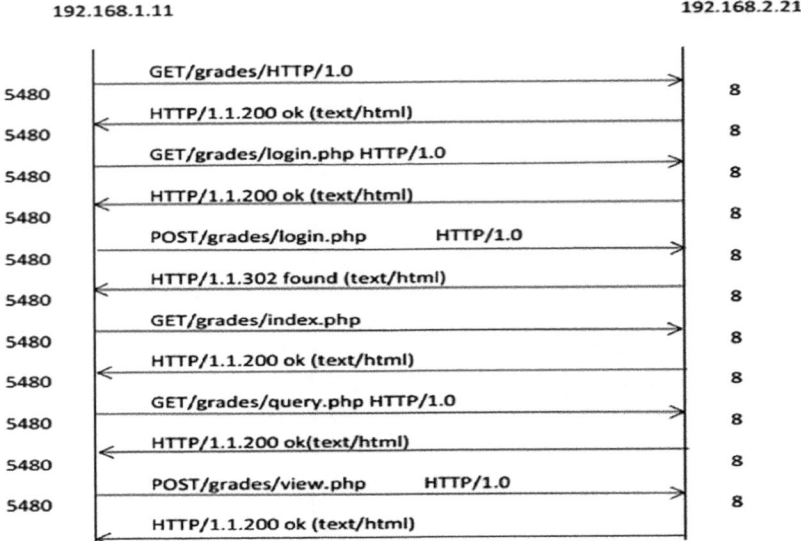

Fig. 2. Protection method of network security using cipher. (Source: [9])

hash function and stated that they also allow high security for the data being routed over the network.

4 Comparative Analysis

Finding from the above review can be described as that the methods of message authentication for the sensor network are compared in this study. The comparative analysis is done as below:

Public key cryptography: This technique is used for securing the key transfer process. In this message communication, session based key is used. This technique has an advantage over the symmetric key algorithms. It allows for secure exchange of the information without require to diverse way to secretly swap the keys.

Elliptic cryptography: This technique is used for key distribution processes. The message communication is being secured with use of RC4 (Rivest Cipher) algorithm. It is generated through properties of elliptic curve equations in traditional methods of generation as the product for the prime numbers.

Secure hashing algorithm: This particular technique is used for messaging the integrity analysis. For the system development, Java in Simulation Time simulation is used. This technique is designed to keep the data secured. It is worked by transformation of data using the hash function.

Key scheduling algorithm: This technique is used for initialization of permutation in the array. It is used as combination of linear plus non-linear transformations and generated 11 sub keys of around 128 bit lengths.

Secret prefix method: This technique is consisted of prepending secret key K to message x before performing the hash operations. MAC(x) = h(k ∥ x). MAC is in secured; a single text MAC pair is contained information equivalent to secret key, independent of the size.

Secret suffix method: This technique is consisted of appending secret key K to message x before performing the hash operations. MAC(x) = h(x∥ k). Off-line collision attack on hash functions are used to get internal collision.

Envelope method: This technique is combined prefix as well as suffix methods. It is consisted of prepending secret key k_1 plus appending key k_2 to message x before performing of hash operations.) MAC (x) = h(k_1 llx ll k_2).

Most of the cryptographic applications uses hash function for example digital signature and others. Hashing is applicable enough for generating random strings in manner to avoid the redundancy of the data those have been stored in the databases. It is widely used in computer graphics that can be helpful in finding the proximity problems and closet pairs in planes. However, the purpose of the encryption is to protect the data from an unauthorized person who is not meant to access the data being transferred over the network and the individual with the cryptographic key can only access the data or information. It can be the most secured way for transferring or exchanging the information over the network emphasizing on four factors including confidentiality, encryption, granular access control, and authentication. Following image can be better descriptive in describing the difference between the encryption or stream cipher and hashing algorithm:

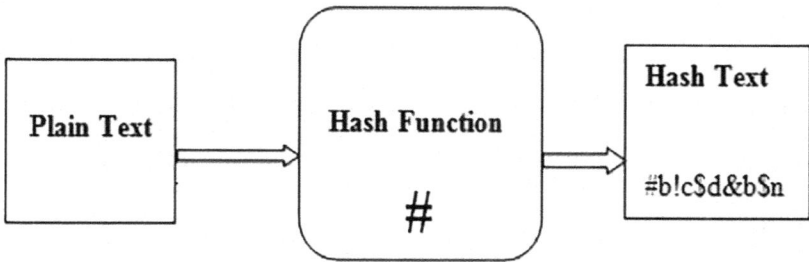

Fig. 3. Comparison between hashing and encryption

Hashing can be applicable in validating the integrity of the content through detecting the modifications and alterations and the output is the hashed output [11, 12]. On the other hand, the encryption encodes data considering the primary purpose of managing the security and the data confidentiality and also requires a cryptographic key that can only be used as the reversible function for regaining the plain text. Encryption is a two-way function including encryption for securing the data and decryption for regaining the access over the data. However, hashing is a one-way function that is utilized for changing the plain text to unique digest that cannot be reversed [13, 14]. Despite of the differences mentioned above, there are certain similarities between the hashing and stream cipher encryption as both are ideal for handling the information and data messages in computing systems.

5 Conclusion

Stream Cipher and Hash Function both are applicable in transforming the data or information into a different format. The attackers are also evolving and adopting new techniques and hence, the future improvements can be difficult for both the technologies. The modern computing system is much portable in implying and updated way for protecting the data and information using these both technologies. In future, using this technology, data protection system can be improved in much easier as well as secure way. Using this technique, people can safe their data from attacker.

References

1. Babita, E.B., Er, G.K.: A review: network security based on cryptography and steganography techniques. Int. J. Adv. Res. Comput. Sci. 8(4), 161 (2017)
2. Guesmi, R., Farah, M.A.B., Kachouri, A., Samet, M.: A novel chaos-based image encryption using DNA sequence operation and Secure Hash Algorithm SHA-2. Nonlinear Dyn. 83(3), 1123–1136 (2016)
3. Jain, A., Buksh, B.: Advance trends in network security with honeypot and its comparative study with other techniques. Int. J. Eng. Trends Technol. 29(6), 304–312 (2015)
4. Perlman, R., Kaufman, C., Speciner, M.: Network Security: Private Communication in a Public World. Pearson Education India, London (2016)
5. Wood, C.A., Uzun, E.: Flexible end-to-end content security in CCN. In: 2014 IEEE 11th Consumer Communications and Networking Conference (CCNC), pp. 858–865. IEEE (2014)
6. Lindteigen, T., Payne, A., Patel, D.: Secure group messaging and data steaming. US Patent Application 14/952,907 (2017)
7. Hwang, T., Gope, P.: Robust stream-cipher mode of authenticated encryption for secure communication in wireless sensor network. Secur. Commun. Netw. 9(7), 667–679 (2016)
8. Haager, J., Sandwith, C., Terrano, J., Saripalli, P.: Systems and methods for security hardening of data in transit and at rest via segmentation, shuffling and multi-key encryption. US Patent 9,990,502 (2018)
9. Pandian, K.S., Ray, K.C.: Dynamic Hash key-based stream cipher for secure transmission of real time ECG signal. Secur. Commun. Netw. 9(17), 4391–4402 (2016)
10. Mathews, M.M., Panchami, V.: Date time keyed-HMAC. In: 2016 Online International Conference on Green Engineering and Technologies (IC-GET), pp. 1–5 (2016)
11. Stallings, W.: Cryptography and Network Security: Principles and Practice, p. 743. Pearson, Upper Saddle River (2017)
12. Puthal, D., Nepal, S., Ranjan, R., Chen, J.: A dynamic prime number based efficient security mechanism for big sensing data streams. J. Comput. Syst. Sci. 83(1), 22–42 (2017)

13. Guesmi, R., Farah, M.A.B., Kachouri, A., Samet, M.: A novel chaos-based image encryption using DNA sequence operation and Secure Hash Algorithm SHA-2. Nonlinear Dyn. **83**(3), 1123–1136 (2016)
14. Lennox, J., Holmberg, C.: Connection-oriented media transport over the transport layer security (TLS). Protocol in the Session Description Protocol (SDP) (No. RFC 8122) (2017)

Interactive Video Gaming with Internet of Things

Yellamma Pachipala$^{(\boxtimes)}$, C. Madhav Bharadwaj, Pakalapati Narendra,
G. Leela Sree, and K. Praveen Reddy

Department of Computer Science and Engineering,
Koneru Lakshmaiah Educational Foundation, Vaddeswaram, Guntur, A.P, India
pachipala.yamuna@gmail.com

Abstract. IoT enabled us to build environments that have potential to transform lifestyle of Humanity. Internet of things (IOT) a collection of networked interacting electronic devices provides the necessary infrastructure and enabling technologies to design and implement smart built environments. In this project it will be demonstrating a proof of concept by making a prototype of a system that can be controlled via game playing for enhancing the game playing experience. In this paper it will be discussing a different approach for video gaming which provides increased interactivity and can be available at a cheap price. In this paper it is focusing on how environments can dynamically interact with us as per the game play. This integration wills a different and more appealing approach for the entertainment industry. The idea came from observing the interactions of electronic items with internet. When we thought what would happen if we send instructions through game instead of sending the instructions manually. The prototype can be implemented with Open Cloud environment and a game engine. Data from game engine will be sent into Arduino.

Keywords: IoT · Arduino · Game engine · Open cloud

1 Introduction

As Internet of things begins to revolutionize with world we saw many projects and advancements being made daily [1]. People started building smart environments that can communicate with other devices via internet. Those are Automated street systems, Water meters, Motion capture systems and many more. Consider that player are playing a video game where we can ride vehicles, climb mountains, cross hot a volcano in each of the above mentioned scenario [2]. It have to different environment conditions that is if player riding a motor at a high speed in the real world [3]. Player can experience air blowing through our face and when player are on mountains it will experience a different climatic conditions

Now way days gaming industry plays a major role in the entertainment environment. An international video gaming industry charming is approximate to be $81.5B in 2014. This is more than two times of the revenue of the international film industry in 2013. In 2015, it was estimated at US$91.5 billion [4] and the demand for new features, new games and new experiences will always be high. Internet of things is a technology

© Springer Nature Switzerland AG 2020
A. P. Pandian et al. (Eds.): ICCBI 2018, LNDECT 31, pp. 436–445, 2020.
https://doi.org/10.1007/978-3-030-24643-3_52

that is allowing humans to communicate with environment [5, 6]. All the electronic gadgets are connected through Internet [6]. It is being used in many industries for the sake of automation and ease of life like; we can operate our electronic gadgets via internet. Some of the examples of using Internet of things in our daily life are switching on microwave when we start from the office, switching the air conditioner in our house when we are present in the office etc.

The most important component of the IOT industry is Relay [6]. The micro controllers like an Arduino, Raspberry Pi are not capable to handle high-powered circuits that are used to handle the house hold electronic items such as lights and fans so to mitigate with this situation Relay can be used [3]. Consider a situation where you have to control a fan from the cloud. The firs step is to find the suitable micro controller. Then the next component which will be needed would be relay to handle the inefficiency of micro controllers in handling high electronic voltages. Another choice is the use of pre-made PCB board with all the circuitry and relays built on, including screw-down terminals for all connections.

In this project we are going to demonstrate a how we can use internet of things can be used in order to increase the effectiveness of video game playing. The paragraph underneath will give a brief of what we will be demonstrating in this project. In this paper demonstrating a workflow of how which can be utilized in video games for increasing interactivity.

2   Literature Survey

As we had some introduction about internet of things now well get a small introduction of video games and where we can implement Internet of things to enhance the gaming experience [4, 5]. Some of the places where we can implement IoT in video games are consider that we are driving a motor cycle in the video game. We will not experience any kind of air and vibrations [7]. We can develop smart fans and chairs that will enhance the game playing experience. If we are playing a horror video game then just imagine how amazing the experience would be when the lights in your gaming room glow according [8].

Research is being held in usage of Internet of things in video games for achieving cheap and better gaming experience in emerging startups and also in Huge Companies for instance Phillips is developing bulbs that can change their properties such as brightness and color to provide a beautiful gaming and cinematic experience [9]. Loop Fit is a product that is developed by loop reality a startup with a integration of virtual reality and internet of things. Loop fit is a product where we will be wearing a VR headset and cycling with a exercising cycle and even though we are stationary [10]. We will find ourselves moving in a virtual world some of the other features are we can connect to other friends online like an online multiplayer game [11].

Much advancement is being made in the field of internet of things [12]. It is making the interactions with environment possible as we can see many projects. It is involving in developing air conditioners, Microwave ovens, Lights that can be controlled with the help of WiFi and not only that facial recognition systems connected to each other to give a person's location etc. [12]. There is a development in gesture recognizing smart environments that reacts based on our gestures such as weaving hands, clapping etc. [13]. Our proposed work is implemented based on a system that can be used to control environments with a cost efficient and home control environmental monitoring system.

3   Proposed Model

The proposed model requires a dedicated cloud that can interact with similar systems and offers the necessary communication protocols which are required to control and monitor the environment with more than just the switching functionality. The model can interact with different devices such as lights, fans, Cycle wheels etc.

3.1   Implementation for Proposed Model

A sample of how to design a proposed model. Consider a video game where the player can jump off from high ledges and ride motorcycles. When designing this video game we have to consider the all the possible states that the player can encounter we should create a data structure for storing speeds and the current action performed by the player. Consider the following flow chart for a very simple game. The velocity of the player is also stored in a separate variable (Fig. 1).

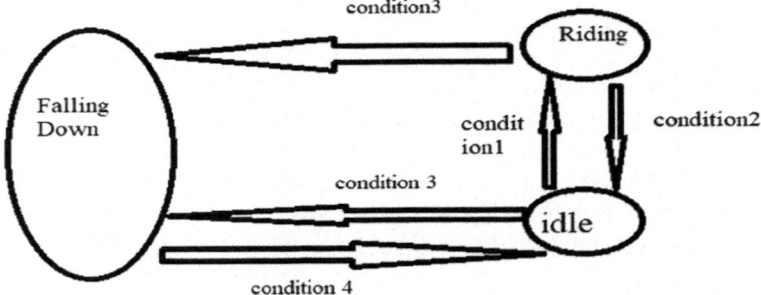

Fig. 1. A sample diagram representing the design of the proposed system

Condition 1: if(isRiding==true)
Condition 2: if(isRiding==false)
Condition 3: if(isfalling==true)
Condition 4: if(isfalling==false)

If any of the both Boolean variables are set to true then the fan will change its speed and that it emits to give a feel of interactivity.

3.2   Algorithm of Proposed Model

(1) Receive data will be from the sensor.
(2) Send the data into a public cloud.

(3) Receive the data from cloud into Arduino.
(4) Write the data on the game as input.
(5) With the received inputs perform the actions specified in the game design document.
(6) Send the some information in the form of variables to cloud.
(7) Receive the data from cloud into Arduino.
(8) Activate the Devices in the environment as per the data received.

The above all steps are consider as workflow of the proposed model is represented in Fig. 2.

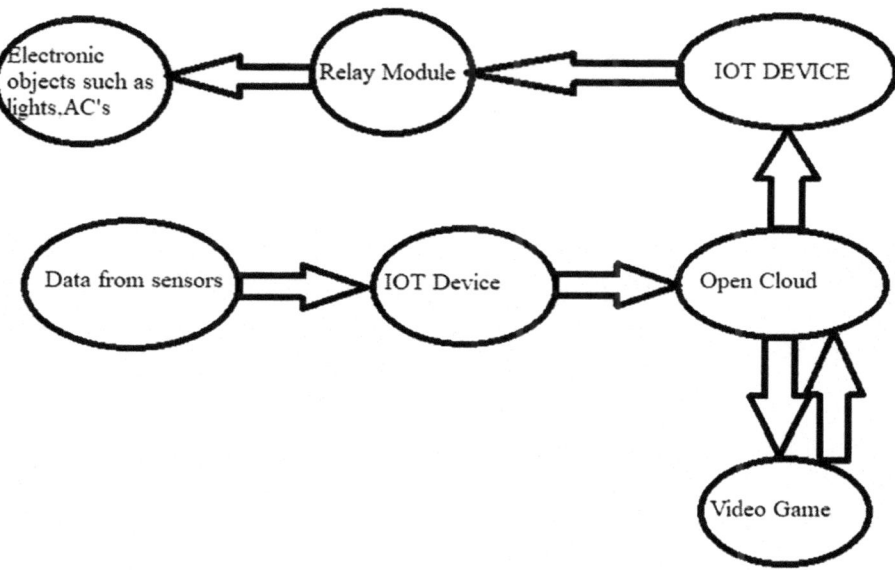

Fig. 2. Workflow of the model

3.3 Working Procedure of the Proposed Models

In Fig. 2 shows the work flow of the proposed model. In Fig. 3 shown the different components are require to implement the model. For implementing this kind of a system we need active internet connection, Arduino and ep8266 or raspberry pi, a public cloud that allows us to update and retrieve data with very low latency which will helps to provide seamless interaction. Each bulb First step is to create a database in one of the public clouds available. Now receive the data from all the sensors available and write the data into cloud database. Now from the cloud database read data the data into the game and as per the data read perform the specified operations mentioned. Another way of environment interaction is getting the data from game and directly writing the data from the game into the cloud database and from cloud database the data should be read into the IoT device and with the help of relay module environment will be controlled as per the received data. A small example of how the environment will be controlled is considering a tube light whose power connection is controlled with Arduino and the power supply is provided as per the value of variables. If the data received from Arduino is false then the light will not get power supply else if the data received is true then light will receive power supply. Consider a light bulb which is to be interacted as per the data from video game let us see how to setup the environment this will be the block diagram of the implementation with Arduino.

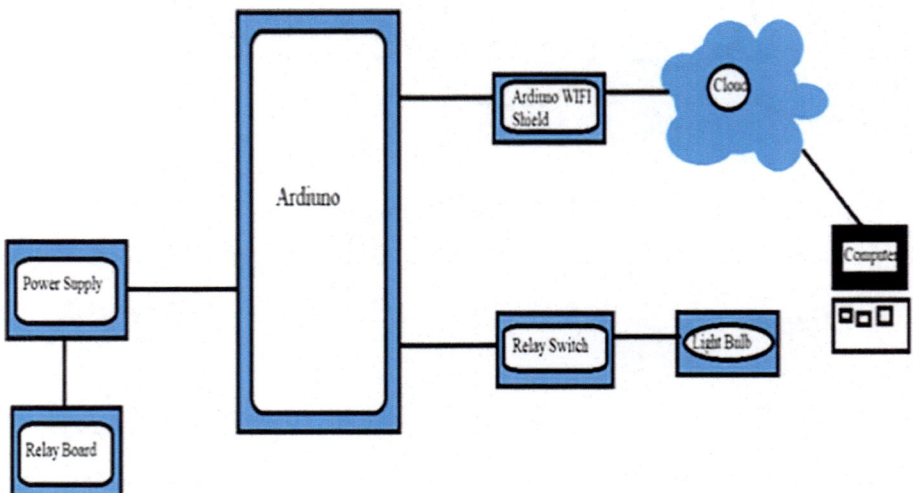

Fig. 3. Architecture of the proposed model

4 Experimental Results and Discussions

After following the above process and execution of the modules are presented. We took input from sensors via Arduino and gave it to a video game which acts as a proof of concept for using IoT in video games for increasing the interaction. In Fig. 4 demonstrate the position of the cube when it is outside the door in video game. Otherwise the LED bulb will glow when the cube didn't enter. It is shown in Fig. 5.

Here the cube is the player and input will be given via keyboard. An LED bulb connected to Arduino will be glowing.

In Fig. 6 shown, the cube is entering in to the door. Whenever the cube will enter the door the LED bulb stops working. This will demonstrate one of the models that we proposed in this project.

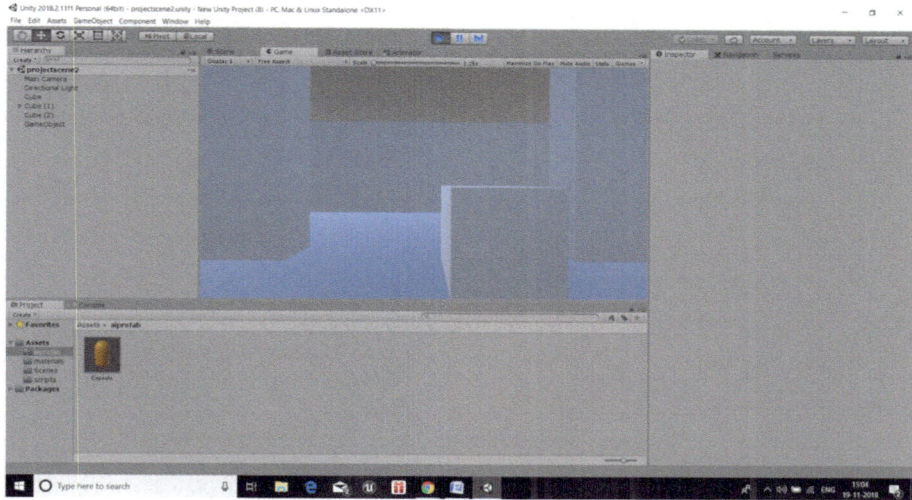

Fig. 4. Demonstrating the position of the cube when it is outside the door in video game

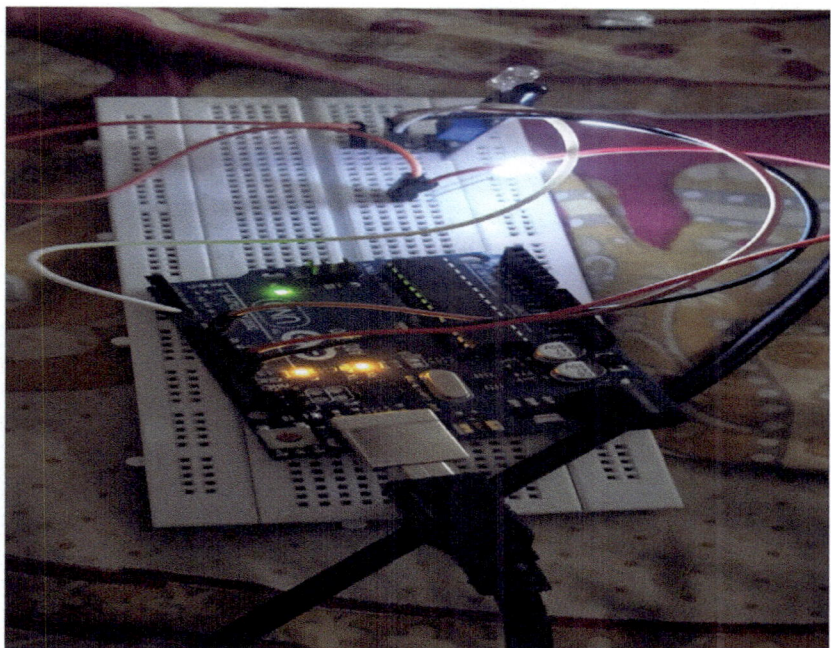

Fig. 5. LED bulb glowing when the cube didn't enter

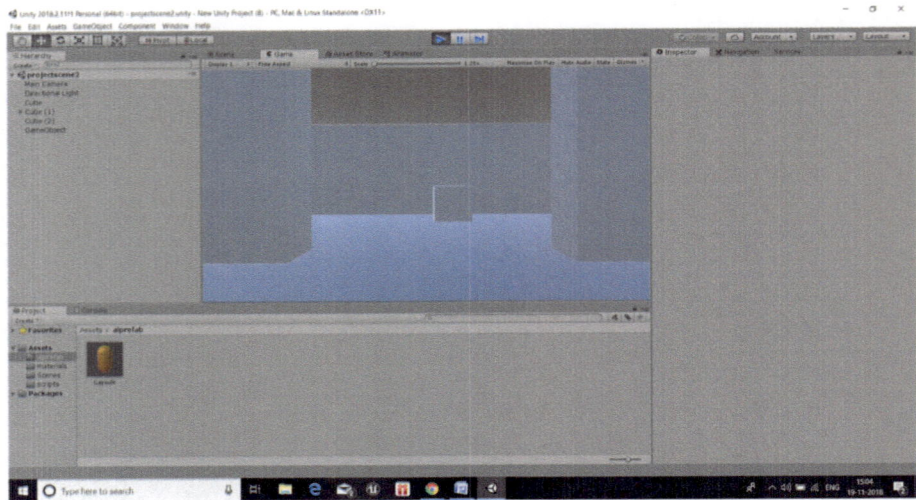

Fig. 6. Cube entering the door

In Fig. 7 shown an implementation of the first mentioned models prototype using open cloud. In this game it the cube should not touch the capsules. The cube should escape the capsules by jumping the cube will jump if the IR sensor detects an intrusion.

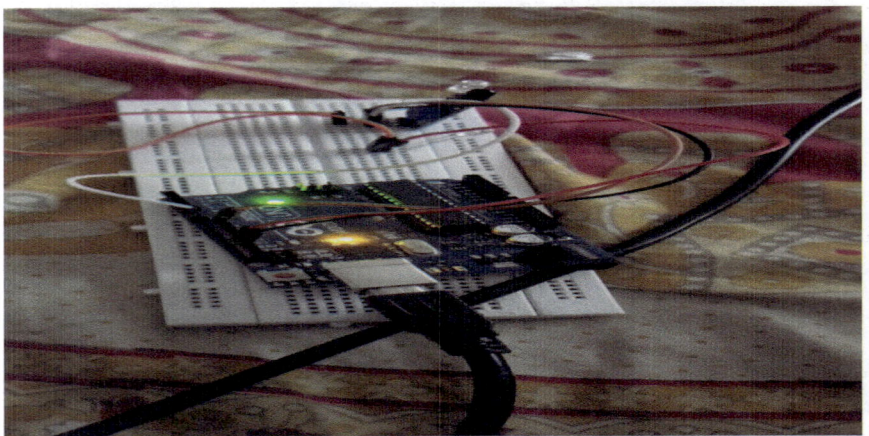

Fig. 7. LED bulb stopped glowing after the cube enters the door

This is the state of game when there is no obstacle with IR Sensor

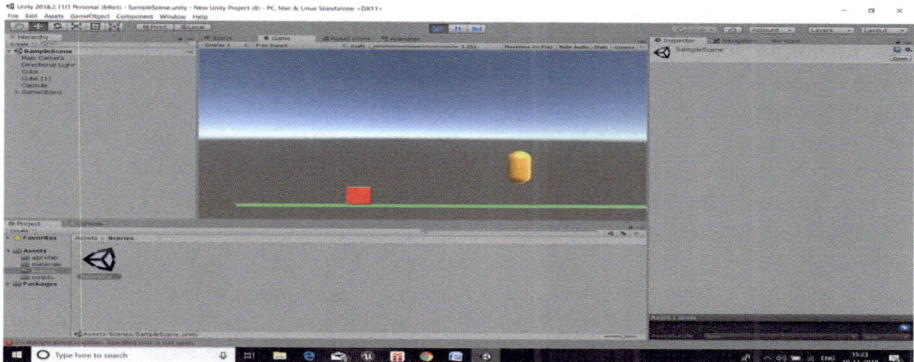

Fig. 8. Cube receives no input from IR Sensor

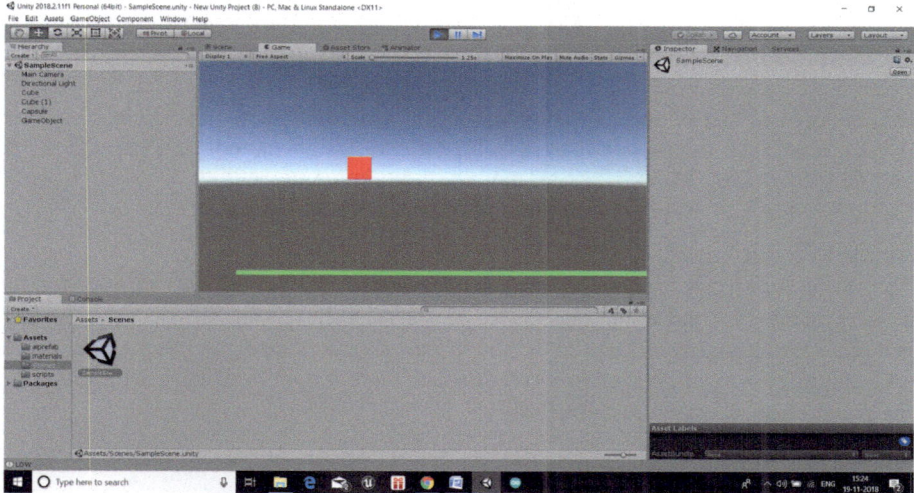

Fig. 9. When cube receives an input from IR Sensor

Discussions

Yet the proposed system is efficient there are some precautions that are to be taken in order to be taken in order to design this kind of systems. First thing to be kept while designing is that different sensors will have different latencies and time delays that are to be used to take inputs. As we have to give inputs from video game to Micro Controller if a same board is used for both input and output the rate at which the micro controller receives input will also have delay which causes the proposed system to give wrong effects at the wrong time. From Figs. 8 and 9, our design was a cube will receive input from Arduino with IR sensor and when the cube receives input it will move into a forward. When the cube enter the hallway a led which is in glowing state that is connected to the Arduino will turn off. For IR sensor we have to transmit on the data with a minimum delay of 50 ms in order to receive the correct input in the above video game that is in Unity3d. But as we send data from Unity3d on the as the light should change its state within a fraction of second by the time Arduino reads data from the serial controller and process it there is at least a delay of 2–3 s. The most important thing to remember is that if complexity of the program that is embedded into the micro controller that give input the performance of the will decrease drastically.

Implementing such kind of system is complex and requires a huge support from the current gaming industry. Because this system has some minimum design requirements on how the electrical system of the house should be designed. It might take time for the existing gaming industry and to adapt to this new technology. When coming to the game experience this experience might be wonderful for experience while playing horror games. As the environments interact according to the variables in the cloud one should make sure that the security of that cloud is not compromised. For some features which require physical motions as input a small extra hardware attachment will do the trick.

5   Conclusion

The technology that is needed to implement the proposed system is already in hand. The main idea of this paper is to organizing the design pattern and strategies and features that might be existing system in providing interactivity in video games. Because of that the developers might find these techniques and strategies to be useful.

Future Enhancements
Once fully automated homes are implemented successfully then we can improve the above system by connecting our video games directly to the micro controller that provides power supply. The whole system can be automated reducing the effort of programming every time. We have to setup such kind of system. If Ipv6 is introduced it might eliminate the need of public cloud.

References

1. Gracanin, D., Handosa, M., Elmonguit, H.G., Matkovic, K.: An approach to user interactions with IoT-enabled spaces
2. Mac Dowell, A., Endler, M.: Internet of things based multiplayer pervasive games: an architectural analysis, vol. 150, pp. 125–138 (2015). https://doi.org/10.1007/978-3-319-19656-5_19
3. David, N., Chima, A., Aronu, U., Obinna, E.: Design of a home automation system using arduino. Int. J. Sci. Eng. Res. 6(6), 795–801 (2015)
4. Shokrnezhad, M., Khorsandi, S.: On the performance improvement of IoT networks using relay-based communications. In: 2016 International Conference on Information Technology (ICIT), Bhubaneswar, pp. 17–22 (2016). https://doi.org/10.1109/icit.2016.017
5. Mu-sigma: How the Internet of Things is Transforming Gaming. Mu Sigma University (2015)

6. Yellamma, P., Bharadeaj, C.M.: Implementing robots in defence through motion capture with mixed reality. Int. J. Eng. Technol. **7**(232), 114–116 (2018)
7. https://internetofthingsagenda.techtarget.com/blog/IoT-Agenda/IoT-gives-the-mobile-gaming-industry-a-facelift
8. Halper, M.: Philips unveils software for Windows and Apple computer games to drive interactive room lighting Published on January 10, 2018 by Contributing Editor, LEDs Magazine, and Business/Energy/Technology Journalist
9. Halper, M. https://www.ledsmagazine.com/articles/2018/01/philips-unveils-software-for-windows-and-apple-computer-games-to-drive-interactive-room-lighting.html
10. Hulivahana, B.: LoopFit: A Virtual Reality Fitness Product by Loop Reality. A Vr affinity VR (2017)
11. Eisenberg, A.: How Internet of Things Will Transform Online. Ignite Outsourcing (2017)
12. Yellamma, P., et al.: Controlling and monitoring home appliances through cloud using IoT. PONTE, **74**(2/1) (2018). https://doi.org/10.21506/j.ponte.2018.2.8
13. Yellamma, P., Narasimham, C.: Hybrid compressed hash based homomorphic ABEncryption algorithm for security of data in the cloud environment. Int. J. Future Revolut. Comput. Sci. Commun. Eng. **3**(11), 604–612 (2017). ISSN 2454-4248

Big Data Challenges and Issues: A Review

Akanksha Mathur[(⊠)] and C. P. Gupta

Computer Science Department, Rajasthan Technical University, Kota, India
mathurakanksha24@gmail.com, cpgupta@rtu.ac.in

Abstract. Data is expanding immensely as well as colossally, multiplying each year. There is no denying the fact that data is and will keep on moulding our lives. Big Data can be thought of as the "development of perpetual information". Big data is pulling in technologists, researchers, and analysts in the last couple of years in different areas of large databases. Big data gathers data from multiple distributed sources in large volumes which makes it a vital issue to process data accurately for better utilization and information quality. Big data poses great challenges in many areas. The paper relates to recent findings in big data science and technology.

Keywords: Big data · Data analytics · Security and privacy · Cloud computing

1 Related Study

Big data is characterized by considering 3V's i.e. volume, variety and velocity, growing to 7v's wherein representation of data cannot be confined to conventional systems. In the 1970s, the term "Big Data" was coined, but rose in 2008. Big data defines a dataset where the data size is beyond the traditional database's capability to record, store, manage and analyze information [1]. Big data has no universally accepted definition of how large it should be for classification as big data. The data volumes are in the range of petabytes (10^{15}), exabytes (10^{18}) and beyond [2]. Data created, collected and arranged in exabyte every year. However, its creation and aggregation is quick and will approach to zeta-byte (10^{21}) in the coming years. A review of big data challenges and issues for data-centric applications expressed in type and significance of information retrieved [3]. Also, this extended too many areas further including cloud computing, IOT (internet of things), social networks, healthcare applications etc. was proven useful. Big data has advancements toward information management and handling challenges like big data analysis, knowledge diversity, knowledge extraction and reducing, integration and cleansing and several other tools for analysis and mining [4]. As business spaces are developing and there is a need to recompile the monetary framework, rethinking connections among producers, merchants, and customers of merchandise and enterprises [6].

© Springer Nature Switzerland AG 2020
A. P. Pandian et al. (Eds.): ICCBI 2018, LNDECT 31, pp. 446–452, 2020.
https://doi.org/10.1007/978-3-030-24643-3_53

1.1 Big Data Challenges and Issues

The most crucial part of big data is information. The appearance of any hazard indicates a need for security and protection during data transition or data storage. Classical security solutions are insufficient with reference to big data to make sure security and privacy. Hence, many security and privacy problems with big data are confidentiality, integrity, visualization and information privacy are expatiate in consequence of literature.

1.1.1 Confidentiality
Confidentiality could be a key measure to handle sensitive data, particularly, having the ability to store and process data whereas assuring confidentiality to assemble data. Confidentiality is imposing a restriction on the information against illegitimate revelation.

1.1.2 Integrity
Data integrity gives assurance against changing information by an unauthorized user. Packet sniffing, password attacks, phishing and pharming, data diddling, the man in the middle attack and session hijacking attacks are the most well-known attacks where integrity is comprised. Integrity is also maintained using data provenance, data trustiness, data loss, and data deduplication.

1.1.3 Visualization
Visualization provides a graphical representation of data that becomes easy to understand and interpret outcomes. This technique is helpful for decision makers to access, evaluate, analyze, comprehend and act on real-time data.

1.1.4 Security and Privacy
Big data contains huge amounts of individual interpersonal data that is voluminous in size and security of private data thus is the greatest challenge (Table 1).

Table 1. Big data challenges and issues

S. no.	Issues	Approaches	Description
1.	Confidentiality	Partial homomorphic encryption [8]	Proposed a secure technique for scrutiny directions, examine totally different routes avoiding data misuse of GPS information, by utilizing halfway homomorphic encryption
		Homomorphic encryption [9]	Utilized homomorphic encryption to give a convention to similitude positioning
		Authentication, Authorization, Access management (AAA) [12]	Information is encoded all through the progress and kept as plaintext
		Property-based protocol [13]	Proposed productive verification scheme for consistent information. Property-based convention for portable clients with information design system, client development procedure, and confirmation method is investigated for affirmation of classification
		Ciphertext-policy attribute-based cryptography scheme (CPABE) [14]	Generated access control scheme running in four stages: framework design, key formation, encryption, and decipherment

(*continued*)

Table 1. (*continued*)

S. no.	Issues	Approaches	Description
2.	Integrity	Feature selection along with integrity [15]	Applied a machine learning technique on a dataset for feature selection
		FPGA [16]	Used an FPGA based hardware for tamper-proof storage while maintaining confidentiality and integrity
		Tamper-resistant hardware [17]	Provide a secure token to avoid any data revelation throughout the execution of a query ensuring a closed execution environment
		Automatic analytic and on-demand data assortment methods [18]	This framework consists of on-demand automatic data collection methods
		Various approaches [19]	Various approaches like data integrity protection, digital signature, data query etc.
		Maintaining security and integrity [20]	An Approach for real-time security verification of streaming data
3.	Visualization	Various approaches to improve visualization [21]	Points out visual noise, large image perception, data loss and rate of image modification
		Toolbox for privacy preservation [22]	Enforced a toolbox for visualizing associate degreed assignment tasks supported an individual's location
4.	Security and privacy	Privacy model and techniques [23]	Explored chance of re-identification of data from various sources and model them for privacy management
		Differentially private data structures [24]	Introduced β-likeness that is a lot of informative and accessible data
		Privacy on their telecommunications platform [25]	Enforced differential privacy by maintaining the utility of the data
		Differential privacy [26]	Implement a personal querying language PINQ [27], to enhance privacy by decreasing adverse entries producing noise to produce optimum results
		PriGen framework [28]	Defines framework "PriGen" for privacy preservation in the cloud
		Reviewed privacy and protection approaches [18]	Approaches like k-anonymity, l-diversity, and t-closeness etc.
		Differentially personal learning [29]	Gathers features associated with different entities and learning useful data
		KNN formula [30]	Enforced a secure KNN within the cloud
		SNN problem [31]	Discusses attacks in existing methods for SNN, and a replacement SNN methodology that approaches to the attacks

1.2   Big Data Analytics Methods

(See Table 2).

Table 2. Big data analytical methods

S. no.	Approaches	Methodology
1.	Text Analytics [32–36]	Prediction of stock market data basically coming from financial news
		Examines text in two forms of data interpretation i.e.
		Entity Recognition (ER) and Relation Extraction (RE)
		Natural language processing technique (NLP)
		Sentiment analysis technique

(*continued*)

Table 2. (*continued*)

S. no.	Approaches	Methodology
2.	Video Analytics [37, 38]	Shared 1 s of high dimension video
		Automatic security and surveillance system
3.	Audio Analytics [39–41]	Audio analytics for treatment of specific medical conditions
		Infant cries and health status was given here
		A comprehensive review of approaches for video categorization
4.	Network Analytics [42, 43]	Large-scale graph analysis, framework supported map-reduce paradigm appeared
		Graph system on synchronous parallel model (BSP)

2 Open Research Challenges in Big Data

2.1 Security and Privacy

Many techniques developed to check personal and private information using security protocols. Still, infrastructure-based aspect for data privacy and management is an issue needed to be resolved.

2.2 Data Fusion and Visualization

Assessment of certain individual group behavior and pattern identification research carried out. An efficient storage and collection of information are required along with solving spatio-temporal problems.

2.3 Cloud Computing

On-demand services not only improve the availability of resources in the cloud environment but also works for cost reduction. Efficient management of data in storing, processing with resources utilization aspects can be considered as one aspect.

2.4 Other Social Media Related

Challenges for example online social network, data integration and performing specified operations for rumor and fake news spread can be the current state of work in big data.

3 Conclusion

Big data is a new data analysis platform that handles multidimensional information on a large-scale for data discovery and higher cognitive processes. Big Data technologies are widely used for data exploitation with the help of large - scale computing infrastructure in social big knowledge analytics tools in many areas, ranging from business intelligence to scientific exploration. This paper is a review of big data challenges, issues and methods in literature focussing on processes and tools to form the next generation computing with big data. The paper also focuses on current research challenges in the field of big data.

References

1. Manyika, J., et al.: Big data: the next frontier for innovation, competition, and productivity, p. 1_137. McKinsey Global Institute, San Francisco (2011)
2. Kaiser, S., Armour, F., Espinosa, J.A., Money, W.: Big data: issues and challenges moving forward. In: 2013 46th Hawaii International Conference on Systems Science, pp. 995–1004 (2013)
3. Agrawal, D., Bernstein, P., Bertino, E.: Challenges and Opportunities with Big Data 2011-1 (2011)
4. Chen, M., Mao, S., Liu, Y.: Big Data: A Survey. Springer, New York (2014)
5. Chen, J., Chen, Y., Du, X., Li, C., Lu, J., Zhao, S., Zhou, X.: Big Data challenge: a data management perspective. Front. Comput. Sci. 7(2), 157–164 (2013)
6. NIST Big Data Public Workinig Group: NIST Special Publication 1500- 4 NIST Big Data Interoperability Framework, Security and Privacy, vol. 4. September 2015
7. Murthy, P., Bharadwaj, A., Subrahmanyam, P., Roy, A., Rajan, S.: Big Data Taxonomy. Cloud Security. Alliance, no. September, p. 33 (2014)
8. Liu, A., et al.: Efficient secure similarity computation on encrypted trajectory data. In: 2015 IEEE 31st International Conference on Data Engineering (ICDE), pp. 66–77 (2015)
9. Chu, Y.W., et al.: Privacy-preserving SimRank over distributed Information network. In: 2012 IEEE 12th International Conference on Data Mining (ICDM) , pp. 840–845 (2012)
10. Bender, G., et al.: Explainable security for relational databases. In: Proceedings of the 2014 ACM SIGMOD International Conference on Management of Data, SIGMOD 2014, pp. 1411–1422. ACM, New York (2014)
11. Meacham, A., Shasha, D.: JustMyFriends: full SQL, full transactional amenities, and access privacy. In: Proceedings of the 2012 ACM SIGMOD International Conference on Management of Data, SIGMOD 2012, pp. 633–636. ACM, New York (2012)
12. Yang, K., Han, Q., Li, H., Zheng, K., Su, Z., Shen, X.: An efficient and fine-grained big data access control scheme with privacy-preserving policy. IEEE Internet Things J. 1–8 (2016)
13. Sudarshan, S., Jetley, R., Ramaswamy, S.: Security and privacy of big data. Studies in Big Data, pp. 121–136 (2015)
14. Jeong, Y., Shin, S.: An efficient authentication scheme to protect user privacy in seamless big data services. Wirel. Pers. Commun. 86(1), 7–19 (2015)
15. Xiao, H., et al.: Is feature selection secure against training data poisoning? In: Proceedings of the 32nd International Conference on Machine Learning (ICML 2015), pp. 1689–1698 (2015)
16. Arasu, A., et al.: Secure database-as-a-service with cipherbase. In: Proceedings of the 2013 ACM SIGMOD International Conference on Management of Data, SIGMOD 2013, pp. 1033–1036. ACM, New York (2013)
17. Lallali, S., et al.: A secure search engine for the personal cloud. In: Proceedings of the 2015 ACM SIGMOD International Conference on Management of Data, SIGMOD 2015, pp. 1445–1450. ACM, New York (2015)
18. Xu, L., Shi, W.: Security theories and practices for big data. Big Data Concepts, Theories, and Applications, pp. 157–192 (2016)

19. Gao, Y., Fu, X., Luo, B., Du, X., Guizani, M.: Handle a framework for investigating data leakage attacks in hadoop. In: 2015 IEEE Global Communications Conference (GLOBECOM), San Diego, CA, pp. 1–6 (2015)

20. Puthal, D., Nepal, S., Ranjan, R., Chen, J.: A dynamic key length based approach for real-time security verification of big sensing data Stream. Lecture Notes in Computer Science, pp. 93–108 (2015)

21. Gorodov, E.Y., Gubarev, V.V.: Analytical review of data visualization methods in application to big data. J. Electr. Comput. Eng. 22 (2013)

22. To, H., et al.: PrivGeoCrowd: a toolbox for studying private spatial Crowdsourcing. In: 2015 IEEE 31st International Conference on Data Engineering (ICDE), pp. 1404–1407 (2015)

23. Lu, R., et al.: Toward efficient and privacy-preserving computing in big data era. IEEE Netw. **28**(4), 46–50 (2014)

24. Cao, J., Karras, P.: Publishing microdata with a robust privacy guarantee. Proc. VLDB Endow. **5**(11), 1388–1399 (2012)

25. Hu, X., et al.: Differential privacy in telco big data platform. Proc. VLDB Endow. **8**(12), 1692–1703 (2015)

26. Proserpio, D., et al.: Calibrating data to sensitivity in private data analysis: a platform for differentially private analysis of weighted dataset

27. McSherry, F.D.: Privacy integrated queries: an extensible platform for privacy-preserving data analysis. In: Proceedings of the 2009 ACM SIGMOD International Conference on Management of Data, pp. 19–30. ACM (2009)

28. Rahman, F., Ahamed, S., Yang, J., Wang, Q.: PriGen: a generic framework to preserve privacy of healthcare data in the cloud. In: Inclusive Society: Health and Wellbeing in the Community, and Care at Home, pp. 77–85 (2013)

29. Jain, P., Thakurta, A.: Differentially private learning with kernels. In: Proceedings of the 30th International Conference on Machine Learning (ICML 2013), pp. 118–126 (2013)

30. Elmehdwi, Y., et al.: Secure k-nearest neighbor query over encrypted data in outsourced environments. In: 2014 IEEE 30th International Conference on Data Engineering (ICDE), pp. 664–675 (2014)

31. Yao, B., et al.: Secure the nearest neighbor revisited. In: 2013 IEEE 29th International Conference on Data Engineering (ICDE), pp. 733–744 (2013)

32. Chung, W.: BizPro: intelligence factors from textual news articles. Int. J. Inf. Manag. **34**(2), 272–284 (2014)

33. Jiang, J.: Information extraction from text. In: Aggarwal, C.C., Zhai, C. (eds.) Mining text Data, pp. 11–41. Springer, New York (2012)

34. Hahn, U., Mani, I.: The challenges of automatic summarization. Computer **33**(11), 29–36 (2000)

35. Liu, B.: Sentiment analysis and opinion mining. Synth. Lect. Hum. Lang. Technol. **5**(1), 1–167 (2012)

36. Feldman, R.: Techniques and applications for sentiment analysis. Commun. ACM **56**(4), 82–89 (2013)

37. Manyika, J., Chui, M., Brown, B., Bughin, J., Dobbs, R., Roxburgh, C., et al.: Big data: the frontier for innovation, competition, and productivity. McKinsey Global Institute (2011). http://www.citeulike.org/group/18242/article/9341321

38. Shockley, R., Smart, J., Romero-Morales, D., Tufano, P.: Gains with "big data," according to a conducted on behalf of SAP (2012)

39. Patil, H.A.: "Crybaby": assess neonatal health status from an infant's cry. In: Neustein, A. (ed.) Advances in Speech Recognition, pp. 323–348. Springer, New York (2010)

40. Hirschberg, J., Hjalmarsson, A., Elhadad, N.: "You're as sick as you sound": using computational approaches to gauge illness and recovery. In: Neustein, A. (ed.) Advances in Speech Recognition, pp. 305–322. Springer, New York (2010)
41. Hu, W., Xie, N., Li, L., Zeng, X., Maybank, S.: A survey on visual content-based video indexing and retrieval. IEEE Trans. Syst. Man Cybern. Part C Appl. Rev. **41**(6), 797–819 (2011)
42. Elser, B., Montresor, A.: An evaluation study of big data frameworks for graph processing. In: Proceedings of IEEE International Conference on Big Data, pp. 60–67. IEEE (2013)
43. Valiant, L.G.: A bridging model for parallelcomputation. Commun. ACM **33**(8), 103–111 (1990). https://doi.org/10.1145/79173.79181
44. Bertino, E., Ferrari, E.: Big Data Security and Privacy. Springer, Berlin (2018)

Set Based Similarity Measure for User Based Collaborative Filtering Recommendation System

K. V. Uma$^{(\boxtimes)}$, M. Deepika, and Vairam Sujitha

Department of Information Technology, Thiagarajar College of Engineering,
Madurai, India
kvuit@tce.edu, deepika.rathnadevi@gmail.com,
vairamsujitha@gmail.com

Abstract. With the advancement of technologies in the modern era, the amount of data that emerges from various sources like online websites, Internet of things, E-commerce etc., keeps on increasing at a larger pace. The data available is too much for a common user to handle. So recommendation system makes efforts to provide right information to the right user at their doorstep and makes it easy for the users. Similarity measure is considered as an important step to determine the accuracy of the recommendation system. A classical collaborative filtering is implemented either by using Pearson correlation coefficient or Cosine similarity which has got its own merits and shortcomings. An enhanced similarity measure is proposed by applying the set based methodology on basic similarity measures and analyze the impact of those various enhanced similarity measures such as set based cosine, set based Pearson correlation coefficient, set based spearman, set based Kendall on the user based collaborative filtering recommendation systems. It is observed that the enhanced similarity measure obtained from set based methodology is more significant than the basic measures.

Keywords: Recommendation systems · Similarity measure ·
Collaborative filtering

1 Introduction

The main objective of recommender systems is to utilize the various sources of collected data inorder to infer the customer interests in buying a product, watching a movie etc. In recommendation system, entity are the users and the product that is recommended is called as an item. Hence, Recommendation analysis is performed based on the history of recommendations between users and items. Because, past interest are the good indicators of future choices. Significant dependencies between the user and item centric activity forms the basic principle of recommendation systems [1, 2]. The correlation between the user and item can be leveraged to make accurate recommendations. The dependencies between the individual item and user will be more accurate for recommendation than that of dependency between the item category and the user. The dependency between the user and item can be learned from the rating matrix between the

© Springer Nature Switzerland AG 2020
A. P. Pandian et al. (Eds.): ICCBI 2018, LNDECT 31, pp. 453–461, 2020.
https://doi.org/10.1007/978-3-030-24643-3_54

user and item. The resulting model will be helpful for targeting the users [12]. If large number of rated items are there for the user, then robust prediction can be done about the future behavior of the user. Different recommender learning models are there such as Content based recommender systems, Collaborative Filtering, Context based recommender systems, Knowledge based recommender systems, Hybrid Recommendation Systems. The principle behind Collaborative filtering methods is that the unspecified ratings by the user for an item can be imputed. The similarity measure can be used to make inferences about incompletely specified values. Collaborative filtering technique is widely used inspite of its shortcoming such as cold start problem and performance degradation due to data scarcity [6]. Basic similarity measures are used mostly which are apparently inadequate for perfect predictions. So to improve the performance of the recommendation system, various methods were developed and compared with the existing similarity measure, thereby to increase the accuracy and efficiency.

2   Related Work

Though collaborative filtering is most popular, it has few shortcomings like performance degradation in a data sparse environment where the rating of items by the users is very much limited, cold start problem in which the preferences data regarding new users is nil so computing recommendations for them is difficult The major problem is the use of basic similarity measures to compute the similarity between the users. Objective is to develop a more advanced similarity measure for user based collaborative filtering algorithm which helps in computing the similarity that exists among the users to produce an accurate user-user similarity matrix, by integrating set based approach with the existing similarity measures like cosine, pearson correlation coefficient, spearman, kendall etc. The advanced similarity measure will not only overcome the drawbacks of existing method but also helps in building an efficient and optimized recommendation system with increased accuracy that overcomes the basic challenges and short comings of existing user based collaborative filtering algorithm.

2.1   Existing Similarity Measures

A similarity measure or similarity function is a real-valued function which quantifies the similarity between two objects under consideration. These measures either takes the large value for similar objects and negative or zero value for dissimilar objects [7].

Cosine Similarity
Cosine similarity is a measure of similarity between two vectors of an inner product space that measures the cosine of the angle between them. When cos angle between the two vectors is 90°, the cos 90 will become zero, and the whole dot product will be zero, that is, they will be orthogonal to each other. The logical conclusion we can infer is that they are very far from each other.

$$Similarity = \cos(\theta) = \frac{(A.B)}{(||A||\,||B||)} \tag{1}$$

Jaccard Similarity

Jaccard similarity coefficient is calculated as the ratio of the intersection of features, to the union of features between two users or items.

$$J(A, B) = \frac{|A \cap B|}{|A \cup B)} \tag{2}$$

Pearson Correlation Coefficient

Another way of finding the aforementioned similarity is to find the correlation between two vectors. Instead of using the distance measures as a way of finding similarity among vectors, we use the correlation between vectors in this approach.

$$r = r_{xy} = \left(\frac{1}{n-1}\right) \sum_{i=1}^{n} \left(\frac{x_i - \bar{x}}{s_x}\right)\left(\frac{y_i - \bar{y}}{s_y}\right) \tag{3}$$

Spearman

Spearman Correlation Similarity is similar to Pearson Correlation Similarity but instead of preference values, it uses ranks. It is given in Eq. 4.

$$\omega_{(a,u)} = \frac{\sum_{i=1}^{n}\left(\text{rank}_{(a,i)} - \overline{\text{rank}_a}\right) * \left(\text{rank}_{(u,i)} - \overline{\text{rank}_u}\right)}{\sigma_a * \sigma_u} \tag{4}$$

Euclidean Distance

Euclidean distance similarity is one of the most common similarity measures which is used to calculate the distance between two points or two vectors. It is given as in Eq. 5.

$$\text{Euclidean Distance}(x, y) = \sqrt{\sum_{i=1}^{x} |x_i - y_i|^2} \tag{5}$$

Adjusted Cosine

Adjusted Cosine is a similarity measure that represents the modified form of vector based similarity where the rating scheme differs for the users. That is some users prefer the items with higher ratings and some others prefer the items with lower rating. To remove this drawback from vector-based similarity, we subtract average ratings for each user from each user's rating for the pair of items in question.

$$\text{sim}(i, j) = \frac{\sum_{u \in U}\left(R_{u,i} - \overline{R_u}\right)\left(R_{u,j} - \overline{R_u}\right)}{\sqrt{\sum_{u \in U}\left(R_{u,i} - \overline{R_u}\right)^2}\sqrt{\sum_{u \in U}\left(R_{u,j} - \overline{R_u}\right)^2}} \tag{6}$$

3   Methodology

Suitable dataset is selected. The dataset is preprocessed and is then converted into a rating matrix, which has username as column and item name as rows with rating as values. This rating matrix is passed as a parameter into set based function. Any one of the default similarity measure is computed and then the set based function for the same is calculated (Fig. 1).

```
Input : Rating Matrix
Output :  User-User Similarity Matrix
Initialize :   corate , i  j,uservector
Function setBased(ratingMatrix)
      uservector <- colnames(ratingMatrix)
      sim_users<- Evaluate any similarity measure
      for I from 1 to length(uservector)
              v1<- store ratingMatrix(x[,i]) as a vector
              for j from 1 to length (uservector)
                      v2<- store ratingMatrix(x[,j])as a vector
                      for k from 1 to length(v1)
                              if (v1[k] >0)
                                      if(v2[k] > 0)
                                              corate <- corate +1
                      setvalue  <- (2 * corate        ) / (length(v1) + length(v2))
                      setbasedvalue<- setvalue *sim_users
```

Fig. 1. Algorithm

A rating matrix is passed as an input to the set based function that computes the user-user similarity matrix, whose dimensions are column length of rating matrix. The two user rating to be compared are stored in two vectors and if both the vectors have non zero values when i == j then increment co-rate by 1. Corated value multiplied by 2 is divided by the sum of length of user vectors that have non zero values gives set value. Set value with default similarity value gives the enhanced set based similarity value.

4   Results

Movie Ratings dataset is used. It consists of 100,000 instances. User-name, Movie name, Ratings are the attributes of Movie Rating dataset. Rating ranges from 1 to 5. R STUDIO is a open source software licensed under GNU General Public License. It is CROSS-PLATFORM and hence supports various operating systems and hardware. It contains 4800 PACKAGES from multiple repositories regarding spatial analysis, data mining etc. RSTUDIO is used for implementation (Figs. 2, 3, 4, 5, 6, 7, 8, 9, 10 and 11).

	Claudia Puig	Gene Seymour	Jack Matthews	Lisa Rose	Mick LaSalle	Toby
Claudia Puig	1.0000000	0.8829944	0.8346223	0.8185759	0.8856583	0.5715966
Gene Seymour	0.8829944	1.0000000	0.8661635	0.8724620	0.9750299	0.5737776
Jack Matthews	0.8346223	0.8661635	1.0000000	0.8505213	0.9289889	0.6908641
Lisa Rose	0.8185759	0.8724620	0.8505213	1.0000000	0.9009699	0.4693768
Mick LaSalle	0.8856583	0.9750299	0.9289889	0.9009699	1.0000000	0.5917263
Toby	0.5715966	0.5737776	0.6908641	0.4693768	0.5917263	1.0000000

Fig. 2. User-user similarity matrix using cosine

User Name	Title	Sum(sim_rating)/sum(similarity)
Claudia Puig	Lady in the water	1.66
Gene Seymour	NIL	NIL
Jack Mathews	Just My Luck	3.46
Lisa Rose	NIL	NIL
Mick LaSalle	NIL	NIL
Toby	Just My Luck	4.32
	Lady in the Water	4.48
	The Night Listener	4.1

Fig. 3. Recommendations generated using cosine

	Claudia Puig	Gene Seymour	Jack Matthews	Lisa Rose	Mick LaSalle	Toby
Claudia Puig	1.00000000	-0.1619553	0.1705437	0.1291143	0.15276304	0.04537303
Gene Seymour	-0.16195527	1.0000000	-0.5360877	-0.1678363	0.34299717	-0.56031553
Jack Matthews	0.17054366	-0.5360877	1.0000000	0.2827785	0.56476374	0.38620063
Lisa Rose	0.12911435	-0.1678363	0.2827785	1.0000000	0.37418788	-0.21942963
Mick LaSalle	0.15276304	0.3429972	0.5647637	0.3741879	1.00000000	-0.06406221
Toby	0.04537303	-0.5603155	0.3862006	-0.2194296	-0.06406221	1.00000000

Fig. 4. User-user similarity matrix using pearson

User Name	Title	Sum(sim_rating)/sum(similarity)
Claudia Puig	Lady in the water	1.66
Gene Seymour	NIL	NIL
Jack Mathews	Just My Luck	3.25
Lisa Rose	NIL	NIL
Mick LaSalle	NIL	NIL
Toby	Just My Luck	3.23
	Lady in the Water	4.12
	The Night Listener	3.87

Fig. 5. Recommendations generated using pearson

	Claudia Puig	Gene Seymour	Jack Matthews	Lisa Rose	Mick LaSalle	Toby
Claudia Puig	1.0000000	0.00000000	0.2760262	0.27602622	0.23354968	-0.14907120
Gene Seymour	0.0000000	1.00000000	-0.5669467	-0.09449112	0.31980107	-0.61237244
Jack Matthews	0.2760262	-0.56694671	1.0000000	0.35714286	0.40291148	0.46291005
Lisa Rose	0.2760262	-0.09449112	0.3571429	1.00000000	0.32232919	0.07715167
Mick LaSalle	0.2335497	0.31980107	0.4029115	0.32232919	1.00000000	0.08703883
Toby	-0.1490712	-0.61237244	0.4629100	0.07715167	0.08703883	1.00000000

Fig. 6. User–user similarity matrix using kendall

User Name	Title	Sum(sim_rating)/sum(similarity)
Claudia Puig	Lady in the water	2.82
Gene Seymour	NIL	NIL
Jack Mathews	Just My Luck	2.76
Lisa Rose	NIL	NIL
Mick LaSalle	NIL	NIL
Toby	Just My Luck	3.43
	Lady in the Water	3.98
	The Night Listener	2.51

Fig. 7. Recommendations generated using kendall

	Claudia Puig	Gene Seymour	Jack Matthews	Lisa Rose	Mick LaSalle	Toby
Claudia Puig	1.00000000	0.0000000	0.2898855	0.20291986	0.37032804	-0.09107651
Gene Seymour	0.00000000	1.0000000	-0.6301260	-0.10502101	0.33541020	-0.65991202
Jack Matthews	0.28988552	-0.6301260	1.0000000	0.51470588	0.50097943	0.70844727
Lisa Rose	0.20291986	-0.1050210	0.5147059	1.00000000	0.42270140	0.09240617
Mick LaSalle	0.37032804	0.3354102	0.5009794	0.42270140	1.00000000	0.03279129
Toby	-0.09107651	-0.6599120	0.7084473	0.09240617	0.03279129	1.00000000

Fig. 8. User-user similarity matrix using spearman

User Name	Title	Sum(sim_rating)/sum(similarity)
Claudia Puig	Lady in the water	2.88
Gene Seymour	NIL	NIL
Jack Mathews	Just My Luck	0.392
Lisa Rose	NIL	NIL
Mick LaSalle	NIL	NIL
Toby	Just My Luck	3.16
	Lady in the Water	4.86
	The Night Listener	1.69

Fig. 9. Recommendations generated using spearman

5   Conclusion

A new enhanced set based similarity measures is proposed for user based collaborative filtering recommendation algorithm where analysis of various basic similarity measures such as cosine, pearson, kendall and spearman along with a mathematical methodology called set based method to build a better user based collaborative filtering recommendation system. And it has been proved that enhanced set based similarity measures has a better performance compared with basic similarity measures used in user based collaborative filtering algorithm. This similarity measures considers both the co-rated

	Claudia Puig	Gene Seymour	Jack Matthews	Lisa Rose	Mick LaSalle	Toby
Claudia Puig	1.0000000	0.8027222	0.6676979	0.7441599	0.8051439	0.4286975
Gene Seymour	0.8027222	1.0000000	0.7874214	0.8724620	0.9750299	0.3825184
Jack Matthews	0.6676979	0.7874214	1.0000000	0.7732012	0.8445353	0.5181481
Lisa Rose	0.7441599	0.8724620	0.7732012	1.0000000	0.9009699	0.3129178
Mick LaSalle	0.8051439	0.9750299	0.8445353	0.9009699	1.0000000	0.3944842
Toby	0.4286975	0.3825184	0.5181481	0.3129178	0.3944842	1.0000000

Fig. 10. User-user similarity matrix using set based cosine

User Name	Title	Sum(sim_rating)/sum(similarity)
Claudia Puig	Lady in the water	3.41
Gene Seymour	NIL	NIL
Jack Mathews	Just My Luck	3.74
Lisa Rose	NIL	NIL
Mick LaSalle	NIL	NIL
Toby	Just My Luck	3.24
	Lady in the Water	4.7
	The Night Listener	3.48

Fig. 11. Recommendations generated using set based cosine

and non-corated items of two users and thereby will provide a better user similarity matrix.

References

1. Yang, Z., Wu, B., Zheng, K.: A survey of collaborative filtering-based recommender systems for mobile internet applications. IEEE Access **4**, 3273–3278 (2016)
2. Riyaz, P.A., Varghese, S.M.: A scalable product recommendations using collaborative filtering in Hadoop for bigdata. In: International Conference on Emerging Trends in Engineering, Science and Technology, ICETEST-2015 (2015)
3. Liu, H., Hu, Z., Mian, A., Tian, H., Zhu, X.: A new user similarity model to improve the accuracy of collaborative filtering. Knowl. Based Syst. **56**, 156–166 (2014)

4. Marques, G., Respício, A., Afonso, A.P.: A mobile recommendation system supporting group collaborative decision making. Proc. Comput. Sci. **96**, 560–567 (2016)
5. Sehgala, S., Chaudhrya, S., Biswasa, P., Jain, S.: A new genre of recommender systems based on modern paradigms of data filtering. In: 2nd International Conference on Intelligent Computing, Communication & Convergence, ICCC 2016 (2016)
6. Sachan, A., Richhariya, V.: Reduction of data sparsity in collaborative filtering based on fuzzy inference rules. Int. J. Adv. Comput. Res. **3**(2), 101 (2013). ISSN (print): 2249-7277 ISSN (online): 2277-7970)
7. Saeed, M., Mansoori, E.G.: A novel fuzzy-based similarity measure for collaborative filtering to alleviate the sparsity problem. Iran. J. Fuzzy Syst. **14**(5), 1–18 (2017)
8. Arsan, T., Köksal, E., Bozkuş, Z.: Comparison of collaborative filtering algorithms with various similarity measures for movie recommendation. Int. J. Comput. Sci. Eng. Appl. (IJCSEA) **6**(3), 1–20 (2016)
9. Aguilar, J., Valdiviezo-Díaz, P., Riofrio, G.: A general framework for intelligent recommender systems. Appl. Comput. Inform. **13**, 147–160 (2017)
10. Kianimajd, A., Ruano, M.G., Carvalho, P., Henriques, J., Rocha, T., Paredes, S., Ruano, A. E.: Comparison of different methods of measuring similarity in physiologic time series. IFAC PapersOnLine **50**(1), 11005–11010 (2017)
11. Bobadilla, J., Ortega, F., Hernando, A., Glez-de-Rivera, G.: A similarity metric designed to speed up, using hardware, the recommender systems k-nearest neighbors algorithm. Knowl. Based Syst. **51**, 27–34 (2013)
12. https://en.wikipedia.org/wiki/Recommender_system

An IOT Based Weather Monitoring System to Prevent and Alert Cauvery Delta District of Tamilnadu, India

R. Murugan[1](✉) ⓘD, R. Karthika devi[2], Anitha Juliette Albert[3]ⓘD,
and Deepak Kumar Nayak[4]

[1] National Institute of Technology Silchar, Silchar 788010, Assam, India
murugan.rmn@gmail.com
[2] Sethu Institute of Technology, Virudhunagar 626115, Tamilnadu, India
[3] Saveetha Engineering College, Chennai 602105, India
[4] Koneru Lakshmaiah Education Foundation, Guntur 522502, Andhra Pradesh, India

Abstract. The Internet of Things (IoT) plays a vital role in medicine, agriculture and smart city and many more. The Cauvery delta district of Tamilnadu, India has affected in the winter season of every year by highest rainfall and cyclone. Hence there is an urgent need for monitoring the weather and other environmental natural disasters. The objective of this paper is to propose an IoT based weather monitoring system to prevent and alert Cauvery delta district of Tamilnadu. This paper proposes a novel architecture to monitor common environmental conditions. This architecture fundamentally utilized the sensors like the temperature, humidity, pressure, light intensity, rain esteem and so on. There are different sorts of sensors present in the model, utilizing which all the previously mentioned parameters can be estimated. It very well may be utilized to screen the temperature or then again moistness of a specific place. With the assistance of temperature and humidity, we can ascertain other information parameters, for example, the dew point. Notwithstanding the previously mentioned functionalities, we can screen the light force, climatic weight, and precipitation esteem and thickness. These sensors are interfaced with Cauvery delta district and display block using serial port of the computer. The proposed system architecture was implemented in MATLAB SIMULINK 2017b. The modules are properly interfaced with the personal computer and tested successfully with software coding environment. The sensor value is fixed a particular threshold value and if there is any change in the reading it displays the abnormality in the monitor, updates in the ThingSpeak is an IoT platform via the internet.

Keywords: Internet of Things · Cauvery delta district ·
Weather monitoring system · ThinkSpeak · Sensors

© Springer Nature Switzerland AG 2020
A. P. Pandian et al. (Eds.): ICCBI 2018, LNDECT 31, pp. 462–469, 2020.
https://doi.org/10.1007/978-3-030-24643-3_55

1   Introduction

Delta is a greek letter. delta means finite increment in mathematics Google. In Geography a delta is a zone of land in which a stream partitions into littler waterways and purges into a bigger waterway. Cauvery Delta is a district of Tamil Nadu state in southern India. It envelops the lower scopes of the Kaveri River and its delta and framed the social country and political base of the Chola Dynasty which controlled the greater part of South India and parts of Sri Lanka and South-East Asia between the ninth and thirteenth hundreds of years AD. The Cauvery delta districts of Tamilnadu are Ariyalur, Nagapattinam, Perambalur, Pudukkottai, Thanjavur, Tiruchirappalli, Karur, Tiruvarur (Shown in Fig. 1).

In every year especially in winter season, these Cauvery delta districts are affected by the cyclone with the highest rainfall. This year in 2018 on November 16, Vedaranyam, the tip of the nose of peninsular India stretching into the Bay of Bengal, was in the eye of a storm when Cyclone Gaja made landfall and swirled through the fertile Cauvery delta in Tamil Nadu. The cyclone left the delta battered like no other in more than half-a-century. The cyclone with high-velocity winds gusting up to 120 kms an hour sheared trees, huts, tiled houses and every other structure in its path. Almost the entire delta spread over Nagapattinam, Tiruvarur and Thanjavur districts, considered the granary of the State, the neighboring Pudukottai and even interior Tiruchi and Dindigul districts staggered under its impact as the cyclone made its way to the Arabian Sea in Kerala [1].

The cyclone swept in wind and water, destroying lakhs of trees, including coconut, banana, cashew, mango, jackfruit, Casuarina, betel vine, eucalyptus, teak and sugarcane on thousands of hectares. The paddy crop of the samba/thaladi seasons was also damaged in some places. Boats and huts of fishermen were destroyed. The Point Calimere Wildlife Sanctuary, a Ramsar site (a wetland of international importance for conservation), was ravaged. Carcasses of blackbuck spotted deer, feral horses and birds were washed on the shores of Karaikal in Puducherry. Scores of villages were wiped out and thousands rendered homeless. The steel roofs of petrol stations, grain storage godowns, and other buildings were blown away. Nearly a lakh tonne of stocks in salt pans in Vedaranyam have washed away. Over 3.41 lakhs houses with thatched or tiled roofs were damaged, according to an official estimate. More than 3.78 lakhs persons were accommodated in over 550 relief centers. Over 92,500 birds and 12,200 heads of cattle perished. Figure 2 shows the damages caused by cyclone Gaja [1]. With the appearance of fast Internet, to an ever-increasing extent people, the world over are interconnected. Internet of Things (IoT) makes this a stride further, and associates not just people, however, electronic gadgets which can talk among themselves. With falling expenses of Wifi empowered gadgets, this pattern will just assemble more force. The fundamental idea behind an IoT is to associate different electronic gadgets through a system and after that recover, the information from these gadgets (sensors) which can be dispersed in any design, transfer them to any cloud benefit where one can examine and process the accumulated data. In the cloud benefit, one can use this information to

Fig. 1. Cauvery delta district of tamilnadu, India.

Fig. 2. The damage caused by cyclone Gaja in their vicinity, at Kodiakkarai in Tamil Nadu's Nagapattinam district on November 22, 2018. Source: PTI.

caution individuals by different means, for example, utilizing a ringer or sending them an email or sending them an SMS and so on [2].

As referenced before, IoT empowers not just Human-to-Human communication, yet in addition Human-to-Device connection and in addition Gadget Device communication. This specific improvement in the state of new roads of associations will affect basically each industry, for example, transportation and coordination, vitality, social insurance and so forth. For instance, on account of vitality, IoT is being connected to make Smart Grids which can recognize and react to changes in nearby and more extensive dimension changes in vitality utilization, which will be an indispensable piece of any countries vitality strategy.

Looking past the previously mentioned vitality model, there are numerous zones of interests where IoT can make an important effect, for example, Smart Homes, which include IoT to increase the level of robotization; Wearable advances, for example, smartwatches and wellness groups; One of the greatest territories of potential in IoT is associated with medicinal services [3].

Numerous worldwide hardware behemoths have just contributed profoundly to an IoT framework. With players like Intel, Rockwell Automation, Siemens, Cisco and General Electric the market is on the cusp of a blast, with examiners foreseeing there will be 26 Billion associated gadgets, more than 4 for every human on the planet, and the business is anticipated to acquire nineteen trillion, in costs reserve funds and benefits with firms like Samsung and Google standing out. With this new technological platform, the weather monitoring system has proposed to prevent and alert the Cauvery delta district of tamilnadu [4].

2 Proposed Architecture

The weather monitoring system architecture is shown in Fig. 3, these architecture divided into two sections. 1. Transmitter module, 2. Receiver module

2.1 Transmitter Module

The transmitter module consists of primary sensors, Arduino UNO R3 board. This system consists of six sensors such as temperature, light, sound, humidity, heat and MEMS sensors.

2.2 Receiver Module

The receiver module consists of wifi modem, smartphone and display channel. The receiver module connected to Arduino board through the comport and its continuously reads the sensors data. The reading data has updated to ThingSpeak channel.

2.3 Interfacing

This IoT system was operating in both wired and wireless modes. To operate in wired mode the Arduino board is connected to serial port and interfaced with display apps like ThingSpeak. To operate in wireless mode the Wi-Fi connection is mandatory. The Arduino connects to the Wi-Fi network using a Wi-Fi module, it creates a Wi-Fi zone using Hotspot of the smartphone. The Arduino board is continuous reads input from these 6 sensors. Then it sends this data to the cloud by sending this data to a particular URL/IP address. Then this act of sending data to IP is repeated after a particular interval of time. For example in this work have sent data every 30 s. This work is very useful since the people can monitor the weather report just by visiting website or URL. In recent days

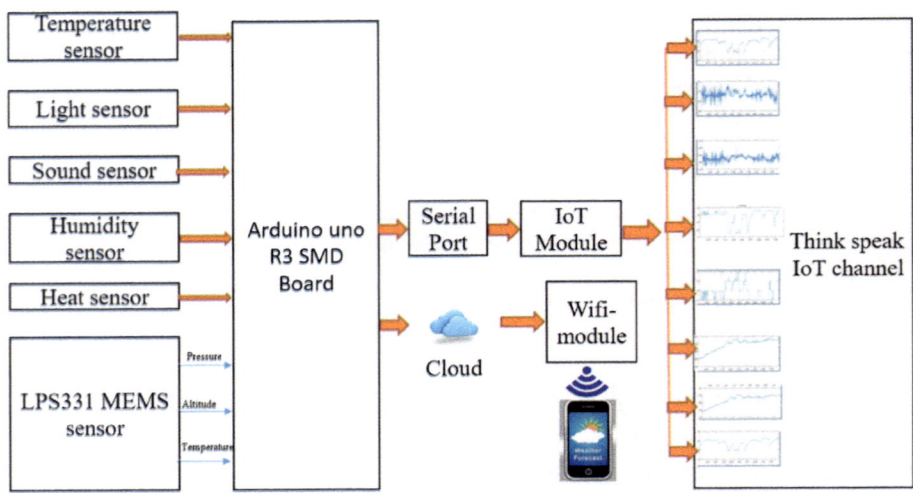

Fig. 3. Proposed architecture.

many IoT apps are also being developed hence the people can monitor or track the weather report through the Android apps or ThinkgSpeak display. There is no need to wait for the metallurgical center weather reports.

2.4   Working

Data Acquisition. Data acquisition is carried out by the sensors that measure the various environmental data and send to the Arduino.

Data Transmission. Typically the data collected by the Arduino is transmitted to the internet using IoT module. Individual sensor's data can be accessed via computer or mobile connection to the internet.

Cloud Processing. The collected Arduino data has sent to cloud and the information has retrieved by wifi module. In this work, we are displaying the results in the website www.thinkspeak.com [12].

3   Implementation

The proposed architecture was implemented (Shown in Figs. 4 and 5) in MAT-LAB SIMULINK 2017b where runs needed on a desktop Intel(R) Core ^{TM}i5–7500 CPU @ 3.40GHz, 8.00 GB RAM 64-bit OS, X64 based processor. The modules are properly interfaced with the personal computer and tested successfully with software coding environment. The sensor value is fixed a particular threshold value and if there is any change in the reading it displays the abnormality in the monitor, updates in the ThingSpeak is an IoT platform via the internet. The obtained channel results of the sensors values were displayed in Fig. 6.

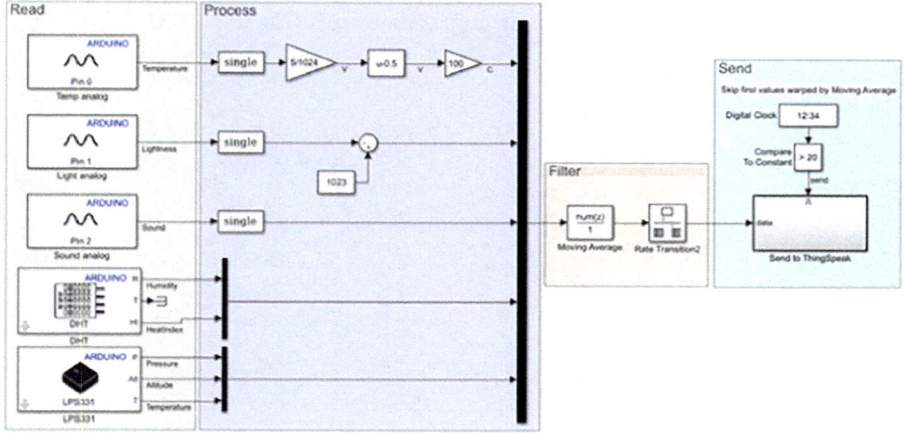

Fig. 4. Transmitter module.

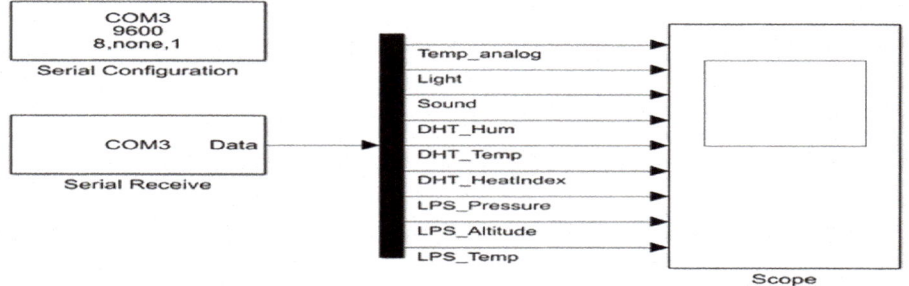

Fig. 5. Receiver module.

4   Result Analysis

The ThingSpeak is an open source IoT application and API to store and recover information from things utilizing the HTTP convention over the Internet or through a Local Area Network. ThingSpeak empowers the production of sensor logging applications. ThingSpeak was initially propelled by ioBridge in 2010 as an administrator in the help of IoT applications. ThingSpeak is a free web benefit that gives you a chance to gather and store sensor information in the cloud and create an Internet of Things applications. The ThingSpeak web benefit gives applications that to examine and imagine MATLAB information, the sensor information can be sent to ThingSpeak from Arduino [5]. ThingSpeak is an IoT stage that utilizes channels to store information sent from applications or gadgets [5]. In the ThingSpeak application, we are created eight channels to monitor the weather. The created channels are temperature, light, sound, humidity, pressure, heat, altitude and temperature that can be shown in Fig. 6.

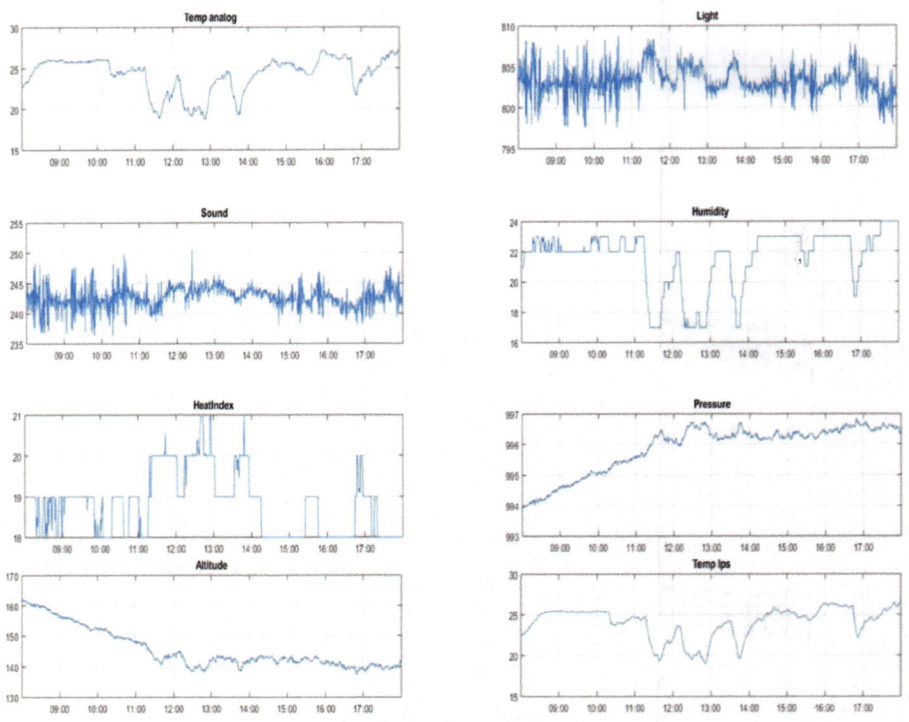

Fig. 6. Channel results of ThingSpeak IoT apps.

5   Conclusion and Future Works

In this paper, an IoT based weather monitoring system was developed and implemented. The performance was analyzed using ThingSpeak an open source IoT application. This system has developed for Cauvery delta district of Tamilnadu, India to automatically aware the weather report of their place. This work helps to alert and save their valuable belongings. In near future, we are going to implement and test this work in any one place of Cauvery delta district with hardware module.

References

1. Shanthi Saravana, B., Veeraraghavalu, R., Alagumariappan, P.: analyzing wind power potential in cauvery delta areas for implementation of renewable energy based standalone pumping system for irrigation. IERI Procedia **5**, 153–160 (2013)
2. https://www.thehindu.com/news/national/tamil-nadu/tamil-nadu-cauvery-delta-districts-where-cyclone-gaja-swirled-through-are-picking-up-the-pieces/article25587103.ece . Accessed 4 Dec 2018
3. https://reliefweb.int/report/india/gaja-cyclone-devastates-hopes-millions-and-millionaires-around-tamil-nadu-delta . Accessed 4 Dec 2018
4. https://www.thehindu.com/news/national/tamil-nadu/destruction-in-the-delta/article25637469.ece . Accessed 4 Dec 2018
5. https://www.thenewsminute.com/article/cyclone-gaja-central-team-visits-thanjavur-district-assesses-damage-92226 . Accessed 4 Dec 2018
6. Noordin, K.A., Chow, C.O., Ismail, M.F.: A low-cost microcontroller-based weather monitoring system. CMU J. **5**(1), 33–39 (2006)
7. Montori, F., Bedogni, L., Bononi, L.: A collaborative internet of things architecture for smart cities and environmental monitoring. IEEE Internet of Things J. **5**(2), 592–605 (2018)
8. Saeed, M.E.S., Liu, Q., Tian, G., Gao, B., Li, F.: remote authentication schemes for wireless body area networks based on the Internet of Things. IEEE Internet of Things Journal. Accepted article (2018)
9. Lu, W., Fan, F., Chu, F., Jing, P., Su, Y.: Wearable computing for Internet of Things: a discriminant approach for human activity recognition. IEEE Internet of Things Journal. Accepted article (2018)
10. Rong-Hua, M.A., Yang, Y.H., Lee, C.Y.: Wireless remote weather monitoring system based on MEMS technologies. Sensors **11**(3), 2715–2727 (2011)
11. Susmitha, P., Sowya Bala, G.: Design and implementation of weather monitoring and controlling system. Int. J. Comput. Appl. **97**(3), 19–22 (2014)
12. https://thingspeak.com/ . Accessed 4 Dec 2018

Summarization of User Reviews on E-Commerce Websites Using Hidden Markov Model

T. Pradeep, M. R. Sai Shibi, P. Manikandan,
Misiriya Shahul Hameed[(✉)], and R. Arockia Xavier Annie

College of Engineering, Department of Computer Science and Engineering,
Anna University, Chennai, India
pradeepthangamuthu94@gmail.com,
saishibi2024@gmail.com, manikandan.prs@gmail.com,
misiriya@gmail.com, annie.benet@gmail.com

Abstract. E-commerce is one of the most important reasons for the growth and sustenance of the internet, with the number of regular users of these sites increasing day by day. The user review forums on e-commerce websites help to get an overview about products of interest. However these reviews are not very organized with respect to whether features of the product, or service provided by the dealers is under discussion. These product reviews can further be positive, negative or a combination of both. This paper discusses a solution to the problem through summarization of product reviews across popular e-commerce websites. The Hidden Markov Model is used to find the component-feature pair, and the SentiWordNet libraries for Weight and Polarity Assignment. After aggregation, the final output is a graph which lists the top six aspects of the product discussed in the review forums along with the weighted polarity representation.

Keywords: Data mining · Natural Language Processing (NLP) ·
Data analytics · Opinion mining

1 Introduction

E-commerce involves business or commercial transactions over the internet and also, transfer of information. Consumer based retail sites, auction or music sites. business exchanges involving trading goods and services between corporations etc. are some e-commerce activities. It's a rapidly growing industry, with the millions of of regular users, thereby keeping the internet busy.

The users of these e-commerce sites make use of the user review forums of a product before they decide to buy them. These forums contain average ratings for the product, giving a quick idea of previous users feedback to the new buyer. But, in order to get a detailed knowledge about each attribute of the product, the user has to read the reviews. The basic problem, a user faces with these reviews is that they are not organized. Certain products may contain thousands of reviews. Among those, some users write about the product, while some people may write about the service provided

A. P. Pandian et al. (Eds.): ICCBI 2018, LNDECT 31, pp. 470–482, 2020.
https://doi.org/10.1007/978-3-030-24643-3_56

(delivery, packaging etc.). There are no means so far to organize these product reviews from all popular e-commerce websites and summarize those reviews for users to easily get a clear idea about the product, without reading reviews from different sites. This is where our proposed system helps. Our project falls under the domain of Data Mining [11] and Natural Language Processing.

1.1   Objective

We wished to develop a system which summarizes the user reviews from all leading e-commerce sites and to display them in a graphical format easily understood by the user. Review summarization involves two main steps:

- Creating component-feature pairs using HMM
- Weighted Polarity assignment to components using Senti-WordNet libraries.

   Given a product name, we need to obtain the user reviews for the product from all leading e-commerce websites, summarize them and produce a graphical representation of the summarized output.

Overall Input: Product name.
Overall Output: Graphical representation of the summarized product reviews.

1.2   Scope

Internet today is a source of various kinds of information, which also aids research purposes. The main benefactors of this kind of research are the E-commerce sites, where data analysis provide the necessary ideas to improve their businesses. The highly competitive e-commerce sites resort to extraction of knowledge from data, using data mining techniques in order to conserve and retain customers [8, 10]. The most important and general data mining tasks done are product search, product recommendation, fraud detection and business intelligence. Even in the presence of a lot of recommender systems (RS) and product comparison softwares already available in the market, there is still no way to get a summarized review based on user reviews. This system is built to help the online shoppers to get a clear idea about the product through a summary.

2   Related Work

Amiya et al. [4] discuss about the implemented working model for television domain, where tools like SentiWordNet, Stanford Core NLP and Hidden Markov Model (HMM) concepts are employed for dictionary and language processor, respectively. POS Tagging of user reviews is performed after which a mapping of the subject and corresponding features is done. A polarity is then assigned to each subject-feature pair. The final results are displayed in the form of pie-charts, bar-charts etc.

   Novotny et al. [2] brings out an enhanced and efficient method for information extraction. An extraction ontology is applied on data segments having multiple data records. They also describe an algorithm for detail-page extraction. Data object instances that are extracted, are processed to achieve tasks such as competitor tracking, market intelligence, tracking of pricing information [2] etc. The three techniques combined here for information extraction are, a pair of algorithms similar to Multifactor Dimensionality Reduction for data record extraction and one more for detail-page extraction [2].

Chainapaporn et al. [3] proposes a Thai herb information extraction process from multiple websites. This employs an JSOUP for parsing, and several template files. Two main phases are mentioned namely, Symptom Name Collection and Treatment Information Extraction. A file of symptom names is generated after parsing multiple websites, and later Thai herb treatment information is extracted. The challenge mentioned here is that every website has non-uniform HTML structures based on complexity. Identification of the Thai herb details and then the symptom contents was hence difficult, due to the lack of uniformity in pattern of representation in the multiple websites. The proposed method of extraction of data from multiple webpages hence proves useful.

3   System Design

The Summarization system contains two phases.

- Review Extraction
- Review Summarization

The first phase extracts the reviews from the various e-commerce websites using web crawling techniques and tools and stores them in the database. All unnecessary contents are omitted without any loss in the required data. Link extractor takes the product name and builds a search query which is used while searching for links in Google search engine. Reviews obtained from different e-commerce websites differ in the format. All the reviews are brought to a single format. Depending on the percentage of people who found the review to be helpful [1], a decision is made whether to take the review to the next level of processing or not. If there are reviews scoring less than minimum threshold of 25%, they are dropped. The output of this stage will be modified JSON objects with an additional Score field of the relevant reviews alone.

After the review extraction phase, the second phase which summarizes the reviews is started. Here, first step involves Sentence splitting. POS tagging helps to identify the components and features. These components then need to be related to the set of opinion words and their modifiers. Hidden Markov Model concepts help to generate the component-feature pair (Fig. 1). The components and features of a sentence form the observable states here. The hidden information that needs to be found out from the observable states, is, whether the components and features are related or not. This is done using conditional probability, which is calculated with two matrices: a matrix to give the probability of a noun being a component, and another to give the probability of an adjective to be a feature to that component.

With the probability matrices, the component feature pairs are identified, using knowledge based training. After component feature pair identification, we find whether

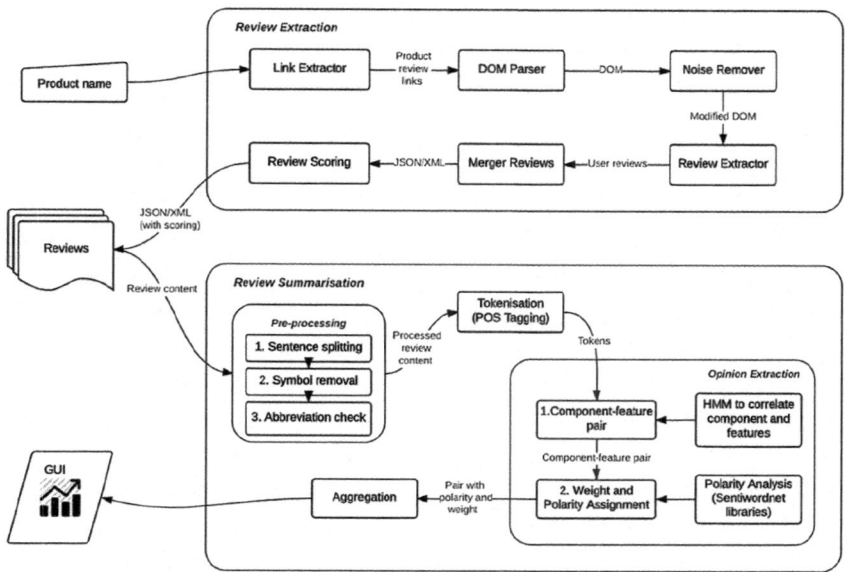

Fig. 1. Overall system architecture

polarity (opinion of the feature on the component) is positive or negative. SentiWordNet [5, 6, 9] is used for this, which assigns to each synset of WordNet three sentiment scores: *positivity, negativity, objectivity*. Finally the output is summarized and visualized as a graph.

4 Implementation of Modules

The list of modules are:

- Review Extraction
- Review Summarization

4.1 Review Extraction

Link Extractor

Input : Name of the product for which reviews are to be summarized.
Output : HTTP links of various e-commerce reviews pages.

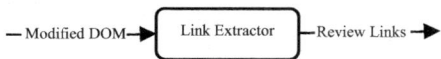

Fig. 2. I/O representation of link extractor

Link extractor takes the product name and builds a search query, which is used to search for links in Google search engine. It is built by appending e-commerce websites name and the word reviews. The query format is (Fig. 2):

PRODUCT NAME + ECOMMERCE SITE NAME + REVIEWS

From the Google search results, unwanted links are removed by text matching. Consider an example product, Samsung Galaxy A7. The Link Extractor creates search queries *Samsung Galaxy A7 Flipkart reviews* [14] and *Samsung Galaxy A7 Amazon reviews* [15] respectively for the two leading online shopping portals in India, Flipkart and Amazon. The first query on Google search engine displays a set of top links. For example:

- www.flipkart.com/samsung-galaxy-a7/p/itme9f3j7bfk2g23
- www.flipkart.com/samsung-galaxy-a7/p/itme4u8trrgepxqh
- http://www.flipkart.com/samsung-galaxy-a7/product-reviews/
  ITME4ZKTGFYWPBRX

The second query produces similar links. In the second set of results, the third result was a link that directed to Samsung's website. Such links are omitted. It is well known that both Flipkart and Amazon have separate pages that contain the product reviews alone for each of their product. In the first set of results for Flipkart, the third link directed to the review page itself, while the other two links directed to the product buying page. Similarly, the second link in the second set of results too directed to the product reviews page. This link would be the optimal link which shall be determined by Text matching, as these links had the string reviews in them.

DOM Parser and Review Extractor

Input : Http links of the e-commerce sites where the product reviews are available.
Output : User reviews containing review content,title and scores.

Jaunt [13], a free Java library for web-scraping & web-automation, is used here. All the links extracted in the previous module are visited. The functions from Jaunt are used to extract the review contents along with the title of the review and the number of users who found the review to be useful. In the previous step of obtaining links, we obtained the following link for reviews on the Flipkart site for the product Samsung Galaxy A7 (Fig. 3):

- http://www.flipkart.com/samsung-galaxy-a7/product-reviews/
  ITME4ZKTGFYWPBRX

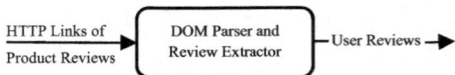

Fig. 3. I/O representation of DOM parser

The first review content on the page obtained is shown as an example here:

Title : Best Phone !! Searching From Long time
Body : I went to Buy Samsung Alpha but Samsung Guy ...... buy this from Samsung Store, 10 days Before.The feeling of un-boxing it and switching it on for the first time is something in itself! ...... u own it, As I am :) One More Imp thing (It Passed Drop Test from 3 Ft :)) If you searching good mob, with 2 GB Ram Good Battry and Dual Sim¿ Go for This..Price Of 30 K + is really worth.
Likes : 88% of 16 users found this review helpful

Merger Reviews

Input : Reviews obtained from the previous module.
Output : JSON object containing review body and review title.

Due to different formats in different e-commerce websites, reviews should be integrated into a single format. The parser helps by iterating for every *hdivitag* and identifies the title, body with the id of the *htitlei*, *hbodyi* tags respectively. This is written to the object. Finally the JSON object contains Title and the Body along with the review number. Say for example, the review from Amazon website is of the following form (Fig. 4):

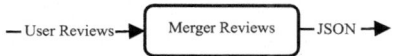

Fig. 4. I/O representation of review merger

"Best Phone !! Searching From Long time"

I went to Buy Samsung Alpha but Samsung Guy introduce my A7, Thanks to him. i had buy this from Samsung Store, 10 days Before. The feeling of un boxing it and switching it on for the first time is .... If you searching good mob, with 2 gb Ram Good Battry and Dual Sim¿ Go for This.. Price Of 30K+is really worth...

This is formatted as,
Review 1:{*Title*:{Best Phone !! Searching From Long time}, *Body*:{I went to Buy Samsung Alpha but Samsung Guy introduce my A7, Thanks to him. i had buy this from Samsung Store, 10 days Before.The feeling of un boxing it and switching it on for the first time is ...... If you searching good mob,with 2 gb Ram Good Battry and Dual Sim¿ Go for This..Price Of 30K+is really worth... }}

This is for one review. This is repeated for *N* such reviews. These are converted to single format and numbered till *N*.

Review Scoring

Input : JSON
Output : Modified JSON with an additional score field.

In order to guide the customers to view the most helpful reviews first, leading e-commerce websites include a feature of *I FIND THIS HELPFUL* option [1] (as in the e-commerce sites Amazon, Flipkart). Reviews are sorted and displayed, based on the percentage of people who found a review to be helpful. This helps customers to read the most helpful reviews first and get a clear idea about the product. Implementing the same concept in our project, the most helpful reviews must be given a higher priority while processing them. Hence, we introduced the score field in the new modified JSON objects. To calculate the score of a specific review, depending on the percentage of people who have found the review to be helpful, the minimum threshold is set as 25%. Reviews scoring less than that value are dropped. Thus, only the applicable reviews are processed and their result is generated and displayed in the final stage, thereby increasing its efficiency. The output will be modified JSON objects with an additional Score field of the relevant reviews alone.

Example: The top review obtained for the product Samsung Galaxy A7 in Amazon website is:

"Stunning design with loads of features to serve your daily needs" by rajesh on 17 March 2015

Bought this phone from Croma in exchange of my old mobile for 25,500.At this price,this is ...... and camera is poorer. 64 people found this helpful - the line taken for processing the score.

5.0 out of 5 stars
Another review for the same product found not so helpful and will be dropped:

"apply emi" by Sriprakash on 29 March 2015
How to apply for EMI on those product

0 people found this helpful - the line taken for processing the score.

4.2   Review Summarization

Sentence Splitting

Input : Review body (Paragraphs)
Output : Reviews as sentences.

After review extraction phase, Review Summarization phase starts. The first step involves Sentence splitting. This is done because, here, we do not process reviews as a whole but as separate sentences. If we process a review as a whole, we may not be able to analyse if the reviewer has got positive or negative views on the particular product.

In some cases, the reviewer may have mixed views on a product. He may like certain features of a product while disliking the other. In such cases, it is better if we process the reviews as separate sentences. We split the reviews into sentences using regular expressions. The usual convention for any sentence is to start with an upper case letter and end with a punctuation. Also, a single space is expected between sentences. But in most reviews, this convention is not followed by the reviewer. Such discrepancies are handled as follows: Check if a punctuation and a space are present before starting the next word. In some cases there may not be any spaces in between sentences. In this case, check if the character next to the punctuation is in upper case (and hence would be the start of the next sentence). It is highly unlikely that the user enters a review with both these discrepancies and hence the above mechanism handles all possible reviews and produce efficient inputs to the next stage of processing.

Tokenization (POS Tagging)

Input : Processed review content
*Output :*Review content with part of speech tagged for each word.

POS tagging is done with the help of Stanford Log-linear Part-Of-Speech Tagger [12]. This software is a Java implementation of the log-linear part-of-speech taggers described in these papers. Our objective is to find the components and features in the review. Components are basic attributes of the product while feature refers to users opinion about the particular component. The following table (Table 1) depicts the form in which the component and feature exist in a review.

Example:
Consider the following sentence, taken from a review.

"The camera is very good."
This sentence is tagged as : *"The/Article camera/noun is/verb very/adverb good/adjective."*

In this sentence, the component is *camera*, feature is *good* and the enhancer is *very*.

Table 1. Mapping

Token	POS
Component	Noun
Feature	Adjective
Enhancer	Adverb

Component-Feature Extraction

Input : A POS tagged review sentence
Output : Component-Feature pair

Here, the relation between the components and the set of opinion words cum modifiers is identified [4]. This is to understand the mapping between the opinion and components. Opinion Extraction process generates pairs of components and relating opinion words with modifiers. With the existence of multiple components, opinion words and modifiers, a component could be related to one or many opinion words. Amiya et al. [4] provides an example, *"The display is huge, sharp and looks really elegant."* In this case, the opinion words related to the component *display* are *huge, sharp* and *elegant*. Similarly, complex sentences can exist (with conjunctions like *and, but,* etc.). This may contain contrasting opinions in a single sentence. To curb this, Hidden Markov Model helps in training the system and relating the feature with the exact component. Figure 5 depicts the need for HMM.

While working with the Hidden Markov Model, the observable states need to be found. This is done using conditional probability. And the conditional probability is calculated with two matrices, as mentioned in Sect. 3. With the probability matrices, the component feature pairs are identified. The probabilities are purely obtained using knowledge based training. Once the training is completed, the unobserved (hidden) states can be obtained from known states. The hidden state is the relation between a component and feature. Baum-Welch Algorithm is used for finding the relation.

Weight and Polarity Assignment

Input : Component along with the feature
Output : Polarity and weight assigned component feature pair

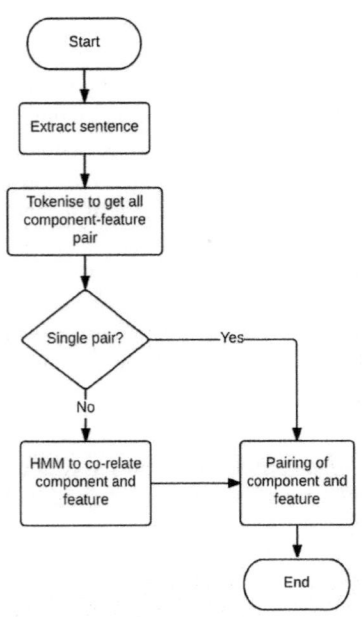

Fig. 5. Hidden Markov model co-relating component-feature pair

Once the component feature pair is identified, the next step is to find whether polarity (opinion of the feature on the component) is positive or negative. This is done with the help of SentiWordNet [5, 6, 9]. Features are passed to the SentiWordNet one by one. Using algorithm 1, the weight of the opinion word is found from SentiWordNet.

Algorithm 1. Algorithm for Weight Assignment

```
1:      p = SUM(AllPositivePolarities)
2:      avgp = AVERAGE(PositivePolarityWeights)
3:      n = SUM(AllNegativePolarities)
4:      avgn = AVERAGE(NegativePolarityWeights)
5:      w = MAX()avgp, avgn
6:      if w = avgp then
7:      Polarity =Positive
8:      else
9:      Polarity =Negative
10:     end if
```

5 Results and Evaluation

The proposed system is a first of its kind product review system, making use of data mining techniques and NLP. The system cannot be tested using any automated techniques, since it would further result in complications depending on the efficiency of the tester software. Thus, the most proper way will be to manually process the reviews for a number of products and compare the results with that obtained by the system. So, *twenty products* are chosen and *fifty random reviews* are taken for each chosen product. These reviews are manually observed and results similar to the ones produced by the system are obtained. Finally, the results obtained in both the ways are tested and the system's efficiency is measured.

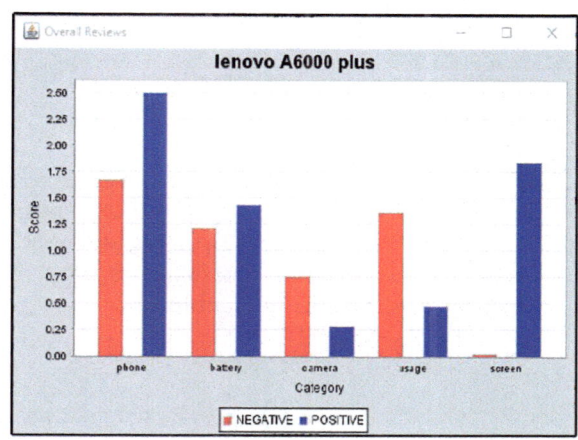

Fig. 6. Output graph indicating weighted polarity of product features

5.1 Output

The output is in the form of two graphs. The first graphical representation (Fig. 6) shows the weighted polarity score for the top six attributes discussed in the review forum by the reviewers. The top six attributes (in Fig. 6, *phone, camera, SIM, screen, quality and apps* are the attributes) are the most frequently repeated key terms pertaining to the product. The red bar and blue bar indicate the positive and negative weighted scores of each attribute. These weighted scores are an outcome of

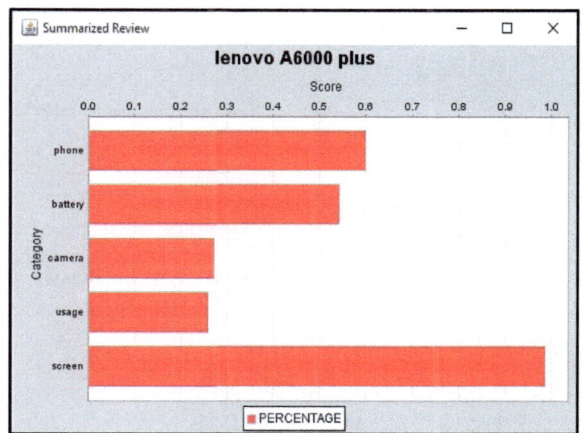

Fig. 7. Output graph indicating product scores

calculations made on the frequency of recurrence and the weightage of the words used to describe the attribute. The second graphical representation, Fig. 7 shows the overall score calculated from the reviews pertaining to each attribute of the product. The bar represents the percentage of positive reviews in each category of the product. This graph gives a quick look on how the users found the product describing each feature of the product.

5.2   Evaluation

Precision and Recall Precision and Recall measures are used here for evaluation [11]. Here we use True Positive, True Negative, False Positive and False Negative values to find the accuracy of the project [7]. In simple terms, high precision shows that we achieved more relevant results than irrelevant, while high recall means that the algorithm has output most of the relevant results.

$$Precision = TP/(TP + FP) \tag{1}$$

$$Recall = TP/(TP + FN) \tag{2}$$

$$Accuracy = (TP + TN)/(TP + TN + FP + FN) \tag{3}$$

TN/True Negative: case was negative and predicted negative
TP/True Positive: case was positive and predicted positive
FN/False Negative: case was positive but predicted negative
FP/False Positive: case was negative but predicted positive

For analysis, we have taken a random set of reviews each particularly describing the attribute *display* shown in Tables 2 and 3 respectively. The system could not identify the sarcasm involved in certain reviews and gives the output as positive for such reviews. This accounts for some comments to fall under a False Positive case. Thus, all the sample test cases are processed for this table (Table 2).

Table 2.   Sample test cases: Attribute - Display

S. no	Review	Predicted output	System output
1	Superior display with sharp colour	Positive	Positive
2	Brilliant display (IPS and AMOLED)	Positive	Positive
3	Display is super for watching cricket matches	Positive	Positive
4	G2 has better display colour combination than S3 Negative	Negative	Positive
5	The 4.8 in. screen is good but motog's 5 in. is better	Negative	Positive
6	Cristal clear display	Positive	Positive
7	Being A-Mold HD display watching video is a bliss	Positive	Positive
8	Best display ever used	Positive	Positive
9	Attrative feature of this phone is its display	Positive	Positive

(*continued*)

Table 2. (*continued*)

S. no	Review	Predicted output	System output
10	Great display for a budget phone	Positive	Positive
11	The phone has such a great display	Positive	Positive
12	The display has great viewing angles	Positive	Positive
13	Amazing screen quality	Positive	Positive
14	High performance touch screen with very good clarity	Positive	Positive
15	Super A-Moled display	Positive	Positive

Evaluation for Display

Here we took 15 sample test cases. Out of 15 test cases there are 13 positive cases and 2 negative cases. Out of them, the number of true and false cases are shown in Table 3 (Fig. 8).

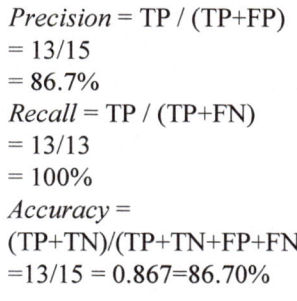

$Precision = TP / (TP+FP)$
$= 13/15$
$= 86.7\%$
$Recall = TP / (TP+FN)$
$= 13/13$
$= 100\%$
$Accuracy =$
$(TP+TN)/(TP+TN+FP+FN)$
$=13/15 = 0.867=86.70\%$

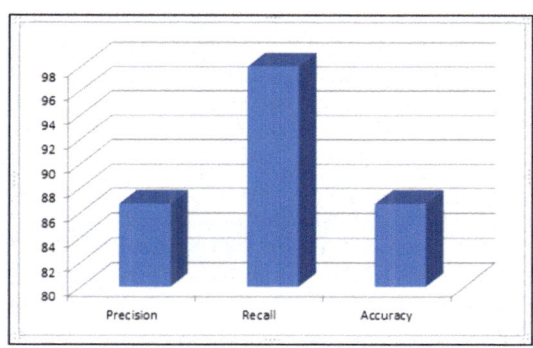

Fig. 8. Evaluation metrics results for display reviews

Table 3. Test case evaluation: Attribute - Display

Cases	Predicted negative	Predicted positive
Negative cases	TN = 0	FP = 2
Positive cases	FN = 0	TP = 13

6   Conclusion and Future Works

A lot of websites are available on the internet which give reviews on most products sought by the customers. Unlike our project, these reviews are made by an editorial review panel. Our work is compared with this review for accuracy. Unlike expert reviews as done in the other websites, ours is one that makes use of, processes and displays the reviews posted by the user and create expert-like review charts defining the scores of each feature of the product. On the basis of these studies it is concluded that the system architecture is very versatile and helps the customers make use of the reviews in a very efficient manner. The system completely works on live datasets. Hence it provides up-to-date information about the product. Hidden Markov Model is used to identify the component-feature pair. This is based on the knowledge based training(Experts knowledge). Polarity of a word is found using SentiWordNet libraries. The system finds the important components and rates them based on the weighted polarity. Finally, the system represents the summarized result using a graph instead of a passage.As of now, the system is provided with a knowledge based training for finding the probability of component-feature existence. Machine learning techniques can be implemented to train the system so that the final output is more efficient and effective. Detecting sarcasm can be implemented for better results from the system as that is lacking at present.

References

1. Chen, L., Chen, G., Wang, F.: Recommender Systems Based on User Reviews: The State of the Art. Springer, Dordrecht (2015). https://doi.org/10.1007/s11257-015-9155-5
2. Novotny, R., Vojtas, P., Maruscak, D.: Information extraction from web pages. In: WI-IAT 2009: Proceedings of the 2009 IEEE/WIC/ACM International Joint Conference on Web Intelligence and Intelligent Agent Technology (2009)
3. Chainapaporn, P., Netisopakul, P.: Thai herb information extraction from multiple websites. In: 4th International Conference Knowledge and Smart Technology (KST) (2012)
4. Tripathy, A.K., Sundararajan, R., Deshpande, C., Mishra, P., Natarajan, N.: Opinion mining from user reviews. In: International Conference on Technologies for Sustainable Development (ICTSD-2015), Feb. 04–06, Mumbai, India (2015)
5. Raut, V.B, Londhe, D.D.: Opinion mining and summarization of hotel reviews. In: 2014 International Conference on Year: 2014 Computational Intelligence and Communication Networks (CICN) (2014)
6. Deepak, T., Singh, J.: The SAFE miner: a fine grained aspect level approach for resolving the sentiment. In: Proceedings of the 2015 Third International Conference on Computer Communication Control and Information Technology (C3IT) (2015)
7. Vidhya, K.A., Soorya, R., Saranavan, N., Geetha, T.V., Singaravelan, M.: Entity resolution for symptom vs disease for top-K treatments. In: 2017 4th International Conference on Advanced Computing and Communication Systems (ICACCS) (2017)
8. Pang, B., Lee, L.: Opinion mining and sentiment analysis. Found. Trends Inf. Retriev. **2**(1–2), 1–135 (2008)
9. Esuli, A., Sebastiani, F.: SentiWordNet: a high-coverage lexical resource for opinion mining. Evaluation **17**(2007), 1–26 (2007)
10. Hu, M., Liu, B.: Mining opinion features in customer reviews. In: AAAI 2004: Proceedings of the 19th National conference on Artificial intelligence, vol. 4, no. 4, pp. 755–760 (2004)
11. Precision and Recall. https://en.wikipedia.org/wiki/Precision_and_recall
12. Stanford Log-linear Part-Of-Speech Tagger. https://nlp.stanford.edu/software/tagger.shtml
13. Jaunt. https://jaunt-api.com/
14. FlipKart. www.flipkart.com/
15. Amazon. https://www.amazon.com/

Multimedia Data Retrieval Using Data Mining Image Pixle Comparison Techniques

D. Saravanan[(⊠)]

Faculty of Operations and IT, ICFAI Business School (IBS), Hyderabad,
The ICFAI Foundation for Higher Education (IFHE),
(Deemed to be university u/s 3 of the UGC Act 1956), Hyderabad, India
Sa_roin@yahoo.com

Abstract. Information extraction is one of the challenging factors for may researchers today. This task more complicated for multimedia data sets because of the quality of the data sets, it gives more challenging factor today. Recording the images convenient even for unprofessional user because of the growing technology allows, every day to produce a large amount of image, audio, video data sets are uploaded by different user community around the world. From this huge content extracting the needed information's are too complex activity. Image extracting done by either text based query or image attribute based retrieval. Extracting the need information from this complex data sets user need additional knowledge about the domain. This gives more attention on this filed. This research paper focuses image retrieval using hierarchical clustering technique. Process divided into two steps initially input data sets need to be trained with help of image pixel value. Second step though image query data's are retrieved from the trained data sets. The proposed technique works well, experimental results also verified this.

Keywords: Image frame · Frame groups · Image attributes
Pixel comparison · Video data mining · Knowledge extraction

1 Introduction

Information extraction done by text based or image based. Traditional text based information retrieval efficiencies are lower than image based retrieval. Image based retrieval gives greater accuracy because of image attribute [1, 2]. Current technology helps this process simple and less time consuming operations. This application used today in various places some of them are cyclone detection, environmental change, climate change, product design and more [3]. For this today most of the organization and research areas they collected huge amount of image data sets for their operations. Image mining are differ from traditional text mining technique, process of extracting the needed information one of the challenging task for the user. Extracting the image data's user need some domain knowledge what type of data need to be retrieved, which type of image attribute used for this operations, data sets are cleaned or not. Before the process get started user need some domain knowledge about the stored image data sets. It gives accurate results with greater accuracy. Number of image mining algorithms and

© Springer Nature Switzerland AG 2020
A. P. Pandian et al. (Eds.): ICCBI 2018, LNDECT 31, pp. 483–489, 2020.
https://doi.org/10.1007/978-3-030-24643-3_57

operations are available today. It bring the need information effectively. But rising the image date sets makes this operations more complicated and more challenging one [4–6]. Based on this work motive in this paper the input video files are converted to static frames, using image properties those frames are analyzed and noisy frames are separated. Using RGB based image pixel property images are stored and retie rived based on the users input image query.

2   Allied Work

Multimedia retrieval needs a special domain knowledge Creation of this data sets are easy for many users but extraction need domain knowledge because of image complex. It requires proper technique or tool without extraction takes more time and user never get needed information. Many research work today motivated for extracting image content from the storage image database. Unlike text retrieval image retrieval considers multiple image attributes properties. Because multimedia data consists of audio, video, time interval, text, motion and more. This properties are consider for any image extraction. Second image extraction requires extra attention before the process get started data based need to be cleaned to reduce searching time.

2.1   Drawback of Existing Image Retrieval System

1. No common technique.
2. Grouping the data sets takes more time.
3. Many existing system works well for specific set of images only.
4. Many existing technique consider single image attribute.
5. Number of output generations are low.

3   Proposed System

Proposed mechanism works based on the image pixel values, multimedia data's are first converted into static images with help of image properties static image values are calculated using histogram technique [7]. After segment the video frames unwanted frames need to be eliminated else the retrieval process takes more time it leads poor output and poor efficiency. After the unwanted frames are eliminated rest of the frames are trained and stored in the separate data base for future operations. The entire process are separated as user side and server side [8]. This technique image extraction get improved and the same more number of output are extracted for given input frame [9]. Experimental explains this process more detailed.

3.1   Improvement of Planned System

1. Given user input query more relevant images are extracted.
2. Time taken reduced.

3. More number of cluster are formed compare with existing techniques.
4. Efficiency improved.

4 Experimental Methodology

4.1 Image Extraction Using Data Mining

Data mining defines extract the new knowledge from the stored data set. In the same video data mining defines finding interesting pattern that are unknown before [10]. Every mining process undergone the preprocessing operations because of noisy data need to be cleaned before it starts extracted. Here the preprocessing operation of video mining shown in the Fig. 1. In stage 1 raw video data are converted into frames, noisy frames are eliminated, cleaned frames are stored in the target data base for further operations [11]. Those data sets are transformed into operations data base for finding knowledge. This extracted knowledge represents to the user community.

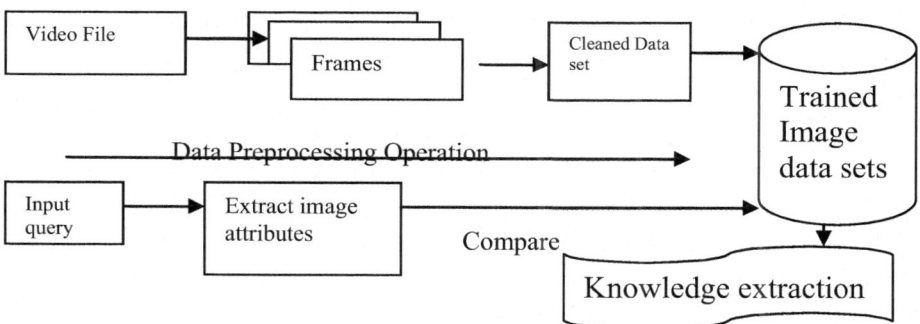

Fig. 1. Proposed architecture

4.1.1 Data Mining in Research Areas

Associations: It helps to find the common factor among the data set. It helps to identify the common factor help to form a new group. This technique helps most of the market analysis helps to find the customer behavior for purchasing the items.

Clusters: It helps to make group based on the identified characters. In image ming it helps to find the image pixel values and form a different cluster. Values are nearer they kept in one group, values are not similar they are all kept in separate groups. This process helps to remove the duplicate frames.

Predications: It helps to identify future occurrence of certain item sets. This process done with help of past analysis. In information extraction it helps even user entered wrong query or wrong input. Information's are extracted based on the past.

Rule induction: Decisions are derived based on rule based operation. After constructing rules, finding the best rule are needed.

The above all are some of the example for data mining used as a tool for extracting the relevant information more accurately.

4.2 Image Segmentation

Motion video are converted into image frames. Every frames image pixel threshold are calculated it helps to remove the unwanted and repeated frames. After this rest of the frames are stored in the separate data based for further operations. This process continue with assign a proper group name for purified data sets. This helps to extract the needed information from specified group sets. It reduces the time taken when user extracting the needed information. All similar items are kept in one group so it increase the retrieval process of the user. This process output shown in the Fig. 2. After this process items are indexed to make users operations more easy [12]. Performance effects of frames Vs time taken shown in the Fig. 6. Experimental results proofs that, time taken for converting more frames in the shorter duration.

4.3 Cleaning and Grouping the Image Data's

After the motion video are converted into to static image data sets, unwanted image data sets to be removed. Some frames are repeated or image frame value may be equal those frames are first removed from the outfitted data sets. Items are grouped together so extraction are very easy [13]. Number of techniques are used to forming the cluster, most of the existing clustering technique based on distance between the data points. This process out put shown in the Figs. 3, 4, 5 and Tables 1 and 2. Cluster formation for image data are differ from text cluster. Image data sets consist of number of attributes such as image pixel, pixel values, time, pixel quality and more. Based on this property information are clustered, because of this reason image cluster formation one of the challenging task for the researchers.

5 Image Mining

Normally in image retrieval the color feature is one of the most utilized visual features. This color feature is strong to background difficulty and independent of size of image and orientation. Even though the overall color feature is very easy to find and can offer logical categorizing power in image recovery, it likes to provide many more fake positives the collection of image is huge. The results of many research recommended that utilizing color layout is a best solution to retrieval of image [14].

5.1 Image Mining Algorithm Steps

The algorithms needed to perform the mining of associations within the context of image. The four major image mining steps are as fallows [15]:

1. Attribute extort: After converted into static image data set each image frames pixel values are calculated and stored I a complete frame value. This values used for duplicate elimination, frame extraction, frame comparison and more.

2. Image discovery and id construction: Every frames now assigned with one unique number helps to differentiate the frame with other frame.
3. Generate Operational data sets: After removing the duplicate frames creates the operational data set for further operation.
4. Apply step 1 to 3 all static image data sets.

6 Experimental Outcomes

Fig. 2. Multimedia video file converted into static file (frames)

Fig. 3. Image frame feature extraction process

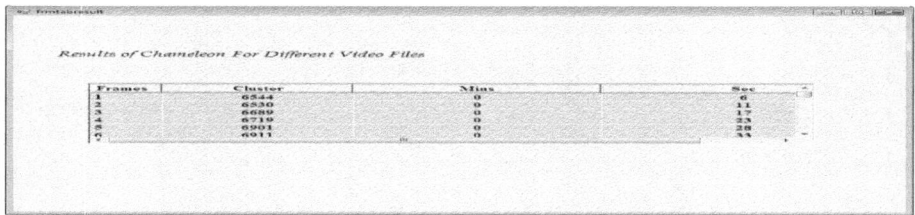

Fig. 4. Frame vs number of cluster formation

Table 1. Image frame clusters vs time takes (millisec)

id	frame	numclus	time
1	0	35204	3:17:55 PM
2	1	35175	3:18:01 PM
3	2	29162	3:18:07 PM
4	3	29762	3:18:14 PM
5	4	30475	3:18:20 PM
6	5	30975	3:18:26 PM
7	6	31351	3:18:33 PM
8	7	31867	3:18:39 PM
9	8	32339	3:18:45 PM
10	9	32427	3:18:52 PM
11	10	32482	3:18:58 PM
12	11	32440	3:19:04 PM
13	12	32842	3:19:11 PM
14	13	32964	3:19:17 PM
(AutoNumber)	0		

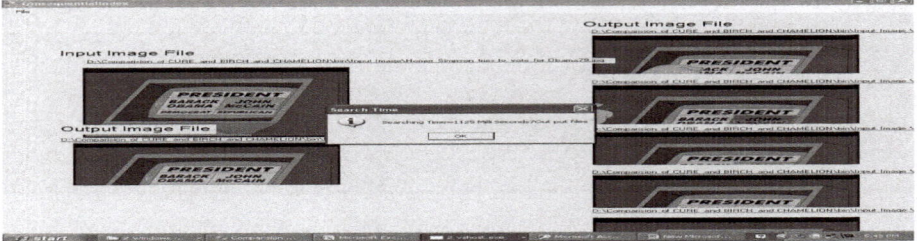

Fig. 5. Image retrieval using RGB technique. 1 input and 5 output

Table 2. Frame count vs time taken

Frame count	Milliseconds	Category
10	726	Cartoon video set
20	765	Cartoon video set
30	786	Cartoon video set
40	814	Cartoon video set
50	828	Cartoon video set
60	860	Cartoon video set

Fig. 6. Proposed technique performance graph no of frame vs time in millisec

7 Conclusion

Image data mining plays very important role in many advanced applications today. Compare to the existing text based retrieval image based retrieval bring more correct and accurate results based on the users input query. Information not only accurate and also brings the output in short span of time. This paper proofs information extraction more effectively than existing techniques. In future works is to expand the work with other image property are combing and bring the more accurate result with reduced time period.

References

1. Fan, J., Zhu, X., Lin, X.: Mining of video database. In: Multimedia Data Mining, Kluwer, Alphen aan den Rijn (2002)
2. Pan, J.-Y., Faloutsos, C.: VideoCube: a new tool for video mining and classification. In: ICADL, Singapore, December 2002
3. Chen, S.C., Shyu, M.L., Zhang, C., Strickrott, J.: Multimedia data mining for traffic video sequence. In: MDM/KDD Workshop 2001, San Francisco, USA (2001)
4. Saravanan, D.: Design and implementation of feature matching procedure for video frame retrieval. Int. J. Control Theory Appl. 9(7), 3283–3293 (2016)
5. Kreuseler, M., Shumann, H.: A flexible approach for visual data mining. IEEE (2001)
6. Fayyad, U.M., Piatetsky-Shapiro, G., Smyth, P., Uthurusamy, R.: Advanced in Knowledge Discovery and Data Mining. AAAI/MIT Press, Seattle (1996)
7. Zhao, W., Wang, J., Bhat, D., Sakiewicz, K., Nandhakumar, N., Chang, W.: Improving color based video shot detection. In: IEEE International Conference on Multimedia Computing and Systems, vol. 2, pp. 752–756 (1999)
8. Ng, R.T., Han, J.: CLARANS: a method for clustering objects for spatial data mining. IEEE Trans. Knowl. Data Eng. 14(5), 1003–1016 (2002)
9. Saravanan, D.: Video content retrieval using image feature selection. Pak. J. Biotechnol. 13 (3), 215–219 (2016)
10. Shyu, M., Xie, Z., Chen, M., Chen, S.: Video semantic event/concept detection using a subspace-based multimedia data mining framework. IEEE Trans. Multimedia 10, 252–259 (2008)
11. Saravanan, D.: Information retrieval using: hierarchical clustering algorithm. Int. J. Pharm. Technol. 8(4), 22793–22803 (2016)
12. Saravanan, D.: Effective video data retrieval using image key frame selection. In: Advances in Intelligent Systems and Computing, pp. 145–155 (2017)
13. Saravanan, D.: Image frame mining using indexing technique. In: Data Engineering and Intelligent Computing, chap. 12, pp. 127–137. Springer, Singapore (2017). ISBN 978-981-10-3223-3
14. Zhang, L., Lin, F., Zhang, B.: A CBIR method based on color-spatial feature. In: The IEEE Region 10 Conference Proceedings, pp. 166–169, September 1999
15. Saravanan, D.: Efficient video indexing and retrieval using hierarchical clustering techniques. In: Advances in Intelligence Systems and Computing, vol. 712, pp. 1–8 (2018). ISBN 978-981-10-8227-6

Generative Adversarial Networks: A Survey of Techniques and Methods

Mohammad Omar Khursheed$^{(\boxtimes)}$, Danish Saeed,
and Asad Mohammed Khan

Department of Computer Engineering, Aligarh Muslim University, Aligarh, India
omarkhu@gmail.com, danishsaeed2@gmail.com,
masadiitr@gmail.com

Abstract. Generative Adversarial Networks (GANs) are a class of deep learning algorithms which are based on two neural networks competing with each other. They are capable of learning representations from data that isn't well annotated and these deep representations can be quite useful in a variety of applications, including image classification, generation, and in deblurring and creating super-resolution images. This paper is an attempt to explore a few different types of GANs, and to discuss issues and solutions in training methods. We also attempt to explore various applications of GANs and the methods we can use these learned deep representations.

Keywords: Neural network · Machine learning · Generative model · Adversarial model · Image generation · Super resolution

1 Introduction

GANs, Generative Adversarial Networks, are currently a swiftly growing topic in the field of Computer Science, especially in field of image generation, and have captivated researchers in recent times. GANs—originally proposed by Ian Goodfellow in 2014 [1] —have two networks, a generator and a discriminator. They are both trained at the same time and compete again each other in a minimax game. The generator tries to make the discriminator think that the images it's producing are real and part of the dataset, while the discriminator attempts to resist these attempts at fooling it.

GANs involve a minimax game between two neural networks, a generator and a discriminator, and the generator, beginning from a noise vector, tries to generate samples that are close enough to the training data that the discriminator becomes unable to distinguish between real and 'fake' images. The generator is given feedback on its performance by the discriminator, which is the only way that the generator can improve, since it has no access to actual images, and begins with a simple noise vector.

In this paper, we attempt to review various advances in the field of Generative Adversarial Networks, as well as explore their architectures. We look at four different types of GANs, three variations as well as Goodfellow et al.'s original architecture. We also give an overview of the current applications of GANs, and the implication that these experiments have in various industries.

© Springer Nature Switzerland AG 2020
A. P. Pandian et al. (Eds.): ICCBI 2018, LNDECT 31, pp. 490–498, 2020.
https://doi.org/10.1007/978-3-030-24643-3_58

2 GAN Architectures

2.1 Vanilla GANs

Overview. In the vanilla implementation (Goodfellow et al.'s [1] original implementation), the two neural networks that are pitted against one another are multilayer perceptron networks and the popular backpropagation and dropout algorithms are used [2]. The idea is to attempt to have the generator learn the distribution p_{vg} of the data x by defining a noise variable $p_{vz}(z)$, which is then mapped to a data space as $G(z; \theta_{vg})$, where G represents a differentiable function that represents the multilayer perceptron being used, and θ_{vg} its parameters. The discriminator is another multilayer perceptron, represented by $D(x; \theta_{vd})$ which outputs a single scalar value $D(x)$, representing the probability of x being from the data rather than from the generator's distribution of p_{vg}. D has the job of ultimately outputting correct labels to both examples generated by G as well as the training samples, and hence the probability of correct labelling must be maximized and G must be trained to minimize log $(1 - D(G(z)))$. The value function $V(G, D)$ is used in a minimax game played by G and D [1].

$$\min_{G} \ \max_{D} V(D, G) = E_{x \sim p_{data}(x)}[\log \ (D(x))] + E_{z \sim p_z(z)}[\log(1 - D(G(z)))]$$

We must optimize while making sure that D is not taken to its ultimate optimal condition within the inner loop, because this would be computationally prohibitive, and in the case of finite datasets, result in overfitting [1]. It is instead that we optimize D for k steps, followed by one step of optimizing G. This leads to D being kept near its optimal condition, and G being slowly improved. Initial learning via the aforementioned minimax game may not provide strong gradients, and a workaround may be to train G to maximize $\log(D(G(z)))$ instead of minimizing $1 - \log(D(G(z))$. This problem early in training is because G is initially a poor generator, and hence D can reject its outputs with very high confidence, leading to the saturation of $\log(1 - D(G(z)))$.

Goodfellow et al., also choose to highlight some important results relating to adversarial nets, which are summarized below. Detailed proofs to these results can be found in [1].

For a fixed generator G, the optimal discriminator D is

$$D_G^*(x) = \frac{p_{data}(x)}{p_{data}(x) + p_{vg}(x)}$$

This is because for the function $y \rightarrow a \log(y) + b \log(y - 1)$ over any $(a, b) \in R^\wedge 2\{0, 0\}$, has its maxima in $[0, 1]$ at $\frac{a}{a+b}$.

The global minimum of the virtual training criteria $C(G) = \max_D V(G, D)$ achieved if and only if $p_{vg} = p_{data}$, where $C(G) = -\log 4$

This is because when $p_{vg} = p_{data}$, it is found that $D_G^*(x) = \frac{1}{2}$.

If G and D have enough capacity and, at each iteration, D is allowed to reach its optimum given G, and p_{vg} is updated to improve

$$\min_{G} \ \max_{D} V(D, G) = E_{x \sim p_{data}(x)} \left[\log D_G^*(x)\right] + E_{x \sim p_{data}(x)} \left[\log\left(1 - D_G^*(x)\right)\right]$$

The training of vanilla GANs is a difficult problem, since it is hard to reach convergence. Various methods are used to improve the chances of this, so that we can generate good samples (Fig. 1).

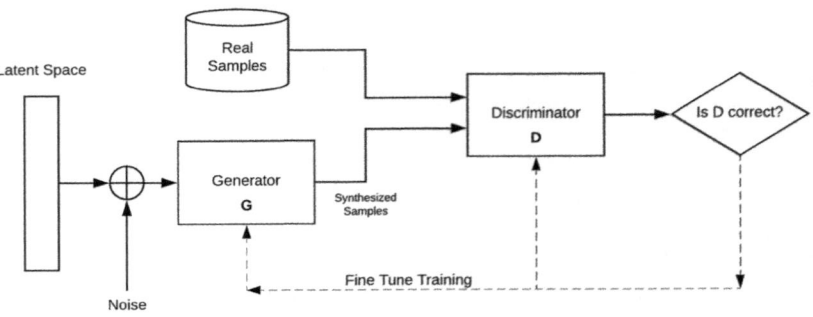

Fig. 1. Vanilla GAN architecture based on paper by Goodfellow et al. [1]

Feature Matching. In feature matching [3], we change the objective of the generator. Instead of directly maximizing the output of the discriminator, we have the generator generate samples that match the statistics of the real data. This is done by training the generator to match output with features that are in an intermediate layer of the discriminator. This is reasonable because this means that we are training the generator to generate samples with features that are characteristic of the real data.

Minibatch discrimination. GANs are prone to mode collapse, which is the state of the generator being parameterized such that it always generates the same point. This happens because each example is processed individually, and hence gradients are unrelated to each other, and hence the generator cannot know when it is supposed to generate more diverse samples [3]. Hence, the discriminator, which when it finds an example that seems to be highly realistic, is unable to recover enough to tell the generator to produce different examples once it knows the sample was from the generator. An obvious strategy to solve this problem is through using data samples from the training set in combination with each other, which intuitively would prevent mode collapse [4]. This is termed minibatch discrimination. By using a measure of closeness, such as L1 distance, between samples, we can essentially continue to discriminate between real and generated samples, but with the network now able to access side information to do so.

2.2 DCGANs

Overview. Deep Convolutional Generative Adversarial Networks, or DCGANs, are an interesting variation on GANs, as they prevent a major problem that happens in vanilla GANs, that of a lack of stability while training. Since there is the attempt to find Nash equilibrium during the training of the vanilla GAN, and each neural net (the discriminator and the generator), attempts to counteract the other's effects, the gradient often oscillates and does not converge. DCGANs, through using certain properties of convolutional neural networks, which are used in the GANs themselves, have been proven to avoid this problem. This was previously found to be infeasible, but due to some improvements in CNN, discussed within the architecture heading below, convolutional neural nets have once more become an option for use in Generative Adversarial Networks. This also helps us get meaningful output, instead of images that do not match the samples at all.

Architecture. The architecture of a DCGAN [6] is markedly different from vanilla GANs in that it does not use multi-layer perceptron networks as its constituents. Instead, the generator is a deconvolutional neural network that generates images from a noise sample, while the discriminator is a convolutional neural network.

There have been many architectural improvements in CNNs that make them suitable for use in GANs. The possibility of replacing the maxpooling layers with strided convolution lets the network learn its own spatial downsampling [5]. The usage of this method in GANs allows the generator to learn its own spatial upsampling, and the discriminator. There is also the relatively new trend of eliminating fully connected layers on top of convolutional features, such as in global average pooling, which increased model stability but reduced the speed with which convergence was reached [7]. By directly connecting convolutional layers to input and output, there is a simulation of a fully connected layer in the input, and the output can be seen as a flattened convolutional layer.

The ReLU [8] function is used in the generator other than in the output layer, where the tanh function is used, and in the discriminator, leaky ReLU [9, 10] is used, which is found to work well, especially for higher resolution modelling. Batch normalization [11] stabilizes the learning process by normalizing the input to have unit mean and zero variance. Training problems often occur due to improper initialization and due to the

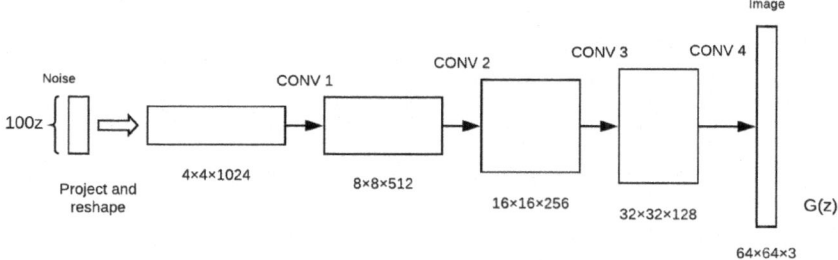

Fig. 2. DCGAN's Generator architecture based on paper by Radford et al. [4]

lack of gradient flow in deeper architectures. Using batch normalization has also been found to prevent mode collapse, and to avoid oscillation in the model, the generator output layer and discriminator input layer do not have batchnorm applied to them. Training, otherwise, is similar to a Vanilla GAN (Fig. 2).

2.3 AC-GANs

Overview. Proposed by Odena, et al. in their 2016 paper [11], AC-GAN (Auxiliary Classifier GAN) is a variant of the vanilla GAN. In GANs, random noise vectors are input to the generator G and outputs and image $X_{fake} = G(z)$. The input to the discriminator D is either from the training set or a generator synthesized image where the output is a probability distribution $P(S \mid X) = D(X)$ over possible image sources [1]. To improve the performance of the GAN, the discriminator and generator were both supplied with side information i.e. class labels for the training data to produce class conditional samples which are of improved quality than those produced by vanilla GANs [11, 13].

To further improve performance of the GAN, the discriminator net can be modified to output class labels for the training data or a set of latent variables for generated samples, using an auxiliary decoder network [3]. This could even be done by pretrained discriminators like image classifiers improving the synthesized images even more [11, 14].

Architecture. The architecture proposed by Odena et al. is not structurally different from that of the vanilla GANs. In AC-GANs, along with the noise z, there's a class

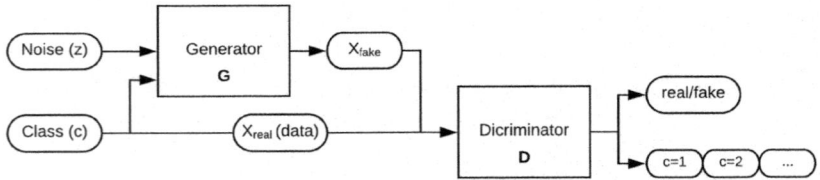

Fig. 3. AC-GAN architecture based on paper by Odena et al. [11]

label $c \sim pc$ that corresponds to each generated sample. The images generate $X_{fake} = G(c, z)$ use both. The discriminator outputs $P(S|X), P(C|X) = D(X)$, the probability distributions over the sources and class labels.

The generator is made up of multiple deconvolution layers that take as input the noise z and class c and transforms it into an image [15]. The discriminator is a deep convolutional neural net having Leaky ReLU nonlinearity [8] (Fig. 3).

The generator is made up of multiple deconvolution layers that take as input the noise z and class c and transforms it into an image [15]. The discriminator is a deep convolutional neural net having Leaky ReLU nonlinearity [8].

In addition to the log-likelihood of the correct source L_S similar to as in vanilla GANs, the objective function also finds out the log-likelihood of correct class L_C:

$$L_S = E[logP(S = real|X_{real})] + E[logP(S = fake|X_{fake})]$$

$$L_C = E[logP(C = c|X_{real})] + E[logP(C = c|X_{fake})]$$

Discriminator D is then trained to maximize $L_S + L_C$ and generator G is trained to maximize $L_C - L_S$ [11].

One of the advantages of AC-GANs is that the modification to the vanilla GAN architecture produces excellent results and seemingly the training process gets stabilized because of it. The addition of class conditional component improves the structure of AC-GANs over vanilla implementation. Early experiments by Odena et al. suggest that on the same model, increasing the number of classes trained on decreased the quality of the synthesized images [11]. Also, large subsets can be separated into smaller datasets in this model and a separate generator and discriminator can be trained for each individual dataset [11].

2.4 InfoGANs

Overview. InfoGAN, proposed by Chen et al. in 2016 [12], is an information theoretic approach to GANs which are able to learn interpretable and meaningful representations. The idea is to maximize the mutual information between a fixed small subset of GAN's noise variables and the observations. Because of the nature of how the noise vector z is input into the generator, the noise will possibly be used in a highly entangled way. So the noise vector is divided into two parts:z, which will be the uncompressed noise source and c, which is called the latent code and points at the structured semantic features of the data distribution. c is the concatenation of all variables $c_1, c_2, \ldots, c_L$ [12].

The generator form $G(z, c)$ is provided with the latent code c and the noise z. We use this to discover the factors in an unsupervised way. Chen et al. propose an information-theoretic regularization because the generator finds a solution $P_G(x|c) = P_G(x)$ and is most likely to ignore the latent code input c. Thus, the mutual information $I(c; G(z, c))$ between $G(z, c)$ and latent codes c must be high [12].

The mutual information can be expressed as:

$$I(X; Y)$$
$$= H(X)$$
$$- H(X|$$
$$= H(Y)$$
$$- H(Y|$$

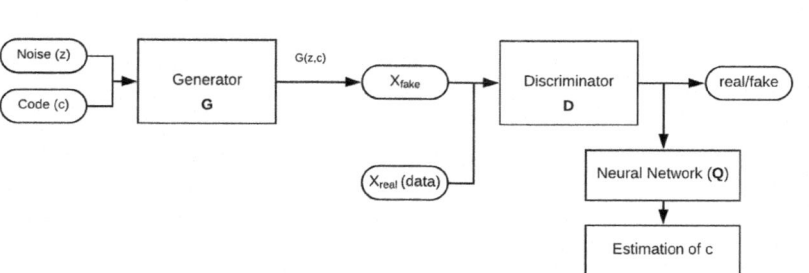

Fig. 4. InfoGAN architecture based on paper by Chen et al. [12]

This expression indicates when Y is observed, uncertainty in X is reduced. This expression equates to 0 when X and Y are independent, and maximizes if they are related by a deterministic invertible function. Given this expression, we require $P_G(c|x)$ to have small entropy given any $x \sim PG(x)$ meaning the generation process should not give rise to loss of information in c (Fig. 4).

Architecture. Chen et al. propose the following regularized expression for GAN model:

$$\min_G \ \max_D V_I(D, G) = V (D, G) - \lambda I(c; G(z, c))$$

Where $V_I(D, G)$ is the minimax game for vanilla GAN [1] written as:

$$\min_G \ \max_D V (D, G) = E_{x \sim P_{data}}[\log D(x)] + E_{z \sim noise}[\log (1 - D(G(z)))]$$

Practically, we obtain a lower bound of $I(c; G(z, c))$ because it is hard to maximise it directly. We define an auxiliary distribution $Q(c|x)$ to approximate $P(c|x)$ the details of which can be found in [12]. This is known as Variational Information Maximization. So, a variational lower bound $L_I(G, Q)$ of the mutual information $I(c; G(z, c))$ is defined which is detailed in [12].

Therefore, including the variational regularization of mutual information, InfoGAN gets defined as:

$$\min_{G,Q} \ \max_D V_{InfoGAN}(D, G, Q) = V (D, G) - \lambda L_I(G, Q)$$

The auxiliary network Q is parameterized as a neural network and shares with D all convolutional layers and only a single fully connected layer to output parameters for the distribution $Q(c|x)$. It has also been observed that $L_I(G, Q)$ converges significantly faster than vanilla GAN objectives means that InfoGAN is a very computationally lightweight add-on to GAN. The additional parameter λ present in InfoGAN can be set to 1 for discrete latent codes, and hence is east to tune. A smaller value of λ is used for latent codes containing continuous variables.

Another interesting point to note is that because of difficulty in training GANs, Chen et al. stabilized their InfoGAN training using techniques based on DC-GANs [4] which suggests that addition of InfoGAN doesn't affect the original GAN training by much amount whatsoever [12].

3   Usage and Potential Industrial Applications

The usage of GANs in general is currently centered on image generation, which is the application described by Goodfellow et al. in their original paper. The generation of realistic images is an application that the architecture of GANs makes them particularly well-suited to, because the adversarial nature leads to even minor implicit features becoming part of the output. Another related application of this architecture is the generation of high resolution images from images that are of a low resolution. Instead of using a mean square error or similar loss function, a ReLU based loss function is used to ensure that spatial coherence is maintained and that similar images are generated from the LR input. Using GANs for face synthesis, i.e., generating different angles of a face for which only a frontal image is available to us, is of significant importance for facial recognition algorithms, and this technique may find its way into face recognition unlocking algorithms in mobile phones.

The usage of GANs for image deblurring has important consequences for the camera and smartphone industry, especially in this day and age, when upon reaching the physical limit in terms of lenses and optical hardware, post-processing via software in the capture of photographs has grown ever more important. A novel application of GANs is that of pattern transfer, from, say handbags to shows, with some improvement, could lead the fashion industry in new and exciting directions.

Since convolutional neural networks are simply an improvement to the vanilla GAN architecture, particularly solving the problem of instability, it is possible to use them as image generators without fear of mode collapse or instability. They have also been used for object detection [16] and for supervised monocular facial reconstruction [17] problems that have significant use in the areas of image processing and medicine, as well as in the development autonomous vehicles.

GANs can be made to generate higher resolution images but they're just blurry version of their lower resolution counterpart and are still not discriminable. AC-GANs are particularly useful in generating images that are not only of higher resolution but are discriminable to a greater extent than low resolution images [11]. Another problem addressed by AC-GAN is the synthesis of only one prototype by the generator that maximally fools the discriminator [1, 3]. They are useful for producing images which are diverse in nature [11]. The AC-GANs framework is also being used in applications involving style transfer i.e. is differentiating the style of a piece of art from its content and applying it to other pieces of art. Furthermore, they can be used to learn distribution of style of many different pieces of art and combine these styles to generate new pieces of art.

In one of the experiments conducted by Chen et al. they were able to separate writing styles from digit shapes in the MNIST dataset. Similar experiments display the versatility of InfoGANs in learning disentangled representations of data without any supervision. Latent codes could bring about variations in style and shapes of images without it being explicitly labelled or fed into the model [12]. This could be very useful in future work especially as it only adds a negligible computation cost over normal GAN.

4   Conclusion and Future Work

The implications that GANs have for the deep learning community are obviously immense, especially in areas related to image generation. There is certainly potential for GANs to be extended for use in other fields, including Natural Language Processing and, even though bottlenecks such as modifying existing architectures to learn well with discrete data do exist, the possibilities of generating meaningful textual content, or combining image synthesis with NLP capabilities, could lead the field in new and interesting directions. Improvement in training methods and architectures can lead to us being able to better represent data and be able to find new applications.

References

1. Goodfellow, I., Pouget-Abadie, J., Mirza, M., Xu, B., Warde-Farley, D., Ozair, S., Courville, A., Bengio, Y.: Generative adversarial nets. In: Advances in Neural Information Processing Systems, pp. 2672–2680 (2014)
2. Hinton, G.E., Srivastava, N., Krizhevsky, A., Sutskever, I., Salakhutdinov, R.: Improving neural networks by preventing co-adaptation of feature detectors. Technical report. arXiv: 1207.0580 (2012)
3. Saliman, T., Goodfellow, I., Zaremba, W., Cheung, V., Radford, A., Chen, X.: Improved techniques for training GANs. In: 30th Conference on Neural Information Processing Systems (NIPS) (2016)
4. Radford, A., Metz, L., Chintala, S.: Unsupervised representation learning with deep convolutional generative adversarial networks. arXiv:1511.06434 (2016)
5. Springenberg, J.T., Dosovitskiy, J., Brox, T., Riedmiller, M.: Striving for simplicity: The all convolutional net. arXiv:1412.6806 (2014)
6. Mordvintsev, A., Olah, C., Tyka, M.: Inceptionism: going deeper into neural networks. http://googleresearch.blogspot.com/2015/06/ inceptionism-going-deeper-into-neural.html. Accessed 17 June 2015
7. Vinod, N., Hinton, G.E.: Rectified linear units improve restricted boltzmann machines. In: Proceedings of the 27th International Conference on Machine Learning (ICML-10), pp. 807–814 (2010)
8. Maas, A.L., Hannun, A.Y., Ng, A.Y.: Rectifier nonlinearities improve neural network acoustic models. In: Proceedings of the ICML, vol. 28 (2013)
9. Xu, B., Wang, N., Chen, T., Li, M.: Empirical evaluation of rectified activations in convolutional network. arXiv:1505.00853 (2015)
10. Ioffe, S., Szegedy, C.: Batch normalization: accelerating deep network training by reducing internal covariate shift. arXiv:1502.03167 (2015)
11. Odena, A., Olah, C., Shlens, J.: Conditional image synthesis with auxiliary classifier gans. arXiv:1610.09585v4 (2017)
12. Chen, X., Duan, Y., Houthooft, R., Schulman, J., Sutskever, I., Abbeel, P.: InfoGAN: interpretable representation learning by information maximizing generative adversarial nets. In: Advances in Neural Information Processing Systems 2016, pp. 2172–2180. arXiv:1606. 03657v1 [cs.LG] (2016)
13. Mirza, M., Osindero, S.: Conditional generative adversarial nets. arXiv:1411.1784 (2014)
14. Nguyen, A., Dosovitskiy, A., Yosinski, J., Brox, T., Clune, J.: Synthesizing the preferred inputs for neurons in neural networks via deep generator networks. In: 30th Conference on Neural Information Processing Systems (NIPS), 2016. Advances in Neural Information Processing Systems, pp. 3387–3395 (2016)
15. Odena, A., Dumoulin, V., Olah, C.: Deconvolution and checkerboard artifacts. Distill 1(10), e3 (2016)
16. Gao, F., Yang, Y., Wang, J., Sun, J., Yang, E., Zhou, H.: A deep convolutional generative adversarial networks (DCGANS)-based semi-supervised methods for object recognition in synthetic aperture radar (SAR) images. Remote Sens. 10, 846 (2018)
17. Bas, A., Huner, P., Smith, W.A.P., Awais, M., Kittler, J.: 3D morphable models as spatial transformer networks. arXiv:1708.07199v1 [cs.CV] (2017)

Natural Language Interface
for Relational Database

Priyanka More[1], Bharti Kudale[2], and Pranali Deshmukh[1(✉)]

[1] Information Technology, Genba Sopanrao Moze CoE, Balewadi, Pune, India
morepriyankad@gmail.com, deshmukhpranali3@gmail.com
[2] Computer Engineering, Genba Sopanrao Moze CoE, Balewadi, Pune, India
gaikwadbharati3@gmail.com

Abstract. This paper is a brief of Natural Language Programming queries to RDBMS structured queries. We propose an approach for building a questioning system through that a user can provide a queries in natural language. This system will interpret the input sentence and process them into SQL queries. The generated queries are run on particular database to gather the expected information in the form of output. Hence it will dramatically simplify the process of handling with large data and making data available for everyone.

1 Introduction

Day by day interaction with computer is increased in the world and result of this is generation of tremendous information. Database systems are designed to store this information but to access this information user should know the knowledge of database languages. It is problematic to non-IT/new users to learn SQL for accessing information. So a simple graphical user interface is required to access information. By using NLP interface the end-user can query the system in native language like English. A natural language Processing interface to a Relational database is a system where the users can access information stored in a relational database by providing requests in English natural language.

In this paper, we propose English Natural Language Interface to Relational Databases (ENLIRDB). This system allows user to extract data from database by replaying to English Natural Language Query (ENLQ) questions. The system used NLP algorithms for translating ENLQ into SQL RDBMS query. The main functionality of this project is that it functions individually of the database language, content and system. It has the capability to enhance its knowledge base through experience.

Advantage is anyone can retrieve information from the database by using this NLP interface. Normally, people are very habitual to work with a search engines. In devices like tablet, palmtop and mobile phone, the display screen is not much wide as a computer or a laptop. Instead, filling a form user can simply type the question similar to the English question.

This system will link the communication gap between users and databases. The, main advantage of ENLIRDB is user has to give query input in English and not required to learn any Structural Query language. ENLIRDB would allow queries to be

A. P. Pandian et al. (Eds.): ICCBI 2018, LNDECT 31, pp. 499–505, 2020.
https://doi.org/10.1007/978-3-030-24643-3_59

defined in English language. This ENLIRDB is mainly suitable for normal users who know English language and there would be no need for them to spend time in learning the database languages for extraction data.

2   Overview

Since a database is having complex architecture, it is very hard for a normal user to carry out operations on database. Therefore, we introduced a system in this user can give queries in natural English language as per their need.

2.1   RDBMS

A database is a collection of data, storing the information of one or more related institutions. Database consists of information two parts.

1. Entities such as employee, departments etc.
2. Relationships between entities, such as employee is working in more than one departments.

A relational database consists of a collection of tables, each of which is assigned a unique name [4].

2.2   NLP

NLP text consists of three phases:

- Morphological analysis.
- Syntactic analysis.
- Semantic analysis.

Natural language processing (NLP)-based interface gives an easy way to interact or write queries in normal English sentences by using a part of-speech (POS) tagging algorithm. The system then groups the given queries as object-categories like banking, social, spatial and similarity-based on previous queries by using POS tagging information POS analysis.

Then queries multiple queries are constructed and provided to the query engine then engine interact with knowledge base and database to verify weather keywords used in query are present in actual table or not. After this step they verify the correctness of each query after final score. To improve query correctness domain ontology is used.

2.3   Architectures Used in ENLIDB Systems [7]

PM System is uses prefixed patterns and guidelines for mapping. This architectonics is easy to implementation but bounded to specific database discipline and patterns. SAVVY is case of PM system.

In syntax-based systems parse tree creation is implemented by using grammar. Syntactic tree is then charted to it's semantic meaning to convert database query. This syntax-based method is used in many systems like LUNAR system. LUNAR systems provides observation to the questions like chemical analysis of mud fragments carried from different locations.

In semantic grammar system the points are identified. Those points are plotted in parse tree with matched words from input. This architecture is used in LADDER system.

Intermediate representation language rules change input into logical query and that logical query is rewritten to SQL query. CHAT-80 system used Prolog language for implementation and it is an interface to geographical databases. The major negative point of this system is, it is useful for particular domain database application.

3 Existing Systems

1. **A rule based approach for NLP query** [4]
2. *Modified word co-occurrence matrix formation module* [5]

In this system Hyperspace Analog to Language matrix (HAL) algorithm is used to find out nouns matched to nouns, adjectives and numeric values calculated from given input.

3. **MARKOV DECISION PROCESS** [2]

This system work as a part of dialogue system. The main functionality of this system is extract list of libraries for keywords used by user in interactions. First it represent all resources with it interacts. Second step is to find out different state including initial state depends upon input values and filter used.

4. A context-free grammar [1]

In this system when any input keyword of particular terminal is matched then it is restored by particular attribute in relational table or operators of SQL.

4 Proposed Systems

In this section we provide a detail concept of the proposed system.

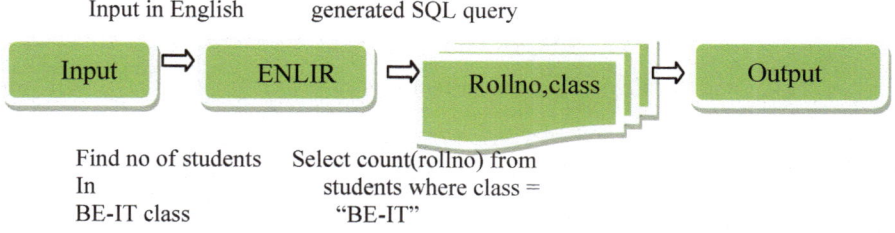

Fig. 1. Natural language interface for relational database

One of the curious and vast area of NLP is the development of a English natural language interface to relational database systems (ENLIDB). In the most recent years NLIDB systems implemented through which users can directly communicate with database in a more suitable and easy way. Figure 1 shows the basic system model of our ENLIDB system. First step collect input in the natural language (English) form from the user. Second step is system uses NLP techniques to converts given input into SQL query. Third step is SQL query directly fire on particular database for getting desired result. The user does not need to learn any languages or no need to worry about translation process SQL query and its syntax.

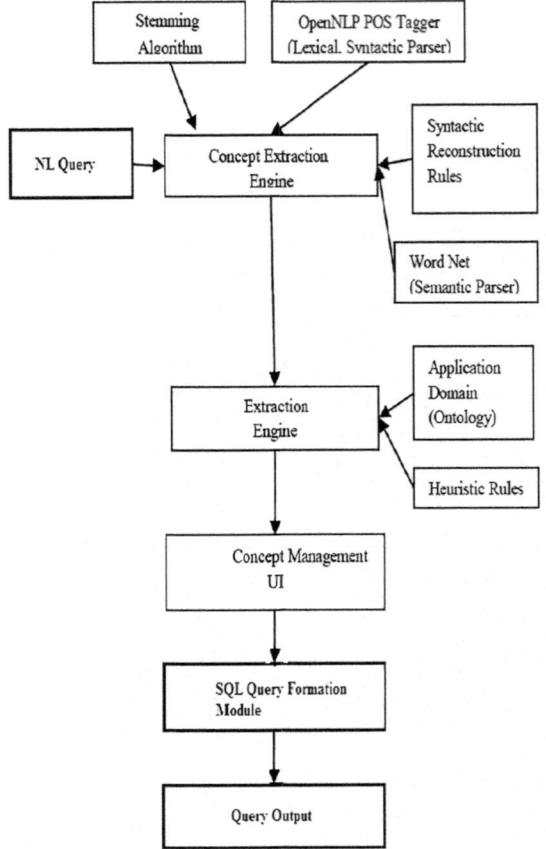

Fig. 2. Proposed architecture of ENLIDB

In Fig. 2 proposed architecture of ENLIDB is explained. There are 5 Modules of this application.

4.1   Concept Extraction Engine

The first module of our project is Concept extraction engine. The main purpose of this engine is understand the natural language query and extract the main key words from query for next module.

4.1.1   Tokenizing and Parsing

Each sentence is further divided into tokens, for each token distinct token id is provided (Fig. 3).

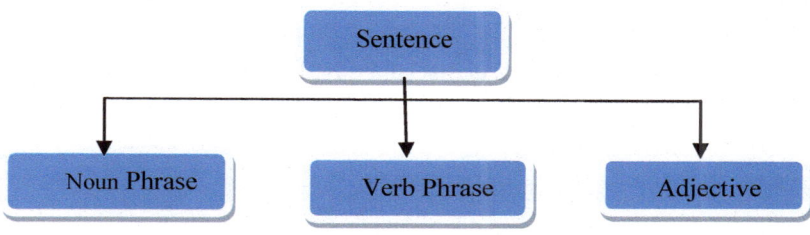

Fig. 3. Tokenizing

4.1.2   Semantic Analysis

It starts by reading all of the words in content to capture the real meaning of any text. It determines the text elements and assigns them to their logical and grammatical role. For example, "destination" and "last stop" technically mean the same thing

4.1.3   Tagging

The lemmas are further tagged according to their part of speech form. For example, the tag will determine the noun in the lemmas and help to find tables and attributes in the database.

4.2   Extraction Engine

It starts by reading all of the words in content to capture the real meaning of any text. It determines the text elements and assigns them to their logical and grammatical role. For example, "destination" and "last stop" technically mean the same thing.

4.3   Concept Management

1. The attribute values are related to their attribute through the relation operators like equal to, is greater than, is less than, etc. For example, Greater than is plotted into ">",
2. Equal to is plotted into "=", etc. In this the comparing attribute has to be written first and the logical operator is then attached to it. After the value which is to be compared is attached.

4.4 SQL Query Formation Module

This is the final stage where actual query is generated and fired on the database management system to get the desired output on GUI.

4.5 Open NLP Libraries

(a) Natural Language Toolkit

This toolkit is used for tasks like tokenization, lemmatization, parsing, POS tagging, etc. This library has tools for almost all NLP tasks.

(b) Stanford CoreNLP

This library is reliable, robust, and accurate NLP platform based on client-server architecture. Written in java, and is accessible through multiple Python wrapper libraries and supports all NLP tasks.

(c) Spacy

This library provides most of standard functionality and is built to be lightning fast. Spacy also really nice interface with all major deep learning frameworks and comes packed with some really good and useful language module.

(d) WordNet

WordNet is a very large lexical database of English language. In this Nouns, verbs, adjectives and adverbs are classified into sets of subjective synonyms, each conveying a specific concept. Synonyms are interconnected by means of conceptual-semantic and lexical relations.

5   Conclusion

In the proposed system user can enter the query in natural language, then the system will translate it into the SQL query and will get the result from the database. Developed system gives correct result for simple queries.

References

1. Bais, H., Machkour, M., Koutti, L.: An independent-domain natural language interface for relational database: case Arabic language. In: IEEE Team of Engineering of Information Systems (2016)
2. Dua, M., Kumar, S., Virk, Z.S.: Hindi language graphical user interface to database management system. In: IEEE 12th Conference on Machine Learning (2013)
3. Mahmud, T., Azharul Hasan, K.M., Ahmed, M., Chak, T.H.C: A rule based approach for NLP based query processing. In: Proceedings of International Conference on Electrical Information and Communication Technology (EICT 2015) (2015)

4. Mahmud, T., Azharul Hasan, K.M., Ahmed, M., Chak, T.H.C.: A rule based approach for NLP based query processing. In: ©2015 IEEE (2015)
5. Mohite, A.: Natural language interface to database using modified co-occurrence matrix technique. In: IEEE International Conference on Pervasive Computing (ICPC) (2015)
6. More, P., Phalnikar, R.: Generating UML diagrams from natural language specifications. IJAIS **1**, 19–23 (2012)
7. Yorozu, Y., Hirano, M., Oka, K., Tagawa, Y.: Electron spectroscopy studies on magneto-optical media and plastic substrate interface. IEEE Trans. J. Magn. Jpn. **2**, 740–741 (1987). (Digests 9th Annual Conference Magnetics Japan, p. 301 (1982))
8. Androutsopoulos, I., Richie, G.D., Thanisch, P.: Natural language interface to databases—an introduction. J. Nat. Lang. Eng. **1**(1), 29–81 (1995)
9. Kucuktunc, O., Gudukbay, U., Ulusoy, O.: A natural language-based interface for querying a video database. IEEE Multimed. **14**, 83–89 (2007)
10. Mahmud, T., Azharul Hasan K.M., Ahmed, M., Chak, T.H.C: A rule based approach for NLP based query processing. In: 2015 IEEE International Conference on Electrical Information and Communication Technologies (EICT) (2015)
11. Fulford, K., Olmsted, A.: Mobile natural language database interface for accessing relational data. In: 2017 IEEE International Conference on Information Society (i-Society) (2017)
12. Gupta, P., Goswami, A.: IQS-intelligent querying system using natural language processing. In: 2017 IEEE International Conference of Electronics, Communication and Aerospace Technology (ICECA) (2017)
13. Surabhi, M.C.: Natural language processing future. In: 2013 IEEE International Conference on Optical Imaging Sensor and Security (ICOSS) (2013)
14. Nicole R.: Database System Concepts, 4th edn. (1997)

Security Analysis on Remote User Authentication Methods

Mukesh Soni[1(✉)], Tejas Patel[2], and Anuj Jain[2]

[1] Smt. S. R. Patel Engineering College, Unjha, India
soni.mukesh15@gmail.com
[2] ITM Universe, Vadodara, India
tjs.patel2@gmail.com, anujjaingit@gmail.com

Abstract. Remote user authentication has been framed to provide security to the users by hiding their details with the help of some technique which can bear all the forgery attacks, but not reveal any details. All the users want CIA (confidentiality, integrity, and availability) to be there if they are working in a network. Various protocols have been proposed and implemented in literature. But the efficient tool is still lacking. Here we find that many of the schemes and protocols suggested are vulnerable to forgery attacks. We have found some problems in secure remote user authentication scheme based on smart card.

Keywords: Attack · Authentication · Remote user · Session key · Smart card

1 Introduction

The authentication process plays a pivotal role in retaining users over a network. There are various methods of authentication that have been prevalent from the times the concept of the network has evolved. If we want a user to respond or give review of a product or a service, or we want a user to log in to our network to gain information, then we have to provide him/her with proper authentication services so as to hold on all the private information shared by the user.

Privacy over a network is a major concern in today's era. Without privacy, a user will never ever think about coming on a network. Keeping this in mind, many schemes and protocols for authentication of the users have been proposed and implemented. Basically, authentication schemes involve verification of username and password of a particular user on the network. But this is a very weak mechanism because it cannot bear forgery attacks for a long period of time.

Remote user authentication provides authentication using smart cards. The scheme of smart card involves the scanning of the card of the user containing all the user details and then asking for password to verify the authenticated user. Smart card, thus, evolved as a safer and better option for traditional method of username and password.

Smart card provides two-way authentication scheme which upholds the privacy of the user to a greater extent. Many protocols work behind the functioning of smart card. With every updated protocol, the functioning behind smart card gets updated but many reports are there from which it is evident that still remote user authentication using smart card is not safe. Figure 1 displays a diagram of verification system. In this

© Springer Nature Switzerland AG 2020
A. P. Pandian et al. (Eds.): ICCBI 2018, LNDECT 31, pp. 506–513, 2020.
https://doi.org/10.1007/978-3-030-24643-3_60

scheme, there are various components such that clients, server, smartcards. A client wishes to get different kind of services and he request to server by a smart card. Server provides the authorization after checking a password and necessary variables. A smart card is utilized to store useful variables.

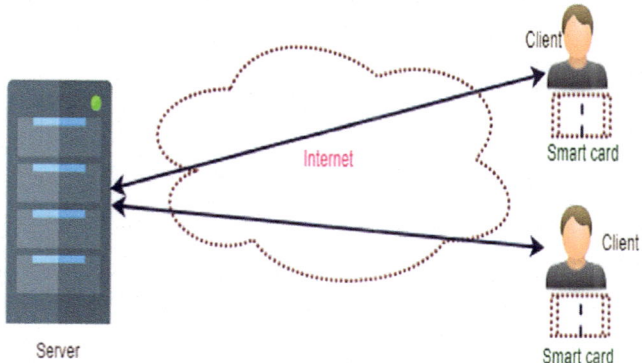

Fig. 1. Authentication system for different users.

In this work, we have discussed the weaknesses of the previously proposed protocols. Further, we will discuss about the forgery attacks that can damage the privacy of the user. Then we will present our proposed protocol, discussing its security and efficiency.

2 Literature Review

In 1981, Lamport suggested an innovative password-based authentication scheme which had cryptography hash function. But it was seen that only high hash overhead and password resetting could not prevent forgery attacks [2]. In 1999, Yang and Shieh proposed two password authentication system with the help of a smart card, based on the discrete logarithmic problem. The discrete logarithmic helped to prevent forgery attacks by the intruder [3].

Hwang and Li, in 2000, presented a verifier-free password confirmation method using a smart card. Its drawbacks include that the users had no opportunity to select or change their passwords, and it was vulnerable to impersonation attack [4]. Later, in 2001, Chein proposed an authentication scheme to achieve mutual authentication. But it was found to be vulnerable to parallel session attack [5]. Further, Ku and Yoon's schemes presented a protocol that required numerous hash operations instead of the costly modular exponentiations. Therefore, though their schemes were a good application in smart card field, the security was at stake. These schemes were vulnerable to security threats such as password guessing, forgery, and denial of service [6].

Continuously researches proposed anonymous password authentication and key exchange (APAKE). After that, Chai et al. [7] found that the previously discussed that APAKE protocol suffered forgery attacks and also proposed a new scheme using smart card. In 2012, Qian et al. [8] proposed a new protocol to extend the usage of general devices. But it suffered from credential sharing problem as discussed in [9]. In further researches, Shin and Kobara [10] proposed S2APA protocol to enhance the security level but later, it was found that it could not resist to a credential forgery attack. Wu et al. [11] proposed a new and efficient APAKE protocol with the help of a tamper-proof device. An adversary can extract any values from this device. However, this method [11] suffers from a denial of service attack.

3   Review of Kumar et al.'s Scheme

In this section, we explain the scheme of Kumar [1] step-by-step to get clear idea of it. There are totally four sections in this method and we have described as follows.

3.1   Registration Procedure

This phase is related to the situation when a user U wants to register himself/herself at the remote server (AS). It is performed over a secure channel.

(1) U chooses his or her name, address, ID_U, PW and performs $R_1 = ID^{PW} \bmod p$.
(2) U sends (name, ID_U, address, R_1) to AS.
(3) After getting the registration request, the AS works out the followings parameters:

- $SID = Red(ID_U)$,
- $C_{ID} = C_K(SID)$,
- $R_2 = R_1^{xs} \bmod p$

(4) SC holds (SID, C_{ID}, R_1, R_2, p, f).
(5) AS gives SC to U.

3.2   Login Procedure

Whenever, U wants to login on the AS, U presents her/his SC to SCR and PIN (personal identification number) to initiate SC. If PIN is invalid again and again, then SC may request a PUK (personal unblocking key). Generally, U gives SC, ID, PW and SCR computes the subsequent computations.

(1) Computes $R_3 = S_{ID} \oplus PW$ and supplies R_3 in SC. It is worked out if U has inserted his or her SC at the first time to SCR. In other scenarios, it avoids this.
(2) Works out $R_1' = ID^{PW} \bmod p$ and compares the performed R_1' and saves R_1. Here, both are equivalent, SC takes PW & continues to the next step, otherwise SCR asks PW again. If valid PW is not inputted again, SCR should enter a PUK.
(3) Computes $C_1 = R_2^{PW-1} \bmod p$ and $C_2 = f(C_1 \oplus T_U)$.
(4) SC sends (ID, C_2, T_U) to AS.

3.3   Verification Procedure

AS collects (ID, C_2, T_U) at T_C & then, performs the following computations to check the validity of (ID, C_2, T_U).

(1) Confirms precise design of ID. If it is not valid, then AS stops (ID, C_2, T_U).
(2) Calculates $S_{ID} = Red(ID)$ & validates $C_{ID} = C_K(S_{ID})$. In case of correct values, it goes to the next step but it will terminate the design in other situations.
(3) After that, it performs $T_C - T_U \leq T_K$ to know validity of (ID, C_2, T_U). If it is passed without any problem, it reaches to the next step. Otherwise, it ends the process in a straight line.
(4) Check, if $C_2 = f(ID^{xs} \oplus T_U) mod\, p$, then the AS takes (ID, C_2, T_U) & continues to the next step. In other cases, it will be terminated by AS.
(5) Then, AS selects r & carries out the subsequent.

- $C_3 = f(ID^{xs} \oplus TS)$
- $S_{key} = f(ID, ID_S, C_3, r)$
- $C_4 = C_3 \oplus S_{key}$
- $C_5 = C_3 \oplus r$

(6) AS sends (C_4, C_5, T_S) to U.
(7) U takes (C_4, C_5, T_S) at T_L, then...

- Performs $T_L - T_S \leq T_U$. In the scenario of more value of the difference, U will not consider it. Else, U performs the followings:
- $C_3^* = f(C_1 \oplus T_S)$
- $K^* = C_3^* \oplus C_4$
- $r* = C_3^* \oplus C_5$
- $S_{key}^* = f(ID, ID_S, C_3^*, r^*)$
- $S_{key}^* = K^*$, then U makes sure that the replying organization is a actual AS and continues to the next step. Else U interferes the connection.

(8) U calculates $C_6 = f\left(C_3^*, S_{key}^*\right)$ & transfers (ID, C_6) to AS.
(9) AS confirms, if $C_6' = f(C_3, S_{key})$, then the AS assures that U will compute the same S_{key}. In other cases, it will be rejected the link.

3.4   Password Change Process

This phase arises when U wants to change his/her password PW with a new password (PW_{new}). This phase has the following steps.

(1) U inserts her/his smart card to the smart card reader and then keys in the PIN to activate the smart card, then inputs her/his identity and the old password PW and then requests to change the password.

(2) Compute $S_{ID}^* = R3 \oplus PW$ and compare the calculated S_{ID}^* and stored S_{ID}, if they are equal the smart card accepts the password PW and proceeds to the next step, otherwise demands the password again. If the password is entered incorrectly multiple times, the smart card may request a PUK (Personal Unblocking Key) code.

(3) SC computes as below.

- $R_3^* = SID \oplus PW_{new}$
- $R_2^* = R_2^{PW^{-1}XPW_{new}}$

(4) SC replaces $R_2 \leftarrow R_2^*$ as well as $R_3 \leftarrow R_3^*$.

4 Security Concerns in Kumar et al.'s Scheme

In this section, we discuss possible security flaws in Kumar's scheme [1] and it makes vulnerable to the computer system technology.

4.1 Smart Card Lost Attack

We assume that SC_i is available to an attacker and thus, she can know $<SID, C_{ID}, R_1, R_2, p, f>$. Now, if an adversary wishes to send a modified request to AS, then she should work out on $C_1 = R_2^{PW^{-1}} mod\, p, C_2 = f(C_1 \oplus T_U)$. Here, an attacker will get C_1' because she knows R_1, p from SC & ID from public domain because an identity is available generally. Now, she can proceed for $C_2' = f(C_1' \oplus T_U')$. After that, she sends a bogus message $<ID, C_2, T_U'>$ to AS. Then, AS functions further as below.

- $S_{ID} = Red(ID)$
- $C_{ID} = C_K(S_{ID})$
- $T_C - T_U' \le T_K$

Above tests are passed with no problem because she has utilized a fresh T_U' in the request. Then, it checks $C_2' = f(ID^{xs} \oplus T_U') mod\, p$. Finally, this test is also cleared and AS will proceed as follows.

- $C_3 = f(ID^{xs} \oplus TS)$
- $S_{key} = f(ID, ID_S, C_3, r)$
- $C_4 = C_3 \oplus S_{key}$
- $C_5 = C_3 \oplus r$

And, AS will send $<C_3, C_4, T_S>$. In this way, a smart card lost attack is possible in the system.

4.2 Password Guessing

It is an attack in which a malicious person guesses PW of U and tries to check weather a guessed password (PW') is true or not. The calculations for the same as below.

- Captures $<ID, C_2, T_U>$ from the public channel.
- Performs $R_1' = ID^{PW'} mod\ p$, $R_2' = R_1'^{xs} mod\ p$, $C_1' = R_2'^{PW'-1} mod\ p$, $C_2' = f(C_1 \oplus T_U)$.
- Compares $C_2 = ?C_2'$.

If both are same, then an attacker concludes with a true password. It means that she has guessed correct password. As a result, she can apply an attack with a guessed password in the system.

4.3 Denial of Service

Attackers intercepts transmitted packets from a public communication channel and tries to change ID in $<ID, C_2, T_U'>$ and retransmits with $<ID', C_2, T_U'>$. After getting it at the receiver point, AS performs as below.

- $S_{ID} = Red(ID)$
- $C_{ID} = C_K(S_{ID})$

However, $C_{ID} \neq C_K(S_{ID})$ and AS stops this communication process and rejects it. But, a sender has sent $<ID, C_2, T_U'>$ correctly but, an attacker has played with ID and forwarded as $<ID', C_2, T_U'>$. Thus, AS rejected it. In this way, a real client will not get service and she should perform a login phase again, which leads to time loss and memory loss.

5 Performance and Security Discussions in Different Protocols

We consider various authentication methods and identify different possible security attacks in them. Next, we also compare performance analysis of these methods to understand their effectiveness in the execution. Table 1 shows security comparison

Table 1. Security comparison among different verification schemes.

Authentication systems	Attacks					
	$\mathcal{A}-1$	$\mathcal{A}-2$	$\mathcal{A}-3$	$\mathcal{A}-4$	$\mathcal{A}-5$	$\mathcal{A}-6$
Kumar [1]	$\checkmark$	$\otimes$	$\checkmark$	$\otimes$	$\otimes$	$\otimes$
Chai et al. [7]	$\checkmark$	$\otimes$	$\checkmark$	$\checkmark$	$\otimes$	$\otimes$
Qian et al. [8]	$\otimes$	$\otimes$	$\otimes$	$\checkmark$	$\otimes$	$\otimes$
Yang et al. [9]	$\otimes$	$\otimes$	$\otimes$	$\checkmark$	$\otimes$	$\otimes$
Shin and Kobara [10]	$\checkmark$	$\otimes$	$\otimes$	$\checkmark$	$\checkmark$	$\otimes$
Wu et al. [11]	$\checkmark$	$\otimes$	$\otimes$	$\checkmark$	$\otimes$	$\otimes$

$\mathcal{A}-1$: Modification Attack
$\mathcal{A}-2$: Impersonation Attack
$\mathcal{A}-3$: Replay Attack
$\mathcal{A}-4$: Man-in-the-middle Attack
$\mathcal{A}-5$: Password Guessing Attack
$\mathcal{A}-6$: Smart Card Lost Attack

among different systems. In Table 1, $\sqrt{}$ denotes that a particular system is protected against that threat and $\otimes$ shows that a specified system is insecure against that attack.

Table 2 represents implementation analysis of authentication methods. Here, *Enc*, *Exp*, and $h(\cdot)$ are cryptographic operations and they are known as an encryption, exponential, and one-way hash function. In this table, we calculate registration cost as well as login and authentication phases. Based on this, we can conclude that which scheme is better in the execution. If an authentication system needs less number of cryptographic functions, it means that that scheme can be implemented quickly and is better compared to other systems.

Table 2. Performance comparison between various verification methods.

Authentication systems	Phases	
	Registration	Login and authentication
Kumar [1]	$1\,Exp$	$6\,h(\cdot) + 4\,Exp$
Chai et al. [7]	$2\,h(\cdot)$	$10\,h(\cdot) + 5\,Exp$
Qian et al. [8]	$1\,Enc + 1\,Exp$	$2\,Enc + 4\,Exp + 6\,h(\cdot)$
Yang et al. [9]	–	$3\,h(\cdot) + 6\,Exp$
Shin and Kobara [10]	$1\,Enc + 1\,Exp$	$2\,Enc + 4\,Exp + 4\,h(\cdot)$
Wu et al. [11]	$2\,h(\cdot)$	$6\,h(\cdot) + 4\,Exp$

6   Conclusion

Security is one of the major issues in wireless network enabled applications. This is because the medium is wireless and results in a lot of packet losses and/or alteration, which leads to certain circumstances. We identified that a mutual authentication scheme is not secured for multiple attacks like password guessing, smart card lost, and denial of service. After that, we do security and performance analysis to know security level and performance of various authentication systems. Then, we notice that none of these schemes is protected fully in this technology world. Therefore, there is a need of proper verification system.

References

1. Kumar, M.: A secure remote user authentication scheme with smart cards. IACR Cryptology ePrint Archive, 2008/331 (2008)
2. Lamport, L.: Password authentication with insecure communication. Commun. ACM **24**(11), 770–772 (1981)
3. Yang, W.H., Shieh, S.P.: Password authentication schemes with smart cards. Comput. Secur. **18**(8), 727–733 (1999)
4. Hwang, M.S., Li, L.H.: A new remote user authentication scheme using smart cards. IEEE Trans. Consum. Electron. **46**(1), 28–30 (2000)
5. Chien, H.Y., Jan, J.K., Tseng, Y.M.: A modified remote login authentication scheme based on geometric approach. J. Syst. Softw. **55**(3), 287–290 (2001)

6. Yoon, E.J., Yoo, K.Y.: More efficient and secure remote user authentication scheme using smart cards. In: 2005 Proceedings of 11th International Conference on Parallel and Distributed Systems, vol. 2, pp. 73–77. IEEE (2005)

7. Chai, Z., Cao, Z., Lu, R.: Efficient password-based authentication and key exchange scheme preserving user privacy. In: International Conference on Wireless Algorithms, Systems, and Applications, pp. 467–477. Springer, Berlin (2006)

8. Qian, H., Gong, J., Zhou, Y.: Anonymous password-based key exchange with low resources consumption and better user-friendliness. Secur. Commun. Netw. 5(12), 1379–1393 (2012)

9. Yang, Y., Lu, H., Liu, J. K., Weng, J., Zhang, Y., Zhou, J.: Credential wrapping: from anonymous password authentication to anonymous biometric authentication. In: Proceedings of the 11th ACM on Asia Conference on Computer and Communications Security, pp. 141–151. ACM (2016)

10. Shin, S., Kobara, K.: Simple anonymous password-based authenticated key exchange (sapake), reconsidered. IEICE Trans. Fundam. Electron. Commun. Comput. Sci. 100(2), 639–652 (2017)

11. Wu, T.Y., Fang, W., Chen, C.M., Wang, E.K.: Efficient anonymous password-authenticated key exchange scheme using smart cards. In: The Euro-China Conference on Intelligent Data Analysis and Applications, pp. 79–87. Springer, Cham (2017)

Human Tracking System Based on GPS and IOT (Internet of Things)

Sonam Priyanka[1], Sandip Dutta[1], and Soubhik Chakraborty[2(✉)]

[1] Department of Computer Science and Engineering,
Birla Institute of Technology Mesra, Ranchi 835215, India
sonampriyankaspl@gmail.com,
sandipdutta@bitmesra.ac.in
[2] Department of Mathematics, Birla Institute of Technology Mesra,
Ranchi 835215, India
soubhikc@yahoo.co.in

Abstract. Human safety is an important issue due to the sudden rise of crimes and accidents. To resolve this issue we propose a Human Tracking System based on GPS (Global Position System) and Internet of Things (IOT) that has dual security features. The proposed system incorporates the available technical know how in sensors (GPS,gravity etc.) to track, monitor and assist a person in different situations [3, 11]. The proposed system may be utilized by children in crowded areas and by anyone who wants to get assistance at the time of distress like patients. Our proposed concept of developing such a device is to create an environment where anyone can be located at any time thus reducing the cost and efforts involved to track them by several agencies. The system will alert the right agencies if a person, like a child or a patient, is in need of assistance.

Keywords: Arduino UNO ·
Ultimate Adafruit GPS (Global Positioning System) shield ·
ESP2866 WiFi module · Arduino IDE software ·
Thing speck cloud server and fritzing software (design circuit diagram)

1 Introduction

Why is it time to take the Internet of Things (IOT) seriously? "The Internet of Things will be five to ten times more impactful in the next decade than the entire internet has been to date" [1] and [2]. So we can define IOT is an internetwork of physical objects embedded with sensors, computers, connectivity and actuator that enables these objects to acquire data, transform it into knowledge, make intelligent decision and generate physical actions to manipulate the environment. This project is based on IOT that works for tracking the human location. We know that today our family/human security has become very important because day by day crime rate is increasing [7]. The proposed device not only identifies or tracks the location but also achieves a complete manufacturing cost of device less than that of any other tracking devices. So any one easily can afford our device and can track our children or any patient (who is not mentally or physically fit). Further literature on IOT can be found in [10–12].

© Springer Nature Switzerland AG 2020
A. P. Pandian et al. (Eds.): ICCBI 2018, LNDECT 31, pp. 514–521, 2020.
https://doi.org/10.1007/978-3-030-24643-3_61

2 System Deployment

2.1 Hardware Component

Our Human Tracking Device is basically tracking the location through GPS Tracker. So our main goal is to read GPS data and send it to the cloud. After that the cloud sends the stored data defining the location through mobile or any other application devices. We are using the following hardware components for making the complete Human tracking device shown in Fig. 1.

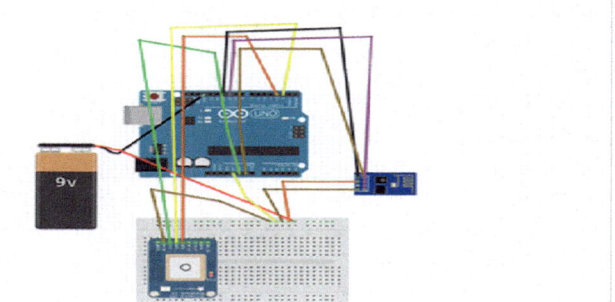

Fig. 1. Hardware component of IOT tools

A. Arduino UNO (Micro controller)
B. Adafruit Ultimate GPS Breakout
C. ESP2866 WiFi
D. Battery

A. Arduino UNO (Micro controller): In Human Tracking Device we are focusing the cost effective product so that we are using Arduino UNO which is also known as tiny computer that we can connect to electrical circuits. This makes it easy to read data from outside and control output and send a command to outside. In this project we are connecting Arduino UNO to Adafruit Ultimate GPS Breakout, ESP2866 WiFi and battery(9 V). The following Tables 1 and 2 showing the Pin connection between devices in Fig. 2.

B. Adafruit Ultimate GPS Breakout: Adafruit Ultimate GPS shield not only reads the Global Positioning System latitude and longitude Data but also is small and portable. This is an excellent GPS being affordable, easy to use, and capable to connect 22 satellites. Table 1 page no 4 shows connection between Arduino Uno.

C. ESP2866 Wi-Fi: It is a low cost WiFi microcontroller and it is used for home automation. Now when we read the GPS Data through Adafruit Ultimate GPS Shield then we have to store and retransmit the data to Cloud (Thing Speak) to client Application (cellular mobile or operating system). We need to connect the ESP2866 WiFi to the internet and to the particular device transmitting to each other shown in Fig. 3. Table 2 shows the connection.

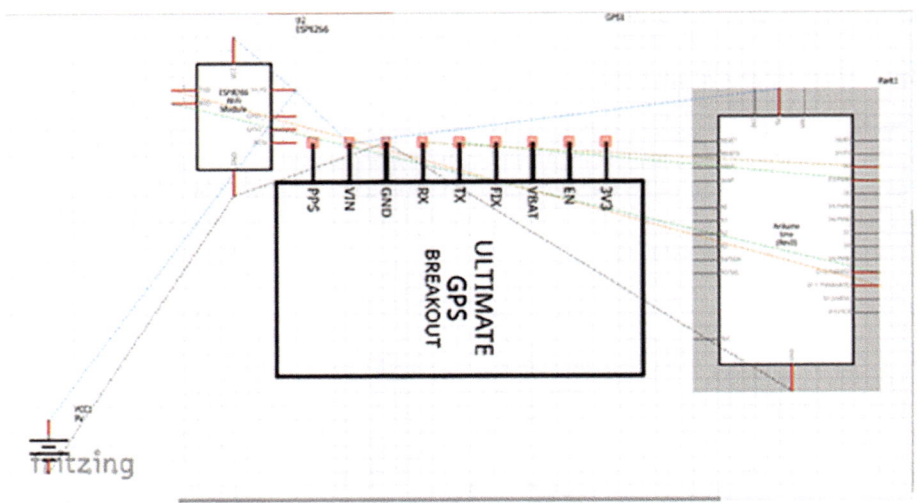

Fig. 2. Circuit diagram of our proposed device

Table 1. Connecting the Adafruit ultimate GPS unit to Arduino UNO

Connecting the Adafruit ultimate GPS unit to Arduino UNO	
GPS Pin	Arduino Pin
Vin	5 V
GND	GND
RX	Pin 2
TX	Pin 3

ESP8266 specifications

- 802.11 b/g/n protocol
- Wi-Fi Direct (P2P), soft-AP
- Integrated TCP/IP protocol stack
- Built-in low-power 32-bit CPU
- SDIO 2.0, SPI, UART

D. Battery: The proposed device requires 9v power supply.

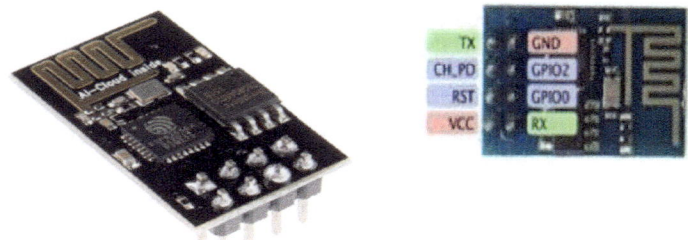

Fig. 3. ESP2866 WiFi module

Table 2. Connecting the ESP2866 WiFi module to Arduino UNO

Connecting the ESP2866 WiFi module to Arduino UNO	
ESP2866 WiFi Pin	Arduino Pin
VCC	5 V
GND	GND
CH_PD	5 V
RX	Pin 10

3   Architecture of Human Tracking System [8]

Figure 4 gives the architecture of the schematic diagram of our processed device showing Arduino Uno working as microcontroller board to connect Ultimate Adafruit GPS shield and ESP2866WiFi through jumper wire through pin modes [6]. The connection description is already given in Tables 1 and 2. This means the architecture of Human Tracking Device working principal is based on the location tracking through GPS navigated by satellites. The transmitted signals of satellite's information of latitude and longitudes reading is done by Ultimate Adafruit GPS shield. That information is retransmitted by WiFi module and stored into the cloud and the data is extracted from cloud through App or any other Application. All the component power

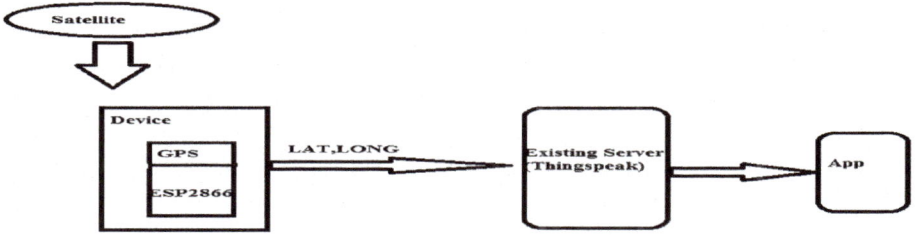

Fig. 4. Architecture of proposed device

supply is provided by DC charge 9 V battery. So the basic concept behind the making of an effecting tracking location device is low cost.

4   Working Principle of Human Tracking System [4]

Our proposed device is based on IOT (internet of things) which has a device communication interconnected with internet. We are using GPS Shield for reading the GPS values. In this device we are using the WiFi module with Arduino UNO processor and GPS Shield that is useful for home Automation/Internet of Things applications.

Our proposed device working on four phase-

(i) Client Device-Tracker Device (ii) Satellite-Data Transmitting (iii) Cloud Server (Thing Speak)- channel where we sending data and storing it. (iv) Cellular Device (Mobile, System)- App are used to show the location of client.

The following diagram (Fig. 5) shows the communication between device and internet. This is the example of working principle human tracking device.

Fig. 5. Example of working principle of human tracking system

5   Software Component

Our proposed device was built using basic programming languages in C++ on Arduino IDE software. To compile Hardware in Sketch book of Arduino IDE we needed a library to install in GitHubs [5]. They are given in the following table (Table 3).

Table 3. Arduino hardware's supportive library [9]

Sl. no.	Hardware/software	Library
1.	GPS Shield (Adafruit ultimate GPS breakout)	Adafruit_GPS
2.	ESP8266 WiFi	All ESP8266_WiFI

6   Results

The Human tracking device gives the results in the following steps:-

- All the hardware components are connected by jumble wiring. After power is supplied all hardware LED lights are blinking which shows that connection is established with the satellites.
- The proposed device GPS is connected with satellite and reads the GPS data in the form of latitude and longitude.
- The transmitted GPS data is retransmitted to the existing server (like Thing Speak)
- We can access the location of any man/woman/child through the existing server through any application devices like mobile app and Google earth etc.

The Figs. 6, 7, 8, 9 depict the pictures of Testing Hardware (Figs. 6 and 9) and Software (Figs. 7 and 8).

Fig. 6. Proposed device

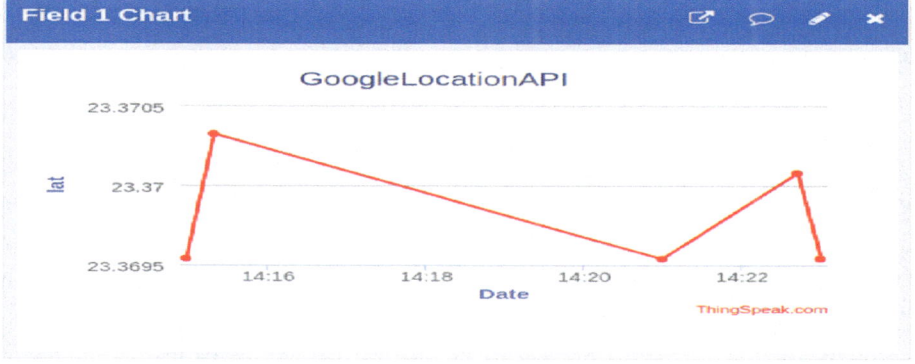

Fig. 7. Graph for latitude value of GPS in Thing Speak

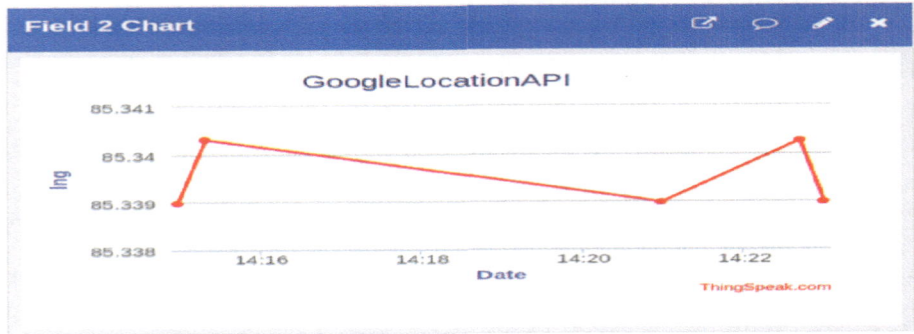

Fig. 8. Graph for longitude value of GPS in Thing Speak

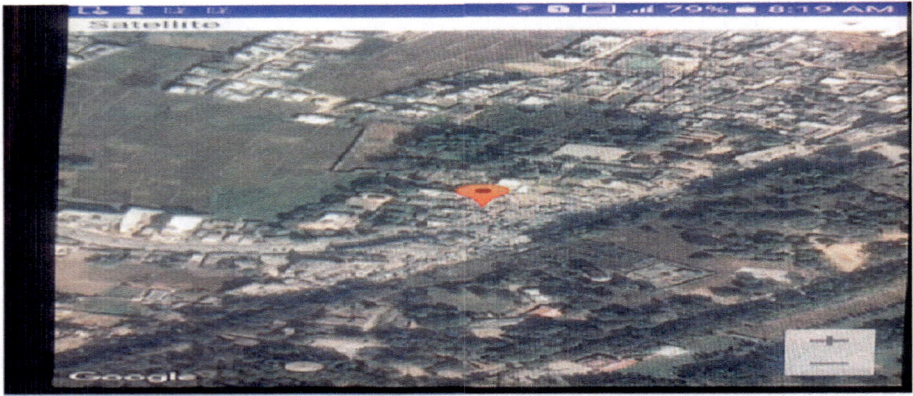

Fig. 9. Access location shown in satellite map

7   Conclusion and Future Works

Our proposed device tracks the location in low cost in a very simple way. So we can make in future smaller devices which can not only track the location with spy camera but do so in low cost also.

Compliance with Ethical Standards.

Conflict of interest All the authors hereby declare that this research did not receive any funding. They further declare that the work is new and that they do not have any conflict of interest.

Ethical Approval This article does not contain any studies with human participants or animals performed by any of the authors.

References

1. www.isro.gov.in. Accessed 4 Dec 2018
2. https://en.wikipedia.org/wiki/Internet_of_things. Accessed 4 Dec 2018
3. http://www.channelfutures.com/security. Accessed 4 Dec 2018
4. Uckelmann, D., Harrison, M., Michahelles, F.: Internet of Things, p. 353. Springer, Berlin (2011)
5. https://github.com/adafruit/Adafruit-Eagle-Library. Accessed 4 Dec 2018
6. Colon, C.R.: An efficient GPS position determination algorithm. Master of Science in Electrical Engineering thesis, Air Force Institute of Technology (1999). https://apps.dtic.mil/dtic/tr/fulltext/u2/a361729.pdf. Accessed 4 Dec 2018
7. Weber, R.H., Weber, R.: Internet of Things, Legal Perspectives, p. 135. Springer, Berlin (2010)
8. Delsing, J.: IOT Automation: Arrowhead Framework, p. 366. CRC Press, Boca Raton (2017)
9. https://learn.adafruit.com/adafruit-ultimate-gps/overview. Accessed 4 Dec 2018
10. Perera, C., Liu, C.H., Jayawardena, S.: The emerging internet of things marketplace from an industrial perspective: a survey. IEEE Trans. Emerg. Top. Comput. 3(4), 585–598 (2015)
11. Chauhuri, A.: Internet of Things, for Things, and by Things. CRC Press, Boca Raton (2018). ISBN: 9781138710443
12. Hassan, Q., Khan, A., Madani, S.: Internet of Things: Challenges, Advances, and Applications. CRC Press, Boca Raton (2017). ISBN: 9781498778510

Review on Mechanisms to Ensure Confidentiality and Integrity of the Data in Cloud Storage

Niharika Srivastav[✉] and Rejo Mathew

Department of IT, Mukesh Patel School of Technology Management and Engineering, NMIMS (Deemed-to-Be) University, Mumbai, India
niharikas869@gmail.com, rejo.mathew@nmims.edu

Abstract. The paper reviews few approaches on securely storing data in the cloud storage. Most of the cloud computing models aim towards ensuring that the data confidentiality and integrity prevails although users of the cloud use internet for using the services provided by the cloud computing system. The first mechanism considered involves checking that data integrity has been preserved or not through use of hash-function and metadata approach. The second mechanism involves making use of SHA-3 hash function to provide confidentiality as well as integrity for the data stored in cloud storage through use of the Mobile Cloud Computing (MCC) services.

Keywords: Cloud computing · Storage · Data integrity and confidentiality · Metadata · Hash function · Mobile Cloud Computing (MCC)

1 Introduction

Cloud computing makes use of the Internetworks, servers that are centralized to organize the several applications, content, etc. that can be accessed by the cloud users. There is a 4-deployment model of the cloud which is comprised of public, private, hybrid and community cloud.

- Public Cloud - Cloud computing occurs outside an organization and any one in the public can access the cloud. Here, files are hosted by the Third Party. [1]
- Private Cloud - Mesh computing is performed within an organization and access to any file is done using a highly secured network.
- Community Cloud - When two or more different organizations have similar policies and similar requirements, they can share the same cloud known as the community cloud.
- Hybrid Cloud - This cloud combines features of the public, private as well as the community cloud. [2]

Numerous models and approaches are proposed to provide solutions to the security problems encountered in the cloud. The main concerns included in all of the cloud models includes dealing with existence of the key components of security i.e. data integrity and confidentiality.

A. P. Pandian et al. (Eds.): ICCBI 2018, LNDECT 31, pp. 522–528, 2020.
https://doi.org/10.1007/978-3-030-24643-3_62

2   Problems

Cloud storage and computing uses the internet which makes the cloud's data vulnerable to unauthorized access, manipulation and theft. Public clouds are highly susceptible to attacks on data confidentiality and integrity [3]. Multi-tenancy is another feature of cloud where data of multiple clients share the same physical space which again raises concerns of data confidentiality and integrity.

Since cloud storage is used by both small and large scale organizations to store highly confidential data, ensuring that there is completely no unauthorized access by any non-organizational member is the biggest concern. Also, because the CSPs have total control of the cloud's content, it is possible that they execute any activity that might be malicious like copying or destroying, manipulating the content, etc. [4].

There are three main concerns for the data stored in cloud-

2.1   Confidentiality-Ensuring Secure Access

Securing access to the protected data and information requires that the user must be authorized to access it. Mechanisms for controlling the access are chosen based on the level of importance of the information that is to be protected.

2.2   Integrity-Protecting Information

Integrity means that the consistency, incorruptibility and accuracy of stored information will be maintained [5].

2.3   Availability-Ensuring the Availability of Content

It is mandatory for the CSPs to guarantee the availability of data without getting affected by any failures of the hardware, malicious physical disks or any server downtime.

3   Solutions

3.1   HAIL

High-availability & integrity layer is an extension of the principles of RAID into the cloud settings, which is quite adversarial. HAIL is a protocol that checks for the integrity of a remote file. HAIL ensures file availability against a strong, mobile adversary by use of proactive cryptographic models [6].

3.2   Integrity in Multi-Cloud Environment by Using Identity-Based Approach

This approach gives demonstration of the proof of concept by testing on a prototype of the application. Block-tag pairs are distributed to different cloud servers and as combiner receives the request for retrieval, the challenge received is distributed between the cloud servers and then the responses from all these cloud servers are collectively sent back to the client.

3.3   PORs: Proofs of Retrievability for Large Files

Proof of retrievability (POR) is a proof by a system of files (the prover) to a particular client (a verifier) that a file F which is the target file is fully recoverable by the client.

3.4   Metadata

Metadata can be seen as small pieces of information that can be attached to various files and other objects in a computing environment. Files in a cloud storage system can conveniently carry custom metadata that relates to security, by controlling access more tightly so that the enterprises using the cloud can have an added confidence that their data storage solution can keep their customer and proprietary information confidential.

3.5   Hash Function

Creation of a fixed length message digest for a message of a particular length is done by the cryptographic hash function. [7] Hash function computes a unique fixed length message digest for every message. Security requirements for hash functions includes them to be 1-way functions and resistant to collision. Collision is caused by the occurrence of same output by the hash function for any given different inputs i.e. Hash (m1) = Hash (m2) (Fig. 1).

3.6   The Solution Using the Combination of Metadata and Hash Function Concepts Can Be Described More Elaborately in the Form of Two Processes-

Encrypted and Hash File along with Metadata Storing in Cloud

- First step is performing encryption of the local file using AES 256 algorithm and then creation of the file's metadata is done.
- For the metadata file, it is needed to record the file's last modified date (in milliseconds), the Length (bytes) of the file and Hash value for the file (which is 32 byte) [6].
- The meta-data and encrypted file are stored in the Cloud.
- SHA 256 is used for hash generation of the local file and after that the hashed file is kept in cloud (Fig. 2).

Checking Integrity by using the Hash Function

- The Metadata of the file, stored earlier, is checked during file retrieval from cloud.
- Next step is retrieval of the encrypted file and using AES-256 algorithm to decrypt the file.
- Then using the hashing algorithm- SHA256, hash function of the decrypted file is generated.

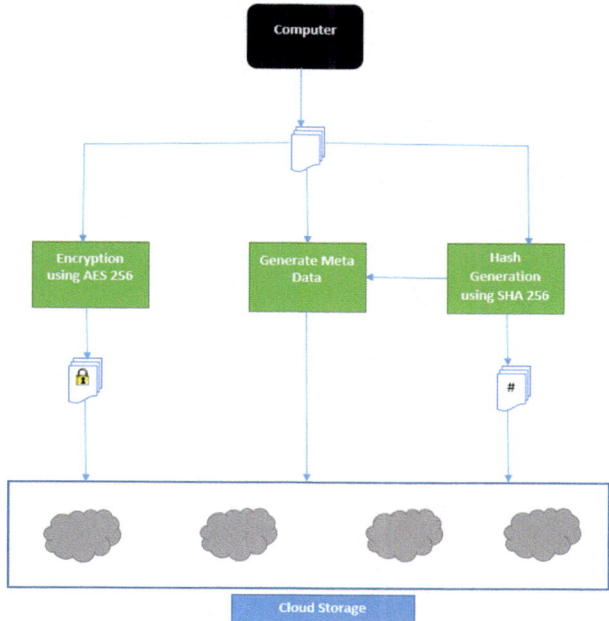

Fig. 1. Encrypted, hash file and metadata stored in cloud

- Next step includes downloading the hash file and making a comparison between both the hash files.
- If both the files have the same hash function and same meta-data value then integrity is preserved otherwise the cloud service has failed to maintain the data integrity of the file [6].

3.7 Ensuring Confidentiality and Integrity in Mobile Cloud Computing (MCC) Using SHA-3 Hash Function

Framework Architecture

The framework includes three main components, namely the mobile device user who is the client, the Cloud Service Provider- CSP and Data Storage Provider - DSP. Mobile device client makes use of any mobile device (like a cellular mobile phone) as a platform through which they access the CSP to perform encryption, decryption, integrity verification of large amount of content, before they store it all on the DSP.

Phases of the framework

Registration

In this phase, user is allowed to create an account with the Cloud Service Provider.

First step involves the mobile user to create an account with CSP. Then second step involves CSP to store the user's account information on the DSP and then the CSP allows the user to generate the public key which is denoted by K(P) and the private key

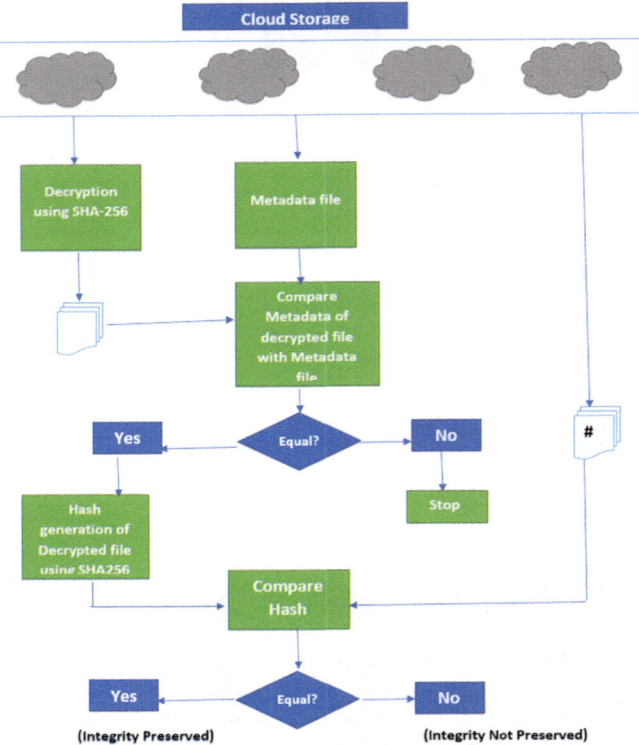

Fig. 2. İntegrity checking

denoted as K(R), by using the ECC algorithm once the registration has been done successfully. Next step includes the CSP to store the generated K(P) on DSP and CSP sends private key K(R) to the mobile client.

Encryption
Encryption phase has following steps-

A. The mobile client uses the Dynamic Keystroke mechanism, which allows for the user's authentication and the client then logs into the CSP [8].
B. The CSP then performs retrieval of the user's K(P) (public key) from DSP.
C. The mobile device client transfer their file.
D. Mobile client gets the option to generate a random Advanced encryption standard key (AES key K(A)).
E. The client is enabled to compute the hash value-V for the client's file and encrypt that file by using AES key K(A) and encrypt the hash value and K(A) by using public key K(P).
F. CSP stores the encrypted file and the hash value and the random AES key – K(A) are on the DSP.

Decryption

The steps included in this phase are-

A. The mobile client uses the Dynamic Keystroke mechanism, which allows for the user's authentication and the client logs into the CSP. [8]
B. CSP performs retrieval of key K(P) from DSP.
C. Client selects the required file from the list of files available in CSP.
D. CSP performs retrieval of the encrypted file and the hash value from DSP.
E. CSP lets the user enter the private key-K(R).
F. User or the client performs decryption of the hash value and the random AES key-K(A) by using the private key, K(R).
G. File is decrypted using K(A).
H. The hash value is used to check the integrity.
I. In the final step, the user makes a request to download the needed file in plaintext form.

4 Review

Approach	Methods used/components	Advantages	Limitations
HAIL [9]	Pseudo random function, Hash-Function and MAC	a. Users are allowed to store data on multiple cloud	a. Applicable only for static data b. Not suitable for a thin client
POR for large files [10]	Cryptographic challenge-response protocol	a. Ensures the possession and retrievability of files on CSP	a. Applicable only with static data b. Increases input/output and transmission cost
Solution using Metadata and Hash-Function	Metadata, SHA-256 hash algorithm, RSA encryption	a. Strong verification for data integrity b. Easy to implement	a. Doesn't guarantee confidentiality b. Metadata management is tedious for large number of files
Framework using SHA-3 for MCC	SHA-3 Hash function, AES encryption-decryption algorithm	a. Minimizes the processing and storage overhead of mobile client and CSP	a. Large number of keys for each integrity check

5 Conclusion

The paper reviewed some approaches for storing data in cloud storage while ensuring that data confidentiality and integrity prevails although users use the internet for using the cloud computing services and discussed about the issues related to confidentiality of data violated due to access by unauthorized users and integrity of content in remote servers.

The paper reviewed some mechanisms for checking integrity of the files, documents etc. stored in the cloud.

References

1. Rao, C., Leelarani, M., Kumar, Y.R.: Cloud: computing services and deployment models. Int. J. Eng. Comput. Sci. **2**, 3389–3392 (2013)
2. Nandgaonkar, S.V., Raut, A.B.: A comprehensive study on cloud computing. Int. J. Comput. Sci. Mob. Comput. **3**(4), 733–738 (2014)
3. Neelima, M.L., Padma, M.: A study on cloud storage. Int. J. Comput. Sci. Mob. Comput. **3**, 966–971 (2014)
4. Aldossary, S., Allen, W.: Data security, privacy, availability and integrity in cloud computing: issues and current solutions. Int. J. Adv. Comput. Sci. Appl. **7**, 485–498 (2016)
5. Al-Saiyd, N.A.: Nada Sail: data integrity in cloud computing security (2014)
6. Vyas, J., Modi, P.: Providing confidentiality and integrity on data stored in cloud storage by hash and meta-data approach. Int. J. Adv. Res. Eng. Sci. Technol. **4**, 38–50 (2017)
7. Tikore, S.V.: Data integrity & confidentiality in cloud storage using hash function and TPA. Int. J. Recent Innov. Trends Comput. Commun. (IJRITCC) **3**, 2738 (2015)
8. Babaeizadeh, M., Bakhtiari, M., Maarof, M.: KeyStroke dynamic authentication in mobile cloud computing (2014)
9. Bowers, K.D., Juels, A., Oprea, A.: HAIL: a high-availability and integrity layer for cloud storage (2012)
10. Aldossary, S., Allen, W.: Data security, privacy, availability and integrity in cloud computing: issues and current solutions (2016)

Review: Cloud Task Scheduling and Load Balancing

N. Manikandan[(✉)] and A. Pravin

School of Computing, Sathyabama Institute of Science and Technology,
Chennai, Tamilnadu, India
macs2005ciet@gmail.com, pravin_ane@rediffmail.com

Abstract. Load Balancing is fundamental for effective activities in conveyed conditions. As Cloud Computing is developing quickly with customers be requesting extra services and superior outcomes, load balancing in favor of the Cloud have turned into an extremely intriguing and imperative research region. Numerous schemes were proposed to give proficient components and schemes to allotting the customer's solicitations to accessible Cloud hubs. These methodologies plan to get better concert of the Cloud as well as offer the client all the extra gratifying and effectual tunes. In this article, we explore the diverse schemes planned to determine the concern of load balancing and task scheduling within Cloud Computing. We talk about with contrast these schemes with give a review of the most recent methodologies in the meadow.

Keywords: Cloud computing · Load balancing · Task scheduling concerns · Challenges and schemes

1 Introduction

Cloud Computing turned out to be extremely prominent over the most recent couple of years. As a component of its services, it gives an adaptable and simple approach to keep and recover information and records. Particularly to make substantial informational collections and documents accessible for the spreading number of clients around the globe. Taking care of such expansive informational indexes require a few systems to advance and streamline activities and give attractive levels of performance to the clients [1]. In this manner, it is imperative to look into a few territories within Cloud near enhance the capacity usage along with the download concert pro the clients. Individual essential concern connected by the meadow is energetic load balancing otherwise task scheduling. Schemes for load balancing are researched intensely in different conditions; in any case, with Cloud situations, some extra difficulties are available with should be tended to. In Cloud Computing the principle concerns include proficiently allotting errands to the Cloud hubs with the end goal that the exertion and demand handling is done as effectively as conceivable [2], while having the capacity to endure the different influencing limitations, for example, heterogeneity and lofty correspondence waits.

Load balancing schemes are delegated dynamic and static transforms. Static transforms are used for the most part reasonable for homogeneous and stable conditions

A. P. Pandian et al. (Eds.): ICCBI 2018, LNDECT 31, pp. 529–539, 2020.
https://doi.org/10.1007/978-3-030-24643-3_63

and can deliver great outcomes in these situations. Be that as it might, they are frequently not malleable and can't synchronize the dynamic transforms to the qualities among the implementation period. Dynamic transforms are further adaptable and mull over different kinds of traits in the frame both proceeding and amid run-period [3]. These schemes may alter to modifies and offer superior outcomes in mixed and dynamic circumstances. In any case, as the circulation qualities turn out to be added impulsive and dynamic. Therefore a part of these schemes might end up wasteful and cause additional overhead than should be expected bringing about a general corruption of the checks concert.

In this article we produce a review of the recent load balancing schemes grew explicitly to ensemble the Cloud Computing situations. We give a diagram of these schemes and examine its possessions. What's more, we look at these schemes dependent on the accompanying possessions: the quantity of characteristics contemplated the general system load, and period arrangement.

2 Task Scheduling in Cloud Computing

Task Scheduling assumes a key job to enhance adaptability and unwavering quality of frameworks in cloud. The primary explanation for scheduling tasks to the assets as per the given period bound [4]. Cloud comprises of various assets that are assorted with one extra by means of a few means and cost of performing tasks in cloud. Utilizing assets of cloud is sole so scheduling of tasks in cloud isn't the same as the conventional approaches for scheduling thus scheduling of tasks in cloud require better thoughtfulness regarding be paid on the grounds that services of cloud relies upon them.

The different types of cloud task scheduling are:

2.1 Cloud Service Scheduling

Cloud Service scheduling depends on client level and framework level [5, 6]. The service arrangement concerns raised between the suppliers and clients are done at client level scheduling. The overseeing of assets with server farm includes the framework level scheduling. A few physical machines remain associated are said to be server farm. The client sends a huge number of tasks and the tasks are allotted to the physical machines in the server farm.

2.2 User-Level Scheduling

For directing the free market activity of cloud assets the market based and activity based schedulers are appropriate. If there should be an occurrence of market based schedulers where the asset is virtualized and conveyed to the client as a service. Service Level Agreement (SLA) incorporates service provisioning in the cloud, where the SLA demonstrates the agreement to be marked among the suppliers and clients expressing the agreement terms, including non-utilitarian necessities determined by Quality-of-service (QoS), commitment and in punishments [7]. The SLA is utilized by novel cloud scheduling [8] having trust screen to give the scheduling quicker secure handling in light of the over flooding client ask. The epic methodology is considered by two variations of heuristics [9] and they are invigorated Annealing and GI-FIFO scheduling.

2.3 Static and Dynamic Scheduling

In the static scheduling where the required information are pre-brought and the task execution at various stages are pipelined. What's more, in dynamic scheduling where the task or occupation parts are not known in advance and though the task execution period may not be known. The scheduling procedures in three level cloud structure dependent on service ask for are

- Resource Providers.
- Service Providers.
- Consumers.

An enormous measure of vitality is squandered by the vast number of cloud computing servers and it produces a carbon dioxide, in this manner the green task scheduling [10, 11] is expected to bring down the vitality use and to diminish the contamination.

2.4 Heuristic Scheduling

The NP difficult concerns are understood by specification, heuristic and estimation techniques. On the off chance that every single conceivable arrangement are identified and thought about one by one, the ideal arrangement is chosen in a specification strategy. If there should be an occurrence of huge occasions, where a count isn't plausible the heuristic handles the examples of finding the best arrangements and sensibly quick. To streamline the arrangement, the rough arrangements are found if there should arise an occurrence of estimation schemes. The make range of the given task is limited while balancing the whole framework load as indicated by the load balancing task plan [12].

These tasks are booked for some predefined period in group mode. The genuine execution period of a substantial number of tasks is known by the clump heuristics. Likewise the heuristics Max-Min [13], Min-Min [14] are utilized by the cluster mode scheduling.

2.5 Real Period Scheduling

The real period scheduling includes a few targets and they are utilized to expand the throughput as opposed to meeting due dates, additionally the normal reaction period will be limited. To amplify the aggregate utility [15] the real period tasks ought to be planned non-preemptively. The benefit Period Utility Function (TUF) and punishment Period Utility Function (TUF) are two diverse utility capacities related in the mean-period for each task.

2.6   Workflow Scheduling

In a Workflow scheduling where the applications are organized as Directed Acyclic Graphs (DAG) [16]. Here the edges indicate the task conditions and the hubs speak to the constituent task. A solitary workflow has some arrangement of task where each task is speaking with another task in the workflow. In the workflow execution the board the workflow scheduling is individual among the real concerns and in the cloud it is empowered to utilize different cloud services to encourage the workflow execution.

3   Concerns in Scheduling Task

3.1   Execution Period

The execution period or CPU period of a specified chore is characterized as the period depleted by the framework performing that task, including the period invested executing run-energy or framework services for its benefit [4]. The instrument used to quantify execution period is usage characterized. It is usage characterized which task, assuming any, is charged the execution period that is devoured by intrudes on handlers and running period for the framework.

3.2   Resource Cost

Resource cost in a cloud can be controlled by asset limit and involved period. Incredible assets prompted greater expense. With cloud computing, business inspires the capacity to get a good deal on equipment as well as programming yet requires spending more on the transfer speed [4]. It is relatively hard to completely abuse the services of cloud computing without fast correspondence channels.

3.3   Performance

Performance is the greatest concern in cloud appropriation. The cloud must give enhanced performance when a client moves to cloud computing foundation. Performance is commonly estimated by capacities of utilizations running on the cloud framework [4]. Poor performance can be caused by absence of legitimate assets viz. circle space, restricted data transfer capacity, bring down CPU speed, memory and system associations and so on.

3.4   Security and Privacy

Security, Performance and Availability are the three greatest concerns in cloud selection [4]. The basic test is the manner by which it tends to security and privacy concerns which happen because of development of information and application on systems, loss of control on information, heterogeneous nature of assets and different security arrangements.

4 Task Scheduling Schemes

The principle motivations behind scheduling schemes are to limit asset starvation and to guarantee reasonableness among the gatherings using the assets. Scheduling manages the concern of choosing which of the exceptional solicitations is to be assigned assets. There are many different scheduling schemes popularly used in task analysis process are [17].

1. Genetic Scheme
2. Greedy Scheme
3. Priority based Job Scheduling Scheme
4. Round Robin
5. Min-Min Scheme
6. Particle Swarm Optimization
7. First Come First Serve Scheme.

5 Load Balancing in Cloud Computing

Load balancing in Cloud is the approach to dispersing workloads above dissimilar computing assets. Cloud load balancing diminishes costs related with records the executive's frameworks and expands convenience of assets and also a kind of load balancing and should not be mistaken for Domain Name System (DNS). At the same time as DNS utilizes programming otherwise equipment to play out the capacity, it utilizes examinations obtainable by different PC arrange organizations.

It also has leeway in excess of DNS as it is capable of exchange loads to servers all inclusive instead of conveying it crosswise over neighborhood servers. In case of a neighborhood main system blackout and it conveys clients to the nearest provincial main system with no interference for the customer.

Load balancing of cloud delivers concerns identifying with Extensible Authentication Protocol (EAP) present amid DNS load balancing. DNS mandates must be authorized one period in every TTL cycle and may get a few hours if exchanging among servers amid a limp or server dissatisfaction. Approaching server movement will remain on steering to the first main system awaiting the point that the TTL lapses and may build a jagged concert as a variety of web access suppliers may achieve the new server before other web access suppliers. Another preferred standpoint is that is enhances reaction period by steering isolated assembly to the finest executing main system farms.

Cloud computing gets points of interest "cost, adaptability and accessibility of service clients". Those favorable circumstances force the interest pro Cloud tunes. The interest brings specialized concerns up in Service Oriented Architectures with Internet of Services (IoS) manner submissions, for example, good accessibility and adaptability. While a noteworthy worry in this concerns, load balancing permits cloud computing to "scale up to expanding requests" through effectively dispensing dynamic neighborhood workload equally over every hubs [18].

For proficient load balancing here should be a few constraints toward assess the load balancing systems by improve asset appropriation for the client requests. Estimation parameters enable us to observe if the specified system is sufficient by adjust the load of the movement on the main system. In this article we deem different load balancing estimation constraints to assess the load balancing methods what are talked about underneath [19]:

- *Throughput:* It is a measure of effort to a PC could achieve in the specified measure of period.
- *Response period:* It is the measure of period utilized to begin satisfying the interest by the client in the wake of enrolling the question.
- *Fault tolerance:* It is the capacity of the load balancing scheme to permits by continue functioning appropriately in a few disappointment state of the framework.
- *Scalability:* Simply the capacity of the scheme to scale its own as indicated by necessary circumstances.
- *Performance:* In general verify of the schemes functioning. It includes the finish of the specified chore beside near realized norms similar to precision, rate and force.
- *Resource utilization:* It is utilized to maintain a mind the use of different assets.

6   Challenges in Cloud Load Balancing

Prior to we might audit the present load balancing techniques for Cloud Computing, we have to distinguish the primary concerns with difficulties included along with influence how the scheme could achieve. At this time we talk about the difficulties to survive tended to while endeavoring to suggest an ideal answer for the concern of load balancing in Cloud Computing [1]. This difficulty is condensed by the accompanying focuses.

6.1   Cloud Nodes for Spatial Distribution

A few schemes are intended to be productive just for an intra net or firmly found hubs where correspondence delays are irrelevant. In any case, it is a test to plan a load balancing scheme that can work for spatially dispersed hubs. This is on the grounds that different elements must be considered, for example, the speed of the system joins among the hubs, the separation between the customer and the task preparing hubs, and the separations between the hubs associated with giving the service. There is a need to build up an approach to organize load balancing component amid all the spatial dispersed hubs whereas having the capacity to viably endure elevated deferrals [20].

6.2   Replication

For a complete replication scheme doesn't consider productive capacity use. This is on the grounds that similar information will be put away in every replication hubs. Complete replication schemes force greater expenses as further stockpiling is required. In any case, halfway replication schemes might spare elements of the informational collections in every hub dependent on every hub's abilities, for example, preparing force and limit [21]. This could prompt better usage, so far it expands the unpredictability of the load balancing schemes as it endeavor to consider the accessibility of the informational collection's elements over the diverse Cloud hubs.

6.3 Complexity in Scheme

Load balancing schemes are liked to be fewer mind boggling as far as usage and activities. The higher usage intricacy would prompt another unpredictable practice what might root a few pessimistic concert concerns. Besides, while the schemes necessitate extra data and superior correspondence for checking and manage, postponements could root other concerns and the effectiveness may plunge. Along these lines, load balancing schemes should be planned in the most straightforward conceivable structures [22].

6.4 Point of Failure

Organizing the load balancing and gathering information concerning the distinctive hubs should be planned in a mode that abstains from have a solitary purpose of disappointment in the scheme. A few schemes (concentrated schemes) may give proficient and viable instruments to understanding the load balancing in a specific example. Be that as it may have the concern of individual organizer for the entire framework. In this crate, in the event that the organizer comes up short, the entire framework could fall flat. At all load balancing scheme should be structured so as to beat this test [23]. Dispersed load balancing schemes appear to give a superior methodology, so far these are substantially further perplexing with necessitate extra synchronization and organize to work effectively.

7 Cloud Load Balancing Schemes

In this article we have examined different schemes in a succession along with toward the last part we will create a relative examination by the considerable number of schemes that we would talk about here [19].

7.1 Ant Colony Optimization Scheme

To take care of the concern of the load balancing creator picks the conduct of the ants amid inquiry of nourishment. As ants has with an exceptionally insightful path for finding the sustenance by the technique for most brief separation creator brought that into the thought [24]. An ant utilizes the guideline of trace deceitful by tumbling the pheromones resting on the land by halting on a few in transits during signal while discharged by pheromone organ. Along these lines the way is refreshed at each new area with the separation esteem which enables the other versatile operators to pick the most limited way among numerous which may outcome the benefit.

7.2   Honey Bee Foraging

This scheme is one of the best schemes for load balancing whereas is enlivened by the conduct of bumble bees. At the same time as bumble bees gone for the inquiry of the sustenance, if it discovers the nourishment they arrive and play out an extraordinary move entitled shake move to advise the eminence and amount of the nourishment to the colony. The move likewise advises the separation from the bee sanctuary to the sustenance. In [25] creator acquires roused from this subterranean insect rule and proposes bumble bee scavenging scheme.

7.3   Genetic Scheme

In [26] the novelist was exposed a new load balancing system utilizing genetic scheme. This scheme endeavors to adjust the load of the main system and furthermore attempt to limit the period spent to satisfy the preparing demands. The creator utilized the hereditary scheme to play out the load balancing through reproducing the scheme in the recreated condition utilizing the cloud Analyst test system. Likewise creator contrasted its reproduction result and the current methods.

7.4   Join Idle Queue

In [27] the novelist was planned a new load balancing scheme knows as Join Idle Queue [JIQ] scheme for apposite load balancing in considerable frameworks. The Join Idle Queue scheme is the improvement of authority of few schemes without connection transparency among the correspondent with the workstations at employment entry. The creator examined the JIQ scheme in substantial framework perimeter and found the viable outcomes. The creator investigated that the JIG scheme constructs 30 crease decreases in the lining transparency when contrasted with the authority of few scheme at standard and heavy load.

7.5   Shortest Job Scheduling Scheme

Shortest job scheduling scheme is important among the amazing schemes for scheduling the occupations on the determination criterion of employment through most brief implementation period determination survive chosen initial. Most brief employment initially had preference of lessened sitting tight period for the procedures which build it an incredible methodology. In [28] the novelist unmistakably pictorised the functioning of the cloud computing with the load balancing by means of the assistance of the most brief activity first scheme. In this article creator reenacted different numeral value of employments through various implementation period and examined the outcomes pro the reproduction. Additionally creator finished an unmistakable end that the most brief activity primary scheme diminishes the normal sitting tight period for the employments.

7.6   Stochastic Hill Scaling

It is the best type of the slope scaling scheme in which is utilized pro the load balancing. Much the same as slope scaling which picks the steepest tough shift stochastic slope scaling picks arbitrarily shape the tough shifts by means of successful likelihood. In [29] creator suggested a delicate computing load balancing technique for balancing the load. Creator utilized the neighborhood streamlining technique stochastic slope moving pro portion by approaching occupations to the main systems or virtual systems.

7.7   Generalized Priority Scheme

In [30] the novelist suggested a load balancing scheme recognized as summed up need scheme. In this scheme the assignments are organized by the dimensions of the assignments to such an extent that the assignment by means of most elevated dimension acquires the most elevated need. Likewise the virtual servers are organized by their MIPS esteem to such an extent that the main system through the most noteworthy MIPS esteem acquires most elevated need. Subsequently the load balancing is done in like manner. Additionally creator made a concise correlation with First Come First Serve (FCFS) and Round Robin (RR) scheduling scheme. The creator utilized a test system for the functioning of summed up need scheme in addition to furthermore for different schemes to employment consequently to create an examination among the suggested scheme and alternate schemes.

8   Conclusion

In this paper, an examination identified with various existing task scheduling and load balancing procedures, difficulties, concerns and schemes in a cloud situation have been introduced. A short portrayal of every scheme strategy has been displayed. The real concern in task scheduling is load balancing, reaction period, asset use along with memory stockpiling. Productive scheduling scheme could be accomplished through consolidating distinctive constraints to offered schemes which would enhance their general concert of cloud environment.

References

1. Al Nuaimi, K., Mohamed, N., Al Nuaimi, M., Al-Jaroodi, J.: A survey of load balancing in cloud computing: challenges and schemes. In: IEEE Second Symposium on Network Cloud Computing and Applications, pp. 137–142 (2012)
2. Randles, M., Lamb, D., Taleb-Bendiab, A.: A comparative study into distributed load balancing schemes for cloud computing. In: Proceedings of the IEEE 24th International Conference on Advanced Information Networking and Applications Workshops, Perth, Australia (2010)
3. Rimal, B.P., Choi, E., Lumb, I.: A taxonomy and survey of cloud computing systems. In: Proceedings of the 5th International Joint Conference on INC, IMS and IDC, IEEE (2009)

4. Kaur, M.: Survey of task scheduling schemes in cloud computing. Int. J. Innov. Res. Comput. Commun. Eng. **5**(4), 7244–7249 (2017)

5. Nallakumar, R., Sengottaiyan, N., Nithya, S.: A survey of task scheduling methods in cloud computing. Int. J. Comput. Sci. Eng. **2**(10), 9–13 (2014)

6. Fei, T.: Resource Allocation and Scheduling Models for Cloud Computing, Paris (2011)

7. Emeakaroha, V.C., Brandic, I., Maurer, M., Breskovic, I.: SLA-Aware application deployment and resource allocation in clouds, IEEE (2011)

8. Daniel, D., Lovesum, S.P.J.: A novel approach for scheduling service request in cloud with trust monitor. IEEE (2011)

9. Boloor, K., Chirkova, R., Salo, T., Viniotis, Y.: Heuristic-based request scheduling subject to a percentile response period SLA in a distributed cloud. IEEE (2011)

10. Mehdi, N.A., Mamat, A., Amer, A., Abdul-Mehdi, Z.T.: Minimum completion period for power-aware scheduling in cloud computing. IEEE (2012)

11. Zhang, L.M., Li, K., Zhang, Y.-Q.: Green task scheduling schemes with speeds optimization on heterogeneous cloud servers. IEEE (2011)

12. Lu, X., Gu, Z.: A load-adaptive cloud resource scheduling model based on ant colony scheme. IEEE (2011)

13. Ming, G., Li, H.: An Improved Scheme Based on Max-Min for Cloud Task Scheduling. Yunnam University, China (2011)

14. Hsu, C.-H., Chen, T.-L.: Adaptive scheduling based on quality of service in heterogeneous environments. IEEE (2010)

15. Liu, S., Quan, G., Ren, S.: On-line scheduling of real-period services for cloud computing. IEEE (2010)

16. Yu, J., Buyya, R.: Workflow scheduling schemes for grid computing. Technical report, GRIDS-TR- 2007-10, Grid Computing and Distributed Systems Laboratory, The University of Melbourne, Australia (2007)

17. Malik, B.H., Amir, M., Mazhar, B., Ali, S., Jalil, R., Khalid, J.: Comparison of task scheduling schemes in cloud environment. Int. J. Adv. Comput. Sci. Appl. **9**(5), 384–390 (2018)

18. Randles, M., Lamb, D., Taleb-Bendiab, A.: A comparative study into distributed load balancing schemes for cloud computing. In: 2010 IEEE 24th International Conference on Advanced Information Networking and Applications Workshops (WAINA). IEEE (2010)

19. Hans, A., Kalra, S.: Comparative study of different cloud computing load balancing techniques. In: International Conference on Medical Imaging, m-Health and Emerging Communication Systems (MedCom), pp. 395–397 (2014)

20. Buyya, R., Ranjan, R., Calheiros, R.N.: Inter cloud: utility-oriented federation of cloud computing environments for scaling of application services. In: Proceedings of the 10th International Conference on Schemes and Architectures for Parallel Processing (ICA3PP), Busan, South Korea (2010)

21. Foster, I., Zhao, Y., Raicu, I., Lu, S.: Cloud computing and grid computing 360-degree compared. In: Proceedings of the Grid Computing Environments Workshop, pp. 99–106 (2008)

22. Grosu, D., Chronopoulos, A.T., Leung, M.: Cooperative load balancing in distributed systems. Concurr. Comput. Practice Exp. **20**(16), 1953–1976 (2008)

23. Ranjan, R., Zhao, L., Wu, X., Liu, A., Quiroz, A., Parashar, M.: Peer- to-peer cloud provisioning: service discovery and load-balancing. In: Cloud Computing—Principles, Systems and Applications, pp. 195–217 (2010)

24. Kumar, R., Sahoo, G.: Load balancing using ant colony in cloud computing. Int. J. Inf. Technol. Convergence Serv. (IJITCS) **3**(5) (2013)

25. Padhy, R.P., Prasad Rao, G.P.: Load balancing in cloud computing systems. Thesis (2011)
26. Dasguptaa, K., Mandalb, B., Duttac, P., Mondald, J.K., Dame, S.: A genetic scheme (GA) based load balancing strategy for cloud computing. In: International Conference on Computational Intelligence Modeling Techniques and Applications (cimta) (2013)
27. Lua, Y., Xiea, Q., Kliotb, G., Gellerb, A., Larusb, J.R., Greenbergc, A.: Join-idle-queue: a novel load balancing scheme for dynamically scalable web services. In: IEEE Conference on Foundations of Computer Science
28. Devi, P., Gaba, T.: Implementation of cloud computing by using short job scheduling. Int. J. Adv. Res. Comput. Sci. Softw. Eng. **3**, 178–183 (2013)
29. Mondal, B., Dasgupta, K., Dutta, P.: Load balancing in cloud computing using stochastic hill climbing-a soft computing approach. Procedia Technol. **4**, 783–789 (2012)
30. Agarwal, A., Jain, S.: Efficient optimal scheme of task scheduling in cloud computing environment. Int. J. Comput. Trends Technol. (IJCTT) **9**(7), 344–349 (2014)

Classification of Encryption Algorithms Based on Ciphertext Using Pattern Recognition Techniques

T. Kavitha[1], O. Rajitha[1], K. Thejaswi[1],
and Naresh Babu Muppalaneni[2(✉)]

[1] Sree Vidyanikethan Engineering College (Autonomous), Tirupathi, India
[2] Department of CSE, National Institute of Technology Silchar,
Silchar, Assam, India
nareshmuppalaneni@gmail.com

Abstract. In digital era security in data communication is a challenging task. For secure data communication between the parties, encryption algorithms are being used. Recently many attacks have been reported for breaking ciphertext using various cryptanalytic techniques. Cryptanalysts are trying to break the cipher without knowing the key. Present day scenario an attacker may not know an encryption algorithm being used by communication entities. Identifying the algorithm itself is a challenging task. Once the encryption algorithm has identified, the cryptanalysts can analyzes the weakness of the encryption algorithm and be able to retrieve the plain text without knowledge of key. Here we present various pattern recognition techniques for identifying encryption algorithm which are used between two parties for communication of data using ciphertext. Thus, we consider the block cipher algorithms DES, IDEA, AES and RC operating in Electronic Code Book (ECB) mode. The classification techniques used are Support Vector Machine (SVM), Bagging (Ba), AdaBoostM1, Neural Network, Naïve Bayesian (NB), Instance Based Learning (IBL), Rotation Forest (RoFo) and Decision Trees to identify the right algorithm for the given ciphertext.

Keywords: Encryption · Ciphertext · Cryptanalysts · Pattern recognition ·
Block ciphers

1 Introduction

Due to a fast increase in the amount of public data transferred through the Internet has more attention for information security, of which the kernel is cryptography. The purpose of Cryptography is for storing sensitive information and for secure communication over public networks. Cryptographic algorithm protects data being attacked by attacker. Cryptanalysis is the process of analyzing and breaking of Ciphers. In Cryptanalysis if ciphertext is available two important tasks can be done, Encryption method Identification and identifying which key is used. For identifying encryption method for the given ciphertext various statistical methods and machine learning techniques can be used. The encrypted file contains Statistical methods which use the

© Springer Nature Switzerland AG 2020
A. P. Pandian et al. (Eds.): ICCBI 2018, LNDECT 31, pp. 540–545, 2020.
https://doi.org/10.1007/978-3-030-24643-3_64

alphabet occurrence frequency, in different methods based on machine learning techniques. Pattern classification task depends on the encryption method identification. The classification algorithms attempt to capture the inferred behaviour of every encryption method from all ciphertexts. Support vector machines are used to identify the Block ciphers which is Proposed in [1]. Methods based on machine learning are also used in crypt analysis for Feistel type block ciphers [2]. Genetic algorithm related methods are used to solve cryptographic systems; it involves finding key used for distinguishing the message [3–5]. Usage of machine learning classifiers is a growing interest of data mining [6]. Performance of classifiers is used to find user profiling.

Here we studied a number of pattern recognition techniques to identify the encryption method, and their accuracy is examined. Here ciphertext is considered as a document characterization problem, for this reason ciphertext is considered as a document and represented by document vector. Usually the ciphertext does not contain any meaning full words and symbols. Subsequence of the bit sequence in a ciphertext is represented by symbols, and sequence of these symbols is words. By using this approach, as a block cipher algorithm problem was identified by encryption method from a ciphertext. So 200 text files were considered and are encrypted with ECB (Electronic Code Book) mode. Every file is encrypted with 200 various keys.

In this paper Support Vector Machine (SVM), Naïve Bayesian (NB), Instance Based Learner (IBL), Neural Networks (MLP), Bagging (Ba), Rotation Forest (RoFo) and AdaBoost are considered and compared based Encrypted data's classification accuracy.

Generally Cryptanalysts does not have an idea for which cryptographic algorithm used when capturing ciphertext. So the cryptanalyst has to identify which cryptographic algorithm used and then decode it through brute force attack, mathematical methods, rainbow table attack and dictionary attack etc. Figure 1 shows how an identification system of cryptographic algorithm works with knowledge of only ciphertext. If ciphertext C is input into the identification system we must know the algorithm.

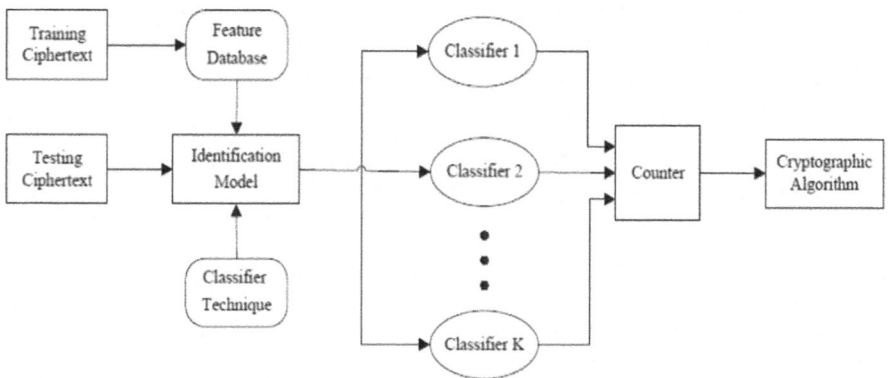

Fig. 1. Model of our identification system Cryptography Algorithm Used

1.1   Various Attacks

Ciphertext Only Attack: here only the encrypted message is available for attacker, but the language is known frequency analysis can be attempted. In this case attacker does not know anything about the contents and must work from ciphertext only.

Known Plain Text Attack: in this both plain text and ciphertext will be available in finding the key. The attacker can guess some parts of the ciphertext. Here the task is to decrypt the rest of the ciphertext blocks using this information. This can be done by determining the key used to encrypt the data.

Chosen Plain Text Attack: This occurs when the attacker gains access to target encryption device, if say it is unattended. The attacker runs various pieces of through the device for encryption. This is compared to plain text to attempt to derive the key.

Chosen Ciphertext Attack: In this the cryptanalyst can choose different ciphertexts to be decrypted and has access to it. This is generally applicable to public key cryptosystems. This involves the attacker selects certain ciphertexts to be decrypted, and then using the results of these decryptions to select subsequent ciphertexts.

2   Methodology

Naive Bayesian: The naive Bayesian classifier is the most straight forward and computationally efficient classifier. It is being used as a best classifier, and it needs small amount training data to estimate parameters. If structure is known then easy to construct and one more advantage is classification process is efficient [7]. This leads to assumption that all features are independent with each other.

$$\text{Similarity (a, b)} = -\sqrt{\sum_{i=1}^{n} f(a_i, b_i)}$$

Support Vector Machines: support vector machines are learning algorithms that analyze the data for classification and regression, it performs linear classification in addition to this it efficiently performs non linear classification. SVM algorithms have been widely applied text and hypertext classification and classification of images and hand written characters are also recognized using SVM's.

Neural Networks: neural network consists of input layer, output layer and hidden layers. Here neuron is taken as a adaptive element with weights, its value modifies based on inputs and outputs the mathematical function is network is tested for random texts with random keys of various lengths [9].

Instance Based Learner: By using specific instances IBL generates classification prediction. Unlike nearest neighbour algorithm, this method normalizes its processes instances, attributes ranges, incrementally and it has a policy for tolerating missing values [10]. The instance based learners applies Euclidean distance function to supply grade matches between given test instance and training instances [11].

Bagging: Bagging is the application of the Bootstrap procedure for machine learning algorithm, typically decision trees. it is a ensemble method that generates individuals by training each classifiers with random redistribution by taking original data with size n, these is known as bootstrap replication of original set. Bootstrap replicate contains 60% of the original set [8, 12].

AdaBoost Algorithms: it is a ensemble algorithm, the idea is create strong classifier from weak classifiers. In this approach each classifier is enough to moderately accurate, this is better than random guessing. In this approach each weak classifier is added in a different level so that it will classify the data. Set of weights is assigned to objects in the data set, so that difficult objects acquire more weight.

Rotation Forest: this algorithm is used for feature extraction and is based on Principal Component Analysis (PCA). Principal component analysis is applied on base classifier by splitting the feature set into k subsets. Here variability information is preserved hence k axis rotation is takes place on base classifier [8].

3   Encryption Method Identification

In this paper we are building a model to predict the block cipher encryption method from the given ciphertext. The algorithms considered are IDEA, DES, AES, and RC2. The block sizes of these algorithms are DES (56-bits), IDEA (64-bits), AES (128, 192, 256 bits) and RC2 (42, 84, 128 bits). 200 text files of fixed size (512 KB) have been used for this study. Bouncy Castel Crypto API library [6] was used to achieve encryption. All these files are encrypted with 200 dissimilar keys to make more difficult to identify. Simulation and Performance Evaluation for identification of cryptographic algorithm mainly focus on some experiments. 200 plaintext and ciphertext pairs were used for this study. For each cryptographic algorithm, we have taken input training ciphertexts as 40 and remaining are divided into 8 groups of 20 ciphertexts for testing. The process of encryption runs in ECB mode. Here for the same cryptographic algorithm the key is fixed. It is clearly observed for final identification of algorithm the size of ciphertext and key plays major role. The following table shows the same key is used for training and testing ciphertext.

Here two different Data Sets are considered; the set P contains 8 classes with different key for different encryption algorithms and 220 input files. The another dataset called Q contains 4 classes for each algorithm and 220 input files. The data are mainly divided into 10 partitions, 9 out of 10 data is used for training and one out of 10 for testing. The process is repeated 12 times; here the overall error rate is reduced. Using the same training set all the classifiers were trained and have been tested for making the classification accuracy.

The simulation results show how accurately classification is done with classifiers. That is Dataset P and Dataset Q. for Dataset P is used for classifiers. In the Table 1, it is observed that RoFo achieves a high accuracy classification (31.90). IBL has the least accuracy classification (11.50) means that only 28 input data out of 220 were classifies correctly.

Table 1. Classification of accuracy of the classifiers

Classifiers	Accuracy	
	Dataset P 8 Classes	Dataset Q 4 Classes
SVM	17.08	32.08
BA	25.83	48.7
AdaBoostM1	20.42	42.23
NN	24.25	45.87
NB	28.33	44.17
IBL	11.50	30.20
RoFo	**31.90**	**54.32**
Decision trees	27.25	43.06

The simulation results of Dataset –Q shows RoFo again performs well than all other classifiers with the accuracy of (54.32) means 130 input data of 220 were classified. Once again the IBL has least classification accuracy (30.20) means only 73 input data are classified.

4   Conclusion

The goal is to determine the best classification algorithm with high accuracy for these four Block ciphers IDEA, DES, AES and RC2 are considered. In this paper eight classifiers (NN, BA, AdaBoostM1 and SVM, NB, IBL, RoFo, and Decision trees) are used for identifying encryption method and its accuracy are evaluated. These simulations are performed using WEKA tool. The different keys are used for different text data. The simulation results shows Rotation Forest (RoFo) classifier has highest classification accuracy for identifying encryption method for ciphered data, where as IBL has least Performance. The accuracy is only 54%, hence new techniques has to develop for identification of encryption method.

References

1. Girish, C.: Classification of modern ciphers. M. Tech thesis, Department of Computer Science and Engineering, Indian Institute of Technology, Kanpur (2002)
2. Brahmaji, M.: Classification of RSA and idea ciphers. M. Tech thesis, Department of Computer Science and Engineering, Indian Institute of Technology, Kanpur (2003)
3. Saxena, G.: Classification of ciphers using machine learning. M. Tech thesis, Department of Computer Science and Engineering, Indian Institute of Technology, Kanpur (2008)
4. Dileep, A.D., Sekhar, C.C.: Identification of block ciphers using support vector machines. In: Proceedings of the International Joint Conference on Neural Networks, Vancouver, BC, Canada, pp. 2696–2701 (2006)

5. Nagireddy, S.: A pattern recognition approach to block cipher identification. M. Tech thesis, Department of Computer Science and Engineering, Indian Institute of Technology, Madras (2008)
6. Mishra, S., Bhattacharjya, A.: Pattern analysis of ciphertext: a combined approach. In: Proceedings of the International Conference on Recent Trends in Information Technology, pp. 393–398 (2013)
7. Chopra, J., Satav, S.: Impact of encryption techniques on classification algorithm for privacy preservation of data. Int. J. Innov. Res. Sci. Eng. Technol. 2(10), 5398–5402 (2013)
8. Cufoglu, A., Lohi, M., Madani, K.: Classification accuracy performance of naïve Bayesian (NB), Bayesian networks (BN), lazy learning of Bayesian rules (LBR) and instance-based learner (IB1)-comparative study. In: 2008 International Conference on Computer Engineering and Systems, pp. 210–215 (2008)
9. Kuncheva, L.I., Rodríguez, J.J.: Classifier ensembles for fMRI data analysis: an experiment. Magn. Reson. Imaging 28, 583–593 (2010)
10. Maeda, Y., Wakamura, M.: Simultaneous perturbation learning rule for recurrent neural networks and its FPGA implementation. IEEE Trans. Neural Netw. Publ. IEEE Neural Netw. Counc. 16, 1664–1672 (2005)
11. Muchoney, D., Williamson, J.: A Gaussian adaptive resonance theory neural network classification algorithm applied to supervised land cover mapping using multitemporal vegetation index data. Network 39, 1969–1977 (2001)
12. Harvey, I.: Cipher Hunting: How to Find Cryptographic Algorithms in Large Binaries. N cipher Corporation Ltd., Cambridge (2001)

Smart Garbage Bin

G. Akshay[1], V. Shivakarthika[1], and M. Dakshayini[2(✉)]

[1] C.N.E, B M S C E, Bangalore, India
{akshayg.scn17,shivakarthika.scn17}@bmsce.ac.in
[2] Department of ISE, B M C E, Bangalore, India
dakshayini.ise@bmsce.ac.in

Abstract. Hospitals generate a lot of thrash and disposal of such waste is one of the important challenge faced by Hospitals. The 80% of hospital waste is harmless trash, 20% of it is harmful and needs to be disposed properly. Waste generated by hospitals contain harmful micro living organisms, which could infect not only patients, but also the staff and the public. To overcome this problem in this paper, an IoT solution system Smart Garbage bin has been developed. This system continuously monitors the waste level in the dustbin, when the level of the waste reaches 90% of the bin, the system sends an alert message to the Hospital Housekeeping supervisor along with the location information. This system will help to locate the filled bin, so that the filled bin can be emptied as early as possible. Implementation results have proved its efficiency.

Keywords: Waste management · Smart bin · Garbage disposal · IoT and networking

1 Introduction

Garbage collection is one of the biggest problem faced by most of the Hospitals in India. The Hospital waste is very sensitive as it may contain Harmful viruses, which may cause harmful diseases to others also. It is important to empty the filled bins as early as possible, but it is not being monitored efficiently in most of the places. In some of the places the bin is emptied every morning irrespective of whether it is filled or not, but in other places it is not monitored at all. The proposed system helps the cleaning system to overcome the irregularities in the collection of garbage from the filled bins.

This work shows the evolution of a smart garbage bin [1] which measures the level of garbage in the garbage bin continuously and then sends alert to the housekeeping supervisor via SMS. The proposed system consists of Ultrasonic sensor [5] to measure the level of garbage, an Arduino Uno [3] to control the system operation, and the GSM module [2] to send the send the warning messages to the housekeeping supervisor with the location information via SMS when the garbage bin is 90% full, so that the garbage bin can be emptied as early as possible.

There are some complexities involved while designing this smart garbage bin. It was important to consider the cost per unit while designing the system [2]. The system should be scalable so that new innovations can easily be integrated into it. It should

© Springer Nature Switzerland AG 2020
A. P. Pandian et al. (Eds.): ICCBI 2018, LNDECT 31, pp. 546–551, 2020.
https://doi.org/10.1007/978-3-030-24643-3_65

provide a user-friendly way to easily setup, monitor and control the setup. The system should also offer some analytical services so that if there is any problem with the system, it can be tracked down. Continues monitoring could also have been done using cloud to store the obtained sensor data, But our aim was to just inform the hospital authority with SMS with the location information when the bin is filled up. In order to avoid the usage of IP address for checking the status via Wi-fi module, GSM Module is used instead of Wi-fi module to check the status. In this paper, we present a smart Garbage Bin to continuously monitor the waste level in the dustbin and send an alert message to the Hospital Housekeeping supervisor along with the location information when the level of the waste reaches 90% of the bin.

This proposed system helps the hospitals to locate the filled bin at various parts of the hospital, so that the filled bin can be emptied as early as possible as it contain possibly harmful micro-living-organisms which could infect not only patients, but also staff and the public if the Bins are not emptied immediately.

Remaining portion of this paper has been arranged as follows. System Model of the proposed work and the algorithm are presented in Sect. 2. Section 3 discusses the detailed implementation of the work. Results obtained are demonstrated in Sect. 4. Section 5 concludes the work.

2 System Model and Design

The main purpose of this proposed work is for helping the hospital management system to provide and maintain the cleanliness at various parts of big hospitals. This work locates the filled dust bin with the help of ultrasonic sensor at different locations, so that the filled bin can be emptied as early as possible as it contain potentially harmful microorganisms.

2.1 Model Design

System model considered for managing and monitoring the waste in the dustbins is shown in Fig. 1. It consists of ultrasonic sensors to measure the garbage level, the information will be sent to Arduino Uno microcontroller, In case when the bins are 90% filled, warning note would be generated and directed to the Housekeeping supervisor through SMS, with the help of GSM module.

The Ultrasonic sensor helps for continuously monitoring the level of the garbage in each Bin. This information is then sent to Arduino Uno for processing. It will compare the garbage level with the threshold level, which was kept by 90% of the Bin. When the level of the garbage in the Bin is more than 90% of the bin, then the notice message would be sent to the Housekeeping supervisor.

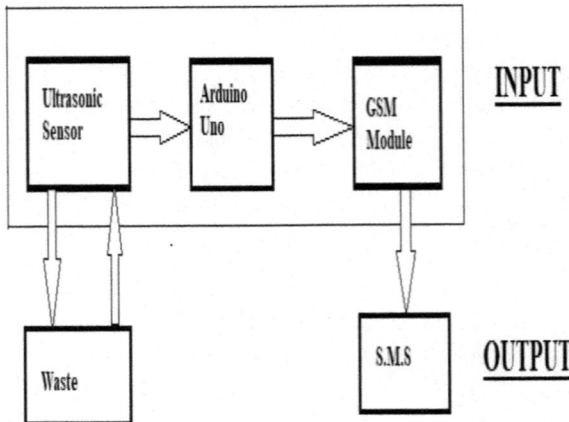

Fig. 1. Garbage waste Monitoring and Management System Model

2.2 Algorithm

[Nomenclature:
    USS – Ultra Sonic Sensor
    Wlevel – Filled level of waste
    Amob – Mobile number of Housekeeping supervisor]

USS(SGB)   Continuously monitors the level of the Wlevel
    if ($W_{level} >= 90\%$)
    {
    W_{level} is High
    Send "Dustbin is full with Location " message to
    A_{mob} indicating W_{level} is High
// Sends the message to Housekeeping supervisor mobile number with location information.
    }
Waste from the respective Dustbin will be cleared

3   Implementation

In this work, as there was chances of this sensor itself getting destroyed while putting the waste into the bin, positioning of the ultrasonic sensor was very challenging and important. The false positioning would also have led to the alert message while putting the waste. The problem could be that it is not feasible to go and collect if only one out of many bins are filled. Only bottom half of the bin should be removed and replaced by empty bin. The cap which contains sensors should be handled carefully. Hence, to make the functioning of the system more effective, lid of the Bin is fitted with an Ultrasonic sensor to check the filled level of the bin as shown in the Fig. 2. The experimental setup shown in Figs. 3 and 4 contains Arduino UNO R3 board, SIM 800A GSM Module – Which is used to send SMS when the bin is filled to 90%, HC-SR04 Ultra-sonic Sensor – Which is used for monitoring level of the waste in the bins.

Fig. 2. Lid with USS

Fig. 3. Experimental setup

Fig. 4. Top view of Experimental setup

4   Results

After completing the connection setup, in order to test the system, start filling the dustbin to which ultrasonic sensor is attached. This sensor continuously monitors the waste level in the bin, when the dustbin is 90% filled up the system sends the message using GSM module to the authority with the location information as shown in Fig. 5 indicating that, the dustbin is full.

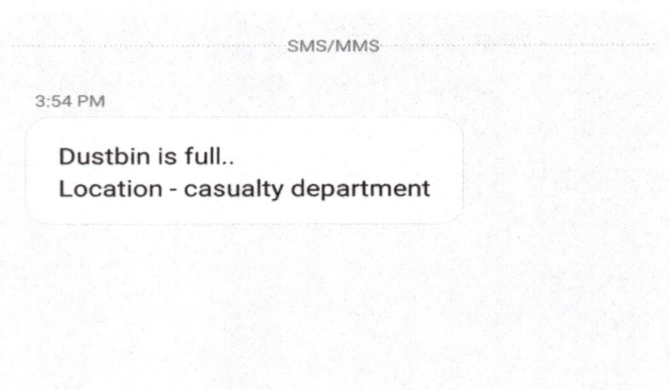

Fig. 5.   Message sent from the proposed system

5   Conclusion

The proposed system is implemented successfully to monitor the waste level in the dustbins of the hospitals and intimate the concerned authority to dispose the same. When the level of the waste reaches 90% of the bin, the system has productively sent an alert message to the Hospital Housekeeping supervisor along with the location information. Thus this system was successful in assisting the hospital management to locate the filled bin and emptied the same as early as possible. This avoids the dangerous wastes of the hospital which contain harmful micro living organisms that can infect not only patients, but also the staff and public if the Bins are not emptied immediately.

6   Future Scope

The proposed system works fine whenever there is message pack as well as the network in the SIM so if the GSM fails the whole system fails so for the backup we can place the WiFi module so that if one fails the other could function and even we can use it for other purpose as we have proposed for hospital waste management we can further enhance it and use it for apartments, public places and even in companies.

References

1. Chaware, S.M., Dighe, S., Joshi, A., Bajare, N., Korke, R.: Smart Garbage Monitoring System using Internet of Things (IOT)
2. Satyamanikanta, S.D., Narayanan, M.: Smart Garbage Monitoring System Using Sensors with RFID Over Internet of Things
3. Ravichandran, S.: Intelligent Garbage Monitoring System Using Internet of Things
4. Dev, A., Jasrotia, M., Nadaf, M., Shah, R.: IoT Based Smart Garbage Detection System
5. Yusof, N.M., Jidin, A.Z., Rahim, M.I.: Smart Garbage Monitoring System for Waste Management

AI BOT: An Intelligent Personal Assistant

A. Shanthini$^{(\boxtimes)}$, Chirag Prasad Rao, and G. Vadivu

SRM Institute of Science and Technology, Kattankulathur, Chennai, India
shanthini.a@ktr.srmuniv.ac.in

Abstract. Intelligent Personal Assistants (IPA) such as Google's Assistant, Apple's Siri and Microsoft's Cortana excels in their natural language user interface. Still there is a need for single bundled IPA with multi-functionalities is increasing day by day. This paper presents the design and development of an artificial Intelligent personal assistant (AIPA), an open end-to-end web service application which receives queries in the form of voice and text and responds with natural language. The paper emphasizes on the implementation of different ways of interacting with a personal assistant and incorporating various functionalities which aren't found in any of the existing virtual assistants with the help of several AI algorithms and open-source tools. Understanding context is important for any AI and increase in the context leads to the effective handling of open-ended requests. Unlike commercial products, this AI bot uses open ended requests than specific tasks to provide multifunctional to artificial brain.

Keywords: Artificial Intelligence · Natural Language
Intelligent Personal Assistant

1 Introduction

Artificial Intelligence (AI) is the simulation of human intelligence exhibited by machines. The term "artificial intelligence" can be understood as a machine imitating "cognitive" functions that associate with the human minds such as "learning" and "problem-solving". AI investigation is partitioned into subfields that emphasis on particular issues or on particular methodologies or on the utilization of a specific device or towards fulfilling specific applications.

Research in AI has discovered how to develop systems with human level intelligence by evolving the extensive academic problems in AI. There are various branches in AI such as logical AI, search, pattern recognition, heuristics, learning from experience, etc. This makes AI have a wide field of applications in various areas like game playing, speech recognition, understanding natural language, computer vision, expert systems, heuristic classification, etc. [1, 2].

Google's Google Now [3], Apple's Siri [4] and Microsoft's Cortana [5] are representing a class of web-service applications known as Intelligent Personal Assistants (IPAs). An IPA is an application which uses the inputs provided by the user such as the user's voice, vision (images), and contextual information.

It uses the received information to supply help with respondent queries in linguistic communication, creating recommendations, and acting actions [17]. These IPAs square

© Springer Nature Switzerland AG 2020
A. P. Pandian et al. (Eds.): ICCBI 2018, LNDECT 31, pp. 552–559, 2020.
https://doi.org/10.1007/978-3-030-24643-3_66

measure the range of how materializing integrated with quickest growing net services. They have recently deployed on well-known platforms reminiscent of iOS [8], Android, and Windows Phone, which is available in mobile devices round the world. In addition, the utilization of these IPAs has been gradually increased with a recent contribution in wearable technologies such as smart watches [6] and smart glasses [7]. Recent projection predicts that the wearables market will be at 485 million annual device shipments by the end of 2018 [8]. This growth in market share, conjoin with the assurance that the design of wearable's is heavily reliant on voice and text inputs, moreover indicate the rapid growth in user demands for IPA services is on the purview.

An AI bot is a machine conversation system which interacts with humans through natural conversational language and performs tasks according to their request. It can perform tasks such as helping the user control his/her home, including lights, appliances, music, that learns the tastes and patterns and provide the best possible desired results intelligently. With the help of several artificial intelligent algorithms such as natural language processing, speech recognition, and reinforcement learning, this personal assistant will help in catering user's queries by providing a human-like conversation and manage everything connected with the user's system.

This Intelligent Personal Assistant is similar to that of Google's Assistant that is bundled with speech recognition, text recognition, and natural language processing algorithms for its functioning. Users can enter queries by talking to the assistant or by text.

2   Literature Survey

There are two pathways of changing complexity nature through this current IPA's back-end [10] in the view of input query. A voice command principally exercises speech recognition on the server-side to execute a command on the mobile/computer device [11]. A voice question boots leverages and complicated Natural language processing (NLP) to deliver a reaction to the consumer. A question such as "What is the temperature outside?" is answered by analysing the sentence and extract the intent of the user from the score with the help of WIT (Wireless Information Terminal). Then it interprets the information using Artificial Intelligence Modelling Language (AIML configured in the WIT) to make a conversation with the user. WIT is a Machine-Learning-Service tool used in IPA implementation for training, understanding and matching the intent of the user queries.

This AI will specifically work towards detecting the different users speaking to it and provide personalized suggestions. It will also use contextual [14] based responses to perform more open-ended tasks, provide more human-like interactivity and make the interactions more lively and conversational. IPA is programed in such a way that it can be accessed by the user from anywhere. User can text anything to AI bot, and it will instantly be relayed to AI server and be processed. Interacting through text feels more appropriate as its less disturbing to people around the user, and also that voice can be more disruptive and text gives you more control of when user want to look at it [12, 13]. Facebook's open-sourced messenger framework is used as a platform for interacting through texts with the IPA.

3 Proposed Methodology

The entire proposal is divided mainly into two separate interface systems. The Speech system is mainly driven by Google's Speech recognition package, using WIT.AI for Natural Language and runs separately from the text interface. The input is the user's speech through the microphone device and the voice formatted input is understood with the help of NLU (Natural Language Understanding), gets processed and returns to the user in form of speech. The corresponding text is also shown on the terminal screen upon execution. The chatbot, on the other hand, uses Facebook's messenger as the underlying framework for connecting and executing the text queries of the users. Users can connect and start conversing over text either by accessing the personalized messenger link generated for the app or by scanning the uniquely generated QR code on their messenger app.

3.1 Voice Interface

There are primarily three key aspects involved when a query is given by the user.

The speech input is first made in the form of voice. The speech input needs to be converted into text [15], where the Googles Speech Recognition and Speech-to-text help in processing it. Then, the knowledge from this text needs to be understood by an NLU system, thereby understanding the intent and then retrieving the answer to the query. This query, which is in the form of text is converted into Speech by (Natural Language Generation) NLG and is thereby returned to the user.

Google's Speech recognition and gTTS (Google Text-to-Speech) packages have been used here for recognizing and converting the speech into text. For understanding and training the intent with the entities, an open source tool developed by Facebook WIT.AI is used. Wit.ai helps to understand the intent of the keywords and match them to their respective entities. This way the entities may be trained and it is auto-generated. Server-side token may be used for accessing it in the program code. WIT identifies the intent in the query based on its trained corpus and extracts a score for the respective entity. Stories and API calls can be configured in WIT in the form of AIML and can also provide branching to user conversations.

Natural Language Processing (NLP) and Natural Language Understanding (NLU) are the two concepts used here for training and understanding the intents in a user's query. When training the bot in WIT, a merge function can be executed to extract the entity and save it to the context object. This way a sentiment can be responded to and acknowledged.

3.2 Conversational Chatbot Using Wit.ai

The messenger bot interface has been developed over Facebook's Messenger framework. Heroku, a cloud platform is used for deploying the code in order to handle the web hooks. Again, there are two main key aspects in the messaging bot: Receiving and Sending messages.

1. **Receiving Texts:** Incoming messages are handled by loading in the JSON POST data and transferred to web hook from Facebook whenever a new messaging event is triggered.
2. **Sending Texts:** In order to send a text message, the basic required things are the recipient's Facebook ID and text message content. A function is written to automatically send the Facebook API with the above information.

The NLU is handled by the WIT works primarily with the help of two main parameters: Intent and Entities. WIT takes in the user's voice or text inputs and returns intents and entities.

Intents: An intent can be simply understood by what the user intends to do. Something like change Temperature or getNews.

Entities: Entities are the variables that contain the details of the user's task. WIT comes with the built-in entity types like location, number, etc. user can also create their own entities.

Depending on the intent and entities that are received from the user, the application takes actions or ask more questions to fill in any missing content that is needed and fulfil the user's request.

The above figure explains the NLP dialog structure as managed by WIT when executing user's message. The way it understands the intent and find the specified entity and provide the conversation story as defined (Fig. 1).

The above figure explains the overview of the Conversation flow upon the user's input in WIT (Fig. 2).

Figure 3 shows the NLU flow of the user's message when passed into WIT for processing and understanding.

3.3 Bot Architecture

All the appliances/devices are connected to the AI central System, where the entire processing is managed with the help of Speech Recognition and NLP Algorithms. This entire System is accessed either by Voice or chatbot/Messenger interface as shown in the above Fig. 4.

4 Output

Figure 5 shows the implementation of how Speech (the voice format) is converted to text and being displayed on the terminal screen along with the speech output. It can be seen sending the user input to the WIT for processing and understanding the intents and then retrieving the response after extracting the respective entity back to the user. It's programmed to always be on "Listening" so as to continue the conversation flow with the current context.

Figure 6 shows the working of the chatbot on Facebook's Messenger.

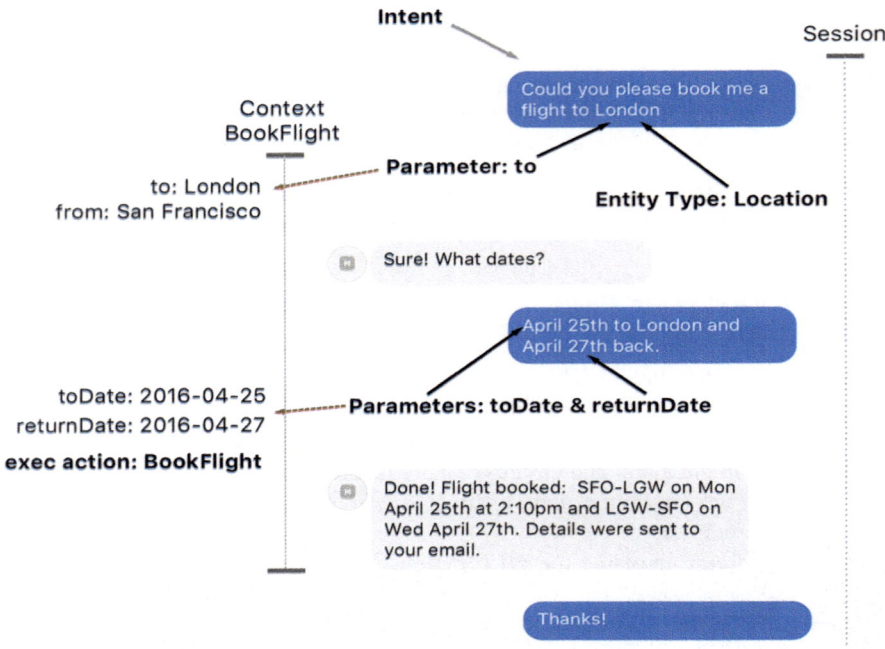

Fig. 1. NLP dialog structure on chatbot

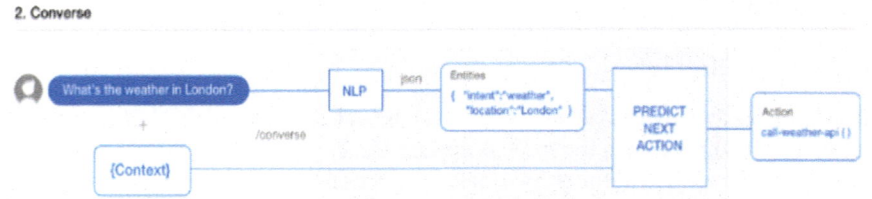

Fig. 2. Conversation Overview

Fig. 3. Overview of Message Understanding by WIT

SYSTEM ARCHITECTURE

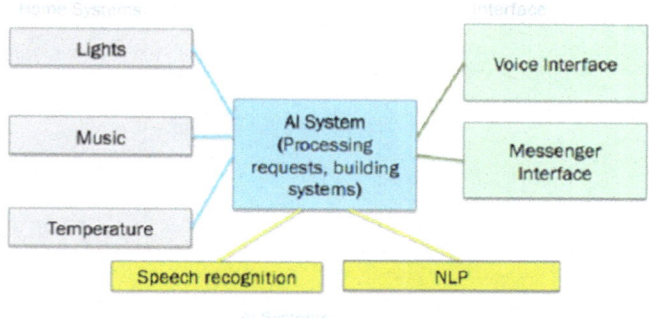

Fig. 4. AI Bot System Architecture

```
●●●                    📁 chatbots — -bash — 80×24
Say something!
You said: what is the weather in chennai'
INFO:requests.packages.urllib3.connectionpool:Starting new HTTPS connection (1):
 api.wit.ai
DEBUG:requests.packages.urllib3.connectionpool:"POST /converse?q=what+is+the+wea
ther+in+chennai%27&session_id=4d6bd95f-2197-4725-875d-02f46606e289 HTTP/1.1" 200
 307
{u'intent': [{u'confidence': 0.9994639558718263, u'value': u'getWeather'}], u'lo
cation': [{u'suggested': True, u'confidence': 0.971465938236801, u'type': u'valu
e', u'value': u'chennai'}]}
INFO:requests.packages.urllib3.connectionpool:Starting new HTTPS connection (1):
 api.wit.ai
DEBUG:requests.packages.urllib3.connectionpool:"POST /converse?session_id=4d6bd9
5f-2197-4725-875d-02f46606e289 HTTP/1.1" 200 100
A moment ! Let me check the weather at chennai
INFO:requests.packages.urllib3.connectionpool:Starting new HTTPS connection (1):
 api.wit.ai
DEBUG:requests.packages.urllib3.connectionpool:"POST /converse?session_id=4d6bd9
5f-2197-4725-875d-02f46606e289 HTTP/1.1" 200 70
INFO:root:ID : 1264527
INFO:root:Clear sky with a temperature of 96.8 F
INFO:requests.packages.urllib3.connectionpool:Starting new HTTPS connection (1):
 api.wit.ai
DEBUG:requests.packages.urllib3.connectionpool:"POST /converse?session_id=4d6bd9
```

Fig. 5.   Speech Recognition

5   Conclusion

This work introduces an Intelligent Personal Assistant, an open-ended application leveraging algorithms such as speech recognition, natural language processing, and text recognition. This IPA is bundled with both voice and text interfaces for interacting and conversing with a user and providing answers to their queries intelligently.

Fig. 6. The Messenger Chatbot

6 Future Work

This work can be extended by implementing Deep Learning techniques to get improvised results. The results can be further analysed and concluded to depict the available techniques which gives us the best results. The Chatbot can be built to provide a well-personalized learning environment that can help the students to have interaction. The Chatbot can analyze the level of the student's understanding of the research or concepts and can promote the exchange of their ideas. Chatbot can also provide more information regarding the new evolving areas of the research which will help the students in their academics.

References

1. Coehoorn, R.M., Jennings, N.R.: Learning about the opponent's preferences to make effective multi-issue negotiation trade-os. In: Proceedings of the 6th International Conference on Electronic Commerce, pp. 59–68. ACM (2004)
2. Chao, L., Tao, J., Yang, M., Li, Y., Wen, Z.: Audio visual emotion recognition with temporal alignment and perception attention. arXiv preprint arXiv:1603.08321 (2016)
3. ABI Research: Wearable Computing Devices, Like Apple iWatch, Will Exceed 485 Million Annual Shipments by 2018. https://www.abiresearch.com/press/wearablecomputing-devices-like-apples-iwatch-will (2013)
4. Hearst, M.A.: 'Natural' search user interfaces. Commun. ACM **54**(11), 60–67 (2011)
5. Siegler, M.G.: Apple's Massive New Data Center Set To Host Nuance Tech. http://techcrunch.com/2011/05/09/applenuance-data-center-deal/
6. Huggins-Daines, D., Kumar, M., Chan, A., Black, A.W., Ravishankar, M., Rudnicky, A.I.: Pocketsphinx: a free, real-time continuous speech recognition system for hand-held devices. In: 2006 IEEE International Conference on Acoustics, Speech and Signal Processing, ICASSP 2006 Proceedings, vol. 1, p. I. IEEE (2006)

7. Povey, D., Ghoshal, A., Boulianne, G., Burget, L., Glembek, O., Goel, N., Hannemann, M., Motlicek, P., Qian, Y., Schwarz, P., Silovsky, J., Stemmer, G., Vesely, K.: The Kaldi speech recognition toolkit. In: IEEE 2011 Workshop on Automatic Speech Recognition and Understanding. IEEE Signal Processing Society (2011)
8. Rybach, D., Hahn, S., Lehnen, P., Nolden, D., Sundermeyer, M., Tuske, Z., Wiesler, S., Schluter, R., Ney, H.: RASR - the RWTH Aachen University Open Source Speech Recognition Toolkit (2011)
9. Seide, F., Li, G., Yu, D.: Conversational speech transcription using context-dependent deep neural networks. In: Interspeech, pp. 437–440 (2011)
10. Ferrucci, D., Brown, E., Chu-Carroll, J., Fan, J., Gondek, D., Kalyanpur, A.A., Adam Lally, J., Murdock, W., Nyberg, E., Prager, J., Schlaefer, N., Welty, C.: Building Watson: an overview of the DeepQA project. AI Mag. 31(3), 59–79 (2010)
11. Bay, H., Tuytelaars, T., Gool, L.V.: Surf: speeded up robust features. In: Computer Vision–ECCV 2006, pp. 404–417. Springer (2006)
12. Hinton, G., Deng, L., Dong, Yu., Dahl, G., Mohamed, A.R., Jaitly, N., Senior, A., Vanhoucke, V., Nguyen, P., Sainath, T., Kingsbury, B.: Deep neural networks for acoustic modeling in speech recognition. Signal Process. Mag. 29, 82–97 (2012)
13. Tckstrm, O., Das, D., Petrov, S., McDonald, R., Nivre, J.: Token and type constraints for cross-lingual part-of-speech tagging. Trans. Assoc. Comput. Linguist. 1, 1–12 (2013)
14. David Forney Jr., G.: The Viterbi algorithm. Proc. IEEE 61(3), 268–278 (1973)
15. Dahl, G.E., Yu, D., Deng, L., Acero, A.: Context-dependent pre-trained deep neural networks for large-vocabulary speech recognition. IEEE Trans. Audio Speech Lang. Process. 20(1), 30–42 (2012)

A Comprehensive Study on Data Management Solutions to Internet of Things (IoT) Based Applications

K. Revathi$^{(\boxtimes)}$ and A. Samydurai

Department of Computer Science and Engineering,
Valliammai Engineering College, Chennai, India
neyadharshini@gmail.com, asamydurai@gmail.com

Abstract. Internet of Things (IoT) brings a major transformation as physical or manual operations into an automatic ones, in terms of its wide variety of applications covering almost all domains from home automation, health monitoring, logistics, structural, environmental monitoring and so on. To facilitate the applications, the data extracted from various sources like sensor, positional and radio frequency identification (RFID) tags have to be maintained and managed properly. Data management in IoT applications is one of the open issues to be determined yet. This paper provides a detailed survey on solutions suggested to handle the data management in IoT applications and also proposes a handy design by exploiting the features of Thingspeak for a remote health monitoring application.

Keywords: Data management · IoT · Sensor · RFID ·
Remote health monitoring and Thingspeak

1 Introduction

Internet of Things (IoT) allows composition of various objects/devices through a common platform to enable efficient communication without much human intervention [13, 14]. IoT is described through two main words: Internet and Things; Internet points towards a combination of network; and Things are generic objects includes wearables such as watches, automobiles, appliances, industrial equipment, animal and even person. The basic idea of IoT is implementing a "smart" system which enables an application from any place, by anyone, through communicate with anything at any time. Although the term IoT was coined by Kevin Ashton in his article on Radio Frequency Identification (RFID) by the year 1999, in recent years alone it captured the attention of all industries and researchers those who like to carry changes in upcoming society, business model and the way of life. The ultimate aim of IoT is to create a better

K. Revathi—Research scholar of Anna University, Chennai and acts as a Junior Research Fellow in Valliammai Engineering College, Chennai under the guidance of Dr. Samydurai A, Associate Professor/CSE in the DST-SERB project numbered EEQ/2017/000485.

© Springer Nature Switzerland AG 2020
A. P. Pandian et al. (Eds.): ICCBI 2018, LNDECT 31, pp. 560–570, 2020.
https://doi.org/10.1007/978-3-030-24643-3_67

world for human beings, by making the objects around us to understand our desire and act according to that without any explicit instructions.

As IoT is viewed as a convergence of a variety of computing and connectivity trends that have been evolving for many decades, it is yet to address so many open challenges such as scalability, robustness, interoperability, creating knowledge, security and privacy concerns. The objects adding up to internet is increasing every day, and it is expected to connect 50 million devices to internet in near future for building any IoT application [2]. A massive amount of heterogeneous objects need to be interconnected for its effective communication by means of exchanging data over internet and thus leads to the proper management of such voluminous data coming from different sources and formats. The term management comprises various processes on data ranging from its collection, preprocessing, unification, manipulation, retention of useful information and also availing security. Data management is considered as one of prime challenge to be focused for attaining better solutions and that will help the IoT development community.

A solution can be made possible, by looking for a way of integrating big data storage and analytics with simple cryptographic algorithms that do agree and act with energy consumption of devices connected together.

The database research group recently renewed their interest in devising new models other than traditional data models in current practice, to facilitate a database for IoT (DoT) which need to address the various aspects such as support for structural, non-structural and semi structural data emitted by the sensing devices, relaxation of the Atomicity, Consistency, Isolation, and Durability (ACID) properties to trade-off consistency and availability, enabling remote storage and analysis by integrating energy efficiency algorithms and security as preliminary design primitives.

The paper is organized as follows: Sect. 2 discusses about various related work carried out for data management in IoT. Section 3 discloses existing efforts as solutions devoted to DoT by considering its characteristics, lifecycle, design requirements and up-to date data models in current scenario. Section 4 reveals the benefit of Thingspeak, an application programming interface (API) for IoT using remote patient monitoring (RPM) as a case study. Section 5 concludes the work through detailed summary of existing work and also addresses the scope for an improvement.

2   Related Work

In general, the IoT applications fall into two categories: closed and wide-open. The things get participating in the network is predicted, then it is known as closed and if those are unpredictable then such kind of applications are called as wide-open ones [6, 7]. The number and type of sources producing data, limitations in sources, structure of the produced data, and the processing requirement (on-demand and offline) are considered as the distinctive design primitives and thus imposes major challenges in the design of effective DoT.

The generation, flow and processing of data aligned with layered architecture of an IoT defined and captured as a life cycle of IoT data [8]. On-demand requests resolved by the device layer itself, whereas voluminous online, offline requests and long-term analysis are handled by cloud server. Data management for device layer has to coordinate with various devices like sensors, actuators, RFID and positional tags that serve in identification, production, collection of data and producing results of online requests via short queries. In device layer, management operations highly dependent on the effective communication carried out among the devices. The data management in server has to deal with massive volume of historical data to make a decision about the devised application.

By taking design primitives and life cycle of data as former consideration, researchers attempted to set comprehensive framework for IoT data management. Some focused on data management in server and derived solutions by devising a distinct data processing layer and use the layer to "cache" fields based on most frequently accessed client's database queries in it [10]. Few focused on data management among sensors in device layer and drawn solutions through query optimization by reducing communication overhead and storage mechanism along with energy efficient techniques to meet the storage and computation constraints in the devices [11].

There exist only two contemporary solutions as database model named, relational database management system (RDBMS) and non-relational (NoSQL) databases. Each addresses only partial aspects of DoT requirements and no one be called as perfect model for DoT [1]. Collaboration models combining the features of RDBMS and NoSQL were also brought to impose the basic requirements of DoT [5]. The classes of NoSQL databases were keenly analyzed for the suitability of DoT [12].

3   Design Challenges with Contemporary Solutions

3.1   Characteristics of IoT Data

IoT applications acts upon a raw data or evaluated information after processing which are retrieved from the sensors. The transformation of sensory input data to an output data predicting or controlling the application can represented as mathematical model through Moore machine [4]. Moore machines are considered as finite state machines with output value which is dependent only on its present state. The mathematical model devoted to represent the sensory data of IoT application is formulated as Table 1 and finite states here are the steps in data processing such as unique identification of data sources, generation clear unified format for data, identification of communication network with enough speed, identification of data processing capacity of server, identification of storage capacity of server, Aggregation of the data, and inference generation and decisions as per application.

The development of database management framework must align with the characteristics of IoT data listed below [2, 3].

1. Volume: Ample sources with the nature of anytime, anywhere, by anyone and anything produces tremendous amount of data
2. Velocity: Data is generated continuously and it is to be continuously
3. Variety: Heterogeneous sources tend to produce different type data as structured, unstructured and semi-structured

4. Veracity: The presence of noise and abnormalities in data as effect of communication failure and lack unification format
5. Validity: The correctness or accuracy of data
6. Volatility: The period about how long the data to be stored

These characteristics are to be met, in the framework proposed for DoT.

Table 1. Mathematical model for IoT data

Representation: $(Q, q0, \Sigma, O, \delta, \lambda)$ where:
- Q is finite set of states
- $q0$ is the initial state
- Σ is the input alphabet is the output alphabet → Sensory Data SD={s1,s2,s3,...sn}
- δ is transition function which maps $Q \times \Sigma \to Q$
- λ is the output function which maps $Q \to O$
- O is the output alphabet → {# - Resources OK, $ - Unified information generated, @ - evaluated information OK, & - Correctness of data}

3.2 Lifecycle of IoT Data

The proposed framework has to be designed, by extending its support for handling both real time and offline data processing found in IoT data life cycle given in Fig. 1.

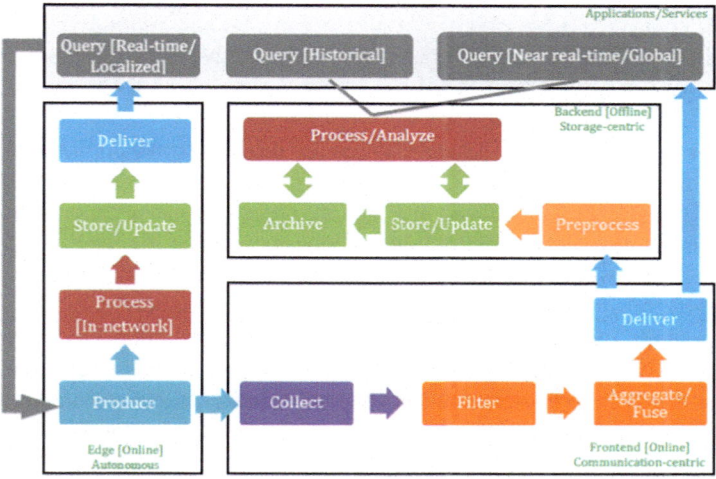

Fig. 1. Life cycle of IoT data and its management operations (Source: Sensors 2013, p. 15585)

In short, online and immediate data requests are handled either by sensor node itself through the deployment of energy harvesting and scavenging techniques or by gateway. The offline and historical data requests are only be solved by cloud server due to its features like availability and capacity to store and analyze the same.

3.3   Requirements of IoT Data

The success of IoT application is predominantly relies on the data created from the sources and processed with qualifying attributes [12]. The qualifying attributes in turn are called as requirements of DoT and are listed below.

- Availability - concerns about the percentage of time a system is operating correctly
- Consistency - associated with transaction specifies the overall integrity and used to enforced via constraints
- Durability - specifies the period of time the data to be valid and committed as an effect of the transaction
- Maintainability - specifies how easily it is upgraded with a change imposes by the environment
- Performance (Read & Write) - specifies the optimization factor for read & write operation
- Recovery Time - reveals the time taken to recover from failures
- Reliability - states the probability of operating without failures for a given period of time
- Robustness - deals with how far the database has the ability to cope with errors while in execution
- Scalability - concerns about ability to support new devices
- Stabilization Time - reveals the time taken for the system to be stabilized after the recovery of failures

By annotating all these attributes carefully in the design, one can bring promising data model for IoT Data.

3.4   Present-Day Solutions

In current scenario, only two categorization of models namely relational and non-relational are available as a solution to any operations and issues associated with the database. These are declared as incompatible ones as it addresses only partial requirements of DoT by thoroughly investigating its features [9]. The features of these contemporary models are captured as Table 2 with the illustration of its advantages and disadvantages in handling IoT data.

Among these two categories of model, non-relational data model grasped the attention of researchers as it provides extended support in scalability and analytics. These are the prominent factors to be addressed for IoT data, the research group turned their attention over NoSQL databases. The promising effort is carried out by investigating the frequent classes of NoSQL databases thoroughly against the DoT requirements deliberated in the Sect. 3.3 [12].

The numerous NoSQL databases with DoT requirements fulfillment is mapped in Fig. 2, that stands as an indicator of the deepest review and reflects the capabilities of NoSQL databases under a five-point scale evaluation. From the listed DoT requirements, scalability, availability and performance are the considered as the primary attributes (or not to be compromised forever in its highest attainment) of the data model as concerned with the nature of IoT devices. In turn, the attribute availability denotes

Table 2. Facts of present data models

Data model	Features	Categories	Merits	Demerits
Relational	Strict schema definition Indexing for fast retrieval Persists data as rows and columns Integrity constraints – Keys Supports view (Virtual Relation) Supports multiple user access controlled by individual user	Classic Client Server a. Single process multithreaded servers – MySQL b. Multiprocess servers – PostgreSQL App resident SQL libraries – SQLite, DeviceSQL Highly distributed SQL environments – MemSQL, MySQL Cluster	Supports structural data Handles massive volume of datasets – Scale up well but expensive	No support for unstructured data/not designed to run efficiently on clusters Difficulty in change management of schema No support for scale out and leads to performance issues
Non-relational (NoSQL)	Horizontal Scalability – scales out well Less and Dynamically modifying Data Schema Memory and Distributed Index	Key-Value DB – DynamoDB, Riak, Redis Column Family Store – Cassandra, Hbase, Hypertable Document DB – MongoDB, OrientDB, CouchDB, Couchbase Graph DB – Neo4j, OrientDB	Designed for distributed data stores for very large scale data needs Provides inexpensive solutions to large data sets – Economical No need for DBA	Maturity level is low No Support from enterprise No support for Analysis and Business Intelligence Need Expertise Administration issues

the possible achievement of reliability, again these are highly dependent on the three factors: robustness, recovery time and stabilization time. Consistency is used to ensure the accuracy of the database while allowing changes by predefined rules and thus avoids the availability of data in different locations as different formats. The attribute durability is intended to persist the committed changes. Consistency and durability are the ACID attributes to be attained at least of moderate level to extend its support for IoT based applications. Likewise the maintainability is considered as a subsidiary attribute which states the flexibility or level of adherence to cope with the technical advancement imposed in future and need not to be attained as high scale from storage perspectives of IoT devices.

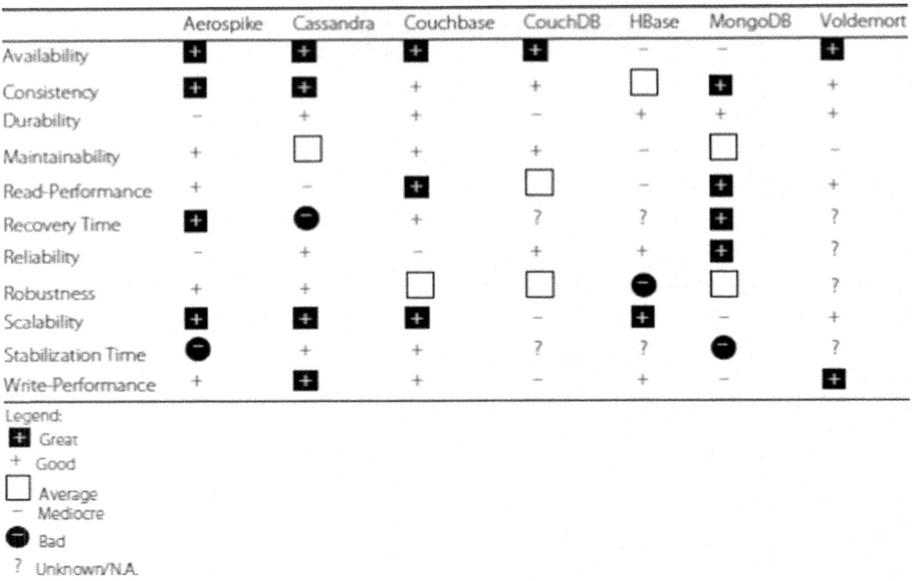

	Aerospike	Cassandra	Couchbase	CouchDB	HBase	MongoDB	Voldemort
Availability	+	+	+	+	–	–	+
Consistency	+	+	+	+	□	+	+
Durability	–	+	+	–	+	+	+
Maintainability	+	□	+	+	–	□	–
Read-Performance	+	–	+	□	–	+	+
Recovery Time	+	●	+	?	?	+	?
Reliability	–	+	–	+	+	+	?
Robustness	+	+	□	□	●	□	?
Scalability	+	+	+	–	+	–	+
Stabilization Time	●	+	+	?	?	●	?
Write-Performance	+	+	+	–	+	–	+

Legend:
+ Great
+ Good
□ Average
– Mediocre
● Bad
? Unknown/N.A.

Fig. 2. Mapping of DoT requirements with NoSQL databases (Source: Journal of Big data, p. 19)

From the Fig. 2, it is observed that the primary quality attributes such as scalability and availability are highly achieved by the databases: Aerospike, Cassandra and Couchbase. With respect to high availability, the dependable attributes like reliability, robustness, recovery and stabilization time is attained in Couchbase as good and average level compared to other two databases. Aerospike offers greater recovery time but bad stabilization time. Unlike to Aerospike, Cassandra provides greater stabilization time but bad recovery time. Couchbase avails good support in yielding consistency and durability comparatively with other two databases. Maintainability is also attained at good level in Aerospike and Couchbase.

No one database is secured with all good for all requirements. By evaluating the attributes as primary and secondary based on the qualities to be enabled in IoT application and through keen investigation, Couchbase database secured greater scale of evaluation for the primary attributes of DoT and provided with average level support for the secondary level of attributes. It has no bad indication in any of the attributes. Couchbase is a memory based, distributed, document-oriented and multi-model database which collects data in JavaScript object notation (JSON) format with flexible schema enable continuous service in place and extend its support on device layer, edge and to cloud server too. It facilitates all the time availability of data and consistent performance over everyday applications. Couchbase would be a better choice of database that promises the quality attributes of IoT data management in a prime level and fine tuning the same in its weaker area brings perfect DoT in real and thus bring excellent solutions.

Few attempts have taken in designing a collaboration model that combines the features of relational, non-relational and file repository to cope with the variety of data produced from the different sources effectively [5].

RDF data storage techniques also used in devising effective DoT over the interactive device network which can process the RDF data faster. These techniques are particularly found to be applicable for various web-based and various mobile applications contributing to IOT community [1].

4   Case Study and Discussion

4.1   Proactive Health Monitoring System

RPM is a golden tool of healthcare industry which increases the access to care constantly well in advance i.e. proactively and reduces the risk and cost involved for later attention and treatment not being in the hospital for long. Nowadays these services are utilized by the society in the name of health or activity trackers, exists as various forms of wearables and lot of research undergone to improve the effectiveness aligned with its usage, life time, accuracy, security and affordability. The most of the RPM based trackers are focused on measuring the cardiac conditions along with activity parameters like steps count, sleeping quality, and calories burnt.

In recent years, some smartphone companies like Apple, HTC and Samsung also delivered smartphones with health care monitoring such as measuring heart rate, calories burned and blood oxygen level by employing a method called photoplethysmography (PPG), which is a 150 years old technique, found to be simple and easy to implement. In PPG, light source like flash and sensor are used and works based on the principle that light entering the body will disseminate in an expectable manner as a result of changes in the blood volume. Through flash, a light shine is made to penetrate into skin and the reflected light which is aligned with the pulse rates are captured through the sensor.

This case study is one of the attempts to bring a low cost activity tracker in place. The entire schematic with workflow of this proposed system is demonstrated as Fig. 3 for enhanced indulgent.

The proposed device uses only heart rate sensor, and from the extracted ECG trace the other vital signs can be extracted through appropriate algorithmic integration at cloud. Thus with reduced circuitry design, it is possible to bring a complete personalized healthcare in place.

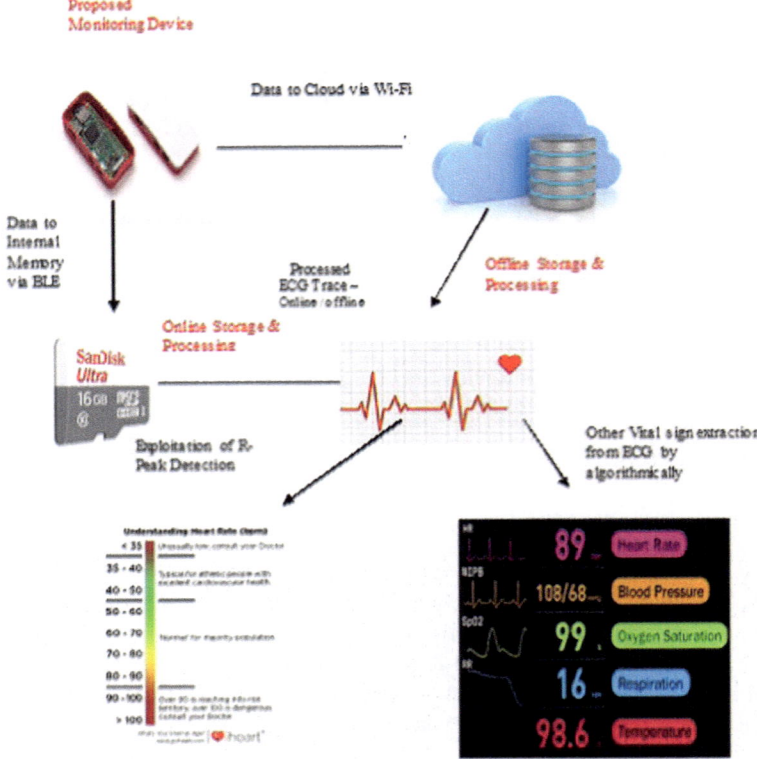

Fig. 3. Schematic structure of proactive health monitoring

4.2   Role and Benefits of Thingspeak

Thingspeak is an open source API and a web service used to store and retrieves the sensor data [15]. Also helps developers to annotate intelligent algorithms to process the data and facilitate visualization techniques in association with MATLAB associated with it. The role of Thingspeak is captured and portrayed as Fig. 4 as below.

Thingspeak sits in cloud environment, acts as an interface between the device layer and application layer, facilitates data storage and retrieval functions through creating channels. The channel can be configured as private and public as per the need of the user. It has eight fields in turn it can collect data arriving from a count of eight sensing devices. It accepts the sensor data gathered in the format extensible markup language (XML) or comma-separated values (CSV). It is able to handle preprocessing of data like noise reduction, obtaining unification of data, aggregating or compressing the data with real time analytics through embodiment of intelligent algorithms through MATLAB. Also facilitates the prominent visualization techniques which is possible through MATLAB and bring the applications to be monitored and to trigger an action in real world based on the processed data arrived from the sensors.

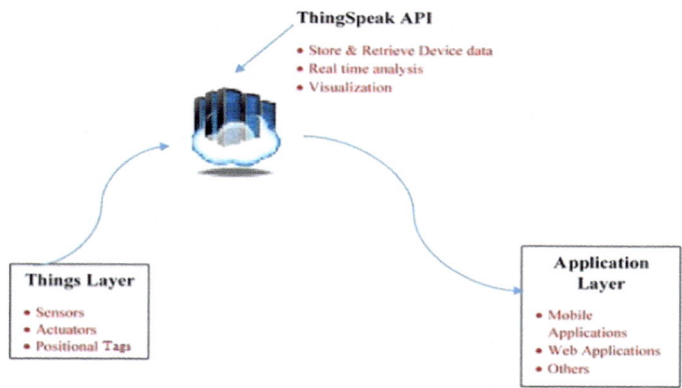

Fig. 4. ThingSpeak in cloud

5 Conclusion

The miniaturization of embedded devices and in built communication technology paves the way for smart IoT applications in use. Open API's like Thingspeak helps the development community in bringing surprising IoT applications. It is good to bring well defined DoT and deploying or associating the same in API's like Thingspeak would be an attractive choice as it takes care of all data pre-processing, processing and visualization in its space and thus offers tremendous instinctive applications ensures potential impacts in environment and society.

Acknowledgement. This work is generously supported by the research grant received from the Science and Engineering Research Board (SERB), Department of Science and Technology (DST), Government of India under the grant scheme (EEQ/2017/000485).

References

1. Cai, H., Xu, B., Jiang, L., Athanasios Vasilakos, V.: IoT-based big data storage systems in cloud computing: perspectives and challenges. IEEE Internet Things J. **4**, 75–87 (2017)
2. Vongsingthong, S., Smanchat, S.: A review of data management in internet of things. KKU Res. J. **20**, 215–240 (2015)
3. Sadiq Ali Khan, M., Jamshed, H., Bano, S., Anwar, M.N.: Big data management in connected world of internet of things. Indian J. Sci. Technol. 10, 29 (2017)
4. Bachhav, B., Parikshit Mahalle, N.: Data management for internet of things: a survey and discussion. Int. Res. J. Eng. Technol. **4**, 1512–1516 (2017)
5. Jiang, L., Da Xu, L., Cai, H., Jiang, Z., Bu, F., Xu, B.: An IoT-oriented data storage framework in cloud computing platform. IEEE Trans. Ind. Inf. **10**, 1443–1451 (2014)
6. Padiya, T., Bhise, M., Rajkotiy, P.: Data management for internet of things. In: IEEE Region 10 Symposium, pp. 62–65 (2015)
7. Trupti Gurav, R., Kudale, A.: DoT (Database for IoT): requirements and selection criteria. Int. J. Comput. Appl. **159**, 39–45 (2017)

8. Abu-Elkheir, M., Hayajneh, M., Ali, N.A.: Data management for the internet of things: design primitives and solution. Sensors **13**, 15582–15612 (2013)

9. Shona, M., Arathi, B.N.: A survey on the data management in IoT. Int. J. Sci. Tech. Adv. **2**, 261–264 (2016)

10. Zhou, Y., De, S., Wang, W., Moessner, K.: Enabling query of frequently updated data from mobile sensing sources. In: IEEE 17th International Conference, Guildford, UK, pp. 946–952 (2014)

11. Lu, T., Fang, J., Liu, C.: A unified storage and query optimization framework for sensor data. In: 12th Web Information System and Application Conference (WISA), pp. 229–234 (2015)

12. Lourenco, J.R., Cabral, B., Carreiro, P., Vieira, M., Bernardino, J.: Choosing the right NoSQL database for the job: a quality attribute evaluation. J. Big Data Anal. **2**, 1–26 (2015)

13. SAP. https://www.sap.com/india/trends/internetof-things.html

14. Wikipedia. https://en.wikipedia.org/wiki/Internetofthings

15. ThingSpeak. https://thingspeak.com

Traffic Prediction Using Decision Tree Classifier in Hive Metastore

D. Suvitha[(⊠)] and M. Vijayalakshmi[(⊠)]

Department of Information Science and Technology, Anna University,
CEG Campus, Chennai, Tamil Nadu, India
suvitha19@gmail.com, vijim@annauniv.edu

Abstract. The capacity to precisely figure the development of road traffic is vital in traffic management and control applications. Urban traffic gridlock is a standout amongst the most extreme issues of regular day to day existence in Metropolitan regions. To manage this issue, variations in rush hour patterns can be exceptionally valuable in helping Intelligent Transport System (ITS) which furnishes drivers with helpful and precise route and travel-time data. In this research work, data storage and analysis is performed in Hive data warehouse. HiveQL Language is used for analyzing the features of traffic data and these analyzed data are processed in SPARK framework. Here with the help of Mlib library, Decision tree classifier algorithm is used for classifying the traffic to low, medium and high traffic range. Using the classifier, abnormality in traffic is detected and it is notified to the user through an android application. Finally the best route with high accuracy is predicted which can be accessed by the drivers. More specifically they can help the user's to choose a much better and less time-consuming path than their original intended route. Experimental result compares forecast exactness of the proposed model with the existing approaches and achieves 95% of accuracy. From the experiments, the results illustrate the proposed approach is unique and robust compared to other models.

Keywords: Decision tree classifier model · Hive warehouse ·
Spark processor · Android app

1 Introduction

Road traffic gridlock is an important drawback in metropolitan region. Roads turn out to be increasingly congested and the traffic moves very gradually and at times does not move at any stretch at all. This causes lot of distress for the individuals who are travelling. The person need to invest his valuable time holding up in roads and this might add to the anxiety. All the previously mentioned issues were a genuine inspiration to do this venture. In order to dwindle this, blocksides of the roads are determined, so that the overcrowded path can be avoided and different path can be picked while migrating from one place to another. The capacity to timely and precisely envision the advancement of traffic plays a fundamental component in traffic authority and restraint applications. In the last few years, significant research studies were accounted for the applications of various empirical and theoretical techniques to predict

© Springer Nature Switzerland AG 2020
A. P. Pandian et al. (Eds.): ICCBI 2018, LNDECT 31, pp. 571–578, 2020.
https://doi.org/10.1007/978-3-030-24643-3_68

the traffic flows. The prime expectation of traffic board methodologies is to deal with road traffic actions up to the most hoisted measure of maintenance to equip decisive and time convincing transportation. With overhauling in technology and the huge distribution of traffic analysis, traffic forecast helps in diminishing stress instead of hanging on in the rush hour gridlock which in turn makes commute experience more enjoyable.

A precise methodology is accommodated for foreseeing the low, medium and high range of possible traffic routes using vehicle count, with the help of that abnormality in traffic route is initiated for suggesting alternative possible paths to the user from that, best path can be predicted using highway name, vehicle count, date and time. The aim of this proposal is to set up a traffic movement with a definitive goal of anticipating the best route to the destination that can be outfitted to the drivers who are utilizing GPS device. Hive data warehouse is used for storing large volume of traffic data's which is integrated with Hadoop Distributed File System (HDFS) for distributed storage. HiveQL Language is used for analyzing the features of traffic data and these analyzed data is stored in metastore which enables SPARK framework to access metadata files. Here with the help of Mlib library, cross validation is performed in machine learning algorithm to give the best algorithm for the dataset which in turn uses Decision tree classifier to predict the best route with high accuracy. Finally this route can be accessed by drivers with the help of android application. In android application the user's input is taken. It is sent as a JSON request to the Hive database server and the possible traffic route with highest probability is calculated and is stored in the database. This possible traffic state is calculated for each and every route for highway in Newyork city. Based on the stored traffic state values, abnormality detection and the best path from source to destination is predicted and finally the output is returned back to the Android device as JSON response.

The remaining flow of this paper is organized in upcoming sections. In Sect. 2, related work is been talked about. Section 3 incorporates the proposed work of Dynamic Traffic Route Prediction Model (DTRPM) Architecture. Section 4 deals with the Methodology. Section 5 approaches with the Dataset Description. Section 6 incorporates Results and Discussion and lastly, Sect. 7 includes the conclusion of the paper.

2   Related Work

Amid past years in Intelligent Transportation System (ITS), different researchers have used diverse strategies to screen traffic events [1, 2] and to anticipate traffic blockage [3]. Cheng et al. proposed automatic incident detection method for urban interstates dependent on geometric conditions and locator areas. These two articles just spotlight on movement recognition rather than the congestion forecasting in this paper. In [3] proposed work used Kalman channel with the Global Positioning System (GPS) area disclosed by drivers to gauge travel time powerfully. The outcomes demonstrated that the forecast exactness of enhanced model is better than the first strategy with Kalman filter. In [4], additionally used Kalman channel yet to anticipate vehicle's future area. Results demonstrated that the proposed technique is better than an expectation strategy. Notwithstanding, both two techniques did not consider best route and vehicle's count prediction. Also, the researchers examined travel time forecast and vehicle's area forecast as opposed to traffic clog forecast in this paper. Also, the analysts in [3] did not think about expressway. In [5], Gray model is used to distinguish traffic occurrences. The analyst utilized genuine guides to analyze the contrast among prediction and reality. Results demonstrated that the proposed technique accomplishes worthy false-caution rate to decide if incident occurred. However, the paper is diverse to our work that traffic occurrences are gathered from PBST to estimate congested driving conditions in next era. In addition, the class and size of gathered information in these papers is comparatively less than this research work. Besides, they didn't use best prediction model which is accomplished in this research work. The vast majority of existing written works utilized a clump of traffic information to anticipate vehicle speed, however it can't accomplish transitory traffic forecast.

3 Proposed Framework

Proposed work is focused on an approach of using big data and machine learning techniques to predict and analyze traffic. With the aid of python's sklearn kit, google maps and hive database integrated with spark, a scalable solution can be implemented for predicting the best route that are impacted by adverse traffic conditions.

In this paper First, we analyze the resident travel behaviors of six features using HiveQL, such as travel time, date, location: (source and destination), vehicle count and Highway. This feature is used to predict high, medium and low traffic ranges for multiple routes accurately using decision tree classifier. Second, optimal route is predicted from the multiple routes using spark SQL which allows relational queries expressed in HiveQL to be executed using Spark.

3.1 Dynamic Traffic Route Prediction Model Architecture

The proposed framework is composed of the following components:

- **Cloud Server:** Download the zip file from the cloud server and extract to .csv file. Validate the observation and transform to hive data warehouse.
- **Hive Database:** Load the Traffic data csv file and create a table to insert the data and finally analyze the traffic data features.
- **HDFS:** Hive uses the HDFS for the distributed storage since it is installed on top of Hadoop.
- **SPARK:** Spark SQL supports intercommunication with Hive metastore, which permits Spark SQL to ingress metadata of Hive tables. Python is used in pyspark for implementing the traffic data with machine learning algorithms. Based on the accuracy results of the machine learning algorithm, best algorithm will be chosen for the traffic data.

- **Android Studio:** The Android application is utilized for getting the info and showing the output to the drivers. The Android application contains a novel key through which the map UI is gotten over the Internet from the Google Maps API. It associates with the Hive server for foreseeing the traffic and the route (Fig. 1).

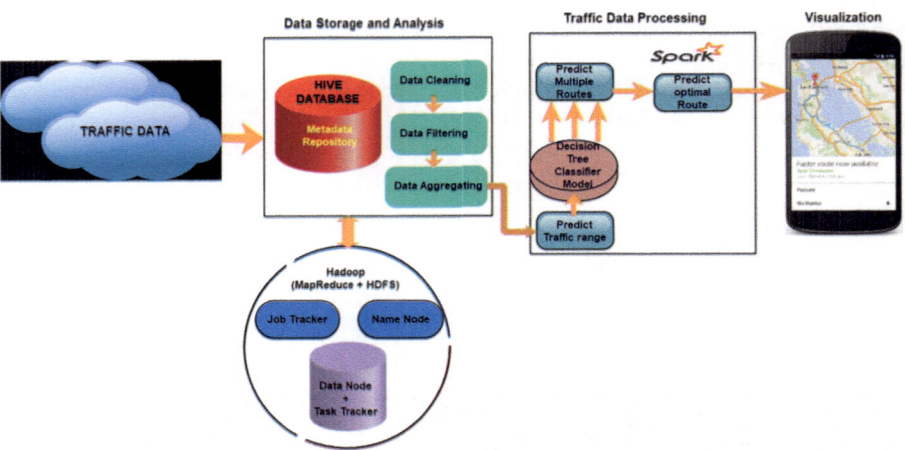

Fig. 1. Dynamic Traffic Route Prediction Model (DTRPM) Architecture

4   Methodology

4.1   Decision Tree Classifier Model

Decision tree is defined in tree structure format. A decision tree is made in two stages:

(a) Tree Structuring Stage

- Training data is continuously partitioned until all the traffic data in every partition is classified into single class.

(b) Tree Skiving Stage

- Dependencies are discarded on statistical noise or anomalies in traffic that might be specific just to the training dataset.

Tree Structuring Stage:

1. *Partition(Data X)*
   if (every coordinates in X are of the clone class) then
   return;
   for each and every attribute Y do
           classify splits on attribute Y;
   Use best split to partition X into X1 and X2;
   Partition(X1);
   Partition(X2);
   ➢ Characteristics of attributes decides the split form.
   ➢ Numerical attributes splits classified as: $Y \leq r$, where r is a real number
   ➢ Categorical attributes splits are classfied as: $Y \in X'$, where X' is a subset of every values of Y

2. *Splitting Index*
   ➢ With the help of splitting index, attribute's different splits are compared
       ○ $Entropy(entropy(K) = -\sum s_l x \log_2(s_l))$
       ○ $Gini\ Index(gini(K) = 1 - \sum s_l^2)$
       (s_l is the relative frequency of class l in K)

3. *The Finest Split*
   ➢ Consider the splitting index as J() and divide X split into X1 and X2
   ➢ The finest split maximizes the upcoming values:
       ○ $J(X) - |X1|/|X| * J(X1) + |X2|/|X| * J(X2)$

Tree Skiving Stage:

- Analyze the tree built which is in initial stage
- Select the subtree with the least calculated error rate
- Error estimation can be classified into 2 stages:
  - Cross-validate the training dataset
  - Utilize nonpartisan dataset

5   Dataset Description

The traffic dataset were collected from NYC Open data https://opendata.cityofnewyork.us/. This dataset was compiled from 2011–2013, traffic counts were recorded on an hourly basis for hundreds of locations. The traffic counts are measured on a particular road (Roadway Name) from one intersection to another (From and To). The data set also specifies a direction of traffic that is being measured (Direction - NB, SB, WB, EB).

6   Results and Discussion

In Fig. 2 Traffic data csv file is loaded in Hive Database. Hive uses the HDFS for the distributed storage since it is installed on top of Hadoop.

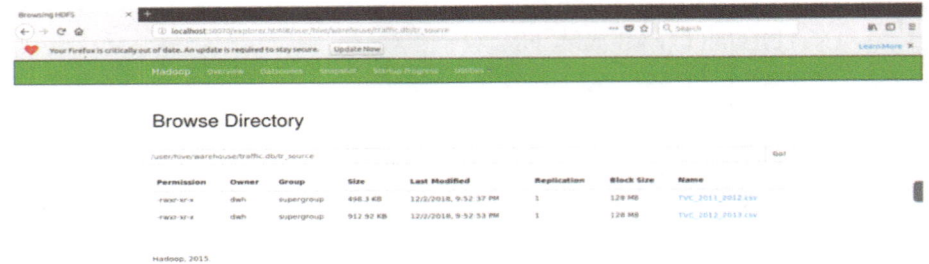

Fig. 2. Processed traffic datasets in HDFS

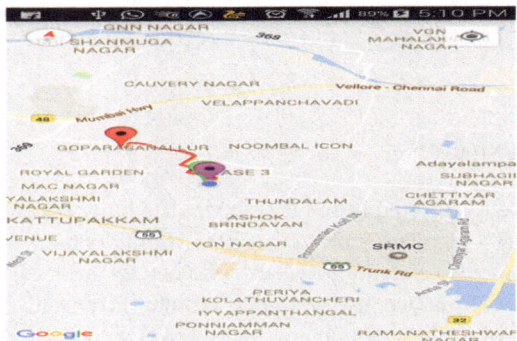

Fig. 3. Data analysis in Apache Hive

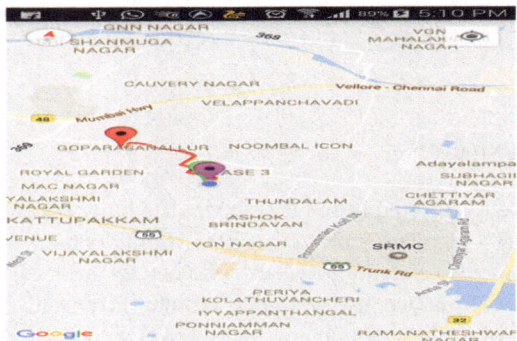

Fig. 4. Route plotting in Android Map

In Fig. 3 map reduce operation is performed for newyork dataset in starting phase. Vehicle count from morning 10:00 AM to 04:00 PM is taken for a particular date 01/09/2012 with highway id, highway name, from and to route (Fig. 4).

Figure 5 shows the traffic density for three days with time data on x-axis and vehicle count data on y-axis plotted. The graph shows the distribution of vehicle count from morning 12 am to afternoon 12 pm.

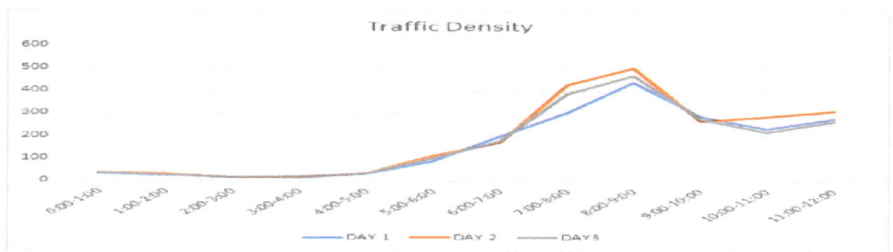

Fig. 5. Traffic density

Figure 6 tells about the precision of the Decision tree classifier model of test dataset. Here the model achieves 95% accuracy, which is higher than the existing models accuracy.

```
In [14]: newyork_data_cross_validate(KNeighborsClassifier(n_neighbors = 3))
Out[14]: 0.6007847533632288

In [15]: newyork_data_cross_validate(GaussianNB())
Out[15]: 0.7705156950672645

In [16]: newyork_data_cross_validate(DecisionTreeClassifier())
Out[16]: 0.9522982062780269

In [17]: newyork_data_cross_validate(RandomForestClassifier())
Out[17]: 0.8681053811659194
```

Fig. 6. Comparison with existing algorithm

7 Conclusion

The proposed work stores large volume of data in hive database and analysis is performed to predict different ranges of traffic in multiple routes from that the best route is predicted with the help of decision tree classifier. Compare to existing algorithm, decision tree classifier achieves high accuracy.

References

1. Milojevic, M., Rakocevic, V.: Short paper: distributed vehicular traffic congestion detection algorithm for urban environments. In: IEE Vehicular Networking Conference (VNC), pp. 182–185. IEEE, December 2013
2. Cheng, Y., Zhang, M., Yang, D.: Automatic incident detection for urban expressways based on segment traffic flow density. J. Intell. Transp. Syst. **19**(2), 205–213 (2015)
3. Ji, H., Xu, A., Sui, X., Li, L.: The applied research of Kalman in the dynamic travel time prediction. In: 2010 18th International Conference on Geoinformatics, pp. 1–5. IEEE, June 2010
4. Feng, H., Liu, C., Shu, Y., Yang, O.W.: Location prediction of vehicles in VANETs using a Kalman filter. Wirel. Pers. Commun. **80**(2), 543–559 (2015)
5. Wang, H.-F.: A freeway automatic incident detection algorithm using grey 840 prediction. M.S. thesis, Department of Computer Science and Information Engineering, National Central University Taoyuan, Taiwan (2003)

Preventing Credit Card Fraud Using Dual Layer Security

Johnson Inbaraj[✉], Mukeshkannan, Arun, and Pradeep Kandasamy

Department of Computer Applications, Kalasalingam Academy of Research
and Education, Krishnankoil, Viruthunagar, India
cajohninba96@gmail.com, smsmukeshkannan@gmail.com,
m.arunvincent@gmail.com, honey.kanda@gmail.com

Abstract. The rise and growth of the E-commerce is obtaining high, thus thanks to that the employment of the credit and revolving credit conjointly been up and still keep growth in day by day and together the dishonest cases caused an explosion within the master card transactions. Because the master card obtaining the foremost in style factor concertedly that impact of this master card fraud being with all of that users. That master card fraud cases makes plenty of losses to the banks and a customers, thus currently the stage of this master card fraud cases is would really like no body cannot offer the protection or answer to the present downside. From the sooner stage there are many varieties of techniques and theories to offer to provide to present to administrator to allow to convey to relinquish the concepts to unravel this downside however nobody will give the answer to the present fraud cases. A transparent understanding on all this approaches will definitely result in economical master card fraud detection system.

In this paper, we just take the step to prevent the credit card transactions, by getting the IP addresses from the first time user and sending OTP to the frequent user by Time Based OTP algorithm.

Keywords: OTP · IP address · Credit card fraud · Fraudulent cases · Master card

1 Introduction

Science and technology is growing day by day, the machines are used to reduce the human works. Before early, people wants to buy a product they have to go to market and buy the products. But now everything is purchased through online. Because people no need to leave home and buy the products. Everything is done just one click, products are arrives to the user house and payment is done through online. Due to this online transaction.

Hacking of data's become more popular. The fraudster hack the details of the credit card holders like account number and PIN. This paper provides the two tier security to the user from these kind of attacks. The first OTP is generated based on the time and the second OTP is generated based on the location and IP/Mac address of the device can be predicted. Through this work fraudulent can be reduced and it provides lots of security to the user during the online payment.

© Springer Nature Switzerland AG 2020
A. P. Pandian et al. (Eds.): ICCBI 2018, LNDECT 31, pp. 579–584, 2020.
https://doi.org/10.1007/978-3-030-24643-3_69

2 Related Works Done

Credit card fraud cases has become a known fallacious cases in inside of the Commerce world. There are several researches and banking sectors take a lot of effort to unravel this fraud cases and build their own concepts to boost the protection for the master card transactions [1]. In our current situation the net looking of multiple product has augmented to a good level extend. Therefore thanks to this all form of users are pushed up to use the debit and credit cards to their buying purpose. So all the fraudsters taking these items as a benefits of their dishonorable services and creating the dishonorable things with the peoples. This work can offer the net page to avoid these form of dishonorable cases and forestall all the peoples from these form of things, for that we have a tendency to are given the twin layer authentications to the users to forestall their account from all the fraudsters by causation OTP supported the cookies to the users [2]. The highest issues of the credit card industry is fraud, so the goal of this paper is detect all type of credit card fraud cases and find the alternative ways to improve the security of this credit card transactions. Many banks or credit card companies using various type of measure to identify the frauds but the proposal of this paper are to have valuable attributes to save the costs and time efficiency. Yet there are many issues for credit card users when they misclassified as fraudulent [3]. The most glorious security for all the net users or online page users is watchword. The passwords are might not be in lexicon as a result of that derived from the user's personal domain therefore we have a tendency to predict the passwords as our own. In this paper, we have a tendency to intend the cookie based mostly virtual watchword authentication protocol that may provide the additional security from all sort of passwords. The info that is in admin facet can mechanically store the user's cookies to their decibel, and whenever the user get into the online page that may mechanically enable them to access the online page supported the cookie based mostly virtual watchword. Therefore this protocol can fight with all kind on-line lexicon attacks and conjointly the dishonorable things [4]. There are a large amount of technological devices used for credit card transaction. Since technology is used in overall transaction devices get into misuse because of that the credit card gets into the frauds way so that it is easy to do fraud on credit card by using these devices, since there is no other way of using credit card for transactions these fraud on credit card is occurring in current world [5]. This paper deals with fraudulent activities occurred in using credit card. The solution given in this paper will not prevent hundred percentage of fraudulent activities occurred on credit cards, but by using this such activities can be reduced at most of it. This paper uses a traditional method named decision tree where every node is connected to the relevant edges which are named in attributes and assigned them with values. This way will be easy to use but the only problem is the process can be done in individual way (Fig. 1).

In this diagram it indicates that the credit card is bring into fraud by the usage of smart devices like smartphones, laptops, tablets, pc's smart home devices. Since in this era time plays a vital role people depend on smart devices for doing their work (Fig. 2).

In this diagram it states that the rate of detection is higher than the rate of permission by this we can assume that detection is one of the major problem which misleads the credit card and use it without the knowledge of permission.

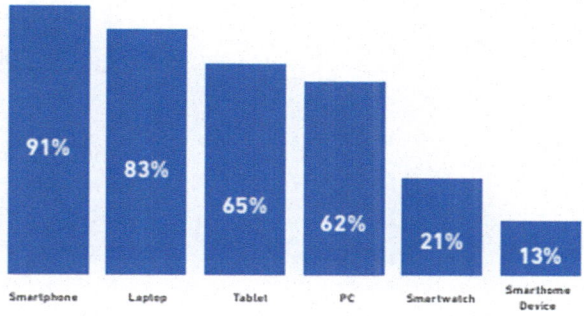

Fig. 1. Fraud occurs in various devices

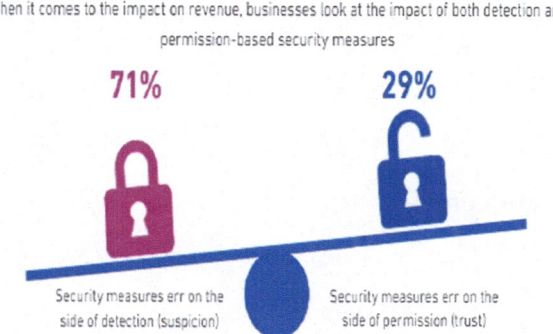

Fig. 2. Detection Vs Permissions

3 Proposed Method

In this paper we have created the web application for the prevention of online credit card transaction fraud using two layer authentication layer mechanism. Two layer in the sense, one is storing IP address of the user to the server database and another one is sending verification code and OTP to the users.

(1) Prevent by Getting IP address: when the user get into the web page through login page, at the time the web application get the IP address of the device which is used by the customer will automatically store in to the servers database and send the verification code to the customer to enter to the next page of the transaction.

(2) Prevent by sending OTP to the correct user: After verify the verification code which is sent to the user, then OTP will be automatically sent to the user's mobile phone number which is stored in the administrators database.

And if user do the same process with same device the first process will be skipped, that is getting card number of the user.

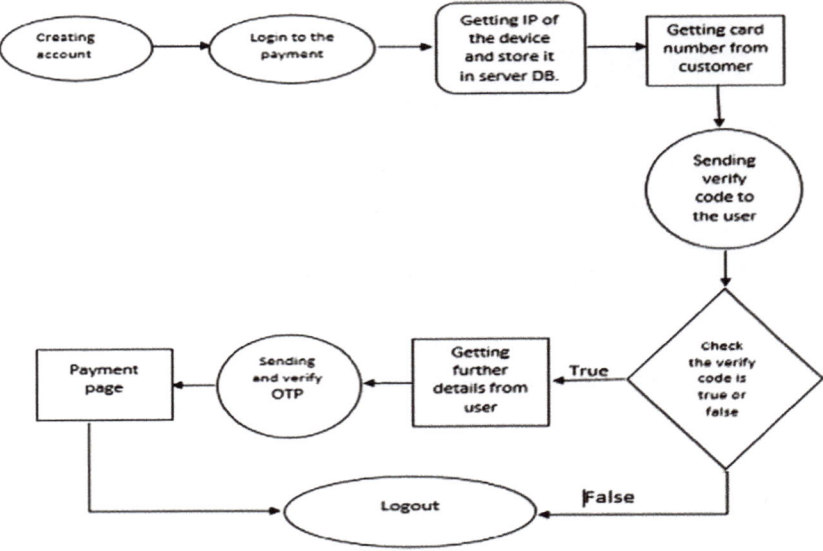

4   System Architecture

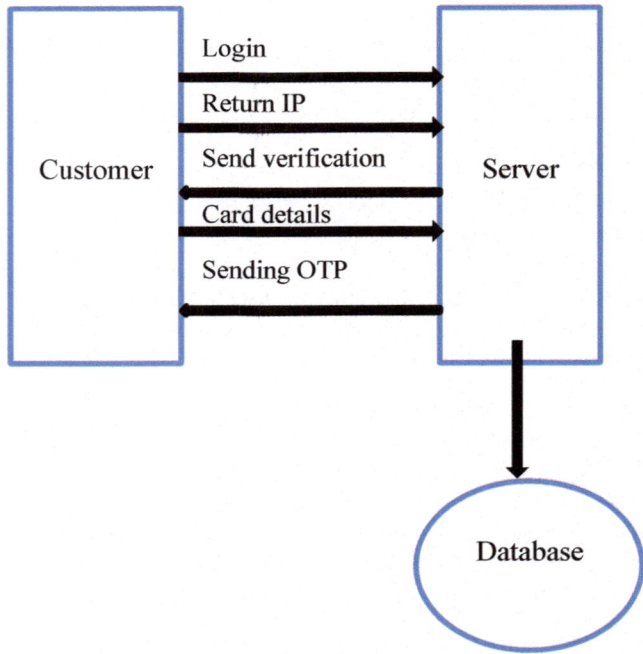

The proposed work is explained below:

(1) At first the new user enter into the web application through the login page, automatically the IP address of the device will be stored to the database. Then the card number will be asked to the user after the entering the card details, the six digit verification code will be sent to the user's mobile phone, after entering the verification code, the validation process will be done, then it links to the next page of the transaction will be displayed to the user.

(2) After completion of entering card details and verification code, the user has to enter the CVV details of the credit card, after submitting the details that will automatically check into the database if it is match into the database it will send the OTP to the user's mobile phone then it will check the OTP, If the entered OTP is correct it will do the further process and move to the next page of transaction.

After completing all the process the user can logout from their account. And if they want to the same transaction through the same device to their account, that will not ask card number of the user because the IP address of the device which they are using is already stored in the database so if they do the same process with the same device it will get the IP address and check it to the database, so if it is already present then it will skip the one step verification in transaction page, so it will ask only the customer name and CVV number of the account holder and then the as usual process will be done in that online transaction.

5 Conclusion

In this our system gives dual security for the credit card users, such a way that fraud detection is minimized and prevented so that the users can use their own credit card safely and securely. This system is only used by the authorized user of credit cards so that other users who are not or unauthorized users cannot use this system. And the main thing of this system is whenever a user use this system he/she should give the correct details which is already present in the bank database or the details which he gave while getting credit card from bank.

References

1. Kulat, A., Kulkarni, R., et al.: Prevention of online transaction frauds using OTP generation based on dual layer security mechanism. Int. Res. J. Eng. Technol. (IRJET), **03**(04) (2016)
2. Delamaire, L., Abdou, H., Pointon, J.: Credit card fraud and detection techniques. Bank Bank Syst. **4**(2), 57–68 (2009)
3. Sood, S.K., Sarje, A.K., Singh, K.: Inverse cookie-based virtual password authentication protocol. Int. J. Netw. Secur. **12**(3), 292–302 (2011)
4. The 2018 Global Fraud and Identity Report. "Experian"
5. Khan, A.A.: Preventing phishing attacks using one time password and user machine identification. Int. J. Comput. Appl. **68**(3), 7–11 (2013)

6. Cha, B.R., Kim, N.H., Kim, J.W.: Prototype analysis of OTP key-generation based on mobile device using voice characteristics. In: Information Science and Application (ICISA) (2011)
7. Parmar, H., Nainan, N., Thaseen, S.: Generation of secure one time password based on image authentication. In: CS & IT-CSCP (2012)
8. Hoffmann, A.O.I., Birnbrich, C.: The impact of fraud prevention on bank-customer relationships: an empirical investigation in retail banking. Int. J. Bank Market. **30**(5), 390–407 (2012)
9. Suman, N.: Review paper on credit card fraud detection. Int. J. Comput. Trends Technol. (IJCTT) **4**(7), 2207–2215 (2013)
10. Renu, Suman: Analysis on credit card fraud detection methods. Int. J. Comput. Trends Technol. (IJCTT) **8**(1), 45–51 (2014)
11. Ghosh, S., Reilly, D.L.: Credit card fraud detection with a neural-network. In: Proceedings of IEEE First International Conference on Neural Networks (2014)
12. Pawar, D., Rabse, S., Paradkar, S., Kaushi, N.: Detection of fraud in online credit card transactions. Int. J. Tech. Res. Appl. **4**(2), 321–323 e-ISSN: 2320-8163

A Network Analysis of the Contributions of Kerala in the Field of Mathematical Research Over the Last Three Decades

K. Reji Kumar[1(✉)] and Shibu Manuel[2]

[1] Department of Mathematics, N. S. S. College, Cherthala, India
rkkmaths@yahoo.co.in
[2] Department of Mathematics, St. Dominic's College, Kanjirappally, India
manuelshibu@gmail.com

Abstract. In this paper we compare the performance made by higher educational institutions in Kerala (a southern state in India) in the field Mathematical research for the period from 1981 to 2015. The entire period is conveniently divided into periods of five years. We use some techniques from social network analysis to compare the performance of institutions over time. His study is based on the research papers published by the faculty of the institutions in the national and international journals which are indexed and listed by MatScinet of American Mathematical Society, which is a prominent abstracting agency in the field of Mathematics. Data available in the data base of MatScinet is used to prepare the network of collaborations of the institutions. These networks are further analyzed and compared to arrive at many valuable conclusions.

1 Introduction

Social network analysis is a branch of network science which deals with the dynamics of the interrelationships of a collection objects. In the context of social network analysis the term society is used in a broad sense. The word society can represent a group of individuals in the context of social sciences, a collection of web pages in the context of World Wide Web, a set of companies in the context of business, or a collection research papers in the context of research collaborations etc. But irrespective of the context the analysis of the network commonly possesses many properties. These properties come under the purview of social network analysis [17].

Social network analysis uses graph theoretic models for representing and studying the properties. Dynamic behaviour is a fundamental characteristic of social networks. Social scientists try to find relationship between the behavior of the society and the structure and topology of the underlying network. They attempt to explain some social phenomena in terms of the observable structural properties in the interconnections in the network. It is useful in many ways. If a phenomenon is associated with some structural properties in a network, one can expect the same phenomenon in some other network having same structural properties. Changes in the structure can affect the behavior of a network in many ways. Explaining the behavioral changes in terms of changes in the structure is an interesting area in the social network analysis.

© Springer Nature Switzerland AG 2020
A. P. Pandian et al. (Eds.): ICCBI 2018, LNDECT 31, pp. 585–590, 2020.
https://doi.org/10.1007/978-3-030-24643-3_70

The formation of internet and World Wide Web shoot up research activities in network science. In modern period network analysis has become an essential area of research in Physical sciences, Biological Sciences and social Sciences.

One of the important characteristics of networks which are studied by network scientists is structure of a network. Every network has its own structure. Network is not a mere collection of nodes and links between them. Dynamic networks have changing structure over time. Network emergence is a property of dynamic networks. Dynamism is another property which is a consequence of emergent behavior of dynamic networks. Dynamism is absent in static networks. Network is usually formed by the autonomous and spontaneous behavior of actors. It provide a character of autonomy to a network. Another important aspect studied is the importance of nodes in the network. Power of nodes in a network is different. It can be measured in many ways. Power can be defined in simple terms as the number of nodes connected to a particular node. In graph theoretic terms it is the degree of a node. In a network it is found that new connections are most likely to be formed between a new node and node having high degree. If two nodes having high degree are not connected in a network, then it is likely that the network may break into at least two sub-networks. Stability is another character of a network which is measured in terms of the rate of change of the state of its node links.

In the next section we discuss various methods, which are widely used by network scientists to find the relative importance of actors in a network.

2   Importance of Nodes in a Network and the Methods to Find It

Graph theory is the backbone of social network analysis. Every network, at a particular instant, can be represented by a graph in which vertices representing the actors in the network and edges between two vertices representing a predetermined relation, which exists between the actors. Graphs are very deeply studied mathematical objects. A vast literature is available which explain many of its structural properties. Social scientists have successfully given interpretations and explanations of these structural properties in the background of social networks. A graph $G = (V, E)$ is a nonempty collection of vertices V and a collection of edges E, which represent the interrelationship among the vertices. The edge set E contains unordered pairs of vertices in V. So elements of the edge set represent reciprocative relations. If the relation is undirected, then directed edges are used in the graph to represent the relation. Corresponding graph is a directed graph. A directed graph is denoted by $D = (V, A)$, where A denote a subset of the set of all ordered pairs of the vertices. We use dynamic graph models to represent dynamic nature of networks.

In a dynamic graph number of vertices and relations among them undergo changes over time. So representation of a dynamic network must specify the time point of the network as well. In the study of the process of spreading of information, identifying the influential spreaders is important [4]. Such information will help us to design methods for boosting of propagation of useful information. In some contexts fast spreaders are dangerous to the whole network. For example, propagation of virus in the process of spreading disease in a society, computer virus spread in complex networks, spreading information in social networks etc. So identifying potential spreaders in a network is a critically important area of research in social network analysis. For deeper analysis of the topic refer [6–8, 13].

There are many other centrality measures which are developed recently. Page-Rank [3], Leader-Rank [7] etc. are examples. These measures are proved to be effective only to some extent, which is context specific. In the paragraphs given below some important node ranking methods are reviewed.

The limitation of this method is that too many nodes are present in each shell even though they have different spreading ability. It is a global metric.

In 2017 Reji Kumar et al. suggested a new ranking method named m-Ranking [11] of nodes. In this method degrees of all nodes are considered to rank nodes of a network. Total power of each node in a network can be calculated giving due consideration to the degree of all nodes and the weights of all edges in the network. Total power is then used to rank the nodes. This method has been used by the authors for the ranking of nodes in a variety of networks [12, 14, 15]. A similar procedure has been developed to identify important nodes in directed networks [16].

Using the new m-ranking method research collaborations among the Indian Institutions conducting mathematical research is studied for the last twenty years [16]. This is the motivation behind the study presented in the following section. This study focuses on the collaborative research conducted by research institutions of Kerala for the last 35 years. We use the data obtained from the database of MatScinet of American Mathematical Society. MatScinet is a major abstracting agency in the field of Mathematics. Articles indexed in their cite contains the names of authors and their institutional affiliation. Each institution is provided with a unique ID. The ID begins with a country code, which helps us to identify the country of the institution. The data is extracted from the database using a Python code and it is further analyzed to make an adjacency matrix in which the row and column represent the institutions in the network. If the institute i and the institute j appear simultaneously in a paper, then the $i j$ entry is increased by 1. Thus we make a weighted adjacency matrix to find the rank of the nodes using another programme written in C.

This method is used in the following section to rank the institutions in the field of mathematical research in Kerala. The rank is then compared to the ranking obtained by some important ranking methods mentioned above.

3   Mathematical Research Collaboration Network in Kerala

In the collaboration network, the institutions are represented by nodes and their collaborations are represented by links. Weight of a link connecting two institutions is the number of papers having coauthors from both institutions. In this paper we study the collaboration of Mathematical research institutions in Kerala for the period from 1981 to 2015. We collected all the papers published in journals which have an MR (Mathematical review of American Mathematical Society) number and at least one author from an institute in Kerala. Using the weighted adjacency matrix of the collaboration matrix we calculate the important institutions using m-Ranking method [11] (Fig. 1).

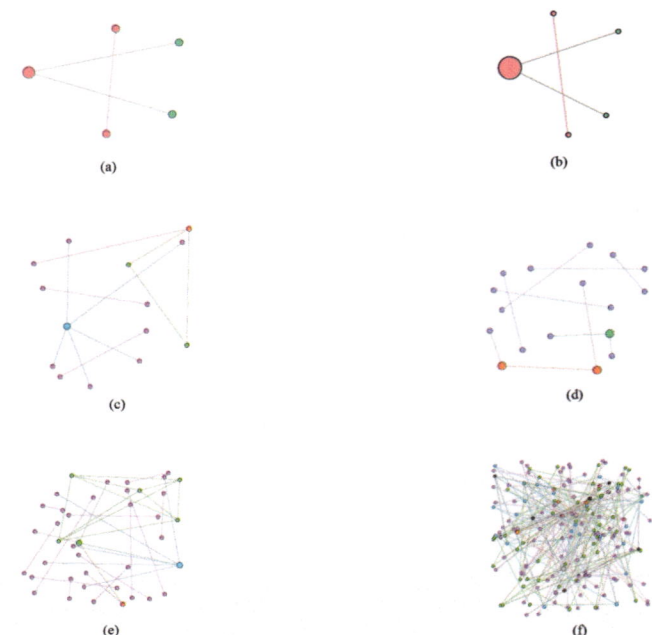

Fig. 1. Network of the periods 1986–90, 1991–95, 1996–2000, 2001–05, 2005–10 and 2011–15 are given in the pictures (a), (b), (c), (d), (e) and (f) respectively

There are 257 institutions which collaborated to publish papers in the network, during the complete period of study. In the first step the papers are categorized into 7 groups according to its year of publication. The groups are 1981–85, 1986–1990, 1991–1995, 1996–2000, 2001–2005, 2006–2010 and 2011–2015. Since some papers of the year 2016 are not yet included in the MR database, we restrict our study up to the year 2015. In 1981–85, there are only 3 institutions which published papers.

In that year no collaboration with other institutions is seen. In the period 1986–90, there are 9 institutions which collaborate each other. In the period 1991–1995, the number of institutions collaborated increased to 14. During the period 1996–2000, 2001–2005, 2006–2010 and 2011–2015 the number of institutions are respectively 20, 25, 49 and 210. We can see a significant increase in the number of institutes during the period 2011–2015 compared to all previous periods.

Using m-Ranking method we rank the institutions in the collaboration network. During the period 1981–85, there are no papers with collaboration of institutions. So rank of all institutions are 0. During the period 1986–90, 6-COCH ranked first with the T value 3.0 and 6-TRCH ranked second with the T value 2.375. The institutions 6-COCH, 6-COCHT-P, 6-KERA, 6-VIKR-AP have papers, but without collaboration. So, their ranks are 0. During the period from 1991 to 1995, 6-COCH and 6-SRIV ranked first with maximum T value 1.75. The first 10 ranks for different periods of study are tabulated in the Tables 1 and 2.

Table 1. Ranking of Institutions for the period from 1981 to 2000.

Rank	1981–85	1986–90	1991–95	1996–2000
1	6-COCH	6-COCH	6-COCHT	6-COCHT
2	6-KERA	6-TRCH	6-SRIV	6-VIKR-AP
3	6-VIKR	6-COCHT	6-CHT-DS	6-VIKR-SE
4	–	6-IIT	6-COCHT-D	6-VIKR
5	–	6-VIKR	6-COCHT-P	6-COCHT
6	–	6-COCH-P	6-PNA-P	6-ANNA
7	–	6-COCHT-P	6-VIKR-AP	6-ANNA-S
8	–	6-KERA	6-VIKR-SD	6-CALI
9	–	6-VIKR-AP	6-COCHT-A	6-COCH
10	–	–	6-COCHT-S	6-IIS

Table 2. Ranking of Institutions for the period from 2001 to 2015

Rank	2001–2005	2006–2010	2011–2015
1	6-COCHT	6-COCHT	6-NIT-M
2	6-COCHT-P	6-CALI-S	6-COCHT
3	6-GCKS	6-BITS-NDM	6-MNPIT-M
4	6-ASOCT	6-IIS-NDM	6-IIIS
5	6-IITND-P	6-IITK-NDM	6-CECTL-M
6	6-CALI	6-MATSCI	6-KSOM
7	6-CALI-S	6-NITC	6-SSYEDC-M
8	6-CALI-P	6-STCT-S	6-NIT-C
9	6-COCH	6-ISI-SMU	6-SILIT
10	6-COCHT-E	6-ALAG	6-NMIE-DM

4 Conclusion

In this paper we present a network based study of collaborative research of the institutions which produce research output in the form of research articles. This study focuses on a long period, which starts from 1981. First article which can be claimed by an institution in Kerala appeared in MR database only in the year 1981. Considering the importance of the institutions that have in the collaboration network, ranks of the institutions are obtained. This helps us to identify the most prominent institution in the network, and the institutions having lesser importance etc. The research work presented here can be a seed for a much elaborate and in depth study of institution collaborations with emphasis to Kerala. Similar studies can be conducted for institutions in India. This research can also be used to predict the future of mathematical research in Kerala.

Acknowledgements. First author would like to acknowledge the research facilities extended by Institute of Mathematical Sciences, Chennai by granting the Associate Visitor-ship. A part of the research is completed during this period. He also likes to acknowledge the financial support of

UGC in the form insert of a major research project No. 40-243/2011(SR). A part of this research is completed during this visit. The second author would like to thank the financial support given to him by UGC in the form of FDP (FIP/12th Plan/KLMG018 TF06).

References

1. Garas, A., Shweitzer, F., Havlin, S.: A k-shell decomposition method for weighted networks. New J. Phys. **4**, 083030 (2012)
2. Zeng, A., Zhang, C.-J.: Ranking spreaders by decomposing complex networks. Phys. Lett. A **377**, 1031–1035 (2013)
3. Brin, S., Page, L.: The anatomy of a large-scale hyper textual web search engine. Comput. New ISDN Syst. **30**, 107–117 (1998)
4. Jaber, L.B., Tamine, L.: Active Micro Bloggers: Identifying Influencers, Leaders in Micro Blogging Networks. Springer, Berlin (2012)
5. Jackson, M.O.: Social and Economic Networks. Princeton University Press, Princeton (2010)
6. Bae, J., Kim, S.: Identifying and ranking influential spreaders in complex networks by neighborhood coreness. Phys. A **395**, 549–559 (2014)
7. Li, Q., Zhou, T., Lu, L.: Identifying influential spreaders by weighted leader rank. Phys. A **404**, 47–55 (2014)
8. Ali, M., Varathan, K.D.: Identification of influential spreaders in online social networks using interaction weighted k-core decomposition method. Phys. A **468**, 278–288 (2017)
9. Bordons, M.: The relationship between the research performance of scientists and their position in co-authorship networks. J. Inform. **9**, 135–144 (2015)
10. Manuel, S., Reji Kumar, K.: An improved k-shell decomposition for complex networks based on potential edge weights. Int. J Appl. Math. Sci. **9**, 163–168 (2016)
11. Manuel, S., Reji Kumar, K., Benson, D.: The m-ranking of nodes in complex networks. In: Proceedings of 9th International COMSNETS 2017 (2017)
12. Reji Kumar, K., Manuel, S: Spreading information in complex networks: a modified method. In: Proceedings of the International Conference on Emerging Technological Trends. IEEE Digital Explore Library (2016)
13. Reji Kumar, K., Manuel, S.: Personal influence on the spreading of information—a network based study. Int. J. Math. Trends Technol. **52**(8) (2017)
14. Reji Kumar, K., Manuel, S., Satheesh, E.N.: Spreading Information in Complex Networks: An Overview and Some Modified Methods, Graph Theory—Advanced Algorithms and Applications. Intechopen, London (2017)
15. Reji Kumar, K., Manuel, S.: A centrality measure for directed networks: m-ranking method. In: Özyer, T., et al. (eds.) Social Networks and Surveillance for Society, Lecture Notes in Social Networks. Springer, Cham (2019)
16. Reji Kumar, K., Manuel, S.: Collaborations of Indian institutions which conduct mathematical research: a study from the perspective of social network analysis. Scientometrics. https://doi.org/10.1007/s11192-018-2898-0
17. Lewis, T.G.: Network Science: Theory and Practice. Wiley, Hoboken (2009)

SoC Based Acoustic Controlled Semi Autonomous Driving System

A. K. Veeraraghavan[1(✉)] and S. Ranga Charan[2]

[1] Department of Electrical and Electronics Engineering,
Sri Sairam Engineering College, Chennai, India
`veeraraghavan.ak.1996@ieee.org`
[2] Department of Automobile Engineering, Hindustan University, Chennai, India
`sr.trans4777@gmail.com`

Abstract. Autonomous driving car provide various features such as Adaptive Cruise Control, Autonomous Emergency Breaking, Lane Detection etc. which all environment mapping using image processing, Lidar, radar, GPS and computer vision. Features like image processing require lot of data processing power and hence can be expensive. In country like where people can afford expensive autonomous car with level 4 or 5 autonomous driving level, and with traffic being highly unreliable, a different system is required which is suitable for Indian road. In this paper, we proposed that is a level 3 autonomous driving vehicle and operates based on the processing of the acoustic, sound and noises from the surrounding. The audio processing is done using Xilinx Zynq FPGA SoC and ICS43432 microphone as hardware, running on convolutional neural network and Fuzzy control. The system on chip (SoC) is the future of embedded system design. An FPGA based SoC platform offers many advantages, including high integration and easy field upgrades of entire systems.

Keywords: FPGA · Fuzzy control · Acoustic controlled driving system · Audio processing · Convolutional neural network · System on chip · Sound classification · Electronic control unit

1 Introduction

The Indian car industry is one of the biggest on the planet. It contributes about 7.1% of India's GDP. All inclusive, the majority of the key car players, for example, Uber, General Motors, Audi, Ford, Honda, Apple, BMW, Nissan, Mercedes-Benz, Toyota, and so forth are contributing critical time and speculation on creating AVs and are now directing test preliminaries in nations, for example, US, Germany, Singapore, UK, Japan, China, and so forth. A survey in 28 nations indicated Indians are the most edgy to lay their hands on self-driving vehicles. Near half said they needed these cars [1] (Fig. 1).

For India, the barrier for Fully Automated Autonomous vehicles is poor street and transport foundation, which prompts most street accidents and deaths in India. Street accidents related deaths in the nation are among the world's most elevated. In 2016 alone, there were 1,50,000 such deaths, which comes to around 400 every day.

© Springer Nature Switzerland AG 2020
A. P. Pandian et al. (Eds.): ICCBI 2018, LNDECT 31, pp. 591–599, 2020.
https://doi.org/10.1007/978-3-030-24643-3_71

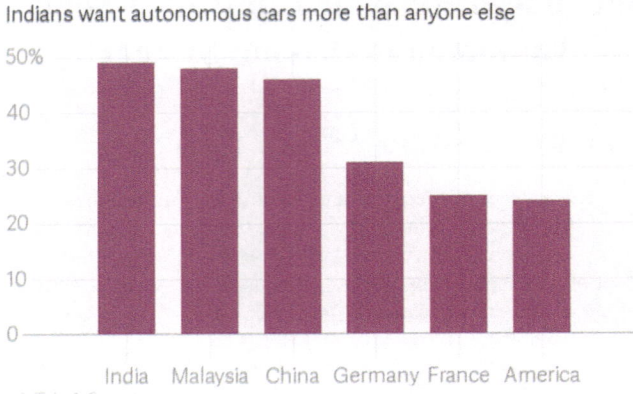

Fig. 1. Survey report

Moreover, roads turned parking lots in the nation's best four urban communities cost a huge $22 billion every year These difficulties have anyway not disheartened Indian car players, for example, Tata, Mahindra and a few tech new businesses from taking a shot at the autonomous vehicles innovation. In addition The Indian Motor Vehicles Act, 1988 and the principles, that control task of vehicles in India, don't as of now permit completely automated frameworks [1]. The laws don't allow notwithstanding testing of autonomous vehicles in India. A human driver should be in compelling control of the vehicle consistently. Considering this point, we decided to develop a Level 3 semi-autonomous system with Advanced Driver Assist feature which is achieved using fuzzy logic based control of electric motor. The proposed technology will open new doors in self driving technology by introducing a new feature called the acoustic controlled driving system similar to Adaptive Cruise Control, Autonomous Emergency Breaking, Lane Detection etc. and introduce audio processing as a new method for automation rather than conventional vision guided systems using LIDAR, radar, GPS and computer vision. This feature will analyze the sounds and noise of vehicles, horn, ambulance, crowd etc. from the surrounding and using signal processing and neural networks provide command and control the Electronic Control Unit which is interfaced to our system using CAN bus. As Indian roads are filled with lot of vehicles, the use a feature that is dependent on sound and noise of will be beneficial in detecting the horn sound, siren sound and give controls and driver assistance, and thus, will be more suitable to Indian roads.

2   Related Work

Each huge automaker is seeking after the tech, anxious to rebrand and remake itself as a "portability supplier" before the possibility of vehicle proprietorship goes done for. Many automakers in the race in developing autonomous cars such as Audi, BMW, Ford, General Motors, Jaguar, Land Rover, Mercedes-Benz, Nissan, Tesla and Volvo.

Not only automobile companies, but also among technology companies like Apple, Microsoft, Uber and Google are planning to work towards Fully Automated Autonomous Vehicles realizing the future market [3]. There are numerous other dynamic research programs concerning self-sufficient vehicles, huge numbers of them highlighting joint efforts among colleges and carmakers. Oxford University, for instance, showed a self-driving Nissan LEAF in 2012. Volkswagen and an examination group from Stanford University have made a driverless Audi sports auto, which has been flashing around US race tracks. In another examination venture financed by the European Union, Volvo effectively drove a caravan of five vehicles that just had a human driver ahead of the pack auto. However, as observed, the technology that most companies development towards Fully Automated Autonomous Vehicles isn't suitable for Indian market. As for India, Volvo plans and mentioning India's plan targeting an all-electric vehicles fleet by 2030 and the Indian government's National Electric Mobility Mission Plan aims to achieve annual sales of electric and hybrid cars of 6 million to 7 million by 2020 [4]. India's streets could at present help semi-Autonomous vehicles, yet further developed vehicles would not be feasible without huge enhancements in infrastructure and control. Automakers have constantly made arrangements for human-operated vehicles to exist together with autonomous vehicles. Therefore, considering this, we in the paper are proposing a system that will provide level 3 semi-autonomous driver's assistance feature based on audio processing.

3   Proposed System

The proposed system mainly focuses of automating of car using audio processing. The system consists of ICS43432 microphone which receives sound from the surrounding and a Xilinx Zynq FPGA SoC. The Xilinx FPGA uses convolutional neural network to execute the audio processing. A fuzzy logic PI controller receives commands from the Xilinx FPGA and the signals of manual systems like steering, braking, acceleration etc., and based on all the input values, it provides output commands to the Electronic Control Unit via the CAN bus interface. The system operation can be classified into 2 portions, hardware operation and the audio processing using the neural network. The block diagram of the proposed system is given below (Fig. 2).

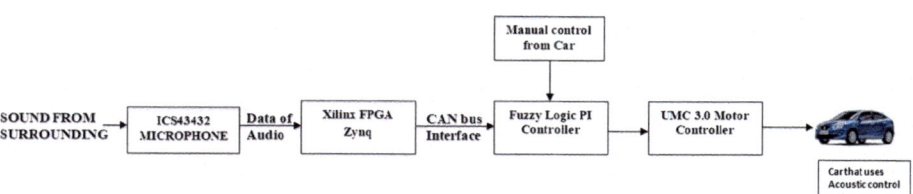

Fig. 2. Block diagram

(a) *Hardware Operation*

The hardware consists of ICS43432 microphone, Xilinx Zynq FPGA, Fuzzy logic PI controller and a UMC motor drive. The ICS43432 microphone is used to receive sound and noise of horn, siren and people from the surrounding and generates digital data based on sound characteristics. The ICS-43432 is a high-performance, low power, omnidirectional MEMS microphone with a bottom port. The ICS43432 bottom-ported digital microphones that output I2S audio as a stream of 24-bit serial words that can be directly read by a controller or processor board with an I2S port. The complete ICS-43432 solution consists of a MEMS sensor, signal conditioning, an analog-to-digital converter, destruction and anti-aliasing channels, control administration, and an industry standard 24-bit I^2S interface. The I^2S interface allows the ICS-43432 to connect directly to digital processors, such as DSPs and microcontrollers, without the need for an audio codec in the system. The reason for choosing ICS-43432 is as it provides digital output and has a high Sound to Noise Ratio High 65 dBA, with a sensitivity of − 26 dB FS. It has a Wide Frequency Response from 50 Hz to 20 kHz and Low Current Consumption of 1.0 mA. It has a High Power Supply Rejection of −80 dB [5] (Fig. 3).

Fig. 3. ICS-43432 microphone

The digital signal generated by ICS43432 microphone is given to Xilinx Zynq FPGA SoC. The FPGA uses Convolutional Neural Network to carry out audio signal processing. The Zynq Xilinx SoC architecture is dual-core or single-core ARM Cortex A9 based processing system. The ARM Cortex-A9 includes on-chip memory, external memory interfaces, and a rich set of peripheral connectivity interfaces. It has Multiprotocol dynamic memory controller of 16-bit or 32-bit interfaces to DDR3, DDR3L, DDR2, or LPDDR2 memories. ECC support in 16-bit mode 1 GB of address space using single rank of 8, 16, or 32-bit-wide memories Static memory interfaces 8-bit SRAM data bus with up to 64 MB [6]. The Zynq Xilinx FPGA supports CAN bus protocol. It has two full CAN 2.0B compliant CAN bus interfaces. The CAN bus feature helps us to interface the FPGA with the Electronic control Unit. The ARM processor is capable of Audio processing and neural networks. This feature enables Zynq FPGA to classify the data sets of the audio by mapping the input audio data with the pre-defined libraries [7].

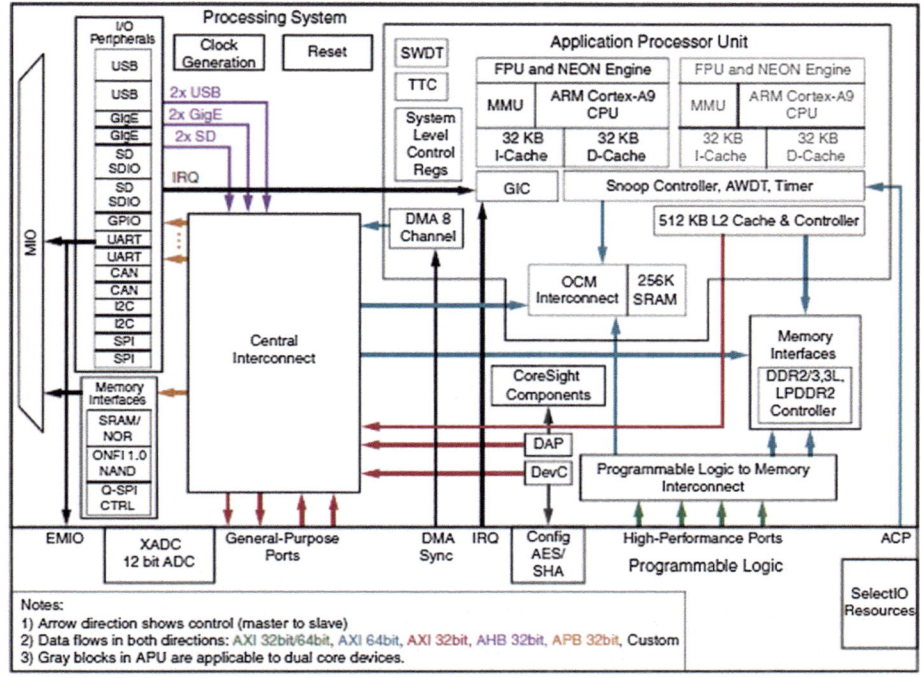

Fig. 4. Zynq Xilinx FPGA architecture

(b) *Neural Network Operation*

The complex nature of the Machine Learning frameworks emerges mainly from the datas. The frameworks assume control over the field of picture acknowledgment and inscribing, with structures like ResNet, GoogleNet shattering benchmarks in the ImageNet rivalry with 1000 classifications of pictures, grouped at about 95% exactness. This was because of a lot of marked data set that were accessible for the Models to prepare on and furthermore quicker PCs with GPU speeding up which makes it simpler to prepare Deep Models. In order to carry out sound classification, we need to classify sounds into a library such that matching of the sound can be done with ease. We have sounds classified into categories such as car, bus, tram etc. where each file will have sounds representing them. The library is endless list that can be updated in future for more accuracy, the more the data set the more accurate will be the detection. We have robust speech recognition systems set up however there is still no universal sound classifier which can enable the system to decipher ordinary sounds, for example, horn or yelling behind us and so on and take actions based on them [8] (Fig. 5).

The issue we look with building a sound classifier is the absence of an expansive data, yet Google's the AudioSet, ESC-50 have a vast accumulation of sound data. The code gives a knowledge into changing over sound data into spectrogram pictures. We utilize library this technique is a standard one for transformation into spectrogram. We utilize a convolutional Neural Network, to characterize the spectrogram images. This is

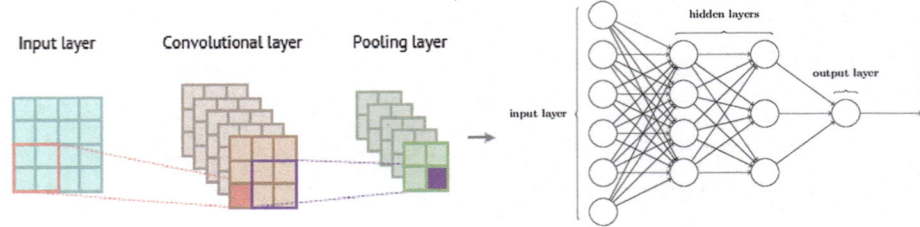

Fig. 5. Convolutional neural network model

because CNNs work better in recognizing various parts of the picture and are additionally great at catching progressive highlights which turn out to be in this way unpredictable with each layer. As the CNNs learn is hierarchical, we can see the underlying couple of layers learn fundamental highlights like different edges which are regular to a wide range of kinds of pictures. This is mainly because CNN has a feature called Pooling Layer, which is a non linear form of down sampling where the image spectrogram is progressively segregated to a minimum value size [9]. audio feature extraction, classifier mapping Spectrogram splits the signal into overlapping segments, windows each segment with the hamming window and forms the output with their zero-padded, N points discrete Fourier transforms. Thus the output contains an estimate of the short-term, time-localized frequency content of the input signal.

(c) *Control of Car*

Each vehicle has a motor driver which provides motor with the pulses to run the motor. The pulses are given according to the acceleration given by the user. Generally, the pulses are given using PWM modulation to the motor to operate with better accuracy. Here, instead of the driver giving the acceleration, the pulses are given based on the audio and data processed by the controller. Based on the data of audio fed to the FPGA and the input signals from the manual system, the pulse signal is given to the Electronic control unit, which controls the motion of vehicle. Mostly for this, UMC Drive 3.0 Universal Motor Controller is used, as shown in Fig. 4, which is used in a lot of electric road cars like Toyota Prius,Tesla Model S, Tesla Model X, Nissan Leaf, Chevrolet Volt and Smart EV. The advantage of this Motor Controller is that it adapts to respective cars drive cycle power stages, inverter drive modes, sensors does not to be changed which greatly reduces the initial conversion cost and its supports sensor-less drive mode too. The UMC provides a CAN bus, 3 phase full bridge control signals as well as a Resolver and inputs encoder. For bus and phase measurements are 4 isolated High Voltage inputs. The Hardware is over-current trip protected and the Digital input, output channels are isolated [10]. As the UMC motor controller supports CAN bus, it can be easily interfaced with the external unit [11] (Fig. 6).

For the control of motor, inputs from the audio signal, speed, acceleration, braking etc. have to be processed. To process these different set of parameters, we are using Fuzzy logic PI controller. The output of the fuzzy system is given as PWM signal. The output values of speed control are determined in the range and states of 00-FF. The components of the fuzzy controller for the DC motor are implemented in VHDL.

Fig. 6. UMC 3.0 Advantics

The model of fuzzy logic is shown in Fig. 7. The input parameters are Xi, which in our case is the inputs from the audio data, the acceleration, braking, speed etc. W$_j$ is the weight associated to each input, which can be assigned based on the priority of the operation of system. For better manual control, the weight associated with the manual parameters such as acceleration, braking speed must be given priority and for higher acoustic automation, input weights for audio datas must be given priority and bi is the baising offered. a $_i$ is the summation of multiplying the input and the weight, which is calculated using

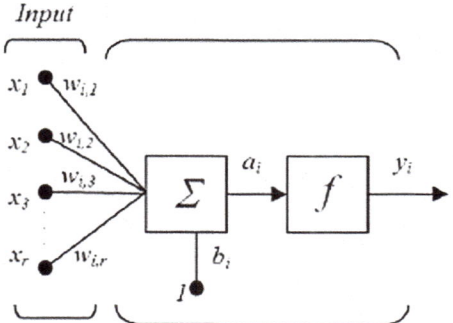

Fig. 7. Model of fuzzy logic

$$a_i \sum X_i W_j + b$$

The output is given as y$_i$ [12]. The simulink modelling of system is shown is Fig. 8. The FPGA and Fuzzy logic PI controller are designed. The various inputs are provided to FPGA and PI Fuzzy controller. Based on the output, feedback is given back for stable running of the system. It is seen from the output graph of the speed vs time graph that the entire controller system helps in achieving a stable speed.

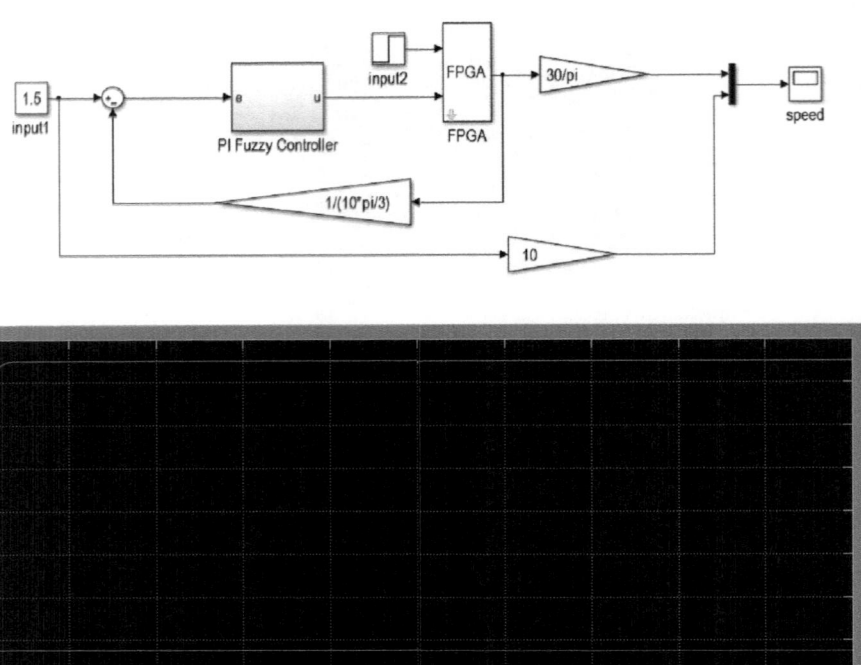

Fig. 8. Simulink model and result

Emulation of legacy instrumentation clusters such as speedometer, RPM, temp and fuel are easily done. The process of controlling the motor remains the same using pulse modulation involving a encoder with more senor based inputs from the motor. In terms of hardware a high voltage capacity relay must interfaced with Xilinx FPGA and PI fuzzy controller to support the currents levels needed to drive the motor.

4 Conclusion

Thus with the proposed design and architecture, we can build and develop a level 3 semi autonomous car based on FPGA and Fuzzy Logic PI controller, which helps us to achieve advanced driver's assistance feature, based on Acoustic controlled driving system, which in itself is a totally new methodology for car automation. The acoustic controlled driving system will be more suitable for Indian roads, as the intensity sound of horns, siren, animals, crowd etc. on Indian roads are very high and by processing these sounds, the car can be suitably controlled. The use of Acoustic Controlled Driving system will open new doors and methods to improve autonomous car technology.

5 Future Scope

- As mentioned, the use of Acoustic Controlled Driving system will open new doors and methods to improve autonomous car technology and can be added as a feature in autonomous vehicles like Adaptive Cruise Control, Autonomous Emergency Breaking, Lane Detection etc.
- With the development of Artificial intelligence, machine learning and processors, in future, a level 5 autonomous car can be developed based in Acoustic Controlled driving system.
- As the system is developed based on programmable SoC, upgrading of system can be achieved easily and with better data sets and design and fabrication of a system exclusively for this application, the efficiency of audio processing and controlling of vehicles can be increased.

References

1. Rosenzweig, J., Bartl, M.: A review and analysis of literature on autonomous driving E-journal: making-of innovation. Mak.-Of Innov. (2015)
2. Kim, T.J.: Automated autonomous vehicles: prospects and impacts on society. J. Transp. Technol. **8**, 137–150 (2018)
3. Bimbraw, K.: Autonomous cars: past, present and future - a review of the developments in the last century, the present scenario and the expected future of autonomous vehicle technology. In: ICINCO 2015-12th International Conference on Informatics in Control, Automation and Robotics, Proceedings, vol. 1, 191–198 (2015). https://doi.org/10.5220/0005540501910198
4. "Driving Mobility Through Autonomy In India" Ipsos Business Consulting
5. Alexandridis, A., Papadakis, S., Pavlidi, D., Mouchtaris, A.: Development and evaluation of a digital MEMS microphone array for spatial audio (2016). https://doi.org/10.1109/eusipco.2016.7760321
6. Xilinx Zynq-7000 SoC Data Sheet: Overview
7. Molanes, R., Amarasinghe, K., Rodriguez-Andina, J.J., Manic, M.: Deep learning and reconfigurable platforms in the internet of things: challenges and opportunities in algorithms and hardware. IEEE Ind. Electron. Mag. **12**, 36–49 (2018). https://doi.org/10.1109/MIE.2018.2824843
8. Valenti, M., Squartini, S., Diment, A., Parascandolo, G., Virtanen, T.: A convolutional neural network approach for acoustic scene classification. In: 2017 International Joint Conference on Neural Networks (IJCNN), Anchorage, AK, pp. 1547–1554 (2017)
9. Piczak, K.J.: Environmental sound classification with convolutional neural networks. In: 2015 IEEE 25th International Workshop on Machine Learning for Signal Processing (MLSP), Boston, MA, pp. 1–6 (2015)
10. o Mueller, P., Ukil, A., Andenna, A.: Intelligent motor control. ABB Review, pp. 27–31 (2010)
11. Kirthika, V., Veeraraghavan, A.K.: Design and development of flexible on-board diagnostics and mobile communication for internet of vehicles. In: International Conference on Computer, Communication, and Signal Processing (ICCCSP), Chennai, pp. 1–6 (2018)
12. Mrs. Abdu-Aljabar, Rana, D.: JDesign and implementation of neural network in FPGA. J. Eng. Dev. **16**(3), 73 (2012). ISSN 1813– 7822

Deep Incremental Learning for Big Data Stream Analytics

Suja A. Alex[1](✉) and J. Jesu Vedha Nayahi[2]

[1] Information Technology, SXCCE, Nagercoil, India
suja@sxcce.edu.in
[2] Computer Science and Engineering, AURC, Tirunelveli, India
vedhaj2000@gmail.com

Abstract. Internet of Things (IoT) consists of many physical devices that generate data continuously and communicate these data among them. This data is continuously evolving large in volume with time varying nature. Processing these data is critical in all applications including monitoring and control applications. The continuous flow of data and time critical decisions make it infeasible to process the data after storage. Hence, the data must be processed on the fly or online. Online learning or incremental learning algorithms are suitable for processing stream data. Traditional machine learning algorithms are not scalable and so, Deep learning would be of great benefit to analyse steam data. This paper provides a survey on the different approaches used for processing data streams. Concept drift is a significant challenge identified and the paper compares the different works to handle concept drift.

Keywords: IoT data · Stream analytics · Concept drift · Deep learning

1 Introduction

Recently, there is a tremendous growth in the number of systems generating huge amount of data streams. Internet of things (IoT), Cyber-Physical Systems, Recommender Systems and Social Networks are some of the applications generating data continuously. In IoT environment, all the devices such as sensors are connected together and its applications include smart city, smart grid and smart health. These devices communicate by exchanging data among them. This IoT network generates large volume of data with high speed. Hence, this high speed data needs to be processed on the go. The nature of the data can be of different kinds, such as structured, semi-structured and unstructured. The difference between IoT data and normal Big Data is that the IoT data is generated by various sensors in the network. The IoT data also has the characteristics such as heterogeneity, noise, variety and rapid growth. Integration of IoT and Big Data solve the problems related to storage, processing and data analysis [1]. This article explores challenges of processing the large IoT data within a short span of time.

Machine learning algorithms are suitable for static data sets. As the world is dynamic, online learning or stream learning algorithm are suitable to process the real

© Springer Nature Switzerland AG 2020
A. P. Pandian et al. (Eds.): ICCBI 2018, LNDECT 31, pp. 600–614, 2020.
https://doi.org/10.1007/978-3-030-24643-3_72

time stream data. Streaming algorithms process examples in one pass with limited memory and limited processing time [2].

Deep Learning is a learning methodology to learn dynamic data streams than machine learning techniques. Deep learning over IoT data provides fast and accurate solution at the right time [3].

The major challenge of IoT data analytics is to handle data of different distributions. During data stream processing, concept change or concept drift must be detected. And, it is hard to quantify the drift rate in non-stationary environment. The objective of this article is to make a survey on the existing literature on IoT data that process the data by detecting concept drift and measuring drift rate or drift magnitude.

2   Data Stream Analytics

2.1   Data Stream Mining

Big data stream is hard to analyze with traditional data mining techniques. It can be handled and processed efficiently using Data stream mining methods. Data stream mining is a process of applying various data mining tasks over a sampled data stream or synopsis created. It includes the tasks such as clustering, classification, association rule mining, frequent item mining and regression. These types of analysis can be done on a continuous data stream.

2.1.1   Classification on Data Streams

Stream data has two significant properties namely infinite length and evolving nature. SyncStream [26] is a classifier which classifies evolving data streams. IBL Streams is another classifier that keeps recent instances while SyncStream dynamically learns a set of prototypes. Identifying a new concept poses a real challenge in stream environment. A CLAss-based Micro (CLAM) classifier ensemble is a more popular data stream classification method. CLAM successfully detects new concept as novel class [27]. While classifying the stream data, concept drift problem occurs. Some of the performance metrics for classification under concept drift are Kappa measure, Temporal Kappa statistics, Memory used, Speed of processing, Confusion matrix and Prequential [28].

2.1.2   Clustering on Data Streams

The challenges in stream clustering are the grouping of the algorithm data in a single pass over the raw data, fast response [24] and calculation of similarity or dissimilarity measure on data. ClipStream approach is used in smart grid environment in order to cluster multiple data streams [29]. Sliding window technique and Data labelling technique are used to group categorical data streams [30]. Due to the presence of concept drift in data streams, it is hard to cluster. Moreover, the estimated cluster may be revealing the true similarity; therefore it is sometimes hard to differentiate outlier and new concept during data stream clustering process [25].

2.1.3   Online Feature Selection

Online feature selection is another challenging task in Data Stream mining. Bayesian framework [31] and Accelerated Particle Swarm Optimization (APSO) [32] can be used for feature selection in data streams. All the features are not necessary. Data Stream can be processed with few optimal features [6]. But finding an optimal feature is an arduous, and this problem is NP-Hard [7]. Some of the online feature selection methods [8] are Group Feature Selection with Streaming Features (GFSSF) [9] and Online Group Feature Selection (OGFS) [10]. GFSSF uses entropy and mutual information for the features. OGFS selects essential feature and reduces redundancy.

2.2   Concept Drift

Concept drift or concept change occurs in the continuous data generated by sensors in IoT. For example, data distribution at the beginning of a stream is not same after some time. Hence, concept drift is a major issue in stream processing and it is necessary to measure the drift. Handling concept drift, using online learning, has been a very challenging task in recent years. Since the drift has time varying nature, the drift can be measured based on the distance between the current distribution and the previous distribution. Some of the used common distance functions are Kullback Leibler (KL) Divergence [4] and Hellinger Distance [5].

Concept drift must be detected to maintain classification performance. It can be achieved for unlabelled data streams and the drift is predicted by the classifier trained with the limited number of labelled data samples. There are three categories of Concept Drift. They are as follows,

1. **Sudden drift:**
   It is also called as concept shift [59]. It arises when concept drifts occur abruptly. It can be seen when the data stream composed of concept c1 changes to concept c2 during a run.
2. **Gradual drift:**
   The concept c1 changes to c2 steadily with an overlapping period.
3. **Incremental drift:**
   It is also called as gradual drift. But, it includes more than two sources [59].
4. **Reoccurring drift:**
   It indicates that the concept which disappeared gets reactivated after some interval of time (or recurring concept).

2.2.1   Concept Drift Detection Techniques

Generally detecting a change in continuous stream data is difficult. But, it is possible to estimate the level of drift that arises using various techniques. The first approach to detect concept drift is based on distance criteria. Kullback-Leibler (KL) distance and Hotelling's T-square test for equal means (H) are likelihood based distance functions for multidimensional data. Secondly, semi-parametric Log-Likelihood (SPLL) is also a likelihood based distance function which detects a concept change in multivariate stream data. The semi-parametric log-likelihood criterion is better than KL and comparable to the Hotelling's T-square test for the non-normalized data and SPLL also performs well on stream data with fewer features [11]. Regional-density estimation is the second way to estimate the drift magnitude. For example, Nearest Neighbour-based density variation identification (NN-DVI) is a Regional-density based concept drift detection method [12]. The third method is based on selection probability. Sampling technique selects the data item based on the probability of recent data item. Reservoir-based sampling finds the probability of most recent items during sampling process. To maximize the selection probability of a newly arrived data item, weighted random sampling (WRS) method is used which assigns a weight to every item. Hence, the item is selected based on probability and its associated weight [13] (Fig. 1).

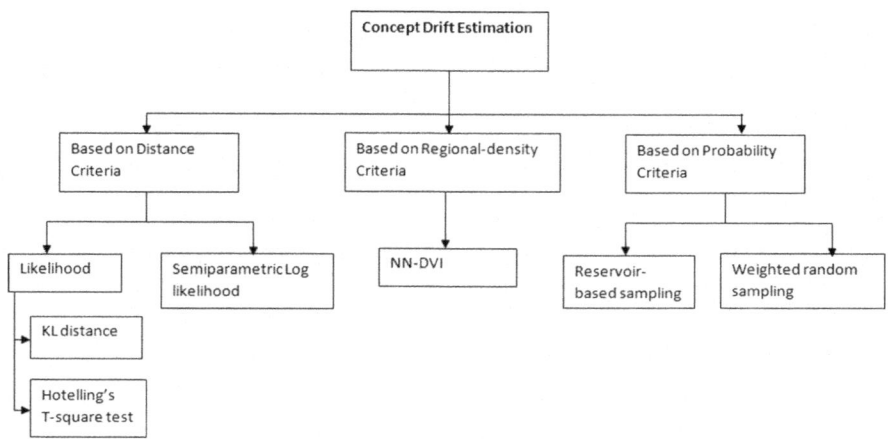

Fig. 1. Concept drift estimation techniques

Sometimes the drift might be asynchronous. PINE framework handles asynchronous drifting concepts in distributed networks [14]. The changes in data streams can also be detected using incremental unsupervised learning called Growing Neural Gas Algorithm (GNG-A). It maintains a continuously updated network as a graph of neurons [15]. An ensemble stream classifier called Knowledge-Maximized Ensemble (KME) is used to handle different types of concept drift. KME can react to multiple types of concept drift by combining the mechanisms of online and chunk-based ensembles [16]. Proactive approach called Detect Abrupt Drift (Detect A) method is used [17] to detect abrupt drift. Another proactive approach called, Competence-based concept detection, requires no prior knowledge of case distribution [18].

The concept drift detection algorithms can be categorized on the basis of adaptive base learners, modifying the training set and ensemble techniques. Adaptive base learners deal with changing concept in training dataset. It can easily adapt to new data with new concept. Adaptive base learning methods can be classified into Decision Tree (DT), Very Fast Decision Tree (VFDT) [45], Hoeffding Window Tree (HWT) and Hoeffding Adaptive Tree (HAT) [46], Concept Drift (CD3) algorithm [47]), k-nearest neighbours [48] and Fuzzy ARTMAP [49]. The Concept-adapting Very Fast Decision Tree (CVFDT) algorithm uses fixed-size window and it must be updated with new data items [73]. The second category is based on Instance Selection and Instance Weighting. The classifier can be built from the selected weight instances. Some of the algorithms of this category are Windowing (FLORA3 [50]) and Weighting techniques (Weighted KNN [51]). Windowing technique handles both gradual and sudden drifts. Gradual drifts need large window size while abrupt (Sudden) drifts need small window size [57]. New training data should be more important than older one and their importance should decrease with time. Hence, Kernel function (SVMs and neural networks) can be used to calculate weights using gradual forgetting function [59]. The third category of concept drift detection algorithms are based on ensemble techniques. Ensemble methods combine multiple classifiers to provide accurate predictions for future data items. It is suitable to detect reoccurring concept drift. Accuracy weighted ensembles (streaming ensemble algorithm [52], Dynamic Weighted Majority (DWM) [53]), Bagging and boosting based methods (Learn++ [54], ADWIN bagging [55]) and Concept locality based approaches (adaptive classifier ensemble (ACE) [56]) are popular ensemble based methods.

Concept drift detection approaches can be broadly classified into Active and Passive approaches. Active approaches can work for detecting abrupt drift and Passive approaches works well for gradual drift or slowly evolving streams [34]. Drift Detection Method (DDM) can detect context changes in the data distribution [74]. It can detect both sudden and gradual drifts. Early Drift Detection Method (EDDM) detects gradual drifts based on the distance between two errors [75]. To compare the distinctness among two windows of data instances, Adaptive Windowing (ADWIN) algorithm is used. This maintains a sliding window with latest data items and drops the old one if recent window is different from the old window [73]. It can detect abrupt changes in the data stream as well. The Weighted Majority Algorithm (WMA) captures all data items from the stream and the classifier is assigned to initial value 1. If the classifier commits error, its weight is reduced to some factor [77]. The changes in single instance can be detected by the algorithm called Exponentially Weight Moving Average (EWMA) [73]. It is also used to estimate confidence interval [42]. ADWIN and EWMA are the examples of forgetting mechanism based Hoeffding Trees. The categorical attributes are classified based on the concept description found by FLORA 3. It classifies the new incoming samples based on the generated concept description [76]. Paired Learners (PL) considers only two learners that detect gradual drift with less time and space [40]. Drift Detection Method based on Hoeffding's inequality (HDDM) can be used to detect drifts when all the incoming data items are labelled [42]. Adaptive-Size Hoeffding Tree (ASHT) takes little time to adapt to new changes in the data stream. The speed of ASHT Bagging can be improved by error change detector and can be added to each classifier [71]. To achieve good classification accuracy with less memory resource, Instance Based (IB3) algorithm can be used. It identifies the best instances for making a classifier [69]. Dynamic Weighted Majority (DWM) is an extended version of WMA. In DWM, each classifier is assigned with some weight. It can add or drop classifiers based on the error committed by the classifiers. If a classifier commits error, its weight is reduced and it can also be dropped from the ensemble. Accuracy Updated Ensemble (AUE) follows incremental fashion but uses batch classifiers [72]. The concept Drift (CD3) detection algorithm considers the data valid and stores it only if the time-stamp of recent data is new; otherwise it drops the old data item [47]. Since all the past instances are not important, SERA is used for selecting best instances based on Mahalanobis distance [68]. Hoeffding Tree with Naive Bayes (HTNB) has influence capacity [72]. Statistical Test of Equal Proportions (STEPD) detects mabrupt drift andny types of concept drift swiftly [58].

Table 1 shows the merits and the future scope of the research in various concept detection methods:

Table 1. Comparison between various concept drift detection techniques

Sl. no	Method	Type of drift	Merit	Future scope
1.	DDM	Abrupt drift and gradual drift	The error rate is based on number of misclassified elements [35]	DDM with incremental decision tree learning algorithm can be used for fast detection. The detection speed can be improved [35]. Very slow gradual changes can be detected [58]
2.	EDDM	Gradual drift	It improves the detection in presence of gradual concept drift with good performance for abrupt concept drift. It can work with slow gradual change type of data stream [36]. It estimates Binomial distribution [71, 74]	It can find a way to determine the values for the parameters of the method. It detects the drift in noisy training data [58]. It can be improved to detect in the presence of noise. Since it can work with slow gradual change type of data stream gradual change, memory space can be improved [36]
3.	ADWIN	Abrupt drift	It is able to track the true centroids more accurately than fixed-window method [37]	Kalman filter can be integrated with ADWIN to give more weight to most recent examples in the window [37]
4.	EWMA	Gradual drift	It monitors false positives and find misclassification rate [38]	It can be extended to classification problem with more than two classes [38]
5.	STEPD	Abrupt drift and Gradual drift	Identify many false positives. It can detect in the presence of noise. Removes stored examples (or concepts) [39]	Other statistical tests can be used to detect concept drift in data streams [39, 58]

(*continued*)

Table 1. (*continued*)

Sl. no	Method	Type of drift	Merit	Future scope
6.	PL	Gradual drift	It maintains and trains only two learners with less time and space [40]	It can be expanded to include single-model methods for concept drift, such as window and Stagger, and other instance generators, such as that used to evaluate AWE [40]
7.	Re-sampling	Gradual drift	It identifies drifts only with variation in the target concept. Robust to noise. The precision and recall rates are good [41]	The re-sampling algorithm can be extended to online [41]
8.	HDDM	Gradual drift	The learning accuracy can be improved [42]	This method can be integrated with other learning algorithms, different weighting schemes for real-world problems [42]
9.	DWCDS	Gradual drift	It detects concept drifts routinely. It is robust to noise. Its classification accuracy is good [43]	DWCDS can be improved to create multiple decision trees [43]
10.	ACED	Gradual drift	It detects concept drift quickly if the window size is small [58]	The drift could be detected for small window size. The misclassifications can be reduced. The algorithm can be improved to detect drifts in the presence of noise [58]
11.	FLORA	Gradual drift	It can work with fixed window size [59]	The dynamic window size can be added [59]
12.	FLORA2	Gradual drift	It dynamically adjusts with the window size. It has better generalization [59]	The recurring concepts can be detected [59]
13.	FLORA3	Reoccurring drift	It can detect recurring concepts [59]	It can handle noisy data [59]

Table 2 shows comparison among various concept drift detection methods based on the performance parameters such as accuracy, time, memory usage, standard deviation and number of instances selection:

Table 2. Performance comparison between concept drift detection methods

Method	Performance parameters				
	Accuracy	Time	Memory	Standard deviation	Number of instances
DDM	–	–	Larger window size leads to more memory beyond the limit [70]	–	–
EDDM	–	–	–	It has low standard deviation [72]	–
EWMA	–	–	It consumes less memory [38], [73]	–	–
CVFDT	Its accuracy is good [73]	–	It needs more memory [73]	–	–
ASHT	–	Needs more time [71]	Needs more memory [71]	–	–
DWM	Less accuracy when it detects gradual concept drift [72]	–	–	Low standard deviation [72]	–
AWE	–	AWE is suited for large data streams when the algorithm improves over time [70], [72]	–	–	–
AUE	The classifiers are updated only if the recent data has high accuracy [72]	–	–	–	–
WMA	Its accuracy is improved when it reaches last concept	–	–	–	–

(*continued*)

Table 2. (*continued*)

Method	Performance parameters				
	Accuracy	Time	Memory	Standard deviation	Number of instances
HTNB	It has low accuracy [72]	–	–	It has low standard deviation [72]	–
PL	–	It requires less time [40].	It requires less space [40]	–	–
IB 3	High classification accuracy [69]	–	Less Storage requirements [69]	–	–
SERA	–	–	–	–	The algorithm may not track drift on the selected instances [68]
FLORA 3	If the coverage is high and the window size is same, then good predictive accuracy can be achieved [67]	–	–	–	–
CD 3	–	If the time-stamp of recent data is same as with previous time-stamp, the old data items are removed [47]	–	–	–

3   Deep Learning

Deep Learning is a new machine learning methodology which is useful for good data representation. Its structure is based on the brain called Artificial Neural Network. DL can be used for processing IoT data. Many IoT applications depend on DL in the recent years since DL models have good accuracy [44].

3.1   Neural Network

Artificial Neural Network (ANN) has many interconnected artificial neurons. Neural network is a machine learning algorithm that solves classification and regression problems. But traditional ANN is not suited for Big Data Analytics, because big stream data has the characteristics such as increased volume, variety and complexity of the data streams that need to be extracted, transformed, analyzed, stored and visualized [19].

3.2   Deep Neural Network

Deep learning is a new methodology to train a neural network for complex big data analytics. Deep Neural Network (DNN) is an extension of the shallow ANN structure [20]. DNN can be used for big data stream analytics. It overcomes the difficulty of training a multilayer neural network. Deep learning models are formed by stacking many layers. It can be formed by either stacking many auto encoders or many Restricted Boltzman Machines (RBM) [21]. RBM is a two layered stochastic NN. It is used for unsupervised learning. It is a basic building block to build a deep neural network. Each RBM is equivalent to one layer of NN. For example, if the multi layer NN has five hidden layers, it is equivalent to five RBMs stacked to form a DNN.

3.3   Deep Auto Encoder

Autoencoder is also a simple neural network. In its architecture, Autoencoder comprises of two parts called encoder and decoder [22]. It is also used for unsupervised learning. The auto encoders can also be stacked to form a deep neural network. In data stream processing, all the data cannot be stored and it might be compressed while processing. Autoencoder performs compression of stream data [23].

4   Issues and Challenges of IoT Data

IoT is a complex environment that has many heterogeneous components. These sensors generate huge amount of data in the real world. Here, the great demands are processing and storage of data [33]. Some of the challenges of IoT data are:

- Managing IoT data
  - The IoT data can be managed using three techniques including Caching, Data partition, Multi-threading [60].
- Securing IoT data
  - Eavesdroppers (example: device manufacturers) gather data about users without their knowledge and they control IoT devices [63].
- Memory efficient architecture
  - Mongo DB architecture is a memory level analytic platform to conduct real-time analysis. When the data size is smaller than the maximum memory of the cluster, the data can be stored in Mongo DB [64].

- Missing value
  - Missing data is a big challenge in IoT applications. Because if the server gets missing data from sensors, its decision making is affected. Probabilistic Matrix Factorization (PMF) is a Bayesian method that recovers incomplete or missing data generated by IoT sensors [65].
- Feature selection
  - Interpreting a raw sensor data into meaningful one is an open challenge. Feature selection finds abstract data or representative dataset from the IoT sensor data [66].
- Storing only necessary data
  - Instead of storing entire data, only important data can be extracted from IoT data and converted it to compatible form suitable for the Data storage system [61].
- Time series values
  - To process huge amount of real-time IoT sensor data for scalable systems, memory and computational cost must be reduced [62].

5   Conclusion

Concept drift, large length and feature selection are the common challenges in upcoming data streams. Online learning algorithm process the data generated in changing environment. Hence, we need a deep neural network framework with incremental learning algorithm to process IoT data and to meet the requirement of non-stationary environment. The performance of framework depends on how the algorithm adapts to concept drift and handle online learning environment. The model's performance can be evaluated on the basis of error rate, accuracy, recall and precision.

References

1. Nicolalde, F.C., Silva, F., Herrera, B., Pereira, A.: Big data analytics in IOT: challenges, open research issues and tools, WorldCIST'18 2018, AISC 746, pp. 775–788 (2018). Springer International Publishing AG, part of Springer Nature 2018
2. Gama, J., Zliobaite, I., Bifet, A., Mykola: A survey on concept drift adaptation. ACM Comput. Surv. (CSUR) **46**(4) (2014)
3. Zhu, D.: Deep, Learning over IoT big data-based ubiquitous parking guidance robot for parking near destination especially hospital, personal and ubiquitous computing. Springer-Verlag London Ltd., part of Springer Nature (2018)
4. Kullback, S., Leibler, R.A.: On information and sufficiency. Ann. Math. Stat. **22**(1), 79–86 (1951)
5. Hoens, T,R., Chawla, N.V., Polikar, R.: Heuristic updatable weighted random subspaces for non-stationary environments. In: IEEE International Conference on Data Mining, ICDM-11, IEEE, pp. 241–250 (2011)
6. Wu, X.D., Yu, K., Wang, H., Ding, W.: Online streaming feature selection. In: Proceedings of the 27th International Conference on Machine Learning, 1159–1166 (2010)
7. Guyon, I., Elisseeff, A.: An introduction to variable and feature selection. J. Mach. Learn. Res. **3**, 1157–1182 (2003)

8. Hu, X., Zhou, P., Li, P., Wang, J., Wu, X.: A survey on online feature selection with streaming features. Front. Comput. Sci. **12**, 479 (2018)
9. Li, H.G., Wu, X.D., Li, Z., Ding, W.: Group feature selection with streaming features. In: Proceedings of the 13th IEEE International Conference on Data Mining, pp. 1109–1114 (2013)
10. Wang, J., Wang, M., Li, P.P., Liu, L.Q., Zhao, Z.Q., Hu, X.G., Wu, X.D.: Online feature selection with group structure analysis. IEEE Trans. Knowl. Data Eng. **27**, 3029–3041 (2015)
11. Kuncheva, L.I.: Change detection in streaming multivariate data using likelihood detectors. IEEE Trans. Knowl. Data Eng. **25**(5) (2013)
12. Liu, A., Lu, J., Liu, F., Zhang, G.: Accumulating regional density dissimilarity for concept drift detection in data streams. Pattern Recognit. **76**, 256–272 (2018)
13. Efraimidis, P.S.: Weighted random sampling over data streams. In: Algorithms, Probability, Networks, and Games, pp. 183–195. Springer, Cham (2015)
14. Ang, H.H., Gopalkrishnan, V., Zliobaite, I., Pechenizkiy, M., Hoi, S.C.H.: Predictive handling of asynchronous concept drifts in distributed environments. IEEE Trans. Knowl. Data Eng. **25**(10) (2013)
15. Bouguelia, M.R., Nowaczyk, S., Payberah, A.H.: An adaptive algorithm for anomaly and novelty detection in evolving data streams. Data Min. Knowl. Disc. **32**, 1597–1633. Springer (2018)
16. Ren, S., Liao, B., Zhu, W.: Li, K.: Knowledge-maximized ensemble algorithm for different types of concept drift. Inf. Sci. **430–431**, 261–281 (2018)
17. Escovedo, T., Koshiyama, A., Abs da Cruz, A., Vellascoa, M.: DetectA: abrupt concept drift detection in non-stationary environments. Appl. Soft Comput. **62**, 119–133 (2018)
18. Lu, N., Zhang, G., Lu, J.: Concept drift detection via competence models. Artif. Intell. **209**, 11–28 (2014)
19. Awad, M., Khanna, R.: Deep Learning in Efficient Learning Machines. Apress, Berkeley, CA, Springer (2015). 978-1-4302-5990-9
20. Geoffrey, E., Osindero, S., Teh, Y.-W.: A fast learning algorithm for deep belief nets. J. Neural Comput. **18**(7), 1527–1554 (2006)
21. Cheng, W., Sun, Y., Li, G. et al.: Jointly network: a network based on CNN and RBM for gesture recognition. Neural Comput. Appli. (2018)
22. Khan, Q.S.U., Li, J., Shuyang, Z.: Training Deep Autoencoder via VLC-Genetic Algorithm. Springer International Publishing AG, Switzerland (2017)
23. Hatcher, W.G., Yu, W.: A survey of deep learning: platforms, applications and emerging research trends. In: Special Section on Human-Centered Smart Systems and Technologies, IEEE (2018)
24. Mansalis, S., Ntoutsi, E., Pelekis, N., Theodoridis, Y.: An Evaluation of Data Stream Clustering Algorithms. Wiley, Hoboken (2018)
25. Abdullatif, A., Masulli, F., Rovetta, S.: Clustering of nonstationary data streams: A survey of fuzzy partitional methods. Wiley (2018)
26. Shao, J., Huang, F., Yang, Q., Luo, G.: Less, robust prototype-based learning on data streams. In: IEEE Transactions on Knowledge & Data Engineering, vol. no. 01 (2018). ISSN: 1041-4347
27. Islam, M.R.: Recurring and novel class detection using class-based ensemble for evolving data stream, advances in knowledge discovery and data mining. PAKDD 2014. Lecture Notes in Computer Science, vol. 8444. Springer (2014)
28. Janardan, Mehta, S.: Concept drift in streaming data classification: algorithms, platforms and issues, ITQM2017. Procedia Comput. Sci. **122**, 804–811 (2017)

29. Laurinec, P., Lucka, M.: Interpretable multiple data streams clustering with clipped streams representation for the improvement of electricity consumption forecasting. Data Min. Knowl. Discov. Springer (2018)
30. Li, Y., Li, D., Wang, S., Zhai, Y.: Incremental entropy-based clustering on categorical data streams with concept drift. Knowl.-Based Syst. **59**, 3–47 (2014)
31. Turkov, P., Krasotkina, O., Mottl, V., Sychugov, A.: Feature selection for handling concept drift in the data stream classification. In: Machine Learning and Data Mining in Pattern Recognition (MLDM). Springer (2016). 978-3-319-41920-6
32. Fong, S., Wong, R., Vasilakos, A.: Accelerated PSO swarm search feature selection for data stream mining big data. IEEE Trans. Serv. Comput. **9**(1), 1–1 (2015)
33. Delicato, F.C., et al.: Resource Management for Internet of Things. Springer Briefs in Computer Science (2017)
34. Ditzler, G., Roveri, M., Alippi, C., Polikar, R.: Learning in nonstationary environments: a Survey. IEEE Comput. Intell. Mag. **10**, 12–25 (2015)
35. Gama, J., Medas, P., Castillo, G., Rodrigues, P.: Learning with Drift Detection, SBIA 2004, LNAI 3171, pp. 286–295. Springer (2004)
36. Baena, M., del Campo- Avila1, J., Fidalgo, R., Bifet, A., Gavalda, R., Morales-Bueno, R.: Early drift detection method. Springer (2005)
37. Bifet, A., Gavalda, R.: Learning from time-changing data with adaptive windowing. Springer (2007)
38. Ross, G.J., Adams, N.M., Tasoulis, D.K., Hand, D.J.: Exponentially weighted moving average charts for detecting concept drift. Springer (2012)
39. de Barros, R.S.M., Hidalgo, J.I.G., de Lima Cabr, D.R.: Wilcoxon rank sum test drift detector, Neurocomputing, January 2018
40. Bach, S.H., Maloof, M.A: Paired learners for concept drift. In: Eighth IEEE International Conference on Data Mining (2008)
41. Harel, M., Crammer, K., El-Yaniv, R., Mannor, S.: Concept drift detection through resampling. In: ICML'14 Proceedings of the 31st International Conference on International Conference on Machine Learning, vol. 32, pp. II-1009-II-1017
42. Frias-Blanco, I., del Campo-Avila, J., Ramos-Jimenez, G., Morales-Bueno, R., Ortiz-Diaz, A., Caballero-Mota, Y.: Online and non-parametric drift detection methods based on Hoeffding's bounds, IEEE (2015)
43. Zhu, Q., Hu, X., Zhang, Y., Li, P., Wu, X.: A double-window-based classification algorithm for concept drifting data streams, IEEE (2010)
44. Mohammadi, M., Al-Fuqaha, A., Guizani, M.: Deep learning for IoT big data and streaming analytics: a survey. IEEE Commun. Surv. Tutor. **20**, 2923–2960 (2018)
45. Domingos, P., Hulten, G.: Mining high-speed data streams. In: KDD, pp. 71–80. ACM, New York (2000)
46. Bifet, A., Gavalda, R.: Adaptive learning from evolving data streams. In: IDA, pp. 249–260 (2009)
47. Black, M., Hickey, R.: Learning classification rules for telecom customer call data under concept drift. Soft Comput. Fusion Found. Methodol. Appl. **8**(2), 102–108 (2003)
48. Alippi, C., Roveri, M.: Just-in-time adaptive classifiers in non-stationary conditions. In: IJCNN, pp. 1014–1019. IEEE, New York (2007)
49. Carpenter, G., Grossberg, S., Markuzon, N., Reynolds, J., Rosen, D.: Fuzzy artmap: a neural network architecture for incremental supervised learning of analog multidimensional maps. TNN **3**(5), 698–713 (1992)
50. Widmer, G., Kubat, M.: Effective learning in dynamic environments by explicit context tracking. In: ECML, pp. 227–243. Springer, Berlin (1993)

51. Alippi, C., Boracchi, G., Roveri, M.: Just in time classifiers: managing the slow drift case. In: IJCNN, pp. 114–120. IEEE, New York (2009)
52. Street, W., Kim, Y.: A streaming ensemble algorithm (SEA) for large-scale classification. In: KDD, pp. 377–382. ACM, New York (2001)
53. Kolter, J., Maloof, M.: Dynamic weighted majority: a new ensemble method for tracking concept drift. In: ICDM, pp. 123–130. IEEE, New York (2003)
54. Kubat, M.: Floating approximation in time-varying knowledge bases. PRL **10**(4), 223–227 (1989)
55. Bifet, A., Holmes, G., Pfahringer, B., Kirkby, R., Gavaldà, R.: New ensemble methods for evolving data streams. In: KDD, pp. 139–148. ACM, New York (2009)
56. Nishida, K., Yamauchi, K., Omori, T.: Ace: adaptive classifiers-ensemble system for concept-drifting environments. In: MCS, pp. 176–185 (2005)
57. Widmer, G., Kubat, M.: Learning flexible concepts from streams of examples: Flora2. In: Proceedings of the 10th European Conference on Artificial Intelligence (ECAI 1992), pp. 463–467 (1992)
58. Nishida, K., Yamauchi, K.: Detecting concept drift using statistical testing. In: International Conference on Discovery Science DS 2007: Discovery Science, pp. 264–269. Springer (2007)
59. Knotek, J., Pereira, W.: Survey on Concept Drift
60. Lu, T., Fang, J., Liu, C.: A Unified storage and query optimization framework for sensor data. In: 12th Web Information System and Application Conference (WISA), pp. 229–234. 11–13 Sept 2015. ISBN: 978-1-4673-9371-3
61. Cerbulescu, C.C., Cerbulescu, C.M.: Large data management in IOT applications. In: 17th International Carpathian Control Conference (ICCC) (2016)
62. A Knowledge-based Approach for Real-Time IoT Data Stream Annotation and Processing
63. Sivaraman, V., Gharakheili, H.H., Vishwanath, A., Boreli, R., Mehani, O.: Network-level security and privacy control for smart-home IoT devices. In: Eight International Workshop on Selected Topics in Mobile and Wireless Computing, IEEE (2015)
64. Chodorow, K.: MongoDB: The Definitive Guide. O'Reilly Media Inc, Newton, MA, USA (2014)
65. Fekade, B., Maksymyuk, T., Kyryk, M., Jo, M.: Probabilistic recovery of incomplete sensed data in IoT, IEEE (2017)
66. Ganz, F., Puschmann, D., Barnaghi, P., Carrez, F.: A practical evaluation of information processing and abstraction techniques for the internet of things. IEEE Internet Things J. **2**, 340–354 (2015)
67. Ryan Hoens, T., Polikar, R., Chawla, N.V.: Learning from streaming data with concept drift and imbalance: an overview. Prog. Artif. Intell. **1**, 89–101 (2011)
68. Chen, S., He, H.: Sera: selectively recursive approach towards nonstationary imbalanced stream data mining. In: IJCNN, pp. 522–529. IEEE, New York (2009)
69. Aha, D.W., Kibler, D., Albert, M.K.: Instance-based learning algorithms. Mach. Learn. **6**, 37–66 (1991)
70. Dongre, P.B., Malik, L.G.: Real time data stream classification and adapting to various concept drift scenarios. In: IEEE International Advance Computing Conference (IACC) (2014)
71. Bifet, A., Holmes, G., Pfahringer, B., Gavalda, R.: Improving adaptive bagging methods for evolving data streams. In: ACML 2009, LNAI 5828, pp. 23–37. Springer-Verlag Berlin Heidelberg (2009)
72. Gonçalves, P.M., de Barros, R.S.M.: RCD: a recurring concept drift framework. Pattern Recognit. Lett. **34**, 1018–1025 (2013)

73. Brzezinski, D.: Mining data streams with concept drift, Thesis (2010). https://doi.org/10. 13140/rg.2.1.4634.6086
74. Gama, J., Medas, P., Castillo, G., Rodrigues, P.: P, Learning with drift detection. In: Bazzan, A., Labidi, S. (eds.) Advances in Artificial Intelligence –SBIA 2004. Lecture Notes in Computer Science, vol. 3171, pp. 66–112. Springer, Berlin, Heidelberg (2004)
75. Baena-Garcı́a, M., Del Campo-Ávila, J., Fidalgo, R., Bifet, A., Gavaldà, R., Morales-Bueno, R., Early drift detection method. In: Internet Workshop on Knowledge Discovery from Data Streams of IWKDDS 2006, vol. 6, Citeseer, pp. 77–86 (2006)
76. Widmer, G., Kubat, M.: Learning in the presence of concept drift and hidden contexts. Mach. Learn. **23**, 69–101 (1996)
77. Blum, A.: Empirical support for winnow and weighted-majority algorithms: results on a calendar scheduling domain. Mach. Learn. **26**, 523 (1997)

Internet of Things Based Reliable Real-Time Disease Monitoring of Poultry Farming Imagery Analytics

S. Balachandar[1(⊠)] and R. Chinnaiyan[2]

[1] Shell India Market Private Limited, Bangalore, India
aaathibala@gmail.com
[2] Department of Information Science and Engineering,
CMR Institute of Technology, Bangalore, India
vijayachinns@gmail.com

Abstract. IOT Industry prediction says 40 Billion connected devices by 2020. Most of the industries already started adopting the "connected" culture for expounding their products or services. Things or Sensors are not new to this world. The real utilization of sensor or things will be determined only by its real usage and increasing the business value. Poultry farming is one of the greatest contributor for our country's GDP (i.e. Agriculture's contribution). This paper explains how effectively technology like IOT, Imagery analytics helps the farmer to control & reduce the bird's mortality rate, recommendation on medicine dosage for birds by remote monitoring. The doctors or bird specialist who leverage advanced analytics (i.e. Imagery analytics) to decide the bird's health and probable recommendation for medicines to control the mortality and disease spreading across poultry. Images taken from the poultry will be stored in an unstructured store with its supporting metadata and data. The algorithmic imagery models will help to identify the deficiencies and recommend the dosages according to the color and stains in different parts of the bird's body.

Keywords: Data lake · Imagery models · Image frames ·
Internet of things (IOT) · Machine learning · Hadoop · Field sensors

1 Introduction

Most of the poultry farm is monitored by the farmer who doesn't know much about the bird's health & medicines and they seek inputs from the Veterinary consultant or doctor who recommend the medicines and dosages based on bird's condition. The availably of doctor's time and time to visit the poultry farm is nearly impossible due to various reasons (e.g. commutation, other business priorities, time to rescue and recover the birds). The images or videos of Poultry farm and birds identity will be collected through CCTV cameras and body camera and other environmental sensor (e.g. Humidity, Temperature and Ammonia) will also ingested into a data lake like Hadoop or Cloud based storages like Amazon S3 or Azure Blob store. the birds' videos and images will be analyzed with individual frames and it will be detected whether it carries different color or blood stain or abnormal droppings (e.g.). The imagery algorithm will

© Springer Nature Switzerland AG 2020
A. P. Pandian et al. (Eds.): ICCBI 2018, LNDECT 31, pp. 615–620, 2020.
https://doi.org/10.1007/978-3-030-24643-3_73

help to predict the disease spread level and recommendation on dosages with alerts and precautionary messages to the farmer will be disseminated through a Mobile Application. The mobile application is a common channel between the farmer and doctor. Doctor will get the imagery model output of disease color and other related issues of the bird and it will be confirmed by the doctor and they might seek additional inputs through mobile app. At the end, farmer receives the right instruction on how to overcome that problem and ensure the birds mortality is under control.

1.1   Data Sources - Poultry

Precision poultry is trending across different countries based on observing, measuring, responding back to farm based on the real-time feeds coming out of the farm.

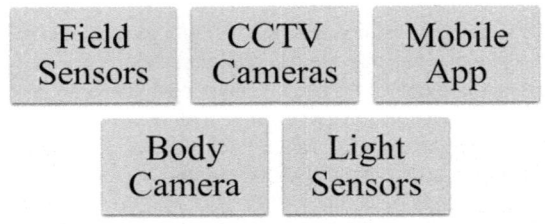

Fig. 1. Data sources of poultry

The components mentioned in Fig. 1 is covered in detail in below Table 1.

Table 1. Data source details

Data Source category	Devices/Channel	Data output	Collection mechanism
Field Sensor	Humidity Sensor	Humidity values ranges based on building type and local temperature. The values will range from 30 to 80% in India	Individual sensor installed in different places of the farm and will be communicating over IOT Edge or Gateway
	Temperature Sensor	Temperature is based on climatic condition and the heating condition in the farm. It will range from 15 to 55 degree Celsius in India	
	Ammonia Gas, Carbon Dioxide Sensor	NH3 (Ammonia) Values ranges from 0 to 10 ppm Carbon Dioxide above 0.001 to 0.03% anything above the range will be considered as abnormal	

(continued)

Table 1. (*continued*)

Data Source category	Devices/Channel	Data output	Collection mechanism
CCTV Cameras	IP Camera of the farm	Stream of images	It will be collected and stored in different intervals (e.g. Minutes/Hour)
Body Camera	Body Camera of the Field Person	Small camera tells the geo position of the bird with its full details (color of the bird, position and its size, etc.)	
Light Sensor	House Light level	Average Lux, Evenness (min/max lux)	Light values will be accessed continuously and stored in data lake
Mobile App	Data Feed and Alert application	Notifications from Veterinary consultant and Inputs and messages from farmer	It will be stored in Mobile App for limited period and the historical data will be available in data lake like Hadoop or Azure or AWS

2   Data Processing and Imagery Analytics

Data collected from above source system will be stored in data lake (block storage like Hadoop (in-house) or cloud storages like Amazon S3 bucket or Azure Blob store) based on the poultry's network communication speed. Data coming from theses sensors and devices will be ingested through message brokers like Kafka or Rabbit MQ (Message Queue) tools which continuously help to do asynchronous communication between the ingestion layer and devices. After the data collected from these sensors and videos or images will be stored independent folders of data lake. The video file will be continuously collected and split into multiple image frames for analytics and detection of any signs for disease or infections. Metadata of the sensors and image metadata like Bird location, data time, height of the bird, etc. will be stored in RDBMS (Relational Database Management System) systems like MySQL or Postgre SQL server. The features of the images (e.g. Segment color, Body portions, blood signs, dropping color, bird head movement, etc.) will be extracted out of the image and the pictures of common diseases like Yoc Sac Infection, E.Coli, Toxcity, Gout, Brooder Pneumonia will be available in the trained data set and the associated parameters of sensor inputs will help to correlate and decide the scores of these images with disease match score.

The data will be trained continuously and it will be tagged by the Veterinary consultant for new disease or suspicious behavior of the bird. The tagged images will be added part of the trained data and new scoring will be calculated and correlated with other sensor readings to determine the disease risk and type of disease that the bird is having at the farm. Based on the score the application will send the alerts and notifications to both Veterinary consultant and the field farmer about this and farmer will immediately take the next action based on the medicine prescribed by the Veterinary consultant. This imagery model will not only help the Veterinary consultant to find the new disease but also save the bird's life by sending the communication and prescription on the fly. It will also save other birds who might get similar infection.

2.1   Literature Survey

Imagery analytics is a new topic for poultry monitoring and disease control. Several published literature and research methodology discussed and referred here.

- Wireless Sensor Network for Cattle Monitoring System - Tsung Ta Wu, Swee Keow Goo, Kae Hsiang Kwong, Craig Michie, Ivan Andonovic, University of Strathclyde, Department of EEE, Glasgow, United Kingdom, [twu/sweegoo/kwong/c.michie/i. andonovic@eee.strath.ac.uk] – it explains the cost effective wireless sensor network technology for monitoring health of cattle farm, also it touches the real-time reporting of animal movement through Implicit Routing Protocol (IRP). Their experiment indicated that how IRP can successfully address the routing path problem.
- Atmospheric ammonia alters lipid metabolism-related genes in the livers of broilers (Gallus gallus)- R.N. Sa, H. Xing, S.J. Luan, H.F. Zhang – it explains how ammonia causes the effects of livestock and animal health. The bird's weight gain and food intake is directly depends to ammonia level in the farm. If it exceeds 50 ppm ($p < 0.5$) it will impact the bird' fat synthesis in its liver.
- Deep Learning Algorithms with Applications to Video Analytics for A Smart City: A Survey, Li Wang, Member, IEEE, and Dennis Sng – It explains the deep learning in object detection, object tracking and face recognition, image classification and how image can be classified with different algorithms like CNN (Convolutional Neural Networks) and Convolutional Hierarchical Recurrent Neural Networks (C-HRNNs) and feature extraction process using Deep ID (J. Lu, V. E. Liong, G. Wang, and P. Moulin, "Joint feature learning for face recognition)

3   Deep Learning Model – Image Classification

As soon as we receive the video stream feeds from the farmer from the plant, the video streams will be converted into different image frames. Each frame can be feed into machine with a sequence number. We can use CNN (Convolutional neural network identifies features to illustrate the object in the image) (Fig. 2).

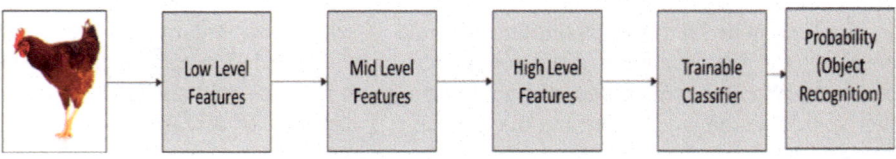

Fig. 2. Deep learning model hierarchy

Features are obtained as vector in different levels. The above example illustrates the learning with hierarchical representation. CNN (Convolutional neural network) model scales with model size and data set size and it can be trained with backpropagation. Every level mentioned above will transforms the images into neuron activations.

Training data set: the initial training data set will be loaded from open source image database which has birds' disease pictures, these are pre-trained and we can use ResNet50 which is one of the best pre-trained models in computer vision. Learning can also do with the help of unsupervised data (e.g. the veterinary consultant tags the picture once they identify the relevant disease for a particular bird). It needs large computation power (e.g. GPU – Graphical Processing Unit) to train a network with in short interval of time.

Understanding of ResNet50: It is 50 layer residual network. Deep convolutional networks lead to series of breakthroughs for image classification. Residual learnings helps to solve the training of neural network becomes difficult while it goes deeper. Residual can be simply understood as subtraction of feature learned from input of that layer.

Testing of the model: the model testing can be done with various deep learning framework and libraries like Tensor flow, Keras, DL4J.

Outcome of the model: the model classifies the birds image in different segments and identify the probable disease (e.g. Avian Flu) with probability level (e.g. 80%) for the bird.

4   Scoring Model

The output of image classifier with a probability values will be correlated with the sensor data set (e.g. Ammonia level, humidity level) of the poultry farm. The scoring model calculates the probability of correlated values and recommend the closest disease title. The scoring model value will be visible to the end user (Veterinary consultant) who can add additional message to the farmer with right prescription tags or links to follow.

5   Conclusion

Technology like IOT and deep learning models are still evolving and the model accuracy will not be gained in the initial stage and it needs constant training of image data sets will be key to prove this experiment will sustain and reliable. The precondition to test this platform is good wireless network to communicate between the farm and doctor or Veterinary consultant. The mobile application will hold good amount FAQ (Frequently Asked Questions) which assists the farmer in poultry farm to monitor the reliability of the sensor readings and values in real-time.

References

1. Deep learning algorithms with applications to video analytics for a smart city: a survey. https://arxiv.org/pdf/1512.03131.pdf
2. The design of chicken growth monitoring system for broiler farm partnership. Int. J. Comput. Sci. Electron. Eng. (IJCSEE) **2**(4) (2014) ISSN 2320–4028
3. Global poultry diagnostics market research report 2017: https://www.qyresearchreports.com/report/global-poultry-diagnostics-market-research-report-2017.htm
4. Kanjilal, D., Singh, D., Reddy, R., Mathew, J.: Smart farm: extending automation to the farm level. Int. J. Sci. Technol. Res. **3**(7), 109–113 (2014)
5. Bang, J., Lee, I., Noh, M., Lim, J., Oh, H.: Design and implementation of a smart control system for poultry breeding's optimal LED environment. Int. J. Control and Automation **7** (2), 99–108 (2014)
6. Corkery, G., Ward, S., Kenny, C., Hemmingway, P.: Incorporating smart sensing technologies into the poultry industry. J. World's Poult. Res. **3**(4), 106–128 (2013)
7. Mahale, R.B., Sonavane, S.S.: Smart poultry farm monitoring using IOT and wireless sensor networks. Int. J. Adv. Res. Comput. Sci. **7**(3), 187–190 (2016)
8. Goud, K.S., Sudharson, A.: Internet based smart poultry farm. Indian J. Sci. Technol. **8**(19), IPL101 (2015)
9. Wireless sensor network for cattle monitoring system: https://www.researchgate.net/publication/220938231_Wireless_Sensor_Network_for_Cattle_Monitoring_System?enrichId = rgreqdde69d721a484dbd0f3f1f81d78c1b88XXX&enrichSource=Y292ZXJQY WdlOzIyMDkzODIzMTtBUzoxMzY5NzUwMjE1MTQ3NTNAMTQwOTY2ODc5 ODkyNA%3D%3D&el=1_x_2&_esc=publicationCoverPdf0
10. Balachandar, S., Chinnaiyan, R.: Centralized reliability and security management of data in internet of things (IoT) with rule builder. Lecture Notes on Data Engineering and Communications Technologies **15**, 193–201 (2018)
11. Balachandar, S., Chinnaiyan, R.: Reliable digital twin for connected footballer. Lecture Notes on Data Engineering and Communications Technologies **15**, 185–191 (2018)
12. Balachandar, S., Chinnaiyan, R.: A reliable troubleshooting model for IoT devices with sensors and voice based chatbot application. Int. J. Res. Appl. Sci. Eng. Technol. **6**(2), 1406–1409 (2018)
13. Chinnaiyan, R., Somasundaram, S.: Evaluating the reliability of component based software systems. Int. J. Qual. Reliab. Manag. **27**(1), 78–88 (2010)
14. Chinnaiyan, R., Somasundaram, S.: Reliability of object oriented software systems using communication variables – a review. Int. J. Softw. Eng. **2**(2), 87–96 (2009)
15. Chinnaiyan, R., Somasundaram, S.: Reliability assessment of component based software systems using test suite - a review. J. Comput. Appl. **1**(4), 34 (2008)
16. Chinnaiyan, R., Somasundaram, S.: An experimental study on reliability estimation of GNU compiler components – a review. Int. J. Comput. Appl. (0975 – 8887) **25**(3), 13–16 (2011)
17. Chinnaiyan, R., Somasundaram, S.: Reliability of component based software with similar software components - a review. i-Manager's J. Softw. Eng. **5**(2), 44–49 (2010)
18. Chinnaiyan, R., Somasundaram, S.: Monte Carlo simulation for reliability assessment of component based software systems. i-Manager's J. Softw. Eng. **4**, 27–32 (2010)
19. Residual learning: https://www.quora.com/What-is-the-deep-neural-network-known-as-%E2%80%9CResNet-50%E2%80%9D

Impacts of Non-linear Effects in DWDM Based Fiber Optic Communication System

S. Esther Jenifa[1(✉)] and K. Gokulakrishnan[2]

[1] Department of ECE, Jayaraj Annapackiam CSI College of Engineering,
Nazareth, India
s.estherjenifa@gmail.com
[2] Department of ECE, Anna University Regional Campus, Tirunelveli, India
gk39762@gmail.com

Abstract. Nowadays, a technology is developed and there is a large amount of bandwidth capacity is necessary to the requirement of users. In this paper, the FWM power is calculated by varying the parameters such as channel spacing, effective area, fiber length and input power. The FWM power is suppressed by using hybrid modulation technique. OSNR is also calculated by varying the dispersion and input power. From this, we observed that the FWM power is reduced by increases channel spacing, fiber length, input power and decreases input power. The result is obtained in Optisystem.

Keywords: FWM four wave mixing ·
DWDM dense wavelength division multiplexing · Equal channel spacing ·
OSNR optical signal to noise ratio · DCF dispersion compensation fiber

1 Introduction

Communication means transmitting information from one place to another with wired or wireless. In fiber-optic communication the information is travel in fiber by sending pulses of light. DWDM is a technology having several no. of wavelengths in a single fiber with independent frequencies of each wavelength. It has the ability to provide unlimited transmission capacity. In this the non-linear effects will occur (i.e., FWM, CPM, SPM) due to refractive index of the optical fiber. This non-linear effects will degrade the system performance and limited the capacity of the fibre. Four Wave Mixing means interaction between the frequencies travel in the fiber and produces another frequencies or creates the sidebands. The generation of new wave is FWM products which will affect the performance of original signal. Habib Ullah Manzoor et al. [4] proposed that the FWM effect was reduced by using circular polarizes alternatively in each channel. Vardges, Vardanyan [2] proposed that the products occurring in the fiber is investigated by optical spectrum and monitor no. of products falling within a channel bandwidth. S. Sugumarani et al. [6] proposed that the FWM is suppressed by placing the optical phase conjugated in mid of the optical fiber link. Gurleen Kaur et al. [3] in this the FWM is suppressed by low level reduction, high level reduction and intermediate level reduction at different distances and different input power levels. RZ pulse generator has two transitions per bit and does not have any

© Springer Nature Switzerland AG 2020
A. P. Pandian et al. (Eds.): ICCBI 2018, LNDECT 31, pp. 621–629, 2020.
https://doi.org/10.1007/978-3-030-24643-3_74

self-clocking. It offers better immunity to fiber nonlinear effects because of high peak power. OSNR is defined as the ratio of optical signal power to the noise power. If the signal power is high compared to noise power then the degree of impairment is low. The requirement of OSNR will be high when it is close to the transmitter. At the receiver, the requirement is low because the signal from transmitter passes through optical amplifier. If the OSNR value is high above the target, then there is proper detection at the receiver. OSNR value is low means the receiver didn't detect or recover the signal.

2   Methodology

This setup is consists of 8 channels with 15 Gbps. The transmitter is consists of CW LASER, PRBS generator, RZ pulse Generator and Hybrid modulator. LASER is act as input source. The prbs generator is used and it selected the bit randomly. The output of prbs generator is binary signal (i.e., 0's or1's). The Hybrid modulator technique is used to reduce the FWM power in DWDM system. The phase modulator introduces the Phase mismatch which added to amplitude modulator. The EDFA is most commonly used in-fiber optical devices which restore the optical signal to its original power level. The combiner is used to combine all channels. Dispersion is occurring in fiber which means spreading of pulse and Dispersion Compensation Fiber (DCF) is used to neglect it. At the receiver, the output is determined (Fig. 1).

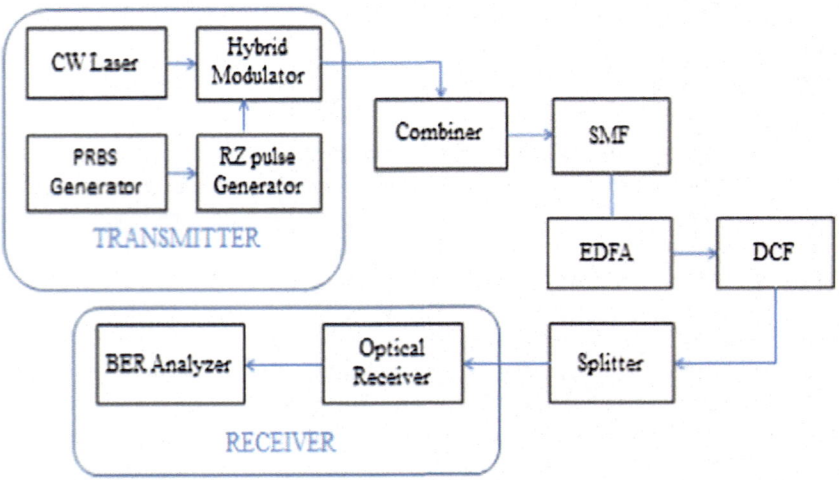

Fig. 1. Reduction of FWM using DWDM simulation setup.

3   Result and Discussion

3.1   Analyzing FWM Power Varying Different Parameter

Case 1

In this case, FWM investigate all the parameters involved in the power analysis and that helps to analyze the performance of overall power system. The different parameters included length of the fiber, affective area and spacing width of channel. By improving the performance of parameters, FWM performance has been degraded. The input power in laser is reduced by increasing the effective area, so the FWM product is minimized (Figs. 2, 3, 4 and 5).

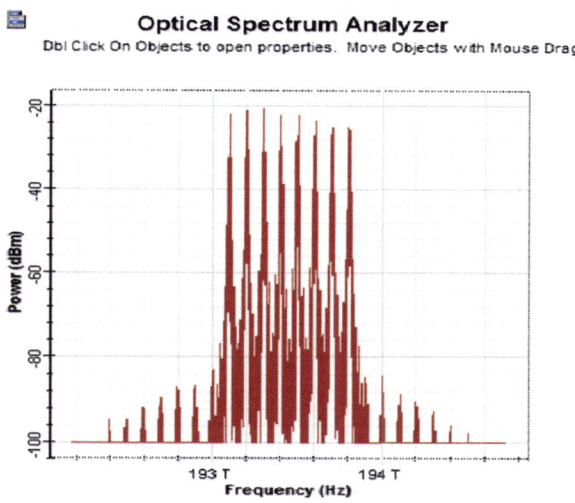

Fig. 2.   Channel spacing = 101 GHz + A_{eff} = 60 + fiber length = 105 km

Case II

In the case 2 situation, FWM investigate the four different parameters and that includes the length of fiber, Area utilization, channel space width and source power. Here by increasing the channel width spacing, there is a chance to reduce the input power and it leads to degrade the performance of FWM power (Figs. 6, 7, 8 and 9).

By comparing both cases, varying the combined of four parameters gives the better result and FWM products is reduced. Reducing the FWM product will increase the system performance.

3.2   OSNR (Optical Signal to Noise Ratio)

Here the OSNR is calculated by varying dispersion and input power. OSNR is defined as the ratio of signal power to the noise power within a valid bandwidth. OSNR value is high means the bit error at receiver is minimized (Tables 1, 2, 3 and 4).

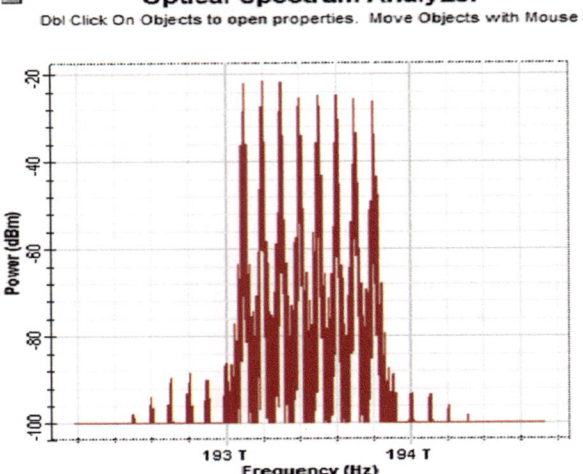

Fig. 3. Channel spacing = 102 GHz + A_{eff} = 65 + fiber length = 106 km

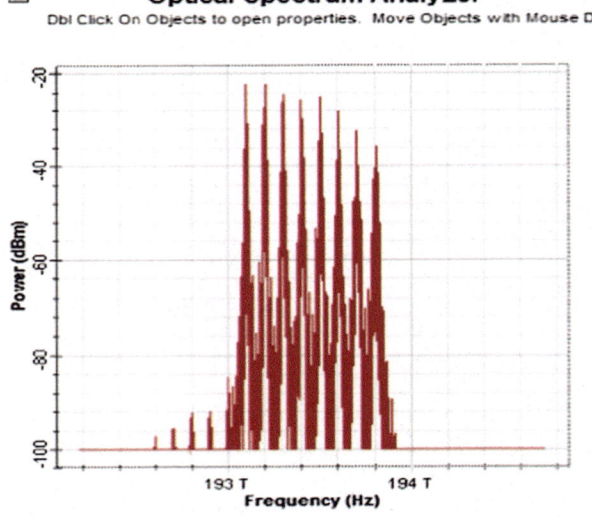

Fig. 4. Channel spacing = 103 GHz + A_{eff} = 70 + fiber length = 107 km

By analyzing these parameter OSNR value is high when different power is used per channel. Simulation software Opti-system 7.0 is used to obtain the results. In the simulation, the input power, bit rate and channel spacing are 1 to 10 dBm, 15 Gbps and 100 GHz, respectively. The results is compared with [7], proposed approach helps to obtain the optimum results for the reduction of FWM power.

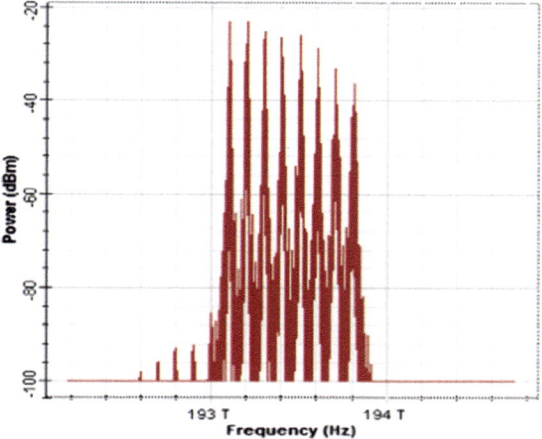

Fig. 5. Channel spacing = 104 GHz + A_{eff} = 75 + fiber length = 110 km

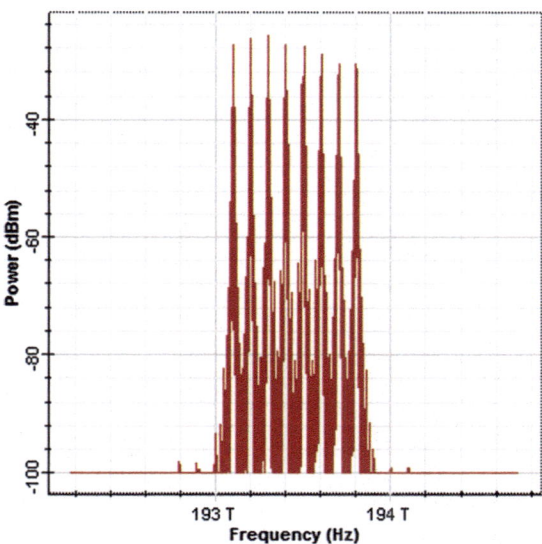

Fig. 6. Channel spacing = 101 GHz + A_{eff} = 60 + fiber length = 105 km + input power = −2 dBm

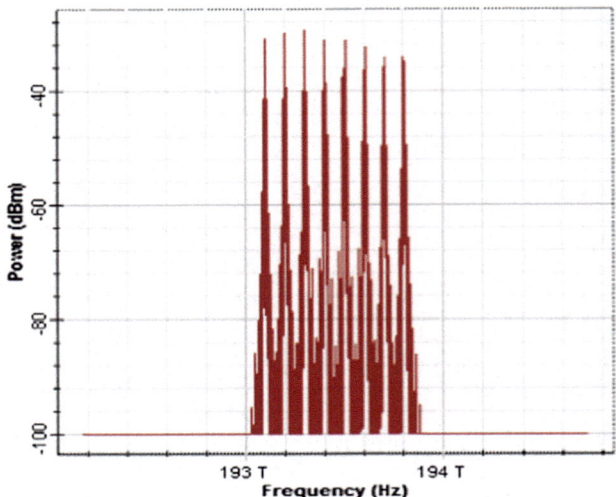

Fig. 7. Channel   spacing = 102 GHz + A_{eff} = 65 + fiber   length = 106 km + input   power = −3 dBm

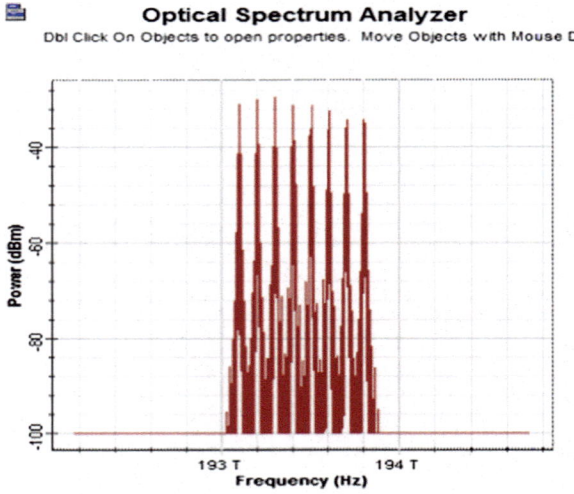

Fig. 8. Channel spacing = 103 GHz + A_{eff} = 70 + fiber length = 107 km + input power =−5 d

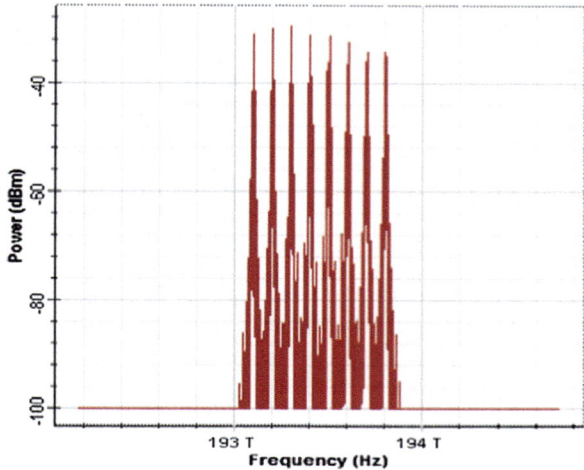

Fig. 9. Channel spacing = 104 GHz + A$_{\text{eff}}$ = 75 + fiber length = 110 km + input power = −8 dBm

Table 1. OSNR when dispersion = 5 ps/nm/km

Frequency (THz)	Signal power (dBm)	Noise power (dBm)	OSNR (dB)
193.1	−8.2111537	−36.892985	28.681831
193.2	−7.2026552	−36.890174	29.687519
193.3	−6.2024774	−36.887366	30.684888
193.4	−5.2040586	−36.884559	31.6805
193.5	−4.198078	−36.884559	32.686481
193.6	−3.2099016	−36.881754	33.671852
193.7	−2.1865034	−36.87895	34.692447
193.8	−1.2089095	−36.876149	35.667239

Table 2. OSNR when Dispersion = 0 ps/nm/km

Frequency (THz)	Signal power (dBm)	Noise power (dBm)	OSNR (dB)
193.1	−8.8124924	−36.892985	28.080493
193.2	−7.2606172	−36.890174	29.629557
193.3	−6.4211215	−36.887366	30.466244
193.4	−4.5963557	−36.884559	32.288203
193.5	−3.8577332	−36.884559	33.026826
193.6	−3.7845329	−36.881754	33.097221
193.7	−2.4625897	−36.87895	34.416361
193.8	−1.0948906	−36.876149	35.781258

Table 3. OSNR when power = −10 dBm

Frequency (THz)	Signal power (dBm)	Noise power (dBm)	OSNR (dB)
193.1	−18.196135	−36.892985	18.69685
193.2	−18.205107	−36.890174	18.685067
193.3	−18.205615	−36.887366	18.681751
193.4	−18.202146	−36.884559	18.682412
193.5	−18.195471	−36.884559	18.689088
193.6	−18.197525	−36.881754	18.684229
193.7	−18.211531	−36.87895	18.667419
193.8	−18.205517	−36.876149	18.670632

Table 4. OSNR when different power per channel

Frequency (THz)	Signal power (dBm)	Noise power (dBm)	OSNR (dB)
193.1	−8.2272219	−36.892985	28.665763
193.2	−7.2136994	−36.890174	29.676475
193.3	−6.1917758	−36.887366	30.69559
193.4	−5.1889422	−36.884559	31.695616
193.5	−4.1859205	−36.884559	32.698638
193.6	−3.2003469	−36.881754	33.681407
193.7	−2.1884771	−36.87895	34.690473
193.8	−1.1998684	−36.876149	35.67628

4   Conclusion

This simulation setup is consists of 8 channels each having 15 Gbps, which is used to calculate the power of four wave mixing and OSNR. The parameters used such as increasing the length of fiber, effective area and channel width spacing, decrease the input power due to the inversely proportional to length, width and effective area and this leads to reduction of FWM power. By reducing FWM power, can increase the system performance with low error rates and high date rate. This will be used for long communication without any interferences.

References

1. Singh, A., Sharma, A.K., Kamal, T.S.: Investigation on modified FWM suppression methods in DWDM optical communication system. Opt. Commun. **282**, 392–395 (2008)
2. Noshada, M., Rostamia, A.: FWM minimization in WDM optical communication systems using the asymmetrical dispersion-managed fibers. Optik, 456–460 (2011)
3. Souza, J.R., Harboe, P.B.: FWM: effect of channel allocation with constant bandwidth, and ultra-fine grids in DWDM system. IEEE Lat. Am. Trans. **9**, 766–773 (2011)

4. Kaler, R., Kaler, R.S.: Investigation of four wave mixing effect at different channel spacing. Optik **123**, 352–356 (2012)
5. Rasheed, I., Abdullah, M., Mehmood, S., Chaudhary, M.: Analyzing the non-linear effects at various power levels and channel counts on the performance of DWDM based optical fiber communication system. In: International Conference on Emerging Technologies, pp. 1–5 (2012)
6. Deshmukh, G., Jagtap, S.: Four wave mixing in DWDM optical system. Int. J. Comput. Eng. Res. **3**, 7–11 (2013)
7. Sharma, V., Kaur, R.: Implementation of DWDM system in the presence of four wave mixing (FWM) under the impact of channel spacing. Optik **124**, 3112–3114 (2013)
8. Sugumarani, S., Arulmozhivarman, P.: Effect of chromatic dispersion on four-wave mixing in WDM systems and its suppression. In: IEEE International Conference (2013)
9. Bhatia, R., Sharma, A.K., Saxena, J.: Optimized alternate polarization and four wave mixing in 60 Gb/s DWDM transmission system. In: IEEE Conferences, pp. 1–4 (2014)
10. Kaur, G., Singh, G.: Analysis to find the best hybrid modulation technique for suppression of four wave mixing. Optik, 25–30 (2015)
11. Vardanyan, V.: Single-fiber optical transmission system with DWDM channels effect on four wave mixing. In: IEEE Conferences, pp. 23–25 (2016)
12. Sabapathi, T., Poovitha, R.: Combating the effects of non-linearities in DWDM system. In: International Conference on Electronics and Communications systems, pp. 38–42 (2017)

Envisagation and Analysis of Mosquito Borne Fevers: A Health Monitoring System by Envisagative Computing Using Big Data Analytics

G. Sabarmathi[1,2(✉)] and R. Chinnaiyan[3]

[1] CMR Institute of Technology, Bangalore, India
sabarganesh@gmail.com
[2] Computer Science, Christ Academy Institute for Advanced Studies,
Bangalore, India
[3] Department of Information Science and Engineering,
CMR Institute of Technology, Bangalore, India
vijayachinns@gmail.com

Abstract. In the recent past, the globalization of economies paved a way to exponential growth of technological expansions in almost all verticals of business and therefore accumulation of huge data became inevitable. The big data is a new paradigm and it has evoked to play a critical role in the advancement of healthcare monitoring system. There is a prospective and subsidy in developing and implementing big data solutions within this terrain. By processing a huge amount of structured, semi-structured and unstructured data from various sources like medical records, pharmaceutical industry, clinical data etc., and assimilate it can offer a persuasive outcome. Envisagative computing is a method used in data analytics process that includes gathering and managing data composed by wide range of biomedical experiments, reviews by expert medical teams, clinical trials and therapeutic trades, then to envisage these data for new pattern and trend evolution, subsequently integrate it with the existing big data tools. Mosquito borne fevers are the prime example of how these data, are an innate aspect of result it generates. In this paper Envisagative computing for mosquito borne disease have been discussed.

1 Introduction

These fevers are breakout by the bite of a disease-ridden mosquito and transfused to humans include Chikungunya, dengue, West Nile virus, Zika virus, and malaria. Traditional approaches to medical investigation have generally focused on the diagnosis of disease states based on the variations in physiology in the form of confined view of certain singular modality of data [2]. Although this approach in understanding diseases is indispensable, research at this level dampers the disparity and associative that define the true underlying medical principles [3]. However, despite the advent of big data concepts, the data captured and gathered from previous epidemic scenarios has remained vastly unexploited and thus intact. Important correlations between

© Springer Nature Switzerland AG 2020
A. P. Pandian et al. (Eds.): ICCBI 2018, LNDECT 31, pp. 630–636, 2020.
https://doi.org/10.1007/978-3-030-24643-3_75

physiological, environmental, climate fluctuations and socio-economical phenomena are concurrently manifest as distinctions across multiple clinical streams. This reveals a strong association among different factors within the system (e.g., interactions between environment and mosquito habitat) there by producing potential cyphers for clinical estimation. Thus, learning and forecasting diseases necessitate an integrated approach where structured and unstructured data stemming from a pile of meteorological and medical modalities are utilized for a more comprehensive perspective of the disease states. By studying patterns exist from these complex data repositories in terms of both characteristic of data itself and taxonomy of data analytics, a meaningful remedial clinical solution can be drawn. Going forward, in this paper Envisagative computing for mosquito borne disease have been discussed. Although Envisagative computing is not a great demonstration of big data analysis in health monitoring system as a whole, the intention is to set forth an ampler model to expedite mosquito borne fever diagnosis more reliable at appropriate time when it acknowledged as time tight measure.

1.1 Big Data Analytics

Big data analysis is the process of the dissimilar formats of information and compiling it into prolific source of collective information. The steps involved in this are identifying the structure, interpreting the data, recognizing the hidden pattern, understanding and analyzing data into recognized procedure. The data related to health care is gathered in different modes. The different analytic tools are used to analysis multi-terabyte health records in the cloud which in turn produces the real insight of the data. Here we have discussed some of such tools as below:

- Yarn: A cluster controlling knowledge that is a key feature of Hadoop second generation.
- Google Big Query: It uses the cloud set up to pile the huge datasets request by facilitating the SQL queries in fraction of time.
- HBase: It is a scattered column-oriented databank built on top of the Hadoop Distributed file system (HDFS).
- Map Reduce: It is the encoding model that allows for huge scalability of data across many servers in a Hadoop cluster. It is distributed into two parts: Map, splits the work to dissimilar nodes in the distributed cluster, and Reduce does the work of collecting the data and produces the outcome in a distinct value.
- Pig: It offers an extraordinary tool of comparable encoding of Map Reduce that is been executed by Hadoop clusters.

2 Envisagative Computing

Envisagative computing is the process of identifying the data by means of correlating and construing the data in the database. The traditional way of curing the patient is by identifying the disease and check for its symptoms. In case of Envisagative computing, we are trying to evaluate the possible cause of that symptoms that will lead to detecting the disease itself. Here we use a term called pseudo-intelligence as part of data analysis which in turn excerpts data from human interaction and machine interaction from those data. It is done with the help of analytical tools which takes the raw data and forecasts at the end, exposes the result in less time. Envisagative computing will capture a disease even before the symptoms starts to rise. When a simulated-Intelligent device does this, we mean that as Envisagative computing.

3   Applying Envisagative Computing

3.1   Case 1

Let us consider the case of dengue, it is a humid and an arboviral sickness, which is spread by Aedes aegypti along with Ae. Albopictus mosquitos and exhibit diverse clinical variety: dengue infection (DF), dengue haemorrhagic fever (DHF), dengue shock syndrome (DSS), and expanded dengue syndrome (EDS) with rare presentations, therefore pretense an investigative paradox. Except we remain alert of distinguishing these descriptions, finding as well as initial instigation of action becomes troublesome. One of the serotypes of a single stranded RNA virus named DENV 1, 2, 3, and 4 are instrumental in prompting infection [4]. Besides mosquito bite, dengue can also be spread by blood transfusion from an infested donor, wounds by diseased sharps, replacement of organs and tissues from diseased patrons, and from infected pregnant mother to her fetus [5–7]. The usual signs and symptoms of dengue are temperature, head and body pain, back pain, retro-orbital pain, indication of blood and rash. Nonspecific or cautioning symptoms are vomiting, faintness, stomach pain, breathing difficulty, dizziness, sweltering and syncope. Added symptoms are coinfections, comorbidities, or difficulties includes diarrhea, throat pain, and neurological manifestations. Later stage of the disease is followed by three phases: feverish, serious, and regaining phase [8]. In the recent years, dengue has traversed terrestrial boundaries and has spread too many new countries. Currently, it is widespread to Southeast Asia and western pacific regions [8]. So, dengue fever has developed a major international communal concern particularly in tropical and subtropical regions disturbing municipal as well as countryside areas. Nevertheless, quite a lot of measures are taken to regulate and prevent it, repeated occurrences have been reported everywhere.

3.2   Case 2

Zika virus is called after its source from Zika through Macaca monkey, Uganda, all through 1947 [9]. The most vital route of Zika virus is Aedes aegypti [10]. The disease can also be spread from mother to fetus through pregnancy or birth. Further means of spreading happens through sexual communication and contaminated blood transfusion.

Clinical presentations of Zika fever resembles "dengue-like" syndrome with low fever, bilateral conjunctivitis, maculopapular rash, headache, retro-orbital pain and arthritis/arthralgia with edema of the minute joins of hands and feet, myalgia, vertigo and asthenia. Rarely, sore throat, cough and loose bowels are experienced. In addition, high fever, chills, rigors, sore throat, hypotension, cervical, submandibular, axillary and/or inguinal lymphadenopathies. Further, intestinal indications like nausea, vomiting, diarrhea, constipation, abdominal pain and aphthous ulcers. Patients with genitourinary symptoms including hematuria, dysuria, perineal pain and hematospermia often have measurable virus constituent part in urine and/or semen. This clinical feature can be wrong for dengue or chikungunya fevers. Some presentations may distinguish Zika fever from chikungunya and dengue fever including more eminent edema of the extremities, less severe headache and milder thrombocytopenia reported in the former. Indeed, Zika fever can be misdiagnosed during the acute stage because of nonspecific signs and symptoms [11].

3.3   Case 3

Chikungunya fever is an acute febrile infection produced by an arthropod-borne alpha virus, Chikungunya virus (CHIKV). It is principally spread to humans through the bite of an infected Aedes species mosquito. During 1950s many countries in Africa and Asia CHIKV initially predicted as a humanoid pathogen [12, 13]. The infection was originally detected as a "dengue-like" illness till test center estimation predicted CHIKV as the basis of disease. Apparently, multiple episodes of fever which lasts for couple of weeks indicate the cause for CHIKV infection.

Most of the cases after a week it develops intensive disintegrating polyarthralgia's. The associated symptoms are rashes, joint pains in wrists, elbows, fingers, knees, and ankles, however remote chances of affecting more-proximal joints. There are reported cases, arthritis with joint swelling is also observed. On the other hand, arthralgias which can persist for months is also a considerate and comprehensive. Although rashes recognized as a symptom it is less common among patients and cannot consider as indication of disease. Other warnings during the course of sickness are headache, weakness, nausea and conjunctivitis; myalgias, even though these are not precise for febrile infections, happen usually. Generally Cervical lymphadenopathy is not witnessed frequently along with o'nyong nyong fever, which is another alphavirus infectious linked with fever and arthralgias [14].

3.4   Case 4

West Nile virus (WNV) occupies notable a place in a genre of vector-borne diseases around the world. The virus called Culex sp.-transmitted was first discovered in Uganda place called West Nile in 1937 and it is well known for triggering small human febrile epidemics in Africa and middle East. When yellow fever outbreaks studied, scientists end up in isolating virus having similar properties of those two flaviviruses called St. Louis encephalitis virus and Japanese B encephalitis virus, and sharing immunological links with these viruses. Even though primary symptom is a fever its pathological studies revealed links to the central nervous system (CNS), portraying its neurotropic behavior. While major symptoms are being fever, stomach ache, nausea, vomiting, forehead pain, myalgias, exanthems, anorexia; diarrhea, angina and lymphadenopathy are not so frequent. During early childhood WNV seems to spread more indicating further symptoms compared with the middle age people. Fever complaints with rare existences of meningitis or encephalitis [15].

In general, the common attributes and classification of dengue fever will be captured along with critical symptoms from earlier cases and fed to the databases.

These data will be utilized more comprehensively to diagnose the patients accurately who has fever with associated symptoms and can instantly recommend him for a blood analysis, otherwise it could be considered as normal fever. The doctor forecast is by means of merely correlating with previous symptoms and also with the help of Envisagative computing; where it computes the outcome based on the various pattern and structure that is been evolved during computing. The Envisagative computing can be performed for the above discussed case studies by executing the functions like

- Dynamic Scan

The Envisagative computing does the scanning of latest updates and gathers data from the various health care sectors, finally compiles all the data and produces outcome with probable form of diseases that is been emerging around the universe. By repeating the scan, a structure can evolve and it will predict the nature of the disease. It can even compute and correlate the relation between fevers so as to produce the final report on classification of fever.

- Individual diagnosis

In this case, while an individual explains the symptoms of his discomfort to the doctor, doctor can effortlessly diagnose his case through Envisagative computing.

- Varied diagnosis

In above examples, the disease is predicted from the existing cases where as here in varied diagnosis, the disease is diagnosed by means of differentiating various factors. For example, while various fevers are manifesting similar kind of symptoms, some

Table 1. Algorithm for Envisagative computing

STEP 1: Framing the structure
1.1: Propose hardware according to applications
1.2: Change to structure picture through Virtualization
1.3: Set up an individual cloud server with captivating data from all health interconnected data granaries
STEP 2: Database Approach
2.1: Storing the Big Data in HDFS* (Hadoop Distributed File System)
2.2: By means of NLP (Natural Language Processing) gather data with similar pattern
2.3: Health hazard gathered Guideline using Map Reduce method
2.4: Accumulate the data in HDFS*
STEP 3: Identify Data
3.1: convert data into meaningful recourse using Analytic tools like Apache
3.2: correlate the anomalies using rational reasoning
3.3: Distinct tree structure made for each fever category by way of multiple correlations and it is been interpreted
STEP 4: Accumulating the Report
4.1: Virtual Intelligence's Fuzzy logic compute is used in outcomes for Abnormal, Dynamic and Indeterminate Forecast
4.2: Reports are tabulated instantly and created to share the details to primary care centres round the worldSource adapted from Srivathsan Ma, Yogesh Arjun Kb [1]

presentations are identified to distinguish Zika fever from chikungunya and dengue fever which includes more eminent edema of the extremities, less severe headache and milder thrombocytopenia reported in theformer. Indeed, Zika fever can be misdiagnosed during the acute stage because of nonspecific signs and symptoms (Table 1).

4 Conclusions

Big data Analysis has the efficiency to provide the technological support to the healthcare sectors in their clinical and other data mining that helps in their decision making. There are issues like security, privacy of data, continuous monitoring of tools and updating technologies that needs proper standard of analytics. By using the advance platform and tools can hasten their evolution in big data analytics healthcare applications. In this paper we try to bring an insight of one case in which this technique can be done this if we work further it can be applied in real time which helps to predict many symptoms early and quick decision making can be made in treatment which in turn saves many lives.

References

1. Srivathsan, M., Yogesh Arjun, K.: Health monitoring system by prognotive computing using big data analytics. Procedia Comput. Sci. **50**, 602–609 (2015)
2. Borckardt, J.J., Nash, M.R., Murphy, M.D., Moore, M., Shaw, D., O'Neil, P.: Clinical practice as natural laboratory for psychotherapy research: a guide to case-based time-series analysis. Am. Psychol. **63**, 77–95 (2008)
3. Celi, L.A., Mark, R.G., Stone, D.J., Montgomery, R.A.: 'Big data' in the intensive care unit: closing the data loop. Am. J. Respir. Crit. Care Med. **187**(11), 1157–1160 (2013)
4. Raghupath, W., et al.: Big data analytics in healthcare: promise and potential. Health Inf. Sci. Syst. **2**, 1–10 (2014)
5. Raghupathi, W., Raghupathi, V.: An overview of health analytics. Working paper (2013)
6. IBM: IBM big data platform for healthcare. Solutions Brief (2012). http://public.dhe.ibm.com/common/ssi/ecm/en/ims14398usen/IMS14398USEN.PDF
7. Bates, D.W., Saria, S., Ohno-Machado, L., Shah, A., Escobar, G.: Big data in health care: using analytics to identify and manage highrisk and high-cost patients. Health Aff. **33**, 1123–1131 (2014)
8. Burghard, C.: Big Data and Analytics Key to Accountable Care Success. IDC Health Insights, Framingham (2012)
9. Laul, A., et al.: Clinical profiles of dengue infection during an outbreak in northern India. J. Trop. Med. **2016** (2016)
10. Dick, G.W., Kitchen, S.F., Haddow, A.J., et al.: Zika virus. I. Isolations and serological specificity. Trans. R. Soc. Trop. Med. Hyg. **46**, 509–520 (1952)
11. Moghadam, S.R.J., Bayrami, S., Moghadam, S.J., Golrokhi, R., Pahlaviani, F.G., SeyedAlinaghi, S.: Zika virus: a review of literature. Asian Pac. J. Trop. Biomed. **6**(12), 989–994 (2016)

12. European Center for Disease Prevention and Control. Zika Virus Epidemic in the Americas: Potential Association with Microcephaly and Guillain-Barré Syndrome. ECDC, Stockholm (2015). http://ecdc.europa.eu/en/publications/Publications/zika-virus-americas-association-with-microcephaly-rapid-risk-assessment.pdf. Accessed 27 Jan 2016
13. Robinson, M.C.: An epidemic of virus disease in Southern Province, Tanganyika territory, in 1952–53. Trans. R. Soc. Trop. Med. Hyg. **49**, 28–32 (1955)
14. Staples, J.E., Breiman, R.F., Powers, A.M.: Chikungunya fever: an epidemiological review of a re-emerging infectious disease. Emerg. Infect. **49**, 942–948 (2009)
15. Sejvar, J.J.: West Nile virus: an historical overview. Ochsner J. **5**(3), 6–10 (2003)
16. Jupp, P.G., McIntosh, B.M.: Chikungunya virus disease. In: Monath, T.P. (ed.) The Arboviruses: Epidemiology and Ecology, vol. II, pp. 137–157. CRC Press, Boca Raton, FL (1988)
17. Harvard Business: How Big Data Impacts Healthcare by Harvard Business Review Analytics Services (2014)
18. Gulati, S., Maheshwari, A.: Atypical manifestations of dengue. Tropical Med. Int. Health **12**(9), 1087–1095 (2007)
19. Tan, F.L., Loh, D.L., Prabhakaran, K., Tambyah, P.A., Yap, H.K.: Dengue haemorrhagic fever after living donor renal transplantation. Nephrol. Dial. Transplant. **20**(2), 447–448 (2005)
20. Pouliot, S.H., Xiong, X., Harville, E., et al.: Maternal dengue and pregnancy outcomes: a systematic review. Obstet. Gynaecol. Surv. **65**(2), 107–118 (2010)
21. Mandal, S.K., Ganguly, J., Sil, K., et al.: Clinical profiles of dengue fever in a teaching hospital of eastern India. Natl. J. Med. Res. **3**(2), 173–176 (2013)
22. Chinnaiyan, R., Somasundaram, S.: Evaluating the reliability of component based software systems. Int. J. Qual. Reliab. Manag. **27**(1), 78–88 (2010)
23. Chinnaiyan, R., Somasundaram, S.: Reliability of object oriented software systems using communication variables – a review. Int. J. Softw. Eng. **2**(2), 87–96 (2009)
24. Chinnaiyan, R., Somasundaram, S.: Reliability assessment of component based software systems using test suite - a review. J. Comput. Appl. **1**(4), 34 (2008)
25. Chinnaiyan, R., Somasundaram, S.: An experimental study on reliability estimation of GNU compiler components – a review. Int. J. Comput. Appl. (0975 – 8887) **25**(3), 13–16 (2011)
26. Chinnaiyan, R., Somasundaram, S.: Reliability of component based software with similar software components - a review. i-Manager's J. Softw. Eng. **5**(2), 44 (2010)
27. Chinnaiyan, R., Somasundaram, S.: Monte Carlo simulation for reliability assessment of component based software systems. i-Manager's J. Softw. Eng. **4**, 27–32 (2010)
28. Sabarmathi, G., Chinnaiyan, R.: Investigations on big data features research challenges and applications. In: IEEE Xplore Digital Library International Conference on Intelligent Computing and Control Systems (ICICCS), IEEE Xplore Digital Library, pp. 782–786 (2018)
29. Sabarmathi, G., Chinnaiyan, R.: Reliable data mining tasks and techniques for industrial applications. IAETSD J. Adv. Res. Appl. Sci. **4**(7), 138–142 (2017)
30. Sabarmathi, G., Chinnaiyan, R.: Big data analytics research opportunities and challenges - a review. Int. J. Adv. Res. Comput. Sci. Softw. Eng. **6**(10), 227–231 (2017)

Session Layer Security Enhancement Using Customized Protocol and Strong Cryptographic Mechanism

Aaditya Jain[✉] and Sunil Nama[✉]

Department of Computer Science and Engineering,
R. N. Modi Engineering College,
Rajasthan Technical University, Kota, India
aadityajain58@gmail.com, nama.sunilraj@gmail.com

Abstract. Cloud computing has been considered as a mechanism to deliver information technology services. Data security is significant aspect of Internet for enterprises of each size and type. Security of Data is must for protective digital privacy measures. However there have been several existing researches that propose security but they have certain limitations. The proposed work introduces security at multiple layers and user defined port to define more secure data transmission protocol in order to enhance the session layer [1] security of network. More over the early session timeout has been also controlled by reducing packet size. Proposed work would be capable to increase the efficiency of data communication along with security. Moreover unauthentic access of data has been restricted using IP validation process.

Keywords: Session layer · IP validation · Port · Socket ·
Cloud · Cryptography

1 Introduction

Cloud computing is considered as a mechanism to deliver information technology services. Here resources have been extracted online with the help of online tools. It is opposed to a direct connection established with a server. With the help of cloud-based storage we can save the files to a remote database instead of keeping them on a local hard disk. We can access the data and softwares running on the cloud as long as we have connectivity with the internet. This is the reason why this computing is known as the cloud computing because the information which is to be accessed is stored on the cloud and it does not require any user to be at a particular place in order to access it. Because of this type of system employees can work from remote location. There are some companies which provide cloud services so that users can store files and applications on remotely located servers, and then access the data with the help of internet [1]. Data security has been considered a significant aspect of Internet for enterprises of each size and type. Security of Data meant for protective digital privacy measures. Such measures would be applied to prevent unauthentic data [2] which access to databases, computer or websites. Data security is also protecting data from corruption. Such security

© Springer Nature Switzerland AG 2020
A. P. Pandian et al. (Eds.): ICCBI 2018, LNDECT 31, pp. 637–644, 2020.
https://doi.org/10.1007/978-3-030-24643-3_76

technologies involve information masking, information removal backups. Encryption [4] has been provided to information security. Digital information is encrypted to cipher text. It becomes unreadable to unauthorized users. At time of authentication, user needs to give a code, password, and biometric data. He may also provide some other form of information in order to verify his identity (Fig. 1).

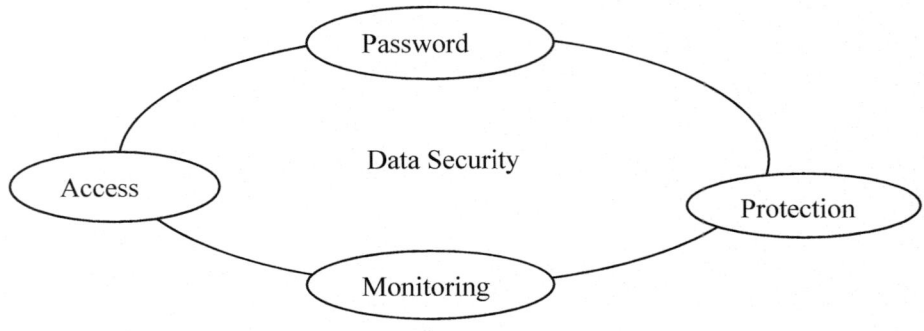

Fig. 1. Data security

2   Literature Review

There have been several researches in field of security. Some of these researches considered IP spoofing, while other discussed threats in peer to peer networks. Lot of researchers have discussed cyber threat to Network security. Research has discussed security technology to handle DDOS attack, IP spoofing. Several researches have proposed encryption mechanism to provide security to information transferred over network. Here in this section the several existing researches made by different authors have been discussed.

In 2013, Abhishek Kumar Bharti [1] used Sensor Nodes and Cryptography to detect session hijacking and IP Spoofing type of attack.

In 2014, Hani Alshamrani [2] wrote research on Internet Protocol Security (IPsec) Mechanisms.

In 2011, Chander Diwakar [3] et al. discussed security threats in peer to peer networks.

In 2014, Haroon Shakirat Oluwatosin [4] did research on Client-Server Model.

In 2014, Ms. Jasmin Bhambure [5] et al. proposed Secure Authentication Protocol in Client – Server Application using Visual Cryptography.

In 2015, Mohan V. Pawar [6] et al. discussed Security of network and Types of Attacks in Network

In 2015, Manjiri N. Muley [7] did study for analysis for exploring the scope of network security techniques in different era.

In 2013, Rupam [8] et al. introduced approach to detect packets using packet sniffing.

In 2013, Sharmin Rashid [9]. et al. proposed some methods of detecting IP Spoofing & also its Preventions.

In 2013, Mukesh Barapatre [10] et al. made a review on Spoofing Attack Detection in Wireless Adhoc Network.

In 2014, Amandeep Kaur [11] et al. did a review on Security Attacks in Mobile Ad-hoc Networks.

In 2014, Md. Waliullah [12] et al. wrote a research on Wireless LAN Security Threats & Vulnerabilities.

In 2014, P. Kiruthika Devi [13] et al. did research on spoofing attack detection & localization in wireless sensor network.

In 2014, Barleen Shinh [14], did a review on Collaborative Black Hole Attack in MANET.

In 2014, Ms. Vidya Vijayan [15] did review on Password Cracking Strategies.

3 Proposed Work

Problem Definition

There were several limitations in existing researches. These researches did not con-sidered security at session layer and transport layer. They just focused on security at application and presentation layer.

Due to limitation of existing security mechanisms there was need to develop a new security [3] system. Chance for decryption without authentication should get reduced.

Need of Proposed Work

There is need to implement IP filter-based security in order to prevent attacker from different network. Here we would also reduce size of packet during transmission using replacement policy.

Session layer security would be enhanced by introducing multilayer security mechanism

1. Here we are using IP filter in order to avoid the unauthenticated packet transmission from server to client.
2. In this, By customizing existing encryption techniques we have to enhance network security.
3. To study the problems of existing security mechanisms and to enhance the level of security in the network.
4. Own socket server and corresponding client will be programmed in order to prevent unauthentic access occurring during the transmission of data.
5. To use more complex key at the time of encryption and decryption by integration of MD5 and multiplicative inverse cryptographic techniques.
6. To form an user interface that helps to established communication between client server.

Algorithm for Multiplicative Inverse

Step 1 Get input aaa as string for encryption and sh as shitting number

Step 2 Take integer shift,i,n and String str,str1="",str2=""

Step 3 Set str=aaa;

Step 4 Convert str to lower case

Step 5 Get lenght of str in n

Step 6 Get array of character from str to ch1 array

Step 7 Take ch3,ch4 as character

Step 8 set shift=sh;

Step 9 set i=0 and increase i by 1 and repeat step 10 until i is less then n

Step 10 if ch1[i] isLetter

      set ch3=(char)(((int)ch1[i]*shift-97)%26+97)

      set str1=str1+ch3

      otherwise

      if(ch1[i]==' ')

      set str1=str1+ch1[i]

Step 11 Caclulation of multiplicative inverse

Step 12 Set q=0,flag=0

Step 13 set i=0 and increase i by i and repeat step 14 until i is less than 26

Step 14 if((((i*26)+1) % shift==0)

      set q=((i*26)+1)/shift and stop the loop

Step15 set array of character from str1 to ch2

Step 16 set i=0 and increment i by 1 and repeat step 17 until i is less than length of str1

Step 17 if ch2[i] .is letter

      ch4=(char)(((int)ch2[i]*q-97) % 26+97);

      set str2=str2+ch4

      otherwise

      if(ch2[i]==' ')

      str2=str2+ch2[i];

Step 18 get str2 as output

Proposed Algorithm for Integration of Multiplicative Inverse with MD5

Step 1 get Input in1 from str2 for encryption

Step 2: set Message Digest md by getting instance MD5;

Step 3: Take array of bytes of message Digest by getting bytes from input string.

Step 4: Set the big integer number from array of bytes of message digest

Step 5: Set string hash text setting value "16"

Step 6 Repeat step 6 until length of hash text is less than 32

Step 7 Concatenate "0" with hash text and set in hash text;

Step 8 Return hash text as output

Process Flow of Proposed Work

Algorithm on Sender side

1. Initialize the port number from receiver for transmission
2. Set the common port number and IP address from sender side
3. Set the file for transmission
4. Perform data compression using replacement table
5. Perform multiplicative encryption of the compressed data using K.
6. Perform MD5 in order to make data more secure.

Algorithm on Receiver Side

1. Wait for the data from sender
2. Receive data from sender
3. Apply md5 to decrypt data
4. Perform multiplicative decryption of data using K.
5. Decompress data using common replacement table
6. Receive the plain data and store in file

4   Results and Analysis

The simulation results represent that the proposed work has reduced the time consumption during data transmission as the size of data get reduce during MD5 encryption. Result of output have been simulated using MATLAB (Fig. 2).

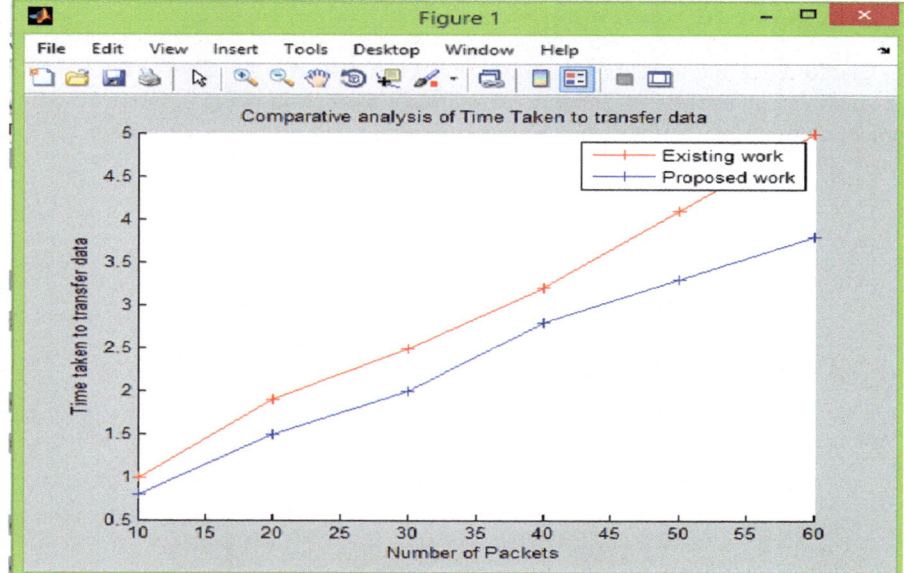

Fig. 2. Comparative analysis of time taken during transmission

Due to reduction in size the probability of error get reduced. Thus proposed work lead to less error as compare to traditional work. Following figure represent the comparative analysis of error rate in case existing and proposed work (Fig. 3).

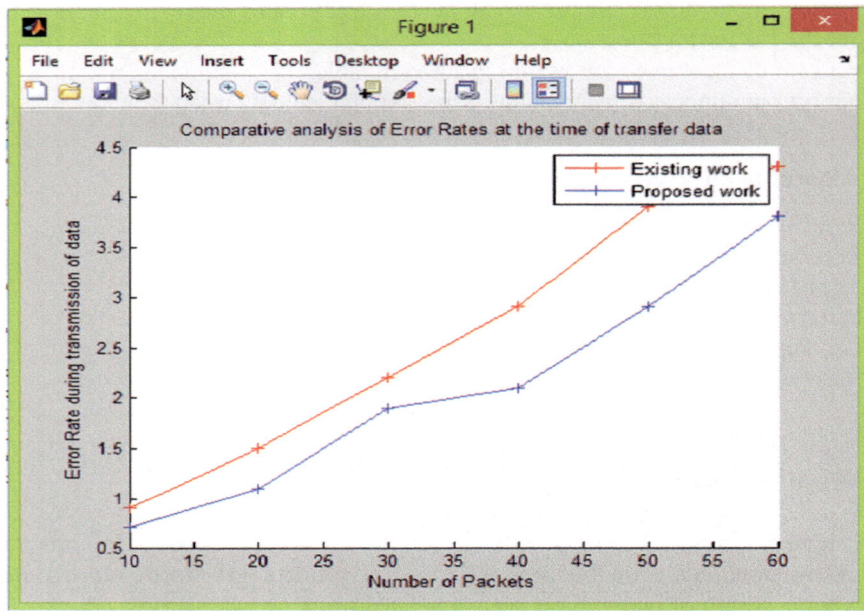

Fig. 3. Comparative analysis of error rates at time of transfer data

Due to reduction in size of data the packet size get reduced. Thus proposed work use small data in packets as compare to traditional work. Following figure represent the comparative analysis of packet size in case existing and proposed work (Fig. 4).

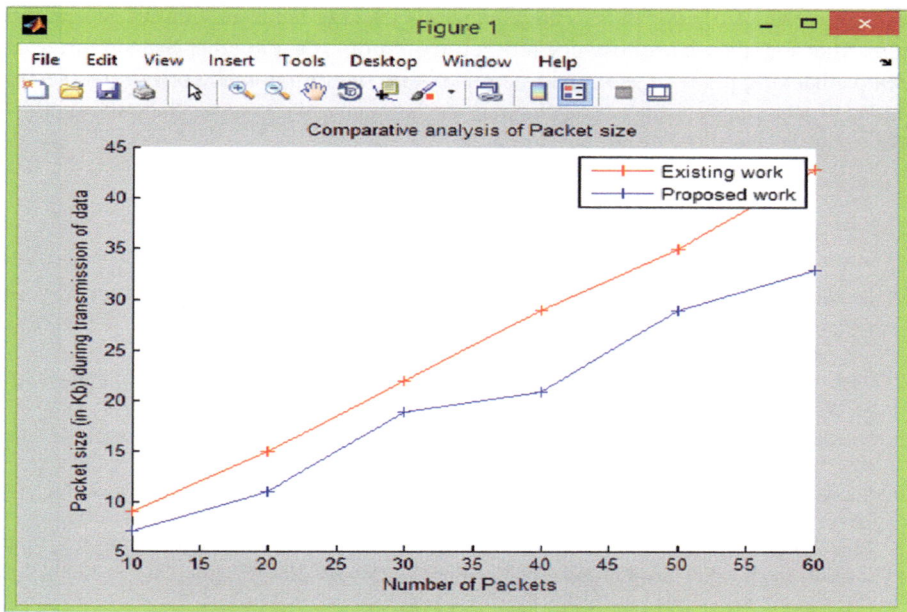

Fig. 4. Comparative analysis of packet size

5 Conclusion

The research concludes that the proposed work has reduced the time consumption during data transmission as the size of data get reduce during MD5 encryption. Due to reduction in size of data the packet size get reduced. Thus proposed work use small data in packets as compare to traditional work.

Due to reduction in size the probability of error get reduced. Thus proposed work lead to less error as compare to traditional work. Moreover the probability of congestion gets reduced and more strong cryptographic mechanism has increased the security.

6 Future Scope

Protection to data from being damaged by the attacker has been considered in this research. In existing research, there was scarcity of strong security system to encrypt data. Thus, transmission of data faced delay because of less powerful security system. So, in order to overcome this hurdle, we have decrease of packet's size. In this way we can shrink the issue of transmission delay. Apart of this, we can also create more powerful security system.

Battle between moral or white hat hackers and non-moral or black hat hackers have many contradictions. That has no ending point. When ethical hacker supports the black hat hacker, he has the knowledge of an organization. They help them to know the security needs of organization. Thus, malicious hackers break in without any permission. After that they harm network for their own profit. On the other hand, Ethical and creative hacking plays an essential role in network security to make sure the protection of data of a company. Ethical hackers help the company to recognize the individuals, to take curative measures to remedy loophole.

References

1. Bharti, A.K.: Detection of session hijacking and IP spoofing using sensor nodes and cryptography. IOSR J. Comput. Eng (IOSR-JCE)e-ISSN: 2278-0661, p-ISSN: 2278-8727 **13** (2), 66–73 (Jul–Aug 2013)
2. Alshamrani, H.: Internet Protocol Security (IPSec) mechanisms. Int. J. Sci. & Eng. Res. **5**(5), 85–87 (May 2014)
3. Diwakar, C., Kumar, S., Chaudhary, A.: Security threats in peer to peer networks. J. Glob. Res. Comput. Sci. **2**(4), 81–84 (2011)
4. Oluwatosin, H.S.: Client-server model. J. Comput. Eng. (IOSR-JCE), **16**(1), 67–71 (Feb 2014)
5. Bhambure, J., Chavan, D., Pallavi, B., Madhuri, L.: Secure authentication protocol in client – server application using visual cryptography. Int. J. Adv. Res. Comput. Sci. Softw. Eng. **4** (2), 556–560 (Feb 2014)
6. Pawar, M.V., Anuradha, J.: Security of network and types of attacks in network. Int. Conf. Intell. Comput., Commun. & Converg., **48,** 503–506 (2015)
7. Manjiri N.M.: Analysis for exploring the scope of network security techniques in different era: a study. Int. J. Adv. Comput. Eng. Netw. **3**(12), 33–36 (Dec 2015)
8. Rupam, Verma, A., Singh, A.: An approach to detect packets using packet sniffing. Int. J. Comput. Sci. & Eng. Surv. (IJCSES) **4**(3), 21–33 (2013)
9. Rashid, S., Paul, S.P.: Proposed methods of IP spoofing detection & prevention, international. J. Sci. & Res. (IJSR) **2**(8), 438–444 (2013)
10. Barapatre, M., Chole, V., Patil, L.: A review on spoofing attack detection in wireless adhoc network. Int. J. Emerg. Trends & Technol. Comput. Sci. **2**(6), 192–195 (November–December 2013)
11. Kaur, A., Singh, A.: A review on security attacks in mobile ad-hoc networks. Int. J. Sci. & Res. **3**(5), 1295–1299 (May 2014)
12. Waliullah, Md, Gan, D.: Wireless LAN security threats & vulnerabilities. Int. J. Adv. Comput. Sci. & Appl. **5**(1), 176–183 (2014)
13. Kiruthika Devi, P., Manavalan, R.: Spoofing attack detection & localization in wireless sensor network. Int. J. Comput. Sci. & Eng. Technol. **5**(09), 877–886 (Sep 2014)
14. Shinh, B., Singh, M.: A review paper on collaborative black hole attack in MANET. Int. J. Eng. & Comput. Sci. **3**(12), 9547–9551 (2014)
15. Vijayan, V., Joy, JP., Suchithra, M.S: A review on password cracking strategies. Int. J. Res. Comput. & Commun. Technol., 8–15 (2014)

Simulation of VANET Using Pointer Base Replacement Mechanism

Aaditya Jain$^{(\boxtimes)}$ and Adarsh Acharya$^{(\boxtimes)}$

Dept. of Computer Science and Engineering, R. N. Modi Engineering College,
Rajasthan Technical University, Kota, India
aadityajain58@gmail.com, aadarshaacharya@gmail.com

Abstract. Wireless networks has performed significant role in daily operations as it has been widely used in military & civilian applications. It is also considered in investigation as well as rescue. It has been used in non- permanent meeting places and airports. Apart from these places, in industrial applications and even in personal area networks, WSNs has been applied. MANET is a temporary wireless network in which is having no fixed infrastructure used. Vehicular ad hoc network has been considered the secondary class of MANET. That has been considered the promising review. It is for the future intelligent transportation environment. The proposed work focuses on the reduction of packet loss during communication. To decrease the packet loss, information would be compressed using pointer. In this mechanism the real data could be replaced by the small short keys available in remote table and encoded data would travel over network. Due to reduce in size of data the success rate of data transmission would increase. There would be minimum chances of packet dropping. More over the probability of congestion get reduced. The use of XOR based encryption has provided quick encryption after compression of data.

Keywords: XOR · MANET · Ad hoc · ITS · VANET · WLAN · IVC

1 Introduction

Wireless networks play a very prominent role in daily transmission which is applied for military & civilian applications, search and rescue, temporary meeting rooms & airports, industrial applications and even in personal area networks. Two kind of wireless networks have existence. First one is the Infrastructure Network and the second is structure-less Network. Infrastructure network contains fixed and wired gateways whereas infrastructure-less network have multi hop wireless nodes. Thus it is lacking the permanent infrastructure. MANET is related to second type. MANET lies in the category of temporary wireless network. In this not permanent structure is applied. So in the MANET, topology alters the frequently as mobile nodes alters without any dependency. It alters their connection to other nodes in short time. Every mobile node works as a router. The traffic transfers to the other nodes. In the case of similarity in the communication range of two mobile nodes, they have the ability to transmission in direct way, or the nodes in between have to forward the packets for them.

© Springer Nature Switzerland AG 2020
A. P. Pandian et al. (Eds.): ICCBI 2018, LNDECT 31, pp. 645–653, 2020.
https://doi.org/10.1007/978-3-030-24643-3_77

Ad hoc Wireless Network
This network has been considered a group of without wire mobile hosts. That is forming not permanent network with lack of the aid of any stand-alone structure or centralized administration [1]. Mobile ad hoc networks are own arranged and own-configured MANETs. Dynamic changes take place in the structure of the network [2]. It is chiefly because of nodes mobility. The network nodes are utilizing the same random access wireless channel. The network nodes cooperate in a friendly way to involve themselves in forwarding of multi-hop. Here the nodes in the network act as hosts. Concurrently, the nodes also performs like routers and the nodes route data between the other nodes in network.

Vehicular ad hoc network
Vehicular ad hoc network is a subordinate class of MANETs. It is a promising approach for future intelligent transportation system (ITS). These networks have lack of permanent infrastructure and hence they depend on the vehicles themselves. It offers the functionality of network. Thus, because of constraints mobility, driver behaviour, and high state mobility, VANETs demonstrate the traits. These characteristics are dramatically varied from several generic MANETs.

VANET has been invented with utilize the MANET principal. It has been employed to transmit the data without wire connection case of vehicles. In 2000, for MANET principal application, the VANET was bringing in use. In the time of 2015, Inter-Vehicle Communication introduced VANET. It comprises many applications such as cooperative awareness message (CAM) and simple hop information dissemination. VANET network has been applied to search location, Vehicular movement, Details of traffic and other related issue of vehicle. The data would be transfer with appropriate way as well as smoothly by this vehicle.

VANET Architectures
VANET has dependency on conception of mobile ad hoc networks. That is capable can communicate using pure cellular or Wireless local area network and ad hoc or hybrid networks. Figure 1 below shows the architecture of VANET.

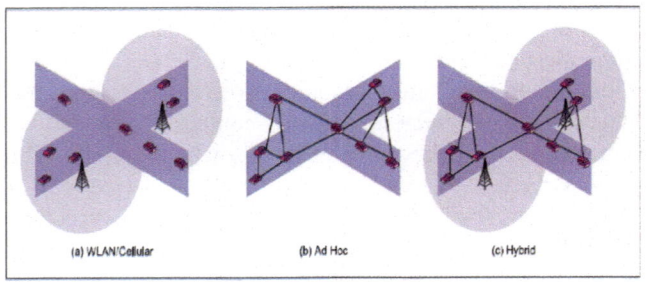

Fig. 1. VANET architectures [25]

VANET based on cellular gateways. WLAN apply the contact points at traffic junctions to attach to net. It is also amass traffic data for objective related to routing. The network structure in this situation is a clean cellular or WLAN structure. Uniform or permanent gateways at the sides of roads have the efficiency to offer the vehicles connection. But these are ultimately not feasible have consideration of structure costs included. In this situation, all vehicles and roadside wireless tolls can establish a mobile ad hoc network. That allows vehicle-to-vehicle transmission. [15]. A hybrid architecture can be shaped with combination of the cellular, WLAN and ad hoc networks jointly for VANET. Hybrid architecture takes helps of vehicles having both WLAN and cellular capabilities as the gateways and mobile network routers. Therefore the vehicles merely with WLAN effectiveness can transmit by multi-hop associates [14, 17].

2 Related Work

VANET assists several applications. That demands transmission in short time. The action of providing Quality-of-Service concentrates on several parameters like Residual Channel Capacity, Faded Signal-to-Noise Ratio, Connection Life Time, End-to-End delays. Handover delays, the packet loss ratios, jitter and Packet Delivery Ratio etc. are also the parameters. On the other hand it is complicated to arrange the Qos because of dynamic topology and specific traits of VANET. After that we will take an overview of the efforts. That has been employed by several investigators for Quality of service provision on VANETs. Here in this section the discussion of several existing researches related to VANET has been made. In 2012, Park Chun et al. [1] explained the Quality of service-approved the IP mobility arrangement. That is for vast vehicles along with several network interfaces. This review proposed a quality-of-service approved Internet Protocol mobility arrangement method for fast vehicles having several without network interfaces.

Regarding to variations in date rate, innovative wireless links are established according the needs of user. The parallel distribution packet tunnels are been applied. That is among an access router and node of mobile. These have been applied for retaining of demanded data rate at the time of progress of mobile node. In 2011, Sharma, et al. [2] A paper is written on the Performance Comparison of VANET Routing Protocols. These reviews have comparison of presentation of 3 routing protocols, whose name are DSDV, AODV and DSR for different parameters. In 2014, SHAH et al. [3] A routing protocols for urban vehicular networks has been proposed. The protocol is known as Unicast routing protocol. They provide the approach and taxonomy. Besides they also offered the challenges related to open research. The present review explained the traits of uncast protocols in city places. A number of parameters are applied to make comparison between reviewed protocols. That also involves crucial study of several perspectives accepted by researchers in investigational protocol valuation. In 2014, Sanjoy Das et al. [4] proposed the Performance Analysis of Routing Protocols for VANETs. The perposed performance analysis has been done together with the Real Vehicular Traces. The paper evaluated the efficiency of several protocols like LAR, AODV, and DSR protocols for VANET. A tool named Vanet Mobi Sim based on an IDM has been applied for formulating the practical mobility traces. As per the network simulator is concerned the network simulator NS-2 has been used. Protocols efficiency has been checked by speeds of variation node.

In 2014, Jabbarpour et al. [5] analysed the performance of V2 V dynamic anchor position-related routing protocols. It has presented a solution for V2V changeable anchor position-related protocols of routing. They have accomplished it by carefully considering the demanding situation of packet routing using VANET. Also it has been probed and assessed technique applying 2 altogether diverse situations i.e., a diversity of vehicle velocities along with densities. Here they estimated the network performance based on an average hold-up routing overhead and delivery ratio of packet etc.In 2013, Gupta [6] Proposed TruVAL abbreviation of Trusted Vehicle Authentication Logic for VANET. This paper introduced a layer form for vehicles substantiation for secure message exchange by the help of trust calculations. Malicious nodes are identified based on this calculated trust.

In 2012, Sutariya et al. [7] A developed AODV routing protocol for VANETS in city scenarios has been presented. The paper has presented a (Improved AODV) routing protocol IAODV. It implements the provision of well-timed and precise information to drivers in V2V communication which is in contrast to AODV protocols in the situation of city VANET.

In 2013, Xu [8] QoS Evaluation of VANET Routing Protocols has been introduced. This paper examined and interpreted the prime quality benchmark among the accepted QoS routing protocols in VANET. Its reason is that the quality of multimedia data broadcast has been considered actually tougher in comparison to the other known wireless networks. The writers have entrusted an appraisal model. They have depicted a parallel base to assess the competency of multimedia quality with separate routing protocols.

In 2013, SafaSadiq et al. [9] described an intelligent vertical handover scheme. That is for audio and video streaming in heterogeneous vehicular networks. The following paper stated an INS system. That relies on function of maximization scoring to apply stages existing wireless network claimants. It has been tried separate input specifications to expand a function of maximization scoring aim of which is to gather data from every network candidate during the selection procedure.

In 2014, Saritha et al. [10] stated a review for channel reservation. They also discussed allocation to grow the quality of service in vehicular transmission. This paper recommended an algorithm for channel allocation which implements channel reuse method and channel borrowing system. They also implement the speed of the vehicle. That is to secure the channel. It is for handoff calls to accentuate QoS in VANET.

In 2014, Dr. Singh [11] made comparative review among UNICAST and multicast routing protocols in several data rates with the use of VANET. The competence of VANET is interpreted with the use of multi QOS metrics. It is impacting the network communication competence, metrics with ratio and estimable delay with 1024 bytes packets. It is with the purpose of inserting uncast routing protocols and multicast routing protocols in metropolitan milieu. Dynamic Source Routing and Ad hoc on demand Distance Vector routing are the uncast routing protocols. The multicast routing protocols are for example Adaptive Demand driven Multicast Routing and On Demand Multicast Routing Protocol.

In 2012, Noori [12] performed the review on Beaconing for VANET-related Vehicle to Vehicle transmission. They considered the probability of Beacon Delivery in Realistic huge sine urban sector with the use of 802.11p. This paper simulated car to car transmission in real life multi-scale metropolitan sector. That is principled on IEEE 802.11p norm. They analyzed the booming possibility of the beacon delivery.

In 2012, Kaur [13] presented paper on competence analysis Of Topology Based Routing Protocols In VANET. This paper standardized and assessed the hands-on and reactive routing protocols that are usually utilized in MANETs. That will certainly apply to VANETs.

In 2012, Singla [14] created a review on performance analysis of routing protocols in VANET with the use of TCP variants On Omnet++ Simulator. This paper probed and analyzed the performance of the routing protocol in VANET. It utilizes TCP variants that are TCPReno, TCP new Reno and TCPTahoe. Optimized Link State Routing along with On-Demand Distance Vector is varying on the foundation of throughput and delay.

In 2013, Spaho [15] wrote paper on performance analysis of AODV routing protocol in VANETS considering multi-flows traffic. This paper assessed and analyzed the AODV routing protocol' competence.

3   Tools and Technology

MATLAB
MATLAB is called as Language of Technical Computing. Within interactive environment it takes as a high level language. MATLAB enables us to can computationally tasks quicker as compare to other different programming languages like C, C++, and FORTRAN. It had introduced environment for organization code, files, and information. It had Interactive tools for different like iterative exploration, design and to resolve problem. For visualize information 2D and 3D graphics function are used. MATLAB is tools that are used to create custom graphical user interfaces.

4   Proposed Work

In proposed work the content have been replaced with short words and XOR based cryptography has made the data secure. This work has been divided in two phases' server side and client side. The server side and client side process flow has been discussed here.

Algorithm on sender side

1. Initialize the port number from receiver for transmission
2. Set the common port number and IP address from sender side
3. Set the file for transmission
4. Perform data compression using replacement table
5. Perform encryption to make data more secure.

Algorithm on receiver side

1. Wait for the data from sender
2. Receive data from sender
3. Apply to decryption on data
4. Decompress data using common replacement table
5. Receive the plain data and store in file

Replacement table at Sender's end

Step 1 Pick the word one by one from file and check their corresponding shortcut in table

Step 2 If no word sound in table there would be no operation Other wise

Step 3 replaces data T Data in XT_Data

Receiver End

Step 1 at the receiver end compare content received one by one with short word in table

Step 2 if word not found perform no operation

Otherwise

Step 3 replaces the short word with actual content

5   Result Analysis

Simulation presented that reduced the size of packet to avoid packet loss during transmission. To minimize packet loss the packet has been compressed using pointer. It also reduces the time taken during data transmission as shown in following Fig. 2.

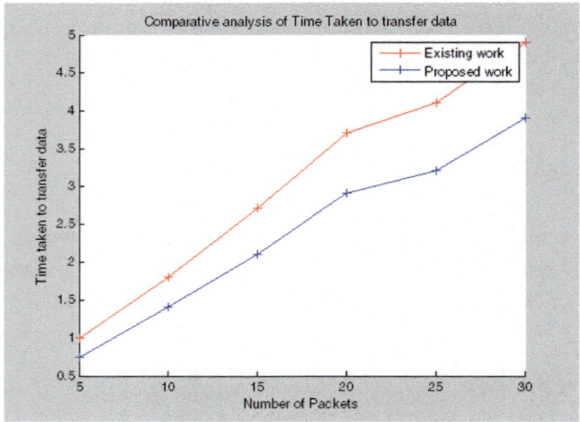

Fig. 2. Time taken to transfer packet

The size of packet is reduced the success rate of data transmission increases. Figure 3 represent the error rate of proposed model is less as compare to traditional.

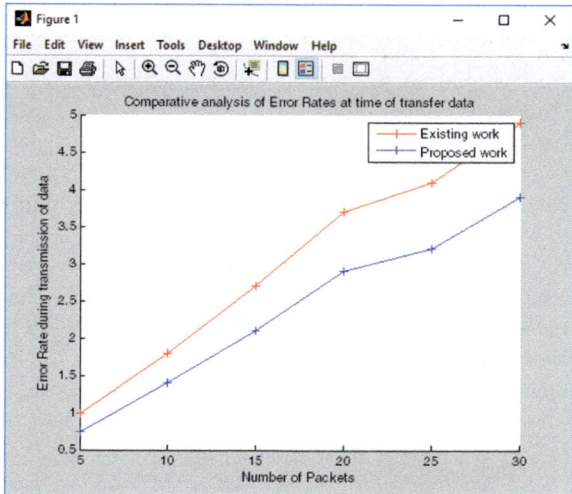

Fig. 3. Probability of errors in tradition and proposed work

More over when the size of packet is reduced by content replacement mechanism the transmission time in secure proposed system is less than time taken by tradition model. The Comparative analysis of transmission time in case of secure and unsecured traditional and proposed work has been shown in Fig. 4.

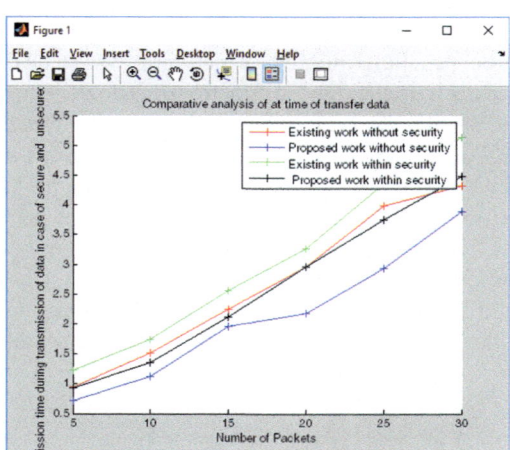

Fig. 4. Comparative analysis of transmission time in case of secure and unsecured traditional and proposed work

6   Conclusion

The proposed work has reduced the size of packet to avoid packet loss during transmission. In order to reduce packet loss the packet is compressed using pointer. In this mechanism the real data is replaced by the small short keys which are available in remote table and secure data travels over network. With reduced size of data the success rate of data transmission increases. There would be minimum chances of packet dropping.

Proposed work also would offer the current security system. Integration of XOR based encryption has enhanced the security. This encrypted content would be too complex for cryptanalyst to decode. In this way presented scheme protects the information on cloud with the help of multiple level of protection.

It has been concluded that in proposed work packet transmission is taking less time as compare to traditional work. The time taken during secure transmission in case of proposed model is less as compare to traditional model.

The proposed work has reduced of packet loss as information has been compressed by pointer. In this mechanism the real data has been exchanged with small short keys available in remote table and encoded data is capable to travel over network. As the size of data is reduced the success rate of data transmission has been increased. There are minimum chances of packet dropping. More over the probability of congestion is reduced. The use of XOR based encryption enabled quick encryption and allowed secure data transmission.

7   Future Scope

Such system would be beneficial in case of network system, Ad hoc system, Cloud environment where data is transmitted using transmission media to remote location. This system would increase the performance. This system would boost up the performance of MANET and VANET. Such system would work on limited bandwidth. The probability of data loss in such system would be minimum. Moreover this system could be useful in case of online commercial system where secure transactions take place. Limited packet transmission would reduce the requirement of costly hardware. In this way the proposed model is cost effective along with better performance and security. There would be less probability of packet dropping and congestion. XOR based encryption has allowed efficient encryption and allowed secure data transmission.

References

1. Park, J.T., Chun, S.M.: QoS-guaranteed IP mobility management for fast moving vehicles with multiple network interfaces (2012)
2. Sharma, N.: Performance comparison of VANET routing protocols (2011)
3. Ali Shah, S.A.: Unicast routing protocols for urban vehicular networks: review, taxonomy, and open research issues (2014)

4. Das, S.: The performance analysis of routing protocols for VANETs with real vehicular traces (2014)
5. Jabbarpour, M.R.: Performance analysis of V2V dynamic anchor position-based routing protocols (2014)
6. Gupta, S.D.: TruVAL: trusted vehicle authentication logic for VANET (2013)
7. Sutariya, D.: An improved AODV routing protocol for VANETs in city scenarios (2012)
8. Xu, S.: QoS evaluation of VANET routing protocols (2013)
9. Sadiq, A.S.: An intelligent vertical handover scheme for audio and video streaming in heterogeneous vehicular networks (2013)
10. Saritha, V.: Approach for channel reservation and allocation to improve quality of service in vehicular communications (2014)
11. Dr. Singh, P.: Comparative study between unicast and multicast routing protocols in different data rates using vanet (2014)
12. Noori, H.: Performed novel study on beaconing for VANET-based vehicle to vehicle communication: probability of beacon delivery in realistic large-scale urban area using 802.11p (2012)
13. Kaur, S.: Performance evaluation of topology based routing protocols in vanet (2012)
14. Singla, R.: Performance evaluation of routing protocols in vanets by using Tcp variants on Omnet++ simulator (2012)
15. Spaho, E.: Performance evaluation of AODV routing protocol in VANETs considering multi-flows traffic (2013)
16. Sommer, C.: Bidirectionally coupled network and road traffic simulation for improved IVC analysis (2011)
17. Boukerche, A.: Wireless communication and the increasing popularity of portable computing devices, wireless and mobile ad hoc networks (2008)
18. Caliskan, M., Graupner, D., Mauve, M.: Proposes a vehicular ad hoc networks (VANET) based on Wireless-LAN IEEE 802.11 (2006)
19. Fracchia, R., Meo, M.: Focuses on inter-vehicular networks providing warning delivery service (2008)
20. Nadeem, T., Dashtinezhad, S., Liao, C., Iftode, L.: Common platform for inter-vehicle communication (2004)
21. Riva, O., Nadeem, T., Borcea, C., Iftode, L.: Ad hoc networks (2007)
22. Wewetzer, C., Caliskan, M., Luebke, A., Mauve, M.: Applications based on vehicular ad-hoc networks (VANETs) rely on information exchanged within the network (2007)
23. Dikaiakos, M.D., Florides, A., Nadeem, T., Iftode, L.: Wireless inter-vehicle communication systems enable the establishment of vehicular ad-hoc networks (VANET) (2007)
24. Acharya, A., Sharma, I.: Bandwidth scalable coding for vehicular networks with infrastructure support, 2016. International Conference on Control, Instrumentation, Communication and Computational Technologies (ICCICCT) (2016)
25. https://encrypted-tbn0.gstatic.com/images?q=tbn: ANd9GcTU9TCSZaDlKSxg7S8kw5AiYw05_21oiCXKDIllFqsYxP7UXZLR
26. Jain, A., Sharma, D.: Approaches to reduce the impact of DOS and DDOS attacks in VANET. IPASJ Int. J. Comput. Sci. (IIJCS) 4(4), April (2016), ISSN: 2321-5992

Splicing Detection Technique for Images

Smita A. Nagtode[1(✉)] and Shrutika A. Korde[2]

[1] Department of Electronics & Telecommunication,
D.M.I.E.T.R., Sawangi (Meghe), Wardha, India
s.nagtode@gmail.com
[2] Electronics & Communication, Department of Electronics &
Telecommunication, D.M.I.E.T.R., Sawangi (Meghe), Wardha, India
shrutika.korde@gmail.com

Abstract. The authenticity of digital image find out by using the Image Forgery Detection technique. In the area of digital forensics and multimedia security it is more important. Photo-editing software, powerful computers, and a device the high resolution images are used for the manipulation. So, it is very difficult for finding the transformation in the digital image by the human eyes. The proposed method to identify splicing forgery for images which are blurred, resized and angular transformed. These images need a special kind of feature extraction namely SWT in order to perform the task. MS-LBP is applied to the SWT for finding key-points.

Keywords: Image splicing · Stationary wavelet transform ·
Singular value decomposition · Multi-Scale Local Binary Pattern

1 Introduction

Image processing is a technique of changing the images into digital images. It performs some kind of action on it like enhancing the image or extracting the data. Image processing is one of the types of signal dispensation. In image processing, input is an image such as photo, video, etc. and output is also the image or attribute related to that image or video. Nowadays, rapidly growing area in technology is image processing.

Following the three steps are involved in the image processing:

- Capturing the image using an image acquisition tool
- Examine the image and manipulating the image
- The last stage is the result/output in which transformed the image [14]

Nowadays, image manipulation has become easier due to various specialized software like Photo editor, Adobe Photoshop, etc., so it expresses as a real. Such manipulation with the real images is called image forgery.

Forgery detection technique is divided in two types: (1) active and (2) passive. In the technique of active forgery detection, some pre-processing operations are required for the images such as digital signatures or digital watermarking. In passive/blind forgery detection technique, the digital images do not require digital signatures or

A. P. Pandian et al. (Eds.): ICCBI 2018, LNDECT 31, pp. 654–660, 2020.
https://doi.org/10.1007/978-3-030-24643-3_78

digital watermarking. It detects the forged regions in the image. Image tampering is the part of the passive forgery detection technique.

Image tampering is a skill which needs to understand the image properties and good visual creativity. One tampers images for various reasons either to enjoy the fun of digital works creating incredible photos or to produce false evidence [8]. Image tampering is classified into three types: copy-move, image splicing, and image retouching.

1.1 Copy-Move

Region duplication forgery is also called as copy-move. Copy-move is the type of image forgery detection technique and this is a common type. In which, some part of an real image is copied, passed it and paste in the particular place of the correlative image (Fig. 1).

Fig. 1. Example of copy-move forgery [15]

1.2 Image Splicing

Image splicing is a fusion of more than one different images also converted to one image to form a duplicate image. In image splicing, cutting/copying some part of the one image and pasted it to another image. So, to detect the tampered region in the image is difficult by the human eye (Fig. 2).

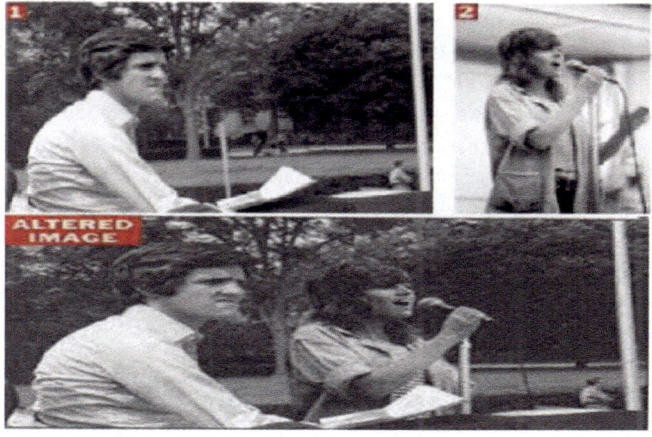

Fig. 2. Example of image splicing [16]

1.3   Image Retouching

Image retouching is the process of changing the original pixel such as enhancing or reducing the features of an image (Fig. 3).

Fig. 3. Example of image retouching [17]

The proposed method to identify splicing forgery for images which are blurred, resized and angular transformed. Images are applied to pre-processing block for conversion of RGB images into YCbCr component. The images are divided into the number of blocks. These images need a special kind of feature extraction namely SWT in order to perform the task. MS-LBP is applied to the SWT for finding key-points.

2   Related Work

Shah and El-Alfy [1] detected image splicing using Multi-Scale LBP and DCT coefficient. Multi-Scale LBP was applied to the images which were divided into numbers of blocks, then computed DCT coefficient as well as the standard deviation. Here, the classifier Support Vector Machine with RBF kernel was predicted the forged and authentic classes of image. The results revealed 97.3% accuracy when applying multi-scale LBP.

Dixit, Naskar et al. [2] used SWT and SVD technique to detect copy-move forgery detection. They negotiated color-based division to implement blur invariance and to decrease the number of the FPR (false positive rate), 8 connected neighborhoods were used for blurring and without blurring the images. The presented method provided higher forgery DA (detection accuracy) comparing with the state-of-the-art.

KANG, WEI [8] detected copy-move image forgery using SVD (singular value decomposition). For algebraic and geometric features extraction SVD (singular value decomposition) provided the method. The proposed method is effective in cases of copy-move image tampering induced by noise and robust next to retouching details.

Hikimi, Hariri et al. [10] detected forgery for an image using LBP, wavelet transform and PCA. All extracted feature was fed into SVM classifier. The proposed method was applied to CASIA TIDA v1.0 and database of Columbia Uncompressed Image Splicing Detection Evaluation. The result showed 97.21% accuracy of CASIA TIDA v1.0 and 95.13% accuracy of Columbia database.

Bayram, Taha Sencar et al. [11] used to transform features of Fourier Mellin, which are invariant to scaling and translation for the detection of copy-move. This method was computationally efficient and being capable forgery detection even in the image which are highly compressed.

TEERAKANOK, UEHARA [3] detected copy-move forgery using Key-point selection and rotation-invariant feature descriptor using SURF and GLCM respectively. Improving the overall accuracy of the system, proposed method needs some threshold value for further analysis.

Fadl, Semary et al. [12] detected copy-move forgery detection using Fast K-means and block frame features as a fast and efficient method, whether without alteration and with alternation modify in a spatial domain. The image was divided into numbers of the block and then extracting the feature for every block. The result of the method was efficient to detect duplicated region under several modifications like JPEG compression, alternation and smoothing environment. The proposed method is 75% faster than other systems.

Shahroudnejad, Rahmati [13] proposed a method to identify tampering regions which were copy-moved under various geometrical transformations. The method detected a large number of matched ASIFT key-points and all pixels were calculated from the duplicate region by utilizing superpixel segmentation and morphological operations. The result of the method was efficient and powerful for copy-move region detection under several transformations and post-processing operation.

George [4] developed the splicing detection technique of tampered blurred images, in this original image and spliced blur image was a different type of blurring.

Tembe, Thombre [5] studied a copy-move forgery detection technique and its classification which are a block-based method and key-point based method.

Sharma, Abrol [7] studied the threat of Digital Image tampering for security as well as different image tampering detection algorithms. Algorithms use various techniques for tamper detection such as Principal Component Analysis (PCA), Discrete Cosine Transform (DCT), Discrete Wavelet Transforms (DWT), and Singular Value Decomposition (SVD). Birajdar, Mankar [6] studied different image tampering detection algorithms for Digital image forgery detection using passive techniques.

3 Proposed Work

Most of the researcher has paid more attention to copy-move forgery. Some techniques are available for the detection of image splicing but accuracy is lower and not considers feature extraction.

Among the proposed approaches is image splicing recognition using multi-scale LBP with DCT but they do not consider any kind of noise or pixel level manipulation [1]. Some author detected copy-move forgery using SWT and SVD technique, in which they do not consider other forms of image region transformations, in copy-move [2]. Some author detected image splicing using the combination of SVD and SVD-DCT but accuracy is lower in that case [9].

In our work, we will identify splicing forgery for images based on MS-LBP with SWT-SVD for blurred images along with resizing and angular shift (Fig. 4).

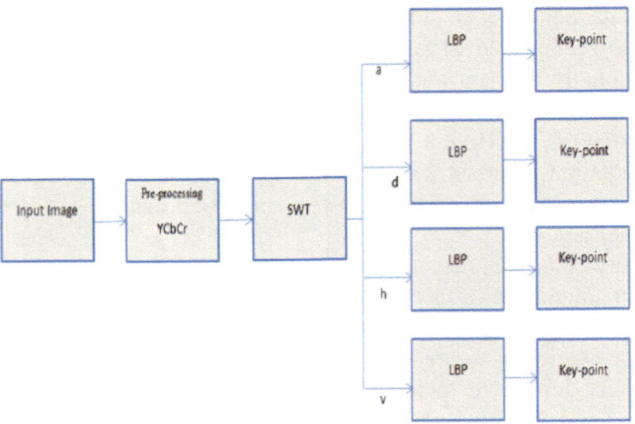

Fig. 4. Proposed model architecture

Pre-processing unit - Images are applied to the pre-processing block, in which RGB images are converted into YCbCr components. YCbCr contains luminance (Y), and the Chrominance (Cb and Cr), where Cb is the Chrominance-blue and Cr is the Chrominance-red color value as a column. The transformation formula is as follows:

$$\begin{bmatrix} Y \\ Cb \\ Cr \end{bmatrix} = \begin{bmatrix} 16 \\ 128 \\ 128 \end{bmatrix} + \begin{bmatrix} 65.485 & 128.553 & 24.966 \\ -37.797 & -74.203 & 112.0 \\ 112.0 & -93.786 & -18.214 \end{bmatrix} \cdot \begin{bmatrix} R \\ G \\ B \end{bmatrix}$$

Pixel are represented in RGB format i.e., 8 bits per sample, where black color represented by 0 and white color represented by 255 respectively, the YCbCr components can be obtain by using above equation.

SWT (Stationary wavelet transform) - The images are divided into the number of blocks. These images need a special kind of feature extraction namely SWT and SVD in order to perform the task. SWT is applied to the input YCbCr images to obtain four subbands, viz A (approximation), V (vertical), H (horizontal), and D (diagonal).

MS-LBP (Multi-Scale Local Binary Pattern) - MS-LBP is applied to each subband (i.e. A (approximation), V (vertical), H (horizontal), and D (diagonal)) to find out the key points in the image (Figs. 5 and 6).

Fig. 5. Input images (a) original image (b) forged image

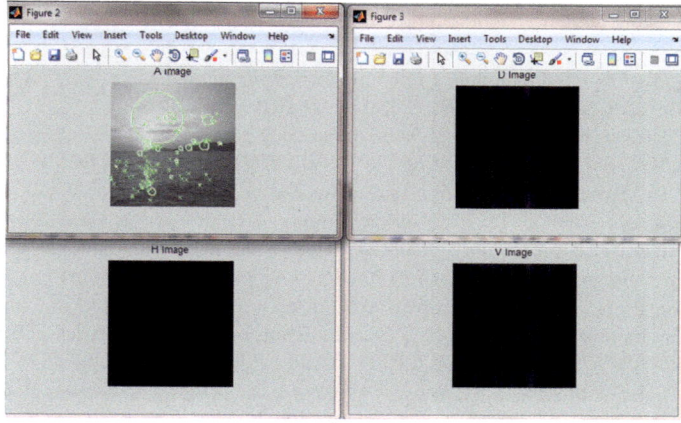

Fig. 6. Result of original image

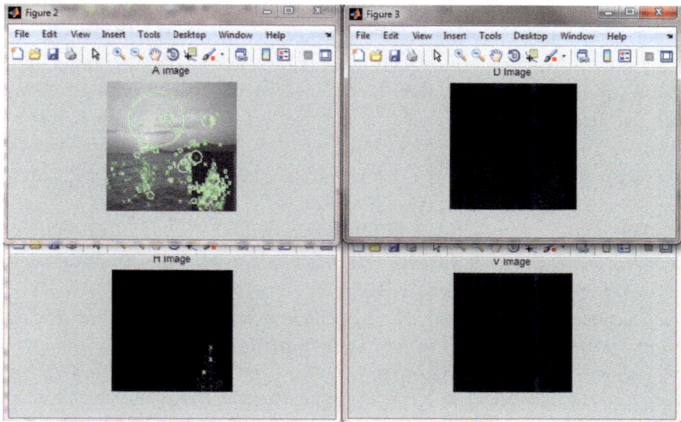

Fig. 7. Result of forged image

4 Conclusion

The authenticity of digital image find out by using the Image Forgery Detection technique. So it must to dig out the image is duplicate or real. The proposed model of this paper is based on Multi-Scale Local Binary Pattern (MS-LBP) with SWT coefficient for the detection image splicing. Images are applied to pre-processing block for conversion of RGB images into YCbCr component. The images are divided into numbers of block. The images need a special kind of feature extraction namely SWT in order to perform the task. MS-LBP is applied to the SWT for finding key-points. We can detect key-point from the image to identify the forged image. For further work we are going to detect blurred images along with resizing and angular shift using SVD, So that this work will give depth analysis for the image under test.

References

1. Shah, A., El-Alfy, E.S.M.: Image splicing forgery detection using DCT coefficients with multi-scale LBP, 978-1-5386-4680-9/18/$31.00 © IEEE (2018)
2. Dixit, R., Naskar, R., Mishra, S.: Blur-invariant copy-move forgery detection technique with improved detection accuracy utilizing SWT-SVD. IET J., ISSN 1751-9659, 2017, 11 Iss. 5, 301–309, © Institute of engineering and technology (2017)
3. Teerakanok, S., Uehara, T.: Copy-move forgery detection using GLCM-based rotation-invariant feature: a preliminary research. 2018 42nd IEEE International Conference on Computer Software & Applications 0730-3157/18/$31.00 © IEEE (2018)
4. Ambili, B., Prof. George, N.: A robust technique for splicing detection in tampered blurred images. International Conference on Trends in Electronics and Informatics (ICEI 2017), 978-1-5090-4257-9/17/$31.00 © IEEE (2017)
5. Tembe, A.U., Thombre, S.S.: Survey of copy-paste forgery detection in digital image forensic. International Conference on Innovative Mechanisms for Industry Applications (ICIMIA 2017), 978-1-5090-5960-7/17/$31.00 © IEEE (2017)
6. Birajdar, G.K., Mankar, V.H.: Digital image forgery detection using passive techniques: a survey. Digit. Invest. 10(3) (2013)
7. Sharma, D., Abrol, P.: Digital image tampering-a threat to security management. Int. J. Ad. Res. Comput. Commun. Eng. 2(10), ISSN (Online): 2278-1021 IJARCCE (2013)
8. Kang, X.B., Wei, S.M.: Identifying tampered region using singular value decomposition in digital image forensics. 2008 International Conference on Computer Science and Software Engineering, © IEEE (2008)
9. Moghaddasi, Z., Jalab, H.A., Md Noor, R.: SVD-based image splicing detection. 2014 International Conference on Information Technology and Multimedia (ICIMU), November 18–20. Putrajaya, Malaysia (2014)
10. Hikimi, F., Hariri, M., GharehBhagi, F.: Image splicing forgery detection using local binary pattern and discrete wavelet transform. 2015 2nd International Conference on Knowledge-Based Engineering and Innovation (KBEI)
11. Bayram, S., Sencar, H.T., Memon, N.: An efficient and robust method for detecting copy-move forgery, 978-1-4244-2354-5/09/$25.00 © IEEE (2009)
12. Fadl, S.M., Semary, N.A., Hadhoud, M.M.: Copy-rotate-move forgery detection based on spatial domain, 978-1-4799-6594-6/14/$31.00 © IEEE (2014)
13. Shahroudnejad, A., Rahmati, M.: Copy-move forgery detection in digital images using affine-SIFT, 978-1-5090-5820-4/16/$31.00 © IEEE (2016)
14. Introduction to image processing website [online]. https://sisu.ut.ee/imageprocessing/book/1
15. https://ai2-s2-public.s3.amazonaws.com/figures/2017-08-08/
    9435988ea91cc60db916beca80782db9bd414a47/2-Figure2-1.png
16. https://ars.els-cdn.com/content/image/1-s2.0-S0923596515001393-gr2.jpg
17. https://static01.nyt.com/images/2009/08/25/technology/personaltech/rikliquify.480.jpg

An Optimized Data Classifier Model Based Diagnosis and Recurrence Predictions of Gynecological Cancer for Clinical Decision Support System

B. Nithya[1(✉)] and V. Ilango[2]

[1] Department of MCA, CMR Institute of Technology, Bangalore, India
`nithya.boopalan@gmail.com`
[2] Department of MCA, CMR Institute of Technology, Bangalore, India
`ilango.v@cmrit.ac.in`

Abstract. Meaningful knowledge can be revealed by applying Soft Computing techniques in healthcare industry. In women-related cancers, the studies that have been done on the diagnosis and recurrence predictions of breast cancers are comparatively more to that of gynecological cancers. While cancer treatments are continuously progressing, there is still a factual risk of relapse after possibly therapeutic treatments. Few types of gynecological cancers are very dangerous and lead to lessening the lifespan of women diagnosed with such type of cancers. To prevent this, an accurate and optimized model is essential in diagnosing and predicting the relapse of gynecological cancers. Given the rising trend in the applications of Machine Learning techniques in various types of cancer predictions, this paper explores on the competent methods and techniques preferred for diagnosing and predicting recurrence in few types of gynecological cancers with the recommendation to achieve optimized results. An advanced, consistent and optimized classifier model for gynecological cancer diagnosis and recurrence predictions is expected as the outcome of this proposed work.

Keywords: Gynecological cancer · Cancer diagnosis ·
Cancer recurrence/relapse · Machine learning · Classification ·
Prediction · Optimization

1 Introduction

The implementation of new ideas into regular medical practice could enhance the treatments received by patients and more prominently it could inspire their survival. Machine Learning techniques have been experienced in several fields and it has been proved that their practice is inevitable, especially in numerous health care applications [6]. The study reveals that as a strong classification approach Machine Learning technique can make the cancer finding fast and can identify the most pertinent data for proper diagnosis and successive clinical studies. In these years, many kinds of cancer predictions were carried out by means of machine learning techniques. The prognostic models created via machine learning can be used with the computational methods for

© Springer Nature Switzerland AG 2020
A. P. Pandian et al. (Eds.): ICCBI 2018, LNDECT 31, pp. 661–669, 2020.
https://doi.org/10.1007/978-3-030-24643-3_79

inference and decision support in healthcare, especially in various types of cancer predictions. This approach is motivating as it is part of a growing trend towards personalized, predictive medicine. Machine learning methods can be used to substantially improve the accuracy of predicting cancer susceptibility, recurrence and mortality [7]. It is mandatory that more attention is required in selecting the dataset for any machine learning exercise. Several studies insisted that a sufficiently large data set that can be partitioned into disjoint training and test sets are subjected to some reasonable form of n-fold cross-validation for smaller data sets. Over the years, a continual progress related to cancer research has occurred concerning the diagnosis and recurrence prediction of cancers that are common in both men and women like lung, skin, liver and stomach cancers [18, 22, 26]. A study with comparative analysis work reviewed about several investigations made to predict cancer prognosis across cancer types and subtypes with functional genomics inputs [10]. These reviews suggested that it is significant to authenticate a method across numerous sets of patients and recommended the importance of knowing relevant features in data. These techniques are most significant for an imperative prediction and it is still extremely limited.

2   Related Work

Recently there had been a great research to diagnose and to predict the relapse of various types of cancers which are common in any human. There are few cancers that mainly affect women: breast, colon, oral, and gynecological cancers. In recent years of research, a few optimized predictive models were established for breast cancer diagnosis and recurrence predictions by means of several prediction algorithms and techniques [2, 5, 13]. There was an investigation [25] on three types of cancer datasets namely breast cancer, prostate and CNS – Central Nervous System cancers. This work applied entropy-based gene selection, discretization for recurrence prediction in these cancers. Gynecological cancers are the most common type of cancers in women after breast cancer. The five main types of gynecological cancer are cervical, ovarian, uterine, vaginal, and vulvar. A few types of gynecological cancers are dangerous and may lead to increased deaths of women who are affected with these type of cancers. There was a study which suggested for a hybrid method for staging prediction of cervical cancer [21]. In this study hybridization included the progression of knowledge-based subnetwork modules with Genetic Algorithms (GA's) by means of rough set theory and ID3. This hybrid method improved the performance of classification, network size and training time, as related to the multilayer perceptron model. Few types of Gynecological cancers are dangerous and the possibility of their relapse should be identified in advance. At present the researches on gynecological cancers are changing the practice of gynecologic oncology. Work done in diagnosis and recurrence predictions in various types of gynecological cancers using machine learning techniques are summarized here with their observations (Table 1).

Table 1. Literature review of gynecological cancer study

Reference	Gyn. cancer type	Types of algorithms used	Datasets used	Model performance and outcome	Observations
[9] Chih-Jen et al. (2014)	Recurrence-Proneness for Cervical Cancer	SVM, C5.0, Extreme Learning Machine - ELM	The data of 168 patients (12 predictor variables), Chung Shan Medical University Hospital, Tumor Registry.	SVM and ELM models used all the 12 predictor variables as inputs. C5.0 model used 2 independent variables (Pathologic T, RT). Classification rates are C5.0 = 96%, SVM = 68% and ELM = 94%	C5.0 is the most suitable and also it can be used to select important independent variables. This work determined that Pathologic Stage, Pathologic T, Cell Type and RT target Summary – were independent prognostic factors to be considered related to the recurrence prediction. Due to the small number of samples, outcomes could not be further analyzed.
[12] Di Wang et al. (2014)	Ovarian Cancer Diagnosis	Genetic Algorithm, Rough Set Incorporated Neural Fuzzy Inference System	Sample of 171 patients (28 features), National University Hospital, Singapore.	GARSINFIS performs better than C4.5 in terms of feature selection. GARSINFIS achieved a greater number of correct findings in all the stages of ovarian cancer diagnosis	Hybrid intelligent system - GARSINFIS derives fuzzy inference rules to diagnose ovarian cancer and identified its stage based on the level of seriousness. To get efficient result the data should consist of DNA gene expressions and images with blood results. But all this information was not entirely available in this work and endorsed for the dataset with this evidence to attain a reliable and accurate predictions.
[4] Bahl et al. (2015)	Predicting Recurrence in Cervical Cancer	K Nearest Neighbor -KNN (for prediction)	The data of 237 patients (12 features) from igcs.org- (82-recurrence, 155 of no recurrence).	The probability of recurrence was high (> 50%) with two features - clinical diameter and MRI Vol.	Determined that the four attributes clinical diameter, MRI volume, PET node and uterine body are the most critical for prediction of recurrence in cervical cancer. The dataset used in this work was relatively small to find a reliable accuracy.

(*continued*)

Table 1. (*continued*)

Reference	Gyn. cancer type	Types of algorithms used	Datasets used	Model performance and outcome	Observations
[3] A.N. J. Huijgens and Mertens (2013)	Recurrence in endometrial cancer	Cox Regression Analysis, Kaplan-Meier method	Samples of 219 patients treated in 2002-10, Orbis Center, Netherlands.	Multiple variate Cox regression analysis revealed that PR and risk profile are identified as independent prognostic factors.	This work decided that new prognostic factors can be used to make the prediction more accurate and efficient. Suggested for further studies to investigate predictive value in clinical outcome.
[15] Hye-Jeong et al. (2013)	Ovarian Cancer Screening	Random Forest, Logistic, Multilayer Perceptron, Bagging, Classification Via Regression, LogitBoost, Multi-Classifier, Simple Logistic, Logistic Regression	176 clinical samples (15 biomarkers), Collected from 2 Hospitals, Korea.	Experiments carried out with different biomarker combinations. 2, 3 and 4 biomarkers were combined using logistic regression, accuracy is 88%.	Determined that logistic regression finds the optimum biomarker combination. 3 marker combinations chosen by logistic regression showed the highest performance. Suggested for the implementation of novel algorithms to improve the performance.
[19] Lin Zhang et al. (2012)	Ovarian Cancer Prediction	Principal Component Analysis – PCA, Elastic net	87 ovarian cancer patients from TCGA repository (http://cancergenome.nih.gov)	PCA, elastic net methods trained only on the gene expression data. Trained both the procedures on the integrated data. Combined methods have shown better results.	Prediction models that combines gene expression and methylation profiles are investigated in this work. Integrative methods improved the prediction performance. This work suggested incorporating biological knowledge such as protein interactions to further improve the performance.
[8] Chi-Chang et al. (2013)	Cervical Cancer Recurrence	Multivariate Adaptive Regression Splines -MARS, C5.0	168 clinical samples (12 predictors) Chung Shan Medical University, Tumor Registry.	The aim is to identify the significant risk factors for the recurrent cervical cancer. Decision tree with 92% accuracy is better than MARS.	This work noticed that p-Stage and p-T as important and independent prognostic factors to find recurrence in cervical cancer. This work concluded that these models were insufficient to provide customized risk assessments.

(*continued*)

Table 1. (*continued*)

Reference	Gyn. cancer type	Types of algorithms used	Datasets used	Model performance and outcome	Observations
[14] D. Sowjanya et al. (2014)	Staging Prediction in Cervical Cancer	Multivariate filter-based model – CFS, J48, SVM, NN, Naïve Bayes, MLP	Dataset of 203 cervical cancer cases (21 Boolean features), Indo-American Cancer Hospital.	Performance of classification algorithms has been analyzed by computing their accuracy, sensitivity and specificity. Accuracy of J48 (93%) is better than other algorithms.	This study aimed at recognizing the most influential risk factors of cervical cancer. Decision tree generated by J48 has selected the attributes that are closely associated with the staging of the cancer. This work achieved better accuracy than any other work considered in their literature study.
[16] Jaree Thongkam, Vatinee Sukmak (2013)	Cervical Cancer Survival Prediction	SVM, C4.5, PART, KNN, AdaBoost, Bagging, Naïve Bayes	2,241 instances (13 variables) 1992–2008, SEER database	The accuracy of bagging is 92%. The average accuracy of PART is better than SVM, C4.5 and K-NN.	Results showed that data filtering with PART can afford optimization of data and improve the performance. Suggested to use hybrid practices to advance accuracy.
[23] R. Vidya and G. M. Nasira (2016)	Cervical Cancer Prediction	Classification and Regression Tree - CART, Random Forest Tree – RFT, K-means with RFT	Dataset of 500 records (61 variables), NCBI - National Centre for Bio-tech. Information	Accuracy was improved by merging the algorithms. CART 83.87% RFT 93.54% RFT with K-mean 96.77%	The study defined the prediction of cervical cancer in two stages i.e. Benign or Malignant. This work suggested for the integration of algorithms for better performance and to reduce prediction errors.
[24] Sharmistha Bhattacharjee et al. (2017)	Ovarian Cancer	Multi-Layer Perceptron (MLP), Decision Tree, KNN, SVM, Ensemble Classifier	215 samples (100 features), FDI-NCI, DB. (https://home.ccr.cancer.gov/ncifdaproteomics/)	Accuracy (98%) and specificity of MLP (NN) model is the maximum than further models considered in this work.	Relative study has been done to find most proper technique under various operational environments and datasets. Suggested MLP as the best method for such findings based on the performance measures like Accuracy, Sensitivity, Specificity and Errors. This dataset had more features.

(*continued*)

Table 1. (*continued*)

Reference	Gyn. cancer type	Types of algorithms used	Datasets used	Model performance and outcome	Observations
[17] Koji Matsuo et al. (2017)	Cervical Cancer Recurrence	Deep Learning, Linear Regression	431 Patients (13 variables) diagnosed during 2008-14, Norris Cancer Center, CA	The deep-learning model had a expressively better prediction related with the linear regression model.	Suggested for the convertion of deep-learning models to a set of decision trees, whose rules may be more relevant for prediction.
[1] Abid Sarwar et al. (2015)	Screening of Cervical Cancer	**15 Algorithms** Bagging, Decision Table, Decorate, Filtered Classifier, J48 graft, END, Multiclass classifier, PART, MBP ANN, Naïve Bayes, Radial basis function network, Random forest, Random subset space, Rotation Forest, Random Committee	2 datasets from Pap smear Benchmark, Herlev Hospital, Denmark (2003–05). Total of 1417 (500 + 917) Pap smear images with 20 features.	Among the algorithms analyzed, for both the datasets END algorithm depicted maximum classification efficiency (77% & 72%) by placing the instances in the respective exact class of diagnosis.	In this work fifteen machine learning algorithms have been used under different platforms over two databases and evaluated their performances for screening the prognosis of cervical cancer. The results proposed that END algorithm can be used to develop a good prognostic aid for screening of cervical cancer. Pap smear images were used as samples in this work.

3   Scope and Importance of This Research Work

Fear of Cancer Recurrence (FCR) has been ranked the largest concern for gynecological cancer survivors. In spite of the attainments of high response rates with surgery followed by chemotherapy, 75% of women ultimately die of complications connected with disease progression [11]. In countries like India a less study has happened to attain an optimized and reliable result of prediction in gynecological cancers as compared with breast cancer. Though a range of prediction models are available for specific gynecological cancers, there is no consistency in accuracy. The dataset used is relatively small in a range of study examined. The accuracy in prediction is the most important issue for determining its usefulness in decision making. The literature study evidenced that feature selection process is so critical and exhibited the importance of selecting the relevant prognostic features for accomplishing accuracy. Optimization in terms of time and space is also a major issue in predicting the consistent results from huge data. An appropriate level of validation for reliability and error rate through computational technique is also necessary. An optimized prediction model for any type of gynecological cancer dataset is essential with novel algorithms to attain high computational efficiency.

4   Proposed Methodology

Machine learning is often used in cancer predictions as it is intended to provide increasing levels of automation in the knowledge engineering process [7]. Most machine-learning problems can be simplified to optimization problems to produce good performance. Through novel algorithms and optimization techniques better computational efficiency could be obtained. An efficient, consistent and accurate classifier model with optimal feature subset is proposed in this work for diagnosis and recurrence prediction of any type of gynecological cancers (Fig. 1).

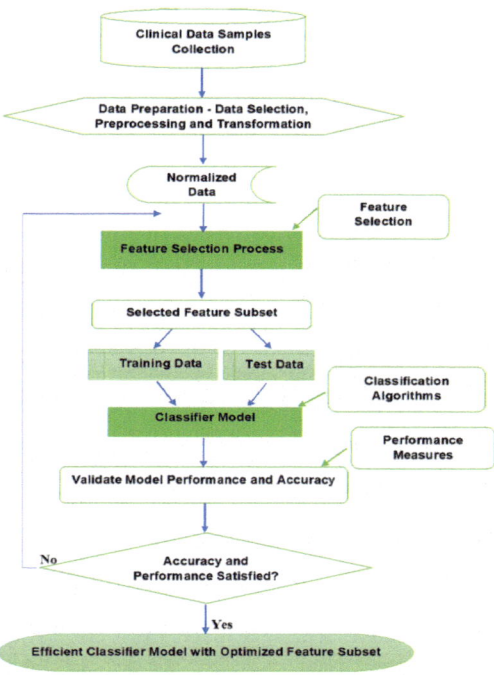

Fig. 1. Proposed system structure

Data preprocessing is vital to avoid the anomalies and noise in data as it prepares data for further processing. Irrelevant and redundant features should be removed through feature selection process to build an effective classifier model. Tools are a big part of feature selection and classification process. Numerous tools and techniques are available to transform data into actionable information to make reliable predictions [6]. The study proved the significance of selecting the relevant algorithms or combining the algorithms through hybrid methods to attain more upgraded results which is an optimized feature subset classifier model. Later the performance of classifier models can be validated using various performance measures [20].

5   Conclusion

Early diagnosis and relapsing malignancies in any type of cancer is important for doctors to alert patients and suggest recommendations. Few types of gynecological cancers and their relapse are dangerous and lead to the diminishing of the life span of women. An advanced and optimized classifier model for gynecological cancer diagnosis and recurrence predictions is anticipated through this work. Computational intelligence with novel algorithms can produce a better performance than traditional methods. Various dataset examination using all relevant predictors for diagnosis and recurrence in gynecological cancer patients will generate results with more accuracy, hence the representation of predicted data will be improved. This will support physicians to plan for appropriate beneficial approaches to cure and to extend disease-free survival of gynecological cancer patients. The scope of the research can be extended to the prediction of all types of gynecological cancers with abundant datasets by signifying the optimization for resources and money.

References

1. Sarwar, A., et al.: Performance evaluation of machine learning techniques for screening of cervical cancer. 2015 IEEE 2nd International Conference on Computing for Sustainable Global Development
2. Ahmad, L.G., Eshlaghy, A.T., Poorebrahimi, A., Ebrahimi, M., Razavi, A.R.: Using three machine learning techniques for predicting breast cancer recurrence. J. Health Med. Inform. **4**, 124 (2013)
3. Huijgens, A.N.J., Mertens, H.J.M.M.: Factors predicting recurrent endometrial cancer. FVV IN OBGYN **5**(3), 179–186 (2013)
4. Bahl, R., et al.: Predicting recurrence in cervical cancer patients using clinical feature analysis. Br. J. Med. Med. Res. **6**(9), 908–917 (2015)
5. Bhardwaj, A., Tiwari, A.: Breast cancer diagnosis using genetically optimized Neural network model. Expert Syst. Appl. **42**(10), 4611–4620 (2015)
6. Nithya, B.: An analysis on applications of machine learning tools, techniques and practices in health care system. Int. J. Adv. Res. Comput. Sci. Softw. Eng. **6**(6), 1–8 (2016)
7. Nithya, B., Ilango, V.: Predictive analytics in health care using machine learning tools and techniques. International Conference on Intelligent Computing and Control Systems, 978-1-5386-2745-7/17/$31.00 © IEEE, 492–499 (2017)
8. Chang, C., et al.: Prediction of recurrence in patients with cervical cancer using MARS and classification. Int. J. Mach. Learn. Comput. **3**, February (2013)
9. Tseng, C.J., et al.: Application of machine learning to predict the recurrence-proneness for cervical cancer. Neural Comput. Appl. **24**(6), 1311–1316. Springer-Verlag, May (2014)
10. Das, J., Gayvert, K.M., Yu, H.: Predicting cancer prognosis using functional genomics data sets. Cancer Inform. **13**(5), 85 (2014)
11. Gupta, D.: Christopher G Lis: role of CA125 in predicting ovarian cancer survival – a review of the epidemiological literature. J. Ovarian Res. **2**, 13 (2009)
12. Wang, D., et al.: Ovarian cancer diagnosis using a hybrid intelligent system with simple yet convincing rules. Appl. Soft Comput. **20**, July. Elsevier (2014)

13. Drukker, C.A.: Optimized outcome prediction in breast cancer by combining the 70-gene signature with clinical risk prediction algorithms. Breast Cancer Res. Treat. **145**(3), 697–705 (2014)
14. Sowjanya Latha, D., et al.: Staging prediction in cervical cancer patients – a machine learning approach. Int. J. Innovative Res. Pract. **2**(2), February (2014)
15. Song, H.J., et al.: Looking for the optimal machine learning algorithm for the ovarian cancer screening. Int. J. Bio-Sci. Bio-Technol. **5**(2), April (2013)
16. Thongkam, J., Sukmak, V.: Cervical cancer survivability prediction models using machine learning techniques. J. Convergence Inf. Technol. (JCIT) **8**(15), October (2013)
17. Matsuo, K., et al.: A pilot study in using deep learning to predict limited life expectancy in women with recurrent cervical cancer. Am. J. Obstet. Gynecol. Dec. (2017)
18. Kourou, K., et al.: Machine learning applications in cancer prognosis and prediction. Elsevier, Comput. Struct. Biotechnol. J. 8–17 (2015)
19. Zhang, L., et al.: An investigation of clinical outcome prediction from integrative genomic profiles in ovarian cancer. Genomic Signal Processing and Statistics, (GENSIPS), IEEE International Workshop, 103–106, 2–4 Dec. 201 (2012)
20. Kurzynski, M., et al.: Evaluating and comparing classifiers: review, some recommendations and limitations. Proceedings of 10th International Conference on Computer Recognition Systems CORES 2017. Springer International Publishing (2018)
21. Mitra, P., Mitra, S.: Staging of cervical cancer with soft computing. IEEE Trans. Biomed. Eng. **47**(7), July (2000)
22. Petalas, P., et al.: Probabilistic neural network analysis of quantitative nuclear features in predicting the risk of cancer recurrence at different follow-up times. Proceedings of the 3rd International Symposium on Image and Signal Processing and Analysis (2003)
23. Vidya, R., Nasira, G.M.: Prediction of cervical cancer using hybrid induction technique: a solution for human hereditary. Indian J. Sci. Technol. **9**(30), August (2016)
24. Bhattacharjee, S., et al.: Comparative performance analysis of machine learning classifiers on ovarian cancer dataset. Third International Conference on Research in Computational Intelligence and Communication Networks, 978-1-5386-1931-5/17/$31.00 © IEEE (2017)
25. Win, S.L., Htike, Z.Z., Yusof, F., Noorbatcha, I.A.: Cancer recurrence prediction using machine learning. IJCSITY **2**(2), May (2014)
26. Wang, X., et al.: A review of cancer risk prediction models with genetic variants. Cancer Inform. **13**(S2), 19–28 (2014). https://doi.org/10.4137/CIN.S13788

Ant Colony Optimization Techniques – An Enhancement to Energy Efficiency in WSNs

Kalpna Guleria[✉] and Anil Kumar Verma

Department of Computer Science Engineering, Thapar Institute of Engineering
and Technology, Patiala, India
guleria.kalpna@gmail.com

Abstract. Wireless sensor networks are resource constraint and set stringent
requirements like highly dynamic nature, application specific behavior, limited
network lifetime and processing capabilities. Limited network lifetime is one of
the major resource constraints which requires energy efficient operation to be
performed. Due to inherently adaptive behavior of wireless sensor networks, it is
highly challenging task to find energy efficient route. It is seen that ant colony
optimization (ACO) based routing protocols are proved as best to find adaptive
and energy efficient routing for the dynamically changing behavior of wireless
sensor networks. This paper proves that ACO based routing protocols consume
less energy and exhibit energy efficient operation.

Keywords: Ant colony optimization (ACO) ·
Wireless sensor networks (WSNs) · Routing protocols · Energy efficiency

1 Introduction

WSNs [1] have a wide variety of applications such as patient health monitoring,
precision agriculture, smart home automation, structural health monitoring, intelligent
highway monitoring in smart cities [2, 3], industrial monitoring/control, earthquake
prediction, surveillance/security and detecting volcano eruptions. ACO make use of
artificial ant agents to search for best path out of available multiple paths between the
nest sites to food site [4–6]. This in turn enhances the performance of WSN routing
protocols.

A set of algorithmic concepts which are used to define heuristic methods or procedures for various optimization problems is called as meta-heuristic. ACO Meta-
heuristic [7–10] is influenced by foraging behavior of actual ant insects. A chemical
substance which is released by ant agents while moving from one point to another is
named as a pheromone. The movement of ants is affected by pheromone value in the
local domain. Information exchange between the ant agents due to pheromone values is
termed as stigmergy [7]. Figures 1(a) and (b) depicts the concept of Ant Colony
Optimized routing by showing that the ants which take small distant paths as compare
to long distant one, reach their destination much early and shows better performance
metrics.

© Springer Nature Switzerland AG 2020
A. P. Pandian et al. (Eds.): ICCBI 2018, LNDECT 31, pp. 670–677, 2020.
https://doi.org/10.1007/978-3-030-24643-3_80

(a)

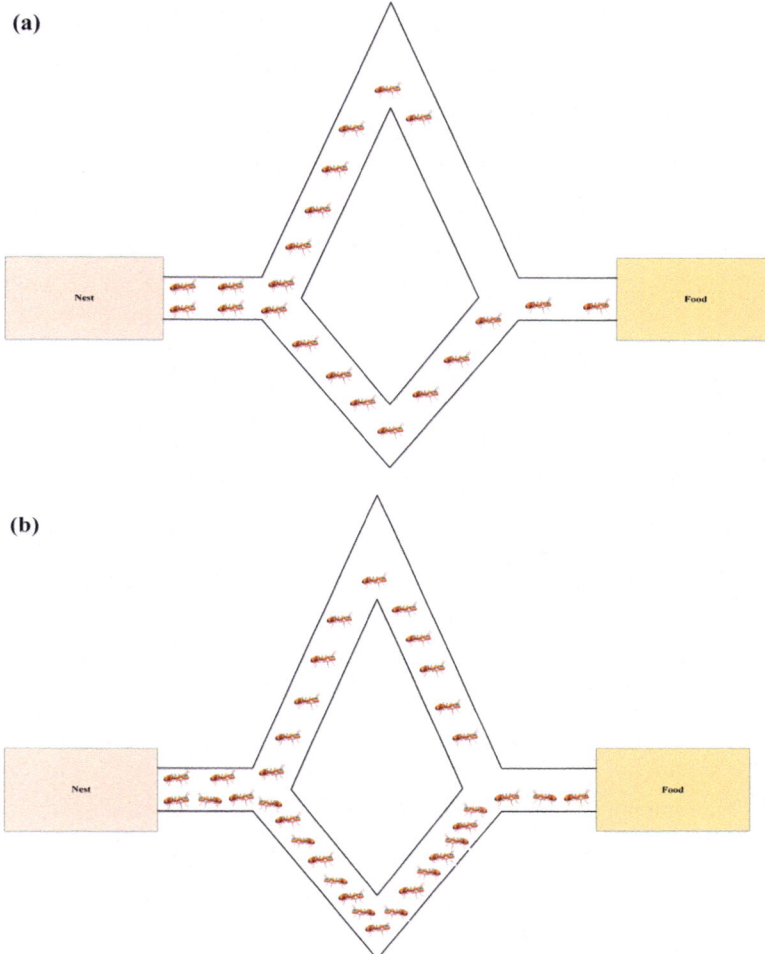

(b)

Fig. 1. (a). Ants opt for a smaller path to reach their food site earlier (b). Ants using smaller path reach back in the nest early than the ants using a long path

2   ACO Based Routing Techniques

2.1   Sensor-Driven Cost-Aware Ant Routing (SC)

Zhang et al. [11] investigated the AntNet routing algorithm for wired networks and thereafter he designed and proposed a routing algorithm for WSN based on AntNet. Three different variables of AntNet: SC (Sensor-driven Cost aware), FF (Flooded Forward), FP (Flooded Piggybacked) were proposed.

In AntNet forward ants search paths randomly and therefore it require time for the selection of good path out of constructed paths. Zhang Y. et al. proposed the solution to the above said problem by defining Sensor-driven Cost-aware routing by making cost estimation initially at network setup time by defining number of hop metric Q_n towards the destination. The cost estimation requires iteratively flooding the message and calculating number of hops by spreading it over the entire network to achieve higher routing updates. The performance of the algorithm is increased by making the forward ants to find out the best direction in which forward ants should move initially.

2.2   Flooded Forward Ant Routing (FF)

To route packets towards destination node, flooding of ants is done in flooded forward [11] method. Ants are required to continually explore the area, during the broadcast. An assumption is made in flooded forward that ants have access to direction and location information of forwarding ants. Stochastic flooding method is used to reduce network control overhead.

2.3   Flooded Piggybacked Ant Routing (FP)

Flooded piggybacked [11] approach is constituted where the data packets are piggy-backed into ant packets to achieve higher rates of routing updates. Flooding of ants is done stochastically as it is done in flooded forward. These three algorithms are compared: FP has very good success rates whereas SC performs poorly on this parameter.

2.4   Energy Efficient Ant-Based Routing (EEABR)

EEABR [12] achieves a significant reduction of communication overhead while path discovery and therefore it results in an increased network lifetime.

EEABR protocol considers the average energy of nodes, remaining energy of nodes, threshold energy of nodes, the number of hops metrics for energy efficient operation. Another point worth mentioning here is that artificial ant agents produced here are fixed in size which means that the size of ant agent is limited. Fixed size forward ants store the nodes traversed during their journey.

Ant generation is done on a periodic basis using a proactive strategy. During its journey, the path traversed by two different forward ants joins at a particular point and then it will form a sub-tree similar to a Steiner tree. Further, the ants will merge as a single ant agent.

Pheromone reinforcement $\Delta\tau$ is calculated as per the equation:

$$\Delta\tau = 1/\left[E_{0\ldots} - \{(E_{min}^k - H_D^k)/(E_{avg}^k - H_D^k)\}\right] \tag{1}$$

Where
E_{avg}^k - Average energy of nodes traversed by ant.
E_{min}^k - Minimum energy/threshold energy of the node.
E_0 - Initial value of node energy.
H_D^k - Total hop count for path.
The Protocol defines pheromone update rule as per the following equation:

$$\tau_{ij} = (1 - \rho)\,\tau_{ij} + (\Delta\tau/\varnothing B_D^k) \qquad (2)$$

Where
Ø - Scaling coefficient
ρ - Pheromone evaporation factor
B_D^k- Total hop count traversed by ants during the return journey.

2.5   An Improved Energy-Efficient Ant-Based Routing (IEEABR)

The IEEABR [13] protocol is an improvement over EEABR. The protocol aims at increasing the network lifetime. Three major enhancements are done over EEABR algorithm like intelligent initialization of routing tables by prioritizing the neighboring nodes, an immediate update about node/link failure, a new mechanism for congestion control by reducing the flooding of ants. These improvements have been significantly done for energy efficient operation for dynamic environments. The above-mentioned enhancements can be further defined as:

Intelligent Initialization of routing table:
There is no standardized routing table initialization method followed in the EEABR algorithm. Initially routing table is set up by a uniform probability distribution, $P_{iD} = 1/N_k$, where P_{iD} is the probability of reaching destination D and Nk is set of neighbors.

Intelligent Update after Network Resources Failures:
The IEEABR protocol deals with the network failure due to link/node failures however EEABR does not. An immediate update about link or node failure is done in the routing table.

A new mechanism for congestion control and self-destruction of control packets:
There is no standardized method which is defined by EEABR whereas IEEABR limits the total no. of forwarding ants launched to 5*N, where N is a number of network nodes. The self-destruction of forwarding ant is done if the number of hops in a cycle is more than half of the already accumulated number of hops.

3   Simulation and Result Analysis

3.1   Evaluation Setup

The Qualnet 6.0 simulator is used to carry out the simulation task. The proposed evaluation model is setup as follows. First, we go to the architecture mode of the simulator. The Terrain size is set to 1500 × 1500 square meters in scenario properties. The subnet 1 properties are set to MAC layer 802.15.4. The network traffic is CBR (Constant Bit Rate). The time for which simulations are performed varies from 100 s to 2 h. The random waypoint mobility model defines the node mobility. Energy of nodes is 25 J/node. Ant ratio considered is 2. The subnet 2 is an IP network and its properties are set as 802.11n. Further, listenable and listening channel properties are set for subnet 1 and subnet 2. In both subnets energy model is set in the physical layer. The battery model for all nodes is residual.

3.2   Results and Analysis

In experimental evaluation, five ACO based routing protocols have been compared for various performance metrics. Paper presents simulation results based on five sets of simulations carried at 20, 40, 60, 80 and 100 s for two hours.

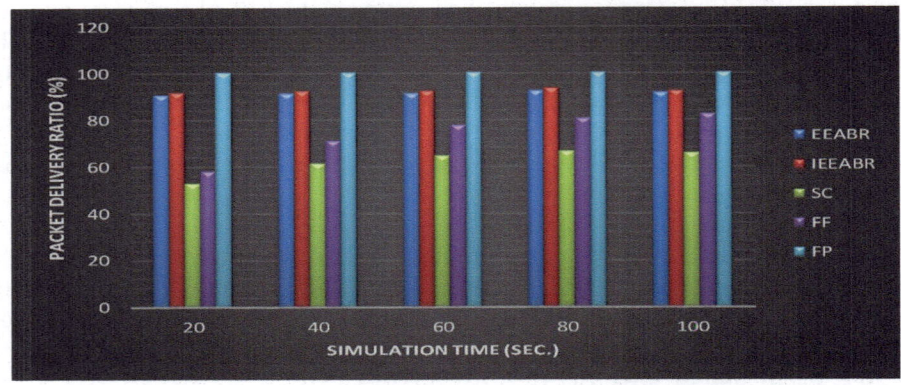

Fig. 2. Packet delivery ratio

Figure 2 represents PDR for FP, FF, SC, EEABR, IEEABR protocols. FP shows highest PDR among all five protocols for all sets of simulations. However, performance of SC is lowest among all. EEABR, IEEABR perform fairly well in comparison to FP. Figure 3 shows that FP attains highest throughput among all. EEABR, IEEABR also exhibit very good throughput whereas SC shows lowest throughput values among all. Figure 4 proves that FF has largest end to end delay. EEABR, IEEABR, SC show small end to end delay. Figure 5 reveals that FP has highest energy consumption among all. EEABR, IEEABR, FF show fairly low energy consumption. Figure 6 exhibits that IEEABR shows higher energy efficiency and EEABR shows 2nd highest energy efficiency among all. However, FP attains lowest energy efficiency among all compared protocols.

It is visualized that EEABR, IEEABR, SC show very less end to end delay whereas FF exhibits highest end to end delay. EEABR, IEEABR, FP shows better throughput in comparison to other protocols. EEABR, IEEABR, FP performs well for packet delivery ratio as compared to SC, FF. Energy consumed is highest in case of FP whereas EEABR and IEEABR consume less energy. IEEABR is most energy efficient and FP is the least. FP protocol can be deployed in applications where 100% packet delivery is of utmost importance and energy consumption is not a constraint. SC protocol also consumes less energy and has less end to end delay but it does not perform well on other performance metrics.

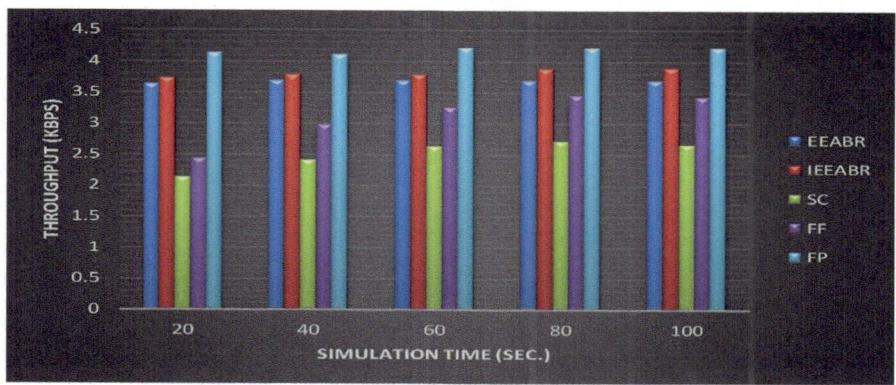

Fig. 3. Throughput

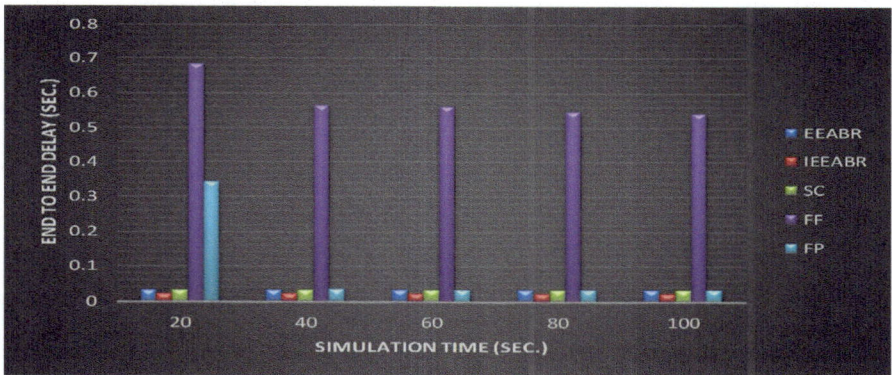

Fig. 4. End to end delay

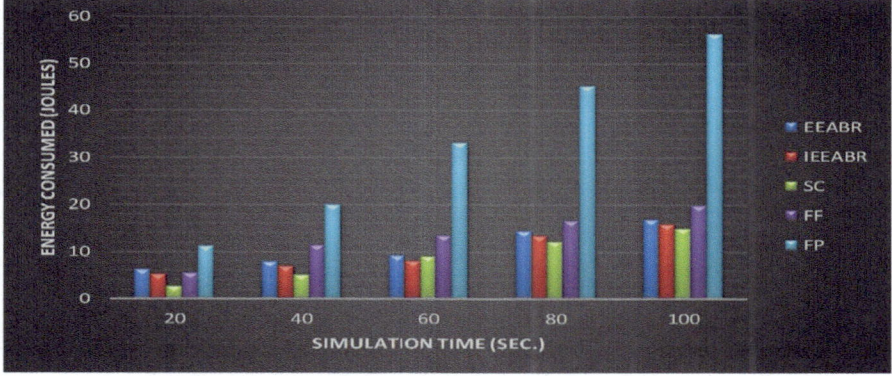

Fig. 5. Energy consumed

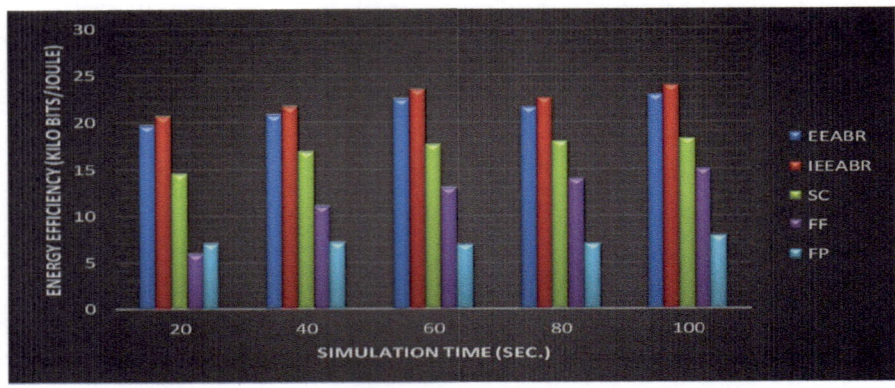

Fig. 6. Energy efficiency

4   Conclusion

ACO based routing protocols can add more contribution in WSN to deal with one of the major resource constraint: energy efficiency. Simulations exhibit that FP consumes highest energy in comparison to SC, FF, EEABR, and IEEABR. IEEABR outperforms SC, FF, FP and EEABR for energy efficient operation. Evaluation of routing algorithms using actual test bed hardware must be motivated. However, hardware cost factor is another challenging aspect.

References

1. Akyildiz, I.F., Su, W., Sankarasubramaniam, Y., Cayirci, E.: A survey on sensor networks. IEEE Commun. Mag. **40**(8), 102–114 (2002)
2. Kumar, S., Verma, A.K.: Position based routing protocols in VANET: A survey. Wireless Pers. Commun. **83**(4), 2747–2772 (2015)
3. Kumar, S., Verma, A.K.: An advanced forwarding routing protocol for urban scenarios in VANETs. Int. J. Pervasive Comput. Commun. **13**(4), 334–344 (2017)
4. Singh, G., Kumar, N., Verma, A.K.: Oantalg: an orientation based ant colony algorithm for mobile ad hoc networks. Wirel. Pers. Commun. **77**(3), 1859–1884 (2014)
5. Wankhade, S.B., Ali, M.S.: Route failure management technique for ant based routing in MANET. Int. J. Sci. Eng. Res. **2**(9), 1–5 (2011)
6. Pankajavalli, P.B., Arumugam, N.: BADSR: an enhanced dynamic source routing algorithm for MANETS based on ant and bee colony optimization. Eur. J. Sci. Res. **53**(4), 576–581 (2011)
7. Di Car, G., Dorgio, M.: Antnet: distributed stigmergetic control for communication networks. J. Artif. Intell. Res. **9**(3), 17–355 (1998)
8. Guleria, K., Verma, A.K.: Comprehensive review for energy efficient hierarchical routing protocols on wireless sensor networks. Wireless Networks, 1–25 (2018)
9. Guleria, K., Verma, A.K.: An energy efficient load balanced cluster-based routing using ant colony optimization for WSN. Int. J. Pervasive Comput. Commun. Nov 24 (2018)

10. Singh, G., Kumar, N., Verma, A.K.: Ant colony algorithms in MANETs: a review. J. Netw. Comput. Appl. **35**(6), 1964–1972 (2012)

11. Zhang, Y., Kuhn, L.D., Fromherz, M.P.: Improvements on ant routing for sensor networks. In: International Workshop on Ant Colony Optimization and Swarm Intelligence 2004 Sep 5, pp. 154–165. Springer, Berlin, Heidelberg

12. Camilo, T., Carreto, C., Silva, J.S., Boavida, F.: An energy-efficient ant-based routing algorithm for wireless sensor networks. In: International Workshop on Ant Colony Optimization and Swarm Intelligence 2006 Sep 4, pp. 49–59. Springer, Berlin, Heidelberg

13. Zungeru, A.M., Seng, K.P., Ang, L.M., Chong Chia, W.: Energy efficiency performance improvements for ant-based routing algorithm in wireless sensor networks. J. Sens. **2013**, 1–17 (2013). https://doi.org/10.1155/2013/759654. Article ID 759654

Overview on Cyber Security Threats Involved in the Implementation of Smart Grid in Countries like India

Venkatesh Kumar[(✉)] and Prince Inbaraj

Department of Electrical and Electronics Engineering,
Karunya Institute of Technology and Sciences, Coimbatore, India
venkateshkumar@karunya.edu, princerhrs@gmail.com

Abstract. Smart grid is an interconnected network of various substations which involves the latest technologies and communication protocols. When a conventional grid gets integrated into a smart grid, along with the various advantages involved, one should consider the security issues related to it. This paper reviews the various challenges and vulnerabilities involved in a smart grid. The various types of cyber-attacks are discussed.

Keywords: Cyber-security · India · Risks · Smart grid · Threat · Types of attack

1 Introduction

To deliver electricity with the help of conventional electric grid to all customers and it will be act as interconnecting network to all electricity providers. Electric stations helps to provide the electric power in high source transmission links. It helps to carry the source from providers to distributors. With recent developments in technologies, renewable-energy technologies like solar and wind power are now challenging the traditional distribution system. This decentralized approach or distributed energy is a one way approach towards power distribution [1, 2].

Now a better method of power transmission and distribution which relies on two way communication approach (that is both data and power), known as the smart grid technology. It helps to produce the smart functions in the electric parameters like measurements meters, appliances and resources in order to achieve efficiency, sustainability, quality, reliability, safety and security as well as to manage distributed generation to consumers [3, 4]. One important concern during an unexpected grid failure, is its security. Any smart grid system must demand these three aspects: (1) how available is the uninterrupted power supply to the user, (2) How integral is the communicated information, and (3) How confidential is the user's data.

In this paper, a summary of the importance of smart grid, challenges, vulnerabilities, objectives and requirements of smart grid is briefed. A couple of attacks has been discussed. The challenges in our country that needs to be rectified are also summarized.

© Springer Nature Switzerland AG 2020
A. P. Pandian et al. (Eds.): ICCBI 2018, LNDECT 31, pp. 678–684, 2020.
https://doi.org/10.1007/978-3-030-24643-3_81

2 Need for Smart Grid

There are several reasons for a power grid to evolve into a smart grid. In traditional grids, the consumers did not participate with the power system. The grids were dominated by central generation which posed many obstacles with energy resource interconnection. They had slow response to power quality issues, were not self-healing and were vulnerable to natural disasters [5, 6].

Whereas, Smart grid enables the consumers to be informed and involved with the power system and can accommodate all generation and storage options with plug-and-play convenience. Power quality is the main priority with quick resolution of issues. They are resilient to attack and natural disasters with rapid restoration capabilities. It protects the assets of the consumers by automatically detecting and responding to the problem, i.e., it is self-healing. Figure 1 summarizes the important aspects required in a smart grid.

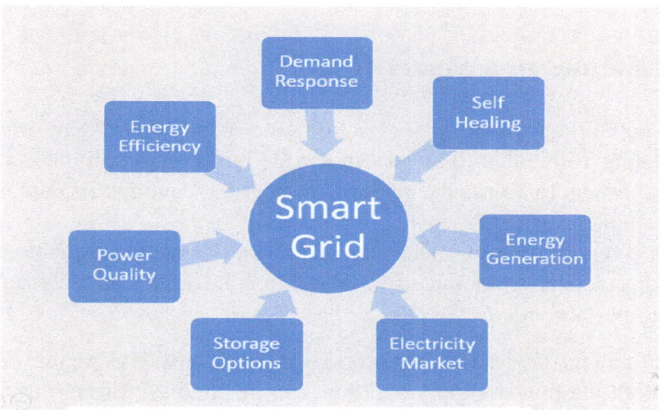

Fig. 1. Need for smart grid.

3 Challenges in Smart Grid

Smart grid is an integration of old system technologies paired with new technologies based on different standards and regulations to support the challenges of smart grid network [7]. The main challenges faced in running a secure smart grid are cyber security, internetworking, interoperability, Lack of performance data, lack of skilled work force and Social Acceptance & user reluctance to change.

A. Cyber security: Though implementing and developing smart grid is important on one side, maintenance and security is equally important. The grid communications are to be properly managed and monitored by the operators and suppliers through the security services. The System cannot be cyber secure if proper policies are not implemented in this area. It takes time to establish a safe grid, but it should be achieved as fast as possible through on-going risk assessment and training. Only with high profile security system with adequate technologies including expertise in communication, monitoring and correction, we can establish a cyber-secure grid [8, 9].

B. Internetworking: The ability of the grid to adapt, communicate, collect and redistribute power from small domestic power sources into the grid, with the acceptance to adopt to the new technologies implemented in the grid.

C. Lack of performance data: The state administration and consumer advocates are not fully convinced with the value of smart grid technologies due to lack of proper analysis of performance data on costs and benefits.

D. Lack of skilled work force: The inability to provide the skilled personals to build, install, operate system and equipment makes it difficult for the administration in solving an inconvenience or operating the grid with efficiency.

E. Societal acceptance and User reluctance to change: It is quite difficult to create an awareness about the electrical infrastructure and how they perform to the customers as they feel vulnerable to adopt into a large network with advanced protocols and electrical advancements.

4   Vulnerabilities in a Smart Grid

Smart grid network various improvisation and capabilities than traditional grid resulting in being vulnerable to attacks in a wide range. Through these vulnerabilities might allow attackers to access the network, tamper the confidential data, and make the service unavailable.

Customer security: Smart meter automatically keeps on collecting data which include various details from the user. These private data can be manipulated and interfere customer activities

(1) More intelligent devices: A smart grid has so many devices that focus on controlled both the power supply and they can be attacked through various nodes in sub, which makes network monitoring and management difficult.

(2) Physical security: As the smart grid consists of more components, it increases the number of insecure components making them more vulnerable

(3) The durability of power systems: As many power systems are short lived IT systems, these equipment can be used to infiltrate various weak security nodes and alter the data.

(4) Different Team's backgrounds: Improper communication between employees causes a lot of wrong decisions which leads to vulnerability in the network

(5) Using Internet Protocol (IP): By IP Protocols in smart grids, it is a big advantage, since it provides wide accessibility between various components. But these devices are more vulnerable to attacks such as IP spoofing, Denial of service (DOS), etc.

(6) More Shareholders: Having many shareholders helps in organizing various types of attacks.

5 Objectives and Requirements of Cyber Security

There are different organizations that do full depth research in the field of cyber security like EPRI and NIST. These bodies design framework and focus on evolving in creating a reliable and safe operation of the grid. One of their main objective is confidentiality, to restrict and constrain access to high priority information like pricing, control commands and utility information. Another objective is integrity, which is restricting the modification of communication and electronic equipment like smart meters to authorized persons. Any corruption in information exchange causing mismanagement in operation like the bad data injection is to be viewed critically. Availability is preventing the adversary from not granting access or control. Attacks like Denial of service (DoS) can tamper the communication and gain access to high priority information [10, 11] (Fig. 2).

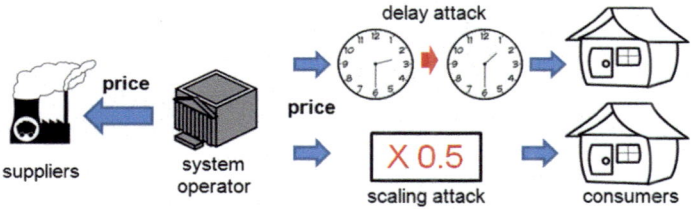

Fig. 2. Real time pricing in smart grid

6 Cyber Attacks

In spite of so many advantages in smart grid, one main major concern is that the technologies involved in the grid are more of the IT-based, which is a threat to the security of the grid. A well secured monitoring of the grid is pretty challenging as the more we include computers and communication nodes the more the hackers try to find week spots in the grid network and gain access to the grid.

In December 2015, the Ukraine power grid was attacked by hackers and were able to access 30 substations and 230 thousand people able left with no electricity for about 6 h. About 73 MWh for electricity was not supplied in the city.

Providing a good cyber security for the nations power grid, helps in preventing Information and power theft, Black outs in the city, unwanted interruption in industries and companies and chaos in normal life. In latest advancements of information sharing such as big data and IoT where data such as online banking information and other valuable information is shared between machine-to-machine, cyber security plays a vital role in protecting the data [12].

7 Types of Cyber Attacks

1. Denial of service: The type of attack where the infiltrator makes the machine or the network resource inaccessible to certain users by disabling the services/hosts connected to the network. They can also target different layers of physical or data link networks. Dos can be classified into volume based attacks, protocol attacks and application layer attacks. Some of them include UDP Flood, SYN flood, NTP Amplification etc (Fig. 3).

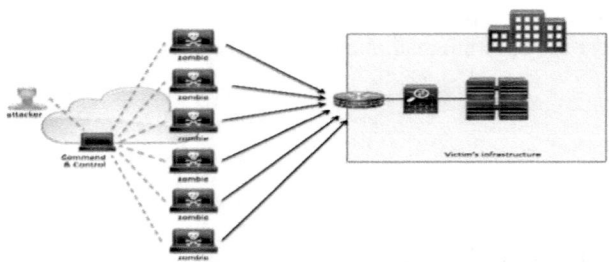

Fig. 3. Denial of service

The main intention of the attacker is to deny or hinder the service to the substations and distribution centers. The damage of this attack can extend up to termination of power supply and cause chaos in midst of consumers.

2. False data injection: The meters and sensors which lack tamper resistance can be altered to give false measurement reports. The reports can be altered bypassing the existing bad measurement detection, which makes the security team unaware of the happening.

As seen in the diagram, attackers try to inject wrong or false data in power system once the configuration of the power system is determined, they can start generating false data [13].

3. Malicious code: This can be spread through viruses. The infected devices can be manipulated by the attacker. A backdoor is created through this, which serves as an entry point to the hacker. Malware infected operating stations may send target commands to DCS server to manipulate the operation. Ex of some these viruses are worms and Trojans.
4. Transmission system attacks: Some of the attacks target the transmission and delivery of power generated. TSA can be classified into Interdiction attacks, Complex network attacks, Substation attacks and CP switching attacks.

8   Why Cyber Security Is a Threat in Countries like India?

The transmission and distribution losses are an average of 26% in India, one among the highest in the world. In some states the losses are as high as 62%. Being one of the poor, careless and weak grid in the world, the country losses money for every unit produced. The Indian power fails in some important aspects like weak reactive power support, metering challenges, overloading of power system equipment and poorly structured distribution network

For the past few years, Information and guidelines are released by the Reserve Bank of India (RBI) and Insurance Regulatory Development Authority of India (IRDAI).

Policy and regulation: India has poor regulation of guidelines for implementing the grid. The present regulations and policies are feasible only for the existing traditional grids. New policies must be regulated which will consider the interest of the consumers, interest of the suppliers and investors.

Improper Infrastructure: It is important that the country should show concern in its grid infrastructure. Only when the power generation and transmission is reliable and secure, the other industries can gloom in, drastically improving the economy of the country. This should start by solving the fundamental problems with the grid.

Lack of awareness: This is one of the main reasons behind the proper implementation of the grid. The consumers have no or very little knowledge about how power is delivered to them. Smart grid and its advantages must be explained to the consumers before installing one. The consumers must have the awareness about the policies and regulations and about the real time pricing. Otherwise we cannot expect an active participation from the consumers.

Another important factor to be considered is the funds granted for implementing smart grid. Huge amount of money is to be invested in smart grids. Consumers must understand and regulate themselves in using energy in a cost effective manner. With the detailed information provided by the smart grid, the consumers will be able to adopt to energy efficient building and appliances.

9   Conclusion

This paper studies the risks involved in integrating the traditional grid to smart grid, mainly how the grid lacks cyber security. The vulnerabilities in the smart grid have been discussed. Even though tradition grids seem to be less vulnerable and safe, they lack so many advantages of the smart grid. A nation's development depends upon the availability and safety of its power resources. Studying the risks and danger involved, makes us stay safer and prepared by taking the right countermeasures to fight against the attackers, and use the power safely and efficiently.

References

1. Rogers, K.M.: An authenticated control framework for distributed voltage support on the smart grid. IEEE Trans. Smart Grid **1**(1), 40–47. IEEE Press, New York (2010)
2. Kosut, O., Jia, L., Thomas, R.J., Tong, L.: Malicious data attacks on the smart grid. IEEE Trans. Smart Grid **2**(4), 645–658. IEEE Press, New York (2011)
3. Aloul, F.: Smart grid security: threats, vulnerabilities and solutions. Int. J. Smart Grid and Clean Energy **1**(1) (2012)
4. Dagle, J.E.: Cyber security of the electric power grid. Proc. IEEE PES General Meeting, Jul. (2009)
5. Amin, S.M.: Securing the electricity grid [Online]. Available: http://massoud-amin.umn.edu/publications/Securing-the-Electricity-Grid.pdf (2010)
6. Amin, S.M., Giacomoni, M.: Smart grid—safe, secure, self-healing. IEEE Power & Energy [Online]. Available: http://magazine.ieee-pes.org/january-february-2012/smart-grid-safe-secure-self-healing/ (2012)
7. Belgi, S.: State of cyber security in Indian electricity industry. India Smart Grid Forum (2013)
8. Ye, Y., Qian, Y.: A survey on smart grid: a communication infrastructures: motivations requirements and challenges. IEEE Commun. Surv. Tutorials **15**(1), 5–20 (2012)
9. Bari, A.: Challenges in the smart grid applications: an overview. Int. J. Distrib. Server Netw. **2014**(1), 1–11 (2014)
10. Ericsson, G.: Cyber security and power system communication—essential parts of a smart grid infrastructure. IEEE Trans. Power Delivery **25**(3), 1501–1507 (2010)
11. Amin, S.M., Wollenberg, B.F.: Toward a smart grid: power delivery for the 21st century. IEEE Power and Energy Mag. **3**(5), 34–41, Sept.-Oct. (2005)
12. Flick, T., Morehouse, J.: Securing the smart grid: next generation power grid security. Syngress (2010)
13. Aouini, I., Azzouz, L.B.: Smart grids cyber security issues and challenges. World Acad. Sci. Eng. Technol. Int. J. Electron. Commun. Eng. **9**(11) (2015)

"IoT Based Medical Diagnosis Expert System Application"

D. Shanthi$^{(\boxtimes)}$, A. Lalitha, and G. Lokeshwari

GCET, Hyderabad, India
Dshanthi01@gmail.com, a.lalitha83@gmail.com,
gandrakoti@yahoo.com

Abstract. Medical diagnosis expert system application is mainly developed to diagnose the probable disease of a user depending upon the symptoms the user has provided. This application also determines the location of the hospitals, clinics and diagnostic centers. The application also offers to deliver the medicines to the user on order to the location of the user.

The main aim of this application is to help the users discover the disease and give them immediate medical care and to do required help or support.

Keywords: Diagnosis · Hospitals · Application · Medical_care

1 Introduction

One of the fastest growing industries now a day is mobile industry. There are many competitors in this area who are doing research and development on new platforms & user experience. The project Medical Diagnosis Expert System is a personal and easy search option for mobile user. This Android based application provides user interface to search options like, information about diseases. The disease names based on symptoms and create awareness to the patients about medical expert system.

An "IoT based Medical diagnosis expert system application", project is to develop an application that provides a user the current insight into symptoms, signs, epidemiology, and treatment of diseases and disorders.

The project provides a strong focus on the clinical diagnosis and patient management tools essential for daily practice. It also provides a full review of all internal medicine and primary care topics. This application is used by patients and doctors. This application is used to provide a strong focus on the clinical diagnosis and patient management tools essential for daily practice. Patient has to go to the doctor even for a small reason like fever, cold and cough which have become very common these days and, going to the doctor each time you suffer from cold or cough is not feasible.

Doctor has to cure the patients with his own knowledge and the doctor cannot get the complete information or assistance on diagnostics. Whenever we require medicines, especially in case of emergency, we have to help ourselves and go to the medical store and get the medicines. The major problem is that, we might not get medicine whenever we want and immediately. So, to solve this issue, we have come up with this proposed system.

© Springer Nature Switzerland AG 2020
A. P. Pandian et al. (Eds.): ICCBI 2018, LNDECT 31, pp. 685–692, 2020.
https://doi.org/10.1007/978-3-030-24643-3_82

Advantages of this App:

Patient need not go to the doctor for very small reasons like fever, cough and cold which would save time and money. Patients and medical practitioners can improve their knowledge about the diagnostics and can get better assistance and help. It provides the diagnostic information based on the symptoms given by the user which is easy for the users to know about the probable disease, its symptoms and the medication and provides the information on care to be taken by the patient.

It can be used by the Doctors (Medical practitioners) for effective diagnosis and assistance. Patients, caregivers, medical professionals can get the information quickly within less time. Whenever we want medicines from medical store, open application, and search medicines, look at the nearest located medical store and order the medicine. We can also find how much it costs and how much quantity is available in the medical store. We can also view the address and contact info of the particular medical store, hospital, clinic and diagnostic centers. And, in case of emergency, we can even call the ambulance.

Reasons how it useful to society:

1. There was no harmful effect on the environment in designing the project and is useful to every use.
2. **Safety:** The application is completely safe as the data is checked for its authenticity and added only by the admin.
3. **Ethics:** The details of the locations provided to the users are added only by the admin. Admin is authenticated by a unique username and password. So the added data is authentic and no other third party can manipulate the data.
4. **Cost:** The cost required for the application is minimal and the admin page and user page is authentic and has a unique username and password.

1.1   Scope of the Project

One of the quickest developing enterprises is software industry. There are numerous rivals here who are doing innovative work on new stages and client encounter. One such innovation is Android from Google which is bolstered for Google telephones. These telephones are portrayed as cutting edge mobiles [As depicted by Google]. Creating application for such cell phones utilizing the open source android SDK is very fascinating. This makes the application call history very simple, productive, adaptable and financial.

"IoT based Medical diagnosis expert system application" has the following modules

- Hospital search
- Diagnosis
- Medical Search
- Admin
- User

Hospital Search

This module allows the user to search for the nearest hospitals, clinics and diagnostic centers in case of any emergency situations and needs, based on the location of the user.

Diagnosis

In this module, application will display the list of symptoms from which the user selects the symptoms based upon the health condition and the probable disease based upon the symptoms is displayed and the user can see the required medication for that disease.

Medical Search

This module allows the medical stores to register themselves so that users can order medicines they require from the nearest medical stores. The users can login using appropriate credentials to order medicines and the medical stores have to login using the appropriate credentials to check the orders. The admin registers the medical stores to ensure that the medical stores are authentic.

Admin

The admin registers the medical stores to ensure that the medical stores are authentic and there is no plagiarism.

User

The users can login using appropriate credentials and can order the medicines they require from the nearest medical stores.

2 Literature Survey

The focus of the project is diagnosing and finding out the disease and symptoms of the patients, the present study is trying to identify the combination of clinical and a laboratory non-invasive variable that can predict the symptoms of diseases in patients in a best way.

A specialist framework can be characterized as "an intelligent computer program that utilizes information and induction methods to take care of issues that are sufficiently troublesome to require huge human mastery for their answers". We can gather from this definition that ability can be exchanged from a human to a computer and afterward put away in the computer in a reasonable frame that clients can call upon the computer for explicit exhortation as required. At that point the framework can make inductions and touch base at an explicit end to give advices and clarifications, if important. The application gives ground-breaking and adaptable intends to get answers for an assortment of issues that frequently can't be managed by other, increasingly conventional and universal strategy. The four fundamental segments of such an application are: a learning base, a deduction motor, an information engineering apparatus, and an explicit UI. A portion of the critical applications incorporate the accompanying: restorative treatment, engineering disappointment examination, choice help, information portrayal, atmosphere determining, basic leadership and learning, and compound process controlling. Master frameworks have applications in numerous areas. They are for the most part suited in circumstances where the master isn't promptly accessible. So as to build up a specialist framework the information must be extricated from space master. This learning is then changed over into a computer program. Learning Engineer plays out the errand of removing the information from the space master (Figs. 1 and 2).

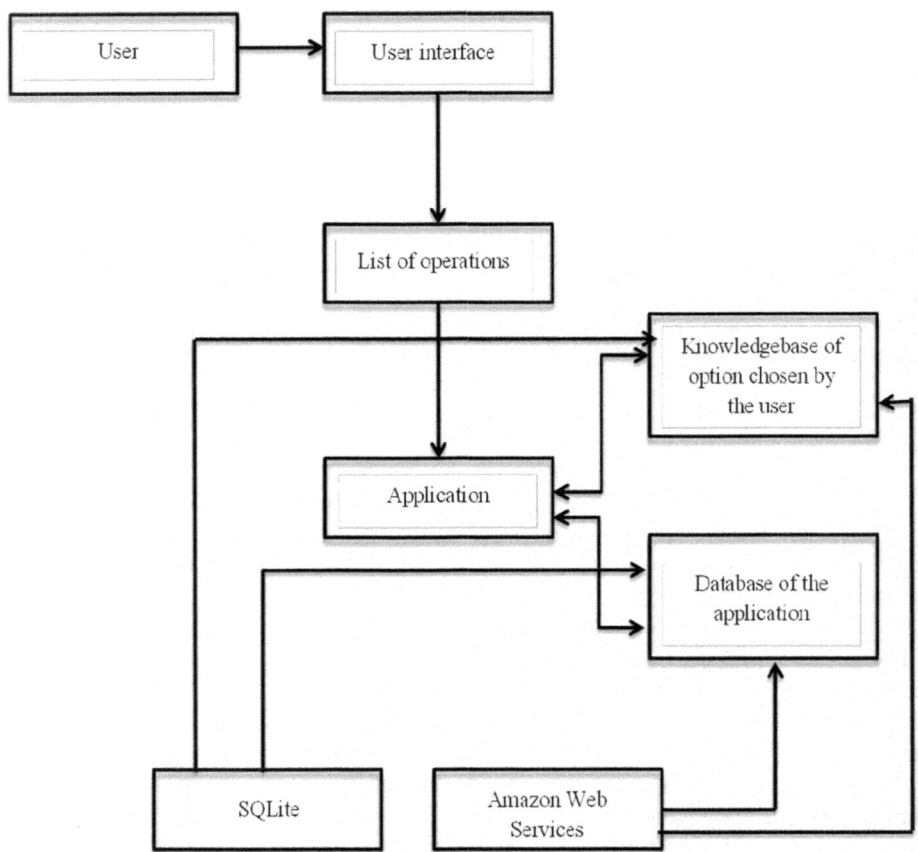

Fig. 1. System architecture

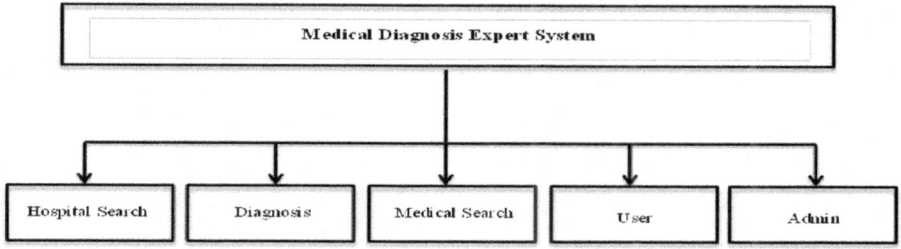

Fig. 2. Module architecture

An example testing cycle:
In spite of the fact that testing shifts between associations, there is a cycle to testing:

Requirements Analysis:
Testing should start in the requirements period of the software development life cycle.

Amid the plan stage, testers work with engineers in figuring out what parts of a structure are testable and under what parameter those tests work.

- **Test Planning:** Test Strategy, Test Plan(s), Test Bed creation.
- **Test Development:** Test Procedures, Test Scenarios, Test Cases, Test Scripts to use in testing software.
- **Test Execution:** Testers execute the software dependent on the plans and tests and report any errors found to the development group.
- **Test Reporting:** Once testing is finished, testers produce metrics and make last reports on their test exertion and regardless of whether the software tested is prepared for discharge.
- **Retesting the Defects**

  Not all errors or imperfections detailed must be settled by a software development group. Some might be caused by errors in designing the test software to coordinate the development or generation condition. A few imperfections can be taken care of by a workaround in the generation condition. Others may be conceded to future arrivals of the software, or the insufficiency may be acknowledged by the business client.

2.1 Test Cases

S.no	Test cases	Input	Description	Result
1	Select medical search	Select user or admin	Displays the page	Pass
2	Select medical search	Select type: User	Type user id and password and user can login and order medicines	Pass
3	Select medical search	Select type: User	Type invalid credentials	Fail
4	Select medical search	Select type: New user registration	Enter all the details and then click on register. It will redirect to the previous page for login	Pass
5	Select medical search	Select type: New user registration	Enter wrong, incomplete or invalid details. It will ask you to enter the complete details	Fail
S.no	Test cases	Input	Description	Result
1	Enter admin username and password	Username: admin Password: admin	Admin logs in if it matches with the database values	Pass
2	Enter invalid admin username and password	Username: abcd Password: abcd	No display	Fail
S.no	Test cases	Input	Description	Result
1	Select disease list	Select type: Disease name	Due to correct input, it will display disease info, symptoms and medication	Pass
S.no	Test cases	Input	Description	Result
1	Select hospital search	Select type: Location	Due to correct input, it will display the hospitals, clinics and diagnostic centres	Pass

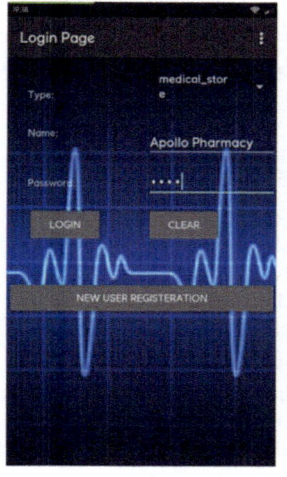
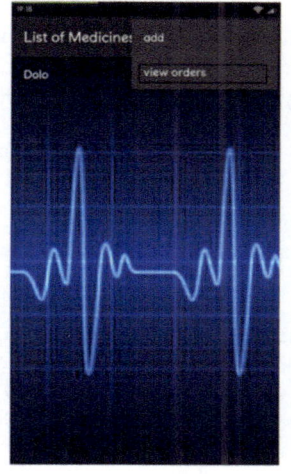
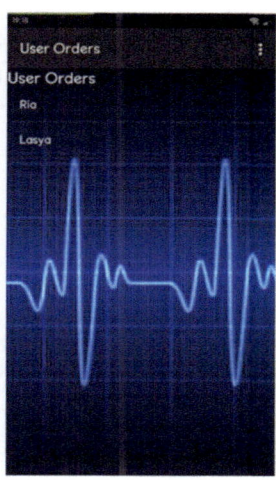

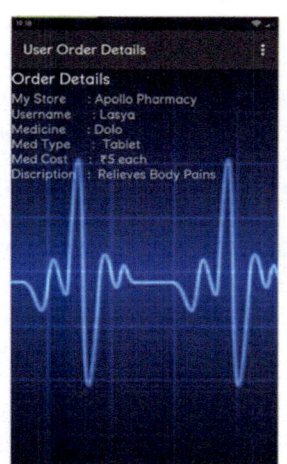
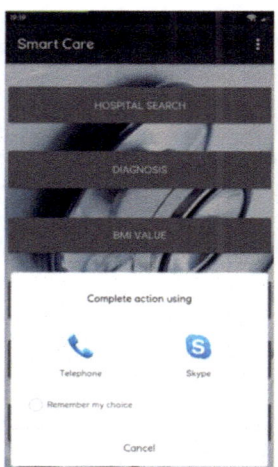
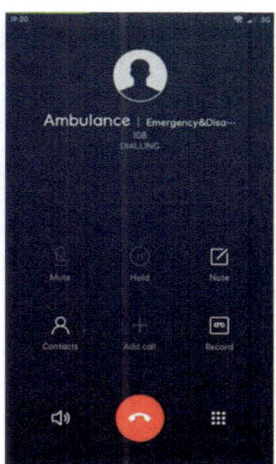

3 Conclusion

- This application is utilized for getting IoT Based Medical Expert System. This application is work in Google mobiles utilizing Android SDK.
- It is a device produced for android stage. The undertaking Medical Expert System is an individual simple look choice for portable client.
- This Android based application gives Good User Interface to seek choices like: Information about diseases, disease names based on symptoms and creates awareness to the patients about medical expert system.

Further enhancements:

The application can be further extended throughout the country and throughout the world. The application can also be updated to contain features like ambulance services and by adding the emergency phone call service facility to it.

References

1. World health organization assesses the world's health systems. 8 December 2010. Who.int. Accessed 6 January 2012
2. WHO country facts: France. Who.int. Accessed 11 November 2013
3. Mitchell, A.: The Divided Path: The German Influence on Social Reform in France After 1870, pp. 252–275. University of North Carolina Press, Chapel Hill, NC (1991)
4. Hildreth, M.L.: Doctors, Bureaucrats & Public Health in France, 1888–1902. Garland Press, New York (1987)
5. Klaus, A.: Every Child a Lion: The Origins of Maternal & Infant Health Policy in the United States & France, 1890–1920. Cornell University Press, Ithaca (1993)
6. Shapiro, A.-L.: Private rights, public interest, and professional jurisdiction: the French public health law of 1902. Bull. Hist. Med. **54**(1), 4–22 (1980)
7. Medical News Today. Accessed 6 January 2012
8. France-prel.indd (PDF). Accessed 6 January 2012
9. Jump up to:[a] [b] (in French) How to choose and declare his referring doctor in France to get maximum health care benefits, Ameli.fr (official web site of the Assurance Maladie)
10. Nombre de lits installés par type d'établissement en hospitalisation complète et de semaine. French Hospital Federation
11. Jump up to:[a] [b] [c] http://www.euro.who.int/document/e83126.pdf. Health Care Systems in Transition – France: WHO
12. L'assurance maladie. Ameli.fr. Accessed 6 January 2012

RGB Component Encryption of Video Using AES-256 Bit Key

N. Geetha$^{(\boxtimes)}$ and K. Mahesh

Department of Computer Applications, Alagappa University, Karaikudi,
Tamilnadu, India
geetha.researchscholar@gmail.com

Abstract. In our lifestyle, the Internet plays an important role. Because sending information through online is more. Especially video files occupy big space on the internet. Here the one thing we all should know, that is security. Encryption is the best job to fulfill the security thirst. The classical ciphers are Substitution, Transposition etc., and the current trendy modern ciphers like a Symmetric key and Asymmetric key have a different variety of algorithms. Symmetric have a block like DES, 3DES, AES and stream cipher like RC4, RC6 etc., Asymmetric have RSA, ECC etc., Here AES-256 algorithm is used for video encryption with RGB Components, The proposed technique can be analyzed with encryption speed, encryption ratio and throughput of the video files.

Keywords: Cryptography · Video encryption · AES-256 · Block cipher · RGB components

1 Introduction

Multimedia is becoming an emerging field in nowadays. The business meeting can be done in video conferencing and every second there is a lot of multimedia information sending through online. But the security mechanisms used for the multimedia data still faces some problems. Cryptography is the one that solves the problems in the security. It has two types, Symmetric key algorithms, and Asymmetric key algorithms. Cryptography has some classical ciphers also such as Substitution ciphers and Transposition codes. In the present world, regular key algorithms and irregular key algorithms can be used for encrypting the multimedia data. When considering the speed of the encryption regular key algorithms are well-matched for multimedia inscribe (Fig. 1).

This paper ordered as follows: Segment 2, describes the associated works executed on videos, Segment 3, Prevailing algorithms which have been used in the video so far. Segment 4, Explanation of the proposed technique. Segment 5, Performance of RGB Components Encryption. Segment 6, Explains about Experimental results and Performance evaluation. Segment 7, Conclusion.

© Springer Nature Switzerland AG 2020
A. P. Pandian et al. (Eds.): ICCBI 2018, LNDECT 31, pp. 693–700, 2020.
https://doi.org/10.1007/978-3-030-24643-3_83

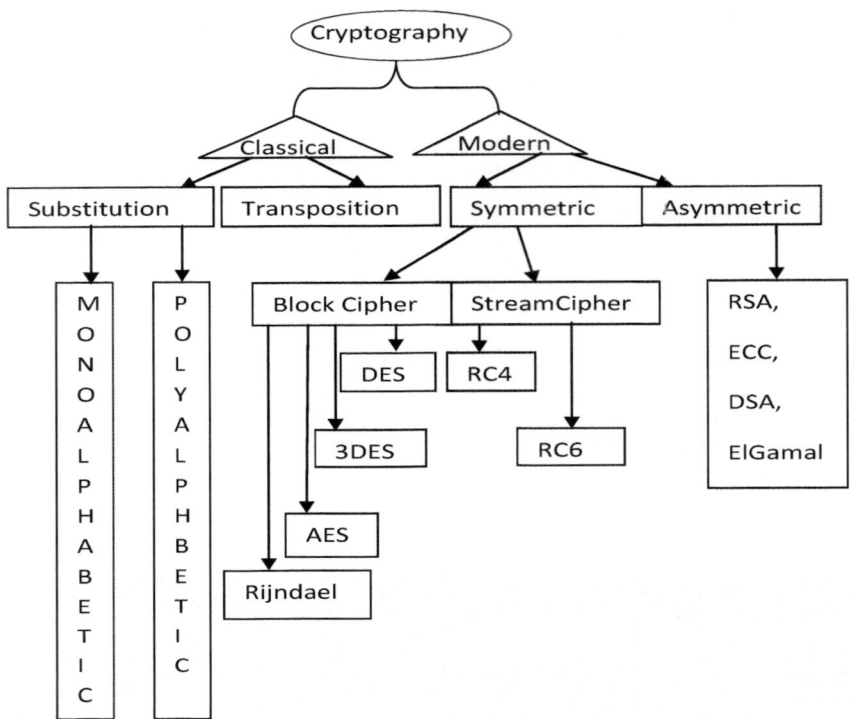

Fig. 1. The life tree of cryptography

2   Literature Review

Jawahar Thakur and Nagesh Kumar (2011), The authors has used the Blowfish algorithm to encrypt the video files. They said compared to AES, Blowfish has better than AES. AES showed meager performance results compared to other algorithms since it needs extra processing power. The processing time of CBC mode had taken more but best suited for big data blocks. OFB shows improved recital than ECB and CBC but necessitates more dispensation time than CFB. Overall time differences between all modes are trifling [1].

Hao Li, Cheng Yang, Jiayin Tian and Jianbo Liu (2016), the authors use three parts in this research, Unpacking, Encryption, Packing. In the Unpack part, the video has been unpacked using the function of FFMPEG, then put the selected payload into the encryption part otherwise take the read frame to packing part directly. The key is retrieved from the key management center and using that key encrypt data use AES. It has two containers, Input format, and Out format container. The pack uses the input format container and packing uses the out format container. The memory usage and time-consuming are more [2].

M S Rohit and R Sanjeev Kunte (2016), In this paper, the author has compared the algorithms of AES, Blowfish and Sattolo's algorithm. The AES and Blowfish algorithms yield very less time related to Sattolo's technique which pulls attention on the multi-faceted tasks and stand-ins done in the Sattolo's technique which consumes more CPU cycles making it extended process. A number of Pixel Change Rate remains same in all the three cases so none of them reveal the identity of original information. The entropy of the AES and Blowfish is higher than the Sattolo's algorithm. The correlation coefficient of Sattolo's scheme is low, hence it draws less suspicion. In Sattolo's Algorithm, the time taken for encryption is high [3].

Jayant Kushwaha and BholaNath Roy (2010), The author mainly uses the correlation among neighboring pixels. The author divides an image into blocks are of n*n size. The double encryption method is proposed in the image file. The key management unit will generate the pair of private-public key. The crypt engine encrypts the image by pixel and by each block using Rijndael algorithm. It has some time consuming [4].

K. Kalaivani and B. R. Sivakumar (2012), The author has offered a collected works evaluation of the state-of-the tools of Multimedia medical information security. Watermarking scheme is not very strong against different types of image operations or bouts. This is clear that these techniques are somewhat complicated to implement in real time and Criminals can use encryption to sheltered communications, or to store impeaching factual on electronic devices. Medical images are delicate. The protection of medical image is a need [5].

KS Tamilkodi and Dr. (Mrs) N Rama (2015), The authors propose the chaotic image encryption for security of image transmission. There are a lot of chaotic image encryptions proposes for a different method of image encryption. The one thing we keep in mind is the security of image transmission. Security analysis for instance Correlation and Histogram analysis were specified in all the reviewed research articles. The security measures of the algorithm like Mean Absolute Error, Entropy Analysis, Deviation Measuring Factors and FIPS Test were merged [6].

Jyoti Singh and Ashok Kumar (2012), The selective encryption can be done using harmony search algorithm. The video is secured over the internet. The entropy increased by the harmony search. The videos like MPEG, AVI, MPEG II are tested through harmony search algorithm. Changing the HMCR, PAR can extra improve the performance of the algorithm. Videos of complex pixels will develop tougher key as the length will be high, large number of transformation and hence more string encryption. In the process of generation of pseudorandom keys if initial key to start is high more robust encryption can be performed as stoutest key will be selected by the harmony search. The field of encryption has widespread prospects and involvement of artificial intelligence has led to generation of hugely extended keys [7].

Nidhi Singhal and J.P.S. Raina (2011), A recital appraisal of AES and RC4 algorithms. The presentation metrics were amount, CPU process time, memory consumption, encryption and decryption time and key size variation. Tests show that the RC4 is debauched and vigor effectual for encryption and decryption. Based on the study done as part of this research, RC4 is works well than AES [8].

Merlyne Sandra Christina C, Karthika M, Vasanthi M, and Vinotha B (2016), In this Work, the author has made six modules, each for exact purpose and will be interlinked to each other. Encryption and decryption modules are the main modules, converting video into the text form and send to the receiver. In decryption, again converting it into the original video. But it depends on the size of the prime. If high security is needed, we have to use big prime numbers [9].

3   Background Algorithms

3.1   Asymmetric Key Algorithms

Tables 1 and 2

Table 1. Asymmetric key algorithms

Developed algorithms	Developed by	Key sizes
RSA	Ron Rivest, Adi Shamir, Leonard Adleman	3072-bit or larger
ECC	Bruce Schneier	160-521 bits
ELGAMAL	Taher Elgamal in 1985	160 bits
DSA	National Institute of Standards and Technology	1024 bits

3.2   Symmetric Key Algorithms

Table 2. Symmetric key algorithms

Developed algorithms	Developed by	Key sizes
Blowfish	Schneier developed block cipher	1-448 bits
DES	In 1977, DES adopted as a U.S. government standard	56 bits
IDEA	Massey and Xuejia developed Block cipher	128 bits
MARS	IBM developed finalist AES	128-256 bits
RC2	Rivest developed Block cipher	1-2048 bits
RC4	Rivest developed Stream cipher	1-2048 bits
RC5	In 1994, Rivest developed and published Block cipher	128-256 bits
RC6	RSA Labs developed finalist AES	128-256 bits
Rijndael	Daemen and Rijmen developed and published in NIST selection for AES	128-256 bits
Serpent	Anderson, Biham, and Knudsen developed finalist AES	128-256 bits
Triple-DES	DES algorithm by three-fold application	168 bits

4   Proposed Work

Read the original video. The individual frames can be loaded into the sytem, then extracting the individual frames. The extracted frames are again extracted by their RGB components. Finally the extracted RGB frames are encrypted by the AES-256 algorithm. Shown in Fig. 2.

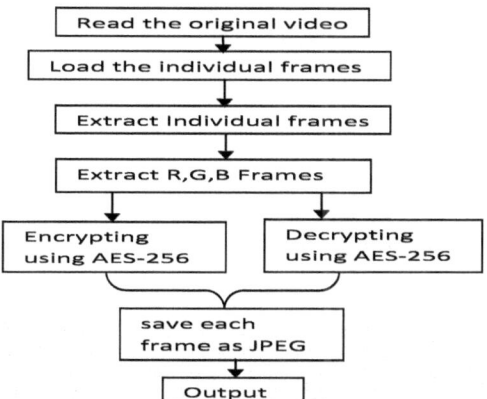

Fig. 2. The flow of proposed video encryption and decryption

4.1   Encryption Process

Step1.   Load the original Mp4 video into the system
Step2.   Convert the video into individual frames
Step3.   Extract the individual frames
Step4.   Extract the R,G,B frames of the video
Step5.   Encrypt R,G,B Component using AES -256 Algorithm.
Step6.   Save the frames as JPEG
Step7.   Recreate the video again

4.2   Decryption Process

Step1.   Load the original Mp4 video into the system
Step2.   Convert the video into individual frames
Step3.   Extract the individual frames
Step4.   Extract the R,G,B frames of the video
Step5.   Decrypt R,G,B Component using AES -256 Algorithm.
Step6.   Save the frames as JPEG
Step7.   Recreate the video again

5   Experimental Results

The Simulation results can be done in by Matlab with a configuration of Intel Pentium Inside, Windows 7 (Fig. 3).

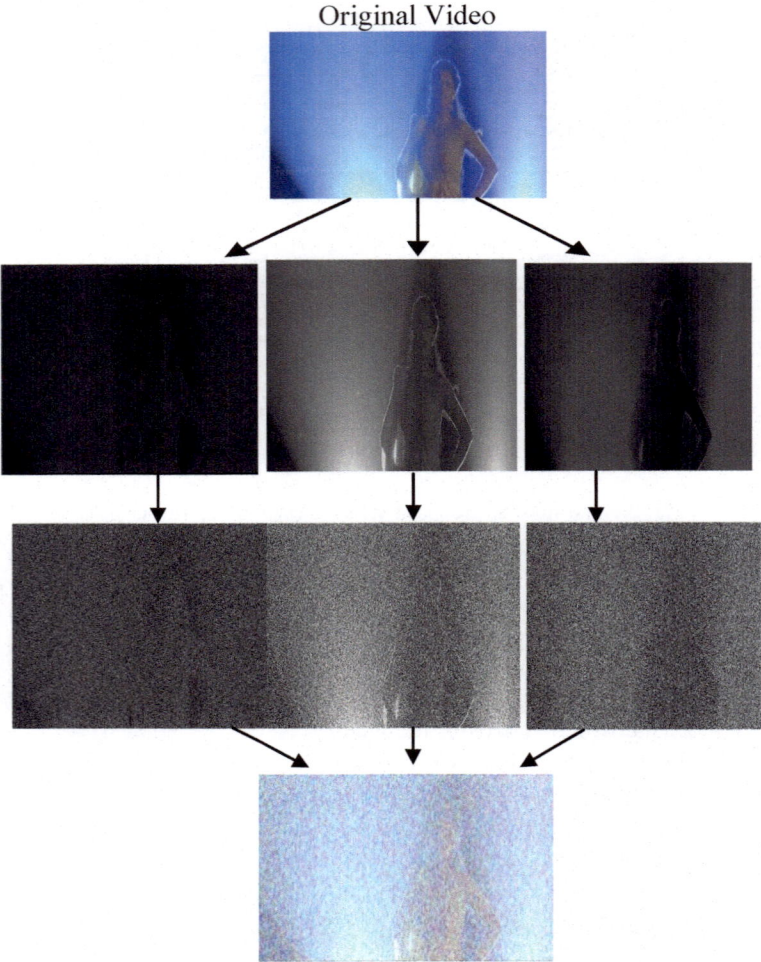

Fig. 3. Encrypted video

6   Performance Evaluation

The performance evaluation can be done in Matlab with a configuration of Intel Pentium Inside, Windows 7. The proposed technique can be compared with RSA algorithm is as follows (Table 3) and graphically shown in Fig. 4.

Table 3. Simulation results

Video source name	Video size	Parameters	AES-256	RSA
Keer.mp4	915 KB	Encrypted size	920 KB	918 KB
		Decrypted size	915 KB	915 KB
		Encrypted Time	03.48 s	02.89 s
		Decrypted Time	04.19	03.34 s
		Encryption Ratio	99.35%	99.67%
		RGB frames Encryption speed	2.34 s	2.47 s
		RGB frames Decryption speed	3.12 s	3.19 s

6.1   Encryption Evaluation of AES-256 Algorithm and RSA Algorithm Is Given Below

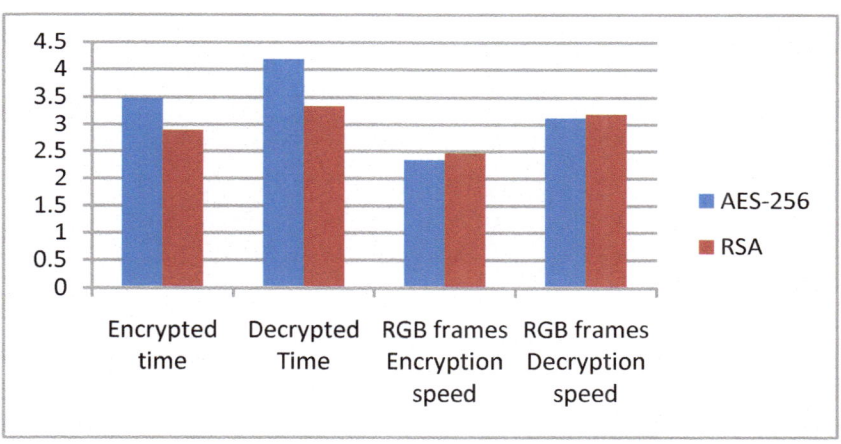

Fig. 4. Encryption evaluation of AES-256 algorithm and RSA algorithm

7   Conclusion

Video encryption is the very challenging tasks for our day today life, mainly for Organizations, Millitary etc. There are many main algorithms are used for securing video files. In this work, The encryption process can be done with RGB components of the video files using AES-256. First load the video and extracting the individual frame of the video. Then again extracting the RGB components of the video and encrypt the RGB frames with AES-256. Finally, convert the saved JPEG frames into video again. This AES-256 is compared with the RSA algorithm. AES-256 is much better than the RSA. In Future, The encryption process will be for multimedia content using cloud environment.

References

1. Thakur, J., Kumar, N.: DES, AES and Blowfish: symmetric key cryptography algorithms simulation based performance analysis. Int. J. Emerg. Technol. Adv. Eng. 1(2), 2250–2459 (2011)
2. Li, H., Yang, C.: A general encryption algorithm for different format videos. Int. J. Secur. Appl. 10, 67–76 (2016)
3. Rohit, M.S., Sanjeev Kunte, R.S.: Performance comparison of AES, blowfish and sattolo's techniques for video encryption. Int. J. Innovative Res. Sci. Eng. Technol. 5(9), May (2016)
4. JayantKushwaha, B.N.: Secure image data by double encryption. Int. J. Comput. Appl. 5(10), 0975–8887 (2010)
5. Kalaivani, K., Sivakumar, B.R.: Survey on multimedia data security. Int. J. Model. Optim. 2(1), Febraury (2012)
6. Tamilkodi, K.S., Rama, N.: A comprehensive survey on perfor -mance analysis of chaotic colour image encryption algorithms based on its cryptographic requirements. Int. J. Inf. Technol. Control and Autom. 5(1/2), April (2015)
7. Jyoti Singh, A.: Selective video encryption using harmony search. Int. J. Innovative Res. Dev. 3(5), May (2012)
8. Singhal, N., Raina, J.P.S.: Comparative analysis of AES and RC4 algorithms for better utilization. Int. J. Comput. Trends and Technol. 177–181, July-August (2011)
9. Merlyne Sandra Christina, C.K.M.: Video encryption and decryption using RSA algorithm. Int. J. Eng. Trends and Technol. 33(7), March (2016)

A Generic Approach for Video Indexing

N. Gayathri[(✉)] and K. Mahesh

Department of Computer Applications, Alagappa University, Karaikudi, India
gayathri.researchscholar@gmail.com

Abstract. From past decades, imaging systems have started to transform the way we communicate the information to the desired users. Most of image processing algorithm works with video promptly. In this way, video processing emerges as an expansion of image processing. Moreover, there are also appears some restricted apparatuses to index, portray, sort out and deal with the image information. A perfectly arranged and viable administration of video archives relies upon the accessibility of indexes. Manual Indexing is not feasible for vast video accumulations. This results in a need for innovative systems and structure that can store, handle, Search, and recover the information from the gigantic media file.

Keywords: Images · Video · Video processing · Index

1 Introduction

Video Processing is one of the rapidly growing technologies. It is a strategy to play out few tasks on a video format, with a goal to achieve an upgraded video or to extract some helpful data from it. Video processing involves noise reduction, Detailed Enhancement, Motion Detection, Frame rate conversion, Aspect ratio Conversion, Color Space Conversion etc.,

Video Processing fundamentally incorporates three stages, they are:

1. Importing the video with the effective use of Digital Photography technique.
2. Analyzing and manipulating the obtained images.
3. Output in which the results can change the video based on their video analysis.

Video:

A Video is defined as a fast movements of images or A group of independent frames. The Nature of the video relies upon the movement of images per minute and the quality of each image frame.

Videos have the some additional Characteristics:

1. It has Significantly number of extravagant content when compared with individual pictures.
2. It contains a tremendous amount of data.
3. Very little prior structure.

These characteristics helps to create a challenging task on indexing and retrieval of videos.

© Springer Nature Switzerland AG 2020
A. P. Pandian et al. (Eds.): ICCBI 2018, LNDECT 31, pp. 701–708, 2020.
https://doi.org/10.1007/978-3-030-24643-3_84

In general, video database has been less in size and indexing and retrieval is based on the keywords that are explained physically. In recent times these databases have turned out to be significantly larger and content based indexing and retrieval are required to create an automatic analysis of videos.

Indexing:

Index is nothing but data structure that tends to sort out the data records available on disk to advance certain sort of retrieval activities. Index enables us to productively retrieve all records that fulfills the search condition present on the inquiry key field of the index. An index is a structure in the database that give quick access to rows and columns in tables to meet certain conditions in a query.

Indexing is disturbed with a compact storing of the extensive gathering of terms and quickly retrieving an arrangement of hopeful terms in fulfilling some property from a vast accumulation of terms.

The Selection of files got from the substance of the video helps to sort out video information that connects to the first video stream. Indexing is the limiting input, ethat enhances the execution particularly in irregular access. It is a dynamic structure. A table can have a few records to fulfill a few inquiries.

Video indexing methodologies can be classified in view of the two fundamental levels of the video content.

1. Perceptual 2. Semantic

When assigning an index to a video archive, it has come out with three major issues. The first is identified with granularity and addresses the query.

(1) What to index?
(2) How to index?
(3) Which is index?

Precedent: A Telephone registry is recorded on the name attribute utilizing lexicographic requesting. Video indexing fills on similar to need in video databases.

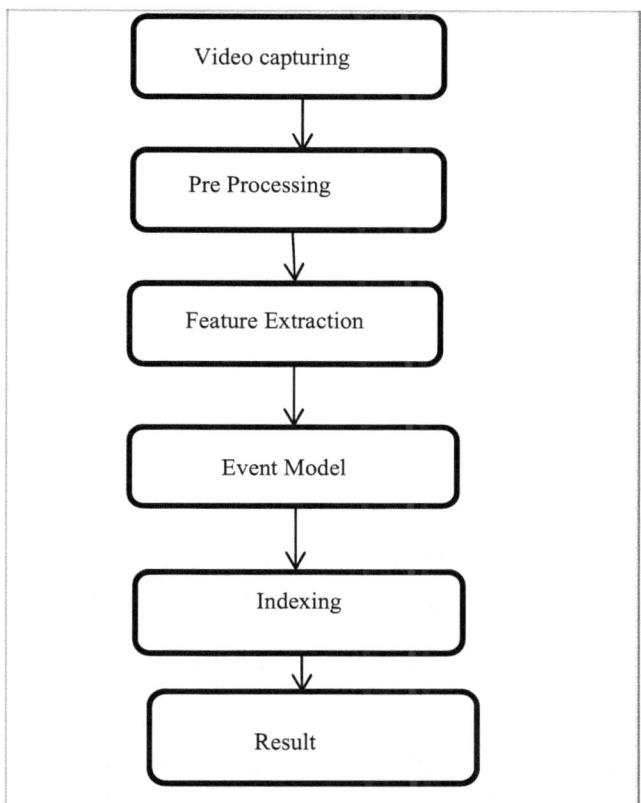

2 Literature Review

Saravanan [1], proposed video catching methodology is an easy process yet the related video recovery is the troublesome procedure, for that procedure the recordings must be indexed. Recovery is the strategy that recovered a video utilizing a client inquiry. The question will be picture or messages rely on the question result yield framework that restored a specific video or picture dependent on that inquiry. In this task we make an indexing for video document by utilizing section based ordering system. Here video will be isolated into a chain of importance which is in storyboards of film making. For example, a various leveled based video seek is made into multi organize deliberation for help the clients to find the particular video portions/outlines intelligently. Here draws out the lessened data transmission and decreased postponements the video through the system of looking and assessing. Author used a method called Segment Based video indexing Technique and Hierarchical Clustering Mechanism.

Kamde et al. [2], entropy is developed for indexing the video with the blend of RGB color esteems and video metadata. The entropy based key frame extraction calculation performs extremely well when the picture foundation is recognizable from the items. Entropy, edge and color features are utilized for video recovery. Database with the videos from various spaces is made and it is seen that mix of different highlights in various ways gives relative outcome. Mix of entropy and edge features can work better in query particular application spaces like news video. Videos recovered using entropy-edge technique may not be comparative in shading and tedious. Thus consolidating entropy and color can enhance the outcome by recovering recordings which are more comparable in shading. Entropy-color strategy is appropriate in applications which requests comparative color videos.

Pandey et al. [3], one of the interesting topics in determining a multimedia database is the association of visual data. From the time when visual media requires expansive measure of capacity and preparing, there is a need to productively index, store and retrieve the visual data from multimedia database. Here introduces an exceptional method for proficient video indexing by observing the video sequence specifically in compressed domain. Unexpected scene changes and unique altering impacts, for example, breaks down, blurs and so forth are effectively distinguished utilizing just DC pictures of I and P outlines and the macro block data of P and B outlines. The proposed conspire is computationally less intricate when contrasted with the traditional strategies, while additionally being hearty to camera or/and protest movement and electric lamp.

3   Proposed Work

1. Preprocessing
2. Feature Extraction
3. Query Processing

1. Preprocessing:
Input Video is taken from the camera which is stored in the form of video file format based on the tool. For example in MATLAB.avi,.mp4. Generally videos are structured according to a descending order of video clips, Scenes, Shots and Frames.

Step 1:Video

Clips

Step 2:Scence

Step 3 : Shots

Step 4:Frame

2. Feature Extraction:

In this feature extraction user have to extract any one of the frame from it. That fetching frame is called a key frame. Analysis the frame and Extract Features from it. Features like Audio Features, Visual Features, RGB Features and Texture Feature, Edge feature, Motion feature.

3. Query Processing

At the point when user don't know how to describe what they need of utilizing words, usually the case that they might want to query based on comparable image. When the user inputs the query image the image features are extracted and then compared with the database. Finally the corresponding result will be displayed.

Query using Matching

1. Exact Match Query
   Requires rigorous match between the query and the video data.
2. Similarity Match Query

This query type didn't allow exact matching. Because of the difficult nature of video data. This query type is more mandatory.

Query Behavior:

1. Browsing Query
2. Iterative Query
3. Location Deterministic Query
4. Statistical Query
5. Tracking Query

4   Experimental Results

Figures 1, 2 and 3

Fig. 1. Frame conversion

Fig. 2. Input image

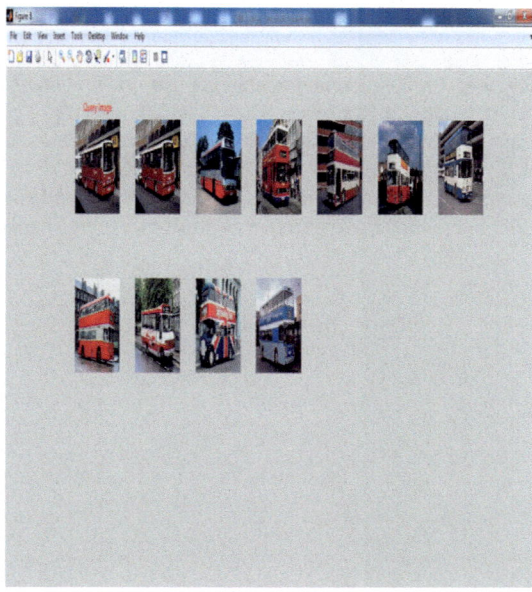

Fig. 3. Output image

5   Performance Analysis

Image	Accuracy	Retrieval time(sec)
Image 1	85.5%	3.8 s
Image 2	86.6%	4 s
Image 3	87%	4 s
Image 4	85.2%	3.8 s
Image 5	86%	3.5 s

6   Conclusion and Future Enhancement

The purpose of this research is to help the budding researchers in the field of video indexing and retrieval. Millions of video files are seen and a large number of them are made each and every day. Anyway there are as yet restricted devices to index, manage, organize and characterize deal with this information. This is just an idea of video indexing and retrieval of relevant videos.

Future Enhancement of this paper is, the user can use the video as a query to retrieve the corresponding videos from the databases. In future we would like to implement efficient methods or algorithm to give accurate results of video for indexing.

References

1. Saravanan, D.: Segment based indexing technique for video data file. 4th International Conference on Recent Trends in Computer Science & Engineering, Pg. No: 12–17 (2016)
2. Kamde, P.M., Shiravale, S., Algur, S.P.: Entropy supported video indexing for content based video retrieval. Int. J. Comput. Appl. **62**(17), January (2013)
3. Pandey, T.P., Bhardwaj, A., Gupta, S.: Robust and efficient method for compressed domain video indexing. IEEE International Conference on Multimedia and Expo, ISBN 0-7695-1198 8/01 (2001)

Energy Aware Routing Applications
for Internet of Things

S. Santiago$^{(\boxtimes)}$ and L. Arockiam

St. Joseph`S College (Autonomous), Trichy, India
ssantiagosj@gmail.com, larockiam@yahoo.co.in

Abstract. Internet is a system of networks and connects millions of users. This structure network moves to interrelated objects called Internet of Things(IoT). It is a new paradim and combines many technologies. This paper proposes energy aware routing application for IoT environment. The proposed technique produces better performance.

Keywords: Energy · Routing · Event rate · Internet of Things

1 Introduction

The century has witnessed an exponential increase in smart devices which are connected to the Internet. The data generated from these devices are useful in making the decision. The devices communicate to the parent node and route the data to the Internt. The data from the devices are captured on the basis of requirements. The event rate or data produced [1] from the devices or services are differnt. The life time of the parent node is estimated based on different parameters such as CPU active, transmission and reception. The ever-increasing number of devices becomes unavoidable requirement in IoT deployment. İn such scenario, the energy of each node plays a vital role. The proposed technique results in considerable energy savings.

2 Related Works

Santiago et al. [2] presented the ways to minimize the energy consumption for IoT networks. The researchers also highlighted lightweight algorithms for the resource-constraint devices. Guerra et al. [3] aimed at fostering independent life of elderly people in the present scenario. Based on the Ambient Assisted Living (AAL)-system, it provided an inexpensive and scarcely intrusive way for providing user localization and identification information.

Ok et al. [4] proposed an energy aware algorithm to select routing paths to forward information. The authors considered shortest path to reach the destination by reducing the hops. Residual energy was considered to route the data. The procedure was repeated for all the routes and the best path was identified. Arockiam et al. [5] suggested energy aware load balancing algorithm for IoT netwoks. The algorithm reduces the radio duty cycle and prolongs the life time of the network.

© Springer Nature Switzerland AG 2020
A. P. Pandian et al. (Eds.): ICCBI 2018, LNDECT 31, pp. 709–714, 2020.
https://doi.org/10.1007/978-3-030-24643-3_85

Risteska et al. [6] introduced a hierarchical distributed approach for health care system which supported three level data managing models such as dew computing, fog computing and cloud computing. The sensors were used to collect the data. Then, the fog computing was responsible for making local decisions. Finally, Data analysis was performed in cloud and solutions were provided for the IoT applications. Several load balancing schemes are also proposed for RPL protocol [7, 8]. Many other routing protocols have been proposed in this regard [9, 10]. However, energy aware routing algorithms for IoT applications are the need of the hour.

3   Proposed Technique

Energy aware load balancing incorporates Un-weighted Pair Group Method using event rate (UPGM) for cluster formation. The goal of this technique is to sustain the life time of individual cum parent nodes. The design process involves cluster formation followed by forwarding of data packets.

The formula used is

$$e(A\ U\ B), x = \frac{|A|.eA, x + |B|.eB, x}{|A| + |B|} \tag{1}$$

where A and B are devices; e is the event rate. eA and eB are the event rate of the devices in A and B (Fig. 1).

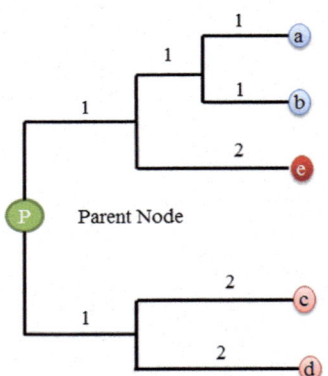

Fig. 1. Un-weighted pair group method

The event rate is considered for the proposed method. The similar events are clustered for a parent. In the given scenario, the data from a device with the same event/time is clustered. For example, the power consumption of node MSP43F1611 when CPU is active 330μA, Low Power Mode 1.1 μA, Transmission 18.8 mA and Reception 17.4 mA. In the proposed method, a node is assigned to a parent according to the event rate. The event rate is assigned at different intervals such as 2, 5, 10, 30 and

60 min per event respectively. Thus, the best parent is utilized for data transmission and cluster is formed based on event rate.

4 Application Scenario

According to the World Population reported by United Nations 2015, the number of older persons aged above 60 years has increased substantially in recent years in most countries. In cities, older people are left in the house to look after themselves with little assistance from neighbours or many older people cannot afford to have maid servants due to financial inabilities. Generally, older people face difficulties such as fall, absence of mind to take pills on time, walking difficulties due to sugar, arthritis, unable to raise on time etc., In such scenario, the proposed technique is implemented.

4.1 Experimental Setup

Sensors are installed in a home where elderly people live alone in an apartment scenario. The sensors collect the data and provide relevant information. Sensor detects the physical environment. The input may be light, temperature, humidity, motion, moisture, pressure, pollution etc., Sensors are deployed to assist elderly people at home. In this experimental study, 10 real time hardware sensors are applied and sample data are uploaded. The sensors are humidity sensor, ultrasonic sensor, temperature sensor, light sensor, flame sensor, rain sensor, heart beat sensor, vibration sensor, sound sensor and gas sensor (Fig. 2).

Fig. 2. Home layout

The experiment is carried out to check the energy consumption cum routing. A laptop is used as a border router which uploads the data to the cloud. The sensors are deployed at different locations to sense the data. The sensors are mostly battery powered. In existing scenario, the parent selection cum transferring of data makes use of the standard objective function. Sensors are deployed at different locations to achieve secured data [11–13]. A home layout is considered for the case study and many researchers have experimented smart assistance [13, 14]. The locations are selected based on the requirements. Flame sensor is placed at kitchen to detect fire. Ultrasonic sensor is kept at the entrance to detect the moment of others. In balcony, the LDR is placed to control the lights outside the house. Humidity sensor is placed at dining hall. Rain sensor is located at the garden to check the rain fall. Temperature sensor is used to check the body temperature cum room temperature.

5   Result and Discussion

The Contiki Cooja simulator is used to carry out the experiment. The hardware sensors listed are connected using the microcontroller and the data plotted in ThingSpeak. The data from the software nodes are stored in simulator for further reference. The data analysis is performed at the cloud. The smart sensors collect and perform initial data processing. Then, the data from the sensors are routed using MSP430 microcontroller and plotted in ThinkSpeak. Node MCU Wi-Fi controller is also used to collect data from the sensors [15].

Table 1. Simulation parameter for assisted living

Parameters	Values
Total number of nodes	30
Hardware nodes	10
Software nodes	20
Node position	Random
Static nods	90%
Dynamic nodes	10%
Operating system	Contiki version 2.7

The simulation parameter is tabulated in Table 1.

The screen shot for the simulation environment is depicted in Figs. 3a and b. The Fig. 3a shows the software sensor nodes simulation environment. Figure 3b shows the radio duty cycle of the simulated environment.

An average of 2500 mAh batteries is connected for real time sensor nodes. Based on the sample data, the life time of sensor nodes are estimated and illustrated in Fig. 4. The simulation results reveal that the life time of sensor nodes is sustained when the technique is implemented.

(a)   **(b)**

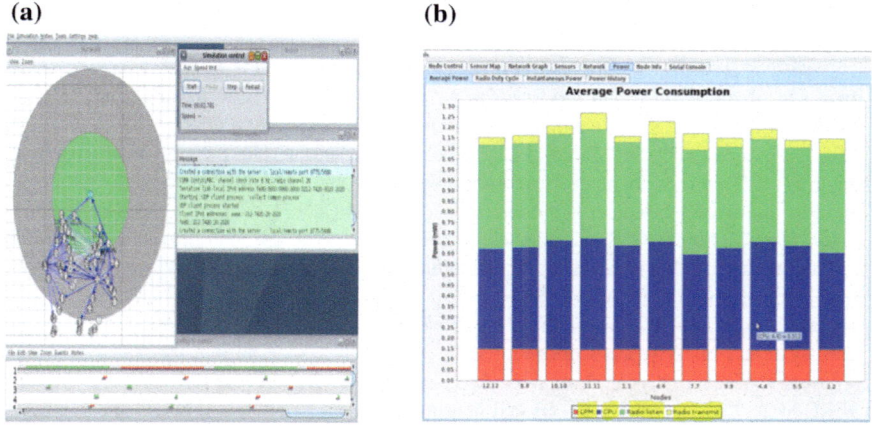

Fig. 3. (a) Simulation environment (b) Power consumption

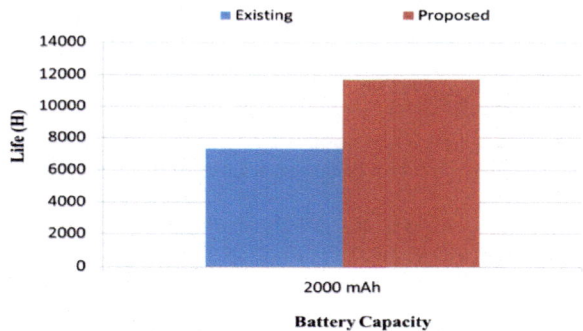

Fig. 4. Output performance

6   Conclusion

The proposed technique is implemented in an IoT elderly home scenario. The IoT devices have different event rate is considered for the deployment. The radio duty cycle is effectively utilized to reduce the energy consumption and assist the elderly people round the clock.

Acknowledge. I thank and acknowledge Mr. Selvaraj for permitting us to deploy the sensor in home environment for this experimental setup.

References

1. Zanella, A., Bui, N., Castellani, A., Vangelista, L., Zorzi, M.: Internet of things for smart cities. IEEE Internet Things J., **1**(1), 22–32 (2014)

2. Santiago, S., Arockiam, L.: Energy efficiency in internet of things: an overview. Int. J. Recent. Trends Eng. Res. **2**(6), 475–482 (2016)
3. Guerra, C., Bianchi, V., Munari, I.D., Ciampolini, P.: CARDEAGate: low-cost, ZigBee-based localization and identification for AAL purposes. In: IEEE International Instrumentation and Measurement Technology Conference, pp. 245–249 (2015)
4. Ok, D., Ahmed, F., Di Marco, P., Chirikov, R., Cavdar, C.: Energy aware routing for Internet of Things with heterogeneous devices. In: International Symposium on Personal, Indoor, and Mobile Radio Communications, pp. 1–5 (2017)
5. Arockiam, L., Santiago, S., Kumar, A.: EALBA: energy aware load balancing algorithm for IoT networks. In: Proceedings of the 2018 International Conference on Mechatronic Systems and Robots, pp. 46–50, ACM (2018)
6. Risteska Stojkoska, B., Trivodaliev, K., Davcev, D.: Internet of things framework for home care systems. In: Wireless Communications and Mobile Computing, pp. 1–8 (2017)
7. Chang, T.-C., Lin, C.-H., Lin, K.C.-J., Chen, W.-T.: Load-balanced sensor grouping for IEEE 802.11 ah networks. In: Global Communications Conference, IEEE pp. 1–6 (2015)
8. Tseng, C.H.: Multipath load balancing routing for internet of things. J. Sens., 1–8 (2016)
9. Hou, J., Jadhav, R., Luo, Z.: Optimization of parent-node selection in RPL-based networks. In: Internet Engineering Task Force (IETF) draft, vol. 1 (2017)
10. Kim, H.-S., Kim, H., Paek, J., Bahk, S.: Load balancing under heavy traffic in RPL routing protocol for low power and lossy networks. IEEE Trans. Mob. Comput. **16**(4), 964–979 (2017)
11. Vithya Vijayalakshmi, V., Arockiam, L.: ARO_VID: a unified framework for data security in internet of things. J. Emerg. Technol. Innov. Res. **1**, 685–690 (2018)
12. Stephen, R., Arockiam, L.: E2 V: techniques for detecting and mitigating rank inconsistency attack (RInA) in RPL based internet of things. J. Phys. Conf. Ser. (JPCS), 1–13 (2018). IOP Publishing
13. Joshitta, R., Arockiam, L.: Anna-a unified framework for data security in healthcare internet of things. Ictact J. Commun. Technol., 1740–1748 (2018)
14. Falco, J., Vaquerizo, E., Lain, L., Artigas, J., Ibarz, A.: AmI and deployment considerations in AAL services provision for elderly independent living. J. Sens. **13**(7), 8950–8976 (2013)
15. Moser, K., Harder, J., Koo, S.G.M.: Internet of things in home automation and energy efficient smart home technologies. In: IEEE International Conference on Systems, Man and Cybernetics, 1260–1265 (2014)

Methods for Finding Brain Diseases Like Epilepsy and Alzheimers

Sasikumar Gurumurthy[1], Naresh Babu Muppalaneni[2(✉)],
and G. Chandra Sekhar[1]

[1] Sree Vidyanikethan Engineering College (Autonomous), Tirupathi, India
[2] Department of CSE, National Institute of Technology Silchar,
Silchar, Assam, India
nareshmuppalaneni@gmail.com

Abstract. In this Paper we have described and discussed about the Epilepsy takes non-recognizable origin in around fifty percentage of the individuals through the disorder. In the further fifty percentage, the disorder could be outlined towards several aspects, together with the Genetic stimulus, Head trauma, Brain conditions, Infectious diseases, Prenatal injury, Developmental disorders, Epilepsy could occasionally connect with developing illnesses, such, for example autism besides neurofibromatosis. Mild cognitive impairment (MCI), grounds additional memory difficulties than ordinary for the people of the similar age group. The bioelectric signals remain verified thru a set of scalp electrodes appropriately located over the head according towards the international 10 to 20 system [6, 7]. Numerous authors approve that the idea of entropy has reached a huge consensus as a pointer of complexity of non-linear signals.

Keywords: Epilepsy · Supervised learning method · Back propagation · Artificial neural networks · Mild cognitive impairment (MCI)

1 Introduction

Epilepsy & Alzheimer's are two kinds of brain diseases happening very fastly around the world. Epilepsy is the kind of process in which a sudden change in electrical activity takes place in brain [9]. Alzheimer's is the process that going to happen in older ages in olden days, but now the process is starting at the early age of life here the person is losing memory it happens because of shrinks happening in brain structure.

The brain disorder can be explained with several aspects, together with the **Genetic stimulus:** Different kinds of epilepsy, which remain considered through the category of seizure for understanding before the slice of a brain that will remains exaggerated, route in family heredity. **Head trauma:** Head trauma for example a consequence of vehicle accident or additional traumatic damage could source the epilepsy. **Brain conditions:** The Brain conditions which create an injury to a brain, such by means of brain tumors otherwise strokes, could causes the epilepsy. **Infectious diseases:** Infectious illnesses, for instance meningitis, AIDS besides viral encephalitis, could cause epilepsy. **Prenatal injury:** Earlier birth, babies remains complex to brain injury that might be triggered by numerous factors, such as a contagion with the mother, underprivileged nutrition

© Springer Nature Switzerland AG 2020
A. P. Pandian et al. (Eds.): ICCBI 2018, LNDECT 31, pp. 715–719, 2020.
https://doi.org/10.1007/978-3-030-24643-3_86

otherwise oxygen shortages. This kind of brain injury could outcome in epilepsy otherwise cerebral palsy. **Developmental disorders:** Epilepsy could occasionally connect with developing illnesses, such, for example autism besides neurofibromatosis. Mild cognitive impairment (MCI), grounds additional memory difficulties than ordinary for the people of the similar age group. The bioelectric signals remain verified thru a set of scalp electrodes appropriately located over the head according towards the international 10 to 20 system [6, 7]. Numerous authors approve that the idea of entropy has reached a huge consensus as a pointer of complexity of non-linear signals.

2   Related Work

In 2007 Srinivasan et al. [1] proposed that by using elman (EN) then probabilistic neural network (PNN) with the time domain feature of EEG signal called approximate entropy (ApEn). In 2007 samanwoy [2] proposed wavelet based neural network based detection for finding the epilepsy. In 2010 pravin et al. [3] proposed sample entropy, wavelet entropy, spectral entropy for findings the epilepsy. Jonathan et al. [4] used adaboost and support vector machine (SVM) for finding the hippocamal segmentation for alzhemeirs detection. srivastava Et al has found alzhemeirs disease by using functional MRI (fMRI). Padilla [5] reported that NMF-SVM of the brain images using this CAD tool. someren et al. reported that alzhemeirs can be found by rhythms in aging Alzhemeirs disease.

3   Methods of Finding Epilepsy

Supervised Learning Method (SVM): This model out perform the traditional statistical methods where the error will only be minimized in this method both the error is minimized and gain is maximized sometimes it is called as maximum classifier. In this we take a raw data set and we do feature extraction and train the model and the evolution of the model (Fig. 1).

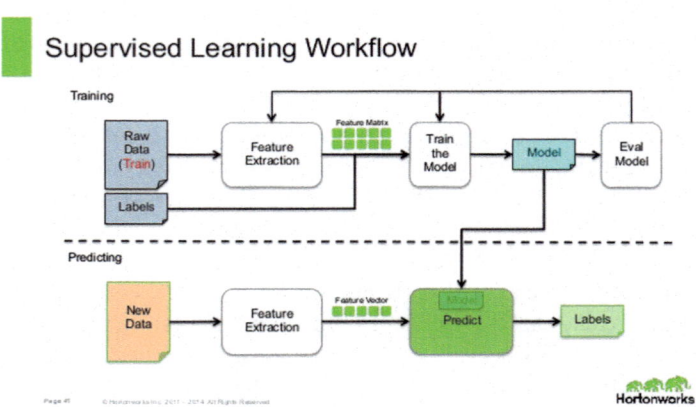

Fig. 1. Supervised learning model

3.1 Back Propagation

Back propagation are the one in which we provide the inputs to the network and expected the output. The process go on weights of the nodes the real output and expected output the error has to be minimum in the process (Fig. 2).

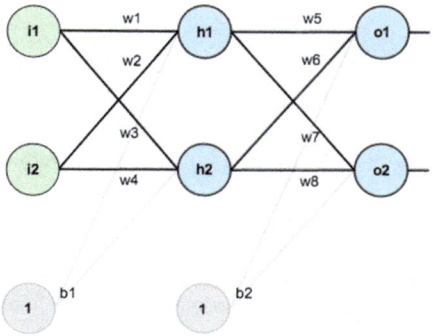

Fig. 2. Basic back propagation

3.2 Artificial Neural Networks (ANN)

It is the one in which each input is associated with weight and summation with a bias is taken and is given to activation function and it produces expected output. There will be a hidden data field between input and output (Fig. 3).

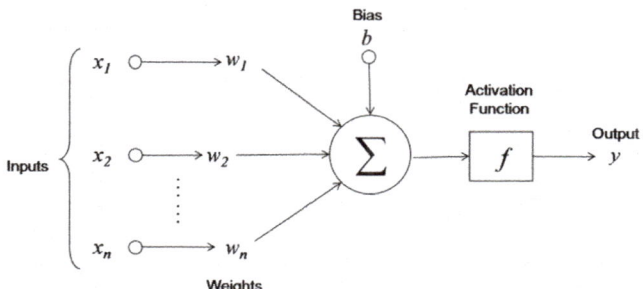

Fig. 3. Artificial neural networks

4 Methods of Finding Alzheimers

4.1 Support Vector Machine

Support vector machine works on MRI, SPECT or PET. In this we have poll based activity. A supervised algorithm analyzes the training data and produces inferred function and a response variable. Here they are two types 1. supervised learning and 2. unsupervised learning. In supervised we are trained and object reorganization is easy. In unsupervised training the object classification is not trained. We will depend on its characteristics [8] (Fig. 4).

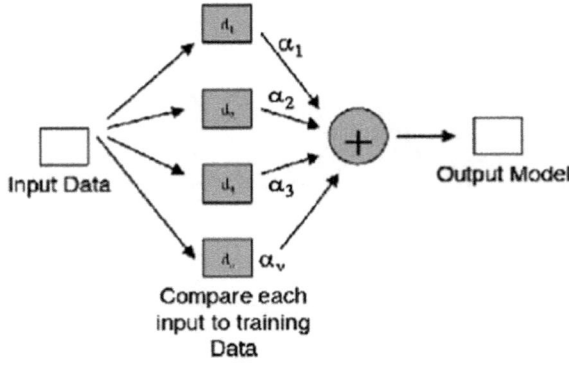

Fig. 4. Support vector machine

4.2 Decision Tree

The decision tree is like tree like structure. In this decision tree there will be a root node and leaves and branches. At each level of the tree there will be certain criteria involved in it. The tree structure success depend on your decision criteria (Fig. 5).

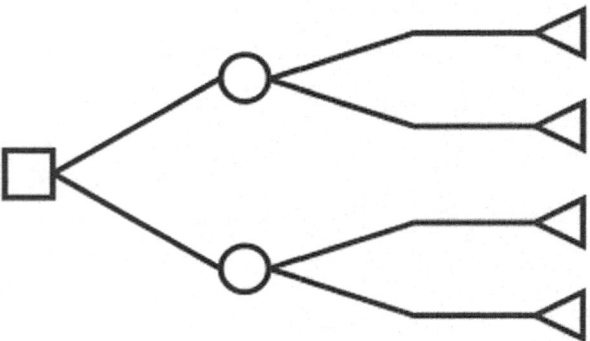

Fig. 5. Decision tree diagram

5   Conclusions

We studied the different ways to find the Epilepsy and causes of epilepsy and also Alzheimer's and methods of finding it.

Acknowledgments. Dr. Sasi kumar gurumoorthy, Dr. M. Naresh Babu and G. Chandra Sekhar would like to thank Department of Science and Technology (DST), Cognitive Science Research Initiative (CSRI). Ref. No.: SR/CSRI/370/2016.

References

1. Morra, J.H., Tu, Z., Apostolova, L.G., Green, A.E., Toga, A.W., Thompson, P.M.: Comparison of AdaBoost and support vector machines for detecting Alzheimer's disease through automated hippocampal segmentation. IEEE Trans. Med. Imaging **29**(1), 30–43 (2010)
2. Padilla, P., López, M., Górriz, J.M., Ramírez, J., Salas-González, D., Álvarez, I.: NMF-SVM based CAD tool applied to functional brain images for the diagnosis of alzheimer's disease. IEEE Trans. Med. Imaging **31**(2), 207–216 (2012)
3. Ulate-Campos, A., Coughlin, F., Gaínza-Lein, M., Fernández, I.S., Pearl, P.L., Loddenkemper, T.: Automated seizure detection systems and their effectiveness for each type of seizure. Seizure **40**, 88–101 (2016)
4. Kumar, A., Singh, T.R.: A new decision tree to solve the puzzle of Alzheimer's disease pathogenesis through standard diagnosis scoring system. Interdiscip. Sci. Comput. Life Sci. **9**(1), 107–115 (2017)
5. Wei, W., Visweswaran, S., Cooper, G.F.: The application of naive Bayes model averaging to predict Alzheimer's disease from genome-wide data. J. Am. Med. Inform. Assoc. **18**(4), 370–375 (2011)
6. Gevins, A.S., Morgan, N.H.: Applications of neural-network (NN) signal processing in brain research. IEEE Trans. Acoust. Speech Signal Process. **36**(I), 1152–1161 (1988)
7. Ullah, I., Hussain, M., Aboalsamh, H.: An automated system for epilepsy detection using EEG brain signals based on deep learning approach. Expert Syst. Appl. **107**, 61–71 (2018)
8. Wena, T., Zhanga, Z.: Effective and extensible feature extraction method using genetic algorithm-based frequency-domain feature search for epileptic EEG multi-classification. Medicine **96**, e6879 (2017)
9. Gigola, S., Ortiz, F., D'Attellis, C.E., Silva, W., Kochen, S.: Prediction of epileptic seizures using accumulated energy in a multiresolution framework. J. Neurosci. Methods **138**, 107–111 (2004)

Ameliorated Domain Adaptation Using Adaptive Classification Algorithm

P. Kola Sujatha, S. Swetha[⊠], and V. Priyadharshini

Department of Information Technology, Madras Institute of Technology,
Chennai, India
pkolasujatha@annauniv.edu,
swecabin@gmail.com, priya.v0096@gmail.com

Abstract. In machine learning, an algorithm usually acquits finer for a particular domain. The work proposes a novel idea of mapping features of source domain to the features of target domain. The model designed for one dataset is applied to another dataset of related domain. The idea throws light upon two subjects namely spam filtering and academic performance analysis. Spam filtering is one of the never-ending problems in the present Internet age. As the spammers guess the type of filter built, there exists a need to design an adaptive filter. The proposed idea is to collect data from a single user mailbox and adapt it to other users. The unseen tokens are estimated using Expectation-Maximization (EM) algorithm, which eliminates zero probability issue. There is an increase in recall value by a minimum of 6.6% from the existing approach. Secondly, the academic performance prediction for school and college students has similar features and hence, the model has been adapted as well. Adaptive mapping is used to map related features between the source and the target. Therefore, the constructed model from the source domain can predict the target data. The proposed classification model shows an increase in accuracy by 7.5% and evaluation accuracy by 75%. The comparison of words in proposed algorithm is in linear time $O(n + m)$, where naïve algorithm is in exponential time $O(n^2)$. Overall, by adapting domains, the need of building multiple models for homologous data is avoided.

Keywords: Domain adaptation · Transfer learning ·
Spam filter · Natural language processing

1 Introduction

The main idea is to develop an algorithm to enable a new dataset to adapt to an already built classifier. For instance, a classifier built for classifying a two-wheeler vehicle can be used for classifying a truck. The features/attributes of motorcycle as well as a truck matches to some extent. Hence, after doing minimal changes, the classifier built for motorcycle can be used for truck. This idea revolves around the concept called transfer learning. In transfer learning, the datasets can either have entirely different features or few matching features. Domain adaptation aims at reducing the effort of building separate models for related problems.

© Springer Nature Switzerland AG 2020
A. P. Pandian et al. (Eds.): ICCBI 2018, LNDECT 31, pp. 720–736, 2020.
https://doi.org/10.1007/978-3-030-24643-3_87

1.1   Research Challenges

Spam Filter. Designing a spam filter is one of the most important challenges faced by the present communication scenario. Spam filter is used to filter unwanted messages in mailboxes, SMS and broadcast messages as well. There are many solutions for spam filtering. Gmail uses an approach of analysing user behaviour towards spam messages. There are other spam filters designed using machine learning, yet many of them are exploited by spammers. Spam filters perform better when designed for a single user. Thus, the requirement of building unique model for each user comes into picture. The foresaid problem or repetition can be solved efficiently by adapting an user model to others, also known as domain adaptation.

Academic Performance Analysis. The academic performance analysis of students can be predicted based on their day-to-day activities as well as by tracking other factors which influence their growth or downfall. A study in Portugal analysed the performance of school students and reason for dropping out of school. The model shows good performance, when built using real-time data. The same study can be extended to college students also. Therefore, by domain adaptation, the model constructed for school can be applied to college data also. The problem statement is taken as means to emphasize the importance of domain adaptation and difficulties during its adaptation phase.

1.2   Scope

Adaptive algorithms can be applied to problems which fall under homologous domains. Related problems can be converged by building a single classification model.

1.3   Machine Learning

Machine learning is one of the fast evolving fields in the field of computational intelligence, to make data driven predictions. It does not rely on static program source code. One of the concepts in machine learning, which is gaining importance in recent days is transfer learning. It is the knowledge of one domain or problem applied to different, but related problem. It is a technique where the domain of one algorithm helps in increasing the accuracy of another. It is the ability of the system to recognize and apply knowledge and skills learnt in previous domains to novel domains. Here, domain refers to the attributes and other tuples pertaining to the problem statement, characterising the attributes. Feature space refers to the collections of features that are used to depict the data. Domain D is the pair of feature space and marginal distribution represented as given in the Eq. 1.1,

$$D = \{\chi, P(\chi)\} \tag{1.1}$$

Marginal distribution $P(\chi)$ is the probability of only X irrespective of Y, where X represents the features and Y represents the class labels. It can also be described as the probability of features irrespective of the class it belongs to in supervised classification.

1.4   Domain Adaptation

Domain adaptation is a combination of machine learning and transfer learning, where an algorithm trained on a specific dataset works for data points outside the dataset. Domain adaptation is applied where multiple source domains are labelled and mapped to a single target domain. In this method, source domain and target domain have few distinguishing features. The source X_S and the target feature space X_T are not distinguishable, where the sample distribution $P(X)$ is distinguishable as represented in Eqs. 1.2–1.4. Domain adaptation can be specified mathematically as,

$$D_S = D_T \tag{1.2}$$

$$P^S(X) \neq P^T(X) \tag{1.3}$$

The task is to build a classifier $C(x)$ such that, $D_T = \{Y, P(Y|X)\}$

$$C(x) : X \to Y \tag{1.4}$$

Domain adaptation defines knowledge that should be extracted from the source and used on the target. There are three approaches in domain adaptation technique. They are:

Reweighting - The source labelled sample is reweighted such that it appears like the target sample.

Iterative approach - This approach iteratively auto labels the dataset. Initially, model C learns from the labelled sample. C automatically labels some target examples. Later, a new model is constructed based on the newly arriving labelled examples.

Search of common representation space - A space where the domains that are close or related to each other are identified. The aim is to obtain the overlapping or related attributes between source and the target domain.

The performance of source labelling is outstanding and can be adapted to target. The capacity to correctly label data coming from target domain, if a model learns from source domain is an issue in domain adaptation. The goal of domain adaptation is to learn 'C' such that it commits little error as possible in D_T. Concept drift is an issue in domain adaptation. It arises due to the changes occurring in the target variable in due course. The changes occur in unpredicted ways, and hence makes it difficult for further analysis. The precision also becomes less accurate as time passes. The mapping function M between the source and target feature space is given as in Eq. 1.5,

$$M : X_S \to X_T \tag{1.5}$$

2   Related Work

Domain adaptation [1] is a combination of transfer learning [9] and machine learning [4]. The main idea is to use an algorithm built for one domain to other domain (same domain or related domain). Source and target domains must have some similar features, but their feature distribution can be different. Existing work has mapped source and target domain values to a feature space by mapping similarities between the source and target domain. One such mapping is Geodesic Flow Kernel (GFK) [3] mapping. GFK aims to predict the class labels for the unlabeled target domain. Dominating features in the source and target domain are mapped by Principal Component Analysis (PCA) [10]. After GFK mapping, the unlabeled target domain is labeled by stream data classification.

An effective spam [6, 18] filter algorithm should eliminate all the spam messages by identifying spam words, without affecting any ham messages. The proposed work uses Naïve bayes (NB) [17, 18] classifier for spam filtering. The aim of this algorithm is to consider no ham message as spam message owing to the fact that, spam word can be in a ham message. But the disadvantage of using NB is that, it computes probability for each word based on the word being spam or ham. NB gives zero probability for unseen words, which will result in zero conditional probability. In order to avoid this problem, existing works have used Laplace smoothing. Proposed work has introduced NB with Expectation Maximization (EM) [15]. EM has two main steps. The first step is expectation step, where it expects the probability of the word to be occurring in spam, whereas the second step is maximization. EM will maximize the probability of the unseen words.

The algorithm proposed performs domain adaptation for spam filtering based on data from different users, considered as domain. Source and target domain datasets are preprocessed in order to get effective results. The preprocessed source dataset is mapped to the target in two steps. First step is mapping the exact feature names using a trie based algorithm [2]. Second step maps the feature labels with synonymous meanings. This mapping is done by WordNet [5, 6, 17] database. Source domain is classified using Naïve bayes with EM and this model is adapted for the target domain. Classified target domain labels and their mappings are persisted in a data store, which learns through re-inforcement learning [13]. The proposed work is tested for real time domains, where the domains adapt well according to the classification algorithm built for student dataset [11] to the college dataset. It shows the domain adaptation between two different, but related domains. Here the datasets are preprocessed by discretisation after performing numeric to nominal conversion. Preprocessed domains' feature labels are mapped to their instances by Instance Weighting [8]. Source domain is subjected to different classification algorithms namely Support Vector Machine (SVM) [12], Naïve Bayes (NB) [15], Random Forest (RF) [7], Best First Tree (BST) [14] and Attribute selection Algorithm (ASC) [10]. The model of source classification is evaluated using target dataset. Among the different classification algorithms, ASC shows the best results, and therefore, it is used for target domain classification.

3 Proposed System - Adaptive System

3.1 Overall System Architecture of Adaptive System (ADAPS)

Figure 1 represents the overall architecture of the proposed system. The proposed work revolves around two problems namely: Academic performance analysis and spam filtering. Each dataset follows a different approach.

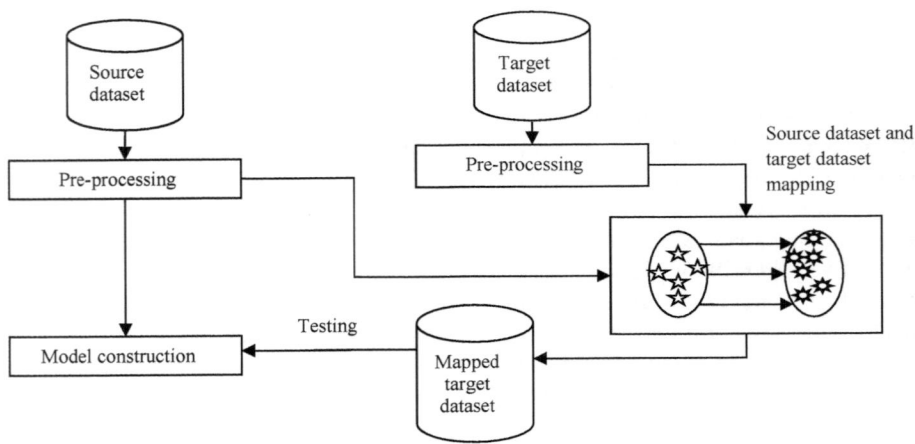

Fig. 1. System architecture of adaptive system (ADAPS)

3.2   Academic Performance Analysis

The academic performance of students can be predicted using psychological and social behaviour. An analysis in Portugal predicts the performance of the student in academics based on these factors. With this scenario as an yardstick, domain adaptation is applied to the problem statement. The performance of school students is considered as source domain and the performance of college students is considered the target. The school student performance is predicted using the data available in UCI Repository contributed during analysis in Portugal. The college student performance data is collected from a real-time survey for college students in India.

3.2.1   Architecture of Academic Performance Analysis

The architecture of academic performance analysis is given in Fig. 2. School students' performance analysis dataset referred as source is pre-processed, transformed and discretized. The pre-processed dataset is classified using different classification algorithms. The best performing model is chosen based on specificity value. Out of the attempted classification algorithms, attribute selected classifier performs better. In college student dataset, EM algorithm is used to estimate missing values, which is an optional step. The feature labels of source and target are mapped in two steps – NLP-based mapping and context-based mapping. The mapped features along with unmapped features are subjected to instance weighting. Instance weighting compares feature values of the two domains.

Individual tuples in the dataset are given importance. The output will be the mapped source and target features. The already built classification model is applied to the mapped target domain. The mapped features are considered, while unmapped features are mapped after manual verification. The approach solves the problem of building different models for related problems.

3.2.2 Modules in Academic Performance Analysis

The proposed system for academic performance analysis consists of modules viz., data

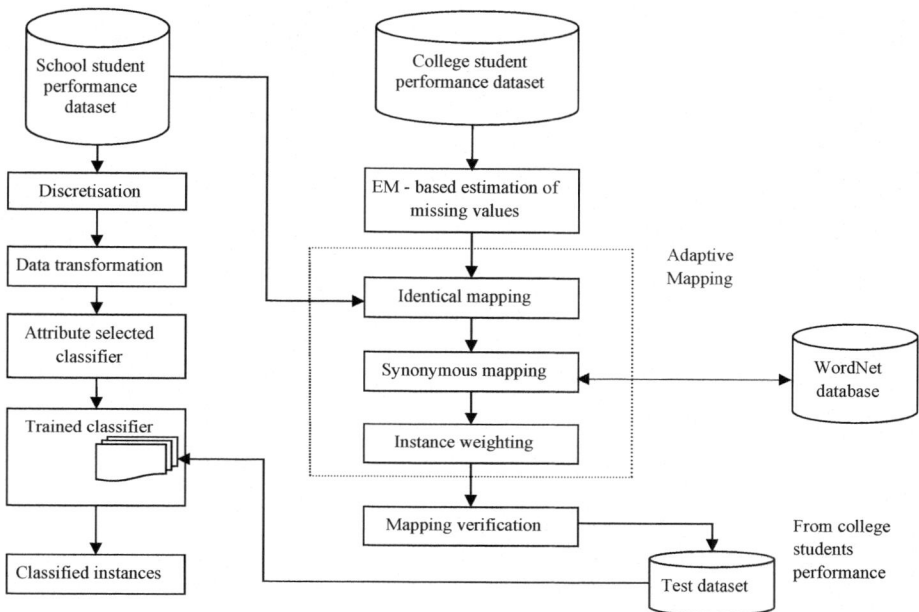

Fig. 2. Architecture of academic performance analysis

transformation, adaptive feature mapping and classification. Each module in the proposed system is discussed below.

Data Pre-processing

The dataset is pre-processed using two steps, discretisation and numeric - nominal conversion. The marks of students are continuous and therefore the continuous values are to be converted to discrete values. Discretisation is performed by splitting the range [a, b] into N - equal width intervals as given in Eq. 3.1.

$$\text{Interval width} = \frac{b - a}{N} \tag{3.1}$$

After discretization, the numeric attributes are converted to nominal attributes. Nominal attributes denote a set of pre-determined class labels, suitable for classification algorithms.

Adaptive Feature Mapping

The identified principal components are mapped to each other by the system itself. Adaptive mapping happens in two phases namely - NLP-based mapping and context-based mapping or Instance weighting

A. **NLP-based Mapping**

In NLP based mapping, the feature labels of source domain and target domain are compared. The source labels with synonymous target labels are identified. The exact match is preferred than approximate matching.

i. **Exact String Matching**

We have proposed a string matching algorithm based on Aho-Corasick algorithm, which constructs a trie for the names of feature vectors in source domain, and checks if the target feature vector name is present in trie. If it is found, it does not check the same node during next iteration.

> **Algorithm: string_matching**
> Input – N_S and N_T (names of feature vectors)
> Output – Mapping of identical feature vector names, N_{Sop} and N_{Top}
> begin
>   Construct a trie for the given N_S
> $N_{Sop} \leftarrow \{\}$
> $N_{Top} \leftarrow \{\}$
>   for each feature name N_{Ti} in N_T
>   do
>   if ($N_{Ti} ==$ leaf node in trie)
>   isLeaf $\leftarrow$ **False**
>   Create a mapping $M : N_{Si} \rightarrow N_{Ti}$
> $N_{Sop} \leftarrow N_{Sop} \cup N_{Si}$
> $N_{Top} \leftarrow N_{Top} \cup N_{Ti}$
>   end for
> end

ii. **Synonymous String Matching**

WordNet, designed and maintained by Princeton University has a lexical database for English, which is stored in ASCII format. Our approach will access the WordNet database to search for the synonym for a given feature name.

B. **Instance Weighting**

The mapping N_{Sop} and N_{Top} are re-checked if they are mapped according to the context. The instances in the dataset are weighed by calculating the number of distinct elements characterising each attribute. For example, gender may contain three distinct elements namely, male, female and transgender. These distinct elements for each

Algorithm: synonymous_matching
Input – N_S, N_T, N_{Sop} and N_{Top}
Output – N_{Sop} and N_{Top}
Assume getWordForm function retrieves wordforms from the synset including duplicates
begin

        File f ← WordNet file
        Synset ← List_all_synsets ( )
        Wordform ← Synset.getWordForm ( )
        Wordform ← remove_duplicates ( )
        for each element in Wordform
        do
          if(Wordform$_i$ == N_{Sm} && Wordform$_i$ == N_{Tn})
                Create a mapping $M : N_{Sm} \rightarrow N_{Tn}$
                $N_{Sop} \leftarrow N_{Sop} \cup N_{Sm}$
                $N_{Top} \leftarrow N_{Top} \cup N_{Tn}$
                $N_S \leftarrow N_S - N_{Sm}$
                $N_T \leftarrow N_T - N_{Tn}$
        end for
end

attribute in source is compared with each attribute in the target. Another method of weighing is to calculate the z-score of the elements in each attribute. Attribute values are normalised using standard method of z-score [7] as in Eq. 3.2.

$$z \text{ score} = \frac{x - \mu}{\sigma} \tag{3.2}$$

where, x be the numerical value of an element, μ is the mean and σ as standard deviation.

The above process standardises the value of the source and target elements. If the source value ranges from 0 to 100, and target ranges from 0 to 20, the values of the source and target can be standardised within a range. The module can also be modified to check for other formulas as well. Thus, it helps the system to identified related features.

Algorithm: instance_weighting
Input – N_{Sop}, N_{Top}, N_S and N_T
Output – N_{Sop} and N_{Top}
begin
    for each mapping N_{Sopi} and N_{Topi} in N_{Sop} and N_{Top}
    do
      flag ← Check_mapping (N_{Sopi}, N_{Topi})
      if (flag == false)
          N_{Sop} ← N_{Sop} − N_{Sopi}
          N_{Top} ← N_{Top} − N_{Topi}
          Create a mapping $M : N_{Sm} \rightarrow N_{Tn}$
          N_{Sop} ← N_{Sop} U N_{Sm}
          N_{Top} ← N_{Top} U N_{Tn}
    end for
    for each element in N_S and N_T
    do
      flag ← Apply_formula (N_{Si}, N_{Ti})
      if (flag)
          Create a mapping $M : N_{Si} \rightarrow N_{Ti}$
          N_{Sop} ← N_{Sop} U N_{Si}
          N_{Top} ← N_{Top} U N_{Ti}
    end for
end

Manual Verification

The system outputs the mapping of principal features identified. The human verifier checks the mapping and remaps the features if there is a mistake. The changes made by the human verifier is updated and stored in the data store. The manual verification is designed as an interactive user interface to enable users with limited domain knowledge to make use of the system.

3.3 Spam Filtering

A spam filter can be designed using machine learning algorithms by collecting information from mail box of users. The message in the mail are converted into tokens. Since there are large number of tokens, the tokens with less frequency are reduced. The messages consist of a set of tokens along with their frequency and the class label to represent ham or spam message. The dataset available will be in the form of tokens, frequencies and class label.

3.3.1   Proposed Approach to Spam Filtering

The algorithm used in the proposed method is naïve bayesian spam filtering. It involves the construction of decision tree [16] using Bayes' theorem [17]. Bayes' theorem is represented as in Eq. 3.3.

$$P(A|B) = \frac{P(B|A)P(A)}{P(B)} \tag{3.3}$$

3.3.2   Algorithm for Naïve Bayesian Spam Filtering with Expectation Maximisation (NBEM)

The algorithm for the naïve bayesian with expectation – maximization (NBEM) built is given below:

Algorithm: nbem_spam_filtering
Input - Bag of words W along with frequency f
Output - Classified tuples
Step 1 - Read input data from file in term-frequency format.
Step 2 - Perform term-frequency conversion for the dataset.
Step 3 - Initialize the classification model along with significant tokens, classification threshold.
Step 4 - Compute the probability of each token W_i

$$P(C = Spam|W_i) = \frac{P(W_i|class = spam)}{P(W_i|class = spam) + P(W_i|class = ham)}$$

Step 5 - Filter the tokens based on significant tokens parameter.
    Step 5.1 - For each token W_i in W,
    Step 5.2 - If (token_count < significant tokens), then add W_i to the list of tokens. Else, skip the token W_i
Step 6 - Merge the two data files into a single data structure.
Step 7 - Repeat steps 8 - 9 until likelihood is positive.
Step 8 - Compute expectation value for the mail based on the current model (E-Step)

$$E = E + log_{10} \left[\frac{P(class = spam)}{P(class = ham)} \right]$$

Step 9 - Maximize the value of likelihood L using Naïve Bayesian conditional probability (M-Step)

$$L = L + \frac{P(W_i|class = spam) \cdot P(class = spam)}{P(W_i|class = ham) \cdot P(class = ham)}$$

Step 10 - If (L > threshold), then message is spam. Else, given message is ham.

3.3.3   Algorithm for Term Frequency Conversion

The algorithm for converting the entries in the dataset to a table containing word ID and frequencies is given below:

Algorithm: term_frequency_conversion
Input - Spam filtering dataset with word ID, frequencies and class labels
Output - Map <Word ID, frequency f>, total number of words, individual count of number of words in spam and ham.
Count - total number of words in bag of words
begin
       for each line in dataset
       do
         if (class is Spam) then
            numSpam++
         if (class is Ham) then
            numHam++
         Split word ID and term frequency
         count++
         if (Word_ID present in Map) then
            Map←<Word_ID.f + prev_f>
         else
            Map ←<Word_ID, f>
       end for
end

3.3.4   Algorithm for Estimation of Unseen Values

The spam filter is trained with a single user mail box, but it is expected to work well for multiple users. Hence, there is a need for a module to estimate the probability of unseen tokens. The algorithm given below illustrates the sequence of steps to be followed to estimate the probability of an unseen token to be a spam.

Algorithm: estimate_unseen_values
Input - Spam and ham probability table
Output - Probability value of unseen tokens
Step 1 - Initialise previous and current likelihood value to zero.
Step 2 - Assign current likelihood to previous likelihood value.
Step 3 - Classify the model using Naïve Bayesian classification algorithm.
Step 4 - Initialise the model by computing the spam and ham probability table.
Step 5 - Initialise i to 1.
Step 6 - for each token W_i in spam probability table,
       Step 6.1 - Compute current likelihood of the token.

$$currentlikelihood += \left(\frac{\text{Prob. of} W_i \text{ in spam prob. table} * \text{Total spam prob.}}{\text{Prob. of} W_i \text{ in ham prob. table} * \text{Total ham prob.}} \right)$$

Step 7 - Compute current likelihood.

$$currentlikelihood \leftarrow log_{10}(currentlikelihood)$$

Step 8 - If (previouslikelihood - currentlikelihood) ≥ 0, then go to step 2. Else, terminate.

4   Analysis and Results Discussion

4.1   Academic Performance Analysis

4.1.1   NLP-Based Mapping

The proposed academic performance analysis involves adaptive mapping, which includes identical and synonymous mapping. The proposed string-mapping algorithm called NLP-based algorithm is implemented using Trie, as a variant of Aho-Corasick [2] algorithm. The time complexity of naïve algorithm is $O(n^2)$, whereas the implemented algorithm shows better complexity $O(n + m)$, where n represents the number of words in the Trie and m represents the maximum height of the trie. The worst case time complexity is given in Fig. 3.

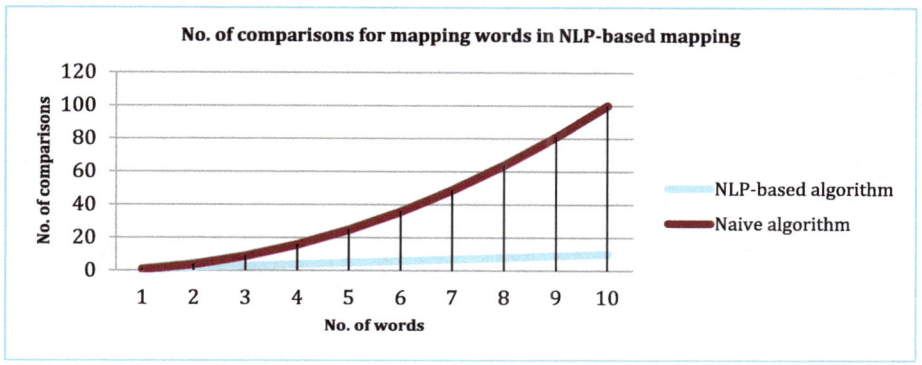

Fig. 3. Big - O time complexities for naïve and NLP-based algorithm

The proposed algorithm outperforms naïve algorithm, in multiple ways. The algorithm is evaluated for elements with varying sizes. The algorithm also avoids the complexity z, which is the number of repeating elements in the given text. The complexity decreases to $O(n + m)$ by a value z from Aho-Corasick algorithm as delineated in Fig. 3.

Synonymous string matching algorithm is evaluated against traditional dictionary-based approach. WordNet [5, 7, 17, 18] is comparatively more efficient than dictionary based approach, as WordNet indexes all words in a hash table. It also allows retrieving words based on the synsets.

4.1.2 Student Performance Prediction (SPP) Vs Academic Performance Prediction (ADAPS)

The student performance prediction analysis [11] is compared with proposed academic performance prediction. Considering the same measures as SPP with same number of iterations, the performance of the proposed algorithm is comparatively higher. The sensitivity rate is tabulated against different classification algorithms in Table 1.

Table 1. Comparison between student performance prediction [11] and proposed ADAPS algorithm (for school level)

Algorithms	NB	SVM	RF	BFT
Sensitivity for student performance prediction	72.9%	64.5%	73.5%	**76.1%**[1]
Sensitivity for ADAPS algorithm	81%	81%	81%	**83.8%**

The Table 2 shows the percentage of correctly classified instances for different types of algorithms which include Naïve Bayesian (NB), Support Vector Machines (SVM), Random Forest (RF), Iterative Dichotomiser 3 (ID3), Attribute Selected Classifier (ASC) and Breadth First Tree (BFT) algorithm. The school students' performance analysis dataset undergoes two steps, which makes it different from the existing SPP [11] approach. Existing approach uses 1-of-C encoding method, whereas the proposed approach uses discretisation and numeric to nominal conversion as the pre-processing step. The foresaid pre-processing steps for the existing system enhances the output of the classifier. Out of all these classifiers, attribute selected classifier performs better and shows lesser error rate.

Table 2. Analysis of student performance with different classification algorithms after applying proposed pre-processing steps

Algorithms	NB	SVM	RF	J48	BFT	ASC
No. of correctly classified instances	530	534	530	539	553	**557**
True positive rate (Sensitivity)	81.2%	80.89%	83.9%	83.05%	85.51%	**85%**
False positive rate (Fall-out)	10.6%	10%	10.8%	9.3%	9.2%	**9.1%**
Precision	80.8%	80.6%	82.4%	82%	**85.3%**	84.1%
F-Measure	81.2%	80.9%	82.4%	82.6%	84.8%	**85.1%**
MCC	70.7%	71%	74.2%	74.1%	**77.7%**	76.9%
ROC area	95%	89.9%	**95.6%**	91.4%	95.5%	94.2%
Relative absolute error	32.02%	97.25%	54.04%	32.28%	44.52%	**31.68%**
Root relative squared error	65.74%	92.38%	64.64%	64.73%	59.60%	**58.15%**

The sensitivity rate or percentage of correctly classified tuples for SPP [11] and ADAPS approach is given in Fig. 4. Decision trees and forests seem to perform well in terms of sensitivity. Sensitivity rate is higher for ASC, as highly correlated or redundant attributes are removed. Dimensionality reduction optimises the performance of the classifier.

4.1.3   Evaluation and Analysis

After constructing a better model, the model is saved. The model is evaluated using test dataset (college performance analysis). The attributes in college dataset are mapped with source dataset. The attributes without mapping are neglected. Therefore, the output is not accurate, but is approximate. The evaluated model shows 75% accuracy for validating using random forest, which shows better results for the particular problem.

Further, the grade is classified into 5 ranges viz., 16–20 (excellent/very good), 14–15 (good), 12–13 (satisfactory), 10–11 (sufficient) and 0–9 (fail). Therefore, five-level classification is performed in school student dataset. The analysis is done based on marks scored in Portuguese subject by students in Portugal. The data is taken from UCI repository.

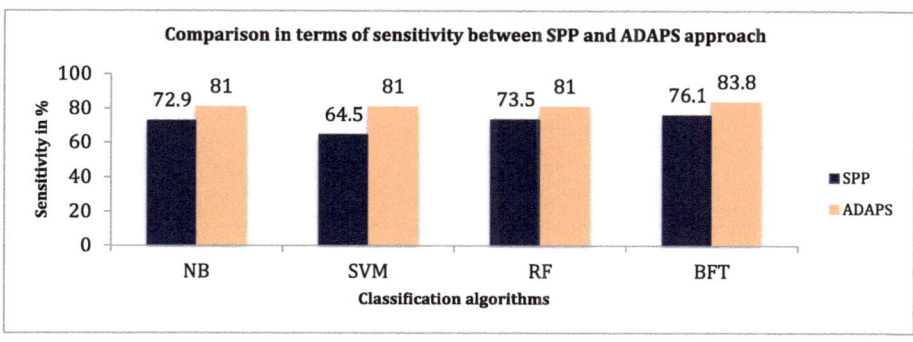

Fig. 4. Sensitivity for different classification algorithms using SPP and ADAPS approach

The classification model shows 99.7% training accuracy after adding an attribute called class label. If the accuracy of the classifier is 100%, procedural programming can solve the problem, and there is no need to implement machine learning concepts for the same. Since, the accuracy is lesser than near ideal accuracy, the inputs are linearly separable. The algorithm is not over-fitting and proves itself better.

To challenge the approach, the attributes causing linear separability are removed. Attribute G3 is removed as it contributes to linear separability of the model. Few of the non-linearly separable attributes are converted to linearly separable attributes after removing G3. Nominal to binary filter is used to convert non-linearity to linearity in datasets. The resulting outcome of classification in ADAPS approach are given in Table 3. Decision table and ASC algorithm shows above 35% accuracy, which is satisfactory. Removing non-linearity in other attributes as well can further increase recall. It can also be enhanced by transforming, standardising, adding or removing dummy attributes.

Table 3. ADAPS approach without attribute G3 (for non-linear data)

Algorithms	Random tree	Decision table	ASC	SVM
No. of correctly classified instances	134	**139**	**139**	2
Sensitivity	33.92%	**35.19%**	**35.19%**	30.89%

4.2   Spam Filtering

The spam filtering system is constructed for a dataset taken from multiple user mail-boxes. Since the number of tokens in the source mailbox is very large, the number of words are reduced. The system is analyzed based on the number of significant tokens. Similar to any classification model, the system also changes in accordance with the above said parameter. The Fig. 5 shows the variation of accuracy in results based on the number of significant tokens. Accuracy is directly proportional to the number of significant tokens. There is a more improvement than the previous level in the accuracy at token number is 10. The graph has been plotted by making classification threshold as a constant, equals 30. The classification threshold can be any positive value.

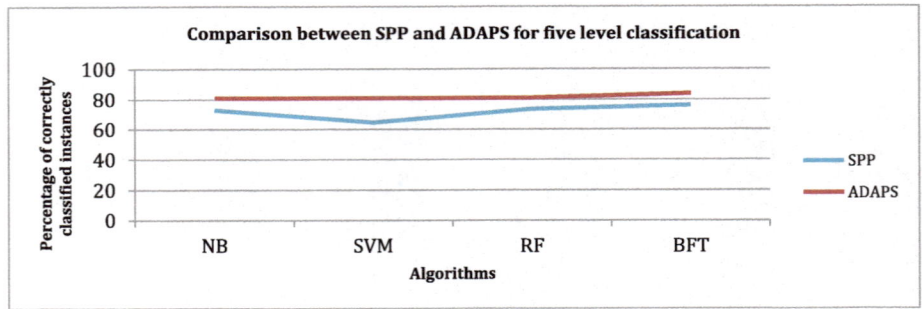

Fig. 5. Comparison in recall between anti-spam filtering [16] and proposed naïve bayesian with EM algorithm

Since the graph has reached better accuracy when the significant tokens are 10, the number of significant tokens is maintained constant with the value of 10. The change in accuracy with varying threshold gives better accuracy when the threshold lies between 0 and 0.2 and all positive values above 0.25. The values are given in Table 4. The spam filtering approach [16] is compared with the proposed EM-based spam filtering approach. The existing system parameters are applied to the proposed system. The outputs show that the proposed algorithm outperforms the existing system in terms of number of significant tokens as shown in Table 5. Thus, the NBEM algorithm predicts labels by learning from multiple user mailboxes and adapts to additional users. The recall rate is also higher and notable.

Table 4. Recall based on classification threshold

Threshold	0.05	0.2	0.25	0.5	1
Recall	**85.5%**	85.2%	44.5%	84.2%	**85.5%**

5   Conclusion

The paper helps in adapting domains of different problems, which is useful for designing a common model for many problems. It requires little human intervention during the early stages of use. The algorithm designed produces better results than the traditional domain adaptation model. In NBEM, the source domain is constructed using multiple user mailboxes. The proposed approach shows a minimum of 6.6% accuracy. If a set of completely unseen tokens arrive, the traditional bayes' theorem shows zero probability. By applying EM, unseen tokens are also assigned a probability. This approach strengthens the conventional naïve bayes approach to suit spam filtering application. Number of significant tokens are limited, so as to reduce the complexity of search in a bag of words and avoid confusion. A word may be present in spam as well as ham. Such words do not change the course of the model.

Domain adaptation is also achieved by constructing a model for school student dataset and then adapting for college student dataset. On detailed analysis of the

Table 5. Comparison in accuracy with anti-spam filtering and proposed naïve bayesian with EM

Number of significant tokens	Accuracy in naïve bayesian anti-spam filtering technique [16]	Accuracy in proposed NBEM
5	85.7%	83%
10	77.4%	**84%**
15	75.8%	83.75%
20	71.4%	83.25%

approach, few conclusions are derived. The proposed NLP-based algorithm reduces the search in the entire Trie, by pruning. It removes the word as soon as it is discovered. This reduces the size of the tree, and therefore the number of comparisons and search time is also decreased. The proposed algorithm is less complex as the creation cost is only O(n*l), where l is the average length of the word.

The proposed pre-processing steps increases the recall of the algorithm, by converting the attributes to nominal attributes. Discrete values are useful in classification algorithms, for fast and distinct classification. WordNet is one of the well constructed lexical database, which enables easy and fast retrieval through its graph structure. The proposed adaptive mapping is different from other, as traditional approach uses statistical model and structural correspondence learning, whereas the proposed approach concentrates upon the attribute structure, type and instances. Each instance in the dataset has a particular weight which contributes to the overall distribution. Z-score standardizes and normalizes the data, which is spread across ranges. Evaluation accuracy of 75% shows that 3 out of 4 instances are classified correctly by the mapped model. At present, z-score, number of distinct elements in the instances and threshold are taken into account. By taking other factors into consideration, the accuracy can be further increased.

6 Future Work

The designed spam filter can be deployed in real-time mail boxes to evaluate the performance and adaptability. It can be verified by customising for an individual user mail box and adapting to other users. The present system categorises words based on the word IDs. But, the system can also be designed to read directly from the mailbox, identify spelling errors and analyse even the misformed word. Student performance designed predicts class labels for college and school students with improved recall and reduced error rate. Nevertheless, the complexity of the trie can also be increased by merging the subtrees using prefixes, which leads to reduction in the height of the tree. In the proposed approach, the elements in the trie are stored in list, where it can be stored in bits, thereby lessening the difficulty. The system can map using NLP-based mapping only, if the source and the target have meaningful attribute names. The NLP-based mapping module can be modified to handle sentences in the place of attribute names. Instance weighting using multiple formulae and context can supplement the system. Additionally, re-inforcement learning can be applied to learn from manual verification made by the user. Algorithms such as Temporal Difference-λ (TD-λ) can be used for reinforcement learning. The system may calculate separate rewards for machine as well as user. The action with higher reward can be considered.

Acknowledgement. There is no participants (humans/animals) are involved in the study.

References

1. Bitarafan, A., Baghshah, M.S., Gheisari, M.: Incremental evolving domain adaptation. IEEE Trans. Knowl. Data Eng. **28**(8), 2128–2140 (2016)
2. Aho, A.V., Corasick, M.J.: String matching: an aid to bibliographic search. Commun. ACM **18**(6), 333–340 (1975)
3. Gong, B., Shi, Y., Sha, F., Grauman, K.: Geodesic flow kernel for unsupervised domain adaptation. In: Proceedings of the 25th IEEE Conference on Computer Vision and Pattern Recognition, pp. 2066–2073 (2012)
4. Taylor, C., Nakhaeizadeh, G.: Learning in dynamically changing domains: theory revision and context dependence issues. In: Machine Learning: ECML-97, vol. 1224, pp. 353–360 (2015)
5. Fellbaum, C.: WordNet: An Electronic Lexical Database. MIT Press, Cambridge, MA (1998)
6. Miller, G.A.: WordNet: A Lexical Database for English. Commun. ACM **38**(11), 39–41 (1995)
7. Abdi, H.: Normalizing data. In: Salkind, N. (ed.) Encyclopedia of Research Design, pp. 1–4. Sage, Thousand Oaks (2010)
8. Jiang, J., Zhai, C.X.: Instance weighting for domain adaptation in NLP. Department of Computer Science University of Illinois at Urbana-Champaign Urbana, IL 61801, USA
9. Asadi, M., Huber, M.: Effective control knowledge transfer through learning skill and representation hierarchies. In: International Joint Conference on Artificial Intelligence (2007)
10. Hall, M.A.: Correlation-based Feature Selection for Machine Learning. University of Waikato, Hamilton, New Zealand (1999)
11. Cortez, P., Silva, A.: Using data mining to predict secondary school student performance. Information Systems/Algorithm R&D Centre University of Minho 4800-058 Guimaraes, Portuga
12. He, Q., Chen, J.-F., He, M., Zhu, R.-X.: The impact of linear transformation on SVM margin. In: IEEE International Conference on Systems, Man, and Cybernetics (2006)
13. Sutton, R.S., Barto, A.G.: Reinforcement Learning: An Introduction. The MIT Press, Cambridge (2012)
14. Rokach, L., Maimon, O.Z.: Data Mining with Decision Trees: Theory and Applications. World Scientific Pub Co Inc, Singapore (2008)
15. Triola, M.F.: Bayes' theorem. http://faculty.washington.edu/tamre/BayesTheorem.pdf
16. Deshpande, V.P., Erbacher, R.F., Harris, C.: An evaluation of naïve bayesian anti-spam filtering techniques. In: Proceedings of the 2007 IEEE Workshop on Information Assurance United States Military Academy, West Point, NY-2022 (2007)
17. Blondel, V.D.: Automatic extraction of synonyms in a dictionary. In: Proceedings in the SIAM Text Mining Workshop (2003)
18. WordNet: An electronic lexical database. MIT Press. https://mitpress.mit.edu/books/wordnet

Fault Tolerance Clustering of Scientific Workflow with Resource Provisioning in Cloud

Ponsy R. K. Sathiabhama, B. Pavithra, and J. Chandra Priya[✉]

Department of Computer Technology, Madras Institute of Technology,
Anna University, Chennai, India
ponsy@mitindia.edu, pavithra3001@gmail.com,
jchandrapriya@gmail.com

Abstract. Cloud computing environment has features that supports the computation of Scientific Workflows and provides easy deployment of workflows. Workflow is predefined ordered tasks executed based on its data dependency and control dependency. Scientific workflow consists of fine and coarse grained computational granularity tasks which are efficiently executed using task clustering. Task clustering is a proficient method which reduces the overhead caused due to execution and it enhances the computational granularity of scientific workflow tasks executed on distributed resources. Transient failures such as task failures in a job can have major impact on the runtime performance of workflow execution. Existing Fault Tolerant Clustering strategies are analyzed in faulty execution environment. Besides task failure, hardware failure is also considered where the virtual machine failure occurs. Scheduling the workflows in cloud computing environment is an NP-Complete problem and it is challenging when tasks failures are considered. To overcome these failures, the workflow scheduling system should be fault tolerant. A clustering algorithm is proposed for resource provisioning to efficiently reduce resource wastage and system overhead. Among various Scientific Workflows, Montage Workflow with different number of tasks is taken for simulation to analyze the proposed work.

Keywords: Scientific workflows · Task clustering · Task failure ·
Fault tolerance · Cloud computing · WorkflowSim

1 Introduction

Cloud computing offers easy access to high performance computing, storage infrastructure through web services. It provides considerable scalability, reliability and configurability along with high performance. Given the recent popularity of cloud computing, and the applicability of cloud computing to the scenario of data and computation intensive scientific workflow applications, there is an increasing demand to investigate scientific cloud workflow systems. Scientific cloud workflow systems support complex e-science applications such as disaster recovery simulation, earthquake modelling, weather forecast, astrophysics and high energy physics and climate modelling.

© Springer Nature Switzerland AG 2020
A. P. Pandian et al. (Eds.): ICCBI 2018, LNDECT 31, pp. 737–751, 2020.
https://doi.org/10.1007/978-3-030-24643-3_88

Workflows are defined as a set of computational tasks and a set of data or control dependencies between them. The workflows are represented as a Directed Acyclic Graph (DAG) in EXtended Markup Language (XML) based formats. Each node of the DAG represents an independent task or an application. The directed edge in the DAG represents an execution dependency among the tasks present in them. These are widely used by the scientific community to analyze and process large amounts of scientific data efficiently.

Workflow technology has been successfully used in business and scientific applications for many years.

- Business Workflows - Business workflows aim to automate and optimize an organization's processes administered by human or computer agents in an administrative context. The main purpose of building a workflow for companies is to enable customers to make better use of its services and gain profits from them. The main concerns are the workflow integrity and security.
- Scientific Workflows - Scientific workflows usually help scientists in streamlining knowledge-intensive activities in their experiments that are domain specific, especially in proving scientific goals or hypotheses. Such scientific processes are exploratory in nature and have new analysis methods that are rapidly evolved from initial ideas and preliminary workflow designs.

A DAG can be used to represent a set of programs where the input, output or execution of one or more programs is dependent on one or more other programs. The data flow and control flow of a workflow, where t is task and d is the data given as input to tasks is shown in Fig. 7. There are two types of dependency the edges can represent: control flow and data flow. Control flow graph made up of tasks and precedence constraints to control the flow. The tasks are the operations and the edges indicating the operations order. The data flow is not only the control, but also about the data that flows between and through the connections from one task to another and that drives the computation of the workflow.

2   Related Works

In [1], Chen et al. proposed the dynamic estimation of parameters where the runtime data are collected during execution of workflow and updating the values for determining the clustering factor. Maximum likelihood estimation for the prior knowledge to adjust parameter estimation is proposed. They consider the overhead-aware DAG (O-DAG) model for task clustering. In [2], Chen et al. proposed the work for increasing the computational granularity for the execution of scientific workflows. They propose two failure modeling frameworks, namely task failure model and job failure model. In [3], Chen et al. proposed balancing methods for addressing the load balance problem for the execution of workflows. Three imbalance metrics namely task runtime variation, task impact factor and task distance variance are measured. In [4], Sahni et al. focused more about the degree of parallelism among the tasks at the level of a workflow while also reducing the system overheads and resource wastage. In [5], Bala et al. proposed a model that predicts the task failures using machine learning methodologies. They make a focus on designing intelligent task failure prediction model for proposing proactive fault tolerance, which is done by predicting task failures for scientific applications. It also presents a Naive Bayes approach with maximum prediction accuracy. In [6], Juve et al. considered different scientific workflows for assessing their performance. Six workflows are considered like astronomy, bioinformatics and earthquake science and gravitational-wave physics. In [7], Calheiros et al. offered an algorithm which uses unused time of provisioned resources and budget excess to replicate tasks. The proposed algorithm increases the possibility of meeting deadlines and reduces the total execution time of applications. They propose strategies for correcting the delays which are caused by undervaluing the execution time of the tasks or performance of the resources used. In [8], Smanchat et al. discussed the taxonomy of scheduling problem and techniques which are proposed based on systematic analysis. The scheduling classifications that are unique to workflow namely, scheduling of the process, task and resource are given. In [9], Mattoso et al. proposed a taxonomy where the issues in dynamic steering of high performance computing are identified for scientific workflows. The main notions are workflow lifecycle, user support for real-time monitoring, interference, analysis and notification by adjusting the workflow execution at runtime. Dynamic workflow features namely adaptive and user-steered workflows are helpful in Large-scale experiments. In [10], Chen et al. proposed a simulator named WorkflowSim, which extended the existing CloudSim simulator by providing a higher layer of workflow management. The simulator assists in workflow optimization techniques by considering system overheads and failures in simulating scientific workflows. In [11], Calheiros et al. (2010) proposed a simulation tool called Cloudsim that supports workflow execution and was the predecessor of the workflowsim tool. It simulates and models cloud computing systems for different application providing environments. In [12], Deelman et al. focuses to achieve scalable and reliable execution of workflows across different computing infrastructures. It supports large-scale applications with I/O intensive, data intensive, computation intensive workflows. It also supports platforms such as Grid and cloud. In [13], Gu et al. proposed analytical cost models and formulated workflow mapping as optimization problems for minimum end-to-end delay and maximum frame rate. A mapping algorithm was offered to minimize the latency based on a recursive critical path optimization procedure.

3 Problem Description

Clustering of scientific workflows is efficient method of executing the workflow. The fault tolerant clustering techniques used in horizontal clustering are utilised and used for proposed hybrid clustering to reduce the increase in execution time caused due to the overheads. The fault tolerant algorithms used are executed in dynamic environment where the parameter values are observed during the runtime of the workflows. Cluster size k is the important factor which is used in the clustering. With the above factors the makespan of the workflow execution is reduced. Along with the transient failure, the hardware failure VM failure is also included.

4 System Model

A workflow is represented in DAG. In DAG representation, the nodes in the graph represent the workflow task and dependencies between the tasks are represented by the edges connecting the nodes, where the dependencies give the order in which the tasks are executed.

The output produced by the tasks in nodes are given as input to another task in node as input in the execution of the workflow. The data-flow dependencies are represented as the input and output files in the workflow. Each task is considered as a program and a set of parameters which is needed to be executed.

A typical example of workflow execution environment is shown in Fig. 8. The host which submits the workflow for execution also prepares worker nodes and an execution site where the jobs are executed separately.

The main components in the system are described as follows:

(a) Workflow Mapper

Workflow Mapper performs a process called Task Clustering. Task Clustering is a process where more than one task is combined together to form a single job based on different characteristics to reduce the system overhead which is caused during the execution of workflow. This type of merging is done by Workflow Mapper. Workflow Mapper is responsible for generating the workflow that is executable in provided environment, which was delivered by the user. It also does the functionalities such as to get required data, identifying the software and to determine the required resources for the execution.

Workflow Mapper generates tasks list and allocates these tasks to an execution site for proper resource. It is done by importing the DAX file and other additional information from metadata such as number of input files, number of child tasks for particular task, file size of the task and so on. This above mentioned data are obtained from Workflow Generator.

(b) Local Execution Engine

Engine executes the jobs in the workflow. The workflow is executed in the order of the dependencies between the tasks. The Clustering Engine considers only the tasks that are free in the workflow engine and also should confirm the constraint required for clustering.

The waiting time required by the job in the execution engine is called as workflow engine delay. It is also given by the time required, when a job submitted from job scheduler till they are executed. The Workflow Engine takes care of the dependencies between the tasks and confirms whether the parent task has been completed before the child task are executed.

(c) Job Scheduler and Local Queue

Job scheduler directs the execution of job on the local and remote resources and also manages individual jobs in the workflow.

The time taken for the workflow engine to submit the job in the local queue and the time taken for the job scheduler to supervise the job running is stated as Queue delay. The queue delay provides efficiency of the job scheduler. The resource required for the execution of jobs in the workflow is an important factor for the calculation of queue delay.

(d) Clustering Engine

The system overhead is greatly reduced by the clustering of tasks based on several factors. This clustering of several tasks as jobs is done by the clustering engine. The job is considered as an entity where it has multiple tasks grouped together which can be executed either in a parallel or in sequential manner.

(e) Workflow Scheduler

Workflow scheduler schedules the job of the workflow to be executed to the worker node where the resources required for the job execution is allocated. The scheduling is done based on the conditions provided by the users.

(f) Failure Generator

It generates failures such as task failure, job failure and VM failure in the execution environment at different layers. The failures are generated after the job execution. The generation of failure is done using different mathematical distributions where the average failure rate considered is provided by the user.

(g) Failure Monitor

The monitor gathers the failure records of the job which include job id, task id and VM id or resource id while the execution of workflow is ongoing. These records are further transferred to the engine where the changes in the scheduling of the job are modified dynamically.

5 System Model

The cloud environment in our system model has a single data center that provides heterogeneous Virtual Machine (VM) resource types, $VT = vt1, vt2, ..., vt_m$ where m is the number of virtual machines Each VM type has a specific configuration with respect to memory, CPU. A static VM start up time is not assigned to all VMs, but is influenced by the start time of the task in the dataset.

6 Fault Tolerance Classification

A fault or failure can be either a hardware defect or a software or programming mistakes. The failure in workflow execution can occur for various reasons namely the variation in the configuration of execution environments, services non-availability or software components non-availability, system running out of memory, overloaded resource conditions and faults in computational and network components.

Based on fault tolerance policies various fault tolerance techniques can be used that can either be task level or workflow level. Reactive policies reduce the effect of failures after the fault has occurred and it does not stop the failure from occurring. There are different techniques are available for fault tolerance as shown in the Fig. 9.

(a) Task-level techniques cover the effects of the failure of tasks in the workflow execution. They are as follows-

(b) Retry - The task that are failed are retried in the same cloud resource. It is the simplest technique in task level.

(c) Check pointing/Restart - For long running applications, this method is most efficient task level fault tolerance technique. When a task fails, the system state can be rolled back to the checkpoint and allowed to continue from that point, rather restarting the application from the beginning.

(d) Replication - The entire workflow is replicated and different replicas are run on different resources till the workflow is completed even when some failures occurs during execution. Replication is a method of maintaining different copies of a data item or object on different resources. Replication adds redundancy in the system.

(e) Job Migration - When task failure occurs during execution, tasks can be migrated to another machine.

(f) Task Resubmission - Whenever a failed task is detected during execution of workflows, the task is resubmitted either to the same or to a different resource at runtime without interrupting the workflow of the system.

The workflow-level techniques deploy the workflow structure such as execution flow to deal with failures or flawed conditions.

(a) Alternate Task - In case of the task failure, different implementation of the task if the previous task failed in the execution of the workflow.

(b) Redundancy - Multiple alternative tasks are executed simultaneously in this technique.

(c) User defined exception handling - During task failure for workflows, the user specifies what treatment should be given for that particular task failure.

(d) Rescue workflow - This technique allows the workflow to proceed, if the task fails and is executed until it becomes impossible to move forward without catering the failed task.

7   Failure Model

To analyze the system overhead Gamma distribution and Weibull distribution are used. Two important parameters namely shape parameter and size parameter along with inter arrival time are considered.

The Gamma distribution is used for modeling the system overhead and task runtime whereas the Weibull distribution is responsible for inter arrival time. This distribution is used to find k where k is the number of tasks in a cluster is estimated using Maximum Likelihood Estimation (MLE).

The system overheads are mainly considered for execution since O-DAG is used. The overheads along with task failures are used. The static execution environment uses the fixed parameter values along with fixed overhead values. In dynamic execution the parameters for calculating MLE is not fixed, because the values are calculated at runtime providing the dynamic execution of the workflows.

This model mainly overlooks the optimal clustering size k* for execution and it is calculated using MLE. The inter-arrival time of the failures is a main factor to calculate k*. The overhead includes system overhead like queue delay, data transfer delay and so on (Fig. 1).

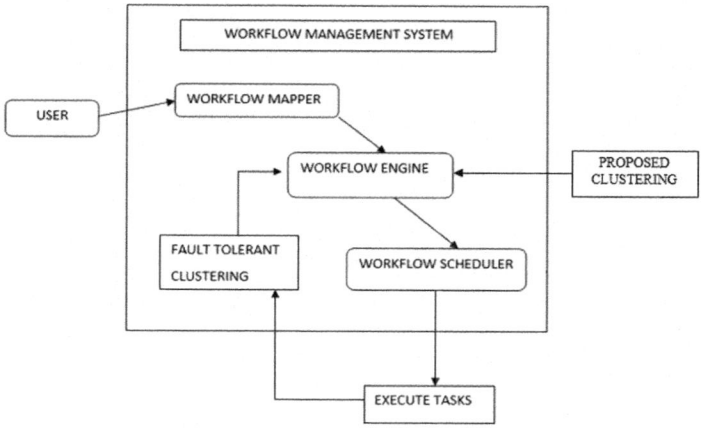

Fig. 1. Architecture of proposed work

Task clustering techniques may be categorized into two broad sorts namely Horizontal and Vertical Clustering. Horizontal clustering tasks are merged at the same horizontal level of the workflow graph, where level or depth of a task t is defined as the longest distance from the entry tasks to task t. Vertical clustering, on the other hand, tasks at sequential levels which have a single parent and single child relationship are merged. Dynamic Clustering (DC) adjusts the clustering factor according to the task failure rates measured from jobs that have already been completed, either successfully or failed which is usually the average value of task failure records.

8   Fault Tolerance Clustering Algorithms

Selective Reclustering (SR) selects the tasks that are failed during execution in a clustered job and clusters them into a new clustered job or considers them as individual jobs. SR involves collecting the task ids in failure records.

Selective Reclustering

Input: W: workflow; N: max number of tasks in a job
Output: Set of clustered tasks executed along with their execution time
Procedure SELECTIVE_RE_CLUSTERING (Jc)
TL←GETTASKS(Jc)
Jnew ←{}
for all tasks t in TL do
if t is failed then
Jnew .add(t)
end if
end for
Wf←Wf+Jnew
end procedure

In Dynamic Reclustering (DR), both the algorithms SR and DC are combined together. This algorithm is used to check whether using both strategies can improve workflow performance. Only tasks which are failed during execution are included into new clustered jobs in DR and the clustering factor k, is also adjusted according to the detected task failure rates.

Dynamic Reclustering

Input: W: workflow; N: max number of tasks in a job

Output: Set of clustered tasks executed along with their execution time

Procedure DYNAMIC_RE_CLUSTERING (J_c)

TL←GETTASKS(J_c)

J_{new} ← {}

for all tasks t in TL do

if t is failed then

J_{new} .add(t)

end if

if J_{new} .size()>k* then

Wf←Wf+J_{new}

J_{new} • {}

end if

end for

Wf←Wf+J_{new}

end procedure

The above algorithm uses retrying technique of fault tolerance mechanisms that are performed during the execution of scientific workflows.

The proposed Hybrid clustering algorithm performs horizontal clustering and also forms clusters when there is a one parent one child relationship between the tasks. Thus when the tasks that have one parent one child relationship are clustered together and executed in the same VM. This algorithm also reduces the resource utilization of VM, where in Horizontal Clustering the tasks with the one parent one child relationship is executed as separate tasks with separate VM allocated for them.

Hybrid clustering Algorithm

Input: : Wf: workflow; N: max number of tasks in a job
Output: Set of clustered tasks executed along with their execution time
Procedure HYBRID_CLUSTERING (Wf,N)
for level <Dep(Wf) do
TL←TASKS_IN_LEVEL(Wf,level)
CL←MERGE(TL,N)
Wf←Wf-TL+CL
end for
end procedure
Procedure Merge(TL,N)
J_c←{}
CL←{}
While TL is not null do
J_c.add(TL.pop(N))
CL.add(J_c)
for all t in TL do
J_c←{t}
while t has only 1 child tchild and tchild has only1 parent do
J_c.add (tchild)
TLmerged← TLmerged + tchild
t←{} tchild
end while
CL.add(J_c)
end for
End while
Return (CL,TLmerged)
end procedure
Procedure RE_CLUSTERING (J_c)
J_{new}←CopyOf J_c
Wf←Wf+J_{new}
end procedure

---

9   Experiments and Results

The code is executed in Workflowsim simulation tool. The fault tolerance algorithm namely horizontal clustering, selective reclustering and dynamic reclustering are executed. The dataset considered is Montage workflow with 25 tasks which is a DAG file in XML format. The factor k is the clustering size of tasks t in workflow execution. Along with 25 tasks, 50, 100, 1000 tasks are considered as datasets. Both clustering algorithms with various cluster sizes namely 2, 3, 5 are considered with the datasets containing different number of tasks are simulated 20 times.

The workflow execution using SR algorithm, here the cluster size does not change itself to produce optimized results. The above Fig. 10 shows the montage workflow execution with cluster size 3 where the fault tolerance occurs in horizontal clustering which is followed by SR algorithm respectively. The following Fig. 2 shows the same dataset execution with cluster size k = 5 using SR algorithm.

```
========== OUTPUT ==========
Cloudlet ID    STATUS    Data center ID    VM ID    Time    Start Time    Finish Time    Depth
    13        SUCCESS          2             0       0.11      14.33         14.44          0
     1        SUCCESS          2             0      27.43      26.37         53.8           1
     0        SUCCESS          2             1      41.23      41.58         82.81          1
     3        SUCCESS          2             2      34.43      98.64        133.07          2
     4        SUCCESS          2             1      33.18      90.58        123.76          2
     2        SUCCESS          2             0      33.04     104.42        137.46          2
     5        SUCCESS          2             0       0.91     151.7         152.61          3
     6        SUCCESS          2             0       1.54     155.47        157.01          4
     8        SUCCESS          2             0      22.43     165           187.44          5
     7        FAILED           2             1      32.39     177.46        209.85          5
    14        SUCCESS          2             0      10.93     232.56        243.49          5
     9        SUCCESS          2             0       1.63     257.6         259.23          6
    10        SUCCESS          2             0       3.18     289.08        292.26          7
    11        SUCCESS          2             0       3.89     320.5         324.38          8
    12        SUCCESS          2             0       4.57     334           338.57          9
BUILD SUCCESSFUL (total time: 1 second)
```

Fig. 2.   Execution of workflow with 25 nodes

10   Horizontal Clustering Results

The following results are for horizontal clustering with SR and DR algorithms. The montage workflow is considered with different set of tasks with 25, 50, 100, 1000 tasks and are executed using horizontal clustering and for fault tolerance clustering SR algorithm is used.

Table 1.   SR in horizontal clustering

Montage workflow	Cluster size_2	Cluster size_3	Cluster size_5
25 tasks	405.1615	428.5755	572.199
50 tasks	515.378	609.399	677.671
100 tasks	781.6425	873.3045	939.918
1000 tasks	3580.583	3054.619	3047.884

The Table 1 shows the execution of montage workflow with different clustering sizes.

Table 2. DR in horizontal clustering

Montage workflow	Cluster size_2	Cluster size_3	Cluster size_5
25 tasks	347.419	378.945	452.657
50 tasks	492.068	558.84	557.967
100 tasks	782.442	705.725	672.767
1000 tasks	3481.6	3096.93	2529.84

The Table 2 shows the execution of montage workflow with different clustering sizes with k value 2, 3, 5 clusters using DR method.

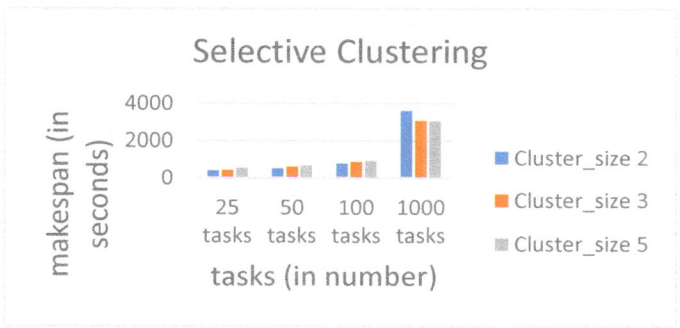

Fig. 3. Graph for SR in horizontal clustering

The graphical representation of the performance of SR clustering, when executing montage workflow as fault tolerant technique while performing horizontal clustering is shown in Figs. 3 and 4

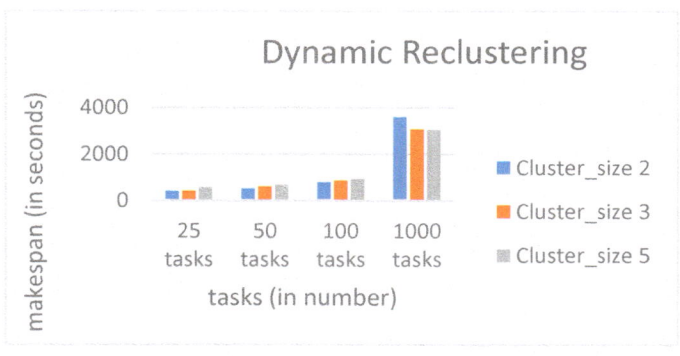

Fig. 4. Graph for DR in horizontal clustering

The graphical representation of the performance of SR clustering, when executing montage workflow as fault tolerant technique while performing hybrid clustering is shown in Fig. 6.

11   Hybrid Clustering Results

The following results are for proposed hybrid clustering with SR and DR algorithms. The same montage workflow dataset is considered with different set of tasks with 25, 50, 100, 1000 tasks and are executed using horizontal clustering and for fault tolerance clustering SR algorithm is used. The Table 3 shows the execution of montage work-flow with different clustering **sizes**.

Table 3. SR in hybrid clustering

Montage workflow	Cluster size_2	Cluster size_3	Cluster size_5
25 tasks	114.831	144.13	213.796
50 tasks	123.147	166.402	231.69
100 tasks	176.348	217.248	306.96
1000 tasks	1487.55	1829.91	2254.309

The Table 4 shows the execution of montage workflow with different clustering sizes with k value 2, 3, 5 clusters. It shows the performance of Hybrid clustering with DR algorithm.

Table 4. DR in hybrid clustering

Montage workflow	Cluster size_2	Cluster size_3	Cluster size_5
25 tasks	143.19	194.04	348.45
50 tasks	159.86	211.99	496.77
100 tasks	247.28	350.9	651.52
1000 tasks	3,067.59	3,556.20	3,299.64

The graphical representation of the performance of SR clustering, when executing montage workflow as fault tolerant technique while performing hybrid clustering is shown in Fig. 5.

The graphical representation of the performance of SR clustering, when executing montage workflow as fault tolerant technique while performing hybrid clustering is shown in Fig. 6.

The Performance of the horizontal clustering algorithm with SR algorithm and DR algorithm for Montage workflow with different tasks are discussed. The proposed hybrid clustering algorithm with SR and DR algorithm are also discussed. In Our proposed work the timespan is reduced considerably and the number of VM required for workflow execution is also reduced.

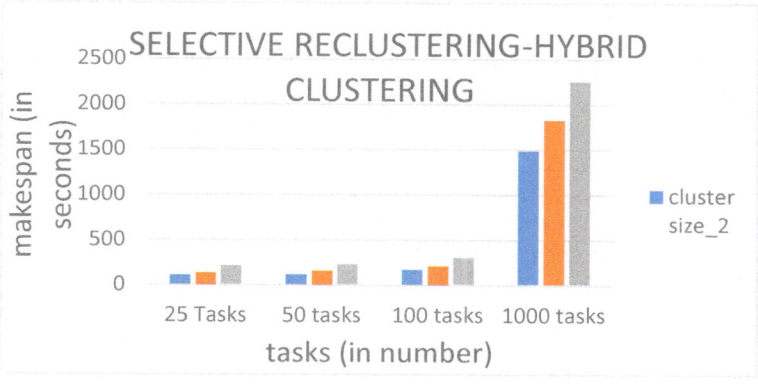

Fig. 5. Graph for SR in hybrid clustering

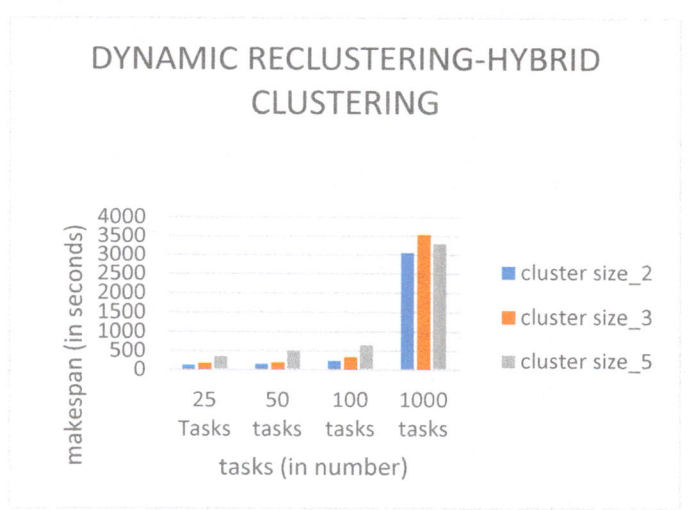

Fig. 6. Graph for DR in hybrid clustering

12   Conclusion and Future Enhancement

Clustering of scientific workflows is efficient method of executing the workflow. The fault tolerant clustering only includes transient failures particularly task failure. Along with the transient failure, the hardware failure VM failure is also included. The hardware failure disrupts the workflow scheduling execution. As future work, different machine learning techniques can be applied for clustering the tasks in workflow and different fault tolerance techniques in task level such as checkpoint, replication and other techniques can be employed for improvement in execution level of workflows.

References

1. Chen, W., Da Silva, R.F., Deelman, E., Fahringer, T.: Dynamic And fault-tolerant clustering for scientific workflows. IEEE Trans. Cloud Comput. **4**(1), 49–62 (2016)
2. Chen, W., Deelman, E.: Fault tolerant clustering in scientific workflows. In: IEEE Eighth World Congress on Services, pp. 10–16 (2012)
3. Chen, W., Da Silva, R.F., Deelman, E., Sakellariou, R.: Balanced task clustering in scientific workflows. In: IEEE 9th International Conference on e-Science, pp. 188–195 (2013)
4. Sahni, J., Vidyarthi, D.P.: Workflow-and-Platform Aware task clustering for scientific workflow execution in Cloud environment. Elsevier Future Gener. Comput. Syst. **64**, 61–74 (2016)
5. Bala, A., Chana, I.: Intelligent failure prediction models for scientific workflows. Elsevier Expert Syst. Appl. **42**, 980–989 (2015)
6. Juve, G., Chervenak, A., Deelman, E., Bharathi, S., Mehta, G., Vahi, K.: Characterizing and profiling scientific workflows. Future Gener. Comput. Syst. **29**, 682–692 (2013)
7. Calheiros, R.N., Buyya, R.: Meeting deadlines of scientific workflows in public clouds with tasks replication. IEEE Trans. Parallel Distrib. Syst. **25**(7), 1787–1796 (2014)
8. Smanchat, S., Viriyapant, K.: Taxonomies of workflow scheduling problem and techniques in the cloud. Future Gener. Comput. Syst. **52**, 1–12 (2015)
9. Mattoso, M., Dias, J., Ocana, K.A.C.S., Ogasawara, E., Costa, F., Horta, F., Silva, V., de Oliveira, D.: Dynamic steering of HPC scientific workflows: a survey. Future Gener. Comput. Syst. **46**, 100–113 (2015)
10. Chen, W., Deelman, E.: WorkflowSim: a toolkit for simulating scientific workflows in distributed environments. In: IEEE 8th International Conference on E-Science, pp. 1–8 (2012)
11. Calheiros, R.N., Ranjan, R., Beloglazov, A., De Rose, C.A.F., Buyya, R.: CloudSim: a toolkit for modeling and simulation of cloud computing environments and evaluation of resource provisioning algorithms. Softw.: Pract. Experience **41**(1), 23–50 (2011)
12. Deelman, E., Vahi, K., Juve, G., Rynge, M., Callaghan, S., Maechling, P.J., Mayani, R., Chen, W., da Silva, R.F., Livny, M., Wenger, K.: Pegasus, a workflow management system for science automation. Future Gener. Comput. Syst. **46**, 17–35 (2015)
13. Gu, Y., Wu, C.Q.: Performance analysis and optimization of distributed workflows in heterogeneous network environments. IEEE Trans. Comput. **65**(4), 1266–1282 (2016)
14. Rimal, B.P., Maier, M.: Workflow scheduling in multi-tenant cloud computing environments. IEEE Trans. Parallel Distrib. Syst. **28**(1), 290–304 (2017)
15. Zhang, F., Cao, J., Hwang, K., Li, K., Khan, S.U.: Adaptive workflow scheduling on cloud computing platforms with iterative ordinal optimization. IEEE Trans. Cloud Comput. **3**(2), 156–169 (2015)
16. Fard, H.M., Prodan, R., Fahringer, T.: A truthful dynamic workflow scheduling mechanism for commercial multicloud environments. IEEE Trans. Parallel Distrib. Syst. **24**(6), 1203–1213 (2013)
17. Zhao, J., Yang, K., Wei, X., Ding, Y., Hu, L., Xu, G.: A heuristic clustering-based task deployment approach for load balancing using Bayes theorem in cloud environment. IEEE Trans. Parallel Distrib. Syst. **27**(2), 305–317 (2016)
18. Zotkiewicz, M., Guzek, M., Kliazovich, D., Bouvry, P.: Minimum dependencies energy-efficient scheduling in data centers. IEEE Trans. Parallel Distrib. Syst. **27**(12), 3561–3575 (2016)

19. Zhu, X., Wang, J., Guo, H., Zhu, D., Yang, L.T., Liu, L.: Fault-tolerant scheduling for real-time scientific workflows with elastic resource provisioning in virtualized clouds. IEEE Trans. Parallel Distrib. Syst. **27**(12), 3501–3517 (2016)
20. Zhao, Y., Li, Y., Raicu, I., Lu, S., Tian, W., Liu, H.: Enabling scalable scientific workflow management in the cloud. Future Gener. Comput. Syst. **46**, 3–16 (2015)

IoT Based Air Quality Monitoring Systems - A Survey

Sumi Neogi, Yugchhaya Galphat, Pritesh Narkar$^{(\boxtimes)}$,
Harsha Punjabi, and Manohar Jethani

Computer Engineering Department, VES Institute of Technology, Mumbai, India
{2015sumi.neogi,yugchhaya.dhote,2015pritesh.narkar,
2015harsha.punjabi,2015manohar.jethani}@ves.ac.in

Abstract. With an increasing level of industrialization, air pollution has become one of the major problems all over the world. The quality of air in the atmosphere is becoming progressively worse due to the emission of harmful gases and other pollutants. With many small, medium and large industries coming up, air pollution has disturbed the entire ecological system and affected lives of humans as well as plants and animals. This creates a need for real-time air quality monitoring systems for micro, small and medium industries so that timely decisions can be taken to avoid environmental degradation. IoT has been proven one of the effective ways for such systems and when merged with cloud computing provides a revolutionary method of management and analysis of data coming from sensors. In this paper, we have done a comparative study of all the existing implementations and various features of the system have been documented.

Keywords: Raspberry Pi · Gas sensors · IoT (Internet of Things)

1 Introduction

Industrial pollution is one of the primary sources of environmental degradation. With the rapid development of the economy and chemical industrial park construction, production activities are increasing frequently leading to an increasing probability of environmental pollution [10]. According to the U.S. Environmental Protection Agency (EPA), 79 million tons of pollution were emitted into the atmosphere in the United States. Also between 1980 to 2016, CO2 emissions increased by 12%. The main countries where worst cases of pollution exist are China, India, Mexico, Indonesia, South Africa, Brazil, and Argentina. According to US EPA, 6 common air pollutants are Particulate Matter (PM), ground-level ozone, Carbon Monoxide (CO), sulfur oxides, nitrogen oxides and lead [1]. Deaths due to air pollution are about 65% in Asia and 25% in India. World Health Organization stated in a report that 9 out of 10 people on the earth breathe contaminated air [12]. In India, most of the industrial regions are

© Springer Nature Switzerland AG 2020
A. P. Pandian et al. (Eds.): ICCBI 2018, LNDECT 31, pp. 752–758, 2020.
https://doi.org/10.1007/978-3-030-24643-3_89

facing poor air quality which adversely affects the health of workers as well as the people staying in the vicinity. The situation is becoming worse day by day. Our capital has been worse hit by air pollution, with 2016 winter recording some of the highest numbers ever [4]. Hence, there is a growing demand for environmental pollution monitoring systems capable of efficiently measuring and analyzing contaminants in the air [11]. Currently, there are a few monitoring systems for air pollution in the world. Though they have improved certain things can be enhanced with time. The most common types are the use of IoT based systems which is gaining popularity over the years. The main component of air pollution monitoring system is the pollutant sensor [6]. The use of a large number of sensors ensures accuracy, reduces the cost and makes data more perfect and systematic for analysis [2]. Section 2 onwards, various implementations and related work have been discussed.

2   Related Work

2.1   IoT Based Air Pollution Monitoring and Predictor System on BeagleBone Black

This system was developed for air pollutants measurement and prediction of pollution level for a smart city. BeagleBone Black is a low cost, high expansion focussed BeagleBoard using a low-cost Sitara XAM3359AZCZ100 Cortex A8 ARM Processor. In this, python programming is used to configure GPIO pins of BeagleBone and also to enable the inbuilt ADC. The advantage of using BeagleBone is that it has inbuilt ADC and data from gas sensors is sensed with the help of inbuilt ADC. The proposed system consists of Beagle Bone connected with gas sensors to detect carbon dioxide, carbon monoxide, and noise. MQ-7 and MQ-11 were used to detect CO and H2 respectively. A GPS module is also used to take the current location and represent pollution concentration at various locations. Microsoft Azure has been used here as a cloud. Data from sensors are put to the cloud database using python SQL and Event hub cloud service [3]. Event hub is a big data streaming platform and event ingestion service capable of receiving and processing millions of events per second. A database is created in the BeagleBone itself in the form of a .csv file, saved for only 1 day and at the end of the day, it is uploaded to the cloud database. Microsoft's Azure Machine Learning Service is further used to predict pollution in the future.

2.2   Air Quality Monitoring System Based on IoT Using Raspberry Pi

The advantage of the system [7] is that it is low cost, low power compact and highly accurate for monitoring the environment. The proposed system can be improved by if pollutants like SO2, NO2 and ground-level ozone were also monitored. The future scope can be that the data obtained can be used to monitor the pattern in pollutants.

2.3   IoT Enabled Air Quality Monitoring Device - A Low-Cost Smart Health Solution

This was a prototype developed for a low-cost indoor air monitoring device. In this system, sensors for light, noxious gases, HCHO, CO2, and temperature were used. The data is transmitted using GSM cellular to the network. Marvell 88MW302 kit was used consisting of SoC which enables easy interfacing to sensors, actuators, and other components.

The data collected by the sensors were stored in the AWS cloud. The user needs user login and password to view the data. The prototype also included a Python Script to access the AWS data. Depending on the threshold limit a notification is sent to the Emergency number or Gmail account. The cost of the overall system was just $170 [1]. This prototype was tested over a couple of hour in Santa Clara. Low values were detected in the early morning and it increased as it enters afternoon.

2.4   Monitoring Pollution: Applying IoT to Create a Smart Environment

This system was proposed to protect the residents living near industrial areas. In this proposed system, the wireless system network consists of Waspmote microcontroller. Waspmote integrates more than 60 different sensors: CO, CO2, NO2, temperature, humidity, radiation, ultraviolet, soil moisture, accelerometer, liquid level etc. This system was proposed to protect the residents living near industrial areas. In this proposed system, the wireless system network consists of Waspmote microcontroller. Waspmote integrates more than 60 different sensors: CO, CO2, NO2, temperature, humidity, radiation, ultraviolet, soil moisture, accelerometer, liquid level etc.

Meshlium, a wireless gateway is used for storing data in the database. It is also used for communication with an access point, Wi-Fi, mobile system and XBee modules. They had used the threshold values for the safe concentration of gases in the environment which are the standards adopted by the UAE government [8]. The gas sensors were used to measure CO2, temperature and air pressure. The other pollutants were categorized into two groups: Air Pollutant 1 (AP1), which measured concentrations of C6H5OH, CH3CH2OH, NH3, H2S, and H2. And Air Pollutant 2 (AP2), which measured the concentrations of C4H10, CH3CH2OH, H2, CH4, and CO. The sensors values are then compared with the predetermined value for further analysis.

2.5   An IoT Based Low-Cost Air Pollution Monitoring System

This is a prototype of environmental air pollution monitoring system which monitors the concentration of major air pollutant gases. In this system, they have used Nucleo F401RE (a 32 bit ARM Microcontroller) with gas sensors which collects the data and sends it to Raspberry Pi 3. They have also used the ESP8266 Serial-to-WiFi module. The ESP8266 has full TCP/UDP stack support so that the data can be transferred as TCP packets from nodes to Raspberry Pi. Here, only two sensors are used MQ-7 for detecting CO and MQ-135 for detecting ammonia, CO2 etc. [9]. They had implemented a TCP server over Raspberry Pi using Node js in javascript. Mongo DB was installed in Raspberry Pi to store the data. Since the data is saved in the JSON format, it is easier to render the webpage with JSON formatted data. After collecting the data, the data is visualized through a MEAN (MongoDB, Express, AngularJS, Node.js) stack created by them. Nginx Web Server is also installed for added functionality.

2.6   Design of Air Quality Meter and Pollution Detector

Here, Arduino Uno R3 board was used since it simplifies the amount of hardware and software implementations. It also provides a wide number of libraries so as to make programming easier. Gas sensors used were DHT11 for ambient temperature and humidity, MQ-2, MQ-7, MQ-135 for detecting smoke, CO and NH3 respectively. The program is written based on C++ in the Arduino IDE. When the gas levels cross a certain level, the people in the vicinity are alerted through a Piezo buzzer. The transmission of data here takes place through Bluetooth since no internet is required. This is a very simple implementation but has a few disadvantages. The data collected from the sensors is transferred through Bluetooth to the serial monitor using the Bluetooth serial monitor app for monitoring [4].

2.7   Design of Air Quality Monitoring System Based on Internet of Things

This system is based on Zigbee, wireless technology for remote monitoring and control. It comprises the web browser, LM3S8962 gateway, and Zigbee coordinator. The temperature measurement is done using DS18B20 sensors using LM3S8962 internal ethernet controller as an ethernet controller. HR911105A is used as a network interface. The LCD5110 screen is the display module used. The module uses STC12C5608AD chip for data acquisition and wireless transmission. Temperature and humidity are measured using AM2305 sensors. PM 2.5 is measured using GP2Y1010AU0F. The gateway is responsible for the collection of temperature and voltage data, LCD control, real-time display, and network transmission. RJ45 network interface module is used which makes use of network interface HR911105A.

The gateway uses ZigBee RF2530 N. The main chip of Zigbee is CC2530. Zigbee gateway module coordinator sets more than one Zigbee router. Hence, the data from this model is aggregated. LCD module of the gateway used is Nokia 5110 LCD screen. To prevent the JTAG port multiplexing causes such as chip lock, one function needs to be called to prevent the failure of JTAG. Yeelink is an open source IoT platform providing access, storage and display of data of the sensors. The system uses the API provided by Yeelink to achieve the function of browsing data on the internet [5] (Table 1).

Table 1. Analysis of various air quality monitoring systems

Related work	Designed for	Approach used	Tools and cloud	Sensors used for	Results	Comments
A. IoT Based Air Pollution Monitoring and Predictor System on BeagleBone Black	Prediction of air pollution level in a smart city	Acquires levels of CO and CO_2 in the air along with GPS location	Python SQL, Event Hub API, Microsoft Azure Cloud	CO, CO_2, Humidity	CO: 35 to 800 ppm. CO_2: 350 to 5000 ppm	No long-term prediction since sensor data was stored for one day
B. Air Quality Monitoring System Based on IoT using Raspberry Pi	Monitoring of air pollution in an area where traffic is high at peak times	Data transmission to the Cloud through Raspberry Pi, Sensors connected to Arduino interfaced with Raspberry Pi	Arduino Board, Raspberry Pi, IBM Bluemix Cloud	PM2.5,CO, CO_2, Temperature, Humidity, Pressure	CO and CO_2: 10 to 10000 ppm. Temperature: 40 to 80 °C	No prediction involved
C. IoT Enabled Air Quality Monitoring Device -A Low-Cost Smart Health Solution	Monitoring indoor air pollution	Communication to the cloud when the threshold of pollutants was crossed	Marvell Board (88MW302), AWS Cloud	Light, Gas, CO_2, HCHO, Temperature	Light: 97–100 mV, HCHO: 321–1432 mV (later using equations it is converted to ppm)	A limited number of pollutants were measured
D. Monitoring Pollution: Applying IoT to Create a Smart Environment	To protect residents living near industrial areas	Smart wireless sensor network. Sensors were connected to the microcontroller, which in turn, were connected to the wireless communication gateway	Waspmote microcontroller, XBee module, the local database of the gateway	C6H5CH3, H2S, CH3CH2OH, NH3, H2, C4H10, CO, CH4	CO_2:350–10000 ppm. C6H5CH3, H2S, NH3:1–30 ppm. CH3CH2OH, C4H10, H2, CO, CH4: 1–100 ppm	Data stored for processing Only monitoring is done, no prevention

(continued)

Table 1. (*continued*)

Related work	Designed for	Approach used	Tools and cloud	Sensors used for	Results	Comments
E. An IOT Based Low-Cost Air Pollution Monitoring System	Monitors concentration of major air pollutant gases	Semiconductor sensor nodes are used with Wi-Fi modules, a base station used is Raspberry Pi	MongoDB	CO, CO_2, SO_2, NO_2	Gases: 229–238 ppm	A number of sensors could have been used to analyze pollutants
F. Design of Air Quality Meter and Pollution Detector	To test the air quality in an open environment and indoor	Arduino U3 board, Bluetooth Serial monitor app is used to transfer data	Bluetooth, no database was used	Temperature, Humidity, NH_3, Smoke, CO	MQ2 smoke: 100–10000 ppm, MQ7 CO: 10–10000 ppm MQ135, NH_3: 10 to 1000 ppm	No CO_2, NO_2 detection No pollution prediction
G. Design of Air Quality Monitoring System Based on Internet of Things	Remote monitoring and controlling the pollution level	ZigBee-wireless technology, ZigBee routers, ZigBee (LM3S8962) communication gateway module	A CC2530 chip of Zigbee, Yealink cloud	Particulate Matter (PM2.5), Temperature, Humidity	Only the pollution level was monitored, no actions were taken	Data is only displayed, No use of data for Prediction No prevention

References

1. Tapashetti, A., Vegiraju, D., Ogunfunmi, T.: IoT-enabled air quality monitoring device: a low-cost smart health solution. In: 2016 IEEE Global Humanitarian Technology Conference (GHTC), pp. 682–685. Seattle, WA (2016)
2. Khot, R., Chitre, V.: Survey on air pollution monitoring systems. In: 2017 International Conference on Innovations in Information, Embedded and Communication Systems (CICS), pp. 1–4. Coimbatore (2017)
3. Desai, N.S., Alex, J.S.R.: IoT based air pollution monitoring and predictor system on Beagle Bone Black. In: 2017 International Conference on Nextgen Electronic Technologies: Silicon to Software (ICNETS2), pp. 367–370. Chennai (2017)
4. Mukhopadhyay, A., et al.: Design of air quality meter and pollution detector. In: 2017 8th Annual Industrial Automation and Electromechanical Engineering Conference (SEMICON), pp. 381–384. Bangkok (2017)
5. Wang, D., Jiang, C., Dan, Y.: Design of air quality monitoring system based on internet of things. In: 2016 10th International Conference on Software, Knowledge, Information Management & Applications (SKIMA), pp. 418–423. Chengdu (2016)
6. Budiarto, A., Febriana, T.: IoT device used for air pollution campaign to encourage cycling habit in inverleith neighborhood. In: 2017 International Conference on Information Management and Technology (Imtech), pp. 356–360. Yogyakarta (2017)
7. Kumar, S., Jasuja, A.: Air quality monitoring system based on IoT using Raspberry Pi. In: 2017 International Conference on Computing, Communication and Automation (ICCCA), pp. 1341–1346. Greater Noida (2017)
8. Alshamsi, A., Anwar, Y., Almulla, M., Aldhoori, M., Hamad, N., Awad, M.: Monitoring pollution: applying IoT to create a smart environment. In: 2017 International Conference on Electrical and Computing Technologies and Applications (ICECTA), pp. 1–4. Ras Al Khaimah (2017)
9. Parmar, G., Lakhani, S., Chattopadhyay, M.K.: An IoT based low-cost air pollution monitoring system. In: 2017 International Conference on Recent Innovations in Signal Processing and Embedded Systems (RISE), pp. 524–528. Bhopal (2017)
10. Xiaojun, C., Xianpeng, L., Peng, X.: IoT-based air pollution monitoring and forecasting system. In: 2015 International Conference on Computer and Computational Sciences (ICCCS), pp. 257–260. Noida (2015)
11. Kim, S.H., Jeong, J.M., Hwang, M.T., Kang, C.S.: Development of an IoT-based atmospheric environment monitoring system. In: 2017 International Conference on Information and Communication Technology Convergence (ICTC), pp. 861–863. Jeju (2017)
12. Gokul, V., Tadepalli, S.: Implementation of a WiFi based plug and sense device for dedicated air pollution monitoring using IoT. In: 2016 Online International Conference on Green Engineering and Technologies (IC-GET), pp. 1–7. Coimbatore (2016)

Cognitive Radio: An NN-PSO Approach Network Selection and Fast Delivery Handover Route 5G LTE Network

Shipra Mishra[✉] and Swapnil Nema

Department of Electronics and Communication, Global Nature Care Sangathan's
Group of Institutions, Jabalpur, India
mishipra2110@gmail.com, swapnil.nema27@gmail.com

Abstract. To move from 4G to 5G technology spectrum deficiency is an important criteria which can be overcome by an efficient technology called Cognitive Radio (CR). This can be achieved by continuous sensing the spectrum band, and detecting the unused frequency bands which would be not used by licensed band and without any unwanted interference to the primary user or licensed user (PU). The demand of spectrum resources has been increased greatly in modern wireless communication system. Network Selection is one of the difficult mechanisms when moving from 4G to 5G technology to address spectrum shortage issue and by achieve Fast delivery handover route with high speed data rate access, and also maintain the QOS. We Proposed NN-PSO method for network selection and Fast Delivery handover route mechanism in order to increase system efficiency by consider more number of SUs in network and providing to their preferences, and respect the criteria of primary network operators, at the same time. An optimization used to reduced unwanted interference suffer by licensed users (PU), due to presence of SU's and also the subscription fees that SUs have to pay for using the licensed network or band (PN). The goal is to provide SUs good network and fast delivery handover route with a high QoS based on the criteria of SU, subject to the interference boundation of each existing network with other channels. The proposed technique Neural Network and Particle swarm optimization would be use to solve the Network selection and optimization problem. Finally, experimental results and numerical parameters represent the effectiveness of the proposed NN-PSO methods to finding a near-optimal solution for network selection.

Keywords: Network selection · Fast delivery route · 5G LTE networks · Particle swarm optimization · Cognitive radio · Neural network

1 Introduction

The ITU report of 2017 represents that 70% of world youth using Internet. The report of ITU's represents that in the past five years, 20% mobile user have added in each year [32]. There would be need of spectrum for various applications in Wireless Communication and also exploit available spectrum in good way. A basic spectrum management technique allows only legal (licensed) users to occupy the available spectrum for

© Springer Nature Switzerland AG 2020
A. P. Pandian et al. (Eds.): ICCBI 2018, LNDECT 31, pp. 759–770, 2020.
https://doi.org/10.1007/978-3-030-24643-3_90

given span of time, but the spectrum scarcity problem led FCC (Federal Communications Commission) to allow unlicensed user for accessing the licensed bands with non-interference approach [1]. In the past years, the demand of mobile data increased 20% annually due to increasing mobile (cell phone) users and various applications. This is evaluate that the data traffic will increase 10 times between 2014 and 2019 and result would be collision of data. Network Selection is based on Resource allocation which is basic criteria for selecting network; and Resource allocation is an important parameter for sharing spectrum in cognitive radio. In which SU (unlicensed user) selected their parameters such as network selection, BW, cost and so on, according to their demands. This huge requirement of mobile data eventuate several challenges which move the research in 5G networks [2]. Here, 5G LTE networks are used to providing significantly high speed with QOS. So, the requirement of spectrum resources is predicted to increase importantly in 5G LTE networks. This need wireless structure designers to suggest an efficient and effective spectrum management technique for improving QOS. Different point of views on 5G architecture are resented in [3, 5] with fundamental technologies such as massive MIMO, micro cells, energy efficient communications, CR etc. The 5G networks will mainly contain of network densification (the act of becoming more dense), i.e., densification over space and frequency. To enhance the available capacity of spectrum band by adding more cell sites, called network densification. Cell sites is mostly positioned in capacity-strained areas for adding more capacity where it is most needed and would be help offload traffic from surrounding sites. There are two terms defined as; Densification over space: defined closed connection of n number of small cell. Densification over frequency: utilizing radio frequency or spectrum in different bands [7]. However, in wireless communication, self-fish spectrum sharing plays an important role in order to achieving hard goals for 5G environment. Spectrum would be also divided between n numbers of devices present in network with multiple applications and also ensures the coverage of 5G networks all over [8]. In addition, this is spectrum efficient as it can use all non-communicated spectrum, can achieve better system capacity, minimum energy consumption, and enhance cell efficiency. Dynamic spectrum access (DSA) has appeared as key for spectrum sharing in an opportunistic way [9, 10]. A cognitive radio network (CRN) used Dynamic Spectrum Access to coexist with a licensed network (primary network) for ease of spectrum allocation [11]. A users have licensed of primary networks (PN) are known as primary users (PUs). A user which can operate licensed band (PU band) with the permission of licensed user (PU) and respect the criteria of PU's, called Secondary user (SU's) or unlicensed user. Unlicensed Spectrum is the open frequency bands to be utilized by an unlimited number of users. There is multiple primary networks incorporates with 5G network. so, there may be possible, that SUs have a problem to selecting one of the primary network for network selection problem with characteristics such as bandwidth, price, and capacity [12, 13]. Most of the time, for accessing the spectrum, Price is a main criterion for selecting any PU network with specific operator. In [14], represents an approach management of interference in cognitive femto cell through joint price based spectrum sharing and power allocation mechanism. In [15] assume that, there is channel between PU (licensed user) and SUs (unlicensed user) and spectrum sharing algorithm is presented price based scheme, through the particle swarm optimization (PSO). The difficulty of sub-carrier sharing through discrete rate allocation to solve price-based spectrum sharing and rate allocation method [16]. The network selection problem solved through PSO technique [17]. But, there is problem of unmanageable nature of the network selection having large processing time in 5G networks, so, it is necessary to introduce better algorithms for determine the network selection difficulties. So that, to improve the previous work, it is necessary to study network selection problem.

This paper represents Related work in Sect. 2, system modal which represent working modal in Sect. 3, proposed work of PSO Intercerebral Neural Network in Sect. 4. Implementation detail and experimental result in Sect. 5; and at last, overall conclusion of work in Sect. 6.

2 Related Work

Andreev [13] represents network selection in 2014 using naturalistic approaches and focused on both network and user centric. An another effective algorithm for selection of network for 5G heterogeneous networks is suggested by Orsino and Araniti [19] in 2015 to an efficiently selected the network with high data rate and user performance. This network is selected according to parameters which consider various measurements related with users present in system, base station, transmitted power, traffic load, and spectral capability. To investigate and evaluate network selection scenario by used uniform framework is presented in [20] with discussing several existing techniques and proposed a gradient-based approach for optimal network selection. Ju and Kang [21] studied for increasing the mutual information between secondary users by availability of the primary service and proposed a single network selection approach in CR. Different way to selection of Spectrum band address in [12], for select a spectrum band in such a way, which have least interference power to licensed band. While satisfying the restriction on delay, and developed a technique by increase transmit power allocation, increase effective capacity with both band selection criteria. Network would also be selected by using a game-theoretic framework which is proposed by Tseng [6] where unlicensed users are players in game and non-cooperative game (where unlicensed user play individual or selfish) is formulated by the network selection problem and a decentralized stochastic learning-based technique was suggested to solving this problem. Author [23] proposed a cross-layer framework and increased efficiency of SUs for considered as Markov decision process, by mutually considering physical layer adaptive modulation, spectrum sensing, processing, quick decision making and coding technique, and frame size scenario. The jointly PSO and genetic algorithm (GA) are used for solving the Optimization problem of 5G heterogeneous network proposed by using price based scheme [1] in 2016.

In short, the above network selection techniques are not suitable for realistic case for good network selection and finding free route where speed, data rate and subscription fee are considered.

3   System Modal

The Basic working structure of network selection of cognitive radio represents in figure. In which, 5G LTE network consisting N number of primary networks, where each primary network consisting m number of secondary users (SUs). Each primary network has maximum number of channels. These channels are available for SU communication based on the PU criteria. First, observed the parameters. Then, the process start with initializing parameters such as, fitness function, population size, initial weight, crossover rate, mutation rate and so on. After that, we applied proposed algorithm Particle Swarm Optimized Intercerebral Neural Network for best network selection with high QOS, by taking 64 iteration in each loop. After that, parameters would be calculating such as mean error, standard error, mean Epoch, standard Epoch, mean weight and standard weight. Finally, finding free route, and selecting network for fast delivery handover route, used for data transferring (Fig. 1).

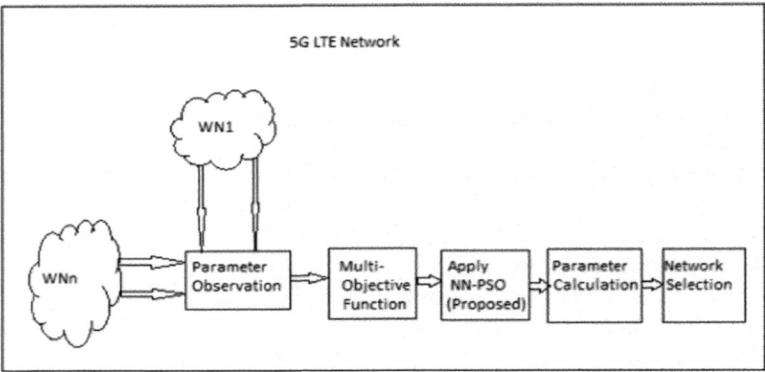

Fig. 1. Basic working structure for network selection and fast delivery Route by using NN-PSO.

4   Proposed Work

4.1   Neural Network

Neural Network is a network for computation and an intelligently compute system; by consisting multiple connected processing elements (nodes) in wireless communication. It is motivated by the human nervous systems. Where neurons represent as nodes and nervous systems are represented as network associated with number of nodes, having an activation functions. Neural networks have different classes, which would be obtained by changing the activation functions of the neurons and the structure of the weighted (w) interconnections between them. Neural Network having one or more layers of nodes (neurons). This is created for the interaction of multiple connected processing elements. A Neural Network can perform any tasks in system either linear network or non-linear network. The desired linear connection between the input and the output represented by approximated straight line, which trained by neural network. In each iteration, the square of the error (also called mean square error) is measured between the actual and desired output, during the training; and applied to a weight algorithm which is updating after each iteration [28]. Due to their parallel structure where each node connected with each other, neural network can continue processing without any problem even when some elements of neural network not work (or failed). Main element (or part) of the neural network is sensor nodes; that's by NN is also referred as sensor network. Data communication between WSN must be an efficient and sensor node must be absorbed less power during processing [29]. In Every sensor node implant multiple sensors; that's by every sensor node is behaving as source of data. In wireless communication, it is not possible to communicate BS by using raw data. Hence, these sensor data streams necessary to be classified. In neural network, a cluster (which is hexagonal shape) also formed by a group of sensor nodes. Each and every node transfer it's own data to a cluster head and then cluster head collect these raw data and sum the data and sends to base station. ART1 and Fuzzy ART categorized under unsupervised neural network models which used for characterized sensor data [29].

- ART1 model: Binary (machine) valued input data.
- Fuzzy ART model: analog data, in which the input data is fuzzy valued.

Another approach using ANNs was developed in [30], and also uses RSSI as the network inputs. The basic Neural Network Neurons work as Received n-inputs, than, multiply each input by it's weight and Apply activation function by sum of results and Output Results.

In this paper, algorithms running many times to train the neural network. We consider hidden layer size of Neural Network is four with size of group 20 and minimum error is 0.01. By taking maximum 64 iteration. The benefits that Neural Network improves their performance by learn from experience and also can deal with incomplete information.

4.2 Particle Swarm Optimization(PSO)

An intelligent optimization which based on population or the moving of particle in swarm is called Particle swarm optimization (PSO). In which, multiple agent (particle) solution coexist and collaborate simultaneously. It was developed in 1995 by Kennedy and Eberhart [31], inspired by the cooperative and social behavior between various particles in swarm [1]. It having multiple agents (*kth* particles) which form a swarm and moving around in search space for searching best result. The location of each and every particle which shows a solution is represented in a vector. The PSO algorithmic procedure starts with an initialize parameter of n random number of particles. Every particle allots with a random position (X_k) and random velocity (V_K) in the search space. Inertia weight (w) of PSO shows,

- Big value w-higher global search ability
- Lower value w-higher local search ability.

PSO is a developmental algorithm. At each iteration, every particle is update or change its velocity (V_k^{new}) and position (X_k^{new}) by using Eqs. (1) and (2). Particle place or position in a swarm found through the Fitness function. It describes best position (called *pbest*) and quality of each particle [25]. The velocity of each particle is compaired continuously with its current value and previous value, until reach the optimum value. If previous value is better than present value than it changed its value with best previous position (*Pbest*) that, best position found by particle itself (fitness value).

- Global best (*gbest*): When every particles in a swarm are considered as best to one particle than, that particle called "*gbest*".
- Local best (*lbest*): if some particle (i.e. only particle in a group) considered as best neighbor of one particle than, that particle called "*lbest*".

PSO used for channel selection in CRN other than using other stochastic algorithms- The PSO has huge advantages compared to others that make using it very attractive:

- Easy to implement.
- Requires a few parameters needed to be adjusted by the user.
- It provides fast convergence and this will save the spectrum sensing process.
- It provides high accuracy.
- It requires less computational time compared to others heuristics.

Suppose, X_k represent the position of the *kth* particle and V_K represent the velocity of *kth* particles. In [26], the author mentioned that they are updated as

$$V_k^{new} = W * V_K + C_1 r_1 (pbest_k - X_k) + C_2 (nbest_k - X_k) \tag{1}$$

$$X_k^{new} = X_k * V_K^{new} \tag{2}$$

Where,

W = inertial weight.
n = number of particles in the swarm.
$c1$ & $c2$ = acceleration constants.
$r1$ & $r2$ = random no. distributed in [0, 1].

In which, W, $c1$, and $c2$ represents the changed of their own previous velocities, personal best position, and its neighbor's best position by the new velocity, respectively. At last, the swarm will converge to the optimal location, as it is driven by individual particle experience and global experience [27].

The proposed 5G LTE network, where, multiple primary networks present and cooperating each other. Where each primary network has a different number of channels with a different capacity are available for the SUs. Each network has some limitations in terms of interference, subscription cost, and capacity. SUs specify their conditions and demands in terms of the maximum data rate and the minimum subscription fee that they are capable to pay. The proposed PSO and NN algorithms use to find a closed-optimal solution based on SU requirements. The goal of this paper is to find best network, free route and select network such that the overall cost for all SUs is minimized and the overall interference acquired to the PUs is also reduced of different primary networks and provide more number of SU's in channel. Our approach different from previous work [1] in the following ways:

- NN-PSO Proposed method requires minimum number of iteration.
- Population size would be increase.
- And best network connection providing with fast deliver handover route.

5   Implementation Details and Experimental Result

5.1   Implementation Details

Simulations are carried out in MATLAB R2018a environment Windows 10 software based computer with 2.5 GHz CPU, for the network selection in 5G LTE network in order to estimate the quality of the solution and the convergence velocity of both NN and PSO algorithms. An Execution of both NN and PSO depends on the selected parameters. After a previous set of analysis, the parameters selected for simulations of NN-PSO algorithms, are given in Table 1 and also compaired with previous consider parameters [1] works.

Table 1. Implementation parameter compared with existing work.

Sr. No.	Parameters	Exiting work	Proposed work	Remark (How much increase or decrease)
1	Number of iteration	3000	64	2936 Decreased
2	Population size (proposed method)	12	30	18 Increased
3	Crossover rate	0.5	0.4	0.1 Decreased
4	Mutation rate	0.03	0.02	0.01 Decreased

5.2   Experimental Result-

Figure 2(a) represents the Number of iteration v/s Fitness Value proceeds by NN-PSO network selection approach iteration proceeds at each loop. Figure 2(b) represent Link parameter v/s Decision Weight where After each looped, signal and decision weight generated as represented in Table 2. Weight before represents initialize random value and each weight after value compaired with weight after values. Key parameters would be Received signal strength Indicator (RSSI), Loss, BW, speed, RTT, Cost.

Figure 3 represents Mean training error and standard Weight deviation. For accurate analysis, the difference between mean to standard deviation must be minimum; ideally it should be 'zero'. We consider expected minimum error 0.01 in System. Calculating standard deviations which show optimum (close) value within one standard

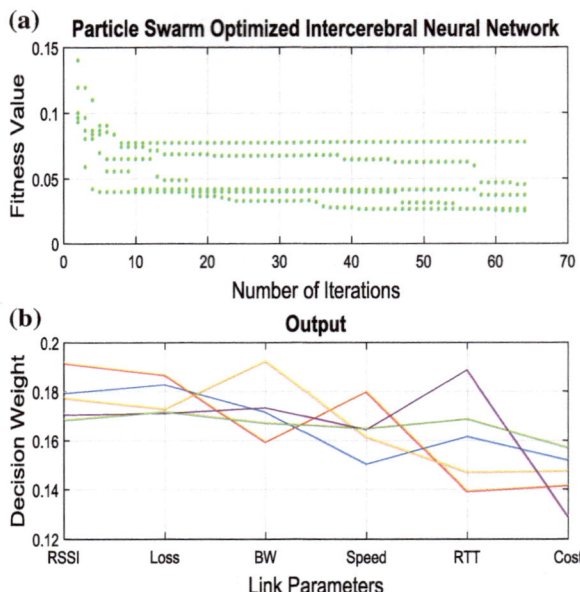

Fig. 2. Particle Swarm Optimized Intercerebral Neural Network.

Table 2. Link parameters v/s decision weight

Link parameter	Weight before	Weight after 1st loop iteration	Weight after 2nd loop iteration	Weight after 3rd loop iteration	Weight after 4th loop iteration	Weight after 5th loop iteration
RSSI	0.1982	0.1799	0.1919	0.1777	0.1712	0.1688
Loss	0.1895	0.1833	0.1870	0.1733	0.1717	0.1722
Band-width	0.1663	0.1722	0.1598	0.1927	0.1740	0.1676
Speed	0.1604	0.1505	0.1801	0.1616	0.1649	0.1651
RTT	0.1514	0.1619	0.1394	0.1471	0.1889	0.1690
Cost	0.1342	0.1522	0.1418	0.1476	0.1293	0.1572

deviation of the mean value. Standard Weight Deviation must be minimum. In Fig. 3, Standard Weight Deviation is 0.005.

Figure 4 represent three dimensional, fast delivery handover routing between each sensor nodes in system. Where, we consider three districts, four cities and six systems. In which, best network route would be find and data would be transferred through the selected route.

The output of the simulation is Mean Epoch, Mean Error, Mean Weight, Standard Epoch, Standard Error and Standard Weight values as represent in Table 3. Where, Epoch represent time. Mean error represent average error of overall simulation signal

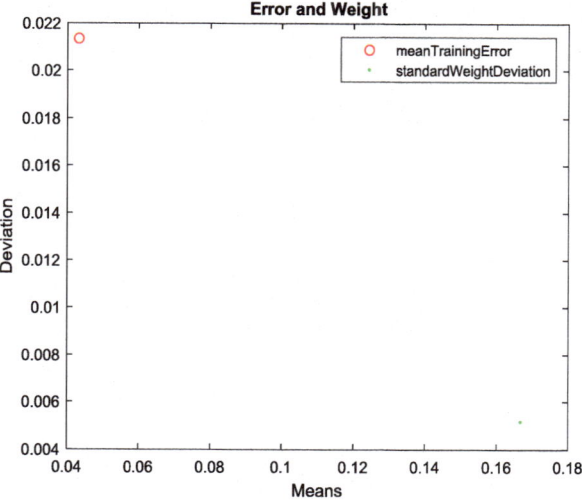

Fig. 3. Error and weight.

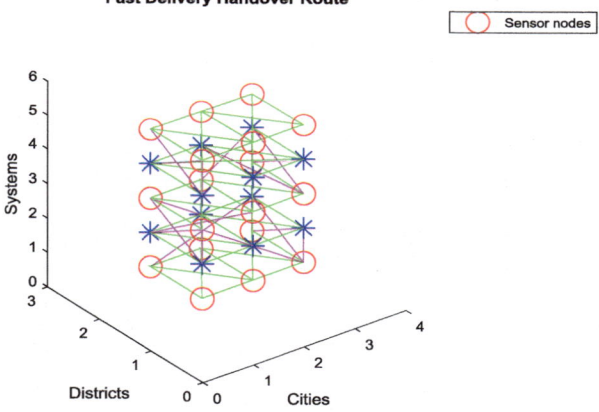

Fig. 4. Fast delivery handover route.

values. And Calculating variance of signal from mean value. After that, calculating standard values of Epoch, Error and Weight, which shows optimum value within one standard deviation of the mean. This process would be applied for finding mean and standard value of Epoch, Error and Weight. Mean error represent average of all error in a vector or set, which is used for measuring accuracy of system. Weight represents balance between exploration and exploitation. In which, overall mean weight is 0.1667 and standard weight is 0.0052.

Table 3. Mean and standard value of epoch, error and weight.

Parameters	Epoch	Error	Weight
Mean	34.3529	0.0431	0.1667
Standard	19.5378	0.0214	0.0052

6   Conclusion

In the wireless environment, QoS is important criteria for Network Selection. In this paper, we studied the network selection problem with fast delivery handover route in 5G LTE networks. We proposed an NN-PSO approach for network selection mechanism based on the criteria of SU's and derived an optimization problem for network selection by reducing the interference to primary networks and provide fast delivery handover route. Here, we solved optimization problem with the Neural Network and PSO, to findout near-optimal solution. We have also found mean and standard values of Epoch, Error and Weight: and also represents two scenarios for performance evaluation with different system settings, SU data rate speed demands, and cost preferences. Then the performance of proposed techniques for selection of network was evaluated under these scenarios. The experimental results represented that the NN-PSO performance and obtain a higher fitness value with less iteration in case of interference reduction, Population size and price related requirement of SU's.

References

1. Hasan, N.U., et al.: Network selection and channel allocation for spectrum sharing in 5G heterogeneous networks. IEEE **4**, 2169–3536 (2016)
2. Chen, S., Zhao, J.: The requirements, challenges, and technologies for 5G of terrestrial mobile telecommunication. IEEE Commun. Mag. **52**(5), 36–43 (2014)
3. Iwamura, M.: NGMN view on 5G architecture. In: Proceedings of IEEE 81st Vehicular Technology Conference (VTC Spring), pp. 1–5. Glasgow, May 2015
4. Droste, H., et al.: The METIS 5G architecture: a summary of METIS work on 5G architectures. In: Proceedings of IEEE 81st Vehicular Technology Conference (VTC Spring), pp. 1–5. Glasgow, Scotland, May 2015
5. Agyapong, P.K., Iwamura, M., Staehle, D., Kiess, W., Benjebbour, A.: Design considerations for a 5G network architecture. IEEE Commun. Mag. **52**(11), 65–75 (2014)
6. Tseng, L.-C., Chien, F.-T., Zhang, D., Chang, R.Y., Chung, W.-H., Huang, C.: Network selection in cognitive heterogeneous networks using stochastic learning. IEEE Commun. Lett. **17**(12), 2304–2307 (2013)
7. Bhushan, N., et al.: Network densification: the dominant theme for wireless evolution into 5G. IEEE Commun. Mag. **52**(2), 82–89 (2014)
8. Ejaz, W., Ibnkahla, M.: Machine-to-machine communications in cognitive cellular systems. In: Proceedings of IEEE International Conference on Ubiquitous Wireless Broadband (ICUWB), pp. 1–5. Montreal, QC, Canada, Oct 2015
9. Rattaro, C., Aspirot, L., Belzarena, P.: Analysis and characterization of dynamic spectrum sharing in cognitive radio networks. In: Proceedings of IEEE International Wireless

Communication Mobile Computing Conference (IWCMC), pp. 166–171. Dubrovnik, Croatia, Aug. 2015

10. Ejaz, W., ul Hasan, N., Lee, S., Kim, H.S.: I3S: intelligent spectrum sensing scheme for cognitive radio networks. EURASIP J. Wireless Commun. Netw. **2013**(1), 1–10 (2013)

11. Ejaz, W., Ul Hasan, N., Kim, H.S.: Distributed cooperative spectrum sensing in cognitive radio for ad hoc networks. Comput. Commun. **36**(12), 1341–1349 (2013)

12. Yang, Y., Aissa, S., Salama, K.N.: Spectrum band selection in delay QoS constrained cognitive radio networks. IEEE Trans. Veh. Technol. **64**(7), 2925–2937 (2015)

13. Andreev, S., et al.: Intelligent access network selection in converged multiradio heterogeneous networks. IEEE Wireless Commun. **21**(6), 86–96 (2014)

14. Ahmad, I., Liu, S., Feng, Z., Zhang, Q., Zhang, P.: Price based spectrum sharing and power allocation in cognitive femto cell network. In: Proceedings of IEEE 78th Vehicular Technology Conference (VTC Fall), pp. 1–5. Las Vegas, NV, USA, Sep. 2013

15. Weng, M.-D., Lee, B.-H., Chen, J.-M.: Two novel price-based algorithms for spectrum sharing in cognitive radio networks. EURASIP J. Wireless Commun. Netw. **2013**, 265 (2013)

16. Fathi, M.: Price-based spectrum sharing and rate allocation in the downlink of multihop multicarrier wireless networks. IET Netw. **3**(4), 252–258 (2014)

17. UlHasan, N., Ejaz, W., Kim, H.S., Kim, J.H.: Particle swarm optimization based methodology for solving network selection problem in cognitive radio networks. In: Proceedings of IEEE Frontiers Information Technology (FIT), pp. 230–234. Islamabad, Pakistan, Dec 2011

18. Hu, S., Wang, X., Shakir, M.Z.: A MIH and SDN-based framework for network selection in 5G HetNet: Backhaul requirement perspectives. In: Proceedings of IEEE International Conference Communication Workshop (ICCW), pp. 37–43. London, UK, June 2015

19. Orsino, A., Araniti, G., Molinaro, A., Iera, A.: Effective RAT selection approach for 5G dense wireless networks. In: Proceedings of IEEE 81st Vehicular Technology Conference (VTC Spring), pp. 1–5. Glasgow, UK, May 2015

20. Wang, Y., Yu, J., Lin, X., Zhang, Q.: A uniform framework for network selection in cognitive radio networks. In: Proceedings of IEEE International Conference Communication (ICC), pp. 3708–3713. London, UK, Jun. 2015

21. Ju, M., Kang, K.: Cognitive radio networks with secondary network selection. IEEE Trans. Veh. Technol. **65**(2), 966–972 (2016)

22. Akhtar, A.M., Wang, X., Hanzo, L.: Synergistic spectrum sharing in 5G HetNets: a harmonized SDN-enabled approach. IEEE Commun. Mag. **54**(1), 40–47 (2016)

23. LeAnh, T., Van Nguyen, M., Do, C.T., Hong, C.S., Lee, S., Hong, J.P.: Optimal network selection coordination in heterogeneous cognitive radio networks. In: Proceedings of IEEE International Conference on Information Network (ICOIN), pp. 163–168. Bangkok, Thailand, Jan. 2013

24. Mitola, J.: Cognitive radio: an integrated agent architecture for software defined radio. IEEE Personal Communications, pp. 13–18 (1999)

25. Poli, R., Kennedy, J., Blackwell, T.: Particle swarm optimization. Swarm Intell. **1**(1), 33–57 (2007)

26. Feng, Y., Teng, G.F., Wang, A.X., Yao, Y.M.: Chaotic inertia weight in particle swarm optimization. In: Innovative Computing, Information and Control, 2007. ICICIC'07. Second International Conference on, pp. 475–475. IEEE, September 2007

27. Mishra, S., Tripathi, G.S.: Comparison of various neural network algorithms used for location estimation in wireless communication. IJAREEIE (2013)

28. Heinzelman, W.B., Chandrakasan, A.P., Balakrishnan, H.: An application – specific protocol architecture for wireless microsensor networks. IEEE Trans. Wirel. Commun. 1(4), 660–670 (2002)
29. Chagas, S.H., Martins, J.B.: An approach to localization scheme of wireless sensor networks based on artificial neural networks and genetic algorithms. IEEE (2012)
30. Kennedy, J., Eberhart, R.: Particle swarm optimization. In: Proceedings of IEEE International Conference on Neural Networks, Perth, Australia. IEEE Service Center, Piscataway, NJ (1995) (in press)
31. ICT Fast and Figure 2017: ITU 15th world telecommunication/ICT indicators symposium (WTIS), Tunisia, 14–16 November 2017

Bluetooth Enabled Electronic Gloves for Hand Gesture Recognition

R. Helen Jenefa[(⊠)] and K. Gokulakrishnan[(⊠)]

Electrical and Electronics Engineering, Anna University Regional Campus,
Tirunelveli, India
helenjenefa@gmail.com, gk39762@gmail.com

Abstract. Human beings have a natural ability to see, listen and interact with their outside surroundings. Unluckily there are some people among us, they are differently talented and do not have the capability to utilize their senses to the greatest extent possible. There are around 300 million hearing and speech impaired people in the world. They depend on other means of communication such as communication by hand gesture called as sign language. This produces a major barrier for people in the deaf and mute communities when they attempt to interact with others, especially in their educational, societal and professional environments. To have communication between speech impaired people and normal people, gesture communication is challenging one. To diminish this communication gap between them, hand gloves are designed which can be easily worn and operated by the hearing and speech impaired person. The flex sensors, accelerometer sensor and contact sensors are fitted in hand gloves, which can measure the bend and movement of hand gesture of the user. These sign signals are converted into a digital signal using ADC. The microcontroller is used to recognize the hand gestures. After recognition of gesture, these signals are sent to the android phone via Bluetooth module. The received text signal at an Android phone is converted into speech signal using Talking display android app. The converted text to speech output is presented with the help of speaker system so that it will be helpful to normal people to understand sign performed by the dumb person. The LCD module is also used to display recognized gesture at microcontroller.

Keywords: Electronic glove · Flex sensors · Accelerometer · Contact sensors · Arduino mega · Gesture recognition

1 Introduction

Communication is inevitable for human life. One cannot think of human life without communication. Communication helps us to understand others. we need to communicate to deal different concerns and problems of daily life. The lack of ability to communicate can lead to a lot of troubles.

Communication is the exchanging the flow of information and opinion from one human being to another; it involve a sender sending a thought, information, or emotion to a recipient. A successful communication requires same basic ability of sender and receiver. Here, thoughts are delivered from sender, which received and an immediate

© Springer Nature Switzerland AG 2020
A. P. Pandian et al. (Eds.): ICCBI 2018, LNDECT 31, pp. 771–777, 2020.
https://doi.org/10.1007/978-3-030-24643-3_91

feedback generated by recipient. But in case of hearing and speech impaired people, when they try to converse with others, they are incapable, While they have not verbal communication and listening power [1]. However they can communicate their thoughts easily by writing it, but suffer a lot while face to face communication. This difficulty can easily diminish by using sign language. Sign language is a special language which usually used by hearing and speech impaired people [7].

2 Sign Language

Sign language is the communication through bodily movements, especially of the hands and arms, used when verbal communication is impractical or not desirable. Wherever vocal contact is impossible, as between speakers of mutually unintelligible languages or when one or more communicators would-be is hearing impaired, this indication language can be used to bridge the break between them. Sign language is used for the hearing impaired express their emotions, contribute to a discussion, learn, and overall live their lives as normal as possible.

A place where group of hearing and speech impaired people exist, sign languages have been developed at that place. Sign Language is not a common language or it is not unique in nature. It may vary from place to place, region to region. Each gesture has a unique meaning. Sign languages generally do not have any relation to the spoken languages of the lands in which they arise. The correlation between sign and spoken languages is complex and varies depending on the country more than the spoken language. There are various sign languages all over the world, namely American Sign Language (ASL) [7], Indian Sign Language (ISL), French Sign Language (FSL), British Sign Language (BSL), Japanese Sign Language (JSL) etc. All of these were developed independently.

Normally, sign language recognition can be achieved in three ways; they were Recognition of alphabets by finger spelling, isolated words, and Recognition of sentences by continuous gesturing. Accordingly, these gestured signs are captured to decide their related meanings which are done by two methods: sensor-based or Glove based and vision-based (Camera). Glove based method uses wearable devices to capture signed gestures; it is very simple and more truthful. But, vision-based technique uses camera to capture the series of images. The second is a more natural technique, that is quiet complex and less truthful.

3 Hardware Description

3.1 Flex Sensors

Flex sensors are nothing but the flexible resistor that is used to detect bend and flex of the object to be identified. It works in both directions. Its resistance value decreases according to the amount it is bent in both directions [5]. When the flex sensor is curved, the sensor produces a resistance output related to the radius of the bend; if radius is small the resistance value is far above the ground. It can be interfaced with the microcontroller [12]. The output from the sensor is analog.

3.2 Accelerometers

The ADXL335 is a 3-axis accelerometer which is thin, tiny, and low power. It measures acceleration with a minimum full-scale range of ± 3 g. It used to measure the static acceleration as well as dynamic acceleration resulting from motion, shock, or vibration.

3.3 Multiplexers

Each Accelerometer has 3 outputs (x, y, z). Five accelerometer has 15 outputs but the Arduino board has only 6 analog pins. so, these sensor outputs should be multiplexed. HEF4067 is the 16 channel multiplexer. It select one input of 16 inputs and allow to flow through the one common pin. Pin diagram of the HEF4067 multiplexer is shown in Fig. 1.

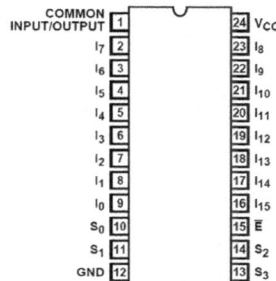

Fig. 1. Pin diagram of HEF4067

3.4 Contact Sensors

Contact sensor which is nothing but a aluminum foil. One of the digital pins and Gnd pin of Arduino Mega are connected to two different aluminum foil. If these two foils are touched it produces the high output otherwise it produces low output. Likewise, it can be used as the Contact sensor.

3.5 Bluetooth Module

The Bluetooth module HC-05 is a MASTER/SLAVE module. By default the factory setting is SLAVE. The Role of the module (Master or Slave) can be configured only by AT COMMANDS. The slave modules cannot begin a connection to another Bluetooth device, but it can accept the connections. Master module can initiate a connection to other Bluetooth devices. The user can use it simply for a serial port replacement to establish connection between microcontroller and other devices.

4 Block Diagram

The block diagram of the electronic gloves is shown in Fig. 2. The block diagram contains ten accelerometers, ten flex sensors, 2 multiplexers, Arduino Mega, LCD and a Bluetooth module. Arduino board has 16 analog pins (A0-A15) only but each accelerometers needs three analog pins to connect with Arduino. So multiplexing is required. Each five accelerometers from dual hand are multiplexed using two HEF4067 which is 16 channel multiplexer. For the Right hand glove: The common input/output pin is connected to the Arduino 's analog pin 5 (A5). Five flex sensors are connected to the analog pin 0 to 5 (A0-A5) through the voltage divider circuit. For the Left hand glove: The common input/output pin is connected to the Arduino's analog pin 11 (A11). Five flex sensors are connected to the analog pin 6 to 10 (A6-A10) through the voltage divider circuit. Five volt power supply is given to the Arduino Mega by either through USB cable or through the power supply circuit. Different sensor readings are assigned to the appropriate signs through program and it dumped to the Arduino Mega board. The Bluetooth module which is connected to the Arduino board will send the data to the mobile phone. The voice and text results are produced through the mobile application in the Android phone. LCD which is connected to the Arduino to show the result.

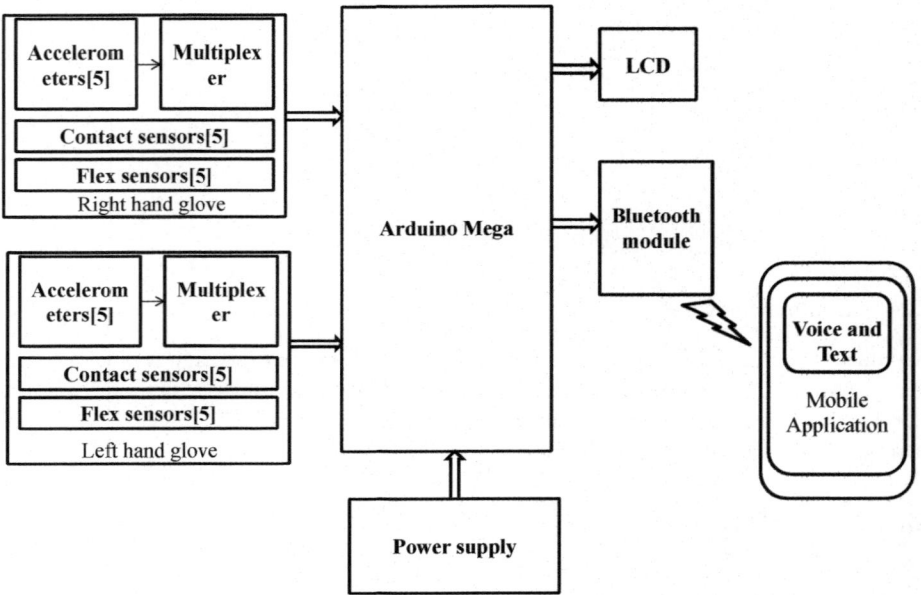

Fig. 2. Block diagram of the system

5 Hardware Setup

Flex sensors and accelerometers and contact sensors are attached to the Glove. Multiplexer and voltage divider circuit also attached with the each glove. LCD and the flex sensors are connected to the Arduino board. Accelerometers are connected to the Arduino board through the multiplexer. The code for converting sign language to text was dumped into Arduino board. The Bluetooth module which is connected to the Arduino board will send the data to the mobile phone. Talking Display Shield Application in the mobile phone is used produce voice and text output. Figure 3 shows the hardware setup of this electronic glove.

Fig. 3. Hardware setup

6 Text and Voice Output

Bluetooth module HC-05 is used to send the resultant data to the mobile. Talking Display shield mobile application is used to get the results through Bluetooth and it display and voicing the output. When the mobile application is opened it shows the all available near by Bluetooth devices. HC-05 Bluetooth device should be chosen manually. After paring, the HC-05 device continuously sends the sensed and recognized gesture data to the mobile phone and this mobile application displays the result and it also produce the audio output of the result. This is shown in Fig. 4.

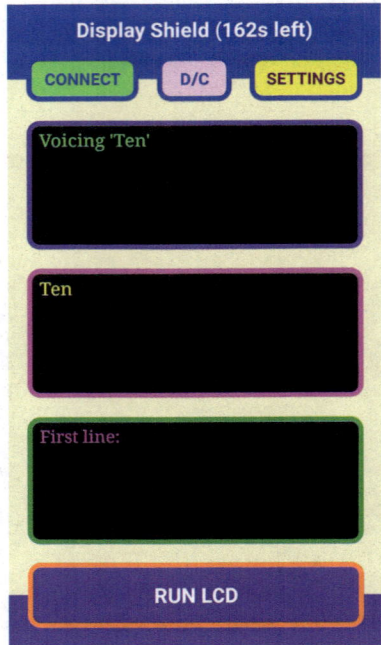

Fig. 4. Text and voice output

7 Conclusion and Future Work

Sign language is one of the useful utensils to simplify the communication between the hearing and speech impaired society and normal communities. Although sign language is created to communicate, the intended person must have the knowledge of the sign language which is not achievable always. Hence this Glove lowers such troubles.

In future work, proposed system can be developed to differentiate the similarity between signs. Moreover it can be focus on converting the sequence of gestures into text i.e. word and sentences.

References

1. Aarthi, M., Vijayalakshmi, P.: Sign language to speech conversion. In: International Conference on Recent Trends in Information Technology (2016)
2. Abhijith Bhakaran, K., Nair, A.G., Deepak Ram, K.: Smart gloves for hand gesture recognition. In: International Conference on Robotic Automation for Humanitarian Application, IEEE (2016)

3. da Silva, A.F., Gonçalves, A.F., Mendes, P.M., Correia, J.H.: FBG sensing glove for monitoring hand posture. In: IEEE Sens.S J. **11**(10) 2442–2448 (2011)
4. Bairagi, V.K.: Gloves based hand gesture recognition using indian sign language. Int. J. Latest Trends Eng. Technol. (2016)
5. Galka, J., Masior, M.: Inertial motion sensing glove for sign language gesture acquisition and recognition. IEEE Sens. J. **16**(16) (Aug 15, 2016)
6. Abhishek, K.S., Qubeley, L.F.: Glove based hand gesture recognition using capacitive touch sensor. In: International Conference on Electronic Devices & Solid state circuits (2016)
7. Elmahgiubi, M., Ennajar, M.: Sign language translator and gesture recognition. IEEE (2015)
8. Praveen Kumar, M., Rajadurai, M., Saidinesh, K.: Hand gesture recognition to translate voice and text. Int. J. Adv. Res. Civ. Struct. Environ. Infrastruct. Eng. Dev. **3**(1) (2015)
9. Zhou, S., Fei, F., Zhang, G., Mai, J.D., Liu, Y.: 2D human gesture tracking and recognition by the fusion of MEMS inertial and vision sensor. IEEE Sens. J. **14**(4), 1160–1170 (2014)
10. Chouhan, T., Panse, A., Voona, A.K., Sameer, S.M.: Smart glove with gesture recognition ability for the hearing and speech impaired. In: Glove with Gesture Recognition Ability for the Hearing and Speech impaired, IEEE (2014)
11. Lavanya, kulapravin, M.S., Mohan, M: Hand gesture and voice conversion system using sign language transcription system. Int. J. Electron. Commun. (2014)
12. Vamsi Praveen, P., Sathya Prasad, K.: Electronic voice to deaf and dumb people using flex sensor. IJIRCCE **4**(8) (2016)
13. Lu, Z., Chen, X., Li, Q., Zhang, X., Zhou, P.: A hand gesture recognition framework and wearable gesture based interaction prototype for mobile devices. IEEE Trans. Hum. Mach. Syst. (2014)

Smart Trolley with Automatic Master Follower and Billing System

R Dhianeswar[1(✉)], M Gowtham[2(✉)], and S Sumathi[1]

[1] Sri Sai Ram Engineering College,
Electronics and Communication Engineering, Chennai, India
dhianesh1612@gmail.com, sumathi.ece@sairam.edu.in
[2] Tata Consultancy Services, Chennai, India
gauthammohan1996@gmail.com

Abstract. In the modern era, every supermarket makes use of shopping trolleys in order to help customers carry products which they wanted to purchase. The customers have to take the products which they wish to purchase and should drop them into the shopping trolley and then they have to reach the billing section. The billing process is quite hectic and it consumes more time and consequently it has created the need for shops to employ more human resources in the billing section, but still this problem is not resolved and some customers even face the problem of pushing the trolley forward when it is loaded with huge quantity. "Smart Shopping Trolley" is going to be a better solution to it where the trolley is made to follow the customer automatically using a "Kinect Sensor". Then, an automatic billing system is also incorporated where the items will be scanned by a Radio Frequency Identification reader and the amount will be summed accordingly so that the final amount is paid without standing in a large queue.

Keywords: Keywords-Kinect · Sensor ·
Radio frequency identification tag and reader

1 Introduction

In this world, each and every day, we see new innovations for the sake of the comfort of people's life and for reducing stress. As everything is getting automated, the retail industry is looking for many automated devices in order to increase the productivity and efficiency. Automation is proving to be one of the valuable tools in the electronic world. Shopping trolley is one of the common things which is seen in all supermarkets. Though these shopping trolleys sound good, they have quite a few issues. Each time, customers have to pull up the trolley and move it to different places in the store. Moreover, the customers have to be cautious while driving the trolley without causing injury to the people standing along the path. At last, they also have to wait in a long queue for getting their items billed. This causes frustration to the customers and lot of their time is wasted. To overcome this problem, the shopping trolley is designed with

© Springer Nature Switzerland AG 2020
A. P. Pandian et al. (Eds.): ICCBI 2018, LNDECT 31, pp. 778–791, 2020.
https://doi.org/10.1007/978-3-030-24643-3_92

the feature of following its master and has an enhanced billing system in it which sums up the amount of the items which are purchased. Thus reducing the shopping and billing time.

2   Literature Survey

2.1   Automated Smart Trolley with Smart Billing Using Arduino

Mobile phone has been one of the most spectacular innovations which so far. It has completely turned the whole world to a newer phase. It has completely changed all the fields of work. Today, the people across the world are able to get information on any field instantly. In this paper, a mobile application is used which displays the products which are available in the market along with their cost. The user can select the required products and once the selection process is over, the products are sorted and displayed based on their location. Radio Frequency Identification (RFID) is a technology which is an alternative to barcode systems. It has many applications and has more accuracy. These tags are used to develop smart shopping cart systems. These shopping carts have the ability to calculate the prices of the products which are dropped into the shopping cart and display the total prices of all the products instantly. Thus they help customers in knowing how much he or she has to pay while shopping and thus save time [1].

2.2   Automated Human Guided Shopping Trolley with Smart Shopping System

Shopping trolleys are one of the common things seen in all supermarkets. Though these shopping trolleys are good to use, they have quite a few issues. These shopping trolleys are abandoned everywhere in supermarkets after being used. It is known to be inconvenience and waste of time for customers who are in rush to search for the desired products in a supermarket. Therefore, an automatic human and line following shopping trolley with a smart shopping system is developed to solve these problems. A line following portable robot is installed under the trolley to lead the users to the items' location that they plan to purchase in the supermarket. On the whole, it presents the hardware and software design of the portable robot [2].

2.3   Intelligent Shopping Cart

RFID technology is used for the identification of the products inside the shopping trolley, thus, saving billing time. The automatic product identification is done by the use of RFID tags in every product and installing RFID reader in the shopping cart. The range of the RIFD reader is maintained in such a way that it does not read the product information which is in other carts nearby [3].

2.4   Fabrication of Automated Electronic Trolley

Shopping trolley is also called a shopping cart which is like a carriage or basket that can be used by customers to transport their purchased products inside grocery shops or supermarkets. The shopping trolley is a cart supplied by supermarkets. In Europe and Canada, supermarkets place a coin lock system on shopping carts in order to encourage the customers to return the carts themselves after use. In addition, the design of shopping trolleys was concerned. This is because poorly designed shopping trolleys can cause potential musculoskeletal injuries from manually pushing or pulling heavy loads. A market survey was conducted where the results showed that most of the users expected the shopping trolley to be energy saving, pulling and pushing motion and adjustable height. The health and safety of users are prioritized when they are shopping in the supermarkets. Furthermore, customers especially parents cannot enjoy during shopping. This is because parents have to take care of children while shopping. It replaces the existing technology [4].

2.5   Smart Trolley in Mega Mall

In view of automating the mall, an automatic microcontroller based TROLLEY is developed. It follows the customer during purchase and it is also programmed to make a proper distance from the customer. The customer has to hold out the barcode of the item in the line of sight of the barcode reader such that it scans the code and sums up the amount. The corresponding amount will be displayed on an LCD display. By using this trolley, customers can save time and buy ample amounts of necessary items within short time [5].

2.6   Shopping Cart Robots

It has a drive system attached to the shopping cart which is used to transmit power for the movement of the cart wheels. The person [6] is followed by the shopping cart by using a stereo camera installed in front of the shopping cart. The person is not required to wear any attached element along with him. But as the recognition is done by colour segmentation, the person is required to wear clothes of a specific colour [7, 8].

2.7   Communicating Shopping Cart

A shopping help system is used which is used to obtain the shopping list from a mobile with the help of the QR code. It intimates regularly about the articles which are present in the shopping cart and continuously communicates with the host computer system. The main components that are used are a Laser range finder, Sonar, and Contact lenses for navigation [9, 10].

2.8   CompaRob – The Shopping Cart Assistance Robot

It is mainly developed in concern with the use of shopping cart by elderly people. The cart uses ultrasonic sensors and radio signals to provide simple and effective person localization. The cart [11] can be connected to a smartphone for better performance. The prototype has been tested in a grocery store, while simulations have been done to analyse its scalability in larger spaces where multiple robots could coexist at a time [12].

3   Proposed Technology

The proposed system is to automate the trolley using a Kinect sensor. The Kinect sensor is used to track the human motion and track accordingly. Kinect is interfaced with Arduino and is used to trigger it. Arduino is programmed to accept the data from the Kinect using Arduino programming. A key is present in the trolley to start automation of the trolley so that it can follow its master. To bill all the items purchased in the shopping mall, a RFID tag along with the reader is used to bill every item put in it. With the help of this setup, customers can easily get the items billed without having to wait in long queues for getting their items billed.

4   Components Used

4.1   Kinect Sensor

Kinect is the one which is made of both hardware and software built in with Kinect sensor accessories. The built in accessories are

  (i) Colour VGA video camera – This is the one which captures the facial expressions and actions of the master by detecting the three colours namely Red, Green and Blue and it is the reason for its name "RGB camera".
 (ii) Depth sensor – It is used to get the view of the room in 3D regardless of the lighting conditions. It has an infrared projector and a monochrome complementary metal oxide semiconductor (CMOS) sensor.
(iii) Multi-array microphone - This is the one which is used to isolate the voices of the players from the noise of the atmosphere.

Both the video and depth sensor cameras have a resolution of 640 × 480-pixel and run at 30 FPS (frames per second). Another important requirement is that there should be about 6 feet (1.8 meters) distance between the person and the Kinect sensor (Fig. 1).

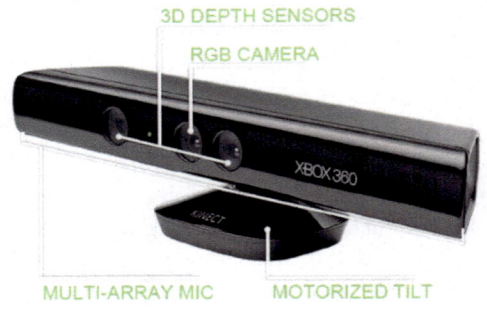

Fig. 1. Kinect Sensor

4.2 RFID Tag and Reader

RFID stands for radio frequency identification. RFID tags are smaller in size that are used in our day to day life for unlocking hotel rooms, entering into cars and for many such applications. These small chips along with an RFID reader form the RFID system. Data are stored in the RFID tag electronically and are retrieved by the RFID reader using EM waves. Tags have small memory in terms of kilobytes. The RFID tag works similar to the Barcode system (Fig. 2).

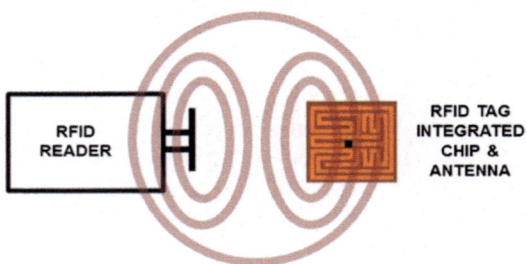

Fig. 2. RFID tag and RFID reader

Advantages of RFID System over Barcode System:

- Line of sight is not required for RFID system.
- Read rate of RFID system is more compared to barcode system.
- The data in RFID tag can be read, written and updated whereas in barcode, the data cannot be changed.

4.3 Software Tools

 (i) *MATLAB software*
(ii) Acquiring Image and Skeletal Data using Kinect

5 Block Diagram

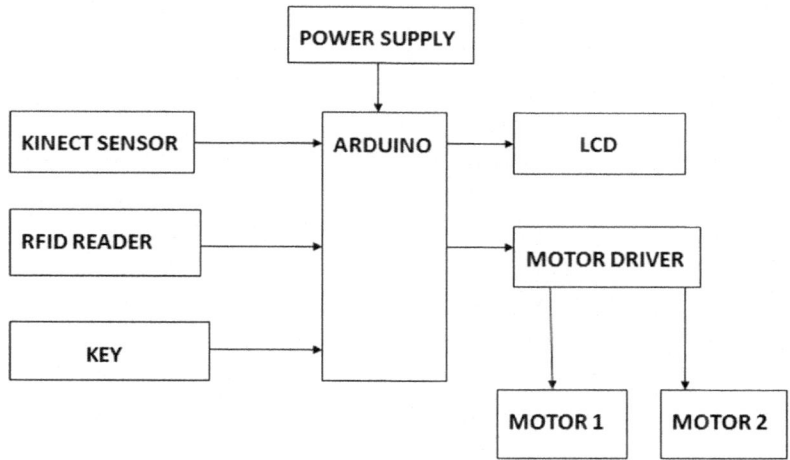

Fig. 3. Block Diagram

6 Flow Chart

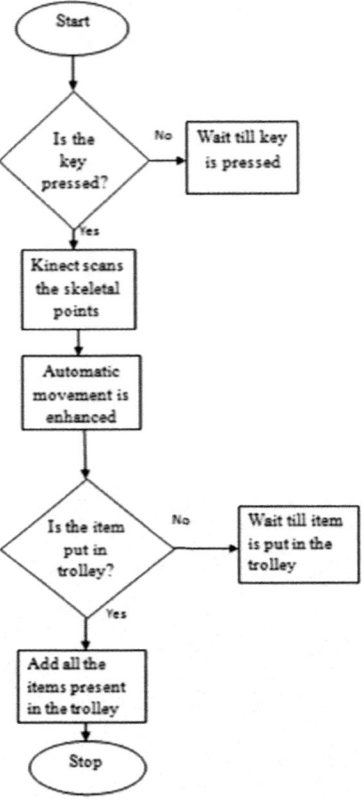

Fig. 4. Flow Chart

7 Description of Block Diagram

7.1 Automation

In this project, Kinect Sensor and L293D Motor Driver are used for automation purpose. Kinect Sensor scans the master of the trolley and L293D Motor Driver enables
the trolley to move according to the master's movement. These two components
exchange their data via Arduino. Once the power supply is switched ON, the Kinect
Sensor is activated (Figs. 3 and 4).

7.2 Enhanced Automation Using Kinect

Kinect Sensor contains an IR projector, RGB (Red-Blue-Green) camera and a Depth Sensor. The IR projector (also known as Depth Projector or IR Emitter) emits IR rays continuously once the supply is switched ON. These IR rays hit the object that is closer to the front of Kinect and gets reflected. The Depth Sensor captures these reflected rays and creates a depth map within the Kinect and sends these data to Depth Sensor's processor using Kinect Driver software. Meanwhile, the RGB camera functions as a web camera by recording a space of room size which includes coloured objects. Thus, RGB camera facilitates the Kinect to track the master. The program used for tracking master is based on 26 Skeletal Points of a human body. After performing all these operations, the Kinect Sensor sends a signal to the Arduino. The Arduino then sends an activation signal to the Motor Driver which enables the Motor Driver to move the trolley. Motor driver is used to drive the motor present in the trolley for automation. Motor driver used is L293D.

7.3 Billing

For billing purpose, RFID Reader, RFID Tags and LCD are used. RFID Tag and Reader are used to calculate the total amount to be paid for the items present in the trolley. Tag is placed in every store product and Reader is used to get the amount of every product by reading the Electronic Product Code present in the RFID Tag. RFID reader has an antenna and a semiconductor chip present inside it. The antenna operates in radio frequency range. RFID tag called as transponder is used to receive the signals from the reader. Active and passive tags are the two types of RFID. Passive RFID tag is generally used for near field communication. RFID Tag also has an antenna and hence both RFID Reader and Tag can act as a transmitter or a receiver. When the customer drops any item into the trolley, the Passive RFID Tag which continuously emits EM waves is received by the Reader. Through this path, the RFID Tag passes on the data to Reader in the form of Electronic Product Code.

The Electronic Product Code is a 96-bit string of data. The first eight bits are used to identify the version of the protocol. The next 28 bits represent the organization that computes the data for the appropriate RFID tag and the number for the organization is assigned by the EPC global consortium. The next 24 bits are an object class which represents the kind of product. The last 36 bits represent a unique serial number for the tag identification. The total electronic product code number can be used for identifying the product accordingly. The Electronic Product Code contains the amount of a product. This data are sent to the LCD via Arduino. Finally, the LCD receives the signals sent by the Reader and displays the total amount to be paid by the customer.

8   Results and Discussions

8.1   Kinect Configuration

8.1.1   Interfacing Kinect

Kinect is interfaced using Kinect driver which is done through Kinect SDK-v1.8. Microsoft Kinect for Windows Software Development Kit (SDK) is used to interface Kinect with the PC through the COM port. Kinect driver consists of Kinect Audio, Kinect camera and Kinect motor. Kinect also supplements other gaming devices (Figs. 5 and 6).

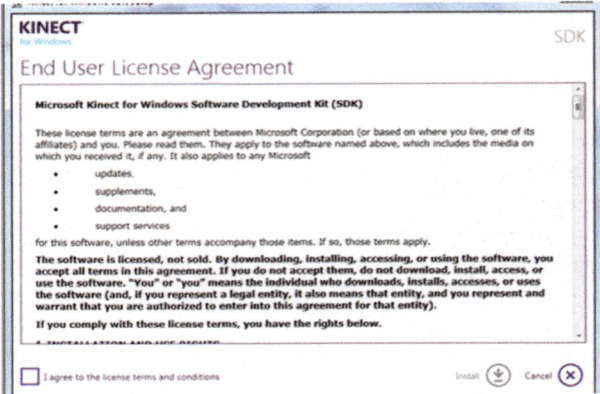

Fig. 5. Kinect installation

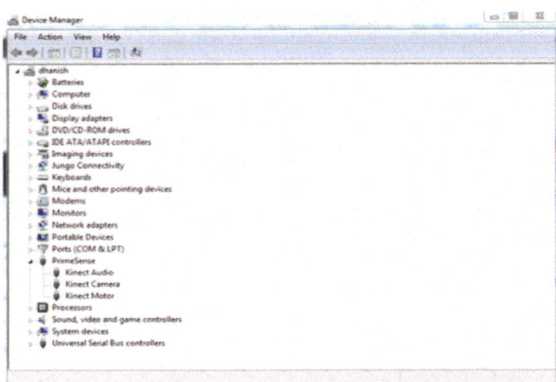

Fig. 6. Kinect Driver

8.2   Results

8.2.1   Kinect Skeletal Imaging

Kinect tracks various skeletal joints in our body. Based on the skeletal image tracked it visualizes our human body structure and tracks our motion. The image given below is the visualization of the human body tracked which is displayed in the computer. It structures the image as a 3 dimensional image using the depth map present in it (Fig. 7).

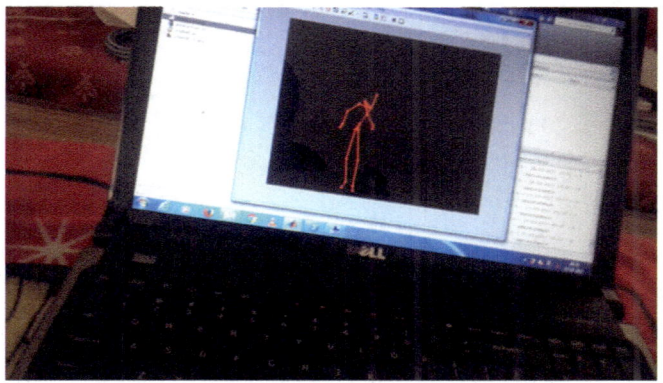

Fig. 7. Skeletal figure

Once the program is run using the MATLAB, the errors are first checked and when all the connections and the program code are verified to be correct, then a window will pop up indicating the generation of the depth map. It will display the video input given

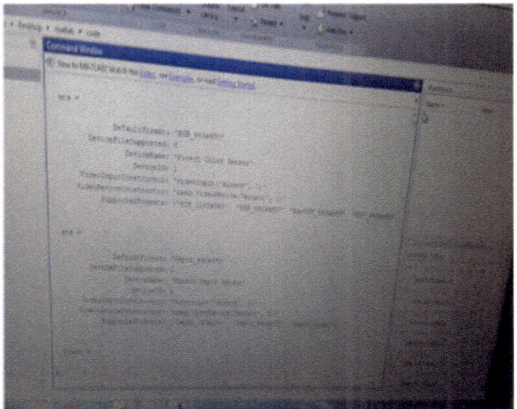

Fig. 8. Video image output in command window

and it also states how that input is got in the programming. Initially the image pixel resolution is given by the RGB camera present in the Kinect and then the depth map is generated. The count values are also displayed to indicate the time interval during which the image is captured (Fig. 8).

Fig. 9. Arduino connections

8.3   Arduino Configuration

8.3.1   Arduino Installation
Arduino version 1.6.7 is used for interfacing with the Kinect in the trolley. Two types of software are generally installed. One for 32 bit software computer and the other for 64 bit computer.

8.3.2   Arduino Setup
Arduino is the microcontroller which is used to interface the Kinect with that of the other devices in the board (Fig. 9).

Fig. 10. Top view of the trolley

Arduino is powered using the COM port in the PC. A motor driver L293D is used to drive the motors present in the wheels. The motor driver is activated by using two 9v batteries. An EM-18 RFID reader is used to detect all the items which are put in the trolley. A LCD is interfaced with the Arduino to display the amount of all the products purchased. A tag is put in every product so that the reader tracks the value of every item purchased by the customer. Kinect is interfaced to the Arduino through the kinect driver and also using MATLAB software.

8.4   Trolley Setup

A trolley is designed with a thick wooden board with four wheels along all the directions. Four DC motors are used to drive the wheels of the trolley and it is driven by the motor driver. Kinect is connected separately to the PC.and powered using the normal AC supply voltage. Arduino is connected using the data cable to the PC. Finally the trolley is made to folow its master until it is in functional mode (Figs. 10 and 11).

Fig. 11. Front view of the trolley

9   Advantages

- Shopping is made easy through automation.
- Customers can avoid standing in long queues for getting their items billed.
- Customers can decide which product to buy based on the amount displayed on the trolley.
- Errors due to the faulty measurement of the items are completely alleviated.
- This is a useful time saving measure for both customers as well as retailers.
- This reduces the tension of standing in long queues.

10   Future Enhancement

There is some scope for further improvement of the project. The mechanism of the robot should be enhanced and designed in a simpler way, to ease the installation under a shopping trolley. A more advanced algorithm should be developed so that the shopping trolley is able to move in a crowded environment and follows the user automatically in any direction. An improved android application reminds the users on the items they need to purchase when they unintentionally pass the good's location. In addition, it can also remind the users who have health problems about the nutrition of products. It allows supermarket staff and users to know the shopping trolley current locations.

11   Conclusion

With the aid of automatic trolley follower using Kinect and the human leading function portable robot, supermarket owners need to purchase only portable robots and can easily install them under shopping trolleys. Users can then enjoy shopping without pushing the shopping trolley themselves. The microcontroller based trolley automatically follows the customer. It also maintains a safe distance between the customer and itself. Meanwhile, the smart shopping system allows users to access the location of the item that they plan to purchase in supermarkets. The location of shopping trolley and items can be easily tracked by using RFID technology. This technology is useful in billing all the items in the trolley and hence prevents customers from standing in long queues.

References

1. Suganya, R., Swarnavalli, N., Vismitha, S.: Mrs. Rajathi, G.M.: Automated smart trolley with smart billing using Arduino. Int. J. Res. Appl. Sci. Eng. Technol. (IJRASET) 4(III), 897–902 (2016)
2. Ng, Y.L.: Automated human guided shopping trolley with smart shopping system. Int. J. Res. Appl. Sci. Eng. Technol. (IJRASET) 3(III), 49–56 (2015)
3. Kumar, R., Gopalakrishna, K., Ramesha, K.: Intelligent shopping cart. Int. J. Eng. Sci. Innov. Technol. (IJESIT) 2(4), 499–507 (2013)
4. Nayak, M., Kamath, K., Karunakara, Lobo, R.J., Anchan, S., Saikrishna, U.: Fabrication of automated electronic trolley. IOSR J. Mech. Civ. Eng. (IOSR-JMCE) 12(3), 72–84 (2015). Ver. II
5. Awati, J.S., Awati, S.B.: Smart trolley in mega mall. In: Int. J. Emerg. Technol. Adv. Eng. (IJETAE) 2(3) (2012)
6. Iyer, J., Dhabu, H., Mohanty, SK.: Smart trolley system for automated billing using RFID and ZIGBEE. Int. J. Emerg. Technol. Adv. Eng. (IJETAE) 5(10) (2015)
7. Nishimura, S., Itou, K., Kikuchi, T., Takemura, H., Mizoguchi, H.: A study of robotizing daily items for an autonomous carrying system—development of person following shopping cart robot. In: Proceedings of the 9th InternationalConference on Control, Automation, Robotics and Vision (ICARCV 2006), pp. 1–6, Singapore, December 2006
8. Nishimura, S., Takemura, H., Mizoguchi, H.: Development of attachable modules for robotizing daily items -person following shopping cart robot. In Proceedings of the IEEE International Conference on Robotics and Biomimetics (ROBIO 2007), pp. 1506–1511, IEEE, Sanya, China, December (2007)

9. Garcia-Arroyo, M., Marin-Urias, L.F., Marin-Hernandez, A., Hoyos-Rivera, G.D.J.: Design, integration, and test of a shopping assistance robot system. In: Proceedings of the 7th Annual ACM/IEEE International Conference on Human-Robot Interaction (HRI 2012), pp. 135–136, ACM, Boston, Mass, USA, March 2012

10. Marin-Hernandez, A., de Jesus Hoyos-Rivera, G., Garc´ıa-Arroyo, M., Marin-Urias, L.F.: Conception and implementation of a supermarket shopping assistant system. In: Proceedings of the 11th Mexican International Conference on Artificial Intelligence (MICAI 2012), pp. 26–31, IEEE, San Luis Potosi, Mexico, November 2012

11. Nagumo, Y., Ohya, A.: Human following behavior of an autonomous mobile robot using light-emitting device. In: Proceedings of the 10th IEEE International Workshop on Robot and Human Interactive Communication (ROMAN 2001), pp. 225–230, Bordeaux, France, September 2001

12. Sales, J., Martí, J.V., Marín, R., Cervera, E., Sanz, P.J.: CompaRob: the shopping cart assistance robot. In: Int. J. Distrib. Sens. Netw. Volume 2016, Article ID 4781280, 15 pages

Biometric Image Processing Recognition in Retinal Eye Using Machine Learning Technique

J. P. Gururaj[1,2(✉)] and T. A. Ashok Kumar[3]

[1] Department of Computer Science, Government First Grade College,
Harihar 577601, Karnataka, India
gururaj_jp@yahoo.com
[2] Garden City University, Bangalore, India
[3] The School of Computational Sciences and Information Technology,
Garden City University, 16th KM, Old Madras Road, Bangalore 560 049,
Karnataka, India

Abstract. Biometrics finds a lot of application in the modern day security aspects whether it may be in industrial sector, defense sector or in domestic sectors. Entry of a person into an organisation in the work environment plays a very important role as unauthorized persons if entered into a premises may create a flutter & inviable to some terrorist attacks. Hence, biometrics of an authorized person into the work environment plays a very important role. There are a large number of biometric methods for identification, viz., physiological & behavioural. Some of them are face, fingerprint, iris, voice, hand, thumb, ear, vein, gait, code, password, signature, pattern, palm print, etc. As there are a number of biometric methods for the automatic detection of human beings, retina scan is being considered as the biometric identity in the research work undertaken by us because of a large number of advantages over the other biometric identities. For biometric recognition of a person using retina scans, a number of methods exists in the literature. In this context, fractal dimension method is one of the method that could be used. Hence, the retinal recognition is performed in our proposed work for identifying the person by developing new approaches using Fractal Dimension methodologies. Hybrid algorithms are to be prepared & finally the developed codes are going to be run which are going to be performed in the Matlab environment, the simulation results are going to be observed and compared with the research work done by alternate scientists & engineers till now in order to build up the amazingness of the works done by us. This article gives the info about the exhaustive summary of the work carried out by us & that is going to be done along with the methodology that is going to be used for solving the desired objective and arrive at the outcome.

Keywords: Biometrics · Retina scan · Physiological · Behavioural · Matlab · Simulation · Security · Eyes

© Springer Nature Switzerland AG 2020
A. P. Pandian et al. (Eds.): ICCBI 2018, LNDECT 31, pp. 792–800, 2020.
https://doi.org/10.1007/978-3-030-24643-3_93

1 Organization of the Paper

The flow of the paper is presented in the following manner. Section 2 gives a detailed info about the research topic. The Retinal Recognition System (RRS) is briefly described in Sect. 3 followed by the different methods that are used for the recognition process in Sect. 4. The motivation that is obtained for doing the research work is presented in Sect. 5. The statement of the problem is also presented here. Section 6 gives the scope of the research work undertaken. Section 7 depicts the research work's objectives, followed by the possible outcome of the work in Sect. 8. A brief review of the literature survey is given in Sect. 9 followed by the proposed research process in Sect. 10. The implementation procedure is presented in Sect. 11. Advantages & future scope of the work undertaken is depicted in Sect. 12. The article finally concludes with the conclusions in Sect. 13 followed by a large number of references.

2 Introduction

Biometrics is the specialty of distinguishing a person by various techniques. Recognizing or checking one distinguish utilizing biometrics is pulling in impressive consideration in this cutting edge computerized world, one the fundamental reason being the security issues in different exceedingly delicate spots. It is the delightful study of programmed distinguishing proof of people that utilizes the extraordinary physical or conduct attributes/qualities of people to remember them. In the present biometric insights innovation, security for frameworks is transforming into progressively increasingly basic need. The scope of structures which have been endangered is regularly expanding and confirmation assumes an essential job as a first line of safeguard against gatecrashers [1].

Cutting edge e-security are in basic need of finding precise, agreeable and savvy options in contrast to passwords and individual personality numbers as monetary misfortunes increment drastically year over year from PC-based misrepresentation including PC hacking and character robberies. Biometric answer the arrangement with those principal inconveniences, in light of the fact that a person's biometric measurements specific and can't be exchanged. Bio-metrics, alternatively, presents a secure approach of authentication and identity, as they're hard to duplicate and use it by someone else other than the owner of the biometric. If biometrics is used together with something you recognize, then this achieves what is called 2-factor authentication. 2-point authentication is an awful lot & has more potential as it requires both additives capturing & identification. Personal identification, recognition is a problem in any bio app [2].

All Biometrics can be classified on the basis of the behavioural & the physiological attributes of human beings. Note that the behavioural traits are examined using the information of how a person behaves which may include how the person makes his/her signatures, voice of a person and dynamics of keystroke. Block diagram of a typical biometric scheme for the recognition purpose is shown in the Fig. 1 [3].

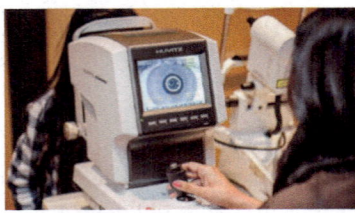

Fig. 1. Biometric device - retina scan machine

In this specific circumstance, in the wake of concentrate the points of interest and downsides of all the biometric techniques, the retina of the human eye was being selected for biometric recognition of a human being in this proposed work as it a complex one. The retinal recognition is going to be performed for identifying the person using the concept of Fractal Dimension in our research work as it finds a lot of applications especially in the medical imaging field that too in the secured areas. In the recent years, biometric authentication using retina scan has gained significant prominence in the security world over all the other methods because of its uniqueness, non-invasiveness, secured approach of recognition with unique identification, no stealing and stability of the human retinal patterns as it is specific to one person only & is shown in the Fig. 2, which is a Biometric device - Retina scan machine [4].

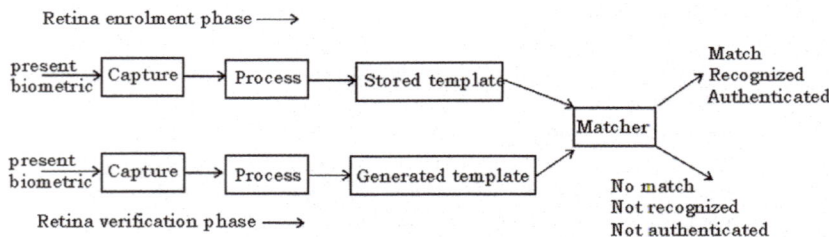

Fig. 2. Block diagram of a typical biometric scheme for the recognition purpose

3   Retinal Recognition System

In the area, the structure of the retinal acknowledgment framework is being exhibited more or less. The structure of human eye is appeared in the Fig. 3 alongside the retinal parts, which is our ROI. The defensive external layer of the eye is called as the sclera. Alternate segments of the eye are cornea, focal point, iris and retina. Retina is the light touchy tissue which shapes the inward side of eye (back end). The optical components of the eye center the picture onto the retina, along these lines starting a progression of substance and electrical occasions inside the retina. Nerve Fibers (NF) inside the retina send electrical signs to the cerebrum, which at that point deciphers these signs as visual pictures. Retina is roughly 0.5 mm thick and spreads the internal side at the back of the eye. A number of blood vessels (BV) are present all over the eyes to provide nutrition to various parts of the eye. 250 μm is the mean dia of the blood capillaries/viens/ arteries & the retina connects all the optic nerves to the brain. One of the important biometric recognition is the usage of the retina, which is a non-contact one & not visible, lying at the back of the eyes, this concept can be used for biometric authentication, identification & recognition of human beings [5].

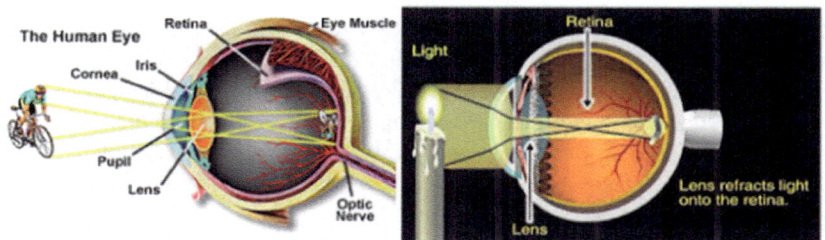

Fig. 3. A front view of the human retina along with its different parts

4 Method Used for Retinal Recognition

For the personal identification of any human beings, recently retinal bio-metrics is playing a very important role in the modern days in this digital technological world. Generally, 3 concepts are used for the biometric authentication using the retina, viz., the Morphological Segmentation, Branching Points Method and the Fractal Dimension Method. The pre-processing techniques for all the 3 retinal identification processes remains the same [10]. The input retinal fundus image is resized for maintain uniform size of the image which is very useful in processing, then the fundus input image is changed into the GS mode for the ease of computation and the contrast of the image is enhanced to effectively differentiate the foreground blood vessel pattern from the background. In our research work, the FD method is used as it is a very popular method compared to the others. A small overview of the Fractal Dimension follows.

Fractal Dimension Method (FD) - Fractal dimension method finds the value of the FD of the viens/capillaries/arteries which will be retrieved from the fundus input image of the retina.

5 Motivation & Problem Statement

Even though, there are quite a number of biometric techniques as mentioned previously, the retina is considered as the efficient method as it the retinal part is at the back end of the eye, it is non-contact and unique for a particular person. On the contrary, retina recognition in the recent days is also playing a very important role in biometric authentication/recognition of human beings. The main advantages of the retina recognition systems are as follows. Retinal scanners have a few focal points over finger printing and voice acknowledgment frameworks [7].

At long last, in this setting in the wake of concentrate the ramifications of each of the biometric procedures, we touched base at the determination of the retina as a standout amongst the best strategy for biometric ID of individuals because of its huge number of focal points referenced previously. Additionally, because of the ebb and flow activity taken up by the focal government in the field of biometrics in all segments, this had additionally persuaded us to take up the examination take a shot at the retina technique for biometric acknowledgment. The main motivation being some of the digital initiatives taken up by the central & state governments in the wake of security issues in the country. This has made us to identify the problem. Consequently in continuation, with energy of this exploration work in the wake of making a through review, we will propose some novel techniques for the programmed acknowledgment of biometrics utilizing retina filters by building up some product calculations in Matlab, the issue at long last, being characterized as the examination issue explanation, "*New approaches in the biometric recognition of humans beings using retinal eye images with the help of Fractal Dimension method*" [8].

6 Scope of the Research Work

The major scope of the proposed research work is mainly of 3 fold, viz., to develop some new approaches to overcome the security and authentication problems faced in many fields using retina scans, to improve the performance of the commonly used existing algorithms, to compare various techniques available for retinal identification based on their accuracy and error rate with the work done by us and to develop some hybrid algorithms for retinal biometric recognition [9].

7 The Research Work's Objectives

The fundamental target of the examination work that will be attempted by me under the direction of my boss is to build up some advanced BMIP algos for the successful biometric recognizable proof of individuals through the retinal piece of eyes & to conquer the difficulties located in the works done by the earlier researchers, at the same time developing some new approaches for retina recognition. Also, consequently, the drawbacks of some of the current existing algorithms that are considered to be greater vital are decided on and the research work is going to be executed to conquer them (overcome them) by improving/enhancing the performance or developing new algorithms by adding some additional parameters, which others have not considered [10].

8 Possible Outcome of the Research Work

The conceivable result of the exploration work is to demonstrate that when the planned calculation/s that will be created in the Matlab condition is run, the programmed acknowledgment of the retina is finished with least computational time prompting whether the individual is being perceived/validated or not in correlation with the work done by the kindred analysts till date thinking about a portion of their downsides, hence upgrading and enhancing the execution of the current calculations and showing that the research work done by us is superior over the others [11].

9   Literature Review

Many works have been proposed on retina biometric human recognition system for secure authentication till date. This section provides the knowledge about the existing technologies, its advantages and disadvantages, work done by other authors till date. Subsequent paragraphs explain how the researchers have contributed to the field of biometric recognition using retina scans. To start with, 100's of research papers were collected from various sources, studied @ length & breadth and its outcomes were studied, analyzed to start with and the problem statement was finalized and defined. In greater part of the work done by the different engineers, creators/scientists [1–12] presented in the earlier sections & in sentences, there were certain disadvantages. Some of the major drawbacks of pre-existing methods mentioned earlier were ....

- Few method fails when there is noise parameters in original picture & becomes sensitive to noise, thus it will produce intricate or additional edges of pictures in the resulting image.
- Some contributions requires excessive computation to achieve high accuracies (computationally quite expensive).
- In some methods, very high compilation time and more memory for storing data is required to get very good recognition process.
- No work is done on expanding the execution and exactness of the current framework.
- False matches in the millions of retina scanned image comparisons were produced.
- The number of computations involved in the recognition process was very high.
- Many methods developed by the authors failed due to noise like mirrored image in the retina photographs & thus the performance index was not upto the mark.
- Full-fledged automation of the retina recognition system was not done, automatic GUI's were not developed.

Some of the drawbacks [1–12] of the works that were carried out by the earlier researchers are taken up in this paper, studied in brief & new algos are surely to be formulated, designed & in order to overcome some of the works done by the earlier researchers. The examination work will be checked through compelling reenactment results in the Matlab-Simulink environment in order to substantiate the research problem undertaken.

10   Proposed Research Methodology

In this section, the methodology for the proposed research work is being presented. The proposed block diagram shown in the Fig. 4 for biometric recognition using retina scans explains the methodology that we are going to adopt in our work to develop an efficient retinal recognition system for humans. In this work, first, the retina recognition system has got 2 levels (arms), viz., the registration arm: the retina enrolment section and the verification arm: the retina recognition phase. Note that in both the above mentioned 2 cases, viz., the enrollment arm and the confirmation arm, the layout of the retina examine picture to be checked is thought about with the set of retina scan templates in different angles which are stored in the retina databases.

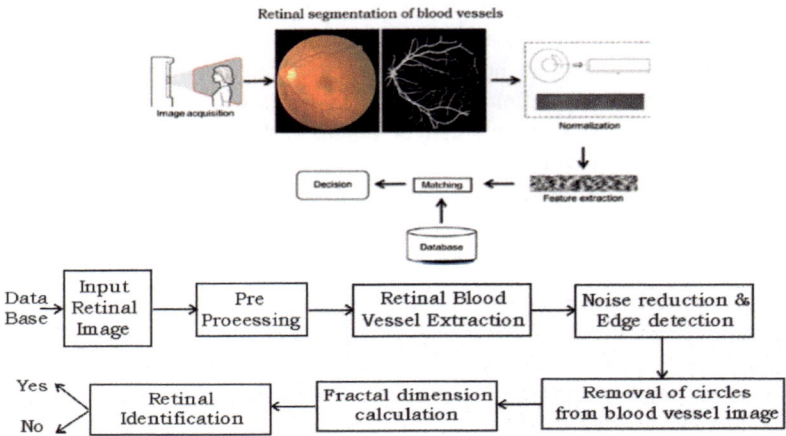

Fig. 4. Block diagram of the proposed retina scan identification technique

11   Implementation

The software implementation results of the automatic biometric retinal identification is going to be carried out using the Matlab tool due to its add-on features and support provided with good tool boxes. Program is created in the Matlab condition as.m file w. r.t. the new approaches that we are going to develop. The developed .m files are going to be run & the results of the simulation is seen, graphically portrayed in the simulation results section. Finally, the conclusions are going to be drawn on the observation of the simulation results & going to be compared with the work done by others. For the simulation purposes, standard retinal databases are going to be used & some images are going to be taken from the eye hospitals across the country. Hardware implementation of the proposed research work may also be thought of if time permits at the end [11].

12   Advantages & Future Scope

Retinal filtering finds a great deal of focal points in the current computerized situation and has been utilized in jails, for ATM character confirmation and the avoidance of welfare misrepresentation. Retinal filtering likewise has medicinal applications. Transferable sicknesses, for example, AIDS, syphilis, intestinal sickness, chicken pox and Lyme ailment just as innate infections like leukemia, lymphoma, and sickle cell weakness, which influence the eyes and different parts of the body can likewise be relieved. One primary preferred standpoint being it is non-contact in nature and henceforth a secured biometric parameter as the retina is at back of the human eyes. Pregnancy additionally influences the soundness of the eyes. Similarly, signs of unending wellbeing conditions, for example, congestive heart disappointment, atherosclerosis & cholesterol issues initially show up in eyes.

13 Conclusions

Research is going to be conducted for the development of novel approaches for secure biometric authentication using retina scans using FD techniques. Different issues w.r.t. the drawbacks of the works done by the various researchers is going to be deeply examined, analyzed, studied in greater depth in more prominent profundity. Currently, the research problem is formulated & defined after an exhaustive survey done on this field. Coding is going to be done in the Matlab environment & the simulation results is going to be observed. Finally, the conclusions are going to be drawn w.r.t. the retinal bio-metric recognition of human beings [1–12].

References

1. Mandelbrot, B.: The Fractal Geometry of Nature. W.H. Freeman and Company, New York (1982)
2. Lopes, R., Betrouni, N.: Fractal and multifractal analysis: A Review. Med. Image Anal. **13**, 634–649 (2009)
3. Blackledge, J., Dubovitskiy, D.: Object detection and classification with applications to skin cancer screening. ISAST Trans. Intell. Syst. **1**(2), 34–45 (2011)
4. Takahashi, T., Kosaka, H., Murata, T., Omori, M., Narita, K., Mitsuya, H., et al.: Application of a multifractal analysis to study brain white matter abnormalities of schizophrenia on T2 weighted magnetic resonance imaging. Psychiatry Res. Neuroimaging **171**, 177–188 (2009)
5. Iftekharuddin, K.: Techniques in fractal analysis and their applications in brain MRI. Med. Imaging Syst.: Technol. Appl., Anal. Comput. Methods **1**, 63–86 (2005)
6. Chaudhuri, B., Sarkar, N.: Texture segmentation using fractal dimension. IEEE Trans. Pattern Anal. Mach. Intell. **7**(1), 72–77 (1995)
7. Oczeretko, E., Jurgilewicz, D., Rogowski, F.: Some remarks on the fractal dimension applications in nuclear medicine. In: Fractals in Biology and Medicine, vol. 3. Birkhäuser, Basel (2002)
8. Belghith, A., Bowd, C., Medeiros, F.A., Balasubramanian, M., Weinreb, R.N., Zangwill, L. M.: Glaucoma progression detection using nonlocal Markov random field prior. J. Med. Imaging **1**(3), 1–10 (2014)
9. Jayasuriya, S.A., Liew, A.W.C., Law, N.F.: Brain symmetry plane detection based on fractal analysis. Comput. Med. Imaging Graph. **37**, 568–580 (2013)
10. Sabaghi, M., Hadianamrei, S.R., Fattahi, M., Kouchaki, M.R., Zahedi, A.: Retinal identification system based on the combination of fourier and wavelet transform. J. Sig. Inf. Process. **3**, 35–38 (2012)

11. Farzin, H., Moghaddam, H.A., Moin, M.S.: A novel retinal identification system. EURASIP J. Adv. Sig. Process **2008**, 280635 (2008)
12. Shahnazi, M., Pahlevanzadeh, M., Vafadoost, M.: Wavelet based retinal recognition. In: 9th International Symposium on Signal Processing and Its Applications (ISSPA), pp. 1–4. Sharjah, Feb 2007

Energy Efficient Dynamic Slot Allocation of Map Reduce Tasks for Big Data Applications

Radha Senthilkumar$^{(\boxtimes)}$ and S. Saranya

Information Technology, MIT, Anna University, Chennai, India
radhasenthilkumar@gmail.com, vsk.saranya@gmail.com

Abstract. Big data applications are very large in size in terms of map and reduce tasks and it takes large time to execute those map reduce based big data applications. Map Reduce is a familiar bulk data processing concept for extensive data computing in cloud environment. The existing Energy Conscious Arrangement is implemented with static map and reduce slot allocation, where there is no room for effective resource utilization. In case of static slot allocation, the pre-determination of map and reduce slots does not employ efficient run time performance and the slots can be strictly under-utilized. To deal with it, the proposed work discovers and enhances the resource utilization and execution time. In the proposed work, dynamic slot allocation technique is accomplished with the existing Energy Conscious Arrangement concept. The technique called Energy Efficient Dynamic Slot Allocation (EEDSA) is accomplished by using dynamic slot allocation technique, which ensures efficient resource utilization. It provides way for slots to be reassigned to map tasks or reduce tasks relying on the requirements. Since slots of the node in the cluster do not deploy any specific functionality, there is no constraint on using map slots in place of reduce slots and vice versa. The proposed technique EEDSA is able to show 10% reduction in execution time for various workloads in comparison with existing Energy Conscious Arrangement technique.

Keywords: Map reduce · Big data · Energy efficient · Resource consumption · Dynamic slot allocation

1 Introduction

Big data applications [1] execute on huge groups within information centers, where the execution time performance becomes the challenge in the execution of high volume workloads. The Map Reduce [2] technique and its open source framework, Hadoop [3], have appeared as the guiding processing framework for big data diagnostics. For arrangement of various Map Reduce jobs, Hadoop initially utilized a FIFO (First In First Out) scheduler. To handle the drawback of waiting time in FIFO, Hadoop then utilized the Fair Scheduler [6]. The recommended Fair scheduler, though, do not consider improving the execution time efficiency when implementing Map Reduce applications. The execution of Map Reduce applications directs to an important factor of the overall expenditure of data centers. The existing Map Reduce scheduling algorithm optimizes

© Springer Nature Switzerland AG 2020
A. P. Pandian et al. (Eds.): ICCBI 2018, LNDECT 31, pp. 801–810, 2020.
https://doi.org/10.1007/978-3-030-24643-3_94

the resource utilization during the execution of big data applications only to a certain level. The recommended scheduling algorithms (FAIR Scheduler) can be integrated and executed by the Hadoop systems. In many situations, execution of big data employs running voluminous workloads at regular intervals. For example, Facebook computes terabytes of data for spam recognition every day. Those workloads make the data centers to utilize job profiling techniques so as to get data for effective resource utilization for each work. Job profiling removes significant performance features of map and reduce tasks for each requesting data centers. Data centers can utilize the facts of extracted work profiles [11] to preprocess latest approximation of jobs' map and reduce phase periods, and make an efficient plan for upcoming implementations.

The existing scheduling algorithms [7] schedule Map Reduce workloads, possessing as the main goal, the reduction of execution time. Most of the existing investigations on the Map Reduce [2] scheduling centres focus on getting better make span (i.e., reducing the time stuck between the starting and the finishing time of an application) of the Map Reduce job's implementation [8, 10]. Still, make span reduction is not inevitably the fine scheme for data centers. Data centers are obligated to carry out executions by their particular time limits, and it is essentially not the best technique to implement the applications as soon as they could, so as to reduce the make span. But the approach is unsuccessful to integrate important optimization chances accessible for data centers to decrease their run time performance. The mainstream of Map Reduce workloads consists of a huge amount of works that do not need quick implementation. With the consideration of this, the resources utilized by the map and reduce jobs when building scheduling results, the data centers could possibly use the resources efficiently and decrease the execution time. The existing resource aware scheduling techniques capture such chances and considerably decrease the Map Reduce energy consumption Most of the big data applications incorporating map reduce concept do not require fast completion. In order to optimize the execution [12] of big data applications, it is essential to consider the resources used up by the tasks. By taking this into consideration, data centers can use their resources effectively and makes scheduling efficiently.

The proposed technique EEDSA (Energy Efficient Dynamic Slot Allocation) ensures effective resource utilization and optimal run time performance. Numerous experiments on a Hadoop platform [5] have been carried out to acquire the time consumption of Map Reduce benchmark dataset such as Terasort, Page Rank, and K Means clustering. The obtained results were used to compare with EMRSA scheduling algorithm.

2   Related Work

2.1   Data Collection

HiBench [5] is a stress testing standard collection for Hadoop and it contains group of Hadoop applications including both simulated micro benchmarks and real big data applications. HiBench is essential for the effective execution of map and reduce tasks of several big data applications over Hadoop framework [10]. It is also used for the enhancement of energy consumption in both hardware and software choices in pro-visions of rate of speed, throughput and resource consumption. ShengSheng Huang et al. [3] estimated and illustrated the Hadoop construction using HiBench, in provi-sions of rate of speed which defines the operation time of a job, throughput which represents the number of odd jobs completed per minute, the Hadoop DFS [Distributed File System], bandwidth, system resource deployment, as well as data contact sample. Currently the benchmark collection consists of eight real time big data applications, categorized into four broad groups. The four groups are micro benchmarks, web search, machine learning and HDFS (Hadoop Distributed File System) benchmark suite. The workloads under four categories are Sort, Wordcount, TeraSort, PageRank, NutchIndexing, Bayesian Classification, K Means Clustering and DFSIO (Distributed File System Input and Output) [5]. The seven applications are directly taken from their open source execution, and the last one is a superior version of the DFSIO (Distributed File System Input Output) benchmark [5] that is extended to evaluate the collective bandwidth delivered by HDFS (Hadoop Distributed File System) [4]. In our proposed work, HiBench is used as the dataset to test Hadoop performance.

2.2   Job Scheduling

Lena Mashayekhy et al. [1] proposed energy sensitive arrangement of map and reduce tasks for big data applications. A map reduce job executes a predetermined number of map and reduce tasks on a cluster which consists of multiple machines. The reduce phase always follows the map phase in the computation of a job. In the map phase, a map slot is allocated with each map task on a machine and each map task takes a segment of the input data and generates key value pairs. The key-value pairs [2], after verified by the reduce phase are then processed by reduce tasks that is allocated to a reduce slot. Therefore, only when the map phase ends, the reduce phase can begin. At last, the distributed file system contains the output produced at the end of the reduce phase. In Hadoop [2], a job tracker process is run by a master node which performs job scheduling and allocates jobs to worker nodes available in the cluster. A task tracker process is executed by each worker node, and the task tracker processes is build up with a predetermined number of map and reduce slots. The status is sent in cyclic to the job tracker by the task tracker to account the number of free slots and the movement of the running tasks.

2.3   Resource Utilization

Map Reduce is a training model which is broadly used for dealing out bulky data exhaustive applications in cluster, cloud background [10]. Most of the scheduling algorithms [9] concentrate on map reduce tasks data locality. Scheduling can be made efficient by using the knowledge of data locality of the intermediate data produced by the map tasks. The knowledge of data locality helps to reduce the intermediate network traffic during the reduce phase and there by speeding the execution of map reduce applications and it results in efficient resource utilization. Time occupied for mapping and reducing the tasks of each job must be known in advance, but this phenomenon is not implemented in the applications. Polo et al. [8] proposed a system for Map Reduce multiple job workloads based on resource aware scheduling technique. This technique focus on developing resource consumption by increasing the utilization of existing task slot to job slot.

2.4 Slot Allocation

YARN [12] solves the ineffectiveness problem of the Hadoop MR Version 1 in the view of resource organization. Instead of using slot, it manages resources into containers. Each host is understood to be accomplished of concurrently managing maximum number of Map tasks and maximum number of Reduce tasks. The containers represents the number of Map slots and Reduce slots, respectively. Cumulating these slots from all the hosts in the cluster, the total number of map slots and reduce slots are worked out. The task of the distribution layer plan is to divide the map slots with the active Map jobs and similarly the number of Reduce slots with the active Reduce jobs. In our proposed work, we consider slots, that are used for map tasks and reduce tasks, which does not hold any specific functionality [5] for execution of two different tasks. And, we prioritize the efficient slots based on its energy consumption using priority queue. Subsequently, during the execution of map tasks in map slots, the reduce slots are idle. Those reduce slots can be used for map task execution efficiently and then the reduce slots are put back in the reduce queue once the map task execution is done.

3   Energy Efficient Dynamic Slot Allocation

3.1 Problem Definition

Big data applications are very large in size in terms of map and reduce tasks and it takes very large time to execute those map reduce based big data applications. To overcome this challenge, we propose a technique EEDSA (Energy Efficient Dynamic Slot Allocation).

3.2 Proposed Work

The first phase of our work is energy focused job arrangement, where the energy consumption by the map and reduce slots were monitored. The most energy efficient slots were prioritized by placing them in the priority queues in ascending order. Job scheduling is generally done to improve the performance of any framework, where the jobs were submitted for execution. The framework that we have taken into consideration is Hadoop, which is the open source implementation for the effective execution of big data applications. The next phase is allotment of map and reduce slots in dynamic manner. Once the slots are prioritized in the priority queues based on their energy utilization, map and reduce slots were placed in their respective queues to provide room for execution. During map reduce execution of tasks, the reduce phase proceeds its execution only after map phase completes its execution. The slot does not hold any specific functionality for execution of map and reduce tasks. Therefore, reduce slots from the reduce queue were utilized for map tasks execution and map slots from the map queue were utilized for reduce tasks execution. In Fig. 1, EEDSA is proposed as architecture diagram. The two priority queues were created (map and reduce) and the slots of the nodes were sorted based on energy utilization and processing time. The slots were placed in the respective queues. The workloads were executed based on dynamic slot allocation for the execution of map and reduce tasks.

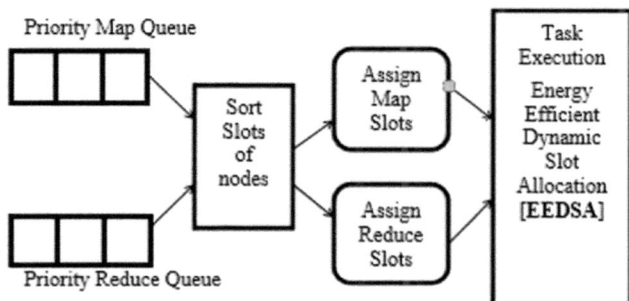

Fig. 1. Framework of proposed work

3.3   Energy Conscious Arrangement

The energy consumption rate is calculated to test the performance of slots by assigning task to a number of slots and then the slot performance is monitored. The performance metrics were acquired by measuring the energy consumption rate, which is the ratio of energy difference (energy consumption by slot during idle and task execution) and the processing time of slots. The energy consumption rate is calculated using energy difference and processing times for both map and reduce tasks. Each and every slot is given a task to execute and its energy usage for a task is obtained during the energy consumption phase. Similarly, all slots are given the same task to execute and their energy usage for the same task are calculated.

After the calculation of energy utilization of the slots in the map phase and reduce phase, sort the slots in the increasing order of energy utilization in the respective queues, map and reduce. In procedure 3.1, the energy utilization calculation is done and the slots are assigned in their respective priority queues. The term 'e' refers to the difference in energy usage between a slot assigned with a task and the idle slot. The term 'p' refers to the processing time of the slot. The slot 'i' refers to the particular map slot and slot 'j' refers to particular reduce slot.

Procedure 3.1: Assigning map and reduce slots to Queues
Input: Slots of nodes in a cluster
Output: Map and Reduce Priority Queues with slots assignment

1 Create map and reduce priority queues
2 Predetermine count of slots for map task
3 Predetermine count of slots for reduce task

4   For each slot in total map slots do
    a. Calculate energy usage = e/p for slot i.
5   For each slot in total reduce slots do
    a. Calculate energy usage = e/p for slot j.
6   Sort the map and reduce slots based on energy usage
7   Assign slots to map and reduce priority queues

The input to the module is the available slots of nodes in a cluster. The empty map and reduce priority queues are created. First, the number of slots for map and reduce tasks are identified. Then, for each and every slot in map phase, its energy usage is calculated based on the energy difference values and the processing times of the slots and repeat the same for reduce phase.

Procedure 3.2: Dynamic Slot Allocation
Input: Status from the task tracker
Output: Dynamic usage of map and reduce slots

1   Calculate used map slots and reduce slots
2   Calculate demand for map reduce slots
3   Find the corresponding load factor
4   If map and reduce slots are adequate then
    no exchange of slots
5   If map and reduce slots are not enough then
    no exchange of slots
6   If map slots are adequate, reduce slots are not adequate then
    get map slots from queue
7   If reduce slots are adequate, map slots are not adequate then
    get reduce slots from queue

4   Experimental Analysis

The execution of the proposed technique were made in order to inspect the performance of the intended algorithm, called EEDSA. The EMRSA scheduling algorithm were evaluated with EEDSA technique to get better execution time. The results of EEDSA were compared with EMRSA by using the Hibench dataset. EEDSA were implemented in four machines in order to improve the execution time. EEDSA performed two types of tasks called small and large.

For small tasks, the outcome of EEDSA was evaluated with EMRSA technique and the same procedure was repeated for large tasks. Three workloads such as Pagerank, K-means clustering and Terasort were used from Hibench dataset.

4.1   Dataset Specification

The performance of EEDSA were analyzed by using HiBench benchmark and the proposed algorithm were executed to improve the performance for both single and four machines, one machine was assigned to be a master, where the name node is configured and the other machine(s) are called slaves, where the data node is configured. In Hadoop, multi node configuration was done to improve the performance, in terms of execution time. In Table 1, the workload specification, represents the number of map tasks and number of reduce tasks, used during map reduce execution.

Table 1. Workload specification

Workload	Map Tasks	Reduce Tasks
(48M, 48R)	48	48
(48M, 64R)	48	64
(64M, 48R)	64	48
(64M, 64R)	64	64

4.2 Experimental Setup

The Operating System used is Ubuntu OS and the version can be 14.04 or 15.04. The installation of Hadoop generally requires large memory needs. i.e. RAM need to be at least 8 GB. The higher the RAM size, the better the Hadoop performance. The platform version of the software is 2.7.1 or Hadoop variant as required. The supporting software package is JAVA JDK and its version is 1.8.0_05. The extensive executions of HiBench workloads were carried out on Hadoop platform to examine the performance of scheduling algorithms [EEDSA, EMRSA]. The outcome of the executions prove that EEDSA is considerably superior in terms of execution time compared with EMRSA.

4.2.1 Terasort

Execuiton of Terasort workloads both small and large task have been carried out on Hadoop Platoform by varying the number of map and reduce taks. It is observed that the execution time of EEDSA is considerably less than EMRSA as depicted in Figs. 2 and 3.

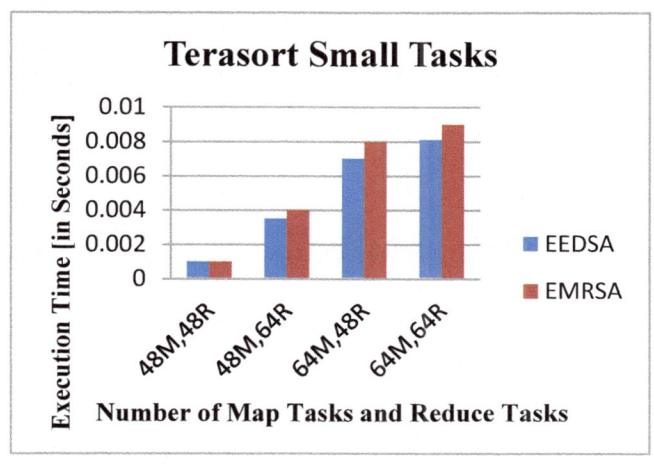

Fig. 2. TeraSort execution with small tasks

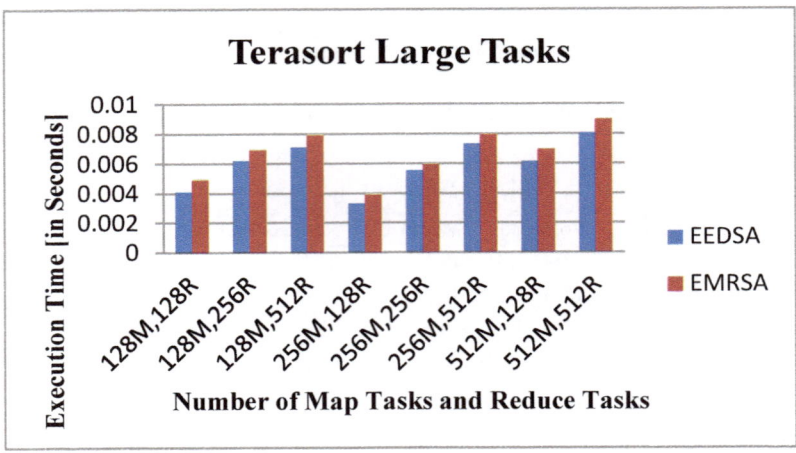

Fig. 3. TeraSort execution with large tasks

4.2.2 Page Rank

The execution time of EEDSA and EMRSA are considered. The obtained results of EEDSA are very close to the optimal solution (Fig. 4).

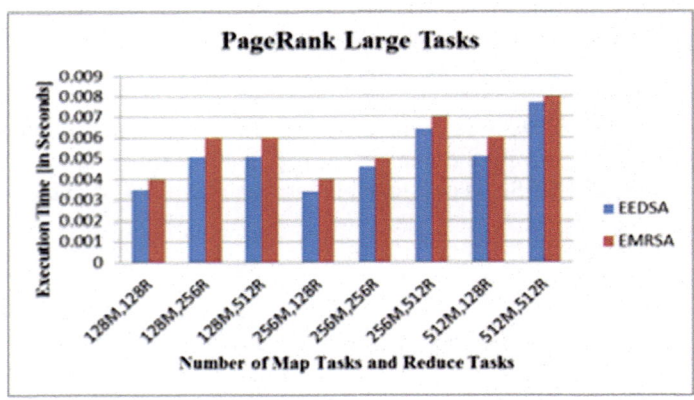

Fig. 4. Page rank large tasks execution.

4.2.3 K-Means Clustering

The execution time of EEDSA and EMRSA are considered. The execution time of K-means clustering were considerably less than the execution time of EMRSA technique. The results show that EEDSA showed considerably less execution time in comparison with EMRSA. For each and every workload, EEDSA is able to achieve the execution time near to the best optimal solution. K-Means clustering execution with large tasks is depicted in Fig. 5 which shows the results of EEDSA.

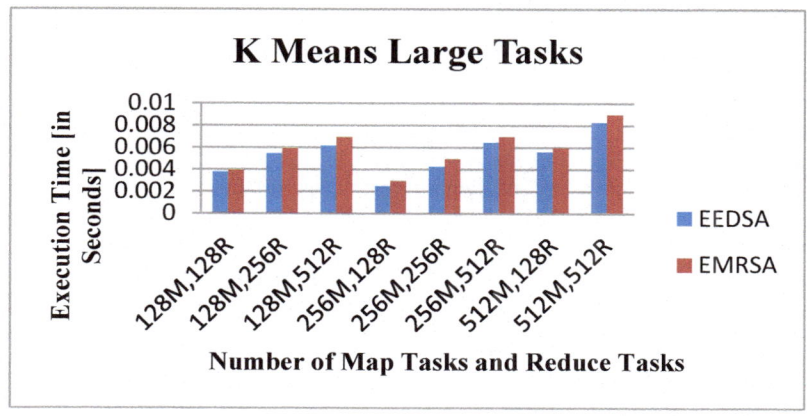

Fig. 5. K-means large tasks execution

5 Conclusion

The Energy Efficient Dynamic Slot Allocation (EEDSA) technique enhances the usage of Map Reduce cluster and execution of workloads fairly. EEDSA is presented in order to ensure fairness for cluster and pools, respectively. In the proposed work, existing Energy Conscious Arrangement technique is integrated with Dynamic Slot Allocation that dynamically assigns map slots and reduce slots to map tasks and reduce tasks alternatively. Dynamic Slot Allocation enhances the resource utilization and reduces execution time considerably. Experimental analysis reveals that EEDSA is able to obtain better performance and also utilize the Hadoop cluster efficiently. The intention is to extend the dynamic slot allocation procedures to heterogeneous atmosphere for many numbers of systems in future.

References

1. Mashayekhy, L., Nejad, M.M.: Energy conscious arrangement of MapReduce jobs. In: IEEE Transactions on Parallel and Distributed Systems, vol. 15, no. 10, pp. 1920–1933, October (2015)
2. Dang, D., Liu, Y.: A crowdsourcing quality estimation technique for big data applications. In: IEEE Transactions on Parallel and Distributed Systems, vol. 11, no. 99, pp. 1–15, February (2015)
3. Huang, S.S., Kodialam, M.S., Kompella, R.R., Lee, M., Mukherjee, S.: Arrangement in MapReduce-like systems for quick deadline time. In: Proceeding IEEE 7th International Conference Computer Communication, vol. 6, no. 4, pp. 774–782, April (2011)
4. Tian, F., Chen, K.: Towards efficient resource delivery for running MapReduce modules in public cloud. Proceeding IEEE International Conference Cloud Computing **16**(7), 155–162 (2011)
5. Huang, S., Huang, J., Dai, J., Xie, T., Huang, B.: The HiBench stress testing benchmark: characterization of MapReduce-based data analysis. Proceeding IEEE **3**(5), 41–51 (2010)

6. Dean, J., Ghemawat, S.: MapReduce: streamlined data computing in large clusters. In: 6th USENIX Symposium Operating System Design Implementation, vol. 15, no. 8, pp. 137–149, March (2004)

7. Salehi, M.A., Radha Krishna, P., Deepak, K.S., Buyya, R.: Preemption-conscious energy administration in virtualized information centers. In: Proceeding IEEE 5th International Conference Cloud Computing, pp. 844–851 (2012)

8. Polo and Ge, R.: Enhancing MapReduce energy optimization for processing exhaustive workloads. In: Proceeding IEEE International Green Computing Conference Workshops, pp. 1–8 (2011)

9. Wang, X., Shen, D., Yu, G., Nie, T., Kou, Y.: A throughput focused task scheduler for enhancing MapReduce performance in job-exhaustive atmospheres. In: Proceeding IEEE 2nd International Conference on Big Data, pp. 171–178 (2013)

10. Ibrahim, S., Jin, H., Lu, L., He, B., Antoniu, G., Wu, S.: Maestro: replica-conscious map development for MapReduce. In: Proceeding 12th IEEE International Symposium Cluster, Cloud Grid Computing, pp. 435–442 (2012)

11. Nejad, M., Mashayekhy, L., Grosu, D.: Truthful greedy algorithms for dynamic virtual machine establishment and allotment in cloud environment. IEEE Trans. Parallel Distrib. Syst. 1, 1 (2014)

12. Kurazumi, S., Tsumura, T., Saito, S., Matsuo, H.: Dynamic computing slots arrangement for i/o exhaustive jobs of big data applications. In: Proceeding IEEE 3rd International Conference Network Computing, pp. 118–92 (2012)

Blockchain Centered Homomorphic Encryption: A Secure Solution for E-Balloting

J. Chandra Priya[(✉)], Ponsy R. K. Sathia Bhama[(✉)], S. Swarnalaxmi,
A. Aisathul Safa, and I. Elakkiya

Department of Computer Technology, MIT, Anna University, Chennai, India
jchandrapriya@gmail.com, ponsydavid@gmail.com

Abstract. .There is a big question arising in today's world of increased technology. If we can bank online and shop online, why we can't vote online? Proficient would confess that affirming the transparency in e-voting, security, and privacy requisites cannot feasibly guarantee through any practically available technology in the imminent future. Even though many security measures have been proposed, it is not extensively implanted, in most of the countries because of the increased security threats. Unlike e-commerce, security of e-voting systems are much more complex. An attack on an e-voting system may cause a legitimate vote to be rejected or a wrong candidate to win. In this research work, a secure online voting system is implemented with ElGamal Elliptic Curve Cryptosystem for vote encryption. The additive homomorphic feature of the Elliptic Curve Cryptosystem is used for computing the results without decrypting each ballot individually. The blockchain is used as a second level of security in the authority side for preserving ballots which utilize SHA-512 for hashing.

Keywords: Blockchain · Homomorphic encryption ·
ElGamal elliptic curve cryptography · E-balloting · SHA-512

1 Introduction

E-balloting with a higher level of security are necessitated to cast votes, making do with the Internet. The online voting methodology is not just increasing the endurance, but it furthermore reduces the comprehensive expenditure on conducting elections. Moreover, there is a chance of inducing the involvement of the citizens to cast their votes by dint of the user-friendly, time-saving procedure, more suitably for the citizens with ailments. The in-depth literature review made on the electronic election technique reveals the demand for a high level of security, thereby preserving the privacy of e-votes [12]. This work is influenced by the requisite for elevating the advancements in the security concern in order to accelerate a wide initiation of online polling all over the world. The secure voting system is designed such that people can open the website and cast their vote at any place and anytime they want and also ensuring confidentiality of voting and verification of ballot integrity and validation of election results.

© Springer Nature Switzerland AG 2020
A. P. Pandian et al. (Eds.): ICCBI 2018, LNDECT 31, pp. 811–819, 2020.
https://doi.org/10.1007/978-3-030-24643-3_95

1.1   ElGamal with Elliptic Curve Cryptography (E²C²)

Elliptic Curve Cryptosystem (ECC) is a public key cryptography system which is formed on discrete logarithms complex of elliptic curves all across finite fields [1]. It encrypts messages appearing at points on an elliptic curve. An elliptic curve is a plane algebraic curve of the general form,

$$y^2 = x^3 + ax + b \tag{1}$$

ECC is used for encryption, the key exchange is based on Diffie-Hellman and ElGamal cryptosystem is used for generating digital signatures. ElGamal is showing multiplicative property of homomorphism. Multiplicative property is computationally expensive and hence ECC is combined with ElGamal to show additive homomorphic property to reduce the computational overhead. Hence, ElGamal ECC was found to be more advantageous that allows the voter to generate a signature for their vote so that the validator can verify whether the vote has been cast by an authorized voter.

1.2   Homomorphic Encryption

Homomorphic encryption is a scheme of encryption technique that accepts the computations can be performed on ciphertexts [3]. Homomorphism makes any encryption algorithm to be more secure as another computation on decryption is not always needed. Henceforth, it reduces the computational overhead in terms of decrypting large number of votes individually [10]. As the additive homomorphic property is used, all the encrypted votes are added and the calculated sum is decrypted to find the results. The additive homomorphism is specified by,

$$E\left(m1\right) + E\left(m2\right) = E\left(m1 + m2\right) \tag{2}$$

where points m1, m2 are cipher texts and E is encryption algorithm. Hence the additive property of ElGamal ECC is used for tallying the individual ballots without decrypting the ballots individually.

1.3   Blockchain

The Blockchain is a digital ledger of transactions which can be automated for recording the data, but eliminates the risk that comes with data being held centrally [2]. A Block in a Blockchain has the valid ballots that are hashed and encoded as a merkle tree [8]. Each block consists of the hash code of the previous block in the chain, and the final block is linked to the initial block of the chain. There are two classification of Blockchain – Public Blockchain and Private Blockchain. Within Public Blockchain anyone can send data to it and also act as a validator. But in Private Blockchain participant and validator access is restricted. In this work, Private Blockchain is used to store the encrypted ballots, thus preventing any modifications in the vote submitted by the voter. The block chain structure is implemented in the authority side as an additional level of security. Blockchain can be viewed as a circular linked list that stores data and the hash value of the precedent block. A merkle tree [11] is a data structure that allows the verification of vote consistency (Fig. 1).

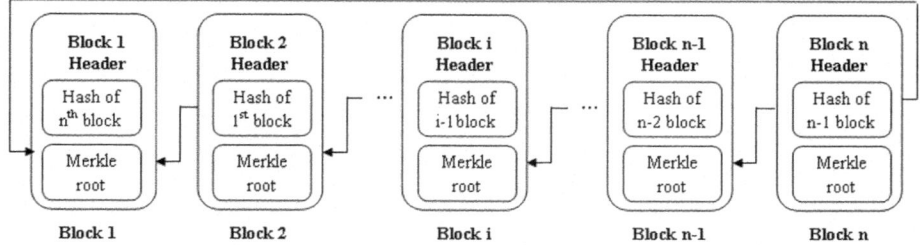

Fig. 1. Structure of blockchain

Hashing is a function that takes input of any length and returns a fixed length output, which makes the blockchain more secured. Table 1 shows that even a small change in the case of the letter can change the hash value drastically. The SHA-512 hashing algorithm [9] is being which consists of the message block of 1024 bits and intermediate hash value of 512 bits. When a hacker tries to attack the blockchain and modify the data, the changes made will results in a totally different hash value of the antecedent block which changes the consecution completely.

Table 1. Comparison of Hash values

Message	Hash
Hello	9b71d224bd62f3785d96d46ad3ea3d73319bfbc2890caadae2dff72519673ca 72323c3d99ba5c11d7c7acc6e14b8c5da0c4663475c2e5c3adef46f73bcdec043
Hello	c2bad2223811194582af4d1508ac02cd69eeeeedeeb98d54fcae4dcefb13cc8 82e7640328206603d3fb9cd5f949a9be0db054dd34fbfa190c498a5fe09750cef

2   Review of Literatures

In paper [1], each vote is encrypted with exponential ElGamal cryptosystem to satisfy the additive homomorphic property. The system depends upon the honesty of authorities and at least one of the authorities is required to be honest. The paper [2] suggests the enhanced version of the standard ElGamal encryption scheme that shows the multiplicative homomorphic property. Generator of the multiplicative group is considered to be a very crucial security factor. The third party to whom the data is submitted is ensured to provide confidentiality. It is said that recovering the private key is highly complex and so it is semantically secured. But, the main characteristics of ElGamal are lost such as digital signature and key exchange. The paper [3] suggests a Secure Privacy Preserving Data Aggregation Scheme (SPPDA) where the homomorphic characteristics of the bilinear ElGamal cryptosystem are used for a privacy preserving computation that ensures the security. An aggregate signature scheme for authentication and integrity in Wireless Body Area Network (WBN) is introduced and proved to be computationally intensive. The paper [4] proposes an optimized realization of an additively homomorphic elliptic curve to be used in the Wireless Sensor Networks. The paper [5] discusses the recording of voting result using Blockchain to be used in the databases of online voting systems to prevent the manipulation of databases. An e-voting system grounded in blockchain structure with Homomorphism exhibited ElGamal encryption and ring digital signature is proposed for across the board electronic voting [6]. The one time ring digital signature confirms the anonymousness of the vote inside the blockchain architecture. The publicly available bulletin boards are out for verification at any instance which henceforth assures the fairness in voting. Moreover, the miner nodes provide sustenance for encrypted ballot counting feasible at a larger scale. The proposed cryptosystem is based on the Identity which conceives the public keys could be instantaneously traced from user IDs in particular phone numbers, email IDs, and social insurance number. Hence the key management strategies of certification-based public key infrastructures (PKI) are made to be simple and can be adapted to achieve the authentication in blockchain. Linear digitally signed data which portrays homomorphism are used in accomplishing linear ciphering upon authenticated data. And the degree of correctness of the ciphering can be publicly verifiable. Even though several schemes for homomorphic signature have been designed in recent pasts, hardly homomorphic signature schemes constructed in identity-based cryptography is very few. A novel identity deployed linear homomorphic digital signature scheme, to avert the drawbacks of the using the public key certificates. This technique has proved to be more secure against existing forgery on robustly chosen message and identity attack on the either side of the heel of the random oracle model. The ID-sited linearly homomorphic digital signature schemes [7] can also be pragmatic in e-business and cloud computing environments. Eventually, the extensive survey of the papers concludes the constraints on how to apply the cryptographic techniques to achieve authentication in blockchain.

3   The Methodology

The online voting system involves: Initialization, Registration, Voting, Tallying and Announcement of results [1] but security must be indulged in all these phases. The entities involved in this system are authorities (5 in number), candidates (max.64), voters and Public Bulletin Board (PBB). Each authority is needed to generate a pair of keys. All the authorities should combinedly generate a public key called Cumulative Public Key (CPK), assuming at least one authority should be honest. Voters are validated by their Voter ID with Voters List Database and registers them. The PBB is a secured broadcast channel that is visible to all the entities where the authorities post the CPK.

ENCRYPTION	**DECRYPTION**
$C_1 = K*G$ where C_1 is in (x,y) form $C_2 = K*M^{-1} \bmod P$	C_2*C_1 is in (x,y) form where $K = x$ $M = C_2*K^{-1} \bmod P$

where M -> Vote
P -> Large Prime Number
K -> Shared Secret Key
(C_1, C_2) -> Cipher Text

3.1   Combined Authority Authentication

Each authority has a unique Authority ID. He/She authenticates to the system using One Time Pad which is sent to their mobile phone or e-mail. On successful authentication, they will be authorized to generate their key pairs (APK, ASK). The multiauthority structure has a protected directory where each of their public components can be posted and is visible only to them (Fig. 2).

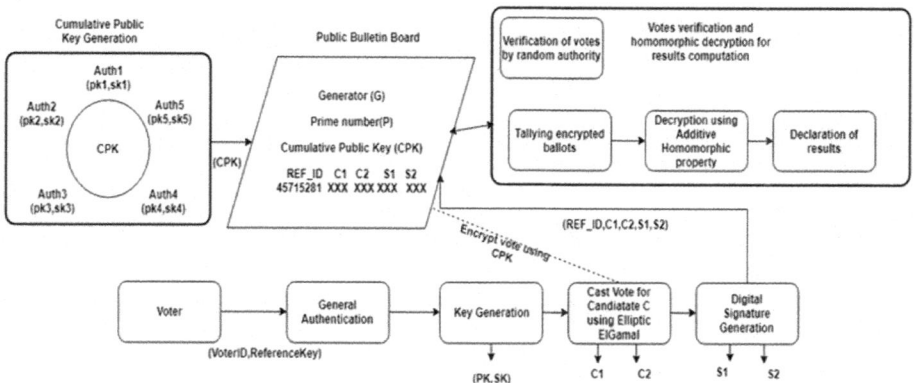

Fig. 2. E^2C^2 blockchain E-ballot architecture

3.2   Voter Authentication and Authorization

The voters should submit their Voter ID for verification, which is authenticated to generate a unique reference key with which they can login into the voting system at the time of election. Furthermore, the system permits them to generate a public-private key combination (PKv, SKv) with the public key components furnished in the bulletin board.

3.3   Key Generation of Authorities and Voters

All the authorities broadcast their public keys to compute the Cumulative Public Key (CPK) which is published in Public Bulletin Board. The key generation follows the ElGamal Elliptic Curve Key Generation scheme which is given in Algorithm 1.

Algorithm 1 Key Generation of ElGamal Elliptic Curve cryptosystem
---
Input: A random number X
Output: Public Key and Secret Key Pair (PK, SK)
Initialization;
Select a number X such that 0<X<P-2
Compute Y such that Y = X.P (ECC product)
Return Y as PK
Return X as SK

3.4   Encrypted Vote Casting

Each Voter aftermath the authentication and key generation process select one Candidate from the list to cast votes which is encrypted with CPK and then voters sign it with their own secret key. Encryption of vote m is done by the selection of a point P_m on the elliptic curve which is encrypted using CPK of the authorities. The exponentiation in ElGamal is replaced by ECC multiplication and multiplication in ElGamal is replaced by ECC addition to reduce computation overhead. The voter needs to generate signature for the vote, which will be verified by the authority to ensure authenticity. The algorithm for encryption of the vote is given in Algorithm 2.

Algorithm 2 Encryption of Vote M_i by Voter V_i

Input: Generator (G), Voter's Secret Key (X), CPK, Vote M
Output: C1, C2 (cipher texts)
Initialization;

1) Compute C1 = ECC.MUL (G, X)
2) Compute T = ECC.MUL (CPK, X)
3) Compute C2 = ECC.ADD (M, T)
4) Return C1, C2

---

where ECC.MUL() and ECC.ADD() are functions that perform point multiplication and point addition on an elliptic curve.

3.5   Vote Authentication Using DSA

The signatures sent by the voters while casting the votes are verified by a random authority. If the verification is found to be false, then the vote is assumed to be a false vote and it is ignored. The authority also verifies whether the particular voter has cast only one vote. If multiple votes are detected, then the vote is ignored.

Algorithm 3 Digital Signature Verification

Input: Generator (G), Voter's Public Key (Y), S1, S2, Hash (H), Order (N)
Output: True / False
Initialization;
1. Compute W = INVERSE (S2, N)
2. Compute V1 = H * W (mod N)
3. Compute V2 = S1 *W (mod N)
4. Compute T1 = ECC.MUL (G, V1)
5. Compute T2 = ECC.MUL (Y, V2)
6. Compute V = ECC.ADD (T1, T2)
7. IF V (mod N) == S1 THEN return
8. ELSE return false

3.6   Homomorphic Decryption for Result Computation

Homomorphic property of Elliptic ElGamal is used for computing the results. Instead of decrypting each ballot individually, the ballots are summed up and decrypted using the additive property of Elliptic ElGamal Cryptography. The voters can verify the results by checking the Public Bulletin Board and verify whether their vote has been modified or not.

4   Security Analysis

4.1   Prevention of Multiple Voting

Each vote casted is established publicly on the Public Bulletin Board. The public board contains only one entry for each voter. Thus, when a voter tries to cast multiple votes, the process is automatically resisted by the system.

4.2   Preserve Privacy of Voters

To ensure privacy of voters, the authority who receives the vote can only verify whether the vote has been casted by the authorized voter or not. The authority cannot see the vote cast by the voter. This verification is done using an ElGamal Digital Signature algorithm. The vote displayed in the PBB is also in an encrypted format.

4.3   End to End Verifiable

To mould the system end to end verifiable, the voter after casting their vote can see the PBB for their reference for ensuring the integrity of their vote.

5   Performance Analysis

To achieve a satisfiable security, RSA requires a key of at least 1024 bits, whereas 160 bit key of ECC gives more security. Figure 3 shows that RSA requires larger key sizes compared to ElGamal-ECC to provide the marked level of security. Also the prime number taken for RSA should be a larger to provide a high level of security, whereas in an ECC lesser number of coordinates are sufficient to provide the same level of security.

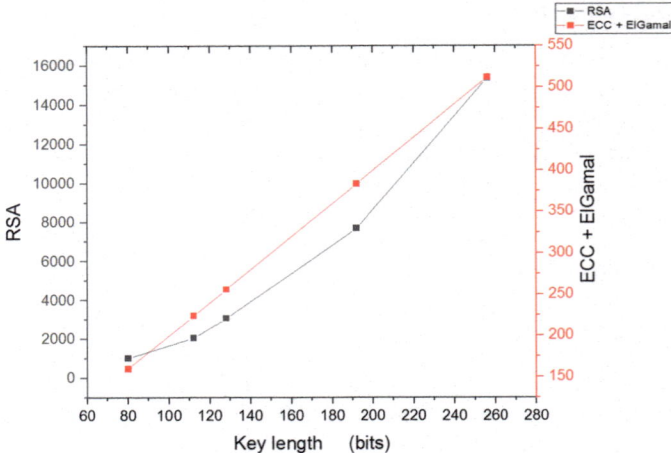

Fig. 3. RSA vs. ECC + ElGamal comparison

6   Conclusion

A secure and privacy preserving online voting is presented with an ElGamal ECC algorithm for encryption and for ensuring authentication through digital signatures. A block chain structure is built to ensure an additional level of security for ballot preservation. The extension of this research work will proceed with an idea of utilizing Blockchain structure further for prediction of poll behavior with classification and deep learning algorithms.

References

1. Yang, X., Yi, X., Nepal, S., Kelarev, A., Han, F.: A secure verifiable ranked choice online voting system based on homomorphic encryption. IEEE Access **6**, 20506–20519 (2018)
2. El Makkaoui, K., Beni-Hssane, A., Ezzati, A.: Cloud-ElGamal: an efficient homomorphic encryption scheme. Intl Conf. Wireless Networks & Mobile Communications (WINCOM), pp. 63–66 (2016)
3. Ara, A., Al-Rodhaan, M., Tian, Y., Al-Dhelaan, A.: A secure privacy-preserving data aggregation scheme based on bilinear ElGamal cryptosystem for remote health monitoring systems. IEEE Access **5**, 12601–12617 (2017)
4. Hanifatunnisa, R., Rahardjo, B.: Blockchain based e-voting recording system design. 11th Intl Conf. Telecommunication Systems Services and Applications (TSSA), pp. 1–6 (2017)
5. Minfeng, F., Wei, C.: Elliptic curve cryptosystem Elgamal encryption and transmission scheme. Intl Conf. Computer Application and System Modeling (ICCASM), pp. V6-51-V6-53 (2010)
6. Wanga, B., Suna, J., Hea, Y., Panga, D., Lua, N.: Large-scale election based on blockchain. Intl Conf. Identification, Information and Knowledge in the Internet of Things, pp. 234–237 (2018)

7. Ekblaw, A., Azaria, A., Halamka, J.D., Lippman, A.: A case study for blockchain in healthcare: medrec prototype for electronic health records and medical research data (2016)
8. Czepluch, J.S., Lollike, N.Z., Malone, S.O.: The use of block chain technology in different application domains. IT University of Copenhagen (2015)
9. Athanasiou, G.S., Michail, H.E., Theodoridis, G., Goutis, C.E.: Optimising the SHA-512 cryptographic hash function on FPGAs. IET Comput. Digital Tech. **8**(2), 70–82 (2014)
10. Yang, Y., Zhang, S., Yang, J., Li, J., Li, Z.: Targeted fully homomorphic encryption based on a double decryption algorithm for polynomials. Tsinghua Science and Technology **19**(5), 478–485 (2014)
11. Macrinici, D., Cartofeanu, C., Gao, S.: Smart contract applications within blockchain technology: a systematic mapping study. In: Telematics Inform. Elsevier **35**(8), 2337–2354 (2018)
12. Feng, Q., He, D., Zeadally, S., Khan, M.K., Kumar, N.: A survey on privacy protection in blockchain system. In: J. Netw. Comput. Appl. **126**, 45–58 (2018)

Hybrid HSW Based Zero Watermarking for Tampering Detection of Text Contents

Fatek Saeed$^{(\boxtimes)}$ and Anurag Dixit

School of Computing Science and Engineering,
Galgotias University, Noida, UP, India
fateksaeed@gmail.com

Abstract. In this paper, the Hybrid Structural component and word length (HSW) based zero watermarking technique is proposed in which the content of text document is not altered for watermark embedding. It involves two steps namely watermark embedding and extraction. From the data owner the plain text is obtained and the watermark is generated. It is based on utilizing the characteristics of text. The text and the watermark are registered to the certifying authority and it is further used for pattern matching which detects the tampering in text document. The proposed embedding algorithm HSW is used for embedding the text. After watermarking, the tampering is applied to the content. The extraction algorithm for HSW is utilized for getting the applied watermark from the text content. For each sub pattern, the corresponding watermarking pattern is extracted from the certified authority. The extracted patterns are compared with the generated pattern. The pattern matching procedure is taken into account for detecting the tampered content

Keywords: Watermarking · Tampering detection · Structural components · Embedding · Extraction

1 Introduction

The watermarking is a type of information that is embedded into a digital media such as image, video, audio, and text. The watermark present in the media files shows the ownership proof of the media file. The watermark in the files may be visible or invisible; it differs based on the visibility of the watermarked content. The media content is watermarked based on the document type. Some classified watermarking types are image watermarking, video watermarking, audio watermarking, text watermarking, database watermarking and so on. The digital watermark in the media files will be robust or fragile. In robust watermarking, the watermark in the watermarked content will not be affected even by altering the original content. In fragile type, the watermark will be affected or destroyed when the original content is altered [1].

The main aim of the watermarking researchers needs to modify or upgrade the watermarking techniques. The area of digital watermarking is required to be concentrated on the precision of watermarking techniques. The techniques that are used for the watermarking need to satisfy some requirements such as, watermark size, the robustness

© Springer Nature Switzerland AG 2020
A. P. Pandian et al. (Eds.): ICCBI 2018, LNDECT 31, pp. 820–826, 2020.
https://doi.org/10.1007/978-3-030-24643-3_96

of fragility, and imperceptibility. Different watermarking techniques are focusing on tamper detection and authentication of digital contents [2].

One of the most challenging types of watermarking the digital media is text watermarking. It's hard to hide watermark information on text contents. The text watermarking techniques are designed to protect the text and to detect the text tampering happening on text documents. It's very difficult to control the tampering over the text contents [3]. The tampering is a process of copying original content or changing the font style and stealing the content from original documents. It's easy to copy and manipulate the original document with the advanced computer technology. Digital watermarking provides different techniques to protect the authentication, validation, and copyrights of the data owner or authorized user [4].

On invisible watermarking, the watermark information in the digital media will not be visible. Different techniques are used in both methods to watermark the media content [5]. The watermarking techniques that are used for tampering detection have many limitations, and they are not valid on all type of text. The watermark will be embedded in the text document make modification on content; it results in reducing the quality, capacity, meaning and values [6].

The geometrical attacks include some malicious process like reformatting, sentence swapping, remove or add words, sentences, paragraphs and shuffling the paragraph. The system attack is held by an attacker with some signal processing tools such as principle component analysis, independent component analysis, clustering, etc. [7].

The semantics of the textual content is used to create watermark on semantic method. The structural watermarking method is a newest watermarking method uses the structure of the textual content to create the watermark. This type of watermarking method never alter any content after embedding the watermark and also this method is not valid on some textual content like web content, mathematical notations with floating point and legal text documents [8].

A novel method HSW is developed in this paper for tampering detection of plain text contents and it utilizes the contents of text itself for its authentication. The tampering attacks such as insertion, deletion and reordering can be avoided and the information authenticity is proved with the help of proposed algorithm.

2 Proposed Algorithm for Tampering Detection

The proposed approach of tampering detection is based on zero watermarking which uses the characteristics of text contents to generate watermark rather than embedding the watermark into the text. The proposed approach is based on structural component and word length and it is applicable for all kinds of documents. It consists of two stages namely text embedding and text extraction. The watermark generation can be done with the data owner and the extraction can be accomplished with the CA. The architecture of proposed tampering detection is shown in Fig. 1.

The proposed algorithm takes text document as an input for tampering detection and the watermark is generated for the text document based on HSW approach. The extracted pattern of watermark is registered with the certifying authority (CA). The attacker may modify the content of the document. In the tampering detection process, the extraction algorithm is applied to extract the watermark and the pattern is matched with the registered pattern of CA. Based on the level of pattern matching the threshold is set to take decision about tampering.

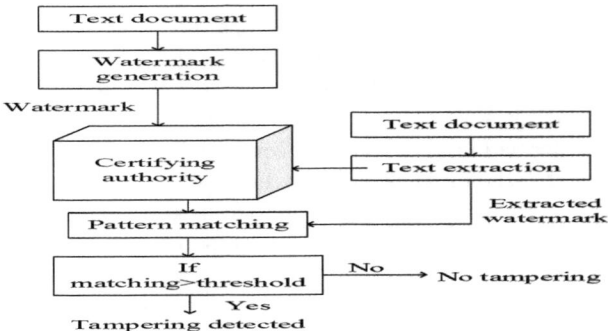

Fig. 1. Overview of proposed algorithm.

2.1 Watermark Generation

The proposed watermarking generation is based on word length and structural components of the text document. Initially, the text partition is formed by considering preposition as a separator. Then the groups are formed by combining the partitions based on the size of the group. In order to form partition and group, the group size and preposition is taken as an input. The frequency of first letter form the double letter word is identified to form the high frequency list. Then the letters are sorted in descending order which places the letter with highest value first. The embedding algorithm is utilized for generating watermark for the text content. The watermark pattern generated is registered with CA which is considered as a trusted third party of digital community. When conflict arises, the extracted watermark pattern is analyzed with the text content to evaluate its authenticity. The CA executed the detection algorithm and it provides response to the data owner who registered the watermark pattern. The text content may be attacked with several possible ways. The proposed algorithm detects tampering for insertion, deletion and reordering attacks.

2.2 Watermark Extraction and Tampering Detection

The extraction algorithm extracts the watermark from the text content. The issue regarding copy right protection can be resolved with CA which keeps the extraction algorithm. Text extraction process is described in Algorithm 1.

Algorithm 1: HSW based text extraction and tampering detection

---

1. Based on the size of the group partition the text extracted.

2. The size of the group is utilized combining the partitions.

3. Consider the double letter word and the word with greater than four letters.

4. Process the consideration for each partition and for each group.

5. Extract the first letter and computes its frequency.

6. The first letter is extracted and combined for each partition if the word length is greater than four.

7. Make the pattern for each partition from step 6.

8. Combine the size of the pattern and extracted first letter from the double letter.

9. Reject the addition information extracted to make unique size.

10. Combine the pattern and the first letter which generates the modified pattern.

11. The extracted pattern is compared with the pattern registered with CA.

12. The score is generated for each pattern based on the pattern similarity.

13. The sore from all patterns is added to get the final score.

14. Set the threshold for the tampering detection.

15. If **(score > threshold)**

16   Tampering detected

17 else

   Document is authenticated.

18 Perform the same process for all kinds of attacks.

In the extraction algorithm, the text document is partitioned based on the preposition which is used in watermark generation. The process of partitioning simplifies the process of pattern extraction. Since the pattern is extracted for each group separately. The partition is also based on the preposition from the same text contents which utilizes the components of text content. The text group is formed by combining the partitions based on the group size. From each group, the double letter word is analyzed and the frequency for the first letter is computed.

The watermark is extracted and compared with the registered pattern from the CA. For each pattern, the score is generated based on the matching probability. Based on the capacity of matching, the score of all patterns is added to get the combined score. The threshold is set to decide that the tampering is high or low. This algorithm efficiently detects the generated watermark from the text content if there is no attack detected. This kind of text document is termed as authentic text without tampering. The proposed algorithm is detects tampering for the attacks such as such as insertion, deletion and reordering. Rather than changing the original contents of text document, the characteristics of text is used. The proposed hybrid algorithm has the advantages of high accuracy since it is based on structural components and word length. In addition to that it also authenticates the original text documents.

2.3   Experimental Results and Analysis

The proposed technique of tampering detection is simulated in MATLAB by varying the number of documents. The absolute score between original watermark pattern and extracted pattern is calculated to evaluate the degree of tampering. The work is evaluated for all possible attacks such as insertion, reordering and deletion. The detection accuracy is considered for the performance evaluation of the proposed method. The detection accuracy for the insertion task is shown in Fig. 2. The accuracy of detection is higher than the existing based on the performance evaluation. The accuracy of insertion attack is 93.5, 92.5, 92, 91.5, 90.8 and 89.5 for the number of documents 20,

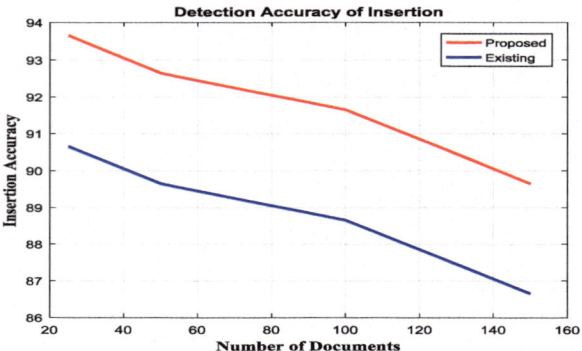

Fig. 2.   Insertion accuracy comparison by varying the number of documents.

40, 60, 80, 100, 120, 140 and 160. For the existing approach, the detection accuracy of insertion task is 90.5, 90, 89.5, 80, 88.5, 87.9 and 87 for the number of documents 20, 40, 60, 80, 100, 120, 140 and 160.

The average accuracy for all detection accuracy for the proposed tampering detection is compared with the detection accuracy of existing approaches. The detection can be performed by varying the number of documents from 20 to 160. The

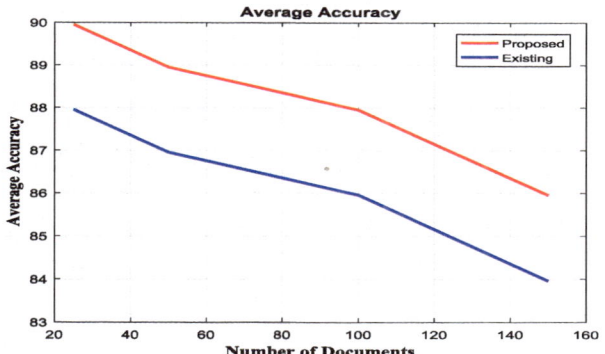

Fig. 3. Average accuracy comparison by varying the number of documents.

numbers of documents are varied as 20, 40, 60, 80, 100, 120,140 and 180 for the detection task. The average accuracy for the proposed detection task is 90, 89.5, 88.8, 88.3, 88, 87.2, 86.3 and 86. The average accuracy for the existing approaches are 88, 87.5, 86.8, 86.3, 86, 85.2, 84.3 and 84 (Fig. 3).

When the numbers of documents are increased, the performance of detection accuracy is degraded. By averaging the detection accuracy of all attacks, the performance is high for the proposed approach and low for the existing approach. The increasing number of documents reduces the overall detection accuracy of both proposed and existing approach. The high detection accuracy obtained by the insertion attack, deletion attacks, reordering attacks and average accuracy shows the efficiency of the proposed approach.

3   Conclusion

Initially, the plain text is obtained from the data owner and the content of the text data is analyzed. Then for the plain text watermark is generated with the novel embedding algorithm Hybrid Structural component and word length based zero watermarking. The text partition is formed and the partition is combined together to form groups depends on the group size. Afterwards, the word or letter from each group is analyzed to detect the maximum occurring list. This list is utilized for the construction of watermark key based on watermark. The watermark is registered with the certifying authority with the original plain text, author name, date and time. The proposed watermarking technique generates the watermark based on the characteristics of plain text rather than embedding the text itself. After watermarking the text may be attacked or un-attacked.

References

1. Murugan, R., Jaseena, K.U., Abraham, J.T.: An invisible watermarking technique for integrity and right protection of relational databases. Int. J. Appl. Eng. Res. **12**(24), 15754–15758 (2017)
2. Bashardoost, M., Rahim, M.S.M., Saba, T., Rehman, A.: Replacement attack: a new zero text watermarking attack. 3D Research **8**(1), 8 (2017)
3. Kamaruddin, N.S., Kamsin, A., Por, L.Y., Rahman, H.: A review of text watermarking: theory, methods, and applications. IEEE Access 6, 8011–8028 (2018)
4. Goyal, L., Raman, M., Diwan, P., Vijay, M., Goyal, L., Raman, M., Diwan, P., Vijay, M.: A robust method for integrity protection of digital data in text document watermarking. Int. J. Sci. Res. Dev 1(6), 14–18 (2014)
5. Alotaibi, R.A., Elrefaei, L.A.: Improved capacity Arabic text watermarking methods based on open word space. J. King Saud Univ.-Comput. Inform. Sci. (2017)
6. Ba-Alwi, F.M., Ghilan, M.M., Al-Wesabi, F.N.: Content authentication of English text via internet using zero watermarking technique and Markov model. Int. J. Appl. Inf. Syst. (IJAIS) **7**(1), 25–36 (2014)
7. Chroni, M., Nikolopoulos, S.D.: Watermarking PDF documents using various representations of self-inverting permutations. arXivpreprint arXiv:1501.02686 (2015)
8. Bhambri, P., Kaur, P.: A novel approach of zero watermarking for text documents. Int. J. Ethics in Eng. Manage. Educ. (IJEEE) **1**(1), 34–38 (2014)
9. Kaur, S., Babbar, G.: A zero-watermarking algorithm on multiple occurrences of letters for text tampering detection. Int. J. Comput. Sci. Eng. **5**(5), 294 (2013)
10. Liu, Y., Zhu, Y., Xin, G.: A zero-watermarking algorithm based on merging features of sentences for Chinese text. J. Chin. Inst. Eng. **38**(3), 391–398 (2015)
11. Ahvanooey, M.T., Dana Mazraeh, H., Tabasi, S.H.: An innovative technique for web text watermarking (AITW). Inform. Secur. J. Glob. Perspect. **25**(4–6), 191–196 (2016)
12. Ahvanooei, M.T., Tabasi, S.H., Rahmani, S.: A novel approach for text watermarking in digital documents by zero-width interword distance changes. DAV Int. J. Sci. **4**(3), 550–558 (2015)
13. Fu, Z., Sun, X., Shu, J., Zhou, L., Wang, J.: Verifiable text watermarking detection to improve security. Int. J. Secur. Appl. **8**(5), 1–10 (2014)

Secure Two-Phase Decentralized Data Deduplication with Assured Deletion in Multi User Environment

Amolkumar N. Jadhav[1(✉)], Pravin B. More[2], Yogesh S. Lubal[2], and Vishal S. Walunj[1]

[1] (Computer Engineering), DYPSOEA, Pune, India
pramolkumar45l@gmail.com, vishal.walunj@dyptc.edu.in
[2] (Computer Science and Engineering), ADCET, Ashta, India
pravinmore9678@gmail.com, yogeshlubal@gmail.com

Abstract. Use of cloud for storage has acquired great importance as number of users and ever increasing growth in generation of data by these users. Even though existence of multiple replicas of files on multiple nodes leads to better availability, on the other hand it unnecessarily drains off valuable storage space on Storage Cloud Service Provider (S-CSP). Data deduplication is a technique that needs to be applied to get rid of these problems. Data deduplication is widely used by S-CSP to reduce amount of storage and save bandwidth by eliminating duplicate copies of same files. Proposed system provides data deduplication at both client and server side thus eliminates target collision. Intra and Inter user data deduplication technique also reduces workload of S-CSP by allowing Deduplication Proxy (DP) to apply intra user data deduplication check. Data deduplication may leads to problem of uncontrolled deletion of files in multi-user environment is also eliminated in this approach.

Keywords: Deduplication · DP · S-CSP · Intra and inter-user deduplication

1 Introduction

Cloud computing allows management of data on remote storage efficiently at low cost in scalable and location independent infrastructure. As data collected from many users at S-CSP is growing continuously, out of which most of the data is repeatedly stored. This leads to wastage of critical resources such as storage and bandwidth.

Deduplication is technique that allows efficient utilization of limited storage space offered by cloud providers. Instead of keeping multiple file copies with same contents, deduplication eliminates redundant files by keeping only one physical copy and referring other redundant files to original. While keeping files on S-CSP, client side data deduplication comes with problem of target collision. Target collision occurs when malicious client uploads a file that does not corresponds to its claimed identifier.

Existing approach does data deduplication by maintaining convergent keys on multiple Key Management Cloud Service Provider (KM-CSPs). Unnecessarily maintaining convergent keys on multiple KM-CSPs requires extra efforts in managing all

© Springer Nature Switzerland AG 2020
A. P. Pandian et al. (Eds.): ICCBI 2018, LNDECT 31, pp. 827–835, 2020.
https://doi.org/10.1007/978-3-030-24643-3_97

KM-CSP nodes. Approach presented in this paper avoids problem of Target Collision by providing hash calculation after client side encryption. This also reduces overhead of maintaining convergent keys at multiple KM-CSPs by storing them on intermediate DP. Deduplication is applied at two different levels, intra-user and inter-user data deduplication. Intra-user data deduplication occurs when files uploaded by the same user are checked for duplicate copies. Inter-user data deduplication occurs when files uploaded by multiple different users are checked for duplicate copies.

Along with these problems approach overcomes problem of uncontrolled deletion of files in multi-user environment. If one user uploads a file and another user uploads same file again then file is not uploaded again instead, a reference is stored to first file. If first user deletes that file then original file might get deleted thus second user's reference now points to file which has been deleted. Unfortunately, this causes loss of file for second user.

2   Related Work

Earlier approaches presented different ways of implementing data deduplication. Method addressed in [2, 3] implements deduplication of data so that in case of public clouds this may efficiently utilize valuable space allowing users to store more and more files even in public clouds.

Approach presented in [1] has to maintain N Key Management Cloud Service Provider (KM-CSPs) for storing convergent keys. Even though this provides better availability, again it increases complexity in maintaining these different nodes.

Similarly implementation of data deduplication in cloud has provided in [2] with the cost of managing N number of deduplication proxies. As number of cloud user increases, a new instance of deduplication proxy needs to be created and maintained. This increases overhead of managing these instances.

3   Proposed System

1. Overcomes Target Collision in Client-Side Deduplication.

Target collision avoided by calculating hash i.e. file identifier of files after encrypting that file. Thus if user uploads file with wrong file identifier, then it will be checked on DP and this upload request is terminated on DP itself and is not forwarded to S-CSP.

2. Overcomes No Encryption in Server-Side Deduplication.

In server-side data deduplication file identifier is calculated on S-CSP. In order to do this file needs to transferred without encryption. This may cause revealing of file contents to malicious user or attacker. As file is encrypted on client-side prior calculating file identifier, this eliminates no encryption problem of server-side data deduplication.

3. Overcomes maintaining Convergent Keys on multiple KM-CSPs.

Instead of storing convergent keys on N number of KM-CSPs this approach stores convergent key on DP. Thus eliminates overhead of maintaining N KM-CSPs. Convergent Keys are stored on S-CSP in encrypted form by public key of users so it preserves Proof of Ownership (PoW).

4. Overcomes problem of uncontrolled deletion in multi-user environment.

Provides controlled deletion of files in multi-user environment. If some file has reference pointing to it then deletion of file does not deletes physical file from S-CSP. Instead it only deletes reference pointing to it. If no reference is pointing then deletion of file cause physical file to be deleted.

Implementation of Secure Two-Phase Decentralized Data Deduplication in Cloud with Assured Deletion in Multi User Environment with the help of proposed system involves three key components:

5. Authorized Clients

To use services provided by the S-CSP, user has to first sign up and agree on terms & conditions offered by the S-CSP. Once they receive their credentials through sign up, they can authorize themselves to access cloud services.

6. Deduplication Proxy (DP)

Intermediate node responsible for carrying out intra-user data deduplication. Authorized clients can request to deduplication proxy for accessing services. Deduplication proxy will check for intra-user data deduplication to reduce some percentage of duplicate check from S-CSP.

7. Storage-Cloud Service Provider (S-CSP)

Storage as a Service (SaaS) is provided by the S-CSP. Authorized client can upload their files to S-CSP but again they are not having direct access to S-CSP. Mandatorily

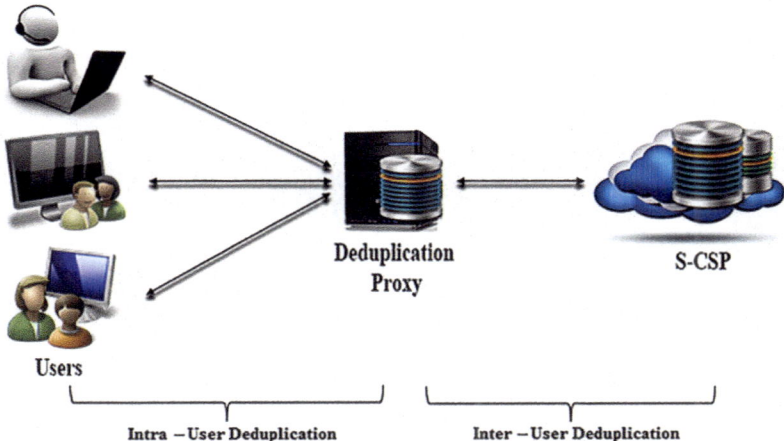

Fig. 1. System design of secure two-phase decentralized data deduplication with assured deletion in multi user environment

they have to go through intermediate deduplication proxy. If intra-user data deduplication check has been false then inter-user data deduplication is implemented at S-CSP.

System setup for secure two phase decentralized data deduplication in cloud
environment is depicted in Fig. 1 with intra and inter-user data deduplication. All
Clients has direct access to DP and indirect access to S-CSP i.e. through DP. All intra-
user data deduplication requests are processed by the DP and they are not forwarded to
S-CSP. If intra-user data deduplication check returns successful then this means that
same file is present on S-CSP and is uploaded by same user. Thus this will only
uploads reference to existing file.

If intra-user data deduplication check returns false then that request will be
forwarded to S-CSP where S-CSP will check for inter-user data deduplication. If this
check returns successful then it will cause upload of reference to existing file uploaded
by some other user.

If both of these data deduplication checks returns false then this implies that file
which client is uploading does not exists on S-CSP, thus upload of original file will
takes place.

3.1   Problem of Uncontrolled Deletion in Multiuser Environment

Problem of uncontrolled deletion in multi user environment is depicted in Fig. 2. User
1 uploads encrypted file1.enc by calculating hash on client-side. This hash is stored on
deduplication proxy for intra-user deduplication and S-CSP for inter-user deduplication. User 2 uploads same file then file will not be uploaded again only reference is
uploaded.

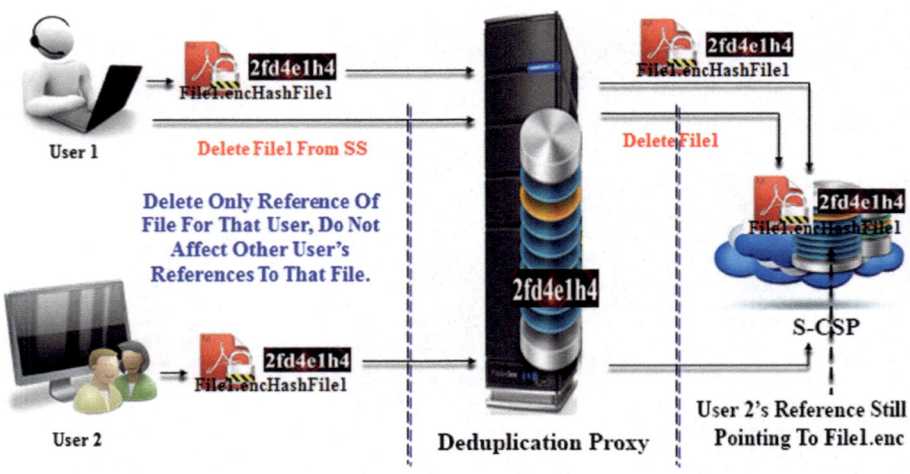

Fig. 2.   Uncontrolled deletion problem in multi user environment

Now if User 1 tries to delete this file, deletion will cause physical file to be deleted.
Thus User 2's file reference will point to file which has been deleted by User 1. So User

2 will not get his file as the file was lost. This approach presents solution for uncontrolled deletion problem in multi user environment.

4   Implementation

Major operations performed by the cloud user are file upload to S-CSP and file download from S-CSP. Before uploading file to S-CSP some prerequisites steps needs to be performed that are depicted in algorithm 3 below (Fig. 3).

Algorithm 1 Preprocessing before uploading file to S-CSP

    Input:
        $F \Longleftarrow$ File to be uploaded
        $K_{cid} \Longleftarrow$ Public encryption key of user
    Output:
        $\{F\}_{hash} \Longleftarrow$ Cipher form of original file
        $F_{id} \Longleftarrow$ Identifier from file contents
        $\{F_{hash}\}_{Kcid} \Longleftarrow$ Encrypted form of key used to encrypt original file, this key encrypted by public key
1: Step 1:
        $F_{hash} \Longleftarrow$ MD5 or SHA1 or SHA256 (file contents)
2: Step 2:
        $\{F\}_{hash} \Longleftarrow$ encrypt (F_{hash}, file contents)
3: Step 3:
        $F_{id} \Longleftarrow$ MD5 or SHA1 or SHA256 ($\{F\}_{hash}$)
4: Step 4:
        $\{F_{hash}\}_{Kcid} \Longleftarrow$ encrypt (K_{cid}, F_{hash})

Fig. 3. Prerequisite steps for upload operation by cloud user

8. Preprocessing Before File Upload

As depicted in algorithm, F represents file to be uploaded by user to S-CSP. First step involves calculation of file hash also called convergent key in baseline approach using either MD5 or SHA1 or SHA256 algorithms. Next this file hash will be used for encryption of file contents which results in cipher form of file to be uploaded. Once file has been encrypted again file hash will be calculated and this file hash is used as File Identifier for data deduplication. Lastly file hash used for encryption of file is again encrypted by using public key of client. Thus it preserves Proof of Ownership (PoW).

4.1   File Upload Operation to S-CSP

Once file has been encrypted, Client will request for file upload operation with file identifier. This request goes to DP where intra-user data deduplication check will be made. If this file is uploaded by the same client previously then Intra-user data deduplication is applied at DP and reference will be uploaded to S-CSP and response to the client has been sent that your reference to file has been uploaded successfully. This process is represented by step 1 and 2 in Fig. 4.

If file does not exists from the same user then upload request will be forwarded to S-CSP where it has been checked for inter-user data deduplication. If same file was uploaded previously by some different user then inter user deduplication is applied at S-CSP which results in upload of reference to file on S-CSP and response to the client

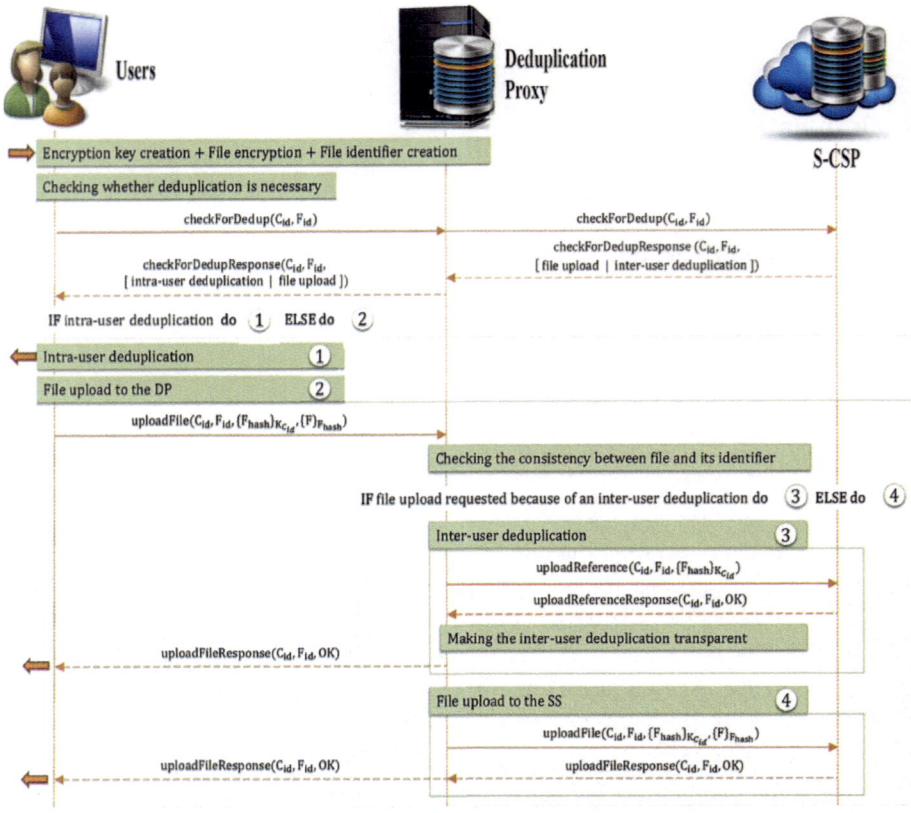

Fig. 4. File upload operation by cloud user

has been sent that your reference to file has been uploaded successfully. This process is represented by step 3 in Fig. 4. If both intra and inter user data deduplication test comes negative then this means that file does not exists on S-CSP. So original file upload will takes place instead of uploading a reference. This process is represented by step 4 in Fig. 4.

Once file upload is successful, file identifier record will be saved on both DP and S-CSP for later intra and inter user data deduplication and response has been sent to the client that your file has been uploaded to S-CSP successfully (Fig. 5).

10. File Download Operation from S-*CSP*
In earlier approach clients request for file download by sending client identifier and file identifier. But as multiple files are having same file identifier store at DP and S-CSP one cannot be able to identify for which file client has sent a request. To resolve this problem instead of sending file identifier to identify files this approach uses file names. Thus client request for file download by sending file name which is unique. So it's easier to identify appropriate requested file.

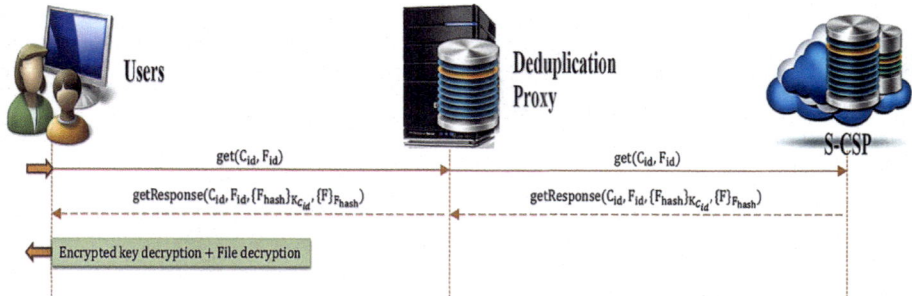

Fig. 5. File download operation by cloud user

Once file download request comes at DP this will forward this request as it is to S-CSP. Next task of S-CSP is to identify whether requested file is original file or a reference pointing to original file. If requested file is original file then that file will be transferred as it is to DP and DP will give this file to client. If requested file is reference pointing to original file then S-CSP will identify to which file this reference is pointing and transfer this file to DP but before transferring it to DP it will rename this file to reference file name. So that client will get a file with the name at the time of uploading this file to S-CSP. File will be transferred in encrypted or cipher form in order to prevent attackers from stealing information to preserve security.

11. Postprocessing After File Download
Once file has been transferred to client this file is in cipher form so needs to be decrypted to get access to original file. Post processing steps performs the same functionality. Input to this algorithm is cipher form of file, encrypted form of key used to encrypt file and private key of client. To get access to key used for encrypting file first it needs decrypt this key by private key of client. This is done in the first step of algorithm. Once key used to encrypt file is available it is possible to decrypted original file and this is done as shown in step 2 of Fig. 6.

Algorithm 2 Post processing after downloading encrypted file from S-CSP

    Input:

        $\{F\}_{hash} \Longleftarrow$ Cipher form of original file

        $\{F_{hash}\}_{Kcid} \Longleftarrow$ Encrypted form of key used to encrypt original file, this key encrypted by public key
of user

        $K_{cid}^{-1} \Longleftarrow$ Private decryption key of user

    Output:

        $F \Longleftarrow$ File to be downloaded

1: Step 1:

        $F_{hash} \Longleftarrow$ decrypt $(K_{cid}^{-1}, \{F_{hash}\}_{Kcid})$

2: Step 2:

        $F \Longleftarrow$ decrypt $(F_{hash}, \{F\}_{hash})$

Fig. 6. Post processing steps after download operation by cloud user

Once file has been decrypted file download request of client is successful. Now client is having access to plain text form of a file which was uploaded by him/her through file upload operation. Thus in general exactly opposite steps are performed in download operation to that of file upload operation.

5 Result Analysis

Result analysis of proposed system is done based on,

(1) *How efficiently it utilizes storage space on S-CSP?*
(2) *How much time it will take to upload a file to S-CSP?*
(3) *How much time it will take to upload a reference to file to S-CSP?*

As same files are not uploaded multiple times this will not over consumes storage space on S-CSP required for storing duplicate copies of files. So as references to file increases it will not increase space used for storing these references. Original files are uploaded based on type of network being used. If system is used in an intra-net this will provide best upload response time. If system is used on internet then type of connection being used effects on upload response time. Uploading of reference to a file which already exists on S-CSP will take very less time in fraction of seconds. Thus upload response time for uploading reference to file is very less thus client will get quick response that your file has been stored on S-CSP successfully (Fig. 7).

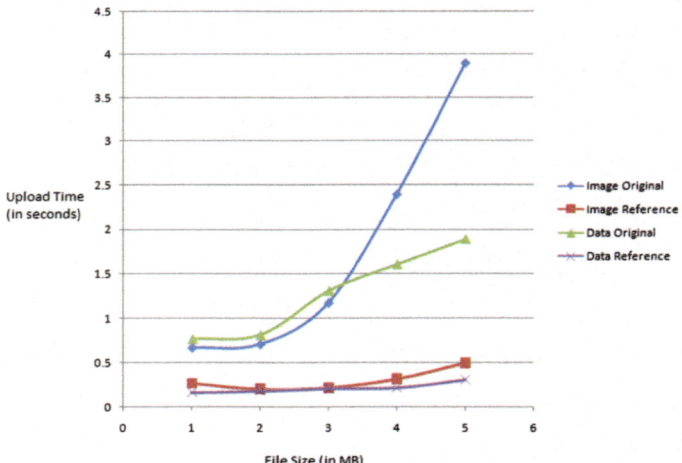

Fig. 7. Ratio of upload response time to different file sizes

Figure 6 depicts Graph for upload time taken by different types of files of different sizes. Upload time for Image & Data files grow as size of files increases. But if same file upload takes place more than one time then time taken for uploading reference is too small and it is slightly same for all image or data files irrespective of their sizes.

6   Conclusion and Future Scope

Approach provides optimal way to perform deduplication of data at two levels one at Deduplication Proxy and second at S-CSP, along with encryption performed by client, prevents different attacks efficiently. Overcomes problems in existing approach of uncontrolled deletion by providing controlled deletion of files with efficient management of references to the files. Distributes load of data deduplication between both Deduplication Proxy and S-CSP by providing intra-user deduplication on Deduplication Proxy and inter-user Deduplication on S-CSP. Overcomes target collision by calculating hash i.e. file identifier of file after file encryption and problem of server-side data deduplication by encrypting file prior to transfer of file to S-CSP.

Deduplication of data is carried out at file level in this approach; one can implement same functionality with block level. Where file is divided into small parts called fragments prior to checking for data deduplication. If file blocks are duplicate then they are not uploaded more than once, thus it will provide better granularity at block level data deduplication as well as this will increase complexity in managing each duplicate blocks of a file.

References

1. Li, J., Chen, X., Li, M., Li, J., Lee, P.P.C., Lou, W.: Secure deduplication with efficient and reliable convergent key management. IEEE Trans. Parallel Distrib. Syst. **25**(6) (2014)
2. Meye, P., Raipin, P., Tronel, F., Anceaume, E.: A secure two-phase data deduplication scheme. IEEE International Conference on High Performance Computing and Communications (HPCC), 2014 IEEE 6th International Symposium on Cyberspace Safety and Security (CSS) and 2014 IEEE 11th International Conference on Embedded Software (2014)
3. Leesakul, W., Townend, P., Xu, J.: Dynamic data deduplication in cloud storage. 2014 IEEE 8th International Symposium on Service Oriented System Engineering (2014)
4. Kokane, Y., Patil, S., Dange, A.: Secure decentralized data deduplication in cloud. Inventi Rapid: Cloud Comput. **3**, 2015 (2015)
5. Russel, D.: Data deduplication will be even bigger in 2010. Gartner, February (2010)
6. Teasley, H.: Your very own cloud [tools & toys]. Spectrum, IEEE **47** (5) (2010). Martini, B., Choo, K.K.R.: Cloud storage forensics: ownCloud as a case study. Digital Investigation. http://dx.doi.org/10.1016/j.diin.2013.08.005 (2013)

People, Technologies, and Organizations Interactions in a Social Commerce Era

Maruti Maurya$^{(\boxtimes)}$ and Milind Gayakwad

Department of Information Technology, Bharati Vidyapeeth
(Deemed to be University) College of Engineering, Pune, Maharashtra, India
marutimaurya14@gmail.com, mdgayakwad@bvucoep.edu.in

Abstract. Social commerce, in today's era is the wealthiest combination of customer oriented technologies and latest commercial features. It has a direct impact on the E-commerce, generating large amount of benefits. It understands both technical and social framework to fulfil the customer's requirements on social media websites. In general, E-Commerce is more beneficial at social commerce. Our results show that customers are more satisfied by the ease of usefulness, thereby enhancing their wish to purchase and trust. For engineers and designers, our results describe the importance of social media commerce in today's era for constructing users trust on social media commerce sites and supporting their wish to buy the trending products.

Keywords: Socio technology · Customer oriented · E-commerce ·
Social commerce

1 Introduction

In media new technologies are giving a serious impact on the society. Households are flushed with these types of changes. Social interactions in the household have been increased by the passage of time. This indeed has increased interactions among the family members leading several conflicts between them. The generation gap has reduced digitally, giving rise to healthy relations, whereas, has led to the privatization within the family. For the same reason, I seek out to investigate the research "How Social Media is giving a severe impact on the social interactions within the households?"

In today's era of E-Commerce (Electronic Commerce) through booming entre-preneurs, we can see a great impact in the online retailer. In this the users buy, do searches, connect with the friend lists are sometimes simply search for the desired products or services. Our aim is to understand E-commerce from the very beginning stage to the factors which various countries supported for internet penetration. ITC's made a new form of commerce known as social commerce which will be analysed in this paper. The purpose is to show that Web 2.0 technology and Social Media have made different ways to communicate, collaborate and share information to a large number of people who are associated with this virtual relationship. The consumer now becomes the social consumer which shares the same passion and taste with other consumers using different social media handles. It not only constraints the buyer of the

© Springer Nature Switzerland AG 2020
A. P. Pandian et al. (Eds.): ICCBI 2018, LNDECT 31, pp. 836–849, 2020.
https://doi.org/10.1007/978-3-030-24643-3_98

product, but helps to increase his/her reach to explore new products or to a 'place' where a consumer can connect with people on the same frequency.

The statistics say, Facebook's monthly active users have increased by 23% from 901 million in March 2012 to 1.11 Billion in March 2013. This increase leads to increase in the number of active mobile users from 488 million to 751 million by March 2013, which represents a 54% hike from the last year. Same is with Google Plus; this is the second largest social media network which grew its monthly active mobile users by 33% in the same year. Similarly, Twitter just like Facebook has set up its users to 44% in the same year.

Predicted Business with s-commerce strategies are in need to consider the mobile device usage in order to fulfil the customer needs while keeping the customer loyalty and trust high, which adds up to one of the important factors for success of online commercial activities. Same is with the m-commerce field in which they need a high social media network to shoot their business up. New mobile social commerce design framework must be completely flexible and should have reliable guidelines which must be accepted by each and every concept of social media business platforms. Formal definition of s-commerce is different from any other type of commerce as all of the commercial transactions depend upon the type of Web Infrastructure and participation of users.

S-commerce linked with Web 2.0 support online transactions and user contribution to help in the acquisition of products and services. The unique infrastructure that a social media handle provides, allows its users to generate data over time.

Say for example, Facebook, Google plus and Twitter are the social media service providers which give their users a specific domain to generate data with the help of Web 2.0 technologies. They are also the example which made social media providers to gain importance in the internet over the years which is a blessing from social commerce.

2 Literature Survey

Recently, social media rapidly changed in brand management and understanding this change is now critical which affected brand management. The paper focuses on existing system and introducing a framework of social networks which impact on brand management. System brands are first for consumers and are to be used in living their own lives [McCracken 1986] and today's challenge is to understand the deep sources and dynamic nature of brands meaning. Author proposed meaning transformation McCracken's model to understand the various brand's meaning. They have presented 20 years of Consumer Research Addressing on Socio cultural, Experiential, Symbolic and Ideological Aspects of Consumption and present Consumer Culture Theory (CCT) which is used for understanding consumption and marketplace values of products. They have discovered three misconceptions about the nature and the orientations of CCT [1].

E-commerce is the evolved form adoption of Web 2.0 technology and its capabilities to understand customer participation and achieve greater economic value in the market. These phenomena are referred as social commerce. Research on the social commerce techniques were done, reviewed and E-Commerce and Web 2.0 techniques were presented. For this, they have used Facebook, Amazon and Starbucks websites.

Social commerce is referring to social, creative and collaborative approach for online marketplaces. The figure shows the author proposed model. The social Commerce model includes phases, namely Commerce, Community, Conversation and Individuals. In conversation phase, we find out the interaction among the users of social media. Fundamental requirements of conversations are Communication, Communities and Connections. Functions of communication are to provide a communication channel between users on the social media. Communities are related to the aggregation of participating groups. The limitations of these papers are identification Web 2.0 and E-Commerce design characteristics [2].

New techniques to recommend friends with similar location preference, location based social network (LBSN) - users, in which friend's online friendship information and the offline user behavior are taken into account. The proposed approach includes Markov chain, cosine similarity based on location clustering and threshold evaluation. In this paper, the authors have focused on relationship perspective to observe the relationship between users of a social network site. Here they describe the buyers and the seller relationship and their roles. It also describes the role of the internet and interfaces. Data is collected from suggested information sources, including suppliers. Suppliers and salespeople are more important than the internet. It directly affects IT adoption.

When we make use of the website for search PEOU, it affects IT adoption because the required information is embedded in the IT. In this the web development is also introduced. In this, firstly, a survey is conducted comparing consumers in China with their U.S. counterparts, and it is shown that a cultural perspective is pertinent and valuable. A recent social media change rapidly. In this, there are social media groups of internet applications that build new technology, that are used to exchange user generated content. A model has been proposed and is the extension of the planned behavior. The author has discussed about the managerial implications [3].

In recent years, social commerce has been a new technology, in which seller and buyer connect to each other using social network. Social Commerce and Social Shopping are the forms of Internet-based which are called as "Social Media" and this allows people to participate in marketing and for selling the products online and provide services in online marketplaces. Applications are needed to merge the Online Shopping and Social Networking into each other. The difference between Social Shopping & Social Commerce is that Social Shopping is connected to the customers and Social Commerce is connected to the sellers. Social Shopping is spreading around the online world. Online user's reviews are important to collect source of information for substituting, consumers, complementing and it's important for understanding customer's requirements. During this investigation has been done to understand Enterprise social networks and to develop models that understands the social psychological processes [4].

The study examines three commonly-used interventions to understand how they influence different user's beliefs and subsequent participation. This tested model has data collected from 366 members. It provides the validated theoretical model that improves the socio-psychological processes governing employees. It also contributes to a more detailed understanding of how and why corporate staffs participate in social networks. It also demonstrates the three commonly used management interventions - Sellers facing challenges of customer related marketing, planning to fulfil the customer requirements and finally building the customer product. In this they have proposed integrated model to describe the components. Structural Equation Modelling with the Partial Least Squares (PLS) is used to analyze valid data of customers. Results show its effect on SC consumer value. This data is collected from Facebook. Results show the benefits of social interaction [5].

In this [6] they propose Collaborative Filtering techniques for the top-n recommendation task they used bi-clustering neighborhood approach. The result shows that the proposed techniques generates a better recommendation that the existing State-of-Art algorithm on sparse data. The performance of the Bicluster Neighborhood (BCN) Framework is evaluated on five real world dataset such as Paypal dataset, lastfm dataset, lastfm friends dataset, and delic bookmarkdatset. Performance of BCN techniques is compared with the three different algorithms, namely SLIM (Sparse Linear methods), Item CF (Collaborative Filtering), and WRMF (Weighted Regulation Matrix Factorization).

They proposed Probabilistic techniques to resolve Diversity-Accuracy Dilemma in the RS (Recommendation System). This proposed recommendation system has two models: First one is that maximization of the accuracy and second one is specified of the recommendation list to the tastes of the users. A recommended technique is based on the Markov model. For this experiment they are using two real datasets such as Movielens and Netflix. These datasets divided into the testing dataset and training dataset [7].

They proposed Topology-based ensemble model for combining users own taste and his trusted users/friends testes. Experimental analysis performed on the Flixster, Epinions, and Ciao for the better results. A proposed technique is not only applicable for Model-Based ensemble, but also Instance-Based ensemble recommendation techniques. Sentiment latent topic model (SLTM) is used for build a connection between sentiments and words [8].

In this [9] they propose emotional aware recommendation approach to incorporate an emotional context into music recommendation. They are using Chinese twitter services dataset. The performance of this technique improved in terms of precision, recall, hit rate and F1 score. A proposed technique is compared with two existing techniques such as Fine grain emotion from micro blogs and Coarse-Gained emotion information. In this they firstly music service is connected to the microblogs services such as Sina Weibo and Twitter for exacting more listing records. In this they improve the accuracy and efficiency of music recommendation techniques.

They proposed techniques for providing personalized service recommendation to individuals user, for this they used trust relationship between services and users. A proposed recommendation technique is based on the collaborative filtering approach such as trust-based service recommendation. TSR techniques are compared with other five approaches relation based TSR, QoS on response rate, QoS on trough output, HITS and PAGERANK [10].

In this [11] they propose techniques for the solving the large scale matrix factorization problems in recommendation system. In this they used optimization techniques such as cyclic coordinate decent approach (CCD) and it is applicable to the large scale problems such as Maximum Entropy Model, NMF problems, Sparse Inverse Covariance Estimation and linear SVM. CCD approach updates the single variable at a time while keeping other fixed. The proposed system is comparing with the ALS, SGD, and CCD++ in large scale data. The dataset used for the experiment is Movielens 10 m, Movielens 1 m, Yahoo Music and Netflix.

They propose improved matrix factorization approach such as Bounded matrix factorization (BMF) approach for rating matrix. They test performance of the proposed algorithm on the real-world dataset such as Netflix and compare state-of-art algorithm with SVD++, SGD, Bias-SVD and ALSWR [12].

3   Research Methology

In recent years, the trend of web service on the Internet is increasing. There is a need for great effect service recommendation organization. Existing methods mainly focus on the characteristics of individual web services (e.g., functional & Non-functional characteristics), they ignore the user's point of view on the service, so they cannot provide a personalized service recommendation. In this paper, network modelling is used to study the relationship of trust between users and web services analytical technology. Based on our knowledge and service network model, we have proposed a collaborative research.

A filtering algorithm called Trust-Based Service Recommendation (TSR) provides recommendations for personalized service. This systematic approach for modelling and analyzing service networks can also be used to recommend other services for the study [6].

The trust region is a term used in mathematical optimization to represent a subset of the region of the objective function approximated using a model function (often quadratic function). If an appropriate model of the objective function is found in the trust region, the region is expanded and conversely, when the approximation is bad the region contracts. Trust region methods are also called restricted step methods.

In Computer Science, clique problem is a computational problem of finding a clique (a subset of vertices, also called a complete sub graph adjacent to one another) graphically. It has several different formulas, depending on what clique and information about that clique should be found. Common formulas for the clique problem include finding the maximum clique (a clique with as many vertices as possible), finding the largest clique in the weighted graph, enumerating all the maximum cliques (in expandable cliques) and solving the decision problem test whether the graph contains larger cliques than the given size.

The problem of clique occurs in the following real-world situation - Consider a social network where the vertices of the graph represent people and the edges of the graphs represent mutual acquaintances. Clique represents a subset of people who know each other and algorithms to find Clique can be used to discover groups of these mutual friends. Clique problem has many applications in Bioinformatics and computational chemistry as well as in social networks.

The first version of the clique problem is difficult. Clique decision problem is NP complete (one of Karp's NP complete problems). The problem of finding the maximum clique is hard to deal with and difficult to approximate with fixed parameters. This is because there is a graph with many maximum cliques exponentially, and we may need an exponential time to list all the maximum cliques. Therefore, much of the theory about the clique problem is focused on identifying special types of graphs that allow more efficient algorithms, or establishing computational difficulties for common problems in various computational models.

The reviews and ratings are taken from students of Bharati Vidyapeeth for testing purpose and around 300 students have participated. The proposed system was provided to them and they were asked to do shopping on the website. They then provided their valuable feedback. The size of the feedback data file is 100.9 KB with 300 records in it.

4 System Architecture

Proposed system architecture is shown below which includes review and rating of websites of the products, recommendation of product and websites as whole to be used, forum and comments of products and five types of websites.

The paper mainly focuses on how online shopping via the website and recommendation affect trust of customers with attitudes. Consumers are the main actor of the marketplace. The role of consumers in the marketplace is to consume and purchase products and services in the market. The difference between the buyer and consumers is that, that the buyers are people whose role are industrial, ultimate and intentional purchasers.

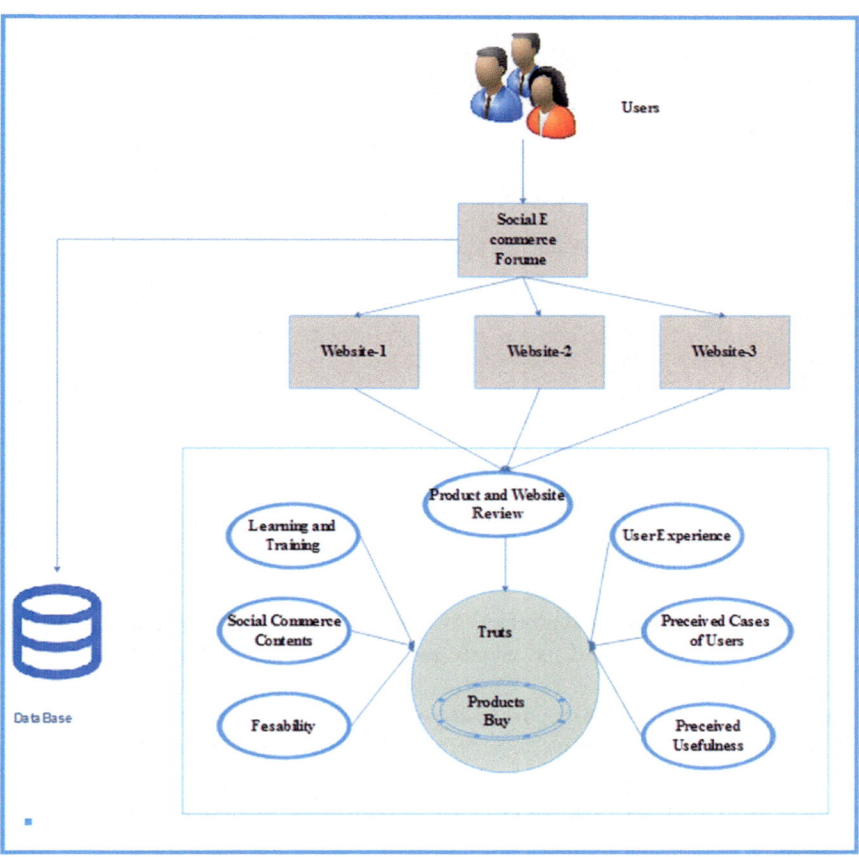

Fig. 1. System architecture

Problem recognition, information search, evaluation, purchase decision, post purchase behavior. Social media strategy below flow chart shows that the social media flow decision making phase. In this several factors are affecting on that the as internal factors and external factors which are leads to social media. External factors are considering popularity and fast growth and it provide low cost solutions for social media. In which phases are including namely external factors, internal factors, activates of social media, expected outcomes, other factors of social media. New farms are going on social media whose follow up same activities of social media which are communication with user of media, announcing of new products these activities are according to the offline marketing (Fig. 1).

5   Mathematical Model

Let S be a solution of problem
   $S = \{I, P, O, Sc, Fc\}$
   Where,
   I = Input of System
   O = Output of System
   P = Process in the System
   Sc = Success case of output of system
   Fc = Failure case of output of system
   • Process:

1. P1 *//Social E-Commerce Forum*

Login to the websites and surf on shopping websites for product that user wants to buy, search the product. The search products and user session details are taken for further use.

2. P2 = {P1}; *//Forum and community page*

Login to the social forum and surf for subjects that are trending and add comments if user wants them.
   Subject is "Website User Experience".
   One of the most frustrating experiences for users of the web is waiting for a page to load for long. With the rise of the mobile devices people are accessing content all over the world or many different platforms.
   Comments:

Comment1:	I am familiar with these kind of bad user experience so, it very frosting for me to find. Is anyone having better user experience website to do surfing?
Comment 2:	Yes, I recommend krusha website for searching and it is very useful and easy to use. They provide good user experience. I am using these websites from last year and I bayed many delicious pastries and they delivered stuff with quality, so it is very trustful website

we use NLP concept sentiment analysis in which every comment is check for positive and negative words to identify negative user and positive user.

3. P3 = {P2}; //*Recommendation and rating page*

Then in recommendation page search every website page for user query and find it by TF-IDF algorithm.

I. Term Frequency (TF):

$$ft = \pm + \log(I + tf_{i,d}) \tag{1}$$

Where,

$tf_{i,d}$ is log terms frequency of terms t in documents d

II. Inverse Documents Frequency(Idf):

Where,

N is the total no of documents in collection so, from Eq. 1 and Eq. 2 is

$$W_{t,d} = \left(1 + \log 1 + tf_{i,d}\right) \log_{10} \frac{N}{dft}$$

Query and location is selected for recommendation and TF = Idf recommend websites. This way every page of all websites is search for user keyword. In all pages check the probabilities of websites reviews and recommend website details to user.

4. P4 = {P3};

User Uses recommendation and visit website for shopping. After shopping it provides ratings for all factors and reviews about product and website are taken and stored them in database. Here Cronbach's alpha is used to find out intercorrelation among the 8 factors.

The standardized Cronbach's alpha can be defined as

$$\alpha_{standardized} = \frac{K \bar{r}}{\left(1 + (K - 1) \bar{r}\right)} \tag{2}$$

Where K is as above and r the mean of the K (K-1)/2 non-redundant correlation coefficients.

Measures to find Internal consistency some internal consistency classes are defined with cronchbach alpha range (Table 1).

Table 1. Cronbach's alpha range table

Cronchbach's alpha	Internal consistency
$0.9 \leq \alpha$	Excellent
$0.8 \leq \alpha < 0.9$	Good
$0.7 \leq \alpha < 0.8$	Acceptable
$0.6 \leq \alpha < 0.7$	Questionable
$0.5 \leq \alpha < 0.6$	Good
$\alpha < 0.5$	Unacceptable

5. P5 = {P4}; //*Result*

To check the probabilities of every parameter, we take all the ratings of eight parameters and show in the way of mean, S.D, Variance, alpha.

Then Trust Cliques Algorithm is used to find Trust. Trust is directly affects the intention to buy.

Sc = When we find class of user, Is it positive or negative.

Fc = When it fail to identify behavior of user.

6 Algorithm

Algorithm: Trust Communities Algorithm:
Input:
Communities size n, iteration limitation l,
User set U, user trust matrix T;
Ensure:
Collection Communities;
Communities = U, n_c = | Communities|;
While, n_c = 1 or number of iteration < l do;
For i = , n_c do;
C_i = the ith Communities;
J = the set of C_1^i s connected vertices;

$$\Delta Call_{max} = 0$$

for each $V_j \in J$ do;

$$C_1^i = C_i \cup V_i;$$

If C_1^i satisfies the first condition of n-trust-clique;
Then;
Calculate ΔCol for the change;
If $\Delta Call > Call_{max}$ then

$$\Delta Call = Call_{max};$$

$$C_i = C_1^i$$

end if;
end if;
end if;
if $\Delta Call_{max} > 0 \vee Random(0, 1) > pro$ then;
delete clique C_i to Communities;;
store current partition result Communities; and the corresponding Coll;
end if;
end if;

$n_c = |\text{Communities}_j|$;
end while;
return optimal Communities with the maximum value of Coll;

The Trust Community Algorithm is required because of Trust is one factor on which all other factors are connected together. If trust has more value then its impact on intention to buy is always be high. This trust factor need to be calculated between users and websites so it help the users to decide from which websites products are good and with good quality. From websites company perspective it will build trust between users and website that will lead to increase in business profits.

So trust community algorithm is made in such a way that it manage to make communities that have High trust, low trust and neurals or average trust. Many users are trust their friends and relatives largely, so they formed groups. This groups are called Communities when they become large in size. The cliques also called as communities, clusters, collection or Groups. In above algorithm Coll is defined as Global Collection of users. It has following given formula:

$$Call = log \left(\frac{\sum_{ci \in Communities} Call_i}{|Communities|} \right) \tag{5}$$

So, those users get coll values similar are put into one community and this way it forms and manages big communities of users.

Process P2 is log into Social Commerce forum, to find details of products like price, vendor, company details and will know the technical aspects with price details of product. Process P3 checks reviews and ratings of product and websites. Also, on shopping website users can add their ratings and reviews on product and websites. For rating and reviews, rating is given in star format that is out of five. One star is extremely poor, two stars are bad, three stars are average, four stars are good and five stars are best.

This is the website module and includes terms - familiarity, learning and training, user experience, perceived ease of use taken from the star rating 1 to 5. 1 means strongly disagree and 5 means strongly agree. These terms are taken from user on websites in star rating and user reviews about websites and its products.

User provides valuable information or opinion about learning and training, user experience which are connected to perceived ease of use and it connects to the perceived ease of use.

Range of Familiarity: 0.455 to 0.873
Range of Learning and training: 0.682 to 0.759
Range of User Experience: 0.649 to 0.849
Range of SCC: 0.627 to 0.759
Range of PEOU: 0.659 to 0.746
Range of PU: 0.629 to 0.840
Range of Trust: 0.555 to 0.778
Reviewpage.jsp (Table 2)

Table 2. Reviewpage.jsp

Familiarity	****	4
Learning and Training	***	3
User Experience	*****	5
Social Commerce Construct	***	3
Perceived case of use	****	4
Perceived Usefulness	****	4
Trust	****	4
Intension to Buy	***	3

Reviews:

This websites has very good UI. It provides best discounts than others

This websites has   Very Good   UI. It has provides   Best
discounts than others.

Product quality is also best.   Best

As we see that user reviews is positive because it has positive words which denote its positivity (Table 3).

Table 3. Word count table

Word	Weight count
Very Good	1
Best	2

This way we assign users class like positives user, negative user and neutral user, all respective user reviews which user has submitted on websites. Positive user are always has high Trust value so their intention to buy products are always high.

After that we also take a note of user search on websites user involvement on social commerce constructs (SCC) and take his/her session time to identify how much user in using SCC.

These are above range in which if values are come then it is satisfactory degree of internal consistency reliability.

Following are examples of reviews given by Users in Social community forums and website reviews are taken into consideration for internal consistency measurement which is shown in below table with factors values for reviews and comments.

Familiarity:

FA_1: I am acquainted with searching information regarding products on the internet.

FA_2: I am acquainted with purchasing products on the internet.

FA_3: I am acquainted with enquiring about product ratings on the web.

Learning and Training:

L_1: I know how to operate a computer and use the web.

L_2: I am well-educated to use the web to buy products online.

L_3: My learning and training about online shopping has been useful.

User Experiences:

UE_1: I am capable to operate a computer.

UE_2: I am capable to use the web.

UE_3: I have used the web from a very long period.

Perceived Ease of Use:

SSC_1: I always make use of portals and blogs for gaining details about a commodity.

SSC_2: I always use customer provided grades about commodities on the web.

SSC_3: I always use customer provided suggestions to buy a commodity on the web.

Social Commerce Constructs:

PE_1: It is very easy to become skilled at using the websites.

PE_2: It is very easy to learn to use the web.

PE_3: The websites that I use for shopping are flexible to interact with.

Perceived Usefulness:

PU_1: Enquiring and purchasing on the web is very useful for me.

PU_2: Enquiring and purchasing on the web makes things easier for me.

Trust:

T_1: Assurance provided by the websites used by me for my last purchase seems trustworthy.

T_2: I can count on the worthiness of the website that I used for my last purchase.

Intention to buy:

IU_1: I am ready to give the details to socio-economic retailers for better shopping experience.

IU_2: I am okay to use my card to buy from a social-commerce retailer (Table 4).

848 M. Maurya and M. Gayakwad

Table 4. Score table

scores:

Perceived Usefulness	Trust	Perceived_ease of_use	User Experience	Learning and Training	Social Commerce Constructs	Familiarty	Intension_to_buy
-0.267405	0.907798	-0.772342	0.797666	0.829146	0.681229	0.245207	0.1513
1.260623	0.1513	0.943973	-0.041982	0.829146	-0.035854	1.226036	0.1513
0.496609	1.664296	0.943973	0.797666	0.829146	1.398312	2.206865	0.1513
-0.267405	0.1513	-0.772342	-1.721279	-1.333843	-0.752937	-0.735622	-1.361696
-1.031419	-0.605198	-1.630499	-1.721279	0.829146	0.681229	0.245207	-0.605198
1.260623	-1.361696	0.085816	0.797666	0.10815	-0.035854	-0.735622	-0.605198
-0.267405	0.907798	1.802131	1.637314	1.550142	0.681229	-1.716451	0.1513
0.496609	-0.605198	0.085816	0.797666	0.10815	-0.752937	-0.735622	0.1513
0.496609	0.907798	0.085816	0.797666	-1.333843	-1.470021	0.245207	0.1513
-1.795433	0.1513	0.943973	-0.881631	1.550142	-1.470021	1.226036	-1.361696
-0.267405	-0.605198	0.943973	-0.881631	-0.612847	0.681229	-0.735622	-0.605198
-1.795433	-1.361696	0.085816	0.797666	-1.333843	-1.470021	-0.735622	1.664296
1.260623	0.907798	0.943973	-0.041982	-1.333843	-1.470021	0.245207	1.664296
-1.031419	0.1513	-1.630499	-0.881631	-0.612847	-0.035854	0.245207	0.1513
1.260623	-1.361696	0.085816	-0.041982	0.829146	1.398312	0.245207	0.907798
-0.267405	-1.361696	0.085816	0.797666	-0.612847	1.398312	-1.716451	-0.605198
1.260623	1.664296	0.943973	-0.041982	0.829146	0.681229	0.245207	1.664296
0.496609	-0.605198	-0.772342	-0.041982	-1.333843	-0.035854	1.226036	-1.361696
-0.267405	0.907798	-1.630499	-1.721279	0.10815	-0.752937	-0.735622	0.907798
-1.031419	-0.605198	-0.772342	0.797666	0.10815	0.681229	0.245207	-1.361696

7 Conclusion

Social network is used increasingly, because of this e-commerce is impacted on social network site. Our findings and results show how increases in purchases intend towards social commerce. From this system results, we can get consumers profiles who buy products online. Results show that Social Commerce Constructs, Communities, Rating and Review of products lead to trust. Trust is established from the Social Commerce Constructs it affects consumer's intentions of buying products. Trust community Algorithm plays major role for focusing on positive users to provide them more positive experience and negative user will get importance because they shows limitations and less quality.

Acknowledgements. The authors are grateful to acknowledge the support of faculty and research scholars of BVDU, COE, PUNE who spent their crucial time while conducting this survey and participated in filled the questionnaire to provide their intellectual feedback related to Social commerce era.

References

1. Huang, Z., Benyoucef, M.: From e-commerce to social commerce: a close look at design features. Electron. Commer. Res. Appl. **12**(4), 246–259 (2013)
2. Stephen, T., Toubia, O.: Deriving value from social commerce networks. J. Mark. Res. **47**(2), 215–228 (2010)
3. Gensler, S., Volckner, F., Liu-Thompkins, Y., Wiertz, C.: Managing brands in the social media environment. J. Interact. Mark. **27**(4), 242–256 (2013)

4. Pentina, I., Gammoh, B.S., Zhang, L., Mallin, M.: Drivers and outcomes of brand relationship quality in the context of online social networks. Int. J. Electron. Commer. **17**(3), 63–86 (2013)
5. Deeter-Schmelz, R., Kennedy, K.N.: Buyer-seller relationships and information sources in an e-commerce world. J. Bus. Ind. Mark. **19**(3), 188–196 (2004)
6. Wu, Z.H., Huang, L.T., Deng, S.G.: Trust based personalized service recommendation a network perspective. J. Comput. Sci. Technol. Jan (2014)
7. Jalili, M., Javari, A.: A probabilistic model to resolve diversity accuracy challenge of recommendation systems. Knowl. Inf. Syst. **44**(3), 609–627 (2015)
8. Rao, Y., Gong, Z., Zhang, N., Zou, H.: Adaptive ensemble with trust networks and collaborative recommendations. Knowl. Inf. Syst. **44**(3), 663–688 (2015)
9. Xu, G., Li, X., Wang, D., Deng, S.: Exploring user emotion in microblogs for music recommendation. Expert Syst. Appl. **42**, 9284–9293 (2015)
10. Wu, Z.H., Huang, L.T., Deng, S.G.: Trust based personalized service recommendation: a network perspective. J. Comput. Sci. Technol. Jan (2014)
11. Dhillon, S., Hsieh, C.J., Yu, H.F., Si, S.: Parallel matrix factorization for recommender systems. Springer (2014)
12. Park, H., Ishteva, M., Kannan, R.S.: Bounded matrix factorization for recommender system. Springer, Knowledge Information System (2014)

Plant Disease Detection System for Agricultural Applications in Cloud Using SVM and PSO

P. Kumar and S. Raghavendran[✉]

Centre for Information Technology and Engineering,
Manonmaniam Sundaranar University,
Abishekapatti, 627012 Tirunelveli, Tamilnadu, India
kumarcite@gmail.com, vsraghavendran@gmail.com

Abstract. In agricultural industry disease in plants is very difficult to identify what type of disease affect and how the disease is to treat the infected plant. Then, the information is store in cloud computing for easily accessing the data for farmers whenever they want and learn about the diseases of the plant and treat it very fast without any loss of plants life. The latest technology e-agriculture service is very important for the development of rural and it contribute the farmers get more knowledge regarding the market place as well as finding varies disease infect the plant by taking image of the affected plant leaves. In this paper, Internet of Things capture the disease affected plant image and it is given to the cloud server then the image is pre-processed with Top hat filter, ostus thresholding Segmentation, Support Vector Machine (SVM) classifier is used and Particle Swarm Optimization technique is perform for task scheduling. By using these techniques execution time and system performance is improved.

Keywords: Support vector machine (SVM) · Ostus thresholding · Top hat filter · Particle swarm optimization (PSO)

1 Introduction

Agriculture industry is important for growing population; today's technology is very helpful to increase the yield of the cultivation. The ecological and environment damage is major hazard for affecting new diseases in plants with the help of fertilizer, pesticide and technological development in which it is increase the yield of the cultivation. Internet of Things such as camera or mobile capture the image of the disease infected plant and Cloud computing are used to access the information from it; what type of disease affect the plant and the already predefined record give the detail of correct pesticide which is used to get high yield. Without any chemical test of infected plant leaf but with the help of image processing technique, the disease can be identified.

The main application of cloud computing is to transfer the huge data from one corner to other corner without any loss of data and with high security [1]. This paper demonstrate that the disease affected leaf is captured by camera and then send to pre-processing and then segmentation by mean based thresholding. SVM classifier

© Springer Nature Switzerland AG 2020
A. P. Pandian et al. (Eds.): ICCBI 2018, LNDECT 31, pp. 850–856, 2020.
https://doi.org/10.1007/978-3-030-24643-3_99

performed better result for getting better result in classification and high accuracy for finding diseases in leaf. [2, 7] describes the image needs more space to save data, so for that cloud computing is used. The overall image of the field or land cultivating plant taken the image by camera to find which type of diseases affect the cultivating land, the disease such as fungal, bacteria etc. Paper [3–6] describes to find the diseases of the leaf either it is affected by fungal, bacteria or other parasite. This image can be detected by the application of image processing; the segmentation K-mean clustering technique is used. [9–11] demonstrate and find the unhealthy leaf from disease affected plant here support vector machine classifier is perform to detect the unhealthy leaf [8, 12]. Particle swarm optimization technique is used for task scheduling for better execution time.

2 Proposed Work

Figure 1 shown as the block diagram of proposed work. Here the image is get from the cultivation land of disease infected leaf that is the leaf may be tree leaf, plant leaf, shrub leaf etc. Then the acquired image is allowed to reduce the noise reduction by using image pre-processing technique as top hat filtering technique. Later the image is segmented through ostus thresholding technique there the edge and region based selection are processed. Then feature extraction is allowed for further technique in that colour, shape, textures of the image are found using GCLM extraction feature technique. Support Vector Network Classifier technique is performed for classify the disease affected leaf or not. By comparing the present data with the already store data in cloud server, there Particle Swam Optimization technique is to find task selection process, this will improve the system performance and execution time. The application orientation of this technique is mainly in agriculture industry.

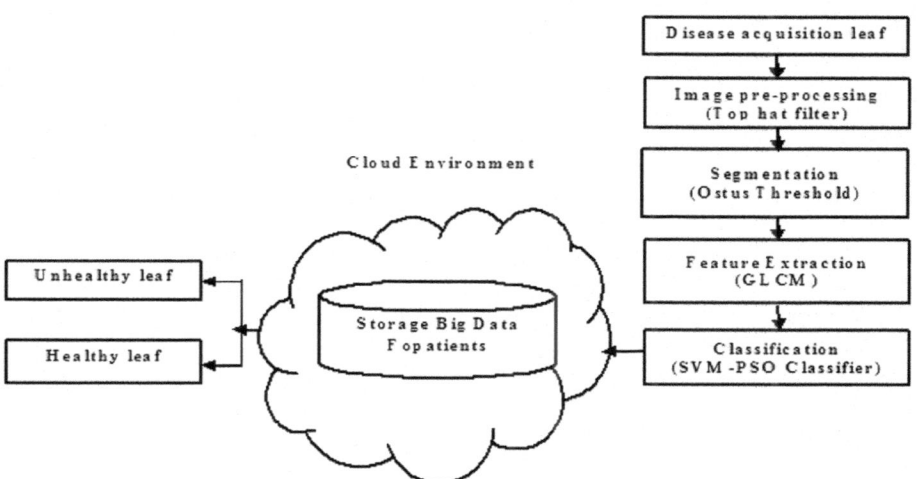

Fig. 1. Block diagram of proposed work

2.1   Image Acquisition

Image acquisition is nothing but the capturing of image from any source file. Here the image is get from the cultivation land or agriculture land. The image of the plant leaf is captured with the help of ordinary camera or mobile camera.

2.2   Image Pre-processing

Before going for any process first need to eliminate the noise of the input image here Top hat filtering technique is used to eliminate the noise of the image. Recover bright object from the dark background by top hat filter.

2.3   Otsus Thresholding

Otsu's method is a global thresholding technique. This technique is mainly used in the histogram of leaf image for searching threshold process. It maximizes "between class variance" of the segmented classes. Otsu proves that Minimizing "within class variance" is same as maximizing "between class variance" of the segmented classes.

2.4   Feature Extraction

A feature is a piece of information that describes a specific item in an image. The visual features of images are analyzed with the help of texture feature of images. One of the popular feature extraction techniques is Gray Level Co-occurrence Matrix (GLCM) which is applied on the images to extract the texture features. It evaluates the pair of pixels with the values obtained from the images. The GLCM is a static geometric device for removing second order texture data from the leaf images. A GLCM is nothing but a matrix here the number of rows and columns is equal to the number of individual gray.

Energy:
Energy is calculated by

$$En = \sqrt{\sum_{x=0}^{m-1}\sum_{y=0}^{n-1}f^2(x,y)} \tag{1}$$

Contrast:
Contrast is calculated by

$$C_{on} = \sum_{x=0}^{m-1}\sum_{y=0}^{n-1}(x-y)^2 f(x,y) \tag{2}$$

Mean:

Mean is calculated by

$$\text{Mean } \mu_i = \frac{1}{N} \sum_{j-1}^{N} f_{ij} \tag{3}$$

Standard Deviation:

Standard deviation is calculated by

$$SD = \sigma_i = \left(\frac{1}{N} \sum_{j-1}^{N} f_{ij} \left(f_{ij} - \mu_i \right)^2 \right)^{\frac{1}{2}} \tag{4}$$

2.5 Support Vector Machine Classifier

Support vector machine is a supervised learning model comes under in the machine learning. Analysis of classification and regression are done by support vector machine algorithm. In which SVM is associated with machine learning algorithm and it produce an output. By comparing with other process among the algorithm SVM algorithm gives better performance for classifications and regressions.

Image classification can be performed by use of SVM. For example, in a given input image for image classification, the task of the classification is to perform whether the input image is a cat or a dog. The input image is given into the SVM which might have gone through some image processing filters so that some features might be extracted such as edges, color and shape.

SVM is localized to compute the nonlinear decision surfaces in n. In this method it consists of projecting the data in high dimension space then where it is considered to remain separate linearly. For determining the non-linear surface of the original space Support Vector Machine algorithm is applied. Especially, the projection could be simulated or formulated using a method called kernel method.

If $x \in n$ is projected in a higher-dimension space H with a non-linear function
$\Phi: n - \rightarrow H$, then $(xi \cdot xj)$ is replaced by $\Phi (xi) \cdot \Phi (xj)$

The kernel function $K (xi, xj) = \Phi (xi) \cdot \Phi (xj)$ and then, the non-linear classifier can be expressed as:

$$f(x; \alpha) = \text{sgn} \sum_{i=1}^{NS} \lambda iyjK(si+x)+b \tag{5}$$

Where the si are the NS support vectors.

2.6 Particle Swarm Optimization (PSO)

PSO is a method of swarm intelligence for finding problems in global optimization. This algorithm copy the characteristics of bird flocking or else fish schooling for getting the foods where it can get a large amount. When a group of birds flying in the sky from one location to other side, only a single bird smell the food and transmit that detail to that group. These techniques minimize the processing time and it will increase the system performance in e-agriculture service by varying the position, fitness and particle search present in the solution space provides best solution.

Algorithm

   i. Initializing the value.
  ii. Finding fitness value.
 iii. Fitness value should be better than pbest.
 iv. pBest should be equal to current value.
  v. Evaluate particle velocity and position.

3   Results and Discussions

Figure 2 describes one of the sample leaf from data set. The image is acquired from data set after pre-processing of an image by top hat filter technique.

Fig. 2. Sample image from data set

- Top hat filtering is the technique used to remove noise of the image.
- Otsus thresholding is a segmentation to detect the region affected area.
- Support vector machine and Particle swam optimization techniques are to be performing the clustering of the image.
- PSO technique is for scheduling the task like as here iteration steps take place that is this technique select which pixel iteration selected for clustering.
- SVM technique classify the disease affected area and unaffected area in the leaf.
- Figure 3 clustering of image in three stages in first stage select the disease affect area, in second stage select the healthy area of the leaf and in third stage disease affected area is concluded.

3.1   Performance of SVM-PSO

Figure 4 describes the performance graph of the SVM-PSO comparing with other classifier techniques.

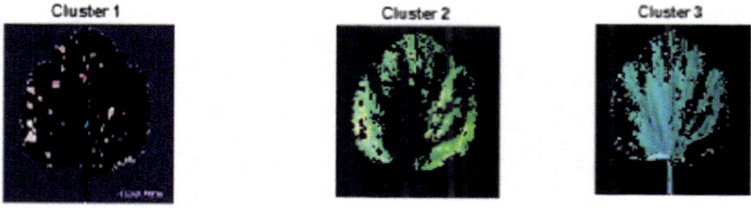

Fig. 3. Clustering using SVM–PSO

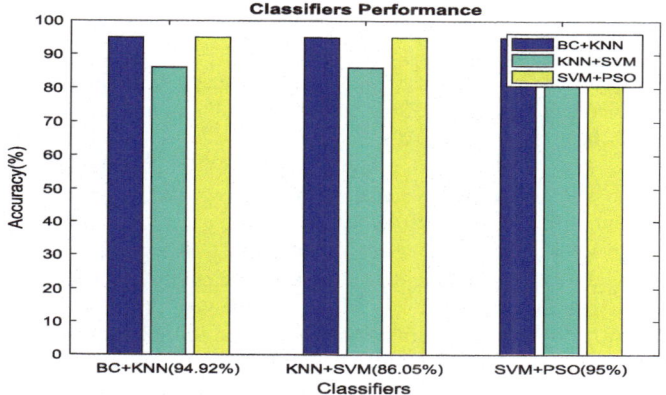

Fig. 4. Performance graph of SVM-PSO with other classifier

- By comparing various classifier techniques with other techniques, SVM-PSO technique is best.
- The accuracy rates of the sample image of various techniques are shown here. SVM-PSO accuracy rate of finding the disease affected area of the leaf is 95%.

4   Conclusion

In agriculture industry fertilizer and pesticide is very important to increase the yield of cultivation but yield is mostly affect by environmental damage and various diseases affect the crops such as vegetable, fruit, cereal crops. The disease may be bacteria, fungi and some other parasite diseases. Algorithm for feature extraction and classification based on image processing techniques. In this work plant leaf is get from data set and find whether the leaf is healthy or unhealthy leaf and compare the leaf to already predefined huge data store in cloud server and which is used to find what type of the leaf and is any disease affect the leaf or not. The accuracy of finding the unhealthy was improved and the execution time and system performance is well performed. In the future work early detection of disease with high possibility and helps to avoid loss of yield using Remote sensing technology or some other modern technology.

References

1. Nandhini, S.A., Hemalatha, R., Radha, S., Indumathi, K.: Web enabled plant disease detection system for agricultural applications using WMSN. Springer Science+Business Media, LLC, part of Springer Nature 2017
2. TongKe, F.: Smart agriculture based on cloud computing and IOT. J. Converg. Inf.
3. Khirade, S.D., Patil, A.B.: Plant disease detection using image processing. In: International Conference on Computing Communication Control and Automation, pp. 768–771, IEEE (2015)
4. Singh, V., Misra, A.K.: Detection of plant leaf diseases using image segmentation and soft computing techniques. Inf. Process. Agric. 4(1), 41–49 (2017)
5. Pujari, J.D., Yakkundimath, R., Byadgi, A.S.: Image processing based detection of fungal diseases in plants. Procedia Computer Science 46 1802–1808 (2015)
6. Aksoy, S., Akçay, H.G., Wassenaar, T.: Automatic mapping of linear woody vegetation features in agricultural landscapes using very high resolution imagery. IEEE Trans. Geosci. Remote. Sens. 48(1), 511–522 (2010)
7. Wang, C., Li, Z., Lv, H., Xu, T.: Application of PSO algorithm based on improved accelerating convergence in task scheduling of cloud computing environment. Int. J. Grid Distrib. Comput. 9(9), 269–280 (2016)
8. Nzanywayingoma, F., Yang, Y.: Analysis of particle swarm optimization and genetic algorithm based on task scheduling in cloud computing environment (IJACSA). Int. J. Adv. Comput. Sci. Appl. 8(1) (2017)
9. Jain, N., Sonali, P.: A machine learning approach: Svm for image classification in cbir. Int. J. Apl. Annovation Eng. Manag. (IJAIEM) 2(4) (2013)
10. Batule, V.B., Chavan, G.U., Sanap, V.P., Wadkar, K.D.: Leaf disease detection using image processing and support vector machine (SVM). J. Res. 2(02) (2016)
11. Srunitha, K., Bharathi, D.: Mango leaf unhealthy region detection and classification. In: Computational Vision and Bio Inspired Computing, pp. 422–436. Springer, Cham (2018)
12. Revathi, P., Hemalatha, M.: Identification of cotton diseases based on cross information gain_deep forward neural network classifier with PSO feature selection. Int. J. Eng. Technol. 5(6), 4637–4642 (2014)

Enhanced Data Security Architecture in Enterprise Networks

V Rashmi Shree$^{(\boxtimes)}$, Zachariah C. F. Antony$^{(\boxtimes)}$,
and N. Jayapandian$^{(\boxtimes)}$

Department of Computer Science and Engineering,
CHRIST (Deemed to be University), Bangalore, India
{rashmi.v,zachariah.antony}@btech.christuniversity.in,
jayapandian.n@christuniversity.in

Abstract. Encryption and storing important information is one of the risky and most challenging tasks. It is the need of the hour in today's fast growing technological transformations that the world is undergoing. A simple Enterprise network is the communication backbone of any organization. It mostly provides better information storage and efficient retrieval, which helps the organization to function smoothly, without having to think twice about their crucial data's security aspects. The information technology paradigm, cloud computing is used to help the organization to focus on its core business. In cloud computing is dealing with many services. That service is used for provide Platform service with infrastructure and software service. This paper, promotes the idea of combining various security and encryption algorithms to connect different enterprise networks using cloud computing, security layer concepts and giving no room for hackers to intrude into the confidential system of data.

Keywords: Cloud computing · Data security · Enterprise network ·
Encryption · Data cleaning · Distributed computing · Hadoop file system ·
Noisy data

1 Introduction

Any organization with a specific set of goals and objectives is known as an enterprise. The communication medium, for a particular organization, to rely upon for, their security, and data retrieval, and database management, and also information management, connecting related computer systems and devices, to provide insights about the organization and its smooth functioning. It can be a LAN or WAN connection, but it depends on the organizations operational requirements. Enterprise network model is a client and server based model. The Enterprise Networks main objective is to connect (combine) users and workgroups [1]. All the users and workgroups that are connected in the network should be able to access and utilize the data easily and efficiently. They must also be able to communicate and provide data. Enterprise network model is a client and server based model. In this model, the function of a server is to store the data and to manage data processing, whereas the client requiring any processes or specific

A. P. Pandian et al. (Eds.): ICCBI 2018, LNDECT 31, pp. 857–864, 2020.
https://doi.org/10.1007/978-3-030-24643-3_100

data will request it from the server. Client-server is a combination of the distributed framework of communication and network processes [2].

A company's process and operations strongly depends on Enterprise computing that denotes the business-oriented information technology. The several components of enterprise computing comprise of enterprise software that includes, relationship and database management. Enterprise computing is able to give a combined software solution to usual issues like, streamlined processes and resource management [3]. Compelled by the combined forces of technological developments and business, enterprise computing is continuously evolving. In an enterprise computing atmosphere, normally the computer systems are spread across a large region. It comprises of huge databases to be maintained. It also engages in frequently fluctuating business requirements. It nurses hundreds to thousands or millions of clients. In addition to this it demands reliability and performance for critical applications. Organizations or enterprises have to adapt flexibly and constantly to impending revolutions in the market. These kinds of alterations and fluctuations may persist in and out of the organization and might even follow at greater occurrences in the future. Addition to this, organizations are required to be extremely dynamic and very resourceful for setting up collaborations with business partners and acquiring new opportunities. Business targets and objectives can be pursued with provisional partnerships. Enterprises or organizations should flexibly be able to adjust and timely adapt. They should be able to exploit technological deviations without demanding the considerable requisite for any re-engineering or redesign. Enterprise computing demands huge flexibility and suitable technological support to accomplish all of this. The applications for which it is mandatory to handle the predominant activities of a particular firm are the core commerce applications. It includes a number of functions, Production, planning, distribution, logistics, sales and marketing, customer relations, financials, and HRM. Future changes can be adapted without any major impact to the enterprises systems. An enterprise network of systems can handle easily any addition or reduction in the number of clients. It is easily configurable and is able to be adapted to assorted environments. It can be effortlessly deployed and allocated to users. Enterprise computing systems are durable as vital information endures over time. They are satisfactorily efficient and responsive, consuming minimum resources and performing well on behalf of clients [4]. It can be maintained and supported easily and fixed with not much of an impact. The method of delivering technology and retrieving information resources that are hosted on the Internet with the help of web-based tools, technology and applications in contrast to connecting and retrieving information directly from the server is referred to as cloud computing. Instead of maintaining files on a privately-owned local storage device, making use of cloud storage will help to save the data on a secluded database. The cloud computing architecture comprises of all the mechanisms and components that make up cloud computing [5]. It consists of the following platforms; they are front end, back end, network and a delivery based on cloud. The front end comprises of client's system and the application software needed to have access to the cloud computing resources. Storage and servers comprise the back end of cloud computing. Internet, intranet and intercloud make up the network component. There exists a central server that will administer the system. It will also help to inspect the traffic and client demands. It is based on the set of rules referred to as protocols and makes use of distinctive software known as middleware. The networked computer systems communicate with each other with the help of middleware. Cloud computing affords certain striking benefits for end users and businesses. Few among those benefits include the pay per use practice in cloud computing, in which end users or clients pay only for the resources that they make use of. Another important aspect of cloud computing is that it provides elasticity as it allows enterprises or organizations to adjust to the workload fluctuations by provisioning when computing resources increase and de-provisioning on demand shrinkage automatically and this is known as elastic computing. Elastic computing helps to alter the usage of computing needs to help meet a changing workload and removes the necessity for enormous investments in the local infrastructure. Cloud computing provides services that can be private, public or hybrid. Private cloud is also stated as enterprise cloud. In private cloud computing the IT services are delivered over a private IT infrastructure or an organization's internal data center and it is dedicated only for the use of one organization and to its internal users. Core resources manage the private cloud. Internal users might or might not be payable for services. Virtual private cloud (VPC) and private cloud are both identical terms. The technical variance between the two is that VPC uses infrastructure provided by some third-party cloud provider while private cloud is built over internal infrastructure.

2 State of Art

Public cloud is a model where the storage and computational aids are provisioned to the use of general public usually over the internet. It is also referred to as a third-party cloud service provider. The usage of public cloud facility is based on "pay as you go" billing method where the services are vended on request. Users only pay for the amount of resources such as storage, bandwidth or CPU cycles, they have consumed. The difference between private cloud and public cloud is that in a private cloud the services are offered behind a firewall. It can be accessed only by the consumers or partners of a particular organization. Hybrid cloud computing model refers to the cloud computing service that combines various cloud service models to provision cloud solutions. It is a mixture of public cloud model and an on-site private cloud service with the planning and coordination between the two. Organizations can test serious workloads or subtle applications on private cloud environment. Infrastructure-as-a-Service is one of the major services in cloud network model. The service offers customers or users the facility to provision certain resources. Such as computing resources, processing, networks, storage resources and other essential resources. These resources are delivered to the customers as a virtual server instance and virtual storage, such as AWS. The consumers are allowed to deploy and run applications, software's and operating systems of their choice on the virtual instances given to them. The users are provided with APIs to help them transfer workloads to the virtual machine. The consumer doesn't have any concern regarding the management or control of the underlying infrastructure of the cloud. Platform-as-a-Service is another service offered by cloud providers. The job of the cloud service provider is to administer the underlying infrastructure of the cloud. This includes the servers, operating system, network and storage structures. The consumer governs the applications and the deployment, development, configuration of it on the cloud platform. Software-as-a-Service is one other service specified by the cloud. It is a type of distribution model. In this service infrastructure, the entire software application is provided to the user. User interface to an application is another option provided. These are offered to the user over the internet and are known as web services. The service provider manages the application as well as the data, therefore the user has the capability to access these services from any location with the help of a computer, laptop, workstations or any smart phone device with internet access. It is available through a thin client interface like a web browser. Since SaaS applications are independent of the platform, the type of operating system it is accessed on is not a problem. The underlying cloud infrastructure is taken care by the service provider comprising of networks, OS, storage, servers and also the software application and the consumer or the client is ignorant of the underlying cloud architecture. Grid computing consists of a distributed set of systems that are connected to each other. It consists of servers or even personal computers that are all linked together over a mutual network using a Wide Area Network. Each system is allocated with an independent task. Each system on the grid is known as a node [6]. In grid computing it is possible to process huge quantity of data with the assistance of a group of computers in the same network that synchronize to resolve a given problem. A grid is capable of handling a large number of resources. Grid's components could be situated at distant locations. Grid computing supports heterogeneity as it hosts software and hardware resources. It consists of data, software components, files, computer systems, networks and other resources. Grid computing has resource sharing properties in which a resource of a grid are part of many organizations and permit other organizations or clients to make use of them. Grid makes up a single virtual computer. It is necessary that a grid computing network is constructed with standard services, interfaces and protocols [7]. The Grid architecture comprises of five distinct layers which offer protocols and services. The fabric layer is responsible for delivering access to computing resources, storage components, network and code source. The central communication and authentication protocol necessary for simple and secure network transactions is handled by the connectivity layer of the Grid architecture. Every Grid transaction has an essential underlying protocol known as Grid Security Protocol, GSI. For performing publication, discovery, monitoring service negotiation, payment for the common operations carried out on individual resources and accounting a protocol is described by the Resource layer. The Grid Resource Access and Management protocol is effective in allocating computational resources. It also helps to examine and regulate the computation on the resources. The GridFTP which is an extension of the FTP (File Transfer Protocol) for grid computing is used for accessing data and transferring data at high speeds [8]. The interaction that occurs across the collections of resources and directory services like Monitoring and Discovery Service is captured by the Collective layer in the Grid architecture. Finally, the Application layer is made up of the user applications which function of virtual organization. The fundamental concept of virtual organization is to create business skill. By doing this they enhance the response to business opportunities. Virtual enterprises are common in the expanse of research and development. The PRODNET Architecture is industrial virtual enterprises have been sustained with the design and implementation of an open infrastructure by the PRODNET II consortium. The unique focus was on the requirements of medium and small sized enterprises [9]. Ineffective struggle to safeguard the enterprise networks helped in developing the concept of SANE that has an ultimate goal of security. Secure Architecture for the Networked Enterprise provides access control that issues minimum opportunity to access resources along with being flexible to not let the network be unused [10]. Implementation should be at the link layer. This is done to thwart lower level layers from undermining it. The information regarding topology and services are hidden from resources that do not have consent to see them. To block the attacker from mapping out the network topology, detecting firewalls, critical servers and other critical data. Modern networks depend on multiple constituents like firewalls, routers, switches etc. If any one component among these is compromised it can cause the devastation of the complete enterprise. Therefore, the goal in SANE is to trust on a central reliable entity where the policy is centrally executed and well-defined.

3   Existing Enterprise Model and Security Challenges

In the existing technology there are complex mechanisms. A standard enterprise network nowadays uses numerous mechanisms concurrently to shield its network. Virtual LANs, Access Control Lists, firewalls and Network Address Translators. The responsibility of security and related policies are distributed among the cases that perform the implementation of these mechanisms. This makes the security fragile and also the configuration is dependent more often on network topology. A possible or usual approach is to put all the security policies in one case. For example, the firewall at the network's entry and exit point, which is the obstruct point in the network is used for this. Consequently, if the attacker breaks through the firewall, then he has access to the entire network. The SANE architecture has centrally conveyed high-level simple policies which are implemented by a single compact mechanism which is within the network [11]. Enterprise network consists of several properties that can be exploited in the effort to make them more secure. Enterprise networks are generally engineered carefully and administered centrally. Simple declarative access control policy can be adopted to allow little privilege to users or clients since most of the systems in the enterprise networks are clients. They usually interact with expected minority of local services. In addition, in an enterprise network, it is possible to assume that authentication is done in the case of hosts and principals. This can be done given the extensively deployed directory service like Lightweight Directory Access Protocol which is used to discover organizations or resources such as devices or files, in a network over the public internet or even a corporate intranet. This procedure will be effective to help expressing policies as meaningful entities such as users and hosts. Enterprise networks are capable of adopting a novel architecture for protection. Due to the increasing and elevated cost of security flops it is crucial to implement a novel security technology for the enterprise networks. Data security is also involved in this enterprise network system [12].

4 Proposed Individual Encryption Architecture

The proposed enterprise model is deals with individual encryption in each module. In this model node 1 is a client machine connected with the enterprise network with cloud environment. The client machine is sending some request to cloud server via enterprise network. The conceptual model of this method is to provide higher security within some limited network system. The concept of enterprise network is to connect several computers and work stations together to allow them to communicate with each other. It helps in accessing the data from the server machine. The number of protocols required for communication is reduced when we use the concept of enterprise networks. Enterprise networks also helps in managing the data in the server efficiently. The concept of enterprise network is connected with multiport system with a cloud environment. This system is seriously facing data security problem, because hacker always tries to hack the client data in network transmission. This kind of network hacking is possible with router or server machine. The computers machines are coupled with the router. The concept of router is focusing on study the data contents that are transmitted from work stations to another workstation or from one workstation to the server. The router makes sure if the beginning node and the ending node to which the data are transferred are on the same network or not. It also checks whether the data that is being transported has to be changed from one type to another. This client machine is connected with the router and transfers the data from client machine to server machine. Similar to that node 2 is an LAN network, the entire LAN network is connected with same cloud server via the enterprise network concept (Fig. 1).

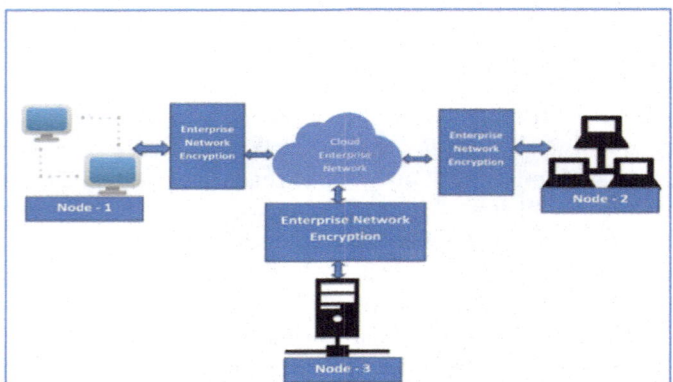

Fig. 1. Enterprise encryption architecture

The concept of LAN is distributed environment, many systems is connected in distributed scenario with a limited access. These kinds of network are also facing security issue, because in distributed environment a greater number of systems connected in single router or single server. The data theft is very easy to these kinds of network connection. Sometime local server is also connected with this enterprise network. Information theft is a process of accessing the others data without their permission. Many types of data theft can happen in the field of financial data, personal data, and various important data. All such data thefts come under the cybercrime. This local server is maintaining distributed system data with in some organization or industry. The propose architecture node 3 is connected with this enterprise network using cloud platform. This level all the nodes are connected with the cloud server, but now question is raised. That means data security is major question in this network environment. The proposed system is providing an idea for individual encryption model. That means individual node is dealing with this security problem with the help of encryption algorithm. This encryption algorithm is encrypting the individual node data and sends it to that encrypted data in server. Then cloud server is stored the encrypted data, if suppose hacker try to access the data from server or network terminal, they got only encryption data. This concept is implemented for the entire node that means distributed network and local server machine is also working with same methodology. The major advantage of this proposed model is providing higher security with a different type of network environment.

5   Conclusion

The concept of enterprise network is discussed. The major focus on data security is implemented and reduces the security attack. This enterprise network is involved in local area network and wide area network. The propose architecture is provide additional security protocol and avoid the data attack from client to server communication. The quality of service is automatically increased because of individual security mechanism. Only problem of this method is taking additional time for data communication, in future direction take this research problem and try to give additional solution for this.

References

1. Yang, S., He, Y., Xu, M., Jiang, Y.: Towards flexible management in enterprise network: an enhanced routing protocol. Int. J. Mach. Learn. Cybernetics. **9**(1), 125–132 (2018)
2. Rice, R.E.: Computer-mediated communication system network data: theoretical concerns and empirical examples. Int. J. Man-Mach. Studies. **32**(6), 627–647 (1990)
3. Van Aken, D., Pavlo, A., Gordon, G.J., Zhang, B.: Automatic database management system tuning through large-scale machine learning. In: International Conference on Management of Data, pp. 1–5. ACM (2017)
4. Jayapandian, N., Rahman, A.M.Z., Gayathri, J.: The online control framework on computational optimization of resource provisioning in cloud environment. Indian J. Sci. Technology. **8**(23), 1–5 (2015)
5. Jayapandian, N., Rahman, A.M.Z., Koushikaa, M., Radhikadevi, S.: A novel approach to enhance multi level security system using encryption with fingerprint in cloud. In: World International Conference on Futuristic Trends in Research and Innovation for Social Welfare (Startup Conclave), pp. 1–5. IEEE (2016)
6. Wang, K., Wang, Y., Hu, X., Sun, Y., Deng, D.J., Vinel, A., Zhang, Y.: Wireless big data computing in smart grid. IEEE Wirel. Communications. **24**(2), 58–64 (2017)

7. Khan, A.A., Rehmani, M.H., Reisslein, M.: Requirements, design challenges, and review of routing and MAC protocols for CR-based smart grid systems. IEEE Commun. Magazine. **55**(5), 206–215 (2017)

8. Nadig, D., Ramamurthy, B., Bockelman, B., Swanson, D.: Identifying anomalies in GridFTP transfers for data-intensive science through application-awareness. In: International Workshop on Security in Software Defined Networks & Network Function Virtualization, pp. 7–12. ACM (2017)

9. Camarinha-Matos, L.M., Afsarmanesh, H.: The PRODNET architecture. In: Working Conference on Virtual Enterprises, pp. 109–126. Springer (1999)

10. Park, J.S., Hwang, J.: Role-based access control for collaborative enterprise in peer-to-peer computing environments. In: International Symposium on Access control models and technologies, pp. 93–99. ACM (2003)

11. Casado, M., Garfinkel, T., Akella, A., Freedman, M.J., Boneh, D., McKeown, N., Shenker, S.: SANE: a protection architecture for enterprise networks. Int. Symp. USENIX Security. **49**(5), 50–55 (2006)

12. Jayapandian, N., Md Zubair Rahman, A.M.J.: Secure deduplication for cloud storage using interactive message-locked encryption with convergent encryption, to reduce storage space. Braz. Arch. Biol. Technol. **61**, 1–13 (2018)

Cloud Virtualization with Data Security: Challenges and Opportunities

Joshua Johnson Abraham$^{(\boxtimes)}$, Christy Sunny$^{(\boxtimes)}$, Anlin Assisi$^{(\boxtimes)}$, and N. Jayapandian$^{(\boxtimes)}$

Department of Computer Science and Engineering,
CHRIST (Deemed to be University), Bangalore, India
{joshua.abraham, christy.sunny,
anlin.assisi}@btech.christuniversity.in,
jayapandian.n@christuniversity.in

Abstract. In recent years, Cloud Computing is emerging as a torrid research area for both academicians and industrialists. It provides effective ways to handle and store the data in advanced system processing applications. Furthermore, it also leverages a radical change in the way the users access and use the available resources. Despite the hype, it also has the challenge of slow data transition from present physical storage to the cloud based platform. This is mainly due to the security challenges associated with the Cloud Computing applications. Hence, data protection has become very critical and always requires an efficient and effective security protocol into the existence. So, the security and reliability of the cloud platform would definitely attract more researchers to this platform. This article discusses an overview of Cloud paradigm and the different virtualization techniques adopted to overcome the security issues associated with the cloud computing platform.

Keywords: Cloud computing · Virtualization · Security · Privacy · Virtual Private Network

1 Introduction

Cloud computing is an emerging technology in the Information Technology industry to deliver an ease to the access of the resources anywhere and at anytime. It has been viewed as a pool of physical and virtual resources through broad area network. This has led to an industrial boom wherein the industries or organization doesn't have to pay for a physical upfront machine, instead virtual machine concept is introduced to reduce the computational cost. Everything, including the storage and computational power would be easily delivered by the cloud provider. As it is clearly evident in the present scenario, top organization and companies are currently utilizing the cloud servers to maintain their resources. Different levels of services are available in the cloud platform, when the customer needs software they use Software as a Service (SaaS). If they need entire platform, they use Platform as a Service (PaaS). Finally if they need infrastructure to execute their project then cloud uses Infrastructure as a Service (IaaS). This is a torrid research topic in both academia and industry. It is considered as a widespread

© Springer Nature Switzerland AG 2020
A. P. Pandian et al. (Eds.): ICCBI 2018, LNDECT 31, pp. 865–872, 2020.
https://doi.org/10.1007/978-3-030-24643-3_101

computing paradigm in which the computational and resources are provided over the internet. Cloud computing method is solely based on the centralized approach, where the cloud server stores the Computing resources and also the datacenter. It also leverages the virtualization technique to efficiently provide the resources among the multiple users. Cloud computing allows the virtualization process for load balancing by means of their dynamic provisioning feature. Hence, Virtualization plays a key role in Cloud computing [1]. This chapter discusses abotu the various security issues linked with the cloud computing platform which is primarily associated with the virtualization techniques to solve the emerging security challenges.

Virtualization is meant for modeling or developing the physical environment similar to the original client or server system. This virtual environment helps to create a virtual server, storage device and network resources for the intended users. It also delivers the capacity to divide the physical resources of a server in a logical manner. In Fig. 1, different kinds of virtualization and methods are discussed. First one is full virtualization, second one is hardware virtualization and the final one is para virtualization method. Full Virtualization is the most commonly used and economical type of virtualization. It is type in which the service requests from a computer are isolated from the computer hardware. The hardware actually intends to permit the software to work on a guest operating system. The concept of this virtualization is mainly to execute the software program without a physical hardware server in different operating systems. Para Virtualization is the another type of virtualization which intensifies the technology of virtualization by checking the new operating system before their installation into the machine. It allows to combine the machine which can undergo certain changes when compared with the hidden hardware. This type of virtualization allows the various operating systems to work on a single set of hardware productively using the available resources like memory and processors. Hardware Virtualization is also known as platform virtualization, designing of a machine will work as an actual computer with any of the available operating system. The software that runs on these machines will be isolated or divided from the concealed hardware present in this type of virtualization. It is used to enhance the virtualization productivity. It involves several hardware elements to help in enhancing the working of a guest operating system.

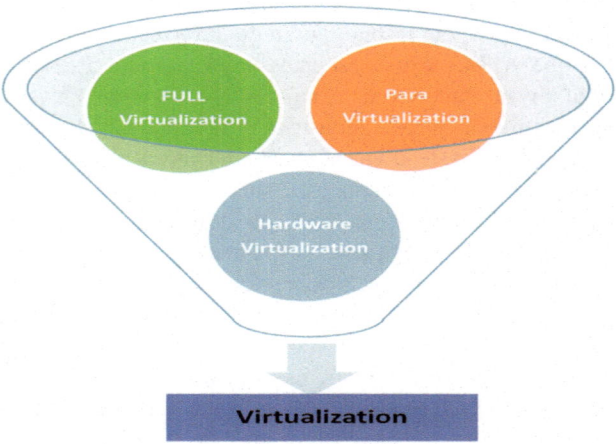

Fig. 1. Virtualization types

2   State-of-the-Art

The word virtualization means the process of changing something that exists in a real form into a virtual version. It is dedicated mainly to create a virtual environment of resources such as device, network, and storage to operate a system in which the framework divides the work into one or more environments. Virtualization deal with different types they are, primary service is storage, memory, network and data virtualization. Then the secondary type is desktop and software virtualization [2]. Virtualization may be defined as the technology that provides a user with the ability to work on different environments using the same computer. For organizations with lesser than 1,000 employees, up to 40% of company expenditure rely on hardware expenses. The following are the advantages of using cloud virtualization, which can easily backup the data. Then the effective information technology management is to reduce the upfront cost. Application virtualization encourages the client to have a remote access for an application from a server. The server stores everything close to the source data and different attributes of the application. However it can be any case running on a nearby workstation through internet. Example of this would be a client who needs to run two unique variants of a similar programming infrastructure. Advancements of application virtualization are facilitated applications and bundled applications. The capacity to run various virtual systems is to control and information management, also it exists together over one physical network. It can be overseen by individual gatherings that can be possibly classified to one another. Network virtualization, dedicates a workspace to develop virtual systems, intelligent switches, switches, firewalls, stack balancer, Virtual Private Network (VPN), and outstanding task at hand security within few days or a week [3].

Desktop virtualization enables the clients' operating system to be remotely put away on a server available in the information center. It enables the client to get to their work area basically, from any area and from various machines. Clients who need to explicate their work frameworks other than Windows Server should have a virtual desktop. Main advantages of work area virtualization are client mobility, portability and simple administration of programming establishment [4]. Virtual storage environment is a stockpiling framework. This storage platform creates different virtual environment and develops a storage copy. It makes overseeing data storage from different sourcs which is then used as a solitary storehouse. Data storage virtualization programming keeps up smooth activities. This is a reliable execution and a ceaseless suite of cutting edge capacities regardless of changes, separate and contrasts in the basic gear. The data storage mechanism can also deal with some encryption and deduplication methods [5].

Virtual Machine Monitor additionally called as the virtual manager embodies to plan the nuts and bolts of virtualization in the distributed computing platform. It is utilized mainly for isolating the physical equipment from its copied parts. This process continually inculcates the System's memory, Input/Output and system activity. The working method of virtual machine is dedicated mainly to identify the originality of computer machine. This virtualization process is involved in execution of operating system, parallel execution of multiple operating systems in single machine. Then the additional layer of this virtualization is named as hypervisor. The product layer is to project and virtualize the assets of a host device to satisfy the client necessities. Physical layer has the ability to run two modes directly, first one is the exposed metal hypervisor and second one is the local machine. The type one hypervisor involves in guiding and accessing the physical equipment. With a type 1 hypervisor, there is no particular working framework to stack as the hypervisor is working with framework [6]. This type 2 method is a direct product application framework. This is a guest machine access link with the CPU and memory access. This method is the example of VMware Fusion and VMware workstation [7]. Figure 2 is to elaborate type 1 and 2 hypervisor model.

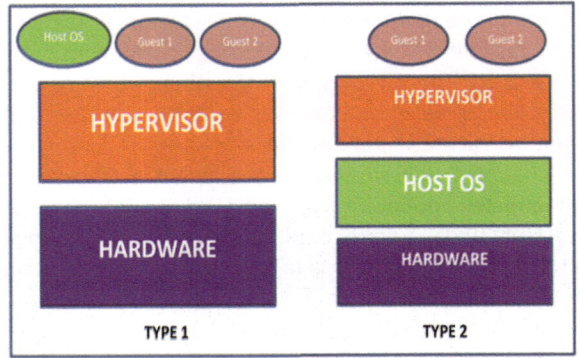

Fig. 2. Hypervisor architecture model

3   Security Issues in Cloud Computing

Security is one of the primary issues, which act as a bottleneck of the emerging cloud computing technology [8]. Cloud computing provides various access control and storage mechanisms. The major problem of this method is losing the control while providing cloud service. One of the challenges is that the user does not have any knowledge on where the data is processed and the data is stored. Another one is the loss of control as the cloud provider has to pay for running a service. Also few of the concerns of security in cloud computing are recovery, privileged user access, data segregation, bug exploitation and privacy. These issues tend to reduce the growth of cloud computing in the Information and Communication industry. Recovery is the issue of data loss when a particular data is not stored properly. This occurs when the user does not store the data in the right manner by which the recovery of data becomes difficult. Privileged user access is the another security issue faced by the cloud computing technology. This enables a second user to have access to files and data that is originally shared by another user. The access given to another user can be misused if the user does not handle the data carefully, which will finally leads to data privacy issues. The data can be breached and can be used by anyone without the knowledge of the original user. Data segregation is another challenge for cloud security which happens due to the unsystematic segregation of data that leads to unknown data storage locations [9]. Then it leads to the loss of data obtained from the user. Bug exploitation leads to severe data loss issues as it can lead to abrupt crash of a program or data transfer. It can lead to data loss and improper storage of data in the storage space. Bugs are caused due to the compatibility issues between the software and hardware, certain errors in the code generates bugs. All these issues lead to data privacy issues for a particular user. Privacy issues leads to hindrance in the development of cloud computing services. Privacy is the main aspect of any technology as the users wants their data to be secure and protected against any threats and attacks [10]. Privacy refers to the revealing of important data stored on the platforms which creates legal liability and loss of reputation. It is one of the main issues which breaches the virtualization security. The set of requirements that needs to be followed are effectiveness, precision, transparency, deplorability, dynamic reaction and accountability.

Effectiveness is the extent to which something is successful in obtaining the desired result in a virtualization environment. The desired result can be successful in accessing the data from anywhere and at any time. Precision refers to refinement of the data accessed without any error. The fact is being exact and accurate while generating the desired result from the user. Transparency is the condition of being devoid of any hidden agendas or conditions while the user tries to access the stored data. It helps in building trust among the users. Deplorability refers to how the virtual machine can be deployed in a user's computer and also if it can be used in various platforms without any hindrance to the operating system. Dynamic reaction is that the user must be readily accessed by the cloud computing service without a delay can help in faster computation than any other resources. The service providers for cloud must be accountable for any data loss caused by the user. The user must also be responsible to store the data as required and while giving the access permissions to other users [11].

4 Improving Security Using Virtualization

4.1 Sandboxing

Sandboxing is one of the unique and rare methodologies present in virtualization. Sandboxing is closely associated with the virtualization and use of the key concepts of virtualization to run this method. Sandboxing basically refers to a process of fooling a computer application or program. It is runs a full-fledged computer platform and analyzes the performance of the individual system. This largely helps in observing the behavior of any vulnerable applications. Since the application is not aware of the tightly coupled virtualized platform that restrains it from replicating acts normally and tries to exhibit its true nature. It is safe to analyze and quarantine an application without harming the existing system [12].

4.2 Server and System Virtualization

Protecting the server information from the physical machine access, this method is generally named as server virtualization. This contains the details on the distinct physical servers, its operating system and system process. Basically server virtualization is used to produce multiple virtual servers from a single physical server. It is a great advantage for organizations to run virtualized servers, it also reduce the cost of the physical server. The management of physical server is tedious when compared with the virtualized server. It can be easily installed, downgraded and upgraded at any point of time without causing a bottleneck in the entire system. A virtual server has effective security as it has own programs to trap worms, viruses and other malevolent programs. The concept of virtualization server is to split the application separately to provide a higher grade of security. This helps in maintaining a server in a secure and safe manner even though the others are attacked by some sort of malware or infection. This method can be used to enhance the disaster recovery process. System virtualization is the process of hardware system equipment and software programming. This virtualization is used to avoid malware virus and creates the special framework. This system virtualization concept is to control and coordinate the entire system architecture. Virtualization concept is to create a dynamic environment and to establish the end to end connection.

4.3   Desktop Virtualization and Infrastructure Security

The virtualization software is also used to create the desktop environment. This environment tends to create logical connection with physical systems. The Information Technology system administrator is dedicated to control the illegal system access and manage the entire system environment. In this, the virtualization concept provide best solution for operating system work place. Compared with other virtualization system this desktop virtualization provides better security and working environment. In the case of virtualized data construction, it enables to access the available resources. Also it provides visibility to properly handle the information. Hence all the activities that happen inside the Virtualized environment are tracked.

4.4   Great Availability and Disaster Recovery

The most important thing of Information technology sector is to keep the data safe and available all the time. Though this cannot be achieved completely, getting back the data after a disaster occurrence is the key aspect of a cloud computing technology. People want their data to be too safe and should always be accessible irrespective of their time and place. Virtualization plays a crucial role in achieving this requirement. The main advantage of virtualization is to reduce the execution time and managing the overall implementation cost for large database. It also reduces the time for implementing or installing many operating systems in a single machine. The major benefit of this is green environment, it reduces the system usage and server implementation. The single server can efficiently manage many client machines and applications. This methodology also reduces the power consumption.

4.5   Hypervisor Security and Server Isolation

The main issue of virtualization is VM jumping or hyper jumping. The same network infrastructure based virtualization server machine helps in preventing malicious and unauthorized access to the clouds resources. This type of environment is creating a loop hole of data attack with the connection of many users in the same network. To avoid these kind of problems, it creates a separate physical link with the virtual local area network connection. This type of situation separates the database server which is to be maintained and also to control the client machines. Hypervisor helps to achieve the creation of separate groups. In this, business database server introduces individual virtual machine concept without the usage of virtualization software. Virtualization is intended to provide permission for running more number of virtual servers in a single server machine. This is a creation of independent virtual server machine with a larger database access.

5   Conclusion

Cloud computing is one of the most significant areas of computer science p[latform. It plays a crucial role in industrial sector and also the way in which different organization leverage their resources in a virtual environment. The aim of this virtualization and cloud computing technology is to reduce the cost of infrastructure, software and maintenance of the traditional physical system. Cloud computing has bought a huge transition in the way the people access and use physical system. This paper discusses about the current trends and growth in virtual cloud computing technologies. It also defines the working principle behind the cloud virtualization and security issues. This article helps to understand the basic cloud virtualization and handling of security challenges in cloud computing environment.

References

1. Xiao, Z., Song, W., Chen, Q.: Dynamic resource allocation using virtual machines for cloud computing environment. IEEE Trans. Parallel Distrib. **24**(6), 1107–1117 (2013)
2. Jain, R., Paul, S.: Network virtualization and software defined networking for cloud computing: a survey. IEEE Commun. Magazine. **51**(11), 24–31 (2013)
3. Han, B., Gopalakrishnan, V., Ji, L., Lee, S.: Network function virtualization: challenges and opportunities for innovations. IEEE Commun. Magazine. **53**(2), 90–97 (2015)
4. Miller, K., Pegah, M.: Virtualization: virtually at the desktop. In: 35th Annual ACM SIGUCCS Fall Conference, pp. 255–260. ACM (2007)
5. Jayapandian, N., Md Zubair Rahman, A.M.J.: Secure deduplication for cloud storage using interactive message-locked encryption with convergent encryption, to reduce storage space. Braz. Arch. Biol. Technol. **61**, 1–13 (2018)
6. Soltesz, S., Pötzl, H., Fiuczynski, M.E., Bavier, A., Peterson, L.: Container-based operating system virtualization: a scalable, high-performance alternative to hypervisors. ACM SIGOPS Oper. Syst. Rev. **41**(3), 275–287. ACM (2007)
7. Vaughan-Nichols, S.J.: Virtualization sparks security concerns. Computer **8**, 13–15 (2008)

8. Jayapandian, N., Rahman, A.M.Z., Radhikadevi, S., Koushikaa, M.: Enhanced cloud security framework to confirm data security on asymmetric and symmetric key encryption. In: Futuristic Trends in Research and Innovation for Social Welfare (Startup Conclave), World Conference, pp. 1–4. IEEE (2016)
9. Jayapandian, N., Rahman, A.M.Z.: Secure and efficient online data storage and sharing over cloud environment using probabilistic with homomorphic encryption. Clust. Comput. **20**(2), 1561–1573 (2017)
10. Losavio, M.M., Chow, K.P., Koltay, A., James, J.: The Internet of things and the smart city: Legal challenges with digital forensics, privacy, and security. Secur. Priv. **1**(3), e23 (2018)
11. Cobb, C., Sudar, S., Reiter, N., Anderson, R., Roesner, F., Kohno, T.: Computer security for data collection technologies. Dev. Eng. **3**, 1–11 (2018)
12. Irazoqui, G., Eisenbarth, T., Sunar, B.: $ A: A shared cache attack that works across cores and defies VM sandboxing–and its application to AES. In: Security and Privacy (SP), 2015 IEEE Symposium on (pp. 591–604). IEEE (2015)

Data Mining Approach for News Inspection on Social Media: A Survey

Yugchhaya Galphat$^{(\boxtimes)}$, Heena Banga$^{(\boxtimes)}$, Isha Dalvi$^{(\boxtimes)}$,
Priya Jethmalani$^{(\boxtimes)}$, and Shraddha Talreja$^{(\boxtimes)}$

Computer Engineering, Vivekanand Education Society's Institute of Technology,
Mumbai, India
{yugchhaya.dhote, 2015heena.banga, 2015isha.dalvi,
2015priya.jethmalani, 2015shraddha.talreja}@ves.ac.in

Abstract. Social Media is the useful servant but a dangerous master, by virtue of it a biggest new-age real world problem arises termed as "Fake News" which represents information that is completely fabricated and is created deliberately to misinform or deceive readers. Also with the advent of social media which on the one hand is low cost, easily accessed and its rapid dissemination of information that lead people to seek out and consume news from social media but on the other hand, it enables the widespread of fake news which has the potential for extremely negative impacts on individuals and society and hence it needs to be stopped. This paper aims to give attention to the fake news problem, its psychological impacts and all the existing approaches to detect fake news. Also an analysis of all the approaches is performed and datasets available for fake news are discussed.

Keywords: Fake news · Social media

1 Introduction

"News" an English word was developed in 14th century as a special use of the plural form of "new" [2]. As name itself implies, its principal meaning is presentation of new information. Often it gives information on Government proclamations, laws, taxes, education, politics, war, public health, the environment, economy, business, fashion, athletic events, entertainment and as well as quirky events. News is indeed an important and only source of seeking information and knowledge about current events. It is also used as a platform to generate opinion for the people. In the technology emerging country, "Internet" became the fastest mode to reach out the population of globe and to spread a word. In the era of Internet, people are mostly using Social media platforms and network sites as the content can be shared freely. People use these platforms for sharing information, entertainment purpose, giving opinions on particular events, gaining knowledge, etc. So, these platforms like Facebook, Twitter, Whatsapp became more powerful mode to reach millions of population across the world. Reports have shown that a majority of adults access their news using digital forms such as social media and traditional search engines rather than using traditional media [1].

© Springer Nature Switzerland AG 2020
A. P. Pandian et al. (Eds.): ICCBI 2018, LNDECT 31, pp. 873–880, 2020.
https://doi.org/10.1007/978-3-030-24643-3_102

2 Psychological Foundations of Fake News

Humans are inherently not very good at discriminating between Fake and Fact. To make people naturally vulnerable to fake news there are two major factors Naive Realism and Confirmation Bias [4]. Also due to echo chamber effect people easily believe in fake news. Social media provides a new model of information creation and consumption for users. On these platforms, mostly people follow like-minded people and because of "news feed" feature they are exposed to selective news which results into receiving only that particular news which promotes their existing views. In this way, users views are polarized resulting into an echo chamber effect [4]. So, these are the psychological and cognitive theories that shows the influential power of fake news.

3 Fake News Detection Systems

In the previous sections, we toss around "Fake News" and its consequences. Now coming to the solution of described problems that is checking the authenticity of news stories. There are numerous techniques for detection but they are mainly classified into two major categories that is Human based approach and Machine based approach.

3.1 Human Based Approach

What help humans to identify between the veracity of the news? Research shows that humans cannot help themselves in finding the deceptive news, the high court judges could possibly distinguish between fake and fact with 50–63% success rate [5, 7]. This research can be explained by several psychological factors:

Social Credibility. It means mostly people consider a source as authentic when they see majority of population considering it as authentic.

Frequency Heuristic. It means mostly users inherently support that news which they read intermittently even if it is deceptive.

Communicative Behaviour. Deceivers are not able to write the appropriate article, since the focus is on verbal communication and due to the lack of knowledge in professional writing humans are not able to recognize the veracity of news.

3.2 Machine Based Approach

To overcome all the flaws associated with above approach, Machine based approach is discovered. This approach is further divided into two major approaches that is Linguistic and Network. Both these approaches are completely different as linguistic approach performs corpora based analysis whereas network approach provides URL based solutions.

Linguistic Approach. Linguistic approach is completely based on knowledge of language and its structure. Although most of the liars manage to curb their language but the main aim of this approach is to contemplate the slip of language. As language has certain features based on which the credibility is judged. Linguistic approach comprises of feature extraction, feature selection and classification task.

Feature Extraction. It is the mechanism of gathering the key linguistic features like POS tags, punctuations etc from news articles. As many of the existing systems depend on feature extraction, all the features mentioned in papers [9] and [10] are:

- Ngrams: It is used for extraction of text using Bag of Words algorithm in which unigrams or bigrams are extracted from the news story and encoded as Term Frequency-Inverse Document Frequency (TF-IDF) vector values.
- Punctuation: Linguistic Inquiry and Word Count (LIWC) is a software which is used to construct the feature set for eleven types of punctuations [9].
- Psycho-linguistic features: To gather psycho-linguistic features, the use of LIWC lexicon to select the appropriate proportions of words is recommended by researchers. LIWC is a tool which cluster single LIWC categories into multiple feature sets [9].
- Automated Readability Index (ARI) [3]: Articles are difficult to understand because of its language complicacy. Hence, feature is used in building the model and is defined as,

$$ARI = 4.71 * \left(\frac{characters_e}{words_e} \right) + 0.5 * \left(\frac{words_e}{sentences_e} \right) - 21.43 \tag{1}$$

where $characters_e$ is the count of characters, $words_e$ is the count of words, $sentences_e$ is the count of sentences in the Eq. (1).

- Syntax: In this, the grammar of the language is checked, with the help of two approaches Context Free Grammar (CFG) and Probabilistic Context Free Grammar (PCFG). In [9], Bennett et al. have used the CFG which is a list of rules that define the set of all wellformed sentences in a language whereas in [10], Conroy et al. have used PCFG in which probability is assigned to each rule and atlast a parse tree is generated assigned with certain probabilities. This approach is used to differentiate rule categories for deception detection with 85–91% accuracy (based upon rule category used) [10].
- Semantic Analysis: It describes the process of understanding natural language–the way that humans communicate–based on meaning and context. It determines the truthfulness of authors by identifying the extent of compatibility of a personal knowledge. The assumption is that since the fake writer does not has previous experience with the particular event or object, then they may end up including contradictions or even leave out important facts that were existent in profiles on related topics [10].

Feature Selection. As there are immense features, feature selection is of main concern. Since all the features are not used simultaneously and to select the most representative and reliable features, feature evaluation, feature subset selection and validation techniques are used [19]. Feature evaluation measures the fitness of features based on some chosen criteria, Feature subset selection finds a combination of the features that maximizes the fitness function and validation is used to determine the number of features that is sufficient for the given data [20].

Classification Task. In this task, all the observations are grouped together based on their characteristics and classification is performed to predict whether article is fake or fact. Some of the popular classifiers used are:

- Naive Bayes Classifier [11]: It works by correlating the use of tokens (typically words, symbols, syntactic or not) with news articles, and then uses Bayes theorem to calculate a probability of an article if fake or fact. Support Vector Machine (SVM): It is a representation of the training data, as points in space categorized by an optimal hyperplane which is as wide as possible. Samples to be evaluated are then mapped into that same space and predicted to belong to a category based on which side they fall.
- Random Forest: It operates by constructing a multitude of decision trees at training time and outputting the class that is the mode of the classes (classification) or mean prediction (regression) of the individual trees.
- Stochastic Gradient Descent: It is a simple and very efficient approach to fit linear models and particularly useful when the number of samples is very large. Gradient Boosting: It involves three elements loss function to be optimized, weak learner to make predictions and additive model to add weak l.
- Bounded Decision Trees: A data of attributes is given together with its classes, a decision tree produces a sequence of rules that can be used to classify the data. Different models were trained using classifiers based on real and fake news articles that are collected from different sources to evaluate the performance of classifier models.

Network Approach. The network approach was designed to integrate the content-based approaches such as the linguistic approach. It collect and compare a wide range of similar and related statement from various sources (network) such as meta tags, linked data, social engagements and social network behaviour. Fact Checking Methods based on network approach:

Linked Data. As web is the great source of information for humans which has improved our daily life. Now, not only humans but machines also use data from web. Linked Data is about using the Web to interlink related data that wasn't previously linked. This process basically depends on querying existing knowledge graphs, or publicly available structured data, such as the Google Relation Extraction Corpus (GREC) or DBPedia ontology [10]. DBPedia, which essentially makes the content of Wikipedia available in RDF (Resource Description Framework). A global web of information (web of data) is created using hyperlinks depending upon the linked data principles stated in [13] as:

- Use Universal Resource Identifiers (URIs) as identifiers for objects and documents.
- Use HyperText Transfer Protocol (HTTP) URIs so that objects can be referred and looked up both by humans as by machines.
- When an URI is requested, it produces useful knowledge, using principles like SPARQL protocol, RDF* and RDF Query Language (SPARQL).
- Link this URI with other URIs, creating structure in the web.

Social Network Behaviour. Identity on social media is not authenticated. News through microblogs and other mass technologies make difficult to find difference between fake and real. Inclusion of hyperlinks and associated metadata results in veracity of assessments. This uses Centering Resonance Analysis (CRA) in order to represent "the content of large sets of text by identifying the most important words that link other words in the network" [10] Sentiment-focused reviews affects online ranking as fake reviews are artificially contributed to distort a ranking.

Online Relevance Score (ORS) [3]. The Jaccard similarity compares a particular query with related news headline search results which were collected from Bing search engine and then defined a term ORS which is given as:

$$ORS(Q, K) = \frac{1}{k} \sum_{i=1}^{k} Jaccard(Q, D_i) \tag{2}$$

In above Eq. (2), Jaccard takes query (Q) and description text (Di) from the top k links returned by the search engine API.

4 Fake News Datasets

Dataset contains over numerous articles tagged as either real or fake. Some of the popular datasets being used are:

- BuzzFeedNews [14]: This dataset is used for testing Linguistic approach and analysis are possible only on text dataset. This dataset consist of titles and links of an actual story or a post that is proclaimed as fake news and mainly covers news on politics, business, tech, entertainment, food, etc.
- LIAR [15]: This dataset also provides analysis only on text-based articles. Its dataset contributes to that news which are considered as real and mainly covers news of political debate, TV ads, interviews, post, etc.
- PHEME [16]: This dataset includes rumoured tweets, collected and annotated within the journalisms. This mainly concentrates on identifying speculation, controversy, misinformation, and disinformation on social media.
- CREDBANK [17]: This is the only dataset which includes social media information and allow users to analyze Twitter data. Though it overlooks multimedia data, it makes researchers an engrossing choice who focuses on fake news detection on social media.
- BS Detector [6]: This is a browser extension which provides datasets. It searches all the links on a webpage as reference of unreliable sources, checking list of domains if compiled manually. Outputs are the labels rather than human annotators.
- Kaggle [3]: This dataset provides a CSV file format for URL links of different fake news articles.

- Signal Media [12]: This dataset is published in conjunction with the Recent Trends to facilitate research on news articles. This dataset contains about 1 million articles from a various sources that include major news outlets and blogs from September 2015.
- Twitter Feed [3]: It extracts the tweets from various twitter handles with the help of subscription to Twitter REST Application.

5   Analysis

After discussing the various approaches and available datasets for the detection of news, an analysis is performed over the results of the work done in following papers (Table 1):

Table 1. Analysis

Related work	Approach	Feature set	Classification algorithms	Datasets	Results
Fake News detection in Social Media	Linguistic, Network	N-gram probabilities, TF-IDF scores, ARI, Syntactic information using Parts Of Speech (POS) tag, ORS	SVM, Random Forest	Twitter Feed (Fake and Real), BBC Database, UCI News Dataset, Kaggle fake news dataset	SVM:-Accuracy = 93.36% Balanced Error Rate (BER) = 0.134 Random Forest:-Accuracy = 94.87% BER = 0.115
Evaluating Machine Learning Algorithms for Fake News Detection	Linguistic	TF-IDF of Bigrams	SVM, Stochastic Gradient Descent, Gradient Boosting, Bounded Decision Trees, Random Forests	Signal Media-11,000 articles	SVM:- Accuracy = 76.2% Stochastic Gradient Descent:-Accuracy = 77.2% Gradient Boosting:-Accuracy = 68.7% Bounded Decision Trees:- Accuracy = 66.1% Random Forests:- Accuracy = 67.6%
Fake News Detection Using Naive Bayes Classifier	Linguistic	N-grams, Syntactic	Naive Bayes	BuzzFeed-2000 articles	True Accuracy = 75.59% Fake Accuracy = 71.73% Total Accuracy = 75.40% Precision = 0.71 Recall = 0.13

6   Conclusion

Mostly people tend to consume news from social media, because of its cheaply available, easily accessible, etc. They use these social media platforms like Facebook, Twitter for online reading of news articles. Fake News problem is growing and making things more complicated. This problem trying to change attitude and opinion of people. It is important that there have some mechanism for detecting fake news, or at the very least, an awareness that not everything which is on social media is true. As this fake news on social media increasing, there comes need to find more accurate ways for detection of false information. This research could be used for comparative studies on different approaches for better results in future.

References

1. Gottfried, J., Shearer, E.: News use across social media platforms 2016. 26 May 2016
2. https://en.m.wikipedia.org/wiki/News
3. Gunasekaran, K., Ganesan, G., Sudarshan, S.R., Balasubramaniam, S.: Fake news detection in social media **1**(1), Article 1 (2016)
4. Shu, K., Sliva, A., Wang, S., Tang, J., Liu, H.: Fake news detection on social media: a data mining perspective. ACM SIGKDD Explor. Newsl. **19**(1), 22–36 (2017)
5. Bond, C.J., DePaulo, B.M.: Accuracy of deception judgments. Pers. Soc. Psychol. Rev. **10**(3), 214–234 (2006)
6. Shu, K., Mahudeswaran, D., Wang, S., Lee, D., Liu, H.: FakeNewsNet: a data repository with news content, social context and dynamic information for studying fake news on social media (2018)
7. Deception detection and rumor debunking for social media
8. Anderson, M.: Social media causes some users to rethink their views on an issue. 7 November 2016
9. Pérez-Rosas, V., Kleinberg, B., Lefevre, A., Mihalcea, R.: Automatic detection of fake news (2017). arXiv preprint arXiv:1708.07104
10. Conroy, N.J., Rubin, V.L., Chen, Y.: Automatic deception detection: methods for finding fake news. In: Proceedings of the Association for Information Science and Technology (2015)
11. Granik, M., Mesyura, V.: Fake news detection using naive Bayes classifier. In: 2017 IEEE First Ukraine Conference on Electrical and Computer Engineering (UKRCON), pp. 900–903. Kiev (2017)
12. Gilda, S.: Evaluating machine learning algorithms for fake news detection. In: 2017 IEEE 15th Student Conference on Research and Development (SCOReD), pp. 110–115. Putrajaya (2017)
13. Valdez, G.G.: Linked data overview and usage in social networks
14. Buzzfeednews: 2017-12-fake-news-top-50. https://github.com/BuzzFeedNews/2017-12-fake-news-top-50
15. Wang, W.Y.: "liar, liar pants on fire": a new benchmark dataset for fake news detection (2017). arXiv preprint arXiv:1705.00648

16. Zubiaga, A., Liakata, M., Procter, R., Hoi, G.W.S., Tolmie, P.: Analysing how people orient to and spread rumours in social media by looking at conversational threads. PLoS One **11**(3), e0150989 (2016)
17. Mitra, T., Gilbert, E.: Credbank: a large-scale social media corpus with associated credibility annotations. In: ICWSM, pp. 258–267 (2015)
18. Parikh, S.B., Atrey, P.K.: Media-rich fake news detection: a survey. In: 2018 IEEE Conference on Multimedia Information Processing and Retrieval (MIPR), pp. 436. Miami, FL (2018)
19. Li, J., Manry, M.T., Narasimha, P.L., Yu, C.: Feature selection using a piecewise linear network. IEEE Trans. Neural Networks **17**(5), 1101–1115 (2006)
20. Exploring linguistic features for deception detection in unstructured text

Knowledge Discovery from Mental Health Data

Shahidul Islam Khan[1,3(✉)], Ariful Islam[1], Taiyeb Ibna Zahangir[2],
and Abu Sayed Md. Latiful Hoque[3]

[1] Department of CSE, International Islamic University Chittagong,
Chittagong, Bangladesh
nayeemkh@gmail.com, akashctg90@gmail.com
[2] Department of Psychiatry, National Institute of Mental Health,
Dhaka, Bangladesh
taiyeb.dmch@gmail.com
[3] Department of CSE, Bangladesh University of Engineering and Technology,
Dhaka, Bangladesh
asmlatifulhoque@cse.buet.ac.bd

Abstract. Mental disorders are increasing rapidly which contributed to intensive mental health-related data. Helpful knowledge can be discovered from this data using data mining techniques. For various mental problems, symptoms are similar which makes diagnoses difficult. Inappropriate diagnosis of the mental disease leads to incorrect treatment that can cause irremediable deterioration of patients. In Bangladesh, 15 million people suffering from mental disease and almost 10% of the people require proper mental treatment. The objective of this research is to examine the performance of classification algorithms to predict mental disorder. In this study, we analyze an authentic and real dataset consisting of 600 treatment records of patients with mental disorders to find the relation between diagnosis and attributes. We collected the dataset from the National Institute of Mental Health, Dhaka ensuring the ethical standard of research on human subjects. We applied two phonetic algorithms and generalization technique to reduce noise in the dataset and to ensure the privacy of patient data. We applied three machine-learning techniques: Random forest, SVM, K-nearest neighbor and compared their performances in diagnosing mental health problems. Experimental results show that Random forest has a better performance than the other algorithms we applied. We have also discovered few rules using association rule mining.

Keywords: Data mining, health informatics · Mental health ·
Classification algorithm · Association rules · Privacy

1 Introduction

According to the World Health Organization (WHO), mental health is a state of well-being within which someone understands his or her own proficiency to deal with the natural stress of life and work effectively, and be able to create a participation in his or her own community [1, 2]. Around the world, the quantity of medical specialty

© Springer Nature Switzerland AG 2020
A. P. Pandian et al. (Eds.): ICCBI 2018, LNDECT 31, pp. 881–888, 2020.
https://doi.org/10.1007/978-3-030-24643-3_103

disorders patient is increasing day by day. Developing countries have a bigger burden of mental disorders than economically advanced countries [3]. According to WHO, more than 450 million people around the world are affected by neuropsychiatric disorders. Treatment of these patients generates huge mental health-related data every day. We can discover many useful and interesting knowledge from this data using data mining.

Data mining involves the identification of unseen patterns from databases using machine learning algorithms. It has applications in many areas including healthcare. Fraud detection in health insurance, risk prediction, medical cost and length of stays in a hospital, etc. are often analyzed using data mining techniques [4–7]. When dealing with health data, privacy preservation is very important. Whenever personally identifiable information or other sensitive information is collected and stored in any form, there exist privacy concerns. A major challenge in health data privacy is to share data among medical practitioners while protecting personally identifiable information [8].

In this paper, we analyzed a dataset of 600 patients with mental disorders to find the relation between diagnosis and attributes. We collected the dataset from the National Institute of Mental Health, Dhaka ensuring the ethical standard of research on human subjects. In Sect. 2, we reviewed some works related to our research. Our methodology is presented in Sect. 3. Sections 4, 5 describes our results and limitations. We conclude in Sect. 6.

2 Related Works

Husain [9] predicts General Anxiety Disorder (GAD) among women using Random Forest classification algorithm. In this research, data collected through a survey where participant filled up a questionnaire consisted of the statement. Only data of female patients were used for this research. The author applied Random Forest to classify the results. The research of R. G. Ramani was based on Schizophrenia affected patients which is one the most common mental disorder [10]. In their paper, they investigated fMRI images of 15 normal controls and 12 Schizophrenia patients by constructing purposeful connectome through image preprocessing techniques and 74 regions of interest (ROI) are classified. Random forest, C4.5, Cost-sensitive classification, and K-Nearest Neighbor are applied to those features found by selection techniques.

In Azar [11] research, a novel study was done that introduces a semi-automated system for preliminary diagnosis of the psychological effected individuals. Here, a patient mental health status was matched with mental illnesses described in DSM-IV-TR. In this study, the semi-automated system was created based on a Genetic Algorithm, and machine learning. Their goal was to ensure, a classifier could match patients symptoms with mental health illnesses.

3   Methodology

A graphical development methodology showed in Fig. 1. We collected data from the National Institute of Mental Health (NIMH) in Dhaka, Bangladesh. These raw data are the prescription report of patients. These reports contain the description of mental illness problems with symptoms, personal and family history records, present condition, ongoing treatments and description of the symptoms and test reports of the patients. We collected 600 patients' data.

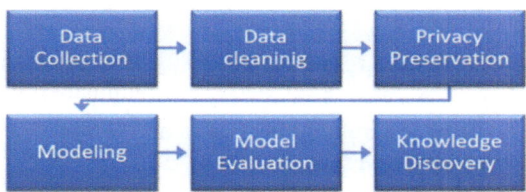

Fig. 1. Block diagram of the process

We converted this data into CSV format. Then we removed all noises. After cleaning, we get 500 data for the data mining process. As our dataset contains categorical values, we used Label Encoder. Label Encoder converts the dataset into numeric values so that machine could understand the data Table 1.

Table 1. Diagnosis after applying Label Encoder

Diagnosis	Diagnosis after encoding
Acute Psychotic Disorder	0
BMD	1
Conversation disorder	2

Our dataset has personal information about every patient. If this private information goes to an intruder, privacy and safety of a patient will be in danger. We used two phonetic algorithms, NameValue and Soundex, for privacy preservation. Another advantage of the phonetic algorithms is that they remove noise from the data.

To preserve the privacy of the patient's name, we used NameValue Algorithm on name column [12]. Some examples showed in Table 2. Here, the name of the patients changed into an encoded format.

Table 2. Applying NameValue algorithm

Patient name	Name value
ANURADHA MISTRY	TJLCESIMNLR
SAHANA	MEJ

To preserve the privacy of the patient's address, we used Soundex Algorithm on the Address column [13]. Some examples showed in Table 3. Here, the name of the patients address changed with encoded values.

Table 3. Applying Soundex algorithm

Address	Address Soundex
BAGERHAT	B263
NARAYANGANJ	N652

We used generalization technique to convert age of the patients to age ranges to enhance the privacy of the patients. The age ranges of patients are shown in Table 4. Our dataset contains patient age from 10 to 70. From generalization concept, we made 12 different age ranges of five intervals.

Table 4. Age generalization

Patient's age	Generalized age
12	10–14
16	15–19
24	20–24

In this paper, we used three Machine Learning Algorithm Random Forest (RF), Support Vector Machine (SVM) and K-Nearest Neighbor. We showed a comparison of these algorithms in the Result Section.

4 Results

In this section, at first, we presented the analyses of which age group of people is affected by Schizophrenia and BMD severely in Fig. 2. We can see that people of age group 25–29 are more affected by BMD & Schizophrenia. Schizophrenia is a chronic and severe mental disorder that affects how a person thinks, feels, and behaves. Bipolar disorder (BMD) is a mental disorder that causes periods of depression and periods of abnormally elevated mood. Depending on its severity, the elevated mood also called mania or hypomania. Environmental and Genetic factors play a vital role. Collecting data from a government facility is very challenging in Bangladesh. It was very tough for us but in the end, we succeeded. For our dataset, we considered Age, Marital status, Gender, All Symptoms as independent attributes. There are 29 diseases in our training dataset marked by the domain experts i.e. doctors. We fixed the disease column as the dependent variable or target class.

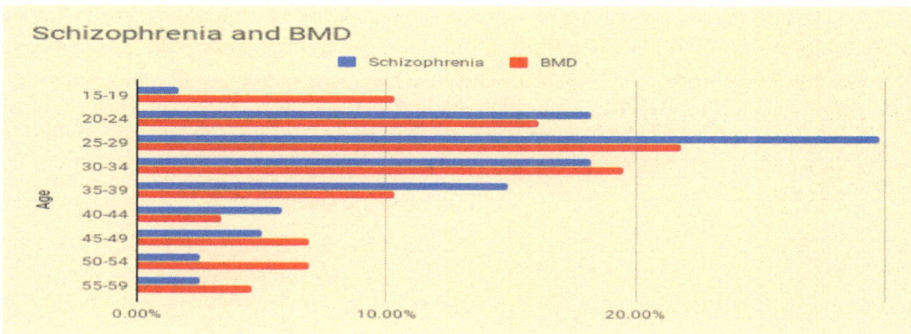

Fig. 2. BMD and Schizophrenia for different age range

We set different parameters of the algorithms as follows:

- Random Forest: n_estimators = 80, max_dept = 10, min_samples_leaf = 2.
- Support Vector Machine: C = 1.0, kernel = linear.
- k-Nearest Neighbor: number of neighbors = 10.

We showed some predicted results and the actual results in Table 5. Here we can see that model able to predict most diseases correctly and few diseases falsely. The reason for false prediction is that symptoms of some diseases are very similar. Another reason is that our collected dataset is not large enough to train our model properly.

Table 5. Result of predicted and actual disease

Random forest		Support vector machine		K nearest neighbor	
Predicted	Actual	Predicted	Actual	Predicted	Actual
27	27	27	27	27	27
2	2	2	2	2	2
26	26	**27**	**26**	26	26
2	**16**	24	24	**27**	**26**
24	24	**27**	**2**	**2**	**26**

Our real data contains 500 patient data. For a good analysis, we need large dataset. Therefore, we also generated a synthetic dataset of 3000 patient records. We applied three Machine Learning Algorithms in Real and Synthetic dataset. Results are shown in Table 6. Here, we can see Random Forest gives the best prediction accuracy.

Table 6. Accuracy of different algorithms

Data	RF	SVM	KNN
Real data	0.857	0.792	0.791
Synthetic data	0.868	0.884	0.854

We also applied our proposed privacy preserved incremental record linkage (PPiRL) framework for the Clustering of our collected patient records [14]. The results of correlation clustering are shown in Table 7. We used gender as the blocking attribute. We found 128 clusters of male patients and 119 clusters of female patients where a cluster contains a maximum of 14 records of the same patients. Correlation Penalty for Male Block was 4560.5 and for Female Block: 2003.6.

Table 7. Results of correlation clustering

Description	Male block	Female block
Number of cluster of patients	128	119
Maximum health records in a single cluster	14	11
Minimum health records in a single cluster	1	1

We showed frequent pattern analysis using association rule mining. For this, we used the Apriori algorithm. Here we set min-confidence = 0.7 and min-support = 0.2. We converted age range as an ordinal attribute as shown in Table 8.

Table 8. Age mapping

Age range	Patient category
0–17	Child and adolescent
18–39	Adult
40 and Above	Senior

We ran Apriori algorithm on the dataset and find some interesting rules. Two of them are presented in Table 9. From first rule, we can say that adult people who are unmarried affected more by Schizophrenia. From second rule, we can say female adults affected more in Bipolar Mode Disorder (BMD). We consulted with domain experts about the findings and they agreed with the results.

Table 9. Association rules

Rule	Confidence	Support
Schizophrenia, Adult, Unmarried	0.90	0.57
BMD, Adult, Female	0.70	0.29

5   Limitation and Future Work

We worked on a small real dataset contains 600 patients data. In the medical reports of the patients, some attributes were missing. For a bigger and complete dataset, prediction results will have greater accuracy. Doctors and record writers should fill up all the attributes offered within the sheet carefully. As our research is one of the pioneering work in the context of mental health in Bangladesh, more analysis needs to be done to predict diagnosing.

6   Conclusions

Around 10% of people in the world are severely affected by mental diseases. In this research, our goal is to discover interesting patterns from the mental health data collected from the National Institute of Mental Health, Dhaka, Bangladesh. We analyze an authentic and real dataset consisting of 600 treatment records of patients with mental disorders to find the relation between diagnosis and attributes. To ensure the privacy of patient data, we applied two phonetic algorithms and a generalization algorithm. We applied three machine-learning techniques: Random forest, SVM, K-nearest neighbor and compared their performances on different measures of accuracy in diagnosing mental health problems. Experimental results show that Random forest has a better performance than the other algorithms we applied. We have also mined a few rules using association rule mining. From our analysis, initial measures can be taken to treat mental patients at an early stage. This model will help the psychiatrist to understand the attributes related to patients with mental disorder.

Acknowledgment. We are grateful to the authorities of the National Institute of Mental Health, Dhaka for permitting us to work on their place and provide us with valuable data.

References

1. World Health Organization political declaration of the high-level meeting of the general assembly on the prevention and control of non-communicable diseases. In: 66th Session of the United Nations General Assembly. WHO, New York (2011)
2. Dooshima, M.P., Chidozie, E.N., Ademola, B.J., Sekoni, O.O., Adebayo, I.P.: A predictive model for the risk of mental illness in Nigeria using data mining. Int. J. Immunol. **6**(1), 5–16 (2018)
3. Bass, J.K., Bornemann, T.H., Burkey, M., Chehil, S., Chen, L., Copeland, J.R., Eaton, W. W., Ganju, V., Hayward, E., Hock, R.S.: A United Nations General Assembly Special Session for mental, neurological, and substance use disorders: the time has come. PLoS medicine **9**(1), e1001159 (2012)
4. Yoo, I., Alafaireet, P., Marinov, M., Pena-Hernandez, K., Gopidi, R., Chang, J.F., Hua, L.: Data mining in healthcare and biomedicine: A survey of the literature. J. Med. Syst. **36**(4), 2431–2448 (2012)
5. Khan, S.I., Hoque, A.S.M.L.: Towards development of national health data warehouse for knowledge discovery. In: Intelligent Systems Technologies and Applications, pp. 413–421. Springer (2016)
6. Khan, S.I., Hoque, A.S.M.L., Ullah, M.: National health data warehouse Bangladesh for remote health monitoring: features, problems and privacy issues. In: Remote Health Monitoring Workshop (2016)
7. Thongkam, J., Xu, G., Zhang, Y., Huang, F.: Toward breast cancer survivability prediction models through improving training space. Expert Syst. Appl. **36**(10), 12200–12209 (2009)

8. Khan, S.I., Hoque, A.S.M.L.: Digital health data: a comprehensive review of privacy and security risks and some recommendations. Comput. Sci. J. Mold. **24**(2), 273–292 (2016)
9. Husain, W., Xin, L.K. Jothi, N.: Predicting generalized anxiety disorder among women using random forest approach. In: 3rd International Conference on Computer and Information Sciences (ICCOINS), pp. 37–42. IEEE, August 2016
10. Geetha, R., Sivaselvi, K.: Data mining technique for identification of diagnostic biomarker to predict Schizophrenia disorder. In: IEEE International Conference on Computational Intelligence and Computing Research (ICCIC), pp. 1–8. IEEE (2014)
11. Azar, G., Gloster, C., El-Bathy, N., Yu, S., Neela, R.H., Alothman, I.: Intelligent data mining and machine learning for mental health diagnosis using genetic algorithm. In: IEEE International Conference on Electro/Information Technology (EIT), pp. 201–206. IEEE, May 2015
12. Khan, S.I., Hoque, A.S.M.L.: Privacy and security problems of national health data warehouse: a convenient solution for developing countries. In: International Conference on Networking Systems and Security (NSysS), pp. 1–6. IEEE, 7 January 2016
13. Soundex System: National Archives (2007)
14. Khan, S.I.: Efficient techniques for privacy preserved incremental record linkage of noisy health data. Ph.D. Thesis, Bangladesh University of Engineering & Technology, Dhaka

Factors Influencing the Scholars Acceptance of Course Recommender System: A Case Study

Zameer Gulzar[1(✉)], L. Arun Raj[1], and A. Anny Leema[2]

[1] BSA Crescent Institute of Science and Technology, Chennai, India
zamir045@gmail.com, arunraj@bsauniv.ac.in
[2] VIT, Vellore, Tamilnadu, India
annyleema@gmail.com

Abstract. Integration of recommender system in an e-learning plays a key role in enhancing its capability which in-turn helps to improve the learning experience of a learner. The primary objective of this research is to assess the hybrid recommender system using technology acceptance model against certain external factors related to the learners and is evaluated for performance efficiency. Therefore, a feedback from the potential users of this recommender system was collected towards using this platform and make an effort to facilitate better technology acceptance. The study included 100 users and out of that 80 filled questionnaire were received after using the course recommender system. The influence of all the three factors on scholars accepting the recommender system for course selection was empirically examined. The results analyzed can assist recommender system developers to maximize user experience as the Implications of this research work are important for researchers, Instructors and Institutions as well.

Keywords: Models · Perceived usefulness · Recommender systems

1 Introduction

To study the usage and the acceptance of information technology remained a focus of many studies in the literature Recommender systems made a tremendous impact on new information system techniques and usually focused on individuals who lack enough of personal experience or skills to assess potential ally large number of available items [1]. Therefore, researchers started to investigate the elements that could have a direct impact on the acceptance of recommendation technology by its users to maximize its popularity. Some of the factors were ease of use, its usefulness, accuracy in suggested items, understanding user's preferences, the intention of reusing system and interaction with the system. Various decision theories and intended models were developed over the last few decades to analyze the learner's behaviour regarding the acceptance of the new innovative technology. The relevance and the aim of this study regarding the acceptance of RS by learners may be treated as technology acceptance. Therefore, Socio-Psychological theories like the Technology Acceptance Model were used in this research.

© Springer Nature Switzerland AG 2020
A. P. Pandian et al. (Eds.): ICCBI 2018, LNDECT 31, pp. 889–895, 2020.
https://doi.org/10.1007/978-3-030-24643-3_104

TAM was developed on the basis of the Theory of Reasoned Action (TRA) and the aim behind using TAM is making predictions on use and embracing latest Information Technology (IT) also systems using features that coerce companies Information System (IS) and their compliance with the work-related needs. The attitudinal models are based on the advantages provided by information systems, excluding the negative characterizes of its use that will lead to the intention of either accepting or rejecting the technological innovation [2]. TAM model emphasis on user beliefs, attitude, and intentions in technology adoptions. Further, more determinants which define the construct of belief: perceived ease of use and user's perceived usefulness. Both these determinants along with intention and attitude form a chain that demonstrates user's adoption of the system [3].

A substantial quantity of research has already been published on users accepting the latest technology and TAM is famous among them. It is being adopted for its simplicity, robustness and its applicability in predicting and explaining the factors affecting the behavioural adoption of users towards an y latest technology [4]. Recommender system based on personality was evaluated using TAM. Their focus was recommending music based on personality while considering emotions and mood of the user [5]. A recommender system where user-centric information is being recommended in terms of courses was developed. They focused on a basic problem which users are facing which searching information is to choose among the variety of choices and by integrating different searching techniques proposed a hybrid framework for suggesting relevant information [6].

Technology enabled learning will generate remarkable results if combined with TAM and aims to create and test socio-technical innovations to support enhanced learning practices. In order to investigate learner's acceptance of recommending learning companion in Facebook, the characteristics of system design and the Facebook usage were considered as externals variables in TAM [7]. Meanwhile, it was analyzed that from 1986 to 2013, that there were 85 scientific publications on TAM which concludes that new constructs have continuously been identified to play key factor to influence the basice variables perceived usefulness and perceived ease of use of TAM [8]. However, all personalized services are not the same. Certain researches indicate that different personalized recommender services likely to have different effects on user's satisfaction. The TAM has been considered as the most significant, economical, and robust model in innovations acceptance behaviour. The TAM framework perhaps is most widely used in the area of Information systems for measuring technology acceptance and therefore, this theoretical model is considered as the base for the present study. TAM Model is used to describe the learner's behavioural intentions on using a technological innovation like course Recommender systems because it explains the essential links between constructs (The Ease of Use of a system and the usefulness of the system) and user or learners Attitude, Intention and the actual usage of the new innovative system.

This study focuses on exploring the possible user acceptance issues on course recommender system using Technology Acceptance Model (TAM). Within the baseline in the form of TAM model, three new external variables will be included as a lack of availability to use a recommender system, past experience, and relevance. A study was conducted using course recommender system as a test bed and a questionnaire appropriate for another recommender system in an educational domain. The main spotlight for this research was to understand the factors and underlying affinity which impact the acceptance and behavioural intention of using recommender system for course selection. As per our knowledge of literature, there is no technology acceptance studies related to recommender approach used in e-learning that include the factor related to the availability of technology in the TAM model. To address this research gap, this paper considers the perceived availability of recommender approach in e-learning systems and actual use of the recommendation technology in such models. The pictorial representation of the fundamental concept of Technology Acceptance Model is can be seen in Fig. 1.

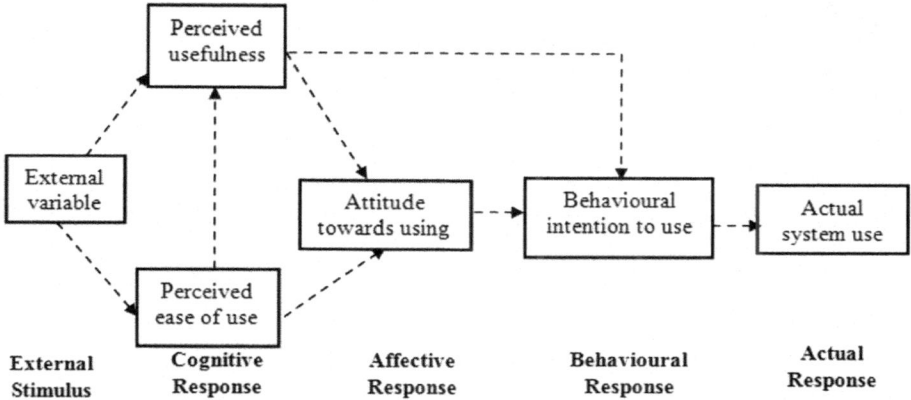

Fig. 1. Fundamental concepts of TAM

1.1 The Objectives of This Study Are as Follows

- To assess the exposure of Recommender System on learners.
- To verify the reasons learners will use Recommender System.
- To identify factors influencing learner's attitude towards use of RS.

2 Research Model and Hypotheses

Higher education institutions are neither exempted nor can afford to lag behind from technological advancements in teaching and learning. This rapid development in higher education is ascribed to the realization that the learning technology will renovate it [9]. The research model presented here used the basic variables of the TAM like the Perceived ease of use, Perceived usefulness, Behavioural intention to use, and Actual system usage. Additionally, three external variables Lack of availability, Relevance, and experience were included in the model. The proposed model was targeted to apply for research scholars based on their expertise in choosing courses and availability of technology for such tasks. The behavioural intention to use technology is almost similar among inexperienced and experienced users as it has already been predicted successfully [10]. It has also been noted that TAM cab applied even before the adoption of new technology [11]. Relevancy is the key to encourage the learner and establishes the fact that the process of learning is relevant as per the learner requirements [12]. By considering various things, this study is very much significant as no previous research has investigated the availability, relevance, and experience towards behavioural intention to use Recommender Systems and validate it with TAM as shown in Fig. 2. Furthermore, this study paves a way to further research on enhancing E-learning systems.

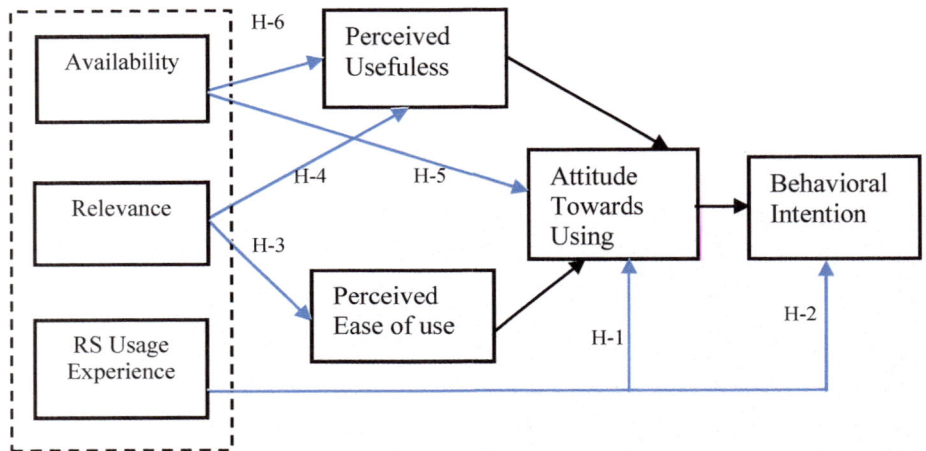

Fig. 2. Modified technology acceptance model

2.1   Added External Constructs

i. Relevance - The perception of an individual with respect to the extent of proposed system being relevant to them.
ii. Experience - The accumulative skills user's gain by getting involved in a particular system is defined as experience.
iii. Availability - The extent to which an individual presume that some technological infrastructure is present to assist the system usage.

This research work is of quantitative in nature and used a manual survey for data collection and the researcher tried the system to give their feedback. The experiment was conducted mostly with researchers and faculty from two different universities in south India. The survey was created to observe the relationship between variables of proposed research model in which course recommender system was introduced to the students. The questionnaire prepared was accessible at the end of the academic year to almost 100 participants. The response percentage recorded was 80%, which means 80 responses were valid and later used for analysis. The hypothesis with respect to external variables is given below.

H-1   RS experience has a positive effect on ATU using RS.
H-2   RS experience has a negative effect on BI of using RS.
H-3   Relevance has a negative effect on PEOU of an RS.
H-4   Relevance has a positive effect on PU of an RS.
H-5   RS availability has a positive effect on ATU.
H-6   RS availability has a positive effect on PU.

3   Result and Discussion

The participant's were research scholars and nearly equal in number with respect to gender, with males 47.46% and females 43%. The participants majority were in the age group of 24 and 45 years, with 63.75% from 23–30, 32.5%, from 31 to 40, and with a little minority of 3.75% above 40 yrs. The respondent has spent much time on course selection procedure with spent almost 4 weeks and 33.75% crossed 8 weeks also. Respondents use 100% traditional methodology of course selection and seem unsatisfied with a partial satisfaction level of 78% respondents. The highest response rate recorded is from Tamilnadu with 75% and the rest of the figures are from Bangalore with a response rate of 25%.

The external variables will serve as a starting point for investigating their impact on TAM construct like BI, ATU, PU, PEOU and Actual Use. The PU and PEOU constructs may not be enough, therefore, external constructs are needed. It was also found that experience and Relevance are the critical factors in determining technology acceptance [13]. Similarly, other studies have shown that relevance affects perceived ease of use. As users get involved or exposed to a particular system and due to accumulative skills they gained while using system is defined as experience [14]. In this study, RS experience is suggested to moderate TAM variables. The summary of the hypothesis proposed concerning external determinant is given in Table 1.

Table 1. Summary of the hypothesis

Hypothesis	Statement	Pearson correlation	Result
H-1	RS usage experience has a positive effect on ATU	−0.213	NO
H-2	RS usage experience has a negative effect on BIU	0.263**	YES
H-3	RS relevance has a negative effect on PEOU	−0.153**	YES
H-4	RS relevance has a negative effect on PU	0.612	NO
H-5	RS availability has a positive effect on ATU	0.289***	YES
H-6	RS availability has a positive effect on PU	0.640***	YES

Significance Level *** = p-Value < 0.01, ** = p-Value < 0.05
YES = Supported NO = Not Supported

In this work, the Technology Acceptance Model (TAM) was modified in order to validate the relationship between core TAM variables and the effect of proposed external variables. Relevance has shown negative relationship with PEOU but at the same time, there is sign of a positive relationship with perceived usefulness which in-turn will affect ATU and BI. RS usage is showing a negative relation with both attitudes towards using and behavioral intention to use. Availability of the recommender system has shown a positive effect on both PU and ATU. Although there is large literature body to examine technology acceptance, this study will enrich the literature by introducing the case study related to recommender systems that have not been covered previously.

4   Conclusion

To interpret the determinants responsible for the successful execution and integration of recommender systems with e-learning systems, there is a need to recognise the main factors that can explain why an individual embrace new technology and other resist to adopt it. This research aims to get the better cognizance of scholar's perception of recommendation technology and the factors that have an effect or influence on the decision for integrating and adopting technology into the learning system. In general, this study tries to measure the behavioural intention to use a recommender system by research scholars and students. The external constructs were adapted that validated the relation between Perceived Usefulness, Perceived Ease of Use, Attitude Towards Usage and in general the impact on Behavioral Intention to Use. Regarding previous constructs, none of the unexpected findings were found and therefore, the current study confirms other findings and empirical evidence based on the Technology Acceptance Model. Further, this study successfully confirms the applicability of TAM on Recommender Systems.

Acknowledgement. The authors are grateful to acknowledge the support of faculty and research scholars of BSACIST who spent their crucial time while conducting this survey and participated in filled the questionnaire to provide their intellectual feedback related to recommender system usage.

Competing Interest: The authors declare that they have no conflict of interest.

Informed Consent: Informed consent was obtained from all individual participants included in the study.

References

1. Selim, H.: An empirical investigation of student acceptance of course websites. Comput. Educ. **40**(4), 343–360 (2003)
2. Pavlou, P.A.: Consumer acceptance of electronic commerce: integrating trust and risk with the technology acceptance model. Int. J. Electron. Commer. **7**(3), 69–103 (2003)
3. Davis, F.D.: User acceptance of information technology: system characteristics, user perceptions and behavioral impacts. Int. J. Man Mach. Stud. **38**(3), 475–485 (1993)

4. Marangunic, N., Granic, A.: Technology acceptance model: a literature review from 1986 to 2013. Univ. Access Inf. Soc. **14**(1), 81–95 (2015)
5. Hu, R., Pu, P.: Acceptance issues of personality-based recommender systems. In: Proceedings of the 3rd ACM Conference on Recommender Systems (RecSys 2009), pp. 221–224 (2009)
6. Zameer, G., Leema, A.A.: A framework for recommender system to support personalization in an e-learning system. Int. J. Web-Based Learn. Teach. Technol. (IJWLTT) **13**(3), 51–68 (2018)
7. Rauniar, R., Rawski, G., Yang, J.: Technology acceptance model (TAM) and social media usage: an empirical study on Facebook. J. Enterp. Inf. Manag. **27**(1), 6–30 (2014)
8. Marangunic, N., Granic, A.: Technology acceptance model: a literature review from 1986 to 2013. Univ. Access Inf. Soc. **14**(1), 81–95 (2015)
9. Farrell, T., Rushby, N.: Assessment and learning technologies: an overview. Br. J. Edu. Technol. **47**(1), 106–120 (2016)
10. Taylor, S., Todd, P.A.: Understanding information technology usage: a test of competing models. Inf. Syst. Res. **6**(2), 144–176 (1995)
11. Shih, H.P.: Extended technology acceptance model of internet utilization behavior. Inf. Manag. **41**(6), 719–729 (2004)
12. Huang, H.-M., Liaw, S.-S.: An analysis of learners' intentions toward virtual reality learning based on constructivist and technology acceptance approaches. Int. Rev. Res. Open Distrib. Learn. **19**(1), 91–115 (2018)
13. Venkatesh, V., Davis, F.D.: A theoretical extension of the technology acceptance model: four longitudinal field studies. Manage. Sci. **46**, 186–204 (2000)
14. Thompson, R., Campeau, D., Higgins, C.: Intentions to use information technologies: an integrative model. J. Organ. End User Comput. (JOEUC) **18**, 25–46 (2006)

Secure User Authentication Using Honeywords

S. Palaniappan$^{(\boxtimes)}$, V. Parthipan, S. Stewart kirubakaran,
and R. Johnson

Computer Science and Engineering, Saveetha School of Engineering,
Saveetha Institute of Medical and Technical Sciences, Chennai, India
{palaniappan, parthipan, stewark.sse}@saveetha.com,
Jaan534275@gmail.com

Abstract. The purpose of the password is to protect the user account from unauthorized usage by the hacker. But in the current situation the field of security also realizes lot of threat to the password even in case if it is hashed. With the rise of hacking technology even the hashed password doesn't provide the required security and also provides the hacker to misuse or exploit the user account without being noticed. The most vulnerable part in this is the misuse of account can be realized only after the user logs and sees the changes in their account usage. And so, the system doesn't yet been improved in safeguarding or detecting the attacks against the database of password which are hashed. Ari Juels and et al. in 2013 [10] discovered the method using honeywords for detecting the password cracking. Honey words are the imitated passwords which are connected with the account of each user. We intended to produce a safer system by creating an authentication using the honey words in the password database. The newly created database contains a combination of both the imitated ones and the original passwords in order to detect whether the attack is happened or not. And hence when the hacker has the password database, he might get confused with the real and fake passwords. Here we make the hacker to fall into our trap by confusing him. Once he tries to enter a false password the administrator will get a notification and the hacker gets identified.

Keywords: Honeywords · Internet security · Honeypot · Hacker detection ·
E-Commerce, etc

1 Introduction

The present password protection system might be defenseless to Distributed Denial of service attacks and hence disturbing the entire system. The stored passwords are more vulnerable and easily hacked by the attacker using online guessing method. Hence, there is not much implementation on security was presented in the current system. In this paper we propose a novel mechanism where we intended to use the Honeywords to identify the attacker by persuading him for making an attack by which we can avoid the DDOS. Initially the user has to enroll with the server and the server in turn produces a set of random passwords called Honey words. The next step is where the user's actual password is encrypted and kept along with the Honey words. The intermediate server

© Springer Nature Switzerland AG 2020
A. P. Pandian et al. (Eds.): ICCBI 2018, LNDECT 31, pp. 896–903, 2020.
https://doi.org/10.1007/978-3-030-24643-3_105

will then invoke the filter to identify the wrong password when the attacker tries to fetch a password from the database by which the DDOS attack can be prevented.

The Honey words generation is based on the information provided by the user and the actual password. Both the honeywords and actual password are encrypted into an unreadable format. We deploy two types of server one is the intermediate server and the other one is the Cloud server for maintaining user account details. When the attacker is aware of the E mail ID information of user, he can easily reset the cloud server password. In our work the main theme is to invite the hacker to perform the attack, so as to make him to fall into ours trap. Once he gains the information of the user the hacker tries to login into the purchase portal, where he has been tracked unknowingly & he is allowed to do purchase. Server identifies the attacker and sends the info to the actual owner and also the server blocks the attacker even before doing transaction from the original account.

2 Literature Survey

With the password stolen from a user the attacker or hacker may be able to imitate as you in front of the device and be able to benefit the know-how or statistics about you. This is called identity fraud. Few people might not agree that the identity theft which is a result of media attention has been increased dramatically in the recent times. The surge in interest is certainly a type of herbal effect or moral panic of extended incidence which continues to be unknown [1].

Despite substantial advances in society and technology, thousands and thousands of humans have their identities stolen. A sensitive statistic is gathered via low-tech methods of physically acquiring it, as well as hello-tech strategies that use numerous technologies to capture a character's statistics. Yet, there are various gear and methods of stopping identity theft that is reachable to the normal client, who either are ignorant of them or simply can't be afflicted. Identity theft is nothing to snicker approximately. With the many methods utilized by identification thieves to thieve individuals' identities, the situation can seem very disheartening. But, doing a few, easy things can alternate it from daunting to conceivable [2].

The vulnerability of social network users to identity robbery after they percentage non-public identification information on-line is exploited. The sharing of info like age, intercourse, deal with and different private facts like pix can assist in organizing an identification. Identity criminals make the most social community users and the weaknesses of social networking websites to accumulate the statistics needed to commit identity theft and identity fraud the usage of this identification information. While there are mechanisms which can lessen the incidence of this crime, data sharing on social networks is voluntary, which, makes it manipulate hard. At the same time as there are numerous ways to respond to this crime, a mix of strategies is possibly to work high-quality, given the pervasive nature of this crime and obstacles supplied with the aid of a couple of jurisdictions. Those problems pertain now not only to the difficulty of making use of criminal sanctions however also to those referring to privacy in a transnational context [3].

Identity theft is a swiftly developing a national trouble. More than seven hundred people had been victimized due to identity theft. Fundamental news stations, magazines, and newspapers are reporting those tales on a normal foundation. The elderly people are colloquially considered an 'easy' target group for criminals, in particular defrauders and schemers. As a greater goal, the elderly need to get hold of good enough protection, which may be performed by a boom in rules and education. The elderly people are focused because its miles assumed, they may be gullible, unsuspecting, and ignorant of scams. Further, its miles assumed that when a lifetime of work they have property to steal. It is critical to educate seniors approximately approaches to shield themselves and decrease capacity harm. However, educating the aged by myself will now not resolve this problem. Regulation needs to be created with the primary goal of deterrence. Growing the penalties in opposition to identity thief ought to do that, especially when the sufferer is a senior. An equally vital objective is to enact law that protects the non-public facts of residents this is publicly to be had because too many people have found out the difficult manner, a touch information in the incorrect fingers can reason economic losses [4].

Erguler et al. [5] stated in their paper about the process of honey word creation; as per their work the password database is designed as the shape of a honey comb where the fake and the actual passwords are kept together. So, only the file which contains the fake password from the password database is noticeable to the hacker whereas the actual password is hidden at the back. However, few drawbacks have been identified after the using the system like less or no process of authentication. Mirante and Justin [6], has given a complete understanding of how the hacker intrudes into the e-commerce web page and get the profile information of users by compromising the credentials.

The internet has seen an enormous growth in the recent times. The net has originally been developed to create a soothing effect among all and to make the customers feel honorable. Instead the dark side of the internet had created a wrong motion and fear among the customers which includes phishing, click on fraud, junk mail, hacking, malware, DOS attacks, invasion into private arena, credit card fraud, insurance ring, violation of virtual belongings, and many others. The dark side of the net have created a wrong impression on the responses and obscured the growth of technologies, public focus, law, litigation, law enforcement etc. Vance [7] on his paper, explored the various types of attacks and the responses style to the same. He had finally delivered a taxonomy of the attack response and aftermath prices.

The popular web social media's such as eHarmony, yahoo and LinkedIn were affected in large by the leaks of password in the recent years. Brown [8] on his work related to the vulnerabilities of weak hashes has suggested few practices that will reduce the damage due to leak of passwords such as using hashes for storing the passwords, prepare the hashes using salt technique, use starching of key or sluggish pseudocode to upsurge the cracking time of the passwords and usage of encryption to the password hashes. He also discussed about the psychology in selecting the passwords. The password selection should be very much strong in such a way that even if the hashes are discovered it should be hard to decrypt the password from the compromised hashes by the hackers.

Cracking a password could be a tough task when we mangle the words and generates others passwords which looks alike. The mangled words further make cracking the password a difficult task if in case they are generated using dictionary-based words. Weir [9] in his paper has discussed about the creation of probabilistic CF grammar for creating a strong password hash. This password hashes are crafted using a training set which contains passwords which are disclosed earlier. With the help of word mangling and the above said grammar he had developed a strong rules which in turn used to generate the password guesses.

3 Honeyword Implementation

Juels along with Ronald L. Rivest [10] devised a solution called honeywords to defend the vulnerabilities against password theft. The main idea in generation of honeywords are to prevent the hackers in gaining the authentication information of the user and to stop them from accessing the hashed password.

Even if the hashed password is compromised by the hacker, he can't access the actual password since the honeywords are used to hash the passwords. we had taken up the honeyword concept and implemented a webpage to make a strong authentication system whereas the administrator of the page creates few forged accounts of the users called honeypots. The Implementation process of honeypot authentication is depicted in Fig. 1

SYSTEM ARCHITECTURE

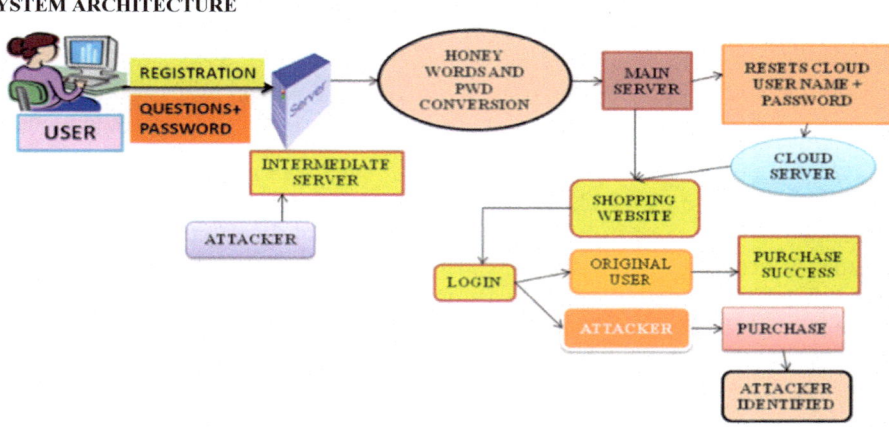

Fig. 1. The Implementation process of honeypot authentication.

An alarm is raised using this honeypot when the hacker gets to solve the hashed password and tries to stole the actual password and then attempt to login onto the forged page. The hacker gets into the trap by performing the above said steps since the forged page doesn't belong to any of the legitimate user.

Each account in turn has multiple combination of passwords (fake and actual). In which Only one password in the combination is genuine, the others are termed as "honeywords". The alarm is raised when the hacker gets to access the password file and attack can be detected reliably.

We have created a scenario that has been framed in such a way that the hackers get copy of the file which contains the username, password linked to the account, salt parameter and other information pertaining to compute the hashed password. The hacker tries to apply the brute force methodology on the hashed password which are recovered. The brute force method can be done on passwords which are short or likely ones. And apply salt technique until he recovers the password which are hashed. The hacker tries to login on to the accounts of user thinking that authentication is the only mechanism used for security

A secondary server is utilized during the process of login to trace the hacker's activity using "honey checker" which differentiates the actual passwords from the faked ones and raise the alarm if any of the honey words are used.

4   Proposed System

We use the honeywords mechanism to prevent unauthorized users from entering and luring the attackers to get information so future attacks can be stopped. Here the shopping server created is designed in such a manner that if a honeyed is used then an alarm will set of the original user and the attacker will be given access to decoy pages. These decoy pages are then used to extract information from the attacker. The system will find the IP and will be sent to the authorized user along with the address given by the unauthorized user in the deliver address box in the decoy pages.

5   User Registration

The user registration module handles the application for creation of new user which in turn allows the user to access the data stored in server. The user must create an account in order to access the server network and the content. The user then be allowed to access their account information. The service of the server could be obtained on request by the user. Once the application is accepted the user details are stored securely on to server database. In this module bank user details are registered with the fields like username, password, and personal details with some set of questions and answers. These details are saved into the server. After proper registration only, the user are allowed to login into the server.

6   Server

In this Server module server will deployed to access the database and web-based application. Server will verify the users and generates honey word for save the users password. In case illegal actions happened, means server will generates alert and intimate it to user. The entire information of the user is stored and monitored by the server and verified on demand from the stored database. Further a connection is established by the server to communicate with the users. The activity of the user is updated by the server on the database. And the user will be authenticated by the server before they access the application. By authenticating the user, the unauthorized access to the application is prevented.

7 Honeyword Generation

The passwords which are generally stored in files are highly unsecured and most of the companies and its users had faced a security threat to such stored passwords. The file though stored in encrypted format are not considered to be secured and lot of chances to be compromised. If hackers try any of the password cracking method and decrypt the file then all of the stored password along with user information will be revealed. So, In this module we deployed honey word creations. That is the user's password and registered questions are combined and then it will generate a key as unknown name.

8 Intermediate Server & Shopping Server Deployment

An intermediate server is a program that handles communications requests to a resource manager program on behalf of a user program. The user program can be referred to as a client of the intermediate server. Here we will generate the Intermediate server to make communication between user and Server. All requests come from the users are first sent to the intermediate server to verify the password and user details. Shopping server is to collect the details from customer and sent to the details to the intermediate server for verification.

9 Password Hacking Process

Hacking is the process of recovering passwords from data that has been stored in or transmitted by a computer system. A common approach is to repeatedly try guesses for the password. Another common approach is to say that you have "forgotten" the password and then change it. Though there are mechanisms already employed to detect such activities, if the hacker is intelligent and cunning then they might find ways to bypass them. The proposed system works on stopping those guessing attacks and prevents access to the hacker to make any changes to the user account.

10 Identification of Attacker

The most important and unique feature of this system is that it not only prevents the attack but alos identifies the attacker. When the attacker gets hold of the password files as the files contains honeywords the password is not known. This forces the attacker to guess the password. If the honeyword is entered repeatedly then the website permits access. But these are the series of fake pages that are generated in order to identify the attacker. The pages appear as the attacker has been granted access to the account. When the purchase is complete the address of delivery given and the Ip address of the attacker is gathered and sent to the alternate e-mail address of the original user. Thus not only preventing the present attack but also by detecting the attacker stops future attacks as well.

11   Conclusion

With the internet age growing strong and everything becoming digital around us security has become a prime issue for its users. In the midst of high-level security mechanisms failing on a regular basis the honeyword mechanism prove that effective security can only be provided at expensive costs is wrong and user details can be secured by being smart. In this respect, we have pointed out that the strength of the honey word system is that it only prevents an attack but it can also be used to track down people who are accessing the system with malicious intent. Though in the outside it may seem that once the advisory guesses the correct password he can easily get in and out easily, the advisory will not be sure if the password entered is a correct password or not. This lack of confidence will lead to him/her making a mistake. In this paper we have showcased what honeywords with the correct type of honeyword generation algorithm can do and also proposed a system where the system can watch over and detect any suspicious activity. It does not stop at detection instead it goes a step further and tracks the source of the problem and collects as much detail as possible about the advisory.

References

1. Copes, H., Kerley, K.R., Huff, R., Kane, J.: Differentiating identity theft: an exploratory study of victims using a national victimization survey. J. Crim. Justice **38**(5), 1045–1052 (2010)
2. Dean, P.C., Buck, J., Dean, P.: Identity theft: a situation of worry. J. Acad. Bus. Ethics. **1**(1), 1–14 (2014)
3. Holm, E.: Social networking and identity theft in the digital society. Int. J. Adv. Life Sci. **6** (3&4), 157–163 (2014)
4. Sylvester, E.L.: Identity theft: are the elderly targeted?
5. Erguler, I.: Achieving flatness: selecting the honeywords from existing user passwords. IEEE Trans. Dependable Secur. Comput. https://doi.org/10.1109/tdsc.2015.2406707
6. Mirante, D., Justin, C.: Understanding Password Database Compromises. Dept. of Computer Science and Engineering Polytechnic Inst. of NYU, Tech. Rep. TRCSE-2013-02 (2013)
7. Vance, A.: If your password is 123456, just make it hackme. The New York Times, vol. 20 (2010)
8. Brown, K.: The dangers of weak hashes. SANS Institute InfoSec Reading Room. Tech. Rep. (2013)

9. Weir, M., Aggarwal, S., de Medeiros, B., Glodek, B.: Password cracking probabilistic context-free grammars. In: Security and Privacy, 30th IEEE Symposium on IEEE, pp. 391–405 (2009)
10. Juels, A., Rivest, R.L.: Honeywords: making password-cracking detectable. In: Proceedings of the 2013 ACM SIGSAC Conference on Computer & Communications Security, pp. 145–160, 4–8 Nov 2013
11. Becker, H.J.: Internet use by teachers: conditions of professional use and teacher-directed student use. Teaching, Learning, and Computing: 1998 National Survey. Report #1
12. Singh, D.: The use of internet among Malaysian librarians. Malays. J. Libr. Inf. Sci. 3(2), 1–10 (1998)
13. Perry, T.T., Anne Perry, L., Hosack-Curlin, K.: Internet use by university students: an interdisciplinary study on three campuses. Internet Res. 8(2), 136–141 (1998)
14. Jefferies, P., Hussain, F.: Using the internet as a teaching resource. Education + Training, 40(8), 359–365 (1998)
15. KantiSrikantaiah, T., Dong, X.: The internet and its impact on developing countries: examples from China and India. Asian Libr. 7(9) 199–209 (1998)
16. Sandelands, E.: Creating an online library to support a virtual learning community. Internet Res. 8(1), 75–80 (1998)
17. Ramesh Babu, B., Gopalakrishnan, S.: Internet: a survey of its use in Chennai. In Malwad, et al. N.M. (ed.) Towards the new information society of Tomorrow: innovations, challenges and impact. Papers presented at the 49th FID (International Federation for Information and Documentation) Conference and Congress, Indian National Scientific Documentation Centre, New Delhi, III-9-13, 11–17 Oct 1998
18. Srikantaiah, T.K., Xiaoying, D.: The internet and its impact on developing countries: examples of China and India. Asian Lib. 7(9), 199–209 (1998)
19. Bavakutty, M., Salih Muhamad, T.K.: Internet services in Calicut University. Conference, Academic Libraries in the Internet Era. In: Proceedings of the 6th National Convention on Academic Libraries in the Internet Era, pp. 37–44. Ahemdabad, India (1999)
20. Kooganurmath, M.M., Jange, S.: Use of internet by social science research scholars: a study in academic libraries in the Internet era. In: National Convention Academic Libraries in the Internet Era, pp. 478–483. Organized by INFLIBNET, Ahemdabad, 18–20 Feb 1999
21. Mahajan, S.G., Patil, S.K.: Internet: its use in university libraries in India. In: National Convention Academic Libraries in the Internet Era, pp. 483–488. Organized by INFLIBNET, Ahmadabad, 18–20 Feb 1999
22. Laite, B.: Internet use survey: analysis. http://www.ship.edu/ ~ bhl/survey/(2000). Accessed 21 May 2004
23. Choo, C.W., Detlor, B., Turnbull, D.: Web Work: Information Seeking and Knowledge Work on the World Wide Web. Kluwer Academic Publishers, Dordrecht, The Netherlands (2000)
24. Branin, J.J., Groen, F.K., Thorin, S.E.: The changing nature of collection management in research libraries. Libr. Resour. & Tech. Serv. 44(1), 23–32 (2000)
25. Password hacking - internet security and ethical hacking. www.insecure.in/password_hacking.asp

A Secured Architecture for IoT Healthcare System

A. Vithya Vijayalakshmi$^{(\boxtimes)}$ and L. Arockiam

Department of Computer Science, St. Joseph's College (Autonomous),
Tiruchirappalli, Tamil Nadu, India
vithyanbalagan@gmail.com, larockiam@yahoo.co.in

Abstract. Internet of Things (IoT) healthcare system, also known as Medical IoT, plays an increasingly important role in improving the health, safety and care of billions of people after it appears. Rather than going to the hospital for help, the health parameters of patients can be monitored remotely, continuously and in real time, then processed and transferred to medical data centers, such as cloud storage, which significantly increases the convenience, efficiency and cost-effectiveness of healthcare. The volume of data handled by Medical IoT devices increases exponentially, leading to increase explosion of sensitive data. The security and privacy of data collected from Medical IoT devices, either during transmission to a cloud or during cloud storage, are major unresolved issues. This paper emphasises on the data security requirements related to data flow in Medical IoT. This paper proposes a secure architecture for IoT healthcare system to ensure data security at all the layers of data flow.

Keywords: Internet of Things · Healthcare system ·
Data security · Cryptography techniques

1 Introduction

Internet of Things is an ideal comprehensive technology Internet and communication technologies are effective. The Internet of Things is a general way of connecting living and non-living on the Internet. Everything in the world is traditionally conceived as smart object, but everything in the world in the IoT paradigm is considered a clever object, and is enabled to communicate physically or virtually with each other through the internet [1]. The Internet of Things research and innovations team defines IoT as "a dynamic global network infrastructure with self-configuring capabilities based on standard and interoperable communication protocols where physical and virtual "things" have identities, physical attributes, and virtual personalities use intelligent interfaces and are seamlessly integrated into the information network" [2]. IoT gets people's approval and connects things and communicates at Anyplace, Anytime with Anyone and Anything. The goal of IoT is to form a smart city, to provide smart healthcare, offer smart transportation etc. There are many possible application areas for IoT. With the rapid growth of the IoT applications, several security and privacy issues are observed [3].

© Springer Nature Switzerland AG 2020
A. P. Pandian et al. (Eds.): ICCBI 2018, LNDECT 31, pp. 904–911, 2020.
https://doi.org/10.1007/978-3-030-24643-3_106

One of the vibrant applications of the Internet of Things is the health system. It plays a vital role in improving people's health and safety. For example, biometric data collection devices in health care usually uses smartphones with sensors or wearable devices. Both are commonly combined with applications for interpreting the signals, processing data and showing users statistics. These apps perform simple processing, which sometimes extends the functionality by transferring the collected data to the cloud to be processed with complex algorithms [4]. In this scenario, security issues can be considered into three main areas: security or as confidentiality, integrity and availability are widely accepted; privacy or proper usage of data; and legal issues, i.e. concerns regarding security related to legislation. In terms of security, three risk points can be identified such as device, data transmission (sensor to smartphone and cloud to smartphone) and cloud storage. To overcome these risks, an architecture has been proposed. This paper proposes a secured architecture for IoT healthcare system to secure the healthcare data.

The rest of the paper is structured as follows, Sect. 2 discusses works related to IoT healthcare architecture, whereas Sect. 3, gives the overview of IoT healthcare system and data security requirements in healthcare system. Section 4 presents the proposed secured architecture for IoT healthcare system. Finally, Sect. 5, presents the research conclusions.

2 Related Works

Teresa Villalba et al. [4] discussed the security issues and challenges related to the IoT healthcare. The objective was to propose an IoT architecture for healthcare. The collection of data from sensors or devices was not perfect to users nowadays, and there were various problems, including integrity, confidentiality and information availability. Security issues related to IoT data transmission and storage in the proposal healthcare system architecture including processing of information and cloud services were analysed. Machorro-Cano et al. [5] discussed literature on the application of the IoT paradigm relating hypertension, overweight and other chronic progressive sicknesses. Smart healthcare platform architecture based on the IoT model was proposed. The proposed architecture was validated by using the information collected from the old people suffering from hypertension and overweight.

Kumar et al. [6] explained a general end-to-end architecture of healthcare system. The current advantages in IoT based healthcare system were summarized. Intel Curie based end-to-end healthcare system design was proposed. Different hardware and sensors used to develop the IoT healthcare system were discussed. Sridhar et al. [7] discussed the limitations of providing the intelligent service. The IoT devices were heterogeneous, including wireless sensors for reduced resources. These devices were hardware- and software- prone and attacks occurs on the network. If it was not secured properly, security problems such as confidentiality and privacy may arise. Intelligent Security Framework for IoT Devices or Cryptography based end-to-end security architecture was proposed. The data were encrypted using lightweight asymmetric cryptography and for securing the end-to-end devices and implements Lattice-based cryptography for securing the device gateway and the cloud services.

3   Overview of IoT Healthcare System

Healthcare is a widespread and significant area of application of the IoT, because the IoT is adapted to improve the quality of service and reduce costs. IoT healthcare parameters such as body temperature, blood pressure and blood glucose level are monitored by different medical sensors or devices. Recent developments in wireless communication, sensors and processing technology drive the use of IoT in healthcare system. Later, wearable, better known as portable body sensors are being developed to monitor patient activities or parameters continuously in real time. In this context, IoT offers a link between the various heterogeneous devices for quick, complete and exact information on health parameters [5].

Assisted living environments help people with disabilities and chronic medical conditions to live on a daily basis. Due to the sensors, the ability to manage data efficiently, real-time living services can be provided to patients. Innovative services such as the collection of vigorous patient data through a medical device network sensors contribute to the use of IoT in health care. It also contributes to the storage and processing of data in the cloud of a medical center and ensures sharing of medical information or ubiquitous access (e.g. health records). Similarly, the challenges of the healthcare sector such as privacy, security, integrity and authentication need to be concentrated for secure healthcare system.

3.1   Data Security Requirements in IoT Based Healthcare System

The importance of data security rises with the adaptations of emerging technologies. Managing large amount of data with respect to storage, ownership, data, and their security has become major areas of concern. Out of which security issues in IoT is growing day by day as they may occur at various levels, when a new technology is implemented. A number of properties such as integrity, confidentiality, privacy, authentication, authorization, availability and non-repudiation, have been measured to ensure data security, services and the entire IoT system. Some of the data security issues faced by IoT are:

A. Data Access Permissions
Different applications have different users; each application will have large number of users. An effective authentication technology should therefore be used to prevent illegal user intervention. Also, the processing and identification of spam and malicious data should be considered [8].

B. Authentication and Identity Management
Authentication and IdM combine processes and technologies to manage and secure access to information and resources while protecting the "things" profile. IdM identifies objects uniquely and authentication involves the validation of the identity between the two communicating parties.

C. Data Integrity
An opponent may modify the data and compromise the integrity of an IoT system. Thus, integrity safeguards that any data received in transit has not been changed [9].

D. Authorization

Authorization allows us to determine if the person or object is allowed to have the resources once identified. Typically, it is implemented using access controls. Authorization and control of access are important in establishing a secure link between a number of devices and services.

E. Data Aggregation

Since node failures are often found in wireless sensor networks, the topology of the network should be able to cure itself. A secure data accumulation method is therefore needed to ensure that reliable data is collected from sensor nodes throughout the network. [10].

F. Data Confidentiality

As data in IoT applications will be related to the physical realm, ensuring data confidentiality is a primary constraint for many cases. For example, bio-sensor data on the bacterial composition of the product used to guarantee the quality required in the food industry [11].

G. User Privacy

Data protection is an important issue for IoT security due to the ubiquitous nature of the IoT environment. Things are connected and data is communicated and exchanged over the Internet, making privacy for users a sensitive topic in many research projects. The key problems are data collection privacy, data sharing and data management and data security.

H. Trust Management

When certain things communicate in an uncertain IoT environment, trust plays an important role in establishing safe interaction. The IoT should take into account two dimensions of trust: trust in the interactions between entities and trust in the user's system.

4   Proposed Secured Architecture for IoT Healthcare System

A huge growth in body wearable sensors has emerged recently as effective tools for healthcare applications and various devices are currently commercially available for a variety of purposes, including fitness, activity awareness and personal healthcare. New clinical applications of such technologies for remote health monitoring systems, which include long-term status recording functionalities and medical access to patient's physiological information. The proposed architecture for IoT healthcare system have a three-layered architecture: Device or sensor layer that includes wearable sensors acts as data acquisition units, such as heart rate, body temperature and blood pressure. The local network layer includes communication and networking, and the service that collects and transmits sensor data. The Back-end services includes the nodes for processing and analysis. Figure 1 shows the secured IoT healthcare system architecture, which includes all the three layers such as data acquisition in environmental monitoring, which is then collected and transferred for the third phase of statistical data analysis.

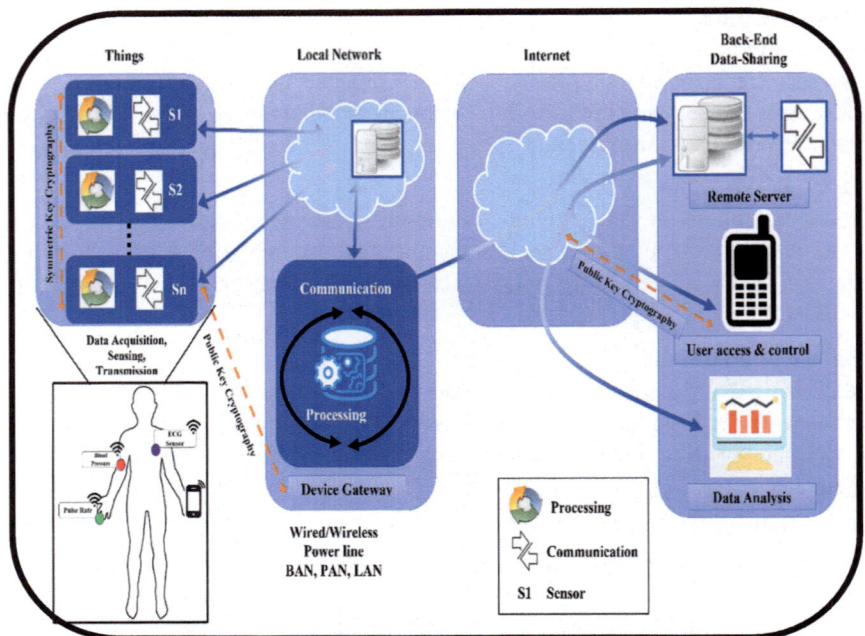

Fig. 1. A secured architecture for IoT healthcare system

(i) Data Acquisition and Sensing layer
In this layer, the data and information collected via peripheral devices (physical equipment) and the physical world is identified, information includes object properties, etc. Environmental conditions, physical RFID reader equipment and all types of Global Positioning System (GPS) sensors and other equipment's are included. Sensors capture and represent the physical world in the digital world are the main element of this layer. Sensor data still needs confidentiality, integrity and authenticity protection, since the storage nodes are very short and the power capacity is too, the public key encryption algorithm cannot be used for security protection.

- Sensors:

    Sensor is a device which detects the environment/events and sends the information to other electronic devices. For example, the data come from the various smart devices such as RFID, barcode labels, actuators and intelligent detection devices are processed and transferred to other devices or local/cloud server.

- Local Processing and Communication:

    The data collected from the sensors are processed in such a way that the attacks or unauthorized users cannot find the source of the data while transferring it through the wireless network.

(ii) Network Layer

The network layer is responsible for the transfer of reliable information through polymerisation from the preliminary processing of information on the conceptual layer. In this level of information transfer, several Internet backbone networks, mobile communications network, satellite television networks, a wireless infrastructure network and communication protocols are necessary. Although it is relatively basic to protect the security of the network, the Man-in-the-Middle attack still exists. For the IoT therefore, security at this level is very important. Either symmetric key cryptography or public key cryptography should be applied for data security in this layer. Public key cryptography will be more suitable for this layer.

- IoT Gateway:

IoT is a local gateway software that acts as an intermediate between the smart objects and cloud computing services also provides security and other functionalities such as data or protocol translation.

(iii) Back-end Data-Sharing Services

This layer allows communication between users to transfer and analyse intelligent object data available on the cloud service with data from other sources. It also allows data to be aggregated and analysed from the IoT data streams. This layer requires a large number of application security structures such as secure third party computing and cloud computing, encryption protocols and strong encryption algorithms, a stronger security system with anti-virus technology. Public key cryptosystem will be appropriate for data protection in this layer.

- Cloud Storage:

One of the features of the IoT is intelligent processing, which means using cloud computing and other intelligent computing technology to analyze and process large amounts of data and information for intelligent object control.

(iv) Cryptography Techniques for Data Security

Data security is a key issue for devices on networks. Security plays a vital role in the field of IoT, where malicious attacks or interference with IoT devices can pose a threat to human life, especially with critical IoT applications. Cryptography is the one of the answers for IoT data security [12]. Cryptography is the most powerful tool available in implementing security services. Cryptography, the art and science of coded or protected communications are only intended for the person who holds the key. Once the communication has been broken down or encoded with a set of instructions, the result is known as the cipher text or cryptogram which usually involves an algorithm and a key. Codes, steganography and ciphers are the types of techniques for cryptography. These various types of ciphers depend on numerical, alphabetical, computer-based, or other scrambling methods and ciphers are the most popular encryption form. Most of these encryption techniques are used in every computer environment.

This research uses cryptography techniques to protect data confidentiality and high security in Internet of Things. The two cryptosystem forms are symmetric key cryptography and asymmetric key cryptography, in which the symmetric cryptosystems use a single key, known as the secret key to encrypting and decrypting data or messages.

Asymmetric cryptosystems use two keys in which one key (the public key) is to encrypt messages or data, while another key (the secret key) to decrypt those messages or data respectively. In this architecture, symmetric key cryptographic encryption is used to secure the data between the sensors and public key cryptographic encryption is used to secure the data in the IoT device gateway and back-end data-sharing services.

5   Conclusion

Nowadays, the most of the devices communicate and send data via wireless network. The importance of data in an intelligent system is increasing. A variety of medical IoT devices and software applications are used to improve the quality of medical services. Ensuring data security and privacy at all stages of data flow is a major challenge in research in the Internet of Things. This paper discussed the internet of things healthcare system and data security of healthcare system. Here, smart healthcare is given great attention; and solution to secure healthcare data at various layers is proposed. A secured architecture is proposed in this paper which efficiently protects healthcare data by applying cryptography techniques.

Acknowledgement. This research was supported by a grant from University Grants Commission (UGC) awarded to A. Vithya Vijayalakshmi under Senior Research Fellowship.

References

1. Patel, S., Singh, N., Pandya, S.: IoT based smart hospital for secure healthcare system. Int. J. Recent Innovation Trends Comput. Commun. **5**(5), 404–408, ISSN: 2321-8169 (2017)
2. Vithya Vijayalakshmi, A., Arockiam, L.: A study on security issues and challenges in IoT. Int. J. Eng. Sci. Manage. Res. **3**(11), 1–9, ISSN: 2349-6193 (2016)
3. Abomhara, M., Koien, G.M.: Security and privacy in the Internet of Things: current status and open issues. In: IEEE International Conference on Privacy and Security in Mobile Systems (PRISMS), pp. 1–8 (2014)
4. Teresa Villalba, M., de Buenaga, M., Gachet, D., Aparicio, F.: Security analysis of an IoT architecture for healthcare. In: ICST Institute for Computer Sciences, Social Informatics and Telecommunications Engineering, LNICST 169, pp. 454–460 (2016)
5. Machorro-Cano, I., Ramos-Deonati, U., Alor-Hernández, G., Sanchez-Cervantes, J.L., Sanchez-Ramirez, C., Rodríguez-Mazahua, L., Segura-Ozuna, M.G.: An IoT-based architecture to develop a healthcare smart platform. Springer International Publishing AG, CCIS 749, pp. 133–145 (2017)
6. Kumar, N.: IoT architecture and system design for healthcare systems. In: IEEE International Conference on Smart Technology for Smart Nation, pp. 1118–1123 (2017)
7. Sridhar, S., Smys, S.: Intelligent security framework for IoT devices - cryptography based end-to-end security architecture. In: Conference on Inventive Systems and Control (ICISC), IEEE, pp. 1–5 (2017)
8. Zhao, K., Ge, L.: A survey on the Internet of Things security. In: IEEE, International Conference on Computational Intelligence and Security (2013)

9. Asghar, M.H., Mohammadzadeh, N., Negi, A.: Principle application and vision in Internet of Things (IoT). In: International Conference on Computing, Communication and Automation, pp. 427–431 (2015)
10. Matharu, G.S., Upadhyay, P., Chaudhary, L.: The Internet of Things: challenges & security issues. IEEE, pp. 54–59 (2014)
11. Santiago, S., Arockiam, L.: Energy efficiency in Internet of Things: an overview. Int. J. Recent Trends Eng. Res. (IJRTER) **2**, 475–482 (2016)
12. Azzawi, M.A., Hassan, R., Bakar, K.A.A. A review on Internet of Things (IoT) in healthcare. Int. J. Appl. Eng. Res. **11**(20), 10216–10221, ISSN: 0973-4562 (2016)

Topic Based Temporal Generative Short Text Clustering

E. S. Smitha[1]([⊠]), S. Sendhilkumar[1], G. S. Mahalakshmi[2],
and S. Krithika Sanju[2]

[1] Department of Information Science & Technology, College of Engineering
Guindy, Anna University, Chennai 600025, India
smithaengoor@gmail.com, ssk_pdy@yahoo.co.in
[2] Department of Computer Science & Engineering, College of Engineering
Guindy, Anna University, Chennai 600025, India
gsmaha@annauniv.edu, krithikhasanju3008@gmail.com

Abstract. In Social network paradigm, Twitter is the most widely used microblog nowadays. In the microblog environment, content or the specific theme for tweets are identified using hashtags but in general, all tweets do not contain hashtags. As the tweets form and spread very fast in microblogs, recommending relevant hashtags to tweets is emerging as a challenging task. This paper proposed an approach for short text clustering using temporal classification approaches. Hence for the batch of tweets both LDA and HDP topic modeling are attempted. In this paper, a hashtag is recommended for each tweet for mapping the topics obtained and the topic with the higher probability is considered as the hashtag of that tweet.

Keywords: Clustering · Hash-tag · Microblog · Semantic analysis ·
Social networks · Topic modeling

1 Introduction

In social network and microblogging sites, a hashtag is a type of metadata tag or keyword phrase used to find specific theme messages or topics required by the users. In general, hashtag is prefixed with symbol "#". Twitter users can use or include hash character # and thereby create hashtags (# is otherwise called as pound or number sign). # is placed before a word and unspaced phrase wherever required (i.e.) at the end or in the main text of the tweet. By searching the hashtag in the twitter search box, it can be easy to find the twitter hashtag which ties the conversation of different users into a single stream. By clicking or tapping on a hashtagged word in any tweet will exhibit all the available tweets that includes that particular hashtag.

In microblog or social networking sites, hashtags can be included wherever in the tweet as the part of 140 characters each i.e., either as prescript (preceding the tweet) or postscript and also within the word of tweet. (e.g. Climate is #hot). The number of hashtags included in the tweet is also important same as the variety of hashtags utilized. At present a tweet with a single hashtag represents a specific conversation. Tweet with two hashtags in most cases represent joining a location to the discussion.

© Springer Nature Switzerland AG 2020
A. P. Pandian et al. (Eds.): ICCBI 2018, LNDECT 31, pp. 912–922, 2020.
https://doi.org/10.1007/978-3-030-24643-3_107

Similarly, hashtags in three numbers in most cases denote "absolute maximum", and if any exceedance of contribution this risks "raising the intensity of the community."

1.1 Motivation and Purpose

In twitter as all the tweets have not assigned with hashtags while searching for each and every tweet under the specific hashtag, only tweets having that hashtag appears in search but corresponding tweets without hashtag does not appear. Hence this problem motivates to assign relevant hashtag to all the tweets. The aim of this work is to include generation of appropriate topics for batch of tweets using LDA and HDP topic model and embed topics for each tweet in that batch. Finally, recommend hashtag for each tweet in that batch using the higher probability topic obtained.

2 Related Works

A novel system to suggest hashtags was proposed by Zhao et al. [1]. A topical translation model for recommending the hashtags of microblog was proposed by Ding et al. [2]. This model assume hashtags and content of tweet as parallel detailing of resource. To facilitate process of translation, integration of latent topical information into translation model carried out.

Wang et al. [3] proposed combined effort of both collaborative intelligent and topical information for hashtag recommendation. It characterizes hashtags based on topic relevance on content models and also forecast hashtag usge preference by collaborative filtering process [4].

Topic modeling, especially Latent Dirichlet Allocation (LDA) [5] is commonly used in news articles and academic abstracts successfully. Mehrotra et al. [6] tried to find some methods to improve topics learned from Twitter content without modifying the basics of LDA. In twitter, Hashtags are used to categorize users posts. But all the tweets may not have Hashtags, so it will hold back the quality of the search results. Godin et al. [6–8], proposed a method for unsupervised and content-based hashtag recommendation for tweets.

Zangerle et al. [9] present an approach for the recommendation of appropriate hashtags and avoid the usage of synonymous hashtags. Zhao et al. [10] empirically compare the content of Twitter with a traditional news medium. Pennacchiotti et al. [11] present a user recommendation system that recommends new friends having similar interests using LDA.

From the literature survey, it is observed that recommending hashtags for all tweets using topic modeling is not an easy problem and the existing methods are not efficient which performs poorly due to many limitations. Therefore, a LDA, HDP-based topic model, where the resultant topics of both can be compared for a tweet. The proposed method can take full advantage of the importance of hashtags in a tweet.

3   Short Text Topic Modeling via Generative Temporal Clustering

The framework of the hashtag recommender system is shown in Fig. 1.

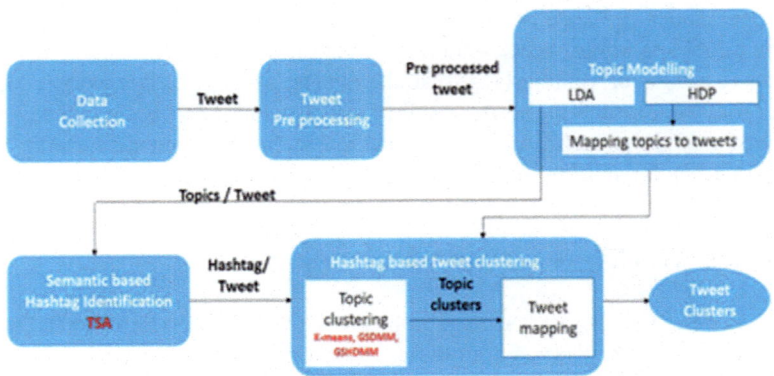

Fig. 1. Short text topic modeling via generative temporal clustering

To effectively recommend relevant hashtags to all tweets in microblogs, topic modeling methods like LDA, HDP are used. Topics are embedded to corresponding tweets and topic with the highest probability for each tweet is considered as the hashtag of that tweet. The main steps of the recommender system includes tweet preprocessing, topic modeling and topic embedding; after topic embedding the key word with highest probability will be recommended.

3.1   Tweet Preprocessing

The twitter data can be collected using the twitter data set or using twitter API. The dataset used for the work is UDI-TwitterCrawl-Aug2012-Tweets.

As real world data are generally inconsistent, noisy-containing errors or outliers and incomplete and also language used by the users is very informal in most of the social media. Users create their own words and spelling, shortcuts; for example 'hapi bday' instead of 'happy birthday'. Thereby, the quality and representation mode of data is very important for data analysis. It will take more amount of time for data preparation and filtering. The main data pre-processing steps are data cleaning, instance selection, normalization, transformation, feature extraction and feature selection. The final training set will obtain after data pre-processing.

Stop words must be removed so that the topics generated by LDA model is not greatly affected since the words present in the dataset forms the base for LDA topic generation also stop words are very less importance which directly affects the results. Along with stop words, punctuation and numbers also must be removed. Domain specific slang words must be eliminated to refine the ground truth. After removing all these deformities from the input dataset, the experimental dataset is obtained.

3.2 Topic Modelling

This module deals with the topic modeling for batch of tweets after it is collected and preprocessed. Latent Dirichlet Allocation (LDA) is an unsupervised learning which may be used to evaluate the multinomial interpretation. It is also a probabilistic generative model. In text modelling, LDA is a means to function Latent Semantic Analysis (LSA). The purpose of LSA is to evaluate the latent structure of concepts, themes or topics in text corpus, which is being captured the reason of the text that is considered to be difficult by "word choice" noise.

The HDP is the same as the LDA but the process involved in HDP is an automatic process where there is no requirement to mention the number of topics and keywords in it.

The parameters of HDP and the process behind generation of topics are explained in this section. Let us assume $|T|$ batch of tweet documents, $|K|$ topics and $|V|$ vocabulary.

1. For k = 1.....K topics: $\emptyset^{(k)} \sim Dirichlet(\beta)$
   Where β is a vector of length $|W|$ which is used to affect how word distribution of topic k, $\emptyset^{(k)}$ is generated. This step indicates that, since having k topics, k different word distributions are obtained.
2. Having k word distributions, generate the tweet documents.
   For each tweet document $t \in T : \theta_t \sim Dirichlet(\alpha)$.

The topics are determined first and have topic distribution θ_t for tweet document t. Hence even a tweet document has a strong single theme so more different topics are covered.

For each word $\omega_i \epsilon t$:

$$z_i \sim Discrete(\theta_t)$$

$$\omega_i \sim Discrete(\emptyset^{(z_i)})$$

Each word has to be assigned to one single topic z_i before each and every word in the tweet document t is written. The topic distribution θ_d is followed by topic assignment of each word w_i. After knowing a word belonging to a topic, determine the word from word distribution $\emptyset^{(z_i)}$.

Here α and β are known as hyper parameters. α, β and ω are known initially whereas $\emptyset, \theta$ and z are unknown initially. Hence ω is the vector of tweet documents word count, θ is the tweet documents topic distribution, $\emptyset$ is the topics word distribution and z is the vector of length $|W|$ representing each topic words

3.3   Topic Embedding and Topic Hashtag Recommendation

After doing the topic modelling for batch of tweets, each topic obtained is embedded to corresponding tweets and n topic with the highest probability for each tweet is considered as the hashtag of that tweet. Hence by doing this all tweet will be processing the hashtag.

The Algorithm I summarizes the overall process used hashtag recommender system. Both LDA and HDP are:

Algorithm I

Input:- Batch of tweets

Topics = Topic mod(n Batch)

$$T_0 \rightarrow Batch\ 1$$

$$T_0 = \{t_{00}, t_{01}, \dots, t_{0\ m-1}\}$$

$$T_1 \rightarrow Batch\ 2$$

$$T_1 = \{t_{10}, t_{11}, \dots, t_{1m-1}\}$$

.

.

.

$$T_{n-1} \rightarrow Batch n$$

$$T_{n-1} = \left\{ t_{n-1\ 0}, t_{n-1\ 1}, \dots, t_{n-1m-2} \right\}$$

where $m = 500$

$$for(i = 0\ to n - 1)do$$

$$bt_i(tpXw) = Top\_mod(i)$$

$$for(j = 1\ to m - 1)do$$

$$embed(tp(w), t_{ij})$$

$$endfor$$

$$t_c = \sum_{q=0}^{n-1} T_q$$

$$t_c = \{T_0 + T_1 + \cdots + T_{n-1}\}$$

$$for(k = 0\ to q)_{t_c}do$$

$$H_T(t_k) = Top_n \left(rank \left(t_k \left(P(t_{p(w)}) \right) \right) \right)$$

$$endfor$$

3.4   Semantic Based Hashtag Identification

Algorithm II

Input:-Topics / tweet.
Output:-Hashtag / tweet.
Methodology:-TSA
Pseudocode:
Input:- Topic words Tw of a tweet
Output:-Semantically relevant HT with probability //finding semantic related word vector for each Tw
//expand Tw using ESA
$<tw0,...,twn> <= ESA(Tw)$
Assign prob(Tw) to every $<tw>$
End for
//adding semmentically relevant HT
Alphab_sort_all(Tw,$<tw>$)
Find unique (Tw,tw) by averaging out probabilities of all occurences rank (Tw,$<tw>$) based on probability
Filter $<Tw,<tw>>$ based on avg_prob_threshold

3.5   Hashtag Based Tweet Clustering

After assigning relevant hashtags to all tweets collected, clustering is done based on certain specific hashtags i.e. the topics based clustering is done and mapping the tweets to the cluster is done. The algorithm used here for clustering is K-Means [12], GSHMM [13] algorithm for LDA and GSHDMM algorithm for HDP.

4   Result and Discussion

4.1   Dataset

The dataset used is UDI-TwitterCrawl-Aug2012-Tweets. It consists of tweets crawled from May 1, 2011 to June 30, 2011, a period of two months. It contains 147909 files. Each file contains at most 500 tweets published by a user. Thus, in total, there are tweets from 147909 users.

4.2   Results & Discussion

Topic Coherence is a means to evaluate topic models and methods that automatically generate topics from a collection of documents, using latent variable models. Each such generated topic consists of words, and the topic coherence is applied to the topmost N words from the topic. It is defined as the average or median of the pairwise word-similarity scores of the words in the topic. The scores are mainly comparative: if topic B incorporates a higher coherence score than topic A, then B is better or more coherent.

The terms of topic coherence are the intrinsic measure UMass and the extrinsic measure UCI, both based on the same high level idea. Both measure compute the sum

$$coherence = \sum i < jscore(w_i, w_j)$$

of pairwise scores on the words $w_1.....w_n$ used to describe the topic, usually the top n words by frequency $p(w|k)$. This measure can be seen as the sum of all edges on complete graph.

Both topic coherence measures UCI and UMass are based on the sum $\sum i <$ js-core (wi, wj) of the pairwise scores of the n top words $w_1.....w_n$ of the topic. Let D(wi) as the count of documents containing the word wi, D(wi, wj) the count of documents containing both words wi and wj, and D the total number or documents in the corpus. The corpus used to compute the counts is specified in a subscript of symbol D. Let, D be the count of documents of the Wikipedia corpus containing the word wi. The Figs. 2 and 3 represent the topic coherence using LDA and HDP respectively. The average value of topic coherence using HDP(0.73) is higher than LDA(−9.39). Figures 5 and 6 shows the result of GSDMM and GSHDMM clustering.

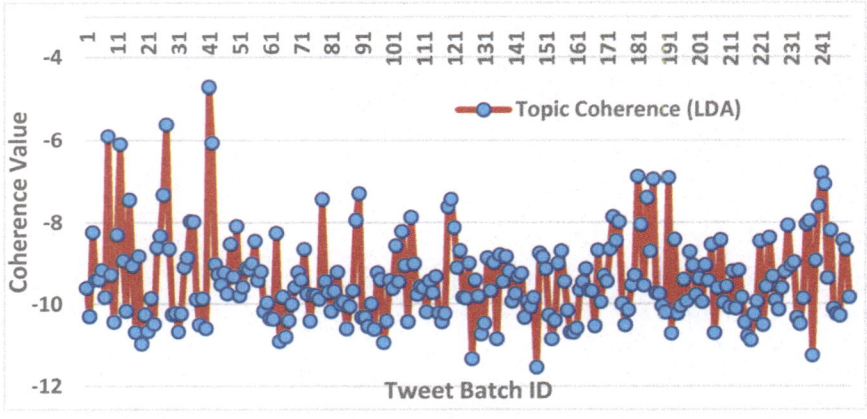

Fig. 2. Topic coherence using LDA

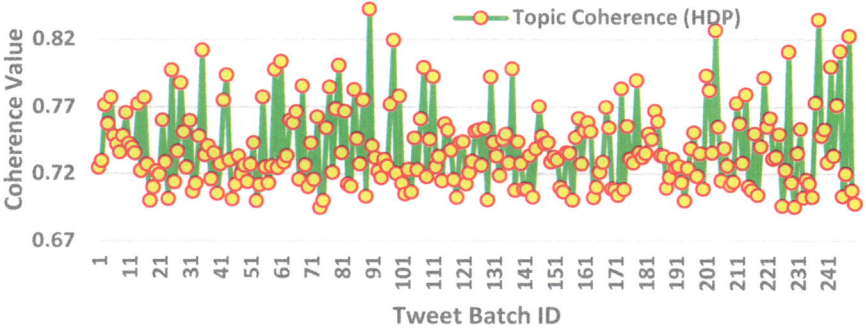

Fig. 3. Topic coherence using HDP

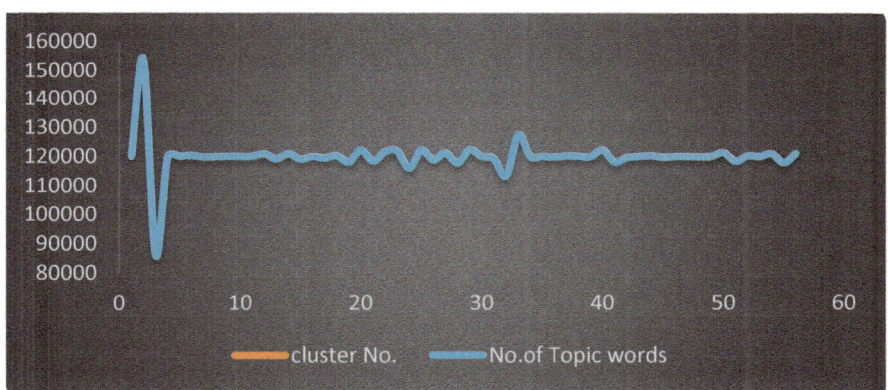

Fig. 4. Output of k Means tweet topic words clustering

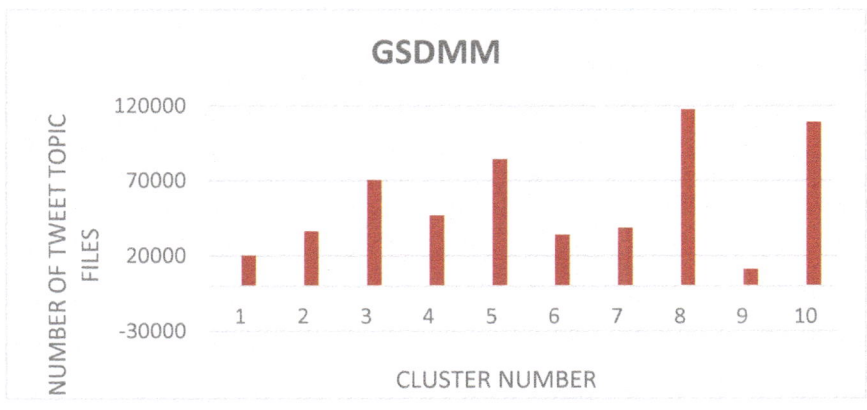

Fig. 5. Output plot of GSDMM clustering

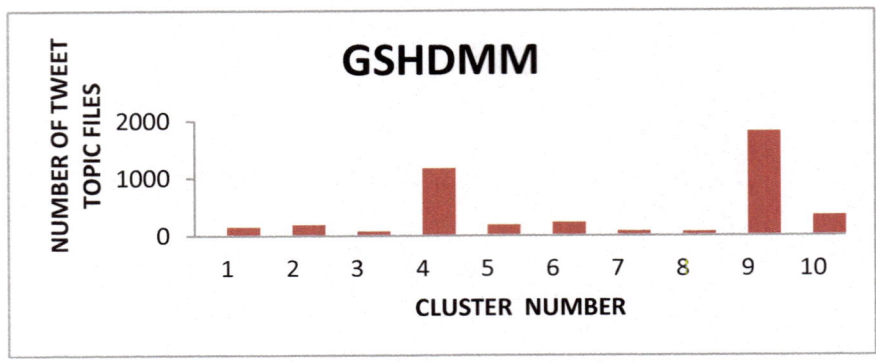

Fig. 6. Output plot of GSHDMM clustering

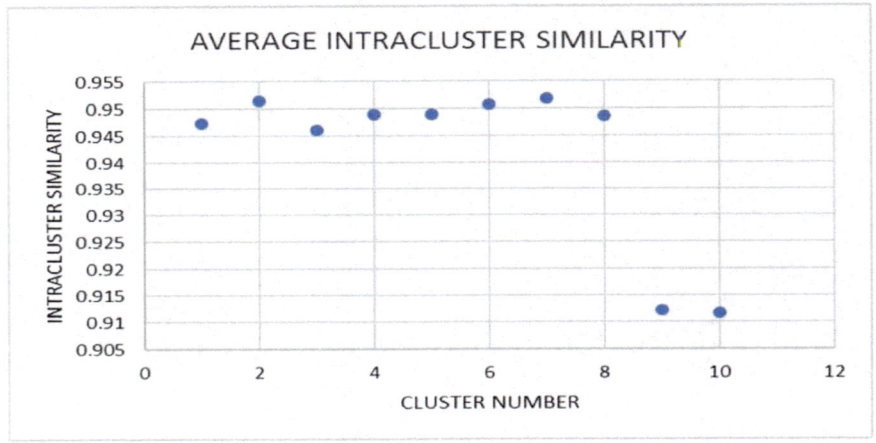

Fig. 7. Result of GSDMM average intra-cluster similarity

The topic models here in the proposed system are evaluated using the metric topic coherence then the recommended hashtags are evaluated by finding the hit matches with the existing hashtag of a tweet, finally average intra-cluster similarity are found for each cluster of GSDMM and GSHDMM clustering. Figures 7 and 8 shows the result of GSDMM and GSHDMM average intra cluster similarity (Fig. 4).

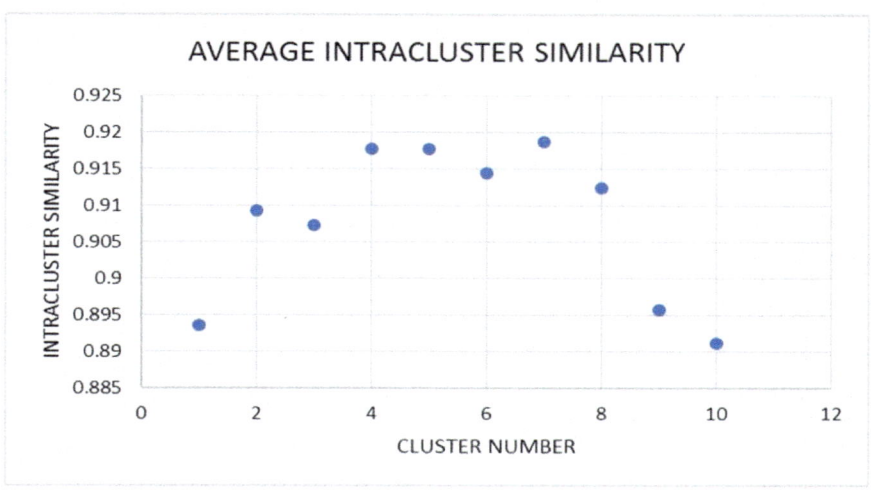

Fig. 8. Result of GSHDMM average intra-cluster similarity

5   Conclusion

Clustering short texts has been attempted much in tweet research. This paper proposed yet another approach for short text clustering using temporal classification. The clustering is generative which is more suitable in bringing the essence of clustering. Earlier works attempted tweet clustering using LDA which demands fixed topics and fixed words per topic. Though a tweet is short containing 150 words at the maximum, generative topic modeling like HDP captures more appropriate cluster topics when processed in batches.

Acknowledgement. The authors are grateful to acknowledge the support of UDI-TwitterCrawl-Aug2012-Tweets.

References

1. Zhao, F., Zhu, Y., Jin, H., Yang, L.T.: A personalized hashtag recommendation approach using LDA-based topic model in microblog environment. Future Gener. Comput. Syst. **65**, 196–206 (2016)
2. Ding, Z., Qiu, X., Zhang, Q., Huang, X.: Learning topical translation model for microblog hashtag suggestion. In: Proceedings of the Twenty-Third International Joint Conference on Artificial Intelligence, AAAI Press, pp. 2078–2084 (2013)
3. Chen, L., Wang, Y., Liang, T., Ji, L., Wu, J.: Data augmented maximum margin matrix factorization for flickr group recommendation. In: Pacific-Asia Conference on Knowledge Discovery and Data Mining, pp. 473–484. Springer, Cham (2014)
4. Bi, B., Tian, Y., Sismanis, Y., Balmin, A., Cho, J.: Scalable topic-specific influence analysis on microblogs. In: Proceedings of the 7th ACM International Conference on Web Search and Data Mining, pp. 513–522. ACM (2014)

5. Mehrotra, R., Sanner, S., Buntine, W., Xie, L.: Improving lda topic models for microblogs via tweet pooling and automatic labeling. In: Proceedings of the 36th International ACM SIGIR Conference on Research and Development in Information Retrieval, pp. 889–892. ACM (2013)

6. Godin, F., Slavkovikj, V., De Neve, W., Schrauwen, B., Van de Walle, R.: Using topic models for twitter hashtag recommendation. In: Proceedings of the 22nd International Conference on World Wide Web, pp. 593–596. ACM (2013)

7. Steyvers, M., Smyth, P., Rosen-Zvi, M., Griffiths, T.: Probabilistic author-topic models for information discovery. In: Proceedings of the Tenth ACM SIGKDD International Conference on Knowledge Discovery and Data Mining, pp. 306–315. ACM (2004)

8. Kywe, S.M., Hoang, T.-A., Lim, E.-P., Zhu, F.: On recommending hashtags in twitter networks. In: International Conference on Social Informatics, pp. 337–350. Springer, Berlin, Heidelberg (2012)

9. Zangerle, E., Gassler, W., Specht, G.: Recommending#-tags in twitter. In: Proceedings of the Workshop on Semantic Adaptive Social Web (SASWeb 2011). CEUR Workshop Proceedings, vol. 730, pp. 67–78, July 2011

10. Zhao, W.X., Jiang, J., Weng, J., He, J., Lim, E.P., Yan, H., Li, X.: Comparing twitter and traditional media using topic models. In: European Conference on Information Retrieval, pp. 338–349. Springer, Berlin, Heidelberg, April 2011

11. Pennacchiotti, M., Gurumurthy, S.: Investigating topic models for social media user recommendation. In: Proceedings of the 20th International Conference Companion on World Wide Web, pp. 101–102. ACM, March 2011

12. Hartigan, J.A., Wong, M.A.: Algorithm AS 136: A k-means clustering algorithm. J. Roy. Stat. Soc. C (Appl. Stat.) **28**(1), 100–108 (1979)

13. Mahalakshmi, G.S., Selvi, G.M., Sendhilkumar, S., 2019. Hierarchical modeling approaches for generating author blueprints. In: Smart Innovations in Communication and Computational Sciences, pp. 411–422. Springer, Singapore

Review of Reinforcement Learning Techniques

Mohit Malpani[(⊠)] and Rejo Mathew

Department IT, Mukesh Patel School of Technology Management
and Engineering, NMIMS (Deemed-to-be University), Mumbai, India
mohitmalpani2@gmail.com, rejo.mathew@nmims.edu

Abstract. The paper with the help of reinforcement learning techniques and its
method helps to find the best techniques that can be used in cyber security to
help defender protect the data against the attackers. The techniques have been
used in a cyber security game and resulted in a game of an unfriendly consecutive decision making problem played between agents i.e. an attacker and a
defender.

Keywords: Cyber security game · Network · Agents · Standard network ·
Game procedure · Reinforcement learning · Neural Network

1 Introduction

Markov game are played amongst two unfriendly agents with help of Q-learning
algorithm. Adverse reinforcement learning is used for playing game with two players
using a Q-learning method [1]. Addressing the concern which may emerge due to an
adversary's incapacity to manage detailed control. Considering defender's concern
about attacker's consequences. We will address these problems in the paper. In cyber
security one network gets attacked by different hackers during same time but depends
on value of data. However in this case we consider only one attacker as well as one
defender playing against each other for one asset [2]. There are 3 different nodes in
networks follows:

(1) First node Start is where attacker is at the origination of each game which has no
    benefit and is attacker's computer [3].
(2) Middle nodes don't have asset values & is between first node and last node which
    attacker needs to hack to access end node.
(3) Last node Data in network is point which has asset value. If the attacker is able to
    reach this node successfully, the game ends with attacker winning the game.

There are two agents in game one attacker and one defender. Both of them have
limited access to the network and having effect on their side of values in nodes network
(strength of attack and level of security). Both try to win as many game as possible to
reach their goal. Intermediate nodes have security flaws that helps attacker to attack.
The end node DATA [4] contains data and where the attacker attacks and the defender
need to make it more secure and tries to defend it (Fig. 1).

© Springer Nature Switzerland AG 2020
A. P. Pandian et al. (Eds.): ICCBI 2018, LNDECT 31, pp. 923–927, 2020.
https://doi.org/10.1007/978-3-030-24643-3_108

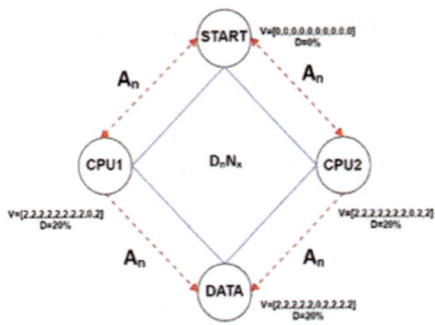

Fig. 1. Attacker and defender architecture

At start of game the attacker is at start node. At each step they choose from their set of actions and from these actions result is decided. The attack value determines on which node attacker has attacked, if value of attack is more than defense value then the attacker wins results in attacker moving to the attacked node and similar thing happens at end node the attacker wins and vice versa. At end of each game winner gets 100 points and loser gets -100 [5].

Step 0   Start node both agents have 0 values so the defender don't need to defend it if attacker tries to attack [6].

Step 1   For CPU2 when attack 2 happens and the defender defends the value of attack is changing at some value from 0 to 1 [7].

Step 2   For the end node when attack 3 happens the defender tries to defend but some data is getting lost and hence value of attack changing from 0 to 1 (Table 1).

2   Reinforcement Learning

Agents decide the actions depending on situation of network and are different for both because complete situation of network is relatively observed for each agent and have to gain different details. There are four techniques involved in the reinforcement learning such as Monte Carlo Learning, Q-Learning, Neural Network and Linear Network.

2.1   Monte Carlo Learning [4]

After every game each agent updates the evaluated reward values of action and state pairs that were used for that game. It updates every state agents participated with same reward values, from:

$$Q_{t+1}(s; a) = Q_t(s; a) + a \ (RQ_t(s; a))$$

Table 1. Actions of attacker and defender

Steps	Attacker's action	Defender's action	Node	Value(Attack)	Value(Defense)
0	–	–	START	[0, 0, 0, 0, 0, 0, 0, 0, 0, 0]	[0, 0, 0, 0, 0, 0, 0, 0, 0, 0]
			CPU1	[0, 0, 0, 0, 0, 0, 0, 0, 0, 0]	[0, 2, 2, 2, 2, 2, 2, 2, 2, 2]
			CPU2	[0, 0, 0, 0, 0, 0, 0, 0, 0, 0]	[2, 0, 2, 2, 2, 2, 2, 2, 2, 2]
			DATA	[0, 0, 0, 0, 0, 0, 0, 0, 0, 0]	[2, 2, 0, 2, 2, 2, 2, 2, 2, 2]
1	Type of attack 2, CPU2	CPU1, Defense Type 1	START	[0, 0, 0, 0, 0, 0, 0, 0, 0, 0]	[0, 0, 0, 0, 0, 0, 0, 0, 0, 0]
			CPU1	[0, 0, 0, 0, 0, 0, 0, 0, 0, 0]	[1, 2, 2, 2, 2, 2, 2, 2, 2, 2]
			CPU2	[0, 1, 0, 0, 0, 0, 0, 0, 0, 0]	[2, 0, 2, 2, 2, 2, 2, 2, 2, 2]
			DATA	[0, 0, 0, 0, 0, 0, 0, 0, 0, 0]	[2, 2, 0, 2, 2, 2, 2, 2, 2, 2]
2	Data, Type of attack3	CPU1, Defense Type 1	START	[0, 0, 0, 0, 0, 0, 0, 0, 0, 0]	[0, 0, 0, 0, 0, 0, 0, 0, 0, 0]
			CPU1	[0, 0, 0, 0, 0, 0, 0, 0, 0, 0]	[2, 2, 2, 2, 2, 2, 2, 2, 2, 2]
			CPU2	[0, 1, 0, 0, 0, 0, 0, 0, 0, 0]	[2, 0, 2, 2, 2, 2, 2, 2, 2, 2]
			DATA	[0, 0, 1, 0, 0, 0, 0, 0, 0, 0]	[2, 2, 0, 2, 2, 2, 2, 2, 2, 2]

a = learning rate (ranging from 0 to 1)
R = reward for wining
Q = estimated reward values
s = current state of world (for attacker node)

Soft-max Learning - The method gives each action in a collection of possible actions, i.e. giving action a chance to get selected. Boltzmann distribution used in algorithm helps to make action selection probabilities.

Upper Confidence Bound 1 (UCB-1) - The method supports its action choice based on their reward values and also on number of previous attempts of that action. At the beginning of the counterfeit, those actions are selected which are not used before. When no actions are left, the action will be chosen that helps in maximizing $V_t(a)$ [5].

Discounted Upper Confidence Bound (Discounted UCB) - It is an alteration to the UCB-1 algorithm [5].

2.2 Q-Learning

It is a model free augmentation learning method and proved that method connect to an ideal policy for a limited collection of actions and state for a sole agent. The Backward Q-learning algorithm helps in instructing the attacker and defender agents [6].

2.3 Neural Network

This network works for defender agent but doesn't work for attackers. It uses an input value for each defense and output value for every action. Results are displayed for every win with 1 and for -1 for loosing, applying only the preferred action.

2.4   Linear Network

This method follows the same technique, but doesn't have any hidden layer which results in direct calculation of output depending on 44 nodes of input layer and thus represents each of the defense and recognition values in the network with common link to the output layer.

3   Results and Discussion

Every win has a value between -1 and 1. If value is -1 that means attacker has won and if value is 1 it means defender has won [7] (Table 2).

Table 2. Comparison of learning techniques

	Monte-Carlo learning	Q-learning
Feature 1 For agents	Monte Carlo is for both agents and is against an opponent with random action selection	It is best for defenders
Feature 2 Algorithm features	Soft-max is fine learning method for the defender against all attackers	Possible to apply algorithm for large problems
Feature 3 Action use advantage	Average % change & standard deviation can be adjusted subject to expert opinion	It is used to find optimal action section policy
Feature 4	Easy to calculate	Speed up learning in finite problems

4   Conclusion

This research work focused in the cyber security game based on different agents includes defender and attacker. It is designed based on Markov game model and that having partial information in which the attacker is trying to attack network by hacking it and the defender is trying to protect the network and it informs the attacker to stop the attacking process. Q-learning is the best algorithm for defender and also worst for attacker agent because of the optimal strategy is less supportive or clear for attacker.

References

1. Auer, P., Cesa-Bianchi, N., Fischer, P.: Finite-time analysis of the multi-armed bandit problem. Ma-chine Learn. **47**, 235–256 (2002)
2. Chung, K., Kamhoua, C., Kwiat, K., Kalbarczyk, Z., Iyer, K.: Game theory with learning for cyber security monitoring. IEEE HASE, pp. 1–8 (2016)
3. Neumann, J.V., Morgenstern, O.: Theory of games and economic behavior. Princeton University Press (2007)

4. Sutton, R.S., Barto, A.G.: Reinforcement learning: an introduction. The MIT press, Cambridge, MA (1998)
5. Garivier, A., Moulines, E.: On upper-confidence bound policies for non-stationary bandit problems. ALT (2008)
6. Wang, Y., Li, T., Lin, C.: Backward q-learning: the combination of sarsa algorithm and q-learning. Eng. Appl. of AI **26**, 2184–2193 (2013)
7. Lin, L.-J.: Reinforcement learning for robots us-ing neural networks. PhD thesis, Carnegie Mellon University (1993)

Machine Learning Algorithms for Diagnosis of Breast Cancer

Richaa Negi[(✉)] and Rejo Mathew

Department of IT, Mukesh Patel School of Technology Management
and Engineering, NMIMS Deemed-to-be University, Mumbai, India
richaanegi24@gmail.com, rejo.mathew@nmims.edu

Abstract. Machine learning is applied on systems for sequence and trend recognition, incomprehensible by humans or traditional programming, as they efficiently utilize the patterns for training the networks to overcome particular challenges. The prevalence of machine learning in medical sciences is devised to decrease the mortality rate of cancer patients owing to its detection at early stages. The objective of this review paper is to compare machine learning algorithms, to be precise, Support Vector Machine (SVM), Random Forest (RF), Bayesian Networks (BN) and k-Nearest Neighbor (kNN) so as to achieve precise detection and classification of breast cancer.

Keywords: Machine learning · Classification · Breast cancer diagnosis ·
Support Vector Machine · K-nearest neighbors · Random Forest ·
Bayesian Networks

1 Introduction

Cancer arises from the uncontrollable development of cells (tumor) in a specific organ. This tumor originates from the tissues present in an organ which then spreads to the surrounding tissues. Breast cancer proves to be one of the main reason of high mortality rate in women around the world.

Although, there is a significant increase in the survival rates due to major improvement in screening and medication, breast cancer proves to be the most invasive cancer in women. By 2020, an estimation of 1.24 lakh women are to be affected by breast cancer as predicted by India's National Health Profile [1].

Machine learning consists of assembling the data, selecting the model and then training and testing the model. The aim of the classification is to appropriately place each observation into its category (Fig. 1).

In supervised learning, the model is trained by a dataset which is defined beforehand so as to achieve the desired outcome. For example, problems on classification and regression. Whereas in unsupervised learning, no such data set exists which denotes that there is no perception of the probable result. For example, problems based upon clustering and associative rule learning algorithms. In this study, we ascertain whether the tumor present is benign or malignant by using four machine learning classifiers [2].

© Springer Nature Switzerland AG 2020
A. P. Pandian et al. (Eds.): ICCBI 2018, LNDECT 31, pp. 928–932, 2020.
https://doi.org/10.1007/978-3-030-24643-3_109

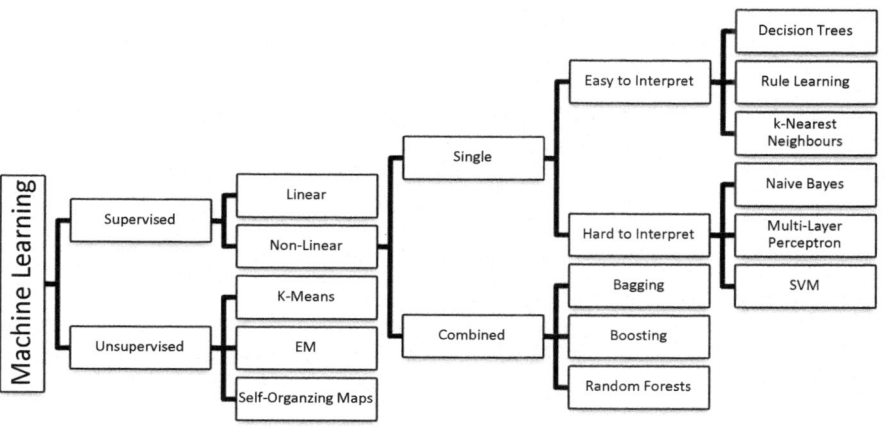

Fig. 1. Taxonomy of machine learning

This review paper entails various machine learning approaches for detection of cancer which are introduced in Sect. 2. Section 3 analyses the algorithms and presents the comparison between them. Sect. 4 concludes the entire paper.

2 Machine Learning Approaches

2.1 Support Vector Machine

This algorithm falls under the category of supervised machine learning and is often used for diagnosis of cancer. SVM generates a linear function which helps in the classification of classes. This linear function, also known as hyper-plane, distinctly splits the classes using specific data points and acts as a decision boundary. The points which contribute to the result of the algorithm are known as the support vectors since it is present in a multi-dimensional space.

The best fitted hyper-plane aims to increase the distance between it and the data element closest to the hyper-plane. This distance is referred as the marginal distance. For example, when the space in which the dataset is present in 2-dimensional, the hyper plane is expressed as ax + by and the objective is to obtain values of a, b and c such that that ax + by < c is valid for the first category and ax + by > c is validated for the second category [6].

Support vectors are a critical element in SVM since they are closest to hyper-plane and even if the data points away from the decision hyper-plane are removed, it won't change the boundary unlike the support vectors.

2.2 Random Forest

Random Forest relates to the concept of ensemble learning which comprises of numerous machine learning algorithms composed to form a final optimal algorithm.

It consists of a combination of numerous decision trees to group a forest since we can run the model a number of times as compared to using one decision tree where the model can be run only once.

Essentially, RF follows a repetitive tactic in which one tree is selected at random from the subset of the present data set. These decision trees might not be the best in general but together they implement and execute the data set very well.

There is a repetition of the aforementioned stages until the count of the number of desired trees is obtained. After the final count of the trees, that classifies the observations into different groups, the categorization of the cases rests on the vote of the majority as taken by the decision trees.

RF is impervious to the undesired disturbance present in the existing data set which presents an increase in the stability when measured to each distinct decision tree. Another reason why RF is favored is due to its capability of managing marginal values of data.

In this case, a tumor can be classified into categories depending upon its features into it being benign or malignant even if the value of the malignant feature is just 15% of the input data [4].

2.3   Bayesian Network

Bayesian Network is operated to aid in estimation and information demonstration of areas that are indeterminate. BN resembles the directed acyclic graph (DAG) which contains numerous nodes. The periphery of the nodes represents the dependency between the conforming nodes (relative to random numerical values) present in the chart [5].

These dependencies are estimated by statistical methods where in each variable has a provisional possibility table which represents the distribution of the likelihood of the variable depending upon its predecessors. These random values are also unassociated of their non-successors concerning their immediate ancestors present in the nodal structure.

The antecedents of a node represents its parents and conditional probability can be acquired from the probability table related to every node [6].

2.4   K-Nearest Neighbors

KNN helps in predicting the properties of the data. Suppose, A and B are two categories present in our data set. A new data point C is added into the data set and is to be classified to fall into either categories. Further, C is categorized depending on its k neighbors comprising of majority data elements. The complete dataset which is used for learning the algorithm is searched for K indistinguishable neighbors [7].

Distance measures such as the Euclidean distance are used to determine which of the K neighbors present in the data set, which is to be studied, are more comparable to the element which is included later on. The Euclidean Distance (ED) is determined by the square root of the sum of the squared differences between a new position (D_1), with coordinates (x_1, y_1) and an existing position (D_2), with coordinates (x_2, y_2), considering all the input attributes for the comparison of the distance between the specific data element and other data elements [8].

$$\text{Euclidean Distance between } D_1 \text{ and } D_2 = \sqrt{\left((x_2 - x_1)^2 + (y_2 - y_1)^2\right)}$$

Algorithm:

1. Select k number of neighbors
2. Calculate the k-nearest neighbors of C according to the Euclidean distance
3. Count the number of data points in each category present in the k-nearest neighbors
4. Allocate the data point (C) to the category with the maximum number of neighbors [9]

3   Comparison Analysis

The four different machine learning algorithms were implemented on the Wisconsin Breast Cancer Data Set where they were computed and analyzed for various models of training and testing. Recall, as a feature, is based on the frequency of accurately predicted outcomes such as when negative observations are rightly anticipated as negative. In this case, for example, it calculates the probability of correctly identifying benign over malignant. The area under a ROC(receiver operating characteristics) graph helps envisage a classifiers functioning by demonstrating the adjustments between its cost and value. In this case, it determines the probabilityof differentiating between benign and malignant [10] (Table 1).

Table 1. Comparison of different methodologies

Properties	Algorithms used			
	SVM [2]	kNN [3]	RF [4]	BN [5]
Training Time	Slow	The training time is slow to compute distance of each instance	The training time is slow to compute distance of each instance	Fast as it handles all the missing data [12]
Training Time for Large Data Set	Increases with increase in extensive memory requirements	Increases because of complexity	Remains constant since it can handle high dimensional spaces	Time remains constant
Recall	High probability	High probability	High probability	Low probability
Area Under ROC	Average chance	Lower chance	Highest chance. Optimum ROC	Higher chance
Average Predictive Accuracy	High	Low	It reduces over fitting for accuracy	Low
Increase in Number of Features	Poor performance	Best performance in medium sized feature space [11]	Average performance	Best performance

4 Conclusion and Future Work

This paper reviews various machine learning algorithms for the detection and analysis of breast cancer. The attained outcomes have proved that the performance of classification varies with respect to the selected method. Results have shown that SVM has achieved an optimal execution with respect to accurateness and distinctness. Whereas, RF comes up with maximum likelihood of appropriately categorizing a tumor.

Additional study can be carried out in this field, for instance, study on the formation of classes of SVM and alteration of applied factors can lead to improved precision. This domain has future possibility of systematic investigation.

References

1. Breast Cancer Among Indian Women. http://anthropology.msu.edu/anp204-us14/2014/07/11/breast-cancer-among-indian-women/
2. Williams, G.: Descriptive and predictive analytics. In: Data Mining with Rattle and R: The Art of Excavation Data for Knowledge Discovery (2011)
3. Chang, R.L.P., Pavlidis, T.: Fuzzy decision tree algorithms. In: IEEE Transactions on Systems and Cybernetics (1977)
4. Kononenko, I.: Machine learning for medical diagnosis: history, state of the art and perspective, vol. 23 (2001)
5. Faltin, F., Kenett, R.: Bayesian networks. In: Encycl. Stat. Qual. Reliab. 1(1), 4 (2007)
6. Bazazeh, D., Shubair, R.: Comparative study of machine learning algorithms for breast cancer detection and diagnosis. In: 2016 5th International Conference on Electronic Devices, Systems and Applications (ICEDSA) (2016)
7. Sarkar, S.K., Nag, A.: Identifying patients at risk of breast cancer through decision trees. In: Int. J. Adv. Res. Comput. Sci. 8, 88–96 (2017)
8. Kurnianingsih, K., Nugroho, L., Widyawan, W., Lazuardi, L., Prabuwono, A.S.: Emergency alert prediction for elderly based on supervised learning. In: 1st International Conference on Biomedical Engineering (IBIOMED) (2016)
9. Amrane, M., Oukid, S., Gagaoua, I., Ensari, T.: Breast cancer classification using machine learning. In: 2018 Electric Electronics, Computer Science, Biomedical Engineering's' Meeting (EBBT) (2018)
10. Sachin, Jayaraj, T., Sanjana, V.G., Darshini, V.P.: A review on neural network and its implementation on breast cancer detection. In: 2016 International Conference on Communication Signal Processing (ICCSP) (2016)
11. Cruz, J.A., Wishart, D.S.: Applications of machine learning in cancer prediction and prognosis. In: Cancer Inform (2006)
12. Sugiyama, M.: Introduction to statistical machine learning, ed. Green, T.: Morgan Kaufman (2006)

Review of Data Security Frameworks for Secure Cloud Computing

Kunaal Umrigar[(⊠)] and Rejo Mathew

Department of IT, Mukesh Patel School of Technology Management
and Engineering, NMIMS Deemed-to-be University, Mumbai, India
kumrigar@gmail.com, rejo.mathew@nmims.edu

Abstract. The most important thing for cloud computing is to offer real time data security for not only small sets of data but in petabytes. Cloud is majorly used because of cost savings and the agile business models it provides. Unfortunately, cloud comes with certain data security challenges. Data security of a user is given the highest priority and hence this paper provides a few methods to mitigate them. The paper discusses layers of CCAF and how addition of excess data validates it. Solutions have been provided to overcome security breach when there is a large increase of data. Another framework is proposed in which each threat is considered individually and a solution is specified to overcome it. The third framework allows providers of cloud management system to define and enforce complex security policies to protect their data. This paper consists of five sections: Introductions, Problems, Solutions, Analysis and Conclusion.

Keywords: Cloud computing Adoption Framework (CCAF) ·
Multidimensional mean failure cost · Framework for secure cloud computing ·
Trojans

1 Introduction

Cloud data security is important because it is important for business owners to make sure that their data remains secure while stored in the cloud. Due to an increase in the number of high-profile hacking cases, cloud computing data security has become one of the major topical issues. In case of a local disaster, both backups could be lost. Cloud security prevents this issue, as data is stored in remote locations thereby protecting the business of data loss.

CCAF is a cloud computing multi layered security framework that integrates three major security technologies: Firewall, encryption, and identity management. CCAF detects and blocks a total of 99.95% viruses and Trojans and can block them for 100 h of continuous attacks. It can also block all SQL injection and thereby provides real time protection to data. CCAF has three layers of security:

1. Firewall and access control
2. Identity management and intrusion prevention
3. Convergent encryption

© Springer Nature Switzerland AG 2020
A. P. Pandian et al. (Eds.): ICCBI 2018, LNDECT 31, pp. 933–941, 2020.
https://doi.org/10.1007/978-3-030-24643-3_110

The focus is on data security while there is an increase of data, the increase can be either from external sources such as attack of viruses or Trojans or they can be from the internal sources if hundreds of terabytes of data is accumulated by users or clients per day. The first step is improve the strength of each layer, since it is difficult for the viruses and Trojans to break different types of security in one attempt. Multidimensional Mean Failure Cost (M^2FC) is another framework presented that identifies security requirements, thereby helping in evaluating cloud security to eliminate the threats. A qualitative risk analysis of systems can be carried out using this model. The solution aims to maintain a trust between the customers and cloud providers. For each stakeholder, the model calculates the losses that the user incurs from security threats and vulnerabilities. It is a model that estimates the security of a system by determining the loss that each stakeholder due to security breaches

Ensuring any customer the integrity of their data is one the major problems in cloud computing. There should be a way for the user to determine if the data which cannot accessed physically is correct or not. Thus, this framework aims to give proof for integrity of the data by providing a scheme. The customer and the cloud provider can agree upon this proof and so this can be integrated at the service level agreement.

2   Problems

2.1   Data Security Issues

Here are some key data security issues that describe the requirement of CCAF:

- **Data tampering:** Occurs when a transaction is modified with unauthorized access.
- **Fabricating User Identities:** When users gain access to data and attempt identity theft by threatening digital signatures with non-repudiation attacks [3].
- **Password-related threats:** deals with cracking and stealing of passwords.
- **Snooping and Data Theft:** During data transmission, the stealing of important personal data such as credit card. Packet sniffers are used to steal information.
- **Scaling the security administration** of multiple Systems poses an extra challenge to manage cloud security as it deals with the problem to provide multiple accesses to multiple applications.
- **Management Interfaces Risks:** The cloud computing services are prone to security risks by insecure interfaces and APIs threats.
- **Security Issues in Virtualization**: Security threats like VM based rootkit attacks occur that infect both client and server machines. The files and registry keys are hidden as a result of this.

2.2   Organisational Challenges

Security and privacy are considered as challenges while adopting a cloud computing network and hence a model is developed that aims at protecting the services and address the issues mentioned below:

Organizational Sustainability: No structured measurement of risk analysis.

Service portability: Cloud portability of all types that supported by FSaaS are dealt with.

Linkage: Different services cannot be connected. Communication between different types of clouds from different vendors is difficult.

Access Problem & Multi-Tenancy Problem:
Several systems find it difficult to access the data stored in different parts of the world. This happens because the environment is not transparent enough and the customer's data is located in data centres of the cloud provider which could be anywhere in the world and hence out of the customer's reach.

IaaS cloud mostly depends on systems where resources are shared between several virtual machines. Sharing of resources means that there could be transfer of malicious activities between machines which carried by one tenant may affect the other tenants too.

2.3   Vulnerabilities in Cloud Computing

Zombie Attack: In cloud, a large number of requests, called zombies flood the virtual machines. Such an attack interrupts availability of Cloud services. Due to this, a large amount of requests cannot be handled by the cloud, this causes Denial of Service.

Man-in-the Middle Attack: If the configuration of the secure socket layer is unsecure then an attacker can access the data exchange.

Metadata spoofing Attack: In this type of attack, the service files that contain Web Services Description Language are modified by an adversary where descriptions about service instances are stored. If it is successful to interrupt the service code, the attack can be successful.

3   Solutions

3.1   Fine Grained Security Model (FGSM)

CCAF security software is implemented using multiple layers of security mechanism that aims at maximizing protection. Implementing different layers of security also means that there is a reduction in the infections by Trojans, virus, worms and unauthorized access and denial of service attacks.

FGSM uses a mark-up language called XACML 3.0 (Extensible Access Control Mark-up Language), based on an XML-schema to define the ports for which secure communications take place with respect to the IP addresses.

Layer description

Layer 1: Firewall

It is the first layer of defence that restricts access to members. Cisco routers and XACML is used together. Cisco routers and networking infrastructure allow us to set the firewall and observe any abnormal activities in the data and thereby maintaining its integrity. XACML is used to strengthen the security system and minimize any errors.

Layer 2: Identity Management

The second layer consists of the intrusion detection system (IDS) and intrusion prevention system (IPS). This layer identifies attacks, intrusions and penetrations, and provides technological model to prevent attacks such as Denial of Service and anti-spoofing. The right person should get the right level of access and that is taken care by the identity management.

Identity management defines the role of users and the privilege and permissions they get.

1. **Users**: They include the ones who can encrypt each key from their block. This step ensures that all the data that users access is stored and protected in the Cloud.
2. **CCAF Server**: The CCAF Server has three functions in particular. First, it authenticates the users during the stages of storage and retrieval.
   Second, it controls the access for the users. Third, the data between users and the cloud is encrypted.
3. **Security Manager**: It stores the metadata that contains encrypted keys, block signatures and also checks whether a user can retrieve a file if authorized.

Layer 3: Encryption:

In this layer, all files that need to be encrypted are integrated. The de duplication of similar files and monitor changes and updates can be reduced in this layer. The behaviour of the fine grain is monitored and early warning is provided if something is abnormal (Fig. 1).

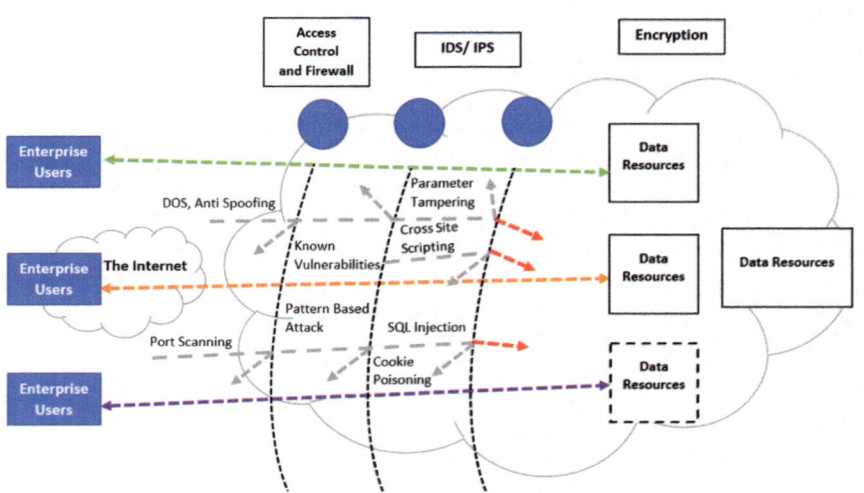

Fig. 1. CCAF offered fine-grained security model

Quarantine Measures:

At the point when the recognized Trojans and viruses are discovered, they are separated and sent to the isolate region instantly. This is done to ensure client security. Solid isolation is employed to identify vulnerabilities in any of the cloud organisations, including blocking unauthorized IPs and attack points/ports.

First the data is backed up safely and then attempts are made to isolate the infected data. If a quarantine action is unsuccessful, the system architect is informed. The files can be kept under "quarantine area" or deleted if chosen.

3.2 Secure Cloud Computing Framework

In this subsection, a security model is described that aims to provide solutions against the threats mentioned in the previous section. Users can be certified by the third party authorities to gain access to the End User Service Portal. After gaining access, the users can use the cloud services which are provided. This framework provides a secure connection to the user to access cloud services.

The End User Service Portal, uses a Virtual Private Network. (VPN) to provide secure access control (Fig. 2).

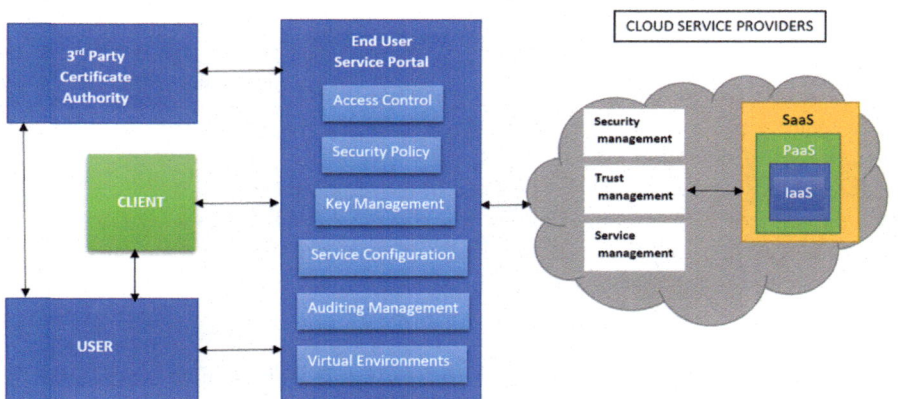

Fig. 2. Cloud computing security model

Services for components in the model are:

- **Client:** They can access the client side, that comprises of the web browser or applications installed by the host via devices like laptops and mobile phones. It is the portal where the users get their personal cloud.
- **End-User Service Portal:** When granted clearance, the certification of the user helps in the issuing of a Single Sign on Access Token (SSAT). This allows sharing of user information between the security policy cloud service providers and the access control component by using XACML.
- **Single Sign-on (SSO):** Many users have multiple accounts with cloud providers and they tend to use the same password at different places, thereby posing a threat to

the security. Therefore the use of SSO technology promises to cut down multiple network and application passwords to one. This enables user to access various different applications in the cloud computing environment using one login, enabling strong authentication at the user level.

- **Security Control:** Access control module supports the providers access control needs. The Role Based Access Control is the most widely accepted model because of its simplicity, flexibility and can easily capture dynamic requirements. It can be used to control the usage and specify certain conditions into authorizations.
- **Security Management:** This component provides privacy specification and security functions. The users are authenticated based on their credentials and characteristics.
- **Trust Management:** Due to the service oriented nature of the cloud, a level of trust should be integrated within the service. We need to establish a bi-directional trust. That is, the users should have some level of trust on the providers so that they can choose their services from them freely, and the providers also need to have some level of trust on the users to release their services to. One approach is to develop a trust management system that includes a generic set of trust negotiation parameters and is integrated with the service, and is bi-directional in nature.
- **Service Monitoring:** High level service performance and availability of data is guaranteed by the monitoring system.

3.3 The M²FC Model

It is a security risk valuation model in which the risk is measured in terms of financial loss per unit of operation time due to security threats.

It assesses the stakeholders cost related to the requirements. Several factors such as the incurred costs of the stakeholder for not meeting a security requirement, the inconsistency between failure cost related to the requirements and the inconsistency between failure impacts as observed by stakeholders are taken it consideration [1].

3.3.1 Framework Methodology
This model aims at providing a solution to each threat individually.

The security steps are:

  i. The framework first identifies a <u>security requirement</u> for each issue.

  ii. Then, for each security requirement, an analysis to identify the probabilities of failure threats is carried out in order to select a <u>dimension element</u> having the highest probabilities when systems fails.

  iii. For each dimension element, the probabilities of failure are analysed to choose security threats having the highest failure probability.

  iv. For each security threat, mitigation problems have been proposed

The model is applied to measure systems and estimate the financial cost for each stakeholder.

3.3.2   Security Requirements

This step identifies that there are three security requirements associated with the security problems mentioned above namely Availability, Integrity and Confidentiality.

The security requirements in the following issues have been identified as:

Virtualization: The malware affects customers' data thereby affecting confidentiality; it also can change the data leading to loss of integrity; and malware can erase the data of the customer thereby affecting availability.

Multi Tenancy Problem: It affects the privacy, access control, accessibility, reliability and accountability of data.

Management Interfaces Risks: Organizations are too reliant on the weak set of APIs and interfaces, hence there is exposure to variety of security issues.

3.3.3   Mitigation

i. **Denial of Service:** Denial of Service indicates that only one VM Machine can consume the system resources, this results in starving the remaining VMs. DoS can be countered by reducing the privileges of the user who is connected to a server. This will help to reduce the DOS attack. Furthermore, one can also limit the resource access to the guests. Technologies currently dealing with virtualization offer a mechanism to limit the resources allocated to each guest machine present in the environment.

ii. **Malicious insiders:** Strict supply chain management should be enforced to counter the malicious virus inside the system. Another effective measure is to specify certain useful human resource requirements in legal contracts, and require the user to be transparent regarding information security.

iii. **Account, service and traffic hijacking:** As a solution, the sharing of account credentials between users and services should be prohibited.

Another effective measure is to implement two-factor authentication techniques wherever possible, and employ user monitoring to detect unauthorized activity.

4   Analysis

See Table 1.

Table 1. Parameter based comparison between the three studied models for secure cloud computing.

Features	Cloud computing adoption framework	Multidimensional mean failure cost model	Secure cloud computing framework
Protection to Trojans and Viruses	• Multi-layered • Contains different security functions at each layer so virus finds it difficult to break into.	• Single layered model • Does not aim at providing solutions to attack on viruses and Trojans.	• Single layered framework • Users with access to the End User Service Portal can use the cloud services.
Use of Extensible Access Control Mark-up Language (XACML)	Yes XACML is used to • strengthen the security system • minimize errors	Not used	Yes XACML is used to • used to share the user information to access control component
Provide Role Based Access Control	Not provided	Not provided	Yes, controls the usage and integrates conditions into the authorizations for gaining access.
Single Sign On	Not allowed	Not allowed	Yes • Enables user to access multiple applications in the cloud computing environment • Ensures user level authentication.
Isolation of threat	Yes • Virus undergoes solid isolation to determine vulnerability • Mitigation of threat	Yes • security threat identified for each issue • Mitigation of threat	No isolation of threat

5   Conclusion and Future Work

This paper addresses the security challenges that exist in cloud computing and a few frameworks have been proposed that identify security and privacy requirements, attacks, threats, concerns and risks associated with the data. The comparison in this paper helps the users select the appropriate methods to assess security risks for their cloud computing environment. Security problems present a major issue in today's time especially in information systems and cloud data centres. Cloud adoption results in the development of organizational challenges where three challenges are identified. The research questions are addressed to deal with these issues. Future research should focus on the rapidly growing domain of cloud computing and studying the security threats and prescribing appropriate measures and solutions to overcome.

References

1. Jouini, M., Rabai, L.B.A.: A security framework for secure cloud computing environments. Int. J. Cloud Appl. Comput. **6**, 32–44 (2016). https://doi.org/10.4018/IJCAC.2016070103
2. Chang, V., Ramachandran, M.: Towards achieving data security with the cloud computing adoption framework. IEEE Trans. Serv. Comput. **9**(1), 138–151 (2016)
3. Ramachandran, M., Chang, V.: Cloud security proposed and demonstrated by cloud computing adoption framework. Int. J. Organ. Collect. Intell. (IJOCI) **6**(3) (2016)
4. Munir, K., Palaniappan, S.: Framework for secure cloud computing. Int. J. Cloud Comput. Serv. Arch. (IJCCSA), **3**(2), 21–35 (2013)
5. Jouinia, M., Rabai, L.B.A.: Comparative study of information security risk assessment models for cloud computing systems In: International Journal of Cloud Applications and Computing (2016)
6. Hale, M., Gamble, R.: SecAgreement: advancing security risk calculations in cloud services. In: Proceedings of 8th IEEE World Congress on Services (2012)
7. Bishop, M.: About penetration testing. IEEE Secur. Priv. **5**(6), 84–87 (2007)
8. Liu, L., Yu, E., Mylopoulos, J.: Security and privacy requirements analysis within a social setting. In: Proceedings of 11th IEEE International Requirements Engineering Conference, pp. 151–161 (2003)
9. Kumar, V., Swetha, M.S., Muneshwara, M.S., Prakash, S.: Cloud computing: towards case study of data security mechanism. Int. J. Adv. Technol. Eng. Res. **2**(4), 1–8 (2012)
10. Behl, A., Behl, K.: An analysis of cloud computing security issues. In: Proceedings of World Congress on Information and Communication Technology, pp. 109–114 (2012)
11. Băsescu, C., Carpen-Amarie, A., Leordeanu, C., Costan, A., Antoniu, G.: Managing data access on clouds: a generic framework for enforcing security policies. In: Proceeding of IEEE International Conference on Advanced Information Networking and Applications (AINA) (2011)
12. Saripalli, P., Walters, B.: QUIRC.: a quantitative impact and risk assessment framework for cloud security. In: Proceedings of the IEEE 3rd International Conference on Cloud Computing (2009)
13. Klems, M., Nimis, J., Tai, S.: A framework for estimating the value of cloud computing. J. Des. E-Bus. Syst. Mark. Serv. Netw. **22**(part 4), 110–123 (2009)

Review of Convolutional Neural Network in Video Classification

Gaurav Talreja$^{(\boxtimes)}$ and Rejo Mathew

Department of IT, Mukesh Patel School of Technology Management
and Engineering, NMIMS (Deemed-to-be) University, Mumbai, India
talrejagaurav98@gmail.com, rejo.mathew@nmims.edu

Abstract. Convolutional Neural Network (CNN), which is a class of deep, Artificial Neural Networks, is most commonly used for image recognition. CNNs have been established as a powerful class for image recognition. But, there hasn't been much research done on the usage of CNNs for video classification. There is very little knowledge on how different CNNs perform for video classification. So, the challenge here is to recognise which CNN architecture is the best for video recognition.

The contribution here is to list down all the different CNN architecture, which were used to classify videos, and compare them with each other based on their performance. The architectures are trained on either the UFC-101 or the sports 1M dataset and their performance is evaluated based on the same. The different groups of CNN architectures which were used for the purpose of video classification are:

- Time Information
- Multiresolution
- Two-stream

Keywords: Convolutional neural networks · Image recognition ·
Video classification · Video recognition · Cnn architecture · Ufc-101

1 Introduction

In recent years, the amount of images and videos on the internet have immensely increased. So there is a need for algorithms which can analyze their semantic content to further improve search, summarization and optimize personal recommendation. CNNs have already been demonstrated as an effective class of models [7] for understanding image content, giving excellent results on image recognition, segmentation, detection and retrieval [4, 8–12].

Currently there is no video classification benchmark which matches the results of the existing image datasets because videos are significantly more difficult to collect, for obvious reasons. So naturally, a question arises: which CNN architecture best utilises and takes advantage of the motion data present in the video? How does motion affect the performance of different CNN architectures? All these questions are answered after evaluating multiple CNN architectures, which have a different approach to process the motion data.

© Springer Nature Switzerland AG 2020
A. P. Pandian et al. (Eds.): ICCBI 2018, LNDECT 31, pp. 942–947, 2020.
https://doi.org/10.1007/978-3-030-24643-3_111

CNNs already require a long amount of time to process and optimize the millions of parameters that are used to parameterize the model and when the videos were taken into consideration, this problem will be further complicated because now the architecture has to work with time as well. To extenuate this issue, the architecture is modified to contain two independent streams of processing. There are two such architectures which modify the architecture to contain two different streams of processing.

In one of those architectures, the architecture will be split into a spatial stream and a temporal stream, results from which will later be fused. In the other architecture, the architecture will be split into a context stream and fovea stream, the results from which will also be later fused.

2 Related Work

The approach to video classification [13] involves three main steps: First, the local visual features of a video will be identified. Next, these features will be bunched into a set-sized video level description. Finally, a classifier is trained to differentiate between the visual classes of interest.

This is where Convolutional Neural Networks come in. CNNs are a biologically inspired class of deep learning models which replace those three stages with a single neural network.

Until recently, CNNs were only applied to relatively small scale images due to the computational constraints, but as the technology has advanced, people have access to really powerful Graphical Processing Units, a.k.a GPUs., and thus these have let CNNs to move up to the network of millions of parameters. This in result has resulted in powerful improvements in image classification, object detection, scene labelling, indoor segmentation, and house number digit classification.

3 Models

Videos are a combination of frames and vary widely in the temporal extent. Thus, videos are not easily able to be processed in a fixed size architecture. In this review, every video is treated as a collection of clips. Each clips contains multiple adjoining frames in time, which extends the connectivity of the network in time domain to learn the spatio-temporal features. The three broad categories which extend the connectivity will be described, in brief, below. Afterwards, the efficiency of the multiresolution architecture is reviewed.

3.1 Time Information Fusion in CNNs [4]

Single-Frame. This is a baseline architecture for this category and this is used for the better understanding of the inputs of static appearance to the classification accuracy. There are multiple convolutional layers with 'd' amount of filters. The final layer is connected to a softmax classifier which gives probabilities for each class label.

Early Fusion. The Early Fusion extension modifies the filters on the first convolutional layer that was present in the single-frame model. This modification helps in combining data throughout an entire time window on a pixel level instantly. This grants the network to explicitly detect local motion direction and speed.

Late Fusion. In the Late Fusion model, two separate single-frame streams are placed at a distance of 15 frames from each other and these two frames are then merged into the first fully connected layer. These two streams have identical parameters. By doing this, none of the two streams can detect any motion, but motion data can be computed by the aforementioned connected layer by analysing the outputs of both separate streams.

Slow Fusion. The Slow Fusion model extends the connectivity of all convolutional layers in time and carries out temporal convolutions, along with spatial convolutions to compute activations. It does so by slowly fusing information throughout the network such that the higher layers would get access to more data in both the space and time dimension. All these models are trained on the Sports-1M dataset. The best model amongst all these is trained on the UFC-101 dataset as well, whose result will be seen later in Sect. 4.

3.2 Multiresolution CNNs

One of the most critical factors while experimenting with different architectures is the runtime performance. Even with the fastest available Graphical Processing Units (GPUs), the processing takes up a huge amount of time. One way to speed up the networks is to minimize the numbers of layers, and neurons in each layers, but due to this, the performance also takes a hit. Instead of reducing the number of layers, the training can be conducted with images of lower resolution [4]. Even though this improves the runtime of the network, there is still a need for images with high frequency in detail to achieve good accuracy.

Fovea and Context stream. A multiresolution architecture in which the processing is divided in to two independent streams over two spatial resolutions is used. The two streams here are the Context stream and the Fovea Stream. The former receives downsampled frames at half the original resolution, while the latter receives the center region of the input, at the original resolution. By doing this, the input dimensionality is effectively halved.

Architecture Changes. Instead of using full sized videos, clips of videos are used. Both the streams will be processed by the same network, and they start at these clips, which are half the spatial size. Thus, the last pooling layer is taken out. The activations from both streams are merged and filled into the first fully connected layer with dense connections. The multiresolution is trained architecture on the Sports-1M dataset and evaluate the results based on that.

3.3   Two-Stream Architecture

Videos can be generally divided into spatial and temporal components. The spatial component contains data about the objects in the video, while the temporal component contains the information about motion of the video.

Spatial Stream ConvNet. The spatial ConvNet works on a single video frame at a time, essentially performing image recognition. The static appearance of the video itself is a clue and can give us a lot of information about the video. Since this ConvNet is essentially performing image recognition, it can be pre-trained on a large image classification dataset [6], like the ImageNet challenge dataset.

Optical Flow ConvNets. The Optical Flow ConvNet forms the temporal recognition stream of this architecture. The pattern of apparent motion caused between two frames due to a movement of camera or an object is called Optical Flow [1]. Optical Flow ConvNets are used to specially process the temporal features of the video. It does so by stacking optical flow displacement fields [2] between multiple consecutive frames. This helps in making the recognition faster as this describes the motion features between video frames explicitly. Many variations of Optical Flow ConvNets can be used which can be achieved by using Optical Flow Stacking, and Trajectory Stacking [5].

Two-stream architecture is used on the UFC-101 dataset and evaluate the result accordingly (Table 1).

Table 1. An overview of the features of all the architectures.

Models	Features
Single-frame	Processes the video frame by frame
Early fusion	Explicitly detects local motion and speed
Late fusion	Motion data is computed by analysing the output of two different streams
Slow fusion	It slowly fuses information throughout the network, and the higher layers get more data in space and time dimension
Multiresolution	Divides the video into two streams, a context stream and a fovea stream to improve the performance
Two-stream	Divides the video into two streams. One stream works on the spatial component while the other works on the temporal component

4   Results

Quantitative Results. The single-frame and the multiresolution models are evaluated on the Sports-1M dataset.

The Single-Frame + Multires Architecture, which is the modified architecture, performed better than the traditional Single Frame architecture. If the Single-Frame architecture is used with just the different streams present in the Multiresolution architecture (i.e the Fovea and Context Stream) separately, there is a drop in performance. So the Single-Frame Fovea Only and Single-Frame Context Only architectures perform worse than the Single-Frame Architecture and the Single-Frame Multires architecture both. The early fusion and the late fusion architectures both perform worse than the Single-Frame architecture, although it is not that significant. However, the slow fusion architecture performs a tad bit better than the Single-Frame architecture if Video Hit is considered. The slow fusion architecture performs the best among the time information fusion architectures. So it becomes a baseline for that category. The slow fusion architecture is trained on the UFC-101 dataset as well. It is observed that the Single-Frame architecture already shows strong performance. The deviations among the different architecture turns out to be insignificant. Looking at the Two-Stream architecture now, which was evaluated on the UFC-101 dataset, the softmax scores were fused using either averaging or SVM. Here, the two-stream models perform better than the Slow Fusion model, which is the baseline for the time information fusion architecture. The two stream model which was fused by SVM performs the best here.

Contributions of Motion. The Slow Fusion, Late Fusion and the Early Fusion architectures are all motion aware architectures. Amongst them, the Slow Fusion architecture performs the best, if compared to the Single-Frame architecture. But the Slow Fusion architecture performed worse than the Two-Stream architecture, which is also a motion-aware architecture. Thus the Two-Stream architecture can be considered as the representative for motion-aware architectures.

Qualitative Analysis. The context stream (i.e the Single-Frame Context Only architecture) learns more of the color features whereas the fovea stream (i.e the Single-Frame Fovea Only architecture) learns high-frequency grayscale filters (Tables 2 and 3).

Table 2. Quantitative analysis of the different CNN architectures which were trained on the Sports-1M dataset.

CNN architecture	Accuracy (Video Hit@1 on the Sports-1M dataset)
Single-frame	59.3
Early fusion	57.7
Late fusion	59.3
Slow fusion	60.9
Single-frame + Multires	60
Single-frame + Fovea	49.9
Single-frame + Context	56

Table 3. Quantitative analysis of different CNN architectures which were trained on the UFC-101 dataset.

CNN architecture	Accuracy (Video Hit@1 on the Sports-1M dataset)
Spatial stream ConcvNet	73.0
Temporal stream ConvNet	83.7
Two-stream model (fusion by averaging)	86.9
Two-stream model (fusion by SVM)	88.0
Slow fusion	65.4

5 Conclusion

The performance of convolutional neural networks in classification of video was studied in this paper. It was found that by using multiresolution architecture, significant improvements to the performance and speed of the network can be made. The results also indicate that a Slow Fusion model persistently performs better than the early and late fusion models. The results also indicate that the networks learn powerful motion features. It was also found that the single-frame model displays strong results, which indicates that local motion cues may not be that important. It was also established the Two-Stream architecture as a representative for the motion-aware architectures.

In future, these results can be used to perform further tests, on different datasets, for video recognition.

References

1. Definition of optical flow on OpenCV.org: https://docs.opencv.org/3.4/d7/d8b/tutorial_py_lucas_kanade.html
2. Simonyan, K., Zisserman, A.: Two stream convolutional networks for action recognition in videos, p. 3 (2014)
3. Convolutional neural networks tutorial in Tensorflow: http://adventuresinmachinelearning.com/convolutional-neural-networks-tutorial-tensorflow/
4. Karpathy, A., et al.: Large scale video classification with convolutional neural networks (2014)
5. Wang, H., Klaser, A., Schmid, C., Liu, C.-L.: Action recognition by dense trajectories. In: Proceedings of CVPR, pp. 3169–3176 (2011)
6. Krizhevsky, A., Sutskever, I., Hinton, G.E.: ImageNet classification with deep convolutional neural networks. In: NIPS, pp. 1106–1114 (2012)
7. LeCun, Y., Bottou, L., Bengio, Y., Haffner, P.: Gradient based learning applied to document recognition. In: Proceedings of the IEEE (1998)
8. Ciresan, D., Giusti, A., Schmidhuber, J., et al.: Deep neural networks segment neuronal membranes in electron microscopy images. In: NIPS (2012)
9. Dollar, P., Rabaud, V., Cottrell, G., Belongie, S.: Behavior recognition via sparse spatio-temporal features. In: International Workshop on Visual Surveillance and Performance Evaluation of Tracking and Surveillance (2005)
10. Farabet, C., Couprie, C., Najman, L., LeCun, Y.: Learning hierarchical features for scene labeling. IEEE Trans. Pattern Anal. Mach. Intell. 35, 1915–1929 (2013)
11. Razavian, A.S., Azizpour, H., Sullivan, J., Carlsson, S.: CNN features off-the-shelf: an astounding baseline for recognition (2014)
12. Sermanet, P., Eigen, D., Zhang, X., Mathieu, M., Fergus, R., LeCun, Y.: OverFeat: integrated recognition, localization and detection using convolutional networks (2013)
13. Wang, H., Ullah, M.M., Klaser, A., Laptev, I., Schmid, C.: Evaluation of local spatio-temporal features for action recognition (2009)

Smart Home Automation Systems

Ritik Hinger[(⊠)] and Rejo Mathew[(⊠)]

Department of IT, Mukesh Patel School of Technology, Management
and Engineering, NMIMS (Deemed-to-be) University, Mumbai, India
hinger.ritik347@gmail.com, rejo.mathew@nmims.edu

Abstract. In today's time the increase in technological advancements have
made people's life easier in aspects of almost everything. Automated systems in
machinery in the industries have been a greatest of them to reduce pressure upon
the people. Smart Homes have helped the human to control every electronic
device with the use of remote controls and smartphones, after successfully
covering the industries these systems have now reached to homes to make living
easy and luxurious. Home automation is a developing technology and has a lot
of scope in the industry where automation systems are installed with self-
learning programs using artificial intelligence. Today there are a lot of different
home automation systems present which works differently. Few of them are
discussed in this paper such as Wi-Fi, GSM, Bluetooth, ZigBee and a comparison is drawn out accordingly to which automation system can be used in the
future.

Keywords: Home automation · Energy conservation · Remotely accessible

1 Introduction

Home automation is a method of controlling the appliances present in the house
remotely. Remotely accessible environment is one in which every appliance present in
the house can be controlled using software application as a junction. There are a lot of
home automation systems which are based on wired communication. Wired communication is not a problem if the entire planning is done at the time of construction of the
property but gets difficult when someone needs to install it elsewhere. Home
automation or Smart homes can be described as the introduction of technology within
home for easy, comfortable, energy efficient and luxurious living [1] for its users. Such
systems are already present in the market and may have some drawbacks in them. The
home automation industry is an evolving industry. This paper will bring off a survey
about all the present home automation systems and compare their features.

Most of the home automation systems in the market today use different wireless
communication [2] techniques to exchange data using Bluetooth, ZigBee, Wi-Fi and
Global System for Mobile Communication (GSM). [3] and interact with the user for
commands.

A. P. Pandian et al. (Eds.): ICCBI 2018, LNDECT 31, pp. 948–952, 2020.
https://doi.org/10.1007/978-3-030-24643-3_112

2   Related Work

2.1   Home Automation System Using GSM

In this, automation systems use global systems for mobile communication to control the devices. This is a text based system [4]. GSM have been used worldwide over a long period of time because of its availability, security and coverage. The control of automation systems is majorly done through SMS [5], by sending attention commands to the respective appliances. The devices send back a message to the user, which makes the interaction of the user [6] and device to be a two way interaction. However this system brings an additional cost for the text messages sent by the users. GSM based automation systems need to have coverage in the devices as it is totally based on SMS service, the commands sent to the appliances may be delayed due to some network issues. There is no user interface which controls the appliances which, sets a drawback for the user for not being able to control or program the devices. The devices do not provide any information to the user and the user has to keep a record of all the devices. One of the drawbacks is that the system requires a computer all the time. The Fig. 1 below shows the GSM system.

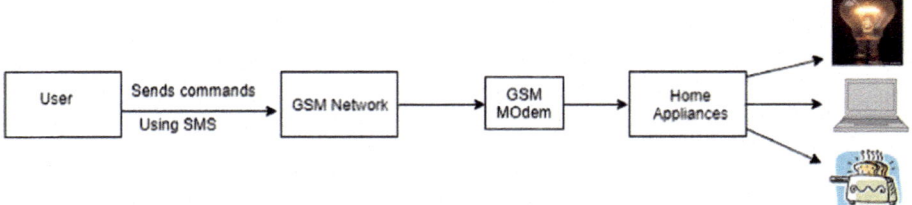

Fig. 1. Communication by GSM systems

2.2   Home Automation System Using Bluetooth

The system uses a mobile phone and the Bluetooth technology. Bluetooth technology is low in cost and is secured. Mobile phones use an interactive interface to communicate with the devices. The Input-Output ports of the Bluetooth software are used to interact with the appliances and operate them. The link between both of the devices and the user is password protected and prevents intruder to access the data. The Bluetooth have a bandwidth of 2.4 GHz, 10 meters of range and transfer rate up to 3 Mbps. This system is fast and economical. It provides feedback to the user and therefore is interactive. It is also capable of detecting any faults in the circuitry. Few of the drawbacks for Bluetooth is the time taken for it to detect and connect to any devices nearby. Also it is confined within its range and will not be helpful if the user gets out of that range. It doesn't provide energy conservation tips and do not have real time access.

2.3   Home Automation Using Phones

Phone based home automation system is an enabling system used to provide a simple framework. The system uses telephone lines for communication [7]. It provides a system for a smart home which provides facilities such as, house-wide wiring, system controller and a common interface. In this type of automation, the system uses a Dual Tone Multi Frequency (DTMF) in the telephone lines. The major drawback of the system is that the number of connections is limited to the number or keys present on the keypad. There are generally 12 keys so there can be 12 devices connected to it.

2.4   Home Automation System Using ZigBee

A wireless communication [8] technology like ZigBee can be used for home automation. It uses voice recognition technology [9] and PIC microcontroller [10]. The voice commands are taken in from the user via mic which is installed on the device and compared to the pre-programmed commands which are stored in the memory and then processed. The system makes use of relays to control the appliances. One of the major setbacks to this type of automation system is the low range of communication which prevents user to access the devices from any remote location outside its range. Also the speech detection sometimes fails to understand the commands given.

2.5   Wireless Automation Systems

Automation systems which use wireless control can be formed by combining all the independent devices to a single co-operative network. This system is made using a combination of Wi-Fi and Bluetooth. The system provides a transparent working of the devices to the user. The appliances are connected to each other and the user using different networking technologies. The system also consists of advanced features like device discovery and connection. The entire interface is designed using Linux platform. The global plug and play system uses many standard protocols for interoperability which is the main advantage of the system. Device discovery is also an advantage for the wireless communication system as it has the ability to share the services.

3   Analysis/Review

(Table 1)

Table 1. Comparison of different home automation systems

System	GSM based	Bluetooth based	Phone based	ZigBee based	Wireless control systems
Primary communication	Text messages, SMS	Bluetooth and AT commands [7]	Phone Lines	ZigBee and AT commands [7]	Radio, infrared or other waves

(*continued*)

Table 1. (*continued*)

System	GSM based	Bluetooth based	Phone based	ZigBee based	Wireless control systems
Real time	No	Yes	No	Yes	Yes
Remote access	Access from anywhere in the world	Restricted to its range of 10 meters	Anywhere with a phone line	Around 10 meters	Depends on the range of spectrum being used
Number of devices	Unlimited	Unlimited	12 due to 12 frequencies DTMF	Unlimited	Unlimited
Speed	Slow due to delivery issues	Fast due to proximity	Fast	Fast	Slow due to interferences
Controller	PIC16F887 microcontroller	Arduino	Microcontroller	Smart Socket	Arduino
User interface	SMS	Bluetooth	Mobile phone	PC or Android phone	Web based application

4 Conclusion and Future Work

Based on all the systems reviewed and the disadvantages and advantages stated, this paper cites the features which an ideal home automation system should have. A home automation system should be accessible to the user everywhere across the globe and should be in real time. Only GSM network provides worldwide connectivity to the user. However GSM networks still need to be worked upon to adopt the internet, data services to operate the automation system remotely from anywhere around the world [11]. The user should have an application to make the use easier and comfortable. An interface should be designed which can be used by all kinds of people [12], only then the home automation system can be of great help to the people.

Future scopes in home automation involves, making homes even smarter and making people's life more luxurious. Artificial Intelligence can be used along with the automation systems to make self-learning programs which learn continuously and adapt with the user's lifestyle. Artificial Intelligence provides opportunities to meet these needs and wishes by intelligent technologies that interact with the user [13]. Energy consumption can be set, based on daily usage. The automation system can also include security systems controlling all sort of cameras, doors, windows etc. The next step of the home automation system can be taking it to higher levels where everything in the house is automated.

References

1. Van Der Werff, M., Gui, X., Xu, W.L.: A Mobile based home automation system, applications and systems. In: 2nd International Conference on Mobile Technology, Guangzhou, p. 5 (2005)
2. Vaidya, B., Patel, A., Panchal, A., Mehta, R., Mehta, K., Vaghasiya, P.: Smart home automation with a unique door monitoring system for old age people using Python, OpenCV, Android and Raspberry pi. In: International Conference on Intelligent Computing and Control Systems (ICICCS) (2017)
3. Dhakar R.: Comparison of various technologies for home automation system. In: Electrical Engineering, vol. IV, issue V. Institute of Technology, Nirma University, Ahmedabad, Gujarat, India (2017)
4. Jivani, N.: Mahesh: GSM based home automation system using app-inventor for android mobile Phone. Int. J. Adv. Res. Electr., Electron. Instrum. Eng. 3(9), 12121–12128 (2014)
5. Alheraish, A.: Design and implementation of home automation system. IEEE Trans. Consum. Electron. 50(4), 1087–1092 (2004)
6. Yuksekkaya, B., Kayalar, A.A., Tosun, M.B., Ozcan, M.K., Alkar, A.Z.: A GSM, internet and speech controlled wireless interactive home automation system (2006)
7. Palaniappan, S., Hariharan, N., Kesh, N.T., Vidhyalakshimi, S.: Angel Deborah S Assistant Prof.,: Home automation systems - a study. In: CSE, SSN College of Engineering, Anna University, Chennai, India Transactions on Consumer Electronics, vol. 52(3), pp. 837–843; 116(11) (2015)
8. Hwang, K., Baek, J.W.: Wireless access monitoring and control system based on digital door lock. IEEE Trans. Consum. Electron. 53(4), 1724–1730 (2007)
9. Baig, F., Baig, S., Khan, M.F.: Controlling home appliance remotely through voice command. Int. J. Comput. Appl. 48(17), 1 – 5 (2012)
10. Liang, N.S., Fu, L.C., Wu, C.L.: An integrated, flexible and Internet based control architecture for home automation system in the internet era. In: IEEE International Conference on Robotics and Automation, vol. 2, pp. 1101–1106 (2002)
11. Deore, R.K., Sonawane, V.R., Satpute, P.H.: Internet of thing based home appliances control. In: International Conference on Computational Intelligence and Communication Networks (CICN) (2015)
12. Javale, D., Mohsin, M., Nandanwar, S., Shingate, M.: Home automation and security system using android ADK. Int. J. Electron. Commun. Comput. Technol. (IJECCT) 3(2), 382–385 (2013)
13. Gottfried Zimmermann Responsive Media Experience Research Group Stuttgart Media University Stuttgart, Germany gzimmermann@acm.org, Tobias Ableitner Responsive Media Experience Research Group Stuttgart Media University Stuttgart, Germany ableitner@hdm-stuttgart.de, Christophe Strobbe Responsive Media Experience Research Group Stuttgart Media University Stuttgart, Germany strobbe@hdm-stuttgart.de

A Survey on Attacks of Bitcoin

Janhvi Joshi[✉] and Rejo Mathew

Department of IT, Mukesh Patel School of Technology Management
and Engineering, NMIMS Deemed-to-be University, Mumbai, India
janhvi.joshi4@gmail.com, rejo.mathew@nmims.edu

Abstract. Bitcoin is a crypto-currency. Bitcoin economy has been gaining immense popularity. Bitcoin mining involves putting new coins into circulation after creating them. Miners utilise electricity on solving cryptographic puzzles. The miners give a proof of bitcoin transactions of the other people. Miners are assumed, to be honest, and have motives to behave well.

This paper looks at the strategies miners use and bring notice to trouble-making and deceitful strategies or the ones which could affect the bitcoin and the security of bitcoin. This paper studies about several recent attacks in which dishonest miners gain an upper reward compares to their corresponding contribution to the network. This paper describes Double Spending Attack, Finney Attack, Vector 76 Attack, >50% Hash-Power Attack, Selfish Mining Attack in detail.

Keywords: Bitcoin · Cryptocurrency · Security threats · Attacks ·
Double spending · Finney · Vector 76 · >50% hash-power · Selfish mining

1 Introduction

Bitcoin is a decentralised electronic payment system based on cryptography, that is, the network which is functioned by the people who use it for making payments among themselves [4, 8]. Bitcoin economy has been growing at an incredibly fast rate [12]. It uses a peer to peer technology. In Bitcoin the owner is completely responsible and has the authority of their own coins. Below the explanation of the working of Bitcoin is provided for the motive of this paper.

The maintenance of a public ledger for all transactions is done in order to keep an account of payments done with units called Bitcoins. When the Bitcoins are created, they are assigned to any of the network node prepared and will enable to spend adequate computing power in order to solve difficult cryptographic puzzles. Proof of Work (PoW) is when the miner's workout cryptographic puzzles to "mine" a block so that it can be added to the blockchain. To every winner, 25 BTC are assigned at an average of every 10 min. This number reduces with time.

Miners, again and again, put together a double SHA-256 hash H2 of a defined data structure being a block header. The winner is decided by a majority vote. Digital signatures enable the possession of any part of the currency. By defining his public key or its hash each obtained quantity of bitcoin recognises its owner. The power to pass on this quantity of bitcoins to the other participants is performed just by the owner of the

© Springer Nature Switzerland AG 2020
A. P. Pandian et al. (Eds.): ICCBI 2018, LNDECT 31, pp. 953–959, 2020.
https://doi.org/10.1007/978-3-030-24643-3_113

relative private key. Transaction fee is defined as the difference between the sum of the inputs and the sum of all output amounts.

The winning miner is the one who would include this transaction block. That miner pays the transaction fee. Every transaction is agreed by every owner of each input number of bitcoins by a separate digital signature agreeing to the passing on to the new owners this money. The blocks build a blockchain. In case of a conflict, an avalanche process is put to use wherein the major part of votes decide the validity of the transaction. This known as the consensus protocol [5]. In the bitcoin network, the propagation is fast: For a node to receive a block, the median time is 6.5 s while the average time being 12.6 s [9].

2   Attacks on Bitcoin

2.1   Double-Spending Attack

Suppose, that the system compromises of an evil-intentioned client C, and a vendor V these are associated through a Bitcoin network [3]. C will try and double-spend the coin she had already passed on to V.

Inorder to perform double spending, C creates two transactions TR_C and TR_V which have the same BTCs though the recipient address of TR_V will have the address of V and the recipient address of TR_C will have the address of C. The attacker C will have to convince the vendor V into accepting a transaction TR_V that V will not be able to redeem later on.

Consider that the major number of the peers in the network receive TR_C, then TR_C has greater chances of being added in a succeeding block but V receives TRv then there is a chance that double spending attack succeeds. Let t_i^V and t_i^C be the time at which node i receives TR_V and TR_C, respectively. Let t_v^V and t_v^C denote the respective times at which V receives TR_V and TR_C. The following steps should meet take place that C can perform a successful double-spending attack.

Step 1—TR_V is added to the wallet of V.
V will not be able to determine if TR_V was broadcasted in the network if it does not get added to the memory pool of V. It is important that $t^V < t^C$ so that TR_V is included in V's wallet; otherwise, V will be adding TR_C in the first place to its memory pool. TR_V is rejected as it arrives later.
Step 2—TR_C gets confirmed to the blockchain.
TR_C cannot appear in succeeding blocks if TR_V is the first to get confirmed in the blockchain. This means that V will not have its BTCs again. In this situation, V will be receiving its BTCs and can utilise the BTCs.
Step 3—V's service time is smaller than the time it takes V to detect misbehaviour. In order for V to successfully detect any misbehaviour by C, the detection time must be smaller than the service time. This is because: The Bitcoin users hold many accounts and are anonymous, there is only limited value in V detecting misbehaviour of C after the user, C has obtained the service as it is very difficult to utilise the Bitcoin address of C is insufficient to identify her.

2.2 Finney Attack

A Finney attack can be successful in two cases: In-case the merchant associated in the trading starts accepting unconfirmed transactions otherwise when the trader waits until a few seconds in order to make sure that everyone agrees that he has paid indeed in the network [4]. The following steps should take place such that a successful Finney attack can take place:

Step 1—The attacker acknowledges the person.
The attacker acknowledges the one he is trying to take advantage of and the one who is going to credit himself.
Step 2—The attacker starts the mining of the second block not using the first one.
The attacker does not use the first transaction. The attacker starts the mining of the second block.
Step 3—The stored first transaction Is used to make purchases once mining of the second block is successful.
A certain amount of time is required for the mining of the second block task and the moment he is successful, he uses the stored first transaction to purchase the good he wanted.
Step 4—The attacker releases the pre-mined block.
The attacker will release the pre-mined block. The invalidation of the former transaction happens even when it shows up in the block table.

2.3 Vector 76 Pre-mining Attack

The Vector 76 attack can take place when the victim will not be able to broadcast blocks to the main network [2, 5]. This type of attack outlines a clear differentiation between full nodes and light node implementations which do not broadcast blocks. The following steps should take place such that a successful vector 76 attack can take place:

Step 1—The attacker has started working on a secret branch of the chain.
The attacker has started working on a secret branch of the chain. Consider the attacker has started a task on a secret branch of the chain. It inserts the transaction tx_1, which the attacker later wants to reverse in this block.
Step 2—Creation of an additional $k - 1$ blocks on top of the one containing tx_1.
In-case a total number of k confirmations. If the defender insists on $\sigma \equiv k$ confirmations. The attacker has to secretly build an additional $k - 1$ blocks on top of the one containing tx_1.
Step 3—The client accepts the attacker's branch of the chain.
If k confirmations are shown to the lightweight client when the honest chain is shorter than the attacker's branch of the chain and if the attacker manages to gain k confirmations for transaction, tx_1, then the client accepts it as the legitimate chain, it being the longest one.
Step 4—The attacker transfers a conflicting transaction tx_2 to the honest network.
The honest network is not informed about the attacker's chain, therefore the attacker transfers a contradictory transaction tx_2 to the honest network. Then the former's

chain will grow long enough for tx_2 to be accepted by all nodes. The honest network would not accept the block containing tx_1.

Inorder to achieve a successful attack, against a similar k-confirmation defender, the network has build up k blocks on top of the transaction.

2.4   >50% Hash-Power Attack

An attacker can control, keep out and alter the ordering of transactions until the time he has control of more than 50% of the network's computing power [1, 7].

The following can take place by a successful attack:

1. When the attacker is in control he can reverse transactions that he sends.
2. He can prevent a few or all transactions from gaining any confirmations.
3. He can prevent some or all other generators from getting any generations.

The following cannot take place by the attacker:

1. He cannot reverse other people's transactions.
2. He cannot prevent transactions from being sent at all.
3. He cannot change the number of coins generated per block.
4. He cannot create coins out of thin air.
5. He cannot send coins that never belonged to him.

It is difficult to change historical blocks. It becomes more difficult the further back you go. It is assumed that no one will attempt this attack since this attack doesn't allow all that much power over the network.

2.5   Selfish Mining Attack

In a selfish mining attack, an attacker keeps its located blocks private and then purposely forks the blockchain [7]. This attack would thus weaken the decentralised nature of the system by leading to a further consolidation of mining power in the attacker's favour [10, 11]. The following steps should take place such that a successful selfish mining attack can take place:

Step 1—Keeping mined blocks private.
The selfish mining pool keeps on mining its blocks private, secretly dividing the blockchain. It keeps on creating a private branch.
Step 2—The honest miners keep on mining.
The honest miners keep on with mining on the shorter public branch.
Step 3—The private branch will not be ahead of the public branch.
The selfish miners command a relatively small portion of the total mining power, their private branch will not be ahead of the public branch for always.
Step 4—Revealing of blocks from the private branch to the public.
Selfish mining carefully discloses blocks from the private branch to the public.
Step 5—Honest miners will switch to the recently disclosed blocks.
The honest miners will switch to the recently disclosed blocks, leaving from the shorter public branch.

Step 6—Previous effort spent on the shorter public branch gets wasted.
The previous effort spent on the shorter public branch gets wasted.
Step 7—The selfish pool collects higher revenues.
The selfish pool collects higher revenues by including a higher fraction of its blocks into the blockchain.

3 Comparison Analysis

Table 1.

Table 1. Analysis of the attacks on bitcoin.

Attacks	Description	Steps	Features
Double-Spending Attack [2]	The attacker will try to double-spend the coin she had already transferred to the vendor inorder to achieve a different service elsewhere	Step 1—TR_V is added to the wallet of V Step 2—TR_C is confirmed in the block chain Step 3—V's service time is smaller than the time it takes V to detect misbehaviour	The attacker can trick the vendor into accepting a transaction that it will not be able to redeem subsequently
Finney Attack [4]	The attack can be successful in two cases: If the merchant associated in the trading accepts unconfirmed transactions OR If the trader waits for a few seconds to ensure that everyone in the network agrees that he has actually paid	Step 1—The attacker acknowledges the person Step 2—The attacker starts the mining of the second block not using the first one Step 3—The stored first transaction Is used to make purchases once mining of the second block is successful Step 4—The attacker releases the pre-mined block	The attacker who is making use of this technique makes two transactions, it is thus a variant of double spending It can only be successful when the attacker mines and keeps control of the contents of the block
Vector 76 Attack [2, 5]	The attack can take place when the victim will not be able to broadcast blocks to the main network This is worked upon by the attacker in secret until he becomes successful	Step 1—The Attacker has started working on a secret branch of the chain Step 2—Creation of an additional $k - 1$ blocks on top of the one containing tx_1 Step 3—The client accepts the attacker's branch of the chain	The attack is a form of pre-mining attack It is easier here to produce less of a lead on the honest network, as it depends upon the network to confirm the double spending payment

(continued)

Table 1. (*continued*)

Attacks	Description	Steps	Features
		Step 4—The attacker transmits a conflicting transaction tx_2 to the honest network	
>50% Hash-Power Attack [1, 7]	This attack can be successful when any attacker that controls more than 50% of the overall network's computing power can control, exclude and modify the ordering of transactions	Step 1—The attacker can reverse transactions that he sends when in control Step 2—He can Prevent some or all transactions from gaining any confirmations Step 3—He can Prevent some or all other generators from getting any generations	The attacker dominates more than 50% of the network's computing power
Selfish Mining Attack [7]	In a selfish mining attack an attacker keeps its located blocks private and then purposely forks the blockchain	Step 1—Keeping mined blocks private the honest miners keep on mining Step 2—The private branch will not be ahead of the public branch Step 3—Revealing of blocks from the private branch to the public Step 4—Honest miners will switch to the recently disclosed blocks Step 5—Previous effort spent on the shorter public branch gets wasted Step 6—The selfish pool collects higher revenues	This attack weakens the decentralised nature of the system by leading to a further consolidation of mining power in the attacker's favour

4 Conclusion and Future Work

In this paper, we looked at a number of different attacks by which some attackers in the bitcoin digital currency can aim to increase their gains while the other miners suffer because of the attacks, such as Double Spending, Finney, Vector 76, >50% Hash-Power, Selfish Mining Attack. The attacks are successful only when certain conditions are fulfilled. This paper specifies the conditions as well as the steps in which these attacks can be successful. Several attacks are just a variant of other attacks which is specified in the paper.

Miners not obeying the protocol, take a chance that their blocks will be turned down by the majority of other network participants and they will not obtain the reward. In this paper, it is shown, that it is not easy to perform the attacks. This paper leads to the questioning of the security of the bitcoin system. It is important that an electronic payment system which is gaining such popularity guarantees against such attacks. It is crucial to find solutions to prevent the attacks on the bitcoin system. This paper raises the collection of future research directions and open questions. Hopefully, this paper will inspire fledgeling researchers towards taking measures about the security issues of Bitcoin systems.

References

1. Joshua, A.K., Davey, I.C., Felten, E.W.: The economics of bitcoin mining, or bitcoin in the presence of adversaries
2. Sompolinsky, Y., Zohar, A.: Bitcoin's security model revisited. In: ArXiv (2016)
3. Karame, G.O., Androulaki, E., Čapkun, S.: Double-spending fast payments in bitcoin. In: CCS '12 Proceedings of the 2012 ACM Conference on Computer and Communications Security (2012)
4. Mukherjee, D., Katragadda, J., Gazula, Y.: Security analysis of bitcoin
5. Tschorsch, F., Scheuermann, B.: Bitcoin and beyond: a technical survey on decentralized digital currencies. In: IEEE Communications Surveys & Tutorials (2015)
6. Conti, M., Sandeep Kumar, E., Lal, C., Ruj, S.: A survey on security and privacy issues of bitcoin. In: IEEE Communications Surveys and Tutorials (2018)
7. Narayanan, A.: The stateless currency and the state: an examination of the feasibility of a state attack on bitcoin
8. Courtois, N., Grajek, M., Naik, R.: The unreasonable fundamental incertitudes behind bitcoin mining. In: ArXiv (2013)
9. Decker, C., Wattenhofer, R.: Information propagation in the bitcoin network. In: IEEE P2P 2013 Proceedings (2013)
10. Eyal, I., Sirer, E.G.: Majority is not enough: bitcoin mining is vulnerable. In: ArXiv (2013)
11. Nakamoto, S.: Bitcoin: a peer-to-peer electronic cash system
12. Bitcoin – Wikipedia. https://en.bitcoin.it/wiki/Introduction

Enotes: Note Taking Tool for Online Learners

R. Sujatha[1(✉)], R. Vijayalakshmi[2], Surya Santhosh Prabhakar[3],
and Prabhakar Krishnamoorthy[4]

[1] School of Information Technology and Engineering,
Vellore Institute of Technology, Vellore, Tamil Nadu, India
r.sujatha@vit.ac.in
[2] Georgia Institute of Technology, Atlanta, USA
vijayalakshmi3@gatech.edu
[3] Georgia State of University, Atlanta, USA
suryaprabhakar@graduate.com
[4] AT&T, Atlanta, USA
pkkrishna@yahoo.com

Abstract. ENotes is a note-taking tool specifically targeted towards online learners. Note-taking in general help students to have a better understanding of what they learn. It also helps the students to improve their writing skills. There are many sophisticated and simple note-taking tools available on the market. Sophisticated tools like Evernotes and OneNote need extra learning for setup and some functionalities are only available as micro transactions. Simple tools like Sticky notes and SimpleApp don't support some basic functionalities required for taking notes such as rich text formatting. ENotes combines the features from both these sophisticated and simple tools to meet the needs of the online learners. This tool is built using Chrome extensions technology in pure JavaScript. It involves no cost, no complex setup, and is user friendly. It covers the basic features like rich text formatting and math formula editor, download, print, search, tool access time, dictionary, speech to text conversion and translations.

Keywords: Note-taking · Open source · Cognitive psychology · Linguistics ·
Working memory · Online learners

1 Introduction

ENotes is a simple and user friendly note-taking tool for online learners. This tool is a Chrome extension which can be installed easily. The prime functionality for any note-taking tool is to provide a good editor and to save the editor content. ENotes provides rich text formatting and math formula editor. It has the ability to save notes to a file, search for specific content, and also to send it to the printer. All these functionalities are implemented in pure JavaScript. Since it is a completely client side programming, it provides better performance. ENotes friendly user interface is shown in Fig. 1. Note-taking helps both in learning and writing. These are the characteristics of "Writing Across the Curriculum" which was a pedagogical movement started in the 1980s. It plays a very vital role in improving cognitive psychology, linguistics, and the teaching

© Springer Nature Switzerland AG 2020
A. P. Pandian et al. (Eds.): ICCBI 2018, LNDECT 31, pp. 960–967, 2020.
https://doi.org/10.1007/978-3-030-24643-3_114

of science. While learning, students grasp the content and take it in simple short format which involves cognitive skills like concentrating, memorizing and recalling. Here the short term memory or working memory plays a major role. This also improves linguistics to understand and reproduce the content. Notes are not only useful for learner, but for teachers for teaching their lessons. Besides learning, note-taking are important for researchers to collect and organize their research information.

Fig. 1. ENotes

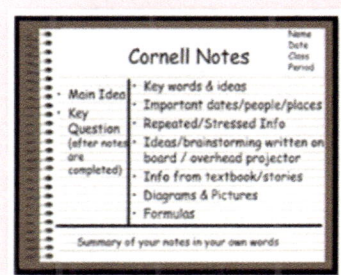

Fig. 2. The Cornell note taking system

It may be tedious task but will be very helpful to recall the information when needed anytime. To make this tedious task easier, there are different methodologies to take notes. Cornell note-taking methodology provide simple layout as shown in Fig. 2 which helps to gather notes efficiently. ENotes uses some of the user interface components from this layout, which will be detailed detail in the design section [1–3].

Fig. 3. Sticky notes

Fig. 4. Simplenote

Knowing the importance of taking notes and don't see any standards to follow, planned to create a tool to support this functionality which will be helpful for learners.

2   Comparisons

Compared ENotes with simple note taking tools like sticky notes (Fig. 3) and simple note (Fig. 4) and also with sophisticated tools like onenote (Fig. 5) and evernote (Fig. 6) popular tools available in the market. Compared and summarized features in Table 1 and benefits in Table 2 of these tools [4–9].

Fig. 5. Onenote   **Fig. 6.** Evernote

Table 1. Feature comparison

Features	Sophisticated applications	Simple applications
Rich text format	Over engineered	Minimum support
Translation	Whole app	Not supported
Print	Supported	Supported in some
Access time	Over engineered	Not supported
Speech to text	Supported in some	Not supported

Table 2. Benefit comparison

Benefits	Sophisticated applications	Simple applications
Setup	Involves learning	Simple
Cost	Depends on feature	No cost
Access	Anywhere	Some anywhere
Ease of use	Involves learning	Simple
Features	Over engineered	Limited

By reviewing the existing tools, compiled the following features that will aid in taking notes.

- Provide rich text editor to type the notes
- Provide math formula editor (currently just gets input without save/clear functionality)
- Display tool access date and time
- Save notes in the editor when the user switches between the tabs (auto save)
- Save notes to the local file
- Clear the notes editor

- Search for the text in the notes editor
- Print the text in the notes editor
- Dictionary lookup
- Speech to text conversion
- Translation

Based on the tool comparison study found that optimization is possible. Chosen to implement ENotes that is going to make a big difference and more successful tool among online learners.

ENotes supports all these basic features with no cost, easy setup, no additional leaning, easy access and independent of any external server or application interfaces.

3 Design

3.1 User Interface

The main design goal is to provide simple, user friendly, robust note-taking tool for online learners with basic note-taking features. As shown in Fig. 1, ENotes provides the simple interface. It has space for editor with scrollbars and buttons to do their functionalities. It gets input required for functionalities using regular html text [10]

3.2 Technology

Since Chrome browser is widely used, chose to develop this tool using Chrome extensions technology. This ENotes Chrome extensions and its flow is shown in Figs. 7 and 8. It is developed completely using web technologies like HTML, CSS and JavaScripts. It does not use any server or external application interfaces. So it provides better performance [11, 14].

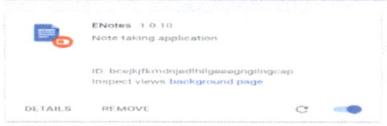

Fig. 7. ENotes chrome extensions

Fig. 8. Chrome extensions flow

A chrome extension contains three environments. They are extensions, background and user environment. In the extensions environment, all the actions take place within the extensions area. Popup.html and popup.js forms this extensions area. The user environment is the browser content environment. Contenscript.js controls the browser environment. Since both evil and good actions can be accomplished using this environment, need to handle this environment wisely and carefully. Background environment is the isolated environment. This environment has access to all Chrome extensions interfaces or API. These environments communicate using messaging and event listeners.

ENotes is heavily based on extension environment. Most of the functionalities are handled by popup,js and popup,html (which is the user interface as shown in Fig. 1). It uses background environment for file handling. Currently dictionary and translation are in the extensions environment. Prefer to move them to the user environment that will be more helpful, while using this environment.

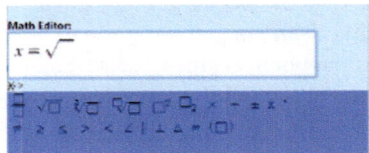

Fig. 9. Text editor using NicEdit   **Fig. 10.** Math formula editor using math

4   Functionalities

4.1   Rich Text Formatting Editor

ENotes editor provides rich text formatting using NicEdit (Fig. 9). NicEdit is extremely lightweight and comes in JavaScripts. So, it can be easily integrated with minimal impact. Rich text formatting like setting font style, alignment, font size, font type, linking to images, changing background and text color are proven and got good feedbacks [12].

4.2   Math Formula Editor

It is really hard to develop math formula editor in JavaScript. I am amazed to see mathquill (Fig. 10) open source formula editor. It is widely used now in educational software. It provides the text box and symbols and so integrated with ENotes [13].

    Enotes tool bar set is shown in Fig. 11. Following contents describe them in detail.

4.3   Tool Access Time

ENotes uses JavaScript Date function to get and format current date and timestamp as show in Fig. 12. It is supporting the format as shown in the screen shot. This can be enhanced to preform display format customization.

4.4   Auto Save

ENotes implements auto-save using chrome storage, it stores and retrieves the editor content when user moves in and out of the ENotes extension. This storage is supported by adding storage to permissions settings in manifest.json.

4.5 Save or Download

ENotes uses document. createElement to create document to download or save to the user local file system. User can specify the file name with html file type to store the content. Else it will use default_name as shown in Fig. 13.

4.6 Search

ENotes uses JavaScript match function to search function as shown in Fig. 14. This can be enhanced further using regex patterns. It shows popup box whether it is able to find the search text in the notes content or not.

4.7 Clear

ENotes uses JavaScript setContent function to empty string to clear the content (Fig. 15).

4.8 Print

ENotes uses window.Print function to print (Fig. 16) the contents of notes editor. It adds the default date and text while printing. It popups the printer selection screen to configure the printer setup. Print dialog is shown in Fig. 17.

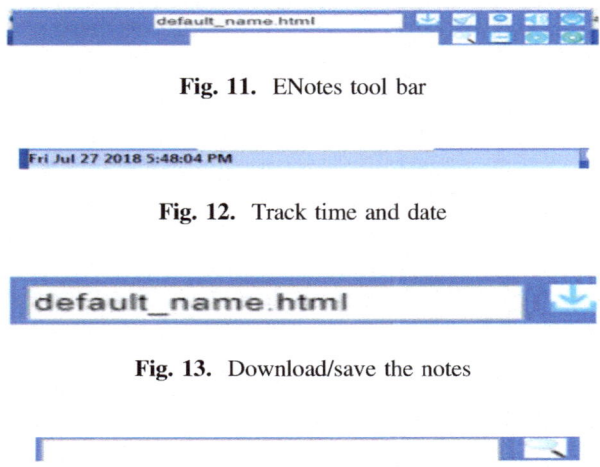

Fig. 11. ENotes tool bar

Fig. 12. Track time and date

Fig. 13. Download/save the notes

Fig. 14. Search the notes

Fig. 15. Clear the notes   **Fig. 16.** Print the notes

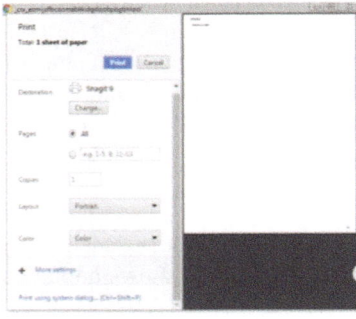

Fig. 17. Print dialog

Fig. 18. Speech to notes   **Fig. 19.** Dictionary

Fig. 20. Notes to speech   **Fig. 21.** Translate

4.9   Speech to Text Conversion

ENotes uses window speech recognition function to convert the speech to note content (Fig. 18). Able to start the speech recognition. But chrome blocks so not able to test the full implementation.

4.10   Dictionary Lookup

ENotes stores dictionary words as (key, value) pairs. Since this is for online classes, dictionary words can be stored locally in a file. Then JavaScript can load this file and implement logic to lookup synonym. ENotes has simple logic storing the dictionary words with in the script, but can be enhanced using better algorithm. Figure 19 represents the dictionary icon.

4.11   Text to Speech Conversion

Enotes uses window's speech synthesis function to convert the speech to text content (Fig. 20). This works pretty good.

4.12   Translation

json format is used to store the translation words. Then using JavaScript function to retrieve them based on the language and content. This can be enhanced like the translation JavaScript available in market such as jQuery.I18n, Globalize and Polyglot. Figure 21 shows the translate icon.

5   Conclusion

Main design issue came across is the security issue that restricts to implement functionalities like email. Since email makes TCP socket connection, need to go through server side programming. So sharing capability is reduced because of this security limit. This can be overcome by researching more on node.js, services and hosting less weight server. It should be bundled in such a way user should not have any impact with setup or access. ENotes is a pilot project. All development and testing in Chrome extension developer mode. Laid the foundation for ENotes tool that fulfills my goal to provide no cost, no setup and feature rich note-taking tool. This can be open sourced so that both design and code can be enhanced to make it more powerful and available to all online users. ENotes debugged using Chrome developer tools and JavaScript alert messages.

References

1. Boch, F., Piolat, A.: Note taking and learning a summary of research. In Writing (2005)
2. Segal, T., Biran, L., Feldman, G., Krikheli, G.: Tegrity Inc system and method for off-line synchronized capturing and reviewing notes and presentations. In: U.S. Patent Application 10/977,257 (2005)
3. Donohoo, J.: Learning how to learn: Cornell notes as an example. J. Adolesc. & Adult Lit. **54**(3) 224–227 (2010)
4. Mueller, P.A., Oppenheimer, D.M.: Technology and note-taking in the classroom, boardroom, hospital room, and courtroom. Trends Neurosci. Educ. **5**(3), 139–145 (2016)
5. van Arnhem, J.P.: Unpacking Evernote: apps for note-taking and a repository for note-keeping. Charlest. Advis. **15**(1), 55–57 (2013)
6. EverNote review. https://www.getapp.com/collaboration-software/a/evernote/reviews/
7. Oldenburg, M.: Excerpts of Using Microsoft OneNote 2010. 6 Sep 2011
8. OneNote review. https://www.pcmag.com/review/346441/microsoft-onenote-web, https://www.slant.co/topics/697/viewpoints/16/ ∼ best-cross-platform-note-taking-app ∼ microsoft-onenote
9. SimpleNote review. http://gettinggeek.com/simplenote-review/, http://www.literatureandlatte.com/forum/viewtopic.php?t=49621
10. Schwartz, M.: Guidelines for Bias-Free Writing. Indiana University Press, 601 N. Morton, Bloomington, IN 47404 (1995)
11. Chrome overview. https://developer.chrome.com/extensions/overview
12. Rich text editor NicEdit. http://nicedit.com
13. Math formula editor. http://mathquill.com/
14. Getting started tutorial. https://developer.chrome.com/extensions/getstarted

Biometric Passport Security Using Encrypted Biometric Data Encoded in the HCC2D Code

Ziaul Haque Choudhury[1]([⊠]) and M. Munir Ahamed Rabbani[2]

[1] Department of Information Technology, B.S. Abdur Rahman Crescent Institute of Science and Technology (Deemed to be University), Chennai, India
ziaulms@gmail.com
[2] School of Computer, Information and Mathematical Sciences,
B.S. Abdur Rahman Crescent Institute of Science and Technology
(Deemed to be University), Chennai, India
rabbani3012@gmail.com

Abstract. In this paper, proposed a biometric encryption in the HCC2D (High capacity 2D barcode) code for national security. The proposed biometric encryption technique is using the AES (Advanced Encryption Standard) algorithm and encoded into the HCC2D code. This technique will enhance the safety and security for a biometric passport from unauthorized access without knowing the passport holder. The proposed method will provide more security for the border crossing and illegal immigrants.

Keywords: Face recognition · Soft biometrics ·
Encrypted data and biometric data in the HCC2D code · e-Passport

1 Introduction

A biometric passport is electronic travel archives that comprise biometric data of the passport bearer that can be naturally perused and handled by a PC validate the citizenship of a person. E-Passport makes utilization of two advances: Radio Frequency Identification (RFID) and Biometrics. Today RFID innovation is progressively being utilized as an anti- forging device in different application territories, for example, stock chains, individual recognizable proof, and access control, financially related accreditations, sensor systems, and so on. Around 20 years back, an authority of the US national government had assessed that among the then 3 million US passport applications got each day in any event 30,000–60,000 identifications are fake and just 1,000 deceitful travel papers have been recognized. As indicated by the reports, 80% illicit drug dealers and an extensive number of militants are supported with phony passports and visas and travel unreservedly everywhere throughout the world [1]. So as to enhance the respectability of the passport, the different nations have been issuing identifications including an RFID chip that contains the passport holder's personal data. Malaysia was the primary nation on the planet to issue e-passport in 1998, and to date, it has been executed in excess of 45 nations.

These Machine Readable Travel Documents (MRTDs) were presented by the International Civil Aviation Organization (ICAO) in its Document 9303 which gives a

© Springer Nature Switzerland AG 2020
A. P. Pandian et al. (Eds.): ICCBI 2018, LNDECT 31, pp. 968–975, 2020.
https://doi.org/10.1007/978-3-030-24643-3_115

lot of tenets and measures for a biometric passport. A biometric passport is the international identification travel document which contains biometric data to replace the conventional passport. This technique contains hybrid documents (i.e., paper, antenna, and embedded chip) that allow wireless communication with the special reader in the border control to authenticate a traveler. The EAC (Extended Access Control) mechanism [2] illustrated the authentication procedure between the e-passport, and an IS (Inspection Terminal).

Since the e-passport or biometric passports are incorporated RFID technology, an access control mechanism is requiring privacy protection. In the existing work found various kinds of security issues on RFID technology. There is a number of threat scenarios are found in the literature which is relevant for the issuance of travel documents in [3]. The relevant issues for the issuing of electronic passports are listed - Data leakage threats, identity theft, tracking, and host listing.

To prevent attacks such as skimming and eavesdropping, it needs to provide a more secure security system. Hence, in this paper, proposed a novel biometric encryption method based on AES (Advanced Encryption Standard) and encoded into the HCC2D (High capacity 2D barcode) code. This technique will enhance the security from various attacks because it's not an active element.

The main novelty and contributions of this paper are:

- AES algorithm is applied for biometric encryption and it will enhance the security.
- The encrypted biometric data's are encoded into the HCC2D code.

The rest of the paper described as follows, in Sect. 2 presents related work. Section 3 elaborated on biometric encryption. In Sect. 4 proposed biometric encryption using the AES algorithm and elaborated on encrypted biometric data encoded in the HCC2D code. In Sect. 5 we conclude this paper.

2   Related Work

A lot of research has just been directed on Machine Readable Travel Documents (MRTD). One of the principal security examinations on the biometric passport was exhibited by Juels, Molnar, and Wagner [4] in 2005. A few downsides in the ICAO standard were distinguished by them which incorporate clandestine scanning and tacking, biometric information spillage, eavesdropping, skimming, and cloning. They likewise discovered certain deformities in the cryptographic structure of the ICAO standard.

M. Lehtonen et al. [5] recommended in 2006 a conceivable arrangement against the clandestine reading and eavesdropping attacks by incorporating the RFID empowered MRTD with the optical memory gadget. The correspondence channel between the peruser and the optical memory gadget is secure since they require an observable pathway association for perusing. The principal disadvantage of this proposed thought is that an equipment change must be done on biometric passports.

There are some menace situations seen in the literature which is important for the issuance of travel documents. The pertinent issues for the issuing of electronic travel papers, for example, identity theft, data leakage threats, tracking, and host listing are examined [6]. Adrian et al. [7] examined two stages of attacks on network and transport layers based on RFID. The attacks on the labels are cloning, spoofing, impersonation, eavesdropping, application layer, unapproved label perusing and label change. In [8] discloses four techniques to attack used by adversaries to comprise the security of a framework that utilizes ISO/IEC 14443 RFID, for example, eavesdropping, skimming cloning and relaying.

Regardless of the way that PC chips and RFID labels have been now used for storing biometric information; they are not suitable for low-cost arrangements. Likewise, The RFID chip is legible, information leakage threats normally without such mindfulness. In the existing, showed that ICAO standard biometric passport based on RFID has countless issues and security infringement issues [9]. Hence to overcome this challenges this paper proposed biometric encryption using the AES algorithm and encoded into the HCC2D code. The HCC2D code has the ability to store the biometric information and it can't be used as an active component, they are significantly more affordable and don't require specific equipment for recuperating data. To be sure, HCC2D codes are efficient, latent read-just segments whose information can't be altered and can be decoded by the explicit gadgets.

3   Biometric Encryption

Biometric encryption employs the physical characteristics of a person as a way to code or decode and provide or deny access to a computer system. Biometric encoding techniques address directly on the clear preference and proposals of the privacy and information protection authorities for applying biometrics to verify the identity instead of identification function alone. The biometric data in itself can't assist as a cryptographic key because of its variableness. Nevertheless, the number of information comprised in a biometric data is large. It is the procedure that firmly bonds a cryptographic key to a biometric data so that the key or biometric can't be retrieved from the stored template. It is only possible to recreate the key only if the right live biometric sample is demonstrated on verification.

The digital key is haphazardly created during enrollment, hence the user unaware of the key. The biometric key is completely independent, and it can always be changed. Later on, a biometric sample is developed and the biometric encryption algorithm consistently and securely holds the key to develop a biometric template and it is also known as a private template. In principle, the biometric key is encrypted with the biometric and the biometric encryption template provides splendid privacy protection. The generated biometric encryption can be stored in a database, smart card or token, etc.

Verification part, the user gives his/her recent biometric sample, which, when employed to the decriminalize biometric encryption template, will let the biometric encryption recall the same key and the biometric assists as a decryption key. At the end of the verification, the biometric sample discarded once again. In another way, the attacker whose biometric data is unlike enough will not be retrieving the key. Hence, these kinds of encryption and decryption are fuzzy because the biometric sample is dissimilar each time; the different encryption key is conventional cryptography.

Afterward, the digital key is retrieved and it can be applied as the basis for any physical or logical application. Hence, biometric encryption is an efficient, secure and privacy-friendly tool for biometric key management. Figure 1 shows the architecture of biometric encryption.

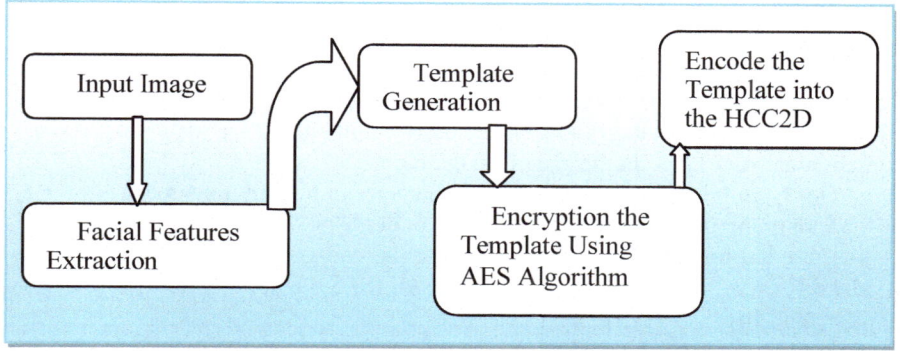

Fig. 1. Biometric encryption architecture.

4   Proposed Biometric Encryption Using AES Algorithm

Encrypted biometric information in the HCC2D code is created by applying the Advanced Encryption Standard (AES) calculation [10]. AES was created by Joan Daemen and Vincent Rijmen, known as Rijndael cipher block. It is generally called symmetric key algorithm the fact that the comparative key is used for encoding and decoding the data. AES is a form of the cipher with different block sizes and key sizes. In AES, 128 bit of block size and 3 types of key sizes are available such as 128 bit, 192 bit and 256 bit. Plain text and key size are selected independently and the size of plaintext, key size decides the number of rounds to be executed. Based on the plaintext and keys there is a minimal of 10 rounds for 128-bit key and 14 round is the maximum for a 256-bit key. We have encrypted the biometric data which contains facial data of an individual and encoded into the HCC2D code. This encoded biometric data is to keep mystery from the intruder. The biometric data has been collected from the FEI database [13] and applied for facial encryption [14]. The encryption of the biometric is applying the AES algorithm and it is illustrated given below.

4.1   Encryption Phase

Two input is required, consider the input image I_1 which is encrypted, and k_1 is the secret key. The k_1 is changed to SHk_1 by applying an SHA 256 algorithm. Therefore, I_1 is encoded into Base 64 string B_1. Hence, B_1 and SHk_1 are enriched into the AES 256 encryption algorithm to create ciphertext C_t.

ALGORITHM (with N = 2):

1. The input biometric data to be read and encode it by applying the base64 standard.
2. The key file to be read and start the encryption using AES 256 bit key by utilizing the hash algorithm, here we considered SHA-256 of the key file.
3. The biometric is encrypted by utilizing the base 64 encoded content and hash produced in step 1 and step 2 severally.
4. Produce another Image I_2 of size (S_1,S_2) with pixel information P_i. Where,
   a. S_1 - character endorse for the key document (Default: 255).
   b. S_2 – Number of characters in the key document
   c. P_i – Pixel information to be occupied (Default: 0)
5. Individual row I_R in the height of image repeat:
   a. Get J_A to be ASCII code of the I_R^{th} character in the key document.
   b. Satisfy the first J_A pixels of the image in the I_R^{th} row with black color. I_R. I_2 $[I_R]$ $[J_A] = 0$ for each I_R, J_A in S_1, S_2 with the end goal that $J_A <$ ASCII(key[I_R])
6. Make N (= 2) Images (S, T) of a similar size (S_1,S_2) and pixel information to such an extent that
   a. For the 1st image S, pixel information is created arbitrarily. It tends to be either 0 (black) or 1 (white).
      I_R. S $[I_R]$ $[J_A]$ = random(0, 1)
   b. 2nd image pixel information T $[I_R]$ $[J_A]$ is characterized such that I_R. T $[I_R]$ $[J_A] = S[I_R]$ $[J_A]$ xor
      I_2 $[I_R]$ $[J_A]$ for each I_R, J_A in (S_1, S_2)
7. The C_t is the encrypted encoding output, the biometric images S and T respectively.

4.2   Decryption Phase

Two inputs are given for decryption phase; C_t is the ciphertext to decrypt and the array of shares k_2, the secret key. Therefore, the original key K_3 image is constructed. The K_2 is then decoded to the k_1 by applying ASCII. The k_1 is changed to SHk_1 by applying an SHA 256 algorithm. The C_t, SHk_1 are fed into the AES 256 decryption algorithm to produce the Base64 encoding of the B_1 image. The B_1 is converted to I_1 output.
   ALGORITHM (with N = 2):

1. Compute the input ciphertext which from the image C_t.
2. Load the Images K_2, K_3 from the input K_2.
3. Produce another Image I_2K_1 of size (S_1, S_2) same as K_2, K_3 with the end goal that
   a. I_2K_1 $[I_R]$ $[J_A] = K_2$ $[I_R]$ $[J_A]$ xor K_3 $[I_R]$ $[J_A]$ for each I_R, J_A in (S_1, S_2)
4. Introduce key K_1 as an array of characters of size same as the height of image I_2K_1 (S_2).
5. For-each row I_R^{th} in the height of image I_2K_1 repeats
   a. Let count = 0
   b. For each pixel J_A of the image in the I_R^{th} row with black color. I_R. Increment count by 1
   c. Find the character K_1I_R by utilizing the ASCII code of the count generated after b. i.e., K_1I_R = char(count)
   d. Set $K_1[I_R] = K_1I_R$

6. With the Key, K_1 initializes the AES 256 Algorithm with a hash (K_1) (SHA-256)
7. Decrypt the ciphertext C_t and save the decrypted base64 encoding as an Image I_1
8. The output I_1 is the decrypted image.

4.3   Sample Input Output for Image Encryption

(Fig. 2).

Input image [13]   Encrypted Data   Decrypted image

Fig. 2. Shows the biometric encryption and decryption

4.4   HCC2D (High Capacity 2D Barcodes)

The HCC2D code [11] is a 2D color standardized tag which is made of a lattice of square shading cells, whose color is chosen from a color palette. The samples of HCC2D codes with 4 and 8 colors. HCC2D code protecting all the Function Patterns, the Format Information and the Version Information characterized in the QR code. Keeping up the structure and the situation of such patterns and basic data permits the HCC2D code to save the solid heartiness to geometric distortions of the QR code. A high limit institutionalized ID named HCC2D (High Capacity Colored 2-Dimensional), which secures the strong quality of QR codes and use colors to construct their data thickness was introduced. The HCCB scanner tag [12] encodes data as triangles and usages distinctive colors, where the shading chose for each triangle is data subordinate. The objective of this movement is to make an HCC2D code encoding the biometric scrambled information. Figure 3 shows encrypted biometric data encoded into the HCC2D code.

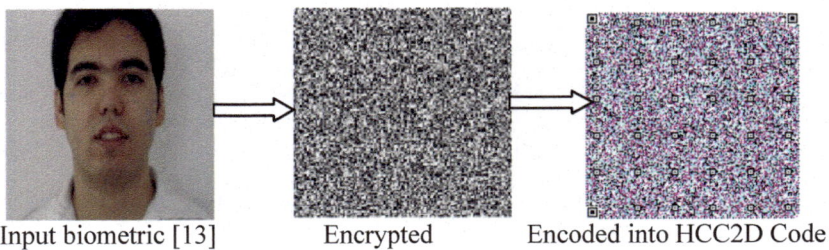

Input biometric [13]   Encrypted   Encoded into HCC2D Code

Fig. 3. Shows the encrypted biometric into the HCC2D code.

5   Conclusions

This paper proposed a novel biometric encryption technique and encoded in the HCC2D code for a secure biometric passport. The biometric data have encrypted using the AES with SHA 256 algorithm in the HCC2D code to secure the data from the legible, data leakage threats. The proposed method will enhance the security in the biometric passport because the encrypted biometric data in the HCC2D code cannot be utilized as an active element, they are substantially less expensive and don't require particular hardware for recovering information. Indeed, HCC2D codes are economical, passive read-only components whose data cannot be modified and can be decoded by the specific devices. We are focusing on secure RFID and enhancement of its security.

Acknowledgment. Authors would like to acknowledge the support of the Artificial Intelligence Laboratory of FEI in São Bernardo do Campo, São Paulo, Brazil, for providing the database for the research work. This work has done in the BSACIST research laboratory, Chennai, India. We thanks to our research department who provided expertise that greatly aided in the research. We thank our research dean and research committee members for assistance in particular technique, methodology and for valuable comments that improved the research work.

References

1. King, B., Zhang, X.: RFID: An anti-counterfeiting tool. In: RFID Security: Techniques, Protocols, and System-On-Chip Design. Springer, Science + Business Media, LLC (2008)
2. Bundesamt fur Sicherheit in der Informationstechnik (BSI), Germany: Advanced security mechanisms for machine readable travel documents–extended Access Control (EAC), version 1.0, TR-03110 (2006)
3. Shivani, K., Aman, D., Riya, B.: The study of recent technologies used in E-passport system. In: IEEE Global Humanitarian Technology Conference - South Asia Satellite (GHTC-SAS), Trivandrum, 26–27 Sep (2014)
4. Juels, A., Molnar, D., Wagner, D.: Security and privacy issues in E-passports. In: Proceedings of the First International Conference on Security and Privacy for Emerging Areas in Communications Networks (SecureComm 2005), Washington, DC, USA, IEEE (2005)
5. Lehtonen, M., Staake, T., Michahelles, F., Fleisch, E.: Strengthening the security of machine readable documents by combining RFID and optical MEMORY devices. In: Presented at Developing Ambient Intelligence: Proceedings of the First International Conference on Ambient Intelligence Development (Amid'06), 2006, Sophia Antipolis, France, Springer, to appear in International Journal of Information Security (IJIS) (2006)
6. Bolle, R.M., Connell, J.H., Pankanti, S., Ratha, N.K., Senior, A.W.: Guide to Biometrics. Springer, New York (2013)
7. Adrian, A., Marius, L.: Biometric Passports (e-passports). 978-1-4244-6363/10/$26.00© IEEE (2010)
8. Tiko, H.: Using NFC enabled android devices to attack RFID systems. www.cs.ru.nl/ bachelors-heses/2018/Tiko_Huizinga___4460898___Using_NFC_enabled_Android_ devices_to_attack_RFID_systems.pdf

9. Juels, A., Molnar, D., Wagner, D.: Security and privacy issues in E-passports. In: Proceedings First International Conference on Security and Privacy for Emerging Areas in Comm. Networks (Securecomm'05), Sept 2005

10. Advanced Encryption Standard (AES). FIPS. https://csrc.nist.gov/csrc/media/publications/fips/197/final/documents/fips-197.pdf. 23 Nov 2001

11. Marco, Q., Giuseppe, F., Italiano. reliability and data density in high capacity color barcodes. Comput. Sci. Inf. Syst. **11**(4), 1595–1615

12. Microsoft Research: High Capacity Color Barcodes. http://research.microsoft.com/en-us/projects/hccb/ (2013)

13. FEI Face Database: https://fei.edu.br/ ~ cet/facedatabase.html

14. Kalubandi, V.K.P., Vaddi, H., Ramineni, V., Loganathan, A.: A novel image encryption algorithm using AES and visual cryptography. In: 2016 2nd International Conference on Next Generation Computing Technologies (NGCT), pp. 808–813. IEEE (2016)

Survey and Analysis of Pest Detection in Agricultural Field

Yugchhaya Galphat[(⊠)], Vedika R. Patange[(⊠)], Pooja Talreja[(⊠)], and Somil Singh[(⊠)]

Vivekanand Education Society's Institute of Technology, Computer Engineering, Mumbai, India
{yugchhaya.dhote, 2015vedika.patange, 2015pooja.talreja, 2015somil.singh}@ves.ac.in

Abstract. Pest infestation is the major problem that our farmers are facing in the agricultural fields. This causes huge damage to the food crops. In order to control the attack of the pests, farmers use pesticides. The excessive use of pesticides turns out to be dangerous to the plants, animals and also to the human beings. It causes various health disorders such as asthma, eye and respiratory tract irritation, skin cancer, etc. In order to decrease the infestation of the pests in the agricultural fields, image analysis techniques are applied to agriculture science and thus provides maximum protection to crops which results in crop management and production. This paper does the survey and analysis of the various image processing algorithms used for pest detection and also the implementation of IOT to detect the pests based on the climatic changes. In addition, this paper is concluded with the analysis of various studies done by the researchers on the techniques and algorithms used for the detection of agricultural pests.

Keywords: Image processing · Internet Of Things · K means clustering · Sensors · Sticky traps

1 Introduction

Agriculture plays an essential role in the economy of India. 54.6% of the population is engaged in agriculture (2011 census). Research in agriculture sector mainly focuses on food quality and increasing the yield of crops. The productivity of crops has been affected by the pest infestation. The Economic Survey states that farm mechanization in the country has to be enhanced. To increase productivity, controlling of pest infestation plays an important role. To manage the pest control, farmers use naked eye method. But this method needs continuous monitoring, which is time-consuming and requires human efforts. Since 1950, the usage of pesticides have increased to 2.5 million short tons annually worldwide but the crop loss has remained relatively constant. In 1992, the World Health Organization (WHO) estimated that three million pesticide poisonings take place every year, causing 2,20,000 deaths. Farmers fail to have precise knowledge about the usage of pesticides. Pesticides and other chemicals used in agriculture field remains hazardous to farmers, causing lung disease, noise-induced

© Springer Nature Switzerland AG 2020
A. P. Pandian et al. (Eds.): ICCBI 2018, LNDECT 31, pp. 976–983, 2020.
https://doi.org/10.1007/978-3-030-24643-3_116

hearing loss, skin diseases, neurological health effects like loss of memory, coordination loss, asthma, cancer and allergies etc. Mildew, pathogen, and other pests destroy more than 40% of the food production. Pathogens, mites, fungi, weeds and insects all refer to pests that reduces crop productivity and profitability. Pests damage the agricultural fields by feeding on crops, such as boll weevil on cotton, codling moth on apples. According to the recent news, the pink bollworm attacks cotton crops in Maharashtra, which eventually affected the economy of the nation and decrease the productivity of cotton crops. Pesticides, insecticides are meant to control pests. There are different types of pesticides used on different pest groups, such as Avicides on birds, bactericides on bacteria, fungicides on fungi, herbicides on plant, insecticides on insects, miticides on mites and so on. These pesticides are not only costly for farmers, but it also causes environmental pollution. So it is necessary to make less use of them. In one study, the United States spent approximately $1.4 billion on crop losses caused by pesticides annually at the beginning of 21st century. The use of technology have been increased in agriculture sector so that to make the farming automatic by converting the manual work of farmer into an automatic way, which reduces the human efforts. The pest control method requires manual work of farmer and the farmer use to spray the pesticides on regular basis. The indiscriminate use of pesticides have adverse effects on farmers health, crops and environment. So there are different technologies being used to solve the pest detection and pesticides spraying like image processing, iot and so on.

2 Literature Survey

2.1 Image Processing

2.1.1 Pest Detection

Miranda [1] proposed a system for the detection of pests and extraction using IP techniques. This study illustrates the implementation of image processing techniques to estimate the density of the pests in the fields, specifically in the rice fields by establishing a detection system that is automated. A network of wireless cameras are set up by the authors along with the sticky traps to capture the insects for image acquisition. For the pre-processing of the image, the RGB image is converted into the grey scale image by using the below formula:

$$I(x, y) = 0.2989 \times R + 0.5870 \times G + 0.1140 \times B \qquad (1)$$

The reference image and input image are used to detect the pests from the image. The following formula is used to determine the difference between the two images in order to detect the pests:

$$O(x, y) = \{255, if\ R(x, y) = I(x, y)\}$$
$$\{I(x, y), if\ R(x, y) \neq I(x, y)\} \qquad (2)$$

Median filter is used by the authors in this study to filter the image. Median filter looks at its nearby neighbour's pixel values to decide whether or not it is representative of its surrounding pixels and replaces it with the median of those values. In the study of Carino et al. [2] there are many devices for pest management and sampling techniques. Light traps and sticky traps are economical and are used to catch the flying insects but they are unable to trap non flying insects. The sweep net is also a quick method and good for sampling arthropods that stays in the rice canopy, but it has a human error of poor catching of arthropods. Visual counting and data recording could have been useful but turned out to be costly.

Pavithra [3] describes an algorithm that was first tested in MATLAB R2011b and then was implemented on FPGA using Verilog. The images are selected from the database and pre processed for the color model to be selected. Saturation color model was selected as the best suited color model. Different color models such as RGB, YCbCr and HCV are used for the pre-processing of images. Then, the segmentation of the pest is carried out wherein the image is partitioned into multiple regions. Noise creates the pixels within an image to produce different intensity values rather than the correct ones and represents unwanted information that deteriorates image quality. As a result of which noise removal technique is used to remove the unnecessary pests. The number of pests were then counted and compared the proposed algorithm with the existing one. Otsu was also tried, but manual thresholding gave the best results. When it comes to processing, outsu is intense, but the time consumed by manual threshold is very less. Erosion and dilation process is done to remove unnecessary pixel noise. According to Van Li [4] binocular stereo method can be used to detect the pest and its location. The difference between the color of the pest and the leaves can be obtained with the help of image segmentation. So, by the combination of binocular stereo and image segmentation, 3D image of the pest can be obtained. But in this paper the results are showing that this method is reliable for depth measurement and the pest size and the color may give different measurements due to the varying noise that occurs in the image.

Bhadane [5] describes a prototype for pest detection on the infested images of different leaves. Images of the leaves that are infested, are captured by digital camera and processed using image segmentation to detect infested parts of a specific plant. İmage segmentation is a process in which a label is assigned to every pixel in an image such that pixels with the same label share certain characteristics such as intensity, color or texture. Object extraction is done by background subtraction. Then segmentation is used to partition the image into multiple regions. Sobel operator is used for these particular type of edge detection. The drawback of this prototype is that if the color of pest and leaf is almost similar then the method of background subtraction cannot identify the pest. The accuracy of identifying a pest depends on the selection of threshold value.

Javed [6] has proposed a methodology wherein there are three steps namely detection, extraction and classification. In the pest detection step, the image is pre processed by reducing the noise from the image and prepare it for further processing. Then, segmentation of the image is performed using k-means clustering. The pest is then classified based on the features that were extracted from the detected pest. To enhance the appearance based intensity variations, Gaussian filter is used. Image pre-processing consists of two steps viz. Smoothening filter and Color transformation. Due to some factors, there might appear sharp edges in the image which can cause local maxima during the segmentation process. In order to get rid of this problem, the sharp edges are converted into the smooth edges Gaussian filter technique. After smoothening, the images will be transformed into RGB by using L*, a* and b* color space. The maximum value of L*, a* and b* can be used as the threshold. For detecting the pest, k-means clustering algorithm is used. In this algorithm, the images are segmented using this method. Some of the researchers were facing some problem with the green colored pests. In this study, the k-means clustering is applied sequentially. This algorithm is decided based on the distance between the two instances. For this, Euclidean distance method is used to decide about any point on the image. To overcome the problem of colors, the image is initially segmented on the basis of number of colors and then on the basis of distance. Finally, classification of the pest is done by cropping the segmented image to get the desired result. Boissard [7] has proposed a system where sticky traps are used. These traps have some kind of smelly materials by which the pests are attracted towards it and get trapped. These traps are useful only for flying pests and not the non flying ones.

Krishnan [8] has proposed a system wherein the early detection of pest is carried out by interpreting the image. Image segmentation and Clustering method is implemented in this study. Clusters can be selected based on a specific condition in the method of clustering. Masking means to identify the green coloured pixels which are set to zero if the value is less than the predefined threshold value. The next step is segmentation where the infected portion of the leaf is extracted and is segmented into a number of patches of equal size. Then the useful segments are obtained and the patches that contain more than 50% of the information is taken into consideration for further analysis. Finally, by using the differential clustering technique, the plant image that ia clustered is subtracted from the acquired image. The required pest image is obtained from this difference image. The number of clusters k as an input parameter ia a drawback of the clustering algorithm. Poor results can be obtained by the wrong choice of k. Our main aim is to detect the starting points of bioagressors attacks and to count these so that we can take necessary actions. The future work of this system would be to modify the algorithm to find the areas that are diseased, in the crops by using softwares that are sophisticated and image acquisition devices that are better.

2.2 Internet Of Things

2.2.1 Effect of Climatic Change

Agrawal [9] has developed a framework wherein Internet Of Things is used for observing the conditions of general climatic parameters. An image processing based solution is used for the automatic leaf diseases detection and classification according to the factors of environment. The system consists of three different fields viz. Embedded system, image processing and wireless networking part for IOT. Results of embedded hardware shows all the recent values of different environmental parameters like temperature, humidity and soil moisture level are continuously displayed on the 20X4 LCD display with its units. Zigbee network is used to send the same data to the PC. Patil [10] has also proposed a technique to detect the infection of leaf. Sugarcane poo is taken as an example for the detection of leaf infection using the four fundamental steps which include picture securing, picture division, leaf locale division and sickness district division.

Shahzadi [11] has developed a framework that is based on IOT. The proposed solution consists of the deployment of sensors in the field such as soil sensors, humidity sensors and temperature sensors in the fields. The collected data is sent to the server by the sensors. The expert system is deployed on the server side which processes the data and sends the notifications to the farmer about the crops. The future work of this system is to deploy the actuators in the fields and to enhance the functionality of the server by using genetic algorithm, artificial neural networks and digital image processing techniques on the server. Based on the research, TongKe [12]. developed a smart agricultural system that was based on cloud computing. In this the load of the resources is balanced and the distribution is carried out with ease with the help of the combination of agricultural information cloud and the Internet Of Things. Chen and Jin [13] proposed Digital Agriculture that was based upon IOT. In this, the information about the temperature, soil and wind is collected by the different sensors and this information is transferred through Zigbee technology.

Mohanraj [14] has proposed a system where the geographical data is used to bring in the weather details. The water table content is also imported to choose the crops that are best suitable. This system is used to map the agricultural field according to the location set by the farmer. The system architecture consists for various modules namely, Reminder, Monitoring plant growth, Watering planner, Crop profit calculator and so on. In the module of calamities check, the watering plan and the setted field is changed according to the weather forecast which is obtained from weather API. It is also used to keep a track on uncommon activities in the field such as sudden firing, etc. Crop water need is calculated using evapotranspiration which is used to calculate the total amount of water that plants consume for a quality growth. Hargreaves and Samani evapotranspiration equation, ETo can be calculated:

$$ETo = 0.0023 * \sqrt{(tmax - tmin)} * (tmin + 17.8) * ExTR \tag{3}$$

The further steps in the algorithm includes calculation of crop factor based on the crop growth, calculation of SAR, determination of the amount of percolation and seepage losses based on soil type. Also, the estimation of effective rainfall of the region based on annual rainfall is carried out. HU [15] proposed the IOT application that consisted Radio Frequency Identification, GPS and smart sensors. RFID was used to record the information and the sensors were used to monitor the field.

Chougule [16] constructed IPM ontology to make farmers aware of integrated pest management. The system is also embedded with the expert system wherein the experts can update the pest symptoms and remedy. Pest ranking description algorithm is used to add the details of pest by the experts. According to this algorithm, vector space model is used to choose the document or text provided by the expert or user for a specific crop, the system ranks these documents respectively. For measuring the temperature, humidity and moisture at fields, the terrestrial sensor network of weather sensors is used. The data that is collected by the sensors is stored in the database servers and can be used by the respective organization or by the agriculture decision support systems. The future work of this system is to extend the proposed algorithms to generate the ontologies in multiple languages in order to make it easy for the farmers to follow the advices.

3   Conclusion

This paper summarize detailed information about the various techniques used for the detection, extraction, classification and the management of the pests in the agricultural field. Image processing techniques along with Internet Of Things proved to be advantageous and helpful for the farmers working in the fields. The system can be more feasible and can be made fully automated by incorporating various techniques such as Machine Learning, Artificial Intelligence, etc. An Android app can also be created to make the farmers aware about the usage of the pesticides.

4   Analysis of Pest Detection

Table 1. Summarizes all the different techniques proposed by the authors.

Table 1. Analysis of pest detection

Proposed system	Technique used	Advantages	Limitations	Results
Detection and extraction of pests	Median filter for enhancement of an image	The system is simple and efficient	Not fully automated	86% of correct recognition ratio and 82% of average accuracy
Pest management	Sticky traps, sweep net, visual counting and data recording	Successful in trapping flying insects	Unable to trap non flying insects, human error, costly.	Amount of insecticides used decreased by 92%
Detection and segmentation of pests	Otsu algorithm, noise removal, erosion and dilation	Thresholding requires less time	11.23% of the original image was infected by the disease	Manual thresholding gave best results
To detect pest and its location	Binocular stereo and image segmentation	The combination of binocular stereo and image segmentation gives a 3D image	Pest size and color gives different measurements	90% average accuracy rate

(*continued*)

Table 1. (*continued*)

Proposed system	Technique used	Advantages	Limitations	Results
To detect pest on infected images of different leaves	Image segmentation, object extraction and sobel operator technique	The image is segmented on number of colors and then on distance	Cannot identify the pests because of similarity in color	94% effectiveness in classifying pest
Pest detection by interpreting image	Image segmentation and clustering algorithm	Reliable for rapid detection of pest	Wrong choice of k may give poor results	97.5% accurate
To observe the conditions of climatic parameters	IOT, Image processing and Zigbee technology	An IP based	Complex visual patterns of diseases and pests	Accuracy upto 96% and correctness above 85%
Deployment of sensors	IOT and genetic algorithm	Farmer is notified about the crops	Attack of insects causes 38% loss of cotton crops	Data is processed by the server and the recommendations are sent to the farmer
Information about weather details and water table contents	Usage of geospatial data	Farmer can keep a timely check about the various parameters of the field	IOT based developed systems not reliable, search engines not helpful	The amount of water intake by the plants per day is calculated using evapotranspiration method

References

1. Miranda, J.L., Gerardo, B.D., Tanguilig III, B.D.: Pest detection and extraction using image processing techniques. Int. J. Comput. Commun. Eng. **3**(3) (2014)
2. Carino, F.A., Kenmore, P.E., Dyck, V.A.: A FARMCOP suction sampler for hoppers and predators in flooded rice fields. In: The International Rice Research Newsletter, vol. 4, ch. 5, pp. 21–22 (1979)
3. Pavithra, N., Murthy, V.S.: An image processing algorithm for pest detection. Perspect. Commun., Embed.-Syst. Signal-Process. (PiCES) **1**(3) (2017). ISSN: 2566-932X
4. Van Li Chunlei Xia Jangmyung Lee: Pusan National University intelligent robot lab: vision-based pest detection and automatic spray of greenhouse plant. In: IEEE International Symposium on Industrial Electronics (ISIE 2009), Seoul Olympic Parktel, Seoul, Korea, 5–8 July 2009

5. Bhadane, G., Sharma, S., Nerkar, V.B.: Early pest identification in agricultural crops using image processing techniques. Int. J. Electr., Electron. Comput. Eng. **2**(2), 77–82 (2013)
6. Javed, M.H., Noor, M.H., Khan, B.Y., Noor, N., Arshad, T.: K-means based automatic pests detection and classification for pesticides spraying, (IJACSA). Int. J. Adv. Comput. Sci. Appl. **8**(11), 236–240 (2017)
7. Boissard, P., Martin, V., Moisan, S.: A cognitive vision approach to early pest detection in greenhouse crops. Comput. Electron. Agric. **62**(2), 81–93 (2008)
8. Krishnan, M., Jabert, G.: Pest control in agricultural plantations using image processing. IOSR J. Electron. Commun. Eng. (IOSR-JECE) **6**(4), 68–74 (2013). e-ISSN: 2278-2834, ISSN: 2278-8735
9. Agrawal, G.H., Galande, S.G., Londhe, S.R.: Leaf disease detection and climatic parameter monitoring of plants using IOT, Department of E & TC Engineering, PREC, Loni, Maharashtra, India
10. Patil, S.B., Bodhe, S.K.: Leaf disease severity measurement using image processing. Int. J. Eng. Technol. **3**(5), 297–301 (2011)
11. Shahzadi, R., Tausif, M., Ferzund, J., Suryani, M.A.: Internet of things based expert system for smart agriculture. Int. J. Adv. Comput. Sci. Appl. **7**(9), 341–350 (2016)
12. TongKe, F.: Smart agriculture based on cloud computing and IoT. J. Converg. Inf. Technol. **8**(2) (2013)
13. Chen, X.-Y., Jin, Z.-G.: Research on key technology and applications for internet of things. Physics Procedia **33**, 561–566 (2011)
14. Mohanraj, I., Ashokumar, K., Naren, J.: Field monitoring and automation using IOT in agriculture domain. In: 6th International Conference on Advances in Computing and Communications, ICACC 2016, 6–8 Sep 2016, Cochin, India
15. Hu, X. SQ. (n.d.): IOT application system with crop growth models in facility agriculture, IEEE
16. Chougule, A., Jha, V. K., Mukhopadhyay, D.: Using IoT for integrated pest management. In: 2016 International Conference on Internet of Things and Applications (IOTA) Maharashtra Institute of Technology, Pune, India 22–24 Jan 2016

Object Detection Using Convex Clustering – A Survey

Madhura P. Divakara, Keerthi V. Trimal[✉], Adithi Krishnan,
and V. Karthik

Department of Computer Science and Engineering,
Vidyavardhaka College of Engineering, Mysuru, Karnataka, India
mpdsimha@gmail.com, keerthivtrimal@gmail.com

Abstract. Clustering is an unsupervised machine learning technique involving the grouping of data points used to classify objects or cases into related groups known as clusters. There are different clustering methods, each with its own advantages and disadvantages. Our main focus in this paper is the newly developed Convex clustering algorithm. Convex clustering uses some hierarchical clustering features while reducing the ability to make false inferences. However, convex clustering is computationally demanding and there are two main obstacles in its path: (a) it is poorly situated on high-dimensional problems and (b) minimal guidance on how to choose penalty weights. Our main objective is to attempt to modify the convex clustering algorithm in order to eliminate the above drawbacks.

Keywords: Convex clustering · Machine learning · Cluster · Object detection

1 Introduction

Object detection is a computer technology for computer vision and image processing that detects instances of semantic objects in digital images and videos of certain classes (such as humans, buildings or cars). Well researched object detection domains include facial detection and pedestrian detection. Object detection has applications in many areas of computer vision, including video surveillance and image collection. It is widely used in computer vision tasks like face detection, face recognition, co-segmentation of video objects. It is also used to track objects, e.g. to track a ball during a football match, to track a cricket bat's movement, to track a person in a video.

Currently, clustering algorithms such as function clustering, k-means clustering, hierarchical clustering, etc. are used for object detection. K-means clustering has several disadvantages, such as data scaling, highly sensitive to initial seeds and noise, data order, unable to handle non-convex clusters of varying density and size. Hierarchical clustering also has disadvantages such as greater time complexity, previous steps, vague termination criteria, etc.

Many convex clustering methods have been proposed and introduced in recent times, but some still need to fix the minimum cluster size, some scale poorly on high-dimensional problems, etc. Our goal is to develop an advanced convex clustering algorithm for object detection and recognition, which at least to some extent alleviates the disadvantages of the algorithm.

© Springer Nature Switzerland AG 2020
A. P. Pandian et al. (Eds.): ICCBI 2018, LNDECT 31, pp. 984–990, 2020.
https://doi.org/10.1007/978-3-030-24643-3_117

2   Related Works

Chen et al. [1] have implemented a proximal distance algorithm to minimize the objective function of convex clustering on graphical processing units ATI and nVidia. It emphasizes the ability of convex clusters to deal with high-dimensional issues. Lashkari et al. [2] have proposed an exemplary probability function that approximates the exact probability resulting in a probabilistic mapping of data points to a set of exemplaries. It reduces the average distance and the mapping costs of information theory. Bilen et al. [3] have established a soft similarity between possible locations of objects in the image and clusters learned in images. It helps to reduce optimization to the global minimum. Chi et al. [4] presented two dividing methods for convex clustering, the first an instance of the Alternating Minimization Algorithm and the second an instance of the Multiplier Algorithm Alternating Direction Method, and examined their complexity.

Pelckmans et al. [5] proposed a convex optimization view for clustering by introducing a shrinkage term that led to data points being merged into a small cluster set. Wang et al. [6] introduced a new clustering method called Sparse Convex Clustering to simultaneously observe clusters and select features. A finite sample error is obtained for the estimator and its consistency of variable selection is determined. Yuan et al. [7] proposed a semi-smooth Augmented Lagrangian method based on Newton for large-scale convex clustering problems with superior performance and scalability. Condat et al. [8] introduced a new convex formulation of data clustering and image segmentation with a fixed number of k-regions and possible penalties for region parameters.

Murray et al. [9] proposed a chromatic object modeling method by optimizing the convex objective function. The weighted kernel number Nc is the optimized feature. Kim et al. [10] have proposed a Q-Learning method where the method incorporates a convex clustering approach to finding a region with some credit assignment property. Collins et al. [11] have implemented a convex term for the regularization of very large image datasets. This term has been applied to spectral cluster algorithms. Tachikawa et al. [12] have used a modified version of the convex clustering method to determine the position of a sound source. Also they have combined sparse coefficients estimation with the convex clustering method.

3   Comparison of Different Convex Clustering in Object Detection

The following Table 1 gives us an idea of the different methods used in the field of convex clustering by different authors. It also illustrates some of our recommendations that we thought could be implemented.

Table 1. Comparison of different convex clustering in object detection

Ref. No	Objective	Concept used	Results/outcome	Advantage	Disadvantage
#1	To minimize the objective function in convex clustering	Separate parameters, accommodate missing data and support prior information on relationships	Successfully implemented with maximum speed achieved on ATI or nVidia GPUs	Allows large scale parallelization. Applicable to high dimensional problems	Execution time is more when moving matrices over slow I/O channels for plotting
#2	To overcome sensitivity to initialization in the gradient descent method of EM algorithm	Find global optimum for initialization using a simple algorithm. Apply rate-distortion theory	A minimum value for global optimum is obtained for cost function and RD function is smooth, convex and has monotonic decrease	Useful in large dataset problems where mixture models fail to give consistent results. Extendable to any proximity data	Clusters split with increase in β value and fluctuation in number of clusters is present
#3	Precision agriculture which Is the modern farming techniques the approach used here	K nearest Neighbours and naïve random tree technique	In this technique it was more productive and more profit and it helped many to plant the right crop at the right time	The use of expert system and IoT is an excellent way to improve the crop rate	The expert system does tell about the herbicide and pesticide recommendation depending on the disease but the use of camera or any image capturing device is used in their paper. Also, the battery consumption can be high
#4	Minimize the objective function of convex clustering	Two splitting methods = ADMM and AMA are introduced and tested on datasets	Both ADMM and AMA lead to less computation time compared to sub gradient method	Storage required and computation time are less for both ADMM and AMA compared to other methods	ADMM did not do as well as AMA because of a large performance gap

(continued)

Table 1. (*continued*)

Ref. No	Objective	Concept used	Results/outcome	Advantage	Disadvantage
#5	Obtain cluster by solving convex optimization problem, efficiently construct solution path	Apply regularization constant as shrinkage term taking Iris dataset for illustration	The method can clearly discriminate the three classes in the dataset from the clustering tree (solution path)	Efficient for very large datasets	Using 2 norm method, the results cannot be interpreted as a form of clustering
#6	To solve the problem of applying convex clustering to a large scale/large dataset	The present semi smooth Newton conjugate gradient method is modified by enforcing appropriate stopping criterion	Efficiency of the algorithm for large datasets is high	Performance is high and the algorithm is extensible compared to existing methods	Not generalized to handle kernel based convex clustering models
#7	To overcome distortion that occurs in convex clustering when uninformative features are included	Assume feature vectors are centered, incorporate adaptive group-lasso penalty into convex clustering objective function	Performs better than k-means algorithm when p >= 500. Improved accuracy of convex clustering by 45%	S-AMA is computationally faster. Convex biclustering can also be extended to sparse convex biclustering	Selection of weights and tuning parameters is difficult when applied practically
#8	To provide a solution to piecewise constant Mumford-Shah Problem	Finite number of candidates is specified. Reformulate the problem by lifting and then apply convex relaxation	Solution to given convex and non-convex problem is found and hence global solution to k means problem is found	Performs better in Colour Image Quantization and Image segmentation	Fails in implementation of parrot case of image segmentation where no rounding procedure is present

(*continued*)

Table 1. (*continued*)

Ref. No	Objective	Concept used	Results/outcome	Advantage	Disadvantage
#9	Modelling chromatic objects such as images etc.	Extract colors, transform colours to colour space, perform convex clustering, set number of clusters, output model	A modelled chromatic object of the given input is obtained	The chromatic model can be used for image retrieval, colour transfer or classification since Nc is the output of clustering	—
#10	Reduce time and memory to learn from datasets in Q-learning	Incorporates convex clustering to find regions with same reward attribution property	The proposed method results in faster convergence than conventional Q-learning methods	Number of iterations required to reach goal state is less. Suitable for visual tracking applications	—
#11	Derive solution that explains given examples in view of the object of interest	Applying convex regularize on stochastic gradient descent approach	Shown that the method is applicable for multi-view spectral clustering apart from increased performance	Highly scalable. Increase in number of examples doesn't need computing full gradient	NMI value for Caltech 250 dataset is less when compared to other methods
	To estimate direction of arrival of sound and distance of a 3D sound source	FISTA to formulate the problem, optimize and post process it, dictionary coherence and generate monopole dictionary, adjust monopoles and apply primal dual splitting	Performance against MUSIC method was much higher	Found that this method could also be applied to special discretization of sound field	Some spurious components remained while tracking distance when experiment was performed in anechoic chamber

4   Conclusion

Clustering is one of the key data mining techniques. Data mining has many applications in different fields, one of which is image processing. Visual information is the most important information that the human brain perceives, processes and interprets. As a computer - based technology, digital image processing has applications in a variety of fields, such as medical/biological image processing, space image processing, etc. After reviewing papers on image processing, image segmentation and clustering, we concluded that hierarchical clustering, clustering of k-means, convex clustering, etc. each has its own advantages and disadvantages. A detailed study of convex clustering papers revealed few disadvantages in convex clustering. Our plan is to create a new algorithm that combines features of various clustering algorithms with convex clustering and to some extent omits its drawbacks.

Acknowledgment. The authors express gratitude towards the assistance provided by Accendere Knowledge Management Services Pvt. Ltd. in preparing the manuscripts. We also thank our mentors and faculty members who guided us throughout the research and helped us in achieving desired results.

References

1. Chen, G.K., Chi, E.C., Ranola, J.M.O., Lange, K.: Convex clustering: an attractive alternative to hierarchical clustering (2015)
2. Lashkari, D., Golland, P.: Convex clustering with exemplar-based models. In: Advances in Neural Information Processing Systems, pp. 1–8 (2007)
3. Bilen, H., Pedersoli, M., Tuytelaars, T.: Weakly supervised object detection with convex clustering. In: Proceedings of the IEEE Computer Society Conference on Computer Vision and Pattern Recognition, pp. 1081–1089 (2015)
4. Chi, E.C., Lange, K.: Splitting methods for convex clustering. J. Comput. Graph. Stat. **24**, 994–1013 (2015)
5. Pelckmans, K., De Brabanter, J., Suykens, J.a.K., De Moor, B.: Convex clustering shrinkage. In: Workshop on Statistics and Optimization of Clustering Workshop (PASCAL) (2005)
6. Wang, B., Zhang, Y., Sun, W.W., Fang, Y.: Sparse convex clustering. J. Comput. Graph. Stat. **27**, 393–403 (2018)
7. Yuan, Y., Sun, D., Toh, K.-C.: An efficient semismooth newton based algorithm for convex clustering. In: Proceedings of the 35th International Conference on Machine Learning, pp. 1–15 (2018)
8. Condat, L.: A convex approach to K-means clustering and image segmentation. In: 11th International Conference on Energy Minimization Methods in Computer Vision and Pattern Recognition, pp. 1–15 (2017)
9. Naila, M., Florent, P., Luca, M., Sandra, S.: Convex clustering for chromatic content modeling, US Patent: US8379974B2 (2012)
10. Kim, S.H., Suh, I.H., Cho, Y.J., K, Y.: Region-based Q-learning using convex clustering approach. In: International Conference on Intelligent Robot and Systems. Innovative Robots for real-world applications. IROS '97, pp. 601–607 (1997)

11. Collins, M.D., Liu, J., Xu, J., Mukherjee, L., Singh, V.: Spectral clustering with a convex regularizer on millions of images. In: Fleet, D., Pajdla, T., Schiele, B., Tuytelaars, T. (eds.) Computer Vision – ECCV 2014. ECCV 2014. Lecture Notes in Computer Science, pp. 282–298 (2014)
12. Tachikawa, T., Yatabe, K., Oikawa, Y.: 3D sound source localization based on coherence-adjusted monopole dictionary and modified convex clustering. Appl. Acoust., Elsevier **139**, 267–281 (2018)

Advanced Load Balancing Min-Min Algorithm in Grid Computing

Menka Raushan[✉], Annmary K. Sebastian[✉], M. G. Apoorva[✉], and N. Jayapandian[✉]

Department of Computer Science and Engineering,
CHRIST (Deemed to be University), Bangalore, India
{kumari.raushan,annmary.sebastian,
apoorva.mg}@btech.christuniversity.in,
jayapandian.n@christuniversity.in

Abstract. Framework figuring has turned into genuine distinctive to old supercomputing situations for creating parallel applications that bridle huge process assets. In any case, the quality acquired in building such parallel Framework mindful applications is over the ordinary parallel registering conditions. It tends to issues like asset disclosure, heterogeneous, adaptation to non-critical failure and assignment programming. Load balancing errand programming inconceivably indispensable downside in cutting edge lattice environment. Load balancing ways is normally utilized for the development of appropriated frameworks. Normally there is a three kind of stages related with Load compromise that is information arrangement, higher psychological process, learning Relocation. Take a gander at the impact of surveying on load assignment by contemplating a fundamental expense in limit. There are three completely hovered tallies to lift which put away the stack ought to be doled out to, pondering the framework action cost among get-togethers. These tallies utilize grouped data trade frameworks and an asset estimation framework to redesign the constrained air framework exactness of load adjusting.

Keywords: Grid computing · Load balancing · Scheduling · Min-Min · Max-Min

1 Introduction

Network programming has its very own troubles because of its inclination of heterogeneousness in agent frameworks, structure, asset providers and asset clients. Matrix Figuring is moreover utilized in application zones like climate expectation, earth investigation, water contamination and multi high-vitality material science. It's difficult to search out partner best asset allotment that limits the calendar length of occupations. Network Processing is considered in light of the fact that the best goals for goals these issues. Since the work of Framework is hyperbolic on a few fields, a few engineers and analysts represent considerable authority in case of every equipment and code required for Lattice plan. Some of the troublesome issues like programming, execution forecast

A. P. Pandian et al. (Eds.): ICCBI 2018, LNDECT 31, pp. 991–997, 2020.
https://doi.org/10.1007/978-3-030-24643-3_118

and asset the board territory unit indispensable in network figuring space. Network registering might be a sort of parallel and conveyed framework that allows the circulation, decision and total of geographically assets powerfully at run time, ability, execution, cost, and client nature of – self-benefit request. Networks practically gather all-inclusive disseminated PCs and information frameworks for making a general supply of registering force and learning. In Framework figuring, the primary concerns are disconnected, and in this way there sources are virtualized. Lattice processing, singular clients will recover PCs and learning, straightforwardly, while not contemplating the circumstance, OS, account organization, and elective subtleties. A key normal for Frameworks is that assets are shared among shifted applications, and hence, the amount of assets possible to some random application to a great degree varies after some time. Load levelling algorithmic principle are two kind static and dynamic, Static [1]. Load levelling calculations designate the undertakings of a parallel program to workstations either the heap at the time hubs are allocated to some errand, or upheld a middle heap of our advanced PC group. The choices related with load balance are made at order time. Various static load levelling systems are round robin algorithmic standard, randomized algorithmic principle, or calculations. And Dynamic load levelling calculations make changes to the dissemination of work among workstations at runtime; they utilize present or ongoing burden information once making appropriation decisions [2]. Multi PCs with dynamic load levelling allot/reallocate assets at runtime bolstered no from the earlier assignment data. Dynamic load level in calculations will offer a genuine enhancement in execution over static calculations. A key normal for Frameworks is that assets are shared among different applications [3]. Grid taking care of has made as the cutting edge parallel and passed on figuring rational. That totals dispersed heterogeneous assets for understanding particular sorts [4]. In that particular points of interest are a huge diverse alliance each with a neighbouring affiliation and booking part in clearing scale. Lattice conditions, the basic structure interfacing them is heterogeneous and data transmission crosswise over assets shifts from relationship with partner. Not constrained to Network, in innumerable present appropriated figuring conditions, the PCs are related by a deferral and data transmission restricted correspondence medium that normally realizes huge postponements between focus point trades and load trade.

2   State of Art

A heap adjustment equation plans to broaden the utilization of assets with lightweight load or inactive assets in this way discharging the assets with noteworthy load. The equation endeavour's to circulate the heap among all the available assets. At indistinguishable time, it intends to lessen the assemble length with the compelling use of assets. In traditional circulated frameworks included uniform and committed assets, stack balance calculations are seriously examined [5]. Anyway these calculations won't function admirably in Framework plan because of its heterogeneousness, quantifiability and independence. This makes stack adjusted programming recipe for matrix figuring harder and a spurring theme for a few specialists [6]. The Non-customary calculations take issue from the conventional calculations in this it produces ideal winds up in a concise measure of your time. There's no best programming recipe for all lattice registering frameworks. Another is to select Partner in Nursing relevant programming equation to use in an exceedingly given lattice surroundings because of the qualities of the errands, machines and system heterogeneousness. Braun et al. have examined the overall execution of 11 heuristic calculations for undertaking programming in lattice processing. They require also gave a recreation premise to scientists to check the calculations. Their outcomes demonstrate that Hereditary equation (GA) performs well in the vast majority of the consequences and along these lines the nearly clear Min-Min recipe performs by GA and in this way the rate of enhancement is moreover awfully little [7]. Go getter Load balance (OLB) doles out the jobs in an exceedingly arbitrary request inside the following available asset while not considering the execution time of the jobs on those assets. So it gives a heap adjusted calendar anyway it creates an extremely poor form range. Maximum Execution Time (MET) relegates employments to the assets bolstered their base expected execution time while not thinking about the supply of the asset and its present load. This equation enhances the fabricate range to some degree anyway it causes a serious load awkwardness [8]. Maximum Consummation Time (MCT) appoints employments to the assets bolstered their base finishing time. Fulfilment time is shown by including normal execution time of work on it asset with the assets readied time. The machine with the base consummation time for that explicit activity is picked. Anyway this recipe considers the errand only each one in turn. Min-Min equation begins with a gathering of every single unfamiliar assignment. The machine that possesses the base fulfilment energy for all employments is picked. At that point the errand with the general least fruition period picked and directed to it asset [9]. Different Load levelling Calculations are offered at present days anyway they contain numerous disadvantages. Such kind of issues will be destroyed by our anticipated unique load levelling rule. Activating Approach in Existing guideline is predicated on Line Length though in anticipated principle its upheld line length and current focal preparing unit length. Decision arrangement in existing standard is done by picked Errand is likewise moved exploitation Occupation Length though in anticipated guideline decision Strategy is additionally released by picked assignment that is relocated essentially dependent on focal handling unit Usage. Real reason for our anticipated principle is Time multifaceted nature that is in anticipated standard time unpredictability is three while in Existing guideline time intricacy is over crisp anticipated standard. Execution time in .Net System exploitation our anticipated guideline is simply too a great deal of speedy contrasted with existing calculations [10]. They contemplated ground-breaking load changing in scattered enrolling structures and a stochastic model was made which ponders haphazardness in deferral. In any case, they don't consider the structure where PCs are orchestrated in various authentic areas. Shah et al. thought about incredible load changing in Network structures utilizing estimation on occupation landing rate. Moreover contemplated dynamic load changing in Network systems, in any case, they don't think about asset easy going quality and how to alter stack in an ideal way among PCs.

3   Problem Statement

Due to the NP-fulfilment nature of the mapping drawback, the created methodologies attempt and notice satisfactory arrangements with shabby esteem considering a few exchange offs and unique cases. Amid this examination, the anticipated calculations are created underneath a gathering of suspicions. The applications to be dead are made out of a gathering of indivisible errands that haven't any reliance among each other, regularly talked as Meta assignment. Errands haven't any due dates or needs identified with them. Appraisals of expected errand execution times on each machine inside the HC suite are prestigious. These appraisals are frequently prepared before an assignment is submitted for execution, or at the time it's submitted. The mapping technique is to be performed statically in an exceedingly clump mode design. The plotter keeps running on a different machine and controls the execution of all employments on all machines inside the suite. Each machine executes one undertaking at any given moment inside the request amid which the errands are allocated (First come back1st Served - FCFS).

Hence assignment programming that is one among the NP Finish issues turns into a spotlight of investigation understudies in lattice processing space. Min-Min equation could be a simple recipe that delivers a timetable that limits the make length than the inverse customary calculations inside the writing. Anyway it neglects to give a heap adjusted calendar. This paper a Heap Adjusted Min-Min (LBMM) equation is arranged that lessens the make length and will expand the asset use [11]. The arranged strategy has two-stages. Inside the first segment the ordinary Min-Min equation is dead and inside the second area the assignments square measure rescheduled to utilize the unutilized assets adequately. Load compromise might be a strategy to support assets, using correspondence, abusing turnout overseeing and to reduce interim through right dispersion of the machine. Network registering might be a copy of circulated figuring that utilizes geologically and scatters assets. To expand execution and power, the Matrix framework needs equipped load compromise calculations for the conveyance of assignments. Load compromise calculations are of 2 sorts, static and dynamic. Our anticipated algorithmic guideline is predicated on unique ways. In perspective of the progression of Matrix enrolling over the Web. There is a little while later a require-ment. This article is an discussed on Network a building where PCs have a place with dispersed genuine locales or get-togethers which are connected with heterogeneous correspondence transmission limits [12]. Address the issue of picking which pack an arriving business ought to be allotted to moreover, how its stack can be passed on among PCs in the get-together to update the execution. The propose calculations which affirmation to discover a stack dispersing over PCs in a get-together that prompts the base reaction time or PC national expense.

4   Proposed Min-Min Load Balancer in Grid

This article is to propose a dynamic load adjustment rule for rising execution of lattice figuring. Amid this there square measure four fundamental advances: recognition computerized PC execution, trading this data between workstations, sagacious new conveyances and making the work development call Genuine data development.

Our proposed framework booking calculation, LBMM, is displayed in Fig. 1. The calculation begins by executing the means in Min-Min technique first. It initially recognizes the errand having least execution time and the asset creating it. Along these lines the undertaking with least execution time is booked first in Min-Min.

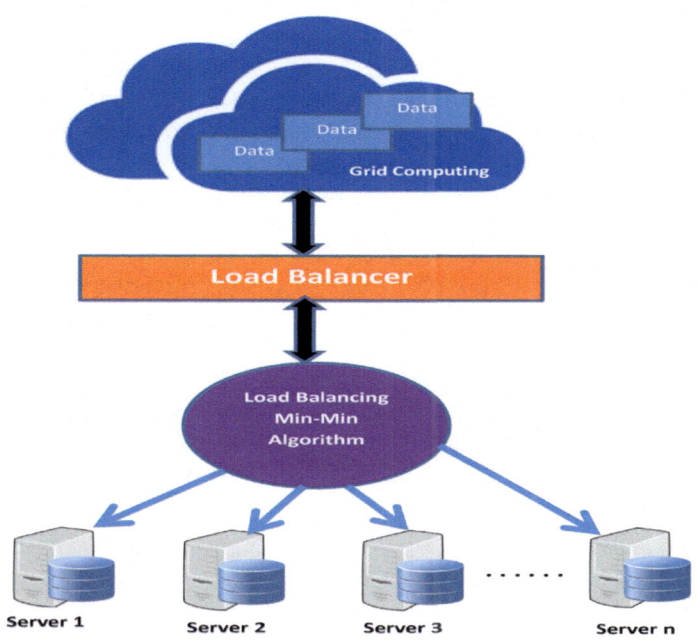

Fig. 1. Propsoed Min-Min load balancer

After that it considers the base culmination time since a few assets are planned with a few errands. Since Min-Min picks the littlest errands first it stacks the quick executing asset more which leaves alternate assets inert. Be that as it may, it is straightforward and produces a decent make span contrasted with different calculations. To assess the power of the arranged algorithmic program, issues having machine heterogeneousness and undertaking heterogeneousness are gathered from various writing and dead for every Min-Min and arranged LBMM algorithmic program. Intuitive programming bundle is created in C++ to execute every calculation. Load adjustment one among the imperative alternatives of Lattice foundation. There are assortment of things, which may affect the network application execution like load evening out, no consistency of assets and asset sharing inside the Lattice environment. In the past area, we uncovered how to scatter stack among a get-together in heterogeneous. Here, portray how to pick which amass the arriving business ought to be doled out. Through this arranged algorithmic principle, we've outline various parts of load evening out algorithmic guideline and presented fluctuated thoughts that show its expansive capacities. Goal of the matrix surroundings is to acknowledge elite registering by ideal use of geologically dispersed and re-enactment of appropriated asset mgmt. what's more, programming for Framework figuring. In this paper, we kept an eye out for the issue of how to logbook and levelling dynamic load in a Network a building containing authoritative social affair related by heterogeneous correspondence data transmissions. This will made tallies which are ensured to discover a heap dissipating over PCs in a party that prompts the base reaction peroid. Pondering correspondence deferral and cost among get-togethers. Through wide beguilement's, we saw that our estimations perform superior to anything static structures like NASH and NASHP under flawless data or a little data trade between times while requiring out and out less correspondence overhead. The reasonability this method is enhanced instigating better execution. For an expansive data trade interim, tallies utilizing the proposed estimation.

Because of the enhancement of Lattice figuring over the Web, there is direct a need for dynamic load changing calculations which consider Framework plan, PC hetero-geneity, correspondence deferral, and asset whim. All these four sections relating to the characteristics of the Framework taking care of condition referenced as of now. The proposed estimations are ensured to discover the stack stream among PCs in a social affair that prompts the base reaction time or cost. The impact of fluctuating cost is analysed which apparently has the capacity to alter the heap among truly stacked and gently stacked PCs. Three appropriated figuring are made to pick which pack the heap ought to be dispatched to, considering correspondence cost among get-togethers. In all calculations, the adjoining scheduler aggregates the data of all its district focus focuses unpredictably.

5 Conclusion

Min-Min and Max-Min calculations square measure material in small scale appropri-ated frameworks. When the measure of the minor assignments is an incredible measure of the huge errands in an extremely meta-undertaking, the Min-Min rule can't plan assignments, appropriately, and along these lines the make length of the framework gets similarly monster. What is more it doesn't give a heap adjusted calendar. To beat the limitations of Min-Min rule, a substitution assignment programing rule, is antici-pated. It's performed in two-stages. This investigation is simply included with the measure of the assets and errand execution time. The investigation will be extra reached out by considering low and high machine no consistency and errand no consistency. Additionally, applying the anticipated standard on real network setting and considering the esteem issue will be elective open downside amid this space. A Heap evening out part has been created that executes in recreated framework environment. It contains connection to totally unique pages: indicate occupations, demonstrate assets, indicate distribution and show standing. Following is that the shot of the record page. Second is work length. Occupation ID is selective for each activity submitted to Matrix. Length is that the normal employment time length given at the season of accommodation. This can be produced once designation of employment to asset. Allotment depends on free asset with regards to request of occupation that has been submitted to asset. This distribution expands the asset usage. Amid this screen capture welcome each Employment ID. Occupation will be moved from sender asset to collector, there's an asset name. Anyway framework application execution remains a test in powerf matrix environment. Assets are regularly submitted to Network and might be pulleα back from Matrix at any minute.

References

1. Cao, J., Spooner, D.P., Jarvis, S.A., Nudd, G.R.: Grid load balancing using intelligent agents. Future Gener. Comput. Syst. **21**(1), 135–149 (2005)
2. Jayapandian, N., Md Zubair Rahman, A.M.J., Gayathri, J.: The online control framework on computational optimization of resource provisioning in cloud environment. Indian J. Sci. Technol. **8**(23), 1–12 (2015)
3. Zhang, J., Guo, H., Hong, F., Yuan, X., Peterka, T.: Dynamic load balancing based on constrained kd tree decomposition for parallel particle tracing. IEEE Trans. Visual Comput. Graph. **24**(1), 954–963 (2018)
4. Garg, S.K., Buyya, R., Siegel, H.J.: Time and cost trade-off management for scheduling parallel applications on utility grids. Future Gener. Comput. Syst. **26**(8), 1344–1355 (2010)
5. Yu, J., Buyya, R.: A taxonomy of workflow management systems for grid computing. J. Grid Comput. **3**(4), 171–200 (2005)
6. Yu, X., Yu, X.: A new grid computation-based Min-Min algorithm. In: Sixth International Conference Fuzzy Systems and Knowledge Discovery. FSKD'09, vol. 1(1), pp. 43–45. IEEE (2009)
7. Herrera, J., Huedo, E., Montero, R.S., Llorente, I.M.: A grid-oriented genetic algorithm. In: International European Grid Conference, pp. 315–322 (2005)
8. Menasce, D.A., Casalicchio, E.: QoS in grid computing. IEEE Internet Comput. **8**(4), 85–87 (2004)
9. Lin, H.C., Raghavendra, C.S.: A dynamic load-balancing policy with a central job dispatcher (LBC). IEEE Trans. Software Eng. **2**, 148–158 (1992)
10. Penmatsa, S., Chronopoulos, A.T.: Price-based user-optimal job allocation scheme for grid systems. In: Proceedings 20th IEEE International Parallel & Distributed Processing Symposium, pp. 396–401 (2006)
11. Reda, N.M., Tawfik, A., Marzok, M.A., Khamis, S.M.: Sort-mid tasks scheduling algorithm in grid computing. J. Adv. Res. **6**(6), 987–993 (2015)
12. Arora, M., Das, S.K., Biswas, R.: A de-centralized scheduling and load balancing algorithm for heterogeneous grid environments. In: Parallel Processing Workshops. Proceedings. International Conference, pp. 499–505. IEEE (2002)

Smart Navigation Application for Visually Challenged People in Indoor Premises

Ganesh Kotalwar[✉], Jigisha Prajapati[✉], Sharayu Patil[✉],
Dilip Moralwar[✉], and Kajal Jewani[✉]

Computer Engineering Department,
Vivekanand Education Society's Institute of Technology, Mumbai, India
{2015ganesh.kotalwar, 2015jigisha.prajapati,
2015sharayu.patil, 2015dilip.moralwar,
kajal.jewani}@ves.ac.in

Abstract. It is very difficult for a visually impaired person to perform its day to day job with ease. Since Mobile applications are largely used among people they have high potential in aiding blind people. In this paper, we are trying to present an application to assist visually disabled. The android application will be using Deep learning object detection and identification techniques such as YOLO, R-CNN etc. The need for navigation help among blind people and a broader look at the advanced technology becoming available in today's world motivated us to develop this project. Technology is something which is there to ease tasks for human beings. Hence, in this project, we use technology to solve the problems of visually impaired people. The project aims to help users in navigation with the use of technology and our engineering profession motivates us to use the technology we have.

Keywords: Object detection · Deep learning · YOLO · R-CNN

1 Introduction

According to the study presented by the World Health Organization (WHO) 2018, approximately 1.3 billion people suffer from some form of vision vitiation. Visually impaired people face many problems in their day-to-day lives. People with disability cannot move from one place to another independently. Traditionally, these persons use guiding sticks for detecting objects in front of them. When visually challenged people navigate in an unfamiliar environment, a system which assists or guide such people is needed [10].

- Furthermore, visually disabled people require assistance in the form of a volunteer to be guided in an unknown environment [8].
- Our objective - To develop an application for blind people.

A. P. Pandian et al. (Eds.): ICCBI 2018, LNDECT 31, pp. 998–1004, 2020.
https://doi.org/10.1007/978-3-030-24643-3_119

2 Ease of Use

2.1 Application Requires

This app is designed in order to make the day to day chores especially for blind people easy. Application of visually challenged peoples with minimal interface complexities. The application should be able to track objects. Since the person is blind, the detected objects would be informed to the user by voice. Additional features like location tracer, the ability to detect logo and expiry date analysis of a product, etc. [5] (Fig. 1).

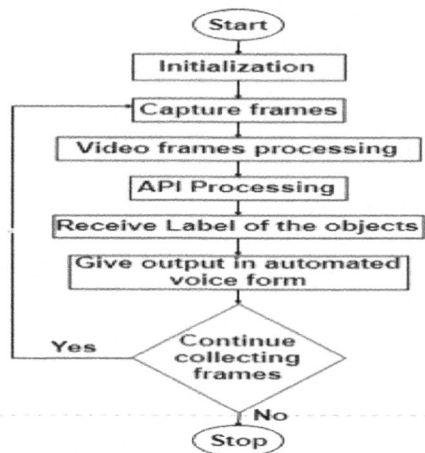

Fig. 1. Application flow diagram

2.2 Relevance of the Project

The application developed can detect the objects in the user's surroundings. It can alert the user of the obstacles in his pathway and this way helps the user to navigate from one place to another saving him from tripping anywhere. It will also solve the problem of keeping a special device or a walking stick. The reason it is more reliable is that it is developed on the Android operating system and Android-based smartphones are very common and highly available almost everywhere [7].

3 Related Work

3.1 Smartphone Application to Assist Visually Impaired People

This paper delineates about the smartphone application for visually impaired people which is based on similar technology as proposed for our project. This application uses various sensory modules which detects the obstacles and thus offers more precise guidance in a certain direction. The technology which we are going to use for object detection would be Deep Learning. The image captured by the camera will be processed in real time and the object would be identified.

…his app is OCR. This is used to scan a document and convert it into
…en converted to speech using TTS. The method includes Text To
…conversion of the scanned image using character recognition techniques.
…of this paper has discussed using technologies which are low cost and are
…As discussed above authors have used sensory MEMS modules for imple-
…ɡ the android application.

…after all the tests were done the following system performance was found. They
…are as follows: The interface of the application was user-friendly, even though the
ambient noise was high the communication did not tamper, just because a visually
challenged person had good hearing abilities, the navigation in indoor premises
required the user to travel slowly. In some areas where the phone signal is weak, it is
required that the speed should be enough to convey proper information. The authors
have also written about how this system differs from the existing systems and how it is
superior to them. They stated that this aiding application with the help of small external
sensory modules proves to be a viable solution [1].

3.2 A Review of Object Detection and Tracking Methods

This paper consists of a typical object tracking framework, generally consisting of three
modules: Object Detection, Object Modeling, and Object Tracking. They interact with
each other during a tracking process. These are discussed in detail in the following
sections as follows:

Object detection, a prerequisite for initializing a tracking process, refers to locate
the object of interest in every frame of a video sequence. There are generally two
approaches of object detection strategies commonly used to initialize a tracking process: manually locating the object in the first frame and let the system detects features,
such as corners, to track the object in the next frame and automatic detection of the
object using predefined features, such as color.

There are many techniques to detect moving objects:

1. Background Subtraction

Background Subtraction is widely used in video sequences having a static background.
The method divides the extracted source i.e. the image into foreground and background. The foreground contains moving objects such as moving people, cars while the
background contains static objects, like road, building, trees, stationary cars, etc. So in
this technique, they have used a reference background image which is first captured
when the objects of interest are not present in the scene. Thus by extracting the current
image frame from its background the object which is moving is found. The resulting
image has values below a predefined threshold in the background area of the current
image except the area occupied by the object.

2. Temporal differencing

Temporal differencing is a method most suitable for situations where the camera is in
motion. It detects objects by taking differences of consecutive frames (two or three),
pixel by pixel. In a moving camera situation, the movement of the camera and the
object are mixed up. Therefore, some researchers proposed to estimated and adjust

camera motion first and then apply the background subtraction method. This method fails to detect the overlapping areas of the moving objects and wrongly detects trailing region of the object, known as ghost region, for a fast moving object.

3. Optical flow

Optical flow is another technique to find moving objects in video frames. It gives a two-dimensional vector field, also called motion field, that represents velocities and directions of each point in consecutive image sequences. The image is segmented into sections due to the discontinuities in the flow. The method, being computationally expensive, has an advantage that it can detect motion in video sequences having a dynamic background.

4. Object detection

It can be done by training a classifier that learns different object views and appearances by means of supervised learning methods. After a classifier is trained, the decision is made on the test region whether it is a target object or not [2].

3.3 Smart Guiding Glasses for Visually Challenged People in Indoor Environment

A multi-sensor fusion based obstacle avoiding algorithm is proposed, which solves one of the major problems that is detecting small obstacles and transparent obstacles e.g. the French door which also includes glass doors, using depth and ultrasonic sensors. Three kinds of auditory cues were developed for the ones who are completely blind so that they are given information about the direction where they can move ahead. Whereas for weak-sighted people, a visual enhancement which leverages the AR (Augmented Reality) technique and integrates the direction that is capable of being traversed and then adopted. The results of various experiments show that smart guiding glasses can efficaciously improve the user's traveling experience in an environment such as a college corridor that is complicated indoor and such everyday places. Thus our system serves a purpose of helping visually impaired people to ease their day to day casual tasks.

The Smart guiding device in the shape of a pair of eyeglasses for blind people for giving guidance efficiently and safely.

Though the ultrasonic method can measure the distance between the objects it cannot determine the accurate direction and suffers through inference problems when the system is tested in a local environment. Laser scanner based method is of high precision and resolution and hence are highly used in mobile robot navigation. However, they are expensive and heavy. Another disadvantage of high power consumption makes them unsuitable for mobile applications. As for camera-based method, there are many choices such as stereo-camera, mono-camera, and RGB-D camera [3].

3.4 Navigation System for Visually Impaired People

This paper is based on a navigation system which comprises an indoor and outdoor positioning of the common object detector system for detecting the position of the user.

navigation systems use GPS for positioning the objects. Unfor-
only be used outside environment basically outdoor environment
loyed radio signals cannot pass through solid walls. Navigation sys-
tdoor environment generally depend upon GPS signals and for indoor
depends upon different methods for finding the position of the user, as GPS
annot be received in indoor premises because it does not cover large walking
nge for the end users. It is possible that in the indoor environment there may
t things that are quite near to each other to be able to get distinguished.

Currently, indoor navigation systems always use radio signal for positioning of the
common objects, which may experience the signal impairment problems, such as Radio
Frequency interference and multipath propagation. A location finding system with
talking assistance is for both navigations of the indoor and outdoor environment. The
System consists of a walking stick with a GSM module for sending a message to the
authorized person at the time of tragedy, RF transmitter and receiver, and sonar sensors.
RFID is used for indoor localization and GPS system is used for outdoor localization.
Thus, this walking stick with GPS system decreases the installing costs of many RFID
tags in outdoor for place identification. "Drishti" is a technique based on GPS which
can switch the system with a simple vocal command from an indoor to an outdoor
environment and vice versa. Authors extend indoor version of Drishti to the outdoor
versions for blind pedestrians with the addition of two ultrasonic transceivers to provide a complete navigation system, which is smaller than a credit card and are attached
on user's shoulder [4].

Table 1. Analysis of various navigation applications for visually challenged people.

Related work	Designed for	Approach used	Tools	Results
A. Smartphone Application to Assist Visually Impaired People	Scanning a document and converting it into text form which is further converted to speech using TTS	OCR and Text To Speech (TTS) and YOLO object detection model	Arduino and Raspberry Pi	Able to help the visually impaired
B. A Review of Object Detection and Tracking Methods	Detection, tracking, and modeling of the object	Background Subtraction and Temporal Differencing	Android mobile camera access.	Objects were identified distinctly
C. Smart Guiding Glasses for Visually Challenged People in Indoor Environment	Determine the depth of the object with respect to users height	Ultrasonic sensor, Depth sensor	RGB-D camera	The stretch between the user and the object is identified
D. Navigation System for Visually Impaired People	Helps to navigate a path for visually impaired	GPS system so as to trace the path	Google API for maps	Path detection helps people for easy navigation

The table given above summarizes all the approaches proposed by different authors (Table 1).

4   Conclusions and Future Works

We started with the motivation and the idea to solve the problems of visually impaired people. We found many methods to implement object detection and found the usage of OpenCV Library and Google Cloud Vision API as the best choice. Our project will be developed on Android and since it is developed by Google, there would be almost no compatibility issues.

Our expected result will be an android application which will be having a user-friendly interface so that a blind person can easily use it for navigation and for different purposes as well. During navigation, the objects that are detected by the application are made known to the user by an automated voice. Thus the user will come to know about the obstacle ahead and can navigate accordingly. It will be also used for knowing the details of a product after it's details are scanned by the camera of the phone. We will be adding a few audio clips of books so that the blind people can hear them in their free time. We will also be using OCR (Optical Character Recognition) for reading and recognizing the printed text and then converting it into automated voice [9].

When we compare and look at the other traditional methods like a modified guiding stick, our application provides more data and functionalities for the visually impaired person using this app. It is a user-friendly application with many functionalities. We will be training our model for various day to day life obstacles in the path of a blind person and provide a huge dataset for the same so that the object recognition is achieved efficiently and successfully. The percentage of recognition will be higher than any other applications.

The proposed system will be compatible with different environments. Users don't need to worry about unfamiliar areas. It will warn users of the obstacles ahead. Feature extraction is one more important feature of this application which has many uses. We are working on to improve the recognition efficiency of this application so that every object is detected accurately and the user won't get any kind of injury or something because of false detection of an object [12].

The point of comparison between various papers is to understand the drawbacks of existing applications. Various points are as follows:

- Smartphone application which is already existing works on manual assistance while our application will be using neural networks to make entire system automatic.
- Smart guiding glasses uses a complex IOT based hardware whose only concept of depth acquisition will be used in our application.
- The object detection and tracking paper helps us to understand
  - Background subtraction for detection
  - Various visual tracking methods for classification
  - Feature-Based tracking.

Hence we are eliminating the use of manual assistance and creating an application entirely by using deep learning. From the above surveys between various papers, we found that the YOLO model for detecting various objects is the best way out of all the ways as it is fast and accurate and also gives results in very less time. Also, the paper suggesting the depth calculation is very useful as the visually impaired person can get the idea of how far the obstacle is from him/her [11].

References

1. Tepelea, L., Gavrilut, I., Gacsadi, A.: Smartphone application to assist visually impaired people. In: 14th International Conference on Engineering of Modern Electric Systems (EMES), Electronics and Telecommunications Department, University of Oradea, Romania (2017)
2. Verma, R.: A review of object detection and tracking methods. In: 2017 IEEE/ACIS 16th International Conference on Computer And Information Science (ICIS), Department of Computer Science and Engineering, Faculty of Engineering, JNV University, Jodhpur, Rajasthan, India
3. Bai, J., Lian, S., Member, IEEE, Liu, Z., Wang, K., Liu, D.: Smart guiding glasses for visually impaired people in indoor environment. In: IEEE Transactions on Consumer Electronics, vol. 63, Issue 3, August 2017
4. Lin, B.-S., Lee, C.-C., Chiang, P.-Y., Jimenez, A., Academic Editor: Simple smartphone-based guiding system for visually impaired people
5. Holton, B., AccessWorld Magazine Correspondent: Smartphone GPS navigation for people with visual impairments
6. Al-Shehabi, M.M., Mir, M., Ali, A.M., Ali, A.M.: An obstacle detection and guidance system for mobility of visually impaired in unfamiliar indoor environments. Int. J. Comput. Electr. Eng. 6(4), 337–341 (2014)
7. Jian, R. (Forest), Lin, Q., Qu, S.: Let blind people see: real-time visual recognition with results converted to 3D audio. In: IEEE Transactions on Consumer Electronics, Civil and Environmental Engineering, Stanford, vol. 21, Issue 26, September 2017
8. Arvai, L.: Mobile phone based indoor navigation system for blind and visually impaired people. In: 19th International Carpathian Control Conference (ICCC), Department of Infocommunication Technologies, Bay Zolton Nonprofit Ltd. for Applied Research, Miskolc, Hungary (2018)
9. Ranaweera, P.S., Madhuranga, S.H.R.: Electronic travel aid system for visually impaired people. Dept. of Electrical and Information Engineering, University of Ruhuna, Sri Lanka
10. Dunai, L.: Obstacle detectors for visually impaired people. Universitat Politècnica de València, Research Center in Graphic Technology, Spain
11. Sharma, T.: NAVI: navigation aid for the visually impaired. PESIT, Bangalore
12. Panchal, A.A.: Character detection and recognition system for visually impaired people. In: 2016 IEEE International Conference on Recent Trends in Electronics, Information & Communication Technology (RTEICT), Department of Electrical Engineering, VJTI, Mumbai, India, 20–21 May 2016

Orisyncrasy - An Ear Biometrics on the Fly Using Machine Learning Techniques

Hitesh Valecha$^{(\boxtimes)}$, Varkha Ahuja$^{(\boxtimes)}$, Labhesh Valechha$^{(\boxtimes)}$,
Tarun Chawla$^{(\boxtimes)}$, and Sharmila Sengupta$^{(\boxtimes)}$

Computer Engineering Department, Vivekanand Education Society's
Institute of Technology, Mumbai, India
{2015hitesh.valecha, 2015varkha.ahuja,
2015labhesh.valechha, 2015tarun.chawla,
sharmila.sengupta}@ves.ac.in

Abstract. Like other biometric using face, iris and finger, ear as a biometric contains a large amount of specific and unique features that allow for human identification. The ear morphology changes slightly after the age of 10 years and the medical studies have shown that significant changes in the shape of the ear happen only before the age of 8 years and after the age of 70 years. It does grow symmetrically in size and begins to bulge downwards as the person ages, but that is a measurable effect. Studies suggest that ear changes only 1.22 mm per year. Although ear and face images can be captured easily from a distance, the ear is unaffected by cosmetics and external entities like spectacles, mask etc. Also, the colour distribution of ear, unlike face, is almost uniform. The position of the ear is almost in the middle of the profile face. Ear data can be captured even without the knowledge of the subject from a distance. Ear biometrics can stand as an excellent example for passive biometrics and does not need much cooperation from the subject, which meets the demand of the secrecy of the proposed system. A digital camera takes the profile face images of the subjects under test from different angles, from which the section of the ear is segmented, preprocessed and feature vectors of the ear are calculated. Further, the feature vectors are compared from the similar outputs developed from the ear images captured at angles by cameras placed at different positions. The feature vectors are then analysed in different test cases which consists of the rotation of face in the same plane, different plane, different light conditions, etc. will be given to machine learning model as input which would be trained to satisfy the recognition of the person. The process, though complicated, would develop a system which would, for a particular person, provide an authenticated ear-based biometric identification system.

Keywords: Authentication · Connected component analysis · Ear biometrics ·
Edge detection · Machine learning · Security

© Springer Nature Switzerland AG 2020
A. P. Pandian et al. (Eds.): ICCBI 2018, LNDECT 31, pp. 1005–1016, 2020.
https://doi.org/10.1007/978-3-030-24643-3_120

1   Introduction

OriSyncrasy is derived from two words - Oricula and Idiosyncrasy where Oricula (in Latin) means ear and Idiosyncrasy means characteristics. Ear biometrics on the fly is the system which automatically identifies the human using ear characteristics without the person's awareness. The ear can become a stable biometrics as it does not vary much with age. It is observed that the change in the ear is much noticeable during the period from four months to eight years old but during the period from eight years to seventy years, it does not vary much.

Each biological feature has its strengths and weaknesses as a biometric like fingerprint has a problem of a fake fingerprint using clays or dummy printing [12]. Iris scanners are expensive and require good lighting conditions [6, 13]. Different facial expressions may increase the false rejection ratio, and ageing also influences an individual's facial components. Voice recognition is not much effective in a noisy environment and suffers from voice spoofing.

The main objective of this paper is to propose a biometric system where identification of the subject is done without the person's awareness. The profile image of a person's face is captured by setting up cameras at different angles, and that image is used for segmentation of the ear part. As ear is remarkably small as compared to face, the ear would lead to a lesser number of computations in future processing. Canny edge detector extracts the helix and antihelix portion of the ear. Then the features of the ear are calculated. These feature vectors are given as training data to the machine learning model and the testing data, i.e., the new feature vectors of the subject, which were calculated at the time of authentication, are passed to the learning model for recognition.

2   Related Work

An ear biometric system was first developed by *Iannarelli* [5] in which measurements were taken manually. But *Burge and Burger* [9] in 1998 developed a system where measurements were taken automatically with the help of a computer, based upon Voronoi diagrams which are taken out from the contours. Recognition system was further developed by *Hurley* et al. [8] using Force field transforms where an image is considered as an array, and there is some mutual attraction between the particles which results in a Gaussian Force Field.

Benzaoui, Hezil, and Boukrouche [7] proposed the use of local descriptors for ear biometrics system which were different and were able to overcome the constraints such as lighting and occlusion. Various descriptors were used such as Local Binary Patterns (LBP) to measure grayscale invariant texture, Local Phase Quantization (LPQ) which are widely used for blur-invariant in texture recognition. LPQ is also used to overcome the relative sensitivity of LBP blur.

Bhanu and Chen [10] proposed a procedure that took helix and antihelix parts of the ear in consideration. The helix and antihelix parts are used for the recognition of an ear, as after the calculation of the helix and antihelix parts, each ear has a unique set of values and hence the ear is used for identifying a person within the considered environment. The procedure is used for 3D matching, as it takes into account the rigid transformation and selected control points and save that information to the database for further analysis and recognition process. Also, iterative closest point (ICP) algorithm is used for alignment and for improvising the results [2]. Root mean square (RMS) is calculated and checked, if it is found to be minimum then the criteria is matched, and the person is authenticated.

3 Proposed Model

The system consists of three stages viz, Segmentation, Edge Detection and Feature Extraction & Recognition. First, the camera detects the profile face, from which the ear portion is segmented. In the Feature Extraction stage, pre-processing is done on the segmented ear image, and then contour edge detection is performed on it. Finally, in the recognition stage, the test image is passed to the system and the person is authorised based on a threshold.

Following are the factors that affect the accuracy in ear recognition:

- Three cameras should be placed in the environment in such a way that the profile face of the person can be captured accurately.
- On the fly biometrics are constrained to a certain number of people and is most suitable for applications such as authenticating people in the board meetings or medical labs.
- Ambient light conditions may also affect the results.
- It would be difficult to segment the ear when some part of the ear is obscured.

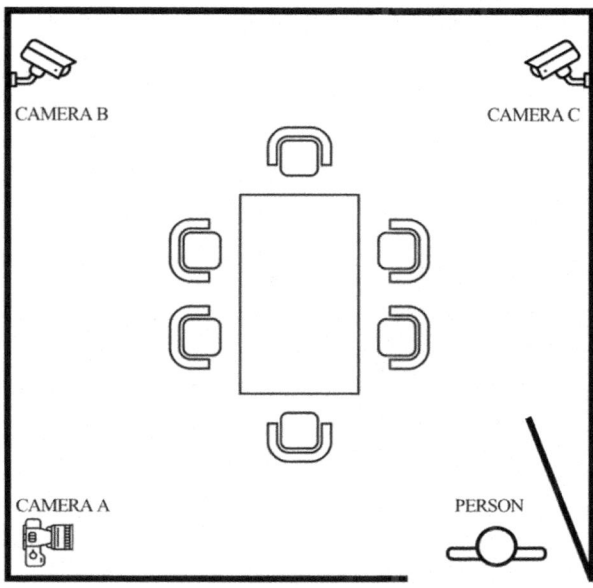

Fig. 1. Illustration of the ideal environment

A. *Controlled Environment*

A high-resolution camera, CAMERA-A, is pointing at the entrance of the room. When a person opens the door, all the cameras will start discreetly surveilling the room (Fig. 1).

The purpose of CAMERA-A is to get the profile shot [1] of the person (head upright) when he enters the room. This angle can get us the sharpest image of the ear, without expecting much tilt or rotation of the ear. Meanwhile, CAMERA-B and CAMERA-C are used for capturing more profile photos for a better verification system.

The profile image will then be normalised, pre-processed, segmented and tested for the authentication of the individual. No action will be taken on the individual if the ear biometric system recognises them as part of the organisation, permitted to be in the room. On the contrary, if the individual doesn't match up with anyone in the database, security will be alarmed instantaneously.

The intruder might also be aware of the presence of such a biometric system and to circumvent it, they could cover their ears by cloth, hair, headphones, jewellery, etc. to trick the system. In this scenario, the system will be inoperative.

In total, there are three cameras to cover the surroundings of the whole environment, but for power management, an additional camera can be placed outside the room to capture anyone entering, while the other cameras could remain switched off. As and when someone approaches for the room and the camera set outside the room detects it, the cameras inside the room will be signalled to switch on for capturing ear image to recognise and authenticate the person entering.

B. *Segmentation*

Segmentation of ear is performed to reduce the redundant data so that the ear could be efficiently processed in the later stages, i.e., feature extraction and feature matching. The added benefit of segmentation is the removal of the background disturbances from the input image which could constrain the feature extraction process and likewise the feature matching process.

Segmentation is done using the Topographic Labelling method which is a relatively lightweight approach [4]. Each pixel is assigned a topographic label based on the intensity gradient around. Each group of labelled points provides a different set of highlighted features. Multiple label groups combined would allow for the selection of specific interest regions, such as the ear. So the labels utilised for ear segmentation are the ridge, convex saddle hill, and convex hill labels.

It then finds the difference of the dilation and erosion of the profile face image and the dot product of this difference is blurred and thresholded.

The connected components of the mask are coloured with different colours and components whose width and height is less than 40px are discarded. On these sub-regions, the difference between triangular erosion and dilation is performed. Finally, the complexity of all the sub-regions is calculated and the sub-region with the highest complexity is found to be the ear of the person (Figs. 2 and 3).

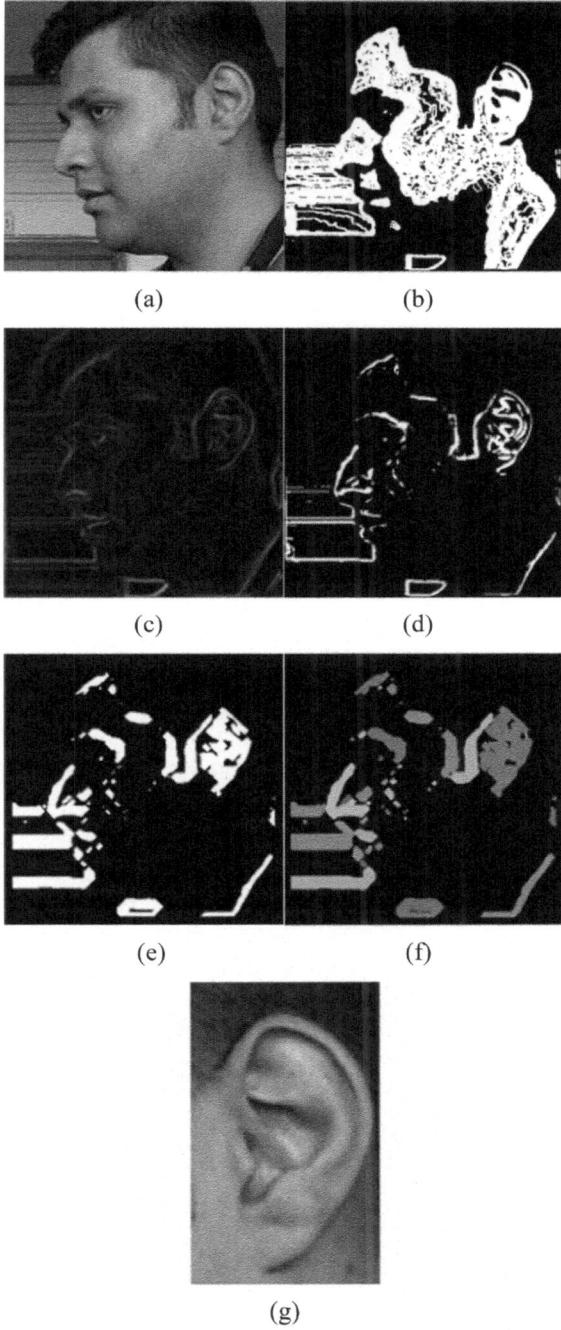

(a)   (b)

(c)   (d)

(e)   (f)

(g)

Fig. 2. Steps of segmenting an ear from the profile image. (a) is the profile face image from CAMERA-A. (b) is the topographical labelling of the image. (c) shows the difference of erosion and dilation and (d) is the dot product of (b) and (c). We get (e) by blurring and rethresholding (d) 10 times. (f) is the colour connected component of the profile image. (g) is the final segmented image from the profile face image.

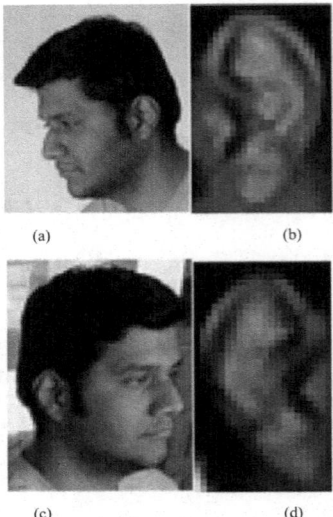

(a)   (b)

(c)   (d)

Fig. 3. Segmentation of ear from different cameras. (a) and (c) are the profile face images from CAMERA-B and CAMERA-C respectively. (b) and (d) are their respective segmented ear images.

C. *Edge Detection*

Canny edge detector is used for edge detection. For the binarization part, the gray values and sigma (standard deviation) is passed to the Canny edge detector in the form of an input array of dimensions same as that of the image (width height) [11]. Now after binarization and detection, the pixels which constitute as the edge are 1, while other pixels are 0 in the output binary image (Fig. 4).

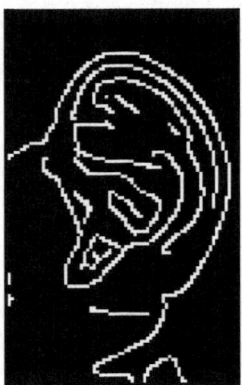

Fig. 4. Edge detected from the segmented ear.

D. *Feature Extraction*

For calculations of the feature vector, helix and antihelix parts of the ear are used (Fig. 5). After extracting edge from the image, max-line and normal-lines are calculated. The term max-line (m) refers to that longest line which can be drawn from both its endpoints on the outer edge of the ear. The normal-lines are the lines which are perpendicular to the max-line and divide the max-line into (n + 1) equal parts, where n is a positive integer. The longest edge can be split into (n + 1) equal parts using the section formula:

$$P \equiv \left(\frac{mx_2 + nx_1}{m+n}, \frac{my_2 + ny_1}{m+n} \right) \tag{1}$$

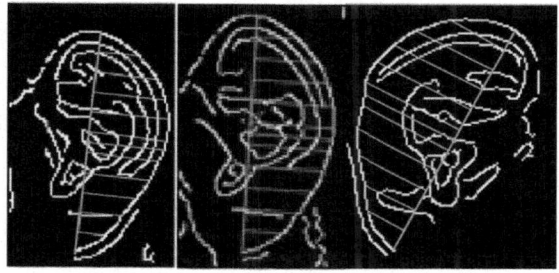

Fig. 5. Extracting feature of the segmented ears from all angles

The midpoint of the max-line (m) is found out, say a [3]. The perpendiculars to the max-line are drawn, and the points where these perpendiculars intersect the outer edge of the ear are noted. The linear distances between the intersection points on the outer ear and the max-line are calculated. Total ten distances were calculated for outer edge and two distances for the inner edge of the ear (Table 1).

Table 1. Extracted features

Parameters	CAMERA-A	CAMERA-B	CAMERA-C
Edge length	450.56	516.99	532.30
Midpoint	(157, 284)	(150, 280)	(209, 290)
P1[a]	(255, 107)	(247, 72)	(217, 33)
P2[a]	(289, 155)	(290, 123)	(286, 58)
P3[a]	(307, 198)	(313, 170)	(334, 86)
P4[a]	(316, 238)	(325, 221)	(370, 122)
P5[a]	(311, 281)	(325, 267)	(394, 166)
P6[a]	(294, 321)	(317, 315)	(398, 218)
P7[a]	(276, 360)	(302, 357)	(409, 266)

(continued)

Table 1. (*continued*)

Parameters	CAMERA-A	CAMERA-B	CAMERA-C
P8[a]	(254, 400)	(283, 405)	(406, 323)
P9[a]	(224, 435)	(252, 452)	(394, 387)
P10[a]	(189, 474)	(220, 495)	(386, 443)
P11[b]	(285, 297)	(291, 289)	(313, 236)
P12[b]	(177, 286)	(223, 284)	(232, 278)

[a]Intersection point on the outer edge of ear
[b]Intersection point on the inner edge of the ear

E. *Recognition*

For the recognition phase, the training data would be as input to the machine learning model (Fig. 6). Now when the person arrives, new values (10 outer + 2 inner) are calculated which would be treated as testing data and passed to the machine learning model for the authentication of a person (Table 2, Fig. 7).

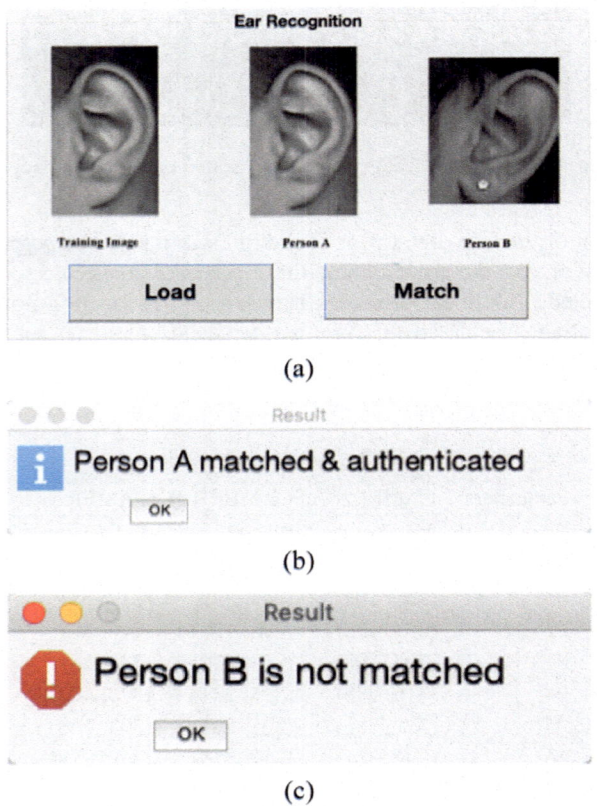

Fig. 6. Authentication of individuals. In (a) we can insert the training and testing images. (b) is the message when training and testing data match, otherwise (c) is the message.

Table 2. Extracted features

Parameters	Training data	Testing data	
		PERSON-A	PERSON-B
Edge length	450.56	451.23	476.03
Midpoint	(157, 284)	(150, 280)	(209, 290)
P1[a]	(255, 107)	(238, 73)	(240, 68)
P2[a]	(289, 155)	(294, 114)	(296, 104)
P3[a]	(307, 198)	(322, 153)	(325, 155)
P4[a]	(316, 238)	(340, 192)	(365, 210)
P5[a]	(311, 281)	(347, 230)	(357, 250)
P6[a]	(294, 321)	(332, 280)	(350, 300)
P7[a]	(276, 360)	(325, 325)	(342, 358)
P8[a]	(254, 400)	(316, 373)	(296, 380)
P9[a]	(224, 435)	(287, 424)	(287, 405)
P10[a]	(189, 474)	(257, 471)	(261, 423)
P11[b]	(285, 297)	(298, 274)	(297, 244)
P12[b]	(177, 286)	(211, 285)	(231, 235)

[a]Intersection point on the outer edge of ear
[b]Intersection point on the inner edge of the ear

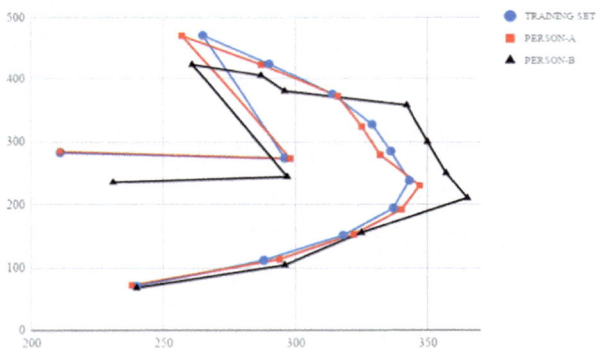

Fig. 7. Analytical graph

The extracted features are stored as the training data. Once anyone enters the environment, their profile face images are captured and the ear features are extracted. Out of the 12 points, if 7 or more than 7 points are matched then the subject is said to be authenticated.

In the above graph, PERSON-A's 10 points are matched with the training set, and hence PERSON-A is authenticated. On the other hand, PERSON-B's only 4 points are matched, which is less than 7 points, and hence PERSON-B is not authenticated.

4 Test Cases

As ear images are taken from different angles and different cameras, the orientation of the images may vary, and there are different test cases and operations to solve those cases viz, normalising the ear image.

A. *Rotation in same plane*

Ear images can be taken while the person is on the move, the rotation of ear image will differ while the person passes the camera (Fig. 8).

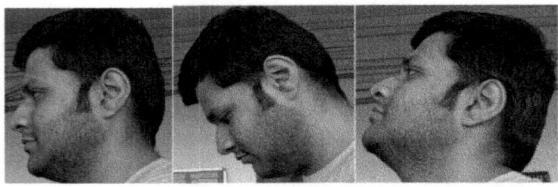

Fig. 8. Rotation of the face in the same plane of the ear

B. *Rotation across different planes*

The ear can be shifted by an angle (15 to 45°) and the image captured by the camera requires skew operations to get the image in the same plane (Fig. 9).

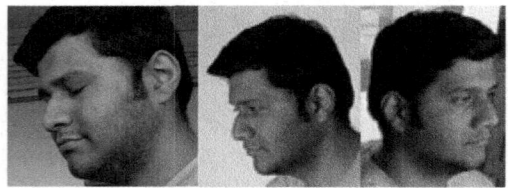

Fig. 9. Rotation not in the same plane of the ear

C. *Light restraint*

The image if taken in limited light will have shade on the ear region. The resulting segmented ear or recognition will not be successful. If the profile face photo is over-exposed, there won't be many details in the ear region. Segmentation and recognition will again not be successful.

D. *Random*

There is a possibility of the person coming in a rush and walking randomly and quickly. In such a case, the camera could capture the images in a blur.

5   Conclusion and Future Scope

This paper demonstrates that automatic identification of a subject can be achieved using the ear's uniqueness. The environment can be set up in places where limited people need to be authenticated. If the person is aware of the proposed system, they can avoid exposing their ear to the camera which results to the failure of the system. The computational time is higher since we're capturing the images from multiple cameras and then applying operation on each image.

The most significant advantage of the system is to identify the person without their knowledge. This feature will change the dynamics of the biometrics, and it can be used in applications which require the subject under test to be unaware of the authentication. Our system does not require any extra machinery for identification and thus, saves a lot of expenses as compared to other biometrics and can bring complete automatism in the biometrics field. Hence in this paper, we have demonstrated that human ear can be used as a unique biometric for passive authentication.

Ear biometric system can have many applications such as access and attendance control, travel control, authorising financial transactions, remote voting, action control, etc. Nowadays airports use a combination of face and iris biometrics to authenticate a person. Instead, a single ear biometric system can be used to verify a person with the same accuracy. Ear lobe has features that can be used for personality analysis - attached lobe, broad lobe, etc.

Acknowledgement. Consent: all the participants are hereby expressing their approval for both the publication of the paper & usage of photographs therein and we are not under any influence. We take the responsibility involved in the publication of our paper.

References

1. Bustard, J.D., Student Member, IEEE, Nixon, M.S., Associate Member, IEEE: Toward unconstrained ear recognition from two-dimensional images. In: IEEE Transactions on Systems, Man, and Cybernetics—Part A: Sytems and Humans, vol. 40, no. 3, May 2010
2. Anwar, A.S., Ghany, K.K.A., Elmahdy, H.: Biometric recognition using 3D ear shape. ICCMIT, Cairo University, Egypt (2015)
3. Yan, P., Bowyer, K.W., Fellow, IEEE: Biometric recognition using 3D ear shape. In: IEEE Transactions on Pattern Analysis and Machine Intelligence, vol. 29, no. 8, August 2007
4. Said, E.H.: Ear segmentation in color facial images using mathematical morphology. North Carolina Agricultural & Technical State University, Greensboro, NC, Ayman Abaza and Hany Ammar, Computer Science and Electrical Engineering Dep., West Virginia University, Morgantown, WV, IEEE (2008)
5. Iannarelli, A.: Ear Identification. Paramount Publishing, Fremont (1989)
6. Ross, A.: Human ear recognition. West Virginia University, and Ayman Abaza, West Virginia High Tech Consortium Foundation, IEEE, November 2011
7. Benzaoui, A., Hezil, N., Boukrouche, A.: Identity recognition based on the external shape of the human ear. Laboratory of PI: MIS, May 1945

8. Hurley, D.J., Nixon, M.S., Carter, J.N.: A new force field transform for ear and face recognition. In: Proceedings of the IEEE International Conference on Image Processing ICIP (2000)

9. Burge, M., Burger, W.: Ear biometrics. In: Biometrics: Personal Identification in Networked Society, pp. 271–286. Springer-Verlag, Berlin, Germany (1998)

10. Bhanu, B., Chen, H.: Human Ear Recognition by Computer. Center for Research in Intelligent Systems, University of California at Riverside, California [Online]. https://doi. org/10.1007/978-1-84800-129-9

11. Prakash, S., Gupta, P.: Ear Biometrics in 2D and 3D. Computer Science and Engineering, Indian Institute of Technology, Kanpur [Online]. https://doi.org/10.1007/978-981-287-375-0

12. Jain, A., Ross, A.A., Nandakumar, K.: Introduction to Biometrics. USA [Online]. https://doi. org/10.1007/978-0-387-77326-1

13. Jain, A., Ross, A.A., Nandakumar, K.: Handbook of Multibiometrics. USA [Online]. https:// doi.org/10.1007/0-387-33123-9

A Comprehensive Study of Various Techniques Used for Flood Prediction

Sagar Madnani[✉], Sahil Bhatia, Kajal Sonawane, Sukhwinder Singh,
and Sunita Sahu

Computer Engineering Department, Vivekanand Education Society's
Institute of Technology, Mumbai, India
{2015sagar.madnani, 2015sahil.bhatia,
2015kajal.sonawane, 2015sukhwindersingh,
sunita.sahu}@ves.ac.in

Abstract. Floods, the naturally occurring hydrological phenomena, caused due to the meteorological events like intense or prolonged rainfall, unusual water overflow of high coastal estuaries on the result of storm surges. On an account of a lot of concrete structures in urban areas, high-intensity rainfall causes urban flooding and as there is no much soil available for water to percolate, this leads to huge drainage problems in urban cities. These types of floods cause harm to houses, buildings, humans, animals, farming land. Flooding leads to contamination of drinking water, spreading of diseases. In recent years, due to the combination of meteorological, hydrological and topographical modeling terminologies, advancement in data collection methods and algorithm analysis, the results of flood forecasting have been improved. In this paper, we have studied different techniques for flood prediction involving Neural Networks, Fuzzy Logic, and GIS-based systems with various algorithms considering different factors. The study shows, on introducing local parameters, increasing the size of acceptable error bounds, and combining different algorithms, better performance of the model is achieved.

Keywords: Flood prediction · Natural hazards · Heavy rains · Drainage

1 Introduction

Flood, one of the natural disaster which occurs over an area or dry land and submerged in water. In other words, the overflow of water in large quantity leads to a flood. Flooding mostly occurs due to the overflow of rivers Flood causes damage to the roads, cars, houses, people and even the natural environment is disturbed. Flooding destroys crops which wipe off trees and vital structures onto land. Every year the people are facing a serious challenge with the drastic nature of flood. The reason behind this involves many parametric and climatic conditions of the region [7]. Some of the floods occur suddenly, some take days or months due to the different size, time duration, and the affected area. In practice, the many areas are flooded easily which are near to the coast, lying on steep slopes.

© Springer Nature Switzerland AG 2020
A. P. Pandian et al. (Eds.): ICCBI 2018, LNDECT 31, pp. 1017–1031, 2020.
https://doi.org/10.1007/978-3-030-24643-3_121

The local overland flood gives a significant threat of unpredictability but short duration flood. It is not right to say, as an area which is not flooded in past, will not be getting flooded in future. Similarly, as an area which is locally flooded, doesn't mean that it will be flooded the next time. While there are areas which can absorb the water into the soil for e.g., Mangrove trees. The river prone flooded areas in many countries are carefully managed using remote sensing GIS-based terminologies [9] & Moderate Resolution Imaging Spectroradiometer (MODIS) based flood mapping techniques [8].

The bursting of banks is prevented using levees, bunds, dams, floodways, reservoirs, retention ponds and weirs. Emergency measures in case of failure of defenses are taken which includes usage of sandbags or portable inflatable tubes [3].

Europe and America face coastal flooding and use seawalls, dikes, millponds as coastal flooding defense. Therefore, there is a high need to predict flood in advance and prevent the huge losses.

2   Factors Causing Flood

There are many causes for the occurrence of floods such as overflowing of rivers, urban drainage basins, channels with steep slides, lack of vegetation, melting snow and ice. The major causes are inadequate drainage, open drains choked with garbage, degraded rivers reduced to drainage, lack of reservoirs and rainwater catchments, missing mangroves that mitigate the impact of torrential rains, inadequate warning, and evacuation systems, low-lying drains that flood during high tide. This section briefly describes the factors responsible for the flood to occur.

2.1   Heavy Rainfall

It is the most severe event which may lead to other numerous hazards, for example, floods can carry away the humans, animals and destroy buildings and infrastructure. The excess rainfall not only destroys these but can also cause degradation of crops. Landslides, which can put human life into tremendous pressure. Droughts will destroy farmlands, fields and enhance the soil erosion, which makes the land infertile. So there is a loss associated with life when rain falls in access or very low. Plants need varying amounts of rainfall to survive [12, 13].

2.2   Drainage

It is the procedure in which water goes from the upper soil layers to the lowermost layers. Most soils do not possess water absorbing capacity, so the drainage of water is important. In nature, bound minerals like sand give speedy voidance, whereas large minerals such as clay will limit avoidance. Drainage also lowers the risk by maintaining the groundwater level so that water cannot be logged or it can result in floods when water overflows through it.

2.3   Tide

The tides are the periodic variation in sea levels that occur primarily because of the gravitational force exerted by the sun and the moon. Tides cause alternate rising and falling of the sea. Tides changes in the sea level, and also makes tidal streams, making a forecast of tides which are important so as to monitor their activities. When the crest, of the ocean or sea wave, reaches a specific location, high tide occurs [3]. As relative water level increases, there is a very less time taken by a hurricane to cause coastal flooding. Flooding currently happens with high tides in several locations due to climate-related sea level rise, land subsidence, and the loss of natural barriers. High tide flooding also called as "nuisance" flooding, is flooding that one forefront to public inconvenience such as road closure.

2.4   Ground Water Level

Groundwater is the water found beneath the earth's surface, within the pores in the soil and in between the spaces in rocks. It goes down deeper in the soil, within the rocks spaces which are known as aquifers. A major source of Groundwater is rainfall. Whenever there is high precipitation, excess water goes down in the earth surface, which results in water level rising causing groundwater flooding. Groundwater flooding occurs in the low-lying areas where there are more permeable rocks. These permeable rocks can be a regional rock in the earth bottoms which lie below less permeable rocks. Groundwater takes more time to dissipate as it moves slowly as compared to surface water.

2.5   Catchment Area

Catchment areas can be defined as an area where the ground water from rainfall or ice flow into a lower elevation to form a single body of water, that may be a lake or a river. Catchment area is the area near the bank of the river from where the water goes into the river. Rainfall water which goes into the catchment area can cause flooding. Also the climatic conditions nearby river and soil absorption capability decides amount of water that can percolate within the earth surface [1].

3   Related Work

Significant works have been done for predicting floods in advance. This section briefly describes the literature survey done in the field of flood prediction.

In [1], authors Tehrany, M., and S. Jones have proposed that the Logistic Regression approach is used as it found very efficient in case of flood hazards. It is used to assess the flood maps which were constructed from seven sets of random points, polygons and considering major 13 flood causative factors. The main advantage of employing this is the set of points which lie between the 500–700 range provides accurate prediction results. The limitation found is using the inventory type as polygon hinders the performance of the algorithm which might be possible because of providing a large amount of data.

In [2], authors Khan, Usman T, et al. have proposed the idea of flood forecasting and mitigation using the fuzzy neural network with the use of data-driven models. The model uses the MATLAB tool with two different terminologies: the first one is the Shuffled Complex Evolution algorithm which finds a solution to minimize the problems and involves refinement which was done with a fmincon function. Now the process consists of three major steps First, an Automated neural pathway strength features selection (ANPSFS) method is selected to optimize input data from a larger dataset. Second, a searching technique is chosen to find out the optimum network architecture. Third, a possibility-theory based fuzzy neural network (FNN) is applied to quantify the uncertainty present in the model parameters and even the outputs by using fuzzy numbers rather than deterministic values. The major advantage of using a data-driven model is that data gathered from real-time, high frequency & flow rate monitoring stations were used to measure the output of the model. Even, there are uncertainties associated like (i) to select the input variables from a huge dataset requires higher analysis. (ii) the architecture selection & its validation. (iii) the parameters involved shown in Fig. 1.

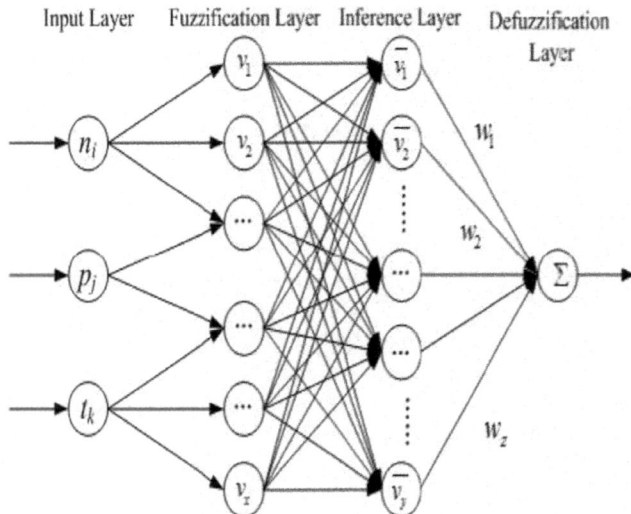

Fig. 1. Neuro-Fuzzy structure (Source: https://www.researchgate.net/figure/The-structure-of-fuzzy-neural-network_fig2_261494763)

In [3], authors S. S. Shahapure et al. designed a model for simulating urban catchment of Navi Mumbai, India. In their work, they have used simulation model which records changes in tidal with the provision of confining ponds. They have used FEM for flood simulation, Geographic Information Systems (GIS) and remote sensing for database preparation. Channel flow and Overland flow are simulated in one dimension by making use of the kinematic wave and diffusion waves approximation of St. Venant's equations and satisfactory results are achieved by comparing their proposed simulated model with already existing models. In this simulation of flooding, an event is shown by a flooded sketch of the channel. This simulated model successfully detects the location of the pond.

Author Liu, Fan, Feng Xu, et al. integrated stacked auto encoders (SAE) along with back propagation neural networks (BPNN) and used deep learning based approach for the prediction of stream flow. This approach combines the benefits of SAE such as powerful feature representation and excellent predicting capacity of BPNN [4]. To balance the data distribution, authors have used K-means clustering which classifies the data into various categories. Then, these classified data are trained by multiple SAE-BP modules which will further reduce the nonlinearity. They have also compared their model with support-vector-machine (SVM) model, the BP neural network model, the RBF neural network model, and extreme learning machine (ELM) model. The experimental results showed that the SAE-BP integrated algorithm has a much better result than other algorithms. The system proposed by the authors predicts stream flow for next 6 h. However further, more outstanding performance can be achieved by considering the imbalance of data distribution and cost sensitive mechanism of the SAE-BP network shown in Fig. 2.

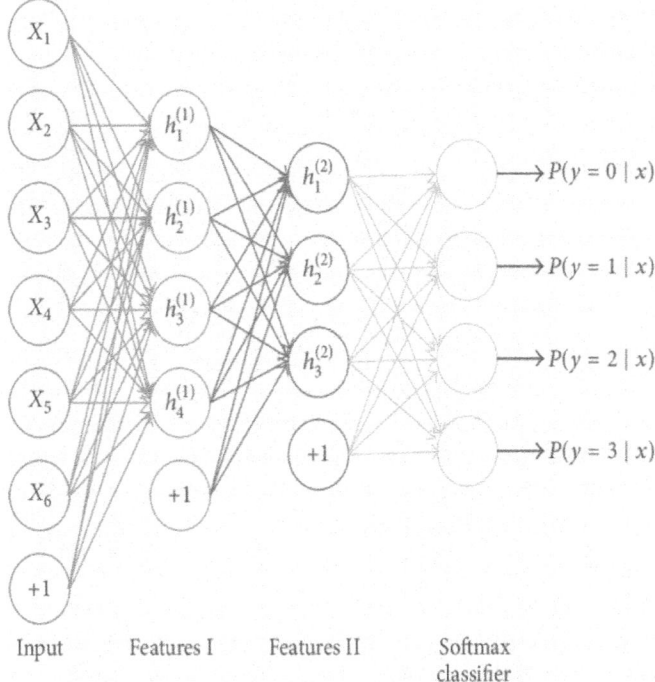

Fig. 2. SAE-BPNN architecture (Source: https://www.researchgate.net/figure/SAE-with-softmax-classifier_fig6_307865936)

Most of the flood forecasting methods are based on statistical data analysis & sometimes, the results obtained are not accurate. In [5], authors Zafar, Sarmad, et al. proposed a GIS model for real-time flood prediction integrating decision support system (DSS) with satellite image processing and hydrological modeling to improve flood forecasting. The study was done on Lower Indus River which covers a large area, and limited gauging stations are available in this area, so to access hydrological data and to improve spatial sampling, radar remote sensing is used. For these satellite images of NASA MODIS, SPOT Imagery was used. The back end of this system takes images as input and these images are processed to extract a river water stage. The first step is image acquisition, then image filtration. Image Filtering is a technique for modifying and enhancing an image. Next step is Lab Color Space Conversion, it describes all colors in 3-dimensional space. It is more enhanced than RGB and CMYK color models. The image is then processed through anisotropic diffusion segmentation and image binarization steps to reduce the noise in an input image. To differentiate the boundaries of objects in an image, morphological operations are used. SVM is used for identifying the stage in river images and classifying the images into different categories. Decision planes separate objects boundary in an image into different class memberships. The SVM work on the basis of the decision boundaries defined by the division planes. The advantage of this system is that it covers the whole spatial range over which the river passes. Inactivity knowledge of employing a GIS system, there's lot more generalization because of the whole knowledge being evolved. Disadvantage of this model is that the user stands to lose a lot of information due to this technique.

Authors Dushyant Patel, Dr. Parekh proposed an Adaptive Neuro-Fuzzy Inference System (ANFIS) model for flood forecasting considering two inputs peak discharge data and rainfall data near the Sabarmati river in Gujarat state in India [6]. The outcome is the predicted flow rate of the Sabarmati River. The proposed architecture has 5 layers, the input layer, the fuzzification layer, inferences process, defuzzification layer, and the final output layer. In layer 1, the input taken are fuzzified, calculation of output is done in layer 4 by applying fuzzy inference rules in layer 2 and 3. Layer 5 predict the final output by combining the results of layer 4 shown in Fig. 3.

Combination of the least square estimate and the gradient descent method is used for optimization of parameters. Results were compared with statistical method Log Pearson Type 3 using parameters Root Mean Square Error (RMSE), Correlation Coefficient (R), Coefficient of Determination (R^2) and Discrepancy Ratio (D). The comparison shows that the ANFIS model can accurately predict the flood as to compare to the statistical method. The advantage of this ANFIS model is that the ANFIS system can handle dynamic, non-linear and noisy data, and can understand the complex relations between the input parameters.

Authors Hallegatte, S. et al. have suggested a model based on climate change along with the associated flood risks [7]. This paper mainly focuses on the stormwater drainage (SWD) system and to adopt a faster transfer of water flow thereby reducing its duration. The SWD networks are spread over more than 100 catchment regions. These catchments examined the inadequacies, assessed design criteria and evaluated a Master Plan for the system. This approach is used to minimize the risks involved in Mumbai Suburban. Hence, rainfall measurements were taken from the PRECIS2 model; a

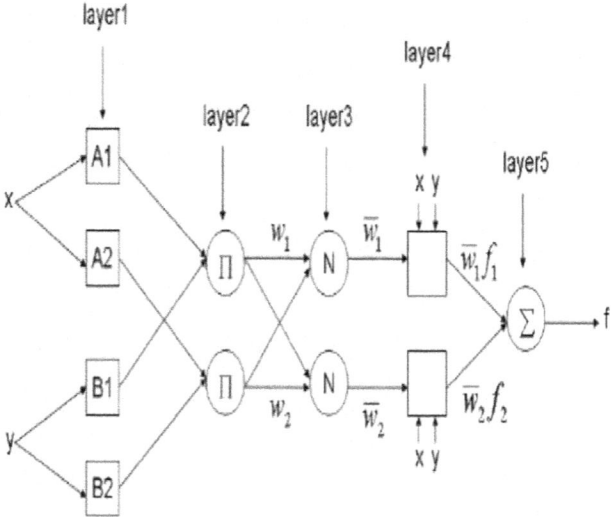

Fig. 3. ANFIS structure with 5 layers (Source: https://www.researchgate.net/figure/Five-layers-of-ANFIS_fig4_259973149)

high-resolution regional climate model (RCM), which depends on the Hadley Centre Coupled Model (HadCM3).

Author R. Brakenridge, et al. proposed a MODIS based approach to detect flood and create the flood inventory maps [8]. MODIS sensor considers having operational application including flood detection characteristics and warning with additional flood prevention or avoidance. To achieve this, it is required to work on strategies for performing the operation. MODIS (Moderate Resolution Imaging Spectroradiometer) sensor abroad USA satellite terra tool. MODIS data offers large coverage and provided by NASA. It is observed that water is passed as a map and combined with other maps. For Flood detection using AI methodologies which mainly based on Deep Learning models shown in Fig. 4.

In [9], author Millan Golic, Shamshirband et al. proposed Precipitation analysis and estimation. Precipitation is an important factor for climate changes. Precipitation Climate Index (PCI) is analyzed in this. PCI is used as a tool to measure hydrological hazards such as flood, tsunami, etc. In this, they proposed a PCI model using soft computing method, namely Support Vector Machines (SVM) and also [9] used the Adaptive Neuro-Fuzzy inference system (ANFIS) for precipitation estimation. The recent technique in this optimization algorithm is the firefly algorithm (FFA). The accuracy of the current model primarily focuses on the proper estimation of the variables associated with it. Wavelet transform (WT) has a number of functions for selection that depends on the analyzed signal. The analysis was performed to fragment the time intervals into its various components, then only the model uses the input for further processing. Data were obtained from the Republic Hydro-Meteorological Service of Serbia.

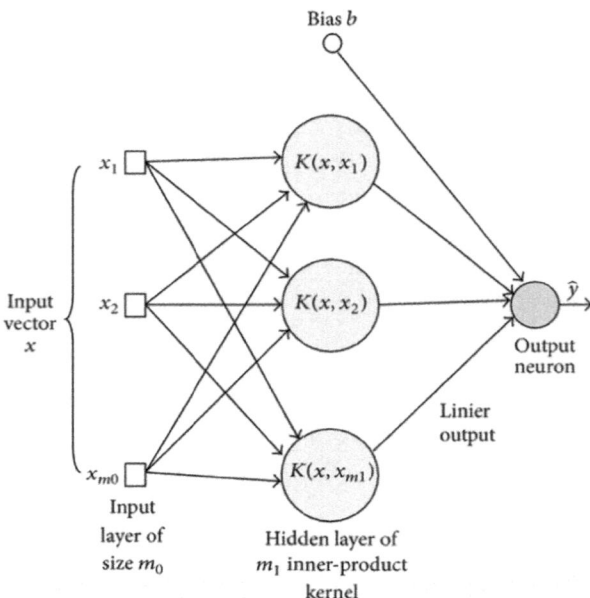

Fig. 4. The network architecture of SVM (Source: https://www.researchgate.net/figure/Architecture-of-SVM_fig10_281287464)

The study administered a scientific technique to form the SVM models for the PCI prediction like SVM-Wavelet, SVM-RBF, and SVM-FFA. The planned SVM-Wavelet model was obtained by combining 2 strategies, that is, the SVM and the wavelet transform.

In [10], author Seal et al. designed a Wireless sensor network (WSN) model for forecasting floods in rivers. This is a statistical model based on simple calculations to give timely warning to the people. With the real-time prediction, the model uses a linear robust multiple variable regression method which is cost efficient, to provide simplicity in cost, with many features having power and speed efficiency. In this way, Prediction accuracy the prime goal of the model is achieved. The proposed model uses theoretically independent parameters which may vary depending upon practical needs. Considering the case of an increasing graph, the water level is calculated using a polynomial and by its nature, the model can predict the future flood line. By comparing the predicted water level with the actual level for a time interval, the model gives the simulation of the flood line, around and below the actual data. It has an advantage of incorporating a wide number of parameters with which the river water-level varies and provide a precise prediction result. This model was simulated in Matlab and results obtained are very close to actual data. The proposed model doesn't have any real hardware implementation and prediction is suitable for only some specific rivers.

In [11], author Nasseri et al. proposed a flood warning system for forecasting rainfall using artificial intelligence. Since input and output parameters relation is nonlinear and non-convex, therefore authors proposed a hybrid model for rainfall forecasting. The model is developed for catchment regions of western suburbs of Sydney, Australia. Back Propagation Neural Network (BPNN) integrated with the Genetic Algorithm is used for training the model. This hybrid model was implemented to forecast rainfall using rainfall hyetograph. The proposed hybrid model gives better results when compared to individual artificial neural algorithms and higher order errors were also reduced. Rain gauging stations taken as input parameters were also reduced with reducing the redundancy in the model. The Authors made use of sensitivity analysis, to reduce the input parameters as well as to identify the time lags. But as rainfall doesn't fall in order, therefore different input gauge stations decrease with increase in time. And results would have been better if cumulative data was used instead of discrete data.

In [12], author R. Usha et al. proposed the use of classification & regression analysis which can be done using Self Organizing Feature Map (SOFM), Support Vector Machine (SVM) algorithms & also makes decision tree from the dataset using Iterative Dichotomiser (ID3) algorithm. The Self-Organised maps help to find relations between the input data and are simple, effective and easy to use. Here, in SOM the high dimensional data needed to be converted to low dimensional data and then the clustering process would be convenient. For handling large-level multi-structures and huge datasets, it consists of two steps. The SOM first reduces a supplied information and then encounters statistical modeling that constitutes a mixture of large variables including the coefficients. Data reduction uses different increasing and schismatic algorithms while the statistical modeling makes the task easier by allocating framework for multilevel structures. For prediction purpose, two different models were used namely (a) The Global Framework Model. (b) Support Vector Regression Kernel Agent.

Moving to next that is SVM which starts the collection with a pair of contrary points near to each other. Coming to the ID3, which uses knowledge or idea so that which attribute goes into a decision node and it doesn't work on continuous numerical data, but on discrete data. So the ID3 which is a supervised learner works on training datasets which make decisions. It uses two approaches to arrive at outcome i.e. Entropy and Gain calculation. Entropy calculates the homogeneity whereas the Gain depends on entropy which evaluates the attribute & gives maximum value from the decision tree. Entropy is calculated as shown in Eq. (3).

$$E(S) = -(p+) * \log_2(p+) - (p-) * \log_2(p-) \tag{3}$$

Where,

S is the set

p+ is the number of elements with positive values

p− represents that of negative values.

Gain is calculated as shown in Eq. (4).

$$\text{Gain}(S, A) = \text{Entropy}(S) - S((|S_v|/|S|) * \text{Entropy}(S_v)) \tag{4}$$

Where,

S is the set & A is the attribute

S_V is the subset of S where attribute A which has value v

|S| is the number of elements in set S and |Sv| is the number of elements in subset Sv.

Hence, the outcome of the model achieves 82% accuracy with all possibilities of attributes.

In [13], author Chatterjee et al. a novel rainfall model has been proposed. For model been employed they used a two-step, a Greedy method is used to find out most promising feature. In the model development phase, K-Means algorithm used on data then each clustered applied Neural Network, and secondly applied Hybrid Neural Network. Data is gathered from Dumdum meteorological station. The Rainfall prediction is divided into two methods, Firstly, by finding different rules that govern rainfall in the specific regional area and the second approach involves expert system to discover a different pattern which are related to rainfall measurements taken. Later they used Artificial Neural Network approach to forecast the rainfall.

In [14], authors Ankita Sharma, Geeta Nijhawan proposed a Back-Propagation Neural network model for rainfall prediction in the Delhi region of India. Rainfall data (seasonal and non-seasonal) have been collected from Indian Meteorological Department IMD website. Input data i.e. humidity, wind speed, and rainfall are considered and normalization is done on input data. Normalization can preserve the connection between the particular knowledge values. Feed Forward Neural Network is implemented with 3 hidden layers. In a feed-forward network, information is passed in the only forward direction. The first layer is the input layer which takes inputs and

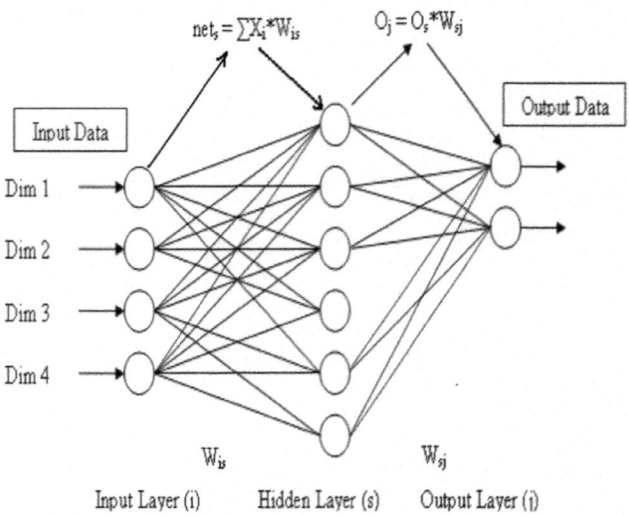

Fig. 5. BPNN architecture (Source: https://www.researchgate.net/Typical-supervised-BPNN-architecture-after-8_fig1_274251820)

computer output and forwards it to the next layer. Output depends on the threshold function when net input is greater than the threshold, the neuron fires shown in Fig. 5.

Backpropagation algorithm is used for training of input data. Matlab tools NFTOOL & NNTOOL is used for training.

NFTOOL (Neural Network Fitting Tool): Used for data fitting problems. It consists of two layer feed forward network trained with the Levenberg-Marquardt algorithm.

NNTOOL (Open network/Data manager): It allows to import, create, use and export neural network and data. It is used for implementing multilayer architecture.

Authors tested three algorithms for multi-layer architecture:

a. Backpropagation algorithm:
   TRAINLM (training function) & Learned (adaptive learning function) gives Best Efficiency with 20 neurons in the hidden layer with MSE 6.58 at 15 epochs.
b. Cascaded Back-propagation
   TRAINLM (training function) & Learned (adaptive learning function) gives Best Efficiency with 20 neurons in the hidden layer with MSE 7.02 at 9 epochs.
c. Layer Recurrent Network
   TRAINLM (training function) & Learned (adaptive learning function) gives Best Efficiency with 20 neurons in the hidden layer with MSE 7.61.

Authors concluded that Back-Propagation algorithm gives the best performance out of 3 algorithms tested with least Mean Square Error (MSE) and accuracy of the model can be increased by increasing the number of neurons in hidden layer and increasing the training data size.

In [15], authors Nikam, et al. proposed a Support Vector Machine technique for rainfall forecasting in the Mumbai region in India with the help of Least Square Method (LSM). It gives better outcome on runoff when compared with the ANN model which primarily focuses on local minima in case of non-linear optimization. Also, the SVM can be applied to river stage forecasting which would be helpful in case of flash flood and found its implementation better than artificial neural network algorithm.

A major advantage of using this model is that its ease of use by redeveloping the SVM and replacing inequality constraints into equality ones. It uses two parameters one is the range of the function which tells about its smoothness of the function and the other is the regularization parameter and gives an idea about the smallest value and flatness. Five years (2007–11) of rainfall data had been used and it was taken from one of the 35 rain gauge stations installed in Mumbai after the major flood attack of 26th July 2005. There was no weather radar data used in the proposed model. 34 measures of rainfall events of 2 years were taken as a variable computation. Twenty-five measures of rainfall events of the year 2010 were used for testing. The model uses all sort of data which cause floods using the generalized version in binary form.

RBF was used which is more compact, the training process was reduced and its overall performance was also enhanced.

The accuracy obtained was quite good enough but involves certain limitations when

(a) the lead time increases.
(b) the prediction values of those of higher rainfall are under prediction.

Although the threshold limits found are forecasted good enough so as to issue warnings and alert (Tables 1 and 2).

Table 1. Comparison of different neural networks based techniques used for flood prediction

Sr. No.	Model used	Characteristics	Advantages	Limitations	Input dataset used
[1]	Logistic Regression	Used for finding flood susceptible maps To find the correlation between factors and inventories	Data of various types such as nominal, categorical can be given to the model No classification required	Size of the data will not affect the performance	Flood Inventory points 13 Flood Causative factors
[2]	Fuzzy Neural Network	The method identifies the dynamic nature of the system A data-driven approach is useful and appropriate for flood modeling	Reduces the complexity of the system Uncertainty is reduced	Test dataset performance was found low To select the input variables from a huge dataset requires higher analysis	Data gathered from real-time, high frequency & FRM stations were used to measure an outcome
[4]	SAE-BP	Stack auto encoders provide better optimization and BPNN supports non-linear relation between input parameters	Powerful feature representation capability Better results than other algorithms Improve the non-linear simulation capability	Not taking full consideration of the imbalance of the data distribution Limited to cost-insensitive approach	Data gathered from previous years 1998–2010 with a sample rate of data as one hour The previous data of 5–7 h was taken from stations and then the runoffs were calculated
[14]	BPNN	Multi-layer feed forward network Errors are propagated backward from the output layer to the previous layer and information in weights of hidden layer are updated	The mathematical formula used in the algorithm can be applied to any network Computation time is reduced if initial weights are small Batch updates of weights exist	Slow and inefficient Suboptimal solutions and solution may change with respect to time. (fuzzy value output)	Rainfall data from IMD Humidity, wind speed

(*continued*)

Table 1. (*continued*)

Sr. No.	Model used	Characteristics	Advantages	Limitations	Input dataset used
[6]	ANFIS	Combines Fuzzy-Logic and Neural Network The method used to implement Non-linear relationship among different input factors	Empirical & Adaptive Model Does not require prior knowledge	Computational cost is high due to a complex structure Difficult to decide the type of membership functions and their initial value	Peak discharge data of river & Rainfall
[15]	SVM	SVM used for precipitation climate change analysis and estimation cost	Gives better result than other methods	It is a complex structure. Its give result in SVM-wavelet and in a different form	Data were obtained from the Republic Hydro-Meteorological Service of Serbia
[12]	SVM, SOM & ID3	Used for classification & regression analysis Statistical modeling makes the task easier by allocating framework for multilevel structures	Simple, convenient & efficient High dimensional data are transformed into low dimensional & then clustering is applied	It takes more time and more storage as high dimensional datasets are used	Outlook, temperature, humidity, windy crop, rainy are all data attributes used

Table 2. Comparison of different GIS-based models used for flood prediction

Sr. No.	Model name	Characteristics	Advantages	Limitations	Input Factors
[3]	GIS Image Processing	Using tidal as input, flood simulation GIS model is developed	The entire area is covered using this technique	Applicable to a 1-D model Hence, the actual area cannot be predicted	GIS for tide input
[5]	GIS Image Processing with DSS	A GIS helps in identifying the state of the river from satellite images and using DSS, useful data can be gathered from raw data to make decisions	The whole spatial range is covered over which the river passes	Using a GIS system, there's tons of generalization because of the huge knowledge being analyzed. Limitation of this system is that the user stands to lose a lot of information due to the generalization of data	Satellite images of Lower Indus River

(*continued*)

Table 2. (*continued*)

Sr. No.	Model name	Characteristics	Advantages	Limitations	Input Factors
[10]	Wireless Sensor Networks (WSN)	WSN is used for real-time prediction Model does not depend on the number of input features or parameters	WSN can be used in a harsh and hostile environment Wireless sensor networks are scalable	Proposed model doesn't have any real hardware implementation Prediction is suitable for only some specific rivers	Statistical data of river water level
[8]	MODIS (Moderate Resolution Imaging Spectroradiometer)	MODIS sensor consider having operational application including flood detection characteristics and warning with additional flood disaster prevention or mitigation	MODIS data offer large coverage and provided by NASA	It involves huge computational power and requires more time for processing	MODIS Data obtained from stations

4   Conclusion

The rising frequencies and intensities of weather events like heavy rainfall, melting of snow ice or heat waves results in severe disasters like floods, droughts, etc. Hence there is a need to make a comparative study of all those which causes such events. This paper briefly describes the various work done for predicting floods. This paper also compared the different techniques by classifying into two main categories: Neural Network based techniques and GIS-based techniques.

This paper also evaluates the effectiveness of these techniques are based on reliability, computation time and accuracy regarding error rates. From this review, we can conclude that Neural Networks approach is a good choice for flood prediction. The study shows that integration of SOM & SVM along with ID3 gives the best possible outcome involving large datasets which are clustered first and then processed. These are found better than others in case of prediction, cost, reliability & computation-wise. Multi-layer feed forward BPNN uses a mathematical formula which can be applied to any network but it may result in a suboptimal solution. GIS-based techniques are efficient, but these techniques generalize the data, due to massive data being generated and the user tends to lose the information. Integration of Stack Auto Encoders with BP Neural Network gives a better result as compared to traditional SVM techniques for short time flood prediction.

References

1. Tehrany, M., Jones, S.: Evaluating the variations in the flood susceptibility maps accuracies due to the alterations in the type and extent of the flood inventory. In: ISPRS International Archives of the Photogrammetry, Remote Sensing and Spatial Information Sciences, pp. 209–214 (2017)
2. Khan, U., He, J., Valeo, C.: River flood prediction using fuzzy neural networks: an investigation on automated network architecture. Water Sci. Technol. **2017**(1), 238–247 (2018)
3. Shahapure, S.S., Eldho, T.I., Rao, E.P.: Flood simulation in an urban catchment of Navi Mumbai City with detention pond and tidal effects using FEM, GIS, and remote sensing. J. Waterw. Port Coast. Ocean Eng. **137**(6), 286–299 (2011)
4. Liu, F., Xu, F., Yang, S.: A flood forecasting model based on deep learning algorithm via integrating stacked auto encoders with BP neural network. Multimedia Big Data (BigMM), IEEE Third International Conference on. IEEE (2017)
5. Zafar, S., Azhar, H.M.S., Tahir, A.: A GIS based hydrological model for river water level detection and flood prediction featuring morphological operations (2018)
6. Patel, D., Parekh, D.F.: Flood forecasting using adaptive neuro-fuzzy inference system (ANFIS). Int. J. Eng. Trends Technol. **12**(10), 510–514 (2014)
7. Hallegatte, S., Ranger, N., Bhattacharya, S., Bachu, M., Priya, S., Dhore, K., Rafique, F., Mathur, P., Naville, N., Henriet, F., Patwardhan, A.: Flood risks, climate change impacts and adaptation benefits in Mumbai (2010)
8. Brakenridge, R., Anderson, E.: MODIS-based flood detection, mapping and measurement: the potential for operational hydrological applications. In: Transboundary floods: reducing risks through flood management, pp. 1–12. Springer, Dordrecht (2006)
9. Gocic, M., Shamshirband, S., Razak, Z., Petkovic, D., Ch, S., Trajkovic, S.: Long-term precipitation analysis and estimation of precipitation concentration index using three support vector machine methods. Advances in Meteorology (2016)
10. Seal, V., Raha, A., Maity, S., Mitra, S.K., Mukherjee, A., Naskar, M.K.: A real time multivariate robust regression based flood prediction model using polynomial approximation for wireless sensor network based flood forecasting systems. In: International Conference on Computer Science and Information Technology, pp. 432–441. Springer, Berlin, Heidelberg (2012)
11. Nasseri, M., Asghari, K., Abedini, M.J.: Optimized scenario for rainfall forecasting using genetic algorithm coupled with the artificial neural network. Expert Syst. Appl. **35**(3), 1415–1421 (2008)
12. Rani, R.U., Krishna, T.R., Reddy, R.K.K.: An efficient machine learning regression model for rainfall prediction. Int. J. Comput. Appl. **115**(23)), 2, 9–10 (2015)
13. Chatterjee, S., Datta, B., Sen, S., Dey, N., Debnath, N.C.: Rainfall prediction using a hybrid neural network approach. Recent Advances in Signal Processing, Telecommunications & Computing (SigTelCom), 2018 2nd International Conference on. IEEE (2018)
14. Sharma, A., Nijhawan, G.: Rainfall prediction using neural network. IJCST **3**(3), 65–69 (2015)
15. Nikam, V., Gupta, K.: SVM-based model for short-term rainfall forecasts at a local scale in the Mumbai urban area, India. J. Hydrol. Eng. **19**(5), 1048–1052 (2013)

Data Analytics in Steganography

T. Vijetha$^{(\boxtimes)}$ and T. Ravinder

Department of Electronics and Communication Engineering,
MLR Institute of Technology, Hyderabad, India
`tummalavijetha4444@gmail.com`

Abstract. Now a day's Digital communication has many advantages like good quality, flexibility in editing, high accuracy and fidelity, data compression, etc. As there is quick or sudden improvement and demand in technology of today's computer network, but this area consists some privacy issues, manipulations of data, and its data authentications. In present modern eras the organisations are having top priority as protection of data, which can be achieved by using digital communication. For this, data can be kept secret or can hide the data as unknown data with an Digital steganography which is an advance technique.

Keywords: Steganography · Watermarking · Digital image processing · PSNR

1 Introduction

Information concealment techniques are necessary for military, intelligence agencies, web banking and privacy. Because the use of web is increasing the data is non inheritable simply thus an individual World Health Organization possesses web will simply get knowledge from web for data whenever that they need. As a lot of and a lot of techniques for concealment data are developed and improved, a lot of and a lot of completely different data police work techniques also are been developed. This has diode to a powerful want for making new techniques for safe guarding direction from hackers. There are numerous knowledge concealment techniques offered for various functions and applications like steganography, cryptography, and watermarking, coated writing is known as Steganography wherever as modifying of information means cryptography such as it becomes in significant to the observers. Watermarking means that introducing of watermark signal into knowledge to get watermark object [1]. Since the rise of the net is one among the foremost vital factors of data technology and communication it's to be supplied with security. In electronic communication the info is reborn into a sequence of bits (ones and zeros) mistreatment special techniques that rely on the kind of knowledge just like the code format is internationally accustomed store text files wherever every characteris drawn by eight bits of knowledge. Digital pictures area unit composed of pixels and these pixels represents the magnitude at a particular location these every pixels area unit drawn by twenty fourbits in RGB format or eight bits within the grey scale pictures. Equally the video files incorporate sequence of pictures. Hence, any sort of information is often drawn by a sequence of 0's and 1's that may be keeping within the memory.

© Springer Nature Switzerland AG 2020
A. P. Pandian et al. (Eds.): ICCBI 2018, LNDECT 31, pp. 1032–1034, 2020.
https://doi.org/10.1007/978-3-030-24643-3_122

2 Software Tool

MATLAB is one of the high level programming language which provides a feasible environment to perform the computations very fast comparing with the programming languages like C, C++ and Fortran. In this proposal MATLAB 2016 version is used to secure the data by dual steganography because MATLAB has an advantage of performing multi-tasking like data manipulations, visualizations, calculation, maths works. This is a programming language which is very easy and sophisticated.

3 Digital Image Processing

The image square measure outlined over 2 dimensions (perhaps more) digital image process could also be modeled within the style of three-D systems. With the quick computers and signal processors on the market within the 2000s, digital image process has become the foremost common style of image process and customarily it's used as a result of it's not solely the foremost versatile methodology, however conjointly the most cost effective. Once the attention perceives a picture on a pc monitor, it's in truly perceiving an oversized assortment of finite color components, or pixels [2]. Therefore, by superimposing these three matrices of "RGB" a full color image will be reconstructed, as shown in Fig. 1.

Fig. 1. The full color image with a 3 matrices

If a picture is qualitied by associate degree by intensity matrix then the cumulative intensity is diagrammatic as a color between the black and white then that will seem to be as a gray scale image, as shown in Fig. 2.

In Fig. 3, a black to white gradient is demonstrated in 4-bits of intensity. Within a RGB range the color RGB image a pixel wise rework is formed to the YUV (luminance, blue chrominance, red chrominance) color house. This house in some cases it is additionally economical than the RGB house that permits higher quantization.

Fig. 2. Gray scale image

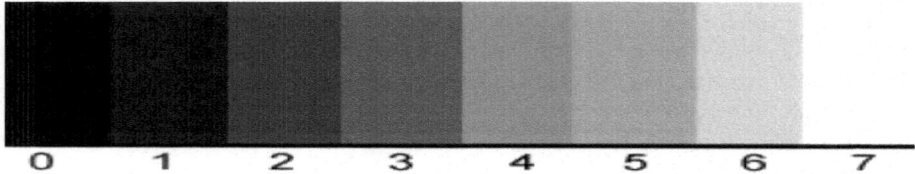

Fig. 3. Bit black to white gradient

4   Conclusion

This paper presents a simple analysis about the concept of dual steganography. In all experiments, for image steganography the typical PSNR is bigger than eighty four dB and for video steganography it will be sixty four. So, the experimental results which are shown here are effective for the planned model.

References

1. Singh, P., Agarwal, S., Pandey, A.: A hybrid DWT-SVD based robust watermarking scheme for color images and its comparative performance in YIQ and YUV color spaces. In: 2013 3rd IEEE International Advance Computing Conference (IACC) 978-1- 4673-4529-3
2. Kaur, S., Bansal, S., Bansal, R.K.: Steganography and classification of image steganography techniques. In: International Conference on Computing for Sustainable Global Development. IEEE (2014). 978-93 -80544-12-0/14
3. Kharrazi, M., Sencar, H.T., Memon, N.: Cover selection for steganographic embedding. In: IEEE International Conference on Image processing, pp. 117–120. Atlanta, USA (2006)

Users' Search Satisfaction in Search Engine Optimization

Ramaraj Palanisamy[(⊠)] and Yifan Liu

Gerald Schwartz School of Business, St. Francis Xavier University,
Antigonish, NS, Canada
rpalanis@stfx.ca

Abstract. This paper identifies the best practices and guidelines for designing an effective and meaningful e-commerce website so that the given site appears at the top in a search engine's search results. A research model for users' search satisfaction is evolved from the literature. The model is empirically tested and validated by obtaining survey-inputs from 101 students from a Canadian University. The factors affecting users' search satisfaction are user characteristics, search characteristics, search engine characteristics, and clicking behavior of advertisements. The data analysis validates the impact of website features, search engine contents and search problems on user's search satisfaction. User's search behavior includes the frequency of different search data input, navigational search and the usage of different devices positively influences user's search satisfaction. Besides, the advertisements clicking negatively influences user's satisfaction. The implications for the theory and practice are given.

Keywords: Search behavior · User's search satisfaction · Search engine optimization · User characteristics

1 Introduction

The purpose of this study is to identify the guidelines and best practices for designing an effective and meaningful e-commerce website so that the given site appears at the top in a search engine's search results. Specifically, this research aims to empirically test a research model to identify the relationships among user characteristics, search characteristics, and user search satisfaction. The study findings help web site designers and marketers to improve the performance of websites in order to enhance the user experience. There are lack of studies in the literature on the user perspective. This study aims to fill this gap by identifying the factors influencing end user search satisfaction.

2 Literature Review

The Fig. 1 shows the proposed research model which is based on user satisfaction metrics which was developed by Kohli and Kumar (2011). Kohli and Kumar (2011) identified several factors that influence user satisfaction, such as response time, URL

© Springer Nature Switzerland AG 2020
A. P. Pandian et al. (Eds.): ICCBI 2018, LNDECT 31, pp. 1035–1045, 2020.
https://doi.org/10.1007/978-3-030-24643-3_124

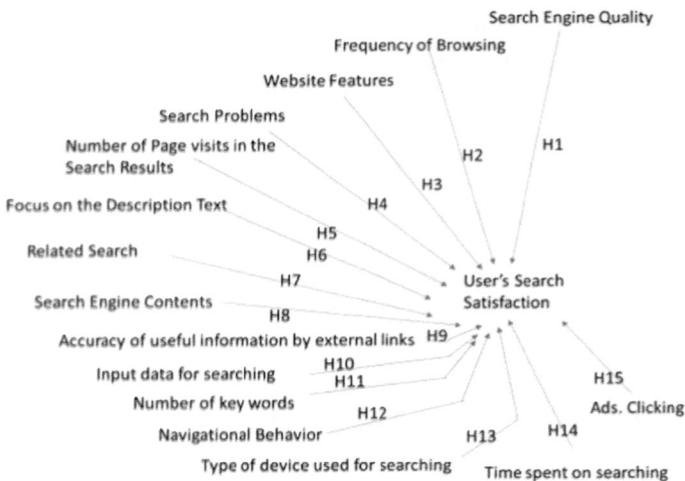

Fig. 1. Research model for user's search satisfaction

correctness, results display, overall impression, and website stability. The various components of the model are reviewed in this section.

2.1 Search Engine Optimization

Search engine optimization (SEO) was defined by Russell (2007) as a process in which search engines find and access the relevant websites. Its purpose is to improve the quality and ranking of a given website and create an effective user experience (Arora and Bhalla 2014). To understand the search engine optimization well, Russell (2007) developed a structure to show how search engine optimization works. The SEO cycle consists of the following phases (Russell 2007): Research/choose key words, Build Website: place keywords, Submit search to engine, Report on search page rankings, Optimize pages, Analyze traffic, conversions, and Maintain Website: reoptimize. In this cycle, SEO specialists do research based on the keywords used by users, location and frequency on a page, and it is the first step to incorporate keywords in the website contents (Gershonowicz 2016). Finding the keywords from the contents is also the simplest way because the users will input their search terms into search engines to describe what they are looking for and specialists can easily locate the keywords which appeal to the users (Crowley 2014). Then the search engine will conduct the research and display the search results via four steps, which include crawling, indexing, searching and ranking (Bai 2013). Based on the results, specialists can optimize the websites by analyzing the problems.

2.2 Search Engine Quality

The search engine quality is measured by the number of websites collected by the search engine, accuracy of search results, display of search results in a sorted order, rich functions such as protect your personal information, relevance of search results, easy to remember URL of search engine, Less ads. in the search page, and the short loading time of search results.

2.2.1   User Characteristics

Education

During the search process, two types of knowledge can be applied to discuss the relationship between users' education and search satisfaction, which are familiarity with the topic or subject and familiarity with the search system. Familiarity with the search system is a kind of advanced knowledge which helps the users to search comfortably. Different search systems have different search facilities for users to choose, if they are familiar with some tools, they can improve their results. Yamin, Ramayah and Wan Ishak (2015) proposed two kinds of dimensions of the user search behavior, one is breadth search query and another is depth search query. Breadth search query can be identified as direct keywords, which focuses on the wide search; depth search query focuses on the complex search such as Boolean search.

Browsing Features

In the search process, users perform different actions to derive their expected results, which can be characterized as many different aspects such as TimeOnPage and TimeOnDomain. TimeOnPage means the time that users spend on a specific webpage while TimeOnDomain shows the visiting time on the domain. Simple aspects of the user web page interactions can be captured and quantified. These features are used to characterize interactions with pages beyond results pages.

User Experience

User experience comprised of the search activities performed by the users, the quality of search results received, and development of knowledge and impression on the search results (Song and Salvendy 2003). Besides, the experience depends on the user friendliness of the search engine, and the support provided by the search engine for query formation. Users gain rich experience when the interactions with the search engine are noteworthy. The experience also depends on the search strategies used for retrieving relevant information. The search strategies include using simple searching key words, phrases, questions, and using browser tools. The intent and purpose of searching is decided by these search inputs and achieved by obtaining a vast amount of relevant information and website links out of searching.

2.2.2   Search Characteristics

Search Purpose

The users use the search engine for the following purposes: navigational (to navigate the web), informational (to retrieve required information), and transactional (to complete a transaction) (Broder 2002). The intent to search the web is expressed by a search query which is used by the search engine to provide relevant and accurate search results (Lewandowski 2008).

Related Search

When the users don't get the required or relevant information from the initial search process, they go for some additional searches by conducting related search (Ohshima 2009). In these instances, instead of simply satisfying with the initial search results, users try with different search key words or phrases; or some times they conduct the related searches by using a different search engine. Thereby, the users try to get relevant and valuable information on the given topic and tend to avoid the irrelevant links or websites in search results.

Related search shows several links about additional information on search results, they are listed at the bottom of each result page and can be reviewed as the search history of other users.

2.2.3   Search Engine Characteristics

On-page Factors

On-page optimization includes all factors that are directly used on a website to achieve ranking on the search engine result pages. Zhang and Dimitroff (2005) identified the elements including page title, meta-content, headings, body text, navigation, and images. Meta description is a kind of short description which appears below the page links. It shows how people and organizations summarize their own web pages (Craven 2000). Keyword analysis and design could be recognized as the most important part in optimizing the webpage, which represents the users' expectation. Yalçın and Köse (2010) gave a basic principle for keywords selection which is that the chosen keywords must be performed by searching to make sure that the website can be found out in top five list, that is, only the valuable keywords should be considered because they can bring the actual positive outcomes.

In designing the web pages, the short and long names, definitions and descriptions of products or services are to be used to form key words for searching. Therefore, the key words could be descriptive instead of single word. Another issue is about keyword density. Martin (2011) defined that density was the percentage of keywords that can be found in the indexed text on a single webpage, many more keywords would lead to confusion and ignorance from web users, and the appropriate number of density is between three percent and five percent.

Content
Website usability (Alghalith 2015) is measured as evaluating items such as page views, time on site, and click paths to determine how user-friendly or user-relevant a web site is. To put it simply, usability enables the users to search the internet in a short time and with high satisfaction (Eisenberg et al. 2008). To improve the users' experience, trust and information credibility were identified by Barnard and Wesson (2003). They regarded the information privacy on the internet as the most important factor, which contains privacy policy, company overview page, feedback, and contact information. One of the main purposes of SEO is to improve user experience during the search thus to increase the popularity of the webpage. The website contents should have maximum number of relevant key words that searchers may likely to use in searching for the site (Bai 2013).

Off-page Factors

Social media
Users can find extra useful information through browsing the social media, and they can also get valuable feedback from others when they want to buy some products online. Social media can also provide promotion information about different products. For the website owners, social media plays an important role in increasing the traffic and value of the site via creating interesting and viral content to attract the attention of potential customers (Zhang and Cabage 2017).

2.2.4 Advertisement Clicking

Users predominantly use searching method to find the required information from the web. This type of mechanism can save much time and cost to some extent because traditional physical trade can be moved to online transaction (Ghose and Yang 2009). One of the main sources of getting information rapidly is from sponsored search, that is, searchers would click on the advertisement links which are listed on the top or the right side of a result page. These advertisement links are sponsored by the advertisers and shown in the search results page (Gupta and Mateen 2014). From the online advertising perspective, the companies increase their budget for sponsored advertisements and in the users' perspective, there is an increase in clicking the sponsored online advertisements (IAB 2013).

2.3 Research Hypotheses

Based on the review of literature, the hypotheses were evolved. The purpose of this research is to find out the best practices for e-commerce website design to enhance users' search satisfaction. Search satisfaction is a major determinant in order to improve the website from users' perspectives, so that the website can be displayed on the top of a search engine's search results page. The evolved hypotheses are given below:

H1.   Higher the search engine usage higher the level of user's search satisfaction.
H2.   High frequency of browsing positively influences the level of user's search satisfaction.

H3.   Enriching website features positively influences the level of user's search satisfaction.

H4.   Less search problems will lead to the higher level of user's search satisfaction.

H5.   More number of visits in search engine result pages will lead to the higher level of user's search satisfaction.

H6.   Adding more description text will lead to the higher level of user's search satisfaction.

H7.   Providing more number of related searches positively influences the higher level of user's search satisfaction.

H8.   Increasing search engine contents positively influences the higher level of user's search satisfaction.

H9.   Increasing the information accuracy by external links positively influences the higher level of user's search satisfaction.

H10.   Increasing the frequency of different data input positively influences the higher level of user's search satisfaction.

H11.   Increasing the number of keywords will positively influences user's search satisfaction.

H12.   Increasing the frequency of navigational behavior positively influences user's search satisfaction.

H13.   Increasing the frequency of different device usage positively influences user's search engine satisfaction.

H14.   Increasing the time spent on searching every day positively influences user's search engine satisfaction.

H15.   Decreasing frequency of advertisement clicking will improve the level of user's search satisfaction.

3   Research Methodology

Considering the research objectives, a questionnaire survey is designed to collect data from undergraduate business students from a Canadian University. The respondents were selected based on their online experience. As most of the students use search engines frequently for university work/assignments and shopping, student respondents were selected. The data was collected through an online survey using Qualtrics survey software. The data was analyzed using SPSS. Overall, the survey had collected 140 responses from business students and 25 responses were from other web users. Too much missing data were found in 64 responses out of 165. As a result, 101 usable responses were obtained. Approximately half of the respondents were male and thereby the sample is gender-balanced.

4   Data Analysis

Pearson correlation analysis was carried out on the conceptual model. Table 1 presents the overall correlations between each independent variable and the dependent variable.

Independent variables	Users search satisfaction	Hypotheses testing
Search Engine use/Quality	0.112**	H1 is validated
Frequency of browsing	−0.041	H2 (data did not support)
Website features	0.262**	H3 is validated
Search problems	−0.149**	H4 is validated
Number of page visits in the search results	0.087	H5 (data did not support)
Focus on the description text	0.208**	H6 is validated
Related search	−0.040	H7 (data did not support)
Search engine content	0.342**	H8 is validated
Accuracy of useful information from links	−0.011	H9 (data did not support)
Input data for searching	0.343**	H10 is validated
Number of key words	0.090	H11 (data did not support)
Navigational behavior	0.300**	H12 is validated
Type of device used for searching	0.243**	H13 is validated
Time spent on searching	0.093	H14 (data did not support)
Ads. Clicking	−0.236**	H15 is validated

**Correlation is significant at the 0.001 level; *Correlation is significant at the 0.001 level

5   Discussion

The managerial implications of the research results are given in this section. The data analysis concludes that better search engine use will positively influences the level of user's search satisfaction. For instance, if users get more results in a sorted order, they will feel more satisfied. A positive correlation shows that web users put much emphasis on their search experience. They have common principles to evaluate the search engines. The results should be in a sorted order with high level of accuracy, the number of target websites provided by search engines should be in a medium level, fewer advertisements, and the loading time should be short. As the Internet grows, search engines have become the main information collector. Web users use search engines more frequently to do many different activities such as online shopping and collecting academic information.

From the user's perspective, most of them spend 4–6 min on the top website and only around 2 min as the tolerance limit for the users to wait or spend on the irrelevant websites. Ryan and Valverde (2003) identified that the length of waiting time could affect the user's online experience. It is possible that more experienced users will have less tolerance for or be more sensitive to delays in downloading the results. The more time spent waiting, the more dissatisfaction will be accumulated.

5.1   Search Characteristics and Search Satisfaction

The data analysis concludes that increasing the frequency of different data input, navigational behavior and different devices used will increase the user's search satisfaction. Web users have a clear tendency to adopt the normal search behavior which is from up to downside. Most web users focus on the first 3–5 links on the first page; the more links clicked, there is a possibility that the accuracy of information may decrease. Website compatibility is another issue that affects the user's judgment. Web users trust more on the desktop or laptop websites than smartphone websites. Al-Khalifa (2014) claimed that the main difference between desktop websites and smartphone websites is the size of the version. Desktop websites have the full version with rich functions while smartphone websites are represented as the tailored version. It only contains the basic functions and services offered by its desktop counterpart. Web users use larger devices such as laptop or desktop than smartphone to ensure the accuracy of information though small devices are convenient and portable.

Users provide inputs for search or "web query". These inputs can be in different forms such as key words, nouns, verbs, noun phrases, web phrases, questions, and composite queries (Bendersky and Croft 2009). The composite queries comprised of the combinations of keywords, nouns, verbs, phrases, and questions. The queries with less key words or shorter phrases are more effective in retrieving accurate results compared to composite and longer queries (Vakkari 2011). That is the main reason why web users rely on the search keyword more than phrases and questions, which is supported by hypothesis 10.

5.2   Search Engine Characteristics and Search Satisfaction

Search problem is a common issue that web users face every day. Bill (2014) listed several "most common" search engine ranking problems, such as slow webpage loading time, 404 errors for broken links, and outdated information. These problems can be a threat to all websites. Basically, no web users want to meet these problems and they will lose patience if problems happen at any time. Reducing the problems means improving the quality of the websites. Web users also show interest in the search engine contents such as blogs, music, images and video, which are displayed on the top of the results page. When users perform a search, they can choose the format of the results.

5.3   Advertisements Clicking and Search Satisfaction

The data analysis concludes that reducing the advertisements will increase the user's search satisfaction. Advertisement is a method that web owners use to make a promotion to compete with other websites, and the location of different advertisements have different purposes. Chen et al. (2011) found that a type of advertisements are located normally to the right and above the organic searching lists. These advertisements often contain a promotion for a specific product. Most of the products are offered at a low price, but the quality is an issue. Web users conceptualize an idea that these advertisements are useless and seldom click on them. Kritzinger and Weideman (2015) claimed that the sharing of real estate space on the user's screen was problematic to users. Web users reduce the frequency of advertisements clicking and view these advertisements as the obstacle (Moxley et al. 2004).

Best Practices for SEO

Based on the research findings, the following best practices for search engine optimization from the user's perspective are given:

1. Original content: It is important that the content be original. Copying an article from other websites has been attempted by many people and usually does not work. The more original content, results in more searches. Although this is not a fully significant correlation because there are many other factors. Website owners cannot expect to get a thousand hits a day with a website with five items or articles. However, a website with a hundred original articles, is virtually assured a thousand visits.
2. Focus on the on-page design: It is the core for each website. Without a clear structure and navigational search, the website can be defined as a failure. It includes many elements such as font design, title, header and external link.
3. Fewer advertisements: Advertisements can attract customers and increase the traffic, but an inappropriate number of advertisements will cause fatigue. When one makes a change on a website, they must be aware that the number of advertisements should be placed in a reasonable number.
4. Regular updates: Updates are rewarded by Google and frequently updated websites show up at the top of the results.
5. HTML errors: Almost all the pages tend to have html errors. A page with clean html code and few errors with a simple structure for search engine robots will be rewarded in the form of a good position.
6. Social networks and forums: In addition to direct visits to get through these means, users can get links to websites, which benefits site owners, allowing them to provide a quote for service and helping them be known in the network. Groups and forums should be related to subject and should not be overly burdensome.

5.4   Future Research

With the rapid growth of mobile search, the evaluation of the websites on mobile devices is becoming a hot topic (Mao et al. 2018). Mobile search traffic will surpass the desktop search in the near future. The future research should focus more on increasing the mobile experience of the users. In the search engine optimization context, the future study should aim for increasing the ranking of mobile websites in mobile searching.

6   Conclusion

In conclusion, the research findings reveal the significant relationships for user's search satisfaction. The conceptual model recommends that website features, search engine contents and search problems have impact on user's search satisfaction. User's search behavior including the frequency of different data input, navigational search and different devices usage have positive effects on user's search satisfaction. Furthermore, advertisement clicking negatively influences the level of user's satisfaction.

From user's perspective, three aspects can be concluded for search engine optimization. One aspect is website content, it should be as original as possible. The contents should be authoritative and could not be imitated easily to increase the websites' competitive advantage; the second aspect is user interaction. As per the discussion in the previous section, social media is essential for web users to have a friendly communication with different websites. Finally and foremost, website design is always a big challenge. The structure should be as simple as possible to increase visual effect with some necessities, navigational search, page title, description text and nontext elements such as image are crucial for an effective website. In the future, search engine optimizers should be more considerate and make the whole website system simplified rather than professional. We believe humanization is the future of websites.

Acknowledgements. This research study has been cleared for ethics compliance with the Tri-Council guidelines (TCPS) and St. Francis Xavier University's ethics policies. Accordingly, St. Francis Xavier University Research Ethics Board approved the project (Letter dated 21 January 2018). The authors would like to thank all the participants who provided their valuable inputs for the completion of this study. The participants were provided adequate information about the study and were encouraged to make their voluntary, informed, and rational decision to participate.

References

Alghalith, N.: Using course-embedded assessment: defining and assessing critical thinking skills of MIS students. J. High. Educ. Theory Pract. **15**, 77–83 (2015)

Al-Khalifa, H.: A framework for evaluating university mobile websites. Online Inf. Rev. **38**(2), 166–185 (2014)

Arora, P., Bhalla, T.: A synonym based approach of data mining in search engine Optimization. Int. J. Comput. Trends Technol. (IJCTT) **12**(4), 201–210 (2014)

Bai, X.: In Google we trust: consumers' perception of search engine optimization and its potential impact on online information search (Order No. 1535565). Available from ABI/INFORM Global (1346703737) (2013)

Barnard, L., Wesson, J.L.: Usability issues for E-commerce in South Africa: An empirical investigation. In: Proceedings of the Annual Conference of the South African Institute of Computer Scientists and Information Technologists (SAICSIT), Pretoria, 17–19 September 2003, pp. 258–267 (2003)

Bendersky, M., Croft, W.B.: Analysis of long queries in a large scale search log. In: Proceedings of the 2009 Workshop on Web Search Click Data, pp. 8–14. ACM, February 2009

Bill, H.: Search Engine Rankings for January 2014. Search Queries (2014). https://www.billhartzer.com/internet-usage/search-engine-rankings-for-january-2014

Broder, A.: A taxonomy of web search. SIGIR Forum **36**(2), 3–10 (2002). 3, Fall 2002

Chen, C.Y., Shih, B.Y., Chen, Z.S., Chen, T.H.: The exploration of internet marketing strategy by search engine optimization: a critical review and comparison. Afr. J. Bus. Manage. **5**(12), 4644–4649 (2011)

Craven, T.C.: Optimize keywords to improve SEO. J. Fin. Plan. **27**(1), 17–21 (2000)

Crowley, M.: Optimize keywords to improve SEO. J. Financ. Plan. **27**(1), 17 (2014)

Eisenberg, B., Quarto-vonTivadar, J., Crossby, B., Davis, L.T.: Always Be Testing: The Complete Guide to Google Website Optimizer. Sybex publisher (2008)

Gershonowicz, J.: Basic tips to transform your website's SEO (2016)

Ghose, A., Yang, S.: An empirical analysis of search engine advertising: sponsored search in electronic markets. Manage. Sci. **55**(10), 1605–1622 (2009)

Gupta, A., Mateen, A.: Exploring the factors affecting sponsored search ad performance. Mark. Intell. Plan. **32**(5), 586–599 (2014)

IAB: Interactive Advertising Bureau (2013). https://www.iab.com

Kohli, S., Kumar, E.: Evolution of user dependent model to predict future usability of a search engine. In: IEEE conference in 2011 World Congress on Information and Communication Technologies, Mumbai, India (2011). https://doi.org/10.1109/WICT.2011.6141259

Kritzinger, W.T., Weideman, M.: Comparative case study on website traffic generated by search engine optimisation and a pay-per-click campaign, versus marketing expenditure. S. Afr. J. Inf. Manag. **17**(1), 1–12 (2015)

Lewandowski, D.: The retrieval effectiveness of web search engines: considering results descriptions. J. Doc. **64**(6), 915–937 (2008)

Mao, J., Liu, Y., Kando, N., Luo, C., Zhang, M., Ma, S.: Investigating result usefulness in mobile search. In: European Conference on Information Retrieval, pp. 223–236. Springer, Cham, March 2018

Martin, M.: Exploring search engine optimization strategies and tactics. Soc. Sci. Res. Netw. (2011). http://papers.ssrn.com/sol3/papers.cfm?abstract_id=1876341

Moxley, D., Blake, J., Maze, S.: Web search engine advertising practices and their effect on library service. Bottom Line **17**(2), 61–65 (2004). https://doi.org/10.1108/08880450410536080

Ohshima, N., Amiri, E., Keshavarz, H.: Resource allocation in grid: a review. In: International Conference on Innovation, Management and Technology Research (ICIMTR 2009). Elsevier, Malaysia (2009)

Russell, K.: Search engine optimization. Comput. World **41**(23), 40 (2007)

Ryan, G., Valverde, M.: Waiting online: a review and research agenda. Internet Res. **13**(3), 195 (2003)

Song, G., Salvendy, G.: Effectiveness of automatic and expert generated narrative and guided instructions for task-oriented web browsing. Int. J. Hum. Comput. Interact. **59**(6), 777–795 (2003)

Vakkari, P.: Comparing Google to a digital reference service for answering factual and topical requests by keyword and question queries. Online Inf. Rev. **35**(6), 928–941 (2011)

Yalçın, N., Köse, U.: What is search engine optimization: SEO? Procedia Soc. Behav. Sci. **9**, 487–493 (2010)

Yamin, F.M., Ramayah, T., Ishak, : W.H.I: Does user search behaviour mediate user knowledge and search satisfaction? Int. J. Econ. Fin. Issues (IJEFI) **5**, 34–39 (2015)

Zhang, J., Dimitroff, A.: The impact of metadata implementation on webpage visibility in search engine results (Part II). Inf. Process. Manage. **41**, 691–715 (2005). https://doi.org/10.1016/j.ipm.2003.12.002

Zhang, S., Cabage, N.: Search engine optimization: comparison of link building and social sharing. J. Comput. Inf. Syst. **57**(2), 148–159 (2017)

Brain Tumour Identification Through MRI Images Using Convolution Neural Networks

N. Jagan Mohana Rao[✉] and B. Anil Kumar

Department of ECE, GMR Institute of Technology, Razam, India
{nallurijagan, anil.revanth}@gmail.com

Abstract. Medical image processing is the highly demanding and emerging field now a day. Gliomas brain tumors are most dangerous and common brain tumors in all the brain tumors. The most efficient imaging approach is the Magnetic resonance imaging (MRI) for accessing tumors, but the is a mostly utilized imaging technique to access these tumors, but the significant data amount is produced by MRI. In the time of plausible, it perverts the manual segmentation. Hence there is a need for novel and automatic segmentation. In this project we proposed an automatic segmentation method predicated on Convolution Neural Networks (CNN) for segmentation process of the MRI images. The utilization of diminutive kernels is to designing an inner architecture and a positive effect against over fitting. We additionally use the normalization intensity approach for pre-processing and it is not common in the segmentation techniques of CNN moreover it proved with the augmentation of data which is use for the segmentation of encephalan in MRI images. Identification of encephalon tumour from encephalon MRI images which is used in MATLAB software.

Index Terms: Brain tumour · Segmentation · Convolution neural networks · Brain tumour identification · Deep learning · Magnetic resonance imaging (MRI)

1 Introduction

The research on encephalon tumor detection and segmentation get highly popular over the past few years. Encephalon tumor, occurs when non typical cells are form within the encephalon. There are two main types of tumors: malignant tumors and benign tumors. Malignant tumor is untypical cell magnification with the possible to grow other components of the body [1]. Medical imaging techniques play an major role in diagnosis and early detection of tumor. Medical Imaging techniques are Computed Tomography (CT), Single-Photon Emission Computed Tomography (SPECT), Positron Emission Tomography (PET), Magnetic Resonance Spectroscopy (MRS) and Magnetic Resonance Imaging (MRI) are all used to provide precious information about shape, size, location and metabolism of encephalon tumors availing in diagnosis [3]. Present treatments which include surgery, chemotherapy or accumulation of them. MRI is subsidiary for tumor in the clinical practice, for acquiring possible MRI images which provides the complementary information. The four MRI modalities utilize in this diagnosis.

A. P. Pandian et al. (Eds.): ICCBI 2018, LNDECT 31, pp. 1046–1053, 2020.
https://doi.org/10.1007/978-3-030-24643-3_125

2 Literature Review

There are different methods utilized for image segmentation. The methods are threshold, region growing, clustering algorithms, classifier methods, artificial neural networks, atlas-guided approaches level set models, morphology predicated and deformable models.

A. Threshold Methods

For image segmentation, the Threshold method is the oldest method. For the two threshold intensity, the segmentation is achieved by grouping all the pixels with the intensity between two thresholds into one class [7]. A process to calculate more than one threshold value is called multi thresholding. In a sequence of image processing operations thresholding is utilized as an initial step [9].

B. Region Growing Method

A great amended method is the Region growing method. For image region extraction, this method predicts the predefined function. This method is established on information intensity or image edges. The utilization rejects the culls in a seed point and extracts all the pixels which connect to predefined seed predicted criteria. Split and merge algorithm which cognates for region growing algorithm but it does not require a seed point. There is a drawback for Region Growing which is called as noise effect.

A different approach is Histogram analysis 2D region growing.

C. Classifier Methods

Supervised Learning Methods are considering as Classifier Methods. This is consider as the pattern apperception approaches which partition a space of the feature which obtained from the image with data utilizing techniques of the predefined labels. The most proximate neighbor classifier is considering as the simple classifier in which every pixel is relegated in same class as the training data. The methodology for generalization is the k-most proximate neighbor. The k-most proximate classifier is denoted as a nonparametric classifier because it does not brands the underlying postulation for statistical structure.

D. Artificial Neural Networks

Artificial Neural Networks (ANNs) are parallel networks of nodes processing which stimulates the biological learning. For computations preformation, every node in the Artificial Neural Network has the capability [17]. Artificial neural Network (ANN) utilizing canny edge detection adaptive threshold relegates the MRI image either as a salubrious encephalon or an encephalon having a tumor. The discrete wavelet transforms with back propagation and neural network (NN) first engage on wavelet transforms to remove features from images, and then implies the technique of primer component analysis to reduce the dimensions of features. Generally, NN predicated methods have engendered copacetic results. For unsupervised method, Artificial Neural Network has been used for clustering approach and deformable models [12].

3   Convolutional Neural Networks

The Convolutional Neural Networks (CNN) is a deep-structured learning model. It learns like the structure of human encephalon in perceiving objects. Moreover, researches additionally show a good performance on CNN-predicated algorithms. CNNs capable of solving many object apperception and biological Image segmentation challenges [1]. CNN can be utilized for patches utilizing kernels; it has leads for taking report context for using the raw data. From the segmentation of encephalon tumor, some research withal observed the utilization of CNNs utilized the Shallow Convolutional Neural Network for two convolution layers which is disunited by max-pooling with 3 stride which follows by one softmax layer and plenary connected (FC) layer. Many of authors estimated the utilization of 3D filters but optate for 2D filters. 3D filters had the favor of 3D nature images, but the computational load had been hiked. Identify the consummate tumor utilize a binary CNN [1]. Then, cellular automata smooth the segmentation, afore a Convolutional Neural Network segregates the multi-class which segregates the tumor sub-region. Each voxel has the ability for removed patches in every plane and also trained the Convolutional Neural Network for the sequence of MRI images. For every Convolutional Neural Network, the previous results in the softmax FC layer with every Convolutional Neural Network are combined and used for training a RF classifier.

4   Proposed Method

There are three main stages for processing: pre-processing, relegation via CNN and post-processing.

A. Pre-Processing

The input MRI images are placed by the inequitableness field distortion. This makes the intensity ranges of the same tissues to vary across the image [10]. To idealize it, we applied some filtering techniques. However, this is not enough to rectification that the intensity dispensation of a tissue type is in a same intensity scale across various samples for the same MRI sequence, which is an explicit or implicit postulation in most segmentation techniques.

B. Relegation via Convolutional Neural Network

CNN were acclimated to procure some more impressive results and win well-kenned challenges and contests [4]. For convolving an image or some signal with kernels of convolutional layers can be habituated to obtain feature maps.

1. At Initial Step: For convergence, initialization is most required. We are initializing the weights in such a way that the variance is constant. This initialization process kenned as Xavier initialization [1]. The gradients & activations are continues in controlled levels or explode and remove the back propagated gradients.
2. Activation Function: This is deals with nonlinear changing the data. Rectifier linear units (ReLU), defined as

$$f(x) = \max\ (0, x) \tag{1}$$

Were come to know get the betterment of results than highly sigmoid classical or the tangent function of hyperbolic and accelerate the training. Then the constant imposing() damaged the flow of gradient and adaption of the consequent weights.

3. Pooling: Pooling is one more paramount context in CNN, which is a form of non-linear down-sampling. It commixes the features of spatiality in feature maps [14]. The redundant feature accumulation engenders may describe more invariant and dense for low number of image changes like nugatory data. Then it completely minimizes the computational load of the sequential stages. For combining the features it has common method like average pooling or max pooling.
4. Regularization: For reducing the over fitting, it will subsidiary. In the FC layer, it use the Dropout methodology.

Step, it eliminates nodes from the network with probability. In this way, it makes all nodes of FC layers to learn improve the representations of the data, preventing nodes from co-adapting to everyone [16]. At testing time, all nodes are used.

5. Loss Function: Loss Function is a function for reduce the data in the training phase. Here the Categorical Cross – Entropy can be chosen.

5   Experimental Setup

A. Data selection

Here, we examine the database of BRATS 2015, 2013 databases and also others [2]. Four sequences are obtainable for each and every patient in BRATS: They are T1-weighted (T1), T2-weighted (T2), T1 with contrast (T1c) & FLAIR. For skull & T1c stripped, each & every input image sample were aligned. The Training set accommodates 20 high graded and 10 low graded tumor MRI samples, with manual segmentations available.

C. Evolution Parameters

Sensitivity: The probability of test outcome will be positive when the infection is present (true positive rate).

$$\text{Sensitivity} = \text{TP} / (\text{TP} + \text{FN}) * 100 \tag{2}$$

Specificity: The probability of test result will be negative when the infection is not present (true negative rate).

$$\text{Specificity} = \text{TN} / (\text{TN} + \text{FP}) * 100. \tag{3}$$

Accuracy: the quality or state of being correct or precise.

$$\text{Accuracy} = (\text{TP} + \text{TN}) / (\text{TP} + \text{TN} + \text{FN} + \text{FP}) * 100. \tag{4}$$

Positive predictive value (PPV): The probability of the infection is present when the test is positive.

$$\text{PPV} = \text{TP} / (\text{TP} + \text{FP}). \tag{5}$$

Dice similarity coefficient (DSC): The DSC measures the overlap between the manual and the automatic segmentation. It is defined as,

$$\text{DSC} = 2\text{TP} / (\text{FP} + 2\text{TP} + \text{FN}). \tag{6}$$

Feature extraction: Feature extraction is transforming an image into its set of features. They are acquired from color, texture and shape. These features are withdrawn for relegation process. Several feature extraction techniques are available. Few are Gabor features, texture feature, decision boundary feature extraction, discriminate analysis [13]. Texture feature is utilized in the system. We developed system used to two techniques for texture feature extraction. They are first order histogram and second order texture feature.

6 Experiment Results

The precision and the performance of the developed system can be tested utilizing confusion matrix. Outcomes of the presage can be True Positive (TP), False Positive (FP), False Negative (FN) and True Negative (TN) [14]. Normal images and tumor images are categorized correctly as True Negative (TN) and True Positive (TP). In some cases, test says no tumor but it has tumor, then it is False Negative (FN) and test says tumor is present but it has no tumor, then it is False Positive (FP) (Figs. 1, 2 and 3; Tables 1 and 2).

Simulation results:

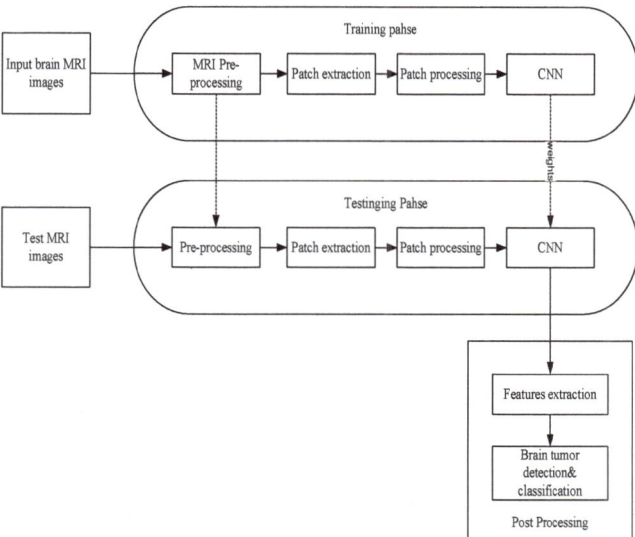

Fig. 1. Proposed method.

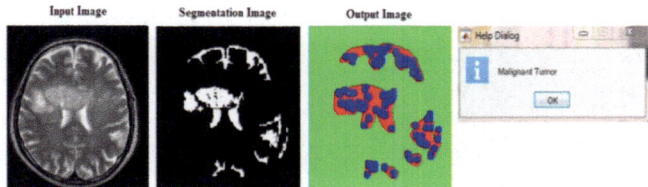

Fig. 2. Example of segmentation for malignant brain tumor.

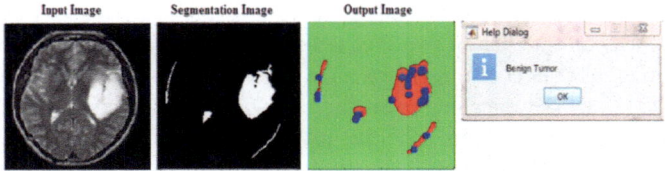

Fig. 3. Example of segmentation for benign brain tumor.

Table 1. Performance measures

S. NO	Evolution parameter	Formulae	Value of result
1	Specificity	TN / (TN + FP)	0.875
2	Sensitivity	TP / (TP + FN)	0.933
3	Accuracy	(TP + TN) / (TP +TN + FN + FP)	0.913
4	Positive predictive value (PPV)	TP / (TP + FP)	0.933
5	Dice similarity coefficient (DSC)	2TP / (FP + 2TP + FN)	0.933

Table 2. Examples for Texture feature parameters measures

S. NO	Parameter	Image 1	Image 2	Image 3	Image 4
1	mean	2.366518e-01	2.474972e-01	2.374861e-01	3.028365e-01
2	Standard Deviation	7.613600e-01	6.705647e-02	1.047619e-01	1.373658e-01
3	Entropy	7.613600e-01	7.513006e-01	7.329509e-01	7.808956e-01
4	RMS	9.327123e-01	9.301261e-01	9.246941e-01	9.389560e-01
5	Variance	3.078092e-03	4.279622e-03	3.307218e-03	4.383120e-03
6	Smoothness	8.976195e-02	8.971268e-02	8.975380e-02	8.970768e-02
7	Kurtosis	3.508794e+00	3.426407e+00	3.113240e+00	2.553089e+00
8	Skewness	8.980265e-02	8.980265e-02	8.980265e-02	8.980265e-02
9	IDM	8.045898e-03	8.041958e-03	8.052602e-03	8.073500e-03
10	Contrast	9.196820e-01	9.408990e-01	9.248281e-01	9.422139e-01
11	Correlation	7.611216e+00	6.956768e+00	6.826727e+00	1.737261e+01
12	Energy	7.048534e-01	5.408277e-01	6.276628e-01	1.417907e+00
13	Homogeneity	-8.899371e-02	1.015106e+00	1.525725e-01	7.420159e-02
14	Tumor	Malignant	Malignant	Benign	Benign

7   Conclusion

The developed technique has been for diagnosis of encephalon tumor from Magnetic Resonance Imaging (MRI) of the encephalon. There are different phases to apperceive the tumor part from the encephalon image. We developed a novel CNN-predicated method for segmentation of encephalon tumors in MRI images. The CNN is constructed over convolutional layers with minute 3*3 kernels to sanction deeper architectures. We achieve better performance of evolution parameters are sensitivity, specificity, PPV, DSC and accuracy.

References

1. Pereira, S., et al.: Brain tumor segmentation using convolutional neural networks in MRI images. IEEE Trans. Med. Imaging **35**(5), 1240–1251 (2016)
2. Menze, B.H., et al.: The multimodal brain tumor image segmentation benchmark (BRATS). IEEE Trans. Med. Imaging **34**(10), 1993–2024 (2015)
3. Işın, A., Direkoğlu, C., Şah, M.: Review of MRI-based brain tumor image segmentation using deep learning methods. Procedia Comput. Sci. **102**, 317–324 (2016)
4. Pan, Y., et al.: Brain tumor grading based on neural networks and convolutional neural networks. In: Engineering in Medicine and Biology Society (EMBC), 2015 37th Annual International Conference of the IEEE. IEEE (2015)
5. Bauer, S., et al.: A survey of MRI-based medical image analysis for brain tumor studies. Phys. Med. Biol. **58**(13), R97 (2013)
6. Jayamal, M.D., et al.: A novel method and tool for detection of tumors and tissue abnormalities. In: Moratuwa Engineering Research Conference (MERCon), 2016. IEEE (2016)
7. Moeskops, P., et al.: Automatic segmentation of MR brain images with a convolutional neural network. IEEE Trans. Med. Imaging **35**(5), 1252–1261 (2016)
8. Ahmmed, R., Hossain, M.F.: Tumor detection in brain MRI image using template based K-means and fuzzy C-means clustering algorithm. In: Computer Communication and Informatics (ICCCI), 2016 International Conference on. IEEE (2016)
9. Chang, P.D.: Fully convolutional neural networks with hyperlocal features for brain tumor segmentation. In: Proceedings MICCAI-BRATS Workshop (2016)
10. Tustison, N.J., et al.: N4ITK: improved N3 bias correction. IEEE Trans. Med. Imaging **29** (6), 1310–1320 (2010)
11. Babu, S.S.: Analysis and evaluation of brain tumour detection from MRI using F-PSO and FB-K means
12. Ravi, A., Sreejith, S.: A review on brain tumor detection using image segmentation. Int. J. Emerg. Technol. Adv. Eng. **5**(6), 60–64 (2015)
13. Raj, J.A., Kumar, S.: An enhanced classifier for brain tumor classification. Int. J. Control. Theory Appl. **9**(26), 325–333 (2016)
14. Naik, J., Patel, S.: Tumor detection and classification using decision tree in brain MRI. Int. J. Comput. Sci. Netw. Secur. (IJCSNS) **14**(6), 87 (2014)
15. Patil, R.C., Bhalchandra, A.S.: Brain tumour extraction from MRI images using MATLAB. Int. J. Electron., Commun. Soft Comput. Sci. Eng. (IJECSCSE) **2**(1), 1 (2012)
16. Cui, Z., Yang, J., Qiao, Y.: Brain MRI segmentation with patch-based CNN approach. In: Control Conference (CCC), 2016 35th Chinese. IEEE (2016)
17. Joshi, D.M., Rana, N.K., Misra, V.M.: Classification of brain cancer using artificial neural network. In: Electronic Computer Technology (ICECT), 2010 International Conference on. IEEE (2010)
18. Gonzalez, R.C., Woods, R.E.: Digital Image Processing, 3rd edn. Prentice-Hall, Upper Saddle River (2009)

An Efficient Worm Detection System Using Multi Feature Analysis and Classification Techniques

B Leelavathi[1(✉)] and Rajesh M Babu[2]

[1] Department of IT, Sri Jayendra Saraswathy CAS, Coimbatore, India
get2leela@rediffmail.com
[2] Department of CSE, Karpagam College of Engineering (Autonomous),
Coimbatore, India

Abstract. Signature based pattern identification is used to detect the worm hole attack in networks. This identification of signature pattern stores the input data and reduce the overhead problem. The malicious node identification based on the packets dropped during the communication between the users. Here applying machine learning approaches to solve the malicious in networks. In this paper, distinguished behaviors of Transferrable Executable headers, API function calls and DLL files are used. The machine learning algorithm, Adaboost ensemble classifier, Naïve Bayesian classifier and Decision tree are applied for improving to detection rate. The performance of each algorithm firstly evaluated to find the most outperformed algorithm each for classifying benign executable and malicious worm executable. And then, these features were combined to detect worm more precisely.

Keywords: Portable executable header · DLLs · API function calls · Machine learning algorithms · Efficient worm detection

1 Introduction

Although the computer solicitation knowledge manipulates at full speed, due to this the safety of computer system creates apprehensive. The security becomes more susceptible than ever and faces more security threats for the reason that the technological development brings the complication of the application surroundings.

Computers in the network cannot be sheltered due to the open characteristic of the internet and it's being short of central control. The malware and worm has stronger reproduction and destructive ability nowadays. A communal way of initiation the occurrences is using malicious behaviors such as worm, virus, Trojan or spyware [1, 2]. Propagation of worm might result in destruction to private users, etc. Spinellis [3] has proved that the discovery of bounded length viruses is an NP-complete problem. Consequently, we must use a diversity of experiential knowledge to recognize malware and increase the accuracy of decisions. An amount of non-signature centered malware discovery technique has been planned recently. These techniques examine several

© Springer Nature Switzerland AG 2020
A. P. Pandian et al. (Eds.): ICCBI 2018, LNDECT 31, pp. 1054–1064, 2020.
https://doi.org/10.1007/978-3-030-24643-3_126

features of malware. The common blockages of such techniques are their extraordinary false positive rate and large dealing out overheads.

The arrangement material of binary executable is exaggerated by its performance. Second executable pursue basic arrangement, such as PE (portable executables) then ELF (executable and linking format). It contains a quantity of headers and sections that express the OS (operating system) loader in what way to map the file into memory. Worm contaminated executable also tail the arrangement, but they have numerous difference with the benign executable in arrangement information. In this research, we obtainable a detection scheme which overwhelms the boundaries of current worm discovery techniques that uses format evidence of PE files to professionally detect zero-day malicious executables. Based on in-depth analysis of the static format information of PE files, we extracted totally 19193 features from arrangement information files. These features were extracted from headers of all major section of PE files and many of them do not supply to the classification accuracy (or even diminish it). We applied feature selection and ranking method to reduce the dimensionality and enhance compactness of the features. When selected features were trained using classification algorithm; the results of our experiments indicate that our scheme achieves about 98.66% of detection rates with less than 0.05% of false positive rates for distinguishing between benign software and malware. The most important is that our method can identify the unknown and new worms.

The rest of the paper is organized as follows. In Section 2 we discuss related works. In Section 3 presents the architecture of our proposed multi feature worm detection approach. Section 4 discusses the experimental results that are performed to regulate which scheme performs better. Finally we present our conclusions in Section 5.

2   Related Works

Research on countermeasures in contradiction of worms has absorbed on both the detection and the containment of worms. An amount of methods have been suggested that are aim to identify worms based on network traffic nature anomalies. Most conspicuously, this performance manifests themselves as a large number of (often failed) construction attempts [4, 5].

The usage of data mining is for data storing process and that helps to identify the patterns [6]. Likewise, machine learning approach helps to resolve the identification of dissimilar patterns in cloud database. So this data mining helps to identify the malicious and produce the better results [7].

3   Proposed Multi Feature Worm Detection System

A structural analysis of multi feature learning framework for worm detection system is proposed. The proposed work framework is given in Fig. 1. It consists of four major parts. It is beginning with the gathering of worm and benign applications from diverse sources. The second part performs the decompresses the collected applications to extract significant features. The third part carries out feature ranking and selection.

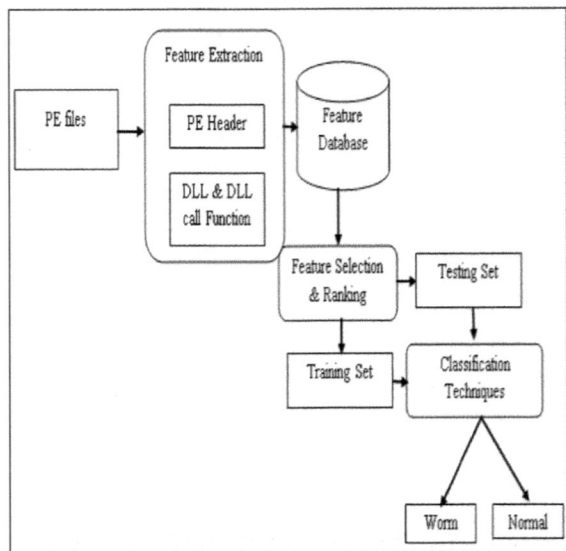

Fig. 1. Function block diagram of the proposed scheme

The last part includes machine learning techniques to classify the applications into normal or worm.

3.1 Dataset Collection

We have collected many benign and worm applications. Our dataset consists of a total of 47875 application programs and it can find 37280 malicious patterns and it resolve the 10595 cleared programs. It does not have any false pattern in the available input dataset. The malevolent instances are collected from several online open sources platforms (Table 1).

Table 1. Description of the collected worm dataset

Name of the malicious worms	Counts
Worm	16,152
Email worm	5,314
Net-worm	1938
P2P-worm	1,573
IRC-Worm	1,545
IM-worm	5379
Total	37280

3.2   PE File Format

PE attitudes for portable executable. It is the innate executable arrangement of Win32. Its requirement is derived rather from the Unix COFF (common object file format). The format evidence of PE file is illustrated in Fig. 2.

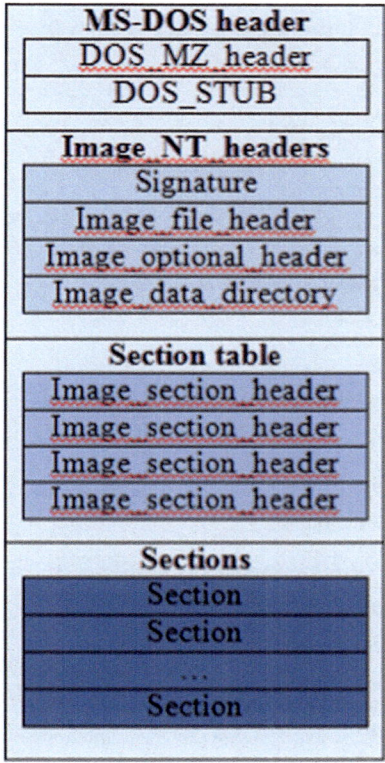

Fig. 2. Structure of PE format

Numerous fields of PE file have no compulsory constraint. There are an amount of dismissed fields and spaces in PE file, so that it has created opportunities for worm's propagation and hide. It also makes the format information of worm and benevolent software show numerous variances.

3.2.1   Feature Extraction

The executable information can without hard derivative in the format of PE file without consecutively programs [8, 9]. We have used IDA pro [10] with implemented python miner for PE parsing table in the executable lists of windows in composed application database. In this research, we applied IDA Pro for static analysis, specifically, to generate PE headers, DLL files and windows API function calls can be extracted.

It excerpts all API function names, PE header evidence & DLL names which is placed within the DLLs which consist in PE file and stored all the mined information in a database. In this study, 37280 raw extracted features of malicious worms and clean programs of 10595 & stored into a database. These features are particularized in Table 2.

Table 2. List of extracted raw features from PE data

S.No	Description	Data type	Quantity
1.	Unique DLL features Binary	Binary	650
2.	Unique API function calls	Binary	18420
3.	PE_DOS header	Integer	21
4.	PE_file header	Integer	6
5.	PE_optional headers	Integer	30
6.	PE_data directories	Integer	14
7.	PE_data directories	Integer	10
8.	PE_import description	Integer	5
9.	PE_export description	Integer	11
10.	PE_resource table	Integer	6
11.	PE_debug info	Integer	7
12.	PE_delay import	Integer	7
13.	PE_TLS table	Integer	6
Total			19293

3.3 Feature Selection and Ranking

The IG is very easily reachable in position method & feature selection [22]. Let the feature vector can be the $x = (f_1, f_2, f_3, f_n)$ of each application, where n denotes the total features. Then the features are nothing but PE headers, API function calls and DLL files.

Let random variable can be considerable as C which signifies the normal or worm application class. $C \in \{normal, worm\}$.

We calculated the IG for each of the application's entire features consistent to the class label C.

The IG for F can be calculated as (Table 3):

$$IG\ (F) = I\ (s_1,\ ....,s_m) - E\ (F)$$

Table 3. List of 20 top ranked features of the PE header

No	Features	Info-Gain score	Rank
1	Size of Initialized Data	0.5867	1
2	Number of Symbols	0.5492	2
3	Certificate Table Size	0.5419	3
4	Export table Size	0.5349	4
5	IATSIZE	0.5310	5
6	Bound Import Size	0.5219	6
7	Load Config Table Size	0.5206	7
8	Base Relocation Table Size	0.5192	8
9	e_cp	0.5132	9
10	CLR Runtime Header Size	0.5086	10
11	Virtual Size	0.4974	11
12	Size of Raw Data	0.4925	12
13	Number of Line Numbers	0.4862	13
14	e_cblp	0.4790	14
15	e_minalloc	0.4650	15
16	Number of Relocations	0.4508	16
17	Address of Funcitons	0.4495	17
18	Address of Name Ordinals	0.4319	18
19	Address of Names	0.4305	19
20	e_ovno	0.4296	20

3.4 Classification Techniques

The machine learning algorithm get an input a training sample $\{(x_1, y_1), (x_2, y_2), ..., (x_n, y_n)\}$ where each x_i belongs to feature vector set of every applications and y_i is a binary class label representing which of two classes the instance x_i belongs. Assumed a new instance, our aspire is to classify the class to which the occurrence fits by using classification approaches. In this work, the efficiency of three different families of classification methods: AdaBoost classifier [12], Naive Bayes (NB) is a statistical classifier [13], Decision Trees (DT) is a predictive model [14] with respect to accuracy have been compared (Table 4).

AdaBoost ensemble classifier: AdaBoost classifier is an efficient ensemble classification algorithm. It takes numerous advantages above other machine learning algorithms and the most important properties of it are powerful classification, less training time and low probability of overtraining. Also, it has several advantages; It is efficient

Table 4. List of the top ranked 30 DLL names

No	DLLs	Description	Info-Gain Score	Rank
1	kernel32.dll	Windows NT BASE API Client DLL	0.6927	1
2	user32.dll	USER API Client DLL	0.6720	2
3	advapi32.dll	Progressive Win32 application programming interfaces	0.6608	3
4	gdi32.dll	GDI Client DLL, complicated in graphical displays.	0.6592	4
5	wininet.dll	Internet Extensions for Win32	0.6514	5
6	comctl32.dll	User Experience Controls Library	0.6498	6
7	shell32.dll	Windows Shell Common DLL	0.6419	7
8	wsock32.dll	Windows Socket 32-Bit DLL	0.6378	8
9	olesut32.dll	OLE 2 - 32 extension	0.6310	9
10	msvbvm50.dll	Visual Basic Virtual Machine	0.6296	10
11	ole32.dll	32-bit OLE 2.0 component	0.6228	11
12	shlwapi.dll	Library for UNC and URL Paths, Registry Entries	0.6190	12
13	ws2_32.dll	Windows Socket 2.0 32-Bit DLL	0.6125	13
14	ntdll.dll	Win32 NTDLL core component	0.6098	14
15	urlmon.dll	OLE32 Extensions for Win32	0.5970	15
16	version.dll	Version Checking and File Installation Libraries	0.5917	16
17	crtdll.dll	Microsoft C Runtime Library	0.5894	17
18	comdlg32.dll	Common Dialogue Library	0.5829	18
19	winmm.dll	Windows Multimedia API	0.5799	19
20	rpcrt4.dll	Remote Procedure Call Runtime	0.5737	20
21	psapi.dll	The process status application programming interface	0.5638	21
22	msvcr100.dll	Microsoft Visual C++ Redistributable DLL	0.5602	22
23	hal.dll	The Windows Hardware Abstraction Layer	0.5593	23
24	mpr.dll	Multiple Provider Router DLL	0.5519	24
25	netapi32.dll	Net Win32 API DLL	0.5419	25
26	avicap32.dll	AVI capture API	0.5329	26
27	rasapi32.dll	Remote Access 16-bit API Library	0.5297	27
28	cygwin1.dll	Cygwin® POSIX Emulation DLL	0.5164	28
29	mscoree.dll	Microsoft .NET Runtime Execution Engine	0.4976	29
30	imagehlp.dll	Windows NT Image Helper	0.4928	30

in cases where great amounts of data are allocated. Millions of instances can be handled by AdaBoost in an reasonably priced time. AdaBoost is more healthy than other machine learning techniques. Also, it provides a threshold which can be rummage-sale to decide the exchange among incorrect positive rate and true positive rate. This feature is used to handle uncertainties and for performing error analysis.

Naive Bayes Classifier: It works on a simple and instinctive concept. It is grounded on the Bayes rule of provisional likelihood. It use of all the attributes enclosed in the data and examine them individually as though they are evenly important and self-governing of each other. This technique works beneath the supposition that one attribute is independent of the other attributes enclosed in the sample. The different characteristics of worm profiles are used to regulate the worm/normal profile likelihood ratio.

Decision Trees: It is an extrapolative machine learning model that make a decision the target value of a new sample based on a variety of attribute values of the available data. The interior knots of a decision tree indicate the different attributes, the undergrowth among the nodes represent the probable values that these attributes can have in the experiential samples, whereas the lethal nodes provides the final value of the reliant variable. The quality that is to be identified is identified as the independent variable, since its value reliable upon the morals of all the additional attributes. The other characteristics, which assistance in forecasting the value of the needy variable, are known as the self-governing variables in the dataset. This analytic support can be especially important to detect worm applications.

4 Experimental Results and Performance Analysis

The evaluation of the offered technique has been used by five different measures namely, precision, F-measure, accuracy, recall &False Positive Rate (FPR) (Tables 5 and 6).

$$\text{Accuracy} = \frac{TP + TN}{TP + FP + FN + TN} \tag{1.1}$$

4.1 Experimental Results

To compare the remaining performance measures like Precision, F-measure & Recall of classifiers, the tabulated experimental results of each time window are plotted as bar charts and shown in Figs. 3 and 4. From the figures, one can infer that all algorithms

Table 5. Performance measures for various classifiers with individual feature set

Classifier	Individual feature set	Precision	Recall	F-Measure	Accuracy	FPR
AdaBoost	PE Header	0.979	0.976	0.968	95.86	0.040
	DLLs	0.989	0.983	0.973	96.53	0.069
	API Functions	0.975	0.971	0.963	95.33	0.080
NB	PE Header	0.958	0.933	0.954	94.13	0.118
	DLLs	0.946	0.933	0.939	94.93	0.115
	API Functions	0.957	0.935	0.927	93.88	0.260
Decision Tree	PE Header	0.936	0.943	0.939	91.73	0.138
	DLLs	0.954	0.910	0.931	90.66	0.102
	API Functions	0.941	0.935	0.939	91.86	0.126

Table 6. Performance measures for various classifiers with combined feature set

Classifier	Combined feature set	Precision	Recall	F-Measure	Accuracy	FPR
AdaBoost	PE Header & DLLs	0.997	0.980	0.988	98.53	0.03
	PE Header & API Functions	0.995	0.980	0.987	98.32	0.09
	All	0.992	0.983	0.987	98.66	0.05
NB	PE Header & DLLs	0.965	0.953	0.959	95.66	0.07
	PE Header & API Functions	0.961	0.955	0.958	95.53	0.06
	All	0.962	0.953	0.971	95.98	0.07
Decision Tree	PE Header & DLLs	0.969	0.937	0.953	92.08	0.33
	PE Header & API Functions	0.974	0.938	0.955	92.44	0.35
	All	0.974	0.934	0.953	92.48	0.12

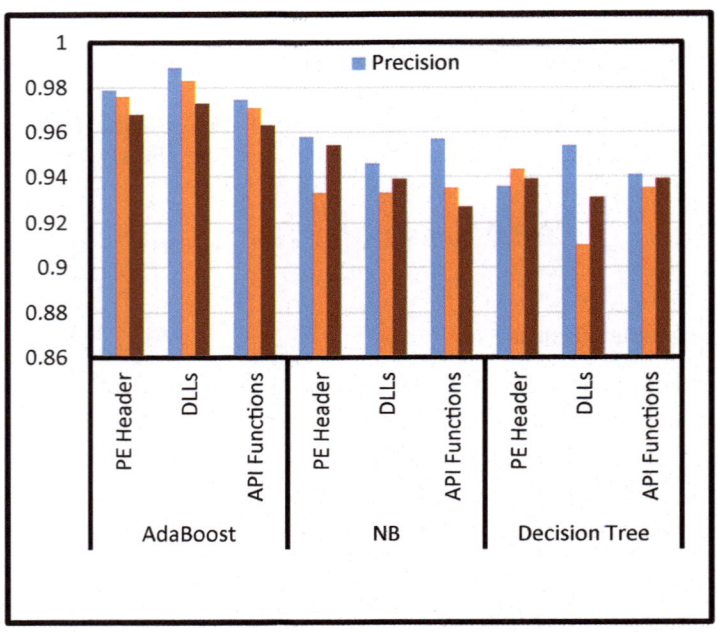

Fig. 3. Comparison of precision, recall & F-measures with individual feature set

have significant improvement in terms of recall, precision & F-measure as combined feature dataset. The Adaboost ensemble classifier achieves good Precision, Recall and F-measure values compared to Naïve Bayesian and decision tree classifiers. Naive Bayesian classifier concludes the likelihood of a class label assumed data using a simple supposition that the attributes are self-governing assumed the class label. The decision tree classifier is very sensitive to noisy data and outliers.

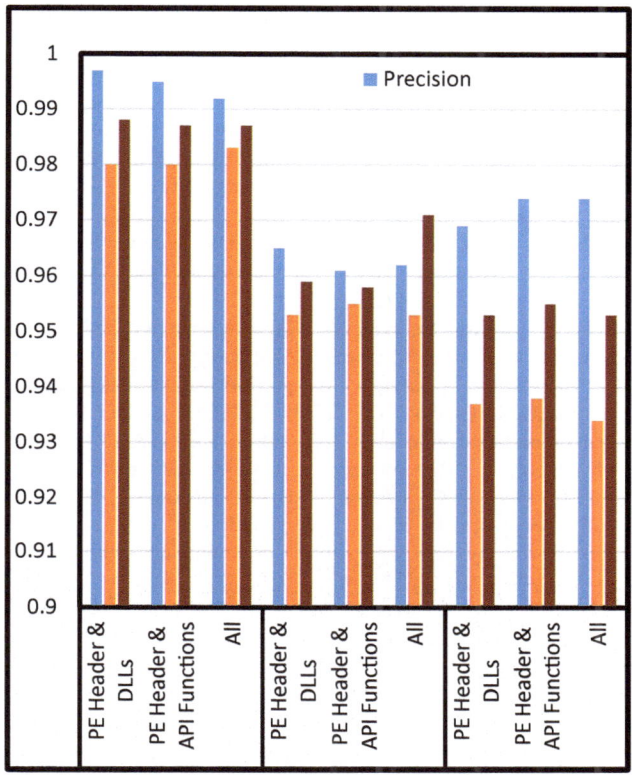

Fig. 4. Comparison of precision, recall & F-measures with combined feature set

5   Conclusion

Automated threats are consider as Worms which consists the massive host number in minimum time amount, making human-based countermeasures futile. Recent days, worms developed the curved sophisticated malware which aids the potential victims in optimized identification at enhanced malware approaches. Here we proposed a innovative multi feature approaches for worm identification. The method is based on structural analysis for characterizing the high level worm executable structure.

References

1. Kienzle, D.M., Elder, M.C.: Recent worms: a survey and trends. In: Proceedings of the 2003 ACM Workshop on Rapid Malcode, pp. 1–10. ACM (2003, October)
2. Menahem, E., Shabtai, A., Rokach, L., Elovici, Y.: Improving malware detection by applying multi-inducer ensemble. Comput. Stat. Data Anal. **53**(4), 1483–1494 (2009)
3. Spinellis, D.: Reliable identification of bounded-length viruses is NP-complete. IEEE Trans. Inf. Theory **49**(1), 280–284 (2003)

4. Ochieng, N., Mwangi, W., Ateya, I.: A tour of the computer worm detection space. Int. J. Comput. Appl. **104**(1) (2014)
5. Aljawarneh, S.A., Moftah, R.A., Maatuk, A.M.: Investigations of automatic methods for detecting the polymorphic worms signatures. Future Gener. Comput. Syst. **60**, 67–77 (2016)
6. So-In, C., Mongkonchai, N., Aimtongkham, P., Wijitsopon, K., Rujirakul, K.: An evaluation of data mining classification models for network intrusion detection. In: Digital Information and Communication Technology and It's Applications (DICTAP), 2014 Fourth International Conference on, pp. 90–94. IEEE (2014, May)
7. Schultz, M.G., Eskin, E., Zadok, F., Stolfo, S.J.: Data mining methods for detection of new malicious executables. In: Security and Privacy, 2001. S&P 2001. Proceedings. 2001 IEEE Symposium on, pp. 38–49. IEEE (2001)
8. Uppal, D., Sinha, R., Mehra, V., Jain, V.: Malware detection and classification based on extraction of API sequences. In: Advances in Computing, Communications and Informatics (ICACCI, 2014 International Conference on, pp. 2337–2342. IEEE (2014, September)
9. Narouei, M., Ahmadi, M., Giacinto, G., Takabi, H., Sami, A.: DLLMiner: structural mining for malware detection. Secur. Commun. Netw. **8**(18), 3311–3322 (2015)
10. Ferguson, J., Kaminsky, D.: Reverse engineering code with IDA Pro. Syngress (2008)
11. Hall, M., Frank, E., Holmes, G., Pfahringer, B., Reutemann, P., Witten, I.H.: The WEKA data mining software: an update. SIGKDD Explorations, vol. 11, no. 1 (2009). Reproduced with permission of the copyright owner. Further reproduction prohibited without permission 4 (2013)
12. Zhang, X.Y., Hou, Z., Zhu, X., Wu, G., Wang, S.: Robust malware detection with Dual-Lane AdaBoost. In: Computer Communications Workshops (INFOCOM WKSHPS), 2016 IEEE Conference on. pp. 1051–1052. IEEE (2016, April)
13. Xiao, L., Chen, Y., Chang, C.K.: Bayesian model averaging of bayesian network classifiers for intrusion detection. In: Computer Software and Applications Conference Workshops (COMPSACW), 2014 IEEE 38th International, pp. 128–133. IEEE (2014, July)
14. Sahu, S., Mehtre, B.M.: Network intrusion detection system using J48 Decision Tree. In: Advances in Computing, Communications and Informatics (ICACCI), 2015 International Conference on, pp. 2023–2026. IEEE (2015, August)

Digital Marketing Framework Strategies Through Big Data

Samee Sayyad[1], Arif Mohammed[2], Vikrant Shaga[3],
Abhishek Kumar[4(✉)] [iD], and K. Vengatesan[5]

[1] School of Engineering, Symbiosis Skill & Open University, Pune, India
samee.syd@gmail.com
[2] Department of MIS, Dhofar University, Salalah, Oman
mdarefl20@gmail.com
[3] Department of Computer Application, Thakur Institute of Management,
Mumbai, India
shagavik@gmail.com
[4] Department of Computer Science, Banaras Hindu University, Varanasi, India
abhishek.maacindia@gmail.com
[5] Department of Computer Engineering, Sanjivani College of Engineering,
Kopargaon, India
vengicse2005@gmail.com

Abstract. In recent times, the amount of data generated by various digital techniques are increasing at an unprecedented rate. These humongous amount of generated data is called "Big Data". It is basically a combination of large data sets that has complex data structure integrated with the difficulties to store, fetch, analyze, transfer, protect and visualize the transferred data. The process involved to retrieve, fetch and analyze the data are named as big data analytics. Big data analytics is more useful in e-business companies to analyse the behavior of customer, systematic analytics and acquiring profit over the competition. This paper aims to study the use of big data in business organizations. The data have been collected from 122 e- business companies located within Maharashtra, India. The outcome of this study is that the big data is basically useful in e-business companies. This paper includes a literature review, sampling, hypothesis testing, expected conclusion and suggestions.

Keywords: Querying · Data curation · Visualization

1 Introduction

Big data images as the buzzword in today's digital environment. Humongous amount of data are generated by different sources such as e-business, online transactions, education, engineering, social media, call centers, medicines, and telecommunication domain. From the past years, generating and exchanging the information has been increasing tremendously in the e-business domain. E- business is recognized as one of the fastest acceptors of big data analytics to know the need of market, customer behavior, face customer retention challenges and increase the profit. E-business applications has gained huge amount of data via online transaction or by different

© Springer Nature Switzerland AG 2020
A. P. Pandian et al. (Eds.): ICCBI 2018, LNDECT 31, pp. 1065–1073, 2020.
https://doi.org/10.1007/978-3-030-24643-3_127

resources. The E-business firms are generally associated with both the structured and unstructured data. Structured data includes name, age, address, date of birth; gender etc. Whereas the unstructured data includes voices, clicks, likes, links, tweets etc. Nowadays, Indian people are mostly engaged in buying and selling the product via internet. The large-scale E-business companies have developed an environment for buying and selling products via online mode. In order to enhance the customer behaviour analytics process, each transaction made by the customer is stored and analyzed.

2   Review of the Literature

Some practitioners and scholars have gone so far as to suggest that Big Data Analytics (BDA) is the "fourth paradigm of science" (Strawn 2012, p. 34). Big Data Analytics (BDA) is a "new paradigm of knowledge assets" (Hagstrom 2012, p. 2), or "the next frontier for innovation, competition, and productivity" (Manyika et al. 2011, p. 1) [5]. Soumya et al. explains the process of data analytics in big data network and that will be helpful to analyse the prediction strategies with lesser outcome [1]. Shahriar Akter et al. [2] gives a brief discussion of business value changes in the big data analytics framework. Said et al. [3] discuss the challenges involved in the electronic commerce and that helps to obtain the variety discussion in statistical areas. Lipsa Sadath et al. [4] focuses on the cyber security crime in the e-commerce application and this helps to meet the requirements of both data mining and e-commerce applications [4].

Objectives

1. To study the new approach in marketing strategies by using big data analysis and large database in E-business applications.
2. To explore the various uses of Big Data concept in E- business transaction and the role of big data in electronic marketing innovation domain.
3. To enumerate the recent trends in E-business.

Hypothesis

H0:   E-business companies does not gain significant benefit due to the implementation of Big Data Analytics framework.

HA: E-business companies benefit due to the implementation of Big Data analytics framework.

Research Methodology:

A)   Primary Data:

Data is collected through structured Questionnaire; some of the data is collected through interviews conducted on the on-site visits and personal observation.

B)   Secondary Data:

The data is collected from Journals, Bulletin, and Newspaper.

C) Tools and Techniques used:

Classifications and tabulation of the data and so are collected from the above mentioned sources is used as per the requirements of the study. The data collected is then analyzed and presented by using the techniques such as:

1. SPSS
2. Microsoft Excel 2007
3. Correlation analysis

D) Period of Study:

The data is collected from 2005 to 2015.

Selection of Sample
Determining Sample size is recognized as the important issue because if the sample data is too large it becomes difficult to process and it consumes more time and money, while samples that are too small may lead to inaccurate results. Hence I restricted my study to "A study of marketing strategies using Big Data in E-business" (With special reference to selected E-business firms in Maharashtra state), its applicability in Maharashtra State only. Hence, out of …437… E-business Companies in Maharashtra, I have decided to take …122 as sample by using the simple random sampling method (purposely for sake of convenience amounting to 27% of the Universe)

Scope of Study
The Scope of this important subject seems to be vast because of the separate data created and stored, which can be utilized as and when required. The study shows that the big data is emerging as a very useful tool in e-business applications, and very few companies have implemented and obtained more number of benefits from big data tool in E-business applications.

Limitation of study
The study is restricted to Maharashtra State only, as the comparative study with other States and Countries will consume more amount of time, money, and energy. We have faced difficulties in collecting data regarding this topic from different sources. Hence we restricted our study to E-business companies in Maharashtra State only.

The study provides a scope for conducting further studies in the areas of:

(1) Increasing big data analytics domain in India
(2) Comparison between marketing strategies obtained by the online shopping portals like Flipkart and Amazon
(3) Case studies on the Big Data implemented companies (Table 1).

From the Graph 1, after taking a survey in Maharashtra got total 122 respondents, major respondents from Mumbai i.e. 61 which is 50% from those 18 i.e. 14.75% respondents using big data analytics and 43 i.e. 35.25% respondents not using big data analytics. A minimum response gets from Aurangabad region i.e. 7 which is 5.74% in that 1 i.e. 0.82% respondent using big data analytics and 6 i.e. 4.92% respondents not using big data analytics. In Pune region total 35 i.e. respondents which are 28.6% from

Table 1. Distribution of company region wise (location) in Maharashtra

Region (location)	A study of marketing strategies in E-business				Total	Percentage
	Using Big Data	Percentage	Not Using Big Data	Percentage		
Mumbai	18	14.75	43	35.25	61	50.00
Pune	12	9.84	23	18.85	35	28.69
Nagpur	4	3.28	6	4.92	10	8.20
Aurangabad	1	0.82	6	4.92	7	5.74
Nashik	1	0.82	8	6.56	9	7.38
TOTAL	36	29.51	86	70.49	122	100

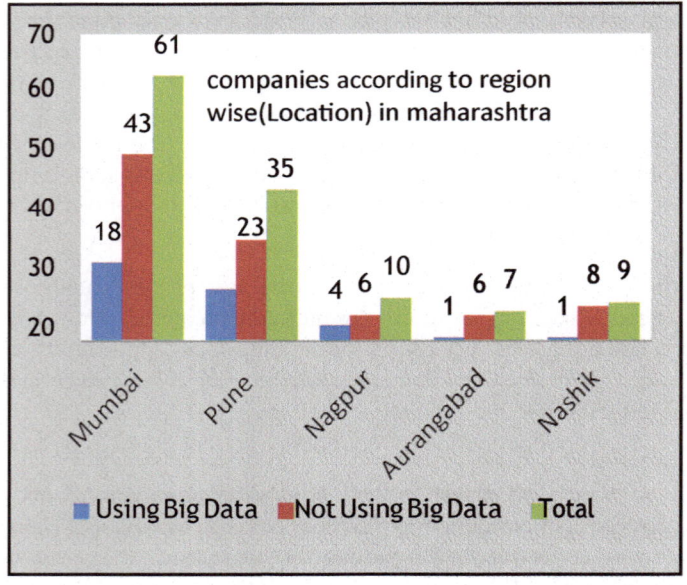

Graph 1. Source Primary Data

those 12 i.e. 9.84% respondents using big data and 23 i.e. 18.85% respondents not using big data in their companies. In Nagpur region gets total 10 respondents which are 8.20% from those 4 i.e. 3.28% respondents using big data analytics and 6 i.e. 4.92% respondents not using big data analytics in their companies. From Nashik region get total 9 responses which are 7.38% in that 1 i.e. 0.82% respondent using big data analytics and 8 i.e. 6.56% companies not using big data analytics (Table 2).

From the Graph 2, out of 122 companies 36 companies are using big data in e-business applications and out of that 30 i.e. 83.33% companies agreed on usefulness of Big data in E-business & 6 i.e. 16.66% companies are not agreed on the usefulness of big data in E- business. While 86 companies not using big data in e- business, from that

Table 2. Association between marketing strategies using big data in E-business and opinion on uses of big data in E-business.

Big data in E-business	Opinion on usefulness of big data in E-business			Chi- square value	p-value
	useful	Not useful	Total		
Using	30	06	36	36.2	P < 0.0001 Significant
Not Using	21	65	86		
Total	51	71	122		

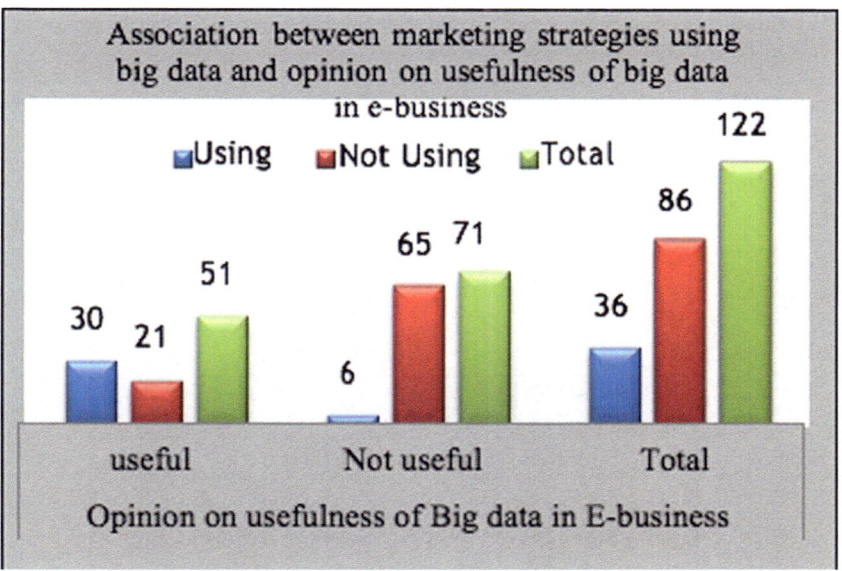

Graph 2. Source Primary Data

21 i.e. 24.41% companies where considered usefulness of big data in e-business and 65 i.e. 75.58% companies where not considered usefulness of big data in e-business. There is statistically significant association between the marketing strategies using Big Data and usefulness of big data in E-business domain (P = 0.0001) (Table 3).

From the Graph 3, out of 122 companies 36 companies are using big data in e-business out of that 14 i.e. 38.88% companies agreed that still it remains as a challenging task to implement big data analytics & 22 companies i.e. 61.11% companies where not agreed that big data analytics is a challenging task in e- business organization. While 86 companies not using big data in e-business framework. Statistically there was no significant association between the use of big data tool in e-business model and the challenging task is to implement the big data analytics in e-business domain (P = 0.086).

Table 3. Association between marketing strategies that utilizes big data in E-business and the challenging task is to implement big data analytics in your e-commerce/e- business framework

Big data in E- business	Challenging task to implements big data analytics in your e- business			Chi- square value	p-value
	Yes	No	Total		
Using	14	22	36	2.91	P = 0.086
Not Using	48	38	86		Not Significant
Total	62	60	122		
Big data in E- business	Company lacking skilled resources for big data analytics technology?			Chi- square value	p-value
	Yes	No	Total		
Using	14	22	36	5.88	P = 0.015
Not Using	54	32	86		Significant
Total	68	54	122		

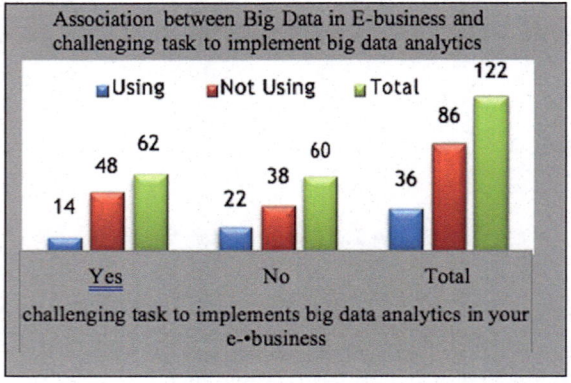

Graph 3. Source Primary Data

From the Graph 4, out of 122 companies only 36 companies were using big data in e-business, and out of that 14 i.e. 38.88% companies are lacking skilled resources & Graph 3: Source Primary Data 22 i.e. 61.11% companies are not lacking skilled resources for implementing big data analytics technology 86 companies where not using big data in e- business out of that 54 i.e. 62.79% companies considered that lacking skilled resources for big data analytics technology & 32 i.e. 37.20% companies were not considered of lacking skilled resources for big data analytics technology.

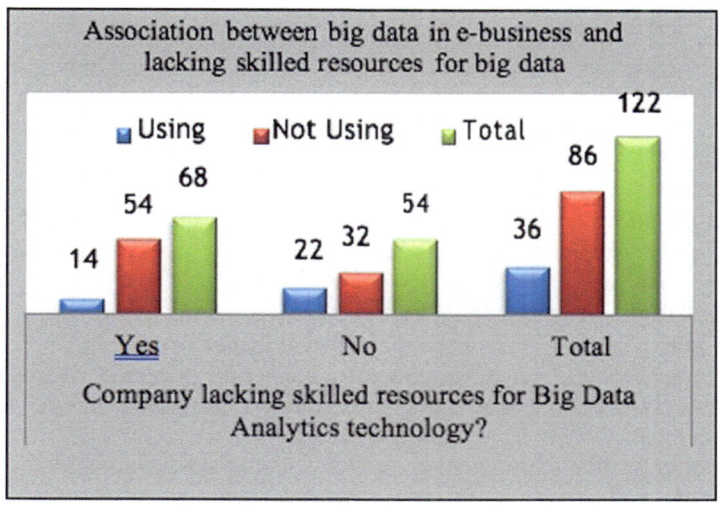

Graph 4. Source Primary Data

There was statistically significant association between the use of big data in e-business and lacking skilled resources for big data (P = 0.015).

3   Conclusions

The survey shows, in Maharashtra state the big data implemented companies are less. Majority of the e-business firm exists in metropolitan cities like Mumbai and Pune region, and from these region maximum companies are found using the big data tool in their E-business applications. In Maharashtra state, maximum companies think and feels that the e-business application have been increased to develop an alternative marketing strategy and it has many advantages over the traditional marketing strategies. All the companies will think positively about the e-business application. From the survey, the major reason behind the non-usage of big data tool is their lack of skilled resources. Other reasons are that many of the marketing companies are not aware of Big Data Analytics technology, non-availability of huge data sources, and the difficulties in shifting companies' database to big data analytics framework. The maximum respondents agree that the big data framework provides a cost-effective mechanism to process the large volume of information available in the organization. The major role of big data in the organization is establishing BI/Analytics platform for the purpose of production reporting, searching, sorting, storing and analysis of the data. After implementing big data Technology, marketing companies tends to achieve major throughput and profit, and their financial performance has been outstanding, and the firm's sales growth has been increased at an unprecedented rate.

Recommendations

(1) Promoting among the academic scholars to learn new technologies like big data and create awareness among the e-business firm owners for ensuring the implementation of big data analytics technology in their traditional marketing strategy.

(2) The big data technology can be included in the college level syllabus so that the students can learn and develop their skilled resources.

(3) Arrange seminars, workshop in the small cities like Aurangabad, Nashik to create websites and implementation the big data tool in the marketing domain of large-scale companies.

(4) Promoting the benefits of big data in e-business through the media like social sites, newspapers, TV channels, journals and magazines.

(5) The companies can reduce the traditional marketing strategies. Big data analytics implements the different marketing strategy for surviving in the global environment.

(6) The company should set its goal, objective, and marketing strategy. According to that they would know the role and tangible benefits of the big data analytics in their company.

(7) The customers can easily access the sites, the products are properly categorized; a secure online transaction can be done to obtain an enhanced customer satisfaction.

Acknowledgement. The authors would like to thank all the participants who provided their valuable inputs for the completion of this study. The participants were provided adequate information about the study and were encouraged to make their voluntary, informed, and rational decision to participate.

References

1. Laxmi, P.S.S., Pranathi, P.S.: Impact of big data analytics on business intelligence-scope of predictive analytics. Int. J. Curr. Eng. Technol. 5(2), Inpressco - International Press Corporation Publication, E-ISSN 2277 – 4106, P-ISSN 2347 – 5161 (2015)

2. Akter, S., Wamba, S.F.: Big data analytics in E-commerce: a systematic review and agenda for future research, electronic market. Int. J. Networked Bus. 26(2), 173–194, ISSN: 1019-6781 (Print) 1422-8890 (Online) (2016)

3. Banks, D.L., Said, Y.H.: Data mining in electronic commerce. Stat. Sci. 21(2), 234–246, published by Institute of Mathematical Statistics (2006)

4. Sadath, L.: Data mining in E-commerce: a CRM platform. Int. J. Comput. Appl. (0975 – 8887) 68(24), 32–37, Scholarly peer reviewed, ISSN: 0123-4560 (2013)

5. Wamba, S.F., Gunasekaran, A., Akter, S., Dubey, R.: Big data analytics and firm performance: effects of dynamic capabilities. Article in J. Bus. Res. 1–11. University of Georgia, Elsevier Publication, ISSN No: 0148-2963 (2016)

6. Erl, T., Khattak, W., Buhler, P.: Big Data Fundamentals –Concepts, Drivers and Techniques. Prentice Hall Publication, ISBN-13: 978-0-13-429107-9 (2015)

7. Kalaivanan, M., Vengatesan, K.: Recommendation system based on statistical analysis of ranking from user. International Conference on Information Communication and Embedded Systems (ICICES), pp. 479–484. IEEE (2013)

8. Kumar, A., Singhal, A., Sheetlani, J.: Essential-replica for face detection in the large appearance variations. Int. J. Pure Appl. Math. **118**(20), 2665–2674
9. Chintal, P.: Study of cyber threats and data encryption techniques to protect national information infrastructure. Int. J. Res. Comput. Commun. Technol. **3**(3) ISSN (Online): 2278 5841, ISSN (Print): 2320–5156 (2014)
10. Sayyad, S.: Analysis of insertion sort in Java. Adv. Comput. Res. **7**(1), ISSN: 0975- 3273, E-ISSN: 0975-9085, Bioinfo Publications (2015)

Author Index

© Springer Nature Switzerland AG 2020
A. P. Pandian et al. (Eds.): ICCBI 2018, LNDECT 31, pp. 1075–1078, 2020.
https://doi.org/10.1007/978-3-030-24643-3

Printed by Printforce, the Netherlands